Errata and Corrigenda number 3 : January 1975 have been entered.
number 4 : January 1976

Errata and Corrigenda number 5 : January 1977 have been entered.
number 6 : January 1978
number 7 : January 1979
8 : January 1980
9 : January 1981

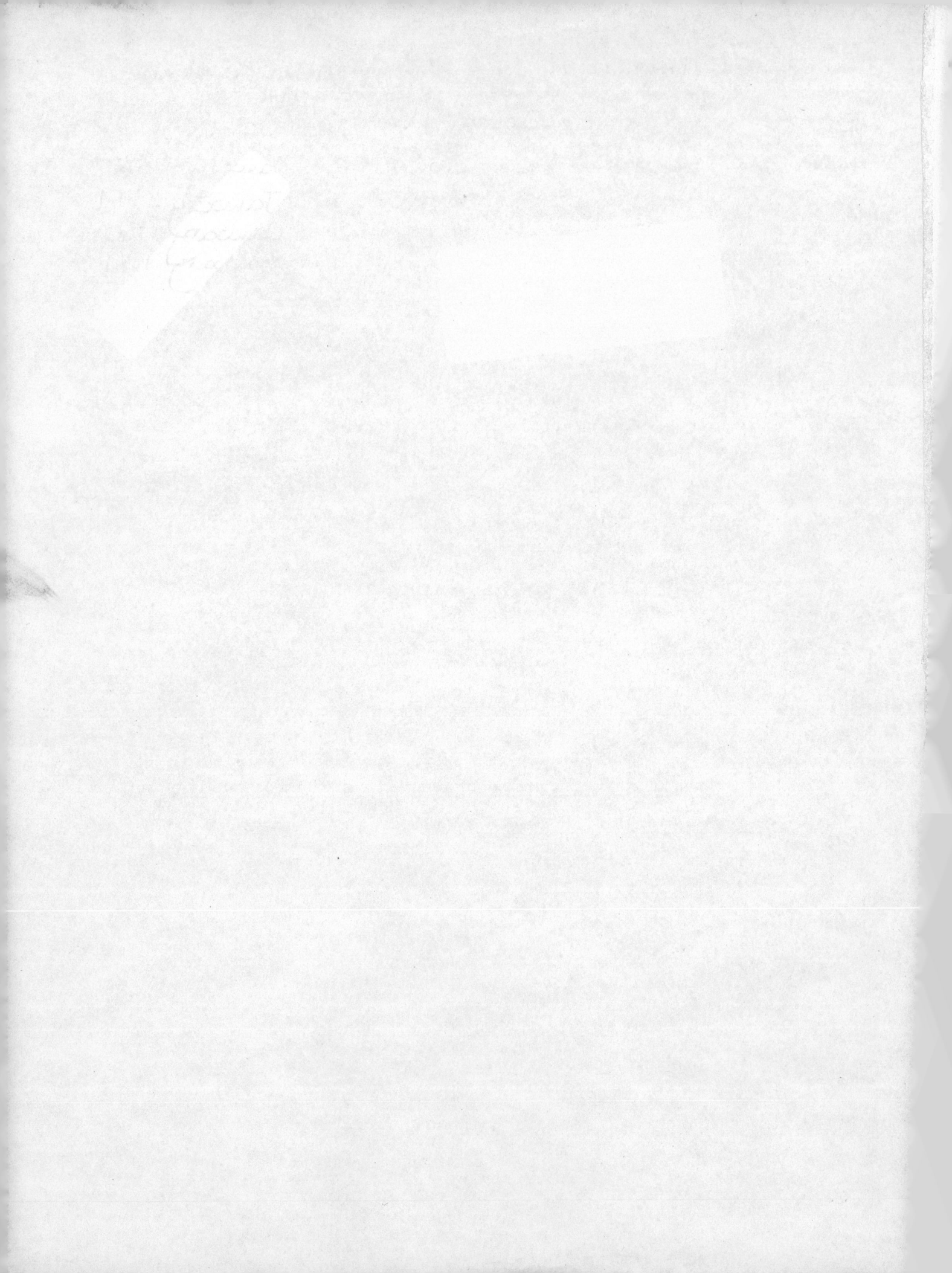

THERMAL CONDUCTIVITY

Nonmetallic Solids

THERMOPHYSICAL PROPERTIES OF MATTER

The TPRC Data Series

A Comprehensive Compilation of Data by the
Thermophysical Properties Research Center (TPRC), Purdue University

Y. S. Touloukian, Series Editor
C. Y. Ho, Series Technical Editor

Volume 1. Thermal Conductivity–Metallic Elements and Alloys
Volume 2. Thermal Conductivity–Nonmetallic Solids
Volume 3. Thermal Conductivity–Nonmetallic Liquids and Gases
Volume 4. Specific Heat–Metallic Elements and Alloys
Volume 5. Specific Heat–Nonmetallic Solids
Volume 6. Specific Heat–Nonmetallic Liquids and Gases
Volume 7. Thermal Radiative Properties–Metallic Elements and Alloys
Volume 8. Thermal Radiative Properties–Nonmetallic Solids
Volume 9. Thermal Radiative Properties–Coatings
Volume 10. Thermal Diffusivity
Volume 11. Viscosity
Volume 12. Thermal Expansion–Metallic Elements and Alloys
Volume 13. Thermal Expansion–Nonmetallic Solids

New data on thermophysical properties are being constantly accumulated at TPRC. Contact TPRC and use its interim updating services for the most current information.

THERMOPHYSICAL PROPERTIES OF MATTER
VOLUME 2

THERMAL CONDUCTIVITY

Nonmetallic Solids

Y. S. Touloukian
Director
Thermophysical Properties Research Center
and
Distinguished Atkins Professor of Engineering
School of Mechanical Engineering
Purdue University
and
Visiting Professor of Mechanical Engineering
Auburn University

R. W. Powell
Senior Researcher
Thermophysical Properties Research Center
Purdue University
Formerly
Senior Principal Scientific Officer
Basic Physics Division
National Physical Laboratory
England

C. Y. Ho
Head of Data Tables Division
and
Associate Senior Researcher
Thermophysical Properties Research Center
Purdue University

P. G. Klemens
Professor and Head
Department of Physics
University of Connecticut
and
Visiting Research Professor
Thermophysical Properties Research Center
Purdue University

IFI/PLENUM • NEW YORK-WASHINGTON • 1970

Library of Congress Catalog Card Number 73-129616

SBN (13-Volume Set) 306-67020-8

SBN (Volume 2) 306-67022-4

IFI/Plenum Data Corporation is a subsidiary of
Plenum Publishing Corporation
227 West 17th Street, New York, N.Y. 10011

Distributed in Europe by Heyden & Son, Ltd.
Spectrum House, Alderton Crescent
London N.W. 4, England

Printed in the United States of America

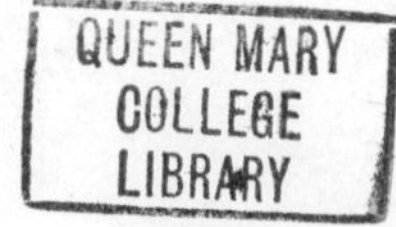

"In this work, when it shall be found that much is omitted, let it not be forgotten that much likewise is performed..."

SAMUEL JOHNSON, A.M.

From last paragraph of Preface to his two-volume *Dictionary of the English Language*, Vol. I, page 5, 1755, London, Printed by Strahan.

Foreword

In 1957, the Thermophysical Properties Research Center (TPRC) of Purdue University, under the leadership of its founder, Professor Y. S. Touloukian, began to develop a coordinated experimental, theoretical, and literature review program covering a set of properties of great importance to science and technology. Over the years, this program has grown steadily, producing bibliographies, data compilations and recommendations, experimental measurements, and other output. The series of volumes for which these remarks constitute a foreword is one of these many important products. These volumes are a monumental accomplishment in themselves, requiring for their production the combined knowledge and skills of dozens of dedicated specialists. The Thermophysical Properties Research Center deserves the gratitude of every scientist and engineer who uses these compiled data.

The individual nontechnical citizen of the United States has a stake in this work also, for much of the science and technology that contributes to his well-being relies on the use of these data. Indeed, recognition of this importance is indicated by a mere reading of the list of the financial sponsors of the Thermophysical Properties Research Center; leaders of the technical industry of the United States and agencies of the Federal Government are well represented.

Experimental measurements made in a laboratory have many potential applications. They might be used, for example, to check a theory, or to help design a chemical manufacturing plant, or to compute the characteristics of a heat exchanger in a nuclear power plant. The progress of science and technology demands that results be published in the open literature so that others may use them. Fortunately for progress, the useful data in any single field are not scattered throughout the tens of thousands of technical journals published throughout the world. In most fields, fifty percent of the useful work appears in no more than thirty or forty journals. However, in the case of TPRC, its field is so broad that about 100 journals are required to yield fifty percent. But that other fifty percent! It is scattered through more than 3500 journals and other documents, often items not readily identifiable or obtainable. Nearly 50,000 references are now in the files.

Thus, the man who wants to use existing data, rather than make new measurements himself, faces a long and costly task if he wants to assure himself that he has found all the relevant results. More often than not, a search for data stops after one or two results are found—or after the searcher decides he has spent enough time looking. Now with the appearance of these volumes, the scientist or engineer who needs these kinds of data can consider himself very fortunate. He has a single source to turn to; thousands of hours of search time will be saved, innumerable repetitions of measurements will be avoided, and several billions of dollars of investment in research work will have been preserved.

However, the task is not ended with the generation of these volumes. A critical evaluation of much of the data is still needed. Why are discrepant results obtained by different experimentalists? What undetected sources of systematic error may affect some or even all measurements? What value can be derived as a "recommended" figure from the various conflicting values that may be reported? These questions are difficult to answer, requiring the most sophisticated judgment of a specialist in the field. While a number of the volumes in this Series do contain critically evaluated and recommended data, these are still in the minority. The data are now being more intensively evaluated by the staff of TPRC as an integral part of the effort of the National Standard Reference Data System (NSRDS). The task of the National Standard Reference Data System is to organize and operate a comprehensive program to prepare compilations of critically evaluated data on the properties of substances. The NSRDS is administered by the National Bureau of Standards under a directive from the Federal Council for Science and Technology, augmented by special legislation of the Congress of the United States. TPRC is one of the national resources participating in the National Standard Reference Data System in a united effort to satisfy the needs of the technical community for readily accessible, critically evaluated data.

As a representative of the NBS Office of Standard Reference Data, I want to congratulate Professor Touloukian and his colleagues on the accomplishments represented by this Series of reference data books. Scientists and engineers the world over are indebted to them. The task ahead is still an awesome one and I urge the nation's private industries and all concerned Federal agencies to participate in fulfilling this national need of assuring the availability of standard numerical reference data for science and technology.

EDWARD L. BRADY
Associate Director for Information Programs
National Bureau of Standards

Preface

Thermophysical Properties of Matter, the TPRC Data Series, is the culmination of twelve years of pioneering effort in the generation of tables of numerical data for science and technology. It constitutes the restructuring, accompanied by extensive revision and expansion of coverage, of the original *TPRC Data Book*, first released in 1960 in loose-leaf format, 11″ × 17″ in size, and issued in June and December annually in the form of supplements. The original loose-leaf *Data Book* was organized in three volumes: (1) metallic elements and alloys, (2) nonmetallic elements, compounds, and mixtures which are solid at N.T.P., and (3) nonmetallic elements, compounds, and mixtures which are liquid or gaseous at N.T.P. Within each volume, each property constituted a chapter.

Because of the vast proportions the *Data Book* began to assume over the years of its growth and the greatly increased effort necessary in its maintenance by the user, it was decided in 1967 to change from the loose-leaf format to a conventional publication. Thus, the December 1966 supplement of the original *Data Book* was the last supplement disseminated by TPRC.

While the manifold physical, logistic, and economic advantages of the bound volume over the loose-leaf oversize format are obvious and welcome to all who have used the unwieldy original volumes, the assumption that this work will no longer be kept on a current basis because of its bound format would not be correct. Fully recognizing the need of many important research and development programs which require the latest available information, TPRC has instituted a *Data Update Plan* enabling the subscriber to inquire, by telephone if necessary, for specific information and receive, in many instances, same-day response on any new data processed or revision of published data since the latest edition. In this context, the TPRC Data Series departs drastically from the conventional handbook and giant multivolume classical works, which are no longer adequate media for the dissemination of numerical data of science and technology without a continuing activity on contemporary coverage. The loose-leaf arrangements of many works fully recognize this fact and attempt to develop a combination of bound volumes and loose-leaf supplement arrangements as the work becomes increasingly large. TPRC's *Data Update Plan* is indeed unique in this sense since it maintains the contents of the TPRC Data Series current and live on a day-to-day basis between editions. In this spirit, I strongly urge all purchasers of these volumes to complete in detail and return the *Volume Registration Certificate* which accompanies each volume in order to assure themselves of the continuous receipt of annual listing of corrigenda during the life of the edition.

The TPRC Data Series consists initially of 13 independent volumes. The initial ten volumes will be published in 1970, and the remaining three by 1972. It is also contemplated that subsequent to the first edition, each volume will be revised, updated, and reissued in a new edition approximately every fifth year. The organization of the TPRC Data Series makes each volume a self-contained entity available individually without the need to purchase the entire Series.

The coverage of the specific thermophysical properties represented by this Series constitutes the most comprehensive and authoritative collection of numerical data of its kind for science and technology.

Whenever possible, a uniform format has been used in all volumes, except when variations in presentation were necessitated by the nature of the property or the physical state concerned. In spite of the wealth of data reported in these volumes, it should be recognized that all volumes are not of the same degree of completeness. However, as additional data are processed at TPRC on a continuing basis, subsequent editions will become increasingly more complete and up to date. Each volume in the Series basically comprises three sections, consisting of a text, the body of numerical data with source references, and a material index.

The aim of the textual material is to provide a complementary or supporting role to the body of numerical data rather than to present a treatise on the subject of the property. The user will find a basic theoretical treatment, a comprehensive presentation of selected works which constitute reviews, or compendia of empirical relations useful in estimation of the property when there exists a paucity of data or when data are completely lacking. Established major experimental techniques are also briefly reviewed.

The body of data is the core of each volume and is presented in both graphical and tabular format for convenience of the user. Every single point of numerical data is fully referenced as to its original source and no secondary sources of information are used in data extraction. In general, it has not been possible to critically scrutinize all the original data presented in these volumes, except to eliminate perpetuation of gross errors. However, in a significant number of cases, such as for the properties of liquids and gases and the thermal conductivity of all the elements, the task of full evaluation, synthesis, and correlation has been completed. It is hoped that in subsequent editions of this continuing work, not only new information will be reported but the critical evaluation will be extended to increasingly broader classes of materials and properties.

The third and final major section of each volume is the material index. This is the key to the volume, enabling the user to exercise full freedom of access to its contents by any choice of substance name or detailed alloy and mixture composition, trade name, synonym, etc. Of particular interest here is the fact that in the case of those properties which are reported in separate companion volumes, the material index in each of the volumes also reports the contents of the other companion volumes.* The sets of companion volumes are as follows:

Thermal conductivity:	Volumes 1, 2, 3
Specific heat:	Volumes 4, 5, 6
Radiative properties:	Volumes 7, 8, 9
Thermal expansion:	Volumes 12, 13

The ultimate aims and functions of TPRC's Data Tables Division are to extract, evaluate, reconcile, correlate, and synthesize all available data for the thermophysical properties of materials with the result of obtaining internally consistent sets of property values, termed the "recommended reference values." In such work, gaps in the data often occur, for ranges of temperature, composition, etc. Whenever feasible, various techniques are used to fill in such missing information, ranging from empirical procedures to detailed theoretical calculations. Such studies are resulting in valuable new estimation methods being developed which have made it possible to estimate values for substances and/or physical conditions presently unmeasured or not amenable to laboratory investigation. Depending on the available information for a particular property and substance, the end product may vary from simple tabulations of isolated values to detailed tabulations with generating equations, plots showing the concordance of the different values, and, in some cases, over a range of parameters presently unexplored in the laboratory.

The TPRC Data Series constitutes a permanent and valuable contribution to science and technology. These constantly growing volumes are invaluable sources of data to engineers and scientists, sources in which a wealth of information heretofore unknown or not readily available has been made accessible. We look forward to continued improvement of both format and contents so that TPRC may serve the scientific and technological community with ever-increasing excellence in the years to come. In this connection, the staff of TPRC is most anxious to receive comments, suggestions, and criticisms from all users of these volumes. An increasing number of colleagues are making available at the earliest possible moment reprints of their papers and reports as well as pertinent information on the more obscure publications. I wish to renew my earnest request that this procedure become a universal practice since it will prove to be most helpful in making TPRC's continuing effort more complete and up to date.

It is indeed a pleasure to acknowledge with gratitude the multisource financial assistance received from over fifty of TPRC's sponsors which has made the continued generation of these tables possible. In particular, I wish to single out the sustained major support being received from the Air Force Materials Laboratory–Air Force Systems Command, the Office of Standard Reference Data–National Bureau of Standards, and the Office of Advanced Research and Technology–National Aeronautics and Space Administration. TPRC is indeed proud to have been designated as a National Information Analysis Center for the Department of Defense as well as a component of the National

*For the first edition of the Series, this arrangement was not feasible for Volume 7 due to the sequence and the schedule of its publication. This situation will be resolved in subsequent editions.

Standard Reference Data System under the cognizance of the National Bureau of Standards.

While the preparation and continued maintenance of this work is the responsibility of TPRC's Data Tables Division, it would not have been possible without the direct input of TPRC's Scientific Documentation Division and, to a lesser degree, the Theoretical and Experimental Research Divisions. The authors of the various volumes are the senior staff members in responsible charge of the work. It should be clearly understood, however, that many have contributed over the years and their contributions are specifically acknowledged in each volume. I wish to take this opportunity to personally thank those members of the staff, research assistants, graduate research assistants, and supporting graphics and technical typing personnel without whose diligent and painstaking efforts this work could not have materialized.

Y. S. Touloukian

Director
Thermophysical Properties Research Center
Distinguished Atkins Professor of Engineering

Purdue University
Lafayette, Indiana
July 1969

Introduction to Volume 2

This volume of *Thermophysical Properties of Matter*, the TPRC Data Series, is among the more comprehensive of all the volumes of the Series. Indeed, it is the result of one of TPRC's oldest data tables programs, initiated in 1960.

The volume comprises three major sections: namely, the front text material together with its bibliography, the main body of numerical data and its references, and the material index.

The text material is intended to assume a role complementary to the main body of numerical data which is the primary purpose of this volume. It is felt that a concise discussion of the theoretical nature of the property under consideration together with a review of predictive procedures and recognized experimental techniques will be appropriate in a major reference work of this kind. The extensive reference citations given in the text should lead the interested reader to sufficient literature for a detailed study. It is hoped, however, that enough detail is presented for this volume to be self-contained for the practical user.

The main body of the volume consists of the presentation of numerical data compiled over the years in a most comprehensive and meticulous manner. The scope of coverage includes most nonmetallic materials of engineering importance which are in the solid state at normal temperature and pressure. The extraction of all data directly from their original sources ensures freedom from errors of transcription. Furthermore, some gross errors appearing in the original source documents have been corrected. The organization and presentation of the data together with other pertinent information on the use of the tables and figures are discussed in detail in the text of the section entitled *Numerical Data.*

While only a very limited number of the materials reported have been critically evaluated, it is planned that the policy of Volume 1 will be extended and that future editions of this volume will include the "recommended values" for an increasing number of materials which can be reasonably well characterized to enable critical analysis.

As stated earlier, all data have been obtained from their original sources and each data set is so referenced. TPRC has in its files all documents cited in this volume. Those that cannot be readily obtained elsewhere are available from TPRC in microfiche form.

The material index at the end of the volume covers the contents of all three companion volumes (Volumes 1, 2, and 3) on thermal conductivity. It is hoped that the user will find these comprehensive indices helpful.

This volume has grown out of activities made possible by TPRC's Founder Sponsors and through the principal support of the Air Force Materials Laboratory—Air Force Systems Command, under the monitorship of Mr. John H. Charlesworth. The limited effort on the critical analysis of the data for the elements and some oxides was made possible through the support of the Office of Standard Reference Data—National Bureau of Standards, under the monitorship of Dr. Howard J. White, Jr. Over the past nine years, many graduate students and research assistants have rendered assistance for long or short periods under the authors' supervision. We wish to acknowledge in chronological order of their association with TPRC, the contributions of Messrs. C. Y. Wang, K. C. Lin, D. Y. Nee, R. L. Feng, J. J. G. Hsia, M. Mangkornkanok, M. Nalbantyan, G. K. Kirjilian, and Mrs. E. K. C. Lee and Mr. K. Y. Wu. The two last mentioned are still at TPRC and participated in the final organization of the tables and figures and the demanding task of checking details. We also wish to acknowledge the benefit of extensive discussions with Dr. J. Kaspar, Senior Staff Scientist, Materials Sciences Laboratory, Aerospace Corporation. He is also a Visiting Research Professor at TPRC.

Inherent to the character of this work is the fact that in the preparation of this volume, we have drawn most heavily upon the scientific literature and feel a debt of gratitude to the authors of the referenced articles. While their often discordant results have caused us much difficulty in reconciling their findings, we consider this to be our challenge and our contribution to negative entropy of information, as an effort is made to create from the randomly distributed data a condensed, more orderly state.

While this volume is primarily intended as a reference work for the designer, researcher, experimentalist, and theoretician, the teacher at the graduate level may also use it as a teaching tool to point out to his students the topography of the state of knowledge on the thermal conductivity of nonmetallic solids. We believe there is also much food for reflection by the specialist and the academician concerning the meaning of "original" investigation and its "information content."

The authors and their contributing associates are keenly aware of the possibility of many weaknesses in a work of this scope. We hope that we will not be judged too harshly and that we will receive the benefit of your suggestions regarding references omitted, additional material groups needing more detailed treatment, improvements in presentation, and, most important, any inadvertent errors. If the *Volume Registration Certificate* accompanying this volume is returned, the reader will assure himself of receiving annually a list of corregenda as possible errors come to our attention.

Lafayette, Indiana
July 1969

Y. S. Touloukian
R. W. Powell
C. Y. Ho
P. G. Klemens

Contents

Numerical Data

Material Index

GROUPING OF MATERIALS AND LIST OF FIGURES AND TABLES

1. ELEMENTS

2. SINGLE OXIDES

*Number marked with an asterisk indicates that recommended values are also reported for this material on separate figure and table of the same number followed by the letter R.

2. SINGLE OXIDES (continued)

3. OXIDE COMPOUNDS

*Number marked with an asterisk indicates that recommended values are also reported for this material on separate figure and table of the same number followed by the letter R.

4. BINARY MIXTURES OF SINGLE OXIDES AND/OR OXIDE COMPOUNDS

4. BINARY MIXTURES OF SINGLE OXIDE AND/OR OXIDE COMPOUNDS (continued)

5. MULTIPLE MIXTURES OF SINGLE OXIDES AND/OR OXIDE COMPOUNDS

6. MIXTURES OF OXIDE AND NONOXIDE

7. IODIDE

8. BROMIDES

9. CARBIDES

10. CHLORIDES

11. FLUORIDES AND THEIR MIXTURES

12. NITRATES

13. NITRIDES

14. PHOSPHATES

15. SULFATES

16. SULFIDES AND THEIR MIXTURES

17. CERMETS

18. MISCELLANEOUS INORGANIC COMPOUNDS AND MIXTURES

19. MINERALS

19. MINERALS (continued)

20. AGGREGATE MIXES

21. MISCELLANEOUS REFRACTORY MATERIALS AND GLASSES

22. POLYMERS

*Number marked with an asterisk indicates that recommended values are also reported for this material on separate figure and table of the same number followed by the letter R.

23. MISCELLANEOUS ORGANIC COMPOUNDS

24. SYSTEMS

25. ANIMAL AND VEGETABLE NATURAL SUBSTANCES AND THEIR DERIVATIVES

25. ANIMAL AND VEGETABLE NATURAL SUBSTANCES AND THEIR DERIVATIVES (continued)

26. FABRICS, YARNS, AND HAIRS

27. PROCESSED COMPOSITES

28. PROCESSED MINERAL SUBSTANCES

29. SLAGS AND SCALES

30. RESIDUES

Theory, Estimation, and Measurement

Notation

A	Cross-sectional area
a	Cube root of atomic volume; Half the focal length of an ellipsoid; Axis of ellipse
b	Amplitude of lattice wave; Axis of ellipse
$\mathbf{b}$	Inverse lattice vector
C	Specific heat per unit volume
$C(\nu)$	Spectral specific heat
C_g	Lattice specific heat per unit volume
c	Defect concentration
D	Thermal diffusivity
E	Voltage drop
$\mathbf{e}$	Unit vector in the direction of polarization
G	Number of atoms in a crystal
h	Planck constant
I	Electric current
i	Suffix denoting the type of carrier
j	Index denoting the possible polarizations of a wave
K	Kelvin
k	Thermal conductivity
k_g	Lattice thermal conductivity
k_i	Intrinsic lattice thermal conductivity
k_r	Thermal conductivity of a reference material
L	Phonon mean free path for boundary scattering
l	Mean free path; Effective length of a specimen
l_g	Phonon mean free path
M	Atomic mass
N	Number of atoms per unit volume
n	Exponent
P	Slope
q	Heat flow per unit time; Wave number
$\mathbf{q}$	Wave vector
r	Radial distance
S	Seebeck coefficient
T	Temperature
ΔT	Temperature difference
t	Time
$t_{1/2}$	Time required for the back-face temperature of a thin-disk specimen to reach half its maximum value
U	Coefficient of equation (13)
$\mathbf{u}$	Displacement of an atom
V	Electrical potential
v	Velocity of lattice waves; Sound velocity
W	Thermal resistivity
W_g	Lattice thermal resistivity
W_i	Intrinsic lattice thermal resistivity
x	Reduced frequency ($x = h\nu/\kappa T$)
$\mathbf{x}$	Equilibrium site of an atom
Δx	Distance difference
α	Index denoting various scattering processes
β	Parameter
γ	Anharmonicity coefficient
δ	Amplitude decrement
ϵ	Local thermal strain
θ	Debye temperature
κ	Boltzmann constant
λ	Wavelength
λ_m	Minimum wavelength of lattice waves corresponding to the maximum wave frequency ν_m
ν	Wave frequency
ν_m	Upper frequency limit of the spectrum of lattice waves
π	Peltier coefficient; Ratio of the circumference of a circle to its diameter
ρ	Electrical resistivity

Theory of Thermal Conductivity of Nonmetallic Solids

1. INTRODUCTION

Heat in solids is conducted by various carriers: electrons, lattice waves (or phonons), magnetic excitations, and, in some cases, electromagnetic radiation. The total thermal conductivity is additively composed of contributions from each type of carrier. It can be shown that

$$k = \frac{1}{3}\sum_{i} C_i v_i l_i \tag{1}$$

where the subscript i denotes the type of carrier, C_i is the contribution of each carrier to the specific heat per unit volume, v_i is the velocity of the carrier (we regard the carrier as a particle; if the carrier is a wave, the appropriate velocity is the group velocity), and l_i is a suitably defined mean free path.

The theory of the thermal conductivity of solids has been the subject of numerous investigations and several review articles [1–4]. It is the purpose of this introductory text to present the major results of the theory only to the extent to which it is needed by the user of these tables: to caution him as to which results are likely to be structure sensitive and thus likely to vary from specimen to specimen, and to help him judge which materials are likely to have similar properties, and thus to guide him in guessing the thermal conductivities of materials which have not been measured. For detailed theoretical developments and discussions, the reader is referred to the aforementioned review articles and to the references given to the individual topics discussed later. Other useful review articles (some are short) and books which contain articles pertaining to this subject are [5–12].

The occurrence of a mean free path in equation (1) opens up the possibility that in some cases one cannot uniquely define the thermal conductivity of a material. This happens whenever a carrier mean free path becomes comparable to the smallest external dimension of the specimen. It happens particularly in dielectric crystals at low temperatures, where the phonon mean free paths are long; it also happens in transparent solids at high temperatures, where photons contribute significantly to the heat transport, provided the specimens are continuous in their optical properties (i.e., single crystals or vitreous solids, but not polycrystalline aggregates). In that case, we can speak of a size-dependent thermal conductivity, though this is only an approximation. The effective thermal conductivity will depend not only on the shortest linear dimension, but also on whether this dimension is parallel or perpendicular to the heat flow. It will also depend on the nature of the external surface: in the former case, on the degree of specularity of the reflection of the carriers, in the latter case, on the ability of the surface to absorb and emit carriers.

While in metals and alloys the free electrons are important carriers of heat, and frequently overshadow the thermal conductivity of the lattice vibrations, the lattice vibrations are the most important carriers in insulating crystals, and in many cases the only carriers. The electrons become important as carriers in semiconductors at higher temperatures, though they may sometimes play a role in other cases, not as carriers, but as a mechanism limiting the mean free path of the lattice waves.

The theory of thermal conductivity of metallic solids, where both the electrons and the lattice waves contribute significantly to the thermal conductivity, has been reviewed in Volume 1.

For purposes of the theory of conduction properties, we distinguish between three temperature regimes: high, intermediate, and low, with rough divisions at temperatures (on the absolute scale) of θ and $\theta/3$, respectively, where θ is the Debye temperature. For our purposes this temperature is related to the upper frequency limit ν_m of the spectrum of lattice waves by $h\nu_m = \kappa\theta$, where h and

κ are the Planck and Boltzmann constants, respectively. Roughly speaking, at high temperatures each atom vibrates independently of its neighbors, and the theories of lattice vibrations simplify. At low temperatures the vibrations are highly correlated and are best described by elastic waves in a continuum, with corresponding simplification. The intermediate temperature regime is somewhat awkward, and theoretical results are obtained by interpolation.

Most solids have Debye temperatures θ around 300 K, but atomically heavy solids have lower θ's (well below 200 K for solids containing gold or lead), while light atom solids (diamond, beryllia, etc.) have much higher Debye temperatures. Another group of solids with low θ's are the solidified rare gases, where binding forces are unusually weak.

2. LATTICE WAVES

The thermal vibrations of atoms in a perfect crystal can be described in terms of vibrations of each atom about its equilibrium site. Although such a simple description would suffice for many purposes, it does not take account of the correlation in the vibrations of neighboring atoms, or of the transfer of vibrational energy from one atomic site to the next. A better description is to regard the thermal vibrations as arising from a superposition of progressive displacement waves (lattice waves), which are the normal modes of the crystal and at the same time the carriers of thermal energy.

The lattice waves occupy a spectrum of frequencies, ν, from the lowest frequencies to some upper limit, ν_m, typically of the order of 10^{13}Hz. At low frequencies these waves are identical to elastic waves in the corresponding elastic continuum; at higher frequencies the atomic structure of the crystal lattice leads to dispersion effects. The corresponding wavelengths range from long waves down to waves of length comparable to atomic dimensions.

The displacement of an atom from its equilibrium site (at $\mathbf{x}$) may thus be written as a superposition of waves, i.e.,

$$\mathbf{u}(\mathbf{x}) = \frac{1}{\sqrt{G}} \sum_{j,\mathbf{q}} \mathbf{e}(\mathbf{q},j) b(\mathbf{q},j) \exp(i\mathbf{q}\cdot\mathbf{x} - i2\pi\nu_j t) \qquad (2)$$

where $\mathbf{q}$ is the wave vector, so that the wavelength $\lambda = 2\pi/q$, q is the wave number, G is the number of atoms in the crystal, j is an index denoting the possible polarizations of a wave of given wave-vector, $\mathbf{e}$ is a unit vector in the direction of polarization, and b is the amplitude of each wave.

The density of $\mathbf{q}$-values (usually derived in terms of idealized boundary conditions) is such that the number of $\mathbf{q}$-values in an element d^3q of $\mathbf{q}$-space is $G(2\pi)^{-3}d^3q$. There are three polarizations to each value of $\mathbf{q}$, and there is an upper limit to the admissible values of $\mathbf{q}$. The physically distinct wave-vectors must lie in a region of $\mathbf{q}$-space, called the fundamental zone, which is defined in terms of the crystal lattice, or rather in terms of the reciprocal lattice of the crystal. It can be shown that the number of physically distinct lattice waves is $3G$, equal to the number of normal modes which a system of G mutually bound atoms should have.

Under certain idealized conditions these displacement waves, equation (2), are the normal modes of the lattice, i.e., the energy content of each wave would be preserved in time. These conditions are:

(a) The interatomic forces are perfectly harmonic (obey Hooke's law).
(b) The lattice is structurally perfect.
(c) The lattice has no external boundaries, or the displacement of the atoms at the boundaries satisfy certain artificial boundary conditions.

Departures from these idealized conditions lead to interchange of energy between these waves and limit their effective mean free path. As seen from equation (1), these processes control the lattice thermal conductivity

$$k_g = \tfrac{1}{3} C_g v l_g \qquad (3)$$

where C_g is the contribution of the lattice waves to the specific heat per unit volume, and v is the velocity of the lattice waves at low frequencies. When λ is large compared with the interatomic spacing, ν is a linear function of $1/\lambda$ and v is constant. At higher frequencies or shorter wavelengths the atomic structure leads to dispersion, i.e., to a departure from the linear relation between ν and $1/\lambda$. In that case the appropriate velocity describing energy transport and entering equation (3) is the group velocity $d\nu/d(1/\lambda)$.

In crystals of complex structure there are also lattice waves describing the relative motion of atoms in the unit cell. These waves, the optical modes, usually have a small group velocity and contribute relatively little to the energy transport. They are often disregarded for purposes of thermal conductivity; however, this neglect may not always be justified.

Because there is some interaction between the different lattice waves, they will tend to be randomly excited, but with an average energy content given by the considerations of statistical mechanics. In thermal equilibrium at temperature T, the lattice specific heat of solids varies as T^3 at very low temperatures, and is independent of T at high temperatures. The contribution to the specific heat per unit volume from waves in the frequency interval ν and $\nu + d\nu$ is given, to a first approximation, by the Debye theory, and is of the form

$$C(\nu)\,d\nu = 9N\kappa\left(\frac{T}{\theta}\right)^3 \frac{x^4 e^x}{(e^x - 1)^2}\,dx \tag{4}$$

when $x = h\nu/\kappa T$ is the reduced frequency and N the number of atoms per unit volume. This holds for $\nu \leq \nu_m$; for $\nu > \nu_m$, $C(\nu) = 0$.

The Debye approximation disregards the dispersion of the high-frequency lattice waves, disregards differences in polarization, and smears out the crystal structure of the solid. It is thus a better approximation at low temperature, when all the important waves are long, than at intermediate and high temperatures. The only concession it makes to the discreteness of the lattice is the choice of a cutoff frequency ν_m or the corresponding minimum wavelength $\lambda_m = v/\nu_m$, where v is an average sound velocity. The cutoff is chosen so that the total number of waves corresponds to the correct number of normal modes ($3N$ per unit volume) which this assembly of atoms ought to have.

The lower limit λ_m of the wavelength is related to a^3, the volume per atom, by $4\pi a^3 = 3\lambda_m{}^3$.

In spite of the obvious inadequacy of the Debye approximation, it is frequently chosen as a basis for discussing thermal conductivity because the theoretical uncertainties in the phonon mean free path usually overshadow errors introduced into the specific heat by the Debye approximation.

3. LATTICE THERMAL CONDUCTIVITY

Since the phonon mean free path is usually a function of frequency ν (it may, of course, also depend on polarization and in extreme cases even on the direction of the wave vector $\mathbf{q}$), the expression (3) for the lattice thermal conductivity has to be generalized to

$$k_g(T) = \frac{1}{3}\int_0^{\nu_m} C(\nu) v l(\nu)\,d\nu \tag{5}$$

where $C(\nu)$ is given by equation (4). The lattice thermal conductivity is thus governed both by the magnitude and the frequency dependence of the mean free path.

The mean free path is limited by several processes which cause interchange of energy among the lattice waves, or scatter phonons from one mode into another. These are broadly grouped into processes due to anharmonicities of the interatomic forces, i.e., departures from Hooke's law, into scattering due to the various kinds of lattice imperfections, and, as a special case of the last, into scattering of phonons by the external boundaries.

The effective scattering probabilities or reciprocals of the mean free path of each group of lattice waves is composed additively of contributions from each process, so that

$$\frac{1}{l(\nu)} = \sum_{(\alpha)} \frac{1}{l_\alpha(\nu)} \tag{6}$$

where α denotes the various processes. If two interaction processes have the same dependence on frequency (ν), then it follows from equations (5) and (6) that the corresponding thermal resistivities follow an additive resistance rule, i.e.,

$$\frac{1}{k_g} = W_g = \sum_{(\alpha)} W_{(\alpha)} \tag{7}$$

where $W_{(\alpha)}$ is the lattice thermal resistivity if only one process (α) were to act. However, if two or more processes have different frequency dependences of $l_{(\alpha)}(\nu)$, then in general

$$W_g > \sum_{(\alpha)} W_{(\alpha)} \tag{7a}$$

The bigger the difference in frequency dependence, the more pronounced will be the deviation from the additive resistance rule.

The interactions due to anharmonicities, which may also be viewed as the scattering by thermal vibrations, lead to a mean free path which depends both on T and ν. The other interactions lead to a mean free path which depends on ν only, since the imperfection structure is practically independent of T, except possibly at very high temperatures.

At high temperatures, $k_g(T)$ is mainly a reflection of the temperature dependence of l, but at low temperatures it depends also on the ν-dependence of $l(\nu)$. This follows because in equation (5)

$$C(\nu) \propto T^3 x^4 e^x (e^x - 1)^{-2} \tag{8}$$

and if $l(\nu) \propto \nu^{-n} \propto T^{-n} x^{-n}$, it follows from equation (5) that

$$k_g \propto T^{3-n} \tag{9}$$

Since different imperfections lead, in the low-frequency limit, to different frequency dependences, or different values of the exponent n, the temperature dependence of the thermal conductivity at low temperatures will differ from case to case and depend upon the nature of the principal imperfections.

For point defects $n = 4$; for long but thin defects (cylinders) $n = 3$; for thin sheets $n = 2$. Dislocations, by virtue of their long-range strain field, do not act as thin cylinders, but scatter with an exponent $n = 1$. Scattering by boundaries, both external boundaries and internal grain boundaries, scatter independently of frequency, with $n = 0$.

Generally speaking, the more extended imperfections have a scattering cross-section which varies less strongly with frequency, and which is relatively more important at low frequencies, and thus more important at low temperatures.

Relation (9) does not hold, strictly speaking, for point defects ($n = 4$) or line defects ($n = 3$), for the integral (5) would then diverge at low frequencies, where $C(\nu)$ of equation (8) varies as ν^2. Relation (9) is then only a rough rule. Point defects or line defects must always be considered in conjunction with another resistive process, such as boundary scattering or intrinsic (anharmonic) scattering.

General reviews of lattice thermal conductivity are given in references [1–4]. Individual research papers on the theory of lattice thermal conductivity include [13–53].

4. INTRINSIC RESISTIVITY AT HIGH TEMPERATURES

At elevated temperatures in good crystals, the major source of thermal resistance is the interchange of energy, or scattering of phonons, due to departures from Hooke's law. A local strain ϵ introduces a fractional change $\gamma\epsilon$ in the local value of the sound velocity. This local change scatters the lattice wave. The coefficient γ, a measure of the anharmonicity, is of order unity (frequently $\gamma \simeq 2$). At high temperatures the thermal strain at neighboring atomic sites is practically uncorrelated. Scattering from each atomic site is thus proportional to $\langle\epsilon^2\rangle$, the mean square thermal strain, which in turn is proportional to T. The intrinsic mean free path thus varies as

$$l_i \propto 1/T \tag{10}$$

so that at high temperatures, where C_g is independent of temperature, k_g varies as $1/T$. This theoretical result, explaining early data on the temperature dependence of the thermal conductivity of insulating crystals [13–15], was derived by Debye [16].

There is, however, a serious theoretical difficulty. The lattice waves have a continuous spectrum of frequencies, all contributing to the specific heat and the thermal conductivity. At high temperatures, the major contribution to the specific heat, according to equation (4), comes from the upper range of the phonon spectrum. However, if $l(\nu)$ increases sufficiently rapidly with decreasing frequency, the major contribution to k would come from lower frequencies. In the Debye model, where every atomic site scatters independently from each other, $l_i(\nu) \propto \nu^{-4}$ and the expression for k diverges at low frequencies. Debye's model, while correctly describing $l_i(\nu)$ at high frequencies, and even producing a good quantitative estimate, fails at low frequencies, where correlations between the displacements of neighboring atomic sites become increasingly important.

To avoid this difficulty, Peierls [17] set up a theory of the anharmonic interaction between lattice waves, which recognizes that the thermal strain which scatters a lattice wave is itself the consequence of other lattice waves. This theory thus treats in detail the interchange of energy which, as a result of cubic anharmonicities (terms in the Hamiltonian which are not quadratic in strain, as Hooke's law would indicate, but cubic), leads to an interchange between groups of three lattice waves. The interchange is such that a phonon in a mode $\mathbf{q}_1$ combines with a phonon in mode $\mathbf{q}_2$ to form a phonon in mode $\mathbf{q}_3$, satisfying certain interference conditions between the frequencies and wave vectors of the participating modes. The frequency conservation condition is

$$\nu_1 + \nu_2 = \nu_3$$

and can also be regarded as energy conservation between the participating phonons. The wave vectors satisfy conservation conditions

$$\mathbf{q}_1 + \mathbf{q}_2 = \mathbf{q}_3 \tag{11a}$$

or

$$\mathbf{q}_1 + \mathbf{q}_2 = \mathbf{q}_3 \pm \mathbf{b} \tag{11b}$$

when $\mathbf{b}$ is one of the inverse lattice vectors of the crystal lattice. The three-phonon processes are accordingly divided into "normal" processes (11a), or "Umklapp" processes (11b), the latter word being derived from the German for "flip-over."

The conservation condition (11a) corresponds roughly to momentum conservation among the phonons. Umklapp processes must be regarded as a combination of three-phonon processes and Bragg reflection. In an elastic continuum only normal processes would occur.

The Peierls' theory is the basis of all subsequent theoretical work. The theory is quite complicated, and has been put into a form suitable for detailed work only by subsequent authors. At low temperatures there is a rigid distinction between the roles of normal and Umklapp processes—only the latter produce thermal resistance, though the former can indirectly influence the effectiveness of other processes. At higher temperatures the distinction between the two types of processes becomes blurred. The theory, since it involves the inverse lattice vector, and thus the details of the crystal structure, becomes quite complicated; however, some rough estimates have been made [3, 27].

The intrinsic phonon mean free path, with some approximations, takes the following form at high temperatures

$$1/l_i(\nu) \propto \nu^2 T \tag{12}$$

so that the divergence difficulties are avoided [3]. The intrinsic thermal resistivity is of the form [3, 27]

$$W_i = U\left(\frac{h}{\kappa}\right)^3 \gamma^2 \frac{1}{Ma}\frac{T}{\theta^3} \tag{13}$$

where M is the atomic mass and a^3 the atomic volume. The numerical coefficient U, typically of the order 1/3, is somewhat uncertain, and depends on the details of the crystal structure.

The major factor controlling the intrinsic lattice conductivity is the Debye temperature θ: solids of high θ will in general have higher values of k_g. Another factor is the crystal structure—solids of complex crystal structure have more inverse lattice vectors and zone boundaries and a greater variety of Umklapp processes; such solids will have a lower thermal conductivity. However, the effect of crystal structure on the intrinsic thermal conductivity is still not well understood from a theoretical point of view.

While the theory predicts the intrinsic thermal resistivity to vary as T at high temperatures, small deviations could arise from several causes. Quartic anharmonicities, i.e., terms in the potential energy quartic in the strain, would lead to four-phonon processes and a contribution to the resistivity proportional to T^2. One would expect the T^2 term to be of lower magnitude by a factor of order $\langle \epsilon^2 \rangle$, which in most solids is about 1 percent at room temperature. A similar effect, also leading to a T^2 term, results from thermal expansion. This would lower the effective value of θ with increasing T, and since W_i varies as $1/\theta^3$, it would lead to an increase in thermal resistivity roughly proportional to T^2. This effect should also be small.

Besides the general review articles, other pertinent references to the theory of intrinsic thermal resistivity at high temperatures are [20, 21, 23, 45, 54–58].

5. INTRINSIC RESISTIVITY AT LOW TEMPERATURES

At low temperatures the vibrations of the lattice cannot be represented as independent vibrations of individual atoms, but the representation of these vibrations in terms of a superposition of lattice waves becomes essential. At this point, the Peierls' theory must be used. The thermal resistance arises from Umklapp processes of type (11b). Since the important thermal waves at low temperatures have low frequencies, these processes occur only as a result of the occasional encounter with a high-frequency wave, of wave vector **q** comparable to **b**/2, where **b** is an inverse lattice vector. Such a wave has a frequency ν_m/β, when β is some parameter, usually of the order of 2, which depends on the details of the zone structure and on the dispersion of the high-frequency phonons. The probability of finding such a phonon decreases exponentially with decreasing temperature, so that

$$W_i \propto e^{-\theta/\beta T} \tag{14}$$

and the intrinsic thermal conductivity increases sharply with decreasing temperature, i.e.,

$$k_i(T) \propto e^{\theta/\beta T} \quad \text{for} \quad T < \theta/6 \tag{14a}$$

Although there are theoretical expressions for the preexponential factor [1, 3, 17], these are somewhat uncertain, and in practice this factor is unimportant in comparison with the theoretical uncertainty relating to the exponent θ/β.

In anisotropic crystals both the preexponential factor and exponent θ/β can depend on the direction of heat flow. Since θ/β depends on the zone structure, it can also be changed by phase transitions, with important consequences to the thermal conductivity.

In spite of the theoretical uncertainties, we can clearly expect, as Peierls had originally predicted

[17], that the thermal conductivity of perfect crystals, which varies roughly as $1/T$ at intermediate and high temperatures, should increase much more rapidly with decreasing temperature at low temperatures, until the phonon mean free path attains macroscopic dimensions. At this point k attains a maximum, and at lower temperatures k decreases as T is further decreased, since l is kept fixed and $k(T)$ is now controlled by the specific heat.

However, this presupposes a structurally perfect crystal. Near the conductivity maximum the thermal conductivity is particularly sensitive to all kinds of crystal imperfections. In some cases, even the mass variation due to the naturally occurring isotopic distribution is enough to appreciably depress the thermal conductivity around its maximum.

In the exponential region also, crystals of higher Debye temperature θ generally have higher thermal conductivities.

The role of normal processes in the low-temperature thermal resistivity has been discussed in [1]. The phonon mean free path for the normal processes has been given in [22, 59, 60].

6. BOUNDARY RESISTANCE

At sufficiently low temperatures, the phonon mean free path is limited by the external boundaries. Under those conditions

$$k(T) = \tfrac{1}{3}C(T)vL \tag{15}$$

where L is of the order of the shortest linear dimension of the specimen. The temperature dependence of thermal conductivity would then parallel the temperature dependence of the specific heat due to lattice waves, and $k(T)$ would vary as T^3 at lowest temperatures.

In many cases $C(T)$ is not only due to lattice waves, but there are additional thermal excitations which contribute to the specific heat, and cause substantial deviations from the T^3 dependence expected for lattice waves. Those excitations would have to be treated as separate carriers according to equation (1). In many cases they have either a low group velocity, or a short mean free path, or both. It is thus possible that crystals having substantial deviations from a T^3 dependence of $C(T)$ may still have a T^3 variation of $k(T)$ at low temperatures.

In a polycrystalline aggregate the phonon mean path L would be given not by the external dimensions, but by the size of the individual crystallites or grains. Such grain sizes, typically of the order of tens of microns, would result in a corresponding depression of $k(T)$, and the extension of the region where $k(T) \propto C(T)$ to higher temperatures than in the case of single crystals. In that case, the temperature range where boundary scattering dominates would include a temperature range where $C(T)$ varies more slowly than T^3. Such cases have been reviewed by Berman [5].

The effective value of L has been calculated by Casimir [61] for the case of heat flow along the axis of a narrow cylinder with perfectly rough surfaces. He assumed that phonons are diffusively scattered at the surface and that there is no effect from the end surfaces. In this case, L depends entirely on the geometry. In many cases, however, phonons arriving at the surfaces may be scattered with a forward bias—this would lengthen the effective values of L correspondingly. This forward bias (often crudely expressed as a fraction specularly reflected) may be a function of phonon frequency. This would make $l(\nu)$ dependent on frequency, and lead to corresponding deviations from the T^3 dependence of the thermal conductivity. Casimir's theory has been extended to apply to short cylinders with smooth surfaces [1, 62, 63]. The factors controlling the degree of specularity of real surfaces for phonons of the relevant wavelength (typically several hundred angstroms) are at present not understood.

The long cylinder is not the only important geometry. Another extreme case is the thin slab, with conduction across the slab. In this case, L is of the order of the slab thickness, and is governed by the acoustic emissivity and reflectivity, respectively, of the surfaces.

Since boundary resistance is a resistive process which is not uniformly distributed throughout the volume of the specimen, one should not really use the additivity relation (6) to obtain the combined effect of boundary scattering and other scattering processes. Deviations from (6) are likely to be significant, but will not change the order of magnitude of the corresponding resistances.

An interesting case, first discussed in reference [3], is the case when the important processes are boundary scattering and "normal" three-phonon processes (11a). The normal processes actually decrease the resistivity by slowing down the rate at which phonon "momentum" is brought to the surface and obliterated. One can define an effective viscosity of the phonon gas in terms of the normal-process mean free path. The heat transport or phonon flow is then analogous to Poiseuille flow in a pipe, while the more usual boundary resistance, as discussed by Casimir [61], corresponds to Knudsen flow. This phenomenon occurs only in very perfect crystals near the conductivity maximum.

7. THERMAL RESISTANCES DUE TO LATTICE IMPERFECTIONS

Nearly all imperfections scatter lattice waves and reduce the lattice thermal conductivity, particularly at low and intermediate temperatures, where the intrinsic phonon mean free path is long. In all cases, one can regard the imperfection as a local change in the velocity of the lattice wave, either because of a change in mass or density, or because of a change in the elastic properties. These local changes in wave velocity act as a perturbation, and scattering is usually calculated by perturbation theory [3, 71], though in the simpler defects it has been possible to arrive at a more self-consistent solution of the wave equation in the presence of a defect through Green's function techniques [87, 88].

According to the simplest perturbation technique, the Born approximation, the perturbation, which is a function of position, can be Fourier analyzed, and the scattering of a wave or phonon from mode $\mathbf{q}$ into $\mathbf{q}'$ involves the $(\mathbf{q} - \mathbf{q}')$th Fourier component of the perturbation. The amplitude of the scattered wave $\mathbf{q}'$ is proportional to that Fourier component, the intensity of scattering to the square of the perturbation, so that total scattering cross sections are generally proportional to the square of the perturbation. This has been reviewed by several authors [1, 3, 4, 7]. The theory bears close resemblance to other theories describing scattering of waves by obstacles.

It follows that, for waves of a particular wavelength, the most important obstacles are those having dimensions comparable to the wavelength. At intermediate and high temperatures, the important defects are thus mainly point defects, while extended defects are more important at lower temperatures.

Another factor to be considered, particularly when the defect scattering cross section varies rapidly with frequency (e.g., point defects and concentrated line defects, where $1/l$ varies as ν^4 and ν^3, respectively), is the interplay with other interaction processes. Not only is it essential to consider other interaction processes such as boundary or anharmonic scattering to remove the low-frequency divergence, but one must consider in detail the role of the "normal" three-phonon processes. These do not reduce the momentum or heat flow of the phonon gas, but they can move momentum from one group of lattice waves to another, and by moving momentum to a group of modes for which the defect scattering is highly effective (such as modes of higher frequency), these processes can contribute indirectly to the thermal resistivity. The exact description of this interplay of processes is mathematically complicated; approximations have been given by Klemens [25] and Callaway [32]; for a recent review see reference [1].

Point defects will depress k_g near the thermal conductivity maximum and, less spectacularly, at higher temperatures. They will tend to obliterate the exponential dependence expected for the intrinsic thermal conductivity (14a), and lead to a thermal conductivity roughly proportional to $1/T$.

At high temperatures, however, point defects will lead to a slower temperature dependence of the thermal conductivity; furthermore, the resistivity due to point defects will not increase linearly with concentration. This is a consequence of the properties of the integral of equation (5), with $C(\nu)$ given by the high-temperature limit of equation (8), i.e., $C(\nu) \propto \nu^2$, and $1/l(\nu)$ being additively composed of point-defect scattering, proportional to ν^4, and intrinsic scattering given by equation (12). In the limit of high temperatures and high defect concentration c, one can show [65]

$$k_g \propto [c(1 - c)]^{-1/2} T^{-1/2} \tag{16}$$

Quantitative estimates of the strength of point defect scattering can be made in terms of the difference between the mass of a normal site and a defect site [25] and in terms of the volume misfit [1, 7].

The magnitude and frequency dependence of the scattering by point defects is also sensitive to short-range order or other types of correlation. The formal theory [122] is analogous to the theory of x-ray scattering, however, there is little information about such correlations in nonmetals and no detailed calculations have been made.

Extended defects which have been treated theoretically include dislocations and stacking faults. In the simple theory the dislocations are treated as fixed in position, scattering because of their long-range strain fields. In nonmetals, where dislocation densities are usually lower, this may not be an adequate model. Whether one must seek a solution in terms of the flutter of dislocations under the applied stress of a low-frequency lattice wave, or in terms of electrostatic interactions between a charged core and other charges in the solid is not clear at present.

Pertinent references to the theory of thermal resistance due to lattice imperfections at high temperatures are [64–69] and at low temperatures include [70–120], of which the references [92–120] deal with point defects of the general type and explain the thermal resistance in terms of phonon-defect resonant scattering processes.

8. AMORPHOUS SOLIDS

In amorphous solids, the mean free path of lattice waves is short; in fact, it is questionable whether one may describe the thermal vibrations in terms of lattice waves. Kittel [121] suggested that the thermal conductivity can be described by equation (3) with a mean free path which is quite short (typically of the order of 10 angstroms) and independent of temperature. This seems to hold at high and intermediate temperatures. At low temperatures, however, when the wavelength of the lattice waves becomes longer, the amorphous solid resembles more and more an elastic continuum, and the randomness of the underlying atomic structure becomes less important. This means that, at low temperatures, l_g should increase with decreasing temperature, as is indeed found. A formal theory of the scattering of elastic waves in such random structures has been given [122]. One interesting consequence is that the role of order is reversed from what one would expect intuitively. The role of order or structural coherence is now to reinforce the scattering process. However, our knowledge of the structure of amorphous material is so rudimentary that we have at the moment no predictive theory. Other references to the theory of the thermal conductivity of amorphous solids include [123–126].

9. RESONANT SCATTERING

So far we have considered defects which scatter passively, that is, scattering is due to local variations in the wave velocity. There are geometrical interferences, leading to reinforcement or destructive interference of the scattered wave. However, it is also possible to have dynamic or time-like interferences. The defect may have internal mechanical resonance frequencies [92–120], or it may have an electronic level structure [127–136]. Lattice waves at or near the corresponding resonance frequencies will be scattered much more strongly than one would expect from passive defects of comparable size. This leads to changes in the temperature dependence of the thermal conductivity, in extreme cases to dips in the conductivity. The theory of mechanical resonances is not well understood. In the case of some paramagnetic impurities whose electronic level structure is strongly strain dependent, the resonant scattering of lattice waves can be calculated from the spin–lattice relaxation. Effects of this type are clearly important in many crystals at low temperatures, particularly in the case of semiconductors with donor or acceptor impurities or radiation damage.

10. OTHER CASES

We have so far discussed the contribution of lattice waves (or phonons) to the thermal conductivity of ordinary nonmetallic solids and the various scattering mechanisms which reduce the phonon mean free path and cause thermal resistances. For other special kinds of solids such as magnetic insulators, solids partially transparent to infrared radiation, and semiconductors which are practically insulators for purposes of thermal conductivity, we have to consider other mechanisms of heat transfer and other mechanisms of resistivity.

In a magnetic insulator at low temperatures, cooperative effects between the magnetic moments arranged in a regular lattice, leading to the concept of spin waves or magnons, can act both as a new mechanism of heat transport and at the same time as a resistive mechanism of phonon transport [137–164]. The exchange energy between neighboring spins of atoms is probably a rough criterion of the upper limit of temperature at which these effects need to be considered. At very low temperatures the spin-wave thermal conductivity of a ferromagnetic insulator is roughly proportional to T^2. Consequently, at sufficiently low temperatures where the mean free paths of both magnons and phonons become boundary limited, frequency independent, and of equal magnitude, the magnon conductivity, varying as T^2, should exceed the phonon conductivity, which varies as T^3 under these conditions. The total thermal conductivity is the sum of the magnon and phonon conductivities.

It is well known that the apparent thermal conductivity of glasses increases very rapidly at high temperatures and that the thermal conductivity versus temperature curves for many translucent crystals turn up at high temperatures after steady decrease and thus exhibit minima in the curve. This is due to the enhanced radiative (photon) heat transfer through the material at high temperatures in addition to the lattice (phonon) thermal conduction. If the material is completely transparent to infrared radiation, there is no interaction between radiation and material. If the material is opaque, heat transfer is entirely by conduction processes. Only for the intermediate case where the material is partially transparent to infrared radiation one can speak of a radiative component of thermal conductivity, which has been the subject of numerous investigations [165–178].

In a material which is partially transparent to infrared radiation, each volume element absorbs a part of the incident radiation and also reradiates radiant energy. A certain amount of energy is therefore transmitted through the material by these radiation and reradiation processes in addition to that conducted by lattice waves. The apparent total thermal conductivity is the sum of the lattice and radiative conductivities. In the limiting case, where the sample is optically thick, i.e., the thickness of the sample is much larger than the mean free path of the photons, the apparent radiative thermal conductivity is proportional to T^3 and inversely proportional to the Rosseland mean extinction coefficient of the material. In the opposite limiting case where the sample is optically thin, the apparent radiative thermal conductivity is proportional to the thickness of the sample as well as to T^3, and is therefore not an intrinsic property of the material.

Finally we must consider the thermal conductivity of semiconductors. In many semiconductors the lattice thermal conductivity is predominant and the electronic component is small enough to be neglected. There are, however, a few cases where the electronic component of thermal conductivity is not negligible at higher temperatures. At sufficiently low temperatures, a typical semiconductor is extrinsic, with only one band, either the conduction or valence band, making a contribution to conduction. At sufficiently high temperatures, it becomes intrinsic and both the conduction and valence bands make comparable contributions. This leads to the ambipolar diffusion of electrons and holes [179–181] and gives rise to the transport of ionization energy in addition to the normal direct transport of energy by the carriers. In the ambipolar diffusion process, electron–hole pairs are created at the hot end of the specimen and absorb energy from the heat source. They move down to the cold end of the specimen under the influence of the temperature gradient and recombine there at the cold end, giving up the ionization energy to the surroundings. The enhancement of the Lorenz function of an intrinsic semiconductor associated with the ambipolar diffusion process can be very significant.

In certain semiconductors the additional heat transport at high temperatures, which cannot be explained in terms of the conduction processes so far mentioned, has been attributed to the exciton contribution to thermal conductivity [182–184]. Excitons are bound electron–hole pairs which remain associated so that they are electrically neutral and can transport excitation energy but no electrical charge. However, the exciton contribution to heat transport is only a rather remote possibility and the experimental evidence is conflicting. The exciton contribution could be significant only if the excitons have sufficiently low excitation energy.

In semiconductors the lattice thermal resistances are caused by the same factors as those for dielectric crystals except for scattering of phonons by electrons [185–194]. However, this phonon–electron scattering is important only at very low temperatures. In the case of semiconductors having high carrier concentrations, the phonons are scattered by electrons (or holes) free to move in some sort of band which is not full [185, 186]. For low carrier concentrations, the phonons are scattered by bound electrons [189–191].

The thermal conductivity of semiconductors has been comprehensively reviewed in [11]. Some less extensive reviews are also available [195–201].

Experimental Determination of Thermal Conductivity

1. INTRODUCTION

In the experimental determination of the thermal conductivity of solids, a number of different methods of measurement are required for different ranges of temperature and for various classes of materials having different ranges of thermal conductivity values. A particular method may thus be preferable over the others for a given material and temperature range, and no one method is suitable for all the required conditions of measurement. The appropriateness of a method is further determined by such considerations as the physical nature of the material, the geometry of samples available, the required accuracy of results, the speed of operation, and the time and funds entailed.

The various methods for the measurement of thermal conductivity fall into two categories: the steady-state and the nonsteady-state methods. In the steady-state methods of measurement, the test specimen is subjected to a temperature profile which is time invariant, and the thermal conductivity is determined directly by measuring the rate of heat flow per unit area and temperature gradient after equilibrium has been reached. In the nonsteady-state methods, the temperature distribution in the specimen varies with time, and measurement of the rate of temperature change, which normally determines the thermal diffusivity, replaces the measurement of the rate of heat flow. The thermal conductivity is then calculated from the thermal diffusivity with a further knowledge of the density and specific heat of the test material.

The primary concern in most methods of measurement is to obtain a controlled heat flow in a prescribed direction such that the actual boundary conditions in the experiment agree with those assumed in the theory. Theoretically, the simplest method to obtain a controlled heat flow is to use a specimen in the form of a hollow sphere with a heater in the center. The heat supplied by the internal heater passes through the specimen in a radial direction without loss. However, in reality it is very difficult to fabricate a spherical heater which produces uniform heat flux in all radial directions. It is also difficult to fabricate spherical specimens and to measure the heat input and the temperature gradient in this experimental arrangement.

A more commonly used method of controlling heat flow in the prescribed direction is the use of guard heaters (combined with thermal insulation in most cases) so adjusted that the temperature gradient is zero in all directions except in the direction of desired heat flow. In most methods of measuring thermal conductivity, a cylindrical specimen geometry ranging from long rod to short disk is utilized, and the heat flow is controlled to be in either the longitudinal (axial) or the radial direction. Thus most methods can be subdivided into longitudinal and radial heat flow methods, as discussed in more detail later.

Experimental study of the thermal conductivity of solids was started in the eighteenth century. Benjamin Franklin in 1753 [202] seems to have been the first to point out the different ability of different materials "to receive and convey away the heat." He observed materials such as metal and wood to be good or poor conductors of heat by the degree of coldness felt when touched. Fordyce [203] pioneered in 1787 to make some experiments on the "conducting powers" of pasteboard and iron. The first steady-state *comparative* method for the measurement of the thermal conductivity of solids was suggested by Franklin and carried out by Ingen-Hausz as reported in 1789 [204]. This method was improved by Despretz as reported in 1822 [205]; Despretz's method was later used by Wiedemann and Franz as reported in 1853 [206] to determine the relative thermal conductivity of a number of metals leading to the postulation of the Wiedemann–Franz law. Since the first steady-state *absolute* method was reported in 1851 by Forbes [207, 208] (see also [209, 210]) and the first nonsteady-state *absolute* method was reported in 1861 by Ångström [211], a number of different methods and their variants have been developed over the years. Several general surveys [212–222] are available for the experimental developments of the methods. The mathematical theories of the methods have been reviewed in several books [223–227].

In the sections that follow, the major methods and the extent of their applicability will be briefly described and discussed. For finer details of experimental designs and techniques, the reader is referred to the references given to the individual methods.

In the category of steady-state methods, we will discuss the longitudinal heat flow method, the Forbes' bar method (which is a quasi-longitudinal heat flow method), the radial heat flow method, the direct electrical heating method, the thermoelectrical method, and the thermal comparator method. In the longitudinal and radial heat flow methods, a distinction is made between absolute and comparative methods according to the means of measuring the heat flow. In an absolute method, the rate of heat flow into a specimen is directly determined, usually by measuring the electrical power input to a heater at one end of the specimen. The rate of heat flow out of a specimen may be measured with a flow calorimeter or boil-off calorimeter. With the latter the rate of heat flow is determined by the boil-off rate of a liquid, such as water, of known heat of vaporization, while with the former it is determined by the flow rate and temperature rise of a circulating liquid, such as water, of known heat capacity. In a comparative method the rate of heat flow is calculated usually from the temperature gradient over a reference sample of known thermal conductivity which is placed in series with the specimen and in which hopefully the same heat flow occurs. The methods are subdivided further according to the various specimen geometries.

In the category of nonsteady-state methods, we will discuss the periodic and the transient heat flow methods. According to the direction of heat flow, each of them is also subdivided into longitudinal and radial heat flow methods. Within the transient heat flow methods, we will discuss also the flash method (which is a variant of the longitudinal heat flow method), the line heat source and probe methods (which are variants of the radial heat flow method), the moving heat source method, and two comparative methods.

2. STEADY-STATE METHODS

A. Longitudinal Heat Flow Methods

In the longitudinal heat flow methods, the experimental arrangement is so designed that the flow of heat is only in the axial direction of a rod (or disk) specimen. The radial heat loss or gain of the specimen is prevented or minimized and evaluated. Under steady-state conditions and assuming no radial heat loss or gain, the thermal conductivity is determined by the following expression from the one-dimensional Fourier–Biot heat-conduction equation [228, 229]:

$$k = \frac{-q\Delta x}{A\Delta T} \tag{17}$$

where k is the average thermal conductivity corresponding to the temperature $\frac{1}{2}(T_1 + T_2)$, $\Delta T = T_2 - T_1$, q is the rate of heat flow, A is the cross-sectional area of the specimen, and Δx is the distance between points of temperature measurements for T_1 and T_2. The different variants of this method are discussed separately below.

a. Absolute Methods

(*i*) *Rod Method.* This method is suitable for good conductors such as beryllium oxide and for all temperatures except for very high temperatures. In fact, this method has been used for most measurements below room temperature. The specimen used is in the form of a relatively long rod so as to produce an appreciable temperature drop along the specimen for precise measurement. A source of heat at a constant temperature is supplied at one end of the rod and flows axially through the rod to the other end where a heat sink at a lower constant temperature is located. The radial heat loss or gain of the rod should be negligible. In order to calculate the thermal conductivity from equation (17), it is necessary to measure the rate of heat flow into and/or out of the rod, the cross-sectional area, the temperatures of at least two points along the rod, and the distance between points of temperature measurements.

For measurements at cryogenic temperatures, radial heat loss does not constitute a serious problem, and thermal insulation and guard heaters are normally not necessary. The measurement is usually made under high vacuum to prevent gas conduction and convection, and a radiation shield surrounding the specimen may be used to minimize radiation losses. The heat is supplied to one end of the specimen by a heating coil of fine resistance wire (which may be wound directly onto the specimen to eliminate contact resistance between heater and specimen) or by a carbon resistor attached to the end. The temperatures may be measured by gas thermometers, vapor-pressure thermometers, thermocouples, resistance thermometers, or magnetic-susceptibility thermometers. General reviews of the low-temperature measurements and experimental techniques have been presented by White [230, 231]. For details of some of the useful low-temperature apparatus the reader may consult references [232–243].

For measurements at high temperatures, heat loss becomes a serious problem because radiant heat transfer increases rapidly with temperature. To prevent radial heat losses, a guard tube surrounding the specimen with controlled guard heaters may be utilized. Insulating powder is usually used to fill the space between the rod specimen and the guard tube, which should have the same temperature distribution along it as does the rod specimen. In fact, as early as 1887, Berget [244, 245] started the use of a guard ring surrounding and with the same temperature distribution as the specimen to prevent heat losses.

The rate of heat flow into the specimen may be determined by measuring the power input to a guarded electrical heater at the free end of the rod specimen [246–248] or by measuring the heat flow out of the specimen with a water-flow calorimeter at the low-temperature end [249], or by both [250–252]. Temperature measurements are made usually with thermocouples. In order to get correct temperature measurements and to minimize heat conduction along thermocouple leads, the thermocouples should be made of fine wires of low-conductivity alloys and the leads from the junction should be along isothermal lines.

This method, as used for measurements at high temperatures, has been comprehensively reviewed and discussed by Laubitz [253] and Flynn [254]. Systematic errors in measurements caused by the effects of heat losses, thermal contact resistance, poor thermocouple contacts, and temperature drift have been analyzed by Bauerle [255].

A variation of this method has been used [256–258] in which the specimen heater is located in a cavity at the center of the rod specimen and a heat sink is at each end. A mean value of the temperature gradient established toward the two ends is used for the thermal conductivity calculation.

(*ii*) *Plate* (*or Disk*) *Method.* This method is suitable for poor conductors and insulators. It is similar to the rod method except for the specimen length-to-width ratio being greatly reduced to a small fraction. This specimen geometry is favorable for measuring poor conductors because the smaller the length to width ratio, the smaller is the ratio of lateral heat losses to the heat flow through the specimen, and the shorter is the equilibrium time. The size of specimen used in various apparatus designed for different kinds of materials varies greatly. For apparatus designed to measure semiconductors, the specimen used may be about 1 cm wide [259], while the apparatus for measuring less homogeneous insulating or refractory materials may require a specimen of over one foot in width [260].

In this method, the thermal conductivity is also given by equation (17). The rate of heat flow may be determined by the electrical power input to a guarded heater [260–262], by a guarded water-flow calorimeter [263], by a boil-off calorimeter [264–267], or by a heat flow meter [268]. Temperature measurements are made generally with thermocouples inserted in the specimen or embedded in grooves on the specimen surfaces depending on the materials tested. Lateral heat losses may be prevented either by utilizing guard heaters or by using a large specimen, of which only a relatively small central area is used for measurement. In the first detailed mathematical analysis of the plate method reported in 1898, Peirce and Willson [269] found already that, if the radius of the specimen is five times larger than that of the central test section whose thickness equals its radius, the temperature at any point within the central test section would not sensibly differ from the temperature at the corresponding point in an infinite disk of the same thickness and same face temperatures. Further mathematical analyses of the errors due to lateral heat loss in guarded hot plate apparatus have been given in [270–272].

Detailed descriptions of recent apparatus for measurements at cryogenic temperatures can be found in the articles collected in [273], and for measurements at high temperatures in [274]. A comprehensive review of the plate method has been given in [275]. A description of the NBS steam calorimeter apparatus and some useful discussions on this method have also been given in [254].

There are two main kinds of experimental arrangements for the absolute plate (or disk) method: the single-plate system and the twin-plate system. The single-plate system [259, 262–269] requires only one specimen, which is placed between a hot plate and a cold plate, while the twin-plate system [260–261] requires two similar specimens to be sandwiched between a hot plate in the middle and two cold plates on the outside. The plate method employing the single-plate system was probably first used by Clément, whose experiment on copper was quoted by Péclet [276] in 1841. Péclet also used this method to measure the thermal conductivity of copper, and both of them obtained erroneous results. Later improvements on this method have been made by Peirce and Willson [269] and Lees [277] among others. The idea of a twin-plate system was developed by Lees [277] in 1898, but he did not actually adopt the twin-plate system for his plate method in the series of measurements reported in [277]. However, he used the twin-plate system in his experiments on the effect of pressure on thermal conductivity reported in 1899 [278]. Great improvement on the plate method employing the twin-plate system was made by Poensgen [261] in 1912, who introduced the guard-ring heater to the system as the prototype of the modern guarded hot-plate apparatus.

b. Comparative Methods

In the earliest steady-state comparative method suggested by Franklin and carried out by Ingen-Hausz [204] as reported in 1789, rods of various metals were coated with wax and heated at one end to a common temperature in a bath of hot water or oil. The wax melted over a greater distance on a rod of better conducting material, and under steady-state conditions the ratio of the conductivities of the rods is roughly proportional to the squares of these distances. The modern comparative methods are the divided-rod (or cut-bar) method and the comparative plate method as discussed below.

(*i*) *Divided-Rod* (*or Cut-Bar*) *Method.* The divided-rod method was originated by Lodge [279] in 1878 and later used by Berget [280], Lees [281], and many others. In this method a reference sample (or samples) of known thermal conductivity is placed in series with the unknown specimen with hopefully the same rate of heat flow through both the reference and the specimen. Under such ideal conditions, the thermal conductivity of the specimen is given by

$$k = k_r \frac{A_r(\Delta T/\Delta x)_r}{A(\Delta T/\Delta x)} \tag{18}$$

where the subscript r designates the reference sample.

This method may be divided into two distinct groups: the "long-specimen" type [280, 282, 283] for measuring the thermal conductivity of good conductors, and the "short-specimen" type [279, 281, 284–287] for measuring poor conductors.

Comparative methods have the advantages of simpler apparatus, easier specimen fabrication, and easier operation. Their disadvantages include additional measurement errors due to the required additional measurements of temperatures and thermocouple separations, difficulty in matched guarding, and lower accuracy due to the additional uncertainty in the conductivity of the reference sample, due to the conductivity mismatch between specimen and reference sample, and due to the interfacial thermal contact resistance. These have been carefully analyzed by Laubitz [253] and Flynn [254]. Flynn [254] has pointed out that the ASTM standard cut-bar method C408-58 [286] is not well designed and the data obtained by using this method can be subject to large errors.

(*ii*) *Plate* (*or Disk*) *Method.* This comparative method is suitable for poor conductors and insulators and is similar to the divided rod method in principle except that the specimen and the reference samples are now flat plates (or disks) sandwiched between a hot and a cold plate. Christiansen [288] was the first to report in 1881 the use of this type of comparative method in which he compared the thermal conductivity of liquids with that of air. Peirce and Willson [269] used this method to measure the thermal conductivity of marble slabs with glass plates as reference material for comparison. Sieg [289] employed the guard ring in his apparatus to prevent lateral heat loss.

c. Combined Method

In using a "combined" method, the apparatus combines the features of both absolute and comparative methods. The rate of heat flow is determined both through a reference sample placed in series with the specimen and simultaneously by a water-flow calorimeter [290–292] or by measuring the electrical power input to a heater [293]. In the measurements reported in [293], a "dual combined" method was employed in which a heater is located at the center of the divided rod between two short specimens with two longer reference samples at the two ends which are cooled by flowing water.

B. Forbes' Bar Method

Forbes' original method [207–210] consists of two separate experiments. The first was termed by Forbes the *statical*, and the second the *dynamical*, or cooling, experiment. In the *statical* experiment a square wrought iron bar with 1.25-inch side and 8 feet long was heated at one end by molten lead or solder at a fixed high temperature, and the steady-state temperature distribution along the bar was determined with the surface of the bar losing heat by convection and radiation to a constant-temperature environment. In the *dynamical* or cooling experiment a similar bar, but only about 20 inches long, was cooled in the same environment from a high uniform temperature and the rate of heat loss was determined. From these two experiments, the thermal conductivity may be computed as follows.

Replacing $\Delta x/\Delta T$ in equation (17) by dx/dT, differentiating the resulting equation with respect to x and rearranging gives

$$k = \frac{1}{A}\frac{dq}{dx}\frac{1}{d^2T/dx^2} \tag{19}$$

The statical experiment provides values for d^2T/dx^2, and the heat loss per unit time per unit length of the bar in the cooling experiment is

$$\frac{dq}{dx} = AC\frac{dT}{dt} \tag{20}$$

where dT/dt is the measured cooling rate and C the specific heat per unit volume.

Hogan and Sawyer [294] have improved this method so that it is not necessary to know the specific heat of the material. They used a thin long rod enclosed in an isothermal furnace. Radial heat loss from the specimen was determined by passing an electric current through the specimen and measuring the electric power required to maintain it at a temperature slightly above that of the furnace. This replaces Forbes' cooling experiment, and it is not necessary to know the specific heat since a steady-state condition is prevailing.

Hogan and Sawyer's method was further improved by Laubitz [295]. In his comprehensive review Laubitz [253] has discussed in detail the generalized Forbes' bar method, including the other major variants currently in use [296–298].

C. Radial Heat Flow Methods

There are several different types of apparatus all employing radial heat flow. The classification is mainly based upon specimen geometry. In the following we will briefly describe the cylindrical, spherical, ellipsoidal, concentric sphere, concentric cylinder, and plate methods. The reader is referred to the references given for individual methods for finer details. A comprehensive review of radial heat flow methods has been made by McElroy and Moore [299].

a. Absolute Methods

(*i*) *Cylindrical Method.* The cylindrical method uses a specimen in the form of a right circular cylinder with a coaxial central hole, which contains either a heater or a heat sink, depending on whether the desired heat flow direction is to be radially outward or inward. The use of this method was first reported by Callendar and Nicolson [300] in 1897 for measuring the thermal conductivity of cast iron and mild steel. The cylindrical specimens used were 5 inches in diameter and 2 feet long with 1-inch coaxial holes heated by steam under pressure. The outside of the cylinder was cooled by water circulating rapidly in a spiral tube. Niven [301] in 1905 also used the radial heat flow method for measurements on wood, sand, and sawdust. His method is close to the so-called hot-wire method developed by Andrews [302] in 1840 and Schleiermacher [303] in 1888 for measurements on gases. Kannuluik and Martin [304] used the hot-wire method for measurements on powders as well as on gases.

In the early experiments and also in many later designs [305–308], end guards are not employed. The effect of heat losses from the ends of the specimen is minimized by using a long specimen and monitoring the electric power within only a small section of the specimen away from the ends.

The guarded cylindrical method employing end guards at both ends of the specimen to prevent axial heat losses was developed by Powell [309] and first reported in 1939 for measurements on Armco iron at high temperatures. In the guarded cylindrical method the specimen is generally composed of stacked disks with a coaxial central hole containing either a heater or a heat sink. Temperatures within the specimen are measured either by thermocouples or by an optical pyrometer. For details of some of the useful apparatus employing the guarded cylindrical method, the reader may consult references [299, 309–314].

The thermal conductivity is calculated from the expression

$$k = \frac{q \ln(r_2/r_1)}{2\pi l(T_1 - T_2)} \tag{21}$$

where l is the length of the central heater and T_1 and T_2 are temperatures measured at radii r_1 and r_2, respectively.

Hoch *et al.* [315] have developed a quasi-radial heat flow method in which a metallic specimen in the form of a disk or short cylinder is heated at its convex cylindrical surface in high vacuum by means of high frequency induction and is losing heat from its flat circular end faces by radiation. In this method the inward flow of heat from the cylindrical surface, at which the heat generation is localized, into the interior of the specimen is, of course, not strictly radial, and the temperature gradient of the flat circular end faces along the radius is related to the thermal conductivity. For measurements on nonmetallic solids such as Al_2O_3 [316], the convex surface of the specimen is covered with a metallic envelope. The theory of this method is improved by Vardi and Lemlich [317], and some of the previously published data are revised.

(*ii*) *Spherical and Ellipsoidal Methods.* In a spherical method, the heater is completely enclosed inside the specimen which is in the form of a hollow sphere. The heat supplied by the internal heater passes through the specimen radially without loss. Theoretically, this method is ideal. However, there are a number of practical difficulties such as difficult fabrication of a spherical heater which produces uniform heat flux in all radial directions, difficult fabrication of spherical specimen, difficult positioning of thermocouples along spherical isotherms, etc., which have prevented this method from being popular. Laws, Bishop, and McJunkin [318] seem the first to have used this method on solids (not loose-filled materials). A detailed description of a modern design may be found in [305]. The thermal conductivity is calculated from the expression

$$k = \frac{q(1/r_1 - 1/r_2)}{4\pi(T_1 - T_2)} \tag{22}$$

The ellipsoidal method is similar, but has some advantages over the spherical method. It was developed by a group of researchers at MIT [319–321]. The major advantage of using a specimen in the form of an ellipsoid instead of a sphere is that the isothermal surfaces near the plane of the minor axes of an ellpsioid are rather flat so that straight thermocouple wires can be used without ill effect. If a is half the focal length of the ellipsoid and T_1 and T_2 are temperatures measured at respectively two radii r_1 and r_2 on the minor axis, the thermal conductivity is determined by the expression

$$k = \frac{q}{8\pi a(T_1 - T_2)} \times \ln\left[\frac{\sqrt{(a^2 + r_2^2)} - a}{\sqrt{(a^2 + r_2^2)} + a} \cdot \frac{\sqrt{(a^2 + r_1^2)} + a}{\sqrt{(a^2 + r_1^2)} - a}\right] \tag{23}$$

Despite the aforementioned advantage, the ellipsoidal method is also rarely used due to the other experimental difficulties common to both the ellipsoidal and spherical methods.

(*iii*) *Concentric Sphere and Concentric Cylinder Methods.* Concentric sphere and concentric cylinder methods are used mainly for measurements on powders, fibers, and other loose-filled materials. The specimen is filled in the space between two concentric spherical (or cylindrical) shells, with the inner sphere (or cylinder) being a heater or a heat sink. In a concentric cylinder apparatus, end guards are usually used to prevent axial heat flow.

A concentric sphere method was first used by Péclet [322] and reported in 1860 with the inner sphere filled with hot water as heater. However, a steady-state condition was not achieved in his pioneering measurements. Later Nusselt [323] succeeded in using this method for measurements on insulating materials with an electric heater installed inside the inner spherical shell. A modern apparatus using a boil-off calorimeter in the inner sphere was described in [324].

A concentric cylinder method was used by Stefan [325] and reported in 1872 for the measurements on gases. It was later adopted for measuring loose-filled materials. Reference [326] describes a modern apparatus employing a guarded boil-off calorimeter inside the inner cylinder. Recently, Flynn and Watson [327] used a concentric cylinder method to measure the high-temperature thermal conductivity of soil.

(*iv*) *de Sénarmont's Plate Method.* de Sénarmont [328–332] in 1847–48 used a radial heat flow plate method to determine the anisotropy in thermal conductivity of crystalline substances. However, this method does not yield absolute values of thermal conductivity, and furthermore the axial heat loss is not prevented.

In his method, a thin plate of the sample was coated with a thin film of white wax; and heat was applied at a central point by means of a hot, thin silver tube tightly fitted in a hole at the center of the plate. The wax melted around the region where heat was supplied and the bounding line of the melted wax was the visible isotherm, the shape of which indicated the variation of thermal conductivity in the different directions.

If the substance is isotropic, the bounding curve of the melted wax is a circle, whereas for anisotropic substances, this curve is elliptical. In such a case, the ratio of the two thermal conductivities k_a and k_b along the two axes a and b of the ellipse is given by the expression

$$\frac{k_a}{k_b} = \left(\frac{a}{b}\right)^2 \tag{24}$$

Powell [333] has modified the method in his simple test for anisotropic materials. In testing gallium, he cooled a slice of crystal locally by means of a piece of solid carbon dioxide and observed the contours of the dew and frost areas which formed around the cooled zone. For testing graphite, he followed de Sénarmont's original method, but the surface of the plate used was covered with frost by precooling instead of being coated with wax.

b. Comparative Methods

(*i*) *Concentric Cylinder Method.* This method has been used for measurements on some special materials such as those that are radioactive or reactive [334–336] and not for ordinary materials, because it does not have any major advantage over the absolute method. A typical apparatus of this kind consists of a cylindrical specimen which is surrounded by a concentric cylindrical reference sample of known thermal conductivity. A coaxial central hole in the specimen contains a heat source, which produces heat flowing radially through both the specimen and the reference sample. The advantage of using this method for measuring radioactive or reactive materials is that the reference sample which encloses the specimen serves also as a means of containment. The thermal conductivity is determined from the expression

$$k = k_r \frac{(T_3 - T_4)\ln(r_2/r_1)}{(T_1 - T_2)\ln(r_4/r_3)} \tag{25}$$

where T_1 and T_2 are two temperatures measured in the specimen at two radii r_1 and r_2, respectively, and T_3 and T_4 in the reference sample respectively at r_3 and r_4.

(*ii*) *Disk Method.* Robinson [337] developed a method, which he termed the "conductive-disk method," for comparative measurements on insulators. This method employs inward radial heat flow from a heater at the circular edge of a disk of suitable conductive reference material sandwiched between two like specimens, which are in turn sandwiched between two circular cold plates at a constant lower temperature. However, the heat flow in this case is not strictly radial, since, as the heat flows radially in the conductive disk toward the center, it flows also from the disk through the specimens to the cold plates. As a result, the steady-state temperature of the disk decreases toward its center, and the rate of decrease depends on the thermal conductivity of the specimens. Robinson obtained an expression for calculating the thermal conductivity of the specimens from the known thermal conductivity and thickness of the disk and from the temperatures of the cold plates and of the disk at its center and at a suitable radius.

D. Direct Electrical Heating Methods

In direct electrical heating methods, the specimen is heated directly by passing an electric current through it. These methods are therefore limited to measurements on reasonably good electrical conductors. Furthermore, they usually yield thermal conductivity in terms of electrical conductivity rather than directly. However, direct electrical heating methods also have certain advantages over other methods, and at high temperatures an increasing number of materials become sufficiently good electrical conductors. Direct electrical heating offers a means of easily attaining very high temperatures, uses simpler apparatus and experimental techniques than other methods at high temperatures, uses relatively small specimens, requires relatively short time to reach equilibrium, and also offers the possibility of concurrent determinations of a number of physical properties on the same specimen. According to specimen geometry, these methods fall into two major categories: cylindrical rod and rectangular bar. They will be briefly discussed below. Comprehensive reviews [338–340] on direct electrical heating methods are available.

The thermoelectrical method to be discussed later involves also the direct passage of an electric current through the specimen. However, in that method the specimen is heated (and cooled) by the Peltier effect which is totally different from the Joulean heating responsible for maintaining the specimen temperature in the direct electrical heating methods discussed here. It is therefore preferable to discuss the thermoelectrical method separately in another section.

a. Cylindrical Rod Methods

The direct electrical heating methods in this category involve heating specimens in the form of rods, thin wires, or tubes by the passage of regulated electric current and measuring potential drops and temperatures for the calculation of thermal conductivity.

There are many different techniques and variants that have been employed over the years since Kohlrausch [341–344] first developed this method. The different variants may be divided into three categories as discussed below.

(*i*) *Longitudinal Heat Flow Method.* In this method the rod is well insulated or guarded to prevent radial heat losses, so that the Joule heat generated in the specimen flows to the two ends. This is the method originally developed by Kohlrausch [341–344]. If the two ends of the rod are held at the same temperature and assuming that in a small temperature range the thermal and electrical conductivities are independent of temperature, the thermal conductivity is given by the simple relation

$$k = \frac{1}{8\rho} \frac{(V_1 - V_3)^2}{(T_2 - T_1)} \tag{26}$$

where ρ is the electrical resistivity, V_1 and V_3 are the electrical potentials at locations 1 and 3 on the specimen which are at equal and opposite distances from the midpoint 2, and T_1 and T_2 are temperatures at locations 1 and 2. This method was first used for actual measurements by Jaeger and Diesselhorst [345]. A variant of it has been used by Mikryukov [338]. The so-called "necked-down-sample method" [346] may also be considered as a longitudinal heat flow method.

(*ii*) *Radial Heat Flow Method.* This method uses a thick rod or tube and allows radial heat transfer. Under steady-state conditions the Joule heat generated in the specimen at regions remote from the ends flows radially to the surface and is then transferred by convection and radiation to the surroundings. This method was first suggested by Mendenhall and applied by Angell [347]. In the case of a cylindrical rod specimen and assuming that in a small temperature range the thermal and electrical conductivities are independent of temperature, the thermal conductivity is given by the simple relation

$$k = \frac{EI}{4\pi l(T_1 - T_2)} \tag{27}$$

where I is the electric current, E is the electrical potential drop over a length l at the central region of the specimen, and T_1 and T_2 are the temperatures at the axis and surface, respectively, of the rod at the central region. These temperatures were too small for precise measurements on metals, but Powell and Schofield [348] used it for poorer conducting carbon and graphite, and they also took account of the variation of thermal and electrical conductivities with temperature.

(*iii*) *Thin-Rod-Approximation Method.* The general form of the present method uses a long thin filament heated electrically in vacuum and allows both longitudinal heat conduction and lateral heat transfer by radiation. The "thin-rod approximation" involves the assumption that the temperatures and potentials in all planes normal to the specimen axis are uniform, i.e., their differences in the radial direction are negligible. Worthing [349] first employed this method for measurements on U-shaped filaments at incandescent temperatures. There are many variants [350–364] of this method, all with more or less different experimental designs, mathematical assumptions, and/or computational techniques.

Taylor, Powell, and co-workers [360, 362–364] at TPRC have made improvements and advancements on this method. They have taken the Thomson effect into account, which had never been done before, and have included the temperature dependence of various physical properties. They used the general equation directly, and their advanced computational techniques have eliminated the need for mathematical approximations and for matching certain experimental conditions.

It seems appropriate to mention the considerable discrepancies which have resulted from the data obtained by various workers, all of whom used different variants of the direct electrical heating method. One of the most recent of the TPRC papers [364] contains an interesting graphical presentation of all the determinations made on tungsten by these methods for the temperature range 1600 to 2800 K. Six of the fourteen groups of workers obtained results lying well above the recommended curve of Powell, Ho, and Liley [365], and one was well below it; the spread being of the order of 50 percent, 80 percent, and 70 percent at 1800, 2200, and 2600 K, respectively. The other seven had results within about 10 percent of the recommended curve, while the curve fitting the new results of [364] was some 3 to 5 percent below the recommended curve.

Earlier reports [339, 362] had contained examples of similar discrepancies for other high-melting-point metals, such as molybdenum, stainless steel, and platinum. The main reasons for these differences include failure to measure accurately small temperature gradients at high temperature, failure to match boundary conditions, errors resulting from simplifying mathematical approximations, and the use of temperature regions in which the thermal conduction term is small compared with the Joulean heating and radiation loss terms.

These have been quoted as examples of current experimental work at the TPRC, which became necessary because of the need to resolve some seriously discordant data and to gain further insight into their causes. The impression must not be given, however, that such discrepancies are confined to metals or to direct electrical heating methods. This is by no means the case, and the literature of heat conduction contains many examples of discordant results for all types of methods used. Titanium carbide, one of the materials dealt with in this volume may be mentioned. The first determinations reported on titanium carbide by Vasilos and Kingery [366] to high temperatures showed the thermal conductivity to decrease from about 0.2 W cm^{-1} C^{-1} at 200 C to 0.1 W cm^{-1} C^{-1} at 500 C and 0.04 W cm^{-1} C^{-1} at 1000 C. Two methods had been used: the divided-rod comparative method for a cube sample up to about 800 C and an ellipsoidal radial-flow method from about 500 to 1100 C. The former method gave results which were greater by 20 percent to 30 percent over their common temperature range. In 1961 Taylor [367] used a better-substantiated radial heat flow method for cylindrical samples of titanium carbide, and found the thermal conductivity to increase linearly from 0.38 W cm^{-1} C^{-1} at 600 C to 0.47 W cm^{-1} C^{-1} at 1600 C.

These two sets of values, differing at about 1000 C by about one order of magnitude and having temperature coefficients of opposite sign naturally aroused interest, and subsequent contributions by Laubitz [368], Hoch and Vardi [369], Powell [370, 371], and the nonsteady-state measurements of Taylor and Morreale [372] all supported the higher values of Taylor [367]. It would seem that the higher thermal conductivity of titanium carbide led to serious errors being associated with the method of Vasilos and Kingery, which were not apparent for substances of lower thermal conductivity. Incidently, had the much simpler measurement of electrical resistivity also been made, the unusually low resultant Lorenz function should have provided warning that abnormal data were being obtained. It might well be added that the inclusion of electrical resistivity measurements on all possible occasions is a simple extra measurement which also serves to provide very useful information about the properties of the material under test and its behavior on temperature cycling.

The foregoing example also indicates that users of the data tables of these volumes should, in the absence of any analysis that has produced a curve of recommended values, tend to be critical of the values presented, until these are seen to be well supported by independent experiments, correlations, or by additional checks such as that of a reasonable Lorenz function.

An additional outcome of the current TPRC investigation has been the development of a method and of equipment capable of determining a large number of high-temperature physical properties [373]. Their multiple-purpose apparatus is the first operational model that can accurately measure the thermal conductivity, electrical resistivity, total and spectral hemispherical emittance, Thomson coefficient, and Lorenz function on one and the same specimen. This apparatus can also measure the specific heat, enthalpy, thermal diffusivity, thermal expansion, Seebeck coefficient, Peltier coefficient, and Richardson coefficient. The merit of obtaining many different physical properties from one and the same specimen so as to permit meaningful quantitative cross-correlations between properties need not be emphasized here.

b. Rectangular Bar Method

This method was developed by Longmire [374] and is a geometrically-deformed variant of the radial heat flow method. The specimen used is in the form of a long rectangular bar. This special specimen geometry enables all temperature measurements to be made on the surface of the specimen. As the specimen is heated electrically in vacuum, the heat loss by radiation establishes a radial temperature gradient, and the temperature at the center line of the wider surface of the rectangular bar will be higher than that at the center line of the narrower surface. From measurements of these two temperatures, the electrical conductivity, and total hemispherical emittance of the bar, the thermal conductivity can be calculated using the equation derived by Longmire.

Longmire's method was improved by Pike and Doar [375–377] both in mathematical analysis and in experimental techniques. They further extended this method to the determination of anisotropy in thermal conductivity.

E. Thermoelectrical Method

The thermoelectrical method was developed by Borelius [378] and reported in 1917 for the combined measurement of the Peltier heat and thermal conductivity of the same material, and is particularly applicable to the measurements on thermoelectric materials.

In this method, the specimen is held between metallic contacts through which a small direct electric current is passed. Peltier heating thus occurs at one end of the specimen and Peltier cooling at the other end, which establishes a temperature gradient along the specimen. Under steady-state conditions, the rate of Peltier heat generation at the hot end is just balanced by the rate of heat conduction from the hot to the cold end. Thus the thermal conductivity can be calculated from the rate of Peltier heat production πI (π being the Peltier coefficient), the temperature difference between the ends ΔT, the cross-sectional area A, and the length l by the expression

$$k = \frac{\pi I l}{A \Delta T} \tag{28}$$

Since $\pi = ST$, S being the Seebeck coefficient, π can be determined by measuring the Seebeck coefficient from the potential difference between the ends after the temperature difference ΔT is established.

When the direct electric current is passed through the specimen, Joulean heating will occur, of course. However, the Joulean heating effect can be made negligibly small in a good thermoelectric material by choosing the current small enough, because the Joule heat production is proportional to I^2 while the Peltier heat production is proportional to I. The Thomson heat effect is generally small.

Borelius' method was used by Sedström [379, 380] for measurements on alloys. Some forty years later, Putley [381] and Harman [382, 383] reinvented this method. A recent apparatus is described in [384].

A transient thermoelectrical method was developed by Hérinckx and Monfils [385]. In this method a direct electric current is passed through the specimen and the time dependence of the resulting potential drop across the specimen is observed. The thermal conductivity can be derived from the shape and asymptote of this potential drop versus time curve provided that the Seebeck coefficient is known.

F. Thermal Comparator Method

The thermal comparator method was developed by Powell [386–389] and is a simple comparative method for the rapid, easy measurement of thermal conductivity.

The essential part of the thermal comparator is an insulated probe with a projecting tip. The probe is integral with a thermal reservoir held at a temperature about 15 to 20 degrees above room temperature. A surface thermocouple is mounted at the tip of the probe and is differentially connected to the thermal reservoir for the measurement of the temperature difference between the reservoir and the tip.

In operation, the probe is gently placed on the surface of the test material. Upon contact of the probe tip of known thermal conductivity k_1 and originally at temperature T_1 with the surface of the test material of thermal conductivity k_2 and at room temperature, T_2, the temperature of the probe tip drops quickly to an intermediate temperature, T, giving the expression

$$T_1 - T = (T_1 - T_2)\left(\frac{k_2}{k_1 + k_2}\right) \tag{29}$$

This temperature difference is registered by the emf reading of the differential thermocouple after a brief transient period (1 to 2 seconds) has elapsed.

From the emf readings of tests on a series of reference samples of known thermal conductivity, a calibration curve is obtained, and the thermal conductivity of an unknown specimen can thus be determined from the emf reading through the calibration curve.

Powell [390] has made a comprehensive review on this method. Some subsequent developments are discussed in [391]. The *thermal comparator* has been developed by TPRC as an instrument [391] for the rapid determination of the thermal conductivity of solids and liquids and is commercially available from The McClure Park Corp., West Lafayette, Indiana.

3. NONSTEADY-STATE METHODS

In nonsteady-state methods, the temperature distribution in the specimen varies with time. The rate of temperature change at certain positions along the specimen is measured in the experiment, and no measurement of the rate of heat flow is required. These methods normally determine the thermal diffusivity, from which the thermal conductivity can be calculated with an additional knowledge of the density and specific heat of the test material. Nonsteady-state methods fall into two major categories, the periodic and the transient heat flow methods, as briefly discussed below. These methods have been comprehensively reviewed by Danielson and Sidles [392], and will be dealt with in Volume 10 of the present TPRC Data Series.

A. Periodic Heat Flow Methods

In periodic heat flow methods, the heat supplied to the specimen is modulated to have a fixed period. The resulting temperature wave which propagates through the specimen with the same period is attenuated as it moves along. Consequently, the thermal diffusivity can be determined from measurements of the amplitude decrement and/or phase difference of the temperature waves between certain positions in the specimen. In most of the periodic heat flow methods, heat flow is in the longitudinal (axial) direction. However, methods with heat flow in the radial direction have also been used.

a. Longitudinal Heat Flow Method

The periodic heat flow method was first developed by Ångström [211, 393] and reported in 1861. In his method a variable heat source capable of producing a sinusoidal temperature variation was attached to the center of a long thin rod specimen, and the temperatures as a function of time at two positions l apart towards the ends of the rod, were measured. From these temperature–time measurements, the velocity, v, and the amplitude decrement, δ, of the temperature wave can be determined for the calculation of thermal diffusivity. This method has been modified and improved by King [394] and others [395–397]. The thermal diffusivity may be calculated from the expression [396]

$$D = \frac{vl}{2 \ln \delta} \tag{30}$$

The Ångström method, which uses a long rod, has its limitations. In some cases, specimens in the form of long rods may not be available, and in other cases, such as in the measurements on poor conductors at high temperatures, heat guarding to prevent lateral heat losses from a long rod may be difficult. Consequently, methods using specimens in the form of a small plate or disk have been developed [398–400].

b. Radial Heat Flow Method

In this method, the specimen in the form of a cylinder is heated by a heat source capable of producing a periodical temperature variation either at the axis or at the circumference, and the radial temperature variations with time are measured. The thermal diffusivity may be calculated from the phase change of the temperature oscillations, or from the amplitude variation of the oscillations with frequency.

Tanasawa [401] used this method in 1935 for the measurements on humid materials. In his method, a sinusoidal temperature was produced on the surface of a cylindrical specimen, and the temperatures at different radial distances were measured for the calculation of thermal diffusivity.

Filippov and his co-workers have further developed a method of this type [402] and used it for the measurements on metals [403] and molten metals [404, 405] at high temperatures.

The nonsteady-state radial heat flow method has also been employed for measurements on insulators [406, 407].

B. Transient Heat Flow Methods

Transient heat flow methods, both longitudinal and radial, were first used by Neumann [408, 409] and reported in 1862. In his method, one end of a bar was heated by a flame until the temperature attained the equilibrium state. The flame was then suddenly removed and the temperatures at two positions along the bar were measured as a function of time. Thermal diffusivity can then be calculated from these measurements. For the measurements on poor conductors he used another method in which a cube or sphere was heated uniformly to a high temperature and then was allowed to cool in the air. The temperatures at the surface and at the center were measured as a function of time.

The modern transient heat flow methods have a wide variety. In the following a number of the major variants are briefly discussed.

a. Longitudinal Heat Flow Method

Similar to the longitudinal periodic heat flow method, the longitudinal transient heat flow method can also be subdivided into two major categories: those using a long rod and those using a small plate (or disk).

Methods in which one end of a long rod, which is initially at uniform temperature, is subjected to a short heating pulse have been developed [410, 411]. There are also methods in which steady heating is provided at one end of a rod and the temperatures as a function of time at two or more positions along the rod are observed [412–414].

Transient heat flow methods in which the specimen used is in the form of a small plate or disk have been developed by a number of workers [415–418].

b. Flash Method

Although the flash method is a variant of the longitudinal transient heat flow method using a small thin disk specimen geometry, it has a very special feature which makes it a class of its own. In the "flash" method, a flash of thermal energy is supplied to one of the surfaces of a disk specimen within a time interval that is short compared with the time required for the resulting transient flow of heat to propagate through the specimen. This method was developed by Parker, Jenkins, Butler, and Abbott [419] and reported in 1961.

In use, a heat source such as flash tube or laser supplies a flash of energy to the front face of a thin disk specimen and the temperature as a function of time at the rear face is automatically recorded. The thermal diffusivity is given from the thickness of the specimen, l, and a specific time, $t_{1/2}$, at which the back face temperature reaches half its maximum value by the expression

$$D = 1.37\, l^2/\pi^2 t_{1/2} \tag{31}$$

Other expressions for the calculation of thermal diffusivity have also been used.

Subsequent improvements on this method have been made [420, 421] by the application of corrections for the finite pulse-time effect and the radiation-loss effect.

c. Radial Heat Flow Method

As mentioned before, a radial heat flow method was used by Neumann [408, 409] for measurements on poor conductors. His specimens were of spherical shape.

In modern apparatus, specimens in the form of cylinders are used. A long cylindrical specimen, hollow or solid, which is initially at uniform temperature, is heated either at the axis or at the outer surface and the temperatures as a function of time at different radial distances are measured. In the methods developed by Ginnings [422] and by Cape, Lehman, and Nakata [423], cylindrical specimens were continuously heated at the outer surface.

Specimens in the form of hollow disks stacked on an axial heater with outer disks as end guards have been used by Carter, Maycock, Klein, and Danielson [424].

Although the line heat source and probe methods are also radial transient heat flow methods, they are quite different from other methods and will be discussed in a separate section below.

d. Line Heat Source and Probe Methods

The line heat source method was originally developed by Stalhane and Pyk [425] in 1931 and used for measurements on ceramic materials [426]. This method is suitable for the measurements on loose-filled materials such as powders.

In this method, a long thin heater wire which serves as a line heat source was embedded in a large specimen initially at uniform temperature. The heater is then turned on, which produces constant heat, q, per unit length and time, and the temperature at a point in the specimen is recorded as a function of time. The thermal conductivity is given by the expression

$$k = \frac{q}{4\pi(T_2 - T_1)} \ln\frac{t_2}{t_1} \tag{32}$$

where $(T_2 - T_1)$ is the temperature difference at two times t_1 and t_2. Subsequently, this method was also developed by van der Held and his co-workers [427, 428] and others.

The probe method is a more practical line heat source method in which the heat source is enclosed inside a probe for protection and for easy insertion into a sample. This method was developed by Hooper and his co-workers [429, 430] and others. Blackwell [431, 432] has derived theoretical treatments for practical departures from a true line source, and in the discussion of a paper [433] dealing with the use of a probe method in connection with the routing of electric power cables, he advocated the use of very small thermistors as an alternative to thermocouples.

e. Moving Heat Source Method

The moving heat source method was developed by Rosenthal and his co-workers [434–436] and involves the establishment of a quasi-steady-state temperature distribution in a long tubular-shaped specimen heated by a moving localized heat source of constant intensity. As the heat source approaches and moves away, each point in the specimen is subjected to a temperature rise and fall. When the heat source passes over the specimen, the temperature at a point remote from the ends is recorded as a function of time. From this record, a curve of the logarithm of the temperature variation with time is made. The thermal diffusivity is given from the velocity of the heat source, v, and the slopes P_r and P_f on the rising and falling portions of the curve at the same temperature by the expression

$$D = \frac{v^2}{P_r + P_f} \tag{33}$$

f. Comparative Method

A comparative method employing transient heat flow was developed by Hsu [437, 438]. In this method, two identical sets of composite blocks are used. Each set consists of a test specimen and a reference sample whose properties are known. Initially the two sets are heated separately to uniform but different temperatures, and then they are suddenly brought into contact, with the two test specimens touching each other. The transient temperature at the contact plane between the test specimen and reference sample corresponding to a certain time is measured, and from this the thermal diffusivity of the specimen can be calculated.

Another transient-heat-flow comparative method has been used by Deem *et al.* [439] for the measurements on irradiated materials. The method of measurement is to place the lower ends of a specimen and a reference sample, which are of the same size and initially at room temperature, in molten tin maintained at a constant elevated temperature and then measure the times required for the tops of them to reach a predetermined intermediate temperature. The ratio of the thermal diffusivities is assumed directly proportional to the ratio of the two times measured for the specimen and the reference material.

References to Text

Review Papers and Books

1. Klemens, P. G., "Theory of the Thermal Conductivity of Solids," in *Thermal Conductivity* (Tye, R. P., ed.), Vol. 1, Chap. 1, Academic Press, London, 1–68, 1969.
2. Klemens, P. G., "Thermal Conductivity of Solids at Low Temperatures," in *Handbuch der Physik* (Flügge, S., ed.), Vol. 14, Springer-Verlag, Berlin, 198–281, 1956.
3. Klemens, P. G., "Thermal Conductivity and Lattice Vibrational Modes," in *Solid State Physics*, Vol. 7, Academic Press, New York, 1–99, 1958.
4. Ziman, J. M., *Electrons and Phonons*, Oxford University Press, 554 pp., 1960.
5. Berman, R., "The Thermal Conductivity of Dielectric Solids at Low Temperatures," *Advan. in Phys.* (Phil. Mag. Suppl.), **2**, 103–40, 1953.
6. Leibfried, G., "Lattice Theory of the Mechanical and Thermal Properties of Crystals," in *Handbuch der Physik*, Vol. 7, 105–324, 1955.
7. Carruthers, P., "Theory of Thermal Conductivity of Solids at Low Temperatures," *Rev. Modern Phys.*, **33**(1), 92–138, 1961.
8. Berman, R., "Heat Conductivity of Non-Metallic Crystals at Low Temperatures," *Cryogenics*, **5**(6), 297–305, 1965.
9. Berman, R., "Heat Conduction in Non-Metallic Crystals," *Sci. Progr. (Oxford)*, **55**, 357–77, 1967.
10. Peierls, R. E., *Quantum Theory of Solids*, Oxford University Press, 229 pp., 1955.
11. Drabble, J. R. and Goldsmid, H. J., *Thermal Conduction in Semiconductors*, Pergamon Press, New York, 235 pp., 1961.
12. Rosenberg, H. M., *Low Temperature Solid State Physics*, Oxford University Press, 420 pp., 1963.

Lattice Thermal Conductivity

13. Eucken, A., "On the Temperature Dependence of the Thermal Conductivity of Nonmetallic Solids," *Ann. Physik*, 4, **34**(2), 185–221, 1911.
14. Eucken, A., "Thermal Conductivity of Some Crystals at Low Temperatures," *Physik. Z.*, **12**, 1005–8, 1911.
15. Eucken, A., "Thermal Conductivity of Some Crystals at Low Temperatures," *Verhandl. Deut. Physik Ges.*, **13**, 829–35, 1911.
16. Debye, P., "Equation of State and the Quantum Hypothesis with an Appendix on Thermal Conduction," in *Vortrage über die kinetische Theorie der Materie und der Elektrizität*, von Planck, M., *et al.* (Mathematische Vorlesungen an der Universität Göttingen: VI.), Teubner, Leipzig and Berlin, 19–60, 1914.
17. Peierls, R. E., "The Kinetic Theory of Heat Conduction in Crystals," *Ann. Physik*, 5, **3**, 1055–1101, 1929; English translation: *OTS*, AEC-TR-1849, 1–67. [TPRC No. 28 528].
18. Peierls, R. E., "Some Typical Properties of Solid Bodies," *Ann. Inst. Henri Poincaré*, **5**, 177–222, 1935.
19. Fröhlich, H. and Heitler, W., "Time Effects in the Magnetic Cooling Method. II—The Conductivity of Heat," *Proc. Roy. Soc. (London)*, **A155**, 640–52, 1936.
20. Pomeranchuk, I., "Thermal Conductivity of Dielectrics at Temperatures Higher than the Debye Temperature," *J. Exptl. Theoret. Phys. (USSR)*, **11**, 246–54, 1941; English translation: *J. Phys. (USSR)*, **4**(3), 259–68, 1941.
21. Pomeranchuk, I., "The Thermal Conductivity of Dielectrics," *Phys. Rev.*, **60**, 820–1, 1941.
22. Pomeranchuk, I., "On the Thermal Conductivity of Dielectrics at Temperatures Lower than that of Debye," *J. Phys. (USSR)*, **6**(6), 237–50, 1942.
23. Pomeranchuk, I., "Heat Conductivity of Dielectrics at High Temperatures," *J. Exptl. Theoret. Phys. (USSR)*, **12**, 419–24, 1942; English translation: *J. Phys. (USSR)*, **7**(5), 197–201, 1943.
24. Akhieser, A. and Pomeranchuk, I., "Heat Conductivity of Salts used in the Magnetic Cooling Method," *J. Phys. (USSR)*, **8**, 216–8, 1944.
25. Klemens, P. G., "The Thermal Conductivity of Dielectric Solids at Low Temperatures (Theoretical)," *Proc. Roy. Soc. (London)*, **A208**, 108–33, 1951.
26. Herpin, A., "The Kinetic Theory of Solids," *Ann. Physik*, **7**, 91–139, 1952.
27. Leibfried, G. and Schlömann, E., "Heat Conduction in Electrically Insulating Crystals," *Nachr. Akad. Wiss. Göttingen, Math.-Physik. Kl.*, **2a**(4), 71–93, 1954; English translation: AEC-tr-5892, 1–36, 1963.
28. Mori, H., "A Quantum-Statistical Theory of Transport Processes," *J. Phys. Soc. Japan*, **11**(10), 1029–44, 1956.
29. Ziman, J. M., "The General Variational Principle of Transport Theory," *Can. J. Phys.*, **34**(12A), 1256–73, 1956.
30. Kubo, R., "Statistical–Mechanical Theory of Irreversible Processes. I. General Theory and Simple Applications to Magnetic and Conduction Problems," *J. Phys. Soc. Japan*, **12**(6), 570–86, 1957.
31. Kubo, R., Yokota, M., and Nakajima, S., "Statistical–Mechanical Theory of Irreversible Processes. II. Response to Thermal Disturbance," *J. Phys. Soc. Japan*, **12**(11), 1203–11, 1957.
32. Callaway, J., "Model for Lattice Thermal Conductivity at Low Temperatures," *Phys. Rev.*, **113**(4), 1046–51, 1959.

33. Keyes, R. W., "Laws of Corresponding States for the Thermal Conductivity of Molecular Solids," *J. Chem. Phys.*, **31**, 452–4, 1959.
34. Callaway, J., "Low-Temperature Lattice Thermal Conductivity," *Phys. Rev.*, **122**(3), 787–90, 1961.
35. Shieve, W. C. and Peterson, R. L., "Correlation Function Calculation of Thermal Conductivity," *Phys. Rev.*, **126**(4), 1458–60, 1962.
36. Hashin, Z. and Shtrikman, S., "Conductivity of Polycrystals," *Phys. Rev.*, **130**(1), 129–33, 1963.
37. Holland, M. G., "Analysis of Lattice Thermal Conductivity," *Phys. Rev.*, **132**(6), 2461–71, 1963.
38. Nettleton, R. E., "Foundation of the Callaway Theory of Thermal Conductivity," *Phys. Rev.*, **132**(5), 2032–8, 1963.
39. Gurzhi, R. N., "Thermal Conductivity of Dielectrics and Ferrodielectrics at Low Temperatures," *Zh. Eksptl. i Teoret. Fiz.*, **46**(2), 719–24, 1964; English translation: *Soviet Physics—JETP*, **19**(2), 490–3, 1964.
40. Luttinger, J. M., "Theory of Thermal Transport Coefficients," *Phys. Rev.*, **135**(6A), A1505–14, 1964.
41. Schieve, W. C. and Leaf, B., "Correlation Function Calculation of the Lattice Thermal Conductivity by Classical Liouville Methods," *Physica*, **30**, 1208–16, 1964.
42. Erdös, P., "Low-Temperature Thermal Conductivity of Insulators Containing Impurities," *Phys. Rev.*, **138**(4A), A1200–7, 1965.
43. Krumhansl, J. A., "Thermal Conductivity of Insulating Crystals in the Presence of Normal Processes," *Proc. Phys. Soc.* (*London*), 5, **85**(547), 921–30, 1965.
44. Krumhansl, J. A. and Guyer, R. A., "Extension of the Relaxation-Time Approximation to Solution of the Phonon Boltzmann Equation," *Bull. Am. Phys. Soc.*, **10**, 530, 1965.
45. Ranninger, J., "Lattice Thermal Conductivity," *Phys. Rev.*, **140**(6A), A2031–46, 1965.
46. Deo, B. and Behera, S. N., "Calculation of Thermal Conductivity by the Kubo Formula," *Phys. Rev.*, **141**(2), 738–41, 1966.
47. Guyer, R. A. and Krumhansl, J. A., "Solution of the Linearized Phonon Boltzmann Equation," *Phys. Rev.*, **148**(2), 766–78, 1966.
48. Guyer, R. A. and Krumhansl, J. A., "Thermal Conductivity, Second Sound, and Phonon Hydrodynamic Phenomena in Nonmetallic Crystals," *Phys. Rev.*, **148**(2), 778–88, 1966.
49. Ranninger, J., "Thermal Conductivity in Nonconducting Crystals," *Ann. Phys.*, **45**, 452–78, 1967.
50. Allen, K. R. and Ford, J., "Lattice Thermal Conductivity for a One-Dimensional, Harmonic Isotopically Disordered Crystal," *Phys. Rev.*, **176**(3), 1046–55, 1968.
51. Nil'sen, Kh. and Shklovskii, B. I., "Nonlinear Thermal Conductivity of Dielectrics in the Region of Viscous Flow of a Phonon Gas," *Fiz. Tverdogo Tela*, **10**(12), 3602–7, 1968; English translation: *Soviet Phys.—Solid State*, **10**(12), 2857–61, 1969.
52. Ranninger, J., "A Simple Microscopic Model for the Lattice Thermal Conductivity," *Ann. Phys.*, **49**(2), 297–308, 1968.
53. Hamilton, R. A. H. and Parrott, J. E., "Variational Calculation of the Thermal Conductivity of Germanium," *Phys. Rev.*, **178**(3), 1284–92, 1969.

Intrinsic Thermal Resistivity at High Temperatures

54. Akhieser, A., "On the Absorption of Sound in Solids," *J. Phys.* (*USSR*), **1**(4), 277–87, 1939.
55. Dugdale, J. S. and MacDonald, D. K. C., "Lattice Thermal Conductivity," *Phys. Rev.*, **98**(6), 1751–2, 1955.
56. Kontorova, T. A., "Thermal Expansion and Thermal Conductivity of Some Crystals," *Zh. Tekh. Fiz.*, **26**(9), 2021–31, 1956; English translation: *Soviet Phys.—Tech. Phys.*, **1**(9), 1959–69, 1956.
57. Lawson, A. W., "On the High Temperature Heat Conductivity of Insulators," *Phys. Chem. Solids*, **3**, 155–6, 1957.
58. Keyes, R. W., "High-Temperature Thermal Conductivity of Insulating Crystals: Relationship to the Melting Point," *Phys. Rev.*, **115**(3), 564–7, 1959.

Normal Processes

59. Landau, L. and Rumer, G., "On the Acoustic Absorption in Solids," *Physik. Z. Sowjetunion*, **11**(1), 18–25, 1937.
60. Herring, C., "Role of Low-Energy Phonons in Thermal Conduction," *Phys. Rev.*, **95**(4), 954–65, 1954.

Boundary Resistance

61. Casimir, H. B. G., "Note on the Conduction of Heat in Crystals," *Physica*, **5**(6), 495–500, 1938.
62. Berman, R., Simon, F. E., and Ziman, J. M., "The Thermal Conductivity of Diamond at Low Temperatures," *Proc. Roy. Soc.* (*London*), **A220**(1141), 171–83, 1953.
63. Berman, R., Foster, E. L., and Ziman, J. M., "Thermal Conduction in Artificial Sapphire Crystals at Low Temperatures. I. Nearly Perfect Crystals," *Proc. Roy. Soc.* (*London*), **A231**(1184), 130–44, 1955.

Thermal Resistance due to Lattice Imperfections at High Temperatures

64. Ambegaokar, V., "Thermal Resistance due to Isotopes at High Temperatures," *Phys. Rev.*, **114**(2), 488–9, 1959.
65. Klemens, P. G., "Thermal Resistance due to Point Defects at High Temperatures," *Phys. Rev.*, **119**(2), 507–9, 1960.
66. Klemens, P. G., "Lattice Thermal Resistance at Normal and High Temperatures due to Point Defects," Westinghouse Research Lab., Research Rept. 929-8904-R3, 1–8, 1961. [AD 277 223]
67. Klemens, P. G., White, G. K., and Tainsh, R. J., "Scattering of Lattice Waves by Point Defects," *Phil.* Mag., **7**(80), 1323–35, 1962.
68. Parrott, J. E., "The High-Temperature Thermal Conductivity of Semiconductor Alloys," *Proc. Phys. Soc.* (*London*), **81**, 726–35, 1963.
69. Abeles, B., "Lattice Thermal Conductivity of Disordered Semiconductor Alloys at High Temperatures," *Phys. Rev.*, **131**(5), 1906–11, 1963.

Thermal Resistance due to Lattice Imperfections at Low Temperatures

70. Pomeranchuk, I., "Thermal Conductivity of Dielectrics at Temperatures Lower than the Debye Temperature,"

J. Exptl. Theoret. Phys. (USSR), **12**, 245–63, 1942; English translation: *J. Phys. (USSR)*, **6**(6), 237–50, 1942.

71. Klemens, P. G., "The Scattering of Low-Frequency Lattice Waves by Static Imperfections," *Proc. Phys. Soc. (London)*, **A68**, 1113–28, 1955.
72. Berman, R., Foster, E. L., and Ziman, J. M., "The Thermal Conductivity of Dielectric Crystals: The Effect of Isotopes," *Proc. Roy. Soc. (London)*, **A237**(1210), 344–54, 1956.
73. Klemens, P. G., "Thermal Resistance due to Isotopic Mass Variation," *Proc. Phys. Soc. (London)*, **A70**(11), 833–6, 1957.
74. Slack, G. A., "Effect of Isotopes on Low-Temperature Thermal Conductivity," *Phys. Rev.*, **105**(3), 829–31, 1957.
75. Berman, R., Nettley, P. T., Sheard, F. W., Spencer, A. N., Stevenson, R. W. H., and Ziman, J. M., "The Effect of Point Imperfections on Lattice Conduction in Solids," *Proc. Roy. Soc. (London)*, **A253**, 403–19, 1959.
76. Carruthers, P., "Electric and Thermal Resistivity of Dislocations," *Phys. Rev. Letters*, **2**, 336–8, 1959.
77. Carruthers, P., "Scattering of Phonons by Elastic Strain Fields and the Thermal Resistance of Dislocations," *Phys. Rev.*, **114**(4), 995–1001, 1959.
78. Klemens, P. G., "Thermal Resistance due to Isotopes and Other Point Defects," *J. Phys. Chem. Solids*, **8**, 345–7, 1959.
79. Sproull, R. L., Moss, M., and Weinstock, H., "Effect of Dislocations on the Thermal Conductivity of LiF," *J. Appl. Phys.*, **30**(3), 334–7, 1959.
80. Callaway, J. and von Baeyer, H. C., "Effect of Point Imperfections on Lattice Thermal Conductivity," *Phys. Rev.*, **120**(4), 1149–54, 1960.
81. Bross, H., "The Effect of Defects on Lattice Thermal Conductivity at Low Temperatures," *Phys. Status Solidi*, **2**(5), 481–516, 1962; English translation: *CFSTI*, NP-TR-963, 1–73, 1962. [TPRC No. 23110]
82. Carruthers, P., "Thermal Conductivity of Solids, III. Modification of Three-Phonon Processes by Isotopic Scattering," *Phys. Rev.*, **126**(4), 1448–52, 1962.
83. Bross, H., Seegar, A., and Haberkorn, R., "Lattice Thermal Resistance of Edge Dislocations," *Phys. Status Solidi*, 3(**6**), 1126–40, 1963.
84. Greig, D., "Lattice Imperfections and the Thermal Conductivity of Solids," *Progr. in Solid State Chemistry*, **1**, 175–208, 1964.
85. Bross, H., Gruner, P., and Kirschenmann, P., "Normal Processes and Their Effect on the Lattice Thermal Conductivity in Solids with Step Dislocations," *Z. Naturforsch.*, **20a**(12), 1611–25 1965.
86. Bross, H., "Effect of Defect Configuration on the Lattice Thermal Conductivity of Dielectrics," *Z. Physik*, **189**(1), 33–54, 1966.
87. Maradudin, A. A., "Theoretical and Experimental Aspects of the Effects of Point Defects and Disorder on the Vibration of Crystals—1," in *Solid State Physics*, Vol. 18, Academic Press, New York, 273–420, 1966.
88. Maradudin, A. A., "Theoretical and Experimental Aspects of the Effects of Point Defects and Disorder on the Vibrations of Crystals—2," in *Solid State Physics*, Vol. 19, Academic Press, New York, 1–134, 1966.
89. Moss, M., "Scattering of Phonons by Dislocations," *J. Appl. Phys.*, **37**(11), 4168–72, 1966.
90. Klemens, P. G., "Phonon Scattering by Cottrell Atmospheres Surrounding Dislocations," *J. Appl. Phys.*, **39**(11), 5304–5, 1968.
91. Ohashi, K., "Scattering of Lattice Waves by Dislocations," *J. Phys. Soc. Japan*, **24**, 437–45, 1968.

Thermal Resistance due to Phonon-Defect Resonant Scattering at Low Temperatures

92. Lifshitz, I. M., "Scattering of Short Elastic Waves in Crystalline Lattices," *Zh. Eksperim. i Teor. Fiz.*, **18**(3), 293–300, 1948.
93. Lifshitz, I. M., "Some Problems of the Dynamic Theory of Non-Ideal Crystal Lattices," *Nuovo Cimento Supplemento*, **3**(4), 716–34, 1956.
94. Pohl, R. O., "Thermal Conductivity and Phonon Resonance Scattering," *Phys. Rev. Letters*, **8**, 481–3, 1962.
95. Callaway, J., "Thermal Resistance Produced by Point Imperfections in Crystals," *Nuovo Cimento*, 10, **29**(4), 883–91, 1963; USAF Rept. AFOSR-64-1235, 1–9, 1963. [AD 442 806]
96. Klein, M. V., "Phonon Scattering by Lattice Defects," *Phys. Rev.*, **131**(4), 1500–10, 1963.
97. Pohl, R. O., "Phonon-Defect Resonance Interaction," in *Proceedings of the Eighth International Conference on Low Temperature Physics*, Butterworths, London, 308–9, 1963.
98. Pohl, R. O., "The Applicability of the Debye Model to Thermal Conductivity," *Z. Physik*, **176**, 358–69, 1963.
99. Takeno, S., "Resonance Scattering of Lattice Waves by Isotopes," *Progr. Theoret. Phys. (Kyoto)*, **29**(2), 191–205, 1963.
100. Takeno, S., "Mass-Difference Scattering and Lattice Thermal Conductivity," *Progr. Theoret. Phys. (Kyoto)*, **30**(1), 144–5, 1963.
101. Wagner, M., "Influence of Localized Modes on Thermal Conductivity," *Phys. Rev.*, **131**(4), 1443–55, 1963.
102. Walker, C. T. and Pohl, R. O., "Phonon Scattering by Point Defects," *Phys. Rev.*, **131**(4), 1433–42, 1963.
103. Baumann, F. C., "Resonance States in Crystals Containing Point Defects," *Bull. Am. Phys. Soc.*, **9**, 644, 1964.
104. Callaway, J., "Theory of Scattering in Solids," *J. Math. Phys.*, **5**(6), 783–98, 1964.
105. McCombie, C. W. and Slater, J., "The Scattering of Lattice Vibrations by a Point Defect. I," *Proc. Phys. Soc. (London)*, **84**(540), 499–509, 1964.
106. Gunther, L., "Lattice-Vibration Effects due to Impurities in an Alkali-Halide," *Phys. Rev.*, **138**(6), A1697–705, 1965.
107. Krumhansl, J. A., "The Scattering of Lattice Vibrations by Vacancy-Type Defects," in *Lattice Dynamics* (Proceedings of the International Conference held at Copenhagen, Denmark, Aug. 5–9, 1963), (Wallis, R. F., ed.), Pergamon Press, London, 523–35, 1965.
108. Krumhansl, J. A. and Matthew, J. A. D., "Scattering of Low-Wavelength Phonons by Point Imperfections in Crystals," *Phys. Rev.*, **140**(5A), A1812–7, 1965.
109. Yussouff, M. and Mahanty, J., "Scattering of Phonons from a Substitutional Impurity," *Proc. Phys. Soc. (London)*, **85**(548), 1223–35, 1965.

110. Klein, M. V., "Phonon Scattering by Lattice Defects. II," *Phys. Rev.*, **141**(2), 716–23, 1966.
111. Seward, W. D. and Narayanamurti, V., "Rotational Degrees of Freedom of Molecules in Solids. I. The Cyanide Ion in Alkali Halides," *Phys. Rev.*, **148**(1), 463–81, 1966.
112. Yussouff, M. and Mohanty, J., "Scattering of Phonons from a Substitutional Impurity in Body-Centered Cubic and Face-Centered Cubic Lattices," *Proc. Phys. Soc. (London)*, **87**(557), 689–701, 1966.
113. Baumann, F. C., Harrison, J. P., Pohl, R. O., and Seward, W. D., "Thermal Conductivity in Mixed Alkali Halides: KCl:Li and KBr:Li," *Phys. Rev.*, **159**(3), 691–9, 1967.
114. Baumann, F. C. and Pohl, R. O., "Scattering of Phonons by Monatomic Impurities in Potassium Halides," *Phys. Rev.*, **163**(3), 843–50, 1967.
115. Benedek, G. and Nardelli, G. F., "Lattice Response Functions of Imperfect Crystals: Effects due to a Local Change of Mass and Short-Range Interaction," *Phys. Rev.*, **155**(3), 1004–19, 1967.
116. Caldwell, R. F. and Klein, M. V., "Experimental and Theoretical Study of Phonon Scattering from Simple Point Defects in Sodium Chloride," *Phys. Rev.*, **158**(3), 851–75, 1967.
117. Radosevich, L. G. and Walker, C. T., "Mass Independence of U-Center-Induced Phonon Resonances in KI," *Phys. Rev.*, **156**(3), 1030–1, 1967.
118. Schwartz, J. W. and Walker, C. T., "Thermal Conductivity of Some Alkali Halides Containing Divalent Impurities. I. Phonon Resonances," *Phys. Rev.*, **155**(3), 959–69, 1967.
119. Pohl, R. O., "Localized Excitations in Thermal Conductivity," in *Localized Excitations in Solids* (Proceedings of the First International Conference on Localized Excitations in Solids, 1967), (Wallis, R. F., ed.), Plenum Press, New York, 434–50, 1968.
120. Peressini, P. P., Harrison, J. P., and Pohl, R. O., "Thermal Conductivity of KCl: Li. Isotope and Electric Field Effects," *Phys. Rev.*, **180**(3), 926–30, 1969.

Thermal Conductivity of Amorphous Solids

121. Kittel, C., "Interpretation of Thermal Conductivity of Glasses," *Phys. Rev.*, **75**(6), 972–4, 1949.
122. Klemens, P. G., "The Thermal Conductivity of Glasses," in *Noncrystalline Solids* (Conference on Noncrystalline Solids, Alfred, New York, 1958), Wiley, New York, 508–30, 1960.
123. Eiermann, K., "Thermal Conductivity of High Polymers," *J. Polymer Science*, Part C, No. 6, 157–65, 1963 (publ. 1964).
124. Klemens, P. G., "Thermal Conductivity in Vitreous Systems," in *Physics of Noncrystalline Solids* (Proceedings of the International Conference, Delft, Netherlands, 1964), North-Holland, Amsterdam, 162–78, 1965.
125. Dreyfus, B., Fernandes, N. C., and Maynard, R., "Low Temperature Thermal Conductivity of Amorphous Solids," in *Proceedings of the Eleventh International Conference on Low Temperature Physics* (St. Andrews, Scotland, 1968), Vol. 1, University of St. Andrews Printing Dept., 589–92, 1968.
126. Dreyfus, B., Fernandes, N. C., and Maynard, R., "Low Temperature Thermal Conductivity of Amorphous Solids," *Phys. Letters*, **26A**(12), 647–8, 1968.

Thermal Resistance due to Phonon-Spin Resonant Scattering at Low Temperatures

127. Morton, I. P. and Rosenberg, H. M., "Scattering of Phonons by Spins at Low Temperatures," *Phys. Rev. Letters*, **8**(5), 200–1, 1962.
128. Orbach, R., "Thermal Resistance of Holmium Ethyl Sulphate," *Phys. Rev. Letters*, **8**(10), 393–6, 1962.
129. Berman, R., Brock, J. C. F., and Huntley, D. J., "The Effect of Spin-Phonon Interactions on the Thermal Conductivity of Lanthanum Cobalt Nitrate," *Phys. Letters*, **3**(6), 310–2, 1963.
130. Orbach, R., "A Note on the Thermal Resistance of Kramers Salts," *Phys. Letters*, **3**(6), 269–70, 1963.
131. Huber, D. L., "A Note on the Thermal Resistivities of Paramagnetic Salts," *Phys. Letters*, **12**(4), 309–10, 1964.
132. McClintock, P. V. E., Morton, I. P., and Rosenberg, H. M., "The Thermal Conductivity of Paramagnetic Crystals at Low Temperatures in Magnetic Fields," in *Proceedings of the International Conference on Magnetism* (Nottingham), Institute of Physics, London, 455–6, 1964.
133. Walton, D., "Phonon Scattering by Paramagnetic Ions and Scattering by Other Defects," *Phys. Rev.*, **151**, 627–8, 1966.
134. McClintock, P. V. E., Morton, I. P., Orbach, R., and Rosenberg, H. M., "The Effect of a Magnetic Field on the Thermal Conductivity of Paramagnetic Crystals: Holmium Ethylsulphate," *Proc. Roy. Soc. (London)*, **A298**(1454), 359–78, 1967.
135. Walton, D., "Study of the Li Ion in KCl Using the Spin-Phonon Interaction," *Phys. Rev. Letters*, **19**, 305–7, 1967.
136. McClintock, P. V. E. and Rosenberg, H. M., "The Effects of a Magnetic Field on the Thermal Conductivity of Paramagnetic Crystals: Cerium Ethylsulphate," *Proc. Roy. Soc. (London)*, **A302** (1470), 419–36, 1968.

Thermal Resistance due to Phonon-Magnon Scattering and Spin-Wave Thermal Conductivity

137. Pomeranchuk, I., "The Thermal Conductivity of the Paramagnetic Dielectrics at Low Temperatures," *J. Exptl. Theoret. Phys. (USSR)*, **11**, 226–45, 1941; English translation: *J. Phys. (USSR)*, **4**(4), 357–74, 1941.
138. Akhiezer, A., "Theory of the Relaxation Processes in Ferromagnetics at Low Temperatures," *J. Phys. (USSR)*, **10**(3), 217–30, 1946.
139. Kittel, C. and Abraham, E., "Relaxation Processes in Ferromagnetism," *Rev. Modern Phys.*, **25**(1), 233–8, 1953.
140. Sato, H., "Thermal Conductivity of Ferromagnetic Substances," *Busseiron Kenkyu*, No. 77, 68–73, 1954.
141. Sato, H., "On the Thermal Conductivity of Ferromagnetics," *Progr. Theoret. Phys. (Kyoto)*, **13**, 119–20, 1955.
142. Dyson, F. J., "General Theory of Spin-Wave Interactions," *Phys. Rev.*, **102**(5), 1217–30, 1956.
143. Akhiezer, A. I. and Shishkin, L. A., "On the Theory of Thermal Conductivity and Absorption of Sound in Ferromagnetic Dielectrics," *Zh. Eksptl. i Teoret. Fiz.*,

34(5), 1267–71, 1958; English translation: *Soviet Phys.—JETP*, **7**(5), 875–8, 1958.

144. Kittel, C., "Interaction of Spin Waves and Ultrasonic Waves in Ferromagnetic Crystals," *Phys. Rev.*, **110**(4), 836–41, 1958.
145. Akhiezer, A. I. and Bar'yakhtar, V. G., "Theory of Heat Conductivity in Ferrodielectric Materials at Low Temperatures," *Fiz. Tverdogo Tela*, **2**(10), 2446–9, 1960; English translation: *Soviet Physics—Solid State*, **2**(10), 2178–82, 1961.
146. Bar'yakhtar, V. G. and Urushadze, G. I., "The Scattering of Spin Waves and Phonons by Impurities in Ferromagnetic Dielectrics," *Zh. Eksptl. i Teoret. Fiz.*, **39**, 355–61, 1960; English translation: *Soviet Phys.—JETP*, **12**, 251–5, 1961.
147. Kaganov, M. I., Tsukernik, V. M., and Chupis, I. E., "Theory of Relaxation Process in Antiferromagnetics," *Fiz. Metal. i Metalloved.*, Akad. Nauk SSSR, Vral. Filial, 10, 797–8, 1960.
148. Bethoux, O., Thomas, P., and Weil, L., "Heat Conductivity of UO_2 at Low Temperatures," *Compt. Rend.*, **253**(19), 2043–5, 1961.
149. Douthett, D. and Friedberg, S. A., "Effects of a Magnetic Field on Heat Conduction in Some Ferrimagnetic Crystals," *Phys. Rev.*, **121**(6), 1662–7, 1961.
150. Luthi, B., "Thermal Conductivity of Yttrium Iron Garnet," *J. Phys. Chem. Solids*, **23**, 35–8, 1962.
151. Sinha, K. P. and Upadhyaya, U. N., "Phonon–Magnon Interaction in Magnetic Crystals," *Phys. Rev.*, **127**(2), 432–9, 1962.
152. Callaway, J., "Scattering of Spin Waves by Magnetic Defects," *Phys. Rev.*, **132**(5), 2003–9, 1963.
153. Douglass, R. L., "Heat Transport by Spin Waves in Yttrium Iron Garnet," *Phys. Rev.*, **129**(3), 1132–5, 1963.
154. Friedberg, S. A. and Harris, E. D., "Heat Transport by Magnons at Low Temperatures," in *Proceedings of the Eighth International Conference on Low Temperature Physics*, Butterworths, London, 302–3, 1963.
155. Kawasaki, K., "On the Behavior of Thermal Conductivity Near the Magnetic Transition Point," *Progr. Theoret. Phys. (Kyoto)*, **29**(6), 801–16, 1963.
156. Callaway, J. and Boyd, R., "Scattering of Spin Waves by Magnetic Defects," *Phys. Rev.*, **134**(6A), A1655–62, 1964.
157. Chupis, I. E., "Theory of Relaxation Processes in a Uniaxial Antiferrodielectric," *Zh. Eksptl. i Teoret. Fiz.*, **46**, 307–19, 1964; *Soviet Phys.—JETP*, **19**, 212–9, 1964.
158. Devyatkova, E. D., Golubkov, A. V., Kudinov, E. K., and Smirnov, I. A., "The Thermal Conductivity of MnTe as Affected by Spin-Phonon Interaction," *Soviet Phys.—Solid State*, **6**(6), 1425–8, 1964.
159. McCollum, D. C., Wild, R. L., and Callaway, J., "Spin-Wave Thermal Conductivity of Ferromagnetic EuS," *Phys. Rev.*, **136**(2A), A426–8, 1964.
160. Stern, H., "Thermal Conductivity at the Magnetic Transition," *J. Phys. Chem. Solids*, **26**(1), 153–61, 1965.
161. Bhandari, C. M. and Verma, G. S., "Scattering of Magnons and Phonons in the Thermal Conductivity of Yttrium–Iron Garnet," *Phys. Rev.*, **152**, 731–6, 1966.
162. Gurevich, L. E and Roman, G. A., "Heat Conductivity of Ferrites at Low Temperatures and the Entrainment of Phonons and Magnons," *Fiz. Tverdogo Tela*, **8**(2), 525–31, 1966; English translation: *Soviet Phys.—Solid State*, **8**(2), 416–20, 1966.
163. Devyatkova, E. D. and Tikhonov, V. V., "Thermal Conductivity and Specific Heat of Yttrium-Calcium Garnets," *Fiz. Tverdogo Tela*, **9**(3), 772–7, 1967; English translation: *Soviet Phys.—Solid State*, **9**(3), 604–8, 1967.
164. Gurevich, L. E. and Roman, G. A., "Thermal Conductivity of Antiferromagnets at Low Temperatures under Conditions of Mutual Magnon and Phonon Drag," *Fiz. Tverdogo Tela*, **8**(9), 2628–32, 1966; English translation: *Soviet Phys.—Solid State*, **8**(9), 2102–5, 1967.

Radiative Thermal Conductivity

165. McCauley, G. V., "Fundamentals of Heat Flow in Molten Glass and in Walls for Use Against Glass," *J. Am. Ceram. Soc.*, **8**(4), 493–504, 1925.
166. Hamaker, H. C., "Radiation and Heat Conduction in Light-Scattering Material," Philips Research Rept., **2**, 55–67, 103–11, 112–125, 420–425, 1947.
167. Preston, F. W., "Meaning of the Term 'Diathermancy' and Heating of Glass in Tank Furnaces," *J. Soc. Glass Technol.*, **31** (142), 134–40, 1947.
168. Genzel, L., "Measurement of the Infrared Absorption of Glass between 20 C and 1360 C," *Glastech. Ber.*, **24**(3), 55–63, 1951.
169. Czerny, M. and Genzel, L., "On the Depth to which Diffuse Radiation Penetrates Glass," *Glastech. Ber.*, **25**(5), 134–9, 1952.
170. Czerny, M. and Genzel, L., "Energy Flux and Temperature Distribution in The Glass Baths of Melting Tanks Resulting from Conduction and Radiation," *Glastech. Ber.*, **25**(12), 387–92, 1952.
171. Geffcken, W., "The Transmission of Heat in Glass at High Temperatures, Part I," *Glastech. Ber.*, **25**(12), 392–6, 1952.
172. van der Held, E. F. M., "The Contribution of Radiation to the Conduction of Heat," *Appl. Sci. Res.*, **A3**, 237–49, 1952.
173. Genzel, L., "Calculation of the Radiation Conductivity of Glasses," *Glastech. Ber.*, **26**(3), 69–71, 1953.
174. Genzel, L., "The Role of Heat Radiation in the Heat Conduction Process," *Z. Physik*, **135**, 177–95, 1953.
175. Monroe, J. E., Jr., "Infrared Transmission and Radiation Conductivity of Alumina from Room Temperature to 1800 C," in *Study of Heat Transfer of Ceramic Materials*, U.S. Office of Naval Research, Metallurgy Branch, NR032–022, Final Rept., Part II, 38–66, 1957.
176. Gardon, R., "A Review of Radiant Heat Transfer in Glass," *J. Am. Ceram. Soc.*, **44**(7), 305–12, 1961.
177. Sparrow, E. M. and Cess, R. D., *Radiation Heat Transfer*, Brooks/Cole, Belmont, Calif., 322 pp., 1966.
178. Condon, E. U., "Radiative Transport in Hot Glass," *J. Quant. Spectrosc. Radiat. Transfer*, **8**, 369–85, 1968.

Ambipolar Diffusion, Exciton Thermal Conductivity, and Thermal Resistance due to Phonon–Electron Scattering in Semiconductors

179. Davydov, B. I. and Shmushkevich, I. M., "Electron Theory of Semiconductors," *Uspekhi Fiz. Nauk*, **24**(1), 21–67, 1940.

180. Price, P. J., "Electronic Thermal Conduction in Semiconductors," *Phys. Rev.*, **95**,596, 1954.
181. Price, P. J., "Ambipolar Thermodiffusion of Electrons and Holes in Semiconductors," *Phil. Mag.*, **46**, 1252–60, 1955.
182. Pikus, G. E., "Thermomagnetic and Galvanomagnetic Effects in Semiconductors, Taking into Account the Variations in the Concentration of Current Carriers. II. Galvanomagnetic Effects in Strong Fields. Electron and Phonon Thermal Conductivity," *Zh. Tekh. Fiz.*, **26**(1), 36–50, 1956; English translation: *Soviet Phys.—Tech. Phys.*, **1**(1), 32–46, 1956.
183. Joffé, A. F., "Heat Transfer in Semiconductors," *Can. J. Phys.*, **34**(12A), 1342–55, 1956.
184. Korolyuk, S. L., "On the Theory of Exciton Thermal Conductivity," *Fiz. Tverdogo Tela*, **4**(3), 790–800, 1962; English translation: *Soviet Phys.—Solid State*, **4**(3), 580–6, 1962.
185. Ziman, J. M., "The Effect of Free Electrons on Lattice Conduction," *Phil. Mag.*, 8, **1**(2), 191–8, 1956.
186. Ziman, J. M., "Corrigendum to 'The Effect of Free Electrons on Lattice Conduction'," *Phil. Mag.*, 8, **2**(14), 292, 1957.
187. Carruthers, J. A., Geballe, T. H., Rosenberg, H. M., and Ziman, J. M., "The Thermal Conductivity of Germanium and Silicon Between 2 and 300 K," *Proc. Roy. Soc.* (*London*), **A238**, 502–14, 1957.
188. Stratton, R., "The Effect of Free Electrons on Lattice Conduction at High Temperature," *Phil. Mag.*, 8, **2**, 422–4, 1957.
189. Keyes, R. W., "Low Temperature Thermal Resistance of *n*-type Germanium," *Phys. Rev.*, **122**(4), 1171–6, 1961.
190. Pyle, I. C., "The Scattering of Phonons by Bound Electrons in a Semiconductor," *Phil. Mag.*, **6**, 609–16, 1961.
191. Griffin, A. and Carruthers, P., "Thermal Conductivity of Solids. IV. Resonance Fluorescence Scattering of Phonons by Donor Electrons in Germanium," *Phys. Rev.*, **131**(5), 1976–95, 1963.
192. Kwok, P. C., "Acoustic Attenuation by Neutral Donor Impurity Atoms in Germanium," *Phys. Rev.*, **149**, 666–74, 1966.
193. Spector, H. N., "Interaction of Acoustic Waves and Conduction Electrons," *Solid State Phys.*, **19**, 291–361, 1966.
194. Mathur, M. P. and Pearlman, N., "Phonon Scattering by Neutral Donors in Germanium," *Phys. Rev.*, **180**(3), 833–45, 1969.
195. ter Haar, D. and Neaves, A., "On the Thermal Conductivity and Thermoelectric Power of Semiconductors," *Advan. in Phys.*, **5**(18), 241–69, 1956.
196. Madelung, O., "Semiconductor," in *Handbuch der Physik*, Vol. 20, Springer-Verlag, Berlin, 1–245, 1957.
197. Appel, J., "Thermal Conductivity of Semiconductors," *Progr. in Semiconductors*, **5**, 141–87, 1960.
198. Keyes, R. W. and Bauerle, J. E., "Thermal Conduction in Thermoelectric Materials," Chap. 5 in *Thermoelectricity: Science and Engineering* (Heikes, R. R. and Ure, R. W., Jr., eds.), Interscience Publishers, New York, 91–119, 1961.
199. Holland, M. G., "Thermal Conductivity," in *Semiconductors and Semimetals* (Willardson, R. K. and Beer, A. C., eds.), Vol. 2 (Physics of III—V Compounds), Academic Press, New York, 3–31, 1966.
200. Probert, S. D. and Thomas, C. B., "The Thermal Conductivity of Semiconductors at Low Temperatures," UKAEA, TRG Report 977 (R/X), 1–34, 1966.
201. Steigmeier, E. F., "Thermal Conductivity of Semiconducting Materials," in *Thermal Conductivity* (Tye, R. P., ed.), Vol. 2, Chap. 4, Academic Press, London, 203–51, 1969.

Experimental Methods

202. Franklin, B., "Meteorological Observations," (written in reply to Cadwallader Colden, Nov. 10, 1753 and read at the Royal Society of London, Nov. 4, 1756), in *The Writings of Benjamin Franklin* (Smyth, A. H., ed.), Vol. III (1750–59), The Macmillan Co., New York, 186–8, 1905.
203. Fordyce, G., "An Account of an Experiment on Heat," *Phil. Trans. Roy. Soc.* (*London*), **77**, 310–7, 1787.
204. Ingen-Hausz, J., "On Metals as Conductors of Heat," *J. de Physique*, **34**, 68, 380, 1789.
205. Despretz, C. M., "On the Conductivity of Several Solid Substances," *Ann. Chim. Phys.*, **19**, 97–106, 1822.
206. Wiedemann, G. and Franz, R., "The Thermal Conductivity of Metals," *Ann. Physik*, **89**, 497–531, 1853.
207. Forbes, J. D., "On the Progress of Experiments on the Conduction of Heat, undertaken at the Meeting of the British Association at Edinburgh, in 1850," *Brit. Assoc. Adv. Sci.*, Rept. Ann. Meeting, **21**, 7–8, 1851.
208. Forbes, J. D., "On Experiments on the Laws of the Conduction of Heat," *Brit. Assoc. Adv. Sci.*, Rept. Ann. Meeting, **22**, 260–1, 1852.
209. Forbes, J. D., "Experimental Inquiry into the Laws of the Conduction of Heat in Bars, and into the Conducting Power of Wrought Iron," *Trans. Roy. Soc. Edinburgh*, **23**, 133–46, 1864.
210. Forbes, J. D., "Experimental Inquiry into the Laws of the Conduction of Heat in Bars. Part II. On the Conductivity of Wrought Iron, Deduced from the Experiments of 1851," *Trans. Roy. Soc. Edinburgh*, **24**, 73–110, 1865.
211. Ångström, A. J., "A New Method of Determining the Thermal Conductivity of Solids," *Ann. Physik*, 2, **114**, 513–30, 1861.
212. Thomson, W. (Lord Kelvin), "Heat," in *Encyclopaedia Britannica*, Vol. 11, 9th Ed., 1880; reprinted in *Mathematical and Physical Papers*, Vol. 3, Cambridge University Press, 113–235, 1890.
213. Preston, T., *The Theory of Heat*, Macmillan and Co., London, 719 pp., 1894; 4th Ed. (J. R. Cotter, ed.), Macmillan and Co., Ltd., London, 836 pp., 1929.
214. Chwolson, O. D., "Thermal Conductivity," in *Traité de Physique* (Translated into French by Davaux, E. and reviewed and augmented by the author), Vol. 3, Chap. VII, Librairie Scientifique A. Hermann et Fils, Paris, 320–408, 1909.
215. Schofield, F. H., "Conduction of Heat," in *A Dictionary of Applied Physics* (Glazebrook, R., ed.), Vol. 1, The Macmillan Co., New York, 429–66, 1922 (reprinted 1950).
216. Ingersoll, L. R., "Methods of Measuring Thermal

Conductivity in Solids and Liquids," *J. Optical Soc. Am.*, **9**, 495–501, 1924.

217. Griffiths, E., "A Survey of Heat Conduction Problems," *Proc. Phys. Soc.* (*London*), **41**, 151–79, 1929.

218. Partington, J. R., "Thermal Conductivity of Solids," in *An Advanced Treatise on Physical Chemistry*, Vol. III, Longmans, Green and Co., London, 410–61, 1952.

219. Seibel, R. D., "Survey and Bibliography on the Determination of Thermal Conductivity of Metals at Elevated Temperatures," Watertown Arsenal Lab. Rept. No. WAL 821/9, 1–65, 1954. [AD 51 228]

220. Kingery, W. D., *Property Measurements at High Temperatures*, John Wiley and Sons, Inc., New York, 416 pp., 1959.

221. Slack, G. A., "Heat Conduction in Solids, Experimental," in *Encyclopaedic Dictionary of Physics* (Thewlis, J., editor-in-chief), Vol. 3, Pergamon Press, Oxford, 601–6, 1961.

222. Tye, R. P. (ed.), *Thermal Conductivity*, Vol. 1 and 2, Academic Press, London, 422 pp. and 353 pp., 1969.

223. Carslaw, H. S. and Jaeger, J. C., *Conduction of Heat in Solids*, Oxford University Press, 1946; 2nd Ed., 510 pp., 1959.

224. Ingersoll, L. R., Zobel, O. J., and Ingersoll, A. C., *Heat Conduction*, McGraw-Hill, New York, 1948; 2nd Ed., University of Wisconsin Press, 325 pp., 1954.

225. Jakob, M., *Heat Transfer*, Vol. 1, John Wiley and Sons, Inc., New York, 758 pp., 1949.

226. Schneider, P. J., *Conduction Heat Transfer*, Addison-Wesley Publ. Co., Cambridge, Mass., 395 pp., 1955.

227. Arpaci, V. S., *Conduction Heat Transfer*, Addison-Wesley Publ. Co., Reading, Mass., 550 pp., 1966.

228. Biot, J. B., *Traité de Physique*, Vol. 4, Paris, 669, 1816.

229. Fourier, J. B. J., *The Analytical Theory of Heat*, Gauthier-Villars, Paris, 1822; English translation by Freeman, A., Cambridge University Press, 466 pp., 1878; Republication of the English translation, Dover Publications, New York, 1955.

230. White, G. K., *Experimental Techniques in Low Temperature Physics*, Oxford University Press, 1959; 2nd Ed., 1968.

231. White, G. K., "Measurement of Solid Conductors at Low Temperatures," in *Thermal Conductivity* (Tye, R. P., ed.), Vol. 1, Chap. 2, Academic Press, London, 69–109, 1969.

232. Lees, C. H., "The Effects of Temperature and Pressure on the Thermal Conductivities of Solids. Part II. The Effects of Low Temperatures on the Thermal and Electrical Conductivities of Certain Approximately Pure Metals and Alloys," *Phil. Trans. Roy. Soc.* (*London*), **A208**, 381–443, 1908.

233. Berman, R., "The Thermal Conductivities of Some Dielectric Solids at Low Temperatures (Experimental)," *Proc. Roy. Soc.* (*London*), **A208**, 90–108, 1951.

234. White, G. K., "The Thermal Conductivity of Gold at Low Temperatures," *Proc. Phys. Soc.* (*London*), **A66**, 559–64, 1953.

235. Mendelssohn, K. and Renton, C. A., "The Heat Conductivity of Superconductors below 1 K," *Proc. Roy. Soc.* (*London*), **A230**, 157–69, 1955.

236. Rosenberg, H. M., "The Thermal Conductivity of Metals at Low Temperatures," *Phil. Trans. Roy. Soc.* (*London*), **A247**, 441–97, 1955.

237. White, G. K. and Woods, S. B., "Thermal and Electrical Conductivities of Solids at Low Temperatures," *Can. J. Phys.*, **33**, 58–73, 1955.

238. Powell, R. L., Rogers, W. M., and Coffin, D. O., "An Apparatus for Measurement of Thermal Conductivity of Solids at Low Temperatures," *J. Res. Nat. Bur. Stand.*, **59**(5), 349–55, 1957.

239. Slack, G. A., "Thermal Conductivity of Potassium Chloride Crystals containing Calcium," *Phys. Rev.*, **105**(3), 832–42, 1957.

240. Williams, W. S., "Phonon Scattering in KCl–KBr Solid Solutions at Low temperatures," *Phys. Rev.*, **119**(3), 1021–4, 1960.

241. Slack, G. A., "Thermal Conductivity of CaF_2 MnF_2, CoF_2, and ZnF_2 Crystals," *Phys. Rev.*, **122**(5), 1451–64, 1961.

242. Berman, R., Bounds, C. L., and Rogers, S. J., "The Effects of Isotopes on Lattice Heat Conduction. II. Solid Helium," *Proc. Roy. Soc.* (*London*), **A289**(1416), 46–65, 1965.

243. Jericho, M. H., "The Lattice Thermal Conductivity of Silver Alloys between 4 K and 0.3 K," *Phil. Trans. Roy. Soc.* (*London*), **A257**, 385–407, 1965.

244. Berget, A., "Measurement of the Thermal Conductivity of Mercury, of Its Absolute Value," *Compt. Rend.*, **105**, 224–7, 1887.

245. Berget, A., "Thermal Conductivity of Mercury and Certain Metals," *J. Phys.* (*Paris*), 2, **7**, 503–18, 1888.

246. "The Physical Society's Exhibition. No. III," Engineer, **159**, 68–70, 1935.

247. Armstrong, L. D. and Dauphinee, T. M., "Thermal Conductivity of Metals at High Temperatures. I. Description of the Apparatus and Measurements on Iron," *Can. J. Phys.*, **A25**, 357–74, 1947.

248. Ditmars, D. A. and Ginnings, D. C., "Thermal Conductivity of Beryllium Oxide from 40 to 750 C," *J. Res. Nat. Bur. Stand.*, **59**(2), 93–9, 1957.

249. Wilkes, G. B., "An Apparatus for Determining the Thermal Conductivity of Metals," *Chem. Met. Eng.*, **21**(5), 241–3, 1919.

250. Powell, R. W., "The Thermal and Electrical Conductivities of Some Magnesium Alloys," *Phil. Mag.*, **27**, 677–86, 1939.

251. Powell, R. W. and Tye, R. P., "High Alloy Steels for Use as a Thermal Conductivity Standard," *Brit. J. Appl. Phys.*, **11**, 195–8, 1960.

252. Larsen, D. C., Powell, R. W., and DeWitt, D. P., "The Thermal Conductivity and Electrical Resistivity of a Round-Robin Armco Iron Sample, Initial Measurements from 50 to 300 C," in *Thermal Conductivity—Proceedings of the Eighth Conference* (Ho, C. Y. and Taylor, R. E., eds.), Plenum Press, New York, 675–87, 1969.

253. Laubitz, M. J., "Measurement of the Thermal Conductivity of Solids at High Temperatures by Using Steady-State Linear and Quasi-Linear Heat Flow," in *Thermal Conductivity* (Tye, R. P., ed.), Vol. 1, Chap. 3, Academic Press, London, 111–83, 1969.

254. Flynn, D. R., "Thermal Conductivity of Ceramics," in

Mechanical and Thermal Properties of Ceramics (Wachtman, J. B., Jr., ed.), NBS Spec. Publ. 303, 63–123, 1969.
255. Bauerle, J. E., "Thermal Conductivity," Section 10.1 in *Thermoelectricity: Science and Engineering* (Heikes, R. R. and Ure, R. W., Jr., eds.), Interscience Publishers, New York, 285–311, 1961.
256. Honda, K. and Simidu, T., "On the Thermal and Electrical Conductivities of Carbon Steels at High Temperatures," *Sci. Repts. Tôhoku Univ.*, 1, **6**, 219–33, 1917.
257. Schofield, F. H., "The Thermal and Electrical Conductivities of Some Pure Metals," *Proc. Roy. Soc. (London)*, **A107**, 206–27, 1925.
258. Powell, R. W., "The Thermal and Electrical Conductivities of Metals and Alloys: I. Iron from 0 to 800 C," *Proc. Phys. Soc. (London)*, **46**, 659–78, 1934.
259. Goldsmid, H. J., "The Thermal Conductivity of Bismuth Telluride," *Proc. Phys. Soc. (London)*, **B69**, 203–9, 1956.
260. ASTM, "Standard Method of Test for Thermal Conductivity of Materials by Means of the Guarded Hot Plate," ASTM Designation: C177-63, in *1967 Book of ASTM Standards*, Part 14, 17–28, 1967.
261. Poensgen, R., "A Technical Method for Investigating the Thermal Conductivity of Slabs of Material," *VDI Zeitschrift*, **56**(41), 1653–8, 1912.
262. Jakob, M., "Measurement of the Thermal Conductivity of Liquids, Insulating Materials and Metals," *VDI Zeitschrift*, **66**, 688–93, 1922.
263. ASTM, "Standard Method of Test for Thermal Conductivity of Refractories," ASTM Designation: C201-47 (1958), in *1967 Book of ASTM Standards*, Part 13, 170–7, 1967.
264. Wilkes, G. B., "Thermal Conductivity, Expansion, and Specific Heat of Insulators at Extremely Low Temperatures," *Refrig. Eng.*, **52**(1), 37–42, 1946.
265. Schröder, J., "A Simple Method of Determining the Thermal Conductivity of Solids," *Philips Tech. Rev.*, **21**(12), 357–61, 1959–60.
266. Schröder, J., "Apparatus for Determining the Thermal Conductivity of Solids in the Temperature Range from 20 to 200 C," *Rev. Sci. Instr.*, **34**(6), 615–21, 1963.
267. ASTM, "Tentative Method of Test for Thermal Conductivity of Insulating Materials at Low Temperatures by Means of the Wilkes Calorimeter," ASTM Designation: C420-62T in *1967 Book of ASTM Standards*, Part 14, 172–9, 1967.
268. ASTM, "Standard Method of Test for Thermal Conductivity of Materials by Means of the Heat Flow Meter," ASTM Designation: C518-67, in *1967 Book of ASTM Standards*, Part 14, 230–8, 1967.
269. Peirce, B. O. and Willson, R. W., "On the Thermal Conductivities of Certain Poor Conductors.—I," *Proc. Am. Acad. Arts and Sci.*, **34**(1), 1–56, 1898.
270. van Dusen, M. S., "The Thermal Conductivity of Heat Insulators," *J. Am. Soc. Heating Vent. Engrs.*, **26**(7), 625–56, 1920.
271. Somers, E. V. and Cyphers, J. A., "Analysis of Errors in Measuring Thermal Conductivity of Insulating Materials," *Rev. Sci. Instr.*, **22**(8), 583–6, 1951.
272. Woodside, W., "Analysis of Errors due to Edge Heat Loss in Guarded Hot Plates," ASTM Spec. Tech. Publ. 217, 49–62, 1957.
273. ASTM. *Thermal Conductivity Measurements of Insulating Materials at Cryogenic Temperatures*, ASTM Spec. Tech. Publ. 411, 118 pp., 1967.
274. Ferro, V. and Sacchi, A., "An Automatic Plate Apparatus for Measurements of Thermal Conductivity of Insulating Materials at High Temperatures," in *Thermal Conductivity—Proceedings of the Eighth Conference* (Ho, C. Y. and Taylor, R. E., eds.), Plenum Press, New York, 737–60, 1969.
275. Pratt, A. W., "Heat Transmission in Low Conductivity Materials," in *Thermal Conductivity* (Tye, R. P., ed.), Vol. 1, Chap. 6, Academic Press, London, 301–405, 1969.
276. Péclet, M. E., "Note on the Determination of the Conductivity Coefficients of Metals for Heat," *Ann. Chim. Physique*, 3, **2**(1), 107–15, 1841.
277. Lees, C. H., "On a Thermal Conductivities of Single and Mixed Solids and Liquids and Their Variation with Temperature," *Phil. Trans. Roy. Soc.* (London), **A191**, 399–440, 1898.
278. Lees, C. H., "Some Preliminary Experiments on the Effect of Pressure on Thermal Conductivity," *Manchester Memoirs*, **43**(8), 1–6, 1899.
279. Lodge, O. J., "On a Method of Measuring the Absolute Thermal Conductivity of Crystals and Other Rare Substances. Part I," *Phil. Mag.*, 5, **5**, 110–7, 1878.
280. Berget, A., "Measurement of the Coefficients of Thermal Conductivity of Metals," *Compt. Rend.*, **107**, 227–9, 1888.
281. Lees, C. H., "On the Thermal Conductivities of Crystals and Other Bad Conductors," *Phil. Trans. Roy. Soc. (London)*, **A183**, 481–509, 1892.
282. van Dusen, M. S. and Shelton, S. M., "Apparatus for Measuring Thermal Conductivity of Metals up to 600 C," *J. Res. Nat. Bur. Stand.*, **12**, 429–40, 1934.
283. Powell, R. W., "The Thermal and Electrical Conductivity of a Sample of Acheson Graphite from 0 to 800 C," *Proc. Phys. Soc. (London)*, **49**, 419–25, 1937.
284. Francl, J. and Kingery, W. D., "Apparatus for Determining Thermal Conductivity by a Comparative Method. Data for Pb, Al_2O_3, BeO, and MgO," *J. Am. Ceram. Soc.*, **37**, 80–4, 1954.
285. Stuckes, A. D. and Chasmar, R. P., "Measurement of the Thermal Conductivity of Semiconductors," *Rept. Meeting on Semiconductors (Phys. Soc., London)*, 119–25 1956.
286. ASTM, "Standard Method of Test for Thermal Conductivity of Whiteware Ceramics," ASTM Designation: C408-58, in *1967 Book of ASTM Standards*, Part 13, 348–52, 1967.
287. Mirkovich, V. V., "Comparative Method and Choice of Standards for Thermal Conductivity Determinations," *J. Am. Ceram Soc.*, **48**(8), 387–91, 1965.
288. Christiansen, C., "Some Experiments on Heat Conduction," *Ann. Physik*, 3, **14**, 23–33, 1881.
289. Sieg, L. P., "An Attempt to Detect a Change in the Heat Conductivity of a Selenium Crystal with a Change in Illumination," *Phys. Rev.*, **6**, 213–8, 1915.
290. Powell, R. W., "The Thermal and Electrical Conductivities of Metals and Alloys: II. Some Heat-Resistant Alloys from 0 to 800 C," *Proc. Phys. Soc. (London)*, **48**, 381–92, 1936.
291. Powell, R. W. and Hickman, M. J., "The Physical Properties of a Series of Steels. 3. Thermal Conductivity

and Electrical Resistivity," Iron and Steel Institute, Special Report No. 24, 242–51, 1939.

292. Powell, R. W. and Tye, R. P., "The Thermal and Electrical Conductivities of Some Nickel–Chromium (Nimonic) Alloys," *The Engineer*, **209**, 729–32, 1960.
293. Sugawara, A., "The Precise Determination of Thermal Conductivity of Pure Fused Quartz," *J. Appl. Phys.*, **39**(13), 5994–7, 1968.
294. Hogan, C. L. and Sawyer, R. B., "The Thermal Conductivity of Metals at High Temperatures," *J. Appl. Phys.*, **23**(2), 177–80, 1952.
295. Laubitz, M. J., "Transport Properties of Pure Metals at High Temperatures. I. Copper," *Can. J. Phys.*, **45**(11), 3677–96, 1967.
296. Watson, T. W. and Robinson, H. E., "Thermal Conductivity of Some Commercial Iron–Nickel Alloys," *ASME J. of Heat Transfer*, Part C, **83**(4), 403–8, 1961.
297. Laubitz, M. J., "The Unmatched Guard Method of Measuring Thermal Conductivity at High Temperatures," *Can. J. Phys.*, **41**(10), 1663–78, 1963.
298. Laubitz, M. J., "The Unmatched Guard Method of Measuring Thermal Conductivity. II. The Guardless Method," *Can. J. Phys.*, **43**(2), 227–43, 1965.
299. McElroy, D. L. and Moore, J. P., "Radial Heat Flow Methods for the Measurement of the Thermal Conductivity of Solids," in *Thermal Conductivity* (Tye, R. P., ed.), Vol. 1, Chap. 4, Academic Press, London, 185–239, 1969.
300. Callendar, H. L. and Nicolson, J. T., "Experiments on the Condensation of Steam. Part I. A New Apparatus for Studying the Rate of Condensation of Steam on a Metal Surface at Different Temperatures and Pressures," *Brit. Assoc. Adv. Sci., Rept. Ann. Meeting*, 418–22, 1897.
301. Niven, C. "On a Method of Finding the Conductivity for Heat," *Proc. Roy. Soc. (London)*, **A76**, 34–48, 1905.
302. Andrews, T., *Proc. Roy. Irish Acad.*, **1**, 465, 1840.
303. Schleiermacher, A., "On the Heat Conduction in Gases," *Ann. Physik Chemie*, **34**(8a), 623–46, 1888.
304. Kannuluik, W. G. and Martin, L. H., "Conduction of Heat in Powders," *Proc. Roy. Soc.*, **A141**, 144–58, 1933.
305. Kingery, W. D., "Thermal Conductivity. VI. Determination of Conductivity of Al_2O_3 by Spherical Envelope and Cylinder Methods," *J. Am. Ceram. Soc.*, **37**, 88–90, 1954.
306. Feith, A. D., "A Radial Heat Flow Apparatus for High-Temperature Thermal Conductivity Measurements," USAEC Rept. GEMP-296, 1–29, 1964.
307. Glassbrenner, C. J. and Slack, G. A., "Thermal Conductivity of Silicon and Germanium from 3 K to the Melting Point," *Phys. Rev.*, **134**(4A), A1058–69, 1964.
308. Banaev, A. M. and Chekhovskoi, V. Ya., "Experimental Determination of the Coefficient of Thermal Conductivity of Solid Materials in the Temperature Range 200–1000 C," *Teplofiz. Vysokikh Temperatur*, **3**(1), 57–63, 1965; English translation: *High Temperature*, **3**(1), 47–52, 1965.
309. Powell, R. W., "Further Measurements of the Thermal Electrical Conductivity of Iron at High Temperatures," *Proc. Phys. Soc. (London)*, **51**, 407–18, 1939.
310. Powell, R. W. and Hickman, M. J., "The Physical Properties of a Series of Steels. Part II. Section IIIc. Thermal Conductivity of a 0.8% Carbon Steel (Steel 7)," *J. Iron Steel Inst. (London)*, **154**, 112–21, 1946.
311. Rasor, N. S. and McClelland, J. D., "Thermal Properties of Materials. Part I. Properties of Graphite, Molybdenum and Tantalum to Their Destruction Temperatures," WADC Tech. Rept. 56-400 Pt I, 1-53, 1957. [AD 118 144]
312. Rasor, N. S. and McClelland, J. D., "Thermal Property Measurements at Very High Temperatures," *Rev. Sci. Instr.*, **31**(6), 595–604, 1960.
313. McElroy, D. L., Godfrey, T. G., and Kollie, T. G., "The Thermal Conductivity of INOR-8 between 100 and 800 C," *Trans. Am. Soc. Metals*, **55**(3), 749–51, 1962.
314. Fulkerson, W., Moore, J. P., and McElroy, D. L., "Comparison of the Thermal Conductivity, Electrical Resistivity, and Seebeck Coefficient of a High-Purity Iron and an Armco Iron to 1000 C," *J. Appl. Phys.*, **37**(7), 2639–53, 1966.
315. Hoch, M., Nitti, D. A., Gottschlich, C. F., and Blackburn, P. E., "New Method for the Determination of Thermal Conductivities between 1000 C and 3000 C," in *Progress in International Research on Thermodynamic and Transport Properties* (Masi, J. F. and Tsai, D. H., eds.), ASME Second Symposium on Thermophysical Properties, Academic Press, New York, 512–8, 1962.
316. Hoch, M., Silberstein, A., and Chapman, H., "Thermal Conductivity of Aluminum Oxide," in *Proceedings of the Fourth Symposium on Thermophysical Properties* (Moszynski, J. R., ed.), ASME, New York, 150–4, 1968.
317. Vardi, J. and Lemlich, R., "Theoretical Approach to a Technique for Measuring Thermal Conductivity at High Temperature," to be published in *J. Appl. Phys* in 1970.
318. Laws, F. A., Bishop, F. L., and McJunkin, P., "A Method of Determining Thermal Conductivity," *Proc. Am. Acad. Arts Sci.*, **41**(22), 455–64, 1906.
319. Adams, M. and Loeb, A. L., "Thermal Conductivity: II. Development of a Thermal Conductivity Expression for the Special Case of Prolate Spheroids," *J. Am. Ceram. Soc.*, **37**(2), 73–4, 1954.
320. Adams, M., "Thermal Conductivity: III. Prolate Spheroidal Envelope Method; Data for Al_2O_3, BeO, MgO, ThO_2, and ZrO_2," *J. Am. Ceram. Soc.*, **37**(2), 74–9, 1954.
321. McQuarrie, M., "Thermal Conductivity: V. High Temperature Method and Results for Alumina, Magnesia, and Beryllia from 1000 to 1800 C," *J. Am. Ceram. Soc.*, **37**(2), p. 84, 1954.
322. Péclet, E., *Traité de la Chaleur*, Vol. I, Paris, 1860.
323. Nusselt, W., "Thermal Conductivity of Thermal Insulators," *VDI Zeitschrift*, **52**(23), 906–12, 1908.
324. "Design Aspects of Plant for Production of Heavy Water by Distillation of Hydrogen," USAEC Rept. NYO-2134, 1957.
325. Stefan, J., "Investigations on the Thermal Conductivity of Gases," *Sitzber Akad.-Wiss. Wien. Math-Naturw. Kl.* IIA, **65**, 45–69, 1872.
326. Kropschot, R. H., Schrodt, J. E., Fulk, M. M., and Hunter, D. J., "Multiple-Layer Insulation," in *Advances in Cryogenic Engineering*, Vol. 5, Plenum Press, 189–97, 1960.

327. Flynn, D. R. and Watson, T. W., "High Temperature Thermal Conductivity of Soils," in *Thermal Conductivity—Proceedings of the Eighth Conference* (Ho, C. Y. and Taylor, R. E., eds.), Plenum Press, New York, 913–39, 1969.
328. de Sénarmont, H., "Memoir on the Conductivity of Crystalline Substances for Heat," *Ann. Chim. Phys.*, 3, **21**, 457–70, 1847.
329. de Sénarmont, H., "Memoir on the Conductivity of Crystalline Substances for Heat," *Compt. Rend.*, 2, **25**, 459–61, 1847.
330. de Sénarmont, H., "Memoir on the Conductivity of Crystalline Substances for Heat," *Ann. Chim. Phys.*, 3, **22**, 179–211, 1848.
331. de Sénarmont, H., "Experiments on the Effects of Mechanical Agents on the Thermal Conductivity of Homogeneous Solids," *Ann. Chim. Phys.*, 3, **23**, 257–67, 1848.
332. de Sénarmont, H., "Thermal Conductivity in Crystallized Substances," *Ann. Physik*, **73**, 191–2, 1848.
333. Powell, R. W., "A Simple Test for Anisotropic Materials," *J. Sci. Instr.*, **30**, 210, 1953.
334. Cohen, I., Lustman, B., and Eichenberg, J. D., "Measurement of the Thermal Conductivity of Metal-Clad Uranium Oxide Rods during Irradiation," *J. Nuclear Materials*, **3**(3), 331–53, 1961.
335. Dumas, J. P., Mansard, B., and Rausset, P., "Uranium Monocarbide Shaping and Irradiation Study. Final Report No. 2, March 1, 1962—December 31, 1963," *United States–Euratom Joint Research and Development Program, Rept.* EURAEC-1179, 1–110, 1964. [English translation of the original French report.]
336. Clough, D. J. and Sayers, J. B., "The Measurement of the Thermal Conductivity of UO_2 under Irradiation in the Temperature Range 150–1600 C," UKAEA Rept. AERE-R-4690, 1–55, 1964.
337. Robinson, H. E., "The Conductive-Disk Method of Measuring the Thermal Conductivity of Insulations," *Bull. Intl. Inst. Refrig. Annexe* 1962–1, 43–50, 1962.
338. Mikryukov, V. E., "*The Thermal Conductivity and Electrical Conductivity of Metals and Alloys*, Metallurgizdat, Moskow, 260 pp., 1959.
339. Powell, R. W., DeWitt, D. P., and Nalbantyan, M., "The Precise Determination of Thermal Conductivity and Electrical Resistivity of Solids at High Temperatures by Direct Electrical Heating Methods," Air Force Materials Laboratory Techn. Rept. AFML-TR-67-241, 1–100, 1967.
340. Flynn, D. R., "Measurement of Thermal Conductivity by Steady-State Methods in Which the Sample is Heated Directly by Passage of an Electric Current," in *Thermal Conductivity* (Tye, R. P., ed.), Vol. 1, Chap. 5, 241–300, 1969.
341. Kohlrausch, F., "On the Thermoelectricity, Heat and Electricity Conduction," *Göttingen Nachr.*, Feb. 7, 1874.
342. Kohlrausch, F., "The Activities at Physical-Technical Institute in the Year 1 February 1897 to 31 January 1898," *Z. Instrumentenkunde*, **18**(5), 138–51, 1898.
343. Kohlrausch, F., "On the Stationary Temperature State of a Conductor Heated by an Electric Current," *Sitz. Berlin Acad.*, **38**, 711–8, 1899.
344. Kohlrausch, F., "On the Stationary Temperature State of an Electrically Heated Conductor," *Ann. Physik*, **1**, 132–58, 1900.
345. Jaeger, W. and Diesselhorst, H., "Thermal Conductivity, Electrical Conductivity, Heat Capacity, and Thermal Power of Some Metals," *Wiss. Abhandl. Physik-tech. Reichsanstalt*, **3**, 269–425, 1900.
346. Holm, R. and Störmer, R., "Measurement of the Thermal Conductivity of a Platinum Probe in the Temperature Range 19–1020 C," *Wiss. Veröff. Siemens-Konzern*, **9**(2), 312–22, 1930.
347. Angell, M. F., "Thermal Conductivity at High Temperatures," *Phys. Rev.*, **33**(5), 421–32, 1911.
348. Powell, R. W. and Schofield, F. H., "The Thermal and Electrical Conductivities of Carbon and Graphite to High Temperatures," *Proc. Phys. Soc. (London)*, **51**(1), 153–72, 1939.
349. Worthing, A. G., "The Thermal Conductivities of Tungsten, Tantalum and Carbon at Incandescent Temperatures by an Optical Pyrometer Method," *Phys. Rev.*, **4**(6), 535–43, 1914.
350. Osborn, R. H., "Thermal Conductivities of Tungsten and Molybdenum at Incandescent Temperatures," *J. Opt. Soc. Am.*, **31**, 428–32, 1941.
351. Krishnan, K. S. and Jain, S. C., "Determination of Thermal Conductivities at High Temperatures," *Brit. J. Appl. Phys.*, **5**, 426–30, 1954.
352. Lebedev, V. V., "Determination of the Coefficient of Thermal Conductivity for Metals in the High Temperature Range," *Phys. Metals Metallog. (USSR)*, **10**(2), 31–4, 1960.
353. Bode, K. H., "A New Method to Measure the Thermal Conductivity of Metals at High Temperatures," *Allgem. Wärmetech.* **10**(6), 110–20, and **10**(7), 125–42, 1961.
354. Gumenyuk, V. S. and Lebedev, V. V., "Investigation of the Thermal and Electrical Conductivity of Tungsten and Graphite at High Temperatures," *Phys. Metals Metallog. (USSR)*, **11**(1), 30–5, 1961.
355. Bode, K. H., "Measurements of the Thermal Conductivity of Metals at High Temperatures," in *Progress in International Research on Thermodynamic and Transport Properties*, (Masi, J. F. and Tsai, D. H., eds.), ASME Second Symposium on Thermophysical Properties, Academic Press, New York, 481–99, 1962.
356. Gumenyuk, V. S., Ivanov, V. E., and Lebedev, V. V., "Determination of the Heat and Electric Conductivity of Metals at Temperatures in Excess of 1000 C," *Instrum. Exper. Techn.*, No. 1, 185–92, 1962.
357. Rudkin, R. L., Parker, W. J., and Jenkins, R. J., "Measurements of the Thermal Properties of Metals at Elevated Temperatures," in *Temperature—Its Measurement and Control in Science and Industry*," Vol. 3, Part 2, 523–34, 1962.
358. Filippov, L. P. and Simonova, Yu. N., "Measurement of Thermal Conductivity of Metals at High Temperatures," *High Temperature*, **2**(2), 165–8, 1964.
359. Platunov, E. S., "Measurement of Heat Capacity and Heat Conductivity of Rod Subjected to Monotonic Heating and Cooling," *High Temperature*, **2**(3), 346–50, 1964.
360. Taylor, R. E., Powell, R. W., Nalbantyan, M., and Davis, F., "Evaluation of Direct Electrical Heating Methods for the Determination of Thermal Conductivity

at Elevated Temperatures," Air Force Materials Laboratory Tech. Rept. AFML-TR-68-227, 1–74, 1968.

361. Bode, K. H., "Possibilities to Determine Thermal Conductivity Using New Solutions for Current-Carrying Electrical Conductors," in *Thermal Conductivity—Proceedings of the Eighth Conference* (Ho, C. Y. and Taylor, R. E., eds.), Plenum Press, New York, 317–37, 1969.
362. Taylor, R. E., Powell, R. W., Davis, F., and Nalbantyan, M., "Evaluation of Direct Electrical Heating Methods," in *Thermal Conductivity—Proceedings of the Eighth Conference* (Ho, C. Y. and Taylor, R. E., eds.), Plenum Press, New York, 339–54, 1969.
363. Taylor, R. E., Davis, F. E., Powell, R. W., and Kimbrough, W. D., "Determination of Thermal and Electrical Conductivity, Emittance and Thomson Coefficient at High Temperatures by Direct Heating Methods," Air Force Materials Laboratory Techn. Rept. AFML-TR-69-277, 1–90, 1969.
364. Taylor, R. E., Davis, F. E., and Powell, R. W., "Direct Heating Methods for Measuring Thermal Conductivity of Solids at High Temperatures," *High Pressures—High Temperatures*, 1, 663–73, 1969.
365. Powell, R. W., Ho, C. Y., and Liley, P. E., "Thermal Conductivity of Selected Materials," National Standard Reference Data Series—National Bureau of Standards NSRDS-NBS 8, 1–168, 1966.
366. Vasilos, T. and Kingery, W. D., "Thermal Conductivity. XI. Conductivity of Some Refractory Carbides and Nitrides," *J. Am. Ceram. Soc.*, **37**, 409–14, 1954.
367. Taylor, R. E., "Thermal Conductivity of Titanium Carbide at High Temperatures," *J. Am. Ceram. Soc.*, **44**(10), 525, 1961.
368. Laubitz, M. J., "On the Series Comparator Methods of Measuring Thermal Conductivity," in *Proceedings of the Black Hills Summer Conference on Transport Phenomena* (1962), S. Dakota School of Mines and Technology, Rapid City, Issued under Office of Naval Research, Contract No. Nonr(G)–00064–62, 8–22, 1962.
369. Hoch, M. and Vardi, J., "Thermal Conductivity of TiC," *J. Am. Ceram. Soc.*, **46**(5), 245, 1963.
370. Powell, R. W., "Correlation of the Thermal and Electrical Conductivity of Metals, Alloys and Compounds," in *Proceedings of the Third Conference on Thermal Conductivity* (1963), Vol. 1, Oak Ridge National Laboratory, 79–112, 1963.
371. Powell, R. W. and Tye, R. P., "The Thermal Conductivities of Some Electrically Conducting Compounds," in *Special Ceramics, 1964* (Popper, P., ed.), Academic Press, London, 243–59, 1965.
372. Taylor, R. E. and Morreale, J., "Thermal Conductivity of Titanium Carbide, Zirconium Carbide and Titanium Nitride at High Temperatures," *J. Am. Ceram. Soc.*, **47**(2), 69–73, 1964.
373. Powell, R. W. and Taylor, R. E., "Multi-Property Apparatus and Procedure for High Temperature Determinations," *Rev. Int. Hautes Temp. Réfract.*, in course of publication in 1970.
374. Longmire, C. L., "Method for Determining Thermal Conductivity at High Temperatures," *Rev. Sci. Instrum.*, **28**(11), 904–6, 1957.
375. Pike, J. N. and Doar, J. F., Union Carbide Research Institute Rept. UCRI-2787, Appendix 1, 1960.
376. Pike, J. N. and Doar, J. F., "High Temperature Thermal Conductivity Measurements. Part 2. The Rectangular Bar Method, Experimental Techniques," Union Carbide Corp. Parma Research Lab. Res. Rept. No. C-10, 1–43, 1961.
377. Pike, J. N. and Doar, J. F., "High Temperature Thermal Conductivity Measurements. Theory of Longmire's Method and the Rectangular Bar Method," Appendix X to Volume II of the Final Report "Research on Physical and Chemical Principles Affecting High Temperature Materials for Rocket Nozzles" (Submitted by Aspinall, S. R.), Union Carbide Research Institute, 1965.
378. Borelius, G., "A Method for the Combined Measurements of Peltier Heat and Thermal Conductivity," *Ann. Physik*, 4, **52**, 398–414, 1917.
379. Sedström, E., "Peltier Heat and Thermal and Electrical Conductivity of Some Solid Metallic Solutions," *Ann. Physik*, 4, **59**, 134–44, 1919.
380. Sedström, E., "On the Knowledge of Gold–Copper Alloys," *Ann. Physik*, **75**, 549–55, 1924.
381. Putley, E. H., "Thermoelectric and Galvanomagnetic Effects in Lead Selenide and Telluride," *Proc. Phys. Soc. (London)*, **B68**, 35–42, 1955.
382. Harman, T. C., "Special Techniques for Measurement of Thermoelectric Properties," *J. Appl. Phys.*, **29**(9), 1373–4, 1958.
383. Harman, T. C., Cahn, J. H. and Logan M. J., "Measurement of Thermal Conductivity by Utilization of the Peltier Effect," *J. Appl. Phys.*, **30**(9), 1351–9, 1959.
384. Calvet, E., Bros, J.-P., and Pinelli, H., "Perfection of an Apparatus for the Measurement of Thermal Conductivities of Solids, under Steady-State Conditions, by Use of the Peltier and Joule Effects," *C. R. Acad. Sci. (France)*, **260**(4), 1164–7, 1965.
385. Hérinckx, C. and Monfils, A., "Electrical Determination of the Thermal Parameters of Semiconducting Thermo-Elements," *Brit. J. Appl. Phys.*, **10**(5), 235–6, 1959.
386. Powell, R. W., "Improvements in and Relating to the Measurement of Thermal Conductivity," British Patent No. 855 658, application date 29 November 1956, complete specification published 7 December 1960.
387. Powell, R. W., "Experiments Using a Simple Thermal Comparator for Measurement of Thermal Conductivity, Surface Roughness and Thickness of Foils or of Surface Deposits," *J. Sci. Instrum.*, **34**, 485–92, 1957.
388. Powell, R. W., "Thermal Conductivity as a Non-destructive-Testing Technique," in *Progress in Non-Destructive Testing*, Vol. 1, Messrs. Heywood & Co. Ltd., 199–226, 1958.
389. Powell, R. W., "Single-End Probe, or Modified Thermal Comparator," British Patent No. 1 036 124, application date 19 January 1962, complete specification published 13 July 1966.
390. Powell, R. W., "Thermal Conductivity Determinations by Thermal Comparator Methods," in *Thermal Conductivity* (Tye, R. P., ed.), Vol. 2, Chap. 6, Academic Press, London, 275–338, 1969.
391. Powell, R. W., DeWitt, D. P., Wolfla, L. H., and Finch, R. A., "An Instrument Embodying the Thermal Comparator Technique for Thermal Conductivity Measure-

ments," in *Temperature Measurements Society—Sixth Conference and Exhibit*, Western Periodicals, Co., North Hollywood, Calif., 233–44, 1969.

392. Danielson, G. C. and Sidles, P. H., "Thermal Diffusivity and Other Nonsteady-State Methods," in *Thermal Conductivity* (Tye, R. P., ed.), Vol. 2, Chap. 3, Academic Press, London, 149–201, 1969.
393. Ångström, A. J., "New Method of Determining the Thermal Conductivity of Bodies," *Phil. Mag.*, **25**, 130–42, 1863.
394. King, R. W., "A Method of Measuring Heat Conductivities," *Phys. Rev.*, **6**(6), 437–45, 1915.
395. Starr, C., "An Improved Method for the Determination of Thermal Diffusivities," *Rev. Sci. Instrum.*, **8**(1), 61–4, 1937.
396. Sidles, P. H. and Danielson, G. C., "Thermal Diffusivity of Metals at High Temperature," *J. Appl. Phys.*, **25**(1), 58–66, 1954.
397. Abeles, B., Cody, G. D., and Beers, D. S., "Apparatus for the Measurement of the Thermal Diffusivity of Solids at High Temperatures," *J. Appl. Phys.*, **31**(9), 1585–92, 1960.
398. Cowan, R. D., "Proposed Method of Measuring Thermal Diffusivity at High Temperatures," *J. Appl. Phys.*, **32**(7), 1363–70, 1961.
399. Hirschman, A., Dennis, J., Derksen, W., and Monahan, T., "An Optical Method for Measuring the Thermal Diffusivity of Solids," in *International Developments in Heat Transfer, Part IV*, ASME, New York, 863–9, 1961.
400. Wheeler, M. J., "Thermal Diffusivity at Incandescent Temperatures by a Modulated Electron Beam Technique," *Brit. J. Appl. Phys.*, **16**(3), 365–76, 1965.
401. Tanasawa, T., "A New Method for the Measurement of the Thermal Constants of Wet Substance (The Second Report)," *Trans. Soc. Mech. Engrs. Japan*, **1**(3), 217–26, 1935.
402. Filippov, L. P. and Pigal'skaya, L. A., "Measurement of the Thermal Diffusivity of Metals at High Temperatures," *High Temperature*, **2**(3), 351–8, 1964.
403. Pigal'skaya, L. A. and Filippov, L. P., "Measurement of the Thermal Diffusivity of Metals at High Temperatures. Part 2. Experimental Method of Alternating Heating in a High-Frequency Furnace," *High Temperature*, **2**(4), 501–4, 1964.
404. Yurchak, R. P. and Filippov, L. P., "Measuring the Thermal Diffusivity of Molten Metals," *Teplofiz. Vysokikh Temperatur*, **2**(5), 696–704, 1964; English translation: *High Temperature*, **2**(5), 628–30, 1964.
405. Yurchak, R. P. and Filippov, L. P., "Thermal Properties of Molten Tin and Lead," *Teplofiz. Vysokikh Temperatur*, **3**(2), 323–5, 1965; English translation: *High Temperature*, **3**(2), 290–1, 1965.
406. van Zee, A. F. and Babcock, C. L., "A Method for the Measurement of Thermal Diffusivity of Molten Glass," *J. Am. Ceram. Soc.*, **34**, 244–50, 1951.
407. Kirichenko, Yu. A., Oleinik, B. N., and Chadovich, T. Z., "Thermal Properties of Polymers," *Inzh-Fiz. Zh.*, **7**(5), 70–5, 1964.
408. Neumann, F., "Experiments on the Thermal Conductivity of Solids," *Ann. Chim. Phys.*, 3, **66**, 183–7, 1862.
409. Neumann, F., "Experiments on the Calorific Conductibility of Solids," *Phil. Mag.*, **25**, 63–5, 1863.
410. Oualid, J., "Determination of the Diffusivity Coefficient of Metals and Semi-Conductors," *J. Phys. Radium*, **22**, 124–6, 1961.
411. Woisard, E. L., "Pulse Method for the Measurement of Thermal Diffusivity of Metals," *J. Appl. Phys.*, **32**(1), 40–5, 1961.
412. Butler, C. P. and Inn, E. C. Y., "Thermal Diffusivity of Metals at Elevated Temperatures," in *Thermodynamic and Transport Properties of Gases, Liquids and Solids*, (Touloukian, Y. S., ed.), *Am. Soc. Mech. Engrs. Symp. Thermal Prop.*, McGraw-Hill, N.Y., 377–90, 1959.
413. Kennedy, W. L., Sidles, P. H., and Danielson, G. C., "Thermal Diffusivity Measurements on Finite Samples," *Adv. Energy Conv.*, **2**, 53–8, 1962.
414. Penniman, F. G., "A Long-Pulse Method of Determining Thermal Diffusivity," *Sol. Energy*, Pt. II, **9**(3), 113–6, 1965.
415. Fitch, A. L., "A New Thermal Conductivity Apparatus," *Am. Phys. Teacher*, **3**, 135–6, 1935.
416. Joffé, A. V. and Joffé, A. F., "Measurement of Thermal Conductivity of Semi-conductors in the Vicinity of Room Temperatures," *Soviet Phys.—Tech. Phys.*, **3**(11), 2163–8, 1958.
417. Plummer, W. A., Campbell, D. E., and Comstock, A. A., "Method of Measurement of Thermal Diffusivity to 1000 C," *J. Am. Ceram. Soc.*, **45**(7), 310–6, 1962.
418. Cutler, M. and Cheney, G. T., "Heat-Wave Methods for the Measurement of Thermal Diffusivity," *J. Appl. Phys.*, **34**(7), 1902–9, 1963.
419. Parker, W. J., Jenkins, R. J., Butler, C. P., and Abbott, G. L., "Flash Method of Determining Thermal Diffusivity, Heat Capacity, and Thermal Conductivity," *J. Appl. Phys.*, **32**(9), 1679–84, 1961.
420. Cape, J. A. and Lehman, G. W., "Temperature and Finite Pulse-Time Effects in the Flash Method for Measuring Thermal Diffusivity," *J. Appl. Phys.*, **34**(7), 1909–13, 1963.
421. Taylor, R. E. and Cape, J. A., "Finite Pulse-Time Effect in the Flash Diffusivity Technique," *Appl. Phys. Letters*, **5**(10), 212–23, 1964.
422. Ginnings, D. C., "Standards of Heat Capacity and Thermal Conductivity," in *Thermoelectricity* (Egli, P. H., ed.), Chap. 20, John Wiley & Sons., New York, 320–41, 1960.
423. Cape, J. A., Lehman, G. W., and Nakata, M. M., "Transient Thermal Diffusivity Technique for Refractory Solids," *J. Appl. Phys.*, **34**(12), 3550–5, 1963.
424. Carter, R. L., Maycock, P. D., Klein, A. H., and Danielson, G. C., "Thermal Diffusivity Measurements with Radial Sample Geometry," *J. Appl. Phys.*, **36**(8), 2333–7, 1965.
425. Stalhane, B. and Pyk, S., "New Method for Measuring the Thermal Conductivity Coefficients," *Tekn. Tidskr.*, **61**(28), 389–93, 1931.
426. Stalhane, B. and Pyk, S., "Determination of the Thermal Conductivity of Ceramic Bodies at High Temperature," *Tekn. Tidskr.*, **64**(48), 445–8, 1934.
427. van der Held, E. F. and van Drunen, F. G., "A Method of Measuring the Thermal Conductivity of Liquids," *Physica*, **15**(10), 865–81, 1949.
428. van der Held, E. F., Hardebol, J., and Kalshoven, J., "The Measurement of the Thermal Conductivity of

Liquids by a Nonstationary Method," *Physica*, **19**(3), 208–16, 1953.

429. Hooper, F. C. and Lepper, F. R., "Transient Heat Flow Apparatus for the Determination of Thermal Conductivities," *Heating, Piping and Air Conditioning*, ASHVE J. Sect. 22(8), 129–34, 1950.
430. Hooper, F. C. and Chang, S. C., "Development of the Thermal Conductivity Probe," *Heating, Piping, and Air Conditioning*, ASHVE J. Sect. 24(10), 125–9, 1952.
431. Blackwell, J. H., "Radial-Axial Heat Flow in Regions Bounded Internally by Circular Cylinders," *Can. J. Phys.*, **31**(4), 472–9, 1953.
432. Blackwell, J. H., "A Transient-Flow Method for Determination of Thermal Constants of Insulating Materials in Bulk. I. Theory," *J. Appl. Phys.*, **25**, 137–44, 1954.
433. Makowski, M. W. and Mochlinski, K., "An Evaluation of Two Rapid Methods of Assessing the Thermal Resistivity of Soil," *Proc. Instn. Elect. Engrs.*, **103A**(11), 453–70, 1956.
434. Rosenthal, D. and Ambrosio, A., "A New Method of Determining Thermal Diffusivity of Solids at Various Temperatures," *Trans. ASME*, **73**(7), 971–4, 1951.
435. Rosenthal, D. and Friedmann, N. E., "Thermal Diffusivity of Metals at High Temperatures," *J. Appl. Phys.*, **25**(8), 1059–60, 1954.
436. Rosenthal, D. and Friedmann, N. E., "The Determination of Thermal Diffusivity of Aluminum Alloys at Various Temperatures by Means of a Moving Heat Source," *Trans. ASME*, **78**(8), 1175–80, 1956.
437. Hsu, S. T., "Theory of a New Apparatus for Determining the Thermal Conductivities of Metals," *Rev. Sci. Instr.*, **28**(5), 333–6, 1957.
438. Hsu, S. T., "Determination of Thermal Conductivities of Metals by Measuring Transient Temperatures in Semi-Infinite Solids," *Trans. ASME*, **79**, 1197–1203, 1957.
439. Deem, H. W., Pobereskin, M., Lusk, E. C., Lucks, C. F., and Calkins, G. D., "Effect of Radiation on the Thermal Conductivity of Uranium – 1.6 Wt.% Zirconium," USAEC Rept. BMI-986, 1–19, 1955.

Numerical Data

Data Presentation and Related General Information

1. SCOPE OF COVERAGE

Presented in this volume are the thermal conductivity data for 5 elements, over 100 different grades of graphites, 53 oxides and oxide compounds, 82 systems of oxide mixtures, 83 nonoxide compounds, 7 systems of mixtures of oxides and nonoxides, 15 organic compounds, 22 kinds of cermets, 10 groups of systems, 8 kinds of refractory materials, over 40 different kinds of glasses, 24 minerals, 16 polymers, 26 animal and vegetable natural substances, 30 processed composites and processed natural and mineral substances, and 10 aggregate mixes, slags, scales, and residues. These data are obtained by processing over 1260 research documents on the thermal conductivity of nonmetallic solids dated from around 1800 to 1967, of which about 590 contain usable data. Materials within each group are arranged in alphabetical order by name, as listed in the *Grouping of Materials and List of Figures and Tables* in the front of the volume. Totally, this volume reports 4627 sets of data on 812 materials, which are listed in the *Material Index* at the end of the volume. The *Material Index* lists also the materials contained in the companion volumes (Volumes 1 and 3) on thermal conductivity.

The ranges of temperatures covered by the thermal conductivity data for some materials are from near absolute zero to past the melting point, though, for most high-temperature materials, the available data are limited to the solid range.

The data for the elements and a number of oxides have been critically evaluated, analyzed, and synthesized, and recommended reference values are presented. This procedure involves critical evaluation of the validity of available data and related information, resolution and reconciliation of disagreements of conflicting data, correlation of data in terms of various affecting parameters, and comparison of the resulting values with theoretical predictions, or with results derived from semi-theoretical relationships, or from generalized empirical correlations. Besides critical evaluation and analysis of the existing data, thermodynamic principles and semi-empirical techniques are employed to fill in gaps and to extrapolate existing data so that the resulting recommended values are internally consistent and cover as wide a range of the controlling parameters as possible. Future editions of this volume will contain recommended values for an increasing number of materials.

2. PRESENTATION OF DATA

The thermal conductivity data and information on test specimens for each material are generally presented in three sections arranged in the following order: Original Data Plot, Specification Table, and Data Table. For the elements and a number of oxides, Graph and Table of Recommended Values is added as a fourth section. Furthermore, for a number of materials for which there exists only a small amount of data, the original data plot may be omitted.

The Original Data Plot is a full-page log-log-scale graphical presentation of the original thermal conductivity data as a function of temperature. When several sets of data are coincident, some of the data sets may be omitted from the plot for the sake of clarity. They are, however, invariably reported in the Data Table and Specification Table.

The Specification Table provides in a concise form the comprehensive information on the test specimens for which the data are reported. The curve numbers in the Specification Table correspond exactly to the numbers which also appear in the Original Data Plot and in the Data Table. The Specification Table gives for each set of data the reference number which corresponds to the number in the list of *References to Data Sources*, the year of publication of the original data, the method of measurement, the temperature range, the reported estimate of error of the data, the specimen designation, and the specimen characterization and test conditions. The information of the last category, which is reported to the extent provided in the original source document, includes the following:

(1) Purity, chemical composition;
(2) Type of crystal, crystal axis orientation, type and concentration of crystal defects;
(3) Microstructure, grain size, pore size and shape, inhomogeneity, additional phases;
(4) Specimen shape and dimensions, method and procedure of fabrication;
(5) Thermal history, heat treatment, mechanical, irradiative, and other treatments;
(6) Manufacturer and supplier, stock number, and catalog number;
(7) Test environment, degree of vacuum or pressure, heat flow direction, strength and orientation of the applied magnetic field;
(8) Pertinent physical properties such as density, porosity, hardness, etc.;
(9) Reference material for a comparative method of measurement;
(10) Form in which the extracted data are presented in the original source document other than raw data points;
(11) Additional information obtained directly from the author.

Unfortunately, in the majority of cases the authors do not report in their research papers all the necessary pertinent information to fully characterize and identify the materials for which their data are reported. This is particularly true for the authors of earlier investigations. Consequently, the amount of information on specimen characterization reported in the Specification Tables varies greatly from specimen to specimen.

In the Data Table, tabular presentation is given for all the data described in the Specification Table and shown or not shown in the Original Data Plot. Many tabular data which are not presented in the original source documents are obtained directly from the authors through private communications. Attempts have often been made to contact the authors of recent publications for tabular data whenever the original data are given in the research paper only in a figure too small to warrant accurate data extraction compatible with the reported accuracy of the measurement. The thermal conductivity data are given in watts per centimeter per degree Kelvin, and the temperatures in degrees Kelvin. For data conversion, the reader is referred to the *Conversion Factors for Thermal Conductivity Units* given later.

The recommended thermal conductivity values for a material are reported in a separate graph and table following the Data Table. The estimated accuracy of the recommended values and special remarks on material characterization and identification are also noted in the table.

3. CLASSIFICATION OF MATERIALS

The classification scheme as shown in the table for nonmetallic elements, compounds, and mixtures

Classification of Materials

Classification		Limits of composition (weight percent)*			
		X_1	$X_1 + X_2$	X_2	X_3
1. Elements		>99.5	—	<0.2	<0.2
2. Mixtures (or solutions) of elements or of elements and compounds	A. Binary	—	≥ 99.5	≥ 0.2	≤ 0.2
	B. Multiple	—	≥ 99.5	>0.2	>0.2
		—	<99.5	≥ 0.2	≤ 0.2
		—	<99.5	>0.2	>0.2
		≤ 99.5	—	<0.2	<0.2
3. Compounds		>95.0	—	<2.0	<2.0
4. Mixtures (or solutions) of compounds	A. Binary	—	≥ 95.0	≥ 2.0	≤ 2.0
	B. Multiple	—	≥ 95.0	>2.0	>2.0
		—	<95.0	≥ 2.0	≤ 2.0
		—	<95.0	>2.0	>2.0
		≤ 95.0	—	<2.0	<2.0

*$X_1 \geq X_2 \geq X_3 \geq X_4 \geq \ldots$

contained in this volume is based strictly upon the chemical composition of the material. This scheme is mainly for the convenience of material grouping and data organization, and is not intended to be used as basic definitions for the various material groups.

4. SYMBOLS AND ABBREVIATIONS USED IN THE FIGURES AND TABLES

In the Specification Tables, the code designations used for the experimental methods are as follows:

C	Comparative method
E	Direct electrical heating method
F	Forbes' bar method
L	Longitudinal heat flow method
P	Periodic or transient heat flow method
R	Radial heat flow method
T	Thermoelectrical method

Other symbols and abbreviations used in the figures and/or tables are as follows:

b.c.c.	Body-centered cubic
c.	Cubic
c.p.h.	Close-packed hexagonal
d	Density
d.	Diamond (crystal structure)
Decomp.	Decomposition
f.c.c.	Face-centered cubic
f.c.t.	Face-centered tetragonal
h.	Hexagonal
I.D.	Inside diameter
k	Thermal conductivity
M.P.	Melting point
monocl.	Monoclinic
NTP	Normal temperature and pressure
O.D.	Outside diameter
orthorh.	Orthorhombic
r.	Rhombohedral
s.c.	Superconducting
Subl.	Sublimation
T	Temperature
t.	Tetragonal
Temp.	Temperature
T.P.	Transition point
Vit.	Vitreous
ρ	Electrical resistivity
μ	Micro
$>$	Greater than
$<$	Less than
$\sim$	Approximately

(3)	Curve number
[4]	Single data point number

5. CONVENTION FOR BIBLIOGRAPHIC CITATION

For the following types of documents the bibliographic information is cited in the sequences given below.

Journal Article:

a. Author(s)—The names and initials of all authors are given. The last name is written first, followed by initials.
b. Title of article—In this volume, the titles of the journal articles listed in the *References to Text* are given, but not of those listed in the *References to Data Sources.*
c. Journal title—The abbreviated title of the journal as used in *Chemical Abstracts* is given.
d. Series, volume, and number—If the series is designated by a letter, no comma is used between the letter for series and the numeral for volume, and they are underlined together. In case series is also designated by a numeral, a comma is used between the numeral for series and the numeral for volume, and only the numeral representing volume is underlined. No comma is used between the numerals representing volume and number. The numeral for number is enclosed in parentheses.
e. Pages—The inclusive page numbers of the article.
f. Year—The year of publication.

Report:

a. Author(s)
b. Title of report—In this volume, the titles of the reports listed in the *References to Text* are given, but not of those listed in the *References to Data Sources.*
c. Name of the responsible organization.
d. Report, or bulletin, circular, technical note, etc.
e. Number
f. Part
g. Pages
h. Year
i. ASTIA's AD number—This is given in square brackets whenever available.

CONVERSION FACTORS FOR UNITS OF THERMAL CONDUCTIVITY

MULTIPLY ↓ by appropriate factor to OBTAIN →	Btu_{IT} hr^{-1} ft^{-1} F^{-1}	Btu_{IT} in. hr^{-1}ft^{-2} F^{-1}	Btu_{th} hr^{-1} ft^{-1} F^{-1}	Btu_{th} in. hr^{-1}ft^{-2} F^{-1}	cal_{IT} sec^{-1} cm^{-1} C^{-1}	cal_{th} sec^{-1} cm^{-1} C^{-1}	$kcal_{th}$ hr^{-1} m^{-1} C^{-1}	J sec^{-1} cm^{-1} K^{-1}	W cm^{-1} K^{-1}	W m^{-1} K^{-1}	mW cm^{-1} K^{-1}
Btu_{IT} hr^{-1}ft^{-1} F^{-1}	1	12	1.00067	12.0080	4.13379×10^{-3}	4.13656×10^{-3}	1.48916	1.73073×10^{-2}	1.73073×10^{-2}	1.73073	17.3073
Btu_{IT} in. hr^{-1}ft^{-2} F^{-1}	8.33333×10^{-2}	1	8.33891×10^{-2}	1.00067	3.44482×10^{-4}	3.44713×10^{-4}	0.124097	1.44228×10^{-3}	1.44228×10^{-3}	0.144228	1.44228
Btu_{th} hr^{-1}ft^{-1} F^{-1}	0.999331	11.9920	1	12	4.13102×10^{-3}	4.13379×10^{-3}	1.48816	1.72958×10^{-2}	1.72958×10^{-2}	1.72958	17.2958
Btu_{th} in. hr^{-1}ft^{-2} F^{-1}	8.32776×10^{-2}	0.999331	8.33333×10^{-2}	1	3.44252×10^{-4}	3.44482×10^{-4}	0.124014	1.44131×10^{-3}	1.44131×10^{-3}	0.144131	1.44131
cal_{IT} sec^{-1}cm^{-1} C^{-1}	2.41909×10^{2}	2.90291×10^{3}	2.42071×10^{2}	2.90485×10^{3}	1	1.00067	3.60241×10^{2}	4.1868	4.1868	4.1868×10^{2}	4.1868×10^{3}
cal_{th} sec^{-1}cm^{-1} C^{-1}	2.41747×10^{2}	2.90096×10^{3}	2.41909×10^{2}	2.90291×10^{3}	0.999331	1	3.6×10^{2}	4.184	4.184	4.184×10^{2}	4.184×10^{3}
$kcal_{th}$ hr^{-1}m^{-1} C^{-1}	0.671520	8.05824	0.671969	8.06363	2.77592×10^{-3}	2.77778×10^{-3}	1	1.16222×10^{-2}	1.16222×10^{-2}	1.16222	11.6222
J sec^{-1}cm^{-1} K^{-1}	57.7789	6.93347×10^{2}	57.8176	6.93811×10^{2}	0.238846	0.239006	86.0421	1	1	1×10^{2}	1×10^{3}
W cm^{-1} K^{-1}	57.7789	6.93347×10^{2}	57.8176	6.93811×10^{2}	0.238846	0.239006	86.0421	1	1	1×10^{2}	1×10^{3}
W m^{-1} K^{-1}	0.577789	6.93347	0.578176	6.93811	2.38846×10^{-3}	2.39006×10^{-3}	0.860421	1×10^{-2}	1×10^{-2}	1	10
$^{-1}$ K^{-1}	5.77789×10^{-2}	0.693347	5.78176×10^{-2}	0.693811	2.38846×10^{-4}	2.39006×10^{-4}	8.60421×10^{-2}	1×10^{-3}	1×10^{-3}	0.1	1

Books:

a. Author(s)
b. Title
c. Volume
d. Edition
e. Publisher
f. Place of publication
g. Pages
h. Year

6. CONVERSION FACTORS FOR THERMAL CONDUCTIVITY UNITS

The conversion factors given in the table on page 44a are based upon the following basic definitions:

1 in.	= 0.0254 (exactly) m*
1 lb	= 0.45359237 kg*
1 cal_{th}	= 4.184 (exactly) J*
1 cal_{IT}	= 4.1868 (exactly) J*
1 $Btu_{th} lb^{-1} F^{-1}$	= 1 $cal_{th} g^{-1} C^{-1}$†
1 $Btu_{IT} lb^{-1} F^{-1}$	= 1 $cal_{IT} g^{-1} C^{-1}$†

*National Bureau of Standards, "New Values for the Physical Constants Recommended by NAS-NRC," *NBS Tech. News Bull.*, **47**(10), 175–7, 1963.

†Mueller, E. F. and Rossini, F. D., "The Calory and the Joule in Thermodynamics and Thermochemistry," *Am. J. Phys.*, **12**(1), 1–7, 1944.

The subscripts "th" and "IT" designate "thermochemical" and "International Steam Table," respectively.

7. CRYSTAL STRUCTURES, TRANSITION TEMPERATURES, AND OTHER PERTINENT PHYSICAL CONSTANTS OF THE ELEMENTS

The table on the following pages contains information on the crystal structures, transition temperatures, and certain other pertinent physical constants of the elements. This information is very useful in data analysis and synthesis. For example, the thermal conductivity of a material generally changes abruptly when the material undergoes any transformation. One must therefore be extremely cautious in attempting to extrapolate the thermal conductivity values across any phase, state, magnetic, or superconducting transition temperature, as given in the table.

No attempt has been made to critically evaluate the temperatures/constants given in the table, and they should not be considered recommended values. This table has an independent series of numbered references which immediately follows the table.

CRYSTAL STRUCTURES, TRANSITION TEMPERATURES, AND OTHER PERTINENT PHYSICAL CONSTANTS OF THE ELEMENTS

Name	Atomic Number	Atomic Weight[a]	Density,[b] kg m^{-3}·10^{-3}	Crystal Structure	Phase Transition Temp., K	Superconducting Transition Temp., K	Curie Temp., K	Néel Temp., K	Debye Temperature at 0 K, K	Debye Temperature at 298 K, K	Melting Point, K	Boiling Point, K	Critical Temp., K
Actinium	89	(227)	10.07[1c]	f.c.c.[2]					124[3]	100[4] (at~50 K)	1323[5]	3200 ± 300[6]	
Aluminum	13	26.9815	2.702[5]	f.c.c.[7]		1.196[5] 1.17[8] 1.18[9]			423 ± 5[3]	390[3]	933.2[3,10]	2723[29]	8650[11] 7740[109]
Americium	95	(243)	11.7[5]	Double c.p.h.[2]							1473[29]	2880[108]	
Antimony	51	121.75	6.684[29]	r.[2] (?) ? (?) ? (?)	367.8[13] (?-?) 690[13] (?-?)	2.6[8] (Sb II, high-pressure modification)			150[3]	200[14]	903.7[13] 903.65[23]	1907 ± 10[3]	2989[15]
Argon	18	39.948	0.0017824[29] (at 273.2 K and 1 atm)	f.c.c.[16]						90[4] (at~45 K)	83.8[17]	87.29[13]	151[15]
Arsenic	33	74.9216	5.73[29] (gray, at 287.2 K) 4.7[29] (black) 2.0[29] (yellow)	r.[7] (gray) c.[5] (yellow)					236[3]	275[18]	1090[13] (35.8 atm) subl. 886[5]	1090[13] (35.8 atm)	
Astatine	85	(210)									573.2[19]	650[20]	
Barium	56	137.34	3.5[29]	b.c.c.[2] (α) ? (β)[13]	648[13,21] (α-β)				110.5 ± 1.8[22]	116[23]	998.2[5]	1910[3]	3663[15] 3920[109]
Berkelium	97	(249)											
Beryllium	4	9.0122	1.85[29]	c.p.h.[2] (α) b.c.c.[2] (β)	1533[24] (α-β)	~6[108] ~8.4[108]			1160[25]	1031[3]	1550[26]	3142 ± 100[3]	6153[15]
Bismuth	83	208.980	9.78[29]	r.[2]		3.9[8] (Bi II, at 25 kbar) 7.2[8] (Bi III, at 27 kbar)			119 ± 2[3]	116 ± 5[3]	544.525[3,111]	1824 ± 8[3]	4620[27]
Boron	5	10.811	2.50[42]	Simple r.[2] (α) r.[2] (β)	1473[2] (α-β)				1315[53]	1362[3]	2573[5]	4050 ± 100[30]	
Bromine	35	79.909	3.119[29]	orthorh.[16]							266.0[17]	331.93[29]	584[15]

[a] Atomic weights are based on $^{12}C = 12$ as adopted by the International Union of Pure and Applied Chemistry in 1961; those in parentheses are the mass numbers of the isotopes of longest known half-life.

[b] Density values are given at 293.2 K unless otherwise noted.

[c] Superscript numbers designate references listed at the end of the table.

Name	Atomic Number	Atomic Weight[a]	Density,[b] $kg\,m^{-3}\cdot 10^{-3}$	Crystal Structure	Phase Transition Temp., K	Superconducting Transition Temp., K	Curie Temp., K	Néel Temp., K	Debye Temperature at 0 K, K	Debye Temperature at 298 K, K	Melting Point, K	Boiling Point, K	Critical Temp., K
Cadmium	48	112.40	8.65[29]	c.p.h.[2] b.c.c.[4] (?)		0.56[5] 0.52[9]			252 ± 48[3]	221[3] 170[4] (b.c.c., at ~ 85 K)	594.18[3,10] Subl. 594.1[13] (at 0.11 mm Hg)	1038[3]	1903[15] 3560[109]
Calcium	20	40.08	1.55[29]	f.c.c.[7] (α) b.c.c.[7] (β)	737[62] (α–β)				234 ± 5[3]	230[3]	1123[19] Subl. 1123[13] (at 0.35 mm Hg)	1765[3]	3267[15]
Californium	98	(251)											
Carbon (amorphous)	6	12.01115	1.8 ~ 2.1[29]								Subl. 3925–3970[5]	4473[5]	
Carbon (diamond)	6	12.01115	3.51[29]	d.[16]					2240 ± 5[31]	1874[3]	> 3823[5]	5100[5]	
Carbon (graphite)	6	12.01115	2.26[29] (α)	h.[2] (α) r.[2] (β)					402 ± 11[3]	1550[3]	Subl. 3925–3970[5]	4473[5]	
Cerium	58	140.12	6.90[29]	f.c.c. (α)[32] Double c.p.h.?[8] (β) f.c.c. (γ)[32] b.c.c. (δ)[32]	103 ± 5[33] (α–β) 263 ± 5[33] (β–γ) 1003[32] (γ–δ)			13[32]	146[3]	138[34]	1077[26]	3972[3]	10400[109]
Cesium	55	132.905	1.873[29]	b.c.c.[2]					40 ± 5[3]	43[23]	301.9[29] Subl. 301.9[13] (at 1.2 μHg)	939[35]	2060[113,114,115] 1900[109]
Chlorine	17	35.453	0.003214[29] (at 273.2 K)	t.[16]						115[4,36] (at ~ 58 K)	172.2[26]	239.10[13]	417[15]
Chromium	24	51.996	7.16[42]	c.p.h.[17,d] (α) b.c.c.[7] (β)	~ 299[17] (α–β)[d]			311[37]	598 ± 32[3]	424[3]	2118[38]	2918 ± 35[3]	
Cobalt	27	58.9332	8.862[42]	c.p.h.[7] (α) f.c.c.[17] (β)	690[39] (α–β)		1400[40]		452 ± 17[3]	386[3]	1765[3,10]	3229[3]	
Copper	29	63.54	8.933[29]	f.c.c.[2]					342 ± 2[3]	310[3]	1356[3,10]	2811 ± 20[41]	8500[11] 8280[109]
Curium	96	(247)	7[42]	Double c.p.h.[8]									
Dysprosium	66	162.50	8.556[42]	c.p.h.[2] (α) b.c.c.[2] (β)	Near m.p.[2] (α–β)			174[43] 83.5[43] (ferro–antiferromag.)	172 ± 35[3]	158[44]	1773[12]	3011[44]	7640[109]

[d]Close-packed hexagonal crystalline modification of chromium may be formed by electrodeposition below 293 K under special conditions of deposition process. This c.p.h. form is unstable and will irreversibly transform into b.c.c. form on heating.

Name	Atomic Number	Atomic Weight[a]	Density[b], $kg\ m^{-3} \cdot 10^{-3}$	Crystal Structure	Phase Transition Temp., K	Superconducting Transition Temp., K	Curie Temp., K	Néel Temp., K	Debye Temperature at 0 K, K	Debye Temperature at 298 K, K	Melting Point, K	Boiling Point, K	Critical Temp., K
Einsteinium	99	(254)											
Erbium	68	167.26	9.06[42]	c.p.h.[2] (α) b.c.c.[2] (β)	1643[2] (α–β)		19[4]	80[4]	134 ± 10[45]	163[44]	1770[26]	3000[3]	7250[109]
Europium	63	151.96	5.245[28]	b.c.c.[7]				~90[4]	127[3]		1099[5]	1971[46]	4600[109]
Fermium	100	(253)											
Fluorine	9	18.9984	0.001695[29] (at 273.2 K and 1 atm)	c.[108] (β-F_2)							53.58[5]	85.24[13]	144[15]
Francium	87	(223)							39[3]		300.2[19]	879[108]	
Gadolinium	64	157.25	7.87[42]	c.p.h.[2] (α) b.c.c.[2] (β)	1535[32] (α–β)		292[40]		170[3]	155 ± 3[3]	1579[19]	3540[3]	8670[109]
Gallium	31	69.72	5.91[29]	orthorh.[4] (α) t.[4] (β)	275.6[13] (α–β) (at 8.86 x 10^6 mm Hg)	1.091[5] 7.2[38] (Ga II, high-pressure modification)			317[3]	240[14] 125[4] (tetra at ~63 K)	302.93[5] 275.6[13] (at 8.86 x 10^6 mm Hg)	2510[3]	7620[27]
Germanium	32	72.59	5.36[29]	d.[7]		5.5[47] (at ~118 kbar) 8.4[108]			378 ± 22[3]	403[3]	1210.6[5]	3100[3]	5642[15]
Gold	79	196.967	19.3[42]	f.c.c.[7]					165 ± 1[3]	178 ± 8[3]	1336.2[3,10] 1336.15[23]	3240[3]	9500[11] 8060[109]
Hafnium	72	178.49	13.28[42]	c.p.h.[48] (α) b.c.c.[48] (β)	2023 ± 20[48] (α–β)	0.16[9] 0.35[108]			256 ± 5[3]	213[23]	2495[19]	4575 ± 150[49]	
Helium	2	4.0026	0.0001785[29] (at 273.2 K and 1 atm)	c.p.h.[16]						30[4] (at ~15 K)	3.45[29] 1.8 ± 0.2[17] (at 30 atm)	4.216[13] 4.22[23]	5.3[15]
Holmium	67	164.930	8.80[29]	c.p.h.[2] (α) b.c.c.[2] (β)	Near m.p.[50] (α–β)		20[4]	132[4]	114 ± 7[45]	161[44]	1734[19]	3228[51]	
Hydrogen	1	1.00797	0.00008987[29] (at 273.2 K and 1 atm)	c.p.h.[16]						116[36] (para., at ~58 K) 105[36] (ortho., at ~53 K)	13.8 ± 0.1[17]	20.39[13] 20.37[23]	33.3[15]
Indium	49	114.82	7.3[29]	f.c.t.[7]		3.4035[5]			108.8 ± 0.3[3]	129[14]	429.76[3,110]	2279 ± 6[3]	4377[15] 7050[109]
Iodine	53	126.9044	4.93[29]	orthorh.[16]						105[4] (at ~53 K)	386.8[29] subl. 298.16[13] (at 0.31 mm Hg)	457.50[29]	785[15]
Iridium	77	192.2	22.5[42]	f.c.c.[7]		0.14[5,9]			425 ± 5[3]	228[3]	2716[3,10]	4820 ± 30[3]	

Name	Atomic Number	Atomic Weight[a]	Density,[b] kg m^{-3} · 10^{-3}	Crystal Structure	Phase Transition Temp., K	Superconducting Transition Temp., K	Curie Temp., K	Néel Temp., K	Debye Temperature at 0 K, K	Debye Temperature at 298 K, K	Melting Point, K	Boiling Point, K	Critical Temp., K
Iron	26	55.847	7.87[28]	b.c.c.-ferromag.[7] (α) b.c.c.-paramag.[7] (β) f.c.c.[7] (γ) b.c.c.[7] (δ)	1183[2] (β-γ) 1673[13] (γ-δ)		1043[40]		457 ± 12[3]	373[3]	1810[19]	3160[20]	[illegible] 6750[123] 9400[109]
Krypton	36	83.80	0.003708[29] (at 273.2 K and 1 atm)	f.c.c.[16]						60[4] (at ~ 30 K)	116.6[5]	119.93[13]	209.4[15]
Lanthanum	57	138.91	6.18[42]	Double c.p.h.[8] (α) f.c.c.[2] (β) b.c.c.[2] (γ)	583[32] (α-β) 1141[32] (β-γ)	4.9[8] (α) 6.3[8] (β)			142 ± 3[52]	135 ± 5[44]	1193[5]	3713 ± 70[3]	10500[109]
Lawrencium	103	(257)											
Lead	82	207.19	11.34[29]	f.c.c.[2]		7.193[5]			102 ± 5[3]	87 ± 1[3]	600.576[3,111]	2022 ± 10[41]	5400[27] 4760[109]
Lithium	3	6.939	0.534[29]	b.c.c.[7]	Martensitic transformation at low temp.[56]				352 ± 17[3]	448[3]	453.7[19]	1599[13]	4150[11] 3720[109]
Lutetium	71	174.97	9.85[29]	c.p.h.[2] (α) b.c.c.[2] (β)	Near m.p.[50] (α-β)				210[54]	116[3]	1923[19]	4140[3]	
Magnesium	12	24.312	1.74[29]	c.p.h.[7]					396 ± 54[3]	330[3]	923[55]	1385[3]	3530[109]
Manganese	25	54.9380	7.43(α)[28] 7.29(β)[28] 7.18(γ)[28]	bcc[93] c.[7] (α) c.[7] (β) f.c.c.[93] (γ) b.c.c.[7] (δ)	1000[13] (α-β) 1374[13] (β-γ) 1410[13] (γ-δ)			95[5]	418 ± 32[3]	363[3]	1517 ± 3[5]	2360[13]	6050[109]
Mendelevium	101	(256)											
Mercury	80	200.59	13.546[29] 14.19[29] (at 234.25 K)	r.[7] (α) b.c.t.-pressure induced structure (β)	Martensitic transformation at low temp.[56]	4.153[5] (α) 3.949[5] (β)			~ 75[58]	92 ± 8[3]	234.28[3,10]	629.73[3,10]	1733[27] 1705[109]
Molybdenum	42	95.94	10.24[42]	b.c.c.[2]		0.92[5,9]			459 ± 11[3]	377[3]	2883[13]	5785 ± 175[3]	17000[11] 16800[109]
Neodymium	60	144.24	7.007[29]	Double c.p.h.[8] (α) b.c.c.[32] (β)	1135[32] (α-β)			8[4] (ordinary) 19[4] (special)	159[3]	148 ± 8[3]	1292[19]	2956[60]	7900[109]
Neon	10	20.183	0.0009002[29] (at 273.2 K and 1 atm)	f.c.c.[16]						60[4] (at ~ 30 K)	24.48[5]	27.23[5] 27.06[23]	44.5[15]

Name	Atomic Number	Atomic[a] Weight	Density,[b] kg m^{-3} · 10^{-3}	Crystal Structure	Phase Transition Temp., K	Superconducting Transition Temp., K	Curie Temp., K	Néel Temp., K	Debye Temperature at 0 K, K	Debye Temperature at 298 K, K	Melting Point, K	Boiling Point, K	Critical Temp., K
Neptunium	93	(237)	20.46[42]	orthorh.[2] (α) t.[2] (β) b.c.c.[2] (γ)	551[2] (α–β) 813[2] (β–γ)				121[3]	163[3]	913.2[5]	4150[3]	
Nickel	28	58.71	8.90[42]	f.c.c.[7]			631[40]		427 ± 14[3]	345[3]	1726[3, 10] 1726 ± 4[61]	3055[63]	6294[15] 11750[109]
Niobium	41	92.906	8.57[42]	b.c.c.[7]		9.13[5] 9.09[8] 9.1[9]			241 ± 13[3]	260[64]	2741 ± 27[3] 2688[65]	4813[66]	19000[109]
Nitrogen	7	14.0067	0.0012506[29]	c.[16] (α) h.[107] (β)	35.62[13] (α–β)					70[4] (at ~35 K)	63.29[5]	77.34[13, 23]	126.2[15]
Nobelium	102	(254)											
Osmium	76	190.2	22.48[29]	c.p.h.[2]		0.655[5] 0.65[8]			500[67]	400[68]	3283 ± 10[69]	5300 ± 100[70]	
Oxygen	8	15.9994	0.001429[29] (at 273.2 K and 1 atm)	b.c. orthorh.[7] (α) r.[7] (β) c.[7] (γ)	23.876 ± 0.01[112] (α–β) 43.818 ± 0.01[112] (β–γ)					250[4] (at ~125 K) 500[36] (at ~250 K)	54.8[5]	90.19[13] 90.18[23]	154.8[15]
Palladium	46	106.4	12.02[28]	f.c.c.[2]					283 ± 16[3]	275[14]	1825[3, 10]	3200[3]	
Phosphorus	15	30.9738	1.82[29] (β) 2.22[29] (γ) 2.69[29] (δ)	h. ?[7] (α) b.c.c.[7] (β) c.[7] (γ) f.c. orthorh.[17] (δ)	196[71] (α–β) 298.16[13] (β–γ) 298.16[13] (β–δ)				193[3] (white) 325[3] (red)	576[3] (white) 800[3] (red)	317.3[5] (white) 1300[72] (black)	553[13]	993.8[15]
Platinum	78	195.09	21.45[29]	f.c.c.[2]					234 ± 1[3]	225 ± 5[3]	2042[3, 10]	4100[3]	8280[15]
Plutonium	94	(242)	19.737[29] (at 298.2 K)	Simple monocl.[2] (α) b.c. monocl.[2] (β) f.c. orthorh.[2] (γ) f.c.c.[2] (δ) b.c.t.[2] (δ') b.c.c.[2] (ϵ)	396.7[73] (α–β) 475[73] (β–γ) 591.4[73] (γ–δ) 729[73] (δ–δ') 757 ± 3[73] (δ'–ϵ)				171[74]	176[74]	912.7[5]	3727[75]	
Polonium	84	(210)	9.3[29] (α) 9.5[29] (β)	Simple c.[7] (α) r.[7] (β)	327 ± 1.5[76] (α–β)				81[3]		527.2[5]	1235[20]	2281[15]
Potassium	19	39.102	0.86[29]	b.c.c.[7]					89.4 ± 0.5[3]	100[3]	336.8[5]	1027[35]	2450[11] 2140[109]
Praseodymium	59	140.907	6.769[29]	Double c.p.h.[8] (α) b.c.c.[2] (β)	1071[32] (α–β)			25[77]	85 ± 1[45]	138[78]	1192 ± 2[79]	3616[80]	8900[109]

Name	Atomic Number	Atomic Weight[a]	Density,[b] kg m^{-3} · 10^{-3}	Crystal Structure	Phase Transition Temp., K	Superconducting Transition Temp., K	Curie Temp., K	Néel Temp., K	Debye Temperature at 0 K, K	Debye Temperature at 298 K, K	Melting Point, K	Boiling Point, K	Critical Temp., K
Promethium	61	(145)		h.[7](α) b.c.c.[120](β)	1185[120](α–β)			6[120]			1353±10[81]	2730[3]	
Protactinium	91	(231)	15.37[42]	b.c.t.[2]		1.4[9]			159[3]	262[3]	1503[5]	4680[3]	
Radium	88	(226)	5[29]						89[3]		973.2[5]	1900[3]	
Radon	86	(222)	0.00973[29] (at 273.2 K and 1 atm)	f.c.c.[7]						400[4] (at∼200 K)	202.2[5]	211[13]	377.16[15]
Rhenium	75	186.2	21.1[42]	c.p.h.[2]		1.698[26]			429±22[3]	275[23]	3453[5]	6035±135[3]	20000[11]
Rhodium	45	102.905	12.45[42]	f.c.c.[7]	possible transformation at 1373–1473 K[57]				480±32[3]	350[3]	2233[3,10,82]	3960±60[3]	
Rubidium	37	85.47	1.53[29]	b.c.c.[2]					54±4[3]	59[23]	312.04[5]	959[35]	2100[113,115,116] 2030[109]
Ruthenium	44	101.07	12.2[29]	c.p.h.[7](α) ? (β) ? (γ) ? (δ)	1308[13,121](α–β) 1473[13,121](β–γ) 1773[13,121](γ–δ)	0.49[5,9]			600[67]	415[3]	2523±10[69]	4325±25[3]	
Samarium	62	150.35	7.54[29]	r.[32](α) b.c.c.[32](β)	1190[32](α–β)		14[8]	106[8]	116[45]	184±4[3]	1345.2[83]	2140[3]	5400[109]
Scandium	21	44.956	3.00[42]	c.p.h.[2](α) b.c.c.[2](β)	1607[2](α–β)				470±80[52]	476[3]	1812[5]	3537±30[3]	
Selenium	34	78.96	4.50[29](α) 4.80[29](β)	monocl.[7](α) h.[7](β) amorphous[7]	304[84,117] (vitrification) 398[13](vit.–β) 423[13](α–β)	7.3[85] (at∼118 kbar)			151.7±0.4[86]	89[36] (at∼45 K) 150[4] (at∼75 K)	490.2[5]	1009[13](Se_6) 958.0[13] ($Se_{4.37}$) 1027[13](Se_2)	1757[15]
Silicon	14	28.086	2.33[42]	d.[7]		7.5[47] (at 118–128 kbar)			647±11[3]	692[3]	1685±2[87]	2753[28]	5159[15]
Silver	47	107.870	10.5[29]	f.c.c.[2]					228±3[3]	221[3]	1234.0[3,13]	2468±15[41]	7460[11]
Sodium	11	22.9898	0.9712[29]	b.c.c.[2]	Martensitic transformation at low temp.[56]				157±1[3]	155±5[3]	371.0[13]	1154[35]	2800[11] 2400[109]
Strontium	38	87.62	2.60[28]	f.c.c.[88](α) c.p.h.[7](β) b.c.c.[7](γ)	488[88](α–β) 878[88](β–γ)				147±1[22]	148[23]	1042[5]	1645[3]	3059[15] 3810[109]
Sulfur	16	32.064	2.07[29](α) 1.96[29](β)	r.[7](α) monocl.[7](β)	368.6[13](α–β)				200[3](β)	527[89](α) 250[89](α, at 40 K)	386.0[5](α) 392.2[5](β) Subl. 368.6[13] (at 0.0047 mm Hg)	717.75[3,10]	1313[15]
Tantalum	73	180.948	16.6[42]	b.c.c.[2]		4.483[5] 4.48[9]			247±13[3]	225[14]	3269[5]	5760±60[3]	22000[11]

Name	Atomic Number	Atomic[a] Weight	Density,[b] kg m^{-3} · 10^{-3}	Crystal Structure	Phase Transition Temp., K	Superconducting Transition Temp., K	Curie Temp., K	Neel Temp., K	Debye Temperature at 0 K, K	Debye Temperature at 298 K, K	Melting Point, K	Boiling Point, K	Critical Temp., K
Technetium	43	(99)	11.50[29]	c.p.h.[2]		8.22[5] 11.2[9]			351[3]	422[3]	2473 ± 50[5]	5300[3]	
Tellurium	52	127.60	6.24[29] (α) 6.00[5] (amorph.)	h.[7] (α) ? (β)[7] amorph.[5]	621[13] (α–β)	3.3[8] (Te II, at 56 kbar)			141 ± 12[3]		722.7[5]	1163 ± 1[3]	2329[15]
Terbium	65	158.924	8.25[29]	c.p.h.[2, 32] (α) b.c.c.[2] (β)	Near m.p.[2] (α–β)		219[90]	230[90]	150[91]	158[44]	1629[19]	3810[3]	
Thallium	81	204.37	11.85[29]	c.p.h.[2] (α) b.c.c.[2] (β)	508.3[5] (α–β)	2.39[5] 2.38[8] 2.37[9]			88 ± 1[3]	96[14]	576.2[19]	1939[92]	3219[15]
Thorium	90	232.038	11.7[42]	f.c.c.[2] (α) b.c.c.[2] (β)	1673 ± 25[93] (α–β)	1.368[5] 1.37[9]			170[94]	100[14]	2023[19]	4500[20]	14550[109]
Thulium	69	168.934	9.32[29]	c.p.h.[2] (α) b.c.c.[2] (β)	Near m.p.[50] (α–β)		22[95] (ferro.–antiferro.)	53[96]	127 ± 1[45]	167[44]	1818[5]	2266[97]	6430[109]
Tin	50	118.69	5.750[29] (α) 7.31[29] (β)	f.c.c.[7] (α) b.c.t.[7] (β) r.[29] (?)	286.2 ± 3[98] (α–β)	3.722[5] (β)			236 ± 24[3] (gray) 196 ± 9[3] (white)	254[3] (gray) 170[14] (white)	505.06[3, 10]	2766 ± 14[3]	8000[11] 9300[109]
Titanium	22	47.90	4.5[29]	c.p.h.[7] (α) b.c.c.[7] (β)	1155[13] (α–β)	0.39[5, 9]			426 ± 5[3]	380[14]	1953[99]	3586[100]	
Tungsten	74	183.85	19.3[29]	b.c.c.[2]		0.011[122]			388 ± 17[3]	312 ± 3[3]	3653[3, 10, 13]	6000 ± 200[3]	23000[11]
Uranium	92	238.03	19.07[28]	orthorh.[7] (α) t.[7] (β) b.c.c.[7] (γ)	37 ± 2[118] (α_0–α) 938[13] (α–β) 1049[13] (β–γ)	0.68[5] (α) 1.80[5] (γ)			200[94]	300[3]	1405.6 ± 0.6[101]	3950 ± 250[102]	12500[27] 12000[109]
Vanadium	23	50.942	6.1[28]	b.c.c.[2]		5.3[5] 5.03[9]			326 ± 54[3]	390[14]	2192 ± 2[61]	3582 ± 42[3]	11200[109]
Xenon	54	131.30	0.005851[29] (at 273.2 K and 1 atm)	f.c.c.[16]							161.2[26]	165.1[13]	289.75[15]
Ytterbium	70	173.04	7.02[42]	f.c.c.[32] (α) b.c.c.[32] (β)	1071[2, 5] (α–β)				118[103]		1097[12]	1970[3]	4420[109]
Yttrium	39	88.905	4.47[29]	c.p.h.[32] (α) b.c.c.[32] (β)	1753[119] (α–β)				268 ± 32[3]	214[104]	1798[119]	3670[105]	8950[109]
Zinc	30	65.37	7.140[29]	c.p.h.[2]		0.875[5] 0.85[9]			316 ± 20[3]	237 ± 3[3]	692.655[3, 110]	1175[106]	2169[15] 2910[109]
Zirconium	40	91.22	6.57[59]	c.p.h.[7] (α) b.c.c.[7] (β)	1135[13] (α–β)	0.546[5] 0.55[9]			289 ± 24[3]	250[14]	2125[19]	4650[20]	12300[109]

REFERENCES

(Crystal Structures, Transition Temperatures, and Other Pertinent Physical Constants of the Elements)

1. Farr, J.D., Giorgi, A.L., and Bowman, M.G., USAEC Rept. LA-1545, 1-13, 1953.
2. Elliott, R.P., Constitution of Binary Alloys, 1st Suppl., McGraw-Hill, 1965.
3. Gschneider, K.A, Jr., Solid State Physics (Sietz, F. and Turnbull, D., Editors), 16, 275-426, 1964.
4. Gopal, E.S.R., Specific Heat at Low Temperatures, Plenum Press, 1966.
5. Weast, R.C. (Editor), Handbook of Chemistry and Physics, 47th Ed., The Chemical Rubber Co., 1966-67.
6. Foster, K.W. and Fauble, L.G., J. Phys. Chem., 64, 958-60, 1960.
7. The Institution of Metallurgists, Annual Yearbook, pp. 68-73, 1960-61.
8. Meaden, G.T., Electrical Resistance of Metals, Plenum Press, 1965.
9. Matthias, B.T., Geballe, T.H., and Compton, V.B., Rev. Mod. Phys., 35, 1-22, 1963.
10. Stimson, H.F., J. Res. NBS, 42, 209, 1949.
11. Grosse, A.V., Rev. Hautes Tempér. et Réfract., 3, 115-46, 1966.
12. Spedding, F.H. and Daane, A.H., J. Metals, 6 (5), 504-10, 1954.
13. Rossini, F.D., Wagman, D.D., Evans, W.H., Levine, S., and Jaffe, I., NBS Circ. 500, 537-822, 1952.
14. deLaunay, J., Solid State Physics, 2, 219-303, 1956.
15. Gates, D.S. and Thodos, G., AIChE J., 6 (1), 50-4, 1960.
16. Gray, D.E. (Coordinating Editor), American Institute of Physics Handbook, McGraw-Hill, 1957.
17. Sasaki, K. and Sekito, S., Trans. Electrochem. Soc., 59, 437-60, 1931.
18. Anderson, C.T., J. Am. Chem. Soc., 52, 2296-300, 1930.
19. Trombe, F., Bull. Soc. Chim. (France), 20, 1010-2, 1953.
20. Stull, D.R. and Sinke, G.C., Thermodynamic Properties of the Elements in Their Standard State, American Chemical Soc., 1956.
21. Rinck, E., Ann. Chim. (Paris), 18 (10), 455-531, 1932.
22. Roberts, L.M., Proc. Phys. Soc. (London), B70, 738-43, 1957.
23. Zemansky, M.W., Heat and Thermodynamics, 4th Ed., McGraw-Hill, 1957.
24. Martin, A.J. and Moore, A., J. Less-Common Metals, 1, 85, 1959.
25. Hill, R.W. and Smith, P.L., Phil. Mag., 44 (7), 636-44, 1953.
26. Moffatt, W.G., Pearsall, G.W., and Wulff, J., The Structure and Properties of Materials, Vol. I, pp. 205-7, 1964.
27. Grosse, A.V., Temple Univ. Research Institute Rept., 1-40, 1960.
28. Lyman, T. (Editor), Metals Handbook, Vol. 1, 8th Ed., American Soc. for Metals, 1961.
29. Lange, N.A. (Editor), Handbook of Chemistry, Revised 10th Edition, McGraw-Hill, 1967.
30. Paule, R.C., Dissertation Abstr., 22, 4200, 1962.
31. Burk, D.L. and Friedberg, S.A., Phys. Rev., 111 (5), 1275-82, 1958.
32. Spedding, F.H. and Daane, A.H. (Editors), The Rare Earths, John Wiley, 1961.
33. McHargue, C.J., Yakel, H.L., and Letter, C.K., ACTA Cryst., 10, 832-33, 1957.
34. Arajs, S. and Colvin, R.V., J. Less-Common Metals, 4, 159-68, 1962.
35. Bonilla, C.F., Sawhney, D.L., and Makansi, M.M., Trans. Am. Soc. Metals, 55, 877, 1962.
36. Rosenberg, H.M., Low Temperature Solid State Physics, Oxford at Clarendon Press, 1965.
37. Arajs, S., J. Less-Common Metals, 4, 46-51, 1962.
38. Edwards, A.R. and Johnstone, S.T.M., J. Inst. Metals, 84 (8), 313-7, 1956.
39. Lagneborg, R. and Kaplow, R., ACTA Metallurgica, 15 (1), 13-24, 1967.
40. Kittel, C., Introduction to Solid State Physics, 3rd Ed., John Wiley, 1967.
41. Kirshenbaum, A.D. and Cahill, J.A., J. Inorg. and Nucl. Chem., 25 (2), 232-34, 1963.
42. Touloukian, Y.S. (Ed.), Thermophysical Properties of High Temperature Solid Materials, MacMillan, Vol. 1, 1967.
43. Griffel, M., Skochdopole, R.E., and Spedding, F.H., J. Chem. Phys., 25 (1), 75-9, 1956.
44. Gschneidner, K.A., Jr., Rare Earth Alloys, Van Nostrand, 1961.
45. Dreyfus, B., Goodman, B.B., Lacaze, A., and Trolliet, G., Compt. Rend., 253, 1764-6, 1961.

46. Spedding, F.H., Hanak, J.J., and Daane, A.H., Trans. AIME, 212, 379, 1958.
47. Buckel, W. and Wittig, J., Phys. Lett. (Netherland), 17 (3), 187-8, 1965.
48. Deardorff, D.K. and Kata, H., Trans. AIME, 215, 876-7, 1959.
49. Panish, M.B. and Reif, L., J. Chem. Phys., 38 (1), 253-6, 1963.
50. Miller, A.E. and Daane, A.H., Trans. AIME, 230, 568-72, 1964.
51. Spedding, F.H. and Daane, A.H., USAEC Rept. IS-350, 22-4, 1961.
52. Montgomery, H. and Pells, G.P., Proc. Phys. Soc. (London), 78, 622-5, 1961.
53. Kaufman, L. and Clougherty, E.V., ManLabs, Inc., Semi-Annual Rept. No. 2, 1963.
54. Lounasmaa, O.V., Proc. 3rd Rare Earth Conf., 1963, Gordon and Breach, New York, 1964.
55. Baker, H., WADC TR 57-194, 1-24, 1957.
56. Reed, R.P. and Breedis, J.F., ASTM STP 387, pp. 60-132, 1966.
57. Hansen, M., Constitution of Binary Alloys, 2nd Edition, McGraw-Hill, p. 1268, 1958.
58. Smith, P.L., Conf. Phys. Basses Temp., Inst. Intern. du Froid, Paris, 281, 1956.
59. Powell, R.W. and Tye, R.P., J. Less-Common Metals, 3, 202-15, 1961.
60. Yamamoto, A.S., Lundin, C.E., and Nachman, J.F., Denver Res. Inst. Rept., NP-11023, 1961.
61. Oriena, R.A. and Jones, T.S., Rev. Sci. Instr., 25, 248-51, 1954.
62. Smith, J.F., Carlson, O.N., and Vest, R.W., J. Electrochem. Soc., 103, 409-13, 1956.
63. Edwards, J.W. and Marshal, A.L., J. Am. Chem. Soc., 62, 1382, 1940.
64. Morin, F.J. and Maita, J.P., Phys. Rev., 129 (3), 1115-20, 1963.
65. Pendleton, W.N., ASD-TDR-63-164, 1963.
66. Woerner, P.F. and Wakefield, G.F., Rev. Sci. Instr., 33 (12), 1456-7, 1962.
67. Walcott, N.M., Conf. Phys. Basses Temp., Inst. Intern. du Froid, Paris, 286, 1956.
68. White, G.K. and Woods, S.B., Phil. Trans. Roy. Soc. (London), A251 (995), 273-302, 1959.
69. Douglass, R.W. and Adkins, E.F., Trans. AIME, 221, 248-9, 1961.
70. Panish, M.B. and Reif, L., J. Chem. Phys., 37 (1), 128-31, 1962.
71. Bridgman, P.W., J. Am. Chem. Soc., 36 (7), 1344-63, 1914.
72. Slack, G.A., Phys. Rev., A139 (2), 507-15, 1965.
73. Sandenaw, T.A. and Gibney, R.B., J. Phys. Chem. Solids, 6 (1), 81-8, 1958.
74. Sandenaw, T.A., Olsen, C.E., and Gibney, R.B., Plutonium 1960, Proc. 2nd Intern. Conf. (Grison, E., Lord, W.B.H., and Fowler, R.D., Editors), 66-79, 1961.
75. Mulford, R.N.R., USAEC Rept. LA-2813, 1-11, 1963.
76. Goode, J.M., J. Chem. Phys., 26 (5), 1269-71, 1957.
77. Cable, J.W., Moon, R.M., Koehler, W.C., and Wollan, E.O., Phys. Rev. Letters, 12 (20), 553-5, 1964.
78. Murao, T., Progr. Theoret. Phys. (Kyoto), 20 (3), 277-86, 1958.
79. Grigor'ev, A.T., Sokolovskaya, E.M., Budennaya, L.D., Iyutina, I.A., and Maksimona, M.V., Zhur. Neorg. Khim., 1, 1052-63, 1956.
80. Daane, A.H., USAEC AECD-3209, 1950.
81. Weigel, F., Angew. Chem., 75, 451, 1963.
82. Nassau, K. and Broyer, A.M., J. Am. Ceram. Soc., 45 (10), 474-8, 1962.
83. McKeown, J.J., State Univ. of Iowa, Ph.D. Dissertation, 1-113, 1958.
84. Abdullaev, G.B., Mekhtiyeva, S.I., Abdinov, D.Sh., and Aliev, G.M., Phys. Letters, 23 (3), 215-6, 1966.
85. Wittig, J., Phys. Rev. Letters, 15 (4), 159, 1965.
86. Fukuroi, T. and Muto, Y., Tohoku Univ. Res. Inst. Sci. Rept., A8, 213-22, 1956.
87. Olette, M., Compt. Rend., 244, 1033-6, 1957.
88. Sheldon, E.A., and King, A.J., ACTA Cryst., 6, 100, 1953.
89. Eastman, E.D. and McGavock, W.C., J. Am. Chem. Soc., 59, 145-51, 1937.
90. Arajs, S. and Colvin, R.V., Phys. Rev., A136 (2), 439-41, 1964.
91. Roach, P.R. and Lounasmaa, O.V., Bull. Am. Phys. Soc., 7, 408, 1962.
92. Shchukarev, S.A., Semenov, G.A., and Rat'kovskii, I.A., Zh. Neorgan. Khim., 7, 469, 1962.
93. Pearson, W.B., A Handbook of Lattice Spacings and Structures of Metals and Alloys, Pergamon Press, 1958.

94. Smith, P.L. and Walcott, N.M., Conf. Phys. Basses Temp., Inst. Intern. du Froid, 283, 1956.

95. Davis, D.D. and Bozorth, R.M., Phys. Rev., 118 (6), 1543-5, 1960.

96. Aliev, N.G. and Volkenstein, N.V., Soviet Physics - JETP, 22 (5), 997-8, 1966.

97. Spedding, F.H., Barton, R.J., and Daane, A.H., J. Am. Chem. Soc., 79, 5160, 1957.

98. Raynor, G.V. and Smith, R.W., Proc. Roy. Soc. (London), A244, 101-9, 1958.

99. Savitskii, E.M. and Burhkanov, G.S., Zhur. Neorg. Khim., 2, 2609-16, 1957.

100. Argent, B.B. and Milne, J.G.C., Niobium, Tantalum, Molybdenum and Tungsten, Elsevier Publ. Co. (Quarrell, A.G., Editor), pp. 160-8, 1961.

101. Argonne National Laboratory, USAEC Rept. ANL-5717, 1-67, 1957.

102. Holden, A.N., Physical Metallurgy of Uranium, Addison-Wesley, 1958.

103. Lounasmaa, O.V., Phys. Rev., 129, 2460-4, 1963.

104. Jennings, L.D., Miller, R.E., and Spedding, F.H., J. Chem. Phys., 33 (6), 1849-52, 1960.

105. Ackerman, R.J. and Rauh, E.G., J. Chem. Phys., 36 (2), 448-52, 1962.

106. Rosenblatt, G.M. and Birchenall, C.E., J. Chem. Phys., 35 (3), 788-94, 1961.

107. Streib, W.E., Jordan, T.H., and Lipscomb, W.N., J. Chem. Phys., 37 (12), 2962-5, 1962.

108. Samsonov, G.V. (Editor), Handbook of the Physicochemical Properties of the Elements, Plenum Press, 1968.

109. Kopp, I.Z., Russ. J. Phys. Chem., 41 (6), 782-3, 1967.

110. Stimson, H.F., in Temperature, Its Measurement and Control in Science and Industry (Herzfeld, C.M., Ed.), Vol. 3, Part 1, Reinhold, New York, pp. 59-66, 1962.

111. McLaren, E.H., in Temperature, Its Measurement and Control in Science and Industry (Herzfeld, C.M., Ed.), Vol. 3, Part 1, Reinhold, New York, pp. 185-98, 1962.

112. Orlova, M.P., in Temperature, Its Measurement and Control in Science and Industry (Herzfeld, C.M., Ed.), Vol. 3, Part 1, Reinhold, New York, pp. 179-83, 1962.

113. Grosse, A.V., J. Inorg. Nucl. Chem., 28, 2125-9, 1966.

114. Hochman, J.M. and Bonilla, C.F., in Advances in Thermophysical Properties at Extreme Temperatures and Pressures (Gratch, S., Ed.), ASME 3rd Symposium on Thermophysical Properties, Purdue University, March 22-25, 1965, ASME, pp. 122-30, 1965.

115. Dillon, I.G., Illinois Institute of Technology, Ph.D. Thesis, June 1965.

116. Hochman, J.M., Silver, I.L., and Bonilla, C.F., USAEC Rept. CU-2660-13, 1964.

117. Abdullaev, G.B., Mekhtieva, S.I., Abdinov, D.Sh., Aliev, G.M., and Alieva, S.G., Phys. Status Solidi, 13(2), 315-23, 1966.

118. Fisher, E.S. and Dever, D., Phys. Rev., 2, 170(3), 607-13, 1968.

119. Beaudry, B.J., J. Less-Common Metals, 14(3), 370-2, 1968.

120. Williams, R.K. and McElroy, D.L., USAEC Rept. ORNL-TM 1424, 1-32, 1966.

121. Jaeger, F.M. and Rosenbaum, E., Proc. Nederland Akademie van Wetenschappen, 44, 144-52, 1941.

122. Gibson, J.W. and Hein, R.A., Phys. Letters, 12(25), 688-90, 1964.

123. Grosse, A.V., Research Institute of Temple Univ., Report on USAEC Contract No. AT (30-1)-2082, 1-71, 1965.

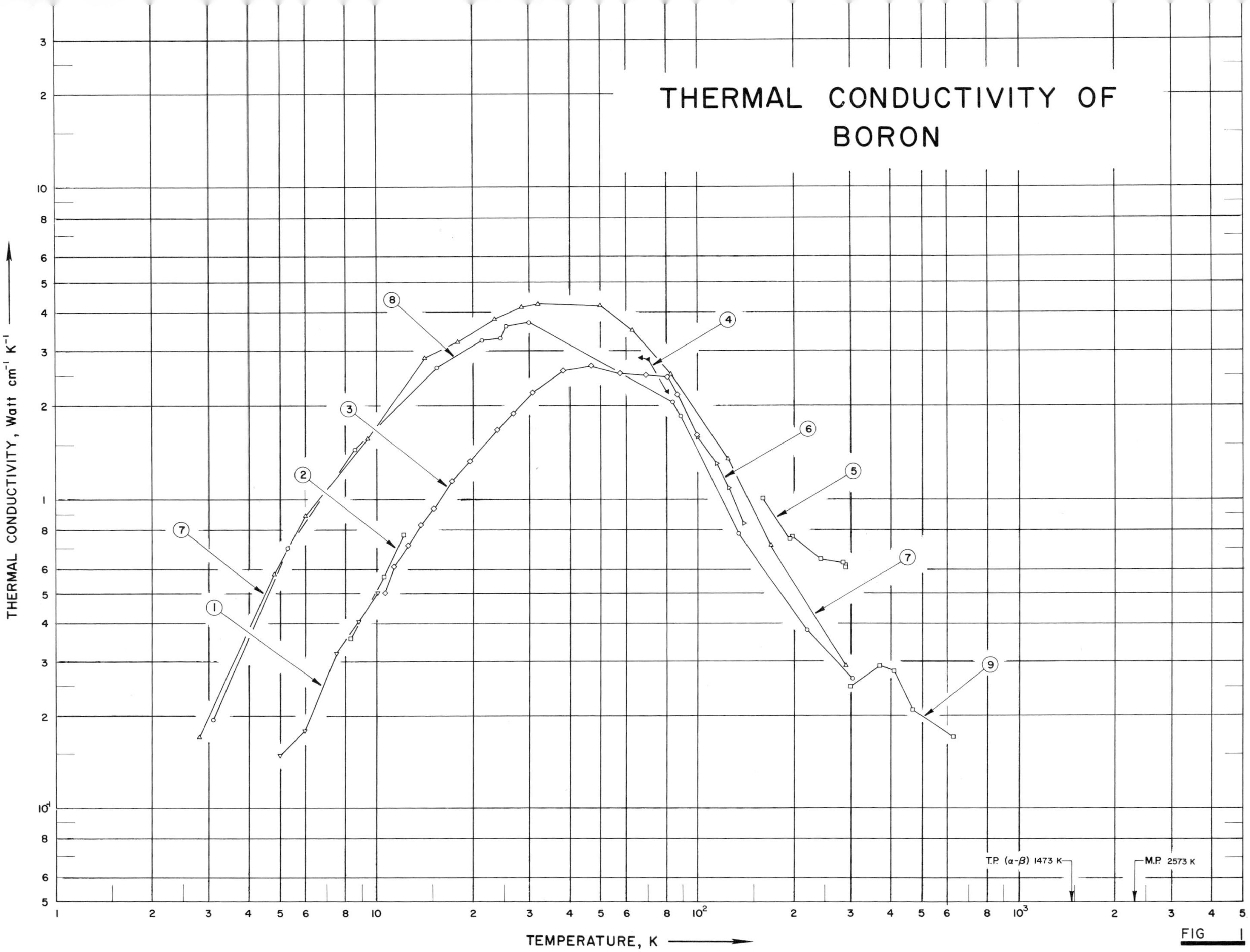

THERMAL CONDUCTIVITY OF BORON

FIG 1

SPECIFICATION TABLE NO. 1 THERMAL CONDUCTIVITY OF BORON

[For Data Reported in Figure and Table No. 1]

Curve No.	Ref. No.	Method Used	Year	Temp. Range, K	Reported Error, %	Name and Specimen Designation	Composition (weight percent), Specifications and Remarks
1	790	L	1963	5-10			99.9 B (by difference), 0.1 C; cylindrical specimen 0.26 cm average diameter 3.8 cm long made from single crystal of the beta-rhombohedral phase, provided by Texaco Experiment Inc.; density 2.342 ± 0.005 g cm^{-3}; electrical resistivity > 5 x 10^6 ohm cm at room temperature; Debye temperature 1219 K.
2	790	L	1963	8-12			Rerun of the above specimen.
3	790	L	1963	10-100			Rerun of the above specimen.
4	790	L	1963	67-80			Rerun of the above specimen.
5	790	L	1963	162-290			Rerun of the above specimen.
6	790	L	1963	100-140			Rerun of the above specimen.
7	776	L	1965	2.8-291		R 4	Major impurities: 10 x 10^{19} Si, 20 x 10^{18} Al, 20 x 10^{18} Mn, 6 x 10^{18} Ti, and 4 x 10^{18} Cu atoms cm^{-3}, also about 0.1% (by volume) of precipitated particles 5-50 μ in diameter (probably of boron nitride, silicon inclusions or small voids); polycrystalline with numerous columnar crystals of β-rhombohedral phase 1 cm long 0.3 cm average diameter; specimen 3.8 cm long 0.7 cm average diameter grown by partially purified boron by General Electric Research Lab.; density 2.33 g cm^{-3}.
8	776	L	1965	3.1-305		R 46	As above but composed of columns 2 cm long 0.1 cm average diameter; specimen 2.6 cm long, 0.6 cm average diameter; provided by Eagle-Picher Research Lab. Miami, Okla. (crystal reference No. M6005CP); grown from the melt by floating zone process.
9	335		1965	300-630			No details reported.
10	1009	R	1959	293, 353			99^{+}B, 0.7 W, and 0.02 total of Ca, Cu, Fe, Mg, and Si; polycrystalline specimen 1 mm in dia and several cm long with a 0.025 mm; tungsten filament at the center amounting to about 0.7% by weight; prepared by the reduction of boron tribromide by hydrogen near the tungsten filament at about 1250 C; data reported as the average for the range 20 to 80 C.

DATA TABLE NO. 1 THERMAL CONDUCTIVITY OF BORON

[Temperature, T, K, Thermal Conductivity, k, Watt $cm^{-1}K^{-1}$]

T	k
CURVE 1	
5.00	0.148
5 97	0.1793
7.50	0.320
8.80	0.405
10.09	0.501
CURVE 2	
8.31	0.357
10.58	0.570
12.21	0.773
CURVE 3	
10.61	0.502
11.44	0.613
12.60	0.712
13.82	0.834
15.07	0.936
17.23	1.115
19.57	1.335
23.95	1.674
26.88	1.893
30.75	2.205
38.06	2.595
46.92	2.688
57.53	2.546
79.23	2.515
80.86	2.476
86.86	2.157
99.74	1.612
CURVE 4	
66.70	2.85
70.36	2.80
80.54	2.219
CURVE 5	
161.8	1.019
194.4	0.747
CURVE 5 (cont.)	
195.6	0.745*
197.1	0.763
242.0	0.648
285.4	0.630
290.0	0.608
290.1	0.616
CURVE 6	
100.6	1.595
116.2	1.304
126.9	1.095
140.9	0.840
CURVE 7	
2.8	0.17
4.8	0.58
6.0	0.89
9.4	1.57
14.2	2.85
18.0	3.21
23.5	3.80
28.5	4.15
32.0	4.25
50.0	4.20
63.0	3.50
83.0	2.52
125.0	1.35
170.0	0.715
291.0	0.29
CURVE 8	
3.1	0.195
5.3	0.700
8.6	1.450
15.5	2.65
21.5	3.25
24.5	3.30
25.5	3.60
30.0	3.70
CURVE 8 (cont.)	
84.0	2.05
89.0	1.85
135.0	0.78
220.0	0.38
305.0	0.265
CURVE 9	
300	0.25
370	0.29
410	0.28
470	0.21
630	0.17
CURVE 10*	
293	0.0126
353	0.0126

*Not shown on plot

FIGURE AND TABLE NO. 1R RECOMMENDED THERMAL CONDUCTIVITY OF BORON

THERMAL CONDUCTIVITY, Watt cm^{-1} K^{-1}

M. P. 2573 K

T. P. 1473 K

TEMPERATURE, K

RECOMMENDED VALUES*
(For Polycrystalline)

T_1	k_1	k_2	T_2
0	0	0	-459.7
1	(0.0150)‡	(0.867)	-457.9
2	(0.0781)	(4.51)	-456.1
3	0.198	11.4	-454.3
4	0.375	21.7	-452.5
5	0.588	34.0	-450.7
6	0.826	47.7	-448.9
7	1.07	61.8	-447.1
8	1.31	75.7	-445.3
9	1.54	89.0	-443.5
10	1.77	102	-441.7
11	1.98	114	-439.9
12	2.19	127	-438.1
13	2.39	138	-436.3
14	2.58	149	-434.5
15	2.76	159	-432.7
16	2.93	169	-430.9
18	3.22	186	-427.3
20	3.46	200	-423.7
25	3.92	226	-414.7
30	4.21	243	-405.7
35	4.30	248	-396.7
40	4.28	247	-387.7
45	4.19	242	-378.7
50	4.04	233	-369.7
60	3.63	210	-351.7
70	3.10	179	-333.7
80	2.63	152	-315.7
90	2.24	129	-297.7
100	1.90	110	-279.7
150	0.910	52.6	-189.7
200	0.525	30.3	- 99.7
250	0.363	21.0	- 9.7
273.2	0.317	18.3	32.0
300	0.276	15.9	80.0
350	0.224	12.9	170.3
400	0.187	10.8	260.3

T_1	k_1	k_2	T_2
500	0.141	8.15	440.3
600	0.113	6.53	620.3
700	(0.0941)	(5.44)	800.3
800	(0.0809)	(4.67)	980.3
900	(0.0708)	(4.09)	1160
1000	(0.0629)	(3.63)	1340
1100	(0.0569)	(3.29)	1520
1200	(0.0518)	(2.99)	1700
1300	(0.0472)	(2.73)	1880
1400	(0.0437)	(2.52)	2060

REMARKS

The recommended values are for high-purity boron. The values that are supported by experimental thermal conductivity data are thought to be accurate to within 5% of the true values near room temperature and 5 to 10% at other temperatures above 80 K. The thermal conductivity near and below the corresponding temperature of its maximum is highly sensitive to small physical and chemical variations of the specimens, and the values below 80 K are intended as typical values for indicating the general trend.

---

* T_1 in K, k_1 in Watt cm^{-1} K^{-1}, T_2 in F, and k_2 in Btu hr^{-1} ft^{-1} F^{-1}.

‡ Values in parentheses are extrapolated.

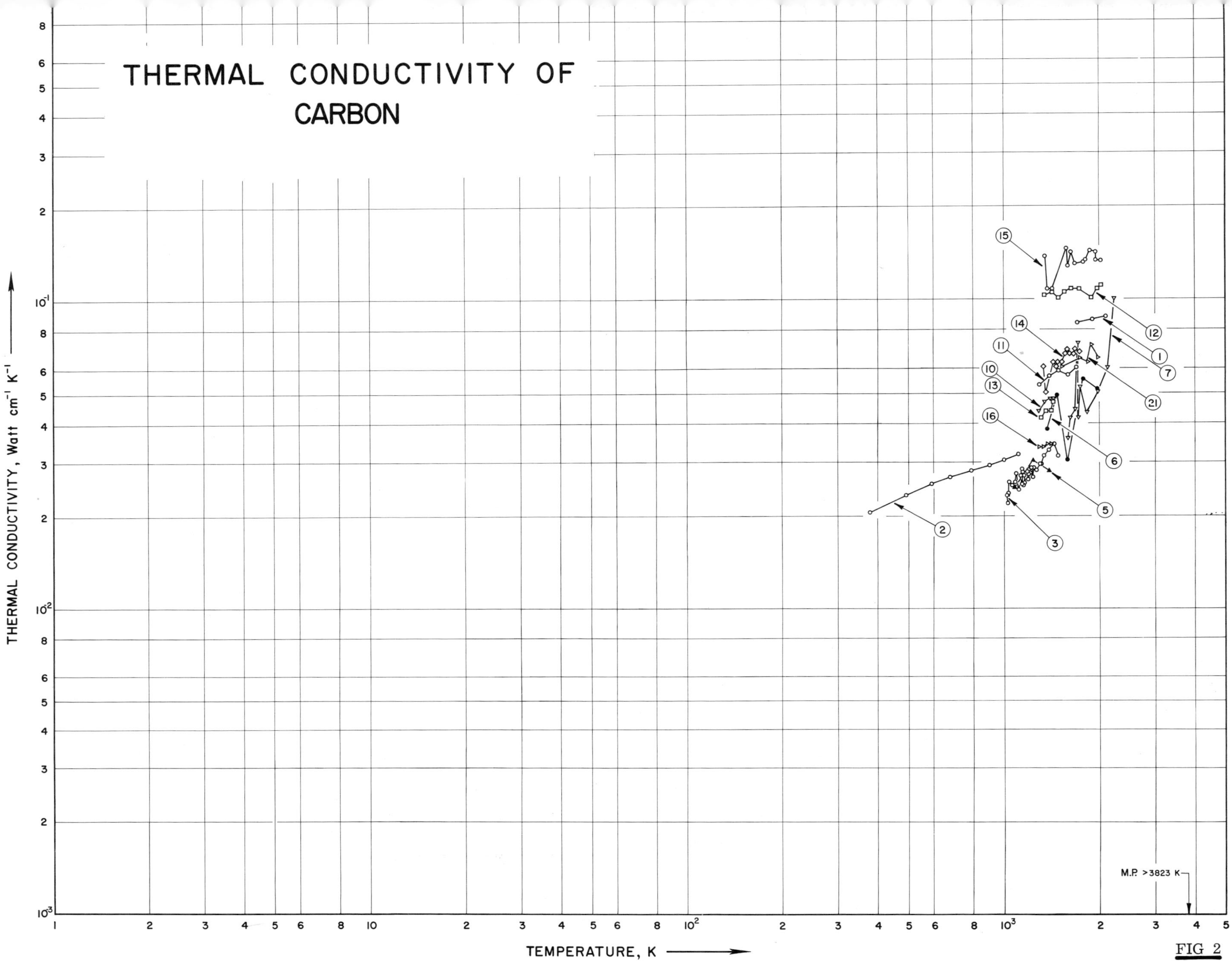

THERMAL CONDUCTIVITY OF CARBON

FIG 2

SPECIFICATION TABLE NO. 2 THERMAL CONDUCTIVITY OF CARBON

[For Data Reported in Figure and Table No. 2]

Curve No.	Ref. No.	Method Used	Year	Temp. Range, K	Reported Error, %	Name and Specimen Designation	Composition (weight percent), Specifications and Remarks
1	16	E	1914	1700-2100			Pure; untreated carbon filament.
2	17	R	1951	377-1107		Petroleum Coke	Petroleum coke electrode; tubular, 6 in. O. D., 1. 5 in. I. D., 2. 25 in. long.
3	18	R	1939	1023-1477		1	80% petroleum coke; 20% lampblack; baked at approx 1100 C.
4	18	R	1939	1033-1290		2	Similar to the above specimen.
5	18	R	1939	1076-1394		3	Similar to the above specimen.
6	18	R	1939	1373-1963		4	Similar to the above specimen.
7	18	R	1939	1598-2238		5	Similar to the above specimen.
8	159	L	1955	373. 2		Lampblack	Compressed under 10 lb in.$^{-2}$ from 0. 375 in. to 0. 25 in. thick; specimen previously used in a high temp neutron absorption experiment.
9	108	L	1920	313, 368		Lampblack	Specimen 0. 476 in. thick; specific gravity 0. 165; prepared from Eagle brand Germantown lampblack.
10	367	R	1963	1293-1433			Prepared by mixing 50 parts 65/100 mesh and 50 parts <2 mesh soft filler (soft Texas coke), and 40 parts soft binder (M-30 pitch); extruded to 0. 5 in. dia; baked for 4 days to 1000 C; heat treated at 1200 C for 10 min; density after baking 1. 55 g cm^{-3}; measured in an argon atmosphere (approx one atmosphere pressure).
11	367	R	1963	1293-1718			The above specimen heat treated at 1500 C for 10 min.
12	367	R	1963	1343-2033			The above specimen heat treated at 1800 C for 10 min.
13	367	R	1963	1318-1438			Prepared by mixing 50 parts 65/100 mesh and 50 parts 200/270 mesh soft filler (soft Texas coke), and 35 parts hard binder (phenol benzaldehyde); extruded to 0. 5 in. dia; baked for 4 days to 1000 C; heat treated at 1200 C for 10 min; density after baking 1. 56 g cm^{-3}; measured in an argon atmosphere (approx one atmosphere pressure).
14	367	R	1963	1338-1738			The above specimen heat treated at 1500 C for 10 min.
15	367	R	1963	1353-2038			The above specimen heat treated at 1800 C for 10 min.
16	367	R	1963	1293-1418			Prepared by mixing 50 parts 100/150 mesh and 50 parts <270 mesh hard filler (phenol formaldehyde), and 48 parts soft binder (M-30 pitch); extruded to 0. 5 in. dia; baked for 4 days to 1000 C; heat treated at 1200 C for 10 min; density after baking 1. 14 g cm^{-3}; measured in an argon atmosphere (approx one atmosphere pressure).
17	367	R	1963	1293-1675			The above specimen heat treated at 1500 C for 10 min.
18	367	R	1963	1298-1938			The above specimen heat treated at 1800 C for 10 min.

SPECIFICATION TABLE NO. 2 (continued)

Curve No.	Ref. No.	Method Used	Year	Temp. Range, K	Reported Error, %	Name and Specimen Designation	Composition (weight percent), Specifications and Remarks
19	367	R	1963	1273-1428			Prepared by mixing 50 parts 100/150 mesh and 50 parts <270 mesh hard filler (phenol formaldehyde), and 43 parts hard binder (phenol benzaldehyde); extruded to 0.5 in. dia; baked for 4 days to 1000 C; heat treated at 1200 C for 10 min; density after baking 1.22 g cm^{-3}; measured in an argon atmosphere (approx. one atmosphere pressure).
20	367	R	1963	1318-1718			The above specimen heat treated at 1500 C for 10 min.
21	367	R	1963	1373-1988			The above specimen heat treated at 1800 C for 10 min.

DATA TABLE NO. 2 THERMAL CONDUCTIVITY OF CARBON

[Temperature, T, K; Thermal Conductivity, k, Watt cm^{-1} K^{-1}]

T	k
CURVE 1	
1700	0.084
1900	0.086
2100	0.088
CURVE 2	
377.2	0.0205
490.2	0.0234
591.2	0.0255
676.2	0.0268
789.2	0.0280
899.2	0.0293
996.2	0.0305
1107.2	0.0318
CURVE 3	
1023.2	0.0233
1024.2	0.0220
1026.2	0.0237
1036.2	0.0258
1061.2	0.0252
1085.7	0.0258
1090.7	0.0275
1116.8	0.0244
1125.2	0.0271
1140.2	0.0254
1141.2	0.0285
1145.2	0.0278
1150.6	0.0253
1154.2	0.0271
1158.0	0.0257
1185.2	0.0278
1191.2	0.0264
1192.2	0.0274
1195.9	0.0283
1209.7	0.0276
1210.7	0.0281
1215.2	0.0287
1236.2	0.0267
1249.9	0.0287
1265.7	0.0282
1300.8	0.0295
CURVE 3 (cont.)	
1311.2	0.0295
1339.3	0.0314
1390.3	0.0326
1443.2	0.0342
1477.2	0.0313
CURVE 4*	
1033.2	0.0244
1071.7	0.0246
1073.2	0.0251
1094.2	0.0249
1111.2	0.0276
1132.2	0.0262
1139.2	0.0291
1150.7	0.0289
1181.7	0.0289
1192.2	0.0283
1193.2	0.0288
1220.2	0.0293
1223.2	0.0293
1249.7	0.0295
1290.2	0.0287
CURVE 5	
1076.2	0.0248
1170.2	0.0277*
1195.2	0.0278*
1247.7	0.0304
1394.2	0.0281
CURVE 6	
1373.2	0.0383
1473.2	0.0494
1583.2	0.0305
1773.2	0.0556
1963.2	0.0519
CURVE 7	
1598.2	0.0356
CURVE 7 (cont.)	
1615.2	0.0418
1678.2	0.0444
1713.2	0.0722
1715.7	0.0418
1741.2	0.0523
1828.2	0.0433
1983.2	0.0506
2133.2	0.0601
2238.2	0.1004
CURVE 8*	
373.2	0.000209
CURVE 9*	
313.2	0.000653
368.2	0.000695
CURVE 10	
1293.2	0.0439
1353.2	0.0469
1408.2	0.0481
1433.2	0.0481
CURVE 11	
1293.2	0.0531
1388.2	0.0565
1493.2	0.0586
1593.2	0.0573
1693.2	0.0607
1718.2	0.0628*
CURVE 12	
1343.2	0.103
1438.2	0.105
1473.2	0.101
1553.2	0.105
1633.2	0.107
1728.2	0.107
CURVE 12 (cont.)	
1883.2	0.101
1983.2	0.108
2033.2	0.111
CURVE 13	
1318.2	0.0418
1338.2	0.0418*
1358.2	0.0439
1373.2	0.0439*
1388.2	0.0439*
1403.2	0.0439
1423.2	0.0469
1438.2	0.0481*
CURVE 14	
1338.2	0.0607
1353.2	0.0502
1403.2	0.0565*
1413.2	0.0628*
1433.2	0.0628
1468.2	0.0607
1478.2	0.0628
1488.2	0.0607*
1523.2	0.0628
1563.2	0.0669
1593.2	0.0690
1613.2	0.0669
1663.2	0.0669
1673.2	0.0690
1703.2	0.0669*
1730.2	0.0678
CURVE 15	
1353.2	0.138
1373.2	0.107
1433.2	0.107
1583.2	0.146
1598.2	0.128
1633.2	0.142
1683.2	0.130
CURVE 15 (cont.)	
1783.2	0.132
1808.2	0.134
1873.2	0.144
1953.2	0.142
1953.2	0.134
2023.2	0.134*
2038.2	0.134
CURVE 16	
1293.2	0.0335
1338.2	0.0335
1383.2	0.0343
1418.2	0.0343
CURVE 17	
1293.2	0.0314
1313.2	0.0314
1358.2	0.0314
1408.2	0.0322
1538.2	0.0356
1583.2	0.0356
1643.2	0.0356
1675.2	0.0356
CURVE 18*	
1298.2	0.0385
1338.2	0.0418
1398.2	0.0418
1443.2	0.0439
1513.2	0.0439
1583.2	0.0439
1623.2	0.0418
1748.2	0.0460
1778.2	0.0439
1853.2	0.0439
1873.2	0.0460
1938.2	0.0460
CURVE 19*	
1273.2	0.0416
1343.2	0.0416
1388.2	0.0418
1428.2	0.0418
CURVE 20*	
1318.2	0.0481
1423.2	0.0481
1538.2	0.0502
1618.2	0.0523
1718.2	0.0565
CURVE 21	
1373.2	0.0552*
1533.2	0.0607
1738.2	0.0649
1843.2	0.0628
1883.2	0.0711
1988.2	0.0649

*Not shown on plot

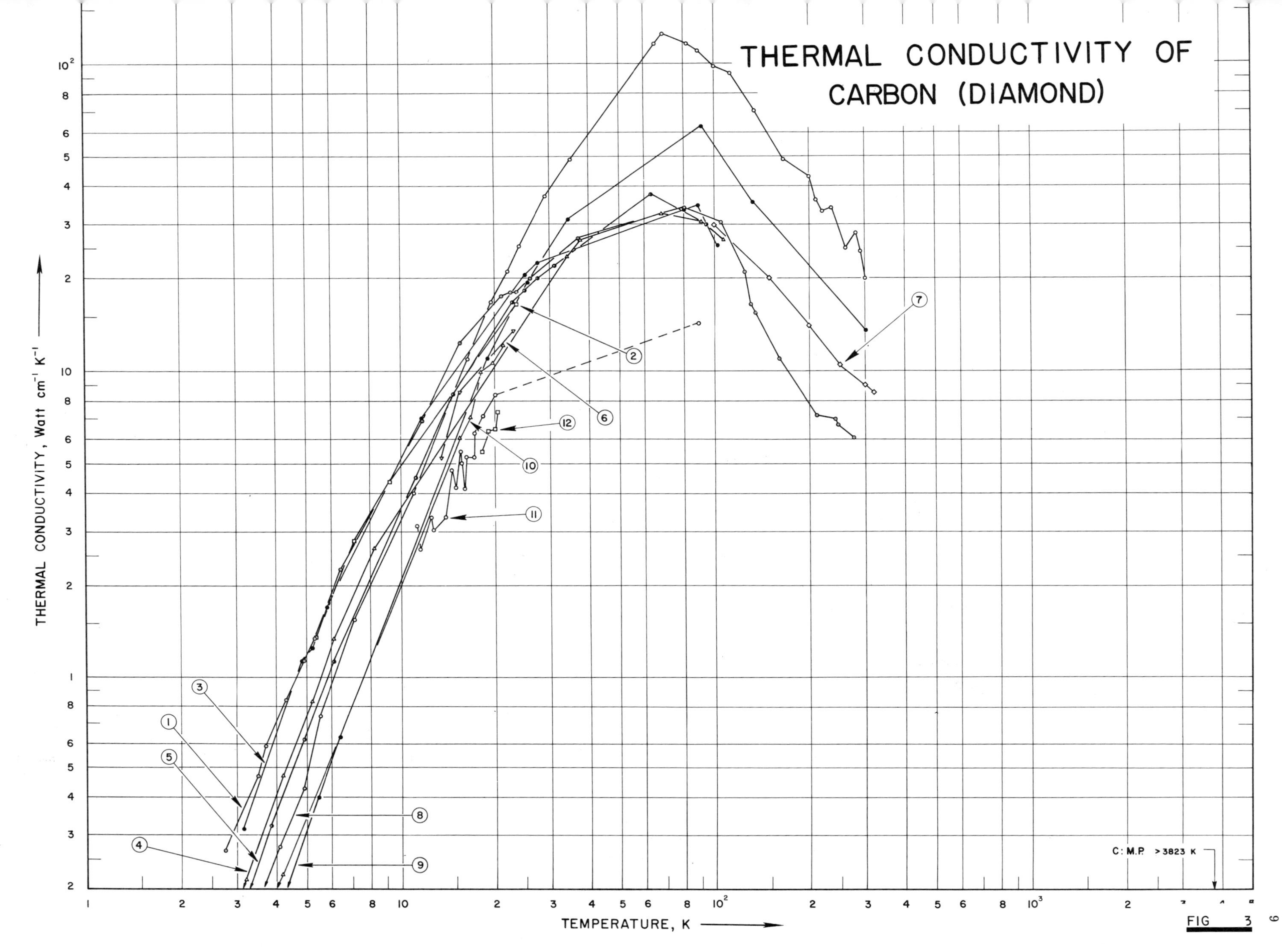
THERMAL CONDUCTIVITY OF
CARBON (DIAMOND)
THERMAL CONDUCTIVITY, Watt cm⁻¹ K⁻¹
TEMPERATURE, K
C: M.P. >3823 K
1
2
3
4
5
6
7
8
9
10
11
12
FIG 3

SPECIFICATION TABLE NO. 3 THERMAL CONDUCTIVITY OF CARBON (DIAMOND)

[For Data Reported in Figure and Table No. 3]

Curve No.	Ref. No.	Method Used	Year	Temp. Range, K	Reported Error, %	Name and Specimen Designation	Composition (weight percent), Specifications and Remarks
1	25	L	1953	2.8-275	5	Type I	Type I stone (gem quality); as classified according to its ultra-violet transparency limit; original dimensions of 3.9 x 3.9 x 10.9 mm; sawn adn ground to other sizes.
2	25	L	1953	5.4-24	5	Type I	Similar to the above specimen (from same stone) except dimensions of 3.9 x 3.9 x 5.8 mm.
3	25	L	1953	3.2-100	5	Type I	Similar to the above specimen (from same stone) except 3.1 x 3.1 mm cross section.
4	25	L	1953	2.4-107	5	Type I	Similar to the above specimen (from same stone) except 1.7 x 1.7 mm cross section.
5	25	L	1953	2.8-94	5	Type I	Similar to the above specimen (from same stone) except dimensions of 1.1 x 1.1 x 6.9 mm.
6	25	L	1953	14-23	5	Type I	Similar to the above specimen (from same stone) except dimensions of 1.1 x 1.1 x 5.2 mm.
7	28	L	1956	100-320		Type I	Type I stone; approx. 1.1 x 1.1 x 11 mm in size; apparatus improved over prior apparatus (Berman, Simon and Ziman, 1953).
8	28	L	1956	2.7-300		Type IIa	Type IIa stone; approx. 0.7 x 1.25 x 10 mm; an electrical insulator.
9	28	L	1956	3.0-300		Type IIb	Type IIb stone; approx. 1.1 x 1.2 x 7 mm; an electrical conductor.
10	231	L	1938	3.0-22			6 mm long; cross sectional area 0.59 mm^2; supplied by I. J. Asscher, Amsterdam, copper wire used for thermal contacts.
11	231	L	1938	11-89			The above specimen remounted; larger copper wire used for thermal contacts.
12	231	L	1938	18-21	20		9 mm long; cross sectional area 0.82 mm^2; supplied by I. J. Asscher, Amsterdam; mercury in copper cups used for thermal contacts.

DATA TABLE NO. 3 THERMAL CONDUCTIVITY OF CARBON (DIAMOND)

[Temperature, T, K; Thermal Conductivity, k, Watt $cm^{-1} K^{-1}$]

T	k
CURVE 1	
2.75	0.267
3.50	0.470
3.70	0.590
4.30	0.840
4.90	1.13
5.30	1.34
6.40	2.25
11.7	6.9
15.5	12.4
21.0	17.5
22.5	18.0
23.5	18.0
26.0	20.0
37.0	27.0
80.0	34.0
105	30.5
125	21.0
130	16.5
135	15.5
160	11.0
210	7.2
240	7.0
245	6.7
275	6.1
CURVE 2	
5.35	1.35
7.1	2.80
9.2	4.35
23.5	16.5
CURVE 3	
3.15	0.315
4.88	1.13
5.20	1.25
5.80	1.70
11.7	7.0
25.0	20.5
27.5	22.5
CURVE 3 (cont.)	
80.0	33.5
89.0	34.5
100.3	25.5
CURVE 4	
2.4	0.105*
2.75	0.144*
3.2	0.215
4.2	0.47
5.2	0.83
6.1	1.33
8.2	2.65
34.0	23.5
37.5	26.5
68.0	32.5
91.0	30.5
107.0	26.7
CURVE 5	
2.75	0.126*
3.85	0.323
4.9	0.62
6.1	1.13
8.2	2.65*
11.2	4.5
14.7	8.4
22.7	16.7
25.0	18.2
27.5	20.0
31.0	22.0
63.0	37.5
94.0	30.0
CURVE 6	
13.5	5.2
15.5	8.5
23.0	13.5
CURVE 7	
100	29.7
150	20.0
200	14.0
250	11.0
300	9.0
320	8.5
CURVE 8	
2.65	0.078*
3.30	0.155*
4.10	0.275
4.9	0.43
5.5	0.74
7.1	1.55
11.0	4.0
16.4	11.0
19.5	16.7
22.0	21.0
24.0	25.5
29.0	37.0
35.0	49.0
65.0	116.0
69.0	125.0
82.0	116.0
89.0	110.0
100.0	98.0
113.0	93.0
135.0	70.5
167.0	49.0
200.0	43.0
210.0	36.0
220.0	33.0
235.0	34.0
260.0	25.0
280.0	28.0
290.0	24.5
300.0	20.0
CURVE 9	
2.95	0.073*
3.90	0.146*
5.45	0.400
6.40	0.63
19.0	11.0
25.5	19.5
34.5	31.0
91.0	63.0
133.0	35.5
300.0	13.5
CURVE 10	
2.99	0.125*
3.59	0.153*
4.21	0.221*
15.6	6.10
16.7	7.09
18.2	9.90
19.8	10.67
21.5	12.1
CURVE 11	
11.4	3.13
11.7	2.63
12.7	3.33
12.8	3.03
14.0	3.33
14.7	4.76
15.1	4.17
15.7	5.46
15.8	5.00
16.1	4.17
16.45	5.26
17.4	5.26
17.45	6.25
18.5	7.14
20.15	8.33
89.4	14.29
CURVE 12	
18.4	5.50
19.2	6.33
20.1	6.49
20.5	7.35

*Not shown on plot

FIGURE AND TABLE NO. 3R RECOMMENDED THERMAL CONDUCTIVITY OF CARBON (Diamond)

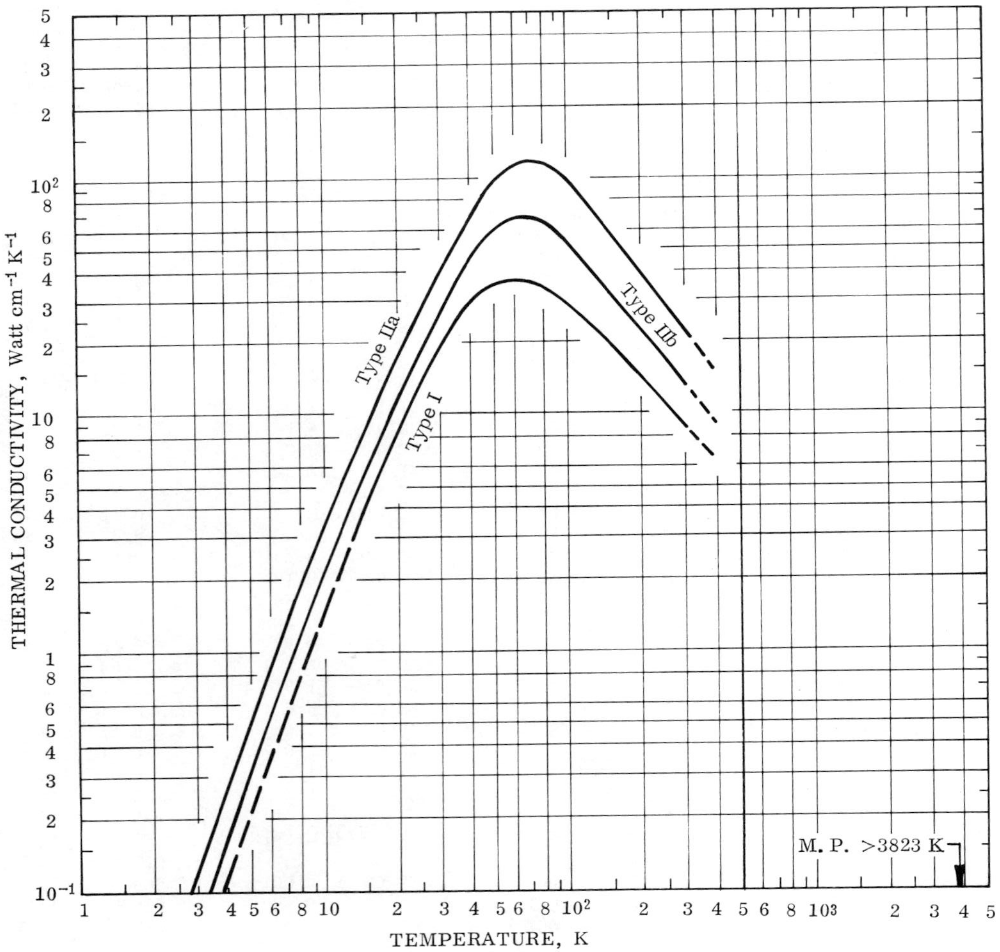

RECOMMENDED VALUES*
(High-purity, high-perfection, water-white diamond).

T_1	Type I k_1	Type I k_2	Type IIa k_1	Type IIa k_2	Type IIb k_1	Type IIb k_2	T_2
0	0	0	0	0	0	0	-459.7
1	(0.00182)‡	(0.105)	(0.00437)	(0.252)	0.00263	0.152	-457.9
2	(0.0142)	(0.820)	(0.0341)	(1.97)	0.0206	1.19	-456.1
3	(0.0471)	(2.72)	0.115	6.64	0.0692	4.00	-454.3
4	(0.111)	(6.41)	0.266	15.4	0.164	9.48	-452.5
5	(0.211)	(12.2)	0.502	29.0	0.313	18.1	-450.7
6	(0.351)	(20.3)	0.836	48.3	0.521	30.1	-448.9
7	(0.540)	(31.2)	1.27	73.4	0.795	45.9	-447.1
8	(0.770)	(44.5)	1.80	104	1.14	65.9	-445.3
9	(1.08)	(62.4)	2.46	142	1.57	90.7	-443.5
10	(1.42)	(82.0)	3.24	187	2.07	120	-441.7
11	(1.82)	(105)	4.09	236	2.63	152	-439.9
12	(2.28)	(132)	5.10	295	3.29	190	-438.1
13	(2.81)	(162)	6.20	358	4.02	232	-436.3
14	3.39	196	7.45	430	4.87	281	-434.5
15	4.03	233	8.77	507	5.77	333	-432.7
16	4.72	273	10.2	589	6.75	390	-430.9
18	6.27	362	13.5	780	8.90	514	-427.3
20	8.00	462	17.1	988	11.4	659	-423.7
25	13.1	757	27.4	1580	18.5	1070	-414.7
30	19.0	1100	39.5	2280	26.9	1550	-405.7
35	24.9	1440	52.5	3030	35.6	2060	-396.7
40	29.7	1720	66.7	3850	44.6	2580	-387.7
45	33.2	1920	80.2	4630	53.0	3060	-378.7
50	35.6	2060	93.0	5370	59.6	3440	-369.7
60	37.5	2170	113	6530	68.0	3930	-351.7
70	37.1	2140	120	6930	69.1	3990	-333.7
80	35.3	2040	117	6760	65.9	3810	-315.7
90	32.8	1900	110	6360	60.3	3480	-297.7
100	30.1	1740	100	5780	54.5	3150	-279.7
150	19.5	1130	60.5	3500	32.6	1880	-189.7
200	14.2	820	40.4	2330	22.6	1310	- 99.7
250	11.1	641	29.7	1720	17.1	988	- 9.7
273.2	10.0	578	26.3	1520	15.3	884	32.0
300	9.00	520	23.1	1330	13.5	780	80.3
350	(7.58)	(438)	(18.6)	(1070)	(11.2)	(647)	170.3
400	(6.52)	(377)	(15.5)	(896)	(9.36)	(541)	260.3

REMARKS

The recommended values are for high-purity high-perfection water-white diamond. The recommended values that are supported by experimental data are thought to be accurate to within 10% of the true values at temperatures above 100 K. The thermal conductivity near and below the corresponding temperature of its maximum is highly sensitive to small physical and chemical variations of the specimens, and the recommended values below 100 K are intended as typical values for indicating the general trend.

* T_1 in K, k_1 in Watt $cm^{-1}K^{-1}$, T_2 in F, and k_2 in Btu $hr^{-1}ft^{-1}F^{-1}$.

‡Values in parentheses are extrapolated.

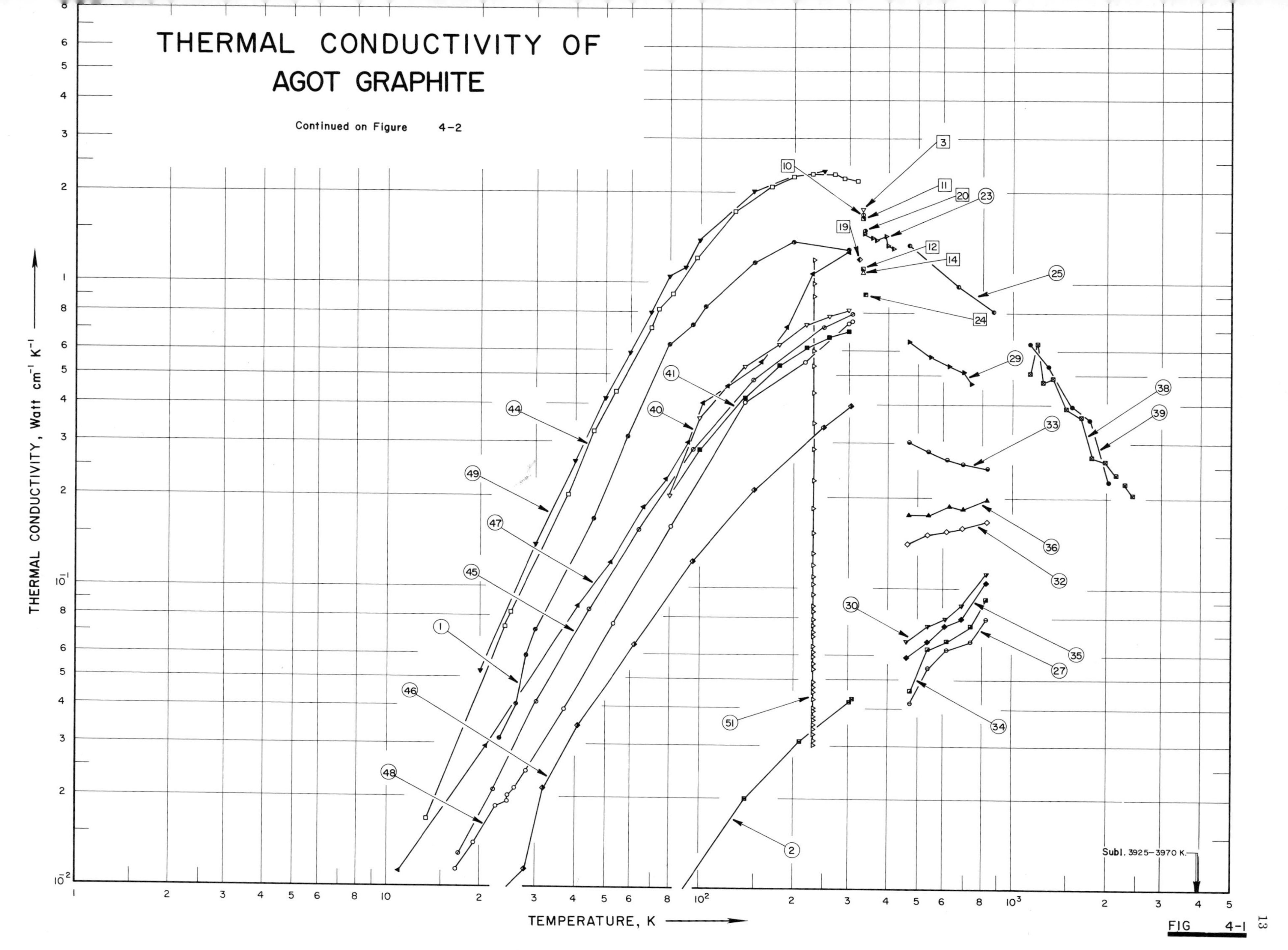
THERMAL CONDUCTIVITY OF
AGOT GRAPHITE
Continued on Figure 4-2
THERMAL CONDUCTIVITY, Watt cm⁻¹ K⁻¹
TEMPERATURE, K
Subl. 3925–3970 K.
FIG 4-1

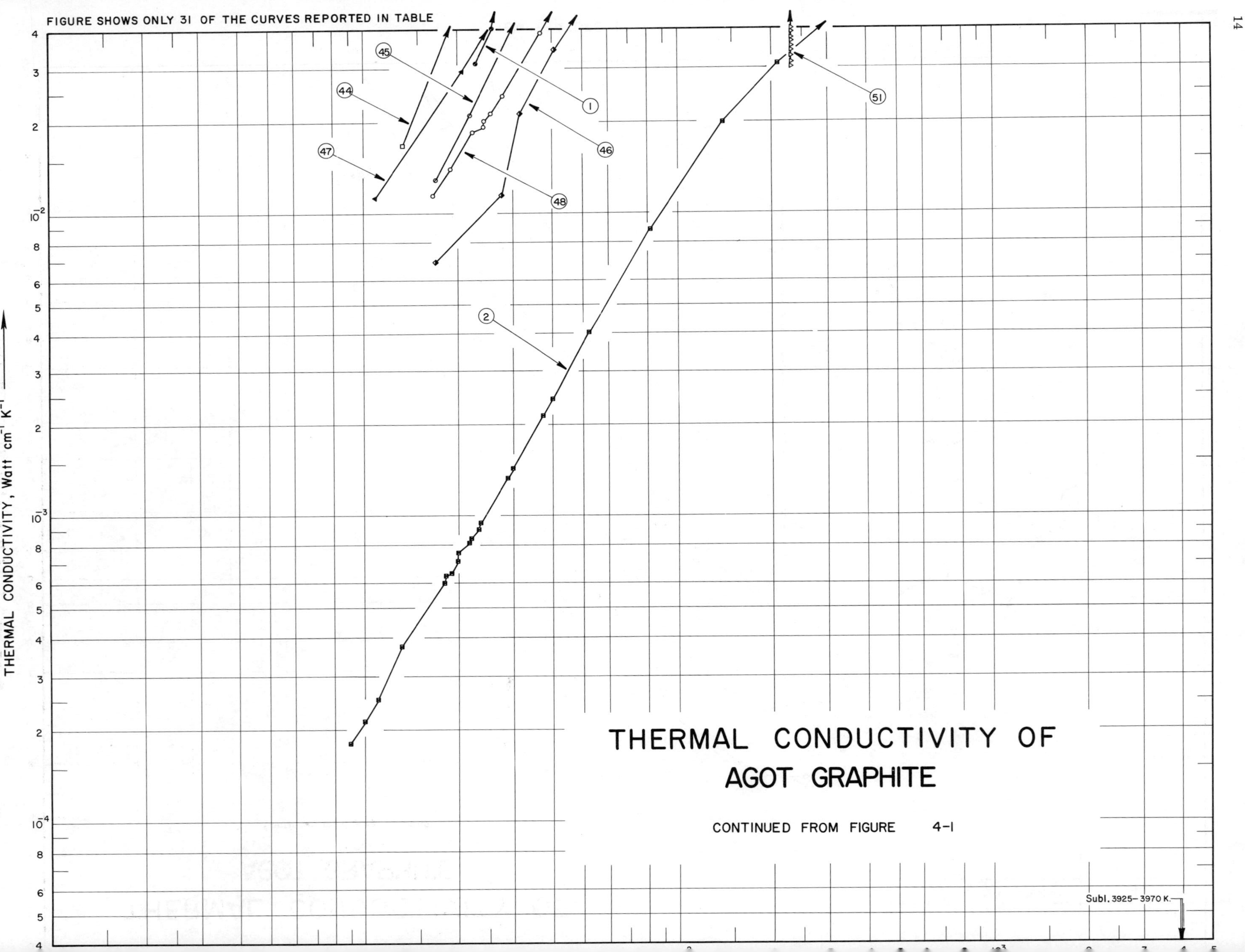
THERMAL CONDUCTIVITY OF
AGOT GRAPHITE
CONTINUED FROM FIGURE 4-1
FIGURE SHOWS ONLY 31 OF THE CURVES REPORTED IN TABLE
Subl. 3925–3970 K.
THERMAL CONDUCTIVITY, Watt cm⁻¹ K⁻¹

SPECIFICATION TABLE NO. 4 THERMAL CONDUCTIVITY OF AGOT GRAPHITE

[For Data Reported in Figure and Table No. 4]

Curve No.	Ref. No.	Method Used	Year	Temp. Range, K	Reported Error, %	Name and Specimen Designation	Composition (weight percent), Specifications and Remarks
1	49	L	1953	23-300	<10	B	Polycrystal; from National Carbon Co.; extruded; bulk density ~1.70 g cm^{-3}; measured perpendicular to the c-axis.
2	163, 53	E	1954	9.0-309	± 5		Index rod made from gas-baked coke (ungraphitized AGOT); extruded; room temp properties: density 1.56 g cm^{-3}, thermoelectric power + 1.3 μvolt K^{-1}, Hall coefficient -0.21 emu, magneto resistivity -0.01 x 10^{-10} emu, electrical resistivity 65.3 milliohm cm, total magnetic susceptibility -3.08 x 10^{-6} c gs unit, orientation factor (ρ_{max}/ρ_{min}) = 1.8.
3	359	L	1949	333.2	5	I	Cylindrical specimen 3.5 in. in dia and 4 in. long; cylinder axis parallel to the axis of extrusion.
4	359	L	1949	333.2	5	I	The above specimen, run No. 2.
5	359	L	1949	333.2	5	I	The above specimen, run No. 3.
6	359	L	1949	333.2	5	I	The above specimen, run No. 4.
7	359	L	1949	333.2	5	I	The above specimen, run No. 5.
8	359	L	1949	333.2	5	I	The above specimen, run No. 6.
9	359	L	1949	333.2	5	I	Measurement of the above specimen to show the "Bashing effect" by striking each end of the cylinder 10 times and 12 times around the circumference with a plastic hammer on a piece of wood on top the cylindrical specimen. (Bashing is the hitting of the specimen hard enough to break crystallites apart but not enough to break the specimen.)
10	359	L	1949	333.2	5	I	The above specimen treated again with 20 blows on each end and 20 blows on the circumference.
11	359	L	1949	333.2	5	I	The above specimen treated again with 10 blows on each end and 20 blows around the circumference but with steel hammer with 2 steel plates at each end of the specimen.
12	359	L	1949	333.2	5	I	Cylindrical specimen 3.5 in. in dia and 4 in. long; cylinder axis at right angle to the extrusion axis.
13	359	L	1949	333.2	5	I	The above specimen measured after striking each end 10 times and 12 times around the circumference with a plastic hammer on a piece of wood on top of the specimen.
14	359	L	1949	333.2	5	I	The above specimen treated again with 20 blows on each end and 20 blows on the circumference with the same hammer.
15	359	L	1949	331.2	5	II	Cylindrical specimen 3.5 in. in dia and 4 in. long; cylinder axis parallel to the extrusion axis.
16	359	L	1949	331.2	5	II	The above specimen, run No. 2.
17	359	L	1949	331.2	5	II	The above specimen, run No. 3.
18	359	L	1949	331.2	5	II	The above specimen, run No. 4.

SPECIFICATION TABLE NO. 4 (continued)

Curve No.	Ref. No.	Method Used	Year	Temp. Range, K	Reported Error, %	Name and Specimen Designation	Composition (weight percent), Specifications and Remarks
19	359	L	1949	324. 2	5	II	Cylindrical specimen 3. 5 in. in dia and 4 in. long; cylinder axis at right angle to the extrusion axis.
20	359	L	1949	337. 2	5	V	Similar to the above specimen but the cylinder axis parallel to the extrusion axis.
21	359	L	1949	338. 2	5	V	Similar to the above specimen but the cylinder axis perpendicular to the extrusion axis.
22	359	L	1949	335. 2	5	VI	Similar to the above specimen but the cylinder axis parallel to the extrusion axis.
23	359	L	1949	335-415	5	VI	Similar to the above specimen.
24	359	L	1949	339. 2	5	VI	Similar to the above specimen but the cylinder axis perpendicular to the extrusion axis.
25	392		1961	446-868			Density 1. 73 g cm^{-3}; grain size >0. 032 in.
26	392		1961	473-737		AK-2	Obtained from Brookhaven pile; density 1. 73 g cm^{-3}; grain size >0. 032 in.
27	392		1961	475-829		AK-1	Similar to the above specimen; irradiated in Brooklyn National Laboratory reactor at 30-50 C by a neutron flux of 1655 megawatt days/adjacent ton.
28	392		1961	474-830		JK-1	Similar to the above specimen except irradiated in Brooklyn National Laboratory reactor at 30-50 C by a neutron flux of 1685 megawatt days/adjacent ton.
29	392		1961	469-836		JK-1	The above specimen annealed at 1400 C for one hr.
30	392		1961	463-829		JK-2	Similar to the above specimen except annealed at 800 C for one hr.
31	392		1961	475-829		LK-1	Obtained from Brookhaven pile; density 1. 73 g cm^{-3}; grain size >0. 032 in. ; irradiated in Brooklyn National Laboratory reactor at 30-50 C by a neutron flux of 1685 megawatt days/adjacent ton.
32	392		1961	465-835		LK-1	The above specimen annealed at 1000 C for one hr.
33	392		1961	469-835		LK-2	Similar to the above specimen except annealed at 1200 C for one hr.
34	392		1961	475-829		CK-1	Obtained from Brookhaven pile; density 1. 73 g cm^{-3}; grain size >0. 032 in. ; irradiated in Brooklyn National Laboratory reactor at 30-50 C by a neutron flux of 1685 megawatt days/adjacent ton.
35	392		1961	464-829		CK-1	The above specimen annealed at 600 C for one hr.
36	392		1961	468-834		PK-1	Similar to the above specimen except annealed at 1100 C for one hr.
37	393	C	1963	324-1069			1 in. dia x 0. 250 in. thick; supplied by National Carbon Co. ; Armco iron used as comparative standard.
38	393	→	1963	1145-2443			1 x 0. 25 x 0. 05 in. ; supplied by National Carbon Co. ; measured in vacuum, the method consists of obtaining the steady-state temp at centers of the narrow and wide faces of specimen by optical pyrometry, specimen electrically heated, thermal conductivity calculated from measured temp, emittance of the specimen, dimensions of the specimen and the Stefan-Boltzmann constant.

SPECIFICATION TABLE NO. 4 (continued)

Curve No.	Ref. No.	Method Used	Year	Temp. Range, K	Reported Error, %	Name and Specimen Designation	Composition (weight percent), Specifications and Remarks
39	393	R	1963	1145-2044			Cylindrical specimen obtained from National Carbon Co.
40	346	R	1956	80-300		AGOT-CSF-MTR	Specimen size 0. 02 x 0. 125 x 1 in.; exposed to 6. 4 x 10^{19} fast neutrons cm^{-2} and 5. 8 x 10^{20} thermal neutron cm^{-2} at 698 K.
41	346	R	1956	80-300		AGOT-CSF-MTR	Similar to the above specimen but exposed to 4. 3 x 10^{19} fast neutrons cm^{-2} and 2. 6 x 10^{20} thermal neutrons cm^{-2} at 933 K.
42	346	R	1956	80-300		AGOT-CSF-MTR	Similar to the above specimen but exposed to 8. 5 x 10^{19} fast neutrons cm^{-2} and 2. 6 x 10^{20} thermal neutrons cm^{-2} at 908 K.
43	346	R	1956	220-300		AGOT-CSF-MTR	Similar to the above specimen but exposed to 4. 9 x 10^{19} fast neutrons cm^{-2} and 1. 5 x 10^{20} thermal neutrons cm^{-2} at 938 K.
44	163, 50, 53	E	1954	14-320	±5	AGOT-KC	Polycrystalline; extruded petroleum coke, pitch bonded; particle size 50 μ; crystallite size 0. 3 μ; specimen size 0. 10 x 0. 03 x 1. 25 in.; density 1. 65 g cm^{-3} at 25 C; thermoelectric power -0. 5 μvolt K^{-1}; Hall coefficient -0. 6 emu; magneto resistivity 5. 4 x 10^{-10} emu; electrical resistivity 6. 2 milliohm cm; total magnetic susceptibility -20. 44 x 10^{-6} cgs unit; orientation factor $(\rho_{max}/\rho_{min}) = 2.0$; measured parallel to the axis of extrusion.
45	163, 50	E	1954	17-308	±5	AGOT-KC	The above specimen exposed to neutron irradiation of 12. 5 MWD/T (megawatt-days per ton) at $<$30 C.
46	163, 50	E	1954	17-305	±5	AGOT-KC	The above specimen exposed to neutron irradiation of 48 MWD/T at $<$30 C.
47	163, 50	E	1954	11-300	±5	AGOT-KC	The above specimen exposed to neutron irradiation of 460 MWD/T at $<$30 C.
48	163, 50	E	1954	17-308	±5	AGOT-KC	The above specimen exposed to neutron irradiation of 1927 MWD/T at $<$30 C.
49	50	E	1956	20-250	±5	AGOT-KC	The virgin specimen before bromination (experiment to show the effect of Br on thermal conductivity of graphite).
50	50	E	1956	10-300	±5	AGOT-KC (Brom-Graphite)	Brominated AGOT-KC graphite; 0. 13 Br.
51	178	L	1950	233. 2		AGOT-KC	AGOT-KC graphite specimen 0. 03 x 0. 125 x 1 in.; irradiated with neutrons of 1927MWD/T; pulse annealed for 1 min; measured under vacuum ($<10^{-6}$ mm Hg) at constant temp of -40 C to show the effect on thermal conductivity of the specimen after being annealed (except the ends) at different temp.
52	178	L	1950	233. 2		AGOT-KC	The above specimen irradiated at 212 MWD/T; both ends annealed.
53	178	L	1950	233. 2		AGOT-KC	The above specimen irradiated at 1927 MWD/T with both ends annealed.

DATA TABLE NO. 4 THERMAL CONDUCTIVITY OF AGOT GRAPHITE

[Temperature, T, K; Thermal Conductivity, k, Watt $cm^{-1} K^{-1}$]

T	k
CURVE 1	
23	0.0310
26	0.0402
28	0.0586
30	0.0711
46	0.167
59	0.314
80	0.628
95	0.732
105	0.837
150	1.17
200	1.38
300	1.30
CURVE 2	
9.0	0.000179
10.0	0.000213
11.1	0.000251
13.2	0.000372
18.1	0.000598
18.2	0.000634
19.0	0.000644
19.9	0.000703
20.0	0.000749
21.7	0.000812
22.0	0.000840
23.3	0.000895
23.6	0.000941
28.8	0.00132
30.0	0.00143
37.4	0.00213
40.0	0.00243
52.5	0.00402
82.0	0.00879
140	0.0199
210	0.0308
302	0.0420
309	0.0431
CURVE 3*	
333.2	1.76
CURVE 4*	
333.2	1.77
CURVE 5*	
333.2	1.72
CURVE 6*	
333.2	1.71
CURVE 7*	
333.2	1.75
CURVE 8*	
333.2	1.75
CURVE 9*	
333.2	1.75
CURVE 10	
333.2	1.69
CURVE 11	
333.2	1.64
CURVE 12	
333.2	1.13
CURVE 13*	
333.2	1.10
CURVE 14	
333.2	1.09
CURVE 15*	
331.2	1.63
CURVE 16*	
331.2	1.64
CURVE 17*	
331.2	1.65
CURVE 18*	
331.2	1.66
CURVE 19	
324.2	1.21
CURVE 20	
337.2	1.51
CURVE 21*	
338.2	1.09
CURVE 22*	
335.2	1.46
CURVE 23	
335.2	1.47
355.2	1.42
366.2	1.40
392.2	1.44
399.2	1.34
412.2	1.32*
415.2	1.31
CURVE 24	
339.2	0.920
CURVE 25	
466.2	1.34
674.2	0.983
868.2	0.812
CURVE 26*	
473.2	1.35
542.2	1.18
619.2	1.04
737.2	0.950
CURVE 27	
475.2	0.0418
541.2	0.0544
621.2	0.0628
736.2	0.0669
829.2	0.0795
CURVE 28*	
474.2	0.0418
620.2	0.0628
830.2	0.0795
CURVE 29	
469.2	0.649
548.2	0.577
631.2	0.540
698.2	0.515
836.2	0.473
CURVE 30	
463.2	0.0669
539.2	0.0753
615.2	0.0795
CURVE 30 (cont.)	
694.2	0.0879
829.2	0.113
CURVE 31*	
475.2	0.0418
541.2	0.0544
621.2	0.0628
736.2	0.0699
829.2	0.0795
CURVE 32	
465.2	0.141
539.2	0.151
620.2	0.155
695.2	0.159
835.2	0.167
CURVE 33	
469.2	0.305
544.2	0.285
624.2	0.268
698.2	0.259
835.2	0.251
CURVE 34	
475.2	0.0460
541.2	0.0628
621.2	0.0669
737.2	0.0753
829.2	0.0920
CURVE 35	
464.2	0.0586
538.2	0.0669
614.2	0.0753
693.2	0.0795
829.2	0.105
CURVE 36	
468.2	0.176
541.2	0.176
623.2	0.188
697.2	0.184
834.2	0.197
CURVE 37*	
324.2	1.67
447.2	1.36
588.2	1.18
697.2	1.02
765.2	0.891
867.2	0.870
1069.2	0.695
CURVE 38	
1145.2	0.510
1206.2	0.636
1252.2	0.477
1360.2	0.494
1483.2	0.393
1669.2	0.368
1803.2	0.272
1999.2	0.264
2153.2	0.238
2301.2	0.222
2443.2	0.205
CURVE 39	
1145.2	0.636
1318.2	0.536
1556.2	0.397
1771.2	0.360
2044.2	0.226
CURVE 40	
80	0.201
100	0.360
CURVE 40 (cont.)	
140	0.534
180	0.628
220	0.732
260	0.780
300	0.816
CURVE 41	
80	0.201*
100	0.285
140	0.423
180	0.536
220	0.615
260	0.669
300	0.696
CURVE 42*	
80	0.188
100	0.272
140	0.410
180	0.519
220	0.598
260	0.657
300	0.673
CURVE 43*	
220	0.590
260	0.657
300	0.686
CURVE 44	
13.5	0.0167
24	0.0732
25	0.0816
38	0.201
46	0.326
54	0.439
70	0.711
74	0.816

*Not shown on plot

DATA TABLE NO. 4 (continued)

CURVE 44 (cont.)

T	k
82	0.920
98	1.21
130	1.72
170	2.09
200	2.26
230	2.30
270	2.30
290	2.22
320	2.18

CURVE 45

T	k
17.1	0.0129
22.0	0.0210
30.2	0.0411
44.4	0.0837
64.4	0.155
95.4	0.285
149.5	0.481
250	0.724
308	0.795

CURVE 46

T	k
17.1	0.00690
27.7	0.0115
31.8	0.0213
41.0	0.0344
62.2	0.0640
95	0.122
150	0.211
250	0.339
305	0.397

CURVE 47

T	k
10.9	0.0113
20.8	0.0293
40.8	0.0866
52.0	0.121
66.5	0.184
78.0	0.227
92.0	0.299
103.0	0.404
124.6	0.460
158.4	0.551

CURVE 47 (cont.)

T	k
190.4	0.718
230	1.080
300	1.289

CURVE 48

T	k
16.7	0.0114
19.1	0.0140
22.5	0.0185
24.4	0.0192
24.5	0.0201
25.7	0.0213
28.0	0.0243
37.0	0.0389
53.3	0.0753
81.0	0.159
140.4	0.406
218.0	0.550
300.0	0.741
308.0	0.753

CURVE 49

T	k
20	0.0523
30	0.138
40	0.259
50	0.418
60	0.586
70	0.795
80	1.05
90	1.13
100	1.38
150	2.01
200	2.26*
250	2.34

CURVE 50*

T	k
10	0.00837
20	0.0460
30	0.121
40	0.230
50	0.368
60	0.502
70	0.690
80	0.920

CURVE 50 (cont.)*

T	k
90	1.05
100	1.21
150	1.76
200	2.0
250	2.09
300	2.09

CURVE 51
T = 233.2K

T_A	k
498.2	0.0300
523.2	0.0311
548.2	0.0326
573.2	0.0341
598.2	0.0351
623.2	0.0364
648.2	0.0374
673.2	0.0391
698.2	0.0400
723.2	0.0405*
748.2	0.0410*
773.2	0.0418*
798.2	0.0427
823.2	0.0435*
848.2	0.0444*
873.2	0.0448*
898.2	0.0452
923.2	0.0469
948.2	0.0490
973.2	0.0531*
998.2	0.0540
1023.2	0.0561
1048.2	0.0569*
1073.2	0.0586
1098.2	0.0611
1123.2	0.0640
1148.2	0.0657*
1173.2	0.0686
1198.2	0.0715
1223.2	0.0753
1248.2	0.0787
1273.2	0.0808*
1298.2	0.0837
1323.2	0.0874

CURVE 51 (cont.)

T_A	k
1348.2	0.0950
1373.2	0.102
1398.2	0.109
1423.2	0.112*
1448.2	0,120
1473.2	0.131
1498.2	0.153
1523.2	0.184
1548.2	0.227
1573.2	0.287
1598.2	0.349
1623.2	0.439
1648.2	0.536
1673.2	0.598
1698.3	0.644*
1723.2	0.699*
1748.2	0.908
1773.2	0.891*
1798.2	0.929*
1823.2	0.996
1848.2	1.05*
1873.2	1.05*
1898.2	1.10*
1935.2	1.20

CURVE 52*
T = 233.2K

T_A	k
298.2	0.0707
323.2	0.0703
348.2	0.0703
373.2	0.0699
398.2	0.0699
423.2	0.0703
448.2	0.0741
473.2	0.0845
498.2	0.0983
523.2	0.111
548.2	0.123
573.2	0.134
598.2	0.142
623.2	0.151
648.2	0.163
673.2	0.174
698.2	0.186
723.2	0.195

CURVE 52 (cont.)*

T_A	k
748.2	0.199
773.2	0.209
798.2	0.217
823.2	0.226
848.2	0.238
873.2	0.246
898.2	0.251
923.2	0.258
948.2	0.270
973.2	0.277
998.2	0.287
1023.2	0.299
1048.2	0.303
1073.2	0.310
1098.2	0.327
1123.2	0.335
1148.2	0.349
1173.2	0.364
1198.2	0.380
1223.2	0.391
1248.2	0.406
1273.2	0.431
1298.2	0.435
1323.2	0.452
1348.2	0.477
1373.2	0.485
1398.2	0.536
1423.2	0.590
1473.2	0.653
1498.2	0.695
1523.2	0.720
1548.2	0.732
1573.2	0.749
1598.2	0.761
1623.2	0.837
1648.2	0.870
1673.2	0.908
1698.2	0.929
1723.2	0.950
1748.2	0.908
1773.2	0.891
1798.2	0.929
1823.2	0.996
1848.2	1.05
1873.2	1.05
1898.2	1.10
1935.2	1.20

CURVE 53*
T = 233.2K

T_A	k
298.2	0.0249
323.2	0.0249
348.2	0.0248
373.2	0.0248
398.2	0.0248
423.2	0.0248
448.2	0.0251
473.2	0.0256
498.2	0.0262
523.2	0.0268
548.2	0.0274
573.2	0.0282
598.2	0.0288
623.2	0.0294
648.2	0.0301
673.2	0.0308
698.2	0.0314
723.2	0.0320
748.2	0.0327
773.2	0.0333
798.2	0.0339
823.2	0.0343
848.2	0.0354
873.2	0.0364
898.2	0.0375
923.2	0.0382
948.2	0.0392
973.2	0.0402
998.2	0.0412
1023.2	0.0427
1048.2	0.0435
1073.2	0.0448
1098.2	0.0464
1123.2	0.0481
1148.2	0.0502
1173.2	0.0519
1198.2	0.0540
1223.2	0.0565
1248.2	0.0590
1273.2	0.0615
1298.2	0.0649
1323.2	0.0682
1348.2	0.0724
1373.2	0.0787
1398.2	0.0862

CURVE 53 (cont.)*

T_A	k
1423.2	0.0971
1448.2	0.112
1473.2	0.138
1498.2	0.173
1523.2	0.204
1548.2	0.239

*Not shown on plot

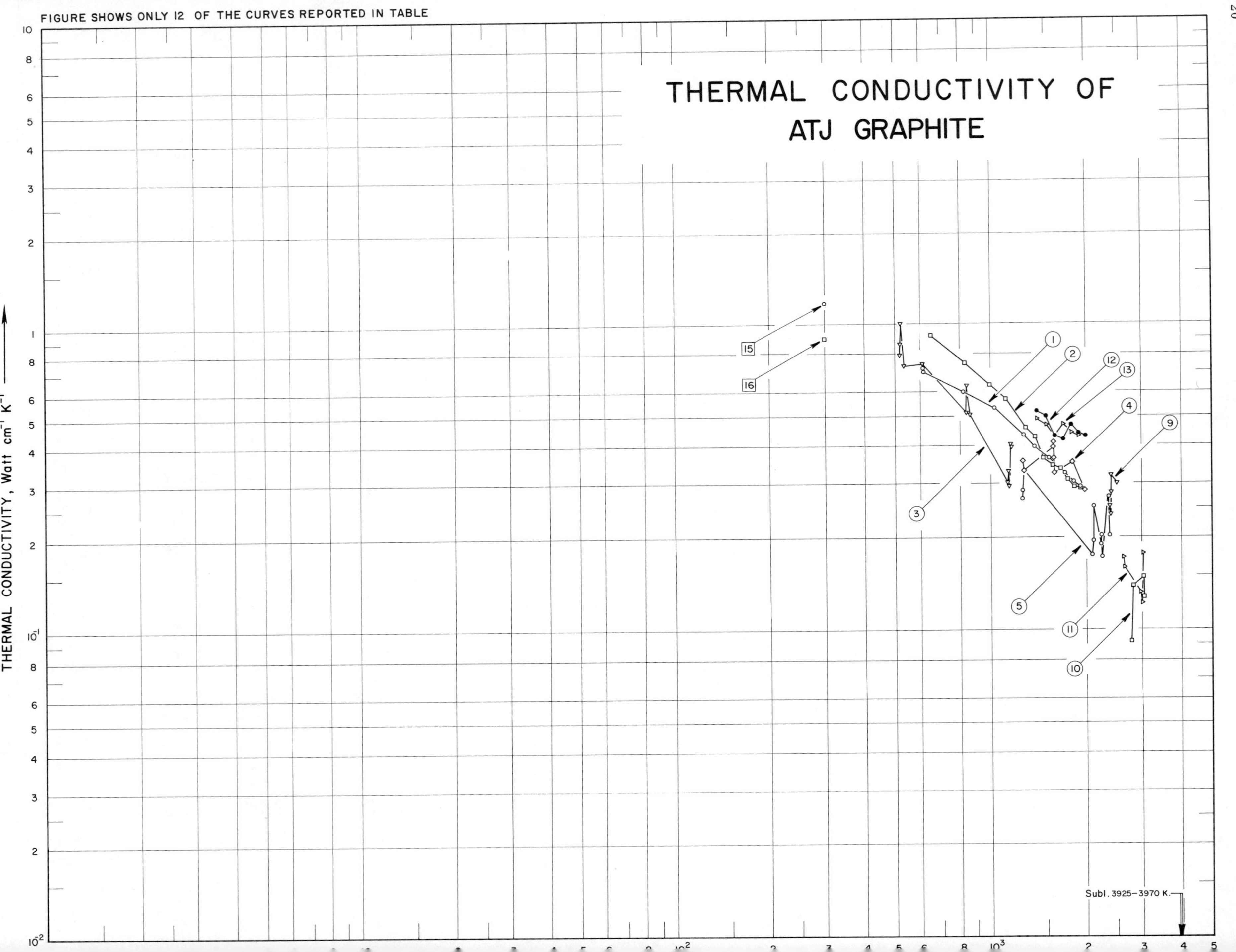

FIGURE SHOWS ONLY 12 OF THE CURVES REPORTED IN TABLE
THERMAL CONDUCTIVITY OF
ATJ GRAPHITE
THERMAL CONDUCTIVITY, Watt cm⁻¹ K⁻¹
Subl. 3925–3970 K.
1
2
3
4
5
9
10
11
12
13
15
16

SPECIFICATION TABLE NO. 5 THERMAL CONDUCTIVITY OF ATJ GRAPHITE

[For Data Reported in Figure and Table No. 5]

Curve No.	Ref. No.	Method Used	Year	Temp. Range, K	Reported Error, %	Name and Specimen Designation	Composition (weight percent), Specifications and Remarks
1	348	R	1960	610-1922			Ash content 0.2%; specimen composed of 15 disks; three of which 1 in. thick, twelve others 0.5 in. thick; each with a dia of 3 ±0.002 in.; maximum grain size ~0.006 in.; made from blocks of ATJ graphite size 9 x 20 x 24 in.; machined perpendicular to grain orientation; siliconized.
2	348	R	1960	650-1929			Similar to the above specimen but machined parallel to grain orientation.
3	243	R	1962	517-1160			Molded and fired; maximum exposure temp 2843 C; specimen 0.75 in. in dia and 0.75 in. long; no deterioration observed after the experiment.
4	243	R	1962	1261-1992			Another run of the above specimen.
5	243	R	1962	1258-2369			Another run of the above specimen.
6	243	R	1962	1425-1769			Another run of the above specimen.
7	243	R	1962	832-837			Another run of the above specimen.
8	243	R	1962	558-560			Another run of the above specimen.
9	243	R	1962	2383-2505			Another run of the above specimen.
10	243	R	1962	2783-3050			Another run of the above specimen.
11	243	R	1962	2622-3022			Another run of the above specimen.
12	343	E	1959	1400-2000			Rectangular bars fabricated by molding; size 1 x 1 x 10 cm; specific gravity 1.74; measured perpendicular to the direction of molding pressure; data averaged from measurements of 4 specimens.
13	343	E	1959	1400-1900			Similar to the above specimens but data averaged from 3 other specimens.
14	343	E	1959	1300-1900			Similar to the above specimens but with specific gravity of 1.69; data averaged from 4 specimens; measured parallel to the direction of molding pressure.
15	338		1962	298.2			Graphite stocks size 9 x 20 x 24 in.; grain size 0.006 in.; bulk density 1.73 g cm^{-3}; electrical resistivity 1100 μohm cm; with grain orientation.
16	338		1962	298.2			Similar to the above but electrical resistivity 1450 μohm cm; across grain orientation.

DATA TABLE NO. 5 THERMAL CONDUCTIVITY OF ATJ GRAPHITE

[Temperature, T, K; Thermal Conductivity, k, Watt $cm^{-1} K^{-1}$]

CURVE 1

T	k
610.4	0.722
614.3	0.697
824.8	0.599
1035.4	0.531
1266.5	0.433
1371.5	0.396
1540.9	0.363
1710.9	0.325
1833.2	0.305
1922.1	0.294

CURVE 2

T	k
649.8	0.917
830.9	0.748
991.5	0.633
1113.7	0.568
1297.1	0.459
1378.2	0.426
1469.8	0.362
1574.8	0.344
1655.4	0.336
1680.9	0.332*
1753.7	0.310
1842.1	0.294
1929.3	0.292

CURVE 3

T	k
516.5	0.793
518.2	0.858
519.3	1.00
524.3	0.733
608.7	0.743
614.3	0.728*
836.5	0.513
840.4	0.629
856.5	0.505
1133.7	0.303
1142.1	0.329
1147.6	0.294
1157.1	0.404
1159.8	0.395

CURVE 4

T	k
1260.9	0.355
1272.1	0.330
1580.4	0.394
1583.2	0.410
1585.9	0.363
1594.3	0.326
1602.6	0.346*
1811.0	0.352
1813.7	0.345*
1813.7	0.353*
1991.5	0.286

CURVE 5

T	k
1258.2	0.267
1263.7	0.286
1266.5	0.332*
2091.5	0.176
2102.6	0.195
2108.2	0.252
2236.0	0.189
2238.7	0.202
2244.3	0.172
2247.0	0.203*
2358.2	0.271
2369.3	0.203

CURVE 6*

T	k
1424.8	0.264
1427.6	0.261
1769.3	0.264
1769.3	0.232

CURVE 7*

T	k
831.5	0.490
833.2	0.577
836.5	0.646

CURVE 8*

T	k
558.2	0.826
559.8	0.795
559.8	0.854
560.4	0.762

CURVE 9

T	k
2383.2	0.253
2391.5	0.238
2394.3	0.281
2397.1	0.219
2505.4	0.201

CURVE 10

T	k
2783.1	0.0923
2811.0	0.139
2811.0	0.139*
3036.0	0.149
3049.8	0.128

CURVE 11

T	k
2622.0	0.172
2641.5	0.160
2972.1	0.133
3002.6	0.123
3022.1	0.177

CURVE 12

T	k
1400	0.520
1500	0.500
1600	0.430
1700	0.420
1800	0.470
1900	0.440
2000	0.430

CURVE 13

T	k
1400	0.489
1500	0.468

CURVE 13 (cont.)

T	k
1600	0.430*
1700	0.470
1800	0.441
1900	0.430

CURVE 14*

T	k
1300	0.430
1400	0.390
1500	0.350
1600	0.370
1700	0.331
1800	0.322
1900	0.330

CURVE 15

T	k
298.2	1.18

CURVE 16

T	k
298.2	0.895

*Not shown on plot

FIGURE AND TABLE NO. 5R RECOMMENDED THERMAL CONDUCTIVITY OF ATJ GRAPHITE

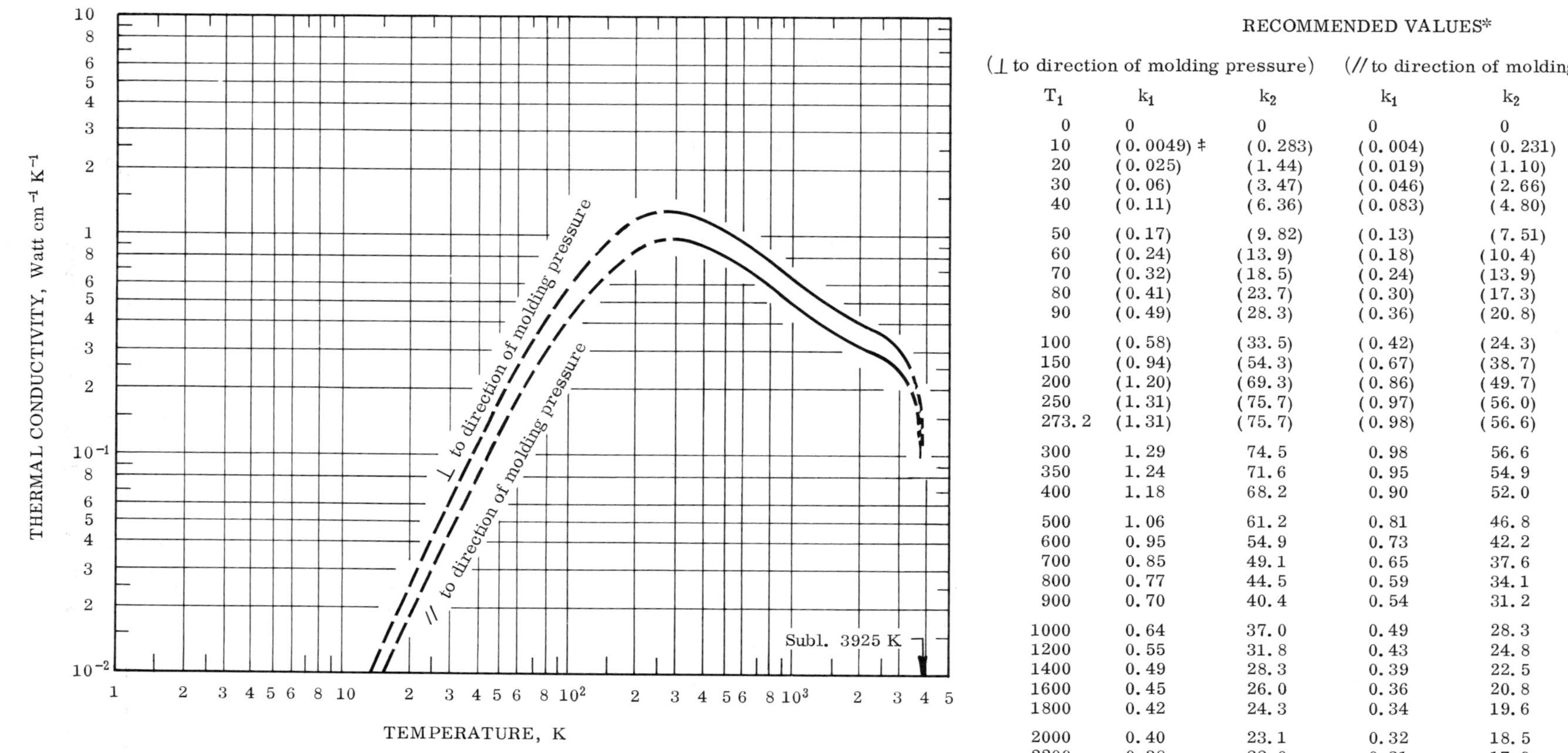

RECOMMENDED VALUES*

	(⊥ to direction of molding pressure)		(// to direction of molding pressure)		
T_1	k_1	k_2	k_1	k_2	T_2
0	0	0	0	0	-459.7
10	(0.0049) ‡	(0.283)	(0.004)	(0.231)	-441.7
20	(0.025)	(1.44)	(0.019)	(1.10)	-423.7
30	(0.06)	(3.47)	(0.046)	(2.66)	-405.7
40	(0.11)	(6.36)	(0.083)	(4.80)	-387.7
50	(0.17)	(9.82)	(0.13)	(7.51)	-369.7
60	(0.24)	(13.9)	(0.18)	(10.4)	-351.7
70	(0.32)	(18.5)	(0.24)	(13.9)	-333.7
80	(0.41)	(23.7)	(0.30)	(17.3)	-315.7
90	(0.49)	(28.3)	(0.36)	(20.8)	-297.7
100	(0.58)	(33.5)	(0.42)	(24.3)	-279.7
150	(0.94)	(54.3)	(0.67)	(38.7)	-189.7
200	(1.20)	(69.3)	(0.86)	(49.7)	- 99.7
250	(1.31)	(75.7)	(0.97)	(56.0)	- 9.7
273.2	(1.31)	(75.7)	(0.98)	(56.6)	32.0
300	1.29	74.5	0.98	56.6	80.3
350	1.24	71.6	0.95	54.9	170.3
400	1.18	68.2	0.90	52.0	260.3
500	1.06	61.2	0.81	46.8	440.3
600	0.95	54.9	0.73	42.2	620.3
700	0.85	49.1	0.65	37.6	800.3
800	0.77	44.5	0.59	34.1	980.3
900	0.70	40.4	0.54	31.2	1160
1000	0.64	37.0	0.49	28.3	1340
1200	0.55	31.8	0.43	24.8	1700
1400	0.49	28.3	0.39	22.5	2060
1600	0.45	26.0	0.36	20.8	2420
1800	0.42	24.3	0.34	19.6	2780
2000	0.40	23.1	0.32	18.5	3140
2200	0.38	22.0	0.31	17.9	3500
2400	0.36	20.8	0.29	16.8	3860
2600	0.35	20.2	0.28	16.2	4220
2800	0.33	19.1	0.27	15.6	4580
3000	0.31	17.9	0.25	14.4	4940
3200	0.28	16.2	(0.23)	(13.3)	5300
3400	(0.25)	(14.4)	(0.21)	(12.1)	5660
3600	(0.20)	(11.6)	(0.17)	(9.82)	6020
3800	(0.113)	(6.53)	(0.095)	(5.49)	6380

REMARKS

ATJ graphite is a pitch-bonded petroleum-coke-base graphite with typical room-temperature density 1.73 g cm^{-3} and is produced by the Carbon Products Division of Union Carbide Corporation. It is formed by molding into blocks and it has very fine grains (0.006 inch maximum). This graphite was previously designated in the development stage as GBH graphite. The uncertainty of the recommended values that are supported by experimental data is probably of the order of ±10 to ±20%, and that of the values obtained by extensive extrapolation is probably twice as great.

* T_1 in K, k_1 in Watt cm^{-1} K^{-1} T_2 in F, and k_2 in Btu hr^{-1} ft^{-1} F^{-1}.

‡ Values in parentheses are extrapolated or estimated.

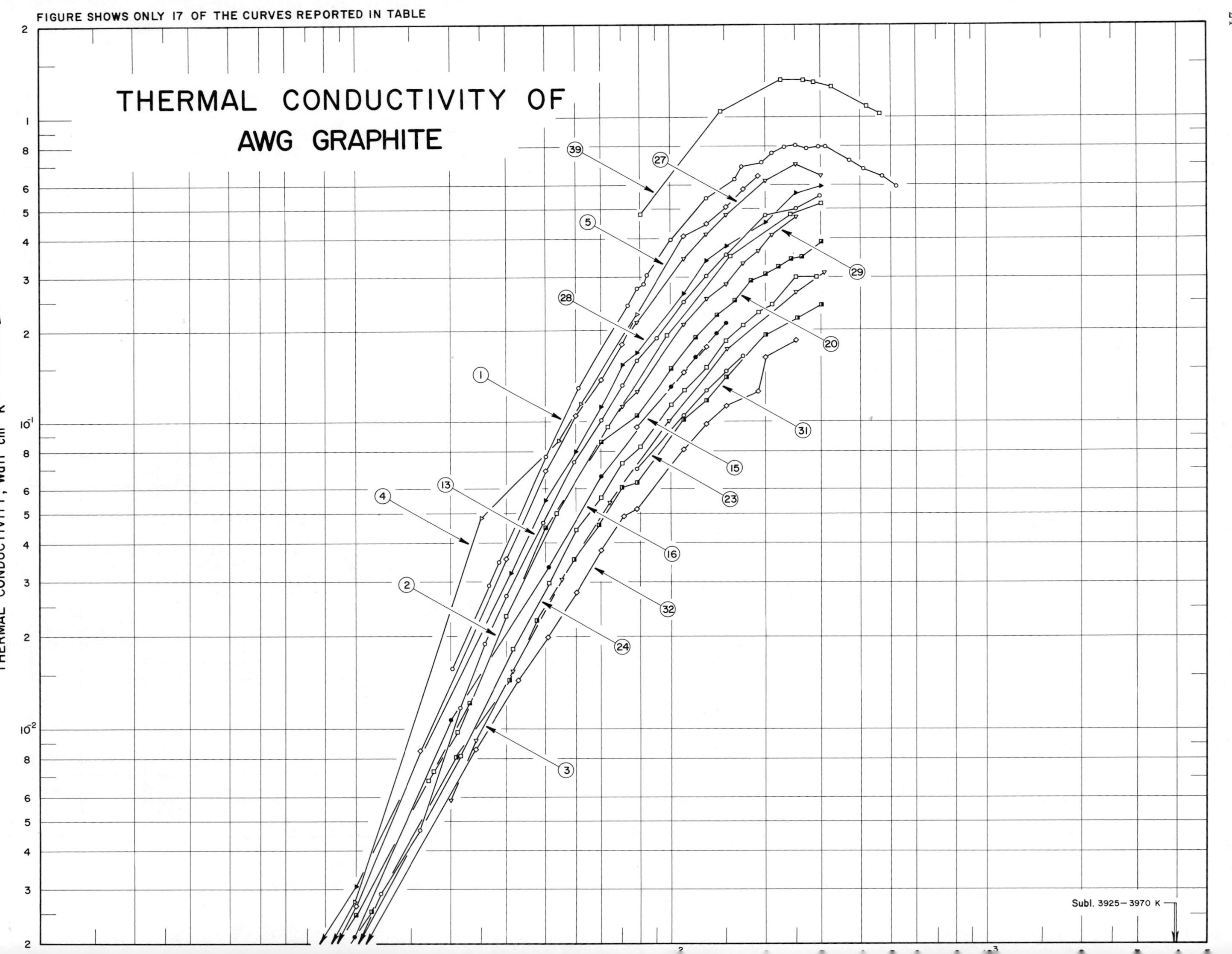
FIGURE SHOWS ONLY 17 OF THE CURVES REPORTED IN TABLE
THERMAL CONDUCTIVITY OF
AWG GRAPHITE
THERMAL CONDUCTIVITY, Watt cm^{-1} K^{-1}
Subl. 3925–3970 K
1
2
3
4
5
13
15
16
20
23
24
27
28
29
31
32
39

SPECIFICATION TABLE NO. 6 THERMAL CONDUCTIVITY OF AWG GRAPHITE

[For Data Reported in Figure and Table No. 6]

Curve No.	Ref. No.	Method Used	Year	Temp. Range, K	Reported Error, %	Name and Specimen Designation	Composition (weight percent), Specifications and Remarks
1	163, 50, 53	E	1954	20-520	±5		Polycrystalline; molded petroleum coke; particle size 25 μ; crystallite size 0.2 μ; density 1.75 g cm^{-3} at 25 C; thermoelectric power +2.3 μvolt K^{-1}; Hall coefficient -0.47 emu; magneto resistivity 1.9 x 10^{-10} emu; electrical resistivity 14.3 milliohm cm; total susceptibility -20.60 x 10^{-6} cgs unit; orientation factor (ρ_{max}/ρ_{min}) = 1.3; measured parallel to the direction of the molding pressure.
2	163, 50	E	1954	17-300	±5		The above specimen exposed to neutron bombardment of 6 MWD/T at <60 C.
3	163, 50	E	1954	20-308	±5		The above specimen exposed to neutron bombardment of 22.7 MWD/T at <30 C.
4	51	E	1956	5.6-78	<2	C-369, No. 1	Specimen 10 mil thick cut from a block of pitch-bonded artificial graphite; electrical resistivity varied from 4.009 to 3.418 milliohm cm at 5.6 to 78 K respectively; irradiated at 103 K by 8.6 Mev protons of 0.65 μah cm^{-2} (micro ampere hour per square centimeter); measured parallel to molding pressure.
5	51	E	1956	6.0-190	<2	C-369, No. 1	The above specimen pulse-annealed for 5 min at 225 K before irradiation; electrical resistivity varied from 3.901 to 2.439 milliohm cm at 6.0 to 190 K respectively.
6	51	E	1956	78-250	<2	C-369, No. 1	The above specimen pulse-annealed at 375 K before irradiation; electrical resistivity varied from 2.968 to 1.924 milliohm cm at 78 to 250 K respectively.
7	51	E	1956	78	<2	C-369, No. 1	The above specimen measured at 78 K after being pulse-annealed at temp ranging from 125 to 375 K.
8	51	E	1956	110	<2	C-369, No. 1	The above specimen measured at 110 K after being pulse-annealed at 225 and 375 K.
9	51	E	1956	130	<2	C-369, No. 1	The above specimen measured at 130 K after being pulse-annealed at 225 and 375 K.
10	51	E	1956	150	<2	C-369, No. 1	The above specimen measured at 150 K after being pulse-annealed at 225 and 375 K.
11	51	E	1956	170	<2	C-369, No. 1	The above specimen measured at 170 K after being pulse-annealed at 225 and 375 K.
12	51	E	1956	190	<2	C-369, No. 1	The above specimen measured at 190 K after being pulse-annealed at 225 and 375 K.
13	51	E	1956	8.6-297	<2	C-369, No. 2	Similar to the above specimen but isothermal-annealed for 2 weeks at 300 K; electrical resistivity varied from 4.139 to 2.372 milliohm cm at 8.6 to 297 K respectively; irradiated by protons of 0.31 μah cm^{-2}.
14	51	E	1956	6.3-78	<2	C-369, No. 3	Similar to the above specimen but electrical resistivity varies from 5.155 to 4.822 milliohm cm at 6.3 to 78 K respectively; irradiated with protons at 103 K of 7.9 μah cm^{-2}.
15	51	E	1956	78-130	<2	C-369, No. 3	The above specimen pulse-annealed for 5 min at 150 K; electrical resistivity varied from 4.895 to 4.736 milliohm cm at 78 to 130 K respectively.
16	51	E	1956	6.0-150	<2	C-369, No. 3	The above specimen pulse-annealed at 175 K; electrical resistivity varied from 5.155 to 4.633 milliohm cm at 6.0 to 150 K respectively.
17	51	E	1956	78-170	<2	C-369, No. 3	The above specimen pulse-annealed at 200 K; electrical resistivity varied from 4.754 to 4.408 milliohm cm at 78 to 170 K respectively.

SPECIFICATION TABLE NO. 6 (continued)

Curve No.	Ref. No.	Method Used	Year	Temp. Range, K	Reported Error, %	Name and Specimen Designation	Composition (weight percent), Specifications and Remarks
18	51	E	1956	6. 0, 78	< 2	C-369, No. 3	The above specimen pulse-annealed at 250 K; electrical resistivity at 6. 0 and 78 K being, respectively, 4. 557 and 4. 291 milliohm cm.
19	51	E	1956	6. 0, 78	< 2	C-369, No. 3	The above specimen pulse-annealed at 300 K; electrical resistivity at 6. 0 and 78 K being, respectively, 4. 179 and 4. 30 milliohm cm.
20	51	E	1956	6. 0-300	< 2	C-369, No. 3	The above specimen pulse-annealed at 375 K; electrical resistivity varied from 4. 295 to 2. 92 milliohm cm at 6. 0 to 300 K respectively.
21	51	E	1956	78	< 2	C-369, No. 3	The above specimen measured at 78 K after being pulse-annealed at temperatures from 125 to 375 K.
22	51	E	1956	5. 5-78	< 2	C-369, No. 4	Similar to the above specimen but not annealed; irradiated at 103 K by protons of 12. 8 μah cm^{-2}; electrical resistivity (before irradiation) varied from 5. 261 to 4. 943 milliohm cm at 5. 5 to 78 K respectively.
23	51	E	1956	78-170	< 2	C-369, No. 4	The above specimen pulse-annealed for 5 min at 200 K; electrical resistivity varied from 4. 935 to 4. 700 milliohm cm at 78 to 170 K respectively.
24	51	E	1956	6. 6-290	< 2	C-369, No. 4	The above specimen pulse-annealed at 375 K; electrical resistivity varied from 4. 362 to 3. 44 milliohm cm at 6. 6 to 290 K respectively.
25	51	E	1956	78	< 2	C-369, No. 4	The above specimen measured at 78 K after being pulse-annealed at temperatures ranging from 125 to 375 K.
26	51	E	1956	78	< 2	Brookhaven	Similar to the above specimen but being exposed to $\sim 10^{18}$ neutrons cm^{-2}.
27	51	E	1956	5. 6-300	< 2	C-376, No. 1	Similar to the above specimen but irradiated at 300 K with an exposure of protons 1. 46 μah cm^{-2}; electrical resistivity vaired from 3. 923 to 1. 964 milliohm cm at 5. 6 to 300 K respectively; not annealed.
28	51	E	1956	5. 2-300	< 2	C-376, No. 2	Similar to the above specimen but being irradiated with an exposure of protons at 2. 47 μah cm^{-2}; electrical resistivity vaired from 4. 083 to 2. 214 milliohm cm at 5. 2 to 300 K respectively.
29	51	E	1956	6. 2-250	< 2	C-376, No. 3	Similar to the above specimen but being irradiated with an exposure of protons at 5. 8 μah cm^{-2}; electrical resistivity varied from 4. 290 to 2. 945 milliohm cm at 6. 2 to 250 K respectively.
30	51	E	1956	6. 4-250	< 2	C-376, No. 4	Similar to the above specimen but being irradiated with an exposure of 9. 3 μah cm^{-2}; electrical resistivity varied from 4. 350 to 3. 308 milliohm cm at 6. 4 to 250 K respectively.
31	51	E	1956	5. 6-250	< 2	C-376, No. 5	Similar to the above specimen but being irradiated with an exposure of 15. 9 μah cm^{-2}; electrical resistivity varied from 4. 465 to 3. 567 milliohm cm at 5. 6 to 250 K respectively.
32	51	E	1956	5. 2-250	< 2	C-376, No. 6	Similar to the above specimen but being irradiated with an exposure of 27. 5 μah cm^{-2}; electrical resistivity varied from 4. 653 to 3. 988 milliohm cm at 5. 2 to 250 K respectively.

SPECIFICATION TABLE NO. 6 (continued)

Curve No.	Ref. No.	Method Used	Year	Temp. Range, K	Reported Error, %	Name and Specimen Designation	Composition (weight percent), Specifications and Remarks
33	51	E	1956	7.6-300	<2	C-381, No. 1	Similar to the above specimen but being irradiated at 423 K with an exposure of 0.95 μah cm^{-2}; electrical resistivity varied from 3.568 to 1.743 milliohm cm at 7.6 to 300 K respectively.
34	51	E	1956	7.2-300	<2	C-381, No. 2	Similar to the above specimen but being irradiated with an exposure of 3.1 μah cm^{-2}; electrical resistivity varied from 3.711 to 1.827 milliohm cm at 7.2 to 300 K respectively.
35	51	E	1956	6.8-300	<2		Similar to the above specimen but being irradiated by protons of 4.96 μah cm^{-2}; electrical resistivity varied from 3.939 to 2.016 milliohm cm at 6.8 to 300 K respectively.
36	51	E	1956	7.6-300	<2		Similar to the above specimen but being irradiated by protons of 11.9 μah cm^{-2}; electrical resistivity varied from 4.186 to 2.304 milliohm cm at 7.6 to 300 K respectively.
37	51	E	1956	6.8-250	<2		Similar to the above specimen but being irradiated by protons of 19.7 μah cm^{-2}; electrical resistivity varied from 4.426 to 2.875 milliohm cm at 6.8 to 250 K respectively.
38	51	E	1956	6.4-300	<2		Similar to the above specimen but being irradiated by protons of 30.2 μah cm^{-2}; electrical resistivity varied from 4.545 to 2.978 milliohm cm at 6.4 to 300 K respectively.
39	53	E	1955	80-460	±5		Made from petroleum coke; molded; specimen size 0.100 x 0.020 x 1.25 in.; room temp properties: density 1.75 g cm^{-3}, thermoelectric power +2.3 μvolt K^{-1}, Hall coefficient -0.47 emu, magneto resistivity 1.9 x 10^{-10} emu, electrical resistivity 14.3 milliohm cm, total magnetic susceptibility -20.6 x 10^{-6} cgs unit, orientation factor (ρ_{max}/ρ_{min}) = 1.0. measured perpendicular to the direction of the molding pressure.

DATA TABLE NO. 6 THERMAL CONDUCTIVITY OF AWG GRAPHITE

[Temperature, T, K; Thermal Conductivity, k, Watt cm^{-1} K^{-1}]

CURVE 1

T	k
20.05	0.0157
26.5	0.0293
28.5	0.0347
40.0	0.0774
51.0	0.130
73.0	0.243
78.0	0.276
82.0	0.285
84.0	0.305
100.0	0.397
130.0	0.544
160.0	0.628
168.0	0.686
195.0	0.711
210.0	0.761
230.0	0.795
250.0	0.808
270.0	0.787
295.0	0.799
310.0	0.799
370.0	0.724
410.0	0.678
470.0	0.640
520.0	0.594

CURVE 2

T	k
17.0	0.00682
17.6	0.00728
21.0	0.00979
30.0	0.0232
43.5	0.0502
63.0	0.0962
97.0	0.192
155.0	0.351
240.0	0.481
300.0	0.523

CURVE 3

T	k
20.0	0.00586
24.0	0.00920
31.5	0.0155
45.0	0.0305
64.0	0.0544
99.0	0.100
151.0	0.172
250.0	0.268
308.0	0.310

CURVE 4

T	k
5.6	0.000962
9.9	0.00272
25.0	0.0485
44.0	0.0866
52.0	0.115
78.0	0.228

CURVE 5

T	k
6.0	0.000991*
10.0	0.00264
16.0	0.00849
30.0	0.0356
40.0	0.0690
50.0	0.105
60.0	0.138
70.0	0.179
78.0	0.228*
110	0.410
130	0.448
150	0.506
170	0.582
190	0.644

CURVE 6*

T	k
78	0.233
110	0.379
130	0.473
150	0.540
170	0.582
190	0.611
210	0.632
230	0.665
250	0.703

CURVE 7* (T = 78K)

T_A	k
0	0.228
125	0.224
150	0.230
175	0.228
200	0.233
225	0.228
250	0.228
275	0.232
300	0.230
325	0.229
350	0.232
375	0.233

CURVE 8* (T = 110K)

T_A	k
225	0.410
375	0.379

CURVE 9* (T = 130K)

T_A	k
225	0.448
375	0.473

CURVE 10* (T = 150K)

T_A	k
225	0.506
375	0.540

CURVE 11* (T = 170K)

T_A	k
225	0.582
375	0.582

CURVE 12* (T = 190K)

T_A	k
225	0.644
375	0.611

CURVE 13

T	k
8.6	0.00138*
12.0	0.00291
16.0	0.00469
21.4	0.0118
25.6	0.0190
30.0	0.0271
39.2	0.0469
49.4	0.0736
60.0	0.101
70.0	0.133
78.0	0.159
90.0	0.188
110.0	0.249
130	0.303
150	0.354
200	0.477
250	0.502
297	0.552

CURVE 14*

T	k
6.3	0.000992
10.0	0.00233
17.0	0.00628
20.0	0.00937
30.0	0.0187
48.0	0.0657
78.0	0.0946

CURVE 15

T	k
78	0.0958
110	0.146
130	0.176

CURVE 16

T	k
6.0	0.000979*
9.9	0.00210
20.0	0.0107
41.0	0.0335
60.0	0.0661
78.0	0.0954*
100.0	0.131
120	0.162
140	0.195
150	0.211

CURVE 17*

T	k
78.0	0.0950
100	0.131
120	0.162
140	0.195
160	0.225
170	0.246

CURVE 18*

T	k
6.0	0.00105
78.0	0.0954

CURVE 19*

T	k
6.0	0.00105
78.0	0.0979

CURVE 20

T	k
6.0	0.00113*
10.0	0.00247
23.0	0.0122
40.0	0.0452
60.0	0.0858
78.0	0.105
100	0.149
120	0.189
140	0.226
160	0.253
180	0.293
200	0.307
220	0.325
240	0.342
260	0.349
300	0.390

CURVE 21* (T = 78 K)

T_A	k
0	0.0946
125	0.0933
150	0.0958
175	0.0954
200	0.0950
225	0.0941
250	0.0954
275	0.0975
300	0.0979
325	0.101
350	0.103
375	0.105

CURVE 22*

T	k
5.5	0.000707
10.0	0.00188
20.0	0.00900
31.0	0.0181
41.0	0.0281
50.0	0.0386
60.0	0.0515
70.0	0.0611
78.0	0.0699

CURVE 23

T	k
78.0	0.0699*
110	0.105
130	0.127
150	0.147
170	0.164

CURVE 24

T	k
6.6	0.000941*
10.0	0.00193*
21.6	0.00816
31.6	0.0182
41.0	0.0298
50.0	0.0444
60.0	0.0565
70.0	0.0732
80.0	0.0828
100	0.114
111	0.127
130	0.151
150	0.184
170	0.208
190	0.230
210	0.244
250	0.300
290	0.300

CURVE 25* (T = 78 K)

T_A	k
0	0.0699
125	0.0674
150	0.0703
175	0.0711
200	0.0699
225	0.0686
250	0.0703
275	0.0724
300	0.0749
325	0.0766
350	0.0778
375	0.0828

*Not shown on plot

CURVE 26*
T = 78 K

T_A	k
0	0.0686
125	0.0682
150	0.0720
175	0.0711
200	0.0707
225	0.0703
250	0.0711
275	0.0736
300	0.0745
325	0.0757
350	0.0782
375	0.0816

CURVE 27

T	k
5.6	0.00120*
10.2	0.00331*
22.4	0.0161*
30.8	0.0389*
41.4	0.0682*
50.4	0.100*
59.4	0.137*
70.0	0.185*
78.0	0.213
110	0.344
130	0.414
150	0.477
200	0.619
250	0.699
300	0.644

CURVE 28

T	k
5.2	0.00109*
10.0	0.00307
31.0	0.0321
40.0	0.0552
50.0	0.0799
60.0	0.112
70.0	0.154
78.0	0.169
110	0.265
130	0.339
150	0.377

CURVE 28 (cont.)

T	k
200	0.452
250	0.565
300	0.594

CURVE 29

T	k
6.2	0.00109*
10.2	0.00264*
21.2	0.0142*
30.8	0.0255*
41.6	0.0435*
49.6	0.0644*
60.0	0.0879*
70.0	0.113
78.0	0.126
110	0.209
130	0.255
150	0.284
170	0.332
190	0.364
210	0.412
250	0.473

CURVE 30*

T	k
6.4	0.00113
10.2	0.00247
20.0	0.00962
30.2	0.0225
41.0	0.0368
50.0	0.0749
60.0	0.0695
70.0	0.0920
78.0	0.0967
110	0.141
130	0.195
150	0.216
170	0.246
190	0.272
210	0.300
250	0.331

CURVE 31

T	k
5.6	0.00086*
11.2	0.00254
20.8	0.00808

CURVE 31 (cont.)

T	k
30.8	0.0145
37.6	0.0226
49.5	0.0355
59.4	0.0464
70.2	0.0607
78.0	0.0632
110	0.102
130	0.117
150	0.141
200	0.193
252	0.220
300	0.244

CURVE 32

T	k
5.2	0.000766*
10.0	0.00175*
24.0	0.00858
32.8	0.0144
40.8	0.0199
50.2	0.0277
60.2	0.0377
71.2	0.0490
78.0	0.0519
110	0.0812
130	0.0983
150	0.113
190	0.126
200	0.162
250	0.185

CURVE 33*

T	k
7.6	0.00164
10.8	0.00350
15.4	0.00770
21.2	0.0156
25.2	0.0239
31.0	0.0377
41.2	0.0686
49.6	0.101
60.0	0.144
70.2	0.194
78.0	0.230
90.0	0.280
110	0.352
130	0.439

CURVE 33 (cont.)*

T	k
150	0.506
200	0.661
250	0.695
300	0.695

CURVE 34*

T	k
7.2	0.00146
10.4	0.00309
16.0	0.00791
20.8	0.0149
27.4	0.0265
30.0	0.0338
40.0	0.0611
50.0	0.0841
59.8	0.133
70.0	0.169
78.0	0.212
90.0	0.260
110	0.331
130	0.400
150	0.464
200	0.623
250	0.690
300	0.695

CURVE 35*

T	k
6.8	0.00133
10.4	0.00284
15.4	0.00674
21.8	0.0147
25.8	0.0207
31.0	0.0302
40.8	0.0531
50.2	0.0803
60.2	0.110
70.3	0.142
78.0	0.169
90.0	0.207
110	0.283
130	0.337
150	0.386
200	0.481
250	0.540
300	0.548

CURVE 36*

T	k
7.6	0.00144
10.4	0.00268
15.0	0.00573
23.0	0.0143
27.0	0.0198
30.0	0.0239
40.8	0.0448
50.2	0.0644
60.0	0.0933
70.0	0.118
78.0	0.138
90.0	0.170
110	0.223
130	0.277
150	0.315
200	0.396
250	0.444
300	0.456

CURVE 37*

T	k
6.8	0.00111
10.8	0.00245
15.2	0.00523
21.2	0.0117
26.4	0.0156
30.6	0.0219
40.8	0.0369
50.0	0.0531
60.0	0.0728
70.6	0.0958
78.0	0.110
90.0	0.134
110	0.174
130	0.216
150	0.248
200	0.331
250	0.367

CURVE 38*

T	k
6.4	0.000912
10.0	0.00215
15.0	0.00444
22.8	0.0109
26.6	0.0143
30.0	0.0179

CURVE 38 (cont.)*

T	k
40.0	0.0290
50.0	0.0444
60.0	0.0615
70.0	0.0816
78.0	0.0929
110	0.143
130	0.178
150	0.206
200	0,267
250	0.308
300	0.317

CURVE 39

T	k
80	0.481
145	1.05
225	1.32
265	1.32
285	1.30
325	1.26
420	1.09
460	1.03

*Not shown on plot

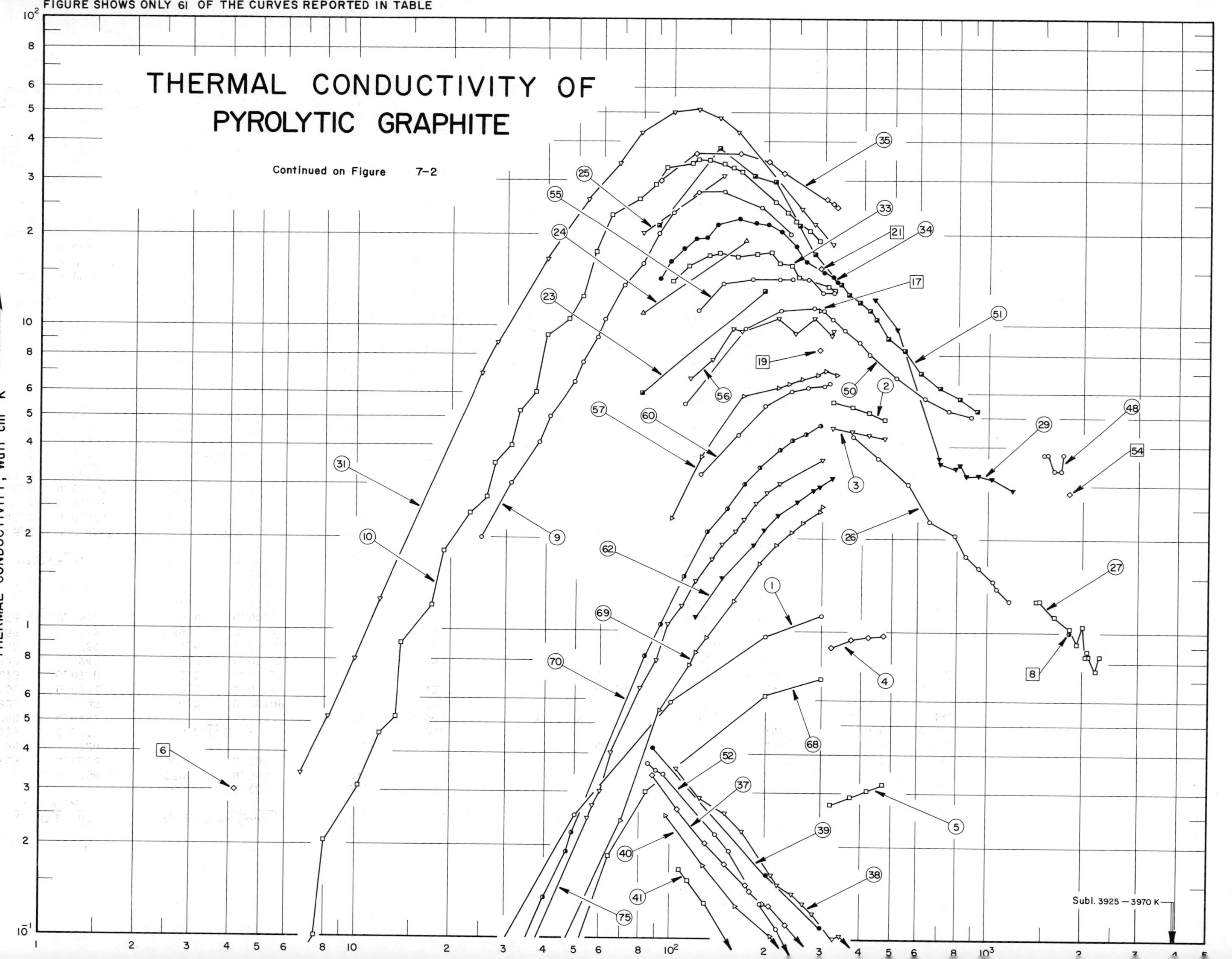
FIGURE SHOWS ONLY 61 OF THE CURVES REPORTED IN TABLE
THERMAL CONDUCTIVITY OF
PYROLYTIC GRAPHITE
Continued on Figure 7-2
THERMAL CONDUCTIVITY, Watt cm^{-1} K^{-1}
Subl. 3925 — 3970 K

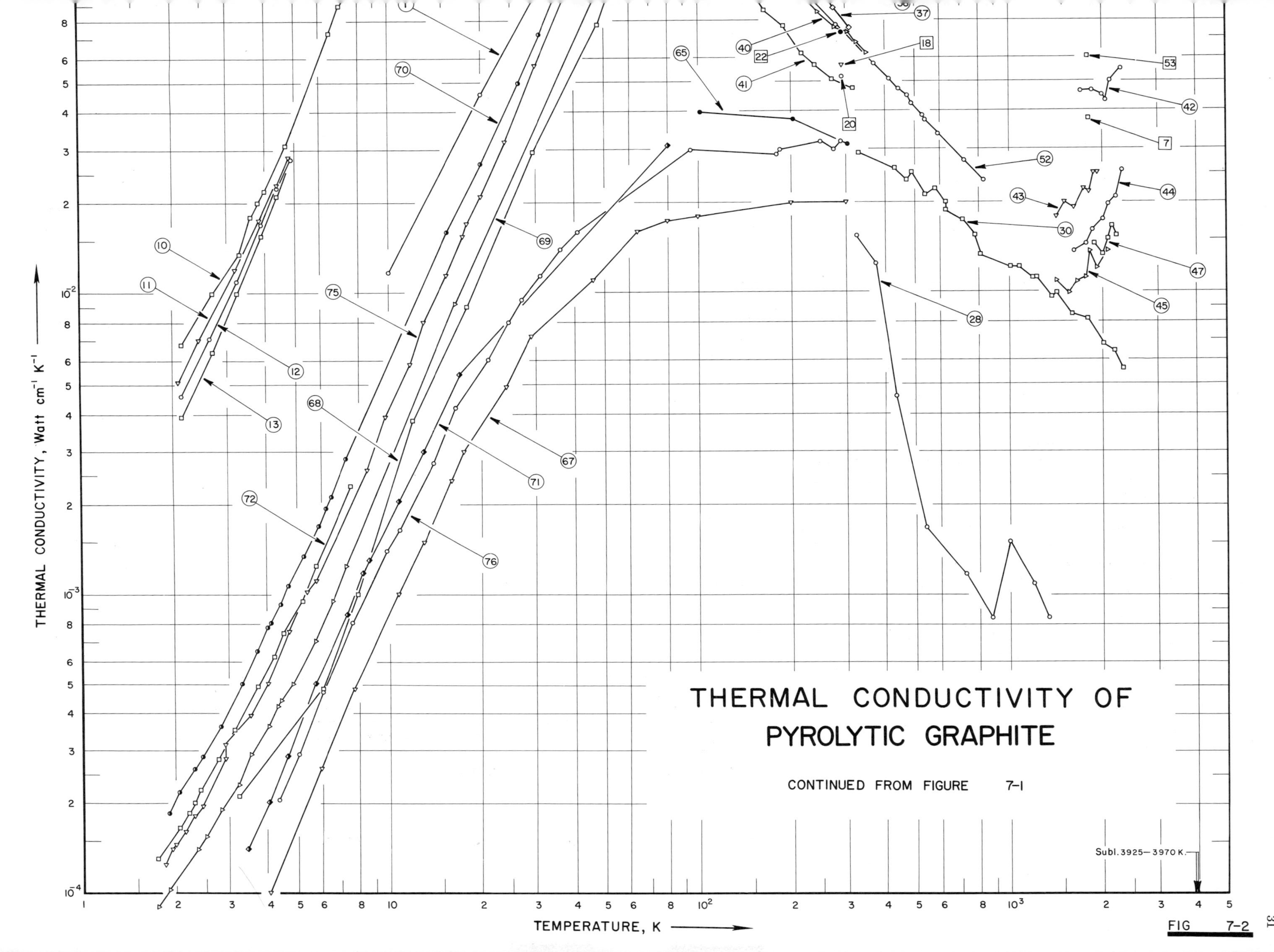
THERMAL CONDUCTIVITY OF
PYROLYTIC GRAPHITE
CONTINUED FROM FIGURE 7-1
Subl. 3925–3970 K.
TEMPERATURE, K
THERMAL CONDUCTIVITY, Watt cm^{-1} K^{-1}

FIG 7-2

SPECIFICATION TABLE NO. 7 THERMAL CONDUCTIVITY OF PYROLYTIC GRAPHITE

[For Data Reported in Figure and Table No. 7]

Curve No.	Ref. No.	Method Used	Year	Temp. Range, K	Reported Error, %	Name and Specimen Designation	Composition (weight percent), Specifications and Remarks
1	50	E	1956	10–300	±5		Obtained by pyrolytic decomposition of a hydrocarbon; no pitch bonding.
2	19	C	1956	323–473		Deposited carbon	Also known as pyrolytic graphite, 99.75 ±0.2 pure with undetectable ash content; deposited from AR grade benzene at 2100 C in a vacuum of 10^{-3} cm Hg; tubular specimen 4.5 cm long, 0.95 cm O.D., and 0.75 cm I.D.; density 1.65 g cm^{-3}; electrical resistivity at 20, 50, 100, 150, and 200 C being, respectively, 245, 230, 215, 200, and 195 μohm cm; measured parallel to axis of tube and parallel to the pronounced layered structure of the specimen.
3	19	C	1956	323–473		Deposited carbon	Similar to the above specimen but deposited at 2000 C; electrical resistivity at 20, 50, 100, 150, and 200 C being, respectively, 380, 360, 330, 305, and 290 μohm cm.
4	19	C	1956	323–473		Deposited carbon	Similar to the above specimen but deposited at 1900 C; electrical resistivity at 20, 50, 100, 150, and 200 C being, respectively, 1.645, 1.585, 1.47, 1.37, and 1.27 milliohm cm.
5	19	C	1956	323–473		Deposited carbon	Similar to the above specimen but deposited at 1800 C; electrical resistivity at 20, 50, 100, 150 and 200 C being, respectively, 3.23, 3.17, 3.065, 2.96, and 2.86 milliohm cm.
6	355	L	1965	4.2			Rectangular block of pyrolytic graphite provided by G. E. Research Lab; reheated to 3500 C after deposition; electrical conductivity in zero magnetic field 7.8 x 10^5 ohm^{-1} cm^{-1}.
7	340	→	1962	1817	12		Pyrolytic graphite obtained from General Electric Co.; 2.3 cm dia x 0.1~0.4 cm thick; k_z determined by using the same method as that for the above specimen.
8	340	→	1962	1817	12		k_r determined simultaneously with the above curve.
9	345	L	1958	25–235		Pyrolytic graphite filament	Specimen of fibrous structures prepared by pyrolysis of methane on a hot carbon wire, measurements made under high vacuum.
10	349	L	1966	2.2–290			Well graphitized and highly heat treated (at 3250 C) pyrolytic graphite; measured in the layer-plane direction; in zero magnetic field.
11	349	L	1966	2.1–4.8			The above specimen measured in a magnetic field of 550 gauss applied in the c-axis direction.
12	349	L	1966	2.2–4.9			The above specimen in a field of 1015 gauss.
13	349	L	1966	2.2–4.9			The above specimen in a field of 2115 gauss.
14	349	L	1966	2.2–3.6			The above specimen in a field of 3805 gauss.
15	349	L	1966	2.2–3.9			The above specimen in a field of 8405 gauss.
16	349	L	1966	3.1–4.8			The above specimen in a field of 12600 gauss.
17	350	P	1965	293.2			Specimen ~0.5 in. long; graphitized at 2910 C; specific heat 0.168 cal g^{-1} C^{-1} at 20 C; measured in the a-axis direction.

SPECIFICATION TABLE NO. 7 (continued)

Curve No.	Ref. No.	Method Used	Year	Temp. Range, K	Reported Error, %	Name and Specimen Designation	Composition (weight percent), Specifications and Remarks
18	350	P	1965	293.2			Similar to the above specimen but about 0.2 in. long; measured in the c-axis direction.
19	350	P	1965	293.2			Specimen ~0.5 in. long; graphitized at 2800 C; specific heat 0.168 cal g^{-1} C^{-1} at 20 C; measured in the a-axis direction.
20	350	P	1965	293.2			Similar to the above specimen but ~0.2 in. long; measured in the c-axis direction.
21	350	P	1965	293.2			Similar to the above specimen but ~0.5 in. long; strain annealed at 3300 C; measured in the a-axis direction.
22	350	P	1965	293.2			Similar to the above specimen but ~0.2 in. long; measured in the c-axis direction.
23	351	L	1965	80, 195			Specimen ~0.2 mm thick; cut from a pyrolytic graphite bar which was made at a deposition temp of 2700 C.
24	351	L	1965	80, 170			Similar to the above specimen but deposition temp 2920 C.
25	351	L	1965	80, 144			Similar to the above specimen but deposition temp 2980 C.
26	361	C	1961	375-1175			Highly regenerative pyrolytic graphite; as deposited; the pyrolytic graphite obtained by passing methane on a graphite slab in the resistance furnace at 2100 C; density 220 g cm^{-3}; measured parallel to the basal planes using dense sintered alumina as a comparative material.
27	361	R	1961	1430-2275			Similar to the above specimen but using another apparatus for higher temp range.
28	361	R	1961	325-1350			Similar to the above specimen but being heat treated for 3 hrs at 2900 C; measured perpendicular to the basal planes.
29	361	C	1961	435-1205			Similar to the above specimen but heat treated for 1 hr at 2900 C; measured parallel to the basal plane using dense sintered alumina as a comparative material.
30	361	R	1961	330-2340			Similar to the above specimen but without heat treatment; measured perpendicular to the basal plane.
31	360	L	1965	6.8-320	±5	No. 1	Pyrolytic graphite specimen size 1 x 5 x 50 mm; the graphite deposition temp 2100 C; annealed under a pressure of 100 bars for 10-15 min at 2800 C; measured parallel to the graphite basal planes.
32	360	L	1965	5.5-320	±5	No. 2	Similar to the above specimen.
33	362	L	1965	100-325	5	AB3	Pyrolytic graphite deposited at 2150 C; annealed at 3000 C; measured parallel to the basal planes.
34	362	L	1965	91-330	5	IFP41	Similar to the above specimen but also hot pressed at 2850 C under 400 Kg cm^{-2}; measured parallel to the basal planes.
35	362	L	1965	91-330	5	IFP56	Similar to the above specimen.
36	362	L	1965	84-318	5	IFPA57	Similar to the above specimen but hot pressed at 2850 C and annealed at 3500 C under 10 Kg cm^{-2} for 0.5 hr; measured parallel to the basal planes.

SPECIFICATION TABLE NO. 7 (continued)

Curve No.	Ref. No.	Method Used	Year	Temp. Range, K	Reported Error, %	Name and Specimen Designation	Composition (weight percent), Specifications and Remarks
37	362	L	1965	88-310	5	IFPA57	The above specimen measured in the c-axis direction.
38	362	L	1965	105-365	5	AB4	Pyrolytic graphite deposited at 2150 C; annealed at 3250 C in induction furnace; measured in the c-axis direction.
39	362	L	1965	88-303	5	AB1	Similar to the AB3 specimen but measured in the c-axis direction.
40	362	L	1965	97-350	5	IFP25	Similar to the IFP41 specimen but measured in the c-axis direction.
41	362	L	1965	107-318	5	IFP25/N4	The above specimen exposed to 2 x 10^{18} fast neutron cm^{-2} at 30 C; measured in the c-axis direction.
42	365	R	1963	1712-2308	12. 5	A1	Specimen 11 cm long, 1. 713 cm O. D., and 1. 465 cm I. D.; pyrolytic graphite deposited from hexane at 1800 C and at a total pressure of 35 cm Hg (partial pressure of hexane 7 cm Hg); the hydrogen carrier gas flows at a rate of 500 cm^3 min^{-1}; the specimen being heat treated for 2 hrs at 2800 C; density 1. 71 g cm^{-3}; data obtained by the first method (direct heating of the graphite tube).
43	365	R	1963	1426-1945		A1	The above specimen measured by the third method (separate heater inserted in the tube).
44	365	R	1963	1637-2313	12. 5	A2	Specimen 11 cm long, 1. 728 cm O. D., and 1. 39 cm I. D.; deposited from hexane at 2100 C and by a method similar to the above; heat treated for 2 hrs at 2800 C; density 2. 21 g cm^{-3}; data obtained by the first method.
45	365	R	1963	1420-2009		A2	The above specimen measured by the third method.
46	365	R	1963	1530-2043	12. 5	B1	Specimen 11 cm long, 1. 75 cm O. D., and 1. 394 cm I. D.; deposited in the same way as the above specimen; heat treated for 1. 25 hrs at 2600 C; density 2. 20 g cm^{-3}; data obtained by the second method (an improvement of the first method to decrease the end contact resistance of the graphite tube).
47	365	R	1963	1895-2231		B1	The above specimen measured by the third method.
48	365	L	1963	1504-1736	30	B1	Thermal conductivity parallel to the basal planes of the above specimen.
49	365	R	1963	1422-2056		B2	Specimen 11 cm long, 1. 75 cm O. D., and 1. 351 cm I. D.; deposited and heat treated in the same way as the above specimen; density 2. 20 g cm^{-3}; data obtained by the third method.
50	369	P	1966	110-880		Specimen 42	Specimen 0. 5 in. long, made from as deposited pyrolytic graphite; annealed at 2900 C for 1 hr in an inert gas atm; thermal conductivity parallel to the deposition plane calculated from measurements of thermal diffusivity, a constant density of 2. 20 g cm^{-3}, and the best fit specific heat data from Magnus 1923, Schlapter and Debrunner 1924, Jacobs and Deem 1956, Wagman et al. 1945, and Rossini et al. 1953.
51	369	P	1966	90-920		Specimen 90	Specimen also 0. 5 in. long supplied by General Electric Co. (structurally more perfect than the above specimen); annealed at 3300 C in inert gas; thermal conductivity parallel to the deposition plane calculated by using the same information as the above specimen.

SPECIFICATION TABLE NO. 7 (continued)

Curve No.	Ref. No.	Method Used	Year	Temp. Range, K	Reported Error, %	Name and Specimen Designation	Composition (weight percent), Specifications and Remarks
52	369	P	1966	85-830		Specimen 90	The above specimen measured perpendicular to the deposition plane.
53	340	→	1962	1808	large	P1	Supplied by General Electric Co.; 2.540 cm dia x 0.238 cm thick; kz determined by using the same method as that for the above specimen.
54	340	→	1962	1808	large	P1	k_r determined simultaneously with the above curve.
55	395	L	1963	122-324		A_2	Specimen approx 0.2 x 1.5 x 8 cm thickness parallel to c-axis; made from pyrolytic graphite (inner layer of sample) deposited at 2150 C from methane atm at 10 cm Hg pressure; in its "as deposited" condition; scattering length 18000 Å; heat flow perpendicular to c-axis.
56	395	L	1963	115-324		A_1	Specimen approx 0.2 x 1.5 x 8 cm, thickness parallel to c-axis; made from pyrolytic graphite deposited at 2180 C from methane atm at 10 cm Hg pressure; in its "as deposited" condition; scattering length 12000 Å; heat flow perpendicular to c-axis.
57	395	L	1963	100-331		A_3	Similar to the above specimen but deposited at 2000 C with scattering length 5000 Å; heat flow perpendicular to c-axis.
58	395	L	1963	116-335		AB_1	Similar to the above specimen but deposited at 2150 C and annealed at 3000 C for 30 min; scattering length 36000 Å; heat flow perpendicular to c-axis.
59	395	L	1963	115-327		AB_2	Similar to the above specimen except scattering length 33000 Å; heat flow perpendicular to c-axis.
60	395	L	1963	125-316		A_2	Specimen approx 0.2 x 1.5 x 8 cm thickness parallel to c-axis; made from pyrolytic graphite (outer layer of sample) deposited at 2150 C and at 10 cm Hg pressure; in its "as deposited" condition; scattering length 5000 Å; heat flow perpendicular to c-axis.
61	395	L	1963	115-327		AB_2	Specimen 0.2 x 1.5 x 8 cm, thickness parallel to c-axis; made from pyrolytic graphite deposited at 2150 C from methane atm at 10 cm Hg pressure; annealed at 3000 C for 30 min; scattering length 36000 Å; measured in vacuum of $<10^{-5}$ mm Hg pressure; heat flow parallel to c-axis.
62	395	L	1963	120-324		N_1	Specimen obtained by sealing AB_2 in an evacuated silica rube and irradiating it in a cooled (~30 C) hollow fuel element in B. E. P. O. at Harwell to an integrated fast neutron dose of about 4 x 10^{18} n. v. t.; heat flow parallel to c-axis; measured in vacuum of $<10^{-5}$ mm Hg pressure.
63	395	L	1963	111-316		N_2	The above specimen annealed in vacuo at 240 C for 70 hrs; heat flow parallel to c-axis; measured in vacuum of $<10^{-5}$ mm Hg pressure.
64	395	L	1963	118-331		N_3	The above specimen annealed in vacuo at 1220 C for 6 hrs; heat flow parallel to c-axis; measured in vacuum of $<10^{-5}$ mm Hg pressure.
65	395	L	1963	104-306		A_1	Specimen 0.2 x 1.5 x 8 cm, thickness parallel to c-axis; made from pyrolytic graphite deposited at 2180 C from methane atm at 10 cm Hg pressure; in its "as deposited" condition; scattering length 12000 Å; measured in vacuum of $<10^{-5}$ mm Hg pressure; heat flow parallel to c-axis.

SPECIFICATION TABLE NO. 7 (continued)

Curve No.	Ref. No.	Method Used	Year	Temp. Range, K	Reported Error, %	Name and Specimen Designation	Composition (weight percent), Specifications and Remarks
66	395	L	1963	89-302		AB_1	Similar to the above specimen except deposited at 2150 C and annealed at 3000 C for 30 min; scattering length 36000 Å; heat flow parallel to c-axis.
67	352	L	1962	4.0-300		PG-0	Specimen 0.7 cm long and having a square cross sectional area of 0.17 cm^2; made from pryolytic graphite deposited on a substrate of commercial graphite in a methane atm at 2250 C and at a total pressure of 20 mm Hg; graphite crystallites shaped like oblate ellipsoids (with rotational symmetry about the c-axis), of minor dia (parallel to c-axis) = 140 Å and major dia (perpendicular to c-axis) = 280 Å; these crystallites with an average angular tilt of 22 degrees from the c-axis formed columnar bundles of 0.1 cm in dia; density of the specimen 2.194 g cm^{-3}; electrical conductivity 1.98 ohm^{-1} cm^{-1} at 298 K; sound velocity 3.4 x 10^5 cm sec^{-1} at 9.8 megacycles sec^{-1} and at 300 K; free from any visible cracks along the [0001] planes; heat flow parallel to the c-axis.
68	352	L	1962	3.2-300		PG-0	Similar to the above specimen but 1.9 cm long with a square cross sectional area of 0.14 cm^2; electrical conductivity 1.85 x 10^3 ohm^{-1} cm^{-1} at 298 K; sound velocity 4.7 x 10^5 cm sec^{-1} at the same conditions as above; heat flow perpendicular to c-axis.
69	353	L	1964	1.8-300	±3	RAY-17	Specimen of pyrolytic graphite in its "as-deposited" condition; manufactured by Raytheon's Adv. Mat. Dept. with a deposition temp of 1700 C; crystallite size 180 Å; density 2.13 g cm^{-3}; cut parallel to the layer plane.
70	353	L	1964	1.9-295	±3	RAY-23	Similar to the above specimen but the deposition temp 2300 C; density 2.22 g cm^{-3} and crystallite size 285 Å; parallel to the layer plane.
71	353	L	1964	3.4-80	±3	RAY-23	Similar to the above specimen but cut perpendicular to the layer plane.
72	353	L	1964	1.8-7.5	±3	RAY-19	Similar to the above specimen but the deposition temp 1900 C; crystallite size 240 Å, and density 2.19 g cm^{-3}; cut parallel to the layer plane.
73	353	L	1964	2.0-6.8	±3	RAY-21	Similar to the above specimen but the deposition temp 2100 C; crystallite size 270 Å, and density 2.20 g cm^{-3}; cut parallel to the layer plane.
74	353	L	1964	3.1-300	±3	RAY-21	Similar to the above specimen but cut perpendicular to the layer plane.
75	353	L	1964	1.9-300	±3	HTM	Specimen of pyrolytic graphite in its "as-deposited" condition; obtained from High Temp. Mat. Inc.; deposition temp ~2100 C; density 2.19 g cm^{-3}; cut parallel to the layer plane.
76	353	L	1964	4.3-290	±3	HTM	Similar to the above specimen but cut perpendicular to the layer plane.

DATA TABLE NO. 7 THERMAL CONDUCTIVITY OF PYROLYTIC GRAPHITE

[Temperature, T, K; Thermal Conductivity, k, Watt cm^{-1} K^{-1}]

T	k
CURVE 1	
10	0.0117
20	0.046
50	0.251
100	0.586
200	0.962
300	1.13
CURVE 2	
323	5.6
373	5.4
423	5.15
473	4.9
CURVE 3	
323	4.6
373	4.45
423	4.35
473	4.25
CURVE 4	
323	0.89
373	0.94
423	0.96
473	0.97
CURVE 5	
323	0.275
373	0.290
423	0.305
473	0.320
CURVE 6	
4.2	0.30
CURVE 7	
1817	0.0377
CURVE 8	
1817	0.996
CURVE 9	
25	2.0
31	3.0
38	4.1
41	5.0
49	6.5
52	7.5
58	9.1
61	10.5
70	13.6
80	16.0
90	20.0
100	23.5
120	27.5
145	27.6
190	24.5
235	20.0
CURVE 10	
2.15	0.0068
2.70	0.0100
3.30	0.0135
3.60	0.0180
3.80	0.020
4.00	0.022
4.70	0.031
6.50	0.073
7.00	0.090
7.50	0.100
8.00	0.205
10.30	0.310
12.00	0.460
13.50	0.520
14.00	0.900
17.50	1.20
19.00	1.80
23.00	2.40
26.00	2.70
CURVE 10 (cont.)	
27.50	3.50
31.00	4.00
33.00	5.20
37.00	6.00
40.00	9.25
47.00	10.50
52.00	12.50
57.00	17.50
64.00	23.0
78.00	26.0
88.0	29.0
95.0	33.0
115.0	34.0
120.0	35.0
130.0	35.0
145.0	34.0
155.0	33.0
165.0	32.0
210.0	25.5
230.0	23.5
245.0	22.0
270.0	20.5
290.0	19.0
CURVE 11	
2.1	0.0051
2.45	0.0070
3.2	0.0120
3.85	0.0175
4.4	0.0230
4.8	0.0285
CURVE 12	
2.15	0.0046
2.65	0.0071
3.25	0.0110
3.90	0.0170
4.40	0.0225
4.90	0.0280
CURVE 13	
2.15	0.0039
2.70	0.0064
3.25	0.0100
3.90	0.0155
4.40	0.0210
4.90	0.0280*
CURVE 14*	
2.15	0.0037
2.75	0.0065
3.60	0.0118
3.60	0.0123
CURVE 15*	
2.15	0.0038
2.75	0.0064
3.25	0.0095
3.70	0.0150
3.90	0.0150
CURVE 16*	
3.1	0.0084
3.4	0.0108
3.5	0.0115
3.9	0.0160
4.8	0.0240
CURVE 17	
293.2	11.3
CURVE 18	
293.2	0.0569
CURVE 19	
293.2	8.28
CURVE 20	
293.2	0.0523
CURVE 21	
293.2	15.5
CURVE 22	
293.2	0.0732
CURVE 23	
80	6.0
195	13.0
CURVE 24	
80	11.0
170	19.0
CURVE 25	
80	20.0
144	31.0
CURVE 26	
375	4.31
450	3.66
560	3.01
655	2.28
790	2.05
855	1.76
940	1.61
1045	1.46
1065	1.38
1175	1.26
CURVE 27	
1430	1.26
1460	1.26
CURVE 27 (cont.)	
1630	1.13
1820	1.03
1925	0.920
2000	1.05
2050	0.837
2065	0.870
2100	0.837
2210	0.753
2275	0.837
CURVE 28	
325	0.0156
375	0.0126
435	0.00456
540	0.00167
725	0.00117
882	0.000711
1000	0.00151
1200	0.00109
1350	0.000836
CURVE 29	
435	12.3
515	9.67
700	3.66
710	3.51
795	3.37
815	3.47
855	3.20
935	3.22
1035	3.14
1205	2.89
CURVE 30	
330	0.0293
431	0.0259
473	0.0238
490	0.0251
540	0.0213
CURVE 30 (cont.)	
580	0.0222
630	0.0200
630	0.0188
715	0.0175
780	0.0156
810	0.0134
1012	0.0122
1080	0.0122
1200	0.0113
1235	0.0113
1380	0.00966
1430	0.0100
1610	0.00849
1800	0.00820
2025	0.00678
2200	0.00644
2340	0.00561
CURVE 31	
6.75	0.34
8.25	0.52
10.0	0.80
12.0	1.25
25.0	6.90
28.0	8.75
40.0	16.5
54.0	26.0
67.5	34.0
79.0	43.0
100.0	50.0
120.0	51.0
140.0	48.0
160.0	43.0
255.0	24.0
280.0	21.5
320.0	18.5
CURVE 32*	
5.5	0.205
6.0	0.22

*Not shown on plot

DATA TABLE NO. 7 (continued)

T	k
CURVE 32 (cont.)*	
6.1	0.24
6.5	0.28
7.75	0.40
9.0	0.625
10.5	0.875
12.0	1.15
15.5	2.20
17.0	2.9
22.5	5.8
23.5	6.1
27.0	7.75
31.0	11.0
35.0	13.0
39.0	15.0
43.0	18.0
58.0	28.5
75.0	38.0
85.0	43.0
92.5	48.0
110	47.0
125	46.0
135	43.0
150	40.0
165	36.0
180	34.0
225	27.0
245	25.0
265	23.0
290	21.0
320	19.5
CURVE 33	
100	14.0
113	15.7
130	17.1
140	17.4
160	16.9
185	17.2
205	17.5
218	16.1
238	15.9
250	14.5
310	13.5
325	13.1
CURVE 34	
91	14.3
98	16.3
108	17.9
118	19.2
128	19.4
138	21.5
163	22.5
183	21.7
200	21.5
220	20.4
245	18.2
263	16.4
300	15.0
320	14.5
330	13.9
CURVE 35	
91	20.0
118	26.7
163	26.7
200	24.7
222	21.7
305	16.1
320	15.2
330	14.5
CURVE 36*	
84	31.3
103	37.0
104	37.3
108	37.7
123	38.5
128	37.7
135	37.3
148	35.7
162	33.9
197	28.6
215	25.6
238	23.0
265	20.2
295	18.2
318	16.7
CURVE 37	
88	0.339
106	0.262
107	0.263*
130	0.204
150	0.175
175	0.149
180	0.142
198	0.130
208	0.127
235	0.111
260	0.0990
275	0.0885
310	0.0758
CURVE 38	
105	0.357
125	0.286
150	0.256
170	0.222
210	0.161
220	0.149
245	0.140
265	0.130
285	0.121
330	0.100
348	0.102
365	0.0980
CURVE 39	
88	0.417
203	0.161
303	0.109
CURVE 40	
97	0.250
128	0.172
163	0.127
210	0.102
245	0.0862
278	0.0762
305	0.0735
315	0.0704*
CURVE 40 (cont.)	
325	0.0680
350	0.0625
CURVE 41	
107	0.167
115	0.154
130	0.130
165	0.0870
190	0.0769
218	0.0625
240	0.0571
272	0.0513
318	0.0476
CURVE 42	
1712	0.0469
1864	0.0469
2002	0.0452
2056	0.0435
2131	0.0502
2308	0.0552
CURVE 43	
1426	0.0180
1519	0.0201
1639	0.0192
1750	0.0222
1819	0.0218
1883	0.0251
1945	0.0251
CURVE 44	
1637	0.0138
1780	0.0146
1867	0.0163
2006	0.0176
2093	0.0197
2221	0.0209
2313	0.0255
CURVE 45	
1420	0.0105
1559	0.0100
1657	0.0109
1772	0.0113
1835	0.0138
1924	0.0121
2009	0.0138
CURVE 46*	
1530	0.0117
1667	0.0126
1715	0.0126
1818	0.0130
1899	0.0130
1972	0.0134
2043	0.0138
CURVE 47	
1895	0.0146
2008	0.0134
2092	0.0151
2154	0.0167
2231	0.0155
CURVE 48	
1504	3.77
1553	3.77
1621	3.35
1705	3.35
1736	3.77
CURVE 49*	
1422	0.0121
1545	0.0117
1670	0.0121
1776	0.0138
1843	0.0138
1901	0.0146
2056	0.0155
CURVE 50	
110	5.5
170	9.75
220	11.3
280	11.5
300	11.3
320	10.5
350	9.60
390	8.75
420	8.00
510	6.75
630	5.75
750	5.25
880	5.00
CURVE 51	
90	21.3
140	38.3
180	31.0
210	29.8
250	21.3
280	17.3
300	15.8*
340	13.8
360	12.8
390	12.0
420	11.3
440	10.5
480	9.00
540	8.25
610	7.00
700	6.25
810	5.75
920	5.25
CURVE 52	
85	0.370
90	0.353
95	0.343
140	0.218
155	0.193
195	0.130
220	0.108
235	0.0950
CURVE 52 (cont.)	
280	0.0775
300	0.0725*
325	0.0675*
370	0.0575
415	0.0513
445	0.0475
475	0.0450
490	0.0425
530	0.0387
540	0.0375
595	0.0337
720	0.0275
830	0.0237
CURVE 53	
1808	0.0699
CURVE 54	
1808	2.84
CURVE 55	
121.6	11.24
149.6	13.88
179.9	14.28
218.8	14.28
239.9	14.28
269.2	14.28
298.5	12.90
323.6	12.90
CURVE 56	
114.8	6.67
134.9	7.69
156.7	9.71
166.0	9.52
216.3	10.53
245.5	9.35
281.8	10.53
319.9	9.17
323.6	9.52

*Not shown on plot

DATA TABLE NO. 7 (continued)

T	k
CURVE 57	
100	2.33
124.5	3.70
167.9	5.88
216.3	6.25
234.4	6.45
257.0	6.67
288.4	6.90
305.5	7.14
331.1	6.90
CURVE 58*	
116.1	18.52
134.9	22.22
162.2	23.26
192.8	23.81
211.3	23.81
245.5	22.73
266.1	21.74
309.0	19.61
335.0	17.86
CURVE 59*	
114.8	17.24
136.5	20.83
158.5	21.28
184.1	21.28
204.2	20.83
213.8	20.41
229.1	20.00
245.5	19.23
263.0	18.52
288.4	18.52
309.0	16.00
327.3	14.81
CURVE 60	
124.7	3.23
162.2	4.35
198.6	5.41
239.9	6.06
269.2	6.25
302.0	6.33
316.2	6.45
CURVE 61*	
114.8	16.66
136.5	20.83
158.5	21.27
186.2	21.27
206.5	20.83
213.8	20.40
231.7	19.60
248.3	19.23
288.4	18.51
309.0	16.00
327.3	15.15
CURVE 62	
120.2	1.11
146.2	1.48
182.0	1.89
197.2	2.13
218.8	2.38
251.2	2.63
281.8	2.86
295.1	2.94
323.6	3.13
CURVE 63*	
110.9	2.701
128.8	3.33
144.5	4.00
164.1	4.55
192.8	5.41
216.3	5.71
239.9	6.06
251.2	6.41
260.0	6.45
309.0	6.90
316.2	7.14
CURVE 64*	
117.5	15.38
141.3	16.66
182.0	20.00
211.3	19.23
216.3	19.60
231.7	18.86
CURVE 64 (cont.)*	
245.5	18.51
266.1	18.51
288.4	17.85
309.0	15.38
331.1	14.28
CURVE 65	
103.5	0.04
204.2	0.0377
305.5	0.0312
CURVE 66*	
89.1	0.455
204.2	0.154
302.0	0.100
CURVE 67	
4.0	0.00010
5.9	0.00026
7.6	0.00048
10.7	0.0010
13.0	0.0015
16.0	0.0024
17.5	0.0030
24	0.0049
29	0.0072
46	0.011
64	0.016
80	0.0175
100	0.018
200	0.020
300	0.020
CURVE 68	
3.2	0.00021
6.0	0.00048
7.8	0.0010
12.0	0.0038
18.0	0.0090
29.5	0.0295
48	0.078
64	0.184
CURVE 68 (cont.)	
84	0.30
200	0.62
300	0.70
CURVE 69	
1.75	0.00009
1.90	0.000102
2.35	0.00014
2.50	0.000155
2.80	0.00019
3.20	0.00023
3.50	0.00029
4.00	0.00036
4.30	0.00042
4.4	0.00044
4.8	0.0005
5.7	0.0007
6.5	0.00095
7.25	0.00125
16.5	0.00925
70.0	0.24
92.5	0.55
115.0	0.755
120	0.850
130	0.950
158	1.25
190	1.65
215	1.90
240	2.10
260	2.25
295	2.45
300	2.55
CURVE 70	
1.9	0.000185
2.05	0.000218
2.3	0.00026
2.45	0.000286
2.80	0.00036
3.3	0.00050
3.7	0.00065
4.0	0.000775
4.1	0.00080
4.4	0.000925
CURVE 70 (cont.)	
4.7	0.00108
5.3	0.00135
5.9	0.00170
6.25	0.00195
6.50	0.00213
7.25	0.00285
15.5	0.016
20.0	0.027
26.5	0.050
31.0	0.0725
40.0	0.135
47.0	0.19
49.0	0.22
82.5	0.825
92.5	1.05
110.0	1.50
130	2.1
150	2.5
170	3.0
190	3.4
220	3.9
240	4.2
265	4.4
295	4.7
CURVE 71	
3.4	0.00014
4.0	0.0002
4.6	0.000285
5.7	0.0005
7.25	0.00085
8.25	0.00118
8.50	0.0013
10.8	0.00205
13.0	0.003
17.0	0.0054
80.0	0.031
CURVE 72	
1.75	0.00013
2.05	0.000165
2.2	0.000185
2.3	0.00020
2.4	0.00022
CURVE 72 (cont.)	
2.75	0.00028
3.1	0.00035
3.7	0.00049
4.2	0.00062
4.5	0.00074
5.2	0.00095
5.8	0.00125
7.5	0.00230
CURVE 73*	
2.0	0.000175
2.45	0.00026
2.70	0.00032
3.80	0.00060
4.40	0.000775
4.60	0.00085
5.0	0.0010
6.0	0.00153
6.75	0.0019
CURVE 74*	
3.1	0.000105
3.7	0.000165
4.5	0.000250
5.25	0.000350
5.7	0.000440
6.5	0.000620
9.3	0.00150
11.5	0.00240
12.5	0.00270
15.5	0.00440
18.5	0.0060
21.0	0.0075
80.3	0.0290
300	0.0350
CURVE 75	
1.85	0.000125
1.95	0.000140
2.0	0.000145
2.15	0.00016
2.30	0.00018
2.45	0.000195
CURVE 75 (cont.)	
2.90	0.00028
2.90	0.00031
3.50	0.00039
4.0	0.00050
4.7	0.00075
5.4	0.00103
5.8	0.0011
8.5	0.0026
9.75	0.0039
11.75	0.0058
13.0	0.0080
15.5	0.0115
17.5	0.0155
18.0	0.0170
20.0	0.021
24.0	0.032
30.0	0.057
55.0	0.245
57.0	0.27
60.0	0.30
65.0	0.40
80.0	0.65
90.0	0.80
98	1.05
108	1.2
120	1.45
135	1.7
145	1.9
160	2.1
170	2.3
185	2.6
200	2.8
220	3.0
300	3.6
CURVE 76	
4.3	0.000205
5.0	0.00029
6.0	0.00047
7.5	0.00080
9.8	0.0014
10.8	0.00165
14.0	0.00275
16.5	0.0042
21.0	0.0060

*Not shown on plot

DATA TABLE NO. 7 (continued)

T	k
CURVE 76 (cont.)	
24.5	0.0080
27.0	0.0095
31.0	0.0115
36.0	0.0140
41.0	0.0160
95.0	0.0300
180	0.029
185	0.030
250	0.032
275	0.030
290	0.032

FIGURE AND TABLE NO. 7R RECOMMENDED THERMAL CONDUCTIVITY OF PYROLYTIC GRAPHITE

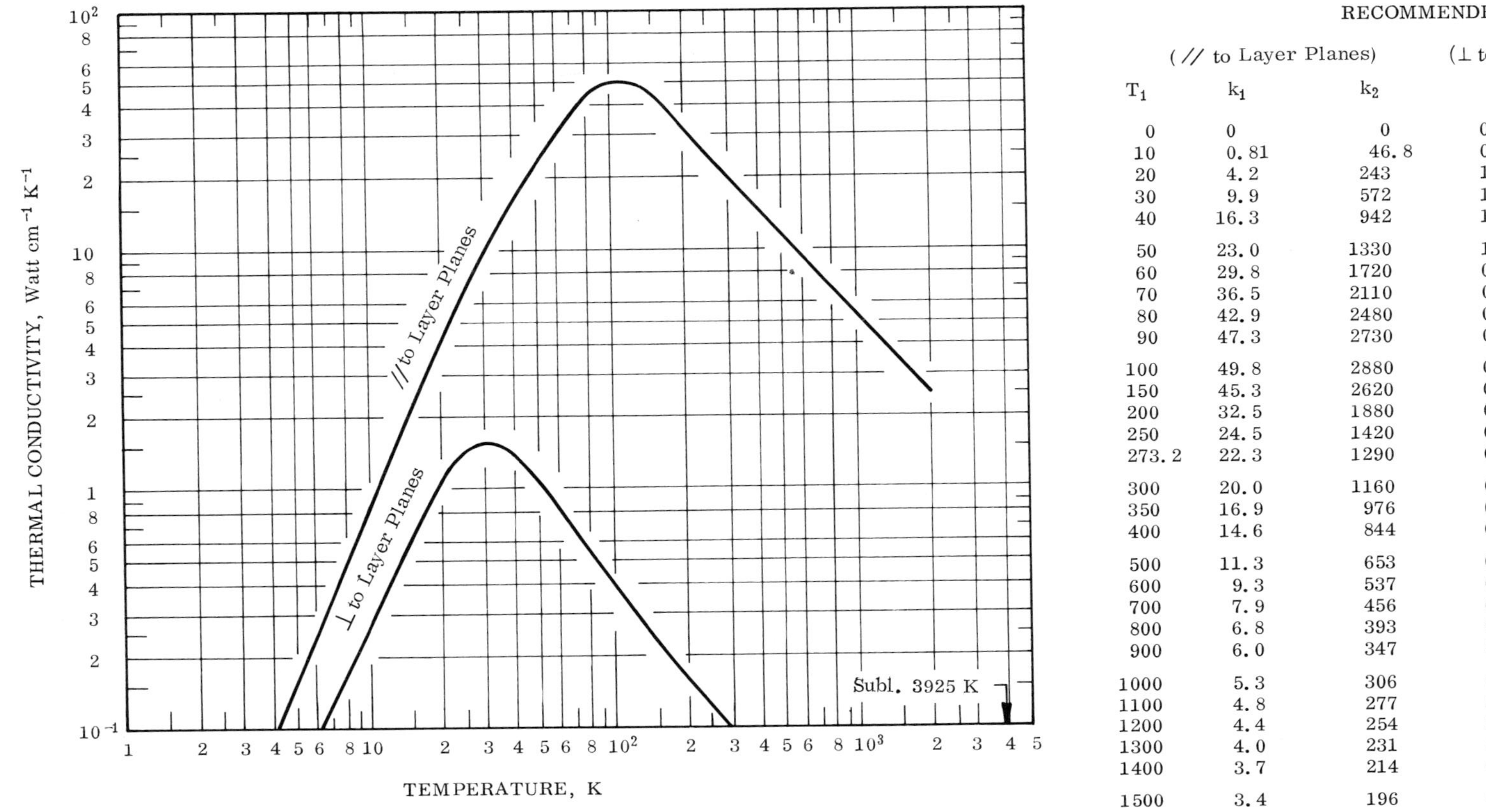

RECOMMENDED VALUES*

T_1	k_1 (// to Layer Planes)	k_2 (// to Layer Planes)	k_1 (⊥ to Layer Planes)	k_2 (⊥ to Layer Planes)	T_2
0	0	0	0	0	-459.7
10	0.81	46.8	0.27	39.0	-441.7
20	4.2	243	1.08	62.4	-423.7
30	9.9	572	1.55	89.6	-405.7
40	16.3	942	1.35	78.0	-387.7
50	23.0	1330	1.03	59.5	-369.7
60	29.8	1720	0.81	46.8	-351.7
70	36.5	2110	0.65	37.6	-333.7
80	42.9	2480	0.54	31.2	-315.7
90	47.3	2730	0.46	26.6	-297.7
100	49.8	2880	0.39	22.5	-279.7
150	45.3	2620	0.23	13.3	-189.7
200	32.5	1880	0.15	8.67	- 99.7
250	24.5	1420	0.116	6.70	- 9.7
273.2	22.3	1290	0.106	6.12	32.0
300	20.0	1160	0.095	5.49	80.3
350	16.9	976	0.080	4.62	170.3
400	14.6	844	0.070	4.04	260.3
500	11.3	653	0.054	3.12	440.3
600	9.3	537	0.044	2.54	620.3
700	7.9	456	0.038	2.20	800.3
800	6.8	393	0.032	1.85	980.3
900	6.0	347	0.028	1.62	1160
1000	5.3	306	0.025	1.44	1340
1100	4.8	277	0.023	1.33	1520
1200	4.4	254	0.021	1.21	1700
1300	4.0	231	0.019	1.10	1880
1400	3.7	214	0.017	0.982	2060
1500	3.4	196	0.016	0.924	2240
1600	3.2	185	0.015	0.867	2420
1700	3.0	173	0.014	0.809	2600
1800	2.8	162	0.013	0.751	2780
1900	2.6	150	0.0125	0.722	2960
2000	2.5	144	0.012	0.693	3140

REMARKS

Pyrolytic graphite is produced by the deposition of carbon from a gaseous hydrocarbon onto a heated surface at high temperature of the order of 2300 K. The uncertainty of the recommended values for the direction parallel to the layer planes at temperatures below 1500 K is probably of the order of ±10 to ±20%, and that at temperatures above 1500 K is probably twice as great. The values for the direction perpendicular to the layer planes are intended as typical values for indicating the general trend.

* T_1 in K, k_1 in Watt cm^{-1} K^{-1}, T_2 in F, and k_2 in Btu hr^{-1} ft^{-1} F^{-1}.

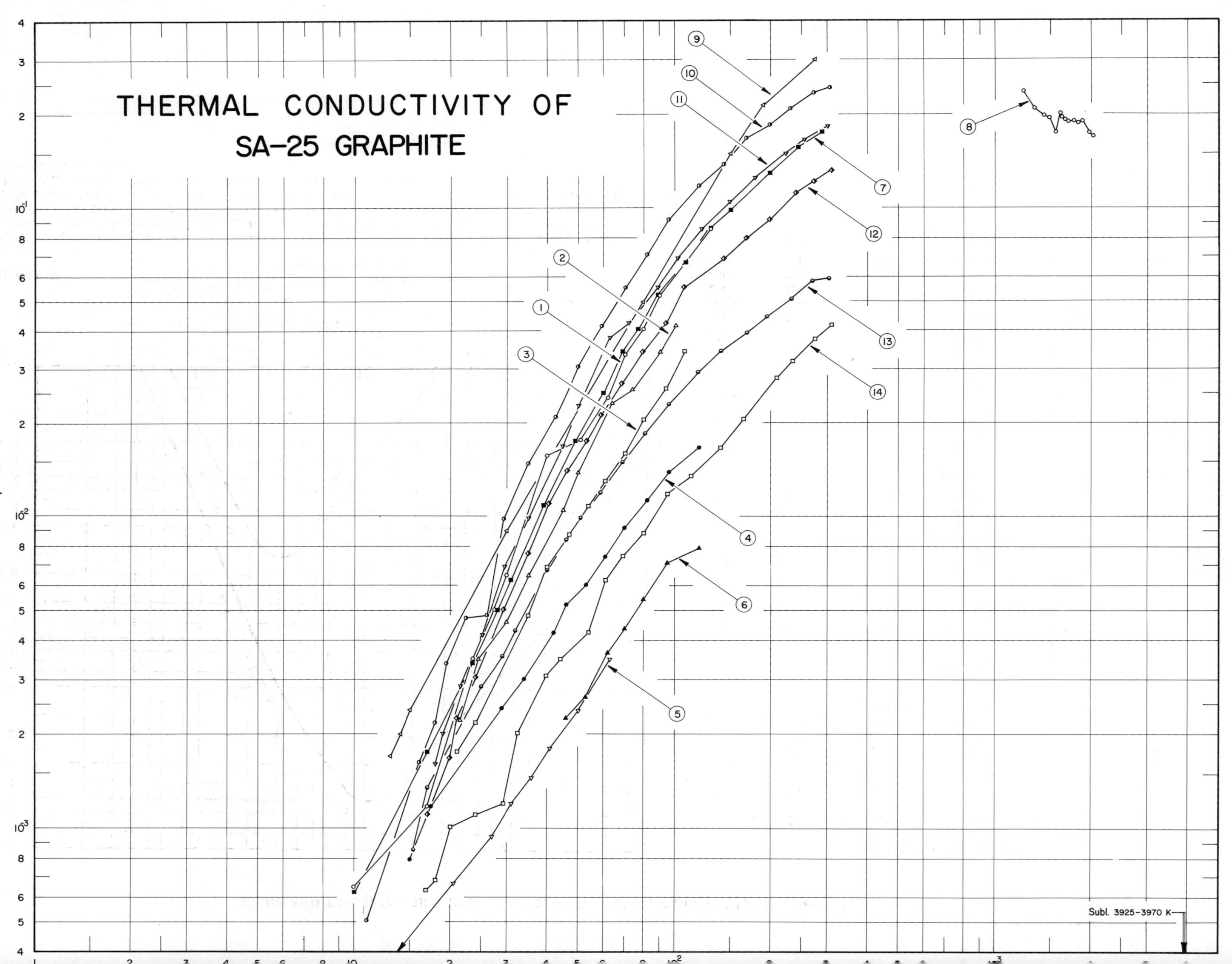
THERMAL CONDUCTIVITY OF
SA-25 GRAPHITE
THERMAL CONDUCTIVITY, Watt cm⁻¹ K⁻¹
Subl. 3925-3970 K

SPECIFICATION TABLE NO. 8 THERMAL CONDUCTIVITY OF SA-25 GRAPHITE

[For Data Reported in Figure and Table No. 8]

Curve No.	Ref. No.	Method Used	Year	Temp. Range, K	Reported Error, %	Name and Specimen Designation	Composition (weight percent), Specifications and Remarks
1	50	E	1956	10-205	±5		Polycrystalline; molded from lampblack; pitch bonded; particle size 0.3 μ, crystallite size 0.05 μ; density 1.55 g cm^{-3} at 25 C.
2	50	E	1956	22-101	±5		The above specimen exposed to neutron bombardment of 12.5 MWD/T at ~30 C.
3	50	E	1956	21-115	±5		The above specimen exposed to neutron bombardment of 22.7 MWD/T at ~30 C.
4	50	E	1956	15-120	±5		The above specimen exposed to neutron bombardment of 146 MWD/T at ~30 C.
5	50	E	1956	13-63	±5		The above specimen exposed to neutron bombardment of 460 MWD/T at ~30 C.
6	50	E	1956	46-112	±5		The above specimen exposed to neutron bombardment of 460 MWD/T at ~30 C (probably second run of the above specimen).
7	53	E	1955	10-290	±5		Made from lampblack; molded; room temp properties: density 1.55 g cm^{-3}, thermoelectric power +9.6 μvolt K^{-1}, Hall coefficient +0.14 emu, magneto resistivity 0.2 x 10^{-10} emu, electrical resistivity 43 milliohm cm, total magnetic susceptibility -21.02 x 10^{-6} cgs unit, orientation factor (ρ_{max}/ρ_{min}) = 1.0.
8	357	L	1959	1246-2045			Emissivity 0.83.
9	345	L	1958	13-275	4		Specimen prepared from lampblack base, molded with a coal-tar pitch binder, measurements made under high vacuo.
10	163	E	1954	10-306			Molded lampblack; density 1.55 g cm^{-3} at room temp; thermoelectric power +9.0 μvolt K^{-1}; Hall coefficient +0.14 emu; magneto resistivity 0.2 x 10^{-10} emu; electrical resistivity 65.3 x 10^{-3} ohm cm; total susceptibility -21.02 x 10^{-6} cgs unit, and orientation factor (ρ_{max}/ρ_{min}) = 1.0.
11	163	E	1954	18-302			The above specimen exposed to neutron irradiation of 12.5 MWD/CT (metawatt days per central metric ton of uranium) at <30 C.
12	163	E	1954	17-310			The virgin specimen exposed to neutron irradiation of 22.7 MWD/CT at <30 C.
13	163	E	1954	15-305			The virgin specimen exposed to neutron irradiation of 146 MWD/CT at <30 C.
14	163	E	1954	17-309			The virgin specimen exposed to neutron irradiation of 460 MWD/CT at <30 C.

DATA TABLE NO. 8 THERMAL CONDUCTIVITY OF SA-25 GRAPHITE

[Temperature, T, K; Thermal Conductivity, k, Watt $cm^{-1} K^{-1}$]

CURVE 1

T	k
10	0.000649
17	0.00172
23.5	0.00335
27.5	0.00502
30	0.00649
40	0.0107
51	0.0176
62	0.0243
70	0.0335
80	0.0402
90	0.0523
130	0.0858
205	0.128*

CURVE 2

T	k
21.5	0.00222
24.5	0.00335
30	0.00460
35	0.00649
45	0.0105
50	0.0138
64	0.0234
74	0.0259
90	0.0343
101	0.0418

CURVE 3

T	k
21	0.00176
24	0.00218
35	0.00481
40	0.00690
47	0.00879
54	0.0107
61	0.0129
70	0.0159
80	0.0205
94	0.0259
115	0.0343

CURVE 4

T	k
15	0.000795
17.5	0.00117
29	0.00243
34	0.00301
42	0.00427
46	0.00523
53	0.00607
61	0.00732
70	0.00920
82	0.0113
96	0.0138
120	0.0167

CURVE 5

T	k
12.5	0.000356*
20.5	0.000669
27	0.000941
31	0.00121
36	0.00146
41	0.00180
50	0.00238
63	0.00343

CURVE 6

T	k
46	0.00226
53	0.00264
62	0.00368
70	0.00439
80	0.00544
95	0.00711
112	0.00795

CURVE 7

T	k
10	0.000628
17	0.00176
23.5	0.00339
28	0.00502
31	0.00628
39	0.0107
49	0.0176

CURVE 7 (cont.)

T	k
60	0.0251
69	0.0343
77	0.0406
89	0.0523
108	0.0669
130	0.0858
150	0.0983
200	0.130
245	0.157
290	0.176

CURVE 8

T	k
1246	0.237
1335	0.209
1420	0.198
1485	0.195
1555	0.175
1605	0.202
1620	0.196
1665	0.193
1710	0.190
1770	0.191
1835	0.187*
1880	0.190
1985	0.175*
2045	0.170*

CURVE 9

T	k
13	0.00170
14	0.00200
15	0.00240
30	0.00900
80	0.0500
150	0.150
190	0.215
275	0.300

CURVE 10

T	k
10.1	0.000506
16.0	0.00163

CURVE 10 (cont.)

T	k
17.9	0.00217
19.4	0.00339
22.3	0.00475
26.1	0.00482
29.4	0.00981
35.0	0.0147
42.5	0.0209
50.8	0.0306
59.0	0.0418
70.0	0.0553
82.4	0.0715
96.2	0.0916
119	0.117
142	0.138
168	0.167
198	0.186
230	0.209
273	0.236
306	0.245

CURVE 11

T	k
18.0	0.00160
19.2	0.00201
21.5	0.00285
25.0	0.00419
29.5	0.00690
35.0	0.00987
44.7	0.0167
50.2	0.0227
63.0	0.0379
72.4	0.0427
89.0	0.0556
103	0.0690
122	0.0854
149	0.105
179	0.126
223	0.149
256	0.166
302	0.185

CURVE 12

T	k
17.0	0.00113*
19.9	0.00167
20.9	0.00226*
23.9	0.00305*
29.3	0.00502
35.0	0.00761
40.5	0.0110*
46.5	0.0140
53.1	0.0176
59.1	0.0213
68.7	0.0269
79.9	0.0343
94.0	0.0427
114	0.0554
142	0.0686
167	0.0803
198	0.0920
240	0.112
273	0.123
310	0.132

CURVE 13

T	k
15.4	0.000858*
17.0	0.00136
25.0	0.00285
29.0	0.00358
31.9	0.00460
40.0	0.00679*
46.0	0.00837*
51.0	0.00991*
58.9	0.0118
68.9	0.0149
81.0	0.0185
96.0	0.0231
118	0.0289
140	0.0343
168	0.0393
195	0.0448
232	0.0510
270	0.0582
305	0.0594

CURVE 14

T	k
16.8	0.000636
18.0	0.000682
20.0	0.00102
24.0	0.00111
29.3	0.00121
32.5	0.00251
39.6	0.00318
44.1	0.00349
54.0	0.00423*
61.0	0.00628
69.1	0.00733
80.0	0.00886*
95.0	0.0117
113.6	0.0134
139	0.0166
164	0.0205
208	0.0281
233	0.0318
275	0.0374
309	0.0418*

*Not shown on plot

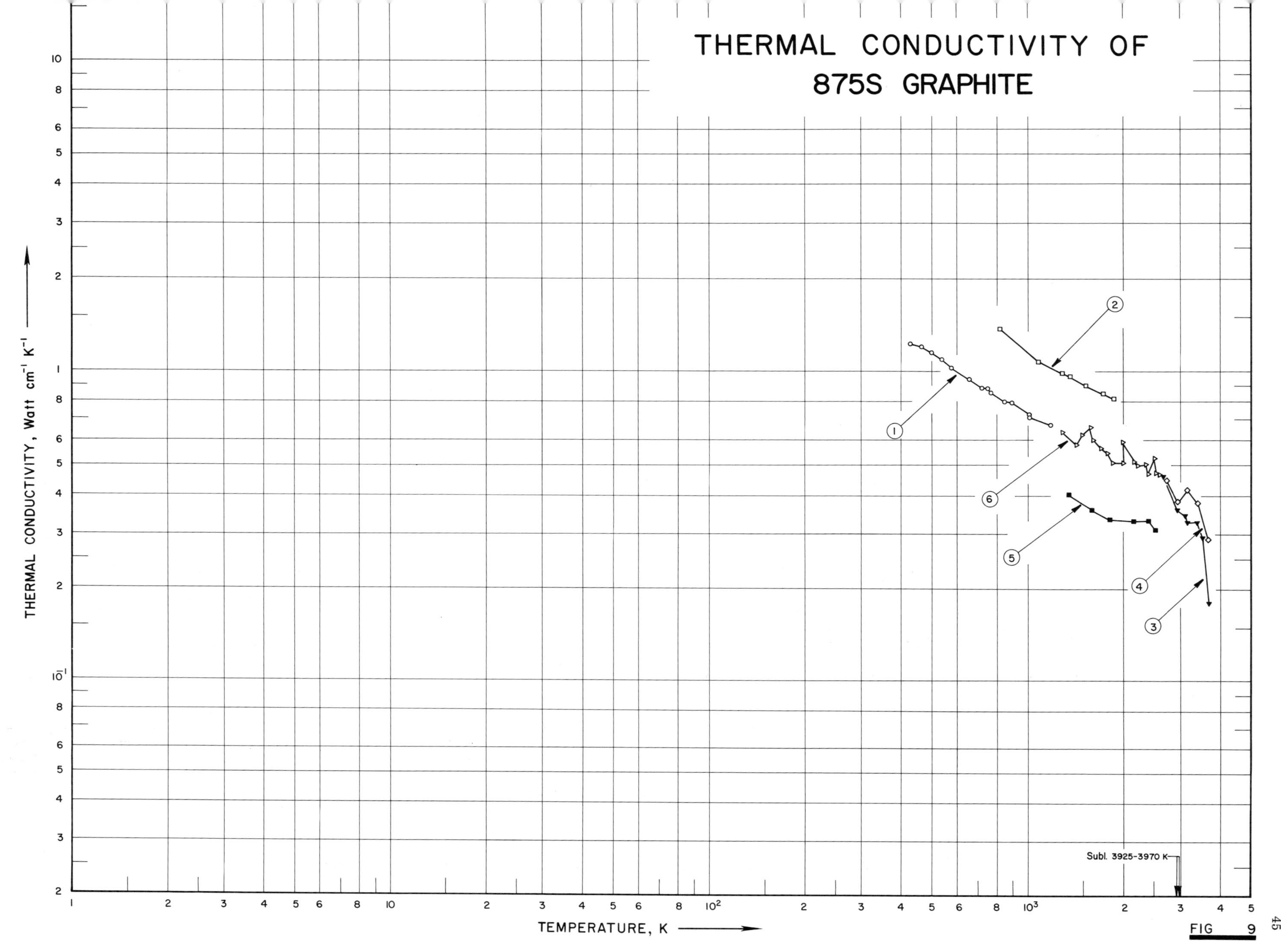
THERMAL CONDUCTIVITY OF
875S GRAPHITE
THERMAL CONDUCTIVITY, Watt cm⁻¹ K⁻¹
TEMPERATURE, K
Subl. 3925-3970 K
1
2
3
4
5
6

FIG 9

SPECIFICATION TABLE NO. 9 THERMAL CONDUCTIVITY OF 875S GRAPHITE

[For Data Reported in Figure and Table No. 9]

Curve No.	Ref. No.	Method Used	Year	Temp. Range, K	Reported Error, %	Name and Specimen Designation	Composition (weight percent), Specifications and Remarks
1	55	C	1956	433-1182			Extruded graphite; density 1.71 g cm^{-3}; measured perpendicular to the direction of extrusion.
2	176	L	1956	820-1865			Grade 7087 graphite from Speer Carbon Co.; density 1.698 g cm^{-3}; measured with heat flow parallel to the axis of extrusion.
3	177	R	1957	2661-3708	±8		Extruded; coarse grain with small voids and fissures; specific gravity 1.63; anisotropy ratio (ratio of electrical resistances measured normal and parallel to the extrusion axis) = 1.19; measured normal to the extrusion axis in inert gas at >150 psi pressure.
4	177	R	1957	2733-3694	±8		Rerun of the above specimen with smaller heat rate.
5	177	R	1957	1351-2527	±5		The above specimen measured with heat flow radially inward.
6	177	R	1957	1289-2600	±5		The above specimen measured after prolonged heating at >2200 C.

DATA TABLE NO. 9 THERMAL CONDUCTIVITY OF 875S GRAPHITE

[Temperature, T, K; Thermal Conductivity, k, Watt $cm^{-1} K^{-1}$]

T	k
CURVE 1	
433.2	1.23
467.6	1.20
503.2	1.15
540.9	1.09
578.7	1.03
658.7	0.943
662.1	0.950*
723.7	0.883
754.8	0.884
772.6	0.852
849.8	0.796
863.2	0.798*
895.4	0.794
1015.9	0.725
1015.9	0.710
1182.1	0.673
CURVE 2	
820.1	1.37
1084.0	1.08
1287.9	0.985
1366.2	0.964
1538.8	0.897
1728.6	0.847
1864.7	0.817
CURVE 3	
2661.0	0.459
2953.2	0.360
3137.1	0.346
3181.0	0.329
3406.0	0.329
3555.4	0.292
3708.2	0.180
CURVE 4	
2733.2	0.447
2960.9	0.384
3183.2	0.419
CURVE 4 (cont.)	
3422.1	0.379
3694.3	0.289
CURVE 5	
1351.0	0.403
1594.8	0.360
1818.2	0.336
2158.7	0.331
2395.4	0.332
2527.1	0.310
CURVE 6	
1288.7	0.639
1427.6	0.582
1483.2	0.625
1588.7	0.663
1610.9	0.604
1705.4	0.568
1783.2	0.545
1855.4	0.509
1999.5	0.509
1999.5	0.594
2166.5	0.512
2222.1	0.498
2355.4	0.504
2394.3	0.470
2505.4	0.528
2533.2	0.474
2599.8	0.466

*Not shown on plot

FIGURE AND TABLE NO. 9R RECOMMENDED THERMAL CONDUCTIVITY OF 875S GRAPHITE

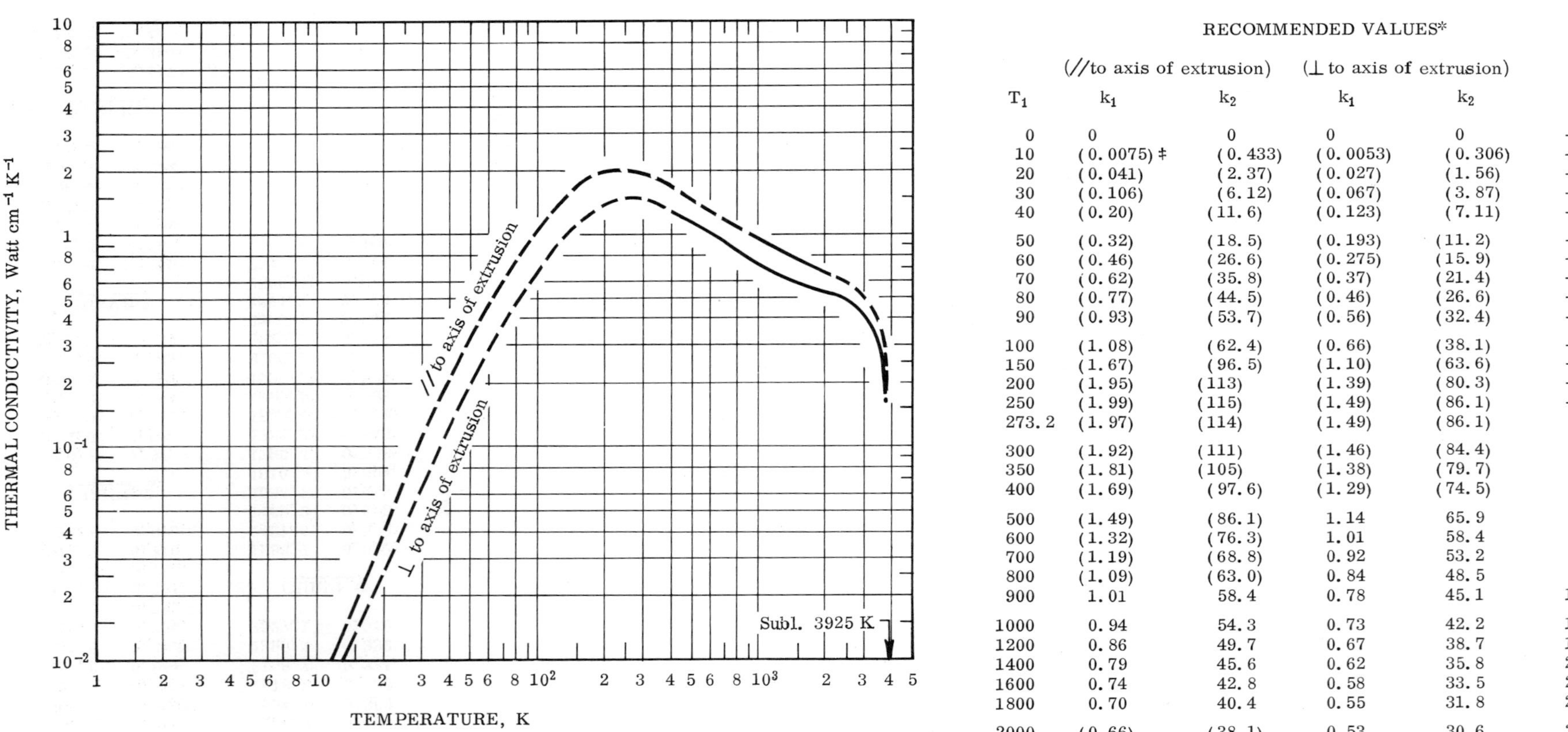

RECOMMENDED VALUES*

T_1	k_1 (// to axis of extrusion)	k_2 (// to axis of extrusion)	k_1 (⊥ to axis of extrusion)	k_2 (⊥ to axis of extrusion)	T_2
0	0	0	0	0	-459.7
10	(0.0075)‡	(0.433)	(0.0053)	(0.306)	-441.7
20	(0.041)	(2.37)	(0.027)	(1.56)	-423.7
30	(0.106)	(6.12)	(0.067)	(3.87)	-405.7
40	(0.20)	(11.6)	(0.123)	(7.11)	-387.7
50	(0.32)	(18.5)	(0.193)	(11.2)	-369.7
60	(0.46)	(26.6)	(0.275)	(15.9)	-351.7
70	(0.62)	(35.8)	(0.37)	(21.4)	-333.7
80	(0.77)	(44.5)	(0.46)	(26.6)	-315.7
90	(0.93)	(53.7)	(0.56)	(32.4)	-297.7
100	(1.08)	(62.4)	(0.66)	(38.1)	-279.7
150	(1.67)	(96.5)	(1.10)	(63.6)	-189.7
200	(1.95)	(113)	(1.39)	(80.3)	- 99.7
250	(1.99)	(115)	(1.49)	(86.1)	- 9.7
273.2	(1.97)	(114)	(1.49)	(86.1)	32.0
300	(1.92)	(111)	(1.46)	(84.4)	80.3
350	(1.81)	(105)	(1.38)	(79.7)	170.3
400	(1.69)	(97.6)	(1.29)	(74.5)	260.3
500	(1.49)	(86.1)	1.14	65.9	440.3
600	(1.32)	(76.3)	1.01	58.4	620.3
700	(1.19)	(68.8)	0.92	53.2	800.3
800	(1.09)	(63.0)	0.84	48.5	980.3
900	1.01	58.4	0.78	45.1	1160
1000	0.94	54.3	0.73	42.2	1340
1200	0.86	49.7	0.67	38.7	1700
1400	0.79	45.6	0.62	35.8	2060
1600	0.74	42.8	0.58	33.5	2420
1800	0.70	40.4	0.55	31.8	2780
2000	(0.66)	(38.1)	0.53	30.6	3140
2200	(0.63)	(36.4)	0.51	29.5	3500
2400	(0.60)	(34.7)	0.49	28.3	3860
2600	(0.57)	(32.9)	0.47	27.2	4220
2800	(0.54)	(31.2)	0.45	26.0	4580
3000	(0.51)	(29.5)	0.42	24.3	4940
3200	(0.47)	(27.2)	0.39	22.5	5300
3400	(0.42)	(24.3)	0.35	20.2	5660
3600	(0.34)	(19.6)	0.28	16.2	6020
3800	(0.19)	(11.0)	(0.16)	(9.24)	6380

REMARKS

875S graphite is a medium-grain (0.032 inch maximum) pitch-bonded petroleum-coke-base graphite with typical room-temperature density 1.67 g cm^{-3}. It is formed by extrusion into rods and is produced by Speer Carbon Company. This graphite was previously designated as 7087 graphite. The uncertainty of the recommended values for the direction perpendicular to the axis of extrusion that are supported by experimental data is probably of the order of ±10 to ±20%, and that of the values obtained by extensive extrapolation is probably twice as great. The values for the direction parallel to the axis of extrusion are intended only for indicating the general trend.

* T_1 in K, k_1 in Watt cm^{-1} K^{-1}, T_2 in F, and k_2 in Btu hr^{-1} ft^{-1} F^{-1}.

‡ Values in parentheses are extrapolated or estimated.

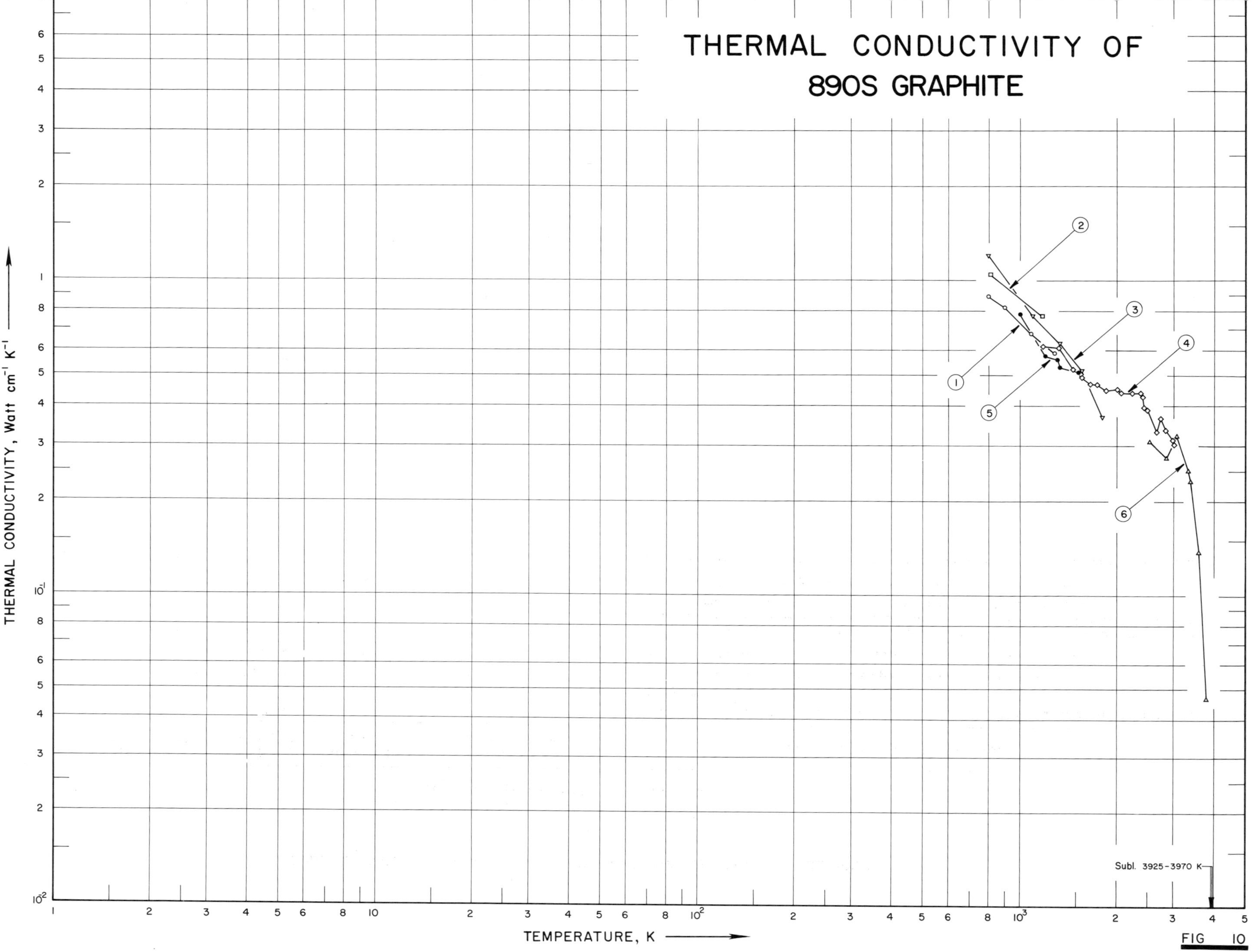
THERMAL CONDUCTIVITY OF
890S GRAPHITE
THERMAL CONDUCTIVITY, Watt cm^{-1} K^{-1}
TEMPERATURE, K
Subl. 3925-3970 K
1
2
3
4
5
6

FIG 10

SPECIFICATION TABLE NO. 10 THERMAL CONDUCTIVITY OF 890S GRAPHITE

[For Data Reported in Figure and Table No. 10]

Curve No.	Ref. No.	Method Used	Year	Temp. Range, K	Reported Error, %	Name and Specimen Designation	Composition (weight percent), Specifications and Remarks
1	48	L	1956	832-1284	±5	1	7 in. dia x 1.5 in. thick; density 1.612 g cm^{-3}; measured with unidirectional heat flow through the disk.
2	48	L	1956	813, 1172	±5	2	Similar to the above specimen but the specimen being heated twice.
3	48	L	1956	798-1809	±5	3	Similar to the above specimen but the specimen being heated three times.
4	177	R	1957	1174-3017	±5		Extruded; very fine grained and uniform; specific gravity 1.67; anisotropy ratio 1.08; measured with heat flow radially inward and normal to the extrusion axis; pyrometer used to measure temp; measured in inert gas at >150 psi pressure.
5	177	R	1957	1002-1533	±5		The above specimen measured by using thermocouples to obtain temp.
6	177	R	1957	2540-3786	±5		The above specimen measured with heat flow radially outward and normal to the extrusion axis.

DATA TABLE NO. 10 THERMAL CONDUCTIVITY OF 890S GRAPHITE

[Temperature, T, K; Thermal Conductivity, k, Watt $cm^{-1} K^{-1}$]

T	k	T	k
CURVE 1		CURVE 5	
831.5	0.891	1002.1	0.781
897.1	0.824	1207.1	0.575
1083.2	0.678	1313.2	0.562
1284.3	0.588	1335.4	0.531
		1533.2	0.514
CURVE 2			
		CURVE 6	
812.6	1.05		
1172.1	0.772	2539.9	0.310
		2853.2	0.275
CURVE 3		3073.1	0.322
		3348.7	0.251
797.6	1.20	3396.0	0.232
1099.8	0.774	3598.2	0.138
1330.4	0.633	3786.0	0.0467
1553.2	0.516		
1808.7	0.367		
CURVE 4			
1173.7	0.616		
1322.6	0.614		
1466.5	0.524		
1557.6	0.493		
1655.4	0.469		
1749.3	0.469		
1860.9	0.448		
2005.4	0.452		
2069.3	0.440		
2244.3	0.440		
2374.9	0.440		
2410.9	0.427		
2437.1	0.395		
2494.3	0.389		
2661.5	0.331		
2744.3	0.365		
2847.1	0.336		
2996.5	0.313		
3016.5	0.303		

FIGURE AND TABLE NO. 10R RECOMMENDED THERMAL CONDUCTIVITY OF 890S GRAPHITE

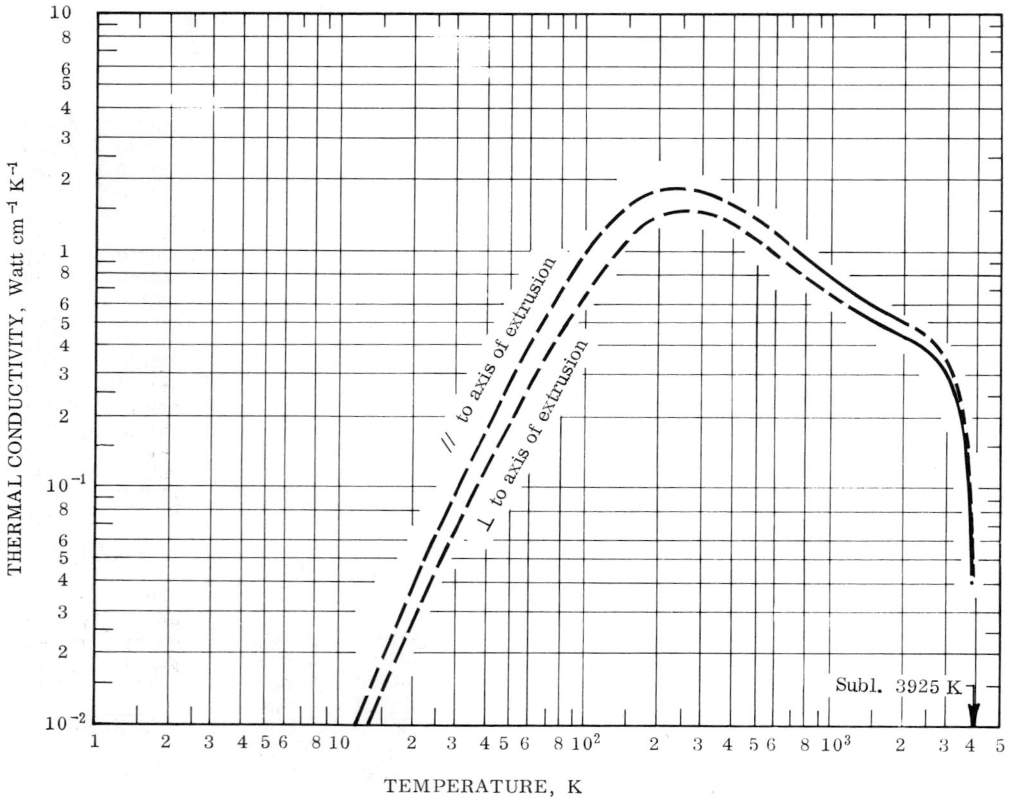

RECOMMENDED VALUES*

T_1	k_1 (// to axis of extrusion)	k_2 (// to axis of extrusion)	k_1 (⊥ to axis of extrusion)	k_2 (⊥ to axis of extrusion)	T_2
0	0	0	0	0	-459.7
10	(0.007) ‡	(0.404)	(0.0054)	(0.312)	-441.7
20	(0.038)	(2.20)	(0.028)	(1.62)	-423.7
30	(0.097)	(5.60)	(0.069)	(3.99)	-405.7
40	(0.18)	(10.4)	(0.127)	(7.34)	-387.7
50	(0.29)	(16.8)	(0.20)	(11.6)	-369.7
60	(0.42)	(24.3)	(0.29)	(16.8)	-351.7
70	(0.56)	(32.4)	(0.38)	(22.0)	-333.7
80	(0.71)	(41.0)	(0.48)	(27.7)	-315.7
90	(0.85)	(49.1)	(0.58)	(33.5)	-297.7
100	(1.00)	(57.8)	(0.68)	(39.3)	-279.7
150	(1.55)	(89.6)	(1.14)	(65.9)	-189.7
200	(1.83)	(106)	(1.42)	(82.0)	- 99.7
250	(1.89)	(109)	(1.52)	(87.8)	- 9.7
273.2	(1.87)	(108)	(1.51)	(87.2)	32.0
300	(1.82)	(105)	(1.48)	(85.5)	80.3
350	(1.71)	(98.8)	(1.40)	(80.9)	170.3
400	(1.59)	(91.9)	(1.32)	(76.3)	260.3
500	(1.38)	(79.7)	(1.16)	(67.0)	440.3
600	(1.21)	(69.9)	(1.01)	(58.4)	620.3
700	(1.08)	(62.4)	(0.90)	(52.0)	800.3
800	0.97	56.0	(0.82)	(47.4)	980.3
900	0.88	50.8	(0.75)	(43.3)	1160
1000	0.81	46.8	(0.69)	(39.9)	1340
1200	0.70	40.4	0.60	34.7	1700
1400	0.63	36.4	0.54	31.2	2060
1600	0.57	32.9	0.50	28.9	2420
1800	0.53	30.6	0.46	26.6	2780
2000	(0.50)	(28.9)	0.44	25.4	3140
2200	(0.48)	(27.7)	0.42	24.3	3500
2400	(0.45)	(26.0)	0.40	23.1	3860
2600	(0.42)	(24.3)	0.38	22.0	4220
2800	(0.39)	(22.5)	0.35	20.2	4580
3000	(0.36)	(20.8)	0.32	18.5	4940
3200	(0.31)	(17.9)	0.28	16.2	5300
3400	(0.25)	(14.4)	0.23	13.3	5660
3600	(0.16)	(9.24)	0.15	8.67	6020
3800	(0.043)	(2.48)	(0.04)	(2.31)	6380

REMARKS

890S graphite is a fine-grain (0.008 inch maximum) pitch-bonded petroleum-coke-base graphite with typical room-temperature density 1.63 g cm^{-3}. It is formed by extrusion into rods and is produced by Speer Carbon Company. This graphite was previously designated as 3474D graphite. The uncertainty of the recommended values at temperatures above 1000 K that are supported by experimental data is probably of the order of ±10 to ±20%, and that of the values obtained by extensive extrapolation is probably twice as great. The values below 1000 K are intended only for indicating the general trend.

* T_1 in K, k_1 in Watt cm^{-1} K^{-1}, T_2 in F, and k_2 in Btu hr^{-1} ft^{-1} F^{-1}.

‡ Values in parentheses are extrapolated or estimated.

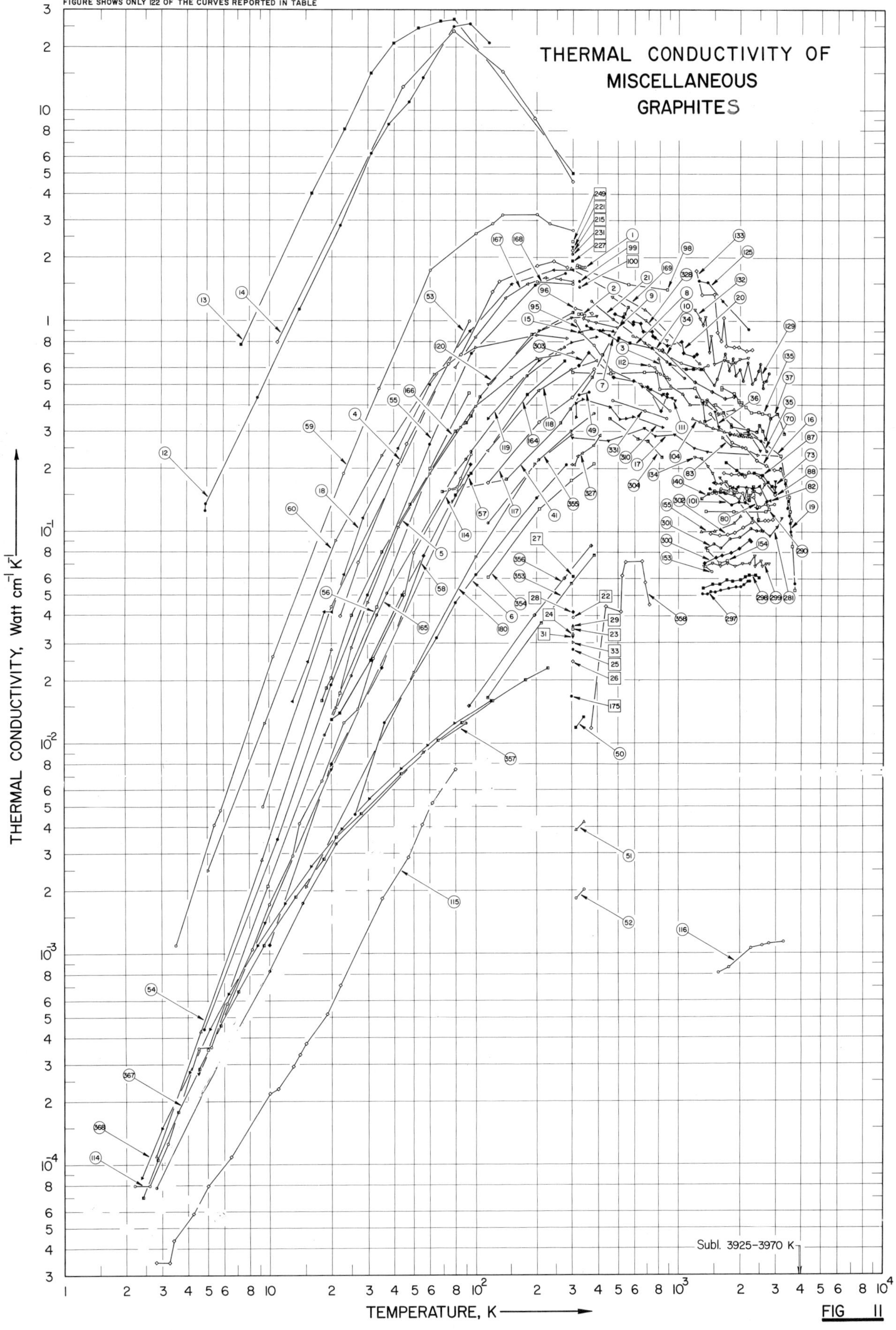
FIGURE SHOWS ONLY 122 OF THE CURVES REPORTED IN TABLE
THERMAL CONDUCTIVITY OF MISCELLANEOUS GRAPHITES
THERMAL CONDUCTIVITY, Watt cm⁻¹ K⁻¹
TEMPERATURE, K
Subl. 3925–3970 K
FIG 11

SPECIFICATION TABLE NO. 11 THERMAL CONDUCTIVITY OF MISCELLANEOUS GRAPHITES

[For Data Reported in Figure and Table No. 11]

Curve No.	Ref. No.	Method Used	Year	Temp. Range, K	Reported Error, %	Name and Specimen Designation	Composition (weight percent), Specifications and Remarks
1	47	L	1944	317-345		No. 1583	Specimen 2.607 cm long and circular cross sectional area 5.068 cm^2; measured lengthwise.
2	47	L	1944	321-344		No. 1583	Similar to the above specimen but only 2.523 cm long and measured crosswise.
3	48	L	1956	789-1869	± 5	GBE	Specimen 7 in. in dia and 1.5 in. thick; density 1.596 g cm^{-3}; measured with unidirectional heat flow through the disk.
4	49	L	1953	22-280	<10	Grade CS, Sample A	Grade CS graphite (conventional coke base, pitch bonded and extruded); polycrystal; form National Carbon Co.; bulk density $\sim$1.70 g cm^{-3}; specimen axis perpendicular to the preferred c-axis orientation.
5	49	L	1953	21-300	<10	Sample C	Polycrystal; natural graphite base, pitch bonded and molded; bulk density $\sim$1.80 g cm^{-3}; specimen axis perpendicular to the preferred c-axis orientation.
6	49	L	1953	26-280	<10	Sample D	Similar to the above specimen but pitch bonded and molded from lampblack; bulk density $\sim$1.65 g cm^{-3}.
7	8	L	1960	351-497	3	Grade RT-0003 (Sample 1)	Specimen cut from a RT-0003 graphite block (National Carbon Co.); density $\sim$1.90 g cm^{-3}; heat flow perpendicular to grain orientation.
8	8	L	1960	500-1294	10	Grade RT-0003 (Sample 2)	Similar to the above specimen.
9	8	L	1960	339-495	3	Grade RT-0003 (Sample 3)	Similar to the above specimen but heat flow parallel to grain orientation; run No. 1.
10	8	L	1960	587-1394	10	Grade RT-0003 (Sample 3)	Second run of the above specimen.
11	8	L	1960	612-1384	10	Grade RT-0003 (Sample 3)	Third run of the above specimen.
12	358, 50	C	1954	4.9-116	5-10	Canadian Natural Graphite	Large crystallite (in the order of 10^{-2} cm); very low ash content; specimen size 0.25 x 0.05 x 0.01 in.; grade AWG graphite used as comparative material.
13	358, 50	C	1954	7.3-300	5-10	Canadian Natural Graphite	Similar to the above specimen.
14	358, 50	C	1954	11-299	5-10	Canadian Natural Graphite	Similar to the above specimen.
15	19	C	1956	323-473		Commercial graphite	High purity; specimen (tubular) 4.5 cm long, 0.95 cm O.D., and 0.75 cm I.D.; density 1.65 g cm^{-3}; electrical resistivity at 20, 50, 100, 150, and 200 C being respectively, 762, 760, 772, 790, and 816 μohm cm; a bar of iron of known thermal conductivity used as comparative material.
16	52	E	1956	300-3710	20	Spektral kohle 1	Large grained artificial graphitized carbon; measured in vacuum.
17	52	E	1956	300-3710	20	Spektral kohle 1	Fine grained artificial graphitized carbon; measured in vacuum.

SPECIFICATION TABLE NO. 11 (continued)

Curve No.	Ref. No.	Method Used	Year	Temp. Range, K	Reported Error, %	Name and Specimen Designation	Composition (weight percent), Specifications and Remarks
18	53	E	1955	13-300	± 5	Natural Ceylon block	Natural Ceylon graphite; size 0.100 x 0.020 x 1.25 in., skew orientation.
19	54	E	1952	3150-3700			Manufactured graphite rod.
20	55	C	1956	484-1227		Grade GBH	Molded graphite; from National Carbon Co.; density 1.75 g cm^{-3}; measured perpendicular to the direction of molding; Armco iron used as comparative material.
21	364	C	1954	373-1073			Polycrystal; bulk density 1.55 g cm^{-3}; porosity 30.2%; dense Al_2O_3 used as comparative material.
22	56		1957	303.2		Korite	Manufactured from Korite petroleum asphalt (from Standard Oil Co., Indiana) and coke prepared from this asphalt; irradiated by exposing to neutrons of 150 MWD/T (mega watt-days per ton).
23	56		1957	303.2		Korite	The above specimen with neutron exposure of 325 MWD/T.
24	56		1957	303.2		Korite	The above specimen with neutron exposure of 830 MWD/T.
25	56		1957	303.2		Korite	The above specimen with neutron exposure of 1100 MWD/T.
26	56		1957	303.2		Korite	The above specimen with neutron exposure of 4270 MWD/T.
27	56		1957	303.2		CSF	Made from Cleves coke (Gulf oil Co.) with standard pitch (Barrett No. 2, medium hard coal tar pitch); purified; exposed to neutrons of 500 MWD/T.
28	56		1957	303.2		CSF	The above specimen with exposure of 1000 MWD/T.
29	56		1957	303.2		CSF	The above specimen with exposure of 1500 MWD/T.
30	56		1957	303.2		CSF	The above specimen with exposure of 2000 MWD/T.
31	56		1957	303.2		CSF	The above specimen with exposure of 2500 MWD/T.
32	56		1957	303.2		CSF	The above specimen with exposure of 3000 MWD/T.
33	56		1957	303.2		CSF	The above specimen with exposure of 3500 MWD/T.
34	176	L	1956	829-1866	5	GBH	Grade GBH graphite from National Carbon Co.; density 1.762 g cm^{-3}; measured with heat flow parallel to the axis of extrusion (should be axis of molding since it was molded).
35	177	R	1957	1220-2700	± 8	GBE	Extruded; extremely coarse grained and fragile; voids and fissures up to 0.125 in. in dia; specific gravity 1.57; anisotropy ratio 1.18; measured normal to the extrusion axis in the heating-up period, in inert gas at > 150 psi pressure.
36	177	R	1957	1510-2507	± 8	GBE	The above specimen in the cooling-down period.
37	177	R	1957	1319-3277	± 8	GBH	Molded; very fine grained and uniform; specific gravity 1.77; anisotropy ratio 0.78; measured normal to the molding pressure, in inert gas at > 150 psi pressure.
38	133	C	1954	353-1093		Grade CS	Cubic specimen 1 x 1 x 1 in.; density 1.55 g cm^{-3}; dense-alumina used as comparative material.

SPECIFICATION TABLE NO. 11 (continued)

Curve No.	Ref. No.	Method Used	Year	Temp. Range, K	Reported Error, %	Name and Specimen Designation	Composition (weight percent), Specifications and Remarks
39	133	C	1954	328-1093		Grade CS	Similar to the above specimen but with cylindrical pores 0.146 cm in dia; porosity 9.8%.
40	133	C	1954	378-1123		Grade CS	Similar to the above specimen but the porosity 19.6%.
41	179	L	1912	352-828	0.1		Specimen 18 mm in dia and 0.79 cm long.
42	106	C	1956	328.7	3	Domestic (Japan) No. 1	Artificial graphite electrode; 80 mm dia x 125 mm long; apparent density 1.501 g cm^{-3}; electrical resistivity 0.00108 ohm cm; copper used as comparative material.
43	106	C	1956	329.7	3	Domestic (Japan) No. 2	Similar to the above specimen but the apparent density 1.520 g cm^{-3}; electrical resistivity 0.00118 ohm cm.
44	106	C	1956	326.7	3	Domestic (Japan) No. 3	Similar to the above specimen but the apparent density 1.533 g cm^{-3}; electrical resistivity 0.00093 ohm cm.
45	106	C	1956	331.7	3	Domestic (Japan) No. 4	Similar to the above specimen but the apparent density 1.59 g cm^{-3}; electrical resistivity 0.00085 ohm cm.
46	106	C	1956	337.2	3	Domestic (Japan) No. 5	Similar to the above specimen but the apparent density 1.586 g cm^{-3}; electrical resistivity 0.00094 ohm cm.
47	106	C	1956	344.7	3	Domestic (Japan) No. 6	Similar to the above specimen but the apparent density 1.591 g cm^{-3}; electrical resistivity 0.00096 ohm cm.
48	106	C	1956	337.2	3	Domestic (Japan) No. 7	Similar to the above specimen but the apparent density 1.60 g cm^{-3}; electrical resistivity 0.00099 ohm cm.
49	108	L	1920	323,363			Solid specimen 1.04 in. thick; specific gravity 1.56.
50	108	L	1920	313,343			Powder (through 20-mesh on 40-mesh) specimen 0.476 in. thick; specific gravity 0.70.
51	108	L	1920	313,343			Powder (through 40-mesh) specimen 0.476 in. thick; specific gravity 0.42.
52	108	L	1920	313,343			Powder (through 100-mesh) specimen 0.476 in. thick; specific gravity 0.48.
53	158	L	1952	9.3-93	2	I	Artificial graphite; made by extrusion which produced a slight anisotropy; crystal size (perpendicular to c-axis) 2000Å; density 1.80 g cm^{-3}; electrical resistivity 1.09 and 0.6 milliohm cm at 90 and 290 K respectively; measured parallel to the axis of extrusion.
54	158	L	1952	2.8-20	2	I	Similar to the above specimen but the density 1.78 g cm^{-3}; electrical resistivity 1.76 and 1.09 milliohm cm at 90 and 290 K respectively; measured perpendicular to the axis of extrusion.
55	158	L	1952	4.8-275	2	II	Similar to the above specimen but the crystal size 1000 Å; density 1.60 g cm^{-3}; electrical resistivity at 4, 20, 90 and 290 K being respectively 2.3, 2.3, 1.7, and 1.08 milliohm cm; measured parallel to the axis of extrusion.

SPECIFICATION TABLE NO. 11 (continued)

Curve No.	Ref. No.	Method Used	Year	Temp. Range, K	Reported Error, %	Name and Specimen Designation	Composition (weight percent), Specifications and Remarks
56	158	L	1952	5-93	2	II	The above specimen measured perpendicular to the axis of extrusion; electrical resistivity at 4, 20, 90, and 290 K being respectively, 3.0, 2.9, 2.2, and 1.35, milliohm cm.
57	158	L	1952	4.5-93	2	III	Similar to the above specimen but the crystal size 300Å; density 1.77 g cm^{-3}; electrical resistivity at 90 and 290 K being 3.01 and 2.33 milliohm cm; measured parallel to the extrusion axis.
58	158	L	1952	10-95	2	III	Similar to the above specimen but the density 1.76 g cm^{-3}; electrical resistivity 3.91 and 2.77 milliohm cm at 90 and 290 K respectively; measured perpendicular to the extrusion axis.
59	158	L	1952	3.5-300	2	IV	Natural graphite; highly anisotropic; crystal size 2000Å; density $\sim$2.25 g cm^{-3}; electrical resistivity 1.16, 1.17, 1.21 and 0.98 milliohm cm at 4, 20, 90, and 290 K respectively; measured perpendicular to the preferred direction of c-axis.
60	158	L	1952	5.0-280	2	IV	The above specimen measured parallel to c-axis; electrical resistivity at 4, 20, 90, and 290 K being respectively 5.3, 5.4, 5.4, and 4.1 milliohm cm.
61	359	L	1949	336.2	5	AGHT, III	Cylindrical specimen; 3.5 in. dia x 4 in. long; cylinder axis parallel to the extrusion axis.
62	359	L	1949	336.2	5	AGHT, III	Similar to the above specimen but the cylinder axis at right angle to the extrusion axis.
63	359	L	1949	339.2	5	AGHT, III	Similar to the above specimen but the direction of cutting the specimen perpendicular to the above specimen.
64	359	L	1949	343.2	5	AGHT, IV	Similar to the above specimen but the cylinder axis parallel to the axis of extrusion.
65	359	L	1949	337.2	5	AGHT, IV	Similar to the above specimen but the cylinder axis perpendicular to the extrusion axis.
66	359	L	1949	344.2	5	AGHT, IV	Similar to the above specimen with the cylinder axis perpendicular to the extrusion axis but the direction of cutting the specimen perpendicular to that of the above specimen.
67	359	L	1949	343.2	5	AGHT, VII	Similar to the above specimen but the cylinder axis parallel to the extrusion axis.
68	359	L	1949	341.2	5	AGHT, VII	Similar to the above specimen but the cylinder axis perpendicular to the extrusion axis.
69	359	L	1949	344.2	5	AGHT, VII	Similar to the above specimen with the cylinder axix perpendicular to the extrusion axis but the direction of cutting of the specimen normal to the above specimen.
70	180	L	1961	1428-3148		AGSR	Graphite rod from National Carbon Company; apparent density 1.54 g cm^{-3} (grade AGSR); heat treated to 3100 C; measured at 1 in. Hg above the atmospheric pressure.
71	180	L	1961	1838.2		AGSR	The above specimen measured at a fixed temperature to show the effect of pressure (approx. from 0 to 60 in. Hg pressure).
72	180	L	1961	2423.2		AGSR	The above specimen measured within the same pressure range but at a higher temperature.

SPECIFICATION TABLE NO. 11 (continued)

Curve No.	Ref. No.	Method Used	Year	Temp. Range, K	Reported Error, %	Name and Specimen Designation	Composition (weight percent), Specifications and Remarks
73	180	L	1961	2973.2		AGSR	The above specimen measured within the same pressure range but at a higher temperature.
74	180	L	1961	1838.2		Lab. prepared rod	Made from soft filler-soft binder mixture particles (200/270 mesh size); heat treated to 3100 C; apparent density 1.58 g cm^{-3}; measured in the same pressure range as the above specimen.
75	180	L	1961	2394.2		Lab. prepared rod	The above specimen measured in the same pressure range at a higher temperature.
76	180	L	1961	2913.2		Lab. prepared rod	The above specimen measured in the same pressure range at a higher temperature.
77	180	L	1961	1829.2		Lab. prepared rod	Made from a soft filler-soft binder mixture; coke (28/35 mesh size) used as filler; very porous; apparent density 1.25 g cm^{-3}; measured in the same pressure range as the above specimen.
78	180	L	1961	2433.2		Lab. prepared rod	The above specimen measured under pressures ranging from 0 to 55.5 in. Hg.
79	180	L	1961	2973.2		Lab. prepared rod	The above specimen measured under pressures ranging from 31 to 55.5 in. Hg.
80	180	L	1961	1473-2933		Test Rod No. 1	Material from National Carbon Co.; graphitized to 3000 C; the pressure within the test chamber kept at 1-2 in. Hg above atmospheric pressure by releasing or admitting argon at various temperature levels.
81	180	L	1961	1773-2523		Test Rod No. 1	Second run of the above specimen.
82	180	L	1961	1478-2968		Test Rod No. 1	Third run of the above specimen.
83	180	L	1961	1643-2433		Test Rod No. 2 (U.B. carbon)	Specimen made from 50 parts of Texas coke (65/100 mesh as the first filler), another 50 parts of Texas coke (200/270 mesh as the second filler), and 40 parts of M-30 coal tar pitch as the binder; extruded, graphitized to 3000 C; apparent density 1.53 g cm^{-3}; measured in argon atmosphere at 1-2 in. Hg above atmospheric pressure.
84	180	L	1961	1513-2933		Test Rod No. 2 (U.B. carbon)	Second run of the above specimen.
85	180	L	1961	1983.2		Test Rod No. 2 (U.B. carbon)	Third run of the above specimen.
86	180	L	1961	1638-2448		Test Rod No. 2 (U.B. carbon)	Fourth run of the above specimen.
87	180	L	1961	1713-2983		Test Rod No. 3 (U.B. carbon)	Specimen made from 100 parts of 200/270 mesh size Texas coke as filler, 50 parts of M-30 coal tar pitch as binder; extruded; heat treated to 3000 C; apparent density 1.57 g cm^{-3}; measured in argon atmosphere at 1-2 in. Hg above the atmospheric pressure.
88	180	L	1961	1663-2993		Test Rod No. 3 (U.B. carbon)	Second run of the above specimen.

SPECIFICATION TABLE NO. 11 (continued)

Curve No.	Ref. No.	Method Used	Year	Temp. Range, K	Reported Error, %	Name and Specimen Designation	Composition (weight percent), Specifications and Remarks
89	180	L	1961	1783-3273		Test Rod No. 4 (U.B. carbon)	Specimen similarly prepared as the above with slight increase in density to 1.58 g cm^{-3}.
90	181		1959	298.2		TS-148	Specimen made by National Carbon Co.; baked to 1425 C; typical impurities after baking 0.15 ash, and 0.042 H; apparent density 1.682 g cm^{-3}; electrical resistivity 1557 μohm cm; measured with grain.
91	181		1959	298.2		TS-148	Similar to the above specimen but electrical resistivity 2594 μohm cm; measure against grain.
92	181		1959	298.2		TS-160	Similar to the above specimen but with 0.13 ash after baking; apparent density 1.685 g cm^{-3}; electrical resistivity 2122 μohm cm; measured with grain.
93	181		1959	298.2		TS-160	Similar to the above specimen but measured against grain; electrical resistivity 3006 μohm cm.
94	181		1959	298.2		TS-160	Similar to the above specimen but baked to 2800 C; apparent density 1.785 g cm^{-3}; electrical resistivity 1842 μohm cm; measured with grain.
95	159, 356	L	1955, 1942	308-903			Glycerine coated; specimen sandwiched between 2 copper disks; the heater being electrically operated.
96	159	L	1955	308, 373			Similar to the above specimen but being sandwiched between 2 silver disks.
97	159	C	1955	323.2			Glycerine coated graphite; boiling water used as heater; brass and steel used as comparative materials.
98	159	L	1955	313-873			Long graphite rod used as specimen; intended to eliminate errors due to uneven flow at heat into and out of the specimen.
99	159	C	1955	323.2		Karbate 2	Commercial impregnated graphite; brass used as the comparative material.
100	159	C	1955	323.2		Karbate 22	Similar to the above specimen.
101	182	R	1961	1422-2422		Sample A	Limited impregnated graphite normal to the extrusion axis; specimen in the form of a short tube with an outer dia of about 3 in. and a wall thickness of about 0.25 in.; experiment performed in helium for temperatures $<$ 1540 C, for temperatures higher than this, argon was used instead.
102	182	R	1961	1367-2255		Sample B	Similar to the above specimen but more fully impregnated.
103	182	R	1961	1417-2255		Sample C	Similar to the above specimen.
104	183	E	1961	1173-2273	<6		Spectrally pure; two thin rods each $\sim$ 1 mm in dia used as the test specimens; annealed in high vacuum at 1700 C for 1 hr; measured in high vacuum.
105	184	P	1961	1193			Measured in a vacuum of 10^{-5} mm Hg; run No. 1.
106	184	P	1961	1185			The above specimen run No. 2.
107	184	P	1961	1185			The above specimen run No. 3.

SPECIFICATION TABLE NO. 11 (continued)

Curve No.	Ref. No.	Method Used	Year	Temp. Range, K	Reported Error, %	Name and Specimen Designation	Composition (weight percent), Specifications and Remarks
108	184	P	1961	1194			The above specimen run No. 4.
109	184	P	1961	1189			The above specimen run No. 5.
110	184	P	1961	1189			The above specimen run No. 6.
111	185	R	1960	653-963		LBR (grade TSP)	Nuclear graphite grade TSP from Nat. Carbon Co.; irradiated with 5 x 10^{20} neutron cm^{-2} at about 315 C.
112	185	R	1960	703-898		LBR (grade TSP)	The above specimen annealed in vacuum at 1000 C for 1 hr.
113	185	R	1960	723-898		LBR (grade TSP)	The above specimen before irradiation and not annealed.
114	339	L	1956	2.2-95		AUG-4	Resin bonded graphite; annealed to 2000 C.
115	339	L	1956	2.8-80		AUG-3	Resin bonded graphite; annealed to 1500 C.
116	161	P	1960	1573-3273		Graphitized Carbon black	99.65 C, 0.27 H, 0.08 O and 0.01 ash; particle size < 1μ; heat treated at 2500 C for 30 min (equivalent to a degree of graphitization of 0.77).
117	15	L	1956	115-385			Polycrystalline; made from 69.14% Kendall coke (soft type carbon), 29.17% medium grade coal tar pitch, and 1.43% Vacwax 80 (from Socony Vacuum Oil Co.); extruded; baked for five days to 1100 C; density after baking 1.49 g cm^{-3}; heat treated again to 2100 C; crystallite dia 98 Å.
118	15	L	1956	115-385			Similar to the above specimen but heat treated to 2200 C; crystallite dia 128 Å.
119	15	L	1956	118-385			Similar to the above specimen but heat treated to 2300 C; crystallite dia 184 Å.
120	15	L	1956	115-385			Similar to the above specimen but heat treated to 2430 C; crystallite dia 290 Å.
121	366		1960	1170-2450		Sample No. 1 (R-0008)	Grade R-0008 (a high quality graphite).
122	366		1960	1170-2600		Sample No. 2 (R-0008)	Similar to the above specimen.
123	366		1960	1115-2725		Sample No. 3 (R-0008)	Similar to the above specimen.
124	340	→	1962	1260-2199	<10	ZT type graphite; G-5, G-9	Thermal conductivity data in the z-direction (k_z) determined simultaneously with thermal conductivity in the r-direction k_r (see next curve) from 4 cylindrical specimens made from ZT type graphite of National Carbon Co.; density 2.00 g cm^{-3}; anisotropy ratio of electrical resistivity ρ(z-direction)/ρ(r-direction) 2.86 at room temperature; the specimens each about 2.54 cm in dia and about 0.3-0.6 cm thick; during measurement the specimens were heated in vacuum by high frequency induction; thermal conductivity determined by equating the heat conduction in specimen to the heat loss by radiation assuming the emissivity of a gray body, the analysis required 2 specimens of different thickness to solve simultaneously for k_z and k_r at a certain temperature.

SPECIFICATION TABLE NO. 11 (continued)

Curve No.	Ref. No.	Method Used	Year	Temp. Range, K	Reported Error, %	Name and Specimen Designation	Composition (weight percent), Specifications and Remarks
125	340	→	1962	1260-2199	<10	ZT type graphite; G-5, G-9	k_r determined simultaneously with the above curve.
126	334	L	1963	1353-2303			Made from soft filler and hard binder carbon; heat treated to 2100 C for 15 min.
127	334	L	1963	1383-2583			The above specimen heat treated to 2400 C for 15 min.
128	335		1963	1170-2340		ZT type; No. 1	Specimen 7 cm long, 1 cm wide and 1 mm thick; made by molding a selected coke-base mixture in one particular direction; impregnated and pressed at high temperatures; all surfaces milled, slightly sandblasted; apparent density 2.15 g cm^{-3}; measured approx. parallel to the grain direction with a tilt angle of 8.1 degrees; room temperature anisotropy in electrical resistivity $\approx$7.0.
129	335		1963	1180-2760		ZT type; No. 1	Second run of the above specimen.
130	335		1963	1180-2400		ZT type; No. 2	Similar to the above specimen but measured perpendicular to the grain direction with a tilt angle of 8.1 degrees.
131	335		1963	1180-2350		ZT type; No. 2	Second run of the above specimen.
132	335		1963	1200-2180		ZT type; No. 3	Similar to the above specimen but measured parallel to the grain direction with a tilt angle of 8.1 degrees.
133	335		1963	1220-2280		ZT type; No. 4	Similar to the above specimen.
134	335		1963	1220-2630		ZT type; No. 5	Similar to the above specimen but with different dimensions of 7 x 6 x 0.1 cm; measured perpendicular to the grain direction with a tilt angle of 8.1 degrees.
135	341	L	1958	1623-2773		H4LM graphite	Specimen 8 in. long and 0.50 in. in dia, density 1.72 g cm^{-3}; heat flow parallel to grain, zero uranium content.
136	341	L	1958	1593-2823		LDH graphite	Specimen 8 in. long and 0.50 in. in dia, density 1.73 g cm^{-3}; heat flow parallel to grain; uranium content 0.125 mg cm^{-3} of carbon.
137	341	L	1958	1623-2823		CK graphite	Specimen 8 in. long and 0.50 in. in dia, density 1.71 g cm^{-3}; heat flow parallel to grain; zero uranium content.
138	341	L	1958	1653-2823		LDC graphite	Specimen 8 in. long and 0.50 in. in dia, density 1.66 g cm^{-3}; heat flow parallel to grain; uranium content 0.250 mg cm^{-3} of carbon.
139	343	E	1959	1300-2200		Boronated graphite	Rectangular specimen 0.1 x 1 x 10 cm (after extrusion and baking at 3273 K); cut from the portion near the parent rod center; the rod being made of a mixture of lamp black, coke, boron carbide, and pitch with a boron content of 1.3% and specific gravity of 1.79.
140	343	E	1959	1300-2465		Boronated graphite	Similar to the above specimen but cut from the central portion of the parent rod.
141	343	E	1959	1300-2200		Boronated graphite	Similar to above specimen but cut from the portion near the center of the parent rod.

SPECIFICATION TABLE NO. 11 (continued)

Curve No.	Ref. No.	Method Used	Year	Temp. Range, K	Reported Error, %	Name and Specimen Designation	Composition (weight percent), Specifications and Remarks
142	344	R	1962	1513-2048			Extruded test rod made from a mixture of 100 parts of soft filler coke particles (200/270 mesh) and 50 parts of M-30 soft binder, baked and heat treated to 3273 K, apparent density 1.54 g cm^{-3}.
143	344	R	1962	1533-1933			The above specimen, run 2.
144	344	R	1962	1653-2393			The above specimen, run 3.
145	344	R	1962	1833-3268			The above specimen, run 4.
146	344	R	1962	1833-2743			The above specimen, run 5.
147	344	R	1962	1683-1963			The above specimen, run 6.
148	344	R	1962	1643-1703			Test rod made from 100 parts of soft coke (50 parts 65/100 mesh and 50 parts 200/270 mesh) and 35 parts of hard binder, baked and graphitized to a temperature of 3273 K prior to testing; apparent density 1.65 g cm^{-3}.
149	344	R	1962	1643-2013			The above specimen, run 2.
150	344	R	1962	1753-2453			The above specimen, run 3.
151	344	R	1962	1683-2093			The above specimen, run 4.
152	344	R	1962	1643-2123			The above specimen, run 5.
153	344	R	1962	1363-1453		U.B. Graphite	Carbon rod sample extruded from a soft filler, soft binder mixture, baked to a temperature of 1273 K, heat treated at 1473 K.
154	344	R	1962	1393-1733		U.B. Graphite	Similar to above specimen except heat treated at 1773 K.
155	344	R	1962	1293-2013		U.B. Graphite	Similar to above specimen except heat treated at 2073 K.
156	344	R	1962	1393-2333		U.B. Graphite	Similar to above specimen except heat treated at 2373 K.
157	344	R	1962	1393-2603		U.B. Graphite	Similar to above specimen except heat treated at 2673 K.
158	344	R	1962	1403-2933		U.B. Graphite	Similar to above specimen except heat treated at 2773 K.
159	344	R	1962	1293-3093		U.B. Carbon	Carbon rod sample extruded from a mixture of soft filler and hard binder, baked and graphitized to a temperature of 3373 K.
160	344	R	1962	1603-3073		U.B. Carbon	The above specimen, run 2.
161	344	R	1962	1413-3103		U.B. Carbon	The above specimen, run 3.
162	344	R	1962	1513-3153		U.B. Carbon	The above specimen, run 4.
163	344	R	1962	1953-2953		U.B. Carbon	The above specimen, run 5.
164	345	L	1958	20-273		C-15	Specimen prepared from petroleum-coke base, molded with coal-tar pitch binder, baked at 2673 K, equivalent bromine residue 0.75 weight percent; measurements made under high temperatures.

SPECIFICATION TABLE NO. 11 (continued)

Curve No.	Ref. No.	Method Used	Year	Temp. Range, K	Reported Error, %	Name and Specimen Designation	Composition (weight percent), Specifications and Remarks
165	345	L	1958	18.5-273		C-15	Similar to the above specimen but baked at 2873 K and with an equivalent bromine residue of 0.5%.
166	345	L	1958	18-300		C-15	Similar to the above specimen but baked at 3073 K and with an equivalent bromine residue of 0.25%.
167	346	R	1956	80-300		CSF-MTR	Virgin 10 mil sample.
168	346	R	1956	200, 300		CSF-MTR	Virgin 20 mil sample.
169	296		1958	293-1273			Impervious graphite.
170	338		1948	1273, 1873		EBP	Rectangular block; 24 x 20 x 6 in.; molded; baked; cut at an angle to give both against and with the grain orientation.
171	338		1948	1273, 1873		AUC	Rod; 12 in. in dia; extruded; baked; specially cut to give an across grain orientation.
172	338		1948	1273, 1873		CS-312	Similar to the above specimen.
173	338		1948	1273, 1873		C-18	Rectangular block; 24 x 20 x 6 in.; molded; baked; cut at angle to give both against and with the grain orientation.
174	338		1948	1273, 1873		L-117	Rod; 3 in. in dia; extruded; baked; specially cut to give an across grain orientation.
175	338		1955	298.2		Porous-40	Molded; baked at 1000 C; specially cut to give with the grain orientation.
176	338		1955	298.2		Porous-60	Similar to the above specimen.
177	338		1955	298.2		255	Molded; baked.
178	338		1955	298.2		CS-112	Rod; 1.125 in. in dia; extruded; baked; specially cut to give with the grain orientation.
179	338		1955	1355-2303		CS-312	Similar to the above specimen but the dia, 12 in.
180	338		1956	15.2-296		CEQ	Rectangular block; 6 x 5 x 3 in.; molded; baked; specially cut to give with the grain orientation.
181	338		1962	298.2		AGSR	Graphite stocks 1-2.75 in. in dia; grain size 0.016 in.; bulk density 1.58 g cm^{-3}; electrical resistivity 839 μohm cm; with grain orientation.
182	338		1962	298.2		AGSR	Similar to the above but electrical resistivity 1500 μohm cm; across grain orientation.
183	338		1962	298.2		AGSR	Graphite stocks 3-5.75 in. in dia; grain size 0.03; bulk density 1.58 g cm^{-3}; electrical resistivity 864 μ ohm cm; with grain orientation.
184	338		1962	298.2		AGSR	Similar to the above but electrical resistivity 1280 μohm cm; across grain orientation.
185	338		1962	298.2		AGSR	Graphite stocks 6-12 in. in dia; grain size 0.06 in.; bulk density 1.57 g cm^{-3}; electrical resistivity 885 μohm cm; with grain orientation.
186	338		1962	298.2		AGSR	Similar to the above but electrical resistivity 1110 μohm cm; across grain orientation.

SPECIFICATION TABLE NO. 11 (continued)

Curve No.	Ref. No.	Method Used	Year	Temp. Range, K	Reported Error, %	Name and Specimen Designation	Composition (weight percent), Specifications and Remarks
187	338		1962	298.2		AGSR	Graphite stocks 14-35 in. in dia; grain size 0.25 in.; bulk density 1.54 g cm^{-3}; electrical resistivity 965 μohm cm; with grain orientation.
188	338		1962	298.2		AGSR	Similar to the above but electrical resistivity 1130 μohm cm; across grain orientation.
189	338		1962	298.2		AGA	Graphite stocks 35 in. in dia; grain size 0.5 in.; bulk density 1.65 g cm^{-3}; electrical resistivity 1040 μohm cm; with grain orientation.
190	338		1962	298.2		AGA	Similar to the above but electrical resistivity 1090 μohm cm; across grain orientation.
191	338		1962	298.2		AGSX	Graphite stocks 1-2.75 in. in dia; grain size 0.016 in.; bulk density 1.67 g cm^{-3}; electrical resistivity 799 μohm cm; with grain orientation.
192	338		1962	298.2		AGSX	Similar to the above but electrical resistivity 1330 μohm cm; across grain orientation.
193	338		1962	298.2		AGSX	Graphite stocks 3-5.75 in. in dia; grain size 0.03 in.; bulk density 1.69 g cm^{-3}; electrical resistivity 821 μohm cm; with grain orientation.
194	338		1962	298.2		AGSX	Similar to the above but electrical resistivity 1390 μohm cm; across grain orientation.
195	338		1962	298.2		AGSX	Graphite stocks 6-12 in. in dia; grain size 0.06 in.; bulk density 1.71 g cm^{-3}; electrical resistivity 820 μohm cm; with grain orientation.
196	338		1962	298.2		AGSX	Similar to the above but electrical resistivity 1010 μohm cm; across grain orientation.
197	338		1962	298.2		CS	Graphite stocks 1-2.75 in. in dia; grain size 0.016 in.; bulk density 1.68 g cm^{-3}; electrical resistivity 819 μohm cm; with grain orientation.
198	338		1962	298.2		CS	Similar to the above but electrical resistivity 1310 μohm cm; across grain orientation.
199	338		1962	298.2		CS	Graphite stocks 3-18 in. in dia; grain size 0.03 in.; bulk density 1.72 g cm^{-3}; electrical resistivity 860 μohm cm; with grain orientation.
200	338		1962	298.2		CS	Similar to the above but electrical resistivity 1100 μohm cm; across grain orientation.
201	338		1962	298.2		ATL	Graphite stocks 20-24 in. in dia; grain size 0.03 in.; bulk density 1.70 g cm^{-3}; electrical resistivity 890 μohm cm; with grain orientation.
202	338		1962	298.2		ATL	Similar to the above but electrical resistivity 1070 μohm cm; across grain orientation.
203	338		1962	298.2		ATL	Graphite stocks 30-50 in. in dia; grain size 0.03 in.; bulk density 1.78 g cm^{-3}; electrical resistivity 1130 μohm cm; with grain orientation.
204	338		1962	298.2		ATL	Similar to the above but electrical resistivity 1180 μohm cm; across grain orientation.
205	338		1962	298.2		AUC	Graphite stocks 1-8 in. in dia; grain size 0.016 in.; bulk density 1.68 g cm^{-3}; electrical resistivity 790 μohm cm; with grain orientation.
206	338		1962	298.2		AUC	Similar to the above but electrical resistivity 1230 μohm cm; across grain orientation.
207	338		1962	298.2		AUC	Graphite stocks 9-18 in. in dia; grain size 0.03 in.; bulk density 1.69 g cm^{-3}; electrical resistivity 767 μohm cm; with grain orientation.

SPECIFICATION TABLE NO. 11 (continued)

Curve No.	Ref. No.	Method Used	Year	Temp. Range, K	Reported Error, %	Name and Specimen Designation	Composition (weight percent), Specifications and Remarks
208	338		1962	298.2		AUC	Similar to the above but electrical resistivity 1230 μohm cm; across grain orientation.
209	338		1962	298.2		CEQ	Graphite stocks size 6 x 5 x 2.875 in.; grain size 0.008 in.; bulk density 1.55 g cm^{-3}; electrical resistivity 5029 μohm cm; with grain orientation.
210	338		1962	298.2		CDA	Graphite stocks size 6 x 5 x 2.6875 in.; grain size 0.006 in.; bulk density 1.62 g cm^{-3}; electrical resistivity 1072 μohm cm; with grain orientation.
211	338		1962	298.2		CDA	Similar to the above but electrical resistivity 1640 μohm cm; across grain orientation.
212	338		1962	298.2		CDG	Graphite stocks size 12 x 12 x 0.25 to 12 x 12 x 1 in.; grain size 0.016 in.; bulk density 1.36 g cm^{-3}; electrical resistivity 1351 μohm cm; with grain orientation.
213	338		1962	298.2		CDG	Similar to the above but the sizes 15 x 18 x 0.25 to 15 x 18 x 2 in.; bulk density 1.40 g cm^{-3}; electrical resistivity 1522 μohm cm.
214	337		1914	290,373		Pencil Lead Graphite	Cylindrical rod; 0.183 cm in dia 10.4 cm long; specific gravity 2.11; specimen made from a 'Kohinor' pencil lead grade 6H.
215	347	L	1964	298.2		ZTA	Bulk density 1.940 g cm^{-3}; electrical resistivity 6.97 x 10^{-4} ohm cm; with grain.
216	347	L	1964	298.2		ZTA	Bulk density 1.940 g cm^{-3}; electrical resistivity 21.87 x 10^{-4} ohm cm; across grain.
217	347	L	1964	298.2		ZTA	Bulk density 1.924 g cm^{-3}; electrical resistivity 7.24 x 10^{-4} ohm cm; with grain.
218	347	L	1964	298.2		ZTA	Bulk density 1.924 g cm^{-3}; electrical resistivity 21.90 x 10^{-4} ohm cm; across grain.
219	347	L	1964	298.2		ZTA	Bulk density 1.953 g cm^{-3}; electrical resistivity 6.91 x 10^{-4} ohm cm; with grain.
220	347	L	1964	298.2		ZTA	Bulk density 1.953 g cm^{-3}; electrical resistivity 23.18 x 10^{-4} ohm cm; across grain.
221	347	L	1964	298.2		ZTA	Bulk density 1.942 g cm^{-3}; electrical resistivity 6.70 x 10^{-4} ohm cm; with grain.
222	347	L	1964	298.2		ZTA	Bulk density 1.942 g cm^{-3}; electrical resistivity 18.95 x 10^{-4} ohm cm; across grain.
223	347	L	1964	298.2		ZTA	Bulk density 1.955 g cm^{-3}; electrical resistivity 6.87 x 10^{-4} ohm cm; with grain.
224	347	L	1964	298.2		ZTA	Bulk density 1.955 g cm^{-3}; electrical resistivity 22.04 x 10^{-4} ohm cm; across grain.
225	347	L	1964	298.2		ZTA	Bulk density 1.923 g cm^{-3}; electrical resistivity 7.07 x 10^{-4} ohm cm; with grain.
226	347	L	1964	298.2		ZTA	Bulk density 1.923 g cm^{-3}; electrical resistivity 22.67 x 10^{-4} ohm cm; across grain.
227	347	L	1964	298.2		ZTA	Bulk density 1.932 g cm^{-3}; electrical resistivity 7.43 x 10^{-4} ohm cm; with grain.
228	347	L	1964	298.2		ZTA	Bulk density 1.932 g cm^{-3}; electrical resistivity 16.09 x 10^{-4} ohm cm; across grain.
229	347	L	1964	298.2		ZTA	Bulk density 1.92 g cm^{-3}; electrical resistivity 7.76 x 10^{-4} ohm cm; with grain.
230	347	L	1964	298.2		ZTA	Bulk density 1.92 g cm^{-3}; electrical resistivity 16.45 x 10^{-4}ohm cm; across grain.
231	347	L	1964	298.2		ZTA	Bulk density 1.93 g cm^{-3}; electrical resistivity 7.54 x 10^{-4}ohm cm; with grain.
232	347	L	1964	298.2		ZTA	Bulk density 1.93 g cm^{-3}; electrical resistivity 15.84 x 10^{-4}ohm cm; across grain.

SPECIFICATION TABLE NO. 11 (continued)

Curve No.	Ref. No.	Method Used	Year	Temp. Range, K	Reported Error, %	Name and Specimen Designation	Composition (weight percent), Specifications and Remarks
233	347	L	1964	298.2		ZTA	Bulk density 1.95 g cm^{-3}; electrical resistivity 6.66 x 10^{-4} ohm cm; with grain.
234	347	L	1964	298.2		ZTA	Bulk density 1.95 g cm^{-3}; electrical resistivity 15.82 x 10^{-4} ohm cm; across grain.
235	347	L	1964	298.2		ZTA	Bulk density 1.94 g cm^{-3}; electrical resistivity 7.42 x 10^{-4} ohm cm; with grain.
236	347	L	1964	298.2		ZTA	Bulk density 1.94 g cm^{-3}; electrical resistivity 16.18 x 10^{-4} ohm cm; across grain.
237	347	L	1964	298.2		ZTB	Bulk density 1.98 g cm^{-3}; electrical resistivity 6.88 x 10^{-4} ohm cm; with grain.
238	347	L	1964	298.2		ZTB	Bulk density 1.98 g cm^{-3}; electrical resistivity 19.74 x 10^{-4} ohm cm; across grain.
239	347	L	1964	298.2		ZTB	Bulk density 1.97 g cm^{-3}; electrical resistivity 6.96 x 10^{-4} ohm cm with grain.
240	347	L	1964	298.2		ZTB	Bulk density 1.97 g cm^{-3}; electrical resistivity 17.81 x 10^{-4} ohm cm; across grain.
241	347	L	1964	298.2		ZTB	Bulk density 1.99 g cm^{-3}; electrical resistivity 6.43 x 10^{-4} ohm cm; with grain.
242	347	L	1964	298.2		ZTB	Bulk density 1.99 g cm^{-3}; electrical resistivity 21.13 x 10^{-4} ohm cm; across grain.
243	347	L	1964	298.2		ZTC	Bulk density 1.93 g cm^{-3}; electrical resistivity 6.97 x 10^{-4} ohm cm; with grain.
244	347	L	1964	298.2		ZTC	Bulk density 1.93 g cm^{-3}; electrical resistivity 11.97 x 10^{-4} ohm cm; across grain.
245	347	L	1964	298.2		ZTC	Bulk density 1.92 g cm^{-3}; electrical resistivity 7.15 x 10^{-4} ohm cm; with grain.
246	347	L	1964	298.2		ZTC	Bulk density 1.92 g cm^{-3}; electrical resistivity 11.00 x 10^{-4} ohm cm; across grain.
247	347	L	1964	298.2		ZTC	Bulk density 1.94 g cm^{-3}; electrical resistivity 6.90 x 10^{-4} ohm cm; with grain.
248	347	L	1964	298.2		ZTC	Bulk density 1.94 g cm^{-3}; electrical resistivity 13.21 x 10^{-4} ohm cm; across grain.
249	347	L	1964	298.2		ZTD	Bulk density 2.01 g cm^{-3}; electrical resistivity 5.41 x 10^{-4} ohm cm; with grain.
250	347	L	1964	298.2		ZTD	Bulk density 2.01 g cm^{-3}; electrical resistivity 7.88 x 10^{-4} ohm cm; across grain.
251	347	L	1964	298.2		ZTE	Bulk density 1.96 g cm^{-3}; electrical resistivity 8.94 x 10^{-4} ohm cm; with grain.
252	347	L	1964	298.2		ZTE	Bulk density 1.96 g cm^{-3}; electrical resistivity 20.40 x 10^{-4} ohm cm; across grain.
253	347	L	1964	298.2		ZTF	Bulk density 1.99 g cm^{-3}; electrical resistivity 7.31 x 10^{-4} ohm cm; with grain.
254	347	L	1964	298.2		ZTF	Bulk density 1.99 g cm^{-3}; electrical resistivity 20.50 x 10^{-4} ohm cm; across grain.
255	347	L	1964	298.2		ZTF	Bulk density 1.99 g cm^{-3}; electrical resistivity 7.24 x 10^{-4} ohm cm; with grain.
256	347	L	1964	298.2		ZTF	Bulk density 1.99 g cm^{-3}; electrical resistivity 21.48 x 10^{-4} ohm cm; across grain.
257	347	L	1964	298.2		RVA	Bulk density 1.84 g cm^{-3}; electrical resistivity 12.21 x 10^{-4} ohm cm; with grain.
258	347	L	1964	298.2		RVA	Bulk density 1.84 g cm^{-3}; electrical resistivity 15.73 x 10^{-4} ohm cm; across grain.
259	347	L	1964	298.2		RVA	Bulk density 1.825 g cm^{-3}; electrical resistivity 12.25 x 10^{-4} ohm cm; with grain.
260	347	L	1964	298.2		RVA	Bulk density 1.825 g cm^{-3}; electrical resistivity 16.87 x 10^{-4} ohm cm; across grain.

SPECIFICATION TABLE NO. 11 (continued)

Curve No.	Ref. No.	Method Used	Year	Temp. Range, K	Reported Error, %	Name and Specimen Designation	Composition (weight percent), Specifications and Remarks
261	347	L	1964	298.2		RVA	Bulk density 1.842 g cm^{-3}; electrical resistivity 12.34 x 10^{-4} ohm cm; with grain.
262	347	L	1964	298.2		RVA	Bulk density 1.842 g cm^{-3}; electrical resistivity 15.20 x 10^{-4} ohm cm;across grain.
263	347	L	1964	298.2		RVA	Bulk density 1.844 g cm^{-3}; electrical resistivity 12.06 x 10^{-4} ohm cm; with grain.
264	347	L	1964	298.2		RVA	Bulk density 1.844 g cm^{-3}; electrical resistivity 15.65 x 10^{-4} ohm cm; across grain.
265	347	L	1964	298.2		RVC	Bulk density 1.84 g cm^{-3}; electrical resistivity 13.08 x 10^{-4} ohm cm; with grain.
266	347	L	1964	298.2		RVC	Bulk density 1.84 g cm^{-3}; electrical resistivity 16.41 x 10^{-4} ohm cm; across grain.
267	347	L	1964	298.2		RVC	Bulk density 1.84 g cm^{-3}; electrical resistivity 12.71 x 10^{-4} ohm cm; with grain.
268	347	L	1964	298.2		RVC	Bulk density 1.84 g cm^{-3}; electrical resistivity 16.03 x 10^{-4} ohm cm; across grain.
269	347	L	1964	298.2		RVC	Bulk density 1.85 g cm^{-3}; electrical resistivity 13.13 x 10^{-4} ohm cm; with grain.
270	347	L	1964	298.2		RVC	Bulk density 1.85 g cm^{-3}; electrical resistivity 16.75 x 10^{-4} ohm cm; across grain.
271	347	L	1964	298.2		RVD	Bulk density 1.87 g cm^{-3}; electrical resistivity 12.62 x 10^{-4} ohm cm; with grain.
272	347	L	1964	298.2		RVD	Bulk density 1.87 g cm^{-3}; electrical resistivity 21.64 x 10^{-4} ohm cm; across grain.
273	347	L	1964	298.2		RVD	Bulk density 1.87 g cm^{-3}; electrical resistivity 12.52 x 10^{-4} ohm cm; with grain.
274	347	L	1964	298.2		RVD	Bulk density 1.87 g cm^{-3}; electrical resistivity 21.72 x 10^{-4} ohm cm; across grain.
275	347	L	1964	298.2		RVD	Bulk density 1.87 g cm^{-3}; electrical resistivity 12.72 x 10^{-4} ohm cm; with grain.
276	347	L	1964	298.2		RVD	Bulk density 1.87 g cm^{-3}; electrical resistivity 21.54 x 10^{-4} ohm cm; across grain.
277	347	L	1964	298.2		CFW	Bulk density 1.90 g cm^{-3}; electrical resistivity 11.98 x 10^{-4} ohm cm; with grain.
278	347	L	1964	298.2		CFW	Bulk density 1.90 g cm^{-3}; electrical resistivity 12.60 x 10^{-4} ohm cm; across grain.
279	347	L	1964	298.2		CFZ	Bulk density 1.91 g cm^{-3}; electrical resistivity 12.77 x 10^{-4} ohm cm; with grain.
280	347	L	1964	298.2		CFZ	Bulk density 1.91 g cm^{-3}; electrical resistivity 16.08 x 10^{-4} ohm cm; across grain.
281	160	R	1958	1593-3198			Specimen 0.5 in. in dia, 8 in. long; prepared by mixing 100 parts (by weight) of raw Texas coke (calcined for 4 hrs at 1200 C in a baking furnace, crushed and ground) and 40 parts of Medium No. 30 coal tar pitch (supplied by Barrett Co.) for 15 min at 160 C and also 3 parts of extrusion oil (VacWax 80 of Socony Vacuum Co.) mixed again at 150 C for 5 hrs; extruded and baked at 1000 C; graphitized in nitrogen atmosphere at 3100 C for 10 min.
282	160	R	1958	2189-3033			Similar to the above but using Texas coke of 200/270 mesh as raw material and extruded at 8200 psi.
283	160	R	1958	1906-3200			Similar to the above but using Texas coke of 100/150 mesh as raw material and extruded at 6100 psi.
284	160	R	1958	2078-3134			Similar to the above but using Texas coke of 28/35 mesh as raw material and extruded at 4100 psi.

SPECIFICATION TABLE NO. 11 (continued)

Curve No.	Ref. No.	Method Used	Year	Temp. Range, K	Reported Error, %	Name and Specimen Designation	Composition (weight percent), Specifications and Remarks
285	160	R	1958	2068-2815			The above specimen measured in high vacuum chamber.
286	265	R	1960	1403-3273		Graphitized carbon rod	Specimen 1.57 in. in dia; made from 100 parts of filler (50 parts of 100/150 mesh and 50 parts of < 270 mesh phenol formaldehyde) and 43 parts binder; extruded at 11500 psi; graphitized to 3100 C.
287	265	R	1960	1418-3188		Graphitized carbon rod	Specimen made from 100 parts of calcined Texas coke (28/35 mesh), 44 parts of coal tar; extruded and baked to 1200 C; density after baking 1.25 g cm^{-3}; graphitized to 3100 C; measured in argon atmosphere at 1-2 in. Hg above atmospheric pressure.
288	265	R	1960	1503-3073		Graphitized carbon rod	Specimen 1.36 in. in dia; made from 100 parts of filler (50 parts of 65/100 mesh and 50 parts of 200/270 mesh Texas coke) and 40 parts of M-30 coal tar pitch as binder; extruded at 7000 psi; graphitized to 3100 C.
289	265	R	1960	1363-3183		Graphitized carbon rod	Specimen 1.61 in. in dia;made from 100 parts of filler (50 parts of 65/100 mesh and 50 parts 200/270 mesh) and 35 parts of phenol benzaldehyde as binder; extruded at 5300 psi; graphitized to 3100 C.
290	265	R	1960	1373-2773		Graphitized carbon rod	Specimen 1.24 in. in dia; made from 100 parts of filler (50 parts of 100/150 mesh and 50 parts of < 270 mesh phenol formaldehyde) and 48 parts of M-30 coal tar pitch as binder; extruded at 2300 psi; graphitized to 3100 C.
291	367	R	1963	1343-2313			Prepared by mixing 50 parts 65/100 mesh and 50 parts < 200 mesh soft filler (soft Texas coke), and 40 parts soft binder (M-30 pitch); extruded to 0.50 in. dia; baked for four days to 1000 C; density after baking 1.55 g cm^{-3}; heat treated at 2100 C for 10 min; measured in an argon atmosphere (pressure approx. one atm).
292	367	R	1963	1303-2603			The above specimen heat treated at 2400 C for 10 min.
293	367	R	1963	1303-2948			The above specimen heat treated at 2800 C for 10 min.
294	367	R	1963	1353-2303			Prepared by mixing 50 parts 65/100 mesh and 50 parts 200/270 mesh soft filler (soft Texas coke), and 35 parts hard binder (phenol benzaldehyde); extruded to 0.50 in. dia; baked for four days to 1000 C; density after baking 1.56 g cm^{-3}; heat treated at 2100 C for 10 min; measured in an argon atmosphere (pressure approx. one atm).
295	367	R	1963	1383-2583			The above specimen heated at 2400 C for 10 min.
296	367	R	1963	1373-2973			The above specimen heated at 2800 C for 10 min.
297	367	R	1963	1318-2233			Prepared by mixing 50 parts 100/150 mesh and 50 parts < 270 mesh hard filler (phenol formaldehyde); and 48 parts soft binder (M-30 pitch); extruded to 0.50 in. dia; baked for four days to 1000 C; density after baking 1.14 g cm^{-3}; heat treated at 2100 C for 10 min; measured in an argon atmosphere (pressure approx. one atm).

SPECIFICATION TABLE NO. 11 (continued)

Curve No.	Ref. No.	Method Used	Year	Temp. Range, K	Reported Error, %	Name and Specimen Designation	Composition (weight percent), Specifications and Remarks
298	367	R	1963	1323-2473			The above specimen heat treated at 2400 C for 10 min.
299	367	R	1963	1333-2763			The above specimen heat treated at 2800 C for 10 min.
300	367	R	1963	1343-2263			Prepared by mixing 50 parts 100/150 mesh and 50 parts <270 mesh hard filler (phenol formaldehyde), and 43 parts hard binder (phenol benzaldehyde); extruded to 0.50 in. dia; baked for four days to 1000 C; density after baking 1.22 g cm^{-3}; heat treated at 2100 C for 10 min; measured in an argon atmosphere (pressure approx. one atm).
301	367	R	1963	1368-2523			The above specimen heat treated at 2400 C for 10 min.
302	367	R	1963	1438-2893			The above specimen heat treated at 2800 C for 10 min.
303	354	C	1962	323-873	5-10	EY9	Specimen made from Morgan Crucible Co. graphite; cut parallel to the direction of extrusion; density 1.64 g cm^{-3}; electrical resistivity reported as 1.93, 1.71, 1.53, 1.40, and 1.30 milliohm cm at 88, 205, 320, 420, and 545 C, respectively; Armco iron used as the comparative material.
304	354	C	1962	313-828	5-10	EY9	Similar to the above specimen but cut perpendicular to the direction of extrusion; electrical resistivity reported as 2.87, 2.58, 2.21, and 2.05 milliohm cm at 70, 185, 350, and 425 C, respectively.
305	354	C	1962	321-916	5-10	HX10	Specimen made from material of Harwell Graphite Plant; cut parallel to the direction of extrusion; density 1.87 g cm^{-3}; electrical resistivity at 83, 195, 360, and 450 C being respectively, 1.50, 1.30, 1.10, and 1.02 milliohm cm.
306	354	C	1962	321-838	5-10	British Reactor Grade A	Specimen cut parallel to the direction of extrusion; density 1.73 g cm^{-3}; electrical resistivity at 100, 200, 300, 400, and 450 C being respectively, 0.60, 0.53, 0.48, 0.45 and 0.44 milliohm cm.
307	354	C	1962	321-846	5-10	British Reactor Grade A	Similar to the above specimen but cut perpendicular to the direction of extrusion; electrical resistivity at 100, 200, and 300 C being respectively, 1.03, 0.90, and 0.82 milliohm cm.
308	354	C	1962	318-816	5-10	British Reactor Grade A	Similar to the above specimen but cut prallel to the direction of extrusion.
309	354	C	1962	318-823	5-10	British Reactor Grade A	Similar to the above specimen but cut perpendicular to the direction of extrusion.
310	354	C	1962	313-831	5-10	British Reactor Grade Carbon	British Reactor Grade Carbon Stock graphitized to 2100 C; not impregnated; cut parallel to the direction of extrusion; density 1.62 g cm^{-3}; electrical resistivity at 100, 200, 300, 400, and 500 C being respectively 3.10, 2.87, 2.67, 2.49, and 2.33 milliohm cm.
311	354	C	1962	313-798	5-10	British Reactor Grade Carbon	Similar to the above specimen but graphitized to 2300 C; density 1.68 g cm^{-3}; electrical resistivity at 100, 200, 300, 400, and 500 C being respectively, 2.35, 2.08, 1.85, 1.64, and 1.46 milliohm cm.

SPECIFICATION TABLE NO. 11 (continued)

Curve No.	Ref. No.	Method Used	Year	Temp. Range, K	Reported Error, %	Name and Specimen Designation	Composition (weight percent), Specifications and Remarks
312	354	C	1962	313-753	5-10	British Reactor Grade Carbon	Similar to the above specimen but graphitized to 2600 C; density 1.62 g cm^{-3}; electrical resistivity at 100, 200, 300, 400, and 500 C being respectively, 1.17, 1.02, 0.92, 0.85, and 0.80 milliohm cm.
313	354	C	1962	303-901	5-10	British Reactor Grade Carbon	Similar to the above specimen but graphitized to 2820 C; density 1.65 g cm^{-3}; electrical resistivity at 100, 200, 300, 400, 500, 600, and 700 C being respectively, 0.78, 0.71, 0.67, 0.65, 0.64, 0.65, and 0.67 milliohm cm.
314	336		1965	323.2		EY9	Grade EY9 graphite from Morgon Crucible Company; electrical resistivity 1.71 milliohm cm at room temperature.
315	336		1965	323.2		EY9	Similar to the above specimen but electrical resistivity 1.86 milliohm cm at room temperature.
316	336		1965	323.2		EY9	Similar to the above specimen but electrical resistivity 1.89 milliohm cm at room temperature.
317	397		1966	364~2239		JTA; 7-F-12	Measured in the with-the-grain direction.
318	363	P	1965	1575-2400		EY9A	Density 1.76 g cm^{-3}; data calculated from measurements of thermal diffusivity; specific heat data from "Nuclear Graphite" by Nightingale, R.E., Yoshikawa, H.H., and Losty, H.H.W., 1962.
319	363	P	1965	1320-2380		Moderator graphite	Density 1.71 g cm^{-3}; data calculated from measurements of thermal diffusivity; specific heat data from the same source as above.
320	391	E	1964	295-511		ZTA	Prepared from coke L, supplied by Pechiney Company, by extruding into a 10 mm dia bar; the graphite was impregnated once with tar; measured along the a-axis.
321	391	E	1964	298-536		ZTA	Similar to the above specimen measured along the c-axis.
322	391	E	1964	306-714		ZTA	Similar to the above specimen; the dia was a bit smaller and measured along the a-axis.
323	391	E	1964	321-721		ZTA	Similar to the above specimen; measured along the c-axis.
324	391	E	1964	302-598		ZTA	Prepared from coke L, supplied by Pechiney Company, by extruding into 10 mm dia; the graphite was impregnated once with tar; measured along the c-axis.
325	391	E	1964	307-605		ZTA	Similar to the above specimen,except neutron-irradiated at 350 C.
326	391	E	1964	311-506		ZTA	Similar to the above specimen, except neutron-irradiated at 250 C.
327	391	E	1964	393-411		ZTA	Similar to the above specimen, except neutron-irradiated at 150 C.
328	392		1961	469-873		MH4LM	Density 1.90 g cm^{-3}; grain size > 0.032 in.
329	392		1961	471-875		MH4LM	Similar to the above specimen except irradiated in Material Testing Reactor at 475 C by a neutron flux of 3.5×10^{19} nut with energy > 0.1 Mev.

SPECIFICATION TABLE NO. 11 (continued)

Curve No.	Ref. No.	Method Used	Year	Temp. Range, K	Reported Error, %	Name and Specimen Designation	Composition (weight percent), Specifications and Remarks
330	392		1961	471-874		ATL-82-1	Grain size 0.016 to 0.03 in.
331	392		1961	471-873		ATL-82-2	Similar to the above specimen; irradiated in Hanford reactor at 360 to 420 C by a neutron flux of 3.2 x 10^{20} nvt with energy > 0.1 Mev.
332	392		1961	472-874		ATL-82-3	Similar to the above specimen except irradiated in Material Testing Reactor at 475 C by a neutron flux of 3.6 x 10^{19} nvt with energy > 0.1 Mev.
333	392		1961	468-868		R0025-1	Obtained from National Carbon Co.; grain size < 0.016 in.
334	392		1961	468-869		R0025-2	Similar to the above specimen.
335	392		1961	472-867		R0025-3	Similar to the above specimen; irradiated in Testing Reactor at 360 to 420 C by a neutron flux of 3.6 x 10^{19} nvt with energy > 0.1 Mev.
336	392		1961	471-865		R0025-3A	The above specimen annealed at 925 C for 16 hrs.
337	340	→	1962	1671	< 10	ZT type graphite; G3A	Thermal conductivity data in the z-direction (k_z) determined simultaneously with thermal conductivity in the r-direction (k_r, see next curve) from 4 cylindrical specimens made from ZT type graphite of National Carbon Co.; density 1.980 g cm^{-3}; anisotropy ratio of electrical resistivity ρ(z-direction)/ρ(r-direction) = 2.50 at room temperature; the specimens each about 2.537 cm dia x 1.126 cm thick being heated in vacuum by high frequency induction, thermal conductivity determined by equating the heat conduction in specimen to the heat loss by radiation assuming the emissivity of a gray body, the analysis required 2 specimens of different thickness to solve simultaneously for k_z and k_r at a certain temperature.
338	340	→	1962	1671	< 10	ZT type graphite; G3A	k_r determined simultaneously with the above curve.
339	340	→	1962	1671	< 10	ZT type graphite; G7	Similar to the above specimen except with size 2.539 cm dia x 0.287 cm thick and density 1.978 g cm^{-3}; k_z was measured.
340	340	→	1962	1671	< 10	ZT type graphite; G7	k_r determined simultaneously with the above curve.
341	336		1965	673-1173		EY9 graphite	Obtained from Morgon Crucible Co.; electrical resistivity 1790 ~ 1850 μohm cm at room temperature; data reported were mean values.
342	394	R	1965	1367-3311	5-7	CFZ grade	99.74 C, < 0.6 H, 0.19 ash, 0.07 CaO, 0.02 Al_2O_3, 0.04 total sulfur, and < 0.01 sulfide sulfur; specimens 1 in. long, 1 in. O.D. and 0.25 in. I.D.; supplied by Union Carbide Co.; heat flow measured parallel to cylindrical axis; with grain; bulk density (mean value) 1.899 g cm^{-3}; thermal conductivity data calculated from the mean values of 9 specimens (standard deviation 0.0946, 0.0609, 0.0786, 0.0963, and 0.111 at 1366.5, 2199.8, 2755.4, 3033.2, and 3310.9 K, respectively).
343	394	R	1965	1367-3311	5-7	CFZ grade	Similar to the above specimens except bulk density (mean value) 1.908 g cm^{-3}; standard deviation 0.0891, 0.129, 0.0986, 0.0968, and 0.0544 at 1366.5, 2199.8, 2755.4, 3033.2, and 3310.9 K, respectively.

SPECIFICATION TABLE NO. 11 (continued)

Curve No.	Ref. No.	Method Used	Year	Temp. Range, K	Reported Error, %	Name and Specimen Designation	Composition (weight percent), Specifications and Remarks
344	394	R	1965	1367-3311	5-7	CFZ grade	Similar to the above specimens except specimen orientation across grain; bulk density (mean value) 1.896 g cm^{-3}; standard deviation 0.0661, 0.0749, 0.0711, 0.0606, and 0.0526 at 1366.5, 2199.8, 2755.4, 3033.2, and 3310.9 K, respectively.
345	394	R	1965	1367-3311	5-7	CFZ grade	Similar to the above specimens except bulk density (mean value) 1.906 g cm^{-3}; standard deviation 0.0535, 0.0362, 0.0799, 0.0862, and 0.0539 at 1366.5, 2199.8, 2755.4, 3033.2, and 3310.9 K, respectively.
346	394	C	1965	338.7	5	CFZ grade	99.74 C, <0.6 H, 0.19 ash, 0.07 CaO, 0.02 Al_2O_3, 0.04 total sulfur and <0.01 sulfide sulfur; specimens 1 in. in dia and 1 in. long; supplied by Union Carbide Co.; with grain; bulk density (mean value) 1.903 g cm^{-3}; thermal conductivity data from the mean values of 10 specimens (standard deviation 0.0937 at 338.7 K); Armco iron used as comparative material.
347	394	C	1965	338.7	5	CFZ grade	Similar to the above specimens except bulk density (mean value) 1.907 g cm^{-3}; standard deviation 0.108 at 338.7 K.
348	394	C	1965	338.7	5	CFZ grade	Similar to the above specimens except bulk density (mean value) 1.881 g cm^{-3}; standard deviation 0.0317 at 338.7 K.
349	394	C	1965	338.7	5	CFZ grade	Similar to the above specimens except bulk density (mean value) 1.907 g cm^{-3}; standard deviation 0.0288 at 338.7 K.
350	396		1967	1088-3030		Supertemp Pyrolytic graphite	Annealed; electrical conductivity 9.54, 7.73, 6.40, 5.42, 4.72, 4.19, 3.71 and 3.28 x 10^3 $ohm^{-1}cm^{-1}$ at 1088, 1365, 1643, 1920, 2198, 2475, 2753, and 3030 K, respectively.
351	397		1966	372-2205		JTA; 14-G-1	Measured with the grain.
352	397		1966	354-2222		JTA; 14-G-1	Measured across the grain.
353	15	L	1956	115-385			Polycrystalline; prepared from a mix consisted of 100 parts of Kendall coke, 42 parts of medium grade coal tar pitch and 2 parts of Socony Vacuum Oil Co. Vacwax 80, the coke calcined to 1100 C, crushed into powder, passed two times through a small Raymond mill, then the mix made and extruded through a 0.5 in. die, cut into 6 in. long rods; the rods baked for 5 days to reach the top temperature of 1100 C, subsequently heat treated at 1200 C for about 5 min; density 1.49 g cm^{-3}; crystallite dia 37 Å.
354	15	L	1956	115-385			Similar to the above specimen except heat treated at 1750 C and the crystallite dia 61.5 Å.
355	15	L	1956	115-385			Similar to the above specimen except heat treated at 1950 C and the crystallite dia 79 Å.
356	20	F	1944	93-373			Cut from a carbon electrode; supplied by National Carbon Co.; 2.9 cm dia x 32 cm long.

SPECIFICATION TABLE NO. 11 (continued)

Curve No.	Ref. No.	Method Used*	Year	Temp. Range, K	Reported Error, %	Name and Specimen Designation	Composition (weight percent), Specifications and Remarks
357	158	L	1952	2.8-90		Carbon resistor	L.A.B. 33 ohm, 0.5 watt resistor.
358	129	L	1909	373-713			Ordinary carbon supplied by National Carbon Co.; made of petroleum coke.
359	13	C	1953	343.2	± 3	Acheson	Density 1.7 g cm^{-3}; heat flow direction perpendicular to the axis of extrusion; Armco iron used as the comparative standard.
360	13	C	1953	343.2	± 3	Acheson	Similar to the above specimen but heat flow parallel to the axis of extrusion.
361	129	L	1909	373-873		Acheson	Specimen 8.5 in. long and 1 in. in dia.
362	354	C	1962	311-853		Reactor Grade Carbon stock	Heat-treated at 1500 C; density 1.64 g cm^{-3}; Armco iron used as comparative material; heat flow parallel to extrusion.
363	354	C	1962	310-804		Reactor Grade Carbon stock	Heat-treated at 1800 C; density 1.61 g cm^{-3}; electrical resistivity measured parallel to extrusion reported as 3.59, 3.49, 3.37, 3.22, and 3.15 milliohm cm at 61, 148, 328, 447, and 516 C, respectively.
364	106	C	1956	336.7	3	Acheson; 1	Artificial graphite electrode 80 mm in dia, 125 mm long; apparent density 1.40 g cm^{-3}; electrical resistivity 0.00123 ohm cm; copper used as comparative material.
365	106	C	1956	334.2	3	Acheson; 2	Similar to the above specimen but the apparent density 1.399 g cm^{-3}; electrical resistivity 0.00121 ohm cm.
366	326	R	1952	363-873		Acheson 2301	Powder; 99 pure; apparent density 0.69 g cm^{-3}; measured after repeated heating.
367	541	L	1962	2.4-227		Ohmite	Carbon resistor 100 ohm 2 W; specimen 0.58 cm in diameter and 1.38 cm long; measured in vacuum.
368	541	L	1962	2.4-118		Ohmite	Similar to the above specimen 3900 ohm 2 W, 0.58 cm in dia and 1.4 cm long.
369	18	L	1939	318-611		Acheson; 1	Tubular specimen 75 cm long, 2.54 cm O.D., and 0.3 cm I.D.; electrical resistivity ρ(0 C) = 0.00110 ohm cm.
370	18	R	1939	1048-1363	<10	Acheson; 2	Similar to the above specimen but ρ(0 C) = 0.00105 ohm cm.
371	18	R	1939	1723-2713	<10	Acheson; 3	Similar to the above specimen but ρ(0 C) = 0.00077 ohm cm.
372	18	R	1939	1798-3048	<10	Acheson; 4	Similar to the above specimen but ρ(0 C) = 0.00077 ohm cm.
373	18	R	1939	1683-2343	<10	Acheson; 5	Similar to the above specimen but ρ(0 C) = 0.00067 ohm cm.
374	20	F	1944	93-373		Acheson	Specimen 16 cm long, 2.9 cm in dia; cut from Acheson graphite electrode (from National Carbon Co.); specimen axis parallel to the electrode axis.
375	20	F	1944	93-373		Acheson	Similar to the above specimen but cut perpendicular to the electrode axis.
376	175	C	1937	613-1128		Acheson	Specimen 1.47 cm in dia and 20 cm long; machined from an Acheson graphite rod; electrical conductivity 1218, 1369, 1445, 1497, 1515, 1517, 1503, 1476, and 1444 ohm^{-1} cm^{-1} at 0, 100, 200, 300, 400, 500, 600, 700, and 1000 C respectively; Armco iron used as comparative material; measured in vacuum.

SPECIFICATION TABLE NO. 11 (continued)

Curve No.	Ref. No.	Method Used*	Year	Temp. Range, K	Reported Error, %	Name and Specimen Designation	Composition (weight percent), Specifications and Remarks
377	175	C	1937	483-1113		Acheson	Similar to the above specimen.
378	175	L	1937	303-423		Acheson	Specimen 3.85 cm in dia and 38 cm long; measured in air.
379	175	L	1937	313-588		Acheson	Specimen 7.34 cm in dia and 38 cm long; measured in air.
380	342	L	1933	123-973		Acheson	Two cylindrical blocks of graphite 10.2 cm in dia and 17.8 cm long placed in a vertical position end to end with a flat electric heater between them.
381	368	E	1954	1300-2000		Acheson	Long thin rod of Acheson graphite electrically heated in vacuo; electrical resistivity at 470, 600, 800, 1000, 1200, 1400, 1600, 1800, 2000, and 2070 K being, respectively, 0.7, 0.645, 0.606, 0.615, 0.640, 0.675, 0.715, 0.765, 0.8175, and 0.840 milliohm cm.
382	368	E	1954	1200-1450		Acheson	Data for a short rod of Acheson graphite.

DATA TABLE NO. 11 THERMAL CONDUCTIVITY OF ~~GRAPHITES MISCELLANEOUS~~ MISCELLANEOUS GRAPHITES.

[Temperature, T, K; Thermal Conductivity, k, Watt $cm^{-1}K^{-1}$]

T	k
CURVE 1	
316.8	1.85
324.6	1.83
325.4	1.83*
329.5	1.82
330.4	1.82*
337.3	1.81
345.1	1.81
CURVE 2	
320.6	1.08
326.8	1.08*
327.9	1.07*
332.1	1.08
333.8	1.07*
339.5	1.07*
344.3	1.06*
CURVE 3	
789.3	0.666
841.5	0.556
914.8	0.495
1049.8	0.453
1133.2	0.402
1364.3	0.403
1552.1	0.365
1869.3	0.381
CURVE 4	
22	0.0397
27	0.0711
42	0.218
46	0.272
74	0.628
85	0.795
120	1.38
130	1.550
200	1.840
230	1.930
280	1.80

T	k
CURVE 5	
21	0.0146
25	0.0285
30	0.0460
42	0.105
60	0.188
85	0.314
100	0.397
150	0.544
200	0.669
300	0.711
CURVE 6	
26	0.00460
36	0.0126
65	0.0314
80	0.0460
100	0.0628
200	0.146
280	0.209
CURVE 7	
351.2	0.950
369.2	0.925
411.2	0.900
454.2	0.858
497.2	0.808
CURVE 8	
500.2	0.828
566.2	0.787
729.2	0.745
900.2	0.623
1080.2	0.594
1178.2	0.594
1294.2	0.590*

T	k
CURVE 9	
339.2	1.03
357.2	1.00
393.2	0.983
495.2	0.929
CURVE 10	
587.2	0.849
797.2	0.728
1029.2	0.632
1260.2	0.594
1394.2	0.615
CURVE 11*	
612.2	0.837
793.2	0.724
1025.2	0.623
1250.2	0.598
1384.2	0.632
CURVE 12	
4.90	0.126
4.94	0.137
8.8	0.435
14.0	1.130
22.1	2.85
31.0	6.28
37.5	8.56
47.2	11.1
55.2	14.4
78.4	25.5
94.1	25.8
116.0	21.1

T	k
CURVE 13	
7.3	0.774
16.0	4.04
23.0	8.18
30.8	15.1
39.5	20.9
52.0	24.7
67.0	26.6
78.4	27.2
300.0	5.02
CURVE 14	
11.0	0.795
44.0	13.0
78.4	23.9
135.0	15.4
195.0	9.21
299.0	4.60
CURVE 15	
323	0.890
373	0.920
423	0.905
473	0.895
CURVE 16	
300	0.573
460	0.565
670	0.573
770	0.575
800	0.477
1200	0.480
1300	0.420
1510	0.400
3170	0.230
3330	0.175
3450	0.137
3475	0.147
3500	0.125
3550	0.113
3600	0.085
3710	0.053

T	k
CURVE 17	
300	0.280
450	0.270
670	0.310
760	0.285
960	0.290
1190	0.250
1380	0.240
3170	0.230*
3330	0.175*
3450	0.137*
3475	0.147*
3500	0.125*
3550	0.113*
3600	0.085*
3710	0.053*
CURVE 18	
13	0.0159
15.5	0.0247
18.5	0.0418
20	0.0418
23	0.0628
28.5	0.117
42	0.251
60	0.502
94	0.900
150	1.51
240	1.76
300	1.76
CURVE 19	
3150	0.20
3315	0.17
3420	0.13
3450	0.14
3480	0.12
3520	0.11
3545	0.10
3575	0.086
3700	0.057

T	k
CURVE 20	
484.3	1.08
526.5	1.05
534.3	1.04*
587.6	0.966
590.4	0.990
642.6	0.971
659.8	0.893
663.2	0.895*
733.2	0.893
734.8	0.848
759.8	0.834
778.7	0.831*
875.4	0.765
904.3	0.762*
904.3	0.768*
1030.9	0.697
1037.1	0.737*
1061.5	0.703
1196.0	0.678
1226.5	0.696
CURVE 21	
373.2	1.25
473.2	1.02
673.2	0.787
873.2	0.644
1073.2	0.536
CURVE 22	
303.2	0.0393
CURVE 23	
303.2	0.0346
CURVE 24	
303.2	0.0328
CURVE 25	
303.2	0.0279

T	k
CURVE 26	
303.2	0.0246
CURVE 27	
303.2	0.0615
CURVE 28	
303.2	0.0418
CURVE 29	
303.2	0.0358
CURVE 30*	
303.2	0.0335
CURVE 31	
303.2	0.0319
CURVE 32*	
303.2	0.0310
CURVE 33	
303.2	0.0301
CURVE 34	
829	0.730
1203.4	0.514
1454.6	0.464
1686.5	0.431
1866.1	0.440
CURVE 35	
1219.8	0.438
1372.6	0.424
1564.8	0.369*
1808.7	0.322

T	k
CURVE 35 (cont.)	
2005.4	0.292
2213.7	0.292
2520.4	0.270
2699.8	0.242
CURVE 36	
1509.8	0.372
1571.5	0.315
1785.4	0.291
2016.5	0.291*
2193.7	0.298
2507.1	0.244
CURVE 37	
1318.7	0.408
1444.8	0.336
1653.7	0.369
1808.2	0.317
2188.2	0.305
2407.1	0.298
2477.6	0.320
2567.6	0.303
2655.4	0.275
2801.0	0.346
3030.4	0.365
3276.5	0.292
CURVE 38*	
353.2	0.1287
413.2	0.1130
473.2	0.1025
503.2	0.983
563.2	0.879
693.2	0.774
773.2	0.732
803.2	0.711
863.2	0.669
988.2	0.586
1093.2	0.544

*Not shown on plot

DATA TABLE NO. 11 (continued)

T	k
CURVE 39*	
328.2	0.92
393.2	0.837
468.2	0.774
543.2	0.669
653.2	0.586
763.2	0.544
913.2	0.481
1043.2	0.460
1093.2	0.460
CURVE 40*	
378.2	0.628
513.2	0.502
623.2	0.460
713.2	0.397
828.2	0.377
893.2	0.356
963.2	0.356
1023.2	0.314
1068.2	0.293
1123.2	0.293
CURVE 41	
79	0.155
142	0.178
261	0.328
292	0.384
375	0.544
423	0.692
535.5	1.08
555	1.17
CURVE 42*	
328.7	1.13
CURVE 43*	
329.7	1.09
CURVE 44*	
326.7	1.30

T	k
CURVE 45*	
331.7	1.17
CURVE 46*	
337.2	1.30
CURVE 47*	
344.7	1.30
CURVE 48*	
337.2	1.34
CURVE 49	
323.2	0.446
363.2	0.461
CURVE 50	
313.2	0.0119
343.2	0.0134
CURVE 51	
313.2	0.00386
343.2	0.00421
CURVE 52	
313.2	0.00183
343.2	0.00202
CURVE 53	
9.3	0.005
20.0	0.044
93.0	1.0
CURVE 54	
2.8	0.00011
4.6	0.00043
9.2	0.0028
20.0	0.028

T	k
CURVE 55	
4.8	0.00044
11	0.0035
20	0.019
30	0.05
60	0.26
95	0.70
196	1.5
275	1.70
CURVE 56	
5	0.00035
9.8	0.0021
22.0	0.0175
33	0.044
60	0.2
93	0.46
CURVE 57	
4.5	0.00027
9.5	0.0014
20	0.0075
93	0.19
CURVE 58	
10	0.00111
20	0.008
34	0.023
56	0.077
95	0.21
CURVE 59	
3.5	0.0011
5.4	0.0041
5.8	0.0048
10.5	0.026
23.0	0.19
34.0	0.48
60	1.75
100	2.6
120	2.9
135	3.2
200	3.2
230	2.9
300	2.7

T	k
CURVE 60	
5	0.0025
9.5	0.0125
21	0.09
35	0.23
63	0.56
100	0.75
200	0.86
280	0.83
CURVE 61*	
336.2	1.30
CURVE 62*	
336.2	1.00
CURVE 63*	
339.2	0.753
CURVE 64*	
343.2	1.34
CURVE 65*	
337.2	1.05
CURVE 66*	
344.2	0.753
CURVE 67*	
343.2	1.30
CURVE 68*	
341.2	0.962
CURVE 69*	
344.2	0.795

T	k
CURVE 70	
1428.2	0.364
1828.2	0.266
2138.2	0.251
2408.2	0.218
2688.2	0.207
2923.2	0.197
3148.2	0.197

p(in. Hg)	k
CURVE 71* (T = 1838.2K)	
0	0.215
10	0.220
20	0.222
30	0.236
40	0.243
50	0.243
60	0.241
CURVE 72* (T = 2423.2K)	
0	1.90
10	1.99
20	2.02
30	2.06
40	2.06
50	2.06
60	2.07
CURVE 73 (T = 2973.2K)	
0	0.153
10	0.164
20	0.167*
30	0.170*
40	0.170*
50	0.174
60	0.172*

p(in. Hg)	k
CURVE 74* (T = 1838.2K)	
0	0.259
10	0.262
20	0.282
30	0.265
40	0.270
50	0.274
60	0.282
CURVE 75* (T = 2394.2K)	
0	0.255
10	0.257
20	0.260
30	0.263
40	0.266
60	0.273
CURVE 76* (T = 2913.2K)	
0	0.210
10	0.217
20	0.222
30	0.220
40	0.220
50	0.228
60	0.224
CURVE 77* (T = 1829.2K)	
0	0.225
5	0.231
10	0.232
19.25	0.241
31.5	0.245
45	0.246
55.75	0.253
CURVE 78* (T = 2433.2K)	
0	0.198
10	0.205
20	0.209
31.25	0.209

p(in. Hg)	k
CURVE 78 (cont.)*	
45	0.222
55.5	0.227
CURVE 79* (T = 2973.2K)	
31	0.199
45	0.204
55.5	0.207

T	k
CURVE 80	
1473.2	0.153
1533.2	0.153
1608.2	0.161
1768.2	0.157
1788.2	0.163
1888.2	0.155
1938.2	0.155
2068.2	0.153
2103.2	0.142
2168.2	0.161
2188.2	0.146
2328.2	0.146
2423.2	0.132
2483.2	0.132
2663.2	0.136
2733.2	0.134
2933.2	0.136
CURVE 81*	
1773.2	0.161
2073.2	0.155
2223.2	0.163
2443.2	0.142
2523.2	0.142
CURVE 82	
1478.2	0.172
1658.2	0.155
1848.2	0.163
1918.2	0.161
1968.2	0.165
2148.2	0.163

T	k
CURVE 82 (cont.)	
2308.2	0.161
2488.2	0.167
2683.2	0.140
2798.2	0.142
2968.2	0.136*
CURVE 83	
1643.2	0.282
1818.2	0.257
2058.2	0.253
2163.2	0.249*
2433.2	0.232
CURVE 84*	
1513.2	0.285
1763.2	0.259
1993.2	0.251
2158.2	0.226
2473.2	0.222
2633.2	0.213
2753.2	0.197
2933.2	0.201
CURVE 85*	
1983.2	0.245
CURVE 86*	
1638.2	0.247
1908.2	0.224
2058.2	0.215
2448.2	0.209
CURVE 87	
1713.2	0.213
1923.2	0.197
2048.2	0.188
2348.2	0.190
2578.2	0.186
2738.2	0.172
2853.2	0.165
2983.2	0.154*

*Not shown on plot

DATA TABLE NO. 11 (continued)

T	k
CURVE 88	
1663.2	0.190
1813.2	0.188
1973.2	0.188
2108.2	0.184
2213.2	0.192
2353.2	0.192*
2508.2	0.184
2763.2	0.172*
2993.2	0.151
CURVE 89*	
1783.2	0.215
2008.2	0.207
2118.2	0.213
2448.2	0.186
2708.2	0.182
2948.2	0.163
3273.2	0.157
CURVE 90*	
298.2	0.402
CURVE 91*	
298.2	0.268
CURVE 92*	
298.2	0.412
CURVE 93*	
298.2	0.310
CURVE 94*	
298.2	0.658
CURVE 95	
308.2	1.004
333.2	0.879
383.2	0.774
463.2	0.586
643.2	0.502
903.2	0.377

T	k
CURVE 96	
308.2	1.17
373.2	1.09
CURVE 97*	
323.2	1.13
CURVE 98	
313.2	1.80
558.2	1.51
873.2	1.42
CURVE 99	
323.2	1.55
CURVE 100	
323.2	1.46
CURVE 101	
1422.1	0.160
1811.0	0.135
1977.6	0.138
2127.6	0.138
2422.1	0.130
CURVE 102*	
1366.5	0.213
1561.0	0.211
1672.1	0.209
1888.7	0.204
2005.4	0.199
2255.4	0.194
CURVE 103*	
1416.5	0.367
1588.7	0.355
1727.6	0.329
1894.3	0.343
2099.8	0.329
2255.4	0.329

T	k
CURVE 104	
1173.2	0.345
1273.2	0.333
1373.2	0.320
1473.2	0.312
1573.2	0.305
1673.2	0.299
1773.2	0.295
1873.2	0.290
1973.2	0.287
2073.2	0.285
2173.2	0.283
2273.2	0.282
CURVE 105*	
1193.0	0.257
CURVE 106*	
1185	0.422
CURVE 107*	
1185	0.422
CURVE 108*	
1194	0.240
CURVE 109*	
1189	0.362
CURVE 110*	
1189	0.362
CURVE 111	
653.2	0.490
723.2	0.477
823.2	0.448
843.2	0.452
873.2	0.437
933.2	0.433
963.2	0.423

T	k
CURVE 112	
703.2	0.619
753.2	0.602
803.2	0.565
873.2	0.544
898.2	0.531
CURVE 113*	
723.2	0.774
783.2	0.745
828.2	0.686
873.2	0.665
898.2	0.644
CURVE 114	
2.2	0.0000795
2.6	0.0000795
3.2	0.000126
4.5	0.000356
5.2	0.000360
6.2	0.000586
8.2	0.00105
10.0	0.00172
13.0	0.00293
14.0	0.00418
18.0	0.00669
23.0	0.0126
27.0	0.0146
32.0	0.0255
45.0	0.0502
50.0	0.0795
75.0	0.159
82.0	0.163
85.0	0.180
95.0	0.234
CURVE 115	
2.8	0.0000347
3.25	0.0000347
3.4	0.0000439
4.25	0.0000586
5.0	0.0000795
6.5	0.000109
10.0	0.000218
11.0	0.000230
13.0	0.000293

T	k
CURVE 115 (cont.)	
14.0	0.000335
15.0	0.000377
19.0	0.000523
22.0	0.000711
35.0	0.00184
47.0	0.00289
55.0	0.00410
62.0	0.00523
80.0	0.00753
CURVE 116	
1573.2	0.000814
1773.2	0.000860
2273.2	0.00107
2573.2	0.00109
2773.2	0.00112
3273.2	0.00114
CURVE 117	
115	0.172
205	0.331
295	0.435
385	0.492
CURVE 118	
115	0.245
205	0.470
295	0.592
385	0.651
CURVE 119	
115	0.345
205	0.650
295	0.795
385	0.840
CURVE 120	
115	0.500
205	0.900
295	1.04
385	1.06

T	k
CURVE 121*	
1170	0.297
1350	0.305
1450	0.302
1550	0.282
1650	0.300
1750	0.290
1830	0.282
1930	0.275
2040	0.308
2160	0.305
2250	0.284
2350	0.280
2450	0.292
CURVE 122*	
1170	0.322
1225	0.315
1350	0.330
1450	0.337
1520	0.325
1580	0.290
1670	0.292
1760	0.305
1850	0.305
1930	0.285
1960	0.305
2030	0.332
2135	0.267
2160	0.317
2230	0.295
2290	0.310
2355	0.298
2410	0.317
2460	0.310
2600	0.307
CURVE 123*	
1115	0.143
1190	0.367
1250	0.393
1330	0.375
1390	0.340
1455	0.362
1510	0.307
1575	0.305
1655	0.312*

T	k
CURVE 123 (cont.)*	
1790	0.295*
1855	0.293*
1935	0.307*
2010	0.307
2035	0.283*
2110	0.293
2205	0.275
2240	0.287
2300	0.285
2350	0.288*
2395	0.303
2475	0.305
2530	0.295
2580	0.315
2660	0.320
2725	0.315
CURVE 124*	
1260	0.241
1387	0.272
1647	0.267
2199	0.303
CURVE 125	
1260	1.57
1387	1.53
1647	1.25
2199	0.912
CURVE 126*	
1353.2	0.155
1463.2	0.163
1573.2	0.151
1633.2	0.151
1643.2	0.159
1743.2	0.165
1868.2	0.146
1873.2	0.151
1963.2	0.157
2013.2	0.155
2103.2	0.157
2113.2	0.160
2223.2	0.172
2303.2	0.172

T	k
CURVE 127*	
1383.2	0.195
1523.2	0.188
1593.2	0.182
1693.2	0.186
1783.2	0.190
1903.2	0.176
1953.2	0.186
2073.2	0.178
2273.2	0.184
2443.2	0.178
2583.2	0.167
CURVE 128*	
1170	0.48
1230	0.62
1300	0.71
1375	0.555
1470	0.56
1530	0.585
1650	0.64
1740	0.56
1800	0.605
1890	0.545
1950	0.555
2040	0.565
2110	0.63
2220	0.55
2340	0.595
CURVE 129	
1180	0.51*
1240	0.6*
1320	0.59
1490	0.495
1550	0.53
1690	0.7
1760	0.585
1840	0.665
1900	0.55
1990	0.59
2070	0.52
2180	0.635
2280	0.5
2370	0.53
2450	0.58
2500	0.625

*Not shown on plot

DATA TABLE NO. 11 (continued)

T	k
CURVE 129 (cont.)	
2580	0.48
2600	0.515
2680	0.52
2760	0.565
CURVE 130*	
1180	0.50
1310	0.30
1425	0.245
1525	0.225
1590	0.21
1680	0.23
1740	0.20
1880	0.20
2150	0.185
2290	0.195
2400	0.195
CURVE 131*	
1180	0.345
1240	0.365
1340	0.31
1420	0.27
1480	0.24
1570	0.205
1700	0.205
1800	0.19
1870	0.185
2070	0.19
2140	0.175
2240	0.2
2350	0.18
CURVE 132	
1200	1.14
1260	1.025
1320	0.95
1350	1.05
1420	0.71
1440	0.845
1490	0.625
1670	0.655
1740	0.645
CURVE 132 (cont.)	
1810	0.625
1890	0.635
1995	0.66
2180	0.67
CURVE 133	
1220	1.73
1300	1.34
1490	1.35
1540	0.88
1610	0.8
1670	1.04
1700	0.76
1880	0.75
1970	0.75
2150	0.72
2280	0.73
CURVE 134	
1220	0.227
1290	0.224
1380	0.205
1460	0.19
1510	0.17
1620	0.17
1700	0.173
1770	0.164
1860	0.15
1900	0.155*
1980	0.143
2080	0.153*
2240	0.15
2330	0.156
2510	0.152
2630	0.169
CURVE 135	
1623.2	0.476
1623.2	0.486
1863.2	0.455
1973.2	0.420
1993.2	0.415
2008.2	0.404
CURVE 135 (cont.)	
2023.2	0.411
2128.2	0.392
2148.2	0.390*
2263.2	0.369
2473.2	0.370
2568.2	0.366
2608.2	0.365*
2713.2	0.360*
2773.2	0.360
CURVE 136*	
1593.2	0.249
1733.2	0.249
1893.2	0.212
1993.2	0.230
2093.2	0.238
2163.2	0.218
2323.2	0.230
2473.2	0.217
2509.2	0.198
2663.2	0.202
2823.2	0.192
CURVE 137*	
1623.2	0.220
1773.2	0.205
1853.2	0.206
2023.2	0.190
2253.2	0.184
2323.2	0.180
2578.2	0.180
2773.2	0.178
2823.2	0.170
CURVE 138*	
1653.2	0.205
1943.2	0.180
2093.2	0.181
2423.2	0.175
2493.2	0.170
2823.2	0.170
CURVE 139*	
1300	0.180
1400	0.164
1500	0.187
1600	0.163
1700	0.188
1800	0.175
1900	0.182
2000	0.172
2100	0.159
2200	0.165
CURVE 140	
1300	0.143
1400	0.152
1500	0.156
1600	0.155
1700	0.153
1800	0.156*
1900	0.151*
2000	0.150*
2100	0.153*
2220	0.142
2330	0.164
2465	0.114
CURVE 141*	
1300	0.145
1400	0.143
1500	0.141
1600	0.150
1700	0.141
1800	0.137
1900	0.159
2000	0.138
2100	0.146
2200	0.160
CURVE 142*	
1513.2	0.197
1833.2	0.174
2048.2	0.159
CURVE 143*	
1533.2	0.188
1933.2	0.163
CURVE 144*	
1653.2	0.175
2061.2	0.162
2393.2	0.0887
CURVE 145*	
1833.2	0.167
2158.2	0.163
2323.2	0.171
2583.2	0.142
2773.2	0.146
2983.2	0.127
3037.2	0.120
3268.2	0.109
CURVE 146*	
1833.2	0.134
2478.2	0.121
2743.2	0.114
CURVE 147*	
1683.2	0.146
1963.2	0.142
CURVE 148*	
1643.2	0.184
1703.2	0.205
CURVE 149*	
1643.2	0.201
2013.2	0.222
CURVE 150*	
1753.2	0.213
2043.2	0.209
2253.2	0.223
2453.2	0.213
CURVE 151*	
1683.2	0.201
1953.2	0.214
1973.2	0.193
2093.2	0.207
CURVE 152*	
1643.2	0.186
1803.2	0.187
1983.2	0.193
2083.2	0.192
2123.2	0.203
CURVE 153	
1363.2	0.0686
1453.2	0.0649
CURVE 154	
1393.2	0.0816
1593.2	0.0711
1733.2	0.0732
CURVE 155	
1293.2	0.100
1533.2	0.101
1858.2	0.111
2013.2	0.119
CURVE 156*	
1393.2	0.131
1563.2	0.155
1813.2	0.176
2073.2	0.186
2333.2	0.174
CURVE 157*	
1393.2	0.207
1653.2	0.218
1813.2	0.207
2158.2	0.180
2525.2	0.174
2603.2	0.174
CURVE 158*	
1403.2	0.253
1613.2	0.218
1893.2	0.206
2263.2	0.192
2503.2	0.184
2803.2	0.180
2933.2	0.190
CURVE 159*	
1293.2	0.163
1533.2	0.169
1743.2	0.169
2033.2	0.170
2443.2	0.168
2783.2	0.167
2998.2	0.166
3093.2	0.180
CURVE 160*	
1603.2	0.155
1793.2	0.167
2113.2	0.172
2403.2	0.172
2573.2	0.167
3073.2	0.174
CURVE 161*	
1413.2	0.153
1853.2	0.169
2463.2	0.174
2693.2	0.167
2963.2	0.167
3023.2	0.174
3063.2	0.165
3103.2	0.172
CURVE 162*	
1513.2	0.176
1933.2	0.182
2273.2	0.176
2703.2	0.174
2963.2	0.167
3153.2	0.172
CURVE 163*	
1953.2	0.180
2953.2	0.163
CURVE 164	
20	0.0130
22	0.0140
31	0.0250
44	0.0500
80	0.150
95	0.190
180	0.450
273	0.650
CURVE 165	
18.5	0.011
25	0.021
33	0.0400
37	0.0510
80	0.190
90	0.222
95	0.240
180	0.550
273	0.750
CURVE 166	
18	0.0160
19	0.0185
20	0.0205
25	0.0400
35	0.0800
48	0.135
80	0.300
90	0.330
95	0.355
100	0.400*
105	0.440
190	0.860
300	1.10

*Not shown on plot

DATA TABLE NO. 11 (continued)

T	k
CURVE 167	
80	0.605
100	0.845
140	1.30
180	1.52
220	1.62
260	1.59
300	1.56
CURVE 168	
200	1.54
300	1.51
CURVE 169	
293.2	1.782
673.2	1.130
1273.2	0.628
CURVE 170*	
1273.2	0.498
1873.2	0.452
CURVE 171*	
1273.2	0.586
1873.2	0.444
CURVE 172*	
1273.2	0.431
1873.2	0.310
CURVE 173*	
1273.2	0.624
1873.2	0.419
CURVE 174*	
1273.2	0.272
1873.2	0.226
CURVE 175	
298.2	0.0167
CURVE 176*	
298.2	0.0335
CURVE 177*	
298.2	0.448
CURVE 178*	
298.2	1.31
CURVE 179*	
1353.2	0.347
1503.2	0.339
1873.2	0.318
2303.2	0.301
CURVE 180	
15.2	0.00209
50.2	0.0218
100.2	0.0762
195.2	0.211
296.2	0.291
CURVE 181*	
298.2	1.55
CURVE 182*	
298.2	0.866
CURVE 183*	
298.2	1.49
CURVE 184*	
298.2	1.02
CURVE 185*	
298.2	1.47
CURVE 186*	
298.2	1.17
CURVE 187*	
298.2	1.34
CURVE 188*	
298.2	1.15
CURVE 189*	
298.2	1.25
CURVE 190*	
298.2	1.19
CURVE 191*	
298.2	1.62
CURVE 192*	
298.2	0.975
CURVE 193*	
298.2	1.58
CURVE 194*	
298.2	0.933
CURVE 195*	
298.2	1.58
CURVE 196*	
298.2	1.28
CURVE 197*	
298.2	1.58
CURVE 198*	
298.2	0.992
CURVE 199*	
298.2	1.51
CURVE 200*	
298.2	1.18
CURVE 201*	
298.2	1.46
CURVE 202*	
298.2	1.21
CURVE 203*	
298.2	1.15
CURVE 204*	
298.2	1.10
CURVE 205*	
298.2	1.64
CURVE 206*	
298.2	1.05
CURVE 207*	
298.2	1.69
CURVE 208*	
298.2	1.33
CURVE 209*	
298.2	0.259
CURVE 210*	
298.2	1.21
CURVE 211*	
298.2	0.791
CURVE 212*	
298.2	0.958
CURVE 213*	
298.2	0.849
CURVE 214*	
290.2	0.155
373.2	0.159
CURVE 215	
298.2	2.16
CURVE 216*	
298.2	0.870
CURVE 217*	
298.2	1.87
CURVE 218*	
298.2	0.833
CURVE 219*	
298.2	2.19
CURVE 220*	
298.2	0.770
CURVE 221	
298.2	2.25
CURVE 222*	
298.2	0.987
CURVE 223*	
298.2	2.30
CURVE 224*	
298.2	0.849
CURVE 225*	
298.2	2.15
CURVE 226*	
298.2	0.828
CURVE 227	
298.2	1.94
CURVE 228*	
298.2	1.04
CURVE 229*	
298.2	2.01
CURVE 230*	
298.2	0.980
CURVE 231	
298.2	2.09
CURVE 232*	
298.2	1.33
CURVE 233*	
298.2	1.78
CURVE 234*	
298.2	0.946
CURVE 235*	
298.2	1.78
CURVE 236*	
298.2	0.950
CURVE 237*	
298.2	1.96
CURVE 238*	
298.2	0.736
CURVE 239*	
298.2	1.95
CURVE 240*	
298.2	0.741
CURVE 241*	
298.2	1.98
CURVE 242*	
298.2	0.732
CURVE 243*	
298.2	2.03
CURVE 244*	
298.2	0.912
CURVE 245*	
298.2	2.23
CURVE 246*	
298.2	1.32
CURVE 247*	
298.2	1.84
CURVE 248*	
298.2	0.929
CURVE 249	
298.2	2.39
CURVE 250*	
298.2	1.78
CURVE 251*	
298.2	1.47
CURVE 252*	
298.2	0.812
CURVE 253*	
298.2	1.56
CURVE 254*	
298.2	0.339
CURVE 255*	
298.2	1.56
CURVE 256*	
298.2	0.339
CURVE 257*	
298.2	1.10

*Not shown on plot

DATA TABLE NO. 11 (continued)

T	k
CURVE 258*	
298.2	0.941
CURVE 259*	
298.2	1.10
CURVE 260*	
298.2	0.946
CURVE 261*	
298.2	1.11
CURVE 262*	
298.2	0.937
CURVE 263*	
298.2	1.09
CURVE 264*	
298.2	0.937
CURVE 265*	
298.2	1.12
CURVE 266*	
298.2	0.987
CURVE 267*	
298.2	1.14
CURVE 268*	
298.2	1.00
CURVE 269*	
298.2	1.10
CURVE 270*	
298.2	0.967
CURVE 271*	
298.2	1.14
CURVE 272*	
298.2	0.828
CURVE 273*	
298.2	1.17
CURVE 274*	
298.2	0.828
CURVE 275*	
298.2	1.12
CURVE 276*	
298.2	0.833
CURVE 277*	
298.2	1.33
CURVE 278*	
298.2	1.26
CURVE 279*	
298.2	1.34
CURVE 280*	
298.2	1.06
CURVE 281	
1593.2	0.182
1913.2	0.143
2288.2	0.116
CURVE 281 (cont.)	
2578.2	0.101
2793.2	0.0959
3014.2	0.106
3198.2	0.117
CURVE 282*	
2189.2	0.177
2348.2	0.162
2398.2	0.156
2583.2	0.149
2904.2	0.112
3033.2	0.112
CURVE 283*	
1905.7	0.142
2285.2	0.115
2577.2	0.100
2793.2	0.0946
3016.0	0.105
3200.2	0.115
CURVE 284*	
2078.2	0.169
2323.2	0.155
2553.2	0.147
2811.2	0.138
2863.2	0.136
2951.2	0.133
3043.2	0.135
3134.2	0.138
CURVE 285*	
2068.2	0.140
2318.2	0.126
2557.2	0.121
2815.2	0.122
CURVE 286*	
1403.2	0.305
1473.2	0.289
1673.2	0.255
1873.2	0.238
CURVE 286 (cont.)*	
2073.2	0.228
2273.2	0.220
2473.2	0.213
2673.2	0.207
2873.2	0.203
3073.2	0.201
3273.2	0.209
CURVE 287*	
1418.2	0.318
1573.2	0.280
1728.2	0.262
1773.2	0.253
2003.2	0.249
2053.2	0.234
2128.2	0.232
2238.2	0.241
2258.2	0.224
2393.2	0.220
2468.2	0.224
2563.2	0.203
2673.2	0.220
2708.2	0.195
2848.2	0.192
2888.2	0.199
2963.2	0.195
2978.2	0.192
3063.2	0.192
3188.2	0.201
CURVE 288*	
1503.2	0.140
2273.2	0.142
3073.2	0.149
CURVE 289*	
1363.2	0.238
1473.2	0.224
1673.2	0.213
1873.2	0.205
2073.2	0.203
2273.2	0.205
2473.2	0.203
2673.2	0.197
CURVE 289 (cont.)*	
2873.2	0.192
3073.2	0.192
3183.2	0.192
CURVE 290	
1373.2	0.126
1873.2	0.126
2673.2	0.126
2773.2	0.128
CURVE 291*	
1343	0.169
1443.2	0.188
1468.2	0.176
1578.2	0.192
1758.2	0.193
1843.2	0.200
1983.2	0.199
2083.2	0.211
2243.2	0.226
2313	0.227
CURVE 292*	
1303.2	0.224
1333.2	0.213
1483.2	0.236
1498.2	0.230
1653.2	0.236
1813.2	0.246
1858.2	0.246
1938.2	0.239
1963.2	0.241
2033.2	0.244
2098.2	0.232
2223.2	0.243
2333.2	0.241
2463.2	0.230
2543.2	0.230
2603.2	0.228
CURVE 293*	
1303.2	0.253
1353.2	0.253
1453.2	0.259
1513.2	0.245
1558.2	0.245
1588.2	0.248
1678.2	0.251
1833.2	0.245
1918.2	0.253
2103.2	0.253
2193.2	0.249
2238.2	0.251
2468.2	0.238
2608.2	0.241
2723.2	0.245
2753.2	0.238
2863.2	0.238
2948.2	0.234
CURVE 294*	
1353.2	0.155
1453.2	0.161
1563.2	0.151
1628.2	0.151
1643.2	0.159
1743.2	0.165
1853.2	0.146
1873.2	0.153
1963.2	0.157
2013.2	0.155
2103.2	0.157
2113.2	0.161
2233.2	0.172
2303.2	0.172
CURVE 295*	
1383.2	0.193
1523.2	0.188
1588.2	0.180
1688.2	0.186
1783.2	0.188
1903.2	0.176
1958.2	0.185
2078.2	0.178
2278.2	0.184
CURVE 295 (cont.)*	
2448.2	0.180
2583.2	0.169
CURVE 296*	
1373.2	0.207
1383.2	0.199
1513.2	0.209
1518.2	0.205
1643.2	0.200
1743.2	0.192
1753.2	0.192
1913.2	0.180
1933.2	0.188
2063.2	0.192
2093.2	0.182
2258.2	0.190
2273.2	0.184
2453.2	0.186
2513.2	0.174
2563.2	0.184
2663.2	0.174
2753.2	0.174
2843.2	0.172
2843.2	0.180
2973.2	0.169
CURVE 297	
1318.2	0.0510
1383.2	0.0510
1433.2	0.0523
1513.2	0.0523
1608.2	0.0531
1768.2	0.0544
1918.2	0.0552
2003.2	0.0552
2093.2	0.0565
2183.2	0.0586
2233.2	0.0586
CURVE 298	
1323.2	0.0544
1468.2	0.0565
1598.2	0.0565
1723.2	0.0594
CURVE 298 (cont.)	
1848.2	0.0586
2008.2	0.0594
2128.2	0.0615
2228.2	0.0628
2388.2	0.0628
2473.2	0.0607
CURVE 299	
1333.2	0.0690
1473.2	0.0711
1603.2	0.0690
1743.2	0.0711
1893.2	0.0690
2063.2	0.0711
2228.2	0.0711
2353.2	0.0774
2368.2	0.0690
2523.2	0.0732
2573.2	0.0690
2688.2	0.0711
2763.2	0.0711
CURVE 300	
1343.2	0.0711
1523.2	0.0753
1633.2	0.0774
1813.2	0.0795
1983.2	0.0837
2223.2	0.0920
CURVE 301	
1368.2	0.0837
1493.2	0.0879
1613.2	0.0837
1793.2	0.0929
1993.2	0.0941
2263.2	0.103
2413.2	0.101
2523.2	0.101

*Not shown on plot

DATA TABLE NO. 11 (continued)

T	k
CURVE 302	
1438.2	0.0983
1598.2	0.0962
1743.2	0.0971
1883.2	0.0992
2073.2	0.105
2288.2	0.112
2473.2	0.113*
2648.2	0.113
2748.2	0.113
2893.2	0.115
CURVE 303	
323.2	0.619
361.2	0.707
478.2	0.544
593.2	0.527
693.2	0.485
818.2	0.377
873.2	0.460
CURVE 304	
313.2	0.431*
343.2	0.464*
458.2	0.343
623.2	0.276
698.2	0.268
828.2	0.226
CURVE 305*	
321.2	0.954
356.2	0.904
468.2	0.841
633.2	0.653
723.2	0.594
916.2	0.402
CURVE 306*	
321.2	1.71
363.2	2.00
468.2	1.39
656.2	0.920
723.2	0.904
838.2	0.816
CURVE 307*	
321.2	1.18
358.2	1.31
473.2	1.01
606.2	0.736
711.2	0.628
846.2	0.494
CURVE 308*	
318.2	1.83
353.2	2.21
451.2	1.43
593.2	1.20
711.2	1.00
816.2	0.941
CURVE 309*	
318.2	1.20
356.2	1.15
373.2	1.17
468.2	0.887
513.2	0.795
631.2	0.682
696.2	0.653
823.2	0.552
CURVE 310	
313.2	0.351
313.2	0.410
338.2	0.427
458.2	0.393
608.2	0.343
646.2	0.347
688.2	0.351
831.2	0.314
CURVE 311*	
313.2	0.577
313.2	0.615
363.2	0.665
483.2	0.586
623.2	0.494
696.2	0.473
798.2	0.427
CURVE 312*	
313.2	1.04
348.2	1.17
488.2	0.937
613.2	0.711
753.2	0.615
CURVE 313*	
303.2	1.44
356.2	1.67
448.2	1.41
623.2	1.05
718.2	0.858
901.2	0.615
CURVE 314*	
323.2	0.705
CURVE 315*	
323.2	0.695
CURVE 316*	
323.2	0.695
CURVE 317*	
364.3	1.23
383.2	1.19
552.6	1.11
553.7	1.10
769.3	0.952
773.2	0.922
991.5	0.852
993.2	0.884
1375	0.787
1672	0.725
1925	0.737
2239	0.694
CURVE 318*	
1575	0.47
1640	0.46
1710	0.46
1845	0.46
1870	0.46
1910	0.44
2045	0.435
2055	0.45
2085	0.44
2175	0.45
2275	0.44
2400	0.46
CURVE 319*	
1320	0.41
1490	0.43
1640	0.43
1725	0.44
1780	0.44
1825	0.44
1875	0.43
1885	0.44
2070	0.45
2380	0.47
CURVE 320*	
295.2	1.69
306.4	1.59
321.7	1.49
341.5	1.44
365.5	1.39
396.6	1.33
427.2	1.30
463.0	1.28
511.4	1.23
CURVE 321*	
297.5	1.42
309.5	1.27
326.7	1.20
348.9	1.17
376.9	1.13
408.5	1.10
444.9	1.07
CURVE 321 (cont.)*	
489.6	1.03
536.2	0.98
CURVE 322*	
306.2	1.77
326.2	1.71
356.2	1.66
390.2	1.61
433.2	1.55
486.2	1.49
549.2	1.38
619.2	1.30
714.2	1.23
CURVE 323*	
321.2	1.43
359.2	1.35
406.2	1.29
463.2	1.29
533.2	1.19
619.2	1.11
721.2	1.03
CURVE 324*	
302.3	1.44
346.8	1.24
397.0	1.17
447.9	1.11
497.8	1.05
573.2	0.967
598.4	0.941
CURVE 325*	
307.0	0.573
322.2	0.594
342.4	0.586
364.4	0.598
389.8	0.607
421.7	0.623
449.2	0.636
478.0	0.653
512.2	0.640
545.4	0.657
605.2	0.628
CURVE 326*	
310.8	0.431
321.4	0.435
332.8	0.431
346.3	0.439
374.8	0.448
410.8	0.464
446.7	0.481
486.1	0.481
506.0	0.515
CURVE 327	
292.5	0.209
307.5	0.209
318.9	0.230
337.3	0.238
384.3	0.259
410.7	0.289
CURVE 328	
469.2	1.30
676.2	1.01
873.2	0.808
CURVE 329*	
471.2	0.879
674.2	0.820
875.2	0.774
CURVE 330*	
471.2	1.04
672.2	0.795
874.2	0.674
CURVE 331	
471.2	0.423
675.2	0.377
873.2	0.347
CURVE 332*	
472.2	0.481
676.2	0.418
874.2	0.464
CURVE 333*	
468.2	0.849
672.2	0.711
868.2	0.582
CURVE 334*	
468.2	0.849
674.2	0.661
869.2	0.548
CURVE 335*	
472.2	0.410
670.2	0.439
867.2	0.452
CURVE 336*	
471.2	0.649
669.2	0.515
865.2	0.460
CURVE 337*	
1671	0.326
CURVE 338*	
1671	1.28
CURVE 339*	
1671	0.326
CURVE 340*	
1671	1.28
CURVE 341*	
673.2	0.635
773.2	0.56
873.2	0.505
973.2	0.495
1073.2	0.45
1173.2	0.42
CURVE 342*	
1366.5	0.638
2199.8	0.644
2755.4	0.594
3033.2	0.629
3310.9	0.594
CURVE 343*	
1366.5	0.643
2199.8	0.573
2755.4	0.588
3033.2	0.592
3310.9	0.536
CURVE 344*	
1366.5	0.553
2199.8	0.553
2755.4	0.572
3033.2	0.554
3310.9	0.483
CURVE 345*	
1366.5	0.457
2199.8	0.452
2755.4	0.468
3033.2	0.497
3310.9	0.450
CURVE 346*	
338.7	1.13
CURVE 347*	
338.7	1.26
CURVE 348*	
338.7	0.868
CURVE 349*	
338.7	0.901

*Not shown on plot

DATA TABLE NO. 11 (continued)

T	k
CURVE 350*	
1088	1.193
1365	0.906
1643	0.734
1920	0.682
2198	0.682
2475	0.693
2753	0.697
3030	0.703
CURVE 351*	
371.5	1.25
374.8	1.24
492.6	1.16
494.8	1.18
752.1	0.957
760.4	0.945
1015	0.874
1017	0.886
1325	0.770
1633	0.762
1942	0.711
2205	0.710
CURVE 352*	
354.3	0.642
354.3	0.640
529.3	0.604
538.7	0.599
642.1	0.573
648.7	0.561
1025	0.495
1029	0.488
1314	0.429
1617	0.381
1911	0.358
2222	0.343
CURVE 353	
115	0.0165
205	0.0369
295	0.0571
385	0.0775

T	k
CURVE 354	
115	0.061
205	0.128
295	0.176
385	0.212
CURVE 355	
115	0.110
205	0.220
295	0.302
385	0.365
CURVE 356	
93.2	0.0151
195.7	0.0402
273.2	0.0602
373.2	0.0862
CURVE 357	
2.8	0.000078
4.7	0.00022
10.0	0.00083
14.5	0.00175
21	0.0033
56	0.0091
90	0.0125
CURVE 358	
373.2	0.0646
438.2	0.0717
515.2	0.0465
523.2	0.0670
543.2	0.0728
655.2	0.0488
676.2	0.0512
713.2	0.0520
CURVE 359*	
343.2	0.154

T	k
CURVE 360*	
343.2	0.253
CURVE 361*	
373.2	2.10
473.2	1.64
573.2	1.44
673.2	1.29
773.2	1.20
873.2	1.13
CURVE 362*	
311.2	0.0544
311.2	0.0795
341.2	0.0879
456.2	0.0879
582.2	0.0879
676.2	0.0711
753.2	0.0711
853.2	0.0669
CURVE 363*	
310.2	0.167
329.2	0.184
345.2	0.201
465.2	0.192
470.2	0.201
601.2	0.167
609.2	0.176
643.2	0.167
696.2	0.159
706.2	0.146
788.2	0.138
794.2	0.151
804.2	0.142
CURVE 364*	
336.7	1.09
CURVE 365*	
334.2	1.05

T	k
CURVE 366*	
363.2	0.00272
378.2	0.00270
388.2	0.00274
423.2	0.00276
448.2	0.00276
463.2	0.00276
493.2	0.00276
523.2	0.00276
643.2	0.00282
703.2	0.00287
763.2	0.00289
873.2	0.00301
CURVE 367	
2.41	0.0000703
2.85	0.000107
3.58	0.000178
4.54	0.000287
5.78	0.000459
7.05	0.000667
9.38	0.00109
13.4	0.00187
18.3	0.00284
22.1	0.00359
27.9	0.00465
43.5	0.00718
66.4	0.0104
86.3	0.0127
122.2	0.0161
177.8	0.0201
227.0	0.0228
CURVE 368	
2.38	0.0000875
3.00	0.000150
4.06	0.000277
5.12	0.000443
6.32	0.000652
8.75	0.00110
11.8	0.00175
15.9	0.00262
22.3	0.00393
30.5	0.00547

T	k
CURVE 368 (cont.)	
43.4	0.00760
58.6	0.00984
79.6	0.0125
118.3	0.0161
CURVE 369*	
318.2	1.16
323.2	1.18
338.2	1.16
393.2	1.07
413.2	1.04
435.2	1.02
543.2	0.900
568.2	0.858
611.2	0.816
CURVE 370*	
1048	0.513
1083	0.464
1342	0.293
1363	0.345
CURVE 371*	
1723	0.268
1923	0.243
1963	0.126
2173	0.176
2713	0.137
CURVE 372*	
1798	0.241
1848	0.146
1907	0.130
2013	0.176
2360	0.126
2548	0.131
2798	0.117
3048	0.142

T	k
CURVE 373*	
1683	0.341
1773	0.257
1783	0.209
1846	0.264
1886	0.202
2233	0.151
2343	0.136
CURVE 374*	
93.2	2.49
193.2	2.05
273.2	1.75
373.2	1.46
CURVE 375*	
93.2	1.75
193.2	1.40
273.2	1.15
373.2	0.931
CURVE 376*	
613.2	1.07
673.2	1.00
710.7	0.946
798.2	0.904
818.2	0.870
913.2	0.787
943.2	0.778
953.2	0.745
1018	0.728
1076	0.669
1128	0.628
CURVE 377*	
483.2	1.26
555.7	1.15
555.7	1.11
633.2	1.08
633.2	1.05
723.2	0.962
748.2	0.946

T	k
CURVE 377 (cont.)*	
818.2	0.895
898.2	0.837
918.2	0.795
998.2	0.753
1048	0.753
1113	0.711
CURVE 378*	
303.2	1.64
323.2	1.58
353.2	1.53
408.2	1.42
423.2	1.41
CURVE 379*	
313.2	1.61
320.7	1.60
338.2	1.57
358.2	1.52
398.2	1.46
440.7	1.38
505.7	1.26
508.2	1.25
588.2	1.11
CURVE 380*	
123.2	1.78
233.2	1.76
313.2	1.72
353.2	1.70
413.2	1.71
453.2	1.67
503.2	1.63
543.2	1.59
628.2	1.46
763.2	1.28
853.2	1.12
973.2	0.962
CURVE 381*	
1300	0.595

T	k
CURVE 381 (cont.)*	
1500	0.500
1700	0.445
1800	0.430
1900	0.420
2000	0.400
CURVE 382*	
1200	0.660
1320	0.580
1450	0.520

* Not shown on plot

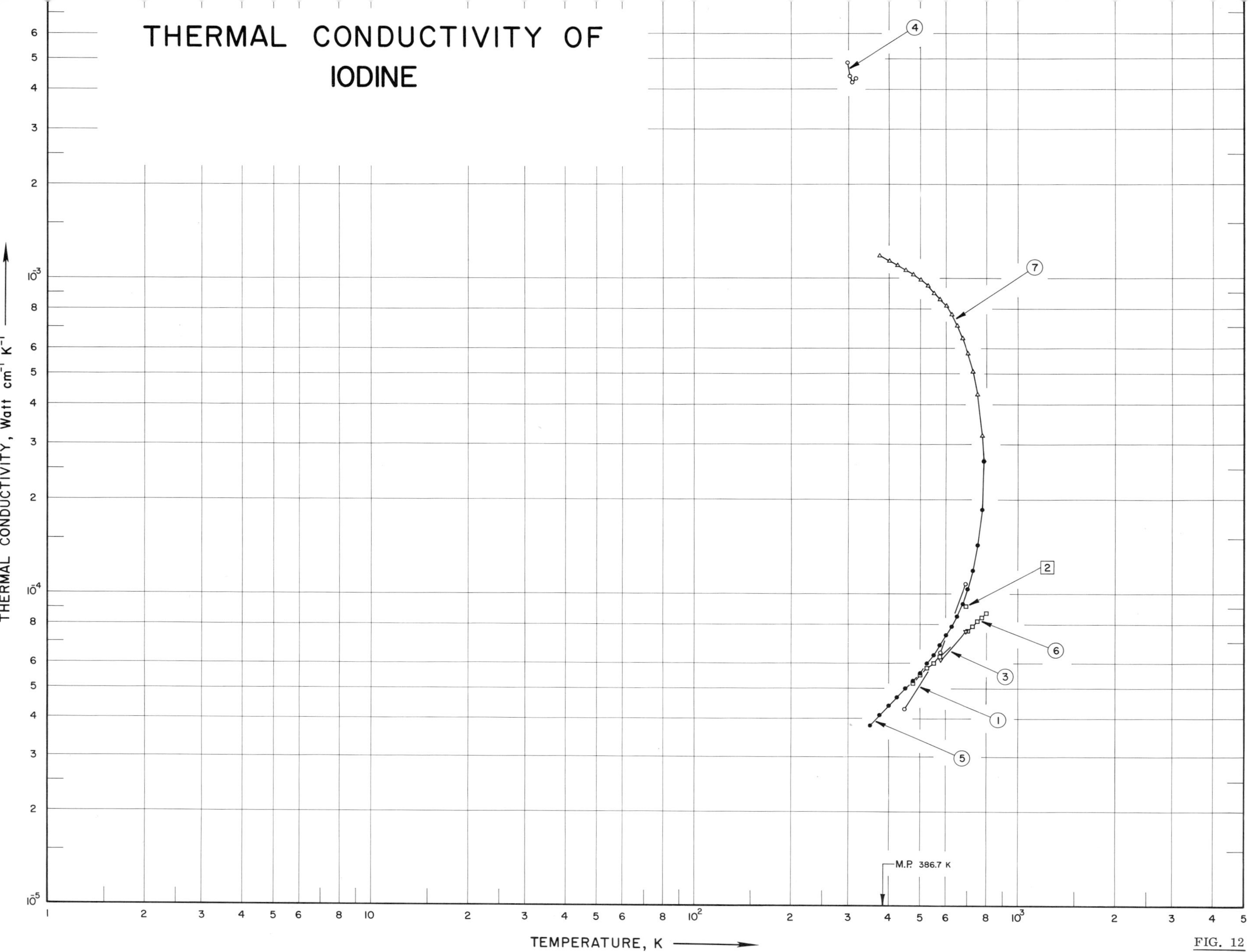
THERMAL CONDUCTIVITY OF
IODINE
THERMAL CONDUCTIVITY, Watt cm⁻¹ K⁻¹
TEMPERATURE, K
M.P. 386.7 K
1
2
3
4
5
6
7

FIG. 12

SPECIFICATION TABLE NO. 12 THERMAL CONDUCTIVITY OF IODINE

[For Data Reported in Figure and Table No. 12

Curve No.	Ref. No.	Method Used	Year	Temp. Range, K	Reported Error, %	Name and Specimen Designation	Composition (weight percent), Specifications and Remarks
1	543	→	1951	447-693	±2.1		In gaseous state; measured at a pressure of 55 mm Hg by the hot-wire method; argon was used for comparison.
2	543	→	1951	693			Similar to above, except measured at a pressure of 120 mm Hg.
3	543	→	1951	579, 693	±1.7		Similar to above, except measured at a pressure of 175 mm Hg.
4	450	C	1923	298-316			Pure iodine; purified by quadruple sublimation; disk specimen 3 cm in dia and 0.27 cm thick; density 4.7 g cm^{-3}; electrical resistivity reported as (data arbitrarily selected from 35 points) 113.2, 106.7, 94.3, 85.2, 67.7, 60.4, 30.9, 19.0, 10.7, 8.6, 6.1, 4.3, 2.9, 1.5, 1.1, and 1.1×10^8 ohm cm at 4.1, 5.0, 6.0, 7.0, 8.2, 9.2, 14.5, 17.3, 20.9, 24.0, 27.3, 31.0, 34.8, 37.7, 41.5, and 42.7 C, respectively; graphite used as comparative material.
5	*			350-785			Saturated vapor; recommended values based on the correlation of Schaefer and Thodos (A.I.Ch.E.Journal, 5, 367-72, 1959).
6	*			350-800			Gas at 1 atm; recommended values based on the correlation of Schaefer and Thodos. (A.I.Ch.E. Journal, 5, 367-72, 1959).
7	*			375-785			Saturated liquid; recommended values based on the correlation of Schaefer and Thodos (A.I.Ch.E. Journal, 5, 367-72, 1959).

DATA TABLE NO. 12 THERMAL CONDUCTIVITY OF IODINE

[Temperature, T, K: Thermal Conductivity, k, Watt $cm^{-1}K^{-1}$]

T	k	T	k	T	k	T	k	T	k	T	k	T	k
CURVE 1		CURVE 4		CURVE 5 (cont.)		CURVE 5 (cont.)		CURVE 6 (cont.)		CURVE 7		CURVE 7 (cont.)	
447	0.0000431	297.6	0.00487	475	0.000053	775	0.000186	525	0.000058	375	1.18	675	0.65
578	0.0000653	302.7	0.00441	500	0.000056	785	0.000265	550	0.000060	400	1.14	700	0.58
693	0.000108	308.6	0.00421	525	0.000060			575	0.000063	425	1.11	725	0.51
		316.1	0.00434	550	0.000064	CURVE 6		600	0.000065**	450	1.07	750	0.43
CURVE 2				575	0.000069			625	0.000068**	475	1.03	775	0.32
		CURVE 5		600	0.000075	350	0.000038**	650	0.000071**	500	0.99	785	0.265
693	0.0000912			625	0.000079	375	0.000041**	675	0.000073**	525	0.95		
		350	0.000038	650	0.000085	400	0.000044**	700	0.000076	550	0.90		
CURVE 3		375	0.000041	675	0.000093	425	0.000047**	725	0.000079	575	0.86		
		400	0.000044	700	0.000104	450	0.000050**	750	0.000082	600	0.82		
579	0.0000619	425	0.000047	725	0.000119	475	0.000052	775	0.000084	625	0.77		
693	0.0000761	450	0.000050	750	0.000143	500	0.000055	800	0.000087	650	0.71		

*Liley, P.E., in course of publication

**Not shown on plot

FIGURE AND TABLE NO. 12R RECOMMENDED THERMAL CONDUCTIVITY OF IODINE

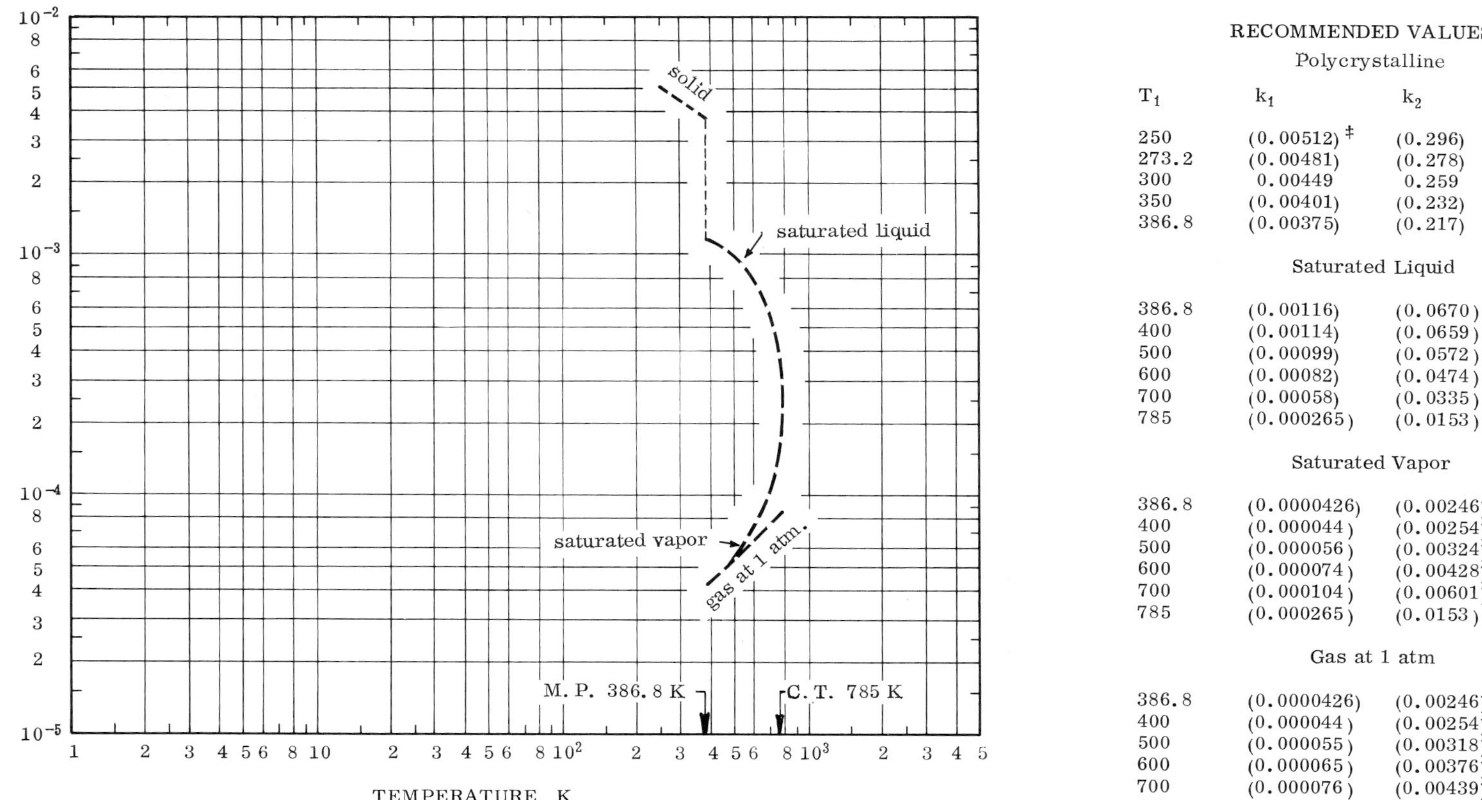

RECOMMENDED VALUES*

T_1	k_1	k_2	T_2
Polycrystalline			
250	(0.00512) ‡	(0.296)	- 9.7
273.2	(0.00481)	(0.278)	32.0
300	0.00449	0.259	80.3
350	(0.00401)	(0.232)	170.3
386.8	(0.00375)	(0.217)	236.5
Saturated Liquid			
386.8	(0.00116)	(0.0670)	236.5
400	(0.00114)	(0.0659)	260.3
500	(0.00099)	(0.0572)	440.3
600	(0.00082)	(0.0474)	620.3
700	(0.00058)	(0.0335)	800.3
785	(0.000265)	(0.0153)	953.3
Saturated Vapor			
386.8	(0.0000426)	(0.00246)	236.5
400	(0.000044)	(0.00254)	260.3
500	(0.000056)	(0.00324)	440.3
600	(0.000074)	(0.00428)	620.3
700	(0.000104)	(0.00601)	800.3
785	(0.000265)	(0.0153)	953.3
Gas at 1 atm			
386.8	(0.0000426)	(0.00246)	236.5
400	(0.000044)	(0.00254)	260.3
500	(0.000055)	(0.00318)	440.3
600	(0.000065)	(0.00376)	620.3
700	(0.000076)	(0.00439)	800.3
785	(0.000087)	(0.00503)	953.3

REMARKS

The recommended values are for high-purity iodine. The values for polycrystalline iodine near room temperature are thought to be accurate to within 10% of the true values. The values for saturated liquid, saturated vapor, and gaseous iodine at 1 atm are estimated and their accuracy is uncertain.

* T_1 in K, k_1 in Watt cm^{-1} K^{-1}, T_2 in F, and k_2 in Btu hr^{-1} ft^{-1} F^{-1}.

‡Values in parentheses are estimated.

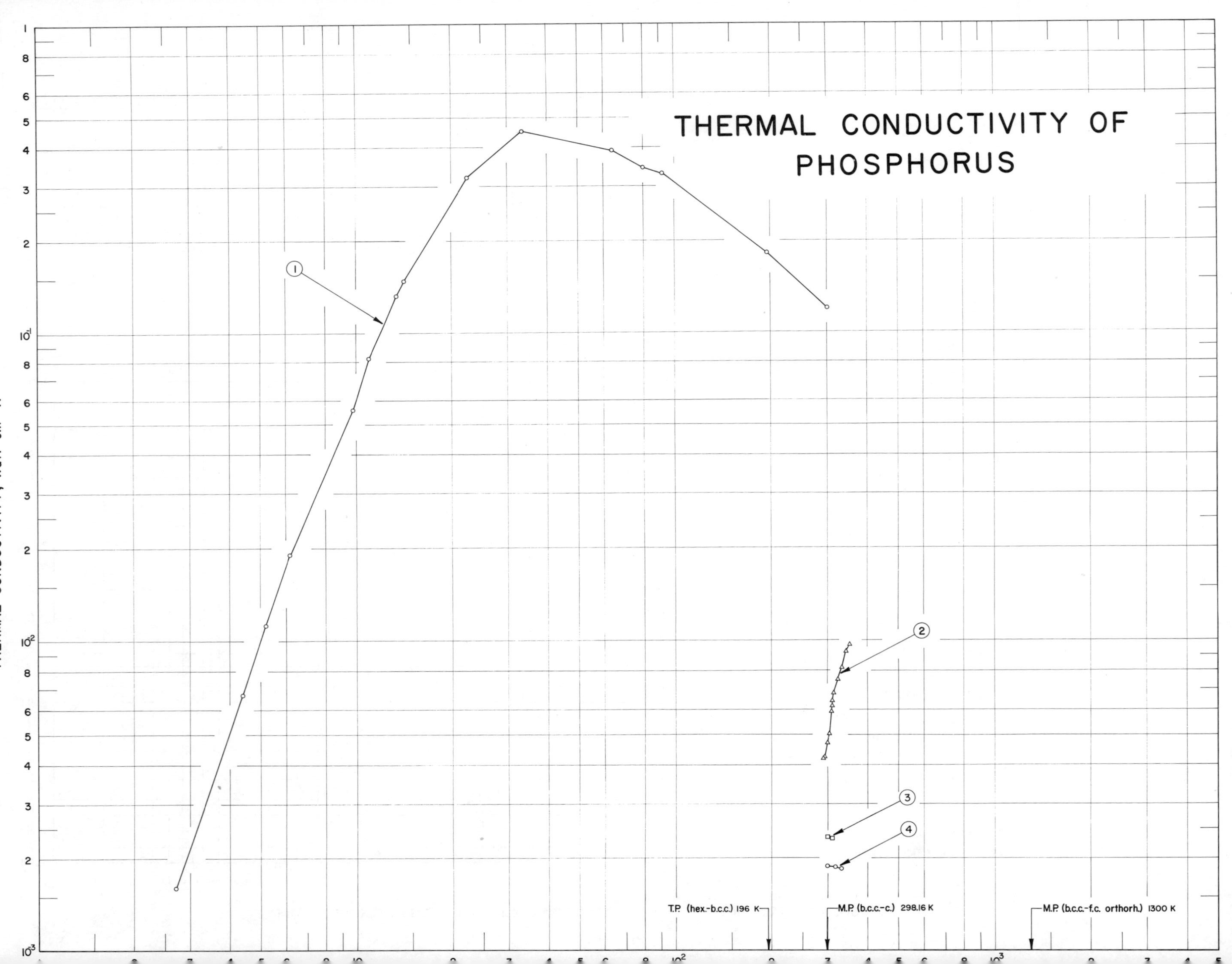

THERMAL CONDUCTIVITY OF
PHOSPHORUS
THERMAL CONDUCTIVITY, Watt cm⁻¹ K⁻¹
T.P. (hex.-b.c.c.) 196 K
M.P. (b.c.c.-c.) 298.16 K
M.P. (b.c.c.-f.c. orthorh.) 1300 K
1
2
3
4

SPECIFICATION TABLE NO. 13 THERMAL CONDUCTIVITY OF PHOSPHORUS

[For Data Reported in Figure and Table No. 13]

Curve No.	Ref. No.	Method Used	Year	Temp. Range, K	Reported Error, %	Name and Specimen Designation	Composition (weight percent), Specifications and Remarks
1	371	L	1965	2.7-300			Prepared from 99.8 pure phosphorus; p-type; polycrystalline with crystal size ~200 μ; orthorhombic (black) phosphorus rod of dimensions 1.2 x 0.35 x 0.35 cm; cut from a larger ingot prepared by R. H. Wentorf of the General Electric Research Lab.; raw material for the ingot from the Fisher Scientific Co.; electrical resistivity 3.1 ohm cm at 300 K; carrier concentration 10^{15} cm^{-3}; heat flow approx in the a-b plane of the crystallites.
2	370	P	1962	289-353	±2		Yellow phosphorus; melting point 44.1 C; includes liquid phase.
3	544	P	1964	299, 311	±1		0.36 dissolved water, < 0.01 other total impurities; solid white α-phase (b.c.c.) phosphorus prepared from chemically pure yellow phosphorus which was treated with warm chromic acid and distilled water until water-white; melting point 44.0 ±0.1 C.
4	544	P	1964	301-331	±2		Similar to the above specimen except in liquid state (the first data point for super-cooled liquid).

DATA TABLE NO. 13 THERMAL CONDUCTIVITY OF PHOSPHORUS

[Temperature, T, K; Thermal Conductivity, k, Watt cm^{-1} K^{-1}]

T	k	T	k	T	k	T	k
CURVE 1		CURVE 1 (cont.)		CURVE 2 (cont.)		CURVE 3*	
2.71	0.0016	64.0	0.390	298.2	0.00471	299.2	0.00234
4.40	0.0067	80.0	0.345	304.2	0.00501	310.7	0.00231
5.20	0.0113	92.0	0.330	306.2	0.00507*		
6.20	0.0190	196.0	0.182	308.2	0.00592	CURVE 4*	
9.80	0.0560	300.0	0.121	309.2	0.00616		
11.0	0.0820			310.2	0.00640	300.6	0.00188
13.5	0.133	CURVE 2		313.2	0.00679	317.7	0.00187
14.2	0.148			323.2	0.00746	331.0	0.00185
22.1	0.320	289.2	0.0042	333.2	0.00818		
33.5	0.450	293.2	0.00424	342.2	0.00920		
				353.2	0.00967		

*Not shown on plot

FIGURE AND TABLE NO. 13R RECOMMENDED THERMAL CONDUCTIVITY OF PHOSPHORUS

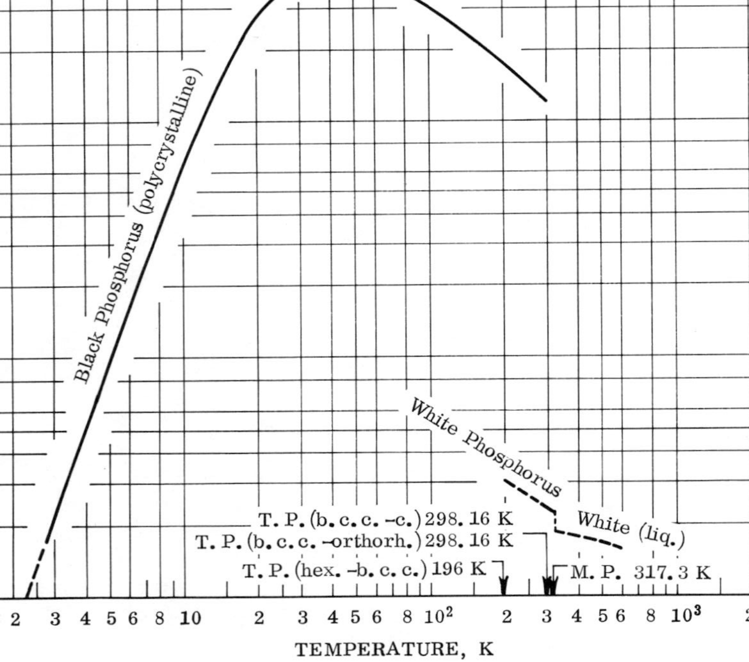

RECOMMENDED VALUES*

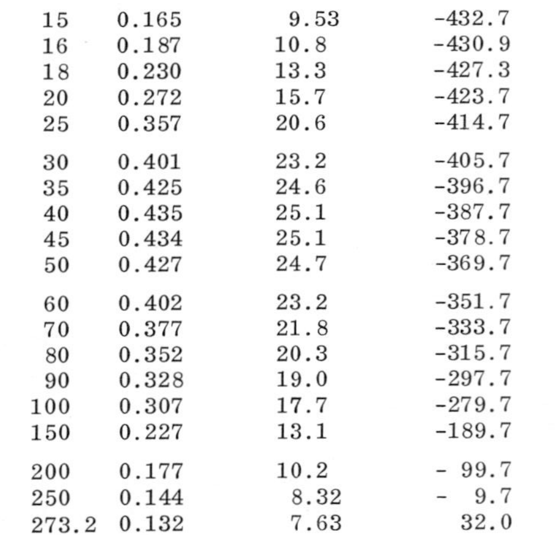

Polycrystalline Black Phosphorus				White Phosphorus			
T_1	k_1	k_2	T_2	T_1	k_1	k_2	T_2
0	0	0	-459.7	0	0	0	-459.7
1	(0.0000796)‡	(0.00456)	-457.9	200	(0.00308)	(0.178)	- 99.7
2	(0.000645)	(0.0373)	-456.1	250	(0.00265)	(0.153)	- 9.7
3	0.00220	0.127	-454.3	273.2	(0.00250)	(0.144)	32.0
4	0.00511	0.295	-452.5	300	0.00235	0.136	80.3
5	0.00998	0.577	-450.7	317.3	(0.00226)	(0.131)	111.4
6	0.0167	0.965	-448.9				
7	0.0255	1.47	-447.1		**Liquid White Phosphorus**		
8	0.0367	2.12	-445.3	317.3	0.00187	0.108	111.4
9	0.0497	2.87	-443.5	350	(0.00183)	(0.106)	170.3
10	0.0653	3.77	-441.7	400	(0.00178)	(0.103)	260.3
11	0.0822	4.75	-439.9	500	(0.00167)	(0.0965)	440.3
12	0.101	5.84	-438.1	600	(0.00157)	(0.0907)	620.3
13	0.122	7.05	-436.3				
14	0.144	8.32	-434.5				
15	0.165	9.53	-432.7				
16	0.187	10.8	-430.9				
18	0.230	13.3	-427.3				
20	0.272	15.7	-423.7				
25	0.357	20.6	-414.7				
30	0.401	23.2	-405.7				
35	0.425	24.6	-396.7				
40	0.435	25.1	-387.7				
45	0.434	25.1	-378.7				
50	0.427	24.7	-369.7				
60	0.402	23.2	-351.7				
70	0.377	21.8	-333.7				
80	0.352	20.3	-315.7				
90	0.328	19.0	-297.7				
100	0.307	17.7	-279.7				
150	0.227	13.1	-189.7				
200	0.177	10.2	- 99.7				
250	0.144	8.32	- 9.7				
273.2	0.132	7.63	32.0				
300	0.121	6.99	80.3				

REMARKS

The recommended values are for high-purity phosphorus. The recommended values that are supported by experimental thermal conductivity data are thought to be accurate to within 5% of the true values near room temperature and 5 to 10% at other temperatures above 150 K. The thermal conductivity near and below the corresponding temperature of its maximum is highly sensitive to small physical and chemical variations of the specimens, and the values below 150 K are intended as typical values for indicating the general trend.

---

*T_1 in K, k_1 in Watt $cm^{-1}K^{-1}$, T_2 in F, and k_2 in Btu $hr^{-1}ft^{-1}F^{-1}$.

‡Values in parentheses are extrapolated.

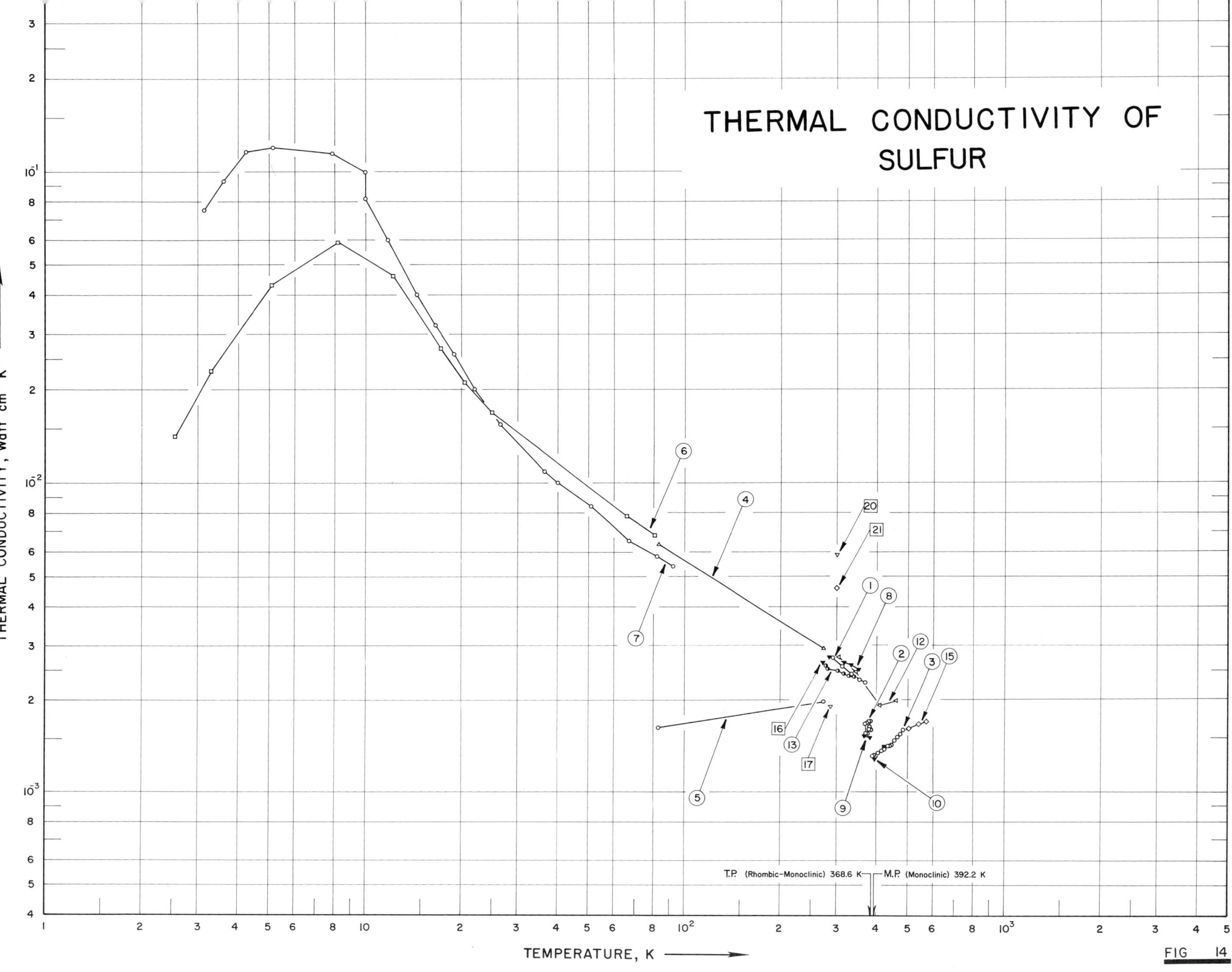
THERMAL CONDUCTIVITY OF SULFUR
THERMAL CONDUCTIVITY, Watt cm⁻¹ K⁻¹
TEMPERATURE, K
T.P. (Rhombic–Monoclinic) 368.6 K
M.P. (Monoclinic) 392.2 K
1
2
3
4
5
6
7
8
9
10
12
13
15
16
17
20
21

FIG 14

SPECIFICATION TABLE NO. 14 THERMAL CONDUCTIVITY OF SULFUR

[For Data Reported in Figure and Table No. 14]

Curve No.	Ref. No.	Method Used	Year	Temp. Range, K	Reported Error, %	Name and Specimen Designation	Composition (weight percent), Specifications and Remarks
1	21	L	1929	293-368			Commercial purity; crystalline variety of rhombic aggregate state; specimen prepared at temp not exceeding 160 C.
2	21	L	1929	368-383			The above specimen in monoclinic aggregate state.
3	21	L	1929	388-483			Commercial purity; in liquid state.
4	22	L	1911	83, 273			Rhombic crystalline sulfur.
5	22	L	1911	83, 273			Amorphous; made by casting boiling sulfur.
6	371	L	1965	2. 6-81		R 124	Single crystal of the stable orthorhombic modification slowly grown from a CS_2 solution at room temp in one week; sulfur starting material >99 pure; rod-shaped sample about 0. 2 cm in dia and 0. 7 cm long; transparent with a yellow color; a few small internal growth flaws visible in the sample; heat flow perpendicular to the c-axis.
7	371	L	1965	3. 2-92		R 127	Similar to the above specimen but heat flow parallel to the c-axis.
8	442, 333	C	1964, 1965	285-354	±1		99. 98 pure; α (rhombic) sulfur; polycrystalline; crown glass plate used as comparative material, the thermal conductivity of which is given as k = 0. 803 + 0. 00054 t with k in kcal m^{-1} hr^{-1} C^{-1} and t in C.
9	333	C	1965	370-389	±1		99. 98 pure; β (monoclinic) sulfur; polycrystalline; same comparative material used as above.
10	333	C	1965	395-428	±1		99. 98 pure; liquid sulfur; same comparative material used as above.
11	443	L	1898	306, 334			4 cm dia x 0. 193 cm thick.
12	444, 445	P	1964	305-460			Data cover both solid and liquid states.
13	446	R	1932	277-353			Rhombic crystal; spherical shell specimen with O. D. 10. 200 cm and I. D. 5. 514 cm; prepared by melting sulfur flowers at 170 C, cast in a brass mould, cooling to complete solification in about 1. 5 hrs, lowering the temp to about 60 C for 30 min, then allowing to cool in the lagging; density 1. 90 g cm^{-3}.
14	446	R	1932	280-358			Rhombic crystal; spherical shell specimen with O. D. 10. 208 cm and I. D. 5. 508 cm; prepared by melting sulfur flowers at 127 C, poured into a brass mould, heated to 135 C, cooled to the melting point in about 5 hrs, completely solidified after another 40 min; density 1. 94 g cm^{-3}.
15	447	P	1959	460-574	<±3		Chemically pure; molten specimen contained in a cell made from 2 thick-walled silver tubes; the liquid annulus has an outer dimension of 28. 5 mm dia x 100 mm long and a width of 8 mm; held for 24 hrs at each temp during measurements.
16	448	P	1904	373. 2			10. 8 cm cubic specimen; density 2. 03 g cm^{-3}; thermal conductivity values calculated from measured thermal diffusivity and the specific heat value of 0. 187 cal g^{-1} C^{-1}.

SPECIFICATION TABLE NO. 14 (continued)

Curve No.	Ref. No.	Method Used	Year	Temp. Range, K	Reported Error, %	Name and Specimen Designation	Composition (weight percent), Specifications and Remarks
17	325	C	1892	287.4			Irregular shaped plate specimen 0.0584 cm in thickness; prepared by pressing between 2 microscope slides having plane surfaces; brass used as comparative material.
18	449	L	1912	276.7			Thin plate specimen.
19	451	L	1952	300			Amorphous sulfur.
20	451	L	1952	300			Single crystal; heat flow along one crystal axis.
21	451	L	1952	300			Single crystal; heat flow along another crystal axis perpendicular to the above.

DATA TABLE NO. 14 THERMAL CONDUCTIVITY OF SULFUR

[Temperature, T, K; Thermal Conductivity, k, Watt $cm^{-1}K^{-1}$]

T	k
CURVE 1	
293.2	0.00273
313.2	0.00256
333.2	0.00243
353.2	0.00233
368.2	0.00229
CURVE 2	
368.2	0.00166
369.2	0.00155
373.2	0.00167
375.2	0.00155
377.2	0.00159
377.2	0.00169
379.2	0.00163
381.2	0.00170
383.2	0.00159
CURVE 3	
388.2	0.00131
393.2	0.00132
CURVE 3 (cont.)	
403.2	0.00134
413.2	0.00136
423.2	0.00138
433.2	0.00141
438.2	0.00141
443.2	0.00142
453.2	0.00147
463.2	0.00151
473.2	0.00154
483.2	0.00158
CURVE 4	
83.2	0.00637
273.2	0.00293
CURVE 5	
83.2	0.00162
273.2	0.00197
CURVE 6	
2.55	0.0141
3.30	0.0229
5.10	0.0430
8.20	0.0590
12.20	0.0460
17.20	0.0270
20.50	0.0210
25.0	0.0168
66.0	0.0078
81.0	0.0068
CURVE 7	
3.15	0.075
3.60	0.093
4.25	0.116
5.15	0.120
7.90	0.115
10.00	0.099
10.00	0.082
11.80	0.060
14.50	0.040
CURVE 7 (cont.)	
16.60	0.032
19.0	0.026
22.0	0.020
26.5	0.0155
36.5	0.0109
40.0	0.010
51.0	0.0084
67.0	0.0065
82.0	0.0058
92.0	0.0054
CURVE 8*	
285.03	0.00275
292.61	0.00274
306.38	0.00268
316.76	0.00265
322.26	0.00263
332.26	0.00260
337.27	0.00258
347.23	0.00254
354.80	0.00251
CURVE 9*	
369.65	0.00154
370.93	0.00153
371.41	0.00150
376.17	0.00153
378.38	0.00156
382.13	0.00150
383.10	0.00152
384.16	0.00148
384.75	0.00156
385.38	0.00150
386.92	0.00159
388.11	0.00148
388.67	0.00152
CURVE 10*	
395.21	0.00129
400.65	0.00132
405.29	0.00130
410.71	0.00132
420.57	0.00136
427.71	0.00139
CURVE 11*	
306.2	0.00284
334.2	0.00254
CURVE 12	
305.2	0.00275
346.2	0.00247
409.2	0.00193
460.2	0.00199
CURVE 13	
277.2	0.00259
281.5	0.00253
303.3	0.00249
303.7	0.00250*
304.0	0.00247*
316.6	0.00244
328.0	0.00244*
328.5	0.00241*
328.8	0.00241
340.6	0.00239
CURVE 13 (cont.)	
352.7	0.00233*
352.8	0.00231*
353.3	0.00233*
CURVE 14*	
278.8	0.00261
298.2	0.00256
303.3	0.00255
307.8	0.00254
318.9	0.00252
327.8	0.00247
338.3	0.00243
358.0	0.00238
CURVE 15	
460.1	0.00149*
501.2	0.00161
537.3	0.00166
574.3	0.00169
CURVE 16	
273.2	0.00264
CURVE 17	
287.4	0.00190
CURVE 18*	
276.73	0.00258
CURVE 19*	
300	0.00209
CURVE 20	
300	0.00586
CURVE 21	
300	0.00460

*Not shown on plot

FIGURE AND TABLE NO. 14R RECOMMENDED THERMAL CONDUCTIVITY OF SULFUR

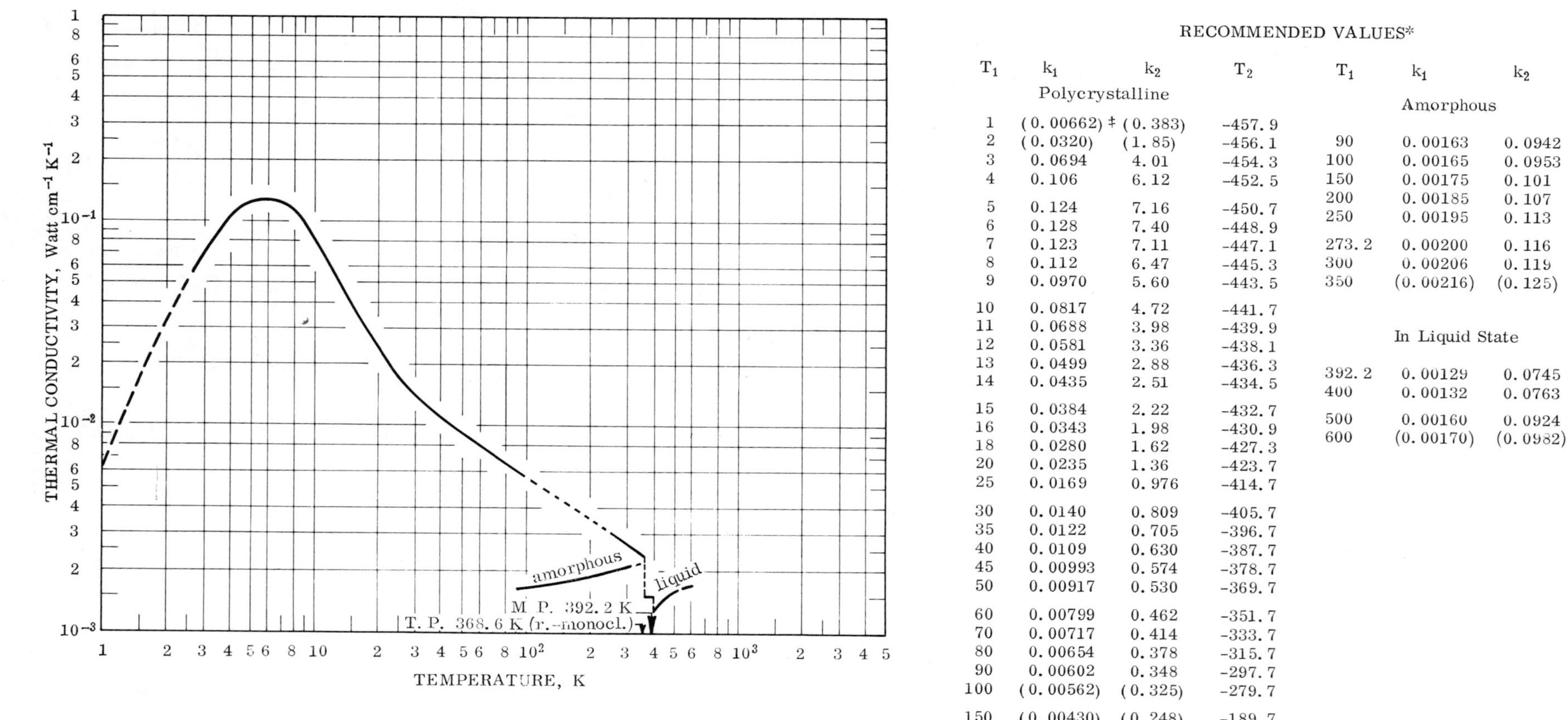

RECOMMENDED VALUES*

Polycrystalline

T_1	k_1	k_2	T_2
1	(0.00662) ‡	(0.383)	-457.9
2	(0.0320)	(1.85)	-456.1
3	0.0694	4.01	-454.3
4	0.106	6.12	-452.5
5	0.124	7.16	-450.7
6	0.128	7.40	-448.9
7	0.123	7.11	-447.1
8	0.112	6.47	-445.3
9	0.0970	5.60	-443.5
10	0.0817	4.72	-441.7
11	0.0688	3.98	-439.9
12	0.0581	3.36	-438.1
13	0.0499	2.88	-436.3
14	0.0435	2.51	-434.5
15	0.0384	2.22	-432.7
16	0.0343	1.98	-430.9
18	0.0280	1.62	-427.3
20	0.0235	1.36	-423.7
25	0.0169	0.976	-414.7
30	0.0140	0.809	-405.7
35	0.0122	0.705	-396.7
40	0.0109	0.630	-387.7
45	0.00993	0.574	-378.7
50	0.00917	0.530	-369.7
60	0.00799	0.462	-351.7
70	0.00717	0.414	-333.7
80	0.00654	0.378	-315.7
90	0.00602	0.348	-297.7
100	(0.00562)	(0.325)	-279.7
150	(0.00430)	(0.248)	-189.7
200	(0.00355)	(0.205)	- 99.7
250	(0.00305)	(0.176)	- 9.7
273.2	0.00287	0.166	32.0
300	0.00269	0.155	80.3
350	0.00242	0.140	170.3
368.6	0.00233	0.135	203.8
368.6	0.00154	0.0890	203.8
392.2	0.00150	0.0867	246.3

T_1	k_1	k_2	T_2
	Amorphous		
90	0.00163	0.0942	-297.7
100	0.00165	0.0953	-279.7
150	0.00175	0.101	-189.7
200	0.00185	0.107	- 99.7
250	0.00195	0.113	- 9.7
273.2	0.00200	0.116	32.0
300	0.00206	0.119	80.3
350	(0.00216)	(0.125)	170.3
	In Liquid State		
392.2	0.00129	0.0745	246.3
400	0.00132	0.0763	260.3
500	0.00160	0.0924	440.3
600	(0.00170)	(0.0982)	620.3

REMARKS

The recommended values are for high-purity sulfur. The recommended values that are supported by experimental thermal conductivity data are thought to be accurate to within 4% of the true values near room temperature and 4 to 10% at other temperatures above 10 K. The thermal conductivity near and below the corresponding temperature of its maximum is highly sensitive to small physical and chemical variations of the specimens, and the values below 10 K are intended as typical values for indicating the general trend.

* T_1 in K, k_1 in Watt $cm^{-1} K^{-1}$, T_2 in F, and k_2 in Btu $hr^{-1} ft^{-1} F^{-1}$.

‡ Values in parentheses are extrapolated.

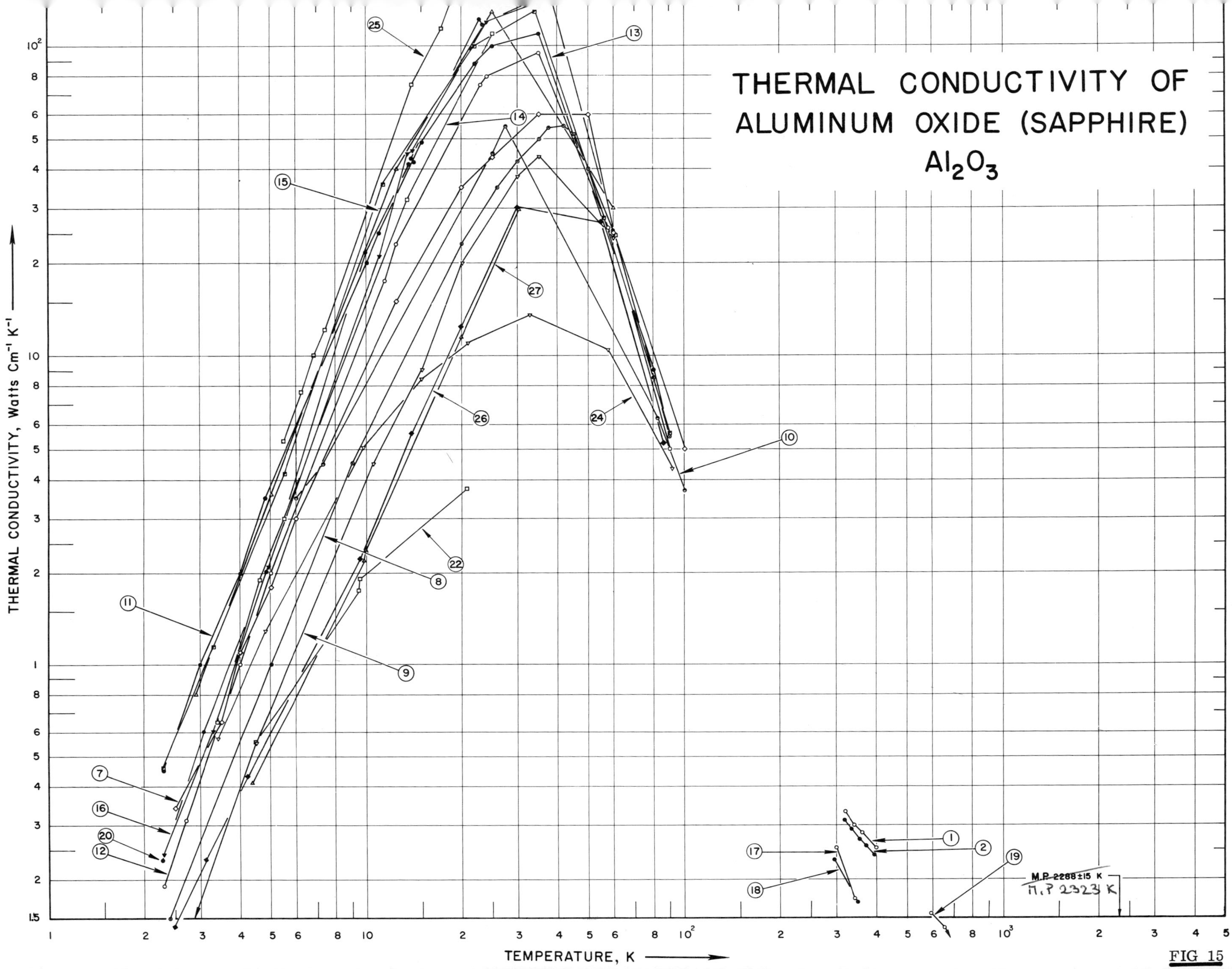

THERMAL CONDUCTIVITY OF ALUMINUM OXIDE (SAPPHIRE) Al_2O_3

THERMAL CONDUCTIVITY, Watts Cm^{-1} K^{-1}

TEMPERATURE, K

M.P. 2288±15 K

M.P. 2323 K

FIG 15

SPECIFICATION TABLE NO. 15 THERMAL CONDUCTIVITY OF ALUMINUM OXIDE (SAPPHIRE) Al_2O_3

[For Data Reported in Figure and Table No. 15]

Curve No.	Ref. No.	Method Used	Year	Temp. Range, K	Reported Error, %	Name and Specimen Designation	Composition (weight percent), Specifications and Remarks
1	42	C	1953	322-399		93B-1	Single crystal; supplied by Linde Air Products Co.; 0.25 in. dia x 0.25 in. long; a-axis parallel to cylinder axis within 3 degrees; heat flow direction parallel to c-axis.
2	42	C	1953	318-394		93C-1	The above specimen measured with heat flow direction within 3 degrees to the a-axis.
3	548, 13	C	1952	343.2			Heat flow direction at 60 degrees to the c-axis; density 4.0 g cm^{-3}.
4	34	C	1943	392-763			Synthetic, colorless specimen 1 cm cube cut from a single crystal with c-axis normal to two opposite faces and parallel to other surfaces; heat flow perpendicular to c-axis.
5	68	C	1954	319-382	±3	93B-2	Single crystal; measured in the direction of c-axis.
6	68	C	1954	316-424	±3	93C-2	Single crystal; measured in the direction of a-axis.
7	69	L	1951	2.5-100		Corundum	Single crystal; prepared by Salford Electrical Instruments Ltd.; specimen dia 3 mm, length 6 cm; rod axis at 36 degrees to the principal axis.
8	70	L	1955	2.4-90		I-b	Pure Al_2O_3 crystal produced by usual commercial techniques; 1.55 mm in dia, 6 cm long; with mat surfaces; rod axis at 36 degrees to the optic axis.
9	70	L	1955	2.3-90		I-c	Similar to the above specimen except the dimensions, 1.02mm in dia, 13 mm long.
10	70	L	1955	6.0-100		II	Pure Al_2O_3 crystal produced by usual commercial techniques; 2.8 mm in dia, 15 mm long; with marked mosaic structure as shown by x-rays.
11	70	L	1955	2.3-90		IIIa	Pure Al_2O_3 crystal; 2.52 mm in dia, 6 cm long; with polished surfaces; annealed at a temperature slightly below the melting point.
12	70	L	1955	2.3-90		IIIb	Pure Al_2O_3 crystal; 2.47 mm in dia, 6 cm long; with mat surfaces; annealed at a temperature slightly below the melting point.
13	70	L	1955	2.3-90		IV	As above but 2.54 mm in dia, 6 mm long; with polished surfaces.
14	70	L	1955	3.4-25		V	As above but with mat surfaces.
15	70	L	1955	2.9-60		VI	As above but with polished surfaces, annealed at a temperature slightly below the melting point.
16	70	L	1955	2.3-35			As above but with mat surfaces.
17	71	C	1951	299, 343			Linde synthetic sapphire; single crystal; heat flow parallel to the optic axis.
18	71	C	1951	296, 350			As above, but heat flow perpendicular to the optic axis.
19	72	R	1955	592-1508		Linde synthetic sapphire	Single crystal; measured with heat flow direction at about 60 degrees with the c-axis.

SPECIFICATION TABLE NO. 15 (continued)

Curve No.	Ref. No.	Method Used	Year	Temp. Range, K	Reported Error, %	Name and Specimen Designation	Composition (weight percent), Specifications and Remarks
20	215		1960	2.3-34		Synthetic sapphire	Single crystal; 2.5 mm dia rod.
21	215		1960	2.5-91		Synthetic sapphire	Single crystal; 2.5 mm dia rod; measured after receiving 3 x 10^7 roentgens γ-ray dose.
22	215		1960	4.5-21		Synthetic sapphire	Single crystal; 2.5 mm dia rod; reactor irradiated.
23	215		1960	3.4-93		Synthetic sapphire	Single crystal; 2.5 mm dia rod; annealed in vacuum at 1000 to 1100 C for about 12 hrs.
24	215		1960	2.8-92		Synthetic sapphire	The above specimen γ irradiated.
25	215		1960	5.5-90		Synthetic sapphire	Single crystal; 2.5 mm dia rod.
26	215		1960	2.5-87		Synthetic sapphire	Single crystal; 2.5 mm dia rod; stretched at 1400 C.
27	215		1960	4.4-31		Synthetic sapphire	The above specimen γ-ray irradiated.

DATA TABLE NO. 15 THERMAL CONDUCTIVITY OF ALUMINUM OXIDE (SAPPHIRE) Al_2O_3

[Temperature, T, K; Thermal Conductivity, k, Watt $cm^{-1}K^{-1}$]

T	k
CURVE 1	
321.8	0.330
342.4	0.299
361.3	0.282
398.6	0.251
CURVE 2	
318.4	0.310
335.9	0.290
355.3	0.268
370.5	0.256
394.0	0.239
CURVE 3*	
343.2	0.297
CURVE 4*	
392.2	0.0331
511.2	0.0387
576.2	0.0423
763.2	0.0602
CURVE 5*	
318.6	0.327
341.7	0.299
361.1	0.277
382.4	0.259
CURVE 6*	
315.6	0.302
335.5	0.289
357.9	0.265
393.5	0.237
424.3	0.219
CURVE 7	
2.5	0.34
3.5	0.65
4.0	1.10
CURVE 7 (cont.)	
5.0	1.80
6.0	3.00
12.5	15.00
20.0	35.00
25.0	44.00
35.0	60.00
50.0	60.00
60.0	25.00
100.0	5.00
CURVE 8	
2.4	0.15
5.0	1.00
9.0	4.50
20.0	23.00
26.0	35.00
30.0	42.50
35.0	50.00
37.5	54.50
42.0	55.00
45.0	52.00
50.0	40.00
60.0	25.00
80.0	8.50
90.0	5.50
CURVE 9	
2.3	0.055*
2.5	0.100*
4.5	0.550
10.5	4.500
15.0	9.000
20.0	20.000
30.0	38.000
35.0	44.000
60.0	24.000
90.0	5.500*
CURVE 10	
6.0	3.50
7.3	4.50
CURVE 10 (cont.)	
25.0	45.00
27.5	55.00
82.5	6.30
100.0	3.70
CURVE 11	
2.3	0.450
3.0	1.00
4.0	2.00
4.8	3.50
10.0	20.00
11.0	25.00
15.0	49.00
22.0	88.00
25.0	100.00
35.0	110.00
80.0	9.00
90.0	5.50*
CURVE 12	
2.3	0.190
2.7	0.310
4.0	1.00
5.0	2.00
11.5	17.50
12.5	23.00
23.0	75.00
24.0	80.00
35.0	95.00
58.0	26.00
90.0	5.00
CURVE 13	
2.3	0.460
3.3	1.150
5.5	4.20
11.3	36.00
22.0	100.00
34.0	130.00
56.0	28.00
90.0	5.60
CURVE 14	
3.4	0.650
4.6	1.90
5.5	3.00
13.5	32.00
25.0	110.00
CURVE 15	
2.9	0.800
5.0	3.60
12.5	40.00
24.0	120.00
25.0	130.00
60.0	30.00
CURVE 16	
2.3	0.240
3.3	0.600
4.9	2.10
11.0	21.00
13.5	45.00
14.0	46.00
24.0	120.00
35.0	140.00
CURVE 17	
299.2	0.251
343.2	0.172
CURVE 18	
296.2	0.230
350.2	0.167
CURVE 19	
591.5	0.155
651.2	0.139
690.2	0.124*
775.2	0.106*
957.2	0.0871*
1073.2	0.0804*
CURVE 19 (cont.)	
1173.2	0.0768*
1257.2	0.0759*
1313.2	0.0761*
1384.2	0.0786*
1439.2	0.0813*
1508.2	0.0849*
CURVE 20	
2.3	0.230
3.1	0.606
4.9	2.041
10.1	21.6
13.8	41.2
14.0	43.1
14.2	42.2
23.1	122.1
23.4	117.9
23.9	120.5*
34.4	141.8
CURVE 21*	
2.5	0.276
4.5	1.79
12.0	23.7
23.2	98.0
32.5	128.5
63.9	27.0
89.5	6.10
90.7	5.81
CURVE 22	
4.5	0.552
9.5	1.75
9.6	1.91
20.7	3.77
CURVE 23*	
3.35	0.633
3.95	0.991
5.22	1.85
CURVE 23* (cont.)	
9.68	5.59
14.5	9.26
20.5	12.3
33.0	14.4
58.0	12.2
93.0	4.55
CURVE 24	
2.78	0.327
3.44	0.571
4.85	1.30
9.8	5.10
15.0	8.40
21.0	11.0
33.0	13.6
58.0	10.5
92.0	4.33
CURVE 25	
5.5	5.30
6.24	7.62
6.89	10.1
7.4	12.2
14.1	75.0
17.3	115.0
18.6	138.0
24.0	189.0
35.5	168.0
37.0	160.0
61.0	24.5
90.1	5.4
CURVE 26	
2.51	0.14
3.18	0.23
4.39	0.43
9.6	2.22
14.1	5.60
19.9	12.4
30.2	30.1
55.2	27.1
86.5	5.25
CURVE 27	
4.4	0.414
9.8	2.20
10.0	2.41
20.0	11.5
30.5	30.2

* Not shown on plot

FIGURE AND TABLE NO. 15R RECOMMENDED THERMAL CONDUCTIVITY OF ALUMINUM OXIDE (SAPPHIRE) Al_2O_3

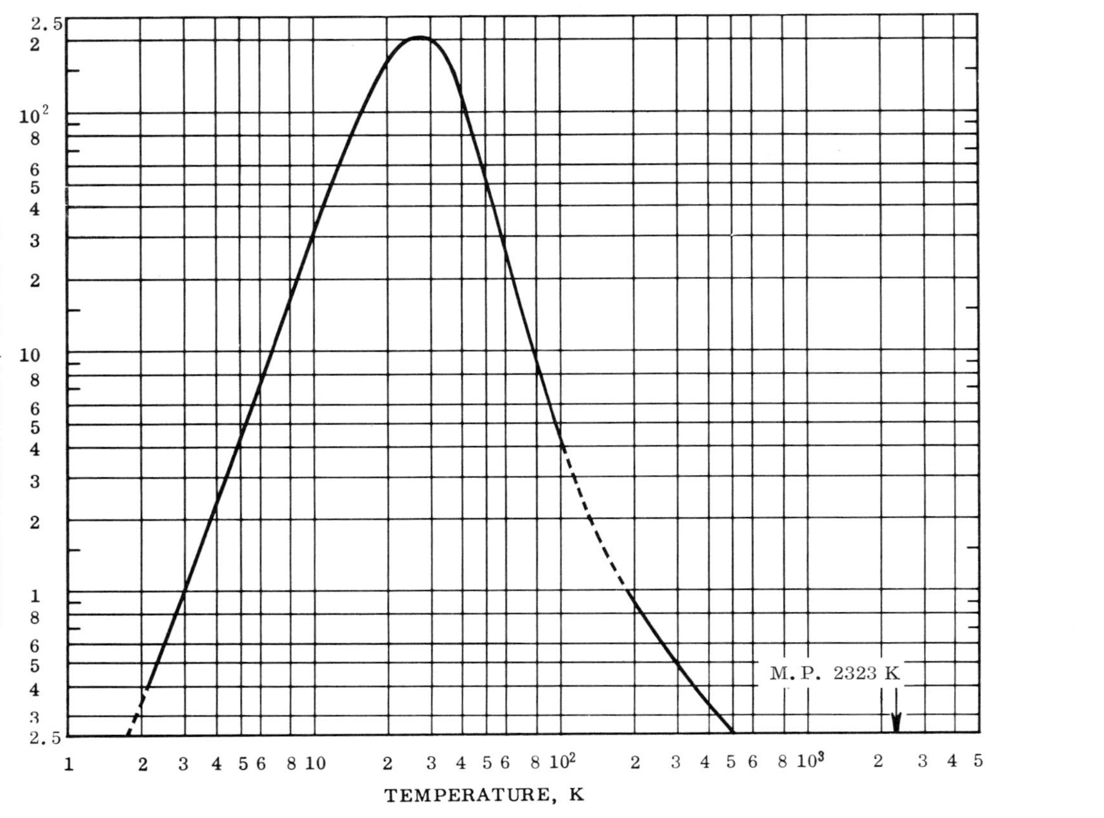

RECOMMENDED VALUES*
(High-purity synthetic sapphire single crystal)

T_1	k_1	k_2	T_2
0	0	0	-459.7
1	(0.039)‡	(2.25)	-457.9
5	4.1	237	-450.7
10	29	1680	-441.7
15	87	5030	-432.7
20	157	9070	-423.7
25	202	11700	-414.7
30	207	12000	-405.7
35	177	10200	-396.7
40	120	6930	-387.7
45	77	4450	-378.7
50	52	3010	-369.7
60	26.5	1530	-351.7
70	15.3	884	-333.7
80	9.6	555	-315.7
90	6.4	370	-297.7
100	4.5	260	-279.7
150	(1.5)	(86.7)	-189.7
200	(0.82)	(47.4)	- 99.7
250	(0.58)	(33.5)	- 9.7
273.2	(0.52)	(30.0)	32.0
300	0.46	26.6	80.3
350	0.38	22.0	170.3
400	0.324	18.7	260.3
450	0.279	16.1	350.3
500	0.242	14.0	440.3
600	0.189	10.9	620.3
700	0.154	8.90	800.3
800	0.130	7.51	980.3
900	0.115	6.64	1160
1000	0.105	6.07	1340

REMARKS

The recommended values are for high-purity synthetic sapphire single crystal with heat flow at 60 degrees to the hexagonal axis. The recommended values that are supported by experimental thermal conductivity data are thought to be accurate to within 10 to 15% of the true values at temperatures above 60 K. The thermal conductivity near and below the corresponding temperature of its maximum is highly sensitive to small physical and chemical variations of the specimens, and the recommended values below 60 K are intended as typical values for indicating the general trend.

---

* T_1 in K, k_1 in Watt cm^{-1} K^{-1}, T_2 in F, and k_2 in Btu hr^{-1} ft^{-1} F^{-1}.

‡ Values in parentheses are extrapolated or interpolated.

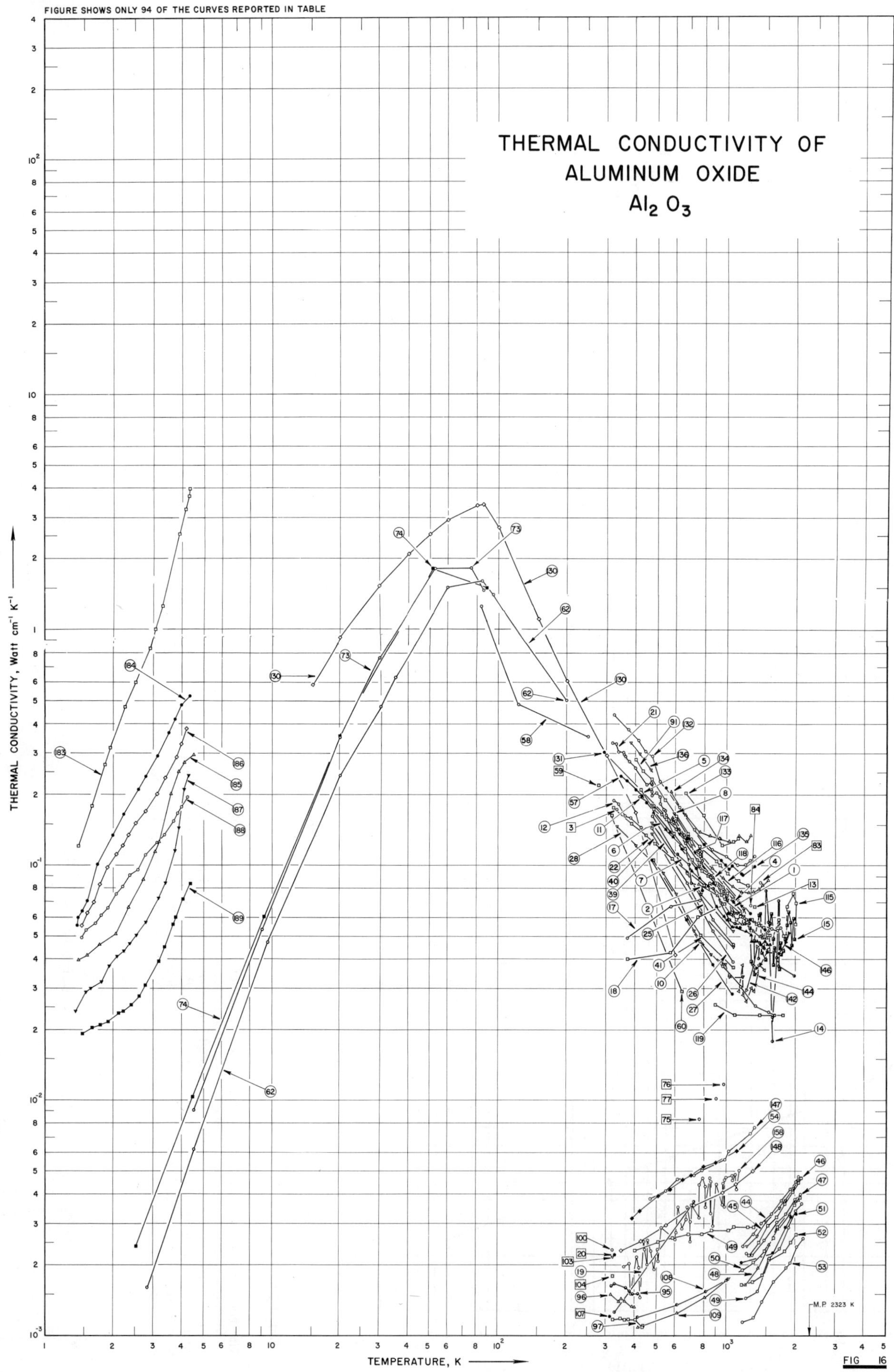
FIGURE SHOWS ONLY 94 OF THE CURVES REPORTED IN TABLE
THERMAL CONDUCTIVITY OF
ALUMINUM OXIDE
Al_2O_3
THERMAL CONDUCTIVITY, Watt cm^{-1} K^{-1}
TEMPERATURE, K
M.P. 2323 K

FIG 16

SPECIFICATION TABLE NO. 16 THERMAL CONDUCTIVITY OF ALUMINUM OXIDE Al_2O_3

[For Data Reported in Figure and Table No. 16]

Curve No.	Ref. No.	Method Used	Year	Temp. Range, K	Reported Error, %	Name and Specimen Designation	Composition (weight percent), Specifications and Remarks
1	1	R	1950	846-1737			Sintered.
2	2	R	1951	758-1535			Pure; polycrystalline; heat-treated.
3	3		1953	333. 2			Maximum moisture absorption = 0. 05%; flexural strength = 33, 000 psi; coefficient of expansion (25-700 C) = 8. 0 x 10^{-6}.
4	4	R	1951	873-1473	±5		Sintered.
5	5	R	1954	423-1553			Hemispherical samples prepared by slipcasting in plaster molds and refiring to 1100 C; 1. 75 in. inside dia and 3. 5 in. outside dia.
6	5	R	1954	483-1553			Similar to the above specimen.
7	5	R	1954	613-1258			Made by slip-casting; hollow cylinder with inside dia 0. 6 in., outside dia 1. 5 in., entire length 18 in., test zone 4 in.
8	5	R	1954	593-1253			Same as the above specimen.
9	302	P	1962	293-1263			Density 3. 04 g cm^{-3}.
10	7	R	1949	769-1631			Sintered; open pores 10. 65%, closed pores 2. 35%; bulk density 3. 48 g cm^{-3}.
11	8	L	1960	419-1067		AP-30	99. 5 Al_2O_3; supplied by McDanel Refractory Porcelain Co.; porosity 25. 69; bulk density 2. 95 g cm^{-3}.
12	9	C	1953	322-425		Hi alumina; 6 Nl-1	No details reported.
13	294	L	1963	1333			Dense alumina; measured by the "Unmatched Guard Method", the given value is the average of 5 runs of two specimens; R. M. S. deviation from average 1. 9%.
14	89	R	1952	1387-1973	5-7	M-2	Prepared by slip casting and firing.
15	89	R	1952	1273-1983	5-7	M-4	Similar to the above specimen.
16	89	R	1952	1336-2028	5-7	M-5	Similar to the above specimen.
17	34	C	1943	373-769		Corundum (African)	Red; single crystal; measured parallel to c-axis; 18-8 stainless steel used as comparative material.
18	34	C	1943	372-985		Corundum (African)	Red; single crystal; measured normal to c-axis; 18-8 stainless steel used as comparative material.
19	79	L	1942	328, 728			Porous; measured in the presence of air atmosphere; density 0. 77 g cm^{-3}.
20	79	L	1942	328. 2			Same as the above specimen but measured in the presence of hydrogen atmosphere.
21	131	C	1954	318-900	<3		1 x 1 x 1 in.; slip-cast from suspension of finely ground material; fired to zero apparent porosity; bulk density 3. 79 g cm^{-3}.

SPECIFICATION TABLE NO. 16 (continued)

Curve No.	Ref. No.	Method Used	Year	Temp. Range, K	Reported Error, %	Name and Specimen Designation	Composition (weight percent), Specifications and Remarks
22	132	R	1958	323-1876	<3		Polycrystal; specimen disks were three annular rings having inside dia 0.625 in., outside dia 3.0 in., and thickness 1 in., with six 0.5 in. thick disks located at each end as guard section.
23	133	C	1954	473-1073	<3		1 x 1 x 1 in.; prepared from Norton Co. 38X 220F alumina by grinding in steel mills, acid treating, and casting in plaster molds from an acid suspension; dense.
24	133	C	1954	473-1073	<3		1 x 1 x 1 in.; prepared from Norton Co. 38X 220F alumina by grinding in steel mills, acid-treating incorporating up to 50 vol.% naphthalene flakes in casting slip, dried for an extended period at 60 to 70 C, then fired at 1830 C for 3 hrs; 12.3% porosity with 0.031 cm dia spherical pores.
25	133	C	1954	473-1073	<3		Same preparation as the above specimen; 23.4% porosity with 0.031 cm dia spherical isometric pores.
26	133	C	1954	473-1073	<3		Same preparation as the above specimen; 30.0% porosity with 0.031 cm dia spherical isometric pores.
27	133	C	1954	473-1073	<3		Same preparation as the above specimen; 44.2% porosity with 0.031 cm dia spherical isometric pores.
28	133	C	1954	333-1183	<3		Same preparation as the above specimen; 48.7% porosity with 0.031 cm dia spherical isometric pores.
29	133	C	1954	473-1073	<3		1 x 1 x 1 in.; prepared from Norton Co. 38X 220F alumina by grinding in steel mills, acid treating and casting in plaster molds from an acid suspension; dense.
30	133	C	1954	473-1073	<3		1 x 1 x 1 in.; prepared from Norton Co. 38X 220F alumina by grinding in steel mills, acid treating, and casting around drill rods firmly set in a plaster mold from an acid suspension; porosity 7.54 with cylindrical pores of dia 0.082 cm; pore axes parallel to heat flow.
31	133	C	1954	428-1093	<3		Same preparation as the above specimen; 11.97% porosity with cylindrical pores of dia 0.082 cm; cylinders parallel to heat flow.
32	133	C	1954	473-1073	<3		Same preparation as the above specimen; 17.95% porosity with cylindrical pores of dia 0.082 cm; cylinders parallel to heat flow.
33	133	C	1954	473-1073	<3		Same preparation as the above specimen; 22.4% porosity with cylindrical pores of dia 0.082 cm; cylinders parallel to heat flow.
34	133	C	1954	473-1073	<3		Same preparation as the above specimen; 4.5% porosity with cylindrical pores of dia 0.146 cm; cylinders parallel to heat flow.
35	133	C	1954	473-1073	<3		Same preparation as the above specimen; 9.75% porosity with cylindrical pores of dia 0.146 cm; cylinders parallel to heat flow.
36	133	C	1954	473-1073	<3		Same preparation as the above specimen; 13.5% porosity with cylindrical pores of dia 0.146 cm; cylinders parallel to heat flow.

SPECIFICATION TABLE NO. 16 (continued)

Curve No.	Ref. No.	Method Used	Year	Temp. Range, K	Reported Error, %	Name and Specimen Designation	Composition (weight percent), Specifications and Remarks
37	133	C	1954	473-1073			Same preparation as the above specimen; 19.75% porosity with cylindrical pores of dia 0.146 cm; cylinders parallel to heat flow.
38	133	C	1954	473-1073			Same preparation as the above specimen; 4.5% porosity with cylindrical pores of dia 0.146 cm; cylinders perpendicular to heat flow.
39	133	C	1954	473-1073			Same preparation as the above specimen; 9.75% porosity with cylindrical pores of dia 0.146 cm; cylinders perpendicular to heat flow.
40	133	C	1954	473-1073			Same preparation as the above specimen; 13.5% porosity with cylindrical pores of dia 0.146 cm; cylinders perpendicular to heat flow.
41	133	C	1954	473-1073			Same preparation as the above specimen; 19.75% porosity with cylindrical pores of dia 0.146 cm; cylinders perpendicular to heat flow.
42	65	C	1958	452-973	2-3	Wesgo Alumina (Al-300)	97.6 Al_2O_3; bulk density 3.70 g cm^{-3}; porosity 6.8%; M.I.T. alumina standard used as comparative material.
43	65	C	1958	452-973	2-3	Wesgo Alumina (Al-300)	2nd run of the above specimen using different alumina as the comparative material.
44	134	R	1952	1193-2093			Porous layer coated cataphoretically on a tungsten wire and then heated to 1600-1700 C.
45	134	R	1952	1248-2083			Porous layer coated cataphoretically on a tungsten wire and then heated to 1600-1700 C.
46	134	R	1952	1268-2143			Porous layer coated cataphoretically on a tungsten wire and then heated to 1600-1700 C.
47	134	R	1952	1243-2143			Porous layer coated cataphoretically on a tungsten wire and then heated to 1600-1700 C.
48	134	R	1952	1223-2093			Porous layer coated cataphoretically on a tungsten wire and then heated to 1600-1700 C.
49	134	R	1952	1233-2143			Porous layer coated cataphoretically on a tungsten wire and then heated to 1600-1700 C.
50	134	R	1952	1193-2123			Porous layer coated cataphoretically on a tungsten wire and then heated to 1600-1700 C.
51	134	R	1952	1173-2043			Porous layer coated cataphoretically on a tungsten wire and then heated to 1600-1700 C.
52	134	R	1952	1173-2043			Porous layer coated cataphoretically on a tungsten wire and then heated to 1600-1700 C.
53	134	R	1952	1193-2193			Porous layer coated cataphoretically on a tungsten wire and then heated to 1600-1700 C.
54	400	R	1966	388-1119	±10	B45F	0.01-0.1 Fe, 0.01-0.1 Na, and 0.01-0.1 Si principal impurities; powder specimen contained in a ~4 in. I.D., 4.5 in. O.D., and ~2.4 in. long container; supplied by Norton Co.; mean particle size 211 μ; volume fraction solid 0.49; pore-free density 3.95 g cm^{-3}.
55	400	R	1966	379-1104	±10	E98	Similar to the above specimen except mean particle size 263 μ, packed to 0.58 volume fraction solid, and pore-free density 3.98 g cm^{-3}.
56	136	C	1959	373-1128	±4		Polycrystalline; containing 0.30 vol.% Cr_2O_3; prepared by mixing calcined Cr_2O_3 and Al_2O_3 in a water suspension and either slip-casting or hydrostatically pressing, and fired at 1800 C to total porosity of 17.2%.

SPECIFICATION TABLE NO. 16 (continued)

Curve No.	Ref. No.	Method Used	Year	Temp. Range, K	Reported Error, %	Name and Specimen Designation	Composition (weight percent), Specifications and Remarks
57	136	C	1959	348-1150	±4		Polycrystalline; preparation same as the above specimen; containing 1. 26 vol. % Cr_2O_3; total porosity 19. 8%.
58	137	P	1957	84-249		AV30	96. 0 Al_2O_3; in vitreous form; 3 mm dia thin rod supplied by McDanel Refractory Porcelain Co.
59	137	P	1957	276		AV30	Same as the above specimen.
60	137	P	1957	317, 639		AV30	Same as the above specimen.
61	316	C	1951	387-693	2		Alumina of known thermal conductivity used as comparative material.
62	158	L	1952	2. 8-200	2		Polycrystalline; a rod of 0. 25 in. dia; obtained from Royal Aircraft Establishment; sintered; density 3. 70 g cm^{-3}.
63	72	R	1955	573-1573			Polycrystalline; prepared from high purity powders; calcined, hydrostatically pressed, and fired.
64	170	R	1953	473. 2			Four specimens with porosity 7. 5 to 22. 3%; heat flow perpendicular to the 0. 82 mm dia cylindrical pores.
65	170	R	1953	773. 2			Four specimens with porosity 7. 5 to 22. 1%; heat flow perpendicular to the 0. 82 mm dia cylindrical pores.
66	170	R	1953	1073. 2			Four specimens with porosity 7. 5 to 22. 2%; heat flow perpendicular to the 0. 82 mm dia cylindrical pores.
67	170	R	1953	473. 2			Four specimens with porosity 4. 6 to 19. 6%; heat flow perpendicular to the 1. 46 mm dia cylindrical pores.
68	170	R	1953	473. 2			Four specimens with porosity 4. 5 to 19. 7%; heat flow parallel to the 1. 46 mm dia cylindrical pores.
69	170	R	1953	773. 2			Four specimens with porosity 4. 3 to 19. 8%; heat flow perpendicular to the 1. 46 mm dia cylindrical pores.
70	170	R	1953	773. 2			Four specimens with porosity 4. 5 to 19. 7%; heat flow parallel to the 1. 46 mm dia cylindrical pores.
71	170	R	1953	1073. 2			Four specimens with porosity 4. 7 to 19. 7%; heat flow perpendicular to the 1. 46 mm dia cylindrical pores.
72	170	R	1953	1073. 2			Four specimens with porosity 4. 7 to 19. 7%; heat flow parallel to the 1. 46 mm dia cylindrical pores.
73	215		1960	4. 5-86			2. 5 mm dia rod; sintered.
74	215		1960	2. 5-89			The above specimen γ-irradiated.
75	216	R	1951	763. 7			No. 8 grain; measured from the inner half of test annulus.
76	216	R	1951	977. 6			Measured from the outer half of the above test annulus.

SPECIFICATION TABLE NO. 16 (continued)

Curve No.	Ref. No.	Method Used	Year	Temp. Range, K	Reported Error, %	Name and Specimen Designation	Composition (weight percent), Specifications and Remarks
77	216	R	1951	901.5			Measured from the entire above annulus.
78	184	→	1961	1318			Polycrystalline; measured in a vacuum of 10^{-5} mm Hg, run No. 1; thermal conductivity was calculated from the measured emissivity data.
79	184	→	1961	1333.5			The above specimen run No. 2.
80	184	→	1961	1284			The above specimen run No. 3.
81	184	→	1961	1289.5			The above specimen run No. 4.
82	184	→	1961	1407.7			The above specimen run No. 5.
83	184	→	1961	1409			The above specimen run No. 6.
84	184	→	1961	1276			The above specimen run No. 7.
85	73	C	1954	573-1473			No details.
86	217	C	1960	354-1171		Gulton HS. B Alumina No. 1	99.3 Al_2O_3, 0.23 Fe_2O_3, 0.22 SiO_2, 0.5 C, 0.05 C, 0.05 CaO, 0.05 TiO_2, 0.02 MgO, 0.01 Na_2O, and 0.07 others; polycrystalline with average grain size supplied by Gulton Ind.; pressed and heat treated; density 3.86 g cm^{-3} and 2.6% porosity; impurity analysis made after heat treatment; MIT alumina standard used as comparative material.
87	217	C	1960	403-1213		Gulton HS. B Alumina No. 2	Same composition,structure,and supplier as the above specimen; average grain size 3 μ; density 3.90 g cm^{-3} and porosity 1.5%; MIT alumina standard used as comparative material.
88	217	C	1960	423-1292		Gulton HS. B Alumina No. 3	The above specimen heat treated for 100 hrs at 1500 C; average grain size 6 μ; density 3.84 g cm^{-3} and porosity 3.0%; MIT alumina standard used as comparative material.
89	217	C	1960	379-1234		Gulton HS. B Alumina No. 4	Same composition,structure,and supplier as the above specimen; average grain size 4 μ; density 3.90 g cm^{-3} and porosity 1.6%; MIT alumina standard used as comparative material.
90	217	C	1960	376-1243		Gulton HS. B Alumina No. 5	Same composition,structure, and supplier as the above specimen; average grain size 10 μ; density 3.91 g cm^{-3} and porosity 1.4%; MIT alumina standard used as comparative material.
91	217	C	1960	403-1241		Norton H. P. Alundum	99.5 Al_2O_3, 0.40 MgO, 0.05 C, 0.01 SiO_2, 0.01 Fe_2O_3, 0.01 CaO, 0.01 Na_2O, and 0.05 others; supplied by Norton Co.; polycrystalline with average grain size 3 μ; hot pressed and heat treated; density 3.97 g cm^{-3} and zero porosity; impurity analysis made after heat treatment; MIT alumina standard used as comparative material.
92	217	C	1960	371-1243		Norton H. P. Alundum	The above specimen heat treated for 100 hrs at 1500 C in helium; average grain size 5 μ; density 3.83 g cm^{-3} and porosity 3.3%; MIT alumina standard used as comparative material.
93	217	C	1960	373-1273		Norton 38-900	99.8 Al_2O_3, 0.05 Fe_2O_3, 0.05 Na_2O, 0.05 C, 0.03 CaO, 0.02 SiO_2, 0.01 MgO, and 0.04 others; polycrystalline with average grain dia 10-11 μ; hot pressed; density 3.89 g cm^{-3} and porosity 1.8%; data corrected to theoretical density; MIT alumina standard used as comparative material.

SPECIFICATION TABLE NO. 16 (continued)

Curve No.	Ref. No.	Method Used	Year	Temp. Range, K	Reported Error, %	Name and Specimen Designation	Composition (weight percent), Specifications and Remarks
94	217	C	1960	373-1273		Norton 38-900	99.3 Al_2O_3, 0.44 MgO, 0.05 Fe_2O_3, 0.05 C, 0.05 Na_2O, 0.03 CaO, 0.02 SiO_2, and 0.04 others; polycrystalline with average grain dia 8-9 μ; hot pressed; density 3.92 g cm^{-3} and porosity 1.1%; data corrected to theoretical density; MIT alumina standard used as comparative material.
95	272	R	1962	318-417	5	FW-5	Prepared from spray-dried alumina powder supplied by the American Cyanamid Co.; Alpha-monohydrate (böhmite) crystal structure; drying loss: 16.4 weight % at 500 F, and 27.0 % at 1800 F; true density of solid (monohydrate) 2.45 g cm^{-3}; B.E.T. surface area 362 m^2g^{-1} and pore volume 0.475 cm^3 g^{-1}; particle size distribution: 61% of 90 μ mesh, 32.5% of 60 μ; 23.5% of 45 μ, 11.5% of 20 μ, and 7.0% of 10 μ; measured in vacuum of 10-25 μ of Hg.
96	272	R	1962	315-400	5	FW-4.5	Prepared from the same powder as the above specimen; pellet density 1.010 g cm^{-3}, micro pore volume 0.383 cm^3 g^{-1}, macro pore volume 0.198 cm^3g^{-1}, micro void fraction 0.387, and macro void fraction 0.200; measured in vacuum (pressure 10-25 microns of Hg).
97	272	R	1962	322-437	5	FW-4	Prepared from the same powder as the above specimen; pellet density 0.896 g cm^{-3}, micro pore volume 0.400 cm^3 g^{-1}, macro pore volume 0.308 cm^3 g^{-1}, micro void fraction 0.359; and macro void fraction 0.275; measured in vacuum (pressure 10-25 microns of Hg).
98	272	R	1962	318-444	5	FW-3.5	Prepared from the same powder as the above specimen; pellet density 0.785 g cm^{-3}, micro pore volume 0.416 cm^3 g^{-1}, macro pore volume 0.451 cm^3 g^{-1}, micro void fraction 0.327, and macro void fraction 0.353; measured in vacuum (pressure 10-25 microns of Hg).
99	272	R	1962	318-378	5	FW-3	Prepared from the same powder as the above specimen; pellet density 0.672 g cm^{-3}, micro pore volume 0.434 cm^3 g^{-1}, macro pore volume 0.670 cm^3 g^{-1}, micro void fraction 0.275, and macro void fraction 0.450; measured in vacuum (pressure 10-25 microns of Hg).
100	272	R	1962	322.1	5	FW-5	Same as the above specimen FW-5 but measured in helium at 1 atm. pressure.
101	272	R	1962	322.1	5	FW-4	Same as the above specimen FW-4 but measured in helium at 1 atm. pressure.
102	272	R	1962	322.1	5	FW-3	Same as the above specimen FW-3 but measured in helium at 1 atm. pressure.
103	272	R	1962	322.1	5	FW-5	Same as the above specimen FW-5 but measured in air at 1 atm. pressure.
104	272	R	1962	322.1	5	FW-4	Same as the above specimen FW-4 but measured in air at 1 atm. pressure.
105	272	R	1962	322.1	5	FW-3	Same as the above specimen FW-3 but measured in air at 1 atm. pressure.
106	273	L	1963	293, 313	<2	Powder	Produced at the NBS by ignition of hydrated aluminum chloride in a muffle furnace at 1150 C; density 0.41 g cm^{-3}.
107	273	L	1963	313.2	<2		Powder; same method of production as the above except density 0.46 g cm^{-3}.

SPECIFICATION TABLE NO. 16 (continued)

Curve No.	Ref. No.	Method Used	Year	Temp. Range, K	Reported Error, %	Name and Specimen Designation	Composition (weight percent), Specifications and Remarks
108	273	R	1963	413-1363	<3		Powder; same method of production as the above except density 0.44 g cm^{-3}.
109	273	R	1963	413-1198	<3		Powder; same method of production as the above except density 0.40 g cm^{-3}.
110	274	R	1953	813-1513	<±5		Specimen in prolate spheroid shape; fabricated in MIT Ceramics Laboratory; fired to a total porosity of 6.35-7.11% with a bulk density of 3.66-3.69 g cm^{-3}.
111	274	R	1953	848-1473	<±5		Same material as the above; separate run.
112	274	R	1953	813-1403	<±5		Same material as the above; separate run.
113	274	R	1953	833-1503	<±5		Same material as the above; separate run.
114	274	R	1953	813-1513	<±5		Same material as the above; separate run.
115	274,130	R	1953	1343-2023	<5		0.31 Fe_2O_3, 0.245 O_2, and 0.01 TiO_2; specimen in prolate spheroid shape; slip-cast; fired to zero apparent porosity at 1850 C and had a final total porosity of 5-10%.
116	293	C	1957	533-1508	±4		Single crystal; 99.5$^+$ pure; 0.875 in. cubic specimen; supplied by Linde Co.; data corrected to zero pososity; polycrystalline alumina used as standard.
117	293	C	1957	548-1158	±4	No. 1	Polycrystalline; 99.5$^+$ pure; 0.875 in. cubic specimen; supplied by Baker Chemical Co.; gravimetric porosity 3.78%; microscopic porosity 4%; average grain size 9 μ; data corrected to zero porosity.
118	293	C	1957	583-1473	±4	No. 2	Similar to the above specimen except for gravimetric porosity 12.09%; microscopic porosity 15%; average grain size 17 μ; data corrected to zero porosity.
119	135, 364	R	1957	893-1773	<2		In cylindrical form 30 mm long, inside dia 30 mm, outside dia 60 mm; porosity 22%.
120	135, 364	R	1957	803-1533	<2		Similar to the above specimen except 10% porosity.
121	330	C	1957	298.2	±6		Commercially pure; thermal comparator applied on the machined curved surface of the 1 in. dia bar specimen.
122	330	C	1957	298.2			Thermal comparator loaded with 100 gram weight applied on the plane surface of the specimen.
123	284	C	1958	473-973		Wesgo Al-300	97.6 Al_2O_3; 1 in. cube ground and polished on diamond laps.
124	286	C	1958	347-900		Norton 38-900	One in. cube; grain size distribution ranging from 5 to 9 microns with a peak at 7.5 microns; hot pressed.
125	287	L	1965	683-883		Specimen a	0.1 MgO, 0.1 NiO; specimen 12.7 mm in dia and 12.7 mm long; sintered; 99.5% theoretical density.
126	287	L	1965	723-843		Specimen b	Similar to above specimen.
127	287	R	1965	1163-1643		Specimen c	Similar to above specimen except 24.00 mm in dia and 25 mm long.
128	287	R	1965	1253-1563		Specimen d	Similar to above specimen; slight melting and cracking were found around the center of the specimen after the measurement.

SPECIFICATION TABLE NO. 16 (continued)

Curve No.	Ref. No.	Method Used	Year	Temp. Range, K	Reported Error, %	Name and Specimen Designation	Composition (weight percent), Specifications and Remarks
129	287	R	1965	1723-1993		Specimen e	Similar to above specimen.
130	288		1963	15-600		Lucalox	Polycrystalline; 99.9% Al_2O_3; density 3.98 g cm^{-3}; gas-tight, essentially zero porosity; melting point 2040 C; manufactured by General Electric Co.
131	296		1958	293-1273			Battelle-developed sinterable alumina powder hydrostatically pressed at 100,000 psi, presintered in air at 1800 F for 1 hr and final sintered at 2800 F for 2 hrs; 98.5% μ theoretical density; average crystal size about 2 μ, with some crystals as large as 70 μ.
132	295	C	1960	323-1323		A-1	Commercial single crystal; gravimetric density 3.98 g cm^{-3}; zero porosity; c-axis inclined at 60 degrees to the direction of heat flow; polycrystalline Al_2O_3 and ZrO_2 used as reference materials.
133	295	C	1960	663-1123		A-2	Commercial single crystal; gravimetric density not determined; c-axis inclined at 60 degrees to the direction of heat flow; polycrystalline Al_2O_3 and ZrO_2 used as reference materials.
134	295	C	1960	571-1271		A-3	Commercial single crystal; gravimetric density 3.98 g cm^{-3}; zero porosity; c-axis inclined at 60 degrees to the direction of heat flow; polycrystalline Al_2O_3 and ZrO_2 used as reference materials.
135	295	C	1960	543-1323		A-4	Polycrystalline specimen fabricated by pressing hydrostatically and sintering; grain size 13 μ; gravimetric density 3.97 g cm^{-3}; microscopic porosity 0.5%; gravimetric porosity 0.25%; pore size 1.0 to 1.5 μ; data corrected for the effect of porosity and presented as at zero porosity.
136	295	C	1960	383-1403		A-5	Polycrystalline specimen fabricated by pressing hydrostatically and sintering; grain size 28 μ; gravimetric density 3.86 g cm^{-3}; microscopic porosity 3.3%; gravimetric porosity 3.0%; pore size 4.0 μ; data corrected for the effect of porosity and presented as at zero porosity.
137	295	C	1960	683-1423		A-6	Polycrystalline specimen fabricated by pressing hydrostatically and sintering; grain size 19 μ; gravimetric density 3.48 g cm^{-3}; microscopic porosity 14.0%; gravimetric porosity 12.5%; pore size 6 to 10 μ; data corrected for the effect of porosity and presented as at zero porosity.
138	314	L	1959	389.8	14	Ignited alumina	Compressed powder; supplied by Anachemia Chemicals, Ltd.; specimen in the shape of a disc of 0.182 in. thick and 9 in. dia, pressed at 63 psi; bulk density 1.02 g cm^{-3}; load reduced to 0.5 psi prior to making measurements.
139	314	L	1959	393.2	14	Ignited alumina	Compressed powder; 9 in. dia x 0.145 in. thick; same supplier as the above specimen; pressed at 940 psi; bulk density 1.27 g cm^{-3}.
140	314	L	1959	396.5		Norton Alundum R. R.	Powder; -90 mesh; supplied by Fisher Scientific Co.; disc of 0.189 in. thick and 9 in. dia, pressed at 63 psi; bulk density 1.92 g cm^{-3}; load reduced to 0.5 psi prior to making measurements.

SPECIFICATION TABLE NO. 16 (continued)

Curve No.	Ref. No.	Method Used	Year	Temp. Range, K	Reported Error, %	Name and Specimen Designation	Composition (weight percent), Specifications and Remarks
141	314	L	1959	403.2		Norton Alundum R. R.	Compressed powder; 9 in. dia x 0.166 in. thick; same supplier as the above specimen; pressed 940 psi; density 2.19 g cm^{-3}.
142	317	R	1960	1238-1473		TC 352	99.3 pure; supplied by Gladding McBean; sintered; bulk density 3.8 g cm^{-3} (95% of theoretical value); run No. 6.
143	317	R	1960	1336-1609		TC 352	The above specimen; run No. 7.
144	317	R	1960	1144-1471		TC 352	The above specimen; run No. 14.
145	317	R	1960	1252-1601		TC 352	The above specimen; run No. 15.
146	317	R	1960	1365-1895		TC 352	The above specimen; run No. 16.
147	373	R	1959	473-1348	±20	Al_2O_3D(35-50)	Powder specimen composed of dense crystalline particles; contained in a 1.125 in. dia cylinder; particle dia 0.04 ±0.01 cm; porosity <4%; apparent density 1.82 g cm^{-3}; volume fraction of particles 0.475.
148	373	R	1959	353-377	±20	Al_2O_3B(35-50)	Powder specimen in the form of porous bubbles; contained in a 1.125 in. dia cylinder; bubble dia 0.04; apparent powder density 1.15 g cm^{-3}; volume fraction of bubbles 0.29.
149	373	R	1959	405-1328	±20	Al_2O_3R	Highly porous specimen of a reagent grade alumina contained in a 1.125 in. dia cylinder; apparent density 1.00 g cm^{-3}; volume fraction 0.25.
150	368	R	1954	813-1503	± 5		Hollow prolate spheriodal specimens with inner minor axis ~2 cm, inner major axis ~10 cm, outer minor axis ~4 cm and outer major axis long enough to make the outer and inner surfaces confocal; prepared by slip-casting from suspensions of finely-ground alumina; total porosity 6.35-7.11; bulk density 3.66-3.69 g cm^{-3}; measured with 4 different specimens.
151	368	R	1954	853-1478	± 5		The above specimens; run No. 2.
152	368	R	1954	813-1493	± 5		The above specimens; run No. 3.
153	368	R	1954	832-1498	± 5		The above specimens; run No. 4.
154	368	R	1954	813-1489	± 5		The above specimens; run No. 5.
155	374	C	1965	419-942	0.5	Al-300	Fired alumina; 97.55 Al_2O_3, 1.35 SiO_2, 1.05 CaO, 0.03 Fe_3O_4, 0.02 Na_2O; grain size in the range 0.01 to 0.15 mm with 80% less than 0.10 mm; bulk density 3.75 g cm^{-3}, true density 3.92 g cm^{-3}; porosity 5% of total volume; material not permeable; prepared to a tolerance of ±0.001 in. in the form of a cylinder 1 in. in dia and 1 in. high; manufactured by the Western Gold and Platinum Co.; alumina AL-300 as reference standard (thermal conductivity of the standard determined by a thermal diffusivity method by J. J. Swica, Alfred University).
156	374	C	1965	442-1225	0-2	Al-300	Similar to the above specimen except Pyroceram 9606 as reference standard (made by Corning Glass Works, thermal conductivity determined by NBS).

SPECIFICATION TABLE NO. 16 (continued)

Curve No.	Ref. No.	Method Used	Year	Temp. Range, K	Reported Error, %	Name and Specimen Designation	Composition (weight percent), Specifications and Remarks
157	375	C	1960	373-1273		Wesgo Al-300	Specimen obtained from Western Gold and Platinum Co.; alumina cubes used as reference standard obtained from W. D. Kingery, MIT.
158	401, 402	R	1960	360~1139			Granular specimen contained in 18 in. dia sphere; produced in a pilot fluidized bed calcinizer operated at 672 K at Idaho Chemical Processing Plant, from a liquid feed of the composition (concentration in g mole liter^{-1}); 8.0 NO_3^-, 2.2 Al^{+++}, 1.25 H^+, 0.15 Na^+, and 0.008 Hg^{++}; porosity 4 to 60; bulk density 1.5 to 0.6 g cm^{-3}; measured in the spherical apparatus.
159	401, 402	R	1960	700, 768		AII-1	Granular specimen contained in 12 in. dia x 48 in. long cylinder; same fabrication method, supplier, and physical properties as the above specimen; measured in the cylindrical apparatus.
160	401, 402	R	1960	838, 1038		AII-2	Similar to the above specimen.
161	401, 402	R	1960	975-1255		AII-3	Similar to the above specimen.
162	401, 402	R	1960	789, 889		AII-4	Similar to the above specimen.
163	401, 402	R	1960	821, 1064		AII-5	Similar to the above specimen.
164	401, 402	R	1960	1001-1229		AII-6	Similar to the above specimen.
165	401, 402	R	1960	923-1145		AII-7	Similar to the above specimen.
166	401, 402	R	1960	980-1198		AII-8	Similar to the above specimen.
167	401, 402	R	1960	458-738		B-1	Similar to the above specimen.
168	401, 402	R	1960	765-1379		B-2	Similar to the above specimen.
169	401, 402	R	1960	910-1499		B-3	Similar to the above specimen.
170	401, 402	R	1960	1040-1698		B-4	Similar to the above specimen.
171	401, 402	R	1960	558-933		C-1	Similar to the above specimen.

SPECIFICATION TABLE NO. 16 (continued)

Curve No.	Ref. No.	Method Used	Year	Temp. Range, K	Reported Error, %	Name and Specimen Designation	Composition (weight percent), Specifications and Remarks
172	401, 402	R	1960	479-1023		C-2	Similar to the above specimen.
173	401, 402	R	1960	536-1145		C-3	Similar to the above specimen.
174	401, 402	R	1960	463-959		D-1	Similar to the above specimen.
175	401, 402	R	1960	604-1254		D-2	Similar to the above specimen.
176	401, 402	R	1960	788-1335		D-3	Similar to the above specimen.
177	401, 402	R	1960	763-1534		E-1	Similar to the above specimen.
178	401, 402	R	1960	796-1725		E-2	Similar to the above specimen.
179	401, 402	R	1960	915-1830		E-3	Similar to the above specimen.
180	401, 402	R	1960	945-1614		E-4	Similar to the above specimen.
181	401, 402	R	1960	752-1708		E-5	Similar to the above specimen.
182	401, 402	R	1960	342-595			Granular specimen of the size of -100 mesh contained in 12 in. dia sphere; same fabrication method and supplier as the above specimen.
183	403	L	1967	1.4-4.3			Pure; single crystal; 8 mm dia x 40 mm; specimen axis made an angle of 60 degrees to the crystal axis.
184	403	L	1967	1.4-4.3		1	0.1 MnO_2, 0.0017 Mn; single crystal; 8 mm dia x 40 mm long; specimen axis made an angle of 60 degrees to the crystal axis.
185	403	L	1967	1.4-4.5		1	Above specimen measured in a magnetic field with strength 11800 oersted along the specimen axis.
186	403	L	1967	1.5-4.2		2	0.2 MnO_2; single crystal; 8 mm dia x 40 mm long; specimen axis made an angle of 60 degrees to the crystal axis.
187	403	L	1967	1.4-4.3		2	Above specimen measured in a magnetic field with strength 11800 orested along the specimen axis.
188	403	L	1967	1.5-4.2		3	0.35 MnO_2, 0.0014 Mn; single crystal; 8 mm dia x 40 mm long; specimen axis made an angle of 60 degrees to the crystal axis.

SPECIFICATION TABLE NO. 16 (continued)

Curve No.	Ref. No.	Method Used	Year	Temp. Range, K	Reported Error, %	Name and Specimen Designation	Composition (weight percent), Specifications and Remarks
189	403	L	1967	1.5-4.4	3		Above specimen measured in a magnetic field with strength 11800 orested along the specimen axis.
190	404, 405	P	1960	333-832			Al_2O_3 powder filled into a cylindrical vessel 32.2 mm inside dia and 300 mm long of a foamed refractory fire clay with specific weight 800 kg cm^{-3}; thermal conductivity data calculated from the measurement of thermal diffusivity, specific heat and density.
191	406	L	1965	4.2			Powder; grain size 6 ±2 μ; load applied to specimen 6 kg cm^{-2}; measured in helium under a pressure 3.27 x 10^{-3} μ Hg.
192	406	L	1965	4.2			Similar to the above specimen; measured under pressure in the range 1.00 x 10^{-5} ~20.2 mm Hg.
193	406	L	1965	4.2			Similar to the above specimen except grain size 35 ±5 μ; measured under pressure in the range 1.84 x 10^{-3} ~3.20 x 10^{-2} μ Hg.
194	406	L	1965	4.2			Similar to the above specimen; measured under pressure in the range 2.24 x 10^{-6} ~63.1 mm Hg.
195	406	L	1965	4.2			Similar to the above specimen; measured under pressure in the range 7.33 x 10^{-2} ~94.4 μ Hg.
196	406	L	1965	4.2			Similar to the above specimen except grain size 150 ±10 μ; measured under pressure in the range 1.93 x 10^{-3} ~2.95 x 10^{-2} μ Hg.
197	406	L	1965	4.2			Similar to the above specimen; measured under pressure in the range 1.37 x 10^{-5} ~67.6 mm Hg.
198	406	L	1965	4.2			Powder; grain size 35 μ; measured in vacuum under a load varying in a cycle.
199	406	L	1965	2.5-4.9			Powder; grain size 6 μ; load applied to specimen 6 kg cm^{-2}; measured in a vacuum of 100 μ Hg.
200	406	L	1965	2.5-4.9			Similar to the above specimen except measured in a vacuum of 10 μ Hg.
201	406	L	1965	2.0-4.1			Powder; grain size 35 μ; load applied to specimen 6 kg cm^{-2}; measured in a vacuum of 1 μ Hg.
202	406	L	1965	2.0-4.2			Similar to the above specimen except measured in a vacuum of 10 μ Hg.
203	406	L	1965	2.8-4.8			Powder; grain size 6 μ; load applied to specimen 6 kg cm^{-2}; measured in a vacuum of $<10^{-1}$ μ Hg.
204	406	L	1965	1.9-4.8			Similar to the above specimen except grain size 35 μ.
205	406	L	1965	2.1-4.1			Similar to the above specimen except grain size 150 μ.

SPECIFICATION TABLE NO. 16 (continued)

Curve No.	Ref. No.	Method Used	Year	Temp. Range, K	Reported Error, %	Name and Specimen Designation	Composition (weight percent), Specifications and Remarks
206	407	L	1949	318-412		C-5	No details reported.
207	407	L	1949	317-421			No details reported.
208	408	C	1962	664-1208			Foam specimen; density 0.593 g cm^{-3}; Min-K 1301 (Johns Manville Corp.) used as comparative material.

DATA TABLE NO. 16 THERMAL CONDUCTIVITY OF ALUMINUM OXIDE Al_2O_3

[Temperature, T,K; Thermal Conductivity, k, Watt $cm^{-1}K^{-1}$]

T	k
CURVE 1	
846.2	0.0833
983.2	0.0715
1152.2	0.0623
1381.2	0.0569
1574.2	0.0536
1674.2	0.0527
1737.2	0.0540
CURVE 2	
758.2	0.0828
913.2	0.0741
1050.2	0.0619
1176.2	0.0561
1181.2	0.0657
1313.2	0.0527
1401.2	0.0498
1442.2	0.0506
1448.2	0.0490
1504.2	0.0490
1535.2	0.0527
CURVE 3	
333.2	0.173
CURVE 4	
873.2	0.0837
1073.2	0.0693
1273.2	0.0588
1473.2	0.0535
CURVE 5	
423.2	0.209
483.2	0.172
548.2	0.157
593.2	0.136
663.2	0.117
710.2	0.105
783.2	0.094
833.2	0.090
893.2	0.0837
CURVE 5 (cont.)	
953.2	0.0745
1013.2	0.0753
1053.2	0.0628
1113.2	0.0586
1153.2	0.0607
1228.2	0.0586
1293.2	0.0586
1353.2	0.0565
1393.2	0.0552
1453.2	0.0544
1523.2	0.0502
1553.2	0.0502
CURVE 6	
483.2	0.167
543.2	0.142
588.2	0.132
663.2	0.117
703.2	0.094
763.2	0.0879
833.2	0.0816
903.2	0.0669
973.2	0.0607
1008.2	0.0586
1063.2	0.0544
1113.2	0.0544
1223.2	0.0523
1283.2	0.0502
1298.2	0.0473
1373.2	0.0473
1418.2	0.0460
1473.2	0.0473
1518.2	0.0460
1553.2	0.0460
CURVE 7	
613.2	0.111
688.2	0.100
703.2	0.0837
873.2	0.0795
943.2	0.0690
CURVE 7 (cont.)	
1033.2	0.0586
1093.2	0.0565
1203.2	0.0565
1258.2	0.0565
CURVE 8	
593.2	0.165
683.2	0.132
763.2	0.0983
853.2	0.0900
933.2	0.0782
1013.2	0.0711
1083.2	0.0628
1113.2	0.0628
1193.2	0.0607
1253.2	0.0586
CURVE 9	
293.2	0.251
323.2	0.213
373.2	0.169
423.2	0.142
473.2	0.121
523.2	0.109
573.2	0.0962
673.2	0.0820
773.2	0.0732*
873.2	0.0669*
973.2	0.0628*
1073.2	0.0607*
1173.2	0.0586*
1263.2	0.0569*
CURVE 10	
769.2	0.0473
964.2	0.0372
1155.2	0.0305
1347.2	0.0251
1535.2	0.0238
1631.2	0.0230
CURVE 11	
419.2	0.195
481.2	0.178
524.2	0.164
604.2	0.142
661.2	0.124
743.2	0.105
839.2	0.0904
964.2	0.0803
1067.2	0.0715
CURVE 12	
322.4	0.188
339.2	0.182
360.9	0.162
382.4	0.159
425.2	0.145
CURVE 13	
1333.2	0.0668
CURVE 14	
1387.2	0.0402
1392.2	0.0469
1447.2	0.0502
1463.2	0.0456
1487.2	0.0782
1523.2	0.0573
1595.2	0.0178
1598.2	0.0226
1641.2	0.0531*
1663.2	0.0402
1681.2	0.0707
1690.2	0.0372
1973.2	0.0339
CURVE 15	
1273.2	0.0515*
1275.2	0.0531
1298.2	0.0389
1333.2	0.0343
CURVE 15 (cont.)	
1338.2	0.0477
1393.2	0.0615
1393.2	0.0410*
1403.2	0.0498*
1413.2	0.0460*
1415.2	0.0623
1428.2	0.0423
1448.2	0.0439*
1483.2	0.0473*
1498.2	0.0397
1504.2	0.0490*
1508.2	0.0435
1553.2	0.0561
1573.2	0.0619
1573.2	0.0368*
1573.2	0.0406*
1607.2	0.0485
1608.2	0.0377
1618.2	0.0444
1666.2	0.0406*
1668.2	0.0460*
1703.2	0.0469*
1718.2	0.0364
1723.2	0.0473*
1723.2	0.0510*
1735.2	0.0469
1738.2	0.0540*
1743.2	0.0414
1748.2	0.0498*
1753.2	0.0427*
1773.2	0.0431
1808.2	0.0481
1813.2	0.0477*
1818.2	0.0498*
1843.2	0.0506*
1848.2	0.0569
1853.2	0.0552*
1873.2	0.0632
1908.2	0.0573
1918.2	0.0582*
1930.2	0.0577*
1943.2	0.0456
1953.2	0.0594
1983.2	0.0481
CURVE 16*	
1336.2	0.0464
1443.2	0.0477
1508.2	0.0460
1628.2	0.0368
1753.2	0.0490
1778.2	0.0410
1826.2	0.0418
1943.2	0.0556
1963.2	0.0632
2023.2	0.0464
2028.2	0.0577
CURVE 17	
373.2	0.0490
575.2	0.0669
769.2	0.0707
CURVE 18	
372.2	0.0397
572.2	0.0423
750.2	0.0598
985.2	0.0682
CURVE 19	
328.2	0.00126
728.2	0.00372
CURVE 20	
328.2	0.00222
CURVE 21	
318.2	0.330
330.2	0.327
338.2	0.304
356.2	0.300
357.2	0.295
363.2	0.290
378.2	0.282*
378.2	0.275
CURVE 21 (cont.)	
388.2	0.275*
398.2	0.264*
400.2	0.259
413.2	0.249*
418.2	0.247*
426.2	0.244*
426.2	0.241*
430.2	0.236*
443.2	0.229*
450.2	0.220*
463.2	0.219*
470.2	0.223
473.2	0.220*
473.2	0.215*
473.2	0.210*
473.2	0.204
473.2	0.197
496.2	0.203
498.2	0.198*
508.2	0.197*
516.2	0.195
518.2	0.183
533.2	0.173
543.2	0.170
548.2	0.164
550.2	0.169*
578.2	0.158*
578.2	0.154
583.2	0.156*
596.2	0.155*
600.2	0.154
633.2	0.139*
636.2	0.135*
646.2	0.132*
657.2	0.135
696.2	0.128
696.2	0.120*
708.2	0.116*
720.2	0.115*
743.2	0.111*
756.2	0.113
758.2	0.111*
760.2	0.106*
CURVE 21 (cont.)	
766.2	0.108
770.2	0.112
773.2	0.114
780.2	0.106
786.2	0.103
813.2	0.0962*
818.2	0.0900*
833.2	0.0933*
836.2	0.0891*
853.2	0.0891*
863.2	0.0920*
873.2	0.0866*
900.2	0.0816*
CURVE 22	
323.0	0.175
385.8	0.150
449.2	0.134
490.0	0.123
585.7	0.106
789.8	0.0820
1007.3	0.0675*
1160.0	0.0613*
1345.7	0.0561*
1512.0	0.0550*
1589.8	0.0549*
1694.4	0.0550*
1875.8	0.0556*
CURVE 23*	
473.2	0.213
673.2	0.121
873.2	0.0837
1073.2	0.0669
CURVE 24*	
473.2	0.184
673.2	0.105
873.2	0.0711
1073.2	0.0544

*Not shown on plot

T	k
CURVE 25	
473.2	0.161
673.2	0.0920
873.2	0.0586
1073.2	0.0460
CURVE 26	
473.2	0.144
673.2	0.0837
873.2	0.0523
1073.2	0.0397
CURVE 27	
473.2	0.105
673.2	0.0586
873.2	0.0377
1073.2	0.0293
CURVE 28	
333.2	0.146
403.2	0.121
518.2	0.0941*
533.2	0.0921
603.2	0.0753
723.2	0.0523
853.2	0.0418
973.2	0.0377
993.2	0.0377
1038.2	0.0335
1183.2	0.0335
CURVE 29*	
473.2	0.209
673.2	0.126
873.2	0.0879
1073.2	0.0711
CURVE 30*	
473.2	0.192
673.2	0.115
873.2	0.0795
1073.2	0.0628
CURVE 31*	
428.2	0.205
468.2	0.188
503.2	0.167
523.2	0.161
553.2	0.146
573.2	0.138
623.2	0.123
643.2	0.117
678.2	0.109
713.2	0.100
733.2	0.0962
753.2	0.0900
783.2	0.0879
853.2	0.0795
873.2	0.0753
913.2	0.0669
953.2	0.0649
1013.2	0.0628
1093.2	0.0586
CURVE 32*	
473.2	0.167
673.2	0.100
873.2	0.0669
1073.2	0.0565
CURVE 33*	
473.2	0.159
673.2	0.0941
873.2	0.0628
1073.2	0.0523
CURVE 34*	
473.2	0.201
773.2	0.0962
1073.2	0.0657
CURVE 35*	
473.2	0.189
773.2	0.0904
1073.2	0.0617
CURVE 36*	
473.2	0.182
773.2	0.0854
1073.2	0.0586
CURVE 37*	
473.2	0.168
773.2	0.0816
1073.2	0.0554
CURVE 38*	
473.2	0.184
773.2	0.0887
1073.2	0.0596
CURVE 39	
473.2	0.155
773.2	0.075
1073.2	0.0523
CURVE 40	
473.2	0.136
773.2	0.0628
1073.2	0.0450
CURVE 41	
473.2	0.105
773.2	0.0502
1073.2	0.0366
CURVE 42*	
452.2	0.216
466.2	0.199
507.2	0.190
525.2	0.171
597.2	0.149
604.2	0.138
697.2	0.115
705.2	0.113
743.2	0.106
800.2	0.0946
CURVE 42* (cont.)	
835.2	0.0891
869.2	0.0870
957.2	0.0749
973.2	0.0740
CURVE 43*	
452.2	0.219
466.2	0.205
507.2	0.193
525.2	0.173
597.2	0.157
604.2	0.145
697.2	0.122
705.2	0.118
743.2	0.115
800.2	0.0983
835.2	0.0946
869.2	0.0874
957.2	0.0808
973.2	0.0795
CURVE 44	
1193.2	0.0024
1323.2	0.0027
1443.2	0.0030
1583.2	0.0033
1753.2	0.00375
1913.2	0.0042
2033.2	0.0045
2093.2	0.00475
CURVE 45	
1248.2	0.0024
1373.2	0.00265
1503.2	0.00305
1713.2	0.0035
1833.2	0.00375
1973.2	0.00425
2083.2	0.0046
CURVE 46	
1268.2	0.0022
CURVE 46 (cont.)	
1403.2	0.0025
1533.2	0.00295
1668.2	0.0032
1803.2	0.00365
1963.2	0.0041
2078.2	0.0045
2143.2	0.0047
CURVE 47	
1243.2	0.00225
1303.2	0.0022
1473.2	0.00255
1673.2	0.00305
1833.2	0.0033
2003.2	0.0038
2143.2	0.0040
CURVE 48	
1223.2	0.00165
1393.2	0.00195
1543.2	0.00245
1693.2	0.00285
1903.2	0.0033
2033.2	0.00375
2093.2	0.0038
CURVE 49	
1233.2	0.00145
1373.2	0.00155
1553.2	0.00215
1713.2	0.00235
1893.2	0.0029
2033.2	0.0034
2143.2	0.00355
CURVE 50	
1193.2	0.0019
1333.2	0.00205
1478.2	0.0023
1623.2	0.0026
1678.2	0.0029
1803.2	0.0033*
CURVE 50 (cont.)	
1993.2	0.0036
2123.2	0.00385
CURVE 51	
1173.2	0.00205
1353.2	0.0021*
1478.2	0.0025*
1613.2	0.00255
1823.2	0.0029
1933.2	0.0032
2043.2	0.0033
CURVE 52	
1173.2	0.00165
1303.2	0.0017
1453.2	0.0018
1533.2	0.0021
1703.2	0.00235*
1813.2	0.0023
1923.2	0.0025
2043.2	0.0027
CURVE 53	
1193.2	0.00115
1353.2	0.00120
1453.2	0.00140
1613.2	0.00170
1833.2	0.00195
1913.2	0.00205
2053.2	0.00240
2193.2	0.00260
CURVE 54	
388.2	0.00315
422.2	0.00337
505.2	0.00393
573.2	0.00417
649.2	0.00459
703.2	0.00479
797.2	0.00524
897.2	0.00542
990.2	0.00559*
1119.2	0.00610
CURVE 55	
379.2	0.00411
423.2	0.00433
517.2	0.00518
599.2	0.00568
701.2	0.00621
799.2	0.00667
895.2	0.00712
1002.2	0.00739
1104.2	0.00763
CURVE 56*	
373.2	0.276
393.2	0.258
423.2	0.230
448.2	0.213
648.2	0.125
783.2	0.0962
888.2	0.0732
1128.2	0.0644
CURVE 57	
348.2	0.238
368.2	0.229
403.2	0.209
428.2	0.197
448.2	0.186
570.2	0.138
643.2	0.117
743.2	0.0967
788.2	0.0782
1023.2	0.0657
1150.2	0.0544
CURVE 58	
84.0	1.259*
126.0	0.481
249.0	0.351
CURVE 59	
276.0	0.218
CURVE 60	
317.0	0.163
639.0	0.029
CURVE 61*	
387.2	0.156
498.2	0.138
633.2	0.117
693.2	0.106
CURVE 62	
2.8	0.0016
4.5	0.0062
9.5	0.047
20.0	0.24
30.5	0.47
35.0	0.63
60.0	1.5
85.0	1.6
95.0	1.4
200.0	0.5
CURVE 63*	
573.2	0.159
623.2	0.143
673.2	0.126
723.2	0.112
773.2	0.104
872.0	0.0870
974.9	0.0758
1073.2	0.0686
1173.2	0.0637
1273.2	0.0590
1373.2	0.0556
1463.2	0.0532
1573.2	0.0527

*Not shown on plot

DATA TABLE NO. 16 (continued)

Porosity (%)	k
CURVE 64* (T = 473.2 K)	
7.5	0.168
11.8	0.147
17.75	0.117
22.25	0.0943
CURVE 65* (T = 773.2 K)	
7.25	0.0818
11.75	0.0701
17.8	0.0565
22.1	0.0471
CURVE 66* (T = 1073.2 K)	
7.5	0.0546
12.0	0.0474
17.75	0.0387
22.15	0.0320
CURVE 67* (T = 473.2 K)	
4.6	0.184
9.65	0.155
13.4	0.136
19.6	0.104
CURVE 68* (T = 473.2 K)	
4.5	0.200
9.7	0.189
13.5	0.182
19.7	0.168
CURVE 69* (T = 773.2 K)	
4.25	0.0885
9.64	0.0749
14.75	0.0628
19.75	0.0502

Porosity (%)	k
CURVE 70* (T = 773.2 K)	
4.5	0.0961
9.7	0.0907
14.66	0.0870
19.7	0.0806
CURVE 71* (T = 1073.2 K)	
4.7	0.0600
9.6	0.0529
14.67	0.0449
19.7	0.0372
CURVE 72* (T = 1073.2 K)	
4.7	0.0656
9.7	0.0620
14.75	0.0588
19.7	0.0557

T	k
CURVE 73	
4.5	0.0091
9.01	0.0532
20.0	0.35
30.3	0.76
53.3	1.81
76.0	1.815
86.0	1.46
CURVE 74	
2.52	0.0024
4.45	0.0103
9.2	0.0608
19.9	0.351
52.2	1.80
89.0	1.50

T	k
CURVE 75	
763.7	0.00829
CURVE 76	
977.6	0.0117
CURVE 77	
901.5	0.0102
CURVE 78*	
1318.0	0.0567
CURVE 79*	
1333.5	0.0482
CURVE 80*	
1284.0	0.0583
CURVE 81*	
1289.5	0.0547
CURVE 82*	
1407.5	0.0776
CURVE 83	
1409.0	0.0845
CURVE 84	
1276.0	0.0676
CURVE 85*	
573.2	0.176
873.2	0.0941
1173.2	0.0643
1473.2	0.0536

T	k
CURVE 86*	
354.2	0.234
410.2	0.209
493.2	0.175
551.2	0.153
611.2	0.135
672.2	0.120
741.2	0.106
819.2	0.0929
862.2	0.0874
925.2	0.0791
962.2	0.0753
1032.2	0.0690
1080.2	0.0657
1171.2	0.0602
CURVE 87*	
403.2	0.259
406.2	0.232
466.2	0.215
507.2	0.194
571.2	0.159
625.2	0.138
680.2	0.120
746.2	0.107
800.2	0.100
892.2	0.0862
1036.2	0.0728
1123.2	0.0644
1213.2	0.0640
CURVE 88*	
423.2	0.258
458.2	0.225
486.2	0.195
524.2	0.187
549.2	0.167
583.2	0.157
629.2	0.135
666.2	0.127
798.2	0.0992
917.2	0.0849
1010.2	0.0749
1100.2	0.0678

T	k
CURVE 88* (cont.)	
1179.2	0.0640
1292.2	0.0587
CURVE 89*	
379.2	0.241
389.2	0.227
465.2	0.196
503.2	0.180
542.2	0.164
628.2	0.136
633.2	0.132
712.2	0.117
802.2	0.100
839.2	0.0916
893.2	0.0862
946.2	0.0812
1019.2	0.0736
1036.2	0.0724
1138.2	0.0661
1234.2	0.0619
CURVE 90*	
376.2	0.249
451.2	0.220
489.2	0.196
526.2	0.170
593.2	0.155
660.2	0.136
792.2	0.107
911.2	0.0891
1024.2	0.0782
1154.2	0.0674
1243.2	0.0628
CURVE 91	
403.2	0.282
429.2	0.251
468.2	0.232
537.2	0.189
602.2	0.161
674.2	0.139
762.2	0.119

T	k
CURVE 91 (cont.)	
900.2	0.103
931.2	0.0992
1008.2	0.0920
1125.2	0.0858
1241.2	0.0816
CURVE 92*	
371.2	0.296
423.2	0.249
474.2	0.216
562.2	0.164
630.2	0.140
667.2	0.129
751.2	0.109
810.2	0.0983
895.2	0.0870
955.2	0.0795
1029.2	0.0741
1120.2	0.0686
1243.2	0.0640
CURVE 93*	
373.2	0.295
473.2	0.234
573.2	0.182
673.2	0.143
773.2	0.118
873.2	0.102
973.2	0.0887
1073.2	0.0795
1173.2	0.0703
1273.2	0.0632
CURVE 94*	
373.2	0.272
473.2	0.218
573.2	0.169
673.2	0.134
773.2	0.111
873.2	0.0962
973.2	0.0845
1073.2	0.0757

T	k
CURVE 94* (cont.)	
1173.2	0.0674
1273.2	0.0594
CURVE 95	
317.6	0.00163
328.7	0.00166
368.7	0.00159
380.9	0.00154
394.3	0.00151
416.5	0.00152
CURVE 96	
315.4	0.00149
340.9	0.00140
349.8	0.00144
362.6	0.00140
388.9	0.00133
399.8	0.00132
CURVE 97	
321.5	0.00117
346.5	0.00118
363.2	0.00117
377.6	0.00118
405.4	0.00118
433.2	0.00109
436.5	0.00111
CURVE 98*	
318.2	0.000987
337.6	0.000987
350.9	0.000987
365.4	0.00104
384.3	0.000935
403.2	0.000969
430.9	0.000935
435.4	0.000952
444.3	0.000900

T	k
CURVE 99*	
317.6	0.000675
326.5	0.000675
339.8	0.000658
355.4	0.000692
366.5	0.000675
377.6	0.000640
CURVE 100	
322.1	0.00232
CURVE 101*	
322.1	0.00208
CURVE 102*	
322.1	0.00166
CURVE 103	
322.1	0.00216
CURVE 104	
322.1	0.00178
CURVE 105*	
322.1	0.00163
CURVE 106*	
293.2	0.000970
313.2	0.00100
CURVE 107	
313.2	0.00121
CURVE 108	
413.2	0.00120
613.2	0.00136
813.2	0.00154

*Not shown on plot

DATA TABLE NO. 16 (continued)

T	k
CURVE 108 (cont.)	
1023.2	0.00174
1223.2	0.00195*
1363.2	0.00215*
CURVE 109	
413.2	0.00108
613.2	0.00125
803.2	0.00146
1003.2	0.00173
1198.2	0.00195*
CURVE 110*	
813.2	0.0900
983.2	0.0711
1133.2	0.0628
1253.2	0.0577
1373.2	0.0544
1433.2	0.0536
1513.2	0.0515
CURVE 111*	
848.2	0.0837
973.2	0.0720
1143.2	0.0623
1223.2	0.0586
1343.2	0.0552
1473.2	0.0544
CURVE 112*	
813.2	0.0887
948.2	0.0745
1073.2	0.0678
1093.2	0.0586
1403.2	0.0552
CURVE 113*	
833.2	0.0920
973.2	0.0774
1093.2	0.0669
1198.2	0.0628
CURVE 113* (cont.)	
1333.2	0.0582
1433.2	0.0565
1503.2	0.0552
CURVE 114*	
813.2	0.0937
973.2	0.0774
1083.2	0.0669
1203.2	0.0619
1323.2	0.0586
1423.2	0.0577
1513.2	0.0540
CURVE 115	
1343.2	0.0544*
1443.2	0.0565*
1503.2	0.0544*
1558.2	0.0439
1703.2	0.0586
1773.2	0.0494
1823.2	0.0502
1843.2	0.0665
1968.2	0.0753
2023.2	0.0690
2023.2	0.0561
CURVE 116	
533.2	0.184*
593.2	0.155*
648.2	0.138*
688.2	0.123*
773.2	0.107*
868.2	0.097
963.2	0.088
1073.2	0.081
1173.2	0.077
1253.2	0.0765
1313.2	0.077
1393.2	0.079
1443.2	0.082
1508.2	0.086
CURVE 117	
548.2	0.165*
623.2	0.136
738.2	0.113
818.2	0.094*
888.2	0.079*
1158.2	0.067
CURVE 118	
583.2	0.159
623.2	0.142
673.2	0.127
773.2	0.103
873.2	0.088
978.2	0.077
1073.2	0.069*
1173.2	0.065*
1273.2	0.059*
1383.2	0.056*
1473.2	0.053*
CURVE 119	
893.2	0.0256
1093.2	0.0232
1403.2	0.0231
1593.2	0.0231*
1773.2	0.0231
CURVE 120*	
803.2	0.0581
913.2	0.0628
1093.2	0.0558
1293.2	0.0581
1478.2	0.0418
1533.2	0.0442
1533.2	0.0488
CURVE 121*	
298.2	0.188
CURVE 122*	
298.2	0.187
CURVE 123*	
473.2	0.201
573.2	0.148
673.2	0.123
773.2	0.0987
873.2	0.0870
973.2	0.0761
CURVE 124*	
347.1	0.316
366.5	0.303
477.6	0.230
588.7	0.176
699.8	0.132
810.9	0.107
899.8	0.0941
CURVE 125*	
683.2	0.140
703.2	0.134
773.2	0.123
783.2	0.109
793.2	0.113
883.2	0.100
CURVE 126*	
723.2	0.128
783.2	0.121
803.2	0.117
823.2	0.111
843.2	0.107
CURVE 127*	
1163.2	0.0753
1293.2	0.0669
1343.2	0.0586
1573.2	0.0586
1643.2	0.0586
CURVE 128*	
1253.2	0.0690
CURVE 128* (cont.)	
1503.2	0.0586
1563.2	0.0649
CURVE 129*	
1723.2	0.0586
1753.2	0.0669
1993.2	0.0711
CURVE 130	
15.0	0.586
20.0	0.920
30.0	1.53
40.0	2.09
50.0	2.55
60.0	2.93
80.0	3.37
85.0	3.41
100.0	2.72
150.0	1.11
200.0	0.607
300.0	0.293
400.0	0.167
600.0	0.0418
CURVE 131	
293.2	0.301
673.2	0.130
1273.2	0.0628
CURVE 132	
323.2	0.439
373.2	0.377
413.2	0.340
443.2	0.310
473.2	0.285
513.2	0.226
623.2	0.176
723.2	0.142
863.2	0.117
998.2	0.109
1118.2	0.100
CURVE 132 (cont.)	
1203.2	0.100
1273.2	0.105
1323.2	0.109
CURVE 133	
663.2	0.201
793.2	0.163
953.2	0.121
1073.2	0.126
1123.2	0.130
CURVE 134	
571.2	0.205
663.2	0.163
763.2	0.138
843.2	0.134
923.2	0.130
933.2	0.129
1028.2	0.126
1135.2	0.134
1223.2	0.126
1271.2	0.134
CURVE 135	
543.2	0.213
653.2	0.159*
793.2	0.124
943.2	0.105
983.2	0.100
1023.2	0.096
1053.2	0.095
1153.2	0.092
1183.2	0.090
1323.2	0.095
CURVE 136	
383.2	0.331
418.2	0.297
468.2	0.255
603.2	0.159*
708.2	0.126*
CURVE 136 (cont.)	
953.2	0.079*
1183.2	0.067*
1403.2	0.059*
CURVE 137*	
683.2	0.130
823.2	0.096
893.2	0.087
983.2	0.079
1203.2	0.063
1423.2	0.059
CURVE 138*	
389.8	0.00147
CURVE 139*	
393.2	0.00453
CURVE 140*	
396.5	0.00272
CURVE 141*	
403.2	0.00872
CURVE 142	
1238.2	0.0295
1319.2	0.0377
1373.2	0.0454
1473.2	0.0481*
CURVE 143*	
1336.2	0.0303
1405.2	0.0425
1497.2	0.0402
1537.2	0.0540
1609.2	0.0485
CURVE 144	
1144.2	0.0293
1179.2	0.0378
1181.2	0.0281
1236.2	0.0264
1295.2	0.0303
1325.2	0.0293
1369.2	0.0347
1397.2	0.0379
1471.2	0.0360
CURVE 145*	
1252.2	0.0341
1324.2	0.0349
1445.2	0.0418
1494.2	0.0444
1588.2	0.0422
1601.2	0.0458
CURVE 146	
1365.2	0.0366
1422.2	0.0379
1510.2	0.0429
1702.2	0.0523
1817.2	0.0472
1895.2	0.0519
CURVE 147	
473.2	0.00380
553.2	0.00414
623.2	0.00461
733.2	0.00480
793.2	0.00504
993.2	0.00560
1033.2	0.00610
1281.2	0.00725
1348.2	0.00770
CURVE 148	
353.2	0.00230
429.2	0.00253
553.2	0.00295

*Not shown on plot

DATA TABLE NO. 16 (continued)

T	k
CURVE 148 (cont.)	
733.2	0.00358
973.2	0.00405
1113.2	0.00440
1377.2	0.00500
CURVE 149	
405.2	0.00230
513.2	0.00250
593.2	0.00260
693.2	0.00270
793.2	0.00269
873.2	0.00280
1033.2	0.00280
1093.2	0.00290
1253.2	0.00290
1328.2	0.00290
CURVE 150*	
813.2	0.0883
993.2	0.0703
1143.2	0.0623
1263.2	0.0531
1368.2	0.0544
1433.2	0.0527
1503.2	0.0515
CURVE 151*	
853.2	0.0828
993.2	0.0715
1113.2	0.0619
1223.2	0.0590
1443.2	0.0552
1478.2	0.0544
CURVE 152*	
813.2	0.0879
958.2	0.0736
1083.2	0.0669
1213.2	0.0615
CURVE 152* (cont.)	
1293.2	0.0577
1408.2	0.0548
1493.2	0.0533
CURVE 153*	
833.2	0.0916
983.2	0.0766
1093.2	0.0665
1203.2	0.0628
1333.2	0.0577
1433.2	0.0561
1498.2	0.0544
CURVE 154*	
813.2	0.0941
983.2	0.0778
1083.2	0.0665
1328.2	0.0586
1343.2	0.0586
1423.2	0.0573
1498.2	0.0536
CURVE 155*	
422.2	0.169
462.2	0.156
454.2	0.155
651.2	0.109
789.2	0.092
786.2	0.091
917.2	0.081
924.2	0.081
945.2	0.078
938.2	0.079
CURVE 156*	
442.2	0.1610
679.2	0.1049
1004.2	0.0748
1007.2	0.0735
1225.2	0.0624
CURVE 157*	
373.2	0.252
473.2	0.202
573.2	0.156
673.2	0.121
773.2	0.101
873.2	0.0862
973.2	0.0757
1073.2	0.0669
1173.2	0.0611
1273.2	0.0561
CURVE 158	
359.8	0.00196
378.7	0.00201
386.5	0.00151
409.8	0.00216
417.1	0.00194
424.8	0.00145
437.6	0.00251
443.2	0.00232
454.3	0.00199
455.4	0.00254
471.5	0.00228
488.2	0.00190
503.2	0.00232
505.4	0.00206
506.5	0.00279*
525.9	0.00287
529.8	0.00258*
600.4	0.00256
604.3	0.00273
620.4	0.00350
644.3	0.00284
671.5	0.00310
677.6	0.00350
698.7	0.00303
698.7	0.00251
709.8	0.00358
764.8	0.00317
764.8	0.00436
793.2	0.00464
812.1	0.00427
814.8	0.00350
CURVE 158 (cont.)	
849.8	0.00467
852.1	0.00329
854.8	0.00450
872.1	0.00291
903.7	0.00438
961.5	0.00358
966.5	0.00441
977.1	0.00455
978.2	0.00350
995.4	0.00469
1063.7	0.00478
1068.7	0.00457
1099.3	0.00481
1102.1	0.00419
1139.3	0.00500
CURVE 159*	
699.8	0.00234
767.6	0.00324
CURVE 160*	
837.6	0.00362
1037.6	0.00476
CURVE 161*	
975.4	0.00424
1032.1	0.00511
1144.8	0.00559
1255.4	0.00668
CURVE 162*	
789.3	0.00339
889.3	0.00389
CURVE 163*	
820.9	0.00350
936.5	0.00391
992.1	0.00514
1064.3	0.00583
CURVE 164*	
1000.9	0.00438
1075.4	0.00492
1165.9	0.00595
1228.7	0.00742
CURVE 165*	
922.6	0.00453
1032.1	0.00452
1144.8	0.00452
CURVE 166*	
980.4	0.00453
1087.1	0.00452
1197.6	0.00452
CURVE 167*	
457.6	0.00222
563.2	0.00241
649.8	0.00277
738.2	0.00315
CURVE 168*	
765.4	0.00376
1015.4	0.00441
1225.9	0.00469
1343.2	0.00481
1378.7	0.00505
CURVE 169*	
909.8	0.00393
1123.7	0.00460
1442.6	0.00479
1498.7	0.00519
CURVE 170*	
1039.8	0.00450
1235.9	0.00509
1474.8	0.00538
CURVE 170 (cont.)*	
1628.2	0.00710
1697.6	0.00850
CURVE 171*	
557.6	0.00227
682.6	0.00308
845.4	0.00362
932.6	0.00434
CURVE 172*	
478.7	0.00199
535.4	0.00267
598.7	0.00292
819.8	0.00351
1022.6	0.00405
CURVE 173*	
535.9	0.00289
864.3	0.00372
961.5	0.00421
1108.2	0.00415
1144.8	0.00531
CURVE 174*	
462.6	0.00251
519.8	0.00267
698.2	0.00296
868.2	0.00341
959.3	0.00365
CURVE 175*	
604.3	0.00313
728.7	0.00331
818.2	0.00353
962.1	0.00379
1125.9	0.00400
1254.3	0.00412
CURVE 176*	
787.6	0.00337
905.9	0.00362
1061.5	0.00403
1186.5	0.00429
1334.8	0.00402
CURVE 177*	
762.6	0.00332
829.8	0.00355
895.4	0.00363
1020.9	0.00391
1124.8	0.00396
1533.7	0.00422
CURVE 178*	
795.9	0.00374
874.8	0.00362
979.8	0.00377
1278.2	0.00414
1553.7	0.00427
1724.8	0.00460
CURVE 179*	
914.8	0.00358
1278.7	0.00363
1642.1	0.00443
1830.4	0.00576
CURVE 180*	
944.8	0.00408
1333.7	0.00426
1614.3	0.00466
CURVE 181*	
751.5	0.00388
1352.6	0.00415
1707.6	0.00448

*Not shown on plot

DATA TABLE NO. 16 (continued)

CURVE 182*

T	k
342.1	0.00149
355.4	0.00151
382.6	0.00152
404.8	0.00156
422.6	0.00163
446.5	0.00173
469.3	0.00178
492.1	0.00187
519.3	0.00199
542.1	0.00211
594.8	0.00222

CURVE 183

T	k
1.41	0.120
1.60	0.178
1.83	0.267
1.92	0.313
2.23	0.470
2.48	0.596
2.87	0.830
3.04	0.998
3.27	1.36
3.88	2.53
4.14	3.25
4.29	3.72
4.31	3.99

CURVE 184

T	k
1.38	0.0551
1.39	0.0598
1.46	0.0738
1.53	0.0802
1.70	0.100
1.97	0.133
2.21	0.164
2.57	0.208
2.77	0.238
3.11	0.290
3.47	0.366
3.72	0.419
3.96	0.478
4.32	0.524

CURVE 185

T	k
1.41	0.0394
1.53	0.0414
1.74	0.0458
2.03	0.0506
2.32	0.0661
2.69	0.0867
3.03	0.113
3.16	0.125
3.56	0.200
3.83	0.250
4.06	0.273
4.47	0.294

CURVE 186

T	k
1.45	0.0551
1.53	0.0625
1.63	0.0695
1.74	0.0828
1.87	0.0975
2.05	0.111
2.18	0.121
2.32	0.134
2.49	0.150
2.76	0.169
3.11	0.200
3.37	0.234
3.77	0.288
3.95	0.327
4.17	0.381

CURVE 187

T	k
1.37	0.0239
1.52	0.0286
1.63	0.0299
1.77	0.0319
1.91	0.0372
2.06	0.0409
2.20	0.0431
2.35	0.0463
2.51	0.0501
2.77	0.0569
3.14	0.0719
3.37	0.0836

CURVE 187 (cont.)

T	k
3.70	0.115
3.85	0.143
4.06	0.209
4.27	0.240

CURVE 188

T	k
1.45	0.0492
1.52	0.0531
1.66	0.0566
1.76	0.0614
1.88	0.0656
2.05	0.0750
2.19	0.0807
2.35	0.0902
2.55	0.0951
2.76	0.110
3.10	0.125
3.33	0.134
3.66	0.153
3.81	0.166
3.96	0.177
4.24	0.194

CURVE 189

T	k
1.46	0.0192
1.61	0.0203
1.75	0.0209
1.89	0.0216
2.10	0.0235
2.21	0.0239
2.39	0.0254
2.58	0.0277
2.76	0.0308
3.15	0.0388
3.33	0.0449
3.64	0.0560
3.74	0.0598
4.02	0.0718
4.35	0.0834

CURVE 190*

T	k
333.0	0.00144
363.9	0.00167
402.5	0.00168
445.7	0.00168
506.9	0.00162
526.0	0.00174
575.4	0.00189
627.3	0.00191
666.6	0.00181
672.0	0.00196
733.1	0.00205
832.3	0.00207

CURVE 191*

T	k
4.2	0.125×10^{-6}

CURVE 192* (T = 4.2 K)

p(mm Hg)	k x 10^6
0.0000100	0.132
0.0000638	0.178
0.000716	0.437
0.00638	1.45
0.0832	7.94
0.457	27.2
2.07	53.1
20.2	61.7

CURVE 193* (T = 4.2 K)

p(μ Hg)	k x 10^6
0.00184	0.417
0.00339	0.452
0.00966	0.457
0.0138	0.531
0.0320	0.741

CURVE 194* (T = 4.2 K)

p(mm Hg)	k x 10^6
0.00000224	0.437
0.00000944	0.543
0.0000269	0.684
0.000197	1.35
0.00204	4.52
0.00767	8.61
0.0513	26.6
0.133	48.4
0.692	94.4
1.97	133
12.9	184
63.1	209

CURVE 195* (T = 4.2 K)

p(μ Hg)	k x 10^6
0.0733	0.944
0.355	1.72
13.3	11.8
94.4	44.7

CURVE 196* (T = 4.2 K)

p(μ Hg)	k x 10^6
0.00193	1.41
0.00767	1.43
0.0295	1.60

CURVE 197* (T = 4.2 K)

p(mm Hg)	k x 10^6
0.0000137	2.66
0.0000589	2.82
0.000123	3.09
0.000324	3.55
0.00130	6.53
0.00376	12.7
0.00804	22.1
0.0473	64.6
0.0966	101

CURVE 197 (cont.)*

p(mm Hg)	k x 10^6
0.684	240
1.97	327
5.31	432
67.6	617

CURVE 198* (T = 4.2 K)

Stress(Kg cm^{-2})	k x 10^6
0.288	0.0380
0.395	0.0490
1.14	0.124
2.28	0.234
5.01	0.427
8.47	0.637
5.78	0.568
2.95	0.365
1.17	0.195
0.389	0.0964

CURVE 199*

T	k x 10^6
2.48	3.95
3.15	4.27
3.82	4.59
4.22	4.66
4.87	5.15

CURVE 200*

T	k x 10^6
2.50	0.968
3.11	1.01
3.78	1.10
4.26	1.10
4.87	1.21

CURVE 201*

T	k x 10^6
2.04	1.29
2.56	1.47
3.05	1.79
3.75	1.70
4.14	1.88

*Not shown on plot

DATA TABLE NO. 16 (continued)

T	$k \times 10^6$
CURVE 202*	
2.00	5.42
2.57	5.60
3.09	6.70
3.76	6.55
4.16	7.02
CURVE 203*	
2.77	0.589
3.58	1.06
4.24	1.41
4.79	2.25
CURVE 204*	
1.88	0.592
2.48	1.17
2.81	2.00
3.02	2.48
3.26	2.98
3.54	3.96
3.67	3.96
3.89	5.22
3.96	5.83
4.31	6.03
4.37	6.49
4.61	7.76
4.80	11.6
CURVE 205*	
2.13	2.29
2.58	3.65
3.05	4.92
3.28	6.52
3.60	9.59
4.14	15.6

T	k
CURVE 206*	
317.7	0.183
331.0	0.177
383.1	0.149
412.1	0.141
CURVE 207*	
316.9	0.0548
354.3	0.0540
394.6	0.0515
421.1	0.0485
CURVE 208*	
664	0.00303
750	0.00303
830	0.00317
1011	0.00375
1208	0.00346

* Not shown on plot

FIGURE AND TABLE NO. 16R RECOMMENDED THERMAL CONDUCTIVITY OF ALUMINUM OXIDE Al_2O_3

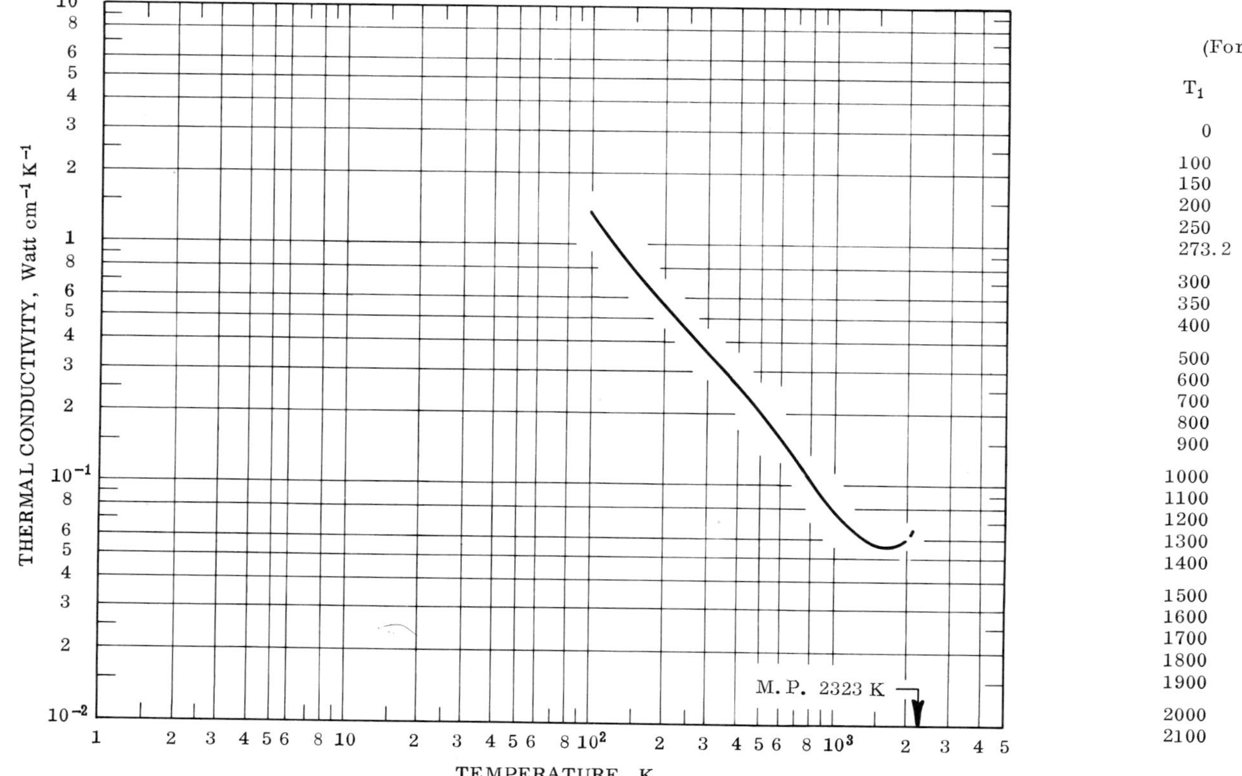

RECOMMENDED VALUES*
(For 99.5% pure, 98% dense, polycrystalline Al_2O_3)

T_1	k_1	k_2	T_2
0	0	0	-459.7
100	1.33	76.9	-279.7
150	0.77	44.5	-189.7
200	0.55	31.8	- 99.7
250	0.434	25.1	- 9.7
273.2	0.397	22.9	32.0
300	0.360	20.8	80.3
350	0.307	17.7	170.3
400	0.264	15.3	260.3
500	0.202	11.7	440.3
600	0.158	9.13	620.3
700	0.126	7.28	800.3
800	0.104	6.01	980.3
900	0.089	5.14	1160
1000	0.0785	4.54	1340
1100	0.0710	4.10	1520
1200	0.0655	3.79	1700
1300	0.0613	3.54	1880
1400	0.0583	3.37	2060
1500	0.0566	3.27	2240
1600	0.0556	3.21	2420
1700	0.0554	3.20	2600
1800	0.0559	3.23	2780
1900	0.0574	3.32	2960
2000	0.0600	3.47	3140
2100	(0.0644) ‡	(3.72)	3320

REMARKS

The recommended values are for 99.5% pure, 98% dense, polycrystalline Al_2O_3. The recommended values are thought to be accurate to within 6% of the true values at temperatures from 500 to 1000 K and 6 to 10% at other temperatures.

* T_1 in K, k_1 in Watt cm^{-1} K^{-1}, T_2 in F, and k_2 in Btu hr^{-1} ft^{-1} F^{-1}.

‡ Values in parentheses are extrapolated.

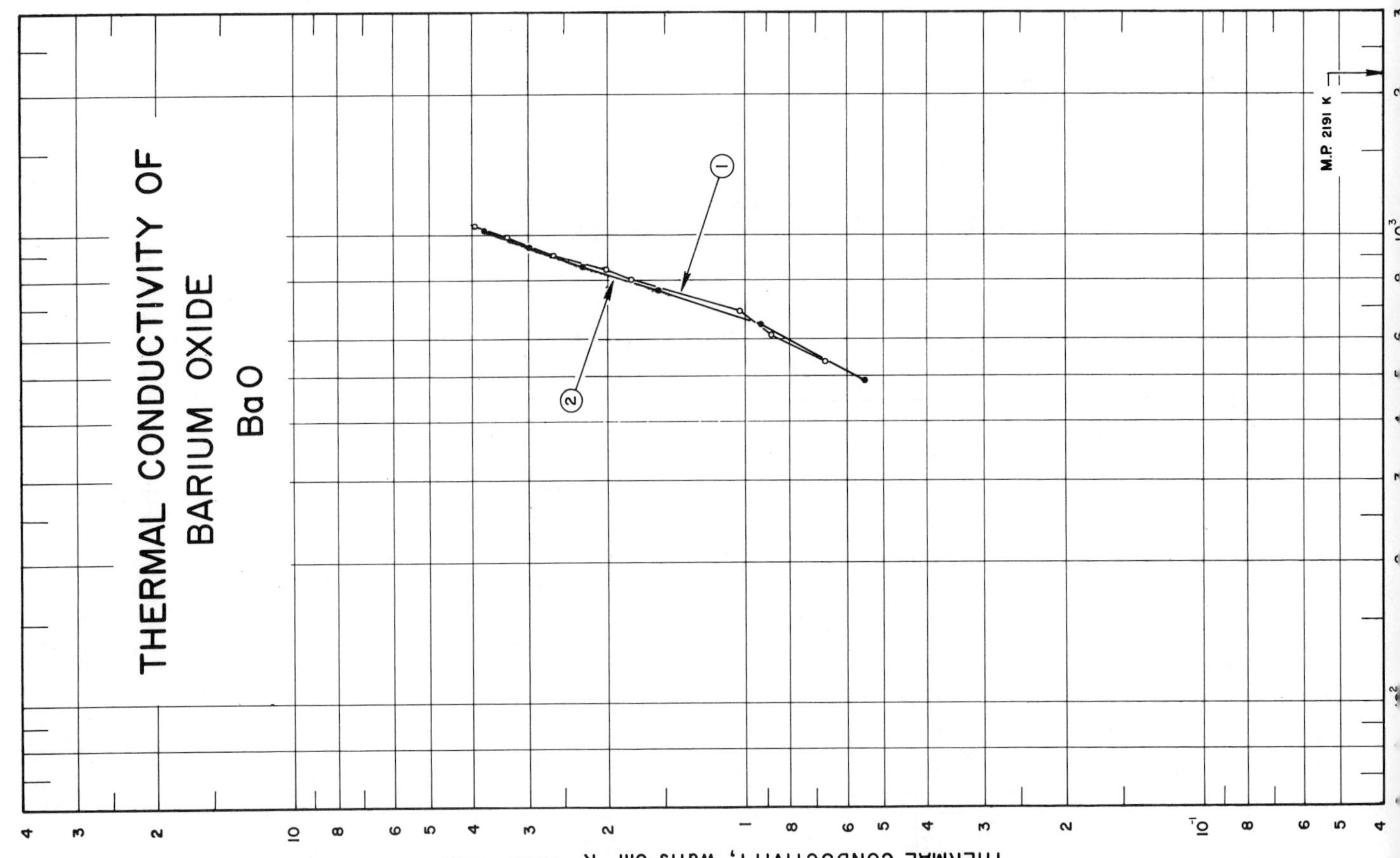
THERMAL CONDUCTIVITY OF
BARIUM OXIDE
BaO
THERMAL CONDUCTIVITY, Watts Cm⁻¹ K⁻¹
M.P. 2191 K
1
2

SPECIFICATION TABLE NO. 17 THERMAL CONDUCTIVITY OF BARIUM OXIDE BaO

[For Data Reported in Figure and Table No. 17]

Curve No.	Ref. No.	Method Used	Year	Temp. Range, K	Reported Error, %	Name and Specimen Designation	Composition (weight percent), Specifications and Remarks
1	75	L	1955	538-1053	< 10	Tube No. 3	Pure; polycrystalline; apparent thermal conductivity (effects due to radiation at high temperatures not considered).
2	75	L	1955	490-1033	< 10	Tube No. 5	Pure; polycrystalline; apparent thermal conductivity.

DATA TABLE NO. 17 THERMAL CONDUCTIVITY OF BARIUM OXIDE BaO

[Temperature, T, K; Thermal Conductivity, k, Watt cm^{-1} K^{-1}]

T	k
CURVE 1	
538.2	0.67
613.2	0.87
690.2	1.03
803.2	1.77
843.2	2.03
908.2	2.65
990.2	3.35
1053.2	3.95
CURVE 2	
490.2	0.55
643.2	0.93
763.2	1.55
853.2	2.27
948.2	3.00
1033.2	3.77

THERMAL CONDUCTIVITY OF BERYLLIUM OXIDE BeO

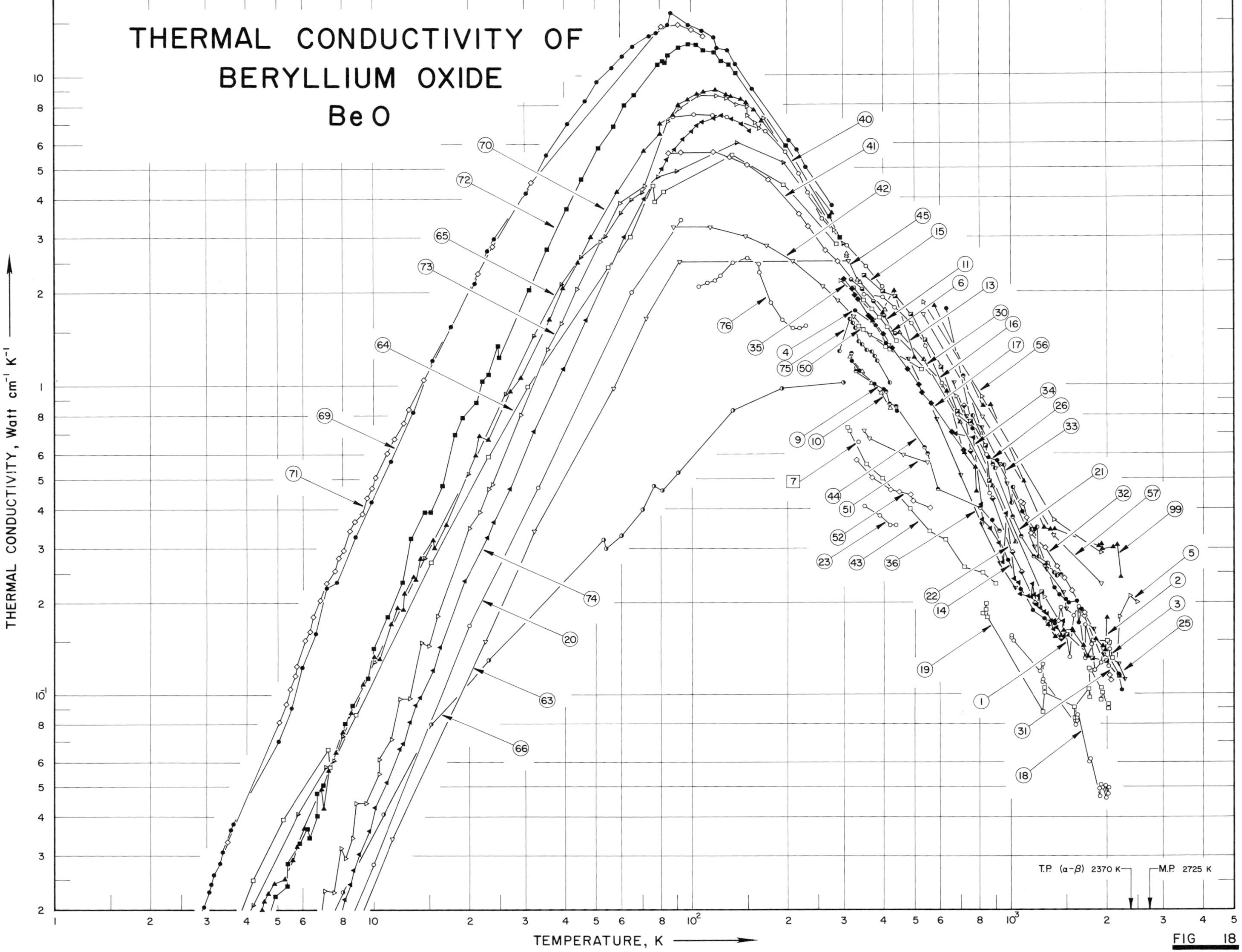

THERMAL CONDUCTIVITY, Watt cm^{-1} K^{-1} ⟶

TEMPERATURE, K ⟶

T.P. ($\alpha-\beta$) 2370 K

M.P. 2725 K

FIG 18

SPECIFICATION TABLE NO. 18 THERMAL CONDUCTIVITY OF BERYLLIUM OXIDE BeO

[For Data Reported in Figure and Table No. 18]

Curve No.	Ref. No.	Method Used	Year	Temp. Range, K	Reported Error, %	Name and Specimen Designation	Composition (weight percent), Specifications and Remarks
1	274, 130	R	1953	1303-2041		M-14	0.01 Fe_2O_3, 0.08 Al_2O_3, and 0.18 MgO; specimen in shape of a prolate spheroid; slip-cast, fired to zero apparent porosity at 1850 C and had a final total porosity of 5-10%.
2	274, 130	R	1953	1303-1998		M-13	Same material as the above; separate run.
3	274, 130	R	1953	1813-2073		M-15	Same material as the above; separate run.
4	141	L	1950	326-407		BeO porcelain	99.9 pure; supplied by Norton Co.; density (25 C) = 2.969 g cm^{-3}; water absorption 0.03%.
5	291	R	1964	1273-2493			Made from UOX-grade BeO powder by isostatically pressing at 7000 psi; the cold compacts crushed and screened through a 20-mesh sieve, cold pressed at about 6500 psi, and then isostatically pressed into a disc at 20,000 psi; sintered in dry hydrogen at 1700 C for approximately 6 hrs and then machined to the final configuration; 2.00 in. outside dia and 0.375 in. inside dia, data corrected to 100% theoretical density (original: 97.8% theoretical density).
6	12	C	1953	322-439		94A-1	Pure; hot-pressed; water absorption 0.03%; density 2.97 g cm^{-3}.
7	3		1953	333.2			65% beryllia in unfired state; max. water absorption = 0.05%; flexural strength 34,000 psi; coefficient of expansion (25-700 C) = 8.4 x 10^{-6}.
8	3		1953	333.2			88% beryllia in unfired state; max. water absorption = 0.05%; flexural strength 34,500 psi; coefficient of expansion (25-700 C) = 8.2 x 10^{-6}.
9	9	C	1953	318-439		273A-1	4811 BeO porcelain supplied by Coors Co.; 0.09% water absorption 0.09%; density 2.90 g cm^{-3}.
10	9	C	1953	315-420		273A-2	Similar to the above specimen.
11	9	C	1953	317-436		94A-3	Pure; hot pressed; water absorption 0.03; density 2.97 g cm^{-3}.
12	13	C	1953	343.2	±3		Hot-pressed; density = 3.0 g cm^{-3}; Armco iron used as comparative material.
13	89	C	1952	370-780			Al_2O_3 used as comparative material.
14	89	R	1952	873-1373			
15	131	C	1954	348-853			1 x 1 x 1 in.; supplied by Bernard Schwartz of MIT; slip-cast from suspensions of finely ground material; fired to zero apparent porosity; bulk density 2.86 g cm^{-3}.
16	131	C	1954	348-853			Second run of the above specimen.
17	138	L	1957	301-712	≤5.4		Impurities (other than carbon) less than 0.2%; fabricated by Norton Co.; hot-pressed and fired at about 1700 C; dia 0.4524 in.; density 2.62 g cm^{-3}.

SPECIFICATION TABLE NO. 18 (continued)

Curve No.	Ref. No.	Method Used	Year	Temp. Range, K	Reported Error, %	Name and Specimen Designation	Composition (weight percent), Specifications and Remarks
18	144	R	1963	1006-2009	5-7	1	Poorly bonded structure; specimen 0.75 in. long, 0.75 in. outside dia and 0.25 in. inside dia; supplied by Zirconium Corp. of America; pressed and sintered; density 3.0 g cm^{-3} at 25 C; specimen broke during experiment.
19	144	R	1963	818-2018	5-7	2	Similar to the above specimen except specimen cracked during experiment.
20	158	L	1952	2.6-93	2		A rod with a square cross section of side 5 mm; sintered; density 2.94 g cm^{-3} (97% of the single crystal value).
21	72		1955	710-1431			Prepared by K. A. P. L.; density 2.78 g cm^{-3}.
22	72		1955	713-1431			Same as the above specimen.
23	407	L	1949	347-436		Commercial	No details reported.
24	220		1947	573-1173		3008-13-3	Extruded from refractory-grade BeO of minus 200 mesh and followed by burning 3 hrs at 3100 F; density 2.818 g cm^{-3}.
25	218	R	1960	665-2290		ORNL-1	Brush SP grade of 0.27 metallic impurities; obtained from ORNL; specimen 3 in. long, 2 in. O.D. and 0.5 in. in I.D.; hot-pressed; average grain size 50 μ; density 2.89 g cm^{-3}.
26	218	R	1960	630-2242		ORNL-2	Similar to the above specimen except density of 2.87 g cm^{-3}.
27	218	R	1960	750-1669		NBC-1	"Pure beryles" of the Nation Beryllia Corp.; slip cast; density 2.72 g cm^{-3}.
28	218	R	1960	970-2256		Al-1	0.07 metallic impurities; supplied by Atomics International; hot pressed at 1700 C and 4000 psi for 4 hrs; density 2.98 g cm^{-3}; average grain size 60 μ.
29	218	R	1960	1175-2293		Al-2	As above, separate run.
30	251	C	1963	450-1039	±4		99.5 BeO, 0.009 Si, 0.005 Al, 0.002 Mo, 0.001 Ca, 0.001 Cr, 0.001 Fe, 0.001 Na, 0.001 Ni, 0.0003 Mn, ≤0.0001 B, Cd, Li, and ≤0.0001 Co, Cu; specimen 2 in. dia by 1 in. thick; cold pressed; firing temperature 1855 K; density 2.87 g cm^{-3}; Armco iron used as comparative material.
31	251	P	1963	1250-2200			Similar to the above specimen; thermal conductivity calculated from the measured thermal diffusivity data, values of density and specific heat were obtained by independent measurements, reported as 2.79, 2.77, 2.74, 2.72, and 2.69 g cm^{-3}, and 2.00, 2.14, 2.27, 2.36, and 2.44 W g^{-1} K^{-1}, respectively, at 1249.8, 1505.4, 1794.3, 1994.3, and 2199.8 K.
32	275	P	1963	1073-2053		BD-98	97% of the theoretical density; manufactured by the Coors Porcelain Co.
33	218	R	1960	761-2131		Al-3	1.0 MgO; hot-pressed at 1700 C and 4000 psi for 4 hrs; specimen 3 in. long, 2 in. O.D. and 0.5 in. I.D.; average grain size 60 μ; density 2.99 g cm^{-3}.
34	316, 274	R	1951	673-1487	±5		99.6$^+$ pure; slip-cast in an acid suspension; density 2.65 g cm^{-3}; total porosity 10%; apparent porosity <1%.

SPECIFICATION TABLE NO. 18 (continued)

Curve No.	Ref. No.	Method Used	Year	Temp. Range, K	Reported Error, %	Name and Specimen Designation	Composition (weight percent), Specifications and Remarks
35	296		1958	293-1273			A commercial beryllia powder was hydrostatically pressed at 10,000 psi then screened through a 40-mesh sieve; hydrostatically pressed again at 100,000 psi, and then presintered at 1255 K for 1 hr in a hydrogen atmosphere, after machining, then sintered at 1811 K for 1 hr in hydrogen; 93.5% of theoretical density.
36	292	C	1961	367-1478		No. 1	Specimen 0.5 x 0.5 x 0.875 in.; fabricated by dry pressing and isostatic compaction; sintered at 1894 K in H_2 for 2 hrs and then heat treated at 2033 K in H_2 for 1 hr; 96-97% of theoretical density; manufactured by Brush Beryllium Co. (Type UOX); sintered aluminum oxide used as reference.
37	292	C	1961	367-1478		No. 2	Same as the above specimen.
38	292	C	1961	367-1367			Average value of the above specimen No. 1 and No. 2 corrected to zero porosity.
39	276	R	1961	700-1700			
40	290	L	1963	88-398	$<3-4$		99 BeO and 1 Al_2O_3; ground cylinder 3.4 cm in dia and 11 cm long; prepared from isostatically pressed bodies using a wax emulsion as binder; calcinated at 1740 C for 2 hrs; density (25 C) 2.89 g cm^{-3}; measured in a vacuum of 10^{-4} mm Hg.
41	290	L	1963	85-408	<4		98 BeO, 1 Al_2O_3, 0.5 MgO + CaO, and 0.5 SiO_2; ground cylinder 3.4 ±0.01 cm in dia and 11 cm long; prepared from isostatically pressed bodies using a wax emulsion as binder; calcinated at 1730 C for 3 hrs; density 2.87 g cm^{-3}; measured in a vacuum of 10^{-4} mm Hg.
42	290	L	1963	88-418	<4		96 BeO, 1 Al_2O_3, 1.5 MgO + CaO, and 1.5 SiO_2; ground cylinder 3.4 ±0.01 cm in dia and 11 cm long; prepared from isostatically pressed bodies using a wax emulsion as a binder; calcinated at 1690 C for 2 hrs; density 2.87 g cm^{-3}; measured in a vacuum of 10^{-4} mm Hg.
43	297	C	1954	308-898		"Triangle" Beryllia; 1	99^+ BeO, main impurity being Al_2O_3; 10.84 x 0.883 x 0.801 cm; cut from the center of a "triangle" beryllia disc, type Y. 1029; fired at 1750 C; baked to about 800 C (to drive off moisture introduced during cutting) before measurement; mean density 1.85 g cm^{-3}; Armco iron used as comparative material.
44	297	C	1954	318-913		"Triangle" Beryllia; 2	99^+ BeO, main impurity being Al_2O_3; 11.41 x 1.006 x 1.002 cm; cut from the center of a "triangle" beryllia disc, type Y. 1033; fired at 1750 C; baked to about 800 C; mean density 2.3 g cm^{-3}; Armco iron used as comparative material.
45	297	C	1954	313-778		Hot-molded beryllia; 3	A rod of hot-molded beryllia; length 9.94 cm, dia 1.00 cm; mean density 2.82 g cm^{-3}; Armco iron used as comparative material.
46	297	C	1954	318-748		Hot-molded beryllia; 4a	The heavier portion of specimen No. 4 of hot-molded beryllia; 1.00 cm dia x 9.95 cm long; mean density 2.80 g cm^{-3}; Armco iron used as comparative material.
47	297	C	1954	358-748		Hot-molded beryllia; 4b	The lighter portion of specimen No. 4 of hot-molded beryllia; 1.00 cm dia x 9.95 cm long; mean density 2.72 g cm^{-3}; Armco iron used as comparative material.

SPECIFICATION TABLE NO. 18 (continued)

Curve No.	Ref. No.	Method Used	Year	Temp. Range, K	Reported Error, %	Name and Specimen Designation	Composition (weight percent), Specifications and Remarks
48	297	C	1954	303-628		Norton's beryllia; 5	Norton's BeO, shipment B1866, Clifton Metal grade; hot-molded; specimen as a strip 0.5 cm wide cut from the center of a disc; density 3.0 g cm^{-3}; Armco iron used as comparative material.
49	319	C	1962	339-568		AOX-BeO(329)	AOX-grade BeO; cylinder 1 in. long by 0.238 in. dia; about 97.5% theoretical density; grain size 20 microns; high density graphite (AGOT) used as comparative material.
50	319	C	1962	335-525		AOX-BeO(329)	Similar to the above specimen but irradiated by 8.6 x 10^{18} neutrons cm^{-2}.
51	319	C	1962	348-548		AOX-BeO(329)	Similar to the above specimen but irradiated by 2.0 x 10^{20} neutrons cm^{-2}.
52	319	C	1962	328-560		AOX-BeO(329)	Similar to the above specimen but irradiated by 3.7 x 10^{20} neutrons cm^{-2}.
53	320	R	1962	1118-1773		Sample 23	Wafer ~2 in. outside dia, 0.375 inside dia, and 0.5 to 2 in. thick; density ~0.9 g cm^{-3}; pore size 0.010 to 0.025 cm; sintered in air at 1565 C for 1.5 hrs.
54	320	R	1962	1133-1833		Sample 23	The above specimen (after the above measurement) sintered in hydrogen for 1.5 hrs at 1590 C to eliminate contained phosphates; density 0.75 g cm^{-3}; pore size 0.01 to 0.025 cm.
55	320	R	1962	1118-1798		Sample 25	Same dimensions as the above specimen; density ~1.0 g cm^{-3}; pore size from 0.01 to 0.02 cm, and sintered at 1566 C for 1.5 hrs.
56	331	R	1963	533-1922			99 pure; specimen of 1 in. in dia and 1 in. long, hot pressed in graphite dies; 98% theoretical density.
57	331	R	1963	533-1922			98 pure; specimen of 1 in. in dia and 1 in. long, cold pressed and fired; 96% of theoretical density.
58	307	R	1954	663-1473	±5		Hollow prolate spheroidal specimen with outer and inner surface confocal, inner minor axis about 2 cm, inner major axis about 10 cm, outer minor axis 4 cm, outer major axis long enough to make both surfaces confocal; specimen prepared by slip casting from suspension of finely ground material fired to a porosity of 9.7%; bulk density 2.7 g cm^{-3}.
59	356		1942	323-658	5-10		Powder.
60	376	C	1959	361-759			Specimen about 1 cm^2 in section and 10 cm in length; density 2.60 g cm^{-3}; Armco iron as reference material.
61	376	C	1959	368-431			Specimen about 1 cm^2 in section and 10 cm in length; density 2.72 g cm^{-3}; irradiated at 5 x 10^{20} neutrons cm^{-2} at 458 K; Armco iron as reference material.
62	376	C	1959	401-803			The above specimen; 2nd run.
63	377	L	1964	4.8-310	<±2	2.2μ No. 2175	Impurities; 0.0800 C, 0.0250 Fe, 0.0050 Al, 0.0050 Ca, 0.0040 Si, and 0.0100 Cu + Ni + Cr + Mg + Na; polycrystalline; 0.77 cm dia x 5 cm long; supplied by Dr. K. D. Reeve; grain size not uniform, about 1 μ to 10 μ; cold-pressed and sintered from "UOX" powder supplied by Brush Beryllium Corp. at 1723 K for 1 to 1.5 hrs; 98% theoretical density.

SPECIFICATION TABLE NO. 18 (continued)

Curve No.	Ref. No.	Method Used	Year	Temp. Range, K	Reported Error, %	Name and Specimen Designation	Composition (weight percent), Specifications and Remarks
64	377	L	1964	2. 2-285	<±2	12 μ No. 2176	Similar to the above specimen except the grain size was 12 μ.
65	377	L	1964	2. 25-300	<±2	35 μ No. 2189	Similar to the above specimen except the grain size was 25 μ tp 45 μ and was sintered at 2023 K.
66	377	L	1964	3. 0-300	±2	35 μ No. 2189	Similar to the above specimen except irradiated at 1 x 10^{20} neutrons cm^{-2} at 75 C.
67	378	L	1962	673-1173	±4	1A	BeO grade I; cylindrical specimen 1. 625 in.dia, 1 in. long; bulk density 2. 95 g cm^{-3}; 98% theoretical density; smoothed data without correction to zero porosity.
68	378	L	1962	673-1173	±4	2A	BeO grade II; cylindrical specimen 1. 625 in. dia, 1 in. long; bulk density 2. 823 g cm^{-3}; 94% theoretical density; smoothed data without correction to zero porosity.
69	409		1965	1. 4-277		Type I	Prepared by sintering under pressure; density 2. 98 g cm^{-3}.
70	409		1965	2. 1-277		Type I	The above specimen irradiated by a dose of 10^{16} ~2 x 10^{16} n >1 MeV/cm^2; identical results obtained with a dose of 3 x 10^6 roentgens.
71	409		1965	1. 6-110		Type I	The above specimen annealed at 500 C for 10 hrs.
72	409		1965	1. 6-292		Type II	Prepared by sintering with the addition of 0. 6 CaO; density 2. 77 g cm^{-3}; no measurable change on irradiation.
73	409		1965	1. 4-281		Type III	Prepared by sintering under pressure; density 2. 75 g cm^{-3}.
74	409		1965	1. 8-284		Type III	The above specimen irradiated by a dose of 10^{16} to 2 x 10^{16} n >1 MeV/cm^2.
75	278	L	1963	291-415	± 10		96 BeO; extruded rod; fired to 1680 C; density 2. 80 g cm^{-3}; using water-ice as coolant.
76	278	L	1963	107-228	± 10		Similar to the above specimen except used liquid nitrogen as coolant.
77	278	L	1963	231-368	± 10		Similar to the above specimen except used solid CO_2 as coolant.
78	410	P	1966	573-1273		UOX grade	0. 5 MgO impurities; unirradiated specimen with grain size 4 microns; density 2. 92 g cm^{-3}; open porosity 0. 1%; thermal conductivity data calculated from the measurements of density, thermal diffusivity and specific heat (specific heat data obtained from High Temp. Materials programs, Part A, "GE-NMPO, GEMP-400A, 1966).
79	410	P	1966	573-1273		UOX grade	Similar to the above specimen except grain size 5 microns; density 2. 905 g cm^{-3}; open porosity 0. 4%; extruded to about 5% in c-axis orientation.
80	410	P	1966	573-1273		UOX grade	Similar to the above specimen except grain size 20 microns; density 2. 930 g cm^{-3}; extruded to about 50% in c-axis orientation; no porosity value was given.
81	410	P	1966	573-1273		UOX grade	0. 5 MgO impurities; unirradiated specimen with grain size 64 microns; density 2. 917 g cm^{-3}; open porosity 0. 2%; thermal conductivity data calculated from the measurements of density, thermal diffusivity and specific heat (specific heat data obtained from High Temp. Materials programs, Part A, "GE-NMPO, GEMP-400A, 1966).

SPECIFICATION TABLE NO. 18 (continued)

Curve No.	Ref. No.	Method Used	Year	Temp. Range, K	Reported Error, %	Name and Specimen Designation	Composition (weight percent), Specifications and Remarks
82	410	P	1966	873-1273		AOX grade	Unirradiated specimen with grain size 8 microns; density 2.904 g cm^{-3}; open porosity 0.4%; see above curve for method of calculation of thermal conductivity.
83	410	P	1966	573-1273		AOX grade	Similar to the above specimen except grain size 17 microns; density 2.866 g cm^{-3}; open porosity 0.1%.
84	410	P	1966	573-1273		AOX grade	Similar to the above specimen except grain size 46 microns; density 2.938 g cm^{-3}; open porosity 0.2%.
85	410	P	1966	573-1273		AOX grade	Similar to the above specimen except grain size 76 microns; density 2.950 g cm^{-3}; open porosity 0.1%.
86	410	P	1966	573-1273		UOX grade	0.5 MgO impurities; similar to the above specimen except grain size 6 microns; density 2.609 g cm^{-3}; open porosity 0.3%.
87	410	P	1966	573-1273		UOX grade	Similar to the above specimen except grain size 4 microns; density 2.453 g cm^{-3}; open porosity 17.9%.
88	410	P	1966	573-1273		UOX grade	Similar to the above specimen except grain size 5 microns; density 2.772 g cm^{-3}; open porosity 0.1%; extruded to about 5% in c-axis orientation.
89	410	P	1966	582-971		UOX grade	0.5 MgO impurities; irradiated specimen with grain size 5 microns; bulk density 2.89 g cm^{-3}; irradiated at 300 C to 2.1 x 10^{20} nvt ($E_n \geq 1$ Mev); thermal conductivity data calculated from the measurements of density, thermal diffusivity and specific heat (specific heat data obtained from High Temp. Materials Programs, Part A, "GE-NMPO, GEMP-400A, 1966).
90	410	P	1966	1066-1272		UOX grade	The above specimen remained at 793 C for approx. 20 hrs.
91	410	P	1966	569-1270		UOX grade	The above specimen annealed at 997 C for 68 hrs.
92	410	P	1966	570-1263		UOX grade	0.5 MgO impurities; irradiated specimen with grain size 4 microns; bulk density 2.90 g cm^{-3}; irradiated at 600 C to 3.1 x 10^{20} nvt ($E_n \geq 1$ Mev); thermal conductivity data calculated as the above specimen.
93	410	P	1966	571-1268		UOX grade	The above specimen annealed at 995 C for 17 hrs; measured with decreasing temp.
94	410	P	1966	576-1281		UOX grade	0.5 MgO impurities; irradiated specimen with grain size 5 microns; bulk density 2.89 g cm^{-3}; irradiated at 800 C to 10 x 10^{20} nvt ($E_n \geq 1$ Mev); thermal conductivity data calculated as the above specimen.
95	410	P	1966	574-1278		UOX grade	The above specimen annealed at 1005 C for 16 hrs; measured with decreasing temp.
96	411		1960	373-1273	$< \pm 5$		0.495 in. dia x 2 in. long; density 94% of theoretical value; data obtained from smoothed curve.
97	412	R	1963	558-565		Specimen 1	$\geq$99.3 commercial high purity BeO; hot pressed; density 2.79 g cm^{-3} ($\geq$ 98% of theoretical value); first run.

SPECIFICATION TABLE NO. 18 (continued)

Curve No.	Ref. No.	Method Used	Year	Temp. Range, K	Reported Error, %	Name and Specimen Designation	Composition (weight percent), Specifications and Remarks
98	412	R	1963	575-2412		Specimen 1	Second run of the above specimen.
99	412	R	1963	576-2222		Specimen 2	Similar to the above specimen.
100	412	R	1963	512-2269		Specimen 1	Commercial high purity BeO; hot pressed; density 2.88 g cm^{-3} (~95% of theoretical value).
101	412	R	1963	533-2227		Specimen 2	Similar to the above specimen.
102	412	R	1963	517-2110		Specimen 1	Commercial high purity BeO; cold pressed; density 2.9 g cm^{-3} (96-97% of theoretical value).
103	412	R	1963	546-2397		Specimen 2	Similar to the above specimen.
104	412	R	1963	563-2163		Specimen 1	Commercial high purity BeO; cold pressed; density 92-94% of theoretical value.
105	412	R	1963	539-2293		Specimen 2	Similar to the above specimen.

DATA TABLE NO. 18 THERMAL CONDUCTIVITY OF BERYLLIUM OXIDE BeO

[Temperature, T, K; Thermal Conductivity, k, Watt cm^{-1} K^{-1}]

T	k
CURVE 1	
1303.2	0.1726
1415.2	0.1736
1438.2	0.1925
1532.2	0.1328
1573.2	0.1803
1633.2	0.1946
1703.2	0.1318
1708.2	0.1690
1793.2	0.1464
1833.2	0.1203
1913.2	0.1276
1963.2	0.1402
2020.7	0.1245
2040.7	0.1402
CURVE 2	
1303.2	0.1695
1373.2	0.1548
1463.2	0.1621
1553.2	0.1642
1565.7	0.1611
1688.2	0.1443
1743.2	0.1333
1847.2	0.1506
1947.2	0.1448
1973.2	0.1391
1998.2	0.1778
CURVE 3	
1813.2	0.1496
1998.2	0.1496
2030.7	0.1475
2073.2	0.1318
CURVE 4	
326.2	1.745
378.2	1.561
407.2	1.372
CURVE 5	
1273.2	0.180*
1313.2	0.170*
1343.2	0.200*
1443.2	0.150*
1443.2	0.210*
1653.2	0.190*
1763.2	0.170*
1823.2	0.155*
1973.2	0.130*
2043.2	0.130*
2173.2	0.140*
2193.2	0.180
2363.2	0.210
2493.2	0.200
CURVE 6	
321.7	2.06
345.6	1.88
368.6	1.76
388.3	1.66
410.0	1.54
439.4	1.39
CURVE 7	
333.2	0.658
CURVE 8*	
333.2	1.140
CURVE 9	
317.7	1.20
334.0	1.12
373.0	1.01
399.4	0.967
439.4	0.824
CURVE 10	
314.5	1.23
342.6	1.11
366.6	1.02
CURVE 10 (cont.)	
392.4	0.941
419.6	0.845
CURVE 11	
317.3	2.19
343.6	2.05
371.2	1.87
401.6	1.68
436.1	1.47
CURVE 12*	
343.2	1.82
CURVE 13	
370.2	1.62
398.2	1.71
433.2	2.02
498.2	1.23
633.2	0.954
718.2	0.607
776.2	0.544
780.2	0.456
CURVE 14	
873.2	0.368
973.2	0.274
1073.2	0.220
1173.2	0.188
1273.2	0.176
1373.2	0.172
CURVE 15	
348.2	2.251
398.2	2.017
435.2	1.946
488.2	1.695
540.2	1.406
608.2	1.138
633.2	0.962*
688.2	0.824
CURVE 15 (cont.)	
748.2	0.766
833.2	0.636
853.2	0.490
CURVE 16	
348.2	1.996
398.2	1.925
435.2	1.787
488.2	1.577
540.2	1.326
608.2	1.046
633.2	0.929*
688.2	0.803
748.2	0.753
833.2	0.594
853.2	0.448
CURVE 17	
301.4	2.19
319.4	2.04
325.8	1.94
332.9	1.879
359.0	1.674
359.4	1.679*
360.1	1.668*
360.5	1.672*
360.5	1.662*
367.6	1.647
397.0	1.454
426.2	1.316
426.6	1.311*
475.2	1.131
475.3	1.124*
514.7	0.995
514.6	0.989*
524.6	0.942
560.5	0.872
560.9	0.872*
652.7	0.704
653.6	0.701*
712.4	0.614
CURVE 18	
1005.9	0.153
1009.3	0.156
1022.6	0.151
1247.1	0.120
1255.4	0.126
1255.4	0.111
1255.4	0.112
1607.6	0.0795
1610.4	0.0822
1612.1	0.0837
1615.4	0.0852
1762.1	0.0600
1785.4	0.0613
1786.5	0.0617*
1909.8	0.0466
1909.8	0.0493
1912.6	0.0508
1920.9	0.0499
1995.9	0.0462
1997.1	0.0503
2001.5	0.0475
2009.3	0.0496
CURVE 19	
817.6	0.185
832.1	0.188
834.3	0.198
842.6	0.179
1287.1	0.0871
1288.7	0.101
1288.7	0.106
1288.7	0.101
1589.8	0.0903
1590.4	0.0838
1591.5	0.0803*
1596.5	0.0909*
1762.6	0.104
1770.4	0.0971
1771.5	0.122
1771.5	0.103
1783.2	0.119
1914.3	0.105
1917.6	0.0953
1923.2	0.102
2002.6	0.0891
2005.9	0.0932
CURVE 19 (cont.)	
2015.9	0.0917*
2017.6	0.0932*
CURVE 20	
2.6	0.00095*
4.5	0.0040*
10.0	0.028
20.0	0.17
33.0	0.47
65.0	2.0
93.0	3.4
CURVE 21	
710.2	1.071
742.2	0.770*
875.2	0.556
1016.2	0.377
1087.2	0.328
1180.2	0.281
1283.2	0.267
1431.2	0.246
CURVE 22	
713.2	0.858
743.2	0.615
871.2	0.435
1015.2	0.292
1092.2	0.251
1183.2	0.199
1290.2	0.177*
1431.2	0.151
CURVE 23	
347.2	0.409
387.9	0.380
419.8	0.357
435.9	0.357
CURVE 24*	
573.2	0.42
723.2	0.35
CURVE 24 (cont.)*	
873.2	0.28
1023.2	0.22
1173.2	0.19
CURVE 25	
665	1.020
721	0.789
810	0.729
922	0.557
971	0.481
1180	0.303
1300	0.258
1470	0.216
1493	0.181*
1695	0.177*
1732	0.158*
1823	0.162
1950	0.136
2008	0.131*
2075	0.130*
2160	0.127
2290	0.112
CURVE 26	
630	1.722
756	0.721
850	0.583
905	0.571
1080	0.391
1182	0.343
1270	0.261*
1360	0.223
1405	0.212
1490	0.205
1503	0.200
1610	0.202
1631	0.171
1648	0.190
1888	0.146*
2046	0.132*
2060	0.130*
2144	0.116
2242	0.103

* Not shown on plot

DATA TABLE NO. 18 (continued)

T	k
CURVE 27*	
750	0.669
851	0.540
952	0.451
989	0.384
1156	0.317
1229	0.251
1365	0.237
1516	0.211
1530	0.190
1636	0.176
1669	0.171
CURVE 28*	
970	0.517
1125	0.373
1141	0.342
1164	0.327
1219	0.344
1371	0.250
1410	0.236
1430	0.239
1470	0.242
1557	0.199
1559	0.204
1575	0.202
1587	0.196
1739	0.162
1745	0.171
1833	0.176
1882	0.158
1921	0.153
1938	0.160
2070	0.144
2120	0.142
2195	0.154
2256	0.160
CURVE 29*	
1175	0.337
1369	0.249
1442	0.260
1490	0.216
1522	0.207
1600	0.209
CURVE 29 (cont.)*	
1610	0.207
1755	0.176
1780	0.184
1920	0.159
1920	0.167
1985	0.158
2065	0.149
2134	0.148
2242	0.151
2293	0.156
CURVE 30	
449.8	1.48
533.2	1.21
694.3	0.762
877.6	0.476
1038.7	0.315
CURVE 31	
1249.8	0.216
1505.4	0.157
1794.3	0.132
1994.3	0.128
2199.8	0.116
CURVE 32	
1073.2	0.420
1083.2	0.418
1118.2	0.374
1153.2	0.335
1283.2	0.305
1398.2	0.262
1473.2	0.239
1553.2	0.218
1698.2	0.186
1698.2	0.180
1843.2	0.158*
1843.2	0.148*
1953.2	0.131*
1953.2	0.127*
2033.2	0.116
2053.2	0.113
CURVE 33	
761	0.797
890	0.540
949	0.551
1006	0.419
1010	0.470
1162	0.346*
1168	0.321*
1181	0.342
1210	0.352
1217	0.281
1301	0.268
1368	0.249
1383	0.243
1393	0.254*
1440	0.242*
1490	0.219*
1501	0.203*
1549	0.213*
1550	0.194*
1564	0.190*
1605	0.202*
1709	0.184*
1720	0.176*
1742	0.174*
1770	0.185*
1775	0.181*
1779	0.176*
1880	0.169*
1920	0.168*
1930	0.174*
1943	0.173*
1979	0.172*
2000	0.173*
2050	0.171*
2080	0.201*
2116	0.184*
2131	0.184*
CURVE 34	
673.2	0.695
768.2	0.680
810.2	0.413
815.2	0.369
860.2	0.347
916.2	0.313
CURVE 34 (cont.)	
942.2	0.270
968.2	0.387
1024.7	0.238
1063.2	0.232
1071.2	0.213
1140.2	0.209
1169.2	0.209
1170.2	0.229
1177.2	0.211
1210.2	0.197
1230.2	0.210
1244.2	0.192
1281.2	0.186
1300.2	0.184
1333.2	0.174
1362.2	0.163
1368.2	0.174
1389.2	0.165
1425.2	0.175*
1440.2	0.155
1455.2	0.174
1463.2	0.155
1487.2	0.155*
CURVE 35	
293.2	2.176
673.2	0.920
1273.2	0.209
CURVE 36	
366.5	1.71*
477.6	1.21
588.7	0.788
699.8	0.511
810.9	0.357
922.1	0.268
1033.2	0.222
1144.3	0.201*
1255.4	0.192*
1366.5	0.185*
1477.6	0.177*
CURVE 37*	
366.5	1.71
477.6	1.19
588.7	0.767
699.8	0.485
810.9	0.339
922.1	0.265
1033.2	0.218
1144.3	0.192
1255.4	0.177
1366.5	0.171
1477.6	0.168
CURVE 38*	
366.5	1.78
477.6	1.24
588.7	0.807
699.8	0.518
810.9	0.360
922.1	0.277
1033.2	0.228
1144.3	0.204
1255.4	0.190
1366.5	0.185
CURVE 39*	
699.8	0.709
810.9	0.554
922.1	0.450
1033.2	0.363
1144.3	0.312
1255.4	0.260
1366.5	0.225
1477.6	0.208
1588.7	0.173
1699.8	0.173
CURVE 40	
88.2	7.32
103.2	7.45
118.2	7.41
133.2	7.32
173.2	6.59
198.2	5.65
CURVE 40 (cont.)	
218.2	4.81
233.2	4.18
258.2	3.45
308.2	2.82
353.2	2.41
398.2	2.09
CURVE 41	
85.2	5.54
93.2	5.65
118.2	5.65
133.2	5.44
151.2	5.13
175.2	4.60
218.2	3.56
233.2	3.24
263.2	2.72
288.2	2.51
363.2	1.88*
408.2	1.67*
CURVE 42	
88.2	3.24
115.2	3.24
148.2	3.03
173.2	2.82
208.2	2.51
258.2	2.09
288.2	1.88
323.2	1.67
363.2	1.46
418.2	1.36
CURVE 43	
308.2	0.73
313.2	0.71
323.2	0.64*
353.2	0.58
398.2	0.50
483.2	0.40
553.2	0.34
628.2	0.32
718.2	0.26
CURVE 43 (cont.)	
818.2	0.25
898.2	0.23
CURVE 44	
318.2	1.27
328.2	1.11
338.2	1.10*
408.2	0.95
418.2	0.87
438.2	0.85
538.2	0.63
548.2	0.60
588.2	0.46
793.2	0.40
913.2	0.34
CURVE 45	
313.2	2.65
313.2	2.61
343.2	2.19
348.2	2.15
403.2	1.75
533.2	1.17*
668.2	0.88*
688.2	0.82*
778.2	0.67*
CURVE 46*	
318.2	2.22
323.2	2.17
338.2	2.05
403.2	1.65
458.2	1.36
543.2	1.09
638.2	0.89
713.2	0.74
748.2	0.70
CURVE 47*	
358.2	1.76
408.2	1.51
493.2	1.19

* Not shown on plot

DATA TABLE NO. 18 (continued)

T	k
CURVE 47 (cont.)*	
633.2	0.85
748.2	0.66
CURVE 48*	
303.2	2.26
343.2	2.05
383.2	1.81
408.2	1.78
443.2	1.55
538.2	1.20
628.2	0.98
CURVE 49*	
339.2	1.799
360.7	1.707
415.7	1.485
425.7	1.414
470.2	1.356
482.2	1.309
537.2	1.213
568.2	1.171
CURVE 50	
334.7	1.552
345.7	1.523
405.7	1.339
420.7	1.318
525.2	1.125
CURVE 51	
348.2	0.711
361.2	0.674
460.2	0.598
548.2	0.565
CURVE 52	
328.2	0.573
368.2	0.502
420.5	0.460
448.2	0.452
486.2	0.444
CURVE 52 (cont.)	
494.2	0.426
560.2	0.402
CURVE 53*	
1118.2	0.0092
1343.2	0.0100
1443.2	0.0105
1558.2	0.0109
1663.2	0.0134
1738.2	0.0184
1773.2	0.0192
CURVE 54*	
1133.2	0.0130
1363.2	0.0138
1473.2	0.0138
1578.2	0.0142
1683.2	0.0146
1833.2	0.0173
CURVE 55*	
1118.2	0.00556
1348.2	0.00736
1453.2	0.00757
1563.2	0.00908
1673.2	0.00996
1748.2	0.0108
1798.2	0.0113
CURVE 56	
533	1.86
811	0.92
1366	0.37
1922	0.29
CURVE 57	
533	1.70
811	0.79
1366	0.33
1922	0.23
CURVE 58*	
663.2	0.703
703.2	0.686
793.2	0.418
795.2	0.372
843.2	0.356
893.2	0.318
923.2	0.276
949.2	0.389
1003.2	0.247
1037.2	0.243
1053.2	0.222
1123.2	0.211
1152.2	0.209
1153.2	0.234
1153.2	0.213
1193.2	0.201
1213.2	0.210
1231.2	0.199
1263.2	0.188
1281.2	0.190
1313.2	0.176
1349.2	0.167
1353.2	0.178
1373.2	0.167
1403.2	0.163
1405.2	0.176
1443.2	0.163
1443.2	0.175
1473.2	0.163
CURVE 59*	
323.2	0.0100
483.2	0.0105
658.2	0.0100
CURVE 60*	
361.2	1.87
374.2	1.79
374.2	1.64
382.2	1.55
419.2	1.38
427.2	1.38
434.2	1.38
435.7	1.30
CURVE 60 (cont.)*	
436.2	1.33
488.2	1.15
497.2	1.11
497.2	1.07
509.2	1.11
514.2	1.05
514.2	1.03
515.2	1.08
527.2	1.06
646.2	0.753
659.2	0.745
707.2	0.619
723.2	0.607
730.2	0.594
759.2	0.577
CURVE 61*	
368.2	1.67
458.2	1.34
482.2	1.30
518.2	1.17
501.2	0.954
431.2	0.661
CURVE 62	
401.2	1.50
401.2	1.45
405.2	1.48
411.2	1.38
418.2	1.41
423.2	1.31
441.2	1.37
483.2	1.23
483.2	1.26
501.2	1.13
505.2	1.27
518.2	1.10
530.2	1.10
605.2	0.900
613.2	0.874
633.2	0.879
689.2	0.774
701.2	0.749
782.2	0.594
801.2	0.594
803.2	0.602
CURVE 63	
4.8	0.0032*
7.2	0.0105*
11.5	0.0340
22.5	0.149
32.0	0.340
57.0	0.975
72.0	1.65
92.0	2.49
310	2.51
CURVE 64	
2.2	0.00500*
2.8	0.00890*
3.5	0.0154*
4.2	0.0250
5.2	0.0392
7.3	0.0580
7.2	0.0660
8.8	0.0860
15.2	0.271
23.0	0.590
30.6	0.990
54.7	2.40
64.5	3.01
76.0	4.40
77.0	3.88
82.0	4.20
135	5.46
195	4.40
285	2.85
CURVE 65	
2.25	0.00425*
2.89	0.00835*
3.5	0.0137*
3.6	0.0153*
4.2	0.0208
5.8	0.0408
7.1	0.0580
7.5	0.0610
8.0	0.0730
10.0	0.128
14.5	0.280
19.95	0.523
CURVE 65 (cont.)	
26.0	0.940
32.0	1.45
39.1	2.13
45.0	2.60
54.5	3.03
60.0	3.59
65.8	3.95
70.8	4.20
79.0	4.70
90.0	4.90
139	6.00
195	5.20
300	2.85
CURVE 66	
3.0	0.00240*
3.5	0.00363*
4.25	0.00490*
4.8	0.00730*
8.0	0.0230
10.7	0.0409
15.2	0.0800
23.1	0.130
54.5	0.300
53.0	0.320
60.5	0.330
70.2	0.399
76.5	0.475
81.0	0.460
91.0	0.525
135	0.830
193	0.975
300	1.02
CURVE 67*	
673.2	0.777
773.2	0.631
873.2	0.526
973.2	0.449
1073.2	0.390
1173.2	0.338
CURVE 68*	
673.2	0.726
773.2	0.595
873.2	0.498
973.2	0.427
1073.2	0.370
1173.2	0.321
CURVE 69	
1.36	0.00238*
1.40	0.00258*
1.66	0.00414*
1.72	0.00462*
2.01	0.00719*
2.32	0.0110*
2.39	0.0117*
2.51	0.0141*
2.58	0.0148*
2.72	0.0173*
2.83	0.0190*
2.95	0.0205
3.06	0.0229
3.10	0.0243
3.16	0.0260
3.30	0.0284
3.37	0.0308
3.57	0.0365
3.64	0.0378
5.04	0.0703
5.53	0.0908
5.98	0.123
6.59	0.159
7.16	0.223
7.66	0.234
8.79	0.327
9.84	0.423
11.4	0.570
13.4	0.820
15.4	1.22
17.6	1.56
20.9	2.15
22.9	2.72
24.0	2.99
30.3	4.19
35.2	5.53
41.2	6.97
46.6	8.30

* Not shown on plot

DATA TABLE NO. 18 (continued)

T	k
CURVE 69 (cont.)	
50.8	9.57
56.5	10.7
61.1	11.7
65.8	12.5
74.1	13.5
78.3	13.8
84.7	14.7
86.5	16.0
98.2	14.7
108.6	14.0
118.0	13.3
120.8	12.3
130.3	12.2
138.0	10.9
157.0	9.00
203.2	6.15
215.8	5.72
228.6	5.05
277.3	3.78
CURVE 70	
2.06	0.00396*
2.08	0.00423*
2.44	0.00575*
2.47	0.00611*
2.73	0.00752*
2.77	0.00804*
3.07	0.0101 *
3.11	0.00964*
3.37	0.0116 *
3.46	0.0122 *
3.61	0.0132 *
3.72	0.0137 *
3.96	0.0157 *
4.10	0.0165 *
4.25	0.0182 *
4.45	0.0200
4.54	0.0214
4.69	0.0227
4.88	0.0243
5.25	0.0251
5.59	0.0290
5.77	0.0322
6.08	0.0368
6.86	0.0493

T	k
CURVE 70 (cont.)	
6.97	0.0426
7.21	0.0569
7.62	0.0646
8.00	0.0753
8.53	0.0877
9.25	0.108
10.0	0.133
10.4	0.131
11.3	0.171
11.8	0.193
12.4	0.191
12.5	0.215
13.4	0.243
13.6	0.239
14.0	0.279
15.3	0.321
15.5	0.301
16.9	0.357
20.9	0.598
21.5	0.690
22.9	0.671
26.8	0.955
29.0	1.07
35.7	1.63
39.5	2.06
44.0	2.50
48.3	3.01
58.5	4.21
71.1	5.70
79.6	6.53
79.6	6.95
87.1	7.46
91.2	8.07
97.3	8.38
104.0	8.71
109.4	8.85
119.7	8.97
128.5	8.71
133.7	8.61
143.2	8.38
148.3	8.15
157.4	7.75
166.7	7.11
277.3	3.57

T	k
CURVE 71	
1.56	0.00335*
1.88	0.00598*
2.47	0.0126 *
2.69	0.0162 *
3.48	0.0334
5.02	0.0813
5.31	0.0935
5.45	0.104
5.69	0.116
5.75	0.124
6.11	0.152
6.31	0.162
6.49	0.180
6.82	0.204
7.11	0.232
7.55	0.255
7.78	0.280
8.02	0.297
8.43	0.343
8.73	0.367
9.20	0.387
9.53	0.437
9.86	0.468
10.2	0.506
11.1	0.605
11.7	0.675
12.5	0.759
12.9	0.843
14.4	1.05
21.4	2.30
23.7	2.82
31.5	4.54
81.1	14.4
91.0	14.6
100.2	14.2
109.9	13.4
CURVE 72	
1.60	0.000951*
1.64	0.00104 *
1.68	0.00113 *
1.75	0.00124 *
1.96	0.00190 *
2.22	0.00248 *
2.30	0.00266 *

T	k
CURVE 72 (cont.)	
2.38	0.00297*
2.62	0.00380*
2.70	0.00413*
2.84	0.00455*
2.90	0.00499*
2.94	0.00522*
3.15	0.00555*
3.24	0.00622*
3.30	0.00686*
3.38	0.00757*
3.44	0.00804*
3.54	0.00861*
3.64	0.00940*
3.72	0.0100 *
3.82	0.0106 *
3.86	0.0114 *
3.95	0.0116 *
3.95	0.0119 *
4.01	0.0129 *
4.16	0.0140 *
4.35	0.0152 *
4.50	0.0163 *
4.58	0.0179 *
4.72	0.0192 *
4.93	0.0222
5.38	0.0239
5.38	0.0284
5.85	0.0328
6.22	0.0367
6.32	0.0341
6.67	0.0402
6.67	0.0476
6.97	0.0506
7.24	0.0575 *
7.52	0.0625 *
7.66	0.0644 *
8.05	0.0769 *
8.13	0.0800
8.59	0.0923
9.57	0.113
10.1	0.142
11.2	0.181
12.3	0.234
13.2	0.324
14.5	0.391
15.2	0.391

T	k
CURVE 72 (cont.)	
16.5	0.478
18.1	0.697
19.1	0.793
21.1	0.887
22.0	1.03
23.0	1.08
24.8	1.24
24.7	1.36
30.9	2.04
35.2	2.75
40.6	3.70
45.4	4.63
51.2	5.83
57.0	6.86
61.7	7.96
66.1	8.65
78.9	10.8
81.7	11.2
83.4	11.0
84.7	11.7
91.6	12.3
98.4	12.7
104.5	12.6
109.9	12.2
118.0	11.9
125.0	11.2
132.4	10.8
138.0	10.2
198.2	5.90
271.6	3.48
292.4	2.99
CURVE 73	
1.41	0.000244*
1.41	0.000269*
1.77	0.000386*
1.81	0.000438*
1.96	0.000525*
2.18	0.000755*
2.24	0.000811*
2.48	0.00105 *
2.79	0.00145 *
2.82	0.00152 *
2.86	0.00165 *
2.86	0.00174 *

T	k
CURVE 73 (cont.)	
2.97	0.00177*
3.24	0.00239*
3.30	0.00250*
3.36	0.00258*
3.43	0.00275*
3.75	0.00342*
3.79	0.00356*
3.98	0.00420*
4.05	0.00439*
4.17	0.00462*
4.40	0.00552*
4.40	0.00647*
4.51	0.00585*
4.56	0.00670*
4.65	0.00569*
4.89	0.00670*
5.18	0.00780*
5.51	0.00950*
5.88	0.0111 *
6.04	0.0124 *
6.34	0.0138 *
6.68	0.0150 *
7.02	0.0232
7.50	0.0229
7.94	0.0318
8.19	0.0294
8.57	0.0342
8.83	0.0443
9.46	0.0443
10.4	0.0553
10.4	0.0617
11.4	0.0716
12.2	0.0975
13.0	0.0975
14.2	0.148
14.9	0.145
15.9	0.182
20.0	0.348
21.6	0.391
23.1	0.466
23.7	0.483
29.0	0.807
35.4	1.32
39.2	1.59
43.9	2.05
52.0	2.92

T	k
CURVE 73 (cont.)	
59.6	3.86
71.5	5.41
84.7	7.03
92.0	7.83
106.7	8.65
121.1	8.59
129.4	8.41
138.4	8.04
150.7	7.94
150.7	7.40
158.1	7.00
164.1	6.71
168.7	7.24
280.5	3.16
CURVE 74	
1.83	0.000308*
1.83	0.000333*
1.92	0.000421*
1.98	0.000447*
2.08	0.000560*
2.17	0.000565*
2.25	0.000659*
2.38	0.000702*
2.58	0.000982*
2.74	0.00106 *
2.83	0.00114 *
2.83	0.00126 *
2.99	0.00130 *
3.23	0.00161 *
3.29	0.00185 *
3.39	0.00177 *
3.42	0.00199 *
3.49	0.00188 *
3.54	0.00205 *
3.62	0.00215 *
3.92	0.00255 *
3.97	0.00271 *
4.06	0.00281 *
4.19	0.00311 *
4.25	0.00327 *
4.43	0.00357 *
4.53	0.00372 *
4.66	0.00399 *
4.73	0.00421 *

*Not shown on plot

DATA TABLE NO. 18 (continued)

T	k
CURVE 74 (cont.)	
4.78	0.00447*
5.09	0.00528*
5.77	0.00730*
6.01	0.00836*
6.18	0.00902*
6.40	0.0102 *
6.75	0.0118 *
6.90	0.0125 *
7.10	0.0135 *
7.26	0.0149 *
7.80	0.0184 *
8.11	0.0217
8.45	0.0242
8.69	0.0268
9.00	0.0303
9.66	0.0361
10.1	0.0428
11.0	0.0518
12.2	0.0662
12.4	0.0693
13.0	0.0813
13.7	0.0927
14.0	0.0980
15.3	0.121
16.2	0.143
19.3	0.236
22.8	0.327
23.8	0.364
26.8	0.466
32.0	0.710
39.5	1.13
46.9	1.63
55.9	2.44*
61.7	3.01
71.0	3.97
81.1	4.99
84.1	5.33
88.7	5.75
94.6	6.22
100.5	6.59
104.7	6.89
107.9	7.05
117.8	7.26
121.6	7.38*
125.0	7.41
129.7	7.33*
CURVE 74 (cont.)	
135.5	7.23
145.5	6.95
153.5	6.61
283.8	2.97*
CURVE 75	
291.1	1.28
313.4	1.64
320.3	1.59
326.5	1.53
333.5	1.38
341.2	1.36
353.5	1.30
366.5	1.28
374.4	1.24
382.5	1.20
391.8	1.13*
401.6	1.13*
414.8	1.02
CURVE 76	
106.7	2.08
112.5	2.14
118.4	2.18
123.1	2.26
135.6	2.48
150.1	2.56
163.1	2.46
163.2	2.31
177.9	1.85
192.7	1.63
208.5	1.53
218.7	1.53
228.3	1.56
CURVE 77*	
231.0	1.47
243.5	1.41
253.3	1.48
265.9	1.58
279.5	1.62
286.6	1.49
295.1	1.49
CURVE 77 (cont.)*	
307.0	1.60
316.3	1.44
324.6	1.32
330.1	1.43
338.3	1.47
348.3	1.33
357.1	1.24
368.1	1.19
CURVE 78*	
573	0.941
673	0.807
773	0.677
873	0.560
973	0.472
1073	0.405
1173	0.351
1273	0.309
CURVE 79*	
573	0.920
673	0.782
773	0.656
873	0.548
973	0.460
1073	0.389
1173	0.330
1273	0.288
CURVE 80*	
573	0.899
673	0.769
773	0.652
873	0.535
973	0.447
1073	0.372
1173	0.326
1273	0.284
CURVE 81*	
573	0.866
673	0.769
CURVE 81 (cont.)*	
773	0.648
873	0.548
973	0.472
1073	0.393
1173	0.343
1273	0.301
CURVE 82*	
873	0.569
973	0.460
1073	0.397
1173	0.343
1273	0.292
CURVE 83*	
573	0.887
673	0.761
773	0.661
873	0.539
973	0.451
1073	0.393
1173	0.351
1273	0.297
CURVE 84*	
573	0.979
673	0.857
773	0.702
873	0.585
973	0.506
1073	0.426
1173	0.364
1273	0.317
CURVE 85*	
573	0.958
673	0.820
773	0.690
873	0.581
973	0.489
1073	0.410
1173	0.355
1273	0.309
CURVE 86*	
573	0.761
673	0.656
773	0.548
873	0.460
973	0.389
1073	0.338
1173	0.292
1273	0.259
CURVE 87*	
573	0.627
673	0.560
773	0.468
873	0.405
973	0.334
1073	0.292
1173	0.251
1273	0.217
CURVE 88*	
573	0.769
673	0.690
773	0.573
873	0.502
973	0.418
1073	0.359
1173	0.309
1273	0.267
CURVE 89*	
582	0.393
584	0.397
673	0.384
774	0.359
872	0.334
971	0.309
CURVE 90*	
1066	0.292
1168	0.259
1272	0.267
CURVE 91*	
569	0.748
669	0.669
766	0.560
863	0.481
1055	0.359
1157	0.313
1270	0.271
CURVE 92*	
570	0.606
668	0.539
769	0.460
864	0.418
867	0.410
968	0.368
1067	0.322
1171	0.284
1263	0.259
CURVE 93*	
571	0.715
714	0.602
811	0.489
952	0.401
1143	0.309
1268	0.263
CURVE 94*	
576	0.690
979	0.422
1076	0.364
1176	0.317
1281	0.276
CURVE 95*	
574	0.748
672	0.677
872	0.510
973	0.422
1072	0.368
1164	0.322
1278	0.284
CURVE 96*	
373.2	1.46
473.2	1.02
573.2	0.82
673.2	0.68
773.2	0.58
873.2	0.50
973.2	0.44
1073.2	0.39
1173.2	0.34
1273.2	0.30
CURVE 97*	
558.2	1.72
562.2	1.66
565.2	1.67
CURVE 98*	
575.2	1.77
575.2	1.76
596.2	1.77
788.2	0.929
831.2	0.956
849.2	0.935
1125.2	0.562
1127.2	0.596
1153.2	0.581
1336.2	0.421
1336.2	0.404
1374.2	0.420
1398.2	0.391
1653.2	0.291
1652.2	0.284
1653.2	0.317
1694.2	0.286
1960.2	0.245
1993.2	0.251
1997.2	0.219
2016.2	0.237
2263.2	0.188
2297.2	0.196
2412.2	0.232

*Not shown on plot

DATA TABLE NO. 18 (continued)

CURVE 99

T	k
576.2	1.85*
576.2	1.83*
576.2	1.81
813.2	0.857
841.2	0.851*
866.2	0.864
1098.2	0.493
1104.2	0.476*
1121.2	0.476*
1274.2	0.350
1293.2	0.336*
1316.2	0.345
1372.2	0.346
1880.2	0.306
1910.2	0.314*
1925.2	0.310
1931.2	0.299
2098.2	0.304
2136.2	0.304*
2161.2	0.307
2222.2	0.242

CURVE 100*

T	k
512.2	1.65
533.2	1.66
788.2	0.655
833.2	0.668
833.2	0.645
1067.2	0.381
1075.2	0.395
1075.2	0.363
1087.2	0.384
1263.2	0.274
1284.2	0.255
1312.2	0.273
1821.2	0.257
1844.2	0.226
1851.2	0.267
1874.2	0.255
2067.2	0.176
2083.2	0.187
2088.2	0.198
2122.2	0.190
2269.2	0.183

CURVE 101*

T	k
533.2	1.63
533.2	1.61
574.2	1.65
817.2	0.770
817.2	0.759
817.2	0.740
1118.2	0.430
1138.2	0.453
1158.2	0.453
1164.2	0.437
1279.2	0.365
1304.2	0.345
1331.2	0.362
1637.2	0.258
1655.2	0.248
1670.2	0.264
1941.2	0.229
1979.2	0.222
2121.2	0.215
2227.2	0.180

CURVE 102*

T	k
517.2	1.50
543.2	1.56
543.2	1.57
768.2	0.760
808.2	0.749
846.2	0.764
1101.2	0.496
1113.2	0.479
1136.2	0.485
1287.2	0.389
1312.2	0.378
1328.2	0.401
1360.2	0.352
1747.2	0.303
1771.2	0.326
1793.2	0.307
1972.2	0.261
2007.2	0.262
2048.2	0.268
2093.2	0.219
2110.2	0.222

CURVE 103*

T	k
546.2	1.72
568.2	1.74
585.2	1.71
797.2	0.708
834.2	0.708
855.2	0.692
1071.2	0.431
1091.2	0.424
1102.2	0.433
1283.2	0.325
1310.2	0.317
1611.2	0.242
1630.2	0.237
1661.2	0.247
1863.2	0.199
1878.2	0.190
1896.2	0.216
2194.2	0.183
2217.2	0.179
2236.2	0.200
2397.2	0.189

CURVE 104*

T	k
563.2	1.86
566.2	1.80
578.2	1.85
802.2	0.813
836.2	0.813
836.2	0.800
864.2	0.816
1061.2	0.418
1106.2	0.395
1106.2	0.431
1294.2	0.325
1319.2	0.309
1326.2	0.322
1670.2	0.248
1683.2	0.232
1717.2	0.247
1950.2	0.244
2002.2	0.244
2053.2	0.244
2163.2	0.190

CURVE 105*

T	k
539.2	1.84
539.2	1.81
563.2	1.75
828.2	0.839
828.2	0.825
1079.2	0.483
1114.2	0.475
1131.2	0.485
1292.2	0.372
1309.2	0.362
1320.2	0.379
1589.2	0.257
1642.2	0.260
1645.2	0.229
1865.2	0.189
1891.2	0.170
1900.2	0.187
2074.2	0.173
2125.2	0.196
2150.2	0.213
2293.2	0.193

*Not shown on plot

FIGURE AND TABLE NO. 18R RECOMMENDED THERMAL CONDUCTIVITY OF BERYLLIUM OXIDE BeO

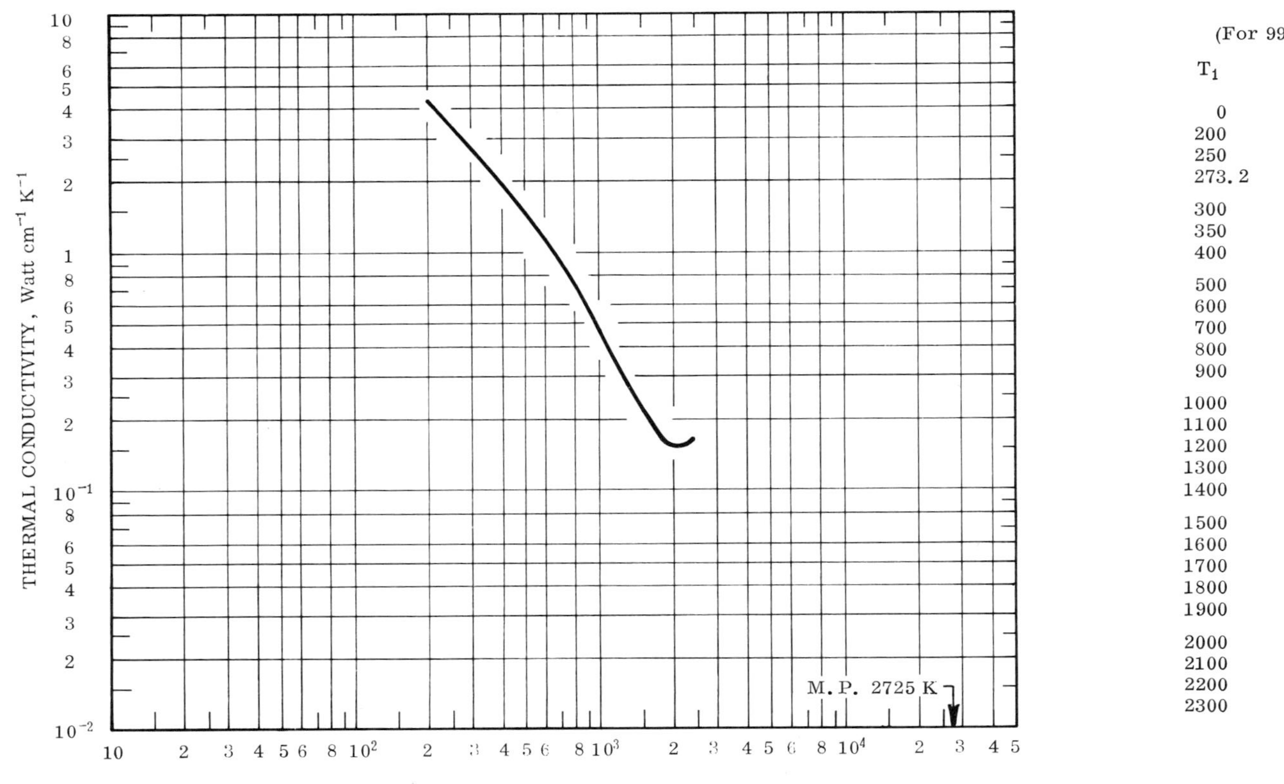

RECOMMENDED VALUES*
(For 99.5% Pure, 98% Dense, Polycrystalline BeO)

T_1	k_1	k_2	T_2
0	0	0	-459.7
200	4.24	245	- 99.7
250	3.34	193	- 9.7
273.2	3.02	175	32.0
300	2.72	157	80.3
350	2.28	132	170.3
400	1.96	113	260.3
500	1.46	84.4	440.3
600	1.11	64.1	620.3
700	0.87	50.3	800.3
800	0.70	40.4	980.3
900	0.57	32.9	1160
1000	0.47	27.2	1340
1100	0.39	22.5	1520
1200	0.33	19.1	1700
1300	0.283	16.4	1880
1400	0.245	14.2	2060
1500	0.215	12.4	2240
1600	0.195	11.3	2420
1700	0.180	10.4	2600
1800	0.167	9.65	2780
1900	0.156	9.01	2960
2000	0.150	8.67	3140
2100	0.150	8.67	3320
2200	0.152	8.78	3500
2300	0.164	9.48	3680

REMARKS

The recommended values are for 99.5% pure, 98% dense, polycrystalline BeO. The recommended values are thought to be accurate to within 8% of the true values at temperatures from 500 to 1000 K and 8 to 15% at other temperatures.

*T_1 in K, k_1 in Watt cm^{-1} K^{-1}, T_2 in F, and k_2 in Btu hr^{-1} ft^{-1} F^{-1}.

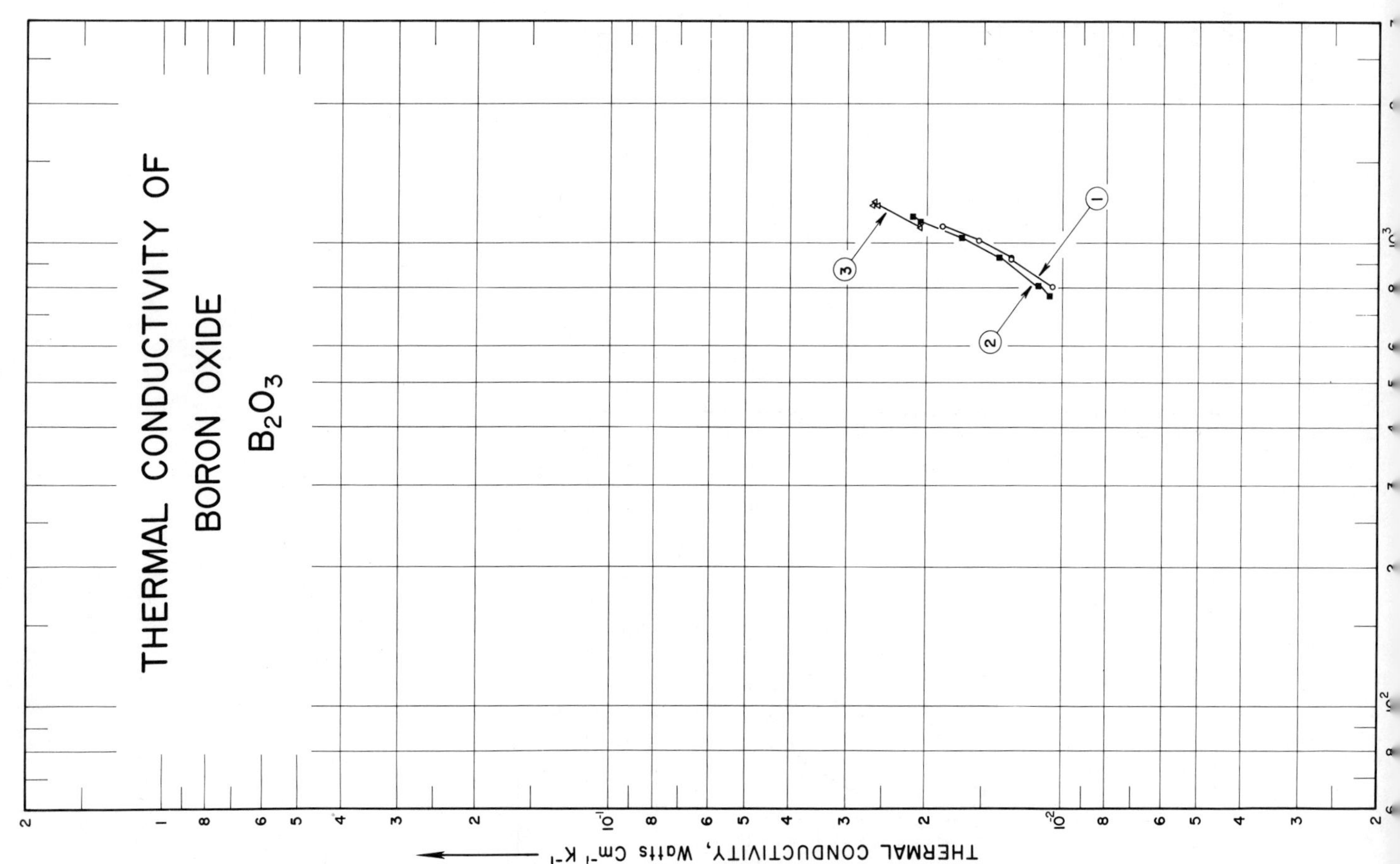

THERMAL CONDUCTIVITY OF
BORON OXIDE
B_2O_3
THERMAL CONDUCTIVITY, Watts Cm^{-1} K^{-1}
1
2
3

SPECIFICATION TABLE NO. 19 THERMAL CONDUCTIVITY OF BORON OXIDE B_2O_3

[For Data Reported in Figure and Table No. 19]

Curve No.	Ref. No.	Method Used	Year	Temp. Range, K	Reported Error, %	Name and Specimen Designation	Composition (weight percent), Specifications and Remarks
1	248	R	1959	809-1098			In liquid state; contained in the annulas of two concentric stainless steel tubes (1. 22 cm radius difference), with a very thin stainless steel cylinder placed midway between the two measuring surfaces.
2	248	R	1959	769-1144			In liquid state; measured after the above specimen was heated to 950 C for two days in the apparatus.
3	248	R	1959	1085-1217			In liquid state; measured after the specimen was heated to 950 C for a period of a week and half in the apparatus.

DATA TABLE NO. 19 THERMAL CONDUCTIVITY OF BORON OXIDE B_2O_3

[Temperature, T, K; Thermal Conductivity, k, Watt $cm^{-1}K^{-1}$]

T	k
CURVE 1	
808.8	0.0106
809.1	0.0106*
809.4	0.0106*
812.6	0.0107*
921.2	0.0130
922.6	0.0130*
925.7	0.0130
925.7	0.0129*
1016.9	0.0154
1017.6	0.0154*
1097.4	0.0185
1097.9	0.0184*
CURVE 2	
768.6	0.0107
770.9	0.0107*
810.4	0.0113
931.3	0.0139
931.3	0.0138*
1024.6	0.0168
1024.9	0.0168*
1117.4	0.0206
1121.6	0.0208*
1140.7	0.0215
1143.7	0.0217*
CURVE 3	
1085.0	0.0207
1091.8	0.0207*
1207.0	0.0258
1207.6	0.0265
1210.7	0.2064*
1211.3	0.0264*
1214.0	0.0261
1217.3	0.0265*

*Not shown on plot

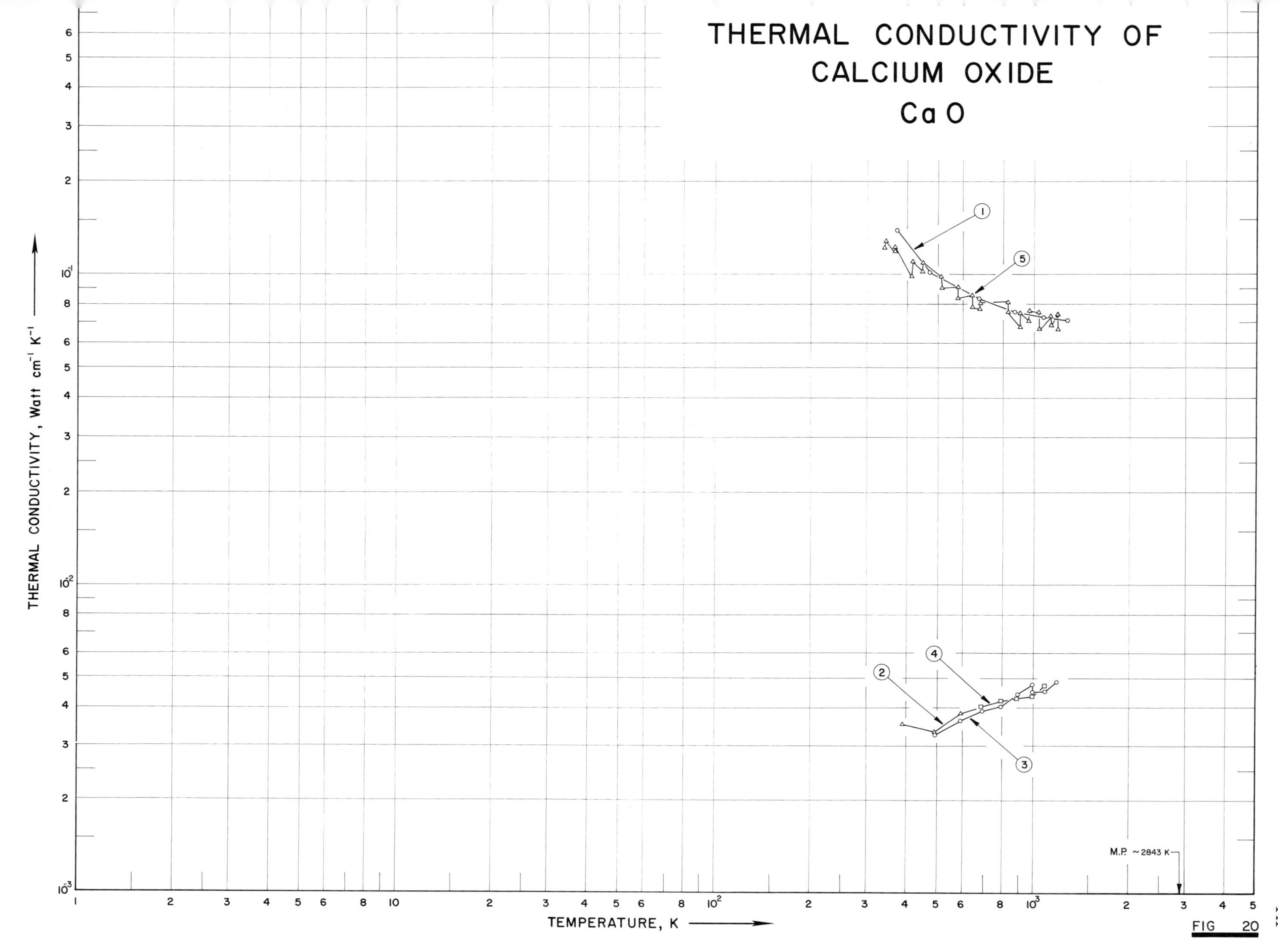
THERMAL CONDUCTIVITY OF
CALCIUM OXIDE
Ca O
THERMAL CONDUCTIVITY, Watt cm⁻¹ K⁻¹
TEMPERATURE, K
M.P. ~2843 K
1
2
3
4
5
FIG 20

SPECIFICATION TABLE NO. 20 THERMAL CONDUCTIVITY OF CALCIUM OXIDE CaO

[For Data Reported in Figure and Table No. 20]

Curve No.	Ref. No.	Method Used	Year	Temp. Range, K	Reported Error, %	Name and Specimen Designation	Composition (weight percent), Specifications and Remarks
1	364	C	1954	373-1273			Polycrystal; prepared by calcining reagent grade calcium carbonate at 1600 C for 2 hr; resulting powder hydrostatically pressed and fired at 1900 C in a zirconia gas-oxygen furnace (oxidizing atmosphere); bulk density 3.03 g cm^{-3}; porosity 8.75%.
2	221	R	1959	390-897	5.9		Packed powdered CaO; 90% of the particles in 0.3-6.6 micron range; density 1.86 g cm^{-3}.
3	221	R	1959	494-1199	5.9		Similar to the above specimen except 1.57 g cm^{-3} density.
4	221	R	1959	693-1197	5.9		Second run of the above specimen.
5	45	C	1953	341-1196			Pure; crystalline cube specimen; prepared from reagent grade calcium carbonate by calcining for 2 hrs at 1600 C; hydrostatically pressed and fired at 1900 C in zirconia gas-oxygen furnace; bulk density 3.03 g cm^{-3}; total porosity 8.75% (true density 3.32).

DATA TABLE NO. 20 THERMAL CONDUCTIVITY OF CALCIUM OXIDE CaO

[Temperature, T, K; Thermal Conductivity, k, Watt cm^{-1} K^{-1}]

T	k
CURVE 1	
373.2	0.139
473.2	0.101
673.2	0.0837
873.2	0.0757
1073.2	0.0728
1273.2	0.0711
CURVE 2	
390.0	0.00353
492.2	0.00330
593.3	0.00382
896.8	0.00439*
CURVE 3	
493.8	0.00326
593.2	0.00361
693.2	0.00390
796.4	0.00405
896.8	0.00445
996.6	0.00480
997.2	0.00451
1099.2	0.00454
1199.2	0.00489
CURVE 4	
692.5	0.00402
692.8	0.00403*
790.6	0.00421
892.2	0.00432
995.2	0.00437
1095.3	0.00475
1097.2	0.00450*
1197.2	0.00488*

T	k
CURVE 5	
341.2	0.123
345.2	0.128
367.2	0.118
367.2	0.123
419.2	0.110
418.2	0.0987
448.2	0.103
448.2	0.109
512.2	0.0983
517.2	0.0900
580.2	0.0908
580.2	0.0837
644.2	0.0858
644.2	0.0787
678.2	0.0874
681.2	0.0808
833.2	0.0812
833.2	0.0757
908.2	0.0749
908.2	0.0678
968.2	0.0703
972.2	0.0761
1042	0.0753
1043	0.0665
1138	0.0736
1141	0.0686
1196	0.0741
1196	0.0661

*Not shown on plot

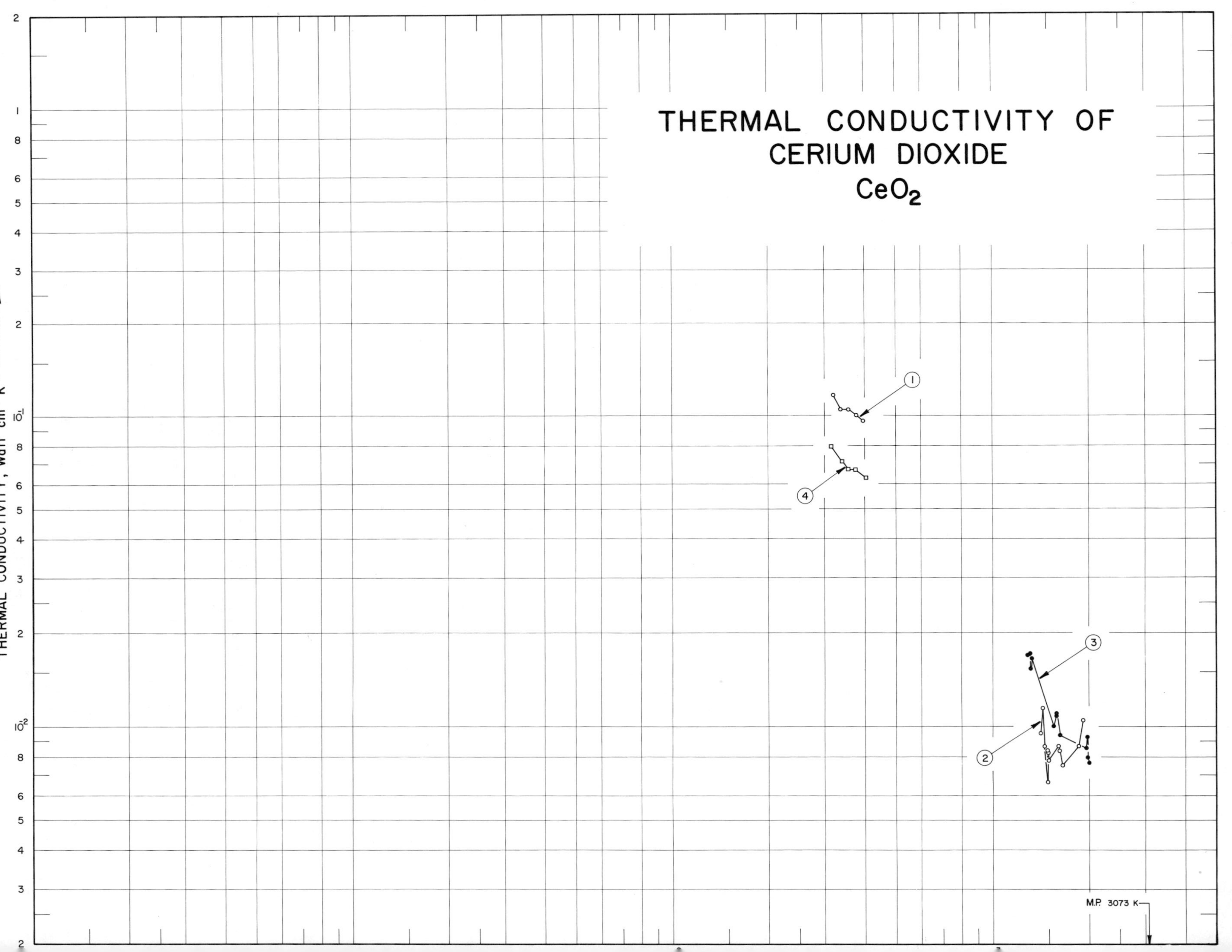
THERMAL CONDUCTIVITY OF
CERIUM DIOXIDE
CeO_2
THERMAL CONDUCTIVITY, Watt cm^{-1} K^{-1}
2
1
8
6
5
4
3
2
10^{-1}
8
6
5
4
3
2
10^{-2}
8
6
5
4
3
2
1
4
3
2
M.P. 3073 K

SPECIFICATION TABLE NO. 21 THERMAL CONDUCTIVITY OF CERIUM DIOXIDE CeO_2

[For Data Reported in Figure and Table No. 21]

Curve No.	Ref. No.	Method Used	Year	Temp. Range, K	Reported Error, %	Name and Specimen Designation	Composition (weight percent), Specifications and Remarks
1	23		1952	320-397		98 A-2	Fired at 1482 C; 0.0042% water absorption; density 6.20 g cm^{-3}; specimen buff color.
2	144	R	1963	1410-1915	5-7	1	0.2 Zr, 0.1 Ca; specimen 0.75 in. long, 0.75 in. O. D. and 0.25 in. I. D.; supplied by Zirconium Corp of America; pressed and sintered; density 6.87 g cm^{-3} at 25 C; specimen melted during test.
3	144	R	1963	1292-2006	5-7		Similar to the above specimen except for size, 3 in. long, 2.5 in. O. D. and 0.75 in. I. D.
4	23		1952	317-404		16 OA-1	2.0 MgO; fired at 1567 C; 0.59% water absorption; density 5.58 g cm^{-3}; specimen buff color.

DATA TABLE NO. 21 THERMAL CONDUCTIVITY OF CERIUM DIOXIDE CeO_2

[Temperature, T, K; Thermal Conductivity, k, Watt cm^{-1} K^{-1}]

T	k
CURVE 1	
320.3	0.117
338.2	0.105
358.5	0.105
378.8	0.100
397.4	0.096
CURVE 2	
1410.4	0.00952
1440.4	0.0115
1456.5	0.00865
1490.4	0.00663
1492.6	0.00837
1496.5	0.00822
1499.8	0.00779
1610.4	0.00865
1627.1	0.00837
1652.6	0.00750
1859.8	0.00865
1914.8	0.0105
CURVE 3	
1292.1	0.0170
1313.7	0.0172
1322.1	0.0154
1327.1	0.0166
1555.9	0.0100
1585.4	0.0108
1588.2	0.0111
1628.2	0.00938
1969.2	0.00851
1977.1	0.00923
1978.2	0.00793
2005.9	0.00765
CURVE 4	
316.5	0.0795
371.4	0.0711
356.4	0.0669
375.3	0.0669
404.1	0.0628

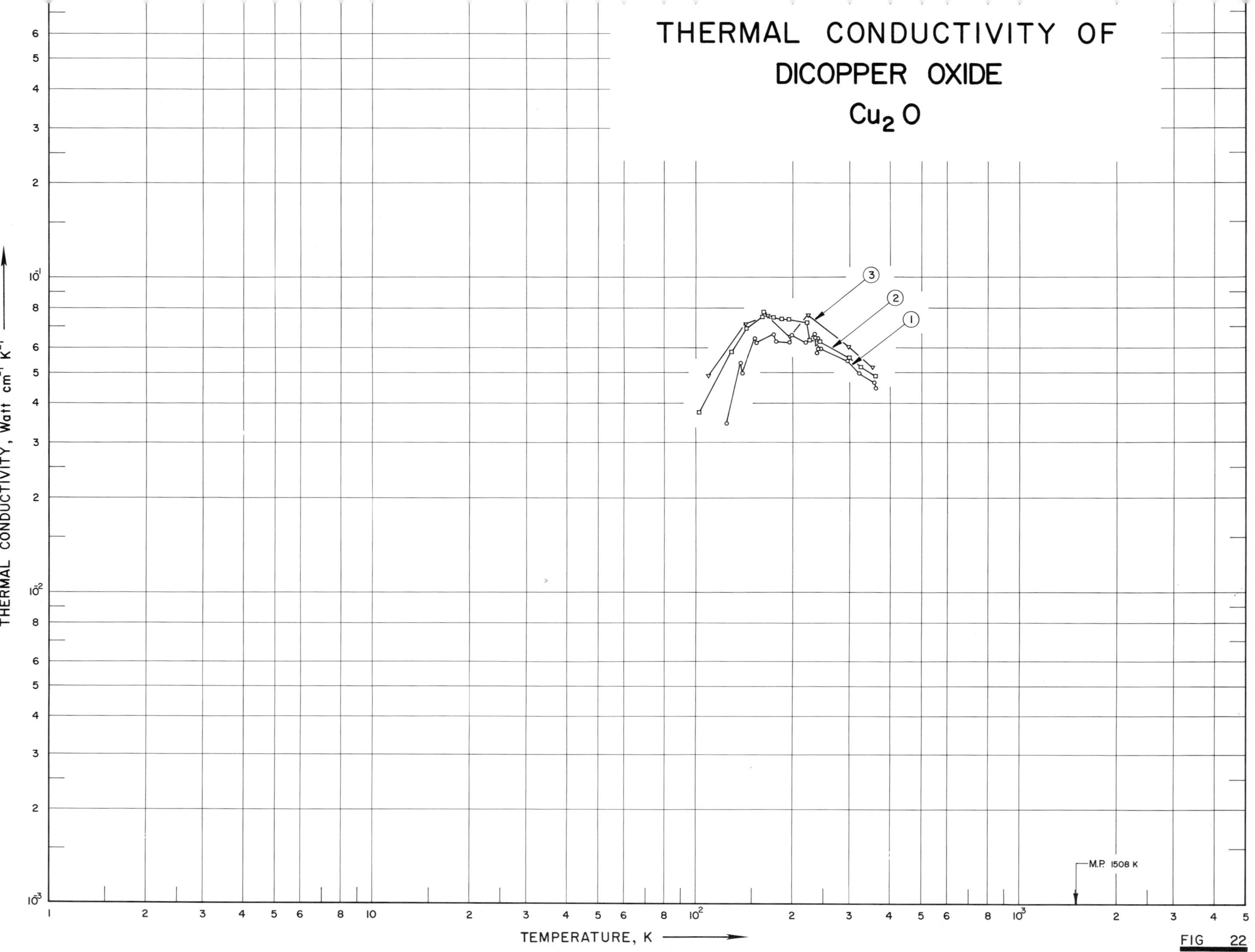
THERMAL CONDUCTIVITY OF
DICOPPER OXIDE
Cu2 O
THERMAL CONDUCTIVITY, Watt cm⁻¹ K⁻¹
TEMPERATURE, K
M.P. 1508 K
1
2
3

FIG 22

SPECIFICATION TABLE NO. 22 THERMAL CONDUCTIVITY OF DICOPPER OXIDE Cu_2O

[For Data Reported in Figure and Table No. 22]

Curve No.	Ref. No.	Method Used	Year	Temp. Range, K	Reported Error, %	Name and Specimen Designation	Composition (weight percent), Specifications and Remarks
1	459	L	1950	125-362	3.3	B	99.9 pure; electrolytic; single crystals containing a polycrystalline interface; specimen prepared by oxidizing machined copper disks of dia 2.86 cm in a furnace at 1020 C for 30 to 210 hrs, flushing nitrogen over the oxide for 1 hr during the slow cooling process; electrical resistivity reported as 47.0, 41.0, 15.9, 4.02, 3.92, 3.25, 3.06, 2.83, 2.42, 1.93, and 1.91 10^3 ohm cm at 206, 208, 234, 276, 277, 288, 293, 295, 306, 336, and 338 K, respectively.
2	459	L	1950	102-360	3.3	E	99.96 pure; electrolytic O.F.H.C.; large single crystal; specimen prepared by oxidizing machined copper disks of dia 2.86 cm in a furnace at 1020 C for 30 to 210 hrs, heated to 1200 C, cooled rapidly.
3	459	L	1950	110-353	3.3	F	Specimen prepared by fusing small pieces of sintered Cu_2O in oxygen in a porcelain ignition capsule.

DATA TABLE NO. 22 THERMAL CONDUCTIVITY OF DICOPPER OXIDE Cu_2O

[Temperature, T, K; Thermal Conductivity, k, Watt $cm^{-1} K^{-1}$]

T	k
CURVE 1	
125.2	0.0346
137.0	0.0536
139.3	0.0498
152.4	0.0642
155.4	0.0623
174.0	0.0662
177.2	0.0627
196.9	0.0625
198.2	0.0657
219.1	0.0624
234.7	0.0662
235.0	0.0645
237.8	0.0578
239.7	0.0644
239.8	0.0595
244.3	0.0594
296.6	0.0544
321.6	0.0497
357.0	0.0465
361.8	0.0448
CURVE 2	
102.3	0.0374
129.8	0.0582
144.2	0.0690
161.9	0.0749
163.2	0.0776
174.1	0.0748
185.0	0.0739
195.0	0.0736
221.8	0.0720
226.6	0.0635
239.5	0.0632*
243.2	0.0628
299.6	0.0558
325.6	0.0524
359.5	0.0486
CURVE 3	
109.6	0.0489
143.9	0.0709
168.0	0.0755
196.9	0.0648
222.8	0.0756
CURVE 3 (cont.)	
236.8	0.0656*
297.0	0.0604
353.2	0.0519

*Not shown on plot

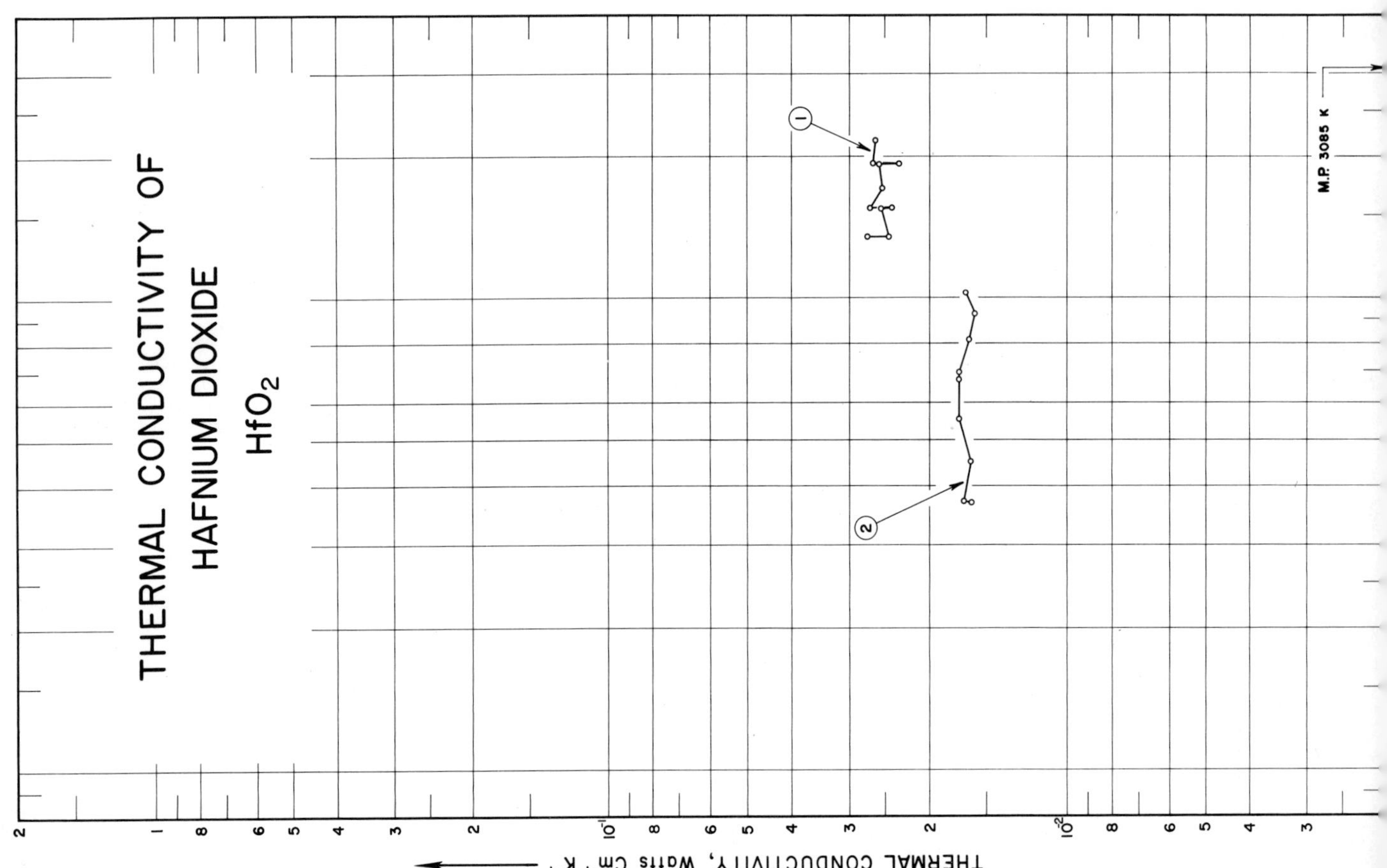
THERMAL CONDUCTIVITY OF
HAFNIUM DIOXIDE
HfO_2
1
2
M.P. 3085 K
THERMAL CONDUCTIVITY, Watts Cm^{-1} K^{-1}

SPECIFICATION TABLE NO. 23 THERMAL CONDUCTIVITY OF HAFNIUM DIOXIDE HfO_2

[For Data Reported in Figure and Table No. 23]

Curve No.	Ref. No.	Method Used	Year	Temp. Range, K	Reported Error, %	Name and Specimen Designation	Composition (weight percent), Specifications and Remarks
1	144	R	1963	1365-2173	5-7	1	2.5 Fe, 0.3 Mg, 0.1 Ca, 0.1 Ti; poorly bonded structure; specimen 0.75 in. long, 0.75 in. O.D. and 0.25 I.D.; pressed and sintered; density 6.87 g cm^{-3} at 25 C; specimen found blistered and partially melted on post inspection.
2	233		1955	368-1033			No details reported.

DATA TABLE NO. 23 THERMAL CONDUCTIVITY OF HAFNIUM DIOXIDE HfO_2

[Temperature, T, K; Thermal Conductivity, k, Watt cm^{-1} K^{-1}]

T	k
CURVE 1	
1364.8	0.0273
1364.8	0.0245
1553.7	0.0255
1565.4	0.0241
1565.4	0.0270
1715.9	0.0254
1727.6	0.0251*
1736.5	0.0252*
1929.8	0.0257
1933.2	0.0232
1937.1	0.0267
2173.2	0.0264
CURVE 2	
368.2	0.0161
373.2	0.0167
453.2	0.0163
558.2	0.0172
673.2	0.0172
698.2	0.0172
818.2	0.0163
928.2	0.0159
1033.2	0.0167

* Not shown on plot

SPECIFICATION TABLE NO. 24 THERMAL CONDUCTIVITY OF INDIUM OXIDE InO

Curve No.	Ref. No.	Method Used	Year	Temp. Range, K	Reported Error, %	Name and Specimen Designation	Composition (weight percent), Specifications and Remarks
1	363	P	1965	1204-1722			Density 6.3 g cm^{-3}; thermal conductivity values calculated from measured thermal diffusivity data and the assumed constant specific heat at 0.2 cal $g^{-1}c^{-1}$.

DATA TABLE NO. 24 THERMAL CONDUCTIVITY OF INDIUM OXIDE InO

[Temperature, T, K; Thermal Conductivity, k, Watt $cm^{-1}K^{-1}$]

T	k
CURVE 1*	
1204	0.05649
1298	0.05675
1334	0.05675
1342	0.05675
1416	0.05675
1469	0.05675
1497	0.05675
1522	0.05834
1560	0.05585
1662	0.05649
1722	0.05649

* No graphical presentation

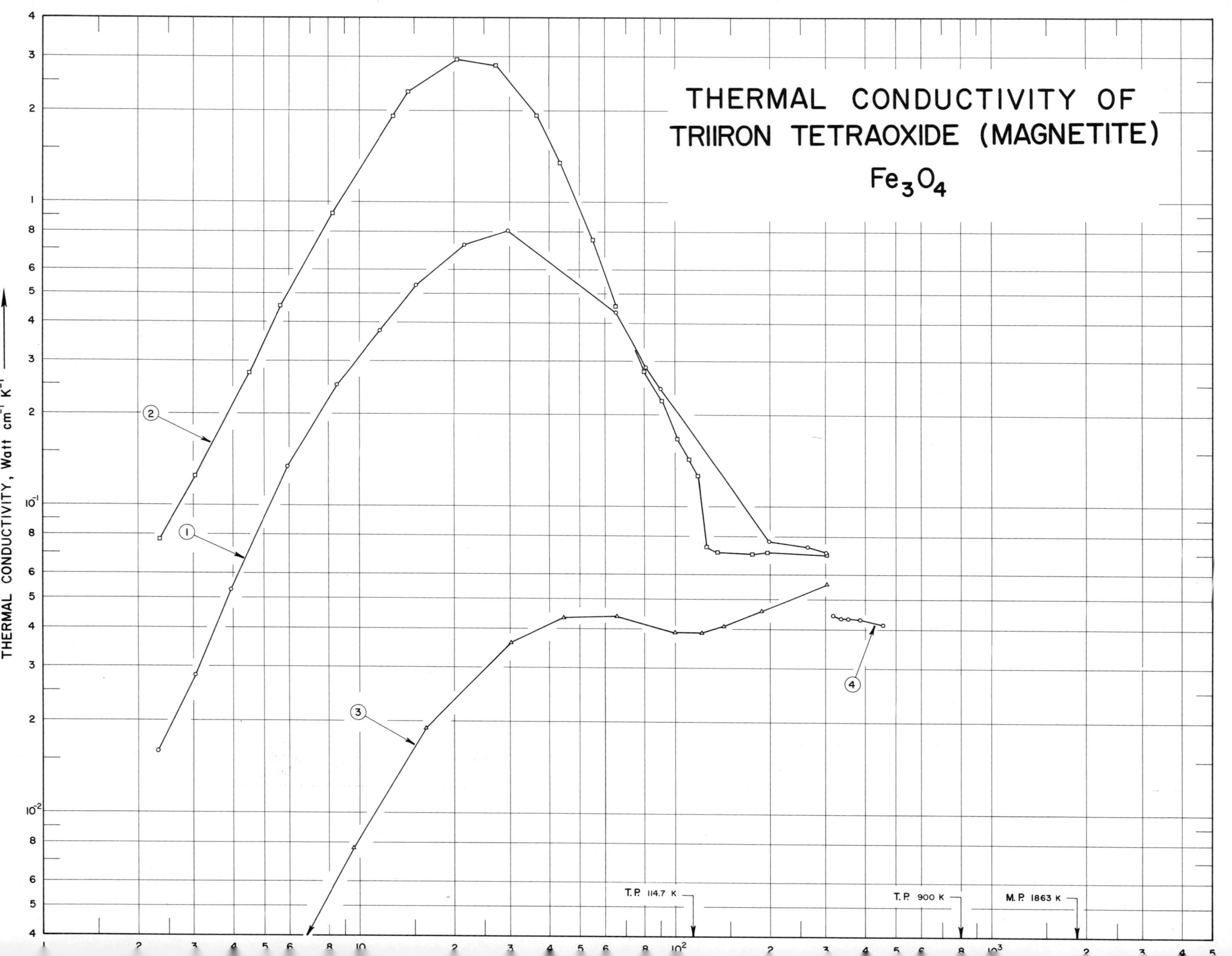
THERMAL CONDUCTIVITY OF
TRIIRON TETRAOXIDE (MAGNETITE)
Fe_3O_4
THERMAL CONDUCTIVITY, Watt cm^{-1} K^{-1}
T.P. 114.7 K
T.P. 900 K
M.P. 1863 K
1
2
3
4

SPECIFICATION TABLE NO. 25 THERMAL CONDUCTIVITY OF TRIIRON TETRAOXIDE (MAGNETITE) Fe_3O_4

[For Data Reported in Figure and Table No. 25]

Curve No.	Ref. No.	Method Used	Year	Temp. Range, K	Reported Error, %	Name and Specimen Designation	Composition (weight percent), Specifications and Remarks
1	300	L	1962	2.3-301		R-37	$10^{18.1}$ atom cm^{-3} Al, $<10^{18.2}$ atom cm^{-3} Be, $<10^{17.9}$ atom cm^{-3} Ca, $10^{17.8}$ atom cm^{-3} Cr, $<10^{17.9}$ atom cm^{-3} K, $<10^{18.1}$ atom cm^{-3} Li, $10^{17.6}$ atom cm^{-3} Mg, $10^{17.1}$ atom cm^{-3} Mn, $10^{18.1}$ atom cm^{-3} Na, $10^{17.6}$ atom cm^{-3} Ni, $<10^{18.0}$ atom cm^{-3} Si, $<10^{17.8}$ atom cm^{-3} Ti, $<10^{17.8}$ atom cm^{-3} V, $<10^{18.0}$ atom cm^{-3} Zn, and $<10^{17.5}$ atom cm^{-3} Zr; single crystal; black luster, opaque specimen 1.26 cm long, 0.54 cm dia; lattice constant 8.397 Å; vacancy concentration 2.6×10^{20} cm^{-3}.
2	300	L	1962	2.3-304		R-57	A section of the above specimen (1.23 cm long, 0.2 cm av. dia) annealed at 1500 K in a CO: 20 CO_2 atmosphere; lattice constant 8.398 Å; vacancy concentration 2×10^{17} cm^{-3}.
3	300	L	1962	3.0-302		R-63 ($Co_{0.07}Fe_{2.93}O_4$)	Similar to specimen R-37 curve No. 1 except 9.9×10^{20} atom cm^{-3} Co; annealed at 1620 K in a 2×10^{-6} atm. O_2; specimen 0.80 cm long, 0.24 cm av. dia; vacancy concentration 8×10^{18} cm^{-3} lattice constant 8.394 Å.
4	68	C	1954	317-453	±3	298 A-1	Single crystal.

DATA TABLE NO. 25 THERMAL CONDUCTIVITY OF IRON OXIDE (MAGNETITE) Fe_3O_4

[Temperature, T, K; Thermal Conductivity, k, Watts $cm^{-1} K^{-1}$]

T	k
CURVE 1	
2.31	0.016
3.04	0.028
3.93	0.053
5.92	0.134
8.49	0.250
11.71	0.376
15.04	0.534
21.54	0.725
29.56	0.808
65.23	0.436
81.02	0.286
90.26	0.245
198.60	0.077
261.63	0.074
300.51	0.071
CURVE 2	
2.34	0.0773
3.03	0.125
4.50	0.274
5.64	0.454
8.27	0.917
12.91	1.92
14.42	2.31
20.51	2.93
27.21	2.80
36.65	1.93
43.32	1.36
54.98	0.751
65.38	0.458
79.85	0.278
90.90	0.223
102.25	0.167
111.07	0.143
118.43	0.127
126.55	0.074
137.47	0.071
175.63	0.070
195.80	0.071
303.67	0.070
CURVE 3	
3.04	0.00081*
4.97	0.00224*
9.60	0.0077
16.26	0.019
30.25	0.036
44.37	0.0435
65.70	0.044
99.55	0.039
120.16	0.039
143.21	0.041
188.29	0.046
302.36	0.056
CURVE 4	
317.1	0.0444
335.7	0.0435
353.9	0.0435
385.6	0.0431
453.2	0.0414

*Not shown on plot

SPECIFICATION TABLE NO. 26 THERMAL CONDUCTIVITY OF LITHIUM OXIDE Li_2O

Curve No.	Ref. No.	Method Used	Year	Temp. Range, K	Reported Error, %	Name and Specimen Designation	Composition (weight percent), Specifications and Remarks
1	3		1953	333.2		β-Spodumene	Maximum water absorption 0.05%; flexural strength 7000 psi.

DATA TABLE NO. 26 THERMAL CONDUCTIVITY OF LITHIUM OXIDE Li_2O

[Temperature, T, K; Thermal Conductivity k, Watt $cm^{-1} K^{-1}$]

T	k
CURVE 1*	
333.2	0.0173

* No graphical presentation

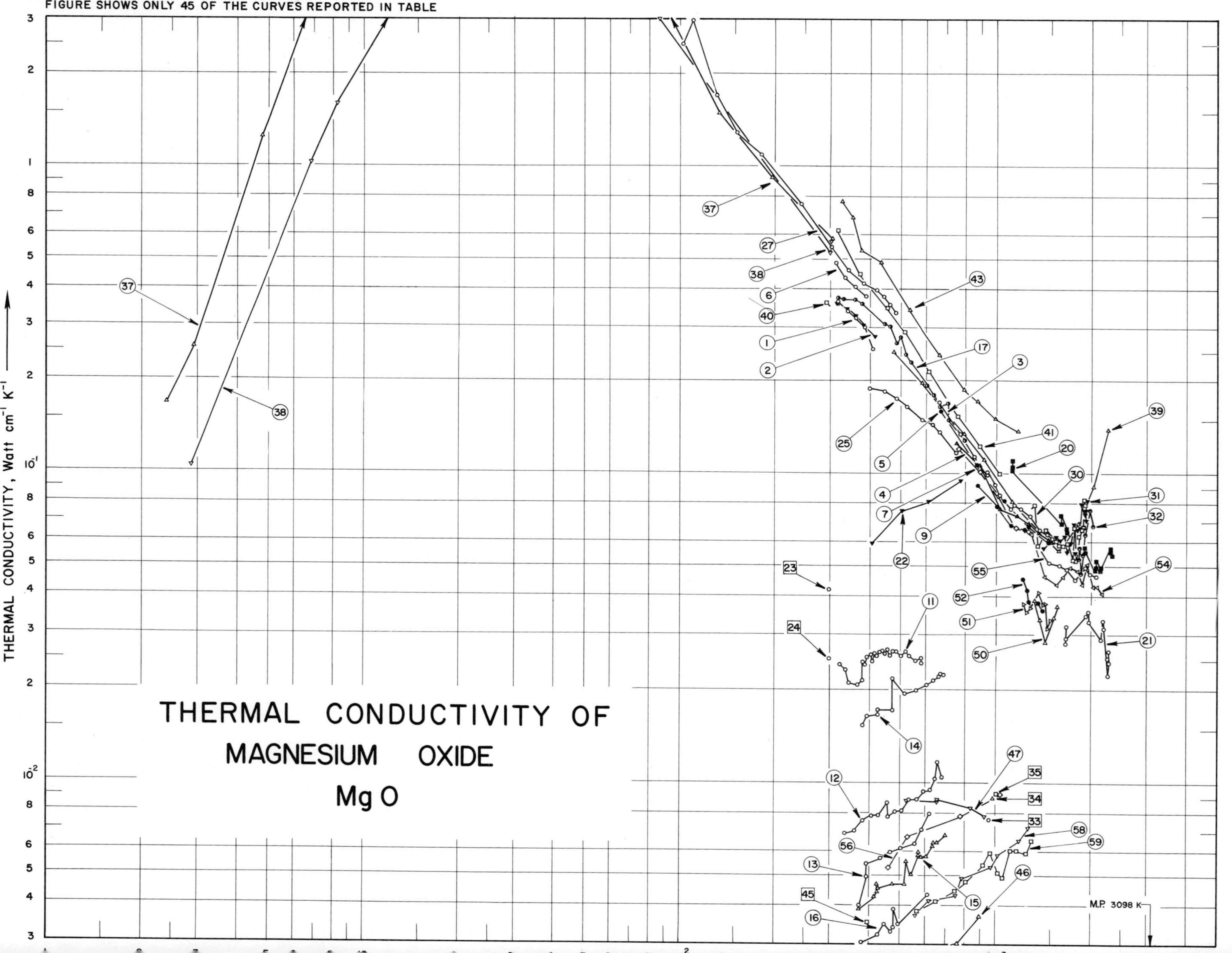

FIGURE SHOWS ONLY 45 OF THE CURVES REPORTED IN TABLE
THERMAL CONDUCTIVITY OF
MAGNESIUM OXIDE
Mg O
THERMAL CONDUCTIVITY, Watt cm⁻¹ K⁻¹
M.P. 3098 K

SPECIFICATION TABLE NO. 27 THERMAL CONDUCTIVITY OF MAGNESIUM OXIDE MgO

[For Data Reported in Figure and Table No. 27]

Curve No.	Ref. No.	Method Used	Year	Temp. Range, K	Reported Error, %	Name and Specimen Designation	Composition (weight percent), Specifications and Remarks
1	3	C	1953	319-411		236A-1	Spectroscopically pure; polycrystalline; supplied by Bell Telephone Laboratories; density (25 C) = 3.21 g cm^{-3}; water absorption = 0.83%.
2	3	C	1953	315-419		236A-2	Same sample as above; separate run.
3	2	R	1951	666-1520			Polycrystalline; 0.30 SiO_2, 0.14 Al_2O_3, 0.35 CaO, 0.05 Fe_2O_3; hydrostatically pressed (30,000 psi); test run No. 1.
4	2	R	1951	751-1517			Same specimen as above; test run No. 2.
5	4	R	1951	673-1473	5		Sintered.
6	23	C	1952	312-388		83A-1	Single crystal.
7	57	R	1950	764-1611		A	Sintered; open pores 0.49%, closed pores 8.44%, total porosity 8.93% bulk density 3.26 g cm^{-3}.
8	57	R	1950	789-1619		A	The second run of the above specimen.
9	57	R	1950	876-1633		B	Sintered; without open pores, closed pores 8.66%, total porosity 8.66%; bulk density 3.27 g cm^{-3}.
10	293	C	1957	707-1692			Single crystal; 99.92 pure, 0.02 SiO_2, 0.05 total Na_2O, CaO, and K_2O, 0.01 Fe_2O_3; data corrected to zero porosity.
11	166	R	1953	324-583			Powder form; 0.64 volume fraction occupied by solid particles; measured in helium.
12	166	R	1953	335-679			Similar to the above specimen except measured in air.
13	166	R	1953	371-620			Similar to the above specimen except measured in argon.
14	239	R	1952	380-683			Powder form; 0.58 fractional volume occupied by the solid particles; measured in helium; gas pressure in range where pressure change does not affect conductivity of powder.
15	239	R	1952	372-694			Similar to the above specimen except measured in air.
16	239	R	1952	378-611			Similar to the above specimen except measured in argon.
17	131	C	1954	318-896			Slip cast from suspensions of finely ground material; fired to zero apparent porosity; bulk density 3.48 g cm^{-3}.
18	131	C	1954	316-898			The second run of the above specimen.
19	240	R	1919	429, 453			Sample size: 12.2 mm I.D., 19 mm O.D., and 8 cm long.
20	144	R	1963	1133-2322	5-7	1	0.3 Fe, 0.3 Si, 0.2 Ca, 0.1 Al; poorly bonded structure; supplied by Zirconium Corp. of America; 0.75 in. long, 0.75 in. O.D., and 0.25 in. I.D.; pressed and sintered; density 3.51 g cm^{-3}; specimen broken and partially melted on post test inspection.
21	144	R	1963	1668-2269	5-7		Similar to the above specimen except dimensions; 3 in. long, 2.5 in. O.D., and 0.75 in. I.D.; specimen found cracked and color changed on post test inspection.

SPECIFICATION TABLE NO. 27 (continued)

Curve No.	Ref. No.	Method Used	Year	Temp. Range, K	Reported Error, %	Name and Specimen Designation	Composition (weight percent), Specifications and Remarks
22	34	C	1943	409-772		Periclase	Synthetic, colorless, isotropic 1 cm cube; cut from a single crystal.
23	152	L	1956	298.2		MgO	Specimen in the form of wafers made of 4 pieces, each 0.75 in. dia x 0.02 in. thick.
24	152	L	1956	298.2		MgO	The above specimen exposed with 3 x 10^{19} epithermal neutrons per cm^2 for 480 megawatt day in the Material Testing Reactor.
25	153	C	1951	400-756	± 5		
26	211	C, R	1954	573-1473			
27	285	P	1962	103-483	± 5	MgO	Single crystal.
28	274	R	1953	653-1513	< 5	Run 1	Fired to total porosity of 8.10-8.93% to a bulk density of 3.25-3.29 g cm^{-3}.
29	274	R	1953	748-1523	< 5	Run 2	Same material as the above; separate run.
30	274	R	1953	1283-1908	<10.7	M-10	0.05 Fe_2O_3, 0.30 SiO_2, 0.14 Al_2O_3, and 0.35 CaO; slip cast; fired to zero apparent porosity at 1850 C; final total porosity 5-10%.
31	274	R	1953	1423-1953	<10.7	M-11	Same material as the above, separate run.
32	274	R	1953	1573-2023	<10.7	M-12	Same material as the above; spearate run.
33	299	R	1958	951.7	±10	Periclase	Compressed granular specimen; porosity 23.5%.
34	299	R	1958	980.7	±10	Periclase	Compressed granular specimen; porosity 15.2%.
35	299	R	1958	1000.7	±10	Periclase	Compressed granular specimen; porosity 13.7%.
36	302	P	1962	313-1263			Density 3.30 g cm^{-3}.
37	300	L	1962	2.4-305		R-38	Single synthetic crystal; transparent, colorless, well-formed and free of visible defects; grown from the melt in an arc furnace using carbon electrodes and a self-crucible technique by R. L. Hansler, Lamp Division, General Electric Co.; specimen 1.11 cm long and 0.28 cm average dia; 4.213 Angstrom lattice constant; impurity (e_{pj} in atms cm^{-3}) given in $\log_{10} e_{pj}$ = 18.9 Al, 18.4 Be, 18.7 Ca, 17.6 Cr; 17.6 Fe, <18.4 K, 17.5 Li, <17.3 Mn, 18.7 Na, <17.2 Ni, 17.9 Si, <17.6 Ti, <17.3 V, <17.8 Zn, <17.7 Zr.
38	300	L	1962	2.9-300		R-14	Similar to above specimen except 1.24 cm long and 0.41 cm average dia; impurity (e_{pj} in atms cm^{-3}) given in $\log_{10} e_{pj}$ = 18.4 Al, 18.4 Be, 18.3 Ca, 17.6 Cr, 17.6 Fe, 18.4 K, 17.5 Li, <17.3 Mn, 18.7 Na, <17.2 Ni, 18.7 Si, <17.6 Ti, <17.3 V, <17.8 Zn, 18.1 Zr.
39	298	R	1961	478-2264			99^+ MgO, 0.5 > Si, 0.3 > Mn; specimen composed of 5 one-in. disks; density 2.98 g cm^{-3}
40	296		1958	293-1273			Sintered; 97% of theoretical density; with an average grain size of about 15 microns.
41	293	C	1957	319-1036	±4	No. 1	Polycrystalline; 99.54 pure, 0.20 SiO_2; 0.12 Al_2O_3, 0.11 total Na_2O, CaO, and K_2O, 0.02 Fe_2O_3, 0.01 TiO_2; gravimetric porosity 4.75%; microscopic porosity 6%; average grain size 8 μ; data corrected to zero porosity; polycrystalline alumina used as comparative material.

SPECIFICATION TABLE NO. 27 (continued)

Curve No.	Ref. No.	Method Used	Year	Temp. Range, K	Reported Error, %	Name and Specimen Designation	Composition (weight percent), Specifications and Remarks
42	293	C	1957	373-1073	± 4	No. 2	Polycrystalline; 99.16 pure, 0.30 SiO_2, 0.14 Al_2O_3, 0.35 total Na_2O, CaO, and K_2O, 0.05 Fe_2O_3, trace TiO_2; gravimetric porosity 13.7%; microscopic porosity 14%; average grain size 12μ; data corrected to zero porosity; same comparative material as above.
43	293	C	1957	328-1168	± 4		Single crystal; 99.92 pure, 0.02 SiO_2, 0.05 total Na_2O, CaO, andK_2O, 0.01 Fe_2O_3; data corrected to zero porosity; same comparative material as above.
44	314	L	1959	395.9			Compressed powder; calcined; manufactured by Fisher Scientific Co.; specimen of 0.152 in. thick and 9 in. dia; pressed at 63 psi; bulk density 0.73 g cm^{-3}; heat flow parallel to the axis of the specimen; load reduced to 0.5 psi prior to making measurements.
45	314	L	1959	398.2			Same as the above specimen except 0.089 in. thick, 1.25 g cm^{-3} bulk density, and 940 psi load.
46	315	R	1948	380-894			Powder; $\leq$200 mesh; porosity 47%; tested in air atmosphere.
47	315	R	1948	525-922			Same as the above specimen except in hydrogen atmosphere.
48	316	C	1951	519-755			Thermal conductivity of alumina determined by ellipsoidal sample used as reference; a crack was discovered in the sample.
49	316	R	1951	665-1673	± 5		Data from 6 runs with 3 specimens.
50	317	R	1960	1332-1565		MgO FPM-1	Fused MgO; 98.9 overall purity; composition: 45% fused MgO with -40 + 60 mesh and 99.0 purity, 15% fused MgO with -80 +100 mesh and 99.0 purity, 35% fused Mgo with -325 mesh and 98.5 purity, and 5% precipitated MgO with 0.02-0.03 μ, and 99.5 purity; bulk density 2.95 g cm^{-3} (83.5% theoretical density); fabricated at JPL; cold pressed at 30,000 psi and sintered at 1783 K for 3 hrs; run No. 11.
51	317	R	1960	1221-1492		MgO FPM-1	The above specimen run No. 12.
52	317	R	1960	1219-1409		MgO FPM-1	The above specimen run No. 20.
53	317	R	1960	1390-1638		MgO FPM-1	The above specimen run No. 21.
54	318	R	1960	1301-2163			Polycrystalline; right cylinder 3 in. long, 2 in. dia with coaxial hole of 0.5 in. dia; density 3.22 g cm^{-3}, supplied by Norton Co.; all data corrected to theoretical density of 3.58 g cm^{-3}; measured during increasing temperatures.
55	318	R	1960	1358-2122			The above specimen measured during decreasing temperatures.
56	373	R	1959	463-1258	±20	MgO D(10-20)	Powder specimen of dense crystalline particles with porosity no larger than 4% and of irregular shape; particle dia 0.140 ± 0.060 cm, powder density 1.60 g cm^{-3}, volume fraction of particles 0.469, central heater power input 4 watts cm^{-1}.
57	373	R	1959	573-1223	±20	MgO D(10-20)	Similar to the above specimen except central heater power input 2 wats cm^{-1}.

SPECIFICATION TABLE NO. 27 (continued)

Curve No.	Ref. No.	Method Used	Year	Temp. Range, K	Reported Error, %	Name and Specimen Designation	Composition (weight percent), Specifications and Remarks
58	373	R	1959	473-1273	±20	MgO D(35-50)	Powder specimen of dense crystalline particles with porosity no larger than 4%, and of irregular shape; particle dia 0.040 ± 0.010 cm; powder density 1.70 g cm^{-3}, volume fraction of particles 0.475.
59	373	R	1959	473-1308	±20	MgO D(100-200)	Powder specimen of dense crystalline particles with porosity no larger than 4%, and of irregular shape, mesh of the sieves limiting particle size 100-200, particle dia 0.011 ± 0.004 cm; powder density 1.54 g cm^{-3}, volume fraction of particles 0.430.
60	368	R	1954	460-1511	± 5		Hollow prolate spheroidal specimen with innor minor axis ~2 cm, inner major axis ~10 cm, outer minor axis ~4 cm and outer major axis long enough to make the outer and inner surfaces confocal; prepared by slip casting from suspension of finely ground material in the M.I.T. Ceramics Lab., fired to a total porosity of 8.10 to 8.93%, bulk density of 3.26 to 3.29 g cm^{-3}.
61	368	R	1954	750-1506	± 5		The above specimen, run 2.
62	400	P	1966	375-810	±11	MgO(E-98)	0.01-0.1 Al, 0.01-0.1 Fe, 0.01-0.1 Si principal impurities; powder specimen contained in a ~4 in. in I.D., 4.5 in. in O.D. and ~24 in. long container; supplied by Norton Co.; as received; volume fraction solid 0.58; thermal conductivity data calculated from the measurement of thermal diffusivity, specific heat and density; mean particle size 268 μ; pore-size density 3.59 g cm^{-3}.
63	400	P	1966	373-842	±11	MgO(E-98)	Similar to the above specimen except specimen packed to 0.64 volume fraction solid.
64	400	R	1966	380-1101	±10	MgO(E-98)	Similar to the above specimen except thermal conductivity data obtained by different method and volume fraction solid 0.58.
65	400	R	1966	369-1099	±10	MgO(E-98)	Similar to the above specimen except specimen packed to 0.61 volume fraction solid.
66	400	R	1966	366-1113	±10	MgO(E-98)	Similar to the above specimen except specimen packed to 0.64 volume fraction solid.
67	400	R	1966	368-1094	±10	MgO(E-98)	Similar to the above specimen except specimen packed to 0.65 volume fraction solid.
68	400	R	1966	371-1095	±10	MgO(E-227)	Similar to the above specimen except mean particle size 369 μ and were packed to 0.61 volume fraction solid; pore-free density 3.58 g cm^{-3}.
69	400	R	1966	378-1022	±10	MgO(E-98)	Similar to the above specimen except mean particle size 268 μ and were packed to 0.58 volume fraction solid; run 34; pore-free density 3.59 g cm^{-3}.
70	400	R	1966	718-1099		MgO(E-98)	Similar to the above specimen except different run, No. 26.

DATA TABLE NO. 27 THERMAL CONDUCTIVITY OF MAGNESIUM OXIDE MgO

[Temperature, T, K; Thermal Conductivity, k, Watt cm^{-1} K^{-1}]

T	k
CURVE 1	
318.5	0.361
341.3	0.338
361.2	0.320
387.5	0.299
410.7	0.252
CURVE 2	
315.4	0.359
340.7	0.343
362.3	0.322
383.2	0.303
418.6	0.278
CURVE 3	
666.2	0.170
773.2	0.134
858.2	0.113
942.2	0.0987
942.2	0.101
993.2	0.0916
1036.2	0.0845
1121.2	0.0761
1150.2	0.0787
1202.2	0.0761
1291.2	0.0724
1382.2	0.0653
1464.2	0.0632
1520.2	0.0615
CURVE 4	
751.2	0.125
890.2	0.106
1042.2	0.0761
1173.2	0.0724
1281.2	0.0686
1380.2	0.0640
1473.2	0.0607
1517.2	0.0598

T	k
CURVE 5	
673.2	0.1608
873.2	0.1079
1073.2	0.0817
1273.2	0.0680
1473.2	0.0601
CURVE 6	
311.7	0.485
334.8	0.435
359.6	0.406
387.6	0.377
CURVE 7	
764.2	0.120
891.2	0.1004
1037.2	0.0849
1158.2	0.0665
1273.2	0.0649
1368.2	0.0649*
1453.2	0.0636*
1544.2	0.0615*
1611.2	0.0590*
CURVE 8*	
789.2	0.106
911.2	0.0929
1022.2	0.0828
1104.2	0.0695
1193.2	0.0636
1267.2	0.0590
1354.2	0.0548
1438.2	0.0523
1493.2	0.0515
1553.2	0.0510
1619.2	0.0510
CURVE 9	
876.2	0.0907
1011.2	0.0774
1120.2	0.0674
1262.2	0.0657

T	k
CURVE 9 (cont.)	
1366.2	0.0602*
1477.2	0.0569*
1536.2	0.0561*
1567.1	0.0577*
1633.2	0.0565*
CURVE 10*	
707.2	0.146
851.2	0.111
956.2	0.103
1055.2	0.0883
1143.2	0.0795
1223.2	0.0745
1299.2	0.0682
1368.2	0.0640
1425.2	0.0615
1454.2	0.0619
1505.2	0.0606
1534.2	0.0577
1591.2	0.0585
1614.2	0.0598
1670.2	0.0590
1692.2	0.0582
CURVE 11	
323.5	0.0239
337.2	0.0232
344.5	0.0209
365.9	0.0206
378.0	0.0213
381.1	0.0244
386.7	0.0241
392.6	0.0253
402.3	0.0258
409.5	0.0246
415.4	0.0260
422.4	0.0256
431.6	0.0263
439.5	0.0265
447.2	0.0260
453.8	0.0267
464.2	0.0256

T	k
CURVE 11 (cont.)	
475.3	0.0265
487.0	0.0265
502.0	0.0256
520.2	0.0265
535.8	0.0256
559.9	0.0247
582.7	0.0251
582.7	0.0244
CURVE 12	
335.3	0.00684
359.4	0.00699
380.2	0.00755
407.9	0.00786
429.1	0.00786
456.4	0.00862
459.0	0.00774
482.4	0.00805
504.3	0.00815
535.2	0.00884
565.1	0.00886
592.5	0.00933
622.4	0.00943
645.8	0.0103
658.4	0.0117
679.0	0.0105
CURVE 13	
371.1	0.00398
392.3	0.00492
393.2	0.00542
435.6	0.00568
466.9	0.00594
502.0	0.00611
557.3	0.00633
584.0	0.00704
620.4	0.00796
CURVE 14	
380.4	0.0154
394.3	0.0164

T	k
CURVE 14 (cont.)	
427.6	0.0166
427.6	0.0171
472.1	0.0173
474.8	0.0216
519.3	0.0194
560.9	0.0199
605.4	0.0207
635.9	0.0214
666.5	0.0221
669.3	0.0224
683.2	0.0224
CURVE 15	
372.1	0.00389
416.5	0.00424
422.1	0.00441
422.1	0.00467
427.6	0.00450
472.1	0.00467
516.5	0.00467
522.1	0.00554
544.3	0.00502
572.1	0.00580
572.1	0.00597
583.2	0.00571
605.4	0.00571
633.2	0.00623
638.7	0.00640
655.4	0.00640
694.3	0.00675
CURVE 16	
377.6	0.00303
424.8	0.00320
449.8	0.00346
466.5	0.00329
477.6	0.00338
480.4	0.00389
494.3	0.00346
610.9	0.00433

T	k
CURVE 17	
318.2	0.371
330.2	0.370
359.2	0.368
378.2	0.358
448.2	0.312
460.2	0.306
483.2	0.265
498.2	0.279
516.2	0.241
536.2	0.228
603.2	0.192
636.2	0.179
659.2	0.165
703.2	0.169
706.2	0.149*
710.2	0.148
786.2	0.128
896.2	0.107*
CURVE 18*	
316.2	0.365
333.2	0.361
360.2	0.330
376.2	0.332
446.2	0.296
463.2	0.290
480.2	0.252
498.2	0.277
513.2	0.229
603.2	0.182
633.2	0.169
663.2	0.157
703.2	0.160
706.2	0.143
716.2	0.142
790.2	0.123
898.2	0.103
CURVE 19	
429.3	0.000795
452.9	0.000753

T	k
CURVE 20	
1133.2	0.110
1133.2	0.105
1133.2	0.104
1605.4	0.0681
1611.5	0.0724
1615.9	0.0737*
1673.7	0.0662
1681.5	0.0664*
1682.1	0.0645
1822.1	0.0535*
1903.7	0.0577
1913.2	0.0577*
1914.3	0.0553
2062.6	0.0488
2069.3	0.0495
2071.5	0.0521
2124.8	0.0485
2130.4	0.0485*
2132.1	0.0498
2310.9	0.0574
2310.9	0.0571*
2310.9	0.0558
2322.1	0.0545
CURVE 21	
1667.6	0.0281
1678.7	0.0290
1685.9	0.0322
1686.5	0.0293*
1939.8	0.0345
1949.8	0.0349*
1954.3	0.0356
1964.8	0.0332
2153.2	0.0289
2168.7	0.0291*
2180.9	0.0332
2189.8	0.0314
2254.8	0.0250
2258.7	0.0222
2260.4	0.0245
2269.3	0.0265

* Not shown on plot

DATA TABLE NO. 27 (continued)

T	k
CURVE 22	
409.2	0.0590
509.2	0.0757
610.2	0.0803
772.2	0.0946
CURVE 23	
298.2	0.0418
CURVE 24	
298.2	0.0251
CURVE 25	
400.2	0.189
448.2	0.185
486.7	0.175
521.2	0.165
589.2	0.148
632.2	0.144
668.2	0.136
756.2	0.116
CURVE 26*	
573.2	0.203
873.2	0.115
1173.2	0.0757
1473.2	0.0611
CURVE 27	
103.2	2.49
111.2	2.97
133.2	1.70
153.2	1.28
183.2	1.09
243.2	0.753
303.2	0.544
343.2	0.460
383.2	0.418
423.2	0.397
443.2	0.377
463.2	0.356
483.2	0.335
CURVE 28*	
653.2	0.171
773.2	0.134
848.2	0.114
943.2	0.102
943.2	0.0996
993.2	0.0920
1023.2	0.0845
1113.2	0.0770
1143.2	0.0795
1193.2	0.0770
1283.2	0.0732
1373.2	0.0661
1458.2	0.0640
1513.2	0.0619
CURVE 29*	
748.2	0.121
893.2	0.107
1033.2	0.0770
1163.2	0.0732
1273.2	0.0703
1373.2	0.0649
1473.2	0.0611
1523.2	0.0598
CURVE 30	
1283.2	0.0686*
1328.2	0.0787
1353.2	0.0586
1448.2	0.0653
1553.2	0.0602
1578.2	0.0586
1643.2	0.0586
1698.2	0.0598
1753.2	0.0527
1773.2	0.0669
1798.2	0.0523
1848.2	0.0628
1873.2	0.0657
1898.2	0.0669
1903.2	0.0816
1908.2	0.0787
CURVE 31	
1423.2	0.0577
1553.2	0.0628
1593.2	0.0607
1653.2	0.0628
1698.2	0.0556
1743.2	0.0598
1768.2	0.0682
1793.2	0.0556
1843.2	0.0661
1858.2	0.0669
1898.2	0.0795
1903.2	0.0736
1923.2	0.0745
1948.2	0.0707
1953.2	0.0816
CURVE 32	
1573.2	0.0628*
1773.2	0.0544*
1813.2	0.0527
1843.2	0.0577
1853.2	0.0473
1863.2	0.0544
1903.2	0.0636
1988.2	0.0753
2023.2	0.0678
CURVE 33	
951.7	0.00761
CURVE 34	
980.7	0.00895
CURVE 35	
1000.7	0.00925
CURVE 36*	
313.2	0.406
373.2	0.326
473.2	0.262
573.2	0.218
CURVE 36 (cont.)*	
673.2	0.188
773.2	0.167
873.2	0.142
973.2	0.126
1073.2	0.109
1173.2	0.0962
1263.2	0.0870
CURVE 37	
2.40	0.167
2.93	0.255
4.81	1.25
6.59	3.05
8.83	7.70
10.39	12.52
13.17	19.61
15.47	24.23
21.40	28.76
27.17	32.37
34.02	28.46
47.02	13.48
49.72	14.96
63.63	8.27
78.19	4.79
88.10	3.46
133.34	1.50
196.93	0.922
304.75	0.584
CURVE 38	
2.88	0.104
6.85	1.03
8.39	1.60
15.02	4.48
15.36	5.23
15.53	5.56
23.44	8.80
24.28	8.91
24.92	9.29
37.70	10.39
52.65	7.81
63.75	6.29
65.10	5.67
65.40	5.67*
83.09	3.40
CURVE 38 (cont.)	
85.19	3.19
86.73	3.036
298.32	0.523
300.30	0.571
CURVE 39	
477.6	0.247
589.8	0.196
703.2	0.154*
795.4	0.134
920.4	0.111
1145.9	0.0801
1361.5	0.0668*
1593.7	0.0562
1816.5	0.0687
2042.6	0.0902
2263.7	0.139
CURVE 40	
293.2	0.360
673.2	0.163*
1273.2	0.0711*
CURVE 41	
319.2	0.619
373.2	0.448
456.2	0.346
516.2	0.290
618.2	0.218
755.2	0.152
890.2	0.121
1036.2	0.0996
CURVE 42*	
373.2	0.360
473.2	0.280
673.2	0.159
873.2	0.113
1073.2	0.088
CURVE 43	
328.2	0.753
353.2	0.686
378.2	0.586
433.2	0.490
533.2	0.342
663.2	0.242
798.2	0.188
873.2	0.172
994.2	0.151
1168.2	0.138
CURVE 44*	
395.9	0.00123
CURVE 45	
398.2	0.00351
CURVE 46	
379.8	0.00218
502.6	0.00270
769.3	0.00301
894.3	0.00368
CURVE 47	
524.8	0.00880
651.0	0.00874
658.7	0.00886
838.7	0.00832
922.1	0.00785
CURVE 48*	
519.2	0.165
679.2	0.135
755.2	0.115
CURVE 49*	
665.2	0.171
715.2	0.146
755.2	0.125
766.2	0.119
CURVE 49 (cont.)*	
775.2	0.135
789.2	0.106
851.2	0.111
859.2	0.113
873.2	0.0912
892.2	0.101
894.2	0.106
912.2	0.0925
942.2	0.0987
946.2	0.100
956.2	0.102
994.2	0.0916
1013.2	0.0774
1023.2	0.0833
1035.2	0.0845
1038.2	0.0837
1043.2	0.0761
1055.2	0.0887
1101.2	0.0690
1121.2	0.0757
1121.2	0.0674
1143.2	0.0795
1153.2	0.0787
1159.2	0.0665
1175.2	0.0720
1196.2	0.0636
1201.2	0.0761
1223.2	0.0745
1263.2	0.0657
1270.2	0.0590
1273.2	0.0644
1281.2	0.0686
1293.2	0.0715
1301.2	0.0682
1356.2	0.0548
1358.2	0.0602
1368.2	0.0640
1369.2	0.0653
1383.2	0.0653
1384.2	0.0640
1426.2	0.0615
1443.2	0.0523
1455.2	0.0635
1456.2	0.0619
1467.2	0.0632
1473.2	0.0606

*Not shown on plot

DATA TABLE NO. 27 (continued)

T	k
CURVE 49 (cont.)*	
1476.2	0.0565
1494.2	0.0515
1505.2	0.0606
1523.2	0.0565
1525.2	0.0615
1537.2	0.0577
1539.2	0.0556
1547.2	0.0573
1555.2	0.0510
1570.2	0.0577
1599.2	0.0586
1615.2	0.0586
1618.2	0.0598
1624.2	0.0510
1639.2	0.0561
1673.2	0.0586
CURVE 50	
1332.2	0.0389
1383.2	0.0337
1443.2	0.0285
1528.2	0.0343
1565.2	0.0372
CURVE 51	
1221.2	0.0379
1251.2	0.0356
1293.2	0.0368
1369.2	0.0412
1407.2	0.0377
1443.2	0.0380
1451.2	0.0316
1492.2	0.0337
CURVE 52	
1219.2	0.0454
1252.2	0.0420
1273.2	0.0386
1288.2	0.0370*
1320.2	0.0368*
1352.2	0.0379*
1357.2	0.0383
1409.2	0.0360
CURVE 53*	
1390.2	0.0343
1478.2	0.0366
1502.2	0.0337
1539.2	0.0326
1603.2	0.0310
1603.2	0.0301
1638.2	0.0314
CURVE 54	
1301.2	0.0635
1438.2	0.0464
1558.2	0.0439
1638.2	0.0464
1713.2	0.0498
1813.2	0.0485
1873.2	0.0439
1948.2	0.0510
2023.2	0.0431
2088.2	0.0431
2163.2	0.0410
CURVE 55	
1358.2	0.0582*
1473.2	0.0515
1593.2	0.0502
1678.2	0.0490
1773.2	0.0452
1903.2	0.0498
1998.2	0.0473
2063.2	0.0464
2122.2	0.0423*
CURVE 56	
463.2	0.00530
633.2	0.00666
783.2	0.00780
1043.2	0.00917
1258.2	0.0101
CURVE 57*	
573.2	0.00582
698.2	0.00634
CURVE 57 (cont.)*	
973.2	0.00760
1223.2	0.00809
CURVE 58	
473.2	0.00338*
563.2	0.00370
623.2	0.00414
753.2	0.00430
788.2	0.00486
973.2	0.00530
1023.2	0.00580
1198.2	0.00650
1273.2	0.00715
CURVE 59	
473.2	0.00330*
573.2	0.00382
653.2	0.00410
748.2	0.00444
811.2	0.00475
918.2	0.00588
973.2	0.00590
1023.2	0.00508
1063.2	0.00490
1123.2	0.00600
1173.2	0.00600
1263.2	0.00589
1308.2	0.00650
CURVE 60*	
460.2	0.171
769.2	0.135
853.2	0.114
943.2	0.102
944.2	0.0996
991.2	0.0929
1033.2	0.0854
1113.2	0.0770
1143.2	0.0795
1193.2	0.0770
1283.2	0.0728
1373.2	0.0661
1453.2	0.0636
1511.2	0.0619
CURVE 61*	
750.2	0.126
883.2	0.107
1035.2	0.0766
1173.2	0.0736
1273.2	0.0690
1373.2	0.0642
1463.2	0.0607
1506.2	0.0598
CURVE 62*	
375.1	0.00433
502.6	0.00502
572.3	0.00552
643.6	0.00604
724.0	0.00661
801.6	0.00689
809.5	0.00666
CURVE 63*	
373.2	0.00635
443.7	0.00717
503.2	0.00786
682.9	0.00981
756.0	0.01005
842.4	0.01018
CURVE 64*	
380.2	0.00420
413.2	0.00431
509.2	0.00529
603.2	0.00567
702.2	0.00632
718.2	0.00658
719.2	0.00643
808.2	0.00690
818.2	0.00673
821.2	0.00698
917.2	0.00739
1016.2	0.00769
1094.2	0.00811
1101.2	0.00801
CURVE 65*	
369.2	0.00495
418.2	0.00535
503.2	0.00628
601.2	0.00681
693.2	0.00766
798.2	0.00828
894.2	0.00874
993.2	0.00884
1099.2	0.00942
CURVE 66*	
366.2	0.00671
431.2	0.00708
568.2	0.00859
622.2	0.00921
711.2	0.01000
716.2	0.01021
814.2	0.01060
915.2	0.01108
1012.2	0.01163
1113.2	0.01158
CURVE 67*	
368.2	0.00720
498.2	0.00892
596.2	0.00997
692.2	0.01086
797.2	0.01158
892.2	0.01176
1094.2	0.01269
CURVE 68*	
371.2	0.00482
433.2	0.00533
528.2	0.00624
628.2	0.00699
699.2	0.00750
724.2	0.00756
799.2	0.00788
799.2	0.00823
897.2	0.00837
898.2	0.00854
898.2	0.00868
998.2	0.00905
CURVE 68 (cont.)*	
998.2	0.00869
1047.2	0.00912
1052.2	0.00886
1087.2	0.00941
1095.2	0.00915
CURVE 69*	
378.2	0.00418
415.2	0.00428
507.2	0.00528
605.2	0.00569
707.2	0.00628
807.2	0.00696
918.2	0.00730
1022.2	0.00781
CURVE 70*	
718.2	0.00657
719.2	0.00665
721.2	0.00648
820.2	0.00680
820.2	0.00700
918.2	0.00742
1015.2	0.00771
1098.2	0.00813
1099.2	0.00802

*Not shown on plot

FIGURE AND TABLE NO. 27R RECOMMENDED THERMAL CONDUCTIVITY OF MAGNESIUM OXIDE MgO

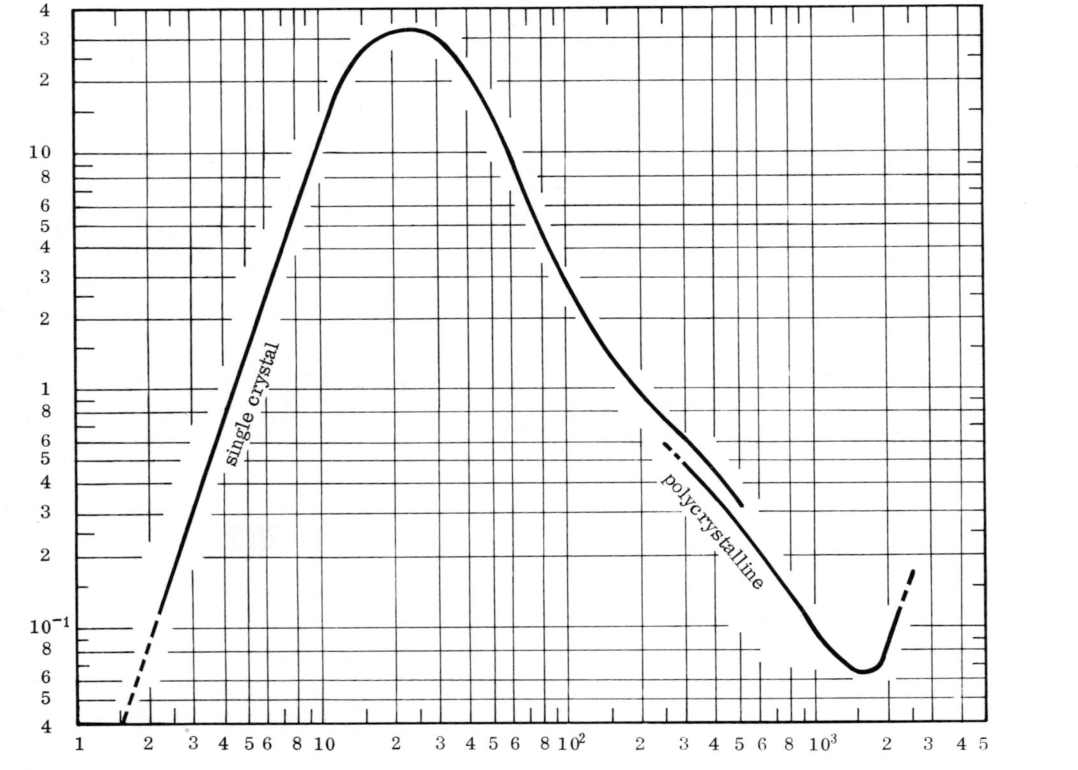

RECOMMENDED VALUES*
(99.96% Pure Single Crystal)

T_1	k_1	k_2	T_2
0	0	0	-459.7
1	(0.012) ‡	(0.693)	-457.9
5	1.55	89.6	-450.7
10	11.7	676	-441.7
15	26.9	1550	-432.7
20	32.5	1880	-423.7
25	33.1	1910	-414.7
30	30.9	1790	-405.7
35	26.5	1530	-396.7
40	21.6	1250	-387.7
50	14.2	820	-369.7
60	9.3	537	-351.7
70	6.3	364	-333.7
80	4.5	260	-315.7
90	3.4	196	-297.7
100	2.7	156	-279.7
150	1.35	78.0	-189.7
200	0.94	54.3	- 99.7
250	0.73	42.2	- 9.7
273.2	0.665	38.4	32.0
300	0.600	34.7	80.3
350	0.507	29.3	170.3
400	0.431	24.9	260.3
450	0.37	21.4	350.3
500	0.32	18.5	440.3

REMARKS

The recommended values for 99.96% pure single crystal are thought to be accurate to within 10 to 15% of the true values at temperatures above 60 K, and the values below 60 K are intended as typical values for indicating the general trend.

The recommended values for 99.5% pure, 98% dense, polycrystalline MgO are thought to be accurate to within 8% of true values at temperatures from 500 to 1000 K and 8 to 15% at other temperatures.

* T_1 in K, k_1 in Watt cm^{-1} K^{-1}, T_2 in F, and k_2 in Btu hr^{-1} ft^{-1} F^{-1}.

‡ Values in parentheses are extrapolated.

TABLE NO. 27R (continued)

99.5% Pure, 98% Dense, Polycrystalline Mgo

T_1	k_1	k_2	T_2
0	0	0	-459.7
250	(0.58)[‡]	(33.5)	- 9.7
273.2	(0.53)	(30.6)	32.0
300	0.484	28.0	80.3
350	0.412	23.8	170.3
400	0.356	20.6	260.3
500	0.269	15.5	440.3
600	0.207	12.0	620.3
700	0.165	9.53	800.3
800	0.134	7.74	980.3
900	0.112	6.47	1160
1000	0.097	5.60	1340
1100	0.085	4.91	1520
1200	0.077	4.45	1700
1300	0.072	4.16	1880
1400	0.068	3.93	2060
1500	0.065	3.76	2240
1600	0.064	3.70	2420
1700	0.064	3.70	2600
1800	0.066	3.81	2780
1900	0.074	4.28	2960
2000	0.085	4.91	3140
2100	0.099	5.72	3320
2200	0.115	6.64	3500
2300	0.132	7.63	3680
2400	(0.150)	(8.67)	3860
2500	(0.170)	(9.82)	4040

[*] T_1 in K, k_1 in Watt cm^{-1} K^{-1}, T_2 in F, and k_2 in Btu hr^{-1} ft^{-1} F^{-1}.

[‡] Values in parentheses are extrapolated.

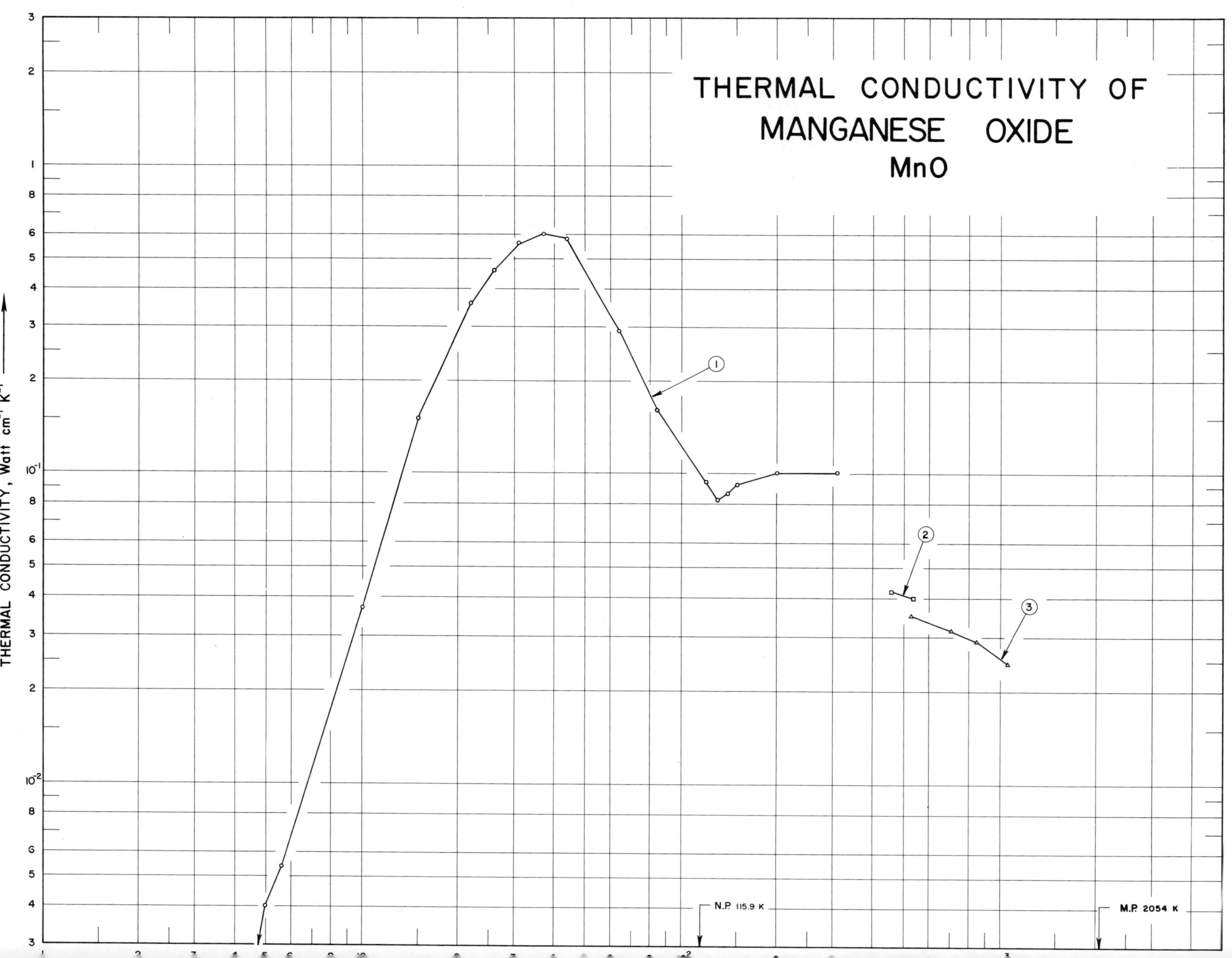

THERMAL CONDUCTIVITY OF
MANGANESE OXIDE
MnO
THERMAL CONDUCTIVITY, Watt cm^{-1} K^{-1}
N.P. 115.9 K
M.P. 2054 K
1
2
3

SPECIFICATION TABLE NO. 28 THERMAL CONDUCTIVITY OF MANGANESE OXIDE MnO

[For Data Reported in Figure and Table No. 28]

Curve No.	Ref. No.	Method Used	Year	Temp. Range, K	Reported Error, %	Name and Specimen Designation	Composition (weight percent), Specifications and Remarks
1	64	L	1958	4-310	8		Single crystal; grown by the Verneuil process; dia 0.5 cm, length 2 cm; unannealed.
2	323	C	1961	458, 538	12		0.5 in. dia and 0.5 in. long; dense, polycrystalline Al_2O_3 used as comparative material; measured in air.
3	323	C	1961	526-1053	15		0.5 in. dia and 0.5 in. long; fused SiO_2 used as comparative material; measured in air.

DATA TABLE NO. 28 THERMAL CONDUCTIVITY OF MANGANESE OXIDE MnO

[Temperature, T, K; Thermal Conductivity, k, Watt $cm^{-1} K^{-1}$]

T	k	T	k
CURVE 1		CURVE 2	
4.0	0.0011*	458.2	0.0420
5.0	0.0040	538.2	0.0400
5.6	0.0054		
10.0	0.037	CURVE 3	
15.0	0.15		
22.0	0.36	526.2	0.0350
26.0	0.46	703.2	0.0315
31.0	0.56	843.2	0.0290
37.0	0.60	1053.2	0.0245
44.0	0.58		
64.0	0.29		
84.0	0.16		
120.0	0.094		
130.0	0.082		
140.0	0.086		
150.0	0.092		
200.0	0.10		
310.0	0.10		

*Not shown on plot

SPECIFICATION TABLE NO. 29 THERMAL CONDUCTIVITY OF TRIMANGANESE TETRAOXIDE Mn_3O_4

Curve No.	Ref. No.	Method Used	Year	Temp. Range, K	Reported Error, %	Name and Specimen Designation	Composition (weight percent), Specifications and Remarks
1	23		1952	318-400		134A	Fired at 1547 K; water absorption 0.024%; density 4.21 g cm^{-3}.

DATA TABLE NO. 29 THERMAL CONDUCTIVITY OF TRIMANGANESE TETRAOXIDE Mn_3O_4

[Temperature, T, K; Thermal Conductivity, k, Watt $cm^{-1}K^{-1}$]

T	k
CURVE 1*	
317.6	0.0343
341.2	0.0351
360.0	0.0356
381.3	0.0356
399.7	0.0347

* No graphical presentation

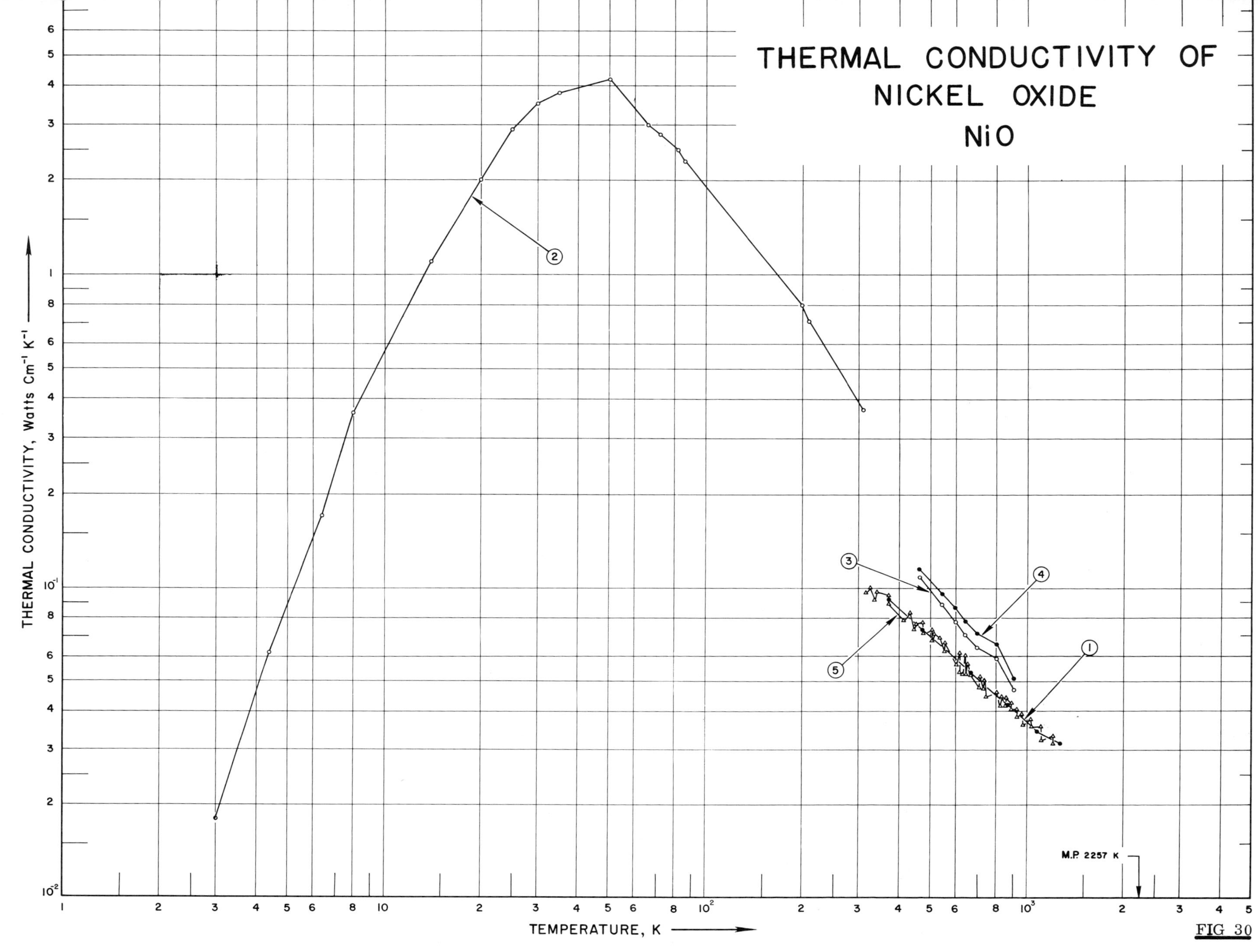
THERMAL CONDUCTIVITY OF
NICKEL OXIDE
NiO
M.P. 2257 K
TEMPERATURE, K
THERMAL CONDUCTIVITY, Watts Cm^{-1} K^{-1}
1
2
3
4
5

FIG 30

SPECIFICATION TABLE NO. 30 THERMAL CONDUCTIVITY OF NICKEL OXIDE NiO

[For Data Reported in Figure and Table No. 30]

Curve No.	Ref. No.	Method Used	Year	Temp. Range, K	Reported Error, %	Name and Specimen Designation	Composition (weight percent), Specifications and Remarks
1	329	C	1953	373-1273			Polycrystal; prepared by calcining chemically pure NiO at 1000 C, hydrostatically pressing, and firing at 1500 C in oxidizing atmosphere; bulk density 5. 05 g cm^{-3}; porosity 25. 7%.
2	64	L	1958	30-310	±8		Single crystal; grown by the Verneuil process; specimen 0. 5 cm dia, 2 cm long.
3	65	C	1958	467-913	±2		Sintered specimen; prepared from calcined (at 1000 C for 1 hr), ball milled (in steel mill with steel balls for 8 hrs) and lubricated (with stearic acid, also used as binder) -325 mesh chemical pure NiO powder; sintered at 1500 C for 4 hrs in an oxidizing atmosphere; bulk density 6. 00 g cm^{-3}; porosity 11. 5%; alumina (Body Al-300) used as standard.
4	65	C	1958	467-913	±2		The above specimen measured with another alumina standard.
5	73	C	1954	318-1209			Bulk density 5. 05 g cm^{-3}; total porosity 32%.

DATA TABLE NO. 30 THERMAL CONDUCTIVITY OF NICKEL OXIDE NiO

[Temperature, T, K; Thermal Conductivity, k, Watt $cm^{-1} K^{-1}$]

T	k
CURVE 1	
373.2	0.0920
473.2	0.0736
673.2	0.0531
873.2	0.0418
1073.2	0.0343
1273.2	0.0314
CURVE 2	
3.0	0.018
4.4	0.062
6.4	0.17
8.0	0.36
14	1.1
20	2.0
25	2.9
30	3.5
35	3.8
50	4.2
66	3.0
72	2.8
82	2.5
86	2.3
200	0.80
210	0.71
310	0.37
CURVE 3	
467.2	0.108
546.2	0.0883
597.2	0.0778
643.2	0.0703
700.2	0.0640
806.2	0.0590
913.2	0.0469
CURVE 4	
467.2	0.116
546.2	0.0962
597.2	0.0862
643.2	0.0782
700.2	0.0711
CURVE 4 (cont.)	
806.2	0.0657
913.2	0.0510
CURVE 5	
318.2	0.0966
326.2	0.100
336.2	0.0914
343.2	0.0975
373.2	0.0949
373.2	0.0886
414.2	0.0787
432.2	0.0833
446.2	0.0732
451.2	0.0764
476.2	0.0771
478.2	0.0715
509.2	0.0736
509.2	0.0677
516.2	0.0711
536.2	0.0690
552.2	0.0624
557.2	0.0669
603.2	0.0562
619.2	0.0616
620.2	0.0533
635.2	0.0628
646.2	0.0602
647.2	0.0525
657.2	0.0569
673.2	0.0527
712.2	0.0475
713.2	0.0519
738.2	0.0473
740.2	0.0502
749.2	0.0446
803.2	0.0461
827.2	0.0418
832.2	0.0447
862.2	0.0418
863.2	0.0444
895.2	0.0427
896.2	0.0403
931.2	0.0408
CURVE 5 (cont.)	
933.2	0.0381
968.2	0.0392
972.2	0.0360
1035.2	0.0377
1038.2	0.0356
1112.2	0.0356
1114.2	0.0323
1208.2	0.0315
1209.2	0.0334

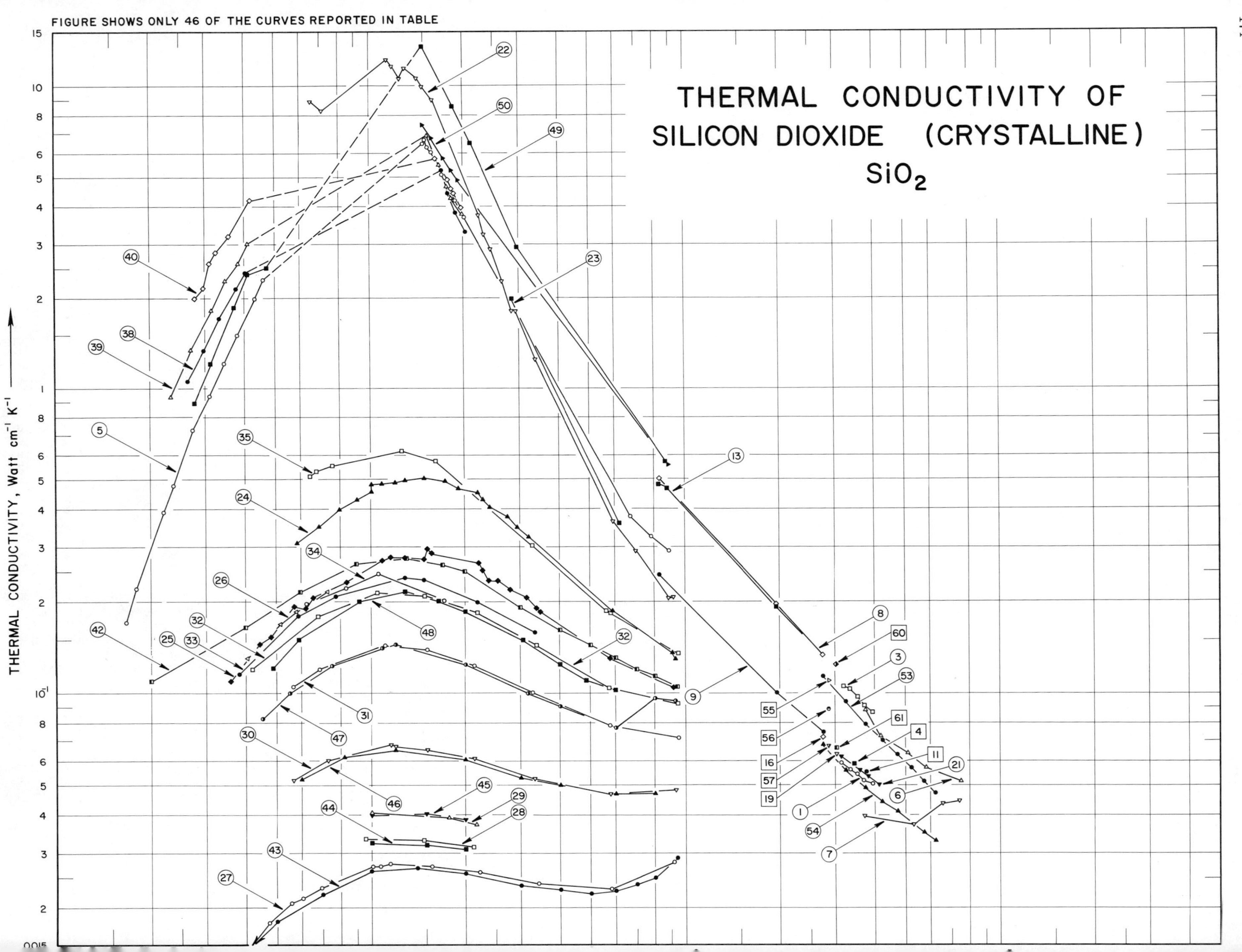

FIGURE SHOWS ONLY 46 OF THE CURVES REPORTED IN TABLE
THERMAL CONDUCTIVITY OF
SILICON DIOXIDE (CRYSTALLINE)
SiO_2
THERMAL CONDUCTIVITY, Watt cm^{-1} K^{-1}
15
10
8
6
5
4
3
2
1
8
6
5
4
3
2
10^{-1}
8
6
5
4
3
2
0.015

SPECIFICATION TABLE NO. 31 THERMAL CONDUCTIVITY OF SILICON DIOXIDE (CRYSTALLINE) SiO_2

[For Data Reported in Figure and Table No. 31]

Curve No.	Ref. No.	Method Used	Year	Temp. Range, K	Reported Error, %	Name and Specimen Designation	Composition (weight percent), Specifications and Remarks
1	23, 35	C	1952	314-394		162A-2	Clear; ground and polished; free from twinning and inclusions; the base of the 0.5 in. cylindrical specimen coated with a special silver paste (No. 45a); c-axis perpendicular to direction of heat flow.
2	23, 35	C	1952	313-390		162B-2	Similar to the above specimen.
3	23, 35	C	1952	318-395		162C-2	Similar to the above specimen except c-axis parallel to direction of heat flow.
4	13	C	1953	343.2	± 3		Measured perpendicular to optical axis; density = 2.6 g cm^{-3}; Armco iron used as the comparative standard.
5	32	L	1937	1.7-89			Length 3.20 cm and dia 0.216 cm, measured perpendicular to the principal axis and in the direction of a binary axis.
6	34	C	1943	373-748		domestic (USA)	1-cm colorless cube; cut from a single crystal; heat flow direction parallel to the c-axis; 18-8 stainless steel used as comparative material.
7	34	C	1943	370-741		domestic (USA)	The above specimen measured with heat flow direction perpendicular to the c-axis.
8	22	L	1911	83-373		1	Single crystal; 3.00 x 3.00 x 2.60 cm; measured parallel to the principal axis in hydrogen.
9	22	L	1911	83-273		1	Above specimen measured perpendicular to the principal axis in hydrogen.
10	22	L	1911	195, 273		1	Same as above.
11	22	L	1911	373.2		1	Same as the above specimen except measured in carbon dioxide.
12	22	L	1911	373.2		1	Same as above.
13	22	L	1911	83-273		2	Single crystal; 3.00 x 3.00 x 2.08 cm; measured parallel to the principal axis in hydrogen.
14	22	L	1911	83-273		2	Same as above.
15	22	L	1911	83, 273		2	Above specimen measured perpendicular to the principal axis in hydrogen.
16	22	L	1911	273.2		2	Same as above.
17	22	L	1911	83-273		3	Single crystal; 3.00 x 3.00 x 2.00 cm; measured perpendicular to the principal axis in hydrogen.
18	22	L	1911	273, 373		3	Same as above.
19	222	L	1959	123-323	± 2～± 5		Single crystal; heat-flow at 90 degrees to the optic axis.
20	68	C	1954	314-415	± 3	162D-1	Natural single crystal of hexagonal type; measured along c-axis.
21	68	C	1954	315-413	± 3	162E-1	The above specimen measured along a-axis.
22	69	L	1951	6.6-92		No. 1	Single crystal; 3.05 x 0.50 x 0.50 cm; measured perpendicular to the principal axis.
23	69	L	1951	28, 62		No. 2	Similar to the above specimen except length 2.15 cm.

SPECIFICATION TABLE NO. 31 (continued)

Curve No.	Ref. No.	Method Used	Year	Temp. Range, K	Reported Error, %	Name and Specimen Designation	Composition (weight percent), Specifications and Remarks
24	69	L	1951	5.8-93			The above specimen No. 1 irradiated in the Harwell pile with a dose of 1.8×10^{18} thermal neutron cm^{-2}.
25	69	L	1951	3.6-92			Same as the above specimen but irradiated with a dose of 2.5×10^{18} thermal neutron cm^{-2}.
26	69	L	1951	4.0-7.3			The above specimen subsequently heated at 100 C for 3 weeks.
27	69	L	1951	3.3-92			The above specimen irradiated with a dose of 29.7×10^{18} thermal neutron cm^{-2}.
28	69	L	1951	9.6-21			The above specimen heated at 300 C for 8 hrs.
29	69	L	1951	10-22			The above specimen heated at 400 C for 6 hrs.
30	69	L	1951	5.7-93			The above specimen heated at 510 C for 6 hrs.
31	69	L	1951	5.6-95			The above specimen heated at 565 C for 6 hrs.
32	69	L	1951	4.2-94			The above specimen heated at 540 C for 60 hrs.
33	69	L	1951	3.8-34			The above specimen heated at 540 C for 677 hrs.
34	69	L	1951	6.3-22			The above specimen heated at 600 C for 1 hr.
35	69	L	1951	6.5-95			The above specimen heated at 700 C for 6 hrs.
36	125	L	1923	298.2	1		Single crystal 0.253 cm thick with heat flow perpendicular to the principal axis of the crystal and measured under pressure of 21 lb $in.^{-2}$.
37	125	L	1923	313.2	1		Single crystal 0.253 cm thick with heat flow parallel to the principal axis of the crystal and measured under pressure of 21 lb $in.^{-2}$.
38	224	L	1938	2.6-20		II A	Length 4.48 cm, dia 0.359 cm; measured perpendicular to the principal axis.
39	224	L	1938	2.3-20		II	Length 4.80 cm, dia 0.454 cm; measured perpendicular to the principal axis.
40	224	L	1938	2.7-20			Length 4.40 cm, dia 0.775 cm; measured perpendicular to the principal axis.
41	225	L	1938	1.7-20			Specimen dia 0.216 cm; heat flow in the direction of the bisector of the angle between the two binary axes.
42	226	L	1950	2-94			Single crystal; irradiated with cumulative irradiation dose $2.4 \times 1.8 \times 10^{18}$ neutrons cm^{-2}.
43	226	L	1950	3.5-94			The above specimen irradiated again with cumulative irradiation dose $19 \times 1.8 \times 10^{18}$ neutrons cm^{-2}.
44	226	L	1950	10-20			The above specimen annealed at 300 C for 8 hrs.
45	226	L	1950	10-20			The above specimen annealed at 400 C for 6 hrs.
46	226	L	1950	6-92			The above specimen annealed at 510 C for 6 hrs.
47	226	L	1950	4.5-93			The above specimen annealed at 565 C for 6 hrs.

SPECIFICATION TABLE NO. 31 (continued)

Curve No.	Ref. No.	Method Used	Year	Temp. Range, K	Reported Error, %	Name and Specimen Designation	Composition (weight percent), Specifications and Remarks
48	226	L	1950	4.9-93			The above specimen annealed at 540 C for 60 hrs.
49	227,308	L	1936	2.8-87		Rod I	Specimen length 5.0 cm, dia 0.308 cm in the first part of the measurement, later cut to 3.033 cm in length and 0.1336 cm in dia when being measured at lower temperatures; heat flow parallel to the c-axis of the crystal.
50	308, 227	L	1935	15-89		Rod II	Specimen 5.58 cm in length and 0.1356 cm in dia cut out of the same crystal as the above specimen Rod I; heat flow parallel to the c-axis of the crystal.
51	308, 227	L	1935	15-20		Rod II	The above specimen heated at 340 C for 8 hrs.
52	308, 227	L	1935	15-20		Rod II	The above specimen again heated at 570 C for 5 hrs.
53	303	L	1940	273-623			Single crystal; less than 0.01 Cu; measured with heat flow direction parallel to the optic axis; specimen 0.250 in. thick and 1.500 in. dia cut from a single crystal about a foot long with well-developed external form, obtained from the Harvard Mineralogical Museum; the whole crystal perfectly transparent except for a few visible fractures; deviation from the intended orientation less than one degree; density 2.652 g cm^{-3}.
54	303	L	1940	273-623			As above but heat flow direction perpendicular to the optic axis.
55	324		1884	285		I	Heat flow direction parallel to the optic axis.
56	324		1884	285		II	Heat flow direction at 45 degrees with the optic axis.
57	324		1884	285		III	Heat flow direction perpendicular to the optic axis.
58	140	C	1926	341,378			Clear quartz (rock crystal) of 1.25 in. in dia and 0.10 in. thick; the faces of the disc were made optically flat and parallel; heat flow direction parallel to the optic axis; aluminum (99.75 pure) used as comparative material.
59	140	C	1926	373,379			As above but heat flow direction perpendicular to the optic axis.
60	325	C	1892	301.8			Quartz disc 1.91 cm dia and 1.005 cm thick; heat flow direction parallel to the optic axis; brass used as comparative material.
61	325	C	1892	302.8			Quartz disc 1.93 cm dia and 0.811 cm thick; heat flow direction perpendicular to the optic axis; brass used as comparative material.
62	326	R	1952	373-723		I	Powder; derived from coarse grain quartz of about 50% grains of 0.3 mm dia, 40% of 0.6 mm and 10% of 1 mm dia; density 0.54 g cm^{-3}.
63	326	R	1952	423-863		II	Powder; derived from coarse grain quartz of 1.0-1.8 mm dia; density 0.44 g cm^{-3}.
64	326	R	1952	378-893		III	Coarse grains of cylindrical form of 3 mm dia and 3-7 mm long; density 0.45 g cm^{-3}.
65	327	R	1958	313-868		Quartz	Powder; grain size 100-200 μ; density 1.35 g cm^{-3}; measured in a vacuum of 5 x 10^{-5} mm Hg.
66	327	R	1958	453-677		Quartz	Above specimen measured in a vacuum of 0.5 mm Hg.

SPECIFICATION TABLE NO. 31 (continued)

Curve No.	Ref. No.	Method Used	Year	Temp. Range, K	Reported Error, %	Name and Specimen Designation	Composition (weight percent), Specifications and Remarks
67	327	R	1958	370-585		Quartz	Above specimen measured under a pressure of 1 atm.
68	328	P	1957	303.2	<7	Quartz; 7	3.49 cm dia and 0.952 cm thick; heat flow perpendicular to optic axis of crystalline quartz; thermal conductivity values calculated from author's experimental data of thermal diffusivity and specific heat.
69	328	P	1957	303.2	<7	Quartz; 8	Similar to the above specimen but the thickness, 0.635 cm.
70	330	C	1957	298.2	±6		1.375 in. dia x 0.25 in. thick; measured with thermal comparator No. 4; heat flow perpendicular to c-axis.
71	330	C	1957	298.2	±6		Similar to the above specimen but measured with thermal comparator No. 5; loaded with 100 g weight.

DATA TABLE NO. 31 THERMAL CONDUCTIVITY OF SILICON DIOXIDE (CRYSTALLINE) SiO_2

[Temperature, T, K; Thermal Conductivity, k, Watt $cm^{-1} K^{-1}$]

T	k
CURVE 1	
313.5	0.0594
333.6	0.0565
351.2	0.0544
369.9	0.0519
393.8	0.0506
CURVE 2*	
312.5	0.0602
336.0	0.0561
352.5	0.0540
370.0	0.0527
389.9	0.0527
CURVE 3	
318.4	0.1060
333.8	0.1030
352.6	0.0971
370.4	0.0908
394.6	0.0866
CURVE 4	
343.2	0.0586
CURVE 5	
1.67	0.171
1.80	0.220
2.20	0.390
2.37	0.477
2.73	0.728
3.10	0.943
3.45	1.22
3.81	1.50
4.33	1.99
4.62	2.30
14.90	6.49
15.11	6.62
15.38	6.29
15.82	6.03
17.17	5.10
18.80	4.22
CURVE 5 (cont.)	
18.89	4.16*
19.07	4.15*
20.12	3.69
67.6	0.379
78.0	0.325
89.4	0.291
CURVE 6	
373.2	0.0887
418.2	0.0728
509.2	0.0640
581.2	0.0573
748.2	0.0515
CURVE 7	
370.2	0.0395
373.2	0.0397*
533.2	0.0370
656.2	0.0435
741.2	0.0444
CURVE 8	
83.2	0.502
195.2	0.197
273.2	0.133
373.2	0.0900*
CURVE 9	
83.2	0.243
195.2	0.100
273.2	0.0746
CURVE 10*	
195.2	0.0996
273.2	0.0719
CURVE 11	
373.2	0.0552
CURVE 12*	
373.2	0.0554
CURVE 13	
83.2	0.481
88.2	0.467
195.2	0.195*
195.2	0.192
273.2	0.133*
CURVE 14*	
83.2	0.477
88.2	0.464
195.2	0.195
195.2	0.200
273.2	0.141
CURVE 15*	
83.2	0.240
273.2	0.0725
CURVE 16	
273.2	0.0717
CURVE 17*	
83.2	0.246
83.2	0.250
195.2	0.103
273.2	0.0735
CURVE 18*	
273.2	0.0728
373.2	0.0567
CURVE 19	
301.2	0.0630
CURVE 20*	
313.9	0.108
333.9	0.105
359.6	0.0954
382.8	0.0891
414.6	0.0816
CURVE 21	
314.5	0.0619
338.2	0.0582*
357.6	0.0561
381.3	0.0536
412.6	0.0502
CURVE 22	
6.56	8.85
7.11	8.28
11.5	12.2
11.9	11.6
12.6	10.6
13.2	11.4
14.3	10.6
14.9	9.93
16.0	8.97
22.4	3.73
23.2	3.25
24.4	2.89
26.5	2.27
28.4	1.82
29.2	1.82
33.7	1.25
59.2	0.365
69.7	0.290
88.7	0.205
91.8	0.206
CURVE 23	
28.3	1.98
61.9	0.358
CURVE 24	
5.85	0.311
6.89	0.352
7.96	0.398
9.06	0.430
10.1	0.458
10.1	0.484
10.9	0.484
12.0	0.489
12.9	0.496
14.8	0.502
17.3	0.493
19.2	0.469
22.0	0.454
22.9	0.430
24.1	0.407
24.8	0.403*
27.4	0.377
29.4	0.348
31.8	0.324
59.2	0.187
90.8	0.137
93.3	0.130
CURVE 25	
3.57	0.110
4.44	0.146
4.84	0.153
5.73	0.193
6.22	0.190
6.56	0.206
8.41	0.232
10.9	0.272
11.7	0.277
12.9	0.277
14.8	0.275
15.2	0.296
15.7	0.288
22.1	0.266
22.7	0.252
23.9	0.234
25.5	0.234
27.9	0.219
31.3	0.207
CURVE 25 (cont.)	
33.6	0.190
34.7	0.185
57.7	0.130
91.8	0.105
CURVE 26	
4.05	0.131
5.18	0.169
5.81	0.186
7.33	0.217
CURVE 27	
3.33	0.0120*
3.73	0.0135*
4.17	0.0151*
4.72	0.0178
5.55	0.0206
6.04	0.0213
6.90	0.0231
10.1	0.0271
10.7	0.0271
11.5	0.0277
15.6	0.0272
22.2	0.0261
34.0	0.0239
58.2	0.0229
91.8	0.0280
CURVE 28	
9.55	0.0333
14.7	0.0330
21.2	0.0314
CURVE 29	
10.1	0.0407
17.7	0.0392
21.6	0.0372
CURVE 30	
5.66	0.0520
7.30	0.0600
11.6	0.0682
12.0	0.0671
15.1	0.0652
21.4	0.0610
33.1	0.0525
57.8	0.0469
93.3	0.0481
CURVE 31	
5.65	0.107
6.89	0.120
11.2	0.143
15.1	0.139
21.3	0.123
32.7	0.100
57.3	0.0785
94.6	0.0715
CURVE 32	
4.23	0.121
6.82	0.178
10.5	0.214
14.9	0.208
21.9	0.183
33.7	0.143
57.3	0.104
94.4	0.0929
CURVE 33	
3.81	0.117
5.88	0.180
7.75	0.208
12.9	0.239
14.8	0.236
21.9	0.199
33.9	0.158

* Not shown on plot

DATA TABLE NO. 31 (continued)

T	k
CURVE 34	
6.28	0.197
8.36	0.221
10.6	0.247
22.2	0.201
CURVE 35	
6.47	0.511
6.76	0.530
7.59	0.551
12.7	0.622
16.1	0.575
22.0	0.451*
32.8	0.303
56.1	0.186
58.1	0.184
94.8	0.136
CURVE 36*	
298.2	0.0615
CURVE 37*	
313.2	0.102
CURVE 38	
2.64	1.07
2.96	1.34
3.33	1.71
3.77	2.16
4.04	2.43
17.08	5.26
17.97	4.44
18.96	3.82
20.32	3.30
CURVE 39	
2.32	0.936
2.68	1.34
3.13	1.82
3.47	2.28
3.83	2.59
4.10	3.02
CURVE 39 (cont.)	
15.40	6.85
15.99	6.08*
16.75	5.49
17.75	4.68
18.27	4.23
19.81	3.76
CURVE 40	
2.78	1.98
2.97	2.17
3.11	2.61
3.26	2.83
3.59	3.20
4.20	4.20
16.34	5.78
17.32	5.21*
17.84	4.90
18.36	4.59
18.71	4.41
19.73	3.95
CURVE 41*	
1.67	0.171
1.80	0.220
2.20	0.390
2.37	0.477
2.73	0.728
3.10	0.943
3.45	1.22
3.81	1.50
4.33	1.99
4.62	2.30
14.90	6.49
15.11	6.62
15.38	6.29
15.82	6.03
17.17	5.10
18.80	4.22
18.89	4.16
19.07	4.16
20.12	3.69
CURVE 42	
2	0.110
4	0.165
6	0.215
9	0.264
13	0.275
17	0.261
20	0.250
30	0.192
40	0.162
50	0.144
60	0.131
70	0.120
80	0.114
94	0.105
CURVE 43	
3.5	0.0125*
5.0	0.0180
7.0	0.0220
10	0.0263
14	0.0269
20	0.0258
30	0.0235
40	0.0229
50	0.0222
60	0.0227
70	0.0238
80	0.0250
94	0.0290
CURVE 44	
10	0.0325
15	0.0320
20	0.0310
CURVE 45	
10	0.0400
15	0.0401
20	0.0385
CURVE 46	
6	0.0525
CURVE 46 (cont.)	
8.2	0.0620
12	0.0655
15.1	0.0655*
20	0.0610
30	0.0530
40	0.0500
60	0.0470
80	0.0470
92	0.0485*
CURVE 47	
4.5	0.0830
5.5	0.100
7.5	0.123
10.8	0.142
12	0.145
20	0.125
31.7	0.100
40	0.0905
60	0.0770
80	0.0965
93	0.0940
CURVE 48	
4.9	0.122
5.9	0.150
9.2	0.200
12.8	0.215
16.5	0.200
20.0	0.185
30.5	0.150
40	0.125
48.5	0.111
60	0.103
80	0.0970*
93.3	0.0938*
CURVE 49	
2.77	0.893
3.12	1.22
3.71	1.87
4.09	2.39
4.74	3.51
CURVE 49 (cont.)	
14.95	13.3
18.55	8.58
21.07	6.49
29.6	2.94
86.9	0.570
CURVE 50	
14.82	7.46
15.85	6.78
17.35	5.80
18.32	5.27
19.24	4.93
19.70	4.90*
88.90	0.56
CURVE 51*	
14.8	6.88
15.1	6.76
17.1	5.62
17.5	5.52
19.8	4.60
20.1	4.53
CURVE 52*	
15.47	6.06
16.87	5.51
19.57	4.54
20.23	4.37
CURVE 53	
273.2	0.114
323.2	0.094
373.2	0.079
423.2	0.070
473.2	0.063
523.2	0.057
573.2	0.0515
623.2	0.0473
CURVE 54	
273.2	0.068
CURVE 54 (cont.)	
323.2	0.056
373.2	0.049
423.2	0.044
473.2	0.041
523.2	0.037*
573.2	0.035
623.2	0.033
CURVE 55	
285	0.110
CURVE 56	
285	0.0887
CURVE 57	
285	0.0667
CURVE 58*	
341.3	0.0929
377.5	0.0778
CURVE 59*	
343.6	0.0540
378.7	0.0485
CURVE 60	
301.8	0.125
CURVE 61	
302.8	0.0661
CURVE 62*	
373.2	0.00178
413.2	0.00178
423.2	0.00178
443.2	0.00178
483.2	0.00184
503.2	0.00186
CURVE 62 (cont.)*	
523.2	0.00188
553.2	0.00190
568.2	0.00201
588.2	0.00209
623.2	0.00218
673.2	0.00230
693.2	0.00243
723.2	0.00259
CURVE 63*	
423.2	0.00126
438.2	0.00126
448.2	0.00132
463.2	0.00142
493.2	0.00151
513.2	0.00159
583.2	0.00218
663.2	0.00293
723.2	0.00343
793.2	0.00397
863.2	0.00795
CURVE 64*	
378.2	0.00136
403.2	0.00146
438.2	0.00167
458.2	0.00167
488.2	0.00205
533.2	0.00230
553.2	0.00238
588.2	0.00259
598.2	0.00293
663.2	0.00368
713.2	0.00427
793.2	0.00544
893.2	0.00711
CURVE 65*	
313.2	0.000289
373.2	0.000335
473.2	0.000356
571.2	0.000418
617.2	0.000448
667.2	0.000515

*Not shown on plot

DATA TABLE NO. 31 (continued)

T	k
CURVE 65 (cont.)*	
713.2	0.000669
811.2	0.000753
863.2	0.000812
868.2	0.000837
CURVE 66*	
453.2	0.00136
533.2	0.00133
593.2	0.00138
608.2	0.00140
677.2	0.00149
CURVE 67*	
370.2	0.00481
389.2	0.00473
408.2	0.00477
461.2	0.00485
490.2	0.00490
533.2	0.00490
585.2	0.00485
CURVE 68*	
303.2	0.0678
CURVE 69*	
303.2	0.0670
CURVE 70*	
298.2	0.067
CURVE 71*	
298.2	0.067

*Not shown on plot

FIGURE AND TABLE NO. 31R RECOMMENDED THERMAL CONDUCTIVITY OF SILICON DIOXIDE (CRYSTALLINE) SiO_2

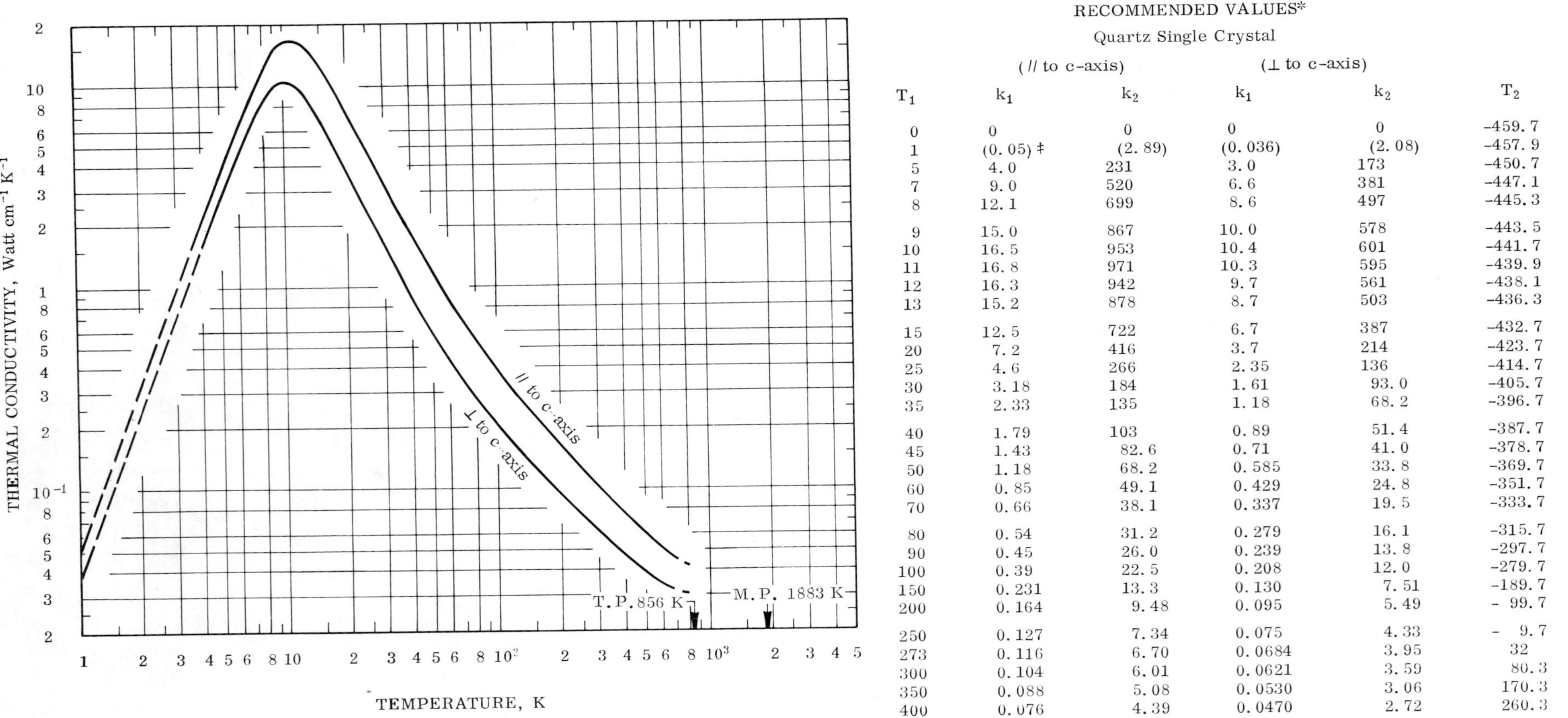

REMARKS

The recommended values are for high-purity quartz single crystal. The recommended values that are supported by experimental data are thought to be accurate to within 5% of the true values at temperatures from 300 to 500 K and 5 to 10% at other temperatures above 20 K. The thermal conductivity near and below the corresponding temperature of its maximum is highly sensitive to small physical and chemical variations of the specimens, and the recommended values below 20 K are intended as typical values for indicating the general trend.

RECOMMENDED VALUES*

Quartz Single Crystal

	(// to c-axis)		(⊥ to c-axis)		
T_1	k_1	k_2	k_1	k_2	T_2
0	0	0	0	0	-459.7
1	(0.05) ‡	(2.89)	(0.036)	(2.08)	-457.9
5	4.0	231	3.0	173	-450.7
7	9.0	520	6.6	381	-447.1
8	12.1	699	8.6	497	-445.3
9	15.0	867	10.0	578	-443.5
10	16.5	953	10.4	601	-441.7
11	16.8	971	10.3	595	-439.9
12	16.3	942	9.7	561	-438.1
13	15.2	878	8.7	503	-436.3
15	12.5	722	6.7	387	-432.7
20	7.2	416	3.7	214	-423.7
25	4.6	266	2.35	136	-414.7
30	3.18	184	1.61	93.0	-405.7
35	2.33	135	1.18	68.2	-396.7
40	1.79	103	0.89	51.4	-387.7
45	1.43	82.6	0.71	41.0	-378.7
50	1.18	68.2	0.585	33.8	-369.7
60	0.85	49.1	0.429	24.8	-351.7
70	0.66	38.1	0.337	19.5	-333.7
80	0.54	31.2	0.279	16.1	-315.7
90	0.45	26.0	0.239	13.8	-297.7
100	0.39	22.5	0.208	12.0	-279.7
150	0.231	13.3	0.130	7.51	-189.7
200	0.164	9.48	0.095	5.49	- 99.7
250	0.127	7.34	0.075	4.33	- 9.7
273	0.116	6.70	0.0684	3.95	32
300	0.104	6.01	0.0621	3.59	80.3
350	0.088	5.08	0.0530	3.06	170.3
400	0.076	4.39	0.0470	2.72	260.3
450	0.067	3.87	0.0423	2.44	350.3
500	0.060	3.47	0.0388	2.24	440.3
600	0.050	2.89	0.0340	1.96	620.3
700	0.0447	2.58	0.0314	1.81	800.3
800	(0.0420)	(2.43)	(0.0306)	(1.77)	980.3

*T_1 in K, k_1 in Watt cm^{-1} K^{-1}, T_2 in F, and k_2 in Btu hr^{-1} ft^{-1} F^{-1}.

‡Values in parentheses are extrapolated.

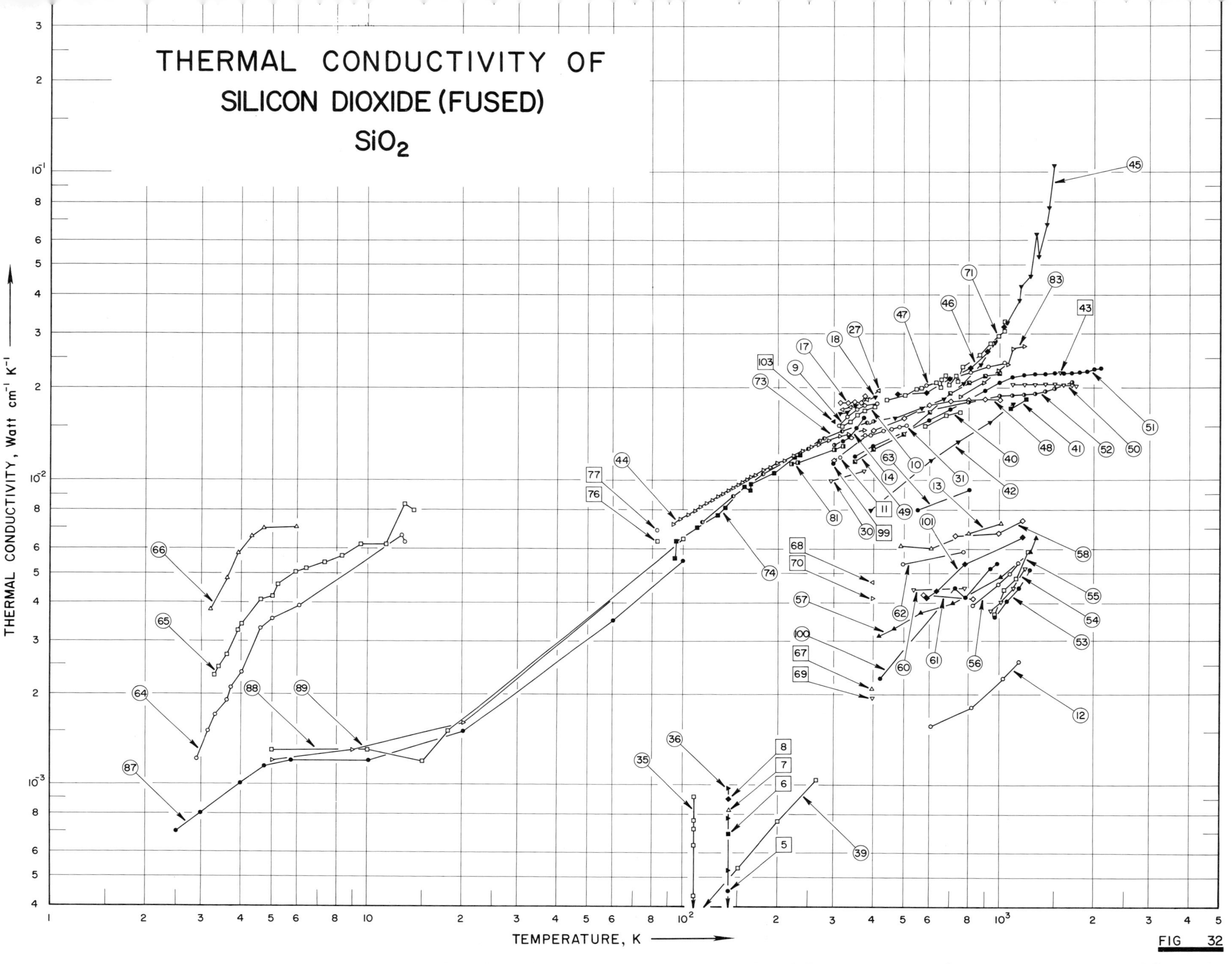
THERMAL CONDUCTIVITY OF
SILICON DIOXIDE (FUSED)
SiO_2
THERMAL CONDUCTIVITY, Watt cm^{-1} K^{-1}
TEMPERATURE, K
FIG 32

SPECIFICATION TABLE NO. 32 THERMAL CONDUCTIVITY OF SILICON DIOXIDE (FUSED) SiO_2

[For Data Reported in Figure and Table No. 32]

Curve No.	Ref. No.	Method Used	Year	Temp. Range, K	Reported Error, %	Name and Specimen Designation	Composition (weight percent), Specifications and Remarks
1	41	L	1948	140.0		Linde Silica	Measured in the presence of helium gas at pressure of 0.0016 mm Hg.
2	41	L	1948	140.3		Linde Silica	Measured in the presence of helium gas at pressure of 0.139 mm Hg.
3	41	L	1948	140.3		Linde Silica	Measured in the presence of helium gas at pressure of 1.304 mm Hg.
4	41	L	1948	140.4		Linde Silica	Measured in the presence of helium gas at pressure of 16.0 mm Hg.
5	41	L	1948	140.6		Linde Silica	Measured in the presence of helium gas at pressure of 51.0 mm Hg.
6	41	L	1948	140.5		Linde Silica	Measured in the presence of helium gas at pressure of 208.0 mm Hg.
7	41	L	1948	140.6		Linde Silica	Measured in the presence of helium gas at pressure of 408.0 mm Hg.
8	41	L	1948	140.6		Linde Silica	Measured in the presence of helium gas at pressure of 597.5 mm Hg.
9	3	L	1953	311-412		53R-1	Silky fused vitreous silica; cylindrical specimen with silky lines parallel to the specimen axis.
10	3	L	1953	317-406		53J-1	Silky fused vitreous silica; cylindrical specimen with silky lines perpendicular to the specimen axis.
11	312	C	1950	314.2		Fused Silica	High optical homogeneity variety "Homosil", manufactured by the W. C. Heraeus Co. of Hanau, Germany; 10 mm^2 cross sectional area and 2-10 mm thick; crystalline quartz used as standard reference.
12	313	L	1959	611-1165		Foamed fused silica	9 in. dia x 1 in. thick; fired; density 46 lb ft^{-3} (0.74 g cm^{-3}).
13	313	L	1959	489-1002		Slip cast fused silica	Similar to the above specimen except density 117 lb ft^{-3} (1.91 g cm^{-3}).
14	42		1953	322-375		53M-1	Vitreous; 0.350 in. in dia and 0.499 in. in length.
15	42		1953	318-408		53P-1	Vitreous; 0.449 in. in dia and 0.498 in. in length.
16	42		1953	319-373		53J-1	Vitreous; 0.251 in. in dia and 0.250 in. in length.
17	42		1953	314-375		53L-1	Vitreous; 0.303 in. in dia and 0.500 in. in length.
18	42		1953	314-406		53N-1	Vitreous; 0.409 in. in dia and 0.500 in. in length.
19	42		1953	319-394		53Q-1	Vitreous; 0.500 in. in dia and 0.500 in. in length.
20	42		1953	318-397		53Q-2	Vitreous; 0.500 in. in dia and 0.499 in. in length.
21	23,35		1952	319-394		53C-2	Clear vitreous silica sample with a clear platinum alloy glaze on its end faces.
22	23,35		1952	318-397		53D-2	Clear vitreous silica sample with a clear silver glaze on its end faces.
23	23,35		1952	316-397		53E-2	Clear vitreous silica sample with silky platinum alloy glaze on its end faces.
24	23,35		1952	314-398		53F-2	Clear vitreous silica sample with silky platinum alloy glaze on its end faces.
25	23,35		1952	318-394		53G-2	Clear vitreous silica sample with silky silver alloy glaze on its end faces.
26	23,35		1952	320-393		53H-2	Clear vitreous silica sample with silky silver alloy glaze on its end faces.

SPECIFICATION TABLE NO. 32 (continued)

Curve No.	Ref. No.	Method Used	Year	Temp. Range, K	Reported Error, %	Name and Specimen Designation	Composition (weight percent), Specifications and Remarks
27	9	C	1953	319-417	± 5	53NI-1	Vitreous silica.
28	9	C	1953	316-407	± 5	53NI-2	Vitreous silica.
29	43	L	1954	132-160			Fine powder contained in a cylindrical container of 1 in. thickness; dried at 380 K for at least 24 hrs; density (25 C) = 6.7 lb ft^{-3} (0.107 g cm^{-3}).
30	44	F	1914	293, 373			Fused silica; 0.332 cm dia x 6.1 cm long; density = 2.17 g cm^{-3}.
31	140	C	1926	342-510			Clear transparent vitreous silica; obtained from Thermal Syndicate; circular plate 1.25 in. in dia, 1 mm thick; surfaces optically flat and parallel to a high degree of accuracy, with air films on the surfaces; density 2.205 g cm^{-3}.
32	140	C	1926	344-510			Similar to the above specimen except with 1.503 mm thickness and 2.204 g cm^{-3} density.
33	140	C	1926	345-509			Similar to the above specimen except with 2.002 mm thickness and 2.203 g cm^{-3} density.
34	140	C	1926	348.3			Similar to the above specimen except with glycerine films on the surfaces and 2.204 g cm^{-3} density.
35	122	R	1953	109.4			Specimen supplied by Linde Air Products Co., packing density 0.0528 g cm^{-3}; measured at constant temperature 109.4 K and at various hydrogen pressures ranging from 0.0053 to 596.5 mm Hg.
36	122	R	1953	140.8			Similar to the above specimen except measured at constant temperature 140.8 K and at various hydrogen pressures ranging from 0.057 to 603 mm Hg.
37	122	R	1953	109.2			Similar to the above specimen except measured at constant temperature 109.4 K and at various nitrogen pressures ranging from 0.0022 to 584 mm Hg.
38	122	R	1953	140.8			Similar to the above specimen except measured at constant temperature 140.8 K and at various nitrogen pressures ranging from 0.939 to 594 mm Hg.
39	139	R	1950	93-265		Silica gel	Density 0.820 g cm^{-3}.
40	79	L	1942	582-752		Silica refractory brick	95.16 SiO_2, 1.46 Al_2O_3, 1.96 CaO, 0.85 Fe_2O_3, 0.08 MgO, 1.57 TiO_2, and 0.21 alkali oxides; cross-section 18 in. x 18 in., prepared from ganister-type quartzite rocks; fired at 1410 to 1420 C for 60 hrs; density 1.81 g cm^{-3} (113 lb ft^{-3}); porosity 22.77%; weight lost on ignition 0.14%.
41	79	C	1942	1091, 1223		Silica refractory brick	Similar to the above specimen but in disc form of dimensions 8 in. dia x 1 in. thick; steel used as comparative material.
42	80	L	1934	393-1101		Star-brand brick	95.9 SiO_2, 1.0 Al_2O_3, 1.0 Fe_2O_3, 2.0 CaO, 0.1 MgO, and 0.1 alkalis; supplied by Harbison-Walker Refractories, Co.; approx. composition; bulk density 1.52 g cm^{-3}; porosity 28.0%.
43	80	L	1934	1568.7		Star-brand brick	The above specimen measured with insulating brick placed between the calorimeter and the lower surface of the brick.

SPECIFICATION TABLE NO. 32 (continued)

Curve No.	Ref. No.	Method Used	Year	Temp. Range, K	Reported Error, %	Name and Specimen Designation	Composition (weight percent), Specifications and Remarks
44	147		1960	93-373		Glass M	Specimen in the form of a pair of 3 in. dia discs; density 2.20 g cm^{-3}; measured in two apparatus for different temperature ranges.
45	72,295	C	1955	368-1481			Specimen obtained from commercial source; cut and polished; data corrected for zero porosity using Loeb's expression (Loeb, A. L., J. Am. Ceram. Soc., 37(2), Pt II, 96-9, 1954.
46	39	L	1960	367-1033			Clear-fused; specimen 3 in. in dia and 0.25 in. in thickness; prepared by Hanovia Chemical Co.; density 2.20 g cm^{-3} at 273.2 K; Armco iron used as comparative material.
47	39	C	1960	367-1033			Same as the above specimen except having low-emissivity aluminum foil discs adjacent to the specimen surface.
48	237	R	1959	300-1000		CQ3	Clear-fused; cylindrical specimen, 0.549 cm dia and 10.2 cm long.
49	237	R	1959	350-1000		CQ3	The third run of the above specimen, CQ3.
50	237	R	1959	1100-1750		CQ3	The fourth run of the above specimen, CQ3.
51	237	R	1959	350-2100		CQ4	Clear-fused; cylindrical specimen,dia 0.478 cm and 8.30 cm long.
52	237	R	1959	300-1700		CQ7	Clear-fused; cylindrical specimen, dia 0.598 cm and 5.34 cm long.
53	238	P	1921	974-1256		Brick No. 1; A_1	95.4 SiO_2, 0.90 Al_2O_3, and 1.68 CaO; brick size: 9 x 4.5 x 2.5 in.; texture very open, and many large and sub-angular rock fragments; bonding of coarse and fine fairly good, although adherence of some of the grains is only fair; abundant large fissures; apparent density 1.75 g cm^{-3}; porosity 24.0%; heat flow in the direction of the length of the brick with thermocouple at a distance of 4.0 cm from the hot face of the specimen; thermal conductivity values calculated from authors measured thermal diffusivity data and the specific heat data of Bradshaw and Emery, (Trans., 19, 84, 1919).
54	238	P	1921	940-1208		Silica brick No. 1; A_2	The above specimen measured with thermocouple at a distance of 5.4 cm from the hot face of the specimen.
55	238	P	1921	968-1229		Silica brick No. 2; A_3	Similar to the above specimen except apparent density 1.80 g cm^{-3} and porosity 22.3%; heat flow in the direction of the length of the brick with thermocouple at a distance of 4.3 cm from the hot face of the specimen.
56	238	P	1921	830-1159		Silica brick No. 2; A_4	The above specimen measured with the thermocouple at a distance of 6.4 cm from the hot face of the specimen.
57	311	L	1959	417-1317		Sample A	Slip cast from fused silica; dried four days at 333 K before being tested; 9 in. in dia and 1 in. thick; unfired; density 1.78 g cm^{-3}.
58	311	L	1959	730-1182		Sample A	The above specimen, 2nd run
59	311	L	1959	393,1282			Same as sample A; unfired.

SPECIFICATION TABLE NO. 32 (continued)

Curve No.	Ref. No.	Method Used	Year	Temp. Range, K	Reported Error, %	Name and Specimen Designation	Composition (weight percent), Specifications and Remarks
60	311	L	1959	537, 778			Same as the above specimen, fired at 1089 K for 3.5 hrs.
61	311	L	1959	578, 829			Same as the above specimen, fired at 1200 K for 3.5 hrs.
62	311	L	1959	499, 775			Same as the above specimen, fired at 1422 K for 3.5 hrs.
63	311	L	1959	553, 805			Same as the above specimen, fired at 1533 K for 3.5 hrs; fine cracks appeared over all surfaces.
64	322		1957	2.9-13.2		Vitreous silica	High purity fused silica; obtained from Corning Works; square cross sectional area 19.8 mm^2; unirradiated; density (determined by hydrostatic weighing) 2.1994 g cm^{-3}; measured by a static method.
65	322		1957	3.30-14.0		Vitreous silica	The above specimen irradiated to 1.71 x 10^{19} fast neutrons cm^{-2}; density after irradiation 2.2412 g cm^{-3}; measured by a static method.
66	322		1957	3.2-6.0		Vitreous silica	The above specimen after an additional exposure to 4.13 x 10^{19} neutrons cm^{-2}; density after second irradiation 2.2602 g cm^{-3}; measured by a static method.
67	314	L	1959	395.4	14		Floated powder; supplied by Fisher Scientific Co.; 0.124 in. thick, 9 in. dia; pressed at 63 psi; -240 mesh; bulk density 1.493 g cm^{-3}; load reduced to 0.5 lb in^{-2} prior to making measurements.
68	314	L	1959	397.0	14		Same as the above specimen except 0.110 in. thick; pressed at 940 psi; bulk density 1.682 g cm^{-3}.
69	314	L	1959	398.7	14		Powder; supplied by Fisher Scientific Co.; 0.107 in. thick, 9 in. dia; pressed at 63 psi; 140 mesh; bulk density 1.552 g cm^{-3}; load reduced to 0.5 lb in^{-2} prior to making measurements.
70	314	L	1959	399.8	14		Same as the above specimen except 0.104 in. thick; pressed at 940 psi; bulk density 1.602 g cm^{-3}.
71	33	C	1954	441-1037			Clear; fused; Armco iron used as comparative material.
72	33	C	1954	445-1065			The above specimen measured with low emissivity foil adjacent to the surface.
73	246	L	1963	116-474	1-3	Fused quartz	$\ell/s = 0.4$ (ℓ, the sample thickness and s, its transverse cross-sectional area); ΔT = 7-10 C (ΔT, the temperature drop across the sample).
74	246	L	1963	94-463	1-3	Fused quartz	$\ell/s = 0.4$, ΔT = 2-4 C.
75	246	L	1963	96-305	1-3	Fused quartz	$\ell/s = 0.2$, ΔT = 4-6 C.
76	22	L	1911	83.2		Quartz glass	Measured in hydrogen.
77	22	L	1911	83.2		Quartz glass	Another run of the above specimen.
78	22	L	1911	195-373		Quartz glass	Measured in carbon dioxide.
79	22	L	1911	195-373		Quartz glass	Another run of the above specimen.
80	222	L	1959	123-323	± 2~± 5		Data obtained from smoothed curve of author's experimental results.

SPECIFICATION TABLE NO. 32 (continued)

Curve No.	Ref. No.	Method Used	Year	Temp. Range, K	Reported Error, %	Name and Specimen Designation	Composition (weight percent), Specifications and Remarks
81	223	R	1928	220-360			Clear fused quartz; hollow cylindrical specimen with one end closed hemispherically; obtained from General Electric Company; heated rapidly to about 800 C, outside surface ground.
82	223	R	1928	490-770			Above specimen measured with another apparatus.
83	223	R	1928	500-1220			Another run of the above specimen.
84	170	R	1953	418-1253		Fused quartz	Ellipsoidal specimen; prepared by grinding from a block of fused quartz.
85	170	R	1953	343-1273			Similar to the above specimen.
86	267	P	1962	297-422			Clear.
87	69	L	1951	2.5-100		Quartz glass	Specimen dia 6.1 mm, length 2.3 cm.
88	69	L	1951	5.0-100		Quartz glass	Specimen dia 7.7 mm, length 2.25 cm.
89	69	L	1951	5.0-100		Quartz glass	Specimen dia 7.4 mm, length 4.6 cm.
90	309	L	1960	50-1100	± 4~5		Fused quartz; five specimens of dimensions 1 x 1 x 1 cm^3 cut out from a single piece obtained from the M.V. Lomonosov Factory in Leningrad; data from three different experimental arrangements.
91	45	C	1953	328-949		Fused quartz	Prepared by grinding.
92	321	P	1961	300-426	± 5	Fused quartz; 1	Tubing; thermal conductivity values calculated from author's measured thermal diffusivity data and values of specific heat C taken from correlated data of Lord, R.C., and Morrow, J.C., J. Chem. Phys. 26, 230-2, 1957.
93	328	P	1957	303.2	<7	Fused quartz; 2	7 cm dia and 2.406 cm thick; thermal conductivity values calculated from author's measured thermal diffusivity and specific heat data.
94	328	P	1957	303.2	<7	Fused quartz; 3	Similar to the above specimen but the thickness, 1.518 cm.
95	328	P	1957	303.2	<7	Fused quartz; 4	Similar to the above specimen but the thickness, 0.0732 cm.
96	328	P	1957	303.2	<7	Fused quartz; 5	Similar to the above specimen except the dimensions 3.49 cm dia x 1.206 cm thick; measured in another apparatus.
97	328	P	1957	303.2	<7	Fused quartz; 6	Similar to the above specimen but the thickness, 0.507 cm.
98	328	P	1957	303.2	<7	Fused quartz; 7	Similar to the above specimen but the thickness, 0.305 cm.
99	415	L	1963	298.2	5	Fused quartz	99.98 pure SiO_2; specimen 0.5 in. in dia and 0.75 in. long; measured in a vacuum of 1.0 x 10^{-5} mm Hg.
100	416	L	1959	420-984	5.6	Slip 18	Specimen 0.25 in. thick; density 1.82 g cm^{-3}; density after firing 1.85-1.95 g cm^{-3}; water absorption 5-6%; porosity 10-14%; particle size: 4-6% greater than 44μ, 23-26% less than 2μ.

SPECIFICATION TABLE NO. 32 (continued)

Curve No.	Ref. No.	Method Used	Year	Temp. Range, K	Reported Error, %	Name and Specimen Designation	Composition (weight percent), Specifications and Remarks
101	416	L	1959	593-1196	5.6	Slip 10	Similar to the above specimen except 0.5 in. thick and particle size: 2-4%, greater than 44 μ 31-34%, less than 2 μ
102	330	C	1957	298.2	±6		3 in. dia x 0.188 in. thick; measured with thermal comparator No. 4; heat flow perpendicular to c-axis.
103	330	C	1957	298.2	±6		Similar to the above specimen but measured with thermal comparator No. 5 loaded with 100 g weight.

DATA TABLE NO. 32 THERMAL CONDUCTIVITY OF SILICON DIOXIDE (fused) SiO_2

[Temperature, T, K; Thermal Conductivity, k, Watt $cm^{-1}K^{-1}$]

T	k
CURVE 1*	
140.03	0.000013
CURVE 2*	
140.27	0.000031
CURVE 3*	
140.30	0.000150
CURVE 4*	
140.38	0.000333
CURVE 5	
140.56	0.000452
CURVE 6	
140.52	0.000692
CURVE 7	
140.61	0.000826
CURVE 8	
140.58	0.000900
CURVE 9	
310.8	0.0151
332.0	0.0161
355.9	0.0170
376.4	0.0175
412.2	0.0177
CURVE 10	
317.1	0.0151
338.6	0.0156
355.4	0.0163
375.6	0.0169
405.6	0.0174
CURVE 11	
314.2	0.0118
CURVE 12	
610.9	0.00156
820.4	0.00180
1033.2	0.00227
1164.8	0.00256
CURVE 13	
488.7	0.00616
608.7	0.00606
797.1	0.00675
1002.1	0.00727
CURVE 14	
322.0	0.0135
336.7	0.0138
353.2	0.0148
374.7	0.0160
CURVE 15*	
318.3	0.0160
334.5	0.0161
353.8	0.0171
371.0	0.0174
408.1	0.0175
CURVE 16*	
318.7	0.0161
337.7	0.0162
352.2	0.0172
373.4	0.0170
CURVE 17	
313.8	0.0179
334.2	0.0179
349.7	0.0180
367.8	0.0179
375.3	0.0189
CURVE 18	
314.0	0.0165
330.9	0.0166
351.9	0.0175
369.1	0.0180*
406.0	0.0187
CURVE 19*	
319.1	0.0159
335.4	0.0159
353.0	0.0167
369.3	0.0173
393.6	0.0175
CURVE 20*	
318.0	0.0151
333.5	0.0159
350.2	0.0167
370.3	0.0176
396.7	0.0179
CURVE 21*	
319.1	0.0159
335.4	0.0159
353.0	0.0167
369.3	0.0173
393.6	0.0175
CURVE 22*	
318.0	0.0151
333.5	0.0159
350.2	0.0167
370.3	0.0176
396.7	0.0179
CURVE 23*	
316.1	0.0153
333.7	0.0156
353.1	0.0165
371.1	0.0167
397.1	0.0178
CURVE 24*	
313.5	0.0159
334.2	0.0159
353.4	0.0172
372.9	0.0169
397.9	0.0174
CURVE 25*	
317.7	0.0151
335.2	0.0155
352.7	0.0164
369.9	0.0170
394.3	0.0181
CURVE 26*	
320.4	0.0154
333.8	0.0160
351.5	0.0165
376.0	0.0165
392.9	0.0172
CURVE 27	
319.1	0.0170
338.6	0.0173
360.1	0.0177*
381.1	0.0183
416.7	0.0197
CURVE 28*	
316.2	0.0165
335.8	0.0172
351.6	0.0179
375.8	0.0185
406.8	0.0201
CURVE 29*	
132.1	0.000187
148.9	0.000209
160.1	0.000224
CURVE 30	
293.2	0.00992
373.2	0.0107
CURVE 31	
341.9	0.0138
376.1	0.0141
392.3	0.0141
403.6	0.0144*
430.7	0.0145
459.6	0.0147
485.5	0.0149
510.2	0.0150
CURVE 32*	
343.6	0.0140
378.6	0.0141
398.3	0.0146
404.4	0.0144
431.7	0.0145
459.6	0.0149
485.1	0.0153
510.4	0.0153
CURVE 33*	
345.4	0.0141
378.0	0.0144
396.6	0.0144
405.0	0.0144
432.2	0.0145
460.8	0.0149
485.6	0.0150
508.5	0.0151
CURVE 34*	
348.3	0.0138

p(mm Hg)	k
CURVE 35 (T = 109.4K)	
0.0053	0.00000544*
0.207	0.0000406*
0.913	0.0000916*
13.5	0.000277
53.5	0.000435
211.5	0.000635
399.0	0.000718
596.5	0.000762
CURVE 36 (T = 140.8K)	
0.057	0.0000209*
1.204	0.000166*
16.0	0.000371*
53.5	0.000528
202.5	0.000780
398.5	0.000908*
603.0	0.000977
CURVE 37* (T = 109.2K)	
0.0022	0.00000460
0.175	0.0000205
1.391	0.0000385
12.0	0.0000678
44.0	0.000101
182.0	0.000148
379.0	0.000174
584.0	0.000191
CURVE 38* (T = 140.8K)	
0.939	0.0000356
2.320	0.0000481
15.0	0.0000812
47.5	0.000112
210.5	0.000155
395.0	0.000171
594.0	0.000179

T	k
CURVE 39	
93.2	0.00030*
100	0.000333*
150	0.000535
200	0.000760
265	0.00104
CURVE 40	
582.2	0.0151
683.2	0.0163
752.2	0.0167
CURVE 41	
1091.2	0.0172
1223.3	0.0184
CURVE 42	
393.0	0.00796
610.4	0.0117
742.1	0.0133
939.3	0.0155
1100.9	0.0176
CURVE 43	
1568.7	0.0223
CURVE 44	
93.2	0.00724
98.2	0.00749
103.2	0.00774
108.2	0.00795
113.2	0.00820
118.2	0.00841
123.2	0.00862
128.2	0.00887
133.2	0.00908
138.2	0.00925
143.2	0.00946
148.2	0.00967
153.2	0.00987
158.2	0.01008
CURVE 44 (cont.)	
163.2	0.0103
168.2	0.0104
173.2	0.0106*
178.2	0.0108
183.2	0.0109*
188.2	0.0111
193.2	0.0113*
198.2	0.0114
203.2	0.0116*
208.2	0.0117
213.2	0.0118*
218.2	0.00120*
223.2	0.0121
228.2	0.0122*
233.2	0.0123*
238.2	0.0125
243.2	0.0126*
248.2	0.0127*
253.2	0.0128
258.2	0.0129*
263.2	0.0130*
268.2	0.0131
273.2	0.0132*
278.2	0.0134*
294.2	0.0134*
288.2	0.0135
293.2	0.0136*
298.2	0.0137*
303.2	0.0138*
308.2	0.0139*
313.2	0.0139
318.2	0.0140*
323.2	0.0136*
328.2	0.0141*
333.2	0.0141
338.2	0.0142*
343.2	0.0143*
348.2	0.0144*
353.2	0.0144*
358.2	0.01450*
363.2	0.0146*
368.2	0.0146*
373.2	0.0146

*Not shown on plot

DATA TABLE NO. 32 (continued)

T	k
CURVE 45	
368.2	0.0151*
468.2	0.0159
568.2	0.0172
668.2	0.0184
773.2	0.0206
871.2	0.0238
968.2	0.0280
1069.2	0.0327
1169.2	0.0385
1173.2	0.0428
1265.2	0.0464
1313.2	0.0628
1339.2	0.0534
1412.2	0.0674
1441.2	0.0766
1481.2	0.105
CURVE 46	
366.5	0.0182
477.6	0.0192
588.7	0.0193
699.8	0.0215
811.0	0.0234
922.1	0.0265
1033.2	0.0318
CURVE 47	
366.5	0.0183*
477.6	0.0192*
588.7	0.0203
699.8	0.0215*
811.0	0.0225
922.1	0.0234
1033.2	0.0242
CURVE 48	
300	0.0117
400	0.0146
500	0.0160
600	0.0177
700	0.0182
800	0.0184
900	0.0185
1000	0.0184
CURVE 49	
350	0.0116
400	0.0127
500	0.0142
600	0.0167
700	0.0192
800	0.0209
900	0.0219
1000	0.0224
CURVE 50	
1100	0.0205
1200	0.0205
1300	0.0205
1400	0.0205
1500	0.0205
1600	0.0204
1700	0.0204
1750	0.0203
CURVE 51	
350	0.0119
400	0.0128
500	0.0142*
600	0.0157
700	0.0170
800	0.0181*
900	0.0196
1000	0.0208
1100	0.0217
1200	0.0220
1300	0.0222
1400	0.0222
1500	0.0223
1600	0.0223
1700	0.0223
1800	0.0225
1900	0.0226
2000	0.0231
2100	0.0232
CURVE 52	
300	0.0131*
400	0.0156
500	0.0163*
CURVE 52 (cont.)	
600	0.0171*
700	0.0174*
800	0.0180
900	0.0187*
1000	0.0188
1100	0.0189
1200	0.0190
1300	0.0192
1400	0.0195
1500	0.0199
1600	0.0203*
1700	0.0208
CURVE 53	
974.2	0.00360
1061.2	0.00406
1156.2	0.00448
1255.7	0.00515
CURVE 54	
940.2	0.00377
1019.7	0.00402
1114.2	0.00448
1208.2	0.00519
CURVE 55	
967.7	0.00368
1044.7	0.00444
1134.2	0.00481
1229.2	0.00590
CURVE 56	
829.7	0.00393
996.7	0.00460
1079.7	0.00498
1159.2	0.00540
CURVE 57	
417.1	0.00310
465.9	0.00331
568.7	0.00374
715.9	0.00400
CURVE 57 (cont.)	
906.5	0.00457*
1014.8	0.00488
1173.2	0.00543*
1258.7	0.00590
1316.5	0.00658
CURVE 58	
730.9	0.00661
996.5	0.00675
1182.1	0.00742
CURVE 59*	
393.2	0.00318
1282.1	0.00602
CURVE 60	
537.1	0.00443
778.2	0.00447
CURVE 61	
578.2	0.00426
828.7	0.00415
CURVE 62	
499.3	0.00535
774.8	0.00588
CURVE 63	
552.6	0.00798
804.8	0.00928
CURVE 64	
2.90	0.00122
3.15	0.00150
3.32	0.00170
3.60	0.00190
3.70	0.00210
4.00	0.00235
4.60	0.00330
5.00	0.00354
CURVE 64 (cont.)	
6.10	0.00390
12.8	0.00660
13.2	0.00631
CURVE 65	
3.30	0.00230
3.40	0.00245
3.60	0.00270
3.90	0.00325
4.00	0.00340
4.60	0.00410
5.00	0.00420
5.20	0.00460
5.92	0.00505
6.40	0.00520
7.30	0.00540
8.30	0.00570
9.50	0.00620
11.50	0.00620
13.2	0.00840
14.0	0.00798
CURVE 66	
3.20	0.00380
3.60	0.00480
3.91	0.00580
4.30	0.00660
4.70	0.00696
5.95	0.00700
CURVE 67	
395.4	0.00209
CURVE 68	
397.0	0.00469
CURVE 69	
398.7	0.00194
CURVE 70	
399.8	0.00417
CURVE 71	
441.2	0.0183
502.2	0.0190
546.2	0.0199
554.2	0.0200*
568.2	0.0200
588.2	0.0202*
631.2	0.0209
654.2	0.0201
657.2	0.0214
659.2	0.0212*
677.2	0.0214*
679.2	0.0220
693.2	0.0205
714.2	0.0209*
714.2	0.0218*
728.2	0.0220
737.2	0.0210
765.2	0.0235
796.2	0.0237*
864.2	0.0256
902.2	0.0258*
907.2	0.0265*
935.2	0.0274*
938.2	0.0280
940.2	0.0282*
959.2	0.0277*
991.2	0.0296
991.2	0.0289*
1025.2	0.0308
1037.2	0.0329
CURVE 72*	
445.2	0.0187
494.2	0.0197
511.2	0.0195
548.2	0.0206
560.2	0.0200
576.2	0.0206
602.2	0.0197
613.2	0.0213
620.2	0.0209
643.2	0.0210
660.2	0.0213
673.2	0.0204
714.2	0.0214
735.2	0.0216
CURVE 72 (cont.)*	
754.2	0.0213
781.2	0.0220
825.2	0.0213
851.2	0.0226
917.2	0.0233
937.2	0.0237
1003.2	0.0243
1065.2	0.0251
CURVE 73	
116	0.00728
144	0.00887
179	0.0105
249	0.0127
270	0.0134
279	0.0136
319	0.0144
332	0.0146*
357	0.0146*
373	0.0148*
381	0.0153
402	0.0156*
418	0.0156*
453	0.0158*
454	0.0162*
474	0.0164*
CURVE 74	
94	0.00561
95	0.00636
112	0.00703
128	0.00770
136	0.00812
157	0.00950
163	0.00929
164	0.00979
195	0.0107
200	0.0113*
205	0.0114*
220	0.0118*
225	0.0119
230	0.0121*
235	0.0122
293	0.0137*
302	0.0142*
CURVE 74 (cont.)	
304	0.0137*
338	0.0145*
344	0.0146*
354	0.0148*
361	0.0149*
370	0.0150*
378	0.0151*
384	0.0151*
392	0.0152*
430	0.0153*
446	0.0159*
463	0.0162*
CURVE 75*	
96	0.00594
108	0.00674
130	0.00803
164	0.00954
226	0.0121
294	0.0138
305	0.0140
CURVE 76	
83.2	0.00636
CURVE 77	
83.2	0.00686
CURVE 78	
195.2	0.0113
273.2	0.0139
373.2	0.0192
CURVE 79*	
195.2	0.0119
273.2	0.0138
373.2	0.0190

* Not shown on plot

DATA TABLE NO. 32 (continued)

T	k
CURVE 80*	
123.2	0.00887
148.2	0.00979
173.2	0.0107
197.2	0.0115
243.2	0.0126
303.2	0.0136
303.2	0.0137
323.2	0.0141
CURVE 81	
220	0.0113
230	0.0114
300	0.0126
320	0.0130
360	0.0134*
CURVE 82*	
490	0.0142
550	0.0146
560	0.0159
700	0.0176
CURVE 83	
500	0.0151*
575	0.0155*
750	0.0188
890	0.0209
990	0.0228
1060	0.0238
1100	0.0270
1220	0.0276
CURVE 84*	
418.2	0.0155
451.2	0.0153
478.2	0.0159
505.2	0.0159
571.2	0.0167
609.2	0.0184
668.2	0.0188
705.2	0.0199
785.2	0.0218
CURVE 84 (cont.)*	
861.2	0.0241
941.2	0.0263
1013.2	0.0297
1074.2	0.0326
1091.2	0.0344
1160.2	0.0393
1215.2	0.0423
1253.2	0.0448
CURVE 85*	
343.2	0.0155
423.2	0.0162
461.2	0.0156
479.2	0.0167
517.2	0.0167
583.2	0.0176
624.2	0.0188
673.2	0.0187
726.2	0.0205
813.2	0.0215
873.2	0.0228
883.2	0.0240
963.2	0.0268
1035.2	0.0302
1091.2	0.0314
1113.2	0.0335
1183.2	0.0377
1243.2	0.0418
1273.2	0.0450
CURVE 86*	
297	0.0145
325	0.0147
362	0.0157
372	0.0163
400	0.0168
413	0.0184
422	0.0182
CURVE 87	
2.5	0.00070
3.0	0.00080
4.0	0.00100
4.75	0.00115
CURVE 87 (cont.)	
5.75	0.00120
10.15	0.00120
20.25	0.00150
60.0	0.00350
100.0	0.00550
CURVE 88	
5.0	0.00130
10.0	0.00130
15.0	0.00120
18.0	0.00150
100.0	0.00650
CURVE 89	
5.0	0.00120
9.0	0.00130
20.25	0.00160
100.0	0.00650*
CURVE 90*	
50	0.00343
100	0.00665
150	0.00925
200	0.0112
250	0.0127
300	0.0138
350	0.0146
400	0.0153
450	0.0159
500	0.0164
550	0.0168
600	0.0172
650	0.0177
700	0.0184
750	0.0193
800	0.0205
850	0.0218
900	0.0236
950	0.0259
1000	0.0291
1050	0.0335
1100	0.0397
CURVE 91*	
328.2	0.0160
328.2	0.0149
378.2	0.0169
378.2	0.0157
440.2	0.0175
440.2	0.0163
503.2	0.0175
505.2	0.0155
624.2	0.0203
626.2	0.0179
723.2	0.0204
723.2	0.0192
815.5	0.0224
817.2	0.0213
890.2	0.0247
890.2	0.0234
949.2	0.0278
949.2	0.0273
CURVE 92*	
300	0.0138
324	0.0146
361	0.0156
370	0.0160
400	0.0167
415	0.0183
426	0.0181
CURVE 93*	
303.2	0.0138
CURVE 94*	
303.2	0.0139
CURVE 95*	
303.2	0.0138
CURVE 96*	
303.2	0.0131
CURVE 97*	
303.2	0.0131
CURVE 98*	
303.2	0.0133
CURVE 99	
298.2	0.014
CURVE 100	
420.38	0.00225
713.16	0.00448
784.83	0.00419
939.27	0.00519
984.27	0.00537
CURVE 101	
592.60	0.00417
646.49	0.00438
782.60	0.00538
1195.94	0.00658
CURVE 102*	
298.2	0.0138
CURVE 103	
298.2	0.0156

* Not shown on plot

FIGURE AND TABLE NO. 32R RECOMMENDED THERMAL CONDUCTIVITY OF SILICON DIOXIDE (FUSED) SiO_2

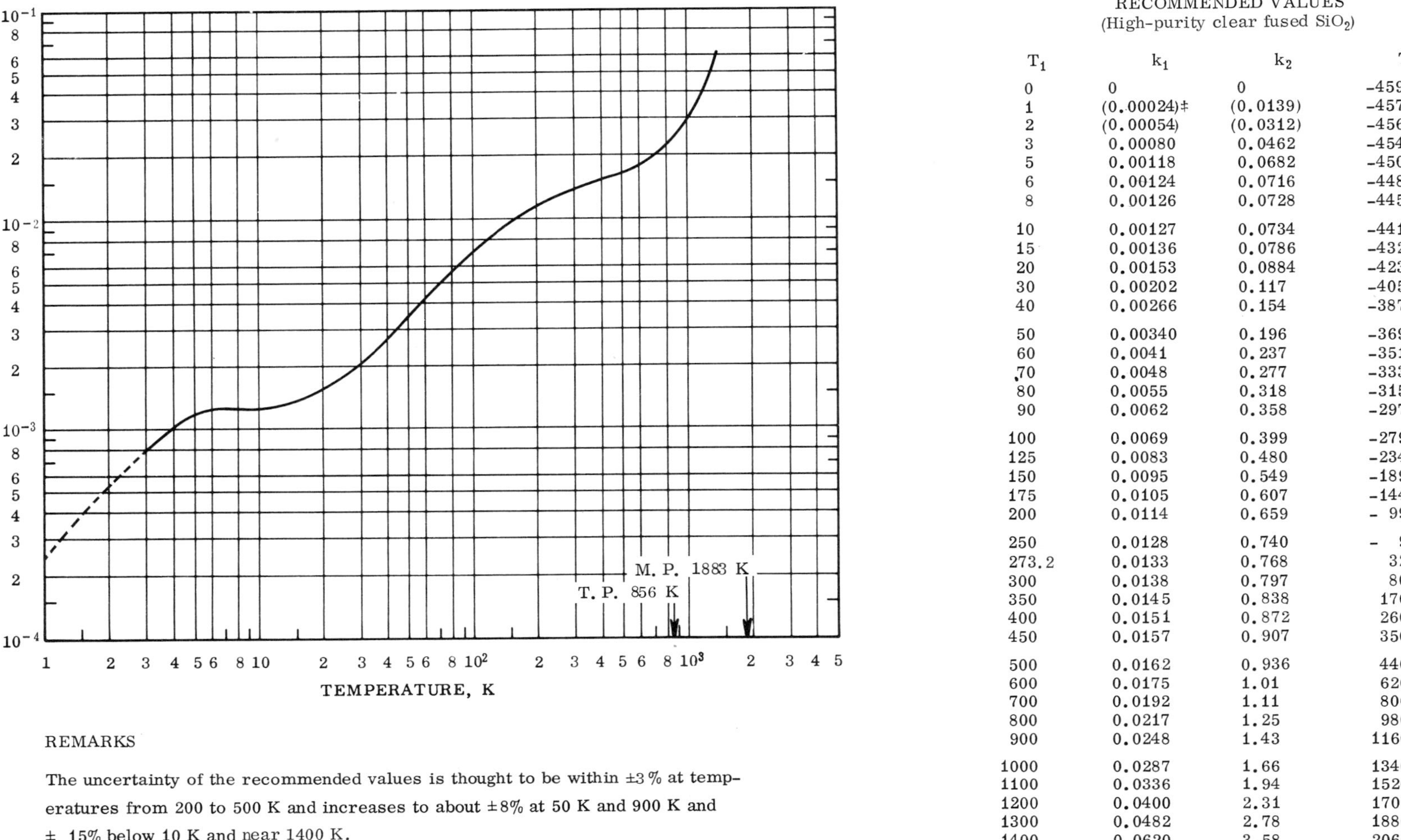

RECOMMENDED VALUES*
(High-purity clear fused SiO_2)

T_1	k_1	k_2	T_2
0	0	0	-459.7
1	(0.00024)‡	(0.0139)	-457.9
2	(0.00054)	(0.0312)	-456.1
3	0.00080	0.0462	-454.3
5	0.00118	0.0682	-450.7
6	0.00124	0.0716	-448.9
8	0.00126	0.0728	-445.3
10	0.00127	0.0734	-441.7
15	0.00136	0.0786	-432.7
20	0.00153	0.0884	-423.7
30	0.00202	0.117	-405.7
40	0.00266	0.154	-387.7
50	0.00340	0.196	-369.7
60	0.0041	0.237	-351.7
70	0.0048	0.277	-333.7
80	0.0055	0.318	-315.7
90	0.0062	0.358	-297.7
100	0.0069	0.399	-279.7
125	0.0083	0.480	-234.7
150	0.0095	0.549	-189.7
175	0.0105	0.607	-144.7
200	0.0114	0.659	- 99.7
250	0.0128	0.740	- 9.7
273.2	0.0133	0.768	32.0
300	0.0138	0.797	80.3
350	0.0145	0.838	170.3
400	0.0151	0.872	260.3
450	0.0157	0.907	350.3
500	0.0162	0.936	440.3
600	0.0175	1.01	620.3
700	0.0192	1.11	800.3
800	0.0217	1.25	980.3
900	0.0248	1.43	1160
1000	0.0287	1.66	1340
1100	0.0336	1.94	1520
1200	0.0400	2.31	1700
1300	0.0482	2.78	1880
1400	0.0620	3.58	2060

REMARKS

The uncertainty of the recommended values is thought to be within ±3% at temperatures from 200 to 500 K and increases to about ±8% at 50 K and 900 K and ± 15% below 10 K and near 1400 K.

*T_1 in K, k_1 in Watt $cm^{-1} K^{-1}$, T_2 in F, and k_2 in Btu $hr^{-1} ft^{-1} F^{-1}$. ‡ Values in parentheses are extrapolated.

SPECIFICATION TABLE NO. 33 THERMAL CONDUCTIVITY OF STRONTIUM OXIDE SrO

Curve No.	Ref. No.	Method Used	Year	Temp. Range, K	Reported Error, %	Name and Specimen Designation	Composition (weight percent), Specifications and Remarks
1	75	L	1955	493-1003	< 10	Tube No. 2	Polycrystalline; prepared from $SrCo_3$; measured under vacuum 10^{-6} mm Hg; apparent thermal conductivity (effects due to radiation at high temperature not considered).

DATA TABLE NO. 33 THERMAL CONDUCTIVITY OF STRONTIUM OXIDE SrO

[Temperature, T, K; Thermal Conductivity, k, Watt $cm^{-1} K^{-1}$]

T	k
CURVE 1*	
493.2	0.50
676.2	0.80
798.2	1.25
893.2	1.76
1003.2	2.50

* No graphical presentation

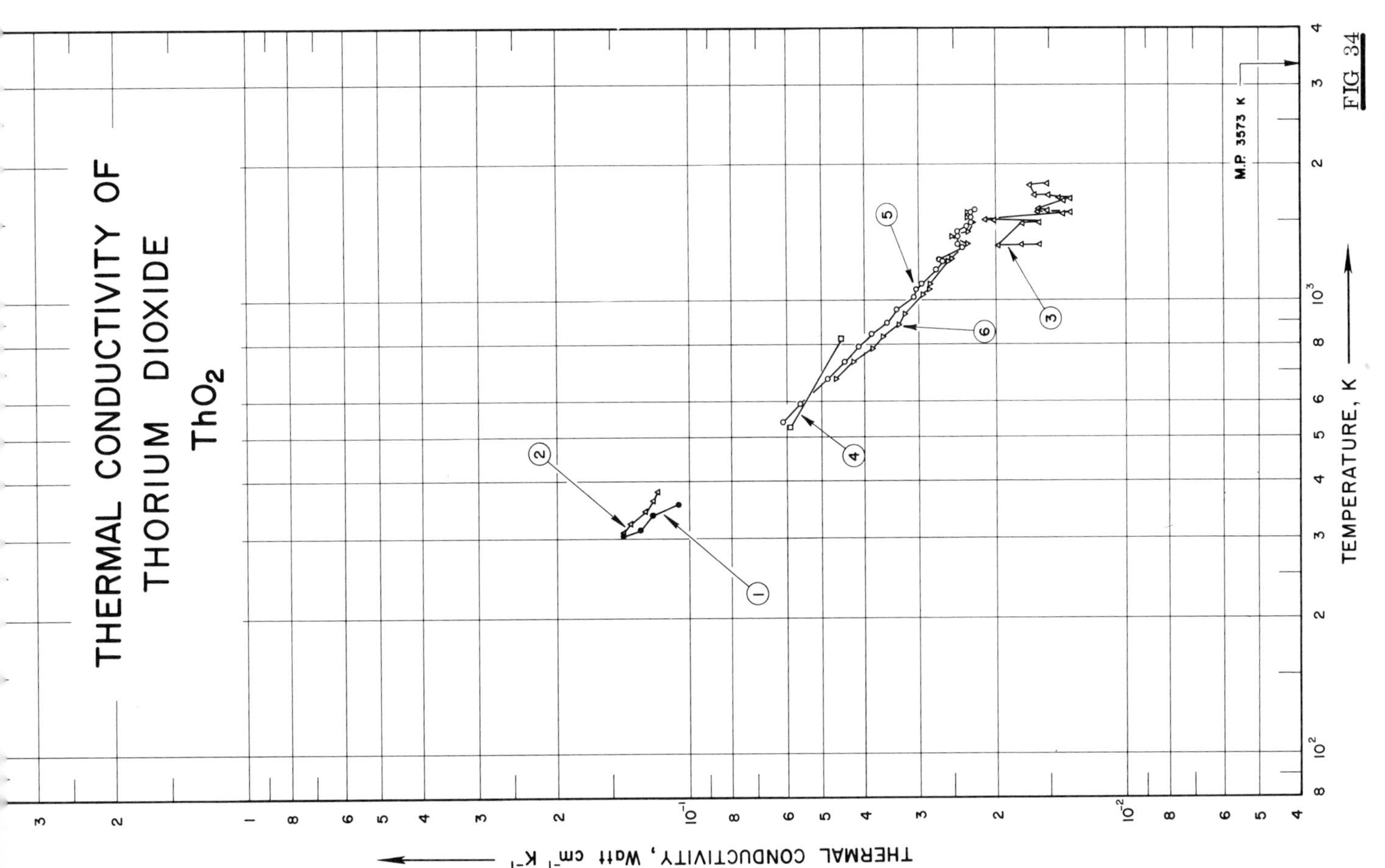
THERMAL CONDUCTIVITY OF
THORIUM DIOXIDE
ThO_2
M.P. 3573 K
TEMPERATURE, K
THERMAL CONDUCTIVITY, Watt cm^{-1} K^{-1}
1
2
3
4
5
6

FIG 34

SPECIFICATION TABLE NO. 34 THERMAL CONDUCTIVITY OF THORIUM DIOXIDE ThO_2

[For Data Reported in Figure and Table No. 34]

Curve No.	Ref. No.	Method Used	Year	Temp. Range, K	Reported Error, %	Name and Specimen Designation	Composition (weight percent), Specifications and Remarks
1	3	C	1953	304-356		239A-1	Spectroscopically pure; supplied by Carbide and Carbon Chemicals Co.; formed by hot pressing at 1790-1820 C; density (25 C) = 9.58 g cm^{-3}; measurements made by using a gold coating on the ends of the cylindrical specimen.
2	3	C	1953	306-379		239A-2	Same specimen as above except using platinum alloy glaze.
3	144	R	1963	1331-1821	5.0-7.0	1	Poorly bonded structure; ground and polished to eliminate all the scratches on the surface of the specimen; 0.75 in. long, 0.75 in. O.D., and 0.25 I.D.; supplied by the Zirconium Corp. of America; density 9.69 g cm^{-3} at 25 C; specimen found broken on post inspection.
4	204	R	1957	527, 824			0.5 CaF_2; specimen consists of 6 discs of average dia 60.03 mm and total height 114.05 mm; hot-pressed at 1500 + 50 C and at pressure of about 100 psi for 30 min; average bulk density 9.37 g cm^{-3}.
5	307	R	1954	543-1593			Specimen in the shape of prolate spheroid prepared by slip casting from suspension of finely ground thoria; total porosity 16.7% and bulk density 8.07 g cm^{-3}; the first run.
6	307	R	1954	538-1593			The above specimen second run.
7	472		1963	1219-2009		a	Precipitated from thorium nitrate and solution of ammonium hydroxide; bulk density 0.26 g cm^{-3}; measured in increasing temperature order.
8	472		1963	1218-2009		a	The above specimen; measured in decreasing temperature order.
9	472		1963	1225-1816		c	Prepared by drying and igniting in air a cotton cloth soaked in thorium nitrate solution; measured in increasing temperature order.
10	472		1963	1520-1816		c	The above specimen; measured in decreasing temperature order.

DATA TABLE NO. 34 THERMAL CONDUCTIVITY OF THORIUM DIOXIDE ThO_2

[Temperature, T, K; Thermal Conductivity, k, Watt $cm^{-1}K^{-1}$]

T	k
CURVE 1	
303.7	0.141
312.6	0.129
336.7	0.121
355.9	0.106
CURVE 2	
306.0	0.141
322.0	0.136
342.1	0.126
361.4	0.122
379.0	0.118
CURVE 3	
1330.9	0.0156
1330.9	0.0172
1330.9	0.0195
1490.9	0.0170
1498.2	0.0156
1502.1	0.0209
1509.3	0.0198
1576.5	0.0138
1580.9	0.0133
1581.5	0.0157
1587.6	0.0149
1597.6	0.0156
1673.7	0.0137
1687.6	0.0133
1691.5	0.0141
1705.9	0.0149
1708.7	0.0160
1801.5	0.0164
1820.9	0.0149
CURVE 4	
526.8	0.0586
824.1	0.0448
CURVE 5	
543.2	0.061
593.2	0.0552
CURVE 5 (cont.)	
673.2	0.048
733.2	0.044
793.2	0.041
843.2	0.0384
893.2	0.0351
953.2	0.0335
1013.2	0.0305
1063.2	0.0301
1093.2	0.0293
1173.2	0.0276
1223.2	0.0264
1243.2	0.0268
1303.2	0.0238
1348.2	0.0243
1393.2	0.0243
1423.2	0.0243
1453.2	0.0234
1493.2	0.0226
1533.2	0.0226
1563.2	0.0226
1593.2	0.0222
CURVE 6	
538.2	0.062
593.2	0.054
673.2	0.046
733.2	0.042
783.2	0.038
833.2	0.036
883.2	0.033
933.2	0.032
1003.2	0.029
1053.2	0.028
1093.2	0.028
1173.2	0.027
1223.2	0.0255
1243.2	0.0251
1303.2	0.0243*
1348.2	0.023
1393.2	0.025
1423.2	0.023
1453.2	0.0225*
1493.2	0.0225
CURVE 6 (cont.)	
1533.2	0.023
1563.2	0.023
1593.2	0.0222*
CURVE 7*	
1219	0.000486
1300	0.000533
1400	0.000618
1500	0.000698
1600	0.000784
1701	0.000889
1800	0.001006
1893	0.00115
1942	0.00128
1980	0.00135
2009	0.00144
CURVE 8*	
1301	0.00105
1400	0.00111
1500	0.00117
1600	0.00123
1700	0.00128
1800	0.00134
1902	0.00139
2009	0.00144
CURVE 9*	
1225	0.000466
1301	0.000539
1400	0.000630
1500	0.000708
1555	0.000746
1600	0.000766
1653	0.000782
1700	0.000783
1760	0.000761
1799	0.000725
1816	0.000702
CURVE 10*	
1520	0.000509
1601	0.000553
1700	0.000613
1752	0.000649
1816	0.000702

*Not shown on plot

FIGURE AND TABLE NO. 34R RECOMMENDED THERMAL CONDUCTIVITY OF THORIUM DIOXIDE ThO_2

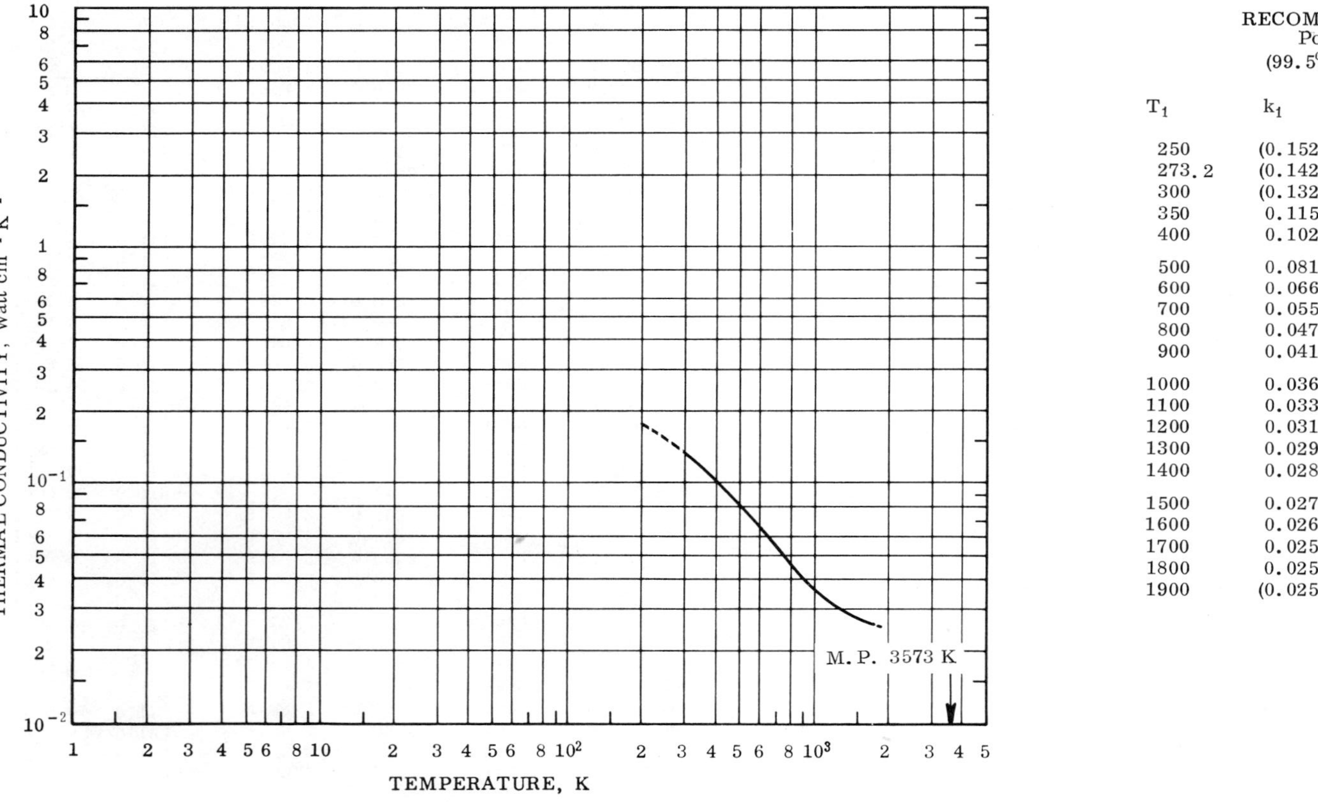

RECOMMENDED VALUES*
Polycrystalline
(99.5% pure, 98% dense)

T_1	k_1	k_2	T_2
250	(0.152)‡	(8.78)	-9.7
273.2	(0.142)	(8.20)	32.0
300	(0.132)	(7.63)	80.3
350	0.115	6.64	170.3
400	0.102	5.89	260.3
500	0.081	4.68	440.3
600	0.066	3.81	620.3
700	0.055	3.18	800.3
800	0.047	2.72	980.3
900	0.041	2.37	1160
1000	0.0368	2.13	1340
1100	0.0336	1.94	1520
1200	0.0312	1.80	1700
1300	0.0296	1.71	1880
1400	0.0284	1.64	2060
1500	0.0273	1.58	2240
1600	0.0266	1.54	2420
1700	0.0259	1.50	2600
1800	0.0254	1.47	2780
1900	(0.0252)	(1.46)	2960

REMARKS

The recommended values are for 99.5% pure, 98% dense, polycrystalline ThO_2. The recommended values that are supported by experimental data are thought to be accurate to within 15% of the true values at temperatures from 350 to 1000 K and 15 to 20% at other temperatures.

* T_1 in K, k_1 in Watt cm^{-1} K^{-1}, T_2 in F, and k_2 in Btu hr^{-1} ft^{-1} F^{-1}.

‡ Values in parentheses are extrapolated.

THERMAL CONDUCTIVITY OF TIN DIOXIDE SnO_2

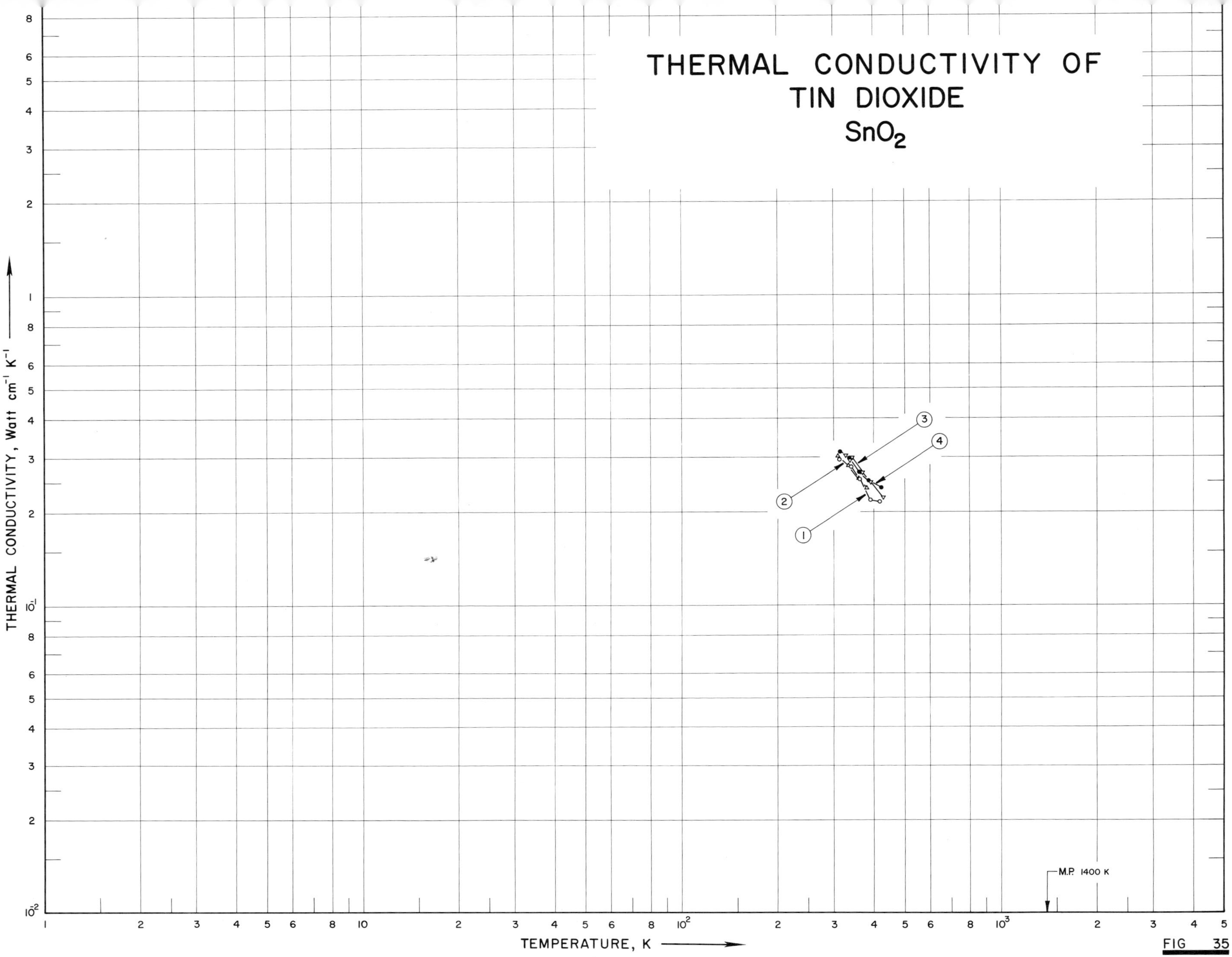

FIG 35

SPECIFICATION TABLE NO. 35 THERMAL CONDUCTIVITY OF TIN DIOXIDE SnO_2

[For Data Reported in Figure and Table No. 35]

Curve No.	Ref. No.	Method Used	Year	Temp. Range, K	Reported Error, %	Name and Specimen Designation	Composition (weight percent), Specifications and Remarks
1	3	L	1953	315-420		166 A-1	98 SnO_2; furnished by Metal and Thermit Corp; density 6.62 g cm^{-3}; water absorption 0.03%.
2	3	L	1953	311-383		166 B-1	98 SnO_2; furnished by Metal and Thermit Corp; density (25 C) = 6.56 g cm^{-3}; water absorption 0.11%.
3	3	L	1953	328-430		47 A-1	98 SnO_2.
4	12, 3	L	1953	315-424		47 A-2	98 SnO_2; calcined at 1092 C; dry pressed at 1500 psi; fired at 1427 C and soaked for 1 hr; density 6.5 g cm^{-3}; water absorption 0.021%.
5	74	C	1954	313.2			97 SnO_2, 0.9 ZnO, 0.1-1.0 Si, 0.05-0.5 Fe, 0.01-0.1 Ca, and < 0.01 other impurities; prepared from -200 mesh tin oxide and zinc oxide powder by ball milling dry for 4 hrs in rubberlined, one-gallon ball mill containing steel balls, pressed in steel die at 6000 to 10,000 psi and sintered in air for 2 to 3 hrs at 1427 C (maximum); apparent porosity 2.6%.

DATA TABLE NO. 35 THERMAL CONDUCTIVITY OF TIN OXIDE SnO_2

[Temperature, T, K; Thermal Conductivity, k, Watt cm^{-1} K^{-1}]

T	k
CURVE 1	
314.6	0.295
340.6	0.279
364.1	0.256
391.4	0.218
420.0	0.216
CURVE 2	
310.5	0.302
335.2	0.281
361.4	0.256
382.8	0.239
CURVE 3	
328.4	0.304
338.1	0.293
344.1	0.299
370.1	0.267
393.3	0.249
429.9	0.221
CURVE 4	
315.0	0.311
338.2	0.295
361.6	0.269
387.4	0.251
423.5	0.238
CURVE 5*	
313.2	0.293

*Not shown on plot

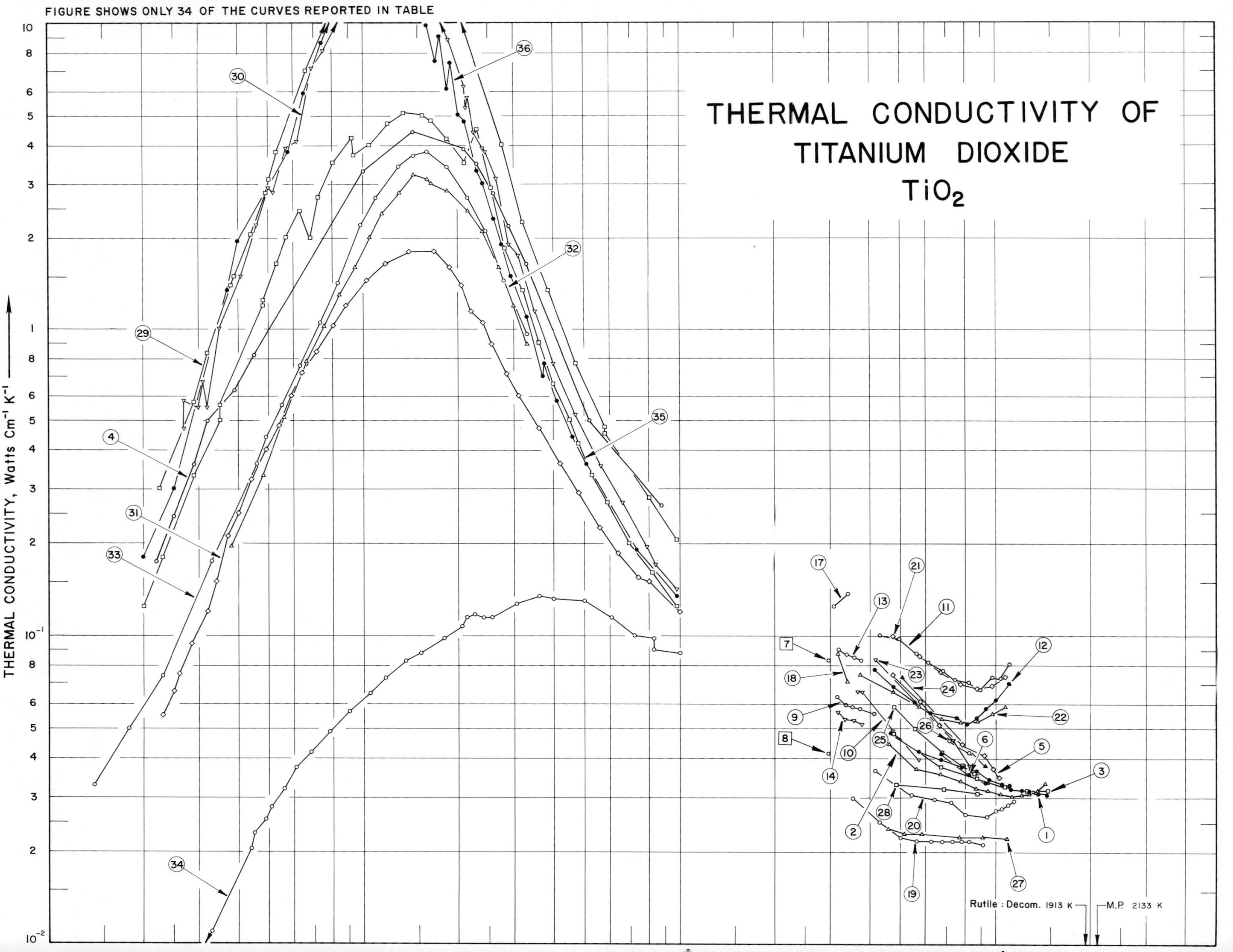
FIGURE SHOWS ONLY 34 OF THE CURVES REPORTED IN TABLE
THERMAL CONDUCTIVITY OF
TITANIUM DIOXIDE
TiO_2
THERMAL CONDUCTIVITY, Watts Cm⁻¹ K⁻¹
Rutile : Decom. 1913 K
M.P. 2133 K

SPECIFICATION TABLE NO. 36 THERMAL CONDUCTIVITY OF TITANIUM DIOXIDE TiO_2

[For Data Reported in Figure and Table No. 36]

Curve No.	Ref. No.	Method Used	Year	Temp. Range, K	Reported Error, %	Name and Specimen Designation	Composition (weight percent), Specifications and Remarks
1	10	R	1952	468-1463			Specimen spheroidal in shape; bulk density 4.11 g cm^{-3}.
2	10	R	1952	463-1448			Similar to the above specimen.
3	364	R	1954	473-1473			Polycrystal; prepared by calcining commercially pure TiO_2 at 1000 C, grinding for 12 hrs in a steel mill, acid leaching slip-casting at a pH of 3.5 with specific gravity 3.5, and then fired at 1700 C in an oxidizing atmosphere; bulk density 4.11 g cm^{-3}; porosity 3.5%.
4	28	L	1956	2.2-88		Rutile	Single crystal; specimen 2 mm in dia and 60 mm long; supplied by Linde Air Products.
5	146		1955	473-1035		I	Polycrystal with crystal size of 15 microns; pressed hydrostatically and sintered for 2 hrs at 1250 C; porosity 2.1%.
6	146		1955	674-1123		II	Polycrystal with crystal size 28 microns; pressed hydrostatically and sintered for 8 hrs at 1450 C; porosity 3.0%.
7	152	L	1956	298.2		$TiO_2$192	Specimen in wafer form made up of 4 pieces of 0.75 in. in dia and 20 mils thick each.
8	152	L	1956	298.2		$TiO_2$192	The above specimen exposed to 5 x 10^{19} epithermal neutrons per cm^2 for 480 megawatt day in the Material Testing Reactor.
9	68	C	1954	317-414	± 3	Rutile; 2176-1	Natural single crystal with tetragonal crystal system; obtained from commercial source; flawless piece cut out and ground to 0.250 in. dia x 0.250 in. long; copper used as comparative material.
10	38	C	1960	363-573		Rutile	Disk specimen; porosity 13.4%; measured in a vacuum of ~ 10^{-1} mm Hg; pure iron used as comparative material; data correct to zero porosity.
11	145		1955	432-1105			Single crystal; heat flow parallel to c-axis.
12	145		1955	419-1102			Single crystal; heat flow perpendicular to c-axis.
13	42	C	1953	319-376		217 A-1	Clear single crystal; heat flow direction parallel to the c-axis.
14	42	C	1953	317-377		217 B-1	Clear single crystal; a-axis parallel to the axis of the cylindrical specimen within 11 degrees; 0.250 ± 0.001 in. in dia and 0.250 ± 0.001 in. in length; heat flow perpendicular to c-axis.
15	42	C	1953	338.2			Same specimen as 217 A-1.
16	42	C	1953	338.2			Same specimen as 217 B-1.
17	71	C	1951	309, 341		Rutile	Single crystal; from Linde Air Products Co. of Tonawanda, N.Y.; measured with heat flow parallel to the optical axis; Pyrex glass used as comparative material.
18	71	C	1951	317, 340		Rutile	Same as above; heat flow perpendicular to the optical axis.

SPECIFICATION TABLE NO. 36 (continued)

Curve No.	Ref. No.	Method Used	Year	Temp. Range, K	Reported Error, %	Name and Specimen Designation	Composition (weight percent), Specifications and Remarks
19	332, 282	R	1955	353-910		Porous sample	Composed of a mixture of 70% Titanox TG and 30% RA10MO, the former being previously calcined at 1550 C for one hr and ground to -100 mesh in a micro-pulverizer; the batch mixed in a porcelain mill using distilled water and flint pebbles; pressed by using 7% binder (composition: 500 g carbowax, 10 g methocel and 1000 cc water) at a pressure of approx. 4400 psi and fired in a Pereny globar kiln at a rate of 120 C per hr and held at 1375 for 30 min; 3.31 sp gr; 11,900 psi mod rupture; 17.1% apparent porosity; 5.6% shrinkage; specimen size: 5.40 cm O.D., 2.8 cm I.D. and 0.5 in. thick, eleven rings stacked to form a cylinder of 5.5 in. high but measurements made only over the centrally-placed rings.
20	332, 282	R	1955	481-1145		Dense titania	Pressed from Titanium Alloy Manufacturing Co. heavy grade titania with 7% binder (same composition as the above); set on c.p. zirconia powder, heated at a rate of 60 C per hr, and held for 20 min at a peak temperature of 1390 C; 3.95 sp gr; 0.1% apparent porosity; 14.0% shrinkage; specimen size: 4.92 cm O.D., 2.61 cm I.D., and 0.5 in. thick; same specimen assembly as the above.
21	293	C	1957	473-1073	±4		99.5^+ pure; single crystal; specimen cubic in shape of 0.875 in. sides; supplied by Linde Co.; cut from fired slugs; heat flow parallel to the c-axis; polycrystalline alumina used as comparative material.
22	293	C	1957	373-1073	±4		Similar to the above specimen except heat flow measured perpendicular to c-axis.
23	293	C	1957	419-1173	±4	No. 1	99.5^+ pure; polycrystalline; specimen cubic in shape of 0.875 in. sides, supplied by Baker Chemical Co.; gravimetric porosity 2.1%; microscopic porosity 2.5%; polycrystalline alumina used as comparative material.
24	293	C	1957	503-923	±4	No. 1	Above specimen, second run.
25	293	C	1957	478-1073	±4	No. 2	99.5^+ pure; polycrystalline; gravimetric porosity 3.0%, microscopic porosity 3%; measured before reheating; same comparative material as above.
26	293	C	1957	508-1015	±4	No. 2	The above specimen measured after reheating; same comparative material as above.
27	293	C	1957	458-1068	±4	No. 3	99.5^+ pure; polycrystalline; gravimetric porosity 5.7%, microscopic porosity 8%; measured before reheating; same comparative material as above.
28	293	C	1957	488-1023	±4	No. 3	The above specimen measured after reheating; gravimetric porosity 3.9%, microscopic porosity 5%; same comparative material as above.
29	310	L	1965	2.3-98		1cO	0.0001 ~0.001 each Ba, Cu, and Si, and perhaps smaller quantities of Al, Ca, and Fe; impurity concentration $10^{17} \sim 10^{18}$ atoms cm^{-3}; single crystal cut from a boule grown by the Verneuil Method; specimen 1 x 4 x 24 mm; supplied by National Lead Co.; specimen axis parallel to the c-axis of the crystal with heat flow in the c direction; oxidized in air for two days at 700 C; specimen surfaces roughened with silicon carbide paper with No. 600 used for the final finish.

SPECIFICATION TABLE NO. 36 (continued)

Curve No.	Ref. No.	Method Used	Year	Temp. Range, K	Reported Error, %	Name and Specimen Designation	Composition (weight percent), Specifications and Remarks
30	310	L	1965	2.7-98		1aO	Similar to the above specimen except for size 1 x 4 x 17 mm and specimen axis parallel to the a-axis of the crystal.
31	310	L	1965	2.3-100		1aOVR	The above specimen 1aO heated at 1175 C in vacuum of about 10^{-5} mm Hg for 27 hrs and then, while still in vacuum, cooled to near room temperature in 1 or 2 min, specimen 1 x 4 x 19 mm; defect concentration 1 x 10^{19} cm^{-3}; electrical resistivity 3.5 ohm cm at 300 K.
32	310	L	1965	3.8-33		1aOVRO	The above specimen 1aOVR reoxidized at 600 C in flowing oxygen for 2 days.
33	310	L	1965	1.4-33		1aOVROO	The above specimen 1aOVRO reoxidized at 700 C for 7 days and at 800 C for 10 days in flowing oxygen.
34	310	L	1965	2.5-100		2aOHR	Cut from the same boule as the above specimen 1cO; specimen axis parallel to the a-axis of the crystal; specimen 1 x 4 x 19mm; heated at 900 C in flowing hydrogen gas for 2 hrs, then the hydrogen flushed out with argon and the furnace cooled to near room temperature in 2 or 3 hrs; defect concentration 1 x 10^{20} cm^{-3}; electrical resistivity 0.35 ohm cm at 300 K.
35	310	L	1965	2.0-98		2aOHRO	The above specimen 2aOHR oxidized at 700 C in flowing oxygen for 10 days.
36	310	L	1965	2.0-98		3aNb	0.1 Nb_2O_5; cut from a boule grown from rutile powder which contained 0.1% Nb_2O_5; specimen 1 x 3.5 x 14 mm; supplied by the National Lead Co.; niobium concentration 1.6 x 10^{19} atoms cm^{-3}; electrical resistivity 3.1 ohm cm at 300 K.
37	474		1958	500-1000			Reduced; electrical resistivity reported as 2.9, 3.3, 1.7, and 1.4 ohm cm at 500, 620, 830, and 1000 K, respectively.
38	468	L	1950	327-403		41B	0.452 in. dia x 0.509 in. long.
39	407	L	1949	322-392			No details reported.

DATA TABLE NO. 36 THERMAL CONDUCTIVITY OF TITANIUM DIOXIDE TiO_2

[Temperature, T, K; Thermal Conductivity, k, Watts $cm^{-1} K^{-1}$]

T	k
CURVE 1	
468.2	0.0487
571.2	0.0425
673.2	0.0399
783.2	0.0377
873.2	0.0366
951.2	0.0343
1055.7	0.0333
1133.2	0.0320
1218.2	0.0318
1293.2	0.0316
1373.2	0.0310
1463.2	0.0308
CURVE 2	
463.2	0.0448
566.2	0.0372
670.7	0.0358
778.2	0.0340
871.2	0.0322
943.2	0.0318
1048.2	0.0310
1128.2	0.0305
1213.2	0.0308
1283.2	0.0310
1365.7	0.0318
1448.2	0.0335
CURVE 3	
473.2	0.0481
673.2	0.0377
873.2	0.0347
1073.2	0.0326
1273.2	0.0318
1473.2	0.0318
CURVE 4	
2.2	0.175
2.5	0.245
2.9	0.360
3.2	0.500
CURVE 4 (cont.)	
3.9	0.630
4.5	0.825
10.0	3.30
14.5	4.45
21.0	3.90
23.0	3.50
26.0	2.80
29.0	2.20
33.0	1.65
52.5	0.500
88.0	0.265
CURVE 5	
473.2	0.0751
580.7	0.0619
668.2	0.0519
784.7	0.0445
923.2	0.0410
983.2	0.0374
1034.5	0.0348
CURVE 6	
674.2	0.0418
820.7	0.0357
943.2	0.0337
1123.2	0.0330
CURVE 7	
298.2	0.0837
CURVE 8	
298.2	0.0418
CURVE 9	
316.7	0.0636
336.2	0.0598
354.4	0.0590
377.5	0.0586
414.0	0.0561
CURVE 10	
363.2	0.066
378.2	0.066
473.2	0.049
573.2	0.040
CURVE 11	
432.2	0.102
496.2	0.0983
565.5	0.0883
616.2	0.0824
681.8	0.0774
743.0	0.0725
820.4	0.0710
892.2	0.0672
975.2	0.0737
1044.9	0.0732
1105.2	0.0812
CURVE 12	
419.2	0.0782
474.6	0.0690
559.2	0.0615
628.2	0.0565
753.2	0.0547
816.8	0.0523
874.2	0.0544
933.2	0.0586
1018.2	0.0628
1102.2	0.0706
CURVE 13	
318.7	0.0904
338.4	0.0874
358.5	0.0858
375.9	0.0837
CURVE 14	
317.0	0.0569
335.3	0.0540
356.9	0.0536
377.0	0.0519
CURVE 15*	
338.2	0.0883
CURVE 16*	
338.2	0.0537
CURVE 17	
309.2	0.126
341.2	0.138
CURVE 18	
317.2	0.0879
340.2	0.0711
CURVE 19	
353.2	0.0299
430.2	0.0251
500.2	0.0224
565.2	0.0218
626.2	0.0217
679.2	0.0218
730.2	0.0217
780.2	0.0217
825.2	0.0217
910.2	0.0213
CURVE 20	
481.2	0.0367
543.2	0.0307
640.2	0.0298
723.2	0.0291
798.2	0.0266
873.2	0.0260
938.2	0.0263
998.2	0.0274
1053.2	0.0278
1101.2	0.0286
1145.2	0.0295
CURVE 21	
473.2	0.100
573.2	0.0866
673.2	0.077
773.2	0.070
873.2	0.068
973.2	0.069
1073.2	0.074
CURVE 22	
373.2	0.075
473.2	0.066
573.2	0.059
673.2	0.054
773.2	0.0527
873.2	0.053
973.2	0.056
1073.2	0.059
CURVE 23	
419.2	0.084*
473.2	0.075*
573.2	0.062*
663.2	0.052*
733.2	0.046
923.2	0.038*
983.2	0.037*
1043.2	0.034*
1173.2	0.033*
CURVE 24	
503.2	0.074*
583.2	0.062*
668.2	0.052*
788.2	0.044*
923.2	0.038
CURVE 25	
478.2	0.059
553.2	0.050
673.2	0.042
CURVE 25 (cont.)	
783.2	0.038
853.2	0.036*
963.2	0.034*
1073.2	0.033*
CURVE 26	
508.2	0.069*
563.2	0.062*
623.2	0.055*
713.2	0.046
823.2	0.042*
908.2	0.038*
1015.2	0.034*
CURVE 27	
458.2	0.024
513.2	0.023
593.2	0.023
768.2	0.0225
903.2	0.0225
1068.2	0.022
CURVE 28	
488.2	0.033
683.2	0.032
878.2	0.031
1023.2	0.0309*
CURVE 29	
2.25	0.30
2.90	0.575
3.2	0.83
3.8	1.40
3.9	1.50
4.4	2.05
4.9	2.80
5.0	3.10
5.3	3.80
6.6	7.0
8.0	11.0*
CURVE 29 (cont.)	
11.1	18.0*
15.0	17.5*
16.2	16.0*
18.0	12.5*
20.0	11.0*
27.5	4.0
32.0	2.25
38.5	1.35
47.0	0.77
58.0	0.475
58.0	0.45
80.0	0.28
98.0	0.205
CURVE 30	
2.7	0.47
2.7	0.58
3.0	0.55
3.1	0.67
3.2	0.55
3.5	1.0
4.1	1.5
4.6	2.2
4.75	2.55
5.0	2.9
5.2	2.8
5.7	3.9
6.2	5.1
6.9	7.1
7.5	8.1
8.6	10.5*
9.2	12.8*
10.0	14.3*
10.2	13.0*
11.1	13.2*
11.6	12.7*
13.0	15.0*
15.0	14.0*
15.5	12.0*
15.7	13.0*
18.7	8.8
21.0	6.2
21.2	5.3

*Not shown on plot

DATA TABLE NO. 36 (continued)

T	k
CURVE 30 (cont.)	
21.5	5.7
22.5	4.4
24.0	3.9
24.5	3.8
26.5	3.1
29.0	1.9
31.0	1.75
35.0	1.15
40.0	0.77
47.0	0.52
56.5	0.355
66.0	0.27
79.0	0.195
84.0	0.17
98.0	0.142
CURVE 31	
2.3	0.055
2.5	0.066
2.6	0.075
2.85	0.094
3.2	0.12
3.4	0.15
3.7	0.21
4.0	0.25
4.4	0.32
4.9	0.40
5.4	0.48
5.9	0.60
6.4	0.72
7.1	0.84
8.0	1.02
8.8	1.2
10.2	1.45
11.8	1.65
14.0	1.8
16.7	1.8
18.7	1.6
20.5	1.4
22.0	1.15
24.0	1.05
25.5	0.89
28.5	0.71
31.0	0.60
CURVE 31 (cont.)	
36.0	0.47
42.0	0.36
48.0	0.29
56.0	0.225
64.0	0.185
74.0	0.155
80.0	0.15
100.0	0.12
CURVE 32	
3.8	0.195
4.8	0.33
5.6	0.51
6.6	0 78
7.5	1.02
8.4	1.3
9.4	1.6
10.5	2.0
11.5	2.4
13.0	2.8
14.5	3.2
16.0	3.1
16.5	3.0
18.5	2.85
21.5	2.45
24.0	2.1
27.0	1.6
30.0	1.2
33.0	0.89
CURVE 33	
1.4	0.033
1.8	0.050
2.3	0.074
3.3	0.175
4.6	0.36
4.9	0.44
5.5	0.56
6.3	0.76
7.3	1.05
8.3	1.42
9.8	2.2
11.0	2.7
CURVE 33 (cont.)	
13.0	3.4
14.5	3.7
16.0	3.8
18.5	3.4
21.5	2.7
24.5	2.1
28.0	1.45
33.0	0.96
CURVE 34	
2.45	0.0058*
2.9	0.0086*
3.3	0.011
4.4	0.0205
4.5	0.023
4.9	0.0255
5.1	0.028
5.6	0.032
6.1	0.0375
6.8	0.042
7.8	0.049
9.0	0.057
10.5	0.065
11.7	0.073
13.6	0.083
15.2	0.088
18.0	0.098
20.5	0.107
21.2	0.115
22.5	0.117
24.0	0.115
25.5	0.115
30.5	0.127
36.0	0.135
40.0	0.132
50.0	0.130
61.0	0.115
72.0	0.10
83.0	0.098
83.0	0.090
100.0	0.088
CURVE 35	
2.0	0.125
2.3	0.18
2.9	0.33
3.5	0.50
3.5	0.56
4.8	1.2
4.8	1.25
5.3	1.65
5.7	2.0
6.3	2.45
6.8	2.0
7.2	2.7
8.0	3.5
9.2	4.2
9.3	3.7
10.5	4.0
12.0	4.7
13.5	5.1
15.5	5.0
16.5	4.8
18.5	4.2
21.0	3.5
23.0	4.5
25.5	2.9
28.0	1.85
32.0	1.35
36.0	0.90
40.0	0.66
45.0	0.50
48.0	0.42
53.0	0.33
59.0	0.27
69.0	0.20
82.0	0.16
98.0	0.125
CURVE 36	
2.0	0.18
2.5	0.30
3.0	0.57*
3.7	1.15*
3.7	1.35
4.0	1.95
5.1	2.8*
CURVE 36 (cont.)	
5.8	3.8
6.5	5.9
7.4	8.6
8.3	11.5
9.1	15.0
10.2	18.0
11.2	21.5
12.3	24.0
12.4	23.0
14.0	18.0
15.5	13.5
16.0	9.8
17.0	7.5
17.5	9.0
18.5	6.1
19.0	7.4
20.0	5.0
21.0	4.8
23.0	3.3
24.0	3.0
26.0	2.3
27.5	1.9
29.5	1.5
33.0	1.1
37.0	0.7
37.5	0.77
41.0	0.58
46.0	0.44
51.0	0.36
58.0	0.27*
73.0	0.19
83.0	0.17*
98.0	0.135
CURVE 37*	
500	0.0478
700	0.0410
900	0.0303
1000	0.0340
CURVE 38*	
327.4	0.0502
354.9	0.0515
384.1	0.0532
402.9	0.0514
CURVE 39*	
312.4	0.0427
339.1	0.0452
353.8	0.0437
392.4	0.0423

* Not shown on plot

FIGURE AND TABLE NO. 36R RECOMMENDED THERMAL CONDUCTIVITY OF TITANIUM DIOXIDE TiO_2

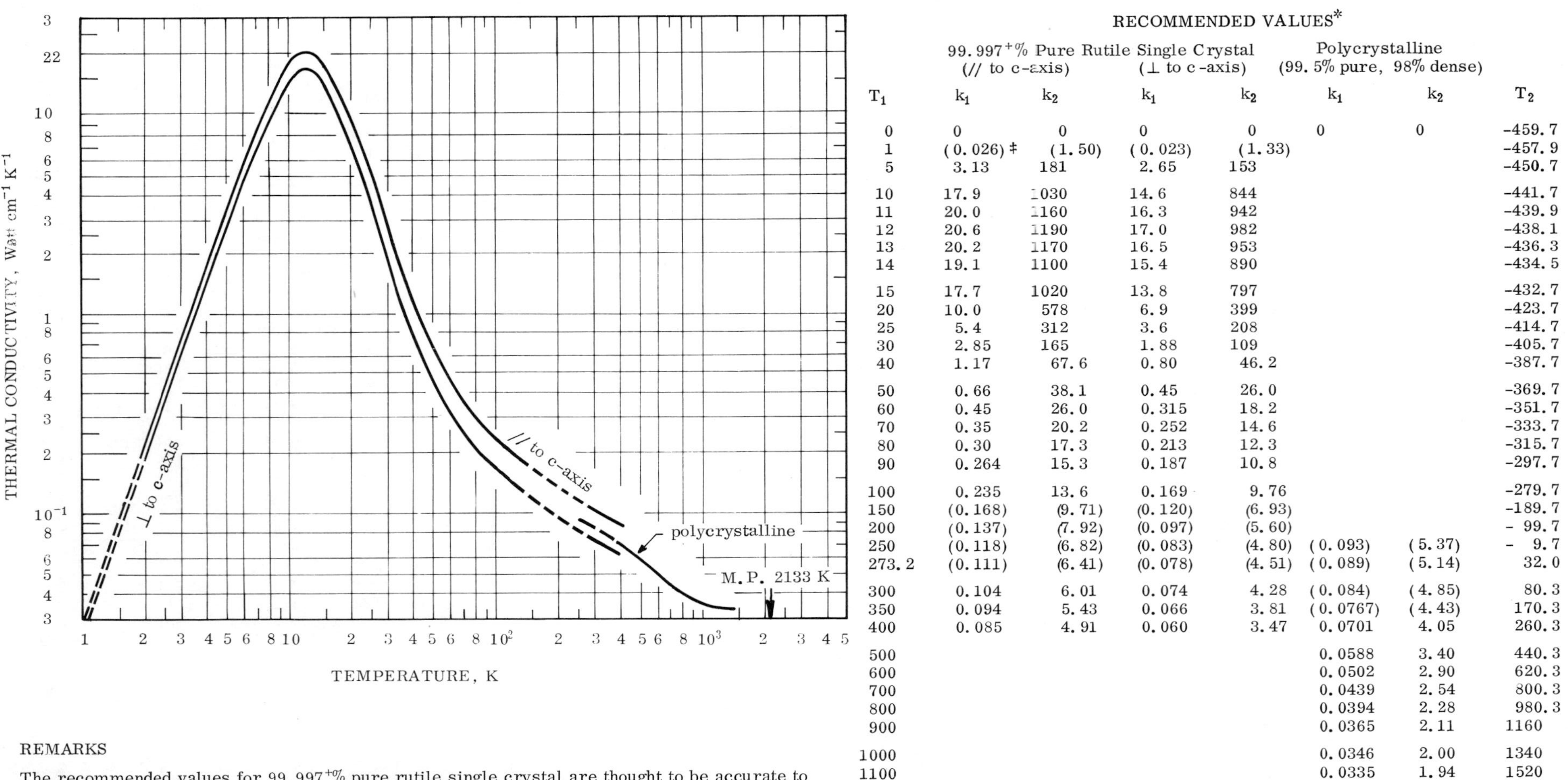

RECOMMENDED VALUES*

T_1	99.997^{+}% Pure Rutile Single Crystal (// to c-axis) k_1	k_2	(⊥ to c-axis) k_1	k_2	Polycrystalline (99.5% pure, 98% dense) k_1	k_2	T_2
0	0	0	0	0	0	0	-459.7
1	(0.026)‡	(1.50)	(0.023)	(1.33)			-457.9
5	3.13	181	2.65	153			-450.7
10	17.9	1030	14.6	844			-441.7
11	20.0	1160	16.3	942			-439.9
12	20.6	1190	17.0	982			-438.1
13	20.2	1170	16.5	953			-436.3
14	19.1	1100	15.4	890			-434.5
15	17.7	1020	13.8	797			-432.7
20	10.0	578	6.9	399			-423.7
25	5.4	312	3.6	208			-414.7
30	2.85	165	1.88	109			-405.7
40	1.17	67.6	0.80	46.2			-387.7
50	0.66	38.1	0.45	26.0			-369.7
60	0.45	26.0	0.315	18.2			-351.7
70	0.35	20.2	0.252	14.6			-333.7
80	0.30	17.3	0.213	12.3			-315.7
90	0.264	15.3	0.187	10.8			-297.7
100	0.235	13.6	0.169	9.76			-279.7
150	(0.168)	(9.71)	(0.120)	(6.93)			-189.7
200	(0.137)	(7.92)	(0.097)	(5.60)			- 99.7
250	(0.118)	(6.82)	(0.083)	(4.80)	(0.093)	(5.37)	- 9.7
273.2	(0.111)	(6.41)	(0.078)	(4.51)	(0.089)	(5.14)	32.0
300	0.104	6.01	0.074	4.28	(0.084)	(4.85)	80.3
350	0.094	5.43	0.066	3.81	(0.0767)	(4.43)	170.3
400	0.085	4.91	0.060	3.47	0.0701	4.05	260.3
500					0.0588	3.40	440.3
600					0.0502	2.90	620.3
700					0.0439	2.54	800.3
800					0.0394	2.28	980.3
900					0.0365	2.11	1160
1000					0.0346	2.00	1340
1100					0.0335	1.94	1520
1200					0.0328	1.90	1700
1300					0.0323	1.87	1880
1400					0.0321	1.85	2060

REMARKS

The recommended values for 99.997^{+}% pure rutile single crystal are thought to be accurate to within 10 to 15% of the true values at temperatures above 30 K. The thermal conductivity near and below the corresponding temperature of its maximum is highly sensitive to small physical and chemical variations of the specimens, and the recommended values below 30 K are intended as typical values for indicating the general trend.

The recommended values for 99.5% pure, 98% dense, polycrystalline TiO_2 that are supported by experimental data are thought to be accurate to within 10% of the true values at temperatures from 400 to 1000 K and 10 to 15% at other temperatures.

* T_1 in K, k_1 in Watt cm^{-1} K^{-1}, T_2 in F, and k_2 in Btu hr^{-1} ft^{-1} F^{-1}.

‡ Values in parentheses are extrapolated or interpolated.

SPECIFICATION TABLE NO. 37 THERMAL CONDUCTIVITY OF TUNGSTEN TRIOXIDE WO_3

Curve No.	Ref. No.	Method Used	Year	Temp. Range, K	Reported Error, %	Name and Specimen Designation	Composition (weight percent), Specifications and Remarks
1	254	E	1955	396-568	3		0.28 ZnO; coarse crystalline structure; prepared by firing H_2WO_3 and ZnO; considerable porosity.

DATA TABLE NO. 37 THERMAL CONDUCTIVITY OF TUNGSTEN TRIOXIDE WO_3

[Temperature, T, K; Thermal Conductivity, k, Watt $cm^{-1} K^{-1}$]

T	k
CURVE 1*	
396.2	0.249
428.2	0.251
487.2	0.254
523.2	0.257
568.2	0.258

* No graphical presentation

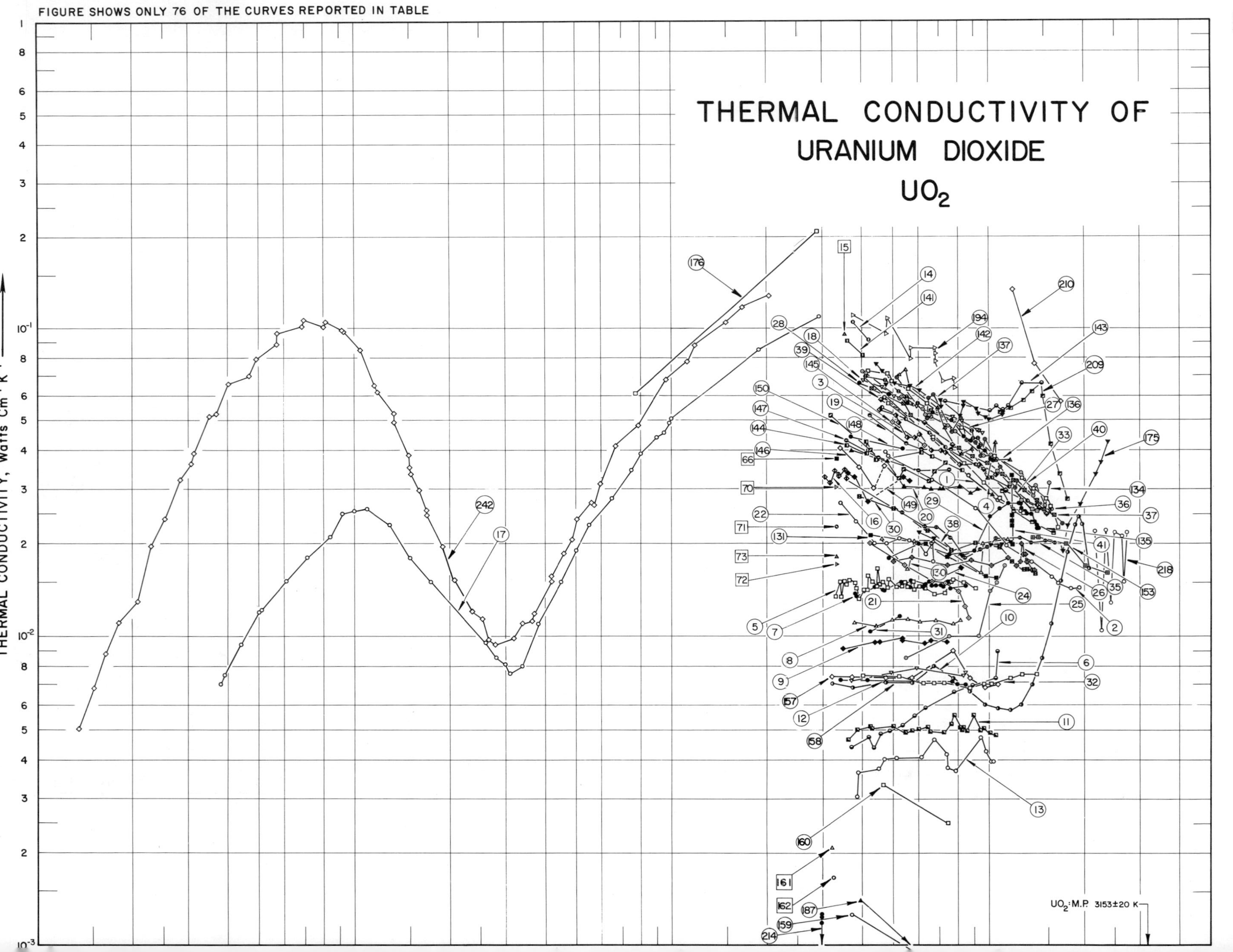

FIGURE SHOWS ONLY 76 OF THE CURVES REPORTED IN TABLE
THERMAL CONDUCTIVITY OF
URANIUM DIOXIDE
UO2
THERMAL CONDUCTIVITY, Watts Cm-1 K-1
UO2: M.P. 3153±20 K

SPECIFICATION TABLE NO. 38 THERMAL CONDUCTIVITY OF URANIUM DIOXIDE UO_2

[For Data Reported in Figure and Table No. 38]

Curve No.	Ref. No.	Method Used	Year	Temp. Range, K	Reported Error, %	Name and Specimen Designation	Composition (weight percent), Specifications and Remarks
1	364	C	1954	473-1273			Polycrystalline; cast from a suspension prepared by Argonne National Laboratory and fired at 1980 C in vacuum; measured in vacuum; bulk density 8. 0 g cm^{-3}; porosity 26. 7%.
2	29	R	1956	476-1941			Specimens made up of 15 pieces of 0. 5 in. I. D. and 3. 0 in. O. D. with a total height of 9 in.; prepared from a mill batch consisting of 500 g of uranium oxide, 1. 75 g of stearic acid, 3. 5 g of methyl cellulose, 2. 5 cm^3 of 0. 5% Aerosol OT in octyl alcohol and 160 cm^3 of water, ground for 16 hrs and dried at 80 C for 24 hrs, then crushed and cold pressed at 20 tons cm^{-2}, again dried for 24 hrs at 80 C after which heated in dry hydrogen up to 1400 C then up to 1500 C in steam finally cooled to room temperature in a hydrogen atmosphere; density 8. 17 g cm^{-3}.
3	30	C	1960	457-1058	±5 (T<673) ±10 (T>673)	No. 4	Oxygen-uranium ratio 2. 01; specimen 0. 625 in. in dia and 2. 242 in. long; prepared from MeW high fired grade (-400 mesh) UO_2 powder with isotropic, coarse, crystalline grains and particles of 14. 9 μ average size; pressure bonded for 3 hrs at 1423 K and 10, 000 psi and then bonded again at 1533 K for 3 hrs at 10, 000 psi; density 10. 1 g cm^{-3} (92% of theoretical density).
4	45	C	1953	486-1298			Pure; crystalline; cubic specimen, supplied by Argonne National Laboratory; prepared by hydrostatical pressing and slowly firing to 1980 C in vacuum; bulk density 8. 0 g cm^{-3}; total porosity 26%.
5	166	R	1953	330-917			In powder form; 63% solid particles by volume; measured in helium.
6	166	R	1953	373-1066			Similar to the above specimen except measured in argon.
7	167	R	1955	379-849			In powder form; 40. 5% void volume; measured in helium at pressures ranging from 59. 3 to 136. 9 psia.
8	167	R	1955	379-819			Similar to the above specimen except measured in a helium and argon mixture of 1. 857 to 1 volumetric ratio at pressures ranging from 47. 3 to 84. 3 psia.
9	167	R	1955	348-742			Similar to the above specimen except measured in a helium and argon mixture of 0. 953 to 1 volumetric ratio at pressures ranging from 39. 4 to 96. 3 psia.
10	167	R	1955	372-846			Similar to the above specimen except measured in a helium and argon mixture of 0. 333 to 1 volumetric ratio at pressures ranging from 46. 3 to 84. 3 psia.
11	167	R	1955	364-1061			Similar to the above specimen except measured in argon at pressures ranging from 44. 3 to 94. 4 psia.
12	167	R	1955	443-904			Similar to the above specimen except measured in nitrogen at pressures ranging from 49. 3 to 83. 3 psia.
13	167	R	1955	385-1035			Similar to the above specimen except measured in a xenon and krypton mixture of 4. 989 to 1 volumetric ratio at pressures ranging from 18. 78 to 74. 3 psia.

SPECIFICATION TABLE NO. 38 (continued)

Curve No.	Ref. No.	Method Used	Year	Temp. Range, K	Reported Error, %	Name and Specimen Designation	Composition (weight percent), Specifications and Remarks
14	136	C	1959	375, 421	±4	$UO_{2.00}$	Bulk density 10.08 g cm^{-3}; porosity 7.5%.
15	136	C	1959	354.2	±4	UO_{2+x}	The above specimen heated in air at 150 to 175 C; composition undetermined.
16	136	C	1959	306-359	±4	$UO_{2.18}$	Bulk density 9.83 g cm^{-3}; porosity 10%.
17	168	L	1961	3.8-294			Prepared by extrusion and sintering; density 9.97 g cm^{-3}; grain 5 to 20 μ in dia; crystal distortion not more than 1000 Å.
18	169	R	1961	403-958		WAPD 22-11	Pressurized Water Reactor Core I production pellets of natural uranium dioxide supplied by Mallinckrodt Chemical Works; molten uranyl nitrate hexahydrate pyrolized to UO_3 and reduced to UO_2 with cracked ammonia to an apparent density of 3 g cm^{-3}, then particles agglomenated with polyvinyl alcohol, blended with sterotex and compacted at 125 psi to a green density of 73-74% of theoretical; sintered for 8 hrs at 1675 C into pellets with density of 10.4 g cm^{-3} (99.9% theoretical density); UO_2 pellets centerless ground on their lateral surfaces and end ground to a height of 0.7976 cm; fuel capsule (containing 15 UO_2 pellets) 0.907 cm I.D., 2.8575 cm O.D.; 11.96 cm nominal length pellets assembled into fuel capsule with zero diametral clearance; fuel chamber atmosphere 1 Kr + 3 X_e by volume at 14.7 psi; irradiated in Material Testing Reactor for 45 days with estimated total exposure of 0.57 x 10^{21} nvt and total burn-up 4.1 x 10^{19} fiss. cm^{-3}; data taken as the effective thermal conductivity (the average thermal conductivity from the fuel center to the inner surface of the cladding of the capsule) versus the temperature at the center of the specimen measured in the Material Testing Reactor (in-pile measurements); this experiment contained a gamma heat capsule, and the gamma heat values used in the thermal conductivity calculations for this test.
19	169	R	1961	423-768		WAPD 22-4	Same supplier, fabrication method, and measuring condition as the above specimen; density 10.27 g cm^{-3} (93.7% theoretical density); diametral clearance 0.0038 cm; fuel chamber filled with helium; irradiated for one day with estimated total exposure 0.43 x 10^{18} nvt and total burn-up 3.6 x 10^{16} fiss. cm^{-3}.
20	169	R	1961	373-1023		WAPD 22-14	Same supplier, fabrication method, and measuring condition as the above specimen; density 10.31 g cm^{-3} (94.1% theoretical density); diametral clearance 0.0064 cm; irradiated for 30 days with estimated total exposure 0.32 x 10^{21} nvt and total burn-up 2.4 x 10^{19} fiss. cm^{-3}.
21	169	R	1961	423-863		WAPD 22-1	Same supplier, fabrication method, and measuring condition as the above specimen; density 10.3 g cm^{-3} (94% theoretical density); diametral clearance 0.0089 cm; fuel chamber filled with helium; irradiated for 15 days with an estimated total exposure 0.8 x 10^{20} nvt and total burn-up 6.5 x 10^{16} fiss. cm^{-3}; no gamma heat measuring capsule in the assembly; gamma heat values assumed in the thermal conductivity calculations for this test.

SPECIFICATION TABLE NO. 38 (continued)

Curve No.	Ref. No.	Method Used	Year	Temp. Range, K	Reported Error, %	Name and Specimen Designation	Composition (weight percent), Specifications and Remarks
22	169	R	1961	343-675		WAPD 22-3	Same supplier, fabrication method, and measuring condition as the above specimen; density 10.15 g cm^{-3} (92.6% theoretical density); diametral clearance 0.020 cm; fuel chamber filled with helium; irradiated for 15 days with estimated total exposure 0.67 x 10^{20} nvt and total burn-up 5.3 x 10^{18} fiss. cm^{-3}.
23	169	R	1961	623, 663		WAPD 22-9	Same supplier, fabrication method, and measuring condition as the above specimen; density 10.28 g cm^{-3} (93.8% theoretical density); diametral clearance 0.020 cm; irradiated for one day at estimated total exposure 0.46 x 10^{18} nvt and total burn-up 3.8 x 10^{16} fiss. cm^{-3}.
24	169	R	1961	623, 823		WAPD 22-2	Same supplier, fabrication method, and measuring condition as the above specimen; density 10.25 g cm^{-3} (93.5 theoretical density); diametral clearance 0.020 cm; fuel chamber evacuated to less than 10^{15} mm; irradiated for one day with estimated total exposure 0.19 x 10^{18} nvt and total burn-up 1.6 x 10^{16} fiss. cm^{-3}.
25	169	R	1961	548-1133		WAPD 22-7	Same supplier, fabrication method, and measuring condition as the above specimen; density 10.41 g cm^{-3} (95% theoretical density); diametral clearance 0.033 cm; irradiated for 15 days with an estimated total exposure 0.19 x 10^{21} nvt and total burn-up 1.5 x 10^{19} fiss. cm^{-3}; first run.
26	169	R	1961	503-1403		WAPD 22-7	The above specimen; second run.
27	169	R	1961	413-948		WAPD 22-5	Laboratory-prepared natural UO_2 pellets; pellets preparation: UNH converted to UO_3 in porcelain ware at 200-400 C, milled for 16 hrs in a zircaloy-2 ball mill into 200 mesh particles, reduced with dry hydrogen at 800 C to UO_2 in platinum boats, and then ball-milled using 0.953 cm dia uranium balls in a rubber-lined vessel, granulated with 102% polythylene glycol binder and pressed at 20-4 psi into pellets; sintered green pellets at 1750 C for 14 hrs to a density of 94.1% theoretical density; the UO_2 pellets ground to the same dimension as that of specimen WAPD 22-11; same fuel capsule dimensions as that of specimen WAPD 22-11; pellets put into fuel capsule with 0.0038 cm diametral clearance; fuel chamber filled with helium at 14.7 psi; irradiated in MTR for 38 days with estimated total exposure 0.5 x 10^{21} nvt and total burn-up 3.6 x 10^{19} fiss. cm^{-3}; data taken as the effective thermal conductivity versus the temperature of the centre of the specimen measured in MTR (in-pile measurement); contained a gamma heat capsule, and the gamma heat values used in the thermal conductivity calculations for this test; first run.
28	169	R	1961	393-1033		WAPD 22-5	The above specimen; second run.
29	169	R	1961	488-1448		WAPD 22-15	UO_2 being 1.02% enriched; same supplier, fabrication method, and measuring condition as the above specimen; density 10.55 g cm^{-3}; diametral clearance 0.0089 cm; fuel chamber filled with 1 Kr + 3 Xe by volume at 14.7 psi; irradiated for 15 days with an estimated total exposure 0.14 x 10^{21} nvt and total burn-up 1.6 x 10^{19} fiss. cm^{-3}.

SPECIFICATION TABLE NO. 38 (continued)

Curve No.	Ref. No.	Method Used	Year	Temp. Range, K	Reported Error, %	Name and Specimen Designation	Composition (weight percent), Specifications and Remarks
30	169	R	1961	393-1413		WARD 22-6	Non-stoichiometric $UO_{2.15}$; same supplier, fabrication method, and measuring condition as the above specimen; density 10.59 g cm^{-3}; no diametral clearance; irradiated for 15 days with an estimated total exposure of 0.19 x 10^{21} nvt and total burn-up 1.5 x 10^{19} fiss. cm^{-3}.
31	170	R	1953	425, 528			Ellipsoidal specimen prepared by Knolls Atomic Power Lab; bulk density 6.01 g cm^{-3}, porosity 45%, measured in air.
32	170	R	1953	404-1425			The above specimen measured in vacuo.
33	171	R	1958	1073-1423	±10		Tubular specimen; 9.44 mm O.D., 3.86 mm I.D., and 102 mm long; prepared by hydrostatic pressing of non-stoichiometric $UO_{2.18}$ at 10 tons in.$^{-2}$; sintered in a nitrogen atmosphere at 1400 C for 2 hrs, followed by reduction in hydrogen at 1200 C for 2 hrs; oxygen-uranium ratio of sintered specimen 2.00 ±0.005; grain size 2 to 10 μ; density 10.5 g cm^{-3} (96% theoretical density).
34	411	L	1960	373-873		1000	0.248 in. dia x 3.12 in. long; isostatically pressed at 40,000 psi; irradiated by a neutron flux of 1.14 x 10^{19} nvt; density 10.27 g cm^{-3} (93.7% of theoretical value); electrical resistivity reported as 11.0 x 10^2, 286, 59.9, 39.1, 20.1, 11.8, 8.47, 6.76 and 5.81 ohm cm at 25, 100, 200, 300, 400, 500, 600, 700 and 800 C, respectively (averaged over data from one cycle of heating and cooling periods of measurement); measured in a vacuum of ~10^{-5} mm Hg; data corrected to UO_2 of 100% of theoretical density.
35	172	R	1961	338-1623	± 6	$UO_{2.18}$, No. 1	Non-stoichiometric; specimen consisting of a stack of 7 discs 2 in. in dia and 0.5 in. thick; prepared from $UO_{2.2}$ powder, cold-pressed at 167 psi in a die, then hydrostatically pressed at 20,000 lb in.$^{-2}$ and sintered at 1400 C in nitrogen for 2 hrs; measured in argon atmosphere; grain size after test 1.5 μ.
36	172	R	1961	466-1598	± 6	No. 2	Stoichiometric; specimen consisting of a stack of 7 discs 2 in. in dia and 0.5 in. thick; prepared from $UO_{2.2}$ powder; cold pressed at 167 lb in.$^{-2}$ in a die, then hydrostatically pressed at 20,000 lb in.$^{-2}$ and sintered at 1400 C in nitrogen for two hrs, followed by 2 hrs reduction in hydrogen at the same temperature; measured in 85 nitrogen + 15 hydrogen atmosphere; density 10.50 g cm^{-3}.
37	172	R	1961	430-1613	± 6	No. 3	Stoichiometric; specimen consisting of a stack of 7 discs 2 in. in dia and 0.5 in. thick; prepared from $UO_{2.03}$ + Cranko powder, enriched to 1.28 Co; same treatment and test atmosphere as the above specimen; density 10.48 g cm^{-3}; grain size (after test) 2.5 μ.
38	172	R	1961	605-1196	± 6	$UO_{2.13}$, No. 3	Non-stoichiometric; the above specimen oxidized to $UO_{2.13}$.
39	172	R	1961	463-1555	± 6	No. 4	Stoichiometric; specimen consisting of a stack of 7 discs 2 in. in dia and 0.5 in. thick; prepared from $UO_{2.2}$ powder with 1 mole % TiO_2 added as an aid for sintering; having the same treatment and test atmosphere as specimen No. 2; density 10.42 g cm^{-2}; grain size (after test) 17.5 μ.

SPECIFICATION TABLE NO. 38 (continued)

Curve No.	Ref. No.	Method Used	Year	Temp. Range, K	Reported Error, %	Name and Specimen Designation	Composition (weight percent), Specifications and Remarks
40	172	R	1961	419-1656	±6	No. 5	Stoichiometric; specimen consisting of a stack of 7 discs 2 in. in dia and 0.5 in. thick; prepared from $UO_{2.03}$ powder; cold pressed at 167 psi in a die, then hydrostatically pressed at 20,000 lb in.$^{-2}$, pre-sintered at 1400 C for 1 hr in hydrogen, followed by final sintering at 1700 C in cracked ammonia for 10 hrs; measured in 85 nitrogen + 15 hydrogen atmosphere; density 10.34 g cm^{-3}; grain size (after test) 6.6 μ.
41	172	R	1961	1193		UO_{2+x}	Stoichiometric; specimen No. 3 measured at 920 C with increasing oxygen content $0 \le x \le 0.13$.
42	165	C	1960	333.2	±2	A_1	Total impurities <0.003; single-phase polycrystalline; 1.27 cm in dia and in length, prepared from UO_2 powder obtained from hydrogen reduction of ammonium diuranate; pressed in a hardened-steel die at a pressure of 2800 Kg cm^{-2} and sintered at 1650 C for 2 hrs in a hydrogen atmosphere; heat flow parallel to the direction of pressing; a constant axial pressure of 56 Kg cm^{-2} exerted on the specimen during measurement; bulk density 8.40 g cm^{-3}.
43	165	C	1960	333.2	±2	A_2	Similar to the above specimen except bulk density 8.45 g cm^{-3}, grain dia 4 x 10^{-4} cm, and coefficient of variation 14%.
44	165	C	1960	333.2	±2	B_1	Similar to specimen A_1 except bulk density 9.55 g cm^{-3}, grain dia 4.5 x 10^{-4} cm, and coefficient of variation 36%.
45	165	C	1960	333.2	±2	B_2	Similar to specimen A_1 except bulk density 9.65 g cm^{-3}.
46	165	C	1960	333.2	±2	C_1	Similar to specimen A_1 except bulk density 10.20 g cm^{-3}.
47	165	C	1960	333.2	±2	C_2	Similar to specimen A_1 except bulk density 10.30 g cm^{-3}, grain dia 10.4 x 10^{-4} cm and coefficient of variation 4%.
48	165	C	1960	333.2	±2	D_1	Similar to specimen A_1 except bulk density 10.40 g cm^{-3}, grain dia 9.10 x 10^{-4} cm and coefficient of variation 4.5%.
49	165	C	1960	333.2	±2	D_2	Similar to specimen A_1 except bulk density 10.45 g cm^{-3}.
50	165	C	1960	333.2	±2	E_1	Similar to specimen A_1 except bulk density 10.60 g cm^{-3}, grain dia 17.5 x 10^{-4} cm and coefficient of variation 9%.
51	165	C	1960	333.2	±2	A_T	Similar to specimen A_1 except bulk density 8.9 g cm^{-3}, grain dia 2.4 x 10^{-4} cm and coefficient of variation 25%; measured with heat flow perpendicular to the direction of pressing.
52	165	C	1960	333.2	±2	B_T	Similar to specimen A_T except bulk density 9.25 g cm^{-3}, grain dia 3.8 x 10^{-4} cm, and coefficient of variation 32%.
53	165	C	1960	333.2	±2	C_T	Similar to specimen A_T except bulk density 10.20 g cm^{-3}, grain dia 4.8 x 10^{-4} cm, and coefficient of variation 15%.
54	165	C	1960	333.2	±2	D_T	Similar to specimen A_T except bulk density 10.35 g cm^{-3}, grain dia 8.3 x 10^{-4} cm, and coefficient of variation 20%.

SPECIFICATION TABLE NO. 38 (continued)

Curve No.	Ref. No.	Method Used	Year	Temp. Range, K	Reported Error, %	Name and Specimen Designation	Composition (weight percent), Specifications and Remarks
55	165	C	1960	333.2	±2	A	Single-phase polycrystal; 1.27 cm in dia and in length, prepared from Mallinckrodt Ceramic Grade UO_2; bulk density 9.90 g cm^{-3}; grain dia 14.3 x 10^{-4} cm; coefficient of variation 17%; measured with a constant axial pressure of 56 Kg cm^{-2} exerted on the specimen and heat flow parallel to the direction of pressing.
56	165	C	1960	333.2	±2	B	Similar to specimen A except bulk density 10.00 g cm^{-3}, grain dia 16.7 x 10^{-4} cm, and coefficient of variation 12%.
57	165	C	1960	333.2	±2	C	Single-phase polycrystal; 1.27 cm in dia and in length, commercially manufactured from ammonium diuranate; bulk density 10.50 g cm^{-3}; grain dia 14.3 x 10^{-4} cm; coefficient of variation 10%; measured with a constant axial pressure exerted on the specimen and heat flow parallel to the direction of pressing.
58	165	C	1960	333.2	±2	D	Similar to the above specimen except bulk density 10.70 g cm^{-3}, grain dia 15.4 x 10^{-4} cm and coefficient of variation 13%; total impurity <0.04; first run.
59	165	C	1960	333.2	±2	E	From the same batch as specimen D.
60	165	C	1960	333.2	±2	E	Second run of the above specimen.
61	165	C	1960	333.2	±2	$UO_{2.01}$, A	Polycrystalline; 1.27 cm in dia and in length; first prepared by the hydrogen-sintering method, then progressively oxidized at 900 C in a static gas mixture of helium and air (30 cm Hg of helium to 2 cm Hg of air); annealed in pure helium at 900 C for 4 hrs after each oxidation, then cooled to room temperature over a period of 20 hrs; heat flow parallel to the direction of pressing; a constant axial pressure of 56 Kg cm^{-2} exerted on the specimen during measurement; data corrected to zero porosity.
62	165	C	1960	333.2	±2	$UO_{2.04}$, A	The above specimen re-oxidized and re-measured in the same way.
63	165	C	1960	333.2	±2	$UO_{2.055}$, A	The above specimen re-oxidized and re-measured in the same way.
64	165	C	1960	333.2	±2	$UO_{2.07}$, A	The above specimen re-oxidized and re-measured in the same way.
65	165	C	1960	333.2	±2	$UO_{2.08}$, A	The above specimen re-oxidized and re-measured in the same way.
66	165	C	1960	333.2	±2	$UO_{2.095}$, A	The above specimen re-oxidized and re-measured in the same way.
67	165	C	1960	333.2	±2	$UO_{2.11}$, A	The above specimen re-oxidized and re-measured in the same way.
68	165	C	1960	333.2	±2	$UO_{2.01}$, B	0.1 TiO_2; polycrystalline; 1.27 cm in dia and in length; prepared and treated the same way as the above specimen and measured under the same conditions; density 10.25 g cm^{-3}; data corrected to zero porosity.
69	165	C	1960	333.2	±2	$UO_{2.055}$, B	The above specimen re-oxidized and re-measured in the same way.
70	165	C	1960	333.2	±2	$UO_{2.09}$, B	The above specimen re-oxidized and re-measured in the same way.
71	165	C	1960	333.2	±2	$UO_{2.14}$, B	The above specimen re-oxidized and re-measured in the same way.

SPECIFICATION TABLE NO. 38 (continued)

Curve No.	Ref. No.	Method Used	Year	Temp. Range, K	Reported Error, %	Name and Specimen Designation	Composition (weight percent), Specifications and Remarks
72	165	C	1960	333.2	±2	$UO_{2.21}$, C	Polycrystalline; 1.27 cm in dia and in length; prepared by pressing the green uranium dioxide powder in a hardened steel die at a pressure of 2800 Kg cm^{-}; the pressed powder sintered in an atmosphere of steam at 1400 C for 2 hrs and cooled in steam from the sintering temperature; heat flow parallel to the direction of pressing; a constant axial pressure of 56 Kg cm^{-2} exerted on the specimen during measurement; density 10.6 g cm^{-3}; data corrected to zero porosity.
73	165	C	1960	333.2	±2	$UO_{2.21}$, D	Similar to the above specimen.
74	165	C	1960	333.2	±2	$UO_{2.21}$, E	Similar to the above specimen.
75	165	C	1960	333.2	±2	$UO_{2.66}$, F	Polycrystalline; 1.27 cm in dia and in length; ammonium diuranate powder ignited in air at 650 C for 6 hrs, pressed in a hardened steel die at a pressure of 2800 Kg cm^{-2}, sintered in air at 1500 C for 3 hrs, sintered again at 1500 C in oxygen for 1.5 hrs and cooled in oxygen from sintering temperature; heat flow parallel to the direction of pressing; a constant axial pressure of 56 Kg cm^{-3} exerted on the specimen during measurement; density 8.05 g cm^{-3}; data corrected to zero porosity.
76	165	C	1960	333.2	±2	$UO_{2.66}$, G	Similar to the above specimen.
77	165	C	1960	333.2	±2	$UO_{2.66}$, H	Similar to the above specimen.
78	165	C	1960	333.2	±2	S_3	Total impurities <0.004; 0.736 cm in dia; prepared by cold-pressing hydrogen-sintering method; irradiated in an air-cooled self-serve facility in the Chalk River NRX reactor for 20 min with integrated thermal-neutron flux 6.4×10^{15} n cm^{-2}; density 10.1 g cm^{-3}.
79	165	C	1960	333.2	±2	L_3	Similar to the above specimen.
80	165	C	1960	333.2	±2	S_4	Similar to specimen S_3 except irradiated for 1 hr with integrated thermal-neutron flux 1.9×10^{16} n cm^{-2}.
81	165	C	1960	333.2	±2	L_4	Similar to the above specimen except density 10.0 g cm^{-3}.
82	165	C	1960	333.2	±2	S_5	Similar to specimen S_3 except irradiated for 3 hrs with integrated thermal-neutron flux 6.25×10^{16} n cm^{-2}.
83	165	C	1960	333.2	±2	L_5	Similar to the above specimen.
84	165	C	1960	333.2	±2	S_6	Similar to specimen S_3 except irradiated for 10 hrs with integrated thermal-neutron flux 2.3×10^{17} n cm^{-2}; density 10.0 g cm^{-3}.
85	165	C	1960	333.2	±2	L_6	Similar to the above specimen.
86	165	C	1960	333.2	±2	S_7	Similar to specimen S_3 except irradiated for 31 hrs with integrated thermal-neutron flux 7.55×10^{17} n cm^{-2}; density 10.0 g cm^{-3}.
87	165	C	1960	333.2	±2	L_7	Similar to the above specimen.

SPECIFICATION TABLE NO. 38 (continued)

Curve No.	Ref. No.	Method Used	Year	Temp. Range, K	Reported Error, %	Name and Specimen Designation	Composition (weight percent), Specifications and Remarks
88	165	C	1960	333.2	±2	S_8	Similar to specimen S_3 except irradiated for 31 hrs with integrated thermal-neutron flux 8.9 x 10^{17} n cm^{-2}; density 10.2 g cm^{-3}.
89	165	C	1960	333.2	±2	L_8	Similar to the above specimen except density 10.3 g cm^{-3}.
90	165	C	1960	333.2	±2	S_1	Similar to specimen S_3 except irradiated for 4 days with integrated thermal-neutron flux 2.3 x 10^{18} n cm^{-2}; density 10.0 g cm^{-3}.
91	165	C	1960	333.2	±2	L_1	Similar to the above specimen except density 10.1 g cm^{-3}.
92	165	C	1960	333.2	±2	S_2	Similar to specimen S_3 except irradiated 13 days with integrated thermal-neutron flux 7.4 x 10^{18} n cm^{-2};
93	165	C	1960	333.2	±2	L_2	Similar to the above specimen except density 10.0 g cm^{-3}.
94	165	C	1960	333.2	±2	L_9	Similar to specimen S_3 except irradiated for 13 days with integrated thermal-neutron flux 8.15 x 10^{18} n cm^{-2}; density 10.3 g cm^{-3}.
95	165	C	1960	333.2	±2	L_{10}	Similar to specimen S_3 except irradiated for 38 days with integrated thermal-neutron flux 2.25 x 10^{19} n cm^{-2}; density 10.3 g cm^{-3}.
96	165	C	1960	333.2	±2	S_{11}	Similar to specimen S_3 except irradiated for 120 days with integrated thermal-neutron flux 6.8 x 10^{19} n cm^{-2}; density 10.2 g cm^{-3}.
97	165	C	1960	333.2	±2	L_{11}	Similar to the above specimen.
98	165	C	1960	333.2	±2	A_{1a}	Total impurity <0.004; prepared by cold-pressing hydrogen-sintered method; irradiated in a water cooled "g"-rod in the Chalk River NRX reactor for 30 days with integrated thermal-neutron flux 1.0 x 10^{19} n cm^{-2}; density 10.1 g cm^{-3}.
99	165	C	1960	333.2	±2	A_{2a}	Similar to the above specimen.
100	165	C	1960	333.2	±2	B_{1a}	Similar to the above specimen.
101	165	C	1960	333.2	±2	B_{2a}	Similar to the above specimen.
102	165	C	1960	333.2	±2	A_{1b}	Total impurity <0.004, prepared by cold-pressing hydrogen-sintering method.
103	165	C	1960	333.2	±2	A_{1c}	The above specimen annealed in vacuum for one hr at 400 ±5 C.
104	165	C	1960	333.2	±2	A_{2b}	Similar to specimen A_{1b}.
105	165	C	1960	333.2	±2	A_{2c}	The above specimen annealed in vacuum for one hr at 900 ±5 C.
106	165	C	1960	333.2	±2	B_{1b}	Similar to specimen A_{1b}.
107	165	C	1960	333.2	±2	B_{1c}	The above specimen annealed in vacuum for one hr at 600 ±5 C.
108	165	C	1960	333.2	±2	B_{2b}	Similar to specimen A_{1b}.
109	165	C	1960	333.2	±2	B_{2c}	The above specimen annealed in vacuum for one hr at 700 ±5 C.

SPECIFICATION TABLE NO. 38 (continued)

Curve No.	Ref. No.	Method Used	Year	Temp. Range, K	Reported Error, %	Name and Specimen Designation	Composition (weight percent), Specifications and Remarks
110	165	C	1960	333.2	±2	S_{5a}	Similar to specimen S_5; measured before being irradiated.
111	165	C	1960	333.2	±2	S_{5b}	The above specimen annealed in vacuum for one hr at 600 ±5 C.
112	165	C	1960	333.2	±2	L_{5a}	Similar to specimen L_5; measured before being irradiated.
113	165	C	1960	333.2	±2	L_{5b}	The above specimen annealed in vacuum for one hr at 1000 ±5 C.
114	165	C	1960	333.2	±2	S_{6a}	Similar to specimen S_6; measured before being irradiated.
115	165	C	1960	333.2	±2	S_{6b}	The above specimen annealed in vacuum for one hr at 700 ±5 C.
116	165	C	1960	333.2	±2	L_{6a}	Similar to specimen L_6; measured before being irradiated.
117	165	C	1960	333.2	±2	L_{6b}	The above specimen annealed in vacuum for one hr at 900 ±5 C.
118	165	C	1960	333.2	±2	L_{2a}	Similar to specimen L_2; measured before being irradiated.
119	165	C	1960	333.2	±2	L_{2b}	The above specimen annealed in vacuum for one hr at 800 ±5 C.
120	165	C	1960	333.2	±2	S_{7a}	Similar to specimen S_7; measured before being irradiated.
121	165	C	1960	333.2	±2	S_{7b}	The above specimen annealed in vacuum for one hr at 600 ±5 C.
122	165	C	1960	333.2	±2	L_{7a}	Similar to specimen L_7; measured before being irradiated.
123	165	C	1960	333.2	±2	L_{7b}	The above specimen annealed in vacuum for one hr at 500 ±5 C.
124	165	C	1960	333.2	±2	S_{11a}	Similar to specimen S_{11}; measured before being irradiated.
125	165	C	1960	333.2	±2	S_{11b}	The above specimen annealed in vacuum for one hr at 600 ±5 C.
126	165	C	1960	333.2	±2	S_{11c}	The above specimen annealed again in vacuum for 30 min at 700 C.
127	165	C	1960	333.2	±2	S_{11d}	The above specimen annealed again in vacuum for one hr at 1000 C.
128	165	C	1960	333.2	±2	L_{11a}	Similar to specimen L_{11}; measured before being irradiated.
129	165	C	1960	333.2	±2	L_{11b}	The above specimen annealed in vacuum for one hr at 800 ±5 C.
130	173	C	1962	461-1161	±5	$UO_{2.002}$, E-1	Mixture of single crystal particles with 60% +4 mesh, 30% -10 +20 mesh, 6% -35 +65 mesh, 4% -100 +200 mesh; compacted vibrationally in stainless steel cup with wall thickness 0.029 in.; specimen 8.738 cm in dia and 2.502 cm long; density 9.52 g cm^{-3} (86.8% of theoretical value); measured in argon.
131	173	C	1962	424-1168	±5	$UO_{2.002}$, E-1	Similar to the above specimen; measured in helium.
132	173	C	1962	546-705	±5	A	Polycrystalline; prepared from high purity UO_2; specimen 7.513 cm in dia and 2.256 cm long; die pressed; sintered in hydrogen at 1650-1800 C; fine hairline cracks resulting from fabrication method used; measured in argon; first run.
133	173	C	1962	674-1335	±5	A	Second run.

SPECIFICATION TABLE NO. 38 (continued)

Curve No.	Ref. No.	Method Used	Year	Temp. Range, K	Reported Error, %	Name and Specimen Designation	Composition (weight percent), Specifications and Remarks
134	173	C	1962	403-1552	±5	A	Third run.
135	173	C	1962	747-1614	±5	A	The above specimen cooled to room temperature and measured again in argon; one of the cracks markedly widened.
136	173	C	1962	526-1317	±5	$UO_{2.006}$; D(L)	Polycrystalline; prepared from high purity UO_2; 7.600 cm in dia and 2.222 cm long; machined from large sample, hydrostatically pressed and sintered in hydrogen at 1650-1800 C; density 0.22 g cm^{-3}; specimen axis coincided with long axis of original sample; edge chipped; measured in argon.
137	173	C	1962	514-1705	±5	$UO_{2.006}$; D(I)	Similar to the above specimen except dia 7.104 cm, density 10.19 g cm^{-3} and disc axis transverse to the long axis of original piece; edge unchipped.
138	173	L	1962	363-1007	±5	$UO_{2.002}$; No. 65	Polycrystalline; prepared from high purity UO_2; cylincrical specimen 0.6325 cm in dia and 7.704 cm long; extruded to near final size and shape followed by hydrostatically pressing and sintering in hydrogen at 1650-1800 C; density 9.55 cm^{-3}; measured in vacuum of about 2 x 10^{-5} mm Hg.
139	173	L	1962	364-882	±5	$UO_{2.002}$; No. 68	Similar to the above specimen except dia 0.6350 cm, length 7.622 cm and density 10.06 g cm^{-3}.
140	173	L	1962	378-1031	±5	$UO_{2.002}$; No. 70	Similar to specimen No. 65 except dia 0.6375 cm, length 7.658 cm and density 10.45 g cm^{-3}.
141	173	L	1962	361-799	±5	$UO_{2.002}$; No. 1000	Polycrystalline; prepared from high purity UO_2; cylindrical specimen 0.6299 cm in dia and 7.925 cm long; machined from larger compacts, hydrostatically pressed and sintered in hydrogen at 1650-1800 C; density 10.22 g cm^{-3}; measured in vacuum of about 2 x 10^{-5} mm Hg.
142	173	C	1962	416-1118	±5	$UO_{2.003}$; G	Monocrystalline; 1.137 x 0.793 x 4.464 cm; prepared by commercial arc-fusion process; density 10.89 g cm^{-3}; measured in vacuum of about 2 x 10^{-5} mm Hg.
143	173	C	1962	660-1489	±5	$UO_{2.003}$; G	Similar to the above specimen.
144	173	L	1962	330-886	±5	$UO_{2.002}$; No. 11	Polycrystalline; prepared from high purity UO_2; 0.6375 cm in dia and 7.669 cm long; extruded, hydrostatically pressed and sintered in hydrogen at 1650-1800 C; irradiation dose 1.40 x 10^{18} fiss. cm^{-3}; maximum UO_2 temperature during irradiation less than 100 C; density 10.07 g cm^{-3}; measured in vacuum of about 2 x 10^{-5} mm Hg.
145	173	L	1962	498-886	±5	$UO_{2.002}$; No. 11	The above specimen measured at decreasing temperature.
146	173	L	1962	341-472	±5	$UO_{2.002}$; No. 19a	Similar to specimen No. 11 except dia 0.6350 cm, length 7.675 cm, density 10.32 g cm^{-3} and irradiation dose 4.11 x 10^{18} fiss. cm^{-3}.
147	173	L	1962	357-752	±5	$UO_{2.002}$; No. 19b	Similar to the above specimen.
148	173	L	1962	368-752	±5	$UO_{2.002}$; No. 19b	The above specimen measured at decreasing temperature.
149	173	L	1962	332-565	±5	$UO_{2.002}$; No. 51a	Similar to specimen No. 11 except dia 0.6325 cm, length 7.6634 cm, density 10.37 g cm^{-3} and irradiation dose 1.1 x 10^{19} fiss. cm^{-3}.

SPECIFICATION TABLE NO. 38 (continued)

Curve No.	Ref. No.	Method Used	Year	Temp. Range, K	Reported Error, %	Name and Specimen Designation	Composition (weight percent), Specifications and Remarks
150	173	L	1962	319-879	± 5	$UO_{2.002}$; No. 51b	Similar to the above specimen.
151	173	L	1962	504-879	± 5	$UO_{2.002}$; No. 51b	The above specimen measured at decreasing temperature.
152	173	L	1962	334-1159	± 5	$UO_{2.002}$; No. 51c	Similar to specimen No. 51a.
153	174	R	1961	1106-2385	<±15		42.0 to 42.2 mm O.D., 8 mm I.D., 3.5 to 4.0 mm thick; prepared from Mallinckrodt PWR-grade powder; pressed and sintered in dry hydrogen to 85% of the theoretical density and then ground flat on both faces.
154	269	R	1958	373-1173			15 mm in dia and 20 mm long; sintered in a carbon furnace; run 1.
155	269	R	1958	323-1073			The above specimen; run 2.
156	269	R	1958	323-1073			The above specimen; run 3.
157	269	R	1958	323-1073			The above specimen; run 4.
158	269	R	1958	323-2073			The above specimen; run 5.
159	159	L	1955	373, 713		Ty O_2	Powder; pressing pressure 100 lb in.$^{-2}$; previously used in neutron absorption experiment.
160	159	L	1955	468, 749		Ty O_2	Pressed in a 150 ton press.
161	159	C	1955	323.2	20.0	Ty O_2	Sintered; iron used as comparative material.
162	159	L	1955	323.2	25.0	Ty O_2	Fused; extremely porous; with several cracks and fissures in it; surface sanded flat and coated with glycerine.
163	417	R	1964	328-1146	± 3.8		0.0593 O (excess), 0.03 Ca, 0.0265 Fe, 0.0062 Nb, 0.004 C, 0.0037 Sn, 0.003 N, 0.0006 Al, 0.00056 Mo, 0.0002-0.0011 Ni, 0.0002-0.0014 Cr, 0.0001 Cd, 0.0001 Na, 0.00002 Cu, <0.002 F, <0.001 Si, <0.0005 Pb, <0.0004 Sm, <0.00005 Ag, <0.00005 B, <0.00005 Eu, and <0.00003 Gd; specimen in disk form, prepared by cold pressing nuclear grade depleted UO_2 powder (produced by the thermal decomposition and reduction of ammonium diuranate), sintering in hydrogen at 1850 C for 4 hrs; density 93.4% of theoretical value (10.97 g cm^{-3}); grain dia 10 to 20 μ; oxygen to uranium ratio 2.012 ±0.002; run No. 1, heating; data corrected for core expansion and corrected to theoretical density.
164	417	R	1964	496-874	± 3.8		The above specimen, run No. 1, cooling; data corrected for core expansion and corrected to theoretical density.
165	417	R	1964	327-1074	± 3.8		The above specimen, run No. 2; data corrected for core expansion and corrected to theoretical density.
166	417	R	1964	373-1465	± 3.8		The above specimen run No. 3; data corrected for core expansion and corrected to theoretical density.

SPECIFICATION TABLE NO. 38 (continued)

Curve No.	Ref. No.	Method Used	Year	Temp. Range, K	Reported Error, %	Name and Specimen Designation	Composition (weight percent), Specifications and Remarks
167	417	R	1964	572-1571	±3.8		The above specimen run No. 3; data corrected for core expansion and corrected to theoretical density.
168	417	R	1964	382-1569	±3.8		The above specimen run No. 5; data corrected for core expansion and corrected to theoretical density.
169	417	R	1964	323-874	±3.8		The above specimen run No. 6; data corrected for core expansion and corrected to theoretical density.
170	417	R	1964	216-382	±3.8		The above specimen run No. 7; data corrected for core expansion and corrected to theoretical density.
171	287	L	1965	543-853		Specimen a	Oxygen to uranium ratio 2.003; 12.70 mm in dia and 12.7 mm long; prepared by compacting nuclear grade UO_2 powder (United Nuclear Corp) at 3 to 5 ton cm^{-2}, sintered in hydrogen at 1700 C for 2 hrs; 96% of theoretical density; measured in argon atmosphere; values corrected to zero porosity using simplified Loeb's equation.
172	287	L	1965	553-878		Specimen b	Similar to the above specimen.
173	287	R	1965	1233-2153		Specimen c	Similar to the above specimen except 24.10 mm in dia and 25 mm long; structural changes and slight cracks were observed after the measurement.
174	287	R	1965	1873, 1953		Specimen c'	The above specimen measured with decreasing temperature.
175	287	R	1965	1603-2373		Specimen d	Similar to specimen c except the characteristic structure with columnar grains and flat pores observed to have grown about 3 mm from the center, an area 2 mm thick from the outer surface of the specimen remained unchanged.
176	418		1963	77.4, 288			Single crystal; 8 mm thick.
177	419	→	1963	2153-2713			Polycrystalline; fuel capsules containing 95% dense UO_2 pellets irradiated (20 min) at thermal-performance conditions; data determined indirectly from measurements of the radial temperature profiles in fuel capsules which were established by measurement of the size distribution of equiaxed grains in the UO_2 and translation of the observed grain size into temperature on the basis of out-of-pile measurements of UO_2 grain growth as a function of time and temperature and calculated by using the highest value of grain growth activation energy reported in the literature; thermal conductivity values calculated from slopes of the temperature profiles.
178	420	R	1964	373-949		Bett 69-4	Thin-walled (0.076 cm thick) fuel cylinder, 0.907 cm O.D. and 12 cm long; pressure-bonded between a nickel capsule and a thin inner tube supported by shrink-fitted alumina pellets from inside after pressure-bonding; 21.4 W/o U^{235} in total μ; 51.0×10^{20} atoms cm^{-3} μ^{235}; average fuel density 10.6 g cm^{-3} (97% of theoretical value); total thermal exposure 0.00004×10^{20} - 0.014×10^{20} nvt; thermocouples A pair and B pair used; measured in vacuum; first start up.

SPECIFICATION TABLE NO. 38 (continued)

Curve No.	Ref. No.	Method Used	Year	Temp. Range, K	Reported Error, %	Name and Specimen Designation	Composition (weight percent), Specifications and Remarks
179	420	R	1964	674-966		Bett 69-4	The above specimen; total thermal exposure 0.040 x 10^{20} - 0.35 x 10^{20} nvt; total burn-up 0.4 x 10^{20} fiss. cm^{-3}; first start-up.
180	420	R	1964	649-948		Bett 69-4	The above specimen; total thermal exposure 0.39 x 10^{20} - 0.56 x 10^{20} nvt; total burn-up 0.9 x 10^{20} fiss. cm^{-3}; B pair thermocouple used (usually lower); second start-up.
181	420	R	1964	437-977		Bett 69-4	The above specimen; total thermal exposure 0.57 x 10^{20} - 0.84 x 10^{20} nvt; total burn-up 1.6 x 10^{20} fiss. cm^{-3}; B pair thermocouple used (usually lower); third start-up.
182	420	R	1964	679-979		Bett 69-4	The above specimen; total thermal exposure 0.85 x 10^{20} - 1.1 x 10^{20} nvt; total burn-up 2.4 x 10^{20} fiss. cm^{-3}; B pair thermocouple used (usually lower); fourth start-up.
183	420	R	1964	491-966		Bett 69-4	The above specimen; total thermal exposure 1.2 x 10^{20} - 1.7 x 10^{20} nvt; total burn-up 3.5 x 10^{20} fiss. cm^{-3}; B pair thermocouple used (usually lower); fifth start-up.
184	420	R	1964	404-945		Bett 69-4	The above specimen; total thermal exposure 1.7 x 10^{20} - 2.2 x 10^{20} nvt; total burn-up 5.2 x 10^{20} fiss. cm^{-3}; B pair thermocouple used (usually lower); sixth start-up.
185	420	R	1964	396-942		Bett 69-4	The above specimen; total thermal exposure 2.2 x 10^{20} - 2.6 x 10^{20} nvt; total burn-up 6.6 x 10^{20} fiss. cm^{-3}; A pair and B pair thermocouples used; seventh start-up.
186	420	R	1964	356-693		Bett 69-4	The above specimen; total thermal exposure 15.6 x 10^{20} - 15.9 x 10^{20} nvt; total burn-up 28 x 10^{20} fiss. cm^{-3}; A pair and B pair thermocouples used; 47th start-up.
187	356		1942	356, 713	5-10		Powder; mean temperature values taken.
188	356		1942	362.2	5-10	Bett 69-4	Powder; pressed; mean temperature value taken.
189	411	L	1960	323-1073		65	0.249 in. dia x 3.033 in. long; extruded and isostatically pressed at 40,000 psi; density 9.58 g cm^{-3} (87.4% of theoretical value); electrical resistivity reported 103, 63.7, 38.9, 24.9, 16.6, 8.47, 4.02, 1.89, and 1.01 ohm cm at 25, 100, 200, 300, 400, 500, 600, 700, and 800 C, respectively (averaged over data from 2 cycles of heating and cooling periods of measurement); measured in a vacuum of $\sim 10^{-5}$ mm Hg; data corrected to UO_2 of 100% of theoretical density.
190	411	L	1960	323-1073		68	0.250 in. dia x 3.001 in. long; extruded and isostatically pressed at 40,000 psi; density 10.07 g cm^{-3} (91.9% of theoretical value); electrical resistivity reported 39.1 x 10^2, 592, 105, 33.8, 15.7, 7.09, 2.85, 1.27, and 0.667 ohm cm at 25, 100, 200, 300, 400, 500, 600, 700, and 800 C, respectively (averaged over data from one cycle of heating and cooling periods of measurement); measured in a vacuum of $\sim 10^{-5}$ mm Hg; data corrected to UO_2 of 100% of theoretical density.

SPECIFICATION TABLE NO. 38 (continued)

Curve No.	Ref. No.	Method Used	Year	Temp. Range, K	Reported Error, %	Name and Specimen Designation	Composition (weight percent), Specifications and Remarks
191	411	L	1960	323-1073		70	0. 251 in. dia x 3. 015 in. long; extruded and isostatically pressed at 40, 000 psi; density 10. 45 g cm^{-3} (95. 3% of theoretical value); electrical resistivity reported 47. 9 x 10^2, 649, 105, 30. 8, 12. 9, 6. 62, 4. 00, 2. 76, and 2. 13 ohm cm at 25, 100, 200, 300, 400, 500, 600, 700, and 800 C, respectively (averaged over data from one cycle of heating and cooling periods of measurement); measured in a vacuum of $\sim 10^{-5}$ mm Hg; data corrected to UO_2 of 100% of theoretical density.
192	13	C	1953	343	±3		Specimen in the form of rectangular parallelepiped 0. 1875 x 0. 1875 x 1. 75 in. ; supplied by Argonne National Laboratory; density 10. 2 g cm^{-3}; Armco iron used as comparative material.
193	421	R	1963	1311-2116		Test 34	No details reported.
194	422	P	1965	373-1376		B-15	Single crystal; specimen 0. 635 cm in dia and 0. 076 cm thick; density 10. 97 g cm^{-3}; thermal conductivity data calculated from the measured data of thermal diffusivity (measurements were conducted in an atmosphere of purified argon containing $<$ 0. 0001 O and $<$ 0. 002 H_2O), specific heat (data obtained from Kelley, K.K. , Bulletin 476 U.S. Bureau of Mines, 1949) and density.
195	422	P	1965	299-973		B-29	0. 01 Fe, 0. 001 Cl and 0. 6 ccg of sorbed gases impurities; single crystal; specimen 0. 635 cm in dia and 0. 051-0. 102 cm thick; density ~10. 95 g cm^{-3}; thermal conductivity data calculated same as the above specimen.
196	422	P	1965	371-1279		1000	Polycrystalline; specimen 0. 635 cm in dia and 0. 132 cm thick; density 10. 22 g cm^{-3}; thermal conductivity data calculated same as the above specimen.
197	423		1963	358-617		57	Polycrystalline; having an exposure of 1. 34 x 10^{19} fiss. cm^{-3}.
198	424		1959	373-873		68	Specimen 0. 250 in. in dia and 3. 001 in. long; density 10. 07 g cm^{-3} (91. 9% of theoretical density).
199	425	C	1963	373-1473		G	Single crystal; oxygen to uranium ratio 2. 003; specimen 0. 31 x 0. 45 x 1. 76 in. ; density 99. 4% of theoretical value; unirradiated; stainless steel used as comparative material.
200	425	C	1963	473-1273		G	The above specimen after first irradiation with a dose of 10^{15} nvt; same comparative material as above.
201	425	C	1963	473-1073		G	The above specimen after second irradiation with a dose of 10^{15} nvt (total 2 x 10^{15} nvt); same comparative material as above.
202	425	C	1963	471, 473		51	Single crystal; specimen 0. 25 in. in dia and 3 in. long; extruded, hydrostatically pressed and sintered; 0. 062 atomic % burn-up; density 94. 5% of theoretical value; same comparative material as above.
203	425	C	1963	379-609		57	Similar to the above specimen except 0. 12 atomic % burn-up and density 85. 7% of theoretical value.

SPECIFICATION TABLE NO. 38 (continued)

Curve No.	Ref. No.	Method Used	Year	Temp. Range, K	Reported Error, %	Name and Specimen Designation	Composition (weight percent), Specifications and Remarks
204	426	R	1958	430-803			Prepared by cold pressing and hydrogen sintering natural UO_2 obtained from Mallinckrodt Chemical Works; pellets centerless ground to 0.3535 in. dia; placed in a stainless-steel-304 capsule and irradiated in the Material Testing Reactor (MTR) for one day at a power of 10 megawatt-day; 1 w/g gamma heating in stainless steel capsule assumed; density 93.5 to 95% of theroetical value based on a theoretical density of 10.96 g cm^{-3}; effective thermal conductivity vs center temperature reported.
205	426	R	1958	432-790			The above specimen after being irradiated in the MTR for six days at an additional power of 10 megawatt-day (total 20 megawatt-day).
206	426	R	1958	555-795			The above specimen after being irradiated in the MTR for six days at an additional power of 10 megawatt-day (total 30 megawatt-day).
207	426	R	1958	409-748			The above specimen after being irradiated in the MTR for one day at an additional power of 10 megawatt-day (total 40 megawatt-day); estimated exposure 0.7 x 10^{20} nvt thermal.
208	427	R	1963	834-2518	± 5 at > 1370K ±10 at < 1370K		< 0.08 total impurity; prepared by compacting UO_2 powder at 4000 psi in a 3 in. dia steel die; isostatic pressing at 3000 psi, sintered in a hydrogen atmosphere for 12 hrs at 1700 C, machined into a disk of O. D. ~2 in. and I. D. ~0.38 in. and thickness > 0.5 in.; density 95.63% of theoretical value.
209	428	R	1963	1077~2073			Single crystal; specimen cylindrical in form with I. D. 0.1 in. and O. D. 0.25 in.; data obtained from smoothed curve.
210	429	R	1959	1189-1673		UO_2, No. 1	Specimen 0.630 cm in dia and 0.551 cm long; cut from extruded rods; surface sand-blasted; density 9.91 g cm^{-3} (90.3 ±.5% of theoretical value).
211	429	R	1959	1145-1395		UO_2, No. 2	Specimen 0.633 cm in dia and 0.625 cm long; cut from extruded rods; surface sand-blasted; density 9.87 g cm^{-3} (90.0 ±0.5% of theoretical value).
212	429	R	1962	327-1146			Pressed and sintered to a density of 93.4% of theoretical value.
213	430	→	1966	298.2			2.54 O (calculated); spherical uranium powder obtained from National Lead Co. containing impurities: 0.05 Fe, 0.01 Mg, 0.008 Mo, 0.005 Si, < 0.005 K, < 0.005 P, < 0.005 Ti, < 0.005 Zn, < 0.002 Ca, < 0.001 As, < 0.001 Na, 0.0005 Ni, < 0.0005 Al, < 0.0005 Co, < 0.0005 Sn, 0.0004 Mn, 0.0002 Cu, 0.0001 Pb, traces of Ag, Bi, Cr, Li, Sb, Be, and B; oxidized by spreading over the bottom of a Petri dish and placed in an oven at 150 C; specimen contained in a 0.75 in. dia x 2 in. long stainless steel cylindrical cell; mesh size -70 + 80; thermal conductivity measured by using the transient line source method, the heat source was a 36-gauge constantan wire contained in a 0.025 in. O. D. hypodermic tube soldered along the axis of the cylindrical cell, data calculated from the measured line temperatures at two certain times; measured in nitrogen at 1 atm.

SPECIFICATION TABLE NO. 38 (continued)

Curve No.	Ref. No.	Method Used	Year	Temp. Range, K	Reported Error, %	Name and Specimen Designation	Composition (weight percent), Specifications and Remarks
214	430	→	1966	298.2			Similar to the above specimen; measured in nitrogen under pressures in the range $5.50 \times 10^{-3} \sim 6.31 \times 10^{3}$ mm Hg.
215	430	→	1966	298.2			0.06 O (calculated); same source, fabrication, and measuring method as the above specimen except the mesh size -230 +325; measured in nitrogen at 1 atm.
216	430	→	1966	298.2			1.17 O (calculated); same source, fabrication, and measuring method as the above specimen; mesh size -230 +325; measured in nitrogen under pressures in the range $1.08 \times 10^{-2} \sim 5.188 \times 10^{3}$ mm Hg.
217	431	R	1961	331-2390			Stoichiometric; 0.5 in. in dia, 0.5 in. high pellets; 95% dense, cold pressed and sintered; irradiated in the General Electric Test Reactor cable facility, UO_2 enrichment 17%.
218	431	R	1966	2154-2713			Similar to the above specimen.
219	432	R	1962	499-882			93.4% dense; pressed and sintered.
220	433	R	1962	300-468	±5		93.4% dense.
221	434	R	1962	373-1465			93.4% dense; pressed and sintered.
222	434	R	1962	512-1571			Similar to the above specimen.
223	435	R	1958	352-1399		Type-1	Specimen 15 mm in dia and 10 cm long made up of 20 mm thick pellets; manufactured from ammonium uranate calcined to UO_3, reduced to $UO_{2.00}$ powder, oxidized in air to $UO_{2.17}$, cold pressed and sintered in a carbon-resistance furnace at 1800 C for 1.5 hrs in an atmosphere of H_2 and H_2O; density 10.2 g cm^{-3}; measured with large clearance between the pellets and the can wall.
224	435	R	1958	329-2101		Type-1	Similar to above specimen except measured in aluminum can with argon as filling gas.
225	435	R	1958	325-1306		Type-2	Specimen 15 mm in dia and 10 cm long made up of 20 mm thick pellets; UO_2 powder produced in a similar way as type-1 but giving $UO_{2.08}$; sintered in a molybdenum resistance furnace at 1600 to 1620 C for 8 hrs in an atmosphere of cracked NH_3; pellets centerless grained to an accuracy of 0.001 cm and the ends plane parallel grained; density 10.0 to 10.1 g cm^{-3}; measured with large clearance between the pellets and the can wall.
226	435	R	1958	472-1273		Type-3	Specimen 15 mm in dia and 10 cm long made up of 20 mm thick pellets; same UO_2 powder as for type-2 used, sintered at 1550 C for 8 hrs in an atmosphere of cracked $NH_3 + H_2O$; pellets centerless grained to an accuracy of 0.001 cm and the ends plane parallel grained; density 10.2 to 10.3 g cm^{-3}; measured with 0.1 mm clearance; average of several runs.
227	435	R	1958	470-1273		Type-3	Similar to the above specimen measured without clearance; average of several runs.

SPECIFICATION TABLE NO. 38 (continued)

Curve No.	Ref. No.	Method Used	Year	Temp. Range, K	Reported Error, %	Name and Specimen Designation	Composition (weight percent), Specifications and Remarks
228	436	R	1963	639.8		CC2-GG	UO_2 pellets, obtained by calcining the powder (specific surface 3 m^2 g^{-1}) at 1650 C in hydrogen; 1.47% enriched (μ^{235}/μ_{total}); irradiated in EL3 successively first for 24 days in peripheral position at a constant power of 17.5 megawatts (α = neutron flux/max EL3 flux = 0.52) and then for 3 days (at α = 1); density 10.6 g cm^{-3}; average value of thermal conductivity at mean temperature reported.
229	436	R	1963	640.7		CC2-GG	The above specimen with a total burn-up of 5.0 x 10^{17} fiss. cm^{-3}.
230	436	R	1963	670.7		CC2-GG	The above specimen with a total burn-up of 2 x 10^{18} fiss. cm^{-3}.
231	436	R	1963	680.7		CC2-GG	The above specimen with a total burn-up of 3.4 x 10^{18} fiss. cm^{-3}.
232	436	R	1963	738.2		CC2-GG	The above specimen with a total burn-up of 7.1 x 10^{18} fiss. cm^{-3}.
233	363	P	1965	1252-1697			No details reported.
234	437	L	1962	338-724			UO_2 wafers prepared from ammonium diuranate; after dry ball milling with Al_2O_3 balls and adding 2% "carbowax 20 M" as a binder, wafers pressed and sintered at 1700 C in a dry hydrogen atmosphere for 4 hrs; specimen ground flat with a surface roughness ranging from 15 to 150 μ in.; density 93% of theoretical density.
235	438	C	1959	785-821			0.25 Nb_2O_5; specimen 0.856 ±0.001 in. in dia and 1.000 ±0.001 in. high; prepared from ceramic grade, low-bulk-density UO_2 supplied by Mallinckrodt Chemical Works and 99 pure, 325 mesh Nb_2O_5 supplied by Fansteel Metallurgical Corp; solid solution of U_3O_8 with Nb_2O_5 prepared by dissolving equal weights of the additive oxide and UO_2 in acid, then the additive and UO_2 coprecipitated with ammonium hydroxide, filtered and calcined in air at 600 C; after granulation, pressed at 20 tsi then heated to 1300 C in argon and soaked for 4 hrs, reduced at 1300 C in an atmosphere of about 10 volume % hydrogen - 90 volume % argon for 1 to 1.5 hrs, cooled to 900 C in the same atmosphere and to room temperature in argon; centerless ground; electrical resistivity 2 x 10^3 to 1.1 x 10^4 ohm cm; density 10.34 g cm^{-3} (94.6% of theoretical); Mallinckrodt ceramic grade UO_2 used as comparative material.
236	438	C	1959	684-1093			0.21 Y_2O_3; specimen 0.856 ±0.001 in. in dia and 1.000 ±0.001 in. high; prepared from ceramic grade low-bulk-density UO_2 supplied by Mallinckrodt Chemical Works and 99.9 pure, 325 mesh Y_2O_3 supplied by Rare Earths, Inc.; prepared in the same way as the above specimen; electrical resistivity 330 ohm cm; density 10.53 g cm^{-3} (96.3% of theoretical); same comparative material as above.
237	438	C	1959	455-1498			Similar to the above specimen except 0.84 Y_2O_3, electrical resistivity 74 ohm cm, and density 10.42 g cm^{-3} (95.9% of theoretical).
238	438	C	1959	633-783			0.59 excess; Mallinckrodt ceramic grade UO_2 used as comparative material.

SPECIFICATION TABLE NO. 38 (continued)

Curve No.	Ref. No.	Method Used	Year	Temp. Range, K	Reported Error, %	Name and Specimen Designation	Composition (weight percent), Specifications and Remarks
239	439	R	1962	881-1613		1	Specimen made up of 60% -4 + 6 mesh, 15% -50 + 70 mesh and 25% -400 mesh UO_2; mean density of compressed UO_2 9.69 g cm^{-3} (88% of theoretical), density of sintered UO_2 grains 10.67 g cm^{-3} (97% of theoretical); mean size of elementary grains 10 μ; measured in a helium atmosphere; the data on those of the effective thermal conductivity.
240	439	R	1962	538-1723		2	Similar to the above specimen except mean density of compressed UO_2 9.46 g cm^{-3} (86% of theoretical).
241	439	R	1962	623-1648		3	95.5 UO_2, 1.0665 ZrO_2, 0.7875 MoO_3, 0.5805 Nb_2O_5, 0.5535 CeO_2, 0.504 $SrCo_3$, 0.2565 $BaCo_3$, 0.18 Y_2O_3, 0.171 La_2O_3, 0.1359 Sm_2O_3, 0.135 Pr_6O_{11}, 0.06795 TeO_2, 0.05085 SeO_2, 0.00495 Eu_2O_3, 0.0018 SnO_2, 0.00135 Sb_2O_5, 0.00135 CdO, 0.00090 Ag_2O, 0.00045 Gd_2O_3; specimen granulometry 63.5% -4 + 6 mesh, 15% -50 + 70 mesh, 17% -400 mesh; mean density of compressed UO_2 9.34 g cm^{-3} (87% of theoretical); mean size of elementary grains 10 μ; measured in a helium atmosphere; the data are those of the effective thermal conductivity.
242	440		1964	1.4-204			Single crystal; supplied by the Hanford Laboratories of the General Electric Co.
243	441		1962	523.2			Specimen prepared by Compagnie Industrielle Des Combustibles Atomiques Frittes; extruded rod; density range 10.40-10.60 g cm^{-3}.
244	441		1962	523.2			Specimen prepared by Compagnie Industrielle Des Combustibles Atomiques Frittes; sintered pellet; density range 10.40-10.60 g cm^{-3}.

DATA TABLE NO. 38 THERMAL CONDUCTIVITY OF URANIUM DIOXIDE UO_2

[Temperature, T, K; Thermal Conductivity, k, Watts $cm^{-1}K^{-1}$]

T	k
CURVE 1	
473.2	0.0594
673.2	0.0431
873.2	0.0331
1073.2	0.0276
1273.2	0.0255
CURVE 2	
475.5	0.0395
486.7	0.0384
664.5	0.0320
911.4	0.0260
1140.0	0.0197
1376.9	0.0175
1592.8	0.0156
1677.8	0.0149
1836.1	0.0144
1941.1	0.0145
CURVE 3	
457.2	0.054
513.2	0.052
522.2	0.049
557.2	0.044
589.2	0.044
610.2	0.045
663.2	0.040
713.2	0.040
737.2	0.039
815.2	0.036
937.2	0.036
1058.2	0.033
CURVE 4	
485.7	0.0569
672.2	0.0439
841.7	0.0361
968.2	0.0314
1098.2	0.0280
1298.2	0.0251
CURVE 5	
330.4	0.0135
341.6	0.0150
345.7	0.0134
353.4	0.0151
357.3	0.0147
366.8	0.0152
380.0	0.0149
384.4	0.0143
391.0	0.0132
405.8	0.0142
418.9	0.0142
419.8	0.0158
433.7	0.0151
440.7	0.0149
445.3	0.0145
448.6	0.0166
456.8	0.0147
468.3	0.0150
483.4	0.0144
485.6	0.0153
516.9	0.0145
536.1	0.0146
543.2	0.0146
546.4	0.0150
558.8	0.0145
581.9	0.0151
588.1	0.0142
609.8	0.0146
635.3	0.0143
681.4	0.0138
731.5	0.0139
767.7	0.0151
778.5	0.0153
839.4	0.0150
916.6	0.0143
CURVE 6	
372.6	0.00440
429.6	0.00476
435.4	0.00440
458.4	0.00486
CURVE 6 (cont.)	
490.5	0.00497
537.4	0.00517
584.8	0.00554
636.5	0.00590
782.8	0.00661
892.4	0.00696
1058.6	0.00732
1066.4	0.00897
CURVE 7	
379.3	0.0137
383.2	0.0135
439.3	0.0145
466.5	0.0142
469.8	0.0141
486.5	0.0145*
531.5	0.0150
552.6	0.0146*
570.4	0.0145
572.6	0.0151
580.4	0.0150*
632.1	0.0148
651.5	0.0151
664.3	0.0147
688.7	0.0147
710.9	0.0147
723.2	0.0145
745.4	0.0151
760.9	0.0143
848.7	0.0146
CURVE 8	
379.3	0.0110
444.3	0.0108
498.2	0.0113
559.8	0.0114
609.3	0.0112
684.3	0.0113
769.3	0.0110
818.7	0.0113
CURVE 9	
348.2	0.00912
439.3	0.00954
455.4	0.00957
538.7	0.00987
538.7	0.00974
630.4	0.00952
660.9	0.00971
741.5	0.00961
CURVE 10	
372.1	0.00720
495.9	0.00768
593.7	0.00791
834.8	0.00749
845.9	0.00765
CURVE 11	
363.7	0.00464
385.9	0.00500
425.4	0.00511
429.3	0.00509
508.2	0.00516
549.8	0.00490
574.8	0.00498
604.3	0.00502
645.9	0.00511
654.3	0.00495
724.8	0.00493
765.4	0.00524
780.9	0.00559
818.2	0.00509
828.2	0.00500
838.7	0.00512
859.8	0.00498
898.2	0.00559
942.1	0.00500
970.4	0.00507
1019.3	0.00490
1060.9	0.00483
CURVE 12	
442.6	0.00722
514.8	0.00723
772.1	0.00715
799.3	0.00697
849.3	0.00696
903.7	0.00694*
CURVE 13	
385.4	0.00306
388.7	0.00365
451.5	0.00376
470.9	0.00402
513.2	0.00407
614.8	0.00410
675.4	0.00467
683.7	0.00419
744.3	0.00379
788.7	0.00370
944.3	0.00474
981.5	0.00429
1022.1	0.00396
1034.8	0.00396
CURVE 14	
375.2	0.105
421.2	0.0912
CURVE 15	
354.2	0.095
CURVE 16	
305.7	0.0329
318.2	0.0314
328.2	0.0343
341.2	0.0328
354.2	0.0347
359.2	0.0324
CURVE 17	
3.8	0.0070
3.9	0.0075
4.4	0.0094
5.0	0.0120
5.1	0.0122
6.1	0.0151
7.1	0.0180
8.4	0.0210
9.2	0.0250
10	0.0255
11	0.0260
13	0.0230
15	0.0180
17.5	0.0150
26	0.0095
28	0.0085
30	0.0081
31	0.0076
34	0.0080
38	0.0110
45	0.0150
50	0.0190
55	0.0230
65	0.0280
75	0.0345
80	0.0390
90	0.0440
95	0.0455
99	0.0490
100	0.0505
190	0.0850
294	0.1100
CURVE 18	
403.2	0.068
463.2	0.0645
508.2	0.060
598.2	0.056
663.2	0.054
773.2	0.0516
828.2	0.050
CURVE 18 (cont.)	
923.2	0.0468
938.2	0.0467
958.2	0.0458
CURVE 19	
423.2	0.052
553.2	0.042
618.2	0.0405
768.2	0.0405
CURVE 20	
373.2	0.040
483.2	0.037
543.2	0.0306
628.2	0.0305
663.2	0.030
708.2	0.030
728.2	0.030
813.2	0.0306
881.2	0.029
925.2	0.030
1023.2	0.0288
CURVE 21	
423.2	0.020
543.2	0.017
663.2	0.015
728.2	0.0147
803.2	0.0140
843.2	0.0125
863.2	0.0115
CURVE 22	
343.2	0.027
383.2	0.0235
438.2	0.020
478.2	0.020
523.2	0.0208

* Not shown on plot

DATA TABLE NO. 38 (continued)

T	k
CURVE 22 (cont.)	
563.2	0.0205
593.2	0.020
635.2	0.0185
663.2	0.020
675.2	0.0175
CURVE 23*	
623.2	0.016
663.2	0.017
CURVE 24	
623.2	0.0145
823.2	0.0167
CURVE 25	
548.2	0.0085
753.2	0.010
933.2	0.010
1013.2	0.014
1063.2	0.015
1133.2	0.017
CURVE 26	
503.2	0.0175
576.2	0.018
743.2	0.017
883.2	0.0177
1023.2	0.017
1153.2	0.018
1233.2	0.017
1263.2	0.0175
1273.2	0.020
1293.2	0.0185
1308.2	0.0185
1323.2	0.0175
1353.2	0.0175*
1378.2	0.0165
1403.2	0.0165
CURVE 27	
413.2	0.070
473.2	0.068
523.2	0.0625
553.2	0.060
598.2	0.057
888.2	0.0465
898.2	0.044
928.2	0.044
948.2	0.042
CURVE 28	
393.2	0.0665
488.2	0.061
558.2	0.057
668.2	0.053
768.2	0.050
838.2	0.0445
883.2	0.043
933.2	0.041
978.2	0.040
1018.2	0.0375
1033.2	0.037
CURVE 29	
488.2	0.018
603.2	0.020
753.2	0.0185
903.2	0.019
1013.2	0.0245
1093.2	0.026
1163.2	0.027
1258.2	0.027
1270.2	0.027
1273.2	0.026
1308.2	0.0255
1333.2	0.025
1348.2	0.025*
1373.2	0.025
1403.2	0.0235
1423.2	0.023
1428.2	0.0225
1433.2	0.0225*
1448.2	0.0225
CURVE 30	
393.2	0.0285
503.2	0.026
643.2	0.022
763.2	0.019
913.2	0.020
1083.2	0.0195
1198.2	0.0175
1303.2	0.0165
1333.2	0.0165
1373.2	0.0165*
1393.2	0.0165*
1403.2	0.0165*
1413.2	0.016
CURVE 31	
425.2	0.0104
528.2	0.0117
CURVE 32	
404.2	0.00745
523.2	0.00742
624.2	0.00707
674.2	0.00705
724.0	0.00709
797.0	0.00697
874.2	0.00686
968.2	0.00703
1022.2	0.00701
1174.2	0.00732
1287.2	0.00753
1425.2	0.00753
CURVE 33	
1073.2	0.0341
1173.2	0.0311
1273.2	0.0276
1373.2	0.0259
1423.2	0.0256
CURVE 34*	
373.2	0.0842
473.2	0.0696
573.2	0.0593
673.2	0.0517
773.2	0.0458
873.2	0.0411
CURVE 35	
338.2	0.0335
533.2	0.0251
628.2	0.0226
688.2	0.0226
758.2	0.0209
843.2	0.0180
943.2	0.0192
993.2	0.0197
1068.2	0.0209
1263.2	0.0209
1373.2	0.0201
1503.2	0.0205
1623.2	0.0201
CURVE 36	
466.2	0.0552
550.2	0.0506
653.2	0.0452
734.2	0.0410
913.2	0.0360
1025.2	0.0326
1169.2	0.0305
1338.2	0.0276
1428.2	0.0268
1598.2	0.0259
CURVE 37	
430.2	0.0636
543.2	0.0565
570.2	0.0519
810.2	0.0406
893.2	0.0381
CURVE 37 (cont.)	
998.2	0.0364
1103.2	0.0335
1198.2	0.0318
1308.2	0.0297
1423.2	0.0276
1513.2	0.0259
1613.2	0.0247
CURVE 38	
605.2	0.0218
673.2	0.0209
749.2	0.0192
830.2	0.0184
917.2	0.0197*
1015.2	0.0197
1111.2	0.0197
1185.2	0.0201
1194.2	0.0197*
1196.2	0.0197*
CURVE 39	
463.2	0.0586
550.2	0.0515*
632.2	0.0490
773.2	0.0414
915.2	0.0360*
970.2	0.0347
1137.2	0.0297
1333.2	0.0276*
1454.2	0.0264
1555.2	0.0251
CURVE 40	
419.2	0.0724
481.2	0.0715
496.2	0.0674
559.2	0.0636
641.2	0.0569
825.2	0.0477
923.2	0.0435*
1045.2	0.0385
1277.2	0.0326
CURVE 40 (cont.)	
1390.2	0.0301
1518.2	0.0280
1586.2	0.0264
1656.2	0.0226
CURVE 41 (T = 1193.2)	
x	k
0.00	0.0337
0.00	0.0327*
0.02	0.0264*
0.03	0.0258*
0.04	0.0248
0.05	0.0237
0.06	0.0229
0.07	0.0209*
0.08	0.0211*
0.09	0.0209*
0.10	0.0206*
0.11	0.0203*
0.12	0.0199
0.13	0.0199
CURVE 42*	
333.2	0.0485
CURVE 43*	
333.2	0.050
CURVE 44*	
333.2	0.056
CURVE 45*	
333.2	0.0575
CURVE 46*	
333.2	0.066
CURVE 47*	
333.2	0.067
CURVE 48*	
333.2	0.066
CURVE 49*	
333.2	0.0655
CURVE 50*	
333.2	0.069
CURVE 51*	
333.2	0.054
CURVE 52*	
333.2	0.0605
CURVE 53*	
333.2	0.0645
CURVE 54*	
333.2	0.0645
CURVE 55*	
333.2	0.0695
CURVE 56*	
333.2	0.0685
CURVE 57*	
333.2	0.071
CURVE 58*	
333.2	0.0715

*Not shown on plot

DATA TABLE NO. 38 (continued)

T	k
CURVE 59*	
333.2	0.069
CURVE 60*	
333.2	0.0685
CURVE 61*	
333.2	0.071
CURVE 62*	
333.2	0.064
CURVE 63*	
333.2	0.061
CURVE 64*	
333.2	0.054
CURVE 65*	
333.2	0.0475
CURVE 66	
333.2	0.0375
CURVE 67*	
333.2	0.033
CURVE 68*	
333.2	0.0535
CURVE 69*	
333.2	0.043
CURVE 70	
333.2	0.031
CURVE 71	
333.2	0.0225
CURVE 72	
333.2	0.017
CURVE 73	
333.2	0.018
CURVE 74*	
333.2	0.018
CURVE 75*	
333.2	0.018
CURVE 76*	
333.2	0.018
CURVE 77*	
333.2	0.018
CURVE 78*	
333.2	0.059
CURVE 79*	
333.2	0.060
CURVE 80*	
333.2	0.0565
CURVE 81*	
333.2	0.056
CURVE 82*	
333.2	0.049
CURVE 83*	
333.2	0.0505
CURVE 84*	
333.2	0.0455
CURVE 85*	
333.2	0.0465
CURVE 86*	
333.2	0.044
CURVE 87*	
333.2	0.047
CURVE 88*	
333.2	0.045
CURVE 89*	
333.2	0.046
CURVE 90*	
333.2	0.046
CURVE 91*	
333.2	0.047
CURVE 92*	
333.2	0.032
CURVE 93*	
333.2	0.0465
CURVE 94*	
333.2	0.047
CURVE 95*	
333.2	0.048
CURVE 96*	
333.2	0.0445
CURVE 97*	
333.2	0.047
CURVE 98*	
333.2	0.044
CURVE 99*	
333.2	0.043
CURVE 100*	
333.2	0.041
CURVE 101*	
333.2	0.0435
CURVE 102*	
333.2	0.062
CURVE 103*	
333.2	0.046
CURVE 104*	
333.2	0.062
CURVE 105*	
333.2	0.0545
CURVE 106*	
333.2	0.062
CURVE 107*	
333.2	0.044
CURVE 108*	
333.2	0.062
CURVE 109*	
333.2	0.050
CURVE 110*	
333.2	0.062
CURVE 111*	
333.2	0.062
CURVE 112*	
333.2	0.062
CURVE 113*	
333.2	0.062
CURVE 114*	
333.2	0.062
CURVE 115*	
333.2	0.0575
CURVE 116*	
333.2	0.062
CURVE 117*	
333.2	0.0615
CURVE 118*	
333.2	0.062
CURVE 119*	
333.2	0.058
CURVE 120*	
333.2	0.062
CURVE 121*	
333.2	0.0505
CURVE 122*	
333.2	0.062
CURVE 123*	
333.2	0.0515
CURVE 124*	
333.2	0.062
CURVE 125*	
333.2	0.045
CURVE 126*	
333.2	0.0475
CURVE 127*	
333.2	0.054
CURVE 128*	
333.2	0.062
CURVE 129*	
333.2	0.054
CURVE 130	
461.2	0.0207
554.2	0.0167
583.2	0.0204
670.2	0.0204
945.2	0.0161
963.2	0.0199
1161.2	0.0205
CURVE 131	
424.2	0.0212
637.2	0.0199
747.2	0.0181
980.2	0.0157
1045.2	0.0311*
1069.2	0.0152
1168.2	0.0308*
CURVE 132*	
546.2	0.0542
566.2	0.0535
670.2	0.0487
705.2	0.0447
CURVE 133*	
674.2	0.0469
877.2	0.0392
1081.2	0.0326
1157.2	0.0337
1191.2	0.0299
1308.2	0.0256
1335.2	0.0248
CURVE 134	
403.2	0.0727
413.2	0.0711
529.2	0.0588
536.2	0.0598
550.2	0.0565*
558.2	0.0584
622.2	0.0542

*Not shown on plot

DATA TABLE NO. 38 (continued)

CURVE 134 (cont.)

T	k
634.2	0.0520
653.2	0.0520
661.2	0.0516
749.2	0.0480
790.2	0.0460
965.2	0.0382
1022.2	0.0386
1271.2	0.0337
1393.2	0.0284
1407.2	0.0306
1506.2	0.0264
1552.2	0.0314

CURVE 135

T	k
747.2	0.0403*
803.2	0.0397*
1011.2	0.0292*
1103.2	0.0297
1220.2	0.0258
1371.2	0.0256*
1392.2	0.0236
1424.2	0.0236*
1614.2	0.0218

CURVE 136

T	k
526.2	0.0706
542.2	0.0579*
548.2	0.0733
565.2	0.0632*
692.2	0.0549
733.2	0.0472
821.2	0.0389
887.2	0.0405
997.2	0.0325
1052.2	0.0426
1052.2	0.0374
1172.2	0.0373
1172.2	0.0295*
1317.2	0.0296*

CURVE 137

T	k
514.2	0.0692
638.2	0.0513
646.2	0.0591
670.2	0.0608
763.2	0.0461
815.2	0.0505
843.2	0.0434
905.2	0.0423
910.2	0.0430*
945.2	0.0386
979.2	0.0438
986.2	0.0400*
1037.2	0.0383*
1095.2	0.0354
1210.2	0.0304
1228.2	0.0320
1337.2	0.0292*
1359.2	0.0282*
1385.2	0.0291*
1439.2	0.0268*
1449.2	0.0286*
1623.2	0.0251*
1705.2	0.0233

CURVE 138*

T	k
363.2	0.0767
440.2	0.0679
494.2	0.0622
531.2	0.0618
572.2	0.0527
631.2	0.0505
711.2	0.0455
771.2	0.0390
828.2	0.0357
914.2	0.0368
1007.2	0.0312

CURVE 139*

T	k
364.2	0.0752
440.2	0.0685
528.2	0.0597
586.2	0.0517
656.2	0.0470
751.2	0.0395
821.2	0.0374
882.2	0.0366

CURVE 140*

T	k
378.2	0.0779
431.2	0.0718
463.2	0.0640
604.2	0.0538
661.2	0.0474
747.2	0.0397
881.2	0.0376
1031.2	0.0302

CURVE 141

T	k
361.2	0.0906
402.2	0.0811
460.2	0.0686*
555.2	0.0601*
635.2	0.0553*
691.2	0.0508*
799.2	0.0479*

CURVE 142

T	k
416.2	0.0739*
442.2	0.0714*
443.2	0.0766
461.2	0.0724
476.2	0.0655
501.2	0.0672
563.2	0.0648
606.2	0.0630
633.2	0.0581*

CURVE 142 (cont.)

T	k
703.2	0.0550
704.2	0.0580
813.2	0.0517
834.2	0.0560
837.2	0.0579
917.2	0.0548
924.2	0.0527
980.2	0.0514
1003.2	0.0505
1112.2	0.0520
1118.2	0.0529

CURVE 143

T	k
660.2	0.061*
729.2	0.058
1019.2	0.054
1067.2	0.056
1112.2	0.054
1166.2	0.056
1284.2	0.067
1489.2	0.067

CURVE 144

T	k
330.2	0.0452*
358.2	0.0415
413.2	0.0389
417.2	0.0428
448.2	0.0377
614.2	0.0398
657.2	0.0383
817.2	0.0345
886.2	0.0341*

CURVE 145

T	k
498.2	0.0493
627.2	0.0416
886.2	0.0341*

CURVE 146

T	k
341.2	0.0407
394.2	0.0352
436.2	0.0301
472.2	0.0356

CURVE 147

T	k
357.2	0.0430
450.2	0.0378
602.2	0.0342
752.2	0.0346

CURVE 148

T	k
368.2	0.0442
538.2	0.0404
752.2	0.0346*

CURVE 149

T	k
332.2	0.0345*
344.2	0.0339*
359.2	0.0336
377.2	0.0327
417.2	0.0275
440.2	0.0306*
540.2	0.0327
565.2	0.0317

CURVE 150

T	k
319.2	0.0519
331.2	0.0509*
427.2	0.0397
430.2	0.0393*
521.2	0.0325
541.2	0.0346
651.2	0.0318
673.2	0.0342
736.2	0.0363
755.2	0.0388*

CURVE 150 (cont.)

T	k
776.2	0.0342*
777.2	0.0370*
806.2	0.0371*
806.2	0.0348*
845.2	0.0358*
879.2	0.0349*

CURVE 151*

T	k
504.2	0.0487
520.2	0.0454
879.2	0.0349

CURVE 152*

T	k
334.2	0.0659
338.2	0.0576
349.2	0.0639
353.2	0.0607
1009.2	0.0291
1048.2	0.0270
1076.2	0.0271
1109.2	0.0279
1159.2	0.0262

CURVE 153

T	k
1106.2	0.025*
1385.2	0.021
1441.2	0.021
1766.2	0.020
2011.2	0.017
2385.2	0.016

CURVE 154*

T	k
373.2	0.0758
473.2	0.0440
513.2	0.0428
573.2	0.0500
673.2	0.0610

CURVE 154 (cont.)*

T	k
708.2	0.0613
773.2	0.0587
873.2	0.0270
908.2	0.0250
973.2	0.0427
1006.2	0.0613
1073.2	0.0610
1173.2	0.0360

CURVE 155*

T	k
323.2	0.0450
363.2	0.0563
373.2	0.0567
399.9	0.0400
433.2	0.0357
473.2	0.0400
553.2	0.0497
573.2	0.0493
673.2	0.0370
726.5	0.0337
773.2	0.0450
826.5	0.0663
873.2	0.0510
973.2	0.0117
993.2	0.0108
1073.2	0.0123

CURVE 156*

T	k
323.2	0.0360
363.2	0.0500
373.2	0.0597
399.9	0.0500
473.2	0.0167
573.2	0.0107
673.2	0.00840
773.2	0.00733
873.2	0.00700
973.2	0.00650
1073.2	0.00633

* Not shown on plot

DATA TABLE NO. 38 (continued)

CURVE 157

T	k
323.2	0.00740
373.2	0.00736
473.2	0.00739
573.2	0.00733
673.2	0.00800*
773.2	0.00900
873.2	0.00733
973.2	0.00683
1073.2	0.00700

CURVE 158

T	k
323.2	0.00700
373.2	0.00684
473.2	0.00713
573.2	0.00713
673.2	0.00800
773.2	0.00730
873.2	0.00667
973.2	0.00601
1073.2	0.00587
1173.2	0.00580
1273.2	0.00600
1373.2	0.00700
1473.2	0.00850
1573.2	0.0110
1673.2	0.0150
1773.2	0.0187
1873.2	0.0230
1933.2	0.0245
1973.2	0.0233
2073.2	0.0167

CURVE 159

T	k
373.2	0.00126
713.2	0.000836

CURVE 160

T	k
468.2	0.00335
743.2	0.00251

CURVE 161

T	k
323.2	0.00209

CURVE 162

T	k
323.2	0.00167

CURVE 163*

T	k
328.2	0.07205
350.2	0.07073
354.2	0.07012
364.7	0.06932
380.2	0.06844
394.7	0.06770
427.7	0.06513
447.7	0.06376
468.2	0.06226
499.2	0.06024
532.2	0.05757
556.2	0.05596
594.2	0.05362
621.2	0.05192
648.2	0.05046
664.2	0.04938
697.2	0.04793
709.2	0.04758
716.2	0.04715
746.2	0.04558
773.2	0.04487
802.2	0.04372
832.7	0.04246
857.7	0.04168
882.2	0.04106
885.7	0.04081
923.2	0.03943
966.2	0.03792
1012.2	0.03654
1063.7	0.03500
1103.2	0.03496
1146.2	0.03297

CURVE 164*

T	k
496.2	0.06346
500.2	0.06310
874.2	0.04116

CURVE 165*

T	k
327.2	0.07923
396.2	0.07273
462.7	0.06699
574.2	0.05786
672.7	0.05090
767.2	0.04627
876.7	0.04178
975.7	0.03811
1074.2	0.03496

CURVE 166*

T	k
373.2	0.07195
475.2	0.06317
675.2	0.04936
869.2	0.04103
870.2	0.04062
970.7	0.03733
1071.1	0.03407
1165.2	0.03196
1173.2	0.03185
1279.2	0.02970
1282.2	0.02950
1372.2	0.02819
1465.2	0.02683

CURVE 167*

T	k
571.7	0.05554
870.2	0.04001
870.2	0.03951
871.7	0.03941
1171.2	0.03092
1175.2	0.03073
1377.2	0.02775
1474.7	0.02601
1475.2	0.02549
1571.2	0.02528

CURVE 168*

T	k
382.2	0.06953
383.2	0.06962
473.2	0.06166
474.2	0.06216
569.7	0.05503

CURVE 168 (cont.)*

T	k
571.7	0.05472
673.2	0.04913
673.2	0.04879
774.2	0.04365
774.2	0.04385
874.7	0.03973
874.7	0.03998
973.2	0.03657
973.2	0.03658
1071.2	0.03393
1071.2	0.03388
1173.2	0.03158
1271.2	0.02945
1323.2	0.02871
1372.2	0.02776
1372.2	0.02773
1422.2	0.02694
1423.2	0.02681
1468.2	0.02628
1521.2	0.02557
1523.2	0.02575
1569.2	0.02470
1569.2	0.02469
1569.2	0.02450

CURVE 169*

T	k
323.2	0.07640
376.2	0.07281
376.2	0.07221
576.2	0.05602
576.2	0.05599
670.7	0.05023
670.7	0.05022
670.7	0.05038
874.2	0.04091

CURVE 170*

T	k
216.2	0.06795
217.2	0.06985
234.2	0.07102
252.2	0.06729
255.2	0.07130
324.2	0.07555
325.2	0.07613

CURVE 170 (cont.)*

T	k
377.2	0.07118
377.2	0.07124
378.2	0.06991
378.2	0.06986
379.2	0.07162
380.2	0.06993
382.2	0.07157

CURVE 171*

T	k
543.2	0.0628
563.2	0.0607
583.2	0.0582
643.2	0.0523
693.2	0.0498
803.2	0.0431
853.2	0.0385

CURVE 172*

T	k
553.2	0.0632
593.2	0.0590
603.2	0.0594
678.2	0.0536
753.2	0.0477
878.2	0.0402

CURVE 173*

T	k
1233.2	0.0272
1243.2	0.0297
1283.2	0.0276
1303.2	0.0276
1353.2	0.0230
1363.2	0.0243
1423.2	0.0226
1443.2	0.0243
1453.2	0.0209
1473.2	0.0213
1493.2	0.0209
1513.2	0.0222
1583.2	0.0218
1593.2	0.0201
1613.2	0.0218
1633.2	0.0218
1713.2	0.0205

CURVE 173 (cont.)*

T	k
1733.2	0.0201
1833.2	0.0201
1983.2	0.0259
1993.2	0.0272
2023.2	0.0280
2073.2	0.0285
2123.2	0.0331
2153.2	0.0347

CURVE 174*

T	k
1873.2	0.0234
1953.2	0.0255

CURVE 175

T	k
1603.2	0.0222*
1633.2	0.0222*
1673.2	0.0209*
1713.2	0.0188
1763.2	0.0230
1853.2	0.0251*
1933.2	0.0268
2173.2	0.0335
2243.2	0.0368
2373.2	0.0427

CURVE 176

T	k
77.4	0.061
288	0.210

CURVE 177*

T	k
2153.2	0.0223
2258.2	0.0198
2268.2	0.0105
2270.2	0.017
2271.2	0.0205
2338.2	0.025
2348.2	0.0122
2393.2	0.0192
2408.2	0.0125
2423.2	0.0165
2428.2	0.0175
2488.2	0.0225

CURVE 177 (cont.)*

T	k
2498.2	0.0172
2518.2	0.0155
2538.2	0.0135
2573.2	0.0155
2588.2	0.0167
2613.2	0.0178
2623.2	0.0215
2648.2	0.0122
2668.2	0.0155
2675.2	0.0185
2708.2	0.0198
2713.2	0.022

CURVE 178*

T	k
373.2	0.047
446.2	0.044
518.2	0.043
597.2	0.041
702.2	0.038
769.2	0.038
773.2	0.038
813.2	0.038
862.2	0.038
910.2	0.036
947.2	0.035
948.2	0.035
949.2	0.034

CURVE 179*

T	k
674.2	0.025
753.2	0.027
792.2	0.026
933.2	0.032
941.2	0.032
942.2	0.032
945.2	0.032
946.2	0.030
948.2	0.033
963.2	0.033
966.2	0.032

CURVE 180*

T	k
649.2	0.021
868.2	0.025
913.2	0.027
930.2	0.028
932.2	0.027
938.2	0.028
948.2	0.028
948.2	0.027

CURVE 181*

T	k
437.2	0.018
491.2	0.019
666.2	0.021
820.2	0.023
822.2	0.023
892.2	0.026
901.2	0.026
908.2	0.026
920.2	0.025
921.2	0.026
923.2	0.025
926.2	0.027
928.2	0.026
936.2	0.027
946.2	0.026
975.2	0.026
977.2	0.026

CURVE 182*

T	k
679.2	0.022
813.2	0.023
912.2	0.025
917.2	0.025
920.2	0.026
979.2	0.025

CURVE 183*

T	k
491.2	0.019
507.2	0.020
570.2	0.020
643.2	0.021
666.2	0.022
738.2	0.022

*Not shown on plot

DATA TABLE NO. 38 (continued)

T	k
CURVE 183 (cont.)*	
761.2	0.025
817.2	0.024
857.2	0.023
870.2	0.023
880.2	0.024
889.2	0.023
903.2	0.023
909.2	0.025
913.2	0.027
913.2	0.025
914.2	0.024
918.2	0.023
919.2	0.024
930.2	0.025
954.2	0.025
957.2	0.025
960.2	0.024
963.2	0.026
966.2	0.023
CURVE 184*	
404.2	0.016
480.2	0.020
504.2	0.019
560.2	0.020
631.2	0.021
731.2	0.023
801.2	0.024
835.2	0.023
857.2	0.023
868.2	0.023
879.2	0.023
885.2	0.023
891.2	0.024
895.2	0.023
896.2	0.023
899.2	0.023
902.2	0.023
906.2	0.022
914.2	0.023
916.2	0.022
931.2	0.023
943.2	0.024
944.2	0.023
945.2	0.023

T	k
CURVE 185*	
396.2	0.020
491.2	0.023
556.2	0.024
635.2	0.025
720.2	0.025
801.2	0.025
842.2	0.025
874.2	0.025
911.2	0.025
913.2	0.024
915.2	0.025
915.2	0.025
928.2	0.025
929.2	0.025
934.2	0.025
934.2	0.026
936.2	0.025
938.2	0.025
941.2	0.025
942.2	0.024
CURVE 186*	
356.2	0.011
405.2	0.016
453.2	0.015
496.2	0.015
544.2	0.016
588.2	0.017
617.2	0.016
671.2	0.017
673.2	0.017
686.2	0.017
689.2	0.017
693.2	0.017
CURVE 187	
395.7	0.0014
713.2	0.0008
CURVE 188*	
362.2	0.0014

T	k
CURVE 189*	
323.2	0.0865
373.2	0.0769
473.2	0.0630
573.2	0.0533
673.2	0.0462
773.2	0.0408
873.2	0.0365
973.2	0.0330
1073.2	0.0301
CURVE 190*	
323.2	0.0865
373.2	0.0769
473.2	0.0630
573.2	0.0533
673.2	0.0462
773.2	0.0408
873.2	0.0365
973.2	0.0330
1073.2	0.0301
CURVE 191*	
323.2	0.0865
373.2	0.0769
473.2	0.0630
573.2	0.0533
673.2	0.0462
773.2	0.0408
873.2	0.0365
973.2	0.0330
1073.2	0.0301
CURVE 192*	
343.2	0.0962
CURVE 193*	
1311.2	0.0459
1585.2	0.0256
1866.2	0.0246
1874.2	0.0213
1905.2	0.0167
2005.2	0.0253

T	k
CURVE 193 (cont.)*	
2046.2	0.0206
2060.2	0.0198
2071.2	0.0251
2106.2	0.0210
2116.2	0.0194
CURVE 194	
373.2	0.112
472.2	0.0960
476.2	0.108
565.2	0.0795
576.2	0.0863
672.2	0.0863
675.2	0.0828
675.2	0.0781
713.2	0.0674
772.2	0.0687
775.2	0.0644
837.2	0.0525*
875.2	0.0530*
875.2	0.0500*
974.2	0.0489*
976.2	0.0480*
1009.2	0.0415*
1020.2	0.0429*
1071.2	0.0416*
1075.2	0.0433*
1076.2	0.0448*
1170.2	0.0399*
1173.2	0.0419*
1192.2	0.0390*
1270.2	0.0377*
1271.2	0.0386*
1272.2	0.0343*
1273.2	0.0421*
1303.2	0.0356*
1371.2	0.0305*
1376.2	0.0363*
CURVE 195*	
299.2	0.0809
307.2	0.0938
332.2	0.0976
372.2	0.0752

T	k
CURVE 195 (cont.)*	
474.2	0.0657
574.2	0.0577
674.2	0.0543
774.2	0.0491
875.2	0.0421
973.2	0.0409
CURVE 196*	
371.2	0.0997
452.2	0.0832
532.2	0.0722
606.2	0.0639
697.2	0.0562
787.2	0.0501
906.2	0.0439
1053.2	0.0379
1179.2	0.0340
1279.2	0.0314
CURVE 197*	
358.2	0.0287
378.2	0.0279
395.2	0.0275
408.2	0.0282
419.2	0.0292
471.2	0.0294
501.2	0.0299
513.2	0.0306
421.2	0.0321
431.2	0.0335
442.2	0.0339
573.2	0.0328
617.2	0.0310
CURVE 198*	
373.2	0.071
473.2	0.057
573.2	0.048
673.2	0.042
273.2	0.037
873.2	0.033

T	k
CURVE 199*	
373.2	0.080
473.2	0.069
673.2	0.058
873.2	0.054
1073.2	0.054
1273.2	0.059
1473.2	0.071
CURVE 200*	
473.2	0.058
673.2	0.044
873.2	0.040
1073.2	0.040
1273.2	0.041
CURVE 201*	
473.2	0.069
673.2	0.058
873.2	0.054
1073.2	0.054
CURVE 202*	
471.2	0.0607
473.2	0.0592
CURVE 203*	
379.2	0.0288
393.2	0.0277
394.2	0.0286
396.2	0.0298
408.2	0.0271
411.2	0.0296
413.2	0.0285
429.2	0.0294
501.2	0.0299
527.2	0.0342
544.2	0.0314
553.2	0.0331
577.2	0.0346
609.2	0.0309

T	k
CURVE 204*	
430.2	0.0261
540.2	0.0204
685.2	0.0179
738.2	0.0179
803.2	0.0165
CURVE 205*	
432.2	0.0336
548.2	0.0233
668.2	0.0212
731.2	0.0198
790.2	0.0197
CURVE 206*	
555.2	0.0230
635.2	0.0208
684.2	0.0206
739.2	0.0199
795.2	0.0194
CURVE 207*	
409.2	0.0409
537.2	0.0296
639.2	0.0212
701.2	0.0203
748.2	0.0202
CURVE 208*	
834	0.0464
1124	0.0372
1331	0.0335
1486	0.0315
1600	0.0307
1753	0.0299
1889	0.0297
2001	0.0301
2211	0.0312
2322	0.0319
2518	0.0339

T	k
CURVE 209	
1077.2	0.0531
1173.2	0.0550
1273.2	0.0583
1373.2	0.0624
1459.2	0.0667*
1473.2	0.0598
1573.2	0.0419
1673.2	0.0336
1773.2	0.0279
1873.2	0.0232*
2073.2	0.020*
CURVE 210	
1189	0.1368
1390	0.07674
1673	0.05781
CURVE 211*	
1145	0.1735
1253	0.05598
1395	0.03516
CURVE 212*	
327.2	0.07400
396.2	0.06798
462.7	0.06263
496.2	0.05935
574.2	0.05410
672.7	0.04762
767.2	0.04336
874.2	0.03860
876.7	0.03918
885.7	0.03827
923.2	0.03699
966.2	0.03558
975.7	0.03577
1012.2	0.03431
1063.2	0.03289
1074.2	0.03283
1103.2	0.03192
1146.2	0.03100

*Not shown on plot

DATA TABLE NO. 38 (continued)

T	k
CURVE 213*	
298.2	0.00175
p (mm Hg)	k
CURVE 214 (T = 298.2K)	
0.0055	0.0000573
1.01	0.000167*
24.3	0.000556*
135	0.000925*
794	0.00118
3055	0.00124
6310	0.00126
T	k
CURVE 215*	
298.2	0.00120
p (mm Hg)	k
CURVE 216* (T = 298.2K)	
0.0108	0.0000962
0.0501	0.0000920
1.82	0.000126
24.8	0.000414
162	0.000837
759	0.00108
2818	0.00113
5188	0.00114
T	k
CURVE 217*	
331.2	0.0822
376.2	0.0803
383.2	0.0997
383.2	0.0733
482.2	0.0772
485.2	0.0599
489.2	0.0681

T	k
CURVE 217 (cont.)*	
581.2	0.0509
647.2	0.0542
647.2	0.0466
785.2	0.0436
815.2	0.0387
878.2	0.0428
900.2	0.0390
1052.2	0.0366
1148.2	0.0284
1283.2	0.0311
1314.2	0.0287
1387.2	0.0269
1486.2	0.0259
1597.2	0.0256
1685.2	0.0191
1751.2	0.0227
1822.2	0.0186
1930.2	0.0182
2390.2	0.0179
CURVE 218	
2154.2	0.0221
2263.2	0.0105
2338.2	0.0222
2413.2	0.0129
2494.2	0.0219
2623.2	0.0212
2659.2	0.0151
2713.2	0.0219
CURVE 219*	
499.2	0.05632
532.2	0.05384
556.2	0.05234
594.2	0.05018
621.2	0.04838
648.2	0.04723
664.2	0.04622
697.2	0.04487
709.2	0.04456
716.2	0.04416
746.2	0.04270
773.2	0.04204
802.2	0.04097

T	k
CURVE 219 (cont.)*	
833.2	0.03980
857.2	0.03910
882.2	0.03850
CURVE 220*	
300.2	0.0732
328.2	0.0674
350.2	0.0660
354.2	0.0656
364.2	0.0648
380.2	0.0640
395.2	0.0634
427.2	0.0608
447.2	0.0596
468.2	0.0582
CURVE 221*	
373.2	0.07198
475.2	0.06323
675.2	0.04948
869.9	0.04120
870.2	0.04079
970.7	0.03752
1071.7	0.03428
1164.9	0.03219
1173.2	0.03208
1279.2	0.02994
1282.7	0.02974
1372.2	0.02846
1465.2	0.02712
CURVE 222*	
571.7	0.05563
870.2	0.04018
870.2	0.03968
871.7	0.03958
1171.2	0.03115
1175.2	0.03096
1377.2	0.02802
1474.7	0.02630
1475.2	0.02578
1571.2	0.02558

T	k
CURVE 223*	
352.2	0.0346
461.2	0.0158
652.2	0.0163
739.2	0.0146
890.2	0.0218
989.2	0.0215
1056.2	0.0290
1342.2	0.0261
1352.2	0.0283
1376.2	0.0324
1399.2	0.0302
CURVE 224*	
329.2	0.0071
565.2	0.0072
775.2	0.0070
1071.2	0.0059
1224.2	0.0055
1395.2	0.0063
1575.2	0.0095
1723.2	0.0158
1846.2	0.0220
1957.2	0.0241
2019.2	0.0228
2068.2	0.0192
2101.2	0.0165
CURVE 225*	
325.2	0.0579
346.2	0.0458
379.2	0.0237
393.2	0.0267
456.2	0.0349
526.2	0.0229
607.2	0.0384
685.2	0.0342
742.2	0.0453
1006.2	0.0494
1306.2	0.0556

T	k
CURVE 226*	
472.2	0.0711
667.2	0.0691
859.2	0.0569
859.2	0.0426
1064.2	0.0324
1273.2	0.0262
CURVE 227*	
470.2	0.0661
660.2	0.0633
867.2	0.0521
1053.2	0.0360
1273.2	0.0366
CURVE 228*	
639.8	0.061
CURVE 229*	
640.7	0.055
CURVE 230*	
670.7	0.055
CURVE 231*	
680.7	0.049
CURVE 232*	
738.2	0.048
CURVE 233*	
1252	0.03133
1317	0.02958
1371	0.02931
1405	0.02786
1417	0.02754
1426	0.02838
1445	0.02773
1461	0.02723
1486	0.02735

T	k
CURVE 233 (cont.)*	
1513	0.02594
1547	0.02535
1561	0.02576
1594	0.02576
1618	0.02455
1625	0.02495
1638	0.02576
1649	0.02576
1649	0.02449
1667	0.02489
1667	0.02576
1697	0.02427
CURVE 234*	
338.2	0.0689
338.2	0.0678
338.7	0.0646
339.8	0.0737
465.4	0.0537
465.4	0.0490
465.9	0.0517
465.9	0.0564
594.8	0.0471
595.4	0.0457
595.4	0.0445
720.9	0.0445
721.5	0.0434
723.2	0.0402
723.7	0.0417
CURVE 235*	
785.2	0.0284
821.2	0.0265
821.2	0.0245
CURVE 236*	
684.2	0.0384
714.2	0.0348
714.2	0.0323
933.2	0.0265
978.2	0.0274
978.2	0.0252
1004	0.0238

T	k
CURVE 236 (cont.)*	
1004	0.0266
1024	0.0292
1024	0.0246
1032	0.0251
1035	0.0278
1093	0.0282
1093	0.0258
1093	0.0224
CURVE 237*	
455.2	0.0567
456.2	0.0629
678.2	0.0537
839.2	0.0418
842.2	0.0457
1498	0.0392
1498	0.0441
CURVE 238*	
633.2	0.0448
636.2	0.0485
690.2	0.0423
691.2	0.0451
783.2	0.0303
CURVE 239*	
880.7	0.0170
1033.2	0.0161
1308.2	0.0161
1488.2	0.0200
1613.2	0.0225
CURVE 240*	
538.2	0.0100
873.2	0.00852
1318.2	0.00986
1538.2	0.0127
1673.2	0.0140
1723.2	0.0170

T	k
CURVE 241*	
623.2	0.0190
1008.2	0.0140
1383.2	0.0116
1523.2	0.0119
1598.2	0.0118
1648.2	0.0123
CURVE 242	
1.36	0.00501
1.52	0.00679
1.65	0.00873
1.82	0.0110
2.08	0.0128
2.29	0.0195
2.53	0.0239
2.84	0.0320
3.07	0.0359
3.14	0.0388
3.51	0.0515
3.68	0.0522
4.02	0.0658
4.69	0.0697
4.92	0.0794
5.73	0.0885
5.73	0.0962
6.86	0.101
6.95	0.107
8.00	0.101
8.15	0.105
9.16	0.0982
9.27	0.0968
10.5	0.0847
11.7	0.0650
11.9	0.0617
13.4	0.0526
13.4	0.0492
14.9	0.0383
15.0	0.0350
15.2	0.0334
16.1	0.0296
16.9	0.0255
17.0	0.0247
19.0	0.0195
20.8	0.0153
22.9	0.0134

*Not shown on plot

DATA TABLE NO. 38 (continued)

T	k
CURVE 242 (cont.)	
23.7	0.0121
25.4	0.0113
26.7	0.00973
27.9	0.00935
31.8	0.00980
33.9	0.0110
36.5	0.0112
36.9	0.0118
41.9	0.0150
41.9	0.0156
45.7	0.0185
48.6	0.0205
50.4	0.0239
56.0	0.0270
57.0	0.0266
59.6	0.0310
66.7	0.0414
78.5	0.0483
96.2	0.0684
112.5	0.0778
117.8	0.0879
148.9	0.105
167.5	0.118
204.2	0.128
CURVE 243*	
523.2	0.0699
CURVE 244*	
523.2	0.0678

*Not shown on plot

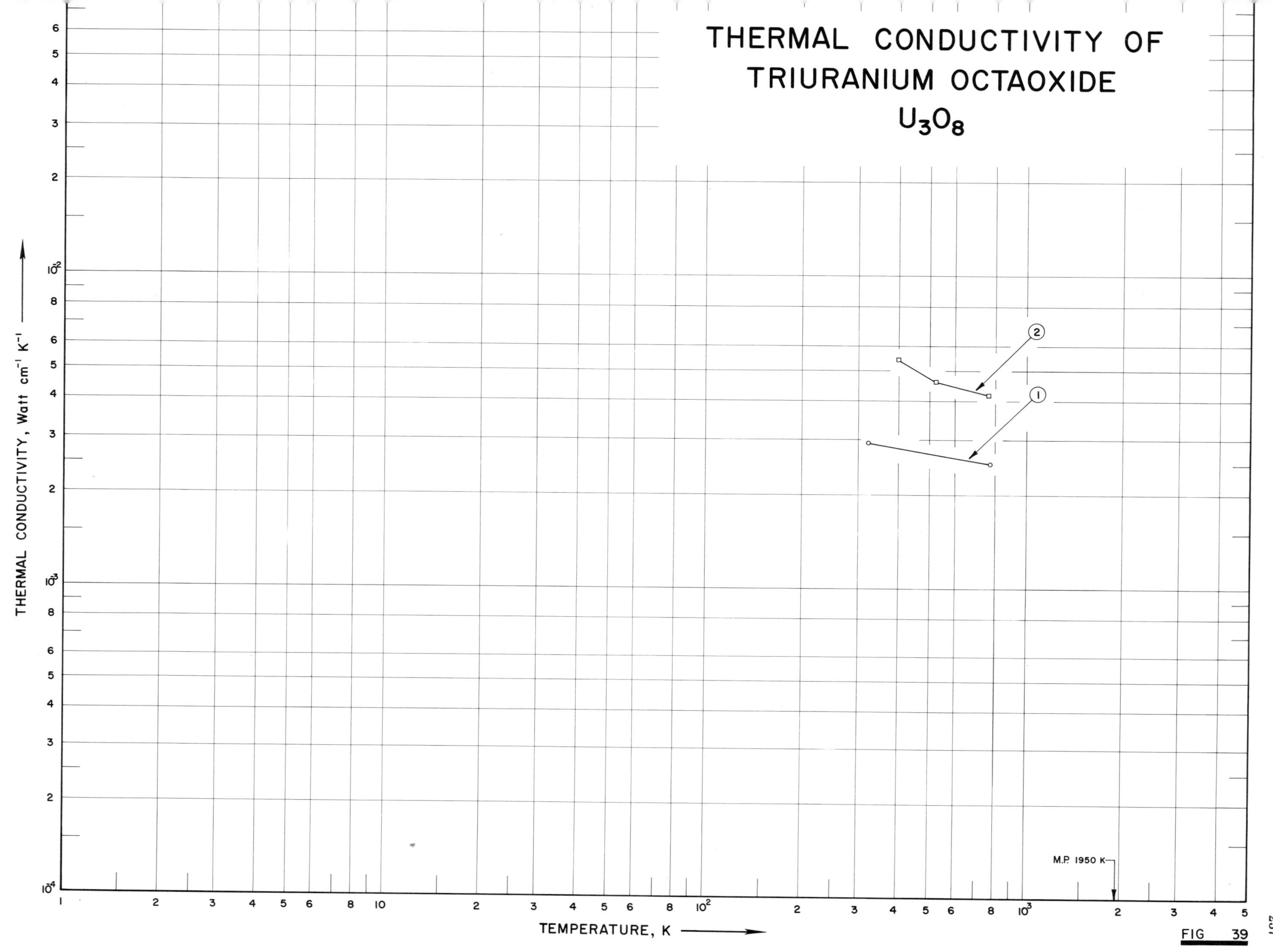
THERMAL CONDUCTIVITY OF
TRIURANIUM OCTAOXIDE
U_3O_8
M.P. 1950 K
TEMPERATURE, K
THERMAL CONDUCTIVITY, Watt cm^{-1} K^{-1}
FIG 39

SPECIFICATION TABLE NO. 39 THERMAL CONDUCTIVITY OF (tri)URANIUM OCTOXIDE U_3O_8

[For Data Reported in Figure and Table No. 39]

Curve No.	Ref. No.	Method Used	Year	Temp. Range, K	Reported Error, %	Name and Specimen Designation	Composition (weight percent), Specifications and Remarks
1	159	L	1955	323, 773			Powder; measured under 100 lb in.$^{-2}$ pressure.
2	159	L	1955	398-763			Specimen 3 in. long, 3 in. dia; pressed in 150 ton press.

DATA TABLE NO. 39 THERMAL CONDUCTIVITY OF (tri)URANIUM OCTOXIDE U_3O_8

[Temperature, T, K; Thermal Conductivity, k, Watt $cm^{-1} K^{-1}$]

T	k
CURVE 1	
323.2	0.00293
773.2	0.00251
CURVE 2	
398.2	0.00544
523.2	0.00460
763.2	0.00418

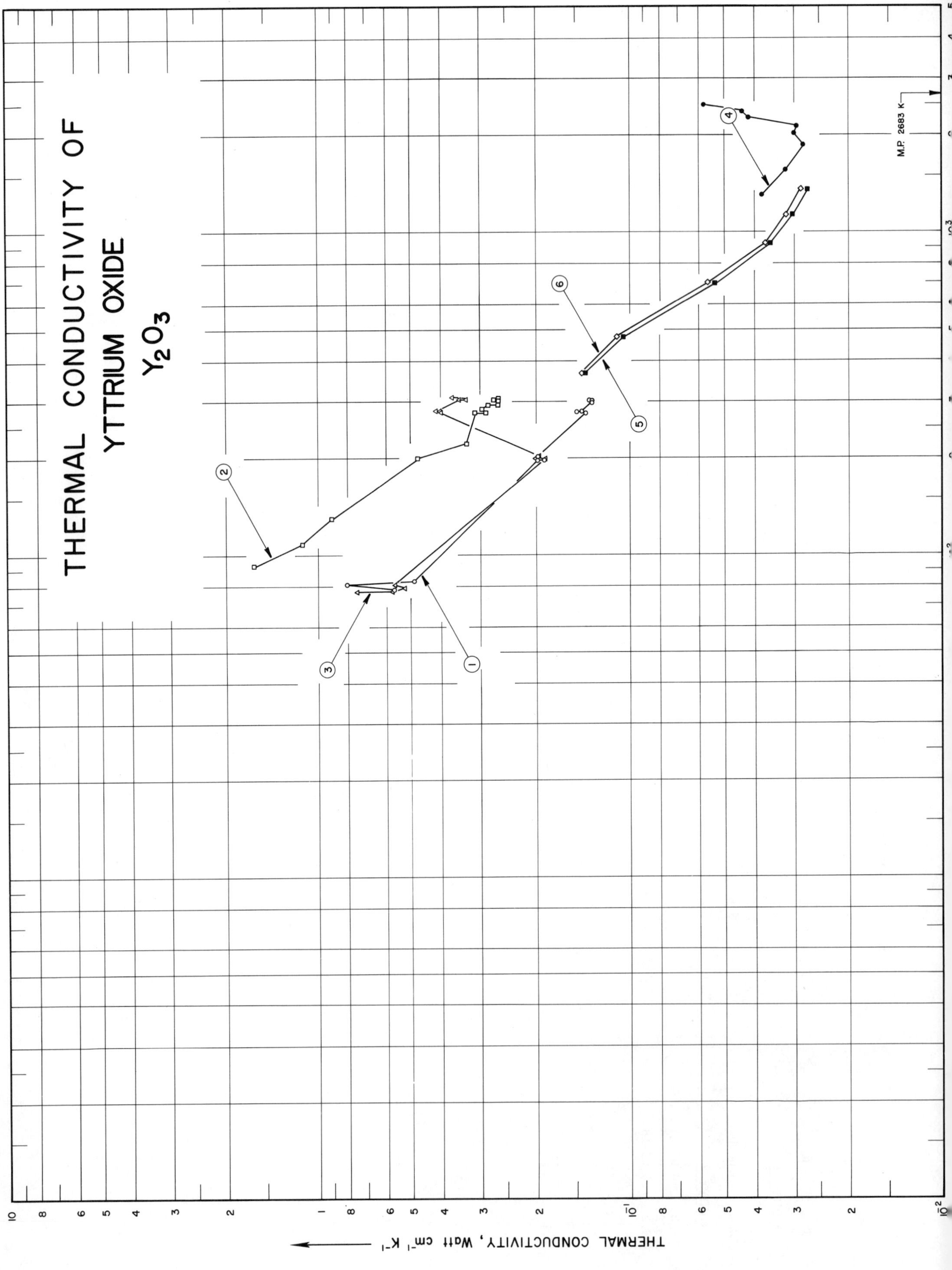

THERMAL CONDUCTIVITY OF
YTTRIUM OXIDE
Y_2O_3
THERMAL CONDUCTIVITY, Watt cm^{-1} K^{-1}
10
8
6
5
4
3
2
1
8
6
5
4
3
2
10^{-1}
8
6
5
4
3
2
10^{-2}
1
2
3
4
5
6
M.P. 2683 K

SPECIFICATION TABLE NO. 40 THERMAL CONDUCTIVITY OF YTTRIUM OXIDE Y_2O_3

[For Data Reported in Figure and Table No. 40]

Curve No.	Ref. No.	Method Used	Year	Temp. Range, K	Reported Error, %	Name and Specimen Designation	Composition (weight percent), Specifications and Remarks
1	460	L	1967	78-304			1.05 Nd_2O_3; single crystal; 0.210 ±0.005 cm dia x 1.040 cm long; prepared by the flame-fusion process, grounded.
2	461	L	1967	93-307			Single crystal; prepared from 99.999$^+$ pure materials by using the flame-fusion technique; density reported as 5.061, 5.055, 5.055, 5.051, 5.045, 5.044, and 5.038 g cm^{-3} at 108, 166, 207, 246, 295, 299, and 324 K, respectively.
3	461	L	1967	77-306			Single crystal; prepared from 99.999$^+$ pure materials by using the flame-fusion technique; Nd^{3+} concentration 2.7 x 10^{20} cm^{-3}; density reported as 5.071, 5.071, 5.069, 5.066, 50.065, 5.057, and 5.055 g cm^{-3} at 132, 137, 166, 198, 217, 296, and 337 K, respectively.
4	291	R	1964	1313-2493			Fabricated by a powder process in which high-purity Y_2O_3 powder (<0.02% total impurities) was cold-pressed in a 3-in. dia steel die at 4000 psi, isostatically pressed at 20,000 psi, and sintered for 19 hrs at 1700 C in a hydrogen atmosphere; the sintered disc were then machined into specimens and guards;disc specimen 2 in. O. D. and 0.375 I. D.; density 93.7% of theoretical and increased to 95.8% of theoretical after measurement; data corrected to 100% of theoretical density;measured in effective thermal conductivity.
5	292	C	1961	367-1367			Fabricated by dry pressing followed by isostatic compaction and then sintered at 1894 C in H_2 for 2 hrs; specimen 1/2 x 1/2 x 7/8 in. rectangular prisms; supplied by Michigan Chemical Co.; 96.3% of theoretical density; commercial stabilized ZrO_2 used as comparative material.
6	292	C	1961	367-1367			The above specimen corrected to zero porosity.

DATA TABLE NO. 40 THERMAL CONDUCTIVITY OF YTTRIUM OXIDE Y_2O_3

[Temperature, T, K; Thermal Conductivity, k, Watt $cm^{-1} K^{-1}$]

T	k
CURVE 1	
78	0.568
81	0.581
83	0.549
196	0.198
197	0.187
201	0.198
276	0.139
278	0.147
279	0.143
299	0.133
302	0.133
304	0.135
CURVE 2	
93	1.64
108	1.14
130	0.909
200	0.478
222	0.332
276	0.313
276	0.289
284	0.299
292	0.285
292	0.265
301	0.265
302	0.274
307	0.265
CURVE 3	
77	0.704
78	0.575
79	0.529
81	0.568
198	0.187
199	0.200
201	0.196
277	0.140
279	0.142
279	0.147
302	0.135
302	0.133
306	0.137
CURVE 4	
1313.2	0.0380
1563.2	0.0320
1853.2	0.0280
2023.2	0.0300
2143.2	0.0295
2273.2	0.0420
2383.2	0.0440
2493.2	0.0580
CURVE 5	
366.5	0.138
477.6	0.105
699.8	0.0537
922.1	0.0357
1144.3	0.0305
1366.5	0.0272
CURVE 6	
366.5	0.143
477.6	0.110
699.8	0.0561
922.1	0.0374
1144.3	0.0318
1366.5	0.0286

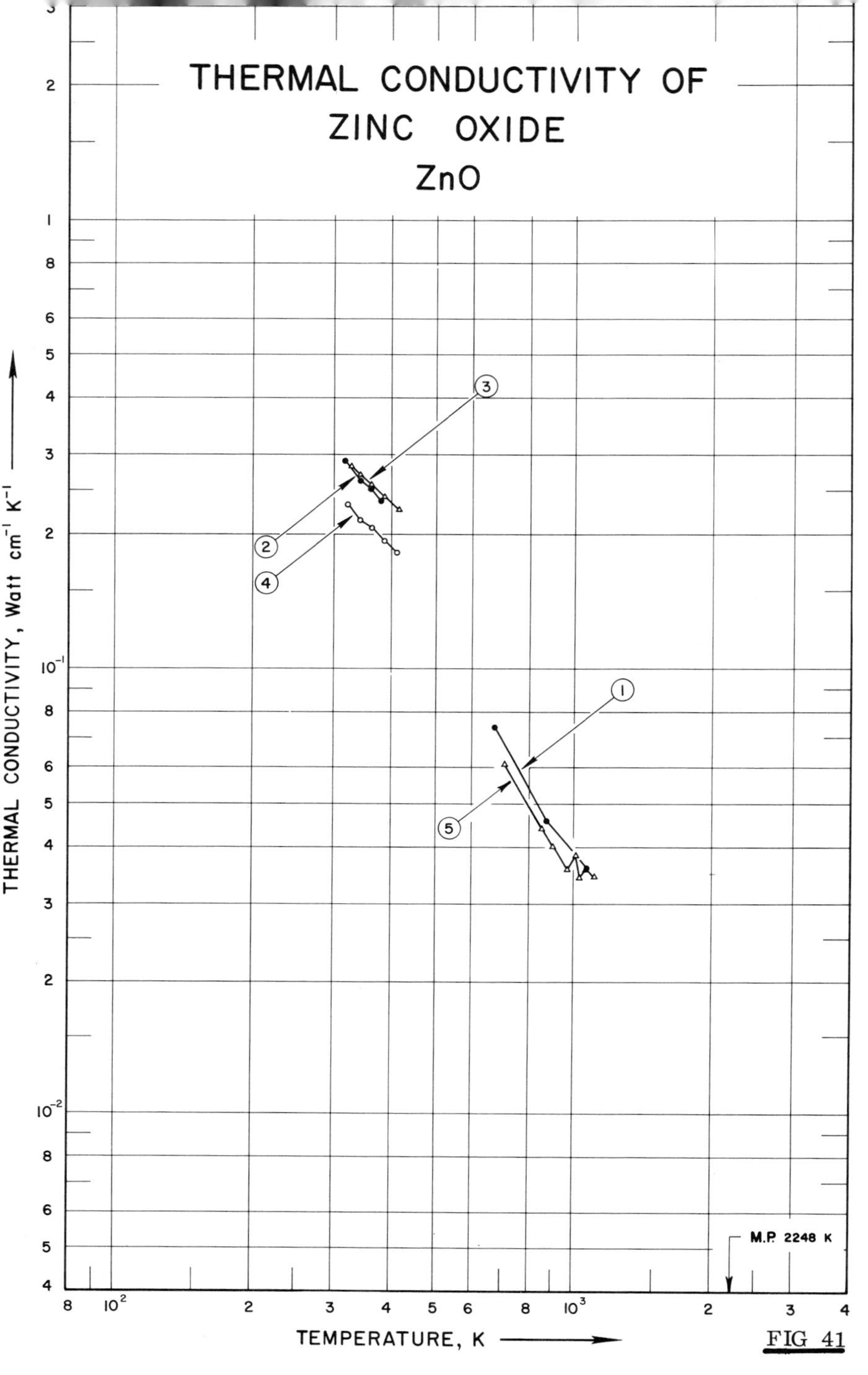
THERMAL CONDUCTIVITY OF
ZINC OXIDE
ZnO
THERMAL CONDUCTIVITY, Watt cm^{-1} K^{-1}
TEMPERATURE, K
M.P. 2248 K
1
2
3
4
5

FIG 41

SPECIFICATION TABLE NO. 41 THERMAL CONDUCTIVITY OF ZINC OXIDE ZnO

[For Data Reported in Figure and Table No. 41]

Curve No.	Ref. No.	Method Used	Year	Temp. Range, K	Reported Error, %	Name and Specimen Designation	Composition (weight percent), Specifications and Remarks
1	364	C	1954	673-1073			Polycrystal; prepared by calcining commercially pure ZnO at 900 C, and slip-casting from a neutral suspension using Daxad No. 23 as a dispersing agent; bulk density 3.72 g cm^{-3}; porosity 34.0%.
2	12		1953	319-380		133A-1	Yellow; prepared by calcining at 1367 K and dry pressing at 15000 psi; fired at 1583 K and soaked for 1 hr; density 5.28 g cm^{-3}.
3	12		1953	328-418		133A-2	Second run of the above specimen.
4	12		1953	322-413		133B-1	Grey; specimen preparation same as the above specimen; fired at 1644 K and soaked for 1 hr; 0.031% water absorption; density 5.20 g cm^{-3}.
5	153	R, C	1951	712-1124			Pure; prepared by slip-casting commercially pure ZnO, using Daxad No. 23 as a dispersant, heated to 1100 C, shaped and then fired at 1300 C; porosity 34%.

DATA TABLE NO. 41 THERMAL CONDUCTIVITY OF ZINC OXIDE ZnO

[Temperature, T, K; Thermal Conductivity, k, Watt $cm^{-1} K^{-1}$]

T	k
CURVE 1	
673.2	0.0741
873.2	0.0460
1073.2	0.0360
CURVE 2	
319.3	0.290
343.0	0.263
361.0	0.252
380.2	0.236
CURVE 3	
327.5	0.283
344.7	0.271
364.5	0.257
387.0	0.242
418.4	0.226
CURVE 4	
322.4	0.232
342.8	0.215
365.6	0.206
386.8	0.194
413.2	0.182
CURVE 5	
712.2	0.0611
860.2	0.0441
910.2	0.0400
978.2	0.0358
1020.7	0.0385
1048.2	0.0344
1078.2	0.0358
1123.5	0.0344

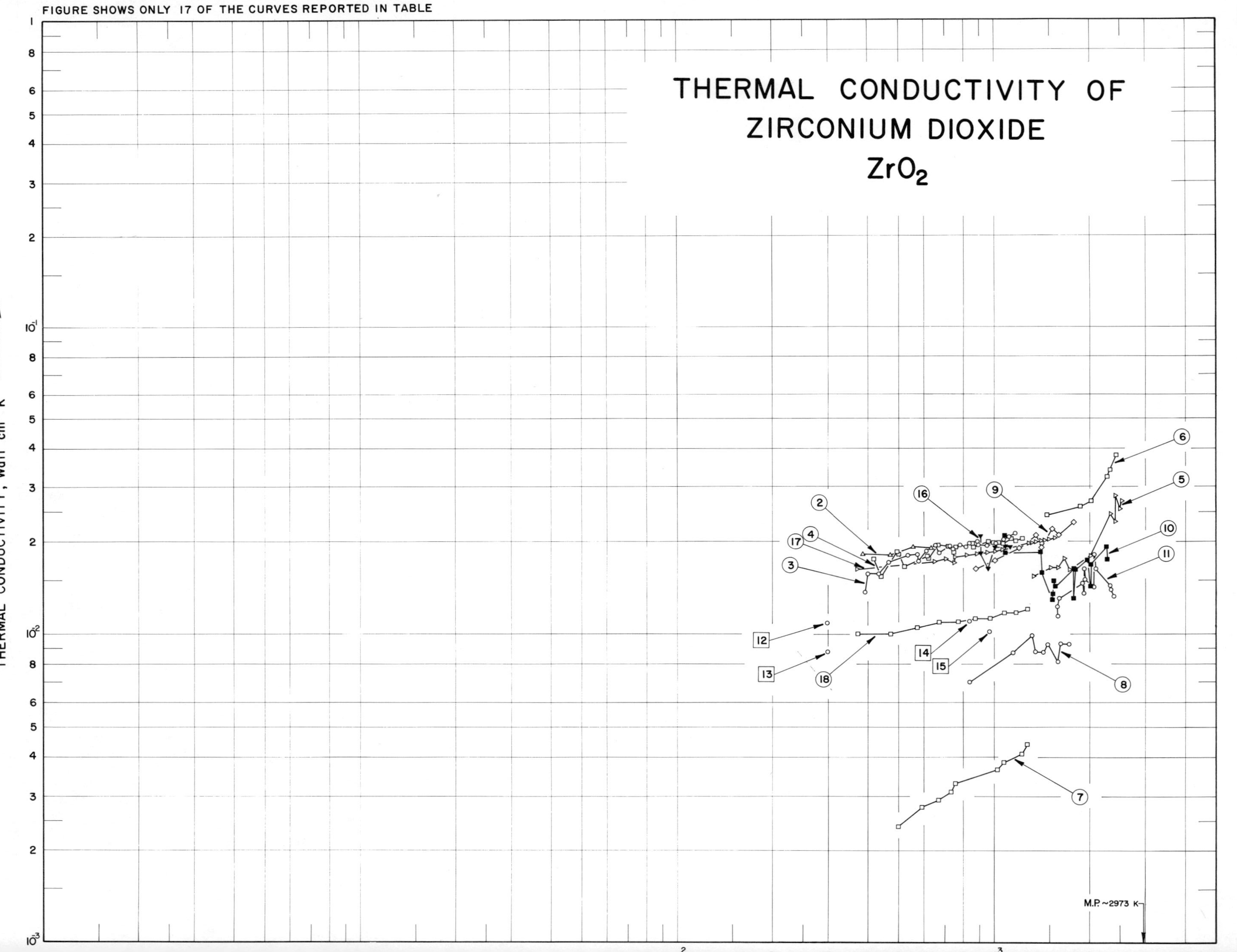

THERMAL CONDUCTIVITY OF
ZIRCONIUM DIOXIDE
ZrO2
FIGURE SHOWS ONLY 17 OF THE CURVES REPORTED IN TABLE
THERMAL CONDUCTIVITY, Watt cm⁻¹ K⁻¹
M.P. ~2973 K

SPECIFICATION TABLE NO. 42 THERMAL CONDUCTIVITY OF ZIRCONIUM DIOXIDE ZrO_2

[For Data Reported in Figure and Table No. 42]

Curve No.	Ref. No.	Method Used	Year	Temp. Range, K	Reported Error, %	Name and Specimen Designation	Composition (weight percent), Specifications and Remarks
1	57	R	1950	766-1233		No. 1	Pure; crystalline; slip-cast; treated four times with 1 N HCl and washed with distilled water; porosity 2.37% (0.25% closed pores, 2.12% open pores); bulk density 5.35 g cm^{-3}.
2	57	R	1950	386-1203		No. 1	The second run of the above specimen.
3	2	R	1951	391-1168		No. 1	The third run of the above specimen.
4	2	R	1951	419-1226		No. 2	Similar to the above specimen except fired at 1550 C; porosity 4.74% (zero open pores); bulk density 5.22 g cm^{-3}.
5	291	R	1964	1343-2523			CaO stabilized ZrO_2; disc shaped specimen 2.00 O.D., 0.375 in. I.D.; density 4.046 g cm^{-3} (66.3% of theoretical).
6	291	R	1964	1473-2423			The above specimen; data corrected to 100% theoretical density.
7	373	R	1959	498-1273	20		Powdered specimen (powder in the form of porous bubbles, 0.040 ± 0.01 cm dia); density 2.02 g cm^{-3}; volume fraction 0.33.
8	135	R	1957	833-1723	<20		Cylindrical specimen 30 mm long, 60 mm O.D., 30 mm I.D.; porosity 40%.
9	135	R	1957	873-1773	<20		Similar to the above except porosity 16%.
10	144	R	1963	1089-2281	5-7		Specimen 0.75 in. long, 0.75 in. O.D., 0.25 in. I.D.; supplied by Zirconium Corp. of America; pressed and sintered; ground and polished to eliminate surface scratches; density 5.63 g cm^{-3}; broke during test.
11	144	R	1963	1592-2374	5-7		Similar to the above specimen except dimensions 3 in. long, 2.5 in. O.D., and 0.75 in. I.D.; cracked and discolored during test.
12	152	L	1956	298.2		550	No details reported.
13	152	L	1956	298.2		550	The above specimen irradiated with an integrated flux of 6×10^{19} epithermal neutrons cm^{-2} above 100 ev for 480 megawatt day in the Material Testing Reactor.
14	228	P	1961	836.2	4.5	C	Specimen 1 in. dia, 0.4 cm thick (cut from a ZrO_2 cylinder); density 4.34 g cm^{-3}; 0.001 in. silver foil placed between the specimen and electrodes.
15	228	P	1961	963.2	4.5	B	Similar to the above specimen (cut from the same ZrO_2 cylinder).
16	228	P	1961	907-1127	4.5	D	Specimen 0.8 in. dia, 0.4 cm thick; density 5.48 g cm^{-3}; 0.001 in. silver foil placed between the specimen and electrodes.
17	307	R	1954	369-1569	5		Hollow prolate spheroidal specimen (inner minor axis ~2 cm, inner major axis ~10 cm, outer minor axis ~4 cm); prepared by slip-casting, fired; total porosity 7.76 to 10.00%; bulk density 5.22 to 5.35 g cm^{-3}.
18	375	C	1960	373-1273		Zirconia;SFCR-50	Specimen supplied by Titanium Alloy Division, National Lead Co.; alumina (Wesgo Al-300) used as standard.

SPECIFICATION TABLE NO. 42 (continued)

Curve No.	Ref. No.	Method Used	Year	Temp. Range, K	Reported Error, %	Name and Specimen Designation	Composition (weight percent), Specifications and Remarks
19	430	→	1966	298.2			Powder specimen contained in a 0.75 in. dia x 2 in. long stainless steel cylindrical cell; mesh size -70 +80; thermal conductivity measured by using the transient line source method, the heat source was a 36-gauge constantan wire contained in a 0.025 in. O.D. hypodermic tube soldered along the axis of the cylindrical cell, data calculated from the measured line temperatures at two certain times; measured in introgen at 1 atm.
20	408	C	1962	553-1180			Foam specimen, density 0.721 g cm^{-3}; Min-K 1301 (Johns Manville Corp.) used as comparative material.
21	463	R	1961	473-1259			Powder specimen contained in a hollow cylinder of 203 mm long and 91 mm internal dia; grain size <0.2 mm; bulk density 2.53 g cm^{-3}.
22	463	R	1961	654-1274			Similar to the above specimen except grain size 0.2~1 mm and bulk density 2.30 g cm^{-3}.
23	463	R	1961	473-1273			Similar to the above specimen except grain size 2~5 mm and bulk density 1.62 g cm^{-3}.

DATA TABLE NO. 42 THERMAL CONDUCTIVITY OF ZIRCONIUM DIOXIDE ZrO_2

[Temperature, T, K; Thermal Conductivity, k, Watt cm^{-1} K^{-1}]

T	k
CURVE 1	
766.2	0.0203
899.2	0.0198
1006.2	0.0196
1090.2	0.0191
1171.2	0.0191
1233.2	0.0190
CURVE 2	
386.2	0.0181
470.2	0.0180
553.2	0.0192
632.2	0.0190
681.2	0.0189*
734.2	0.0195*
793.2	0.0192*
839.2	0.0192*
887.2	0.0192*
961.2	0.0197*
1004.2	0.0198*
1076.2	0.0198*
1097.2	0.0210*
1134.2	0.0204*
1163.2	0.0204*
1203.2	0.0229*
CURVE 3	
391.2	0.0137
399.2	0.0157
434.2	0.0157
465.2	0.0171
535.2	0.0180
573.2	0.0181
577.2	0.0173
612.2	0.0186
655.2	0.0195
671.2	0.0183
725.2	0.0193
744.2	0.0189
782.2	0.0195
815.2	0.0192
836.2	0.0197
CURVE 3 (cont.)	
887.2	0.0200
890.2	0.0195
948.2	0.0194
1012.2	0.0197
1071.2	0.0204
1115.2	0.0208
1168.2	0.0214
CURVE 4	
419.2	0.0175
440.2	0.0155
494.2	0.0185
497.2	0.0182
520.2	0.0166
616.2	0.0179
619.2	0.0178
661.2	0.0195
718.2	0.0193
745.2	0.0184
758.2	0.0192
855.2	0.0197
858.2	0.0191
952.2	0.0199
992.2	0.0198
1014.2	0.0198*
1026.2	0.0198
1064.2	0.0197
1131.2	0.0205
1168.2	0.0201
1226.2	0.0205
CURVE 5	
1343.2	0.0154
1513.2	0.0164
1593.2	0.0164
1663.2	0.0176
1743.2	0.0162
2003.2	0.0179
2103.2	0.0180*
2323.2	0.0246
2413.2	0.0233
2413.2	0.0280
CURVE 5 (cont.)	
2493.2	0.0256
2523.2	0.0270
CURVE 6	
1473.2	0.0244
1873.2	0.0260
2023.2	0.0270
2273.2	0.0322
2333.2	0.0340
2423.2	0.0380
CURVE 7	
498.2	0.00240
593.2	0.00278
669.2	0.00292
733.2	0.00310
755.2	0.00330
1025.2	0.00366
1073.2	0.00386
1233.2	0.00410
1273.2	0.00440
CURVE 8	
833.2	0.00697
1153.2	0.00872
1313.2	0.00988
1353.2	0.00872
1433.2	0.00872
1478.2	0.00930
1593.2	0.00814
1613.2	0.00930
1723.2	0.00930
CURVE 9	
873.2	0.0163
1003.2	0.0174
1203.2	0.0198*
1243.2	0.0198
1363.2	0.0209
1413.2	0.0192
1536.2	0.0221
1613.2	0.0209
1773.2	0.0232
CURVE 10	
1088.7	0.0175*
1088.7	0.0208
1088.7	0.0182
1404.3	0.0185
1409.8	0.0185*
1410.9	0.0159
1533.2	0.0130
1542.6	0.0136
1550.4	0.0149
1553.7	0.0143
1779.8	0.0164
1798.2	0.0131
1801.5	0.0163
1999.3	0.0173
2012.6	0.0144
2019.8	0.0169
2268.7	0.0193
2278.7	0.0176
2280.9	0.0176*
CURVE 11	
1592.1	0.0123
1595.4	0.0115
1599.8	0.0123*
1603.2	0.0131
1912.1	0.0147
1927.6	0.0137
1929.8	0.0164
1936.5	0.0151
2089.8	0.0143
2092.1	0.0182
2099.8	0.0164
2310.9	0.0144
2341.5	0.0140
2373.7	0.0133
CURVE 12	
298.2	0.0109
CURVE 13	
298.2	0.00879
CURVE 14	
836.2	0.0111
CURVE 15	
963.2	0.0102
CURVE 16	
907.2	0.0208
907.2	0.0207*
907.2	0.0181
953.2	0.0163
1016.7	0.0192
1085.2	0.0192
1127.2	0.0192
CURVE 17	
369.2	0.0163
747.2	0.0170
649.2	0.0172
701.2	0.0176
749.2	0.0177
818.2	0.0180
883.2	0.0182
949.2	0.0184
1001.2	0.0186
1061.2	0.0188
1125.2	0.0191*
1189.2	0.0192*
1243.2	0.0194*
1283.2	0.0197
1307.2	0.0197
1325.2	0.0197*
1333.2	0.0197*
1353.2	0.0199
1389.2	0.0200*
1403.2	0.0201
1425.2	0.0201*
1433.2	0.0201*
1447.2	0.0202
1487.2	0.0203
1503.2	0.0205*
1553.2	0.0205
1569.2	0.0206*
CURVE 18	
373.2	0.0100
473.2	0.0100
573.2	0.0105
673.2	0.0109
773.2	0.0109
873.2	0.0113
973.2	0.0113
1073.2	0.0117
1173.2	0.0117
1273.2	0.0121
CURVE 19*	
298.2	0.00190
CURVE 20*	
553	0.00159
700	0.00159
811	0.00216
897	0.00216
939	0.00216
1180	0.00202
CURVE 21*	
473.2	0.00232
653.2	0.00303
862.2	0.00386
1054	0.00452
1259	0.00517
CURVE 22*	
654.2	0.00350
873.2	0.00449
1073	0.00508
1274	0.00564
CURVE 23*	
473.2	0.00306
677.2	0.00394
873.2	0.00501
1073	0.00697
1273	0.00931

*Not shown on plot

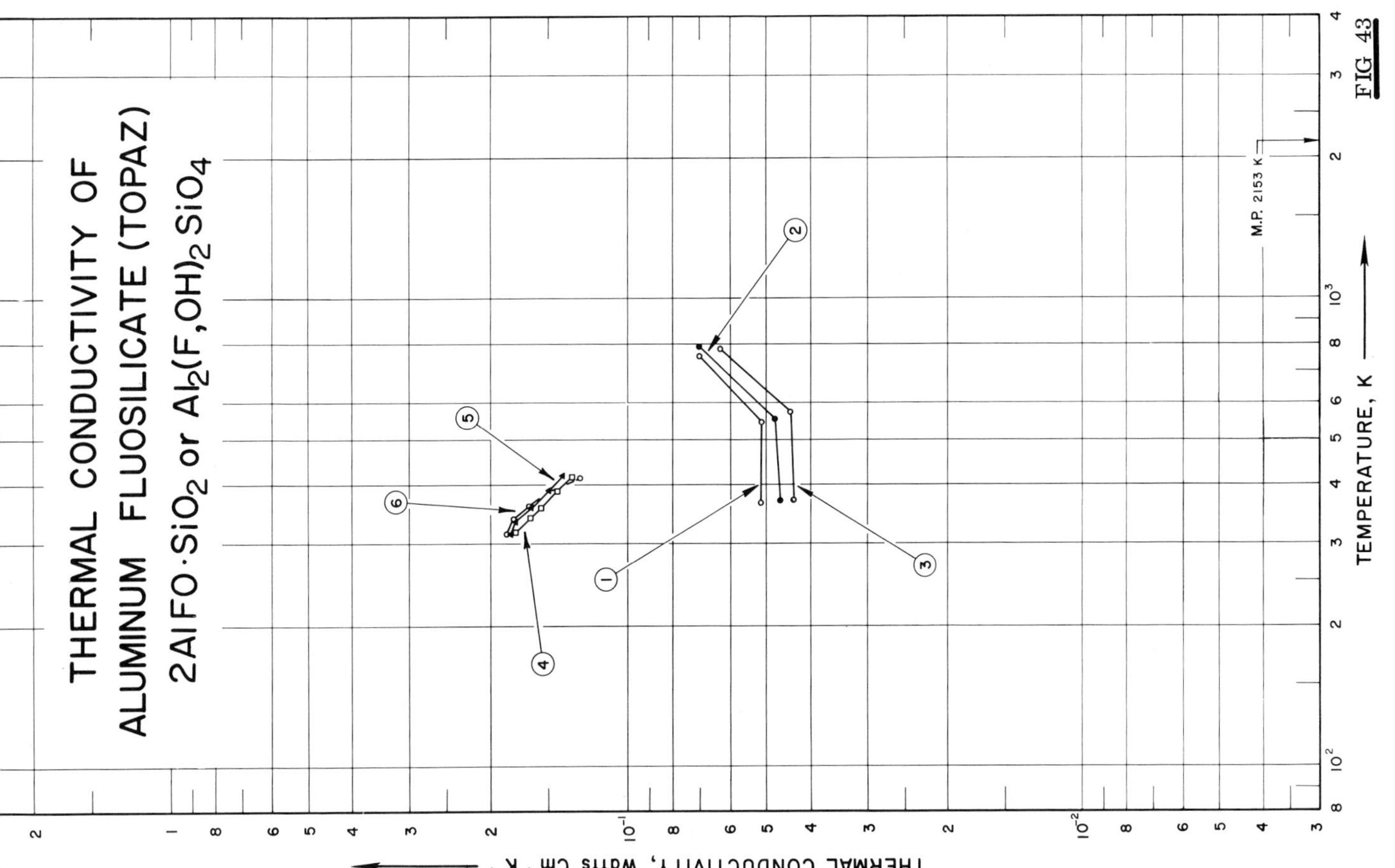
THERMAL CONDUCTIVITY OF
ALUMINUM FLUOSILICATE (TOPAZ)
$2AlFO \cdot SiO_2$ or $Al_2(F,OH)_2SiO_4$
M.P. 2153 K
TEMPERATURE, K
THERMAL CONDUCTIVITY, Watts Cm^{-1} K^{-1}
1
2
3
4
5
6

FIG 43

SPECIFICATION TABLE NO. 43 THERMAL CONDUCTIVITY OF ALUMINUM FLUOSILICATE (TOPAZ) $Al_2(F, OH)_2SiO_2$ or $2AlFO \cdot SiO_2$

[For Data Reported in Figure and Table No. 43]

Curve No.	Ref. No.	Method Used	Year	Temp. Range, K	Reported Error, %	Name and Specimen Designation	Composition (weight percent), Specifications and Remarks
1	34	C	1943	367-757			Domestic; colorless; specimen 1 x 1 x 1 cm cut from a single crystal with its axes a, b and c normal to their corresponding surfaces; heat flow parallel to c-axis; 18-8 stainless steel used as comparative material.
2	34	C	1943	372-793			The above specimen measured with heat flow parallel to b-axis.
3	34	C	1943	373-785			The above specimen measured with heat flow parallel to a-axis.
4	68	C	1954	315-417	±3.0	Brazil topaz; 293A-1	Orthorhombic single crystal (cell dimensions: a = 4.64 Å, b = 8.78 Å, and c = 8.38Å); specimen 0.25 in. in dia and 0.25 in. long; measured in vacuum; heat flow parallel to c-axis; copper used as comparative material.
5	68	C	1954	314-420	±3.0	Brazil topaz; 293B-1	The above specimen measured along b-axis.
6	68	C	1954	315-417	±3.0	Brazil topaz; 293C-1	The above specimen measured along a-axis.

DATA TABLE NO. 43 THERMAL CONDUCTIVITY OF ALUMINUM FLUOSILICATE (TOPAZ) $Al_2(F, OH)_2SiO_2$ or $2AlFO \cdot SiO_2$

[Temperature, T, K; Thermal Conductivity, k, Watt $cm^{-1}K^{-1}$]

T	k
CURVE 1	
367.2	0.0515
549.2	0.0515
757.2	0.0703
CURVE 2	
372.2	0.0469
556.2	0.0481
793.2	0.0703
CURVE 3	
373.2	0.0439
573.2	0.0444
785.2	0.0632
CURVE 4	
315.3	0.177
339.6	0.165
357.5	0.156
388.1	0.144
416.5	0.133
CURVE 5	
314.2	0.182
334.7	0.177
356.4	0.163
388.9	0.150
419.8	0.140
CURVE 6	
314.8	0.185
337.6	0.178
357.4	0.165*
390.0	0.149
416.8	0.128

*Not shown on plot

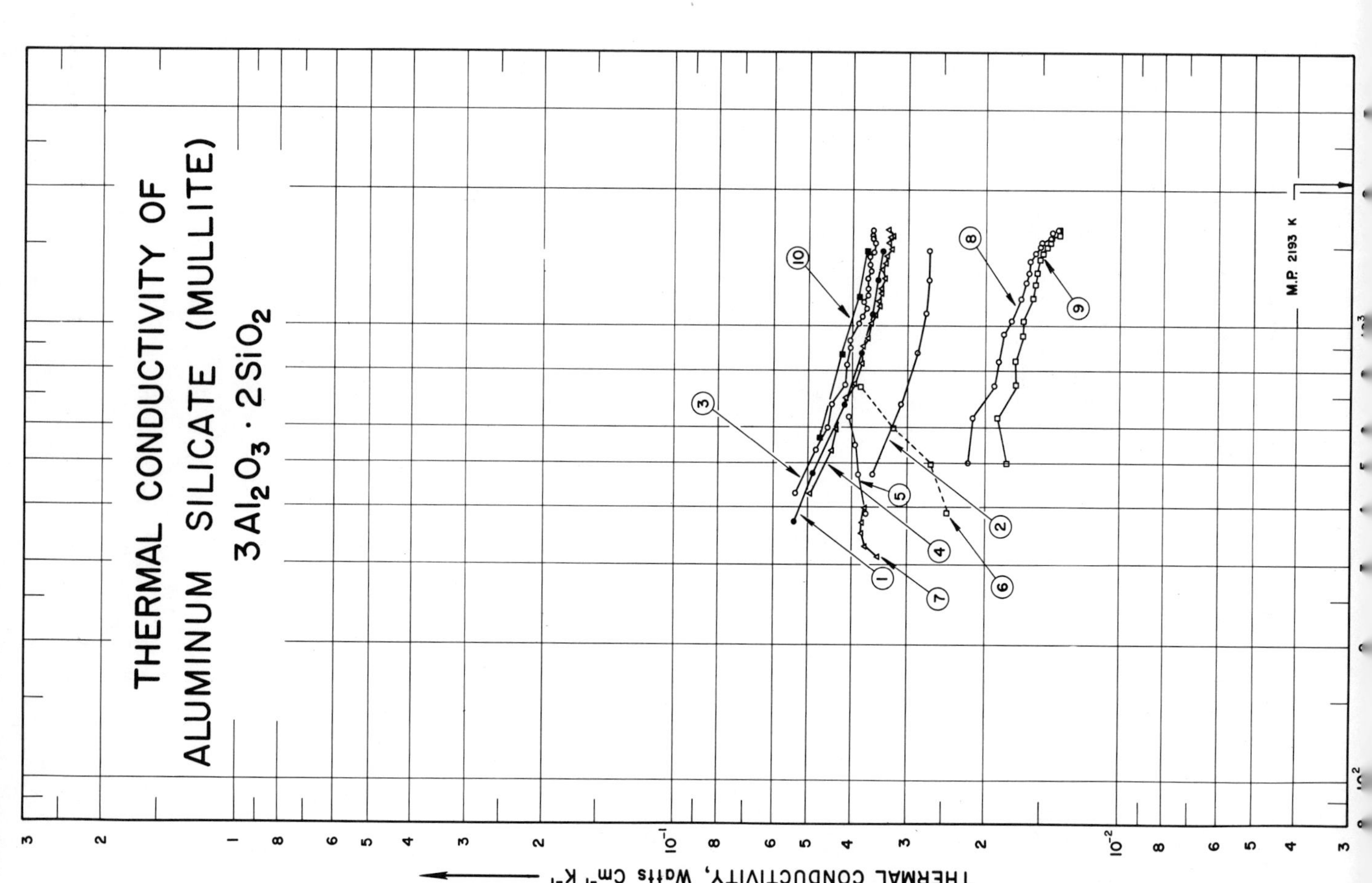
THERMAL CONDUCTIVITY OF
ALUMINUM SILICATE (MULLITE)
3Al2O3 · 2SiO2
M.P. 2193 K
THERMAL CONDUCTIVITY, Watts Cm-1 K-1
1
2
3
4
5
6
7
8
9
10

SPECIFICATION TABLE NO. 44 THERMAL CONDUCTIVITY OF ALUMINUM SILICATE (MULLITE) $3Al_2O_3 \cdot 2SiO_2$

[For Data Reported in Figure and Table No. 44]

Curve No.	Ref. No.	Method Used	Year	Temp. Range, K	Reported Error, %	Name and Specimen Designation	Composition (weight percent), Specifications and Remarks
1	364	C	1954	373-1473		Mullite	69. 0 Al_2O_3, 31. 9 SiO_2; polycrystal; supplied by Babcock and Wilcox Co. ; prepared from pure fused material; ground for 24 hrs in a steel ball mill; particle size 50% < 5 microns; slip cast from slip of pH 3. 0, specific gravity 2. 1 and fired at 1780 C; bulk density 2. 79 g cm^{-3}; porosity 11. 4%; alumina used as comparative material.
2	364	C	1954	473-1473		Mullite	Similar to the above specimen except bulk density 2. 21 g cm^{-3} and porosity 29. 8%.
3	10	R	1952	431-1633		Mullite	Bulk density 2. 79 g cm^{-3}; total porosity 11. 43%.
4	10	R	1952	428-1636		Mullite	Similar to the above specimen.
5	34	C	1943	392-740		Mullite	Specimen 1 cm cube; electrocast; polycrystalline with needles well aligned; measured parallel to c (principal) axis; 18-8 stainless steel used as comparative material.
6	34	C	1943	392-744		Mullite	The above specimen measured normal to c (principal) axis.
7	68	C	1954	315-402	±3. 0	Mullite; 5A2	Measured in vacuum; copper used as comparative material.
8	153	R	1951	502-1652		Mullite No. 2	Crystalline specimen; porosity 28%; fired at 1710 C.
9	153	R	1951	502-1629		Mullite No. 2	2nd run of the above specimen.
10	73	C	1954	573-1473		Mullite	Prepared by slip casting; fired.

DATA TABLE NO. 44 THERMAL CONDUCTIVITY OF ALUMINUM SILICATE (MULLITE) $3Al_2O_3 \cdot 2SiO_2$

[Temperature, T, K; Thermal Conductivity, k, Watt $cm^{-1}K^{-1}$]

T	k
CURVE 1	
373.2	0.0540
473.2	0.0490
673.2	0.0418
873.2	0.0381
1073.2	0.0360
1273.2	0.0351
1473.2	0.0343
CURVE 2	
473.2	0.0360
673.2	0.0310
873.2	0.0285
1073.2	0.0272
1273.2	0.0268
1473.2	0.0268
CURVE 3	
430.7	0.0536
535.7	0.0481
598.2	0.0452
673.2	0.0446
745.7	0.0414
825.7	0.0412
898.2	0.0406
935.7	0.0406
1018.2	0.0387
1053.2	0.0381
1098.2	0.0372
1133.2	0.0377
1175.7	0.0370
1218.2	0.0370
1273.2	0.0370
1328.2	0.0364
1372.2	0.0366
1463.2	0.0360
1533.2	0.0356
1565.7	0.0360
1583.2	0.0360
1633.2	0.0360
CURVE 4	
428.2	0.0496
533.2	0.0446
593.2	0.0435
CURVE 4 (cont.)	
698.2	0.0414
748.2	0.0397
825.7	0.0381
903.2	0.0379
943.2	0.0370
1018.2	0.0364
1053.2	0.0356
1108.2	0.0347
1138.2	0.0349
1185.7	0.0345
1218.2	0.0345
1278.2	0.0339
1335.7	0.0345
1378.2	0.0337
1438.2	0.0335
1473.2	0.0326
1535.7	0.0331
1573.2	0.0328
1583.2	0.0324
1635.7	0.0331
CURVE 5	
392.2	0.0372
477.2	0.0387
563.2	0.0395
740.2	0.0502
CURVE 6	
392.2	0.0243
500.2	0.0264
608.2	0.0321
744.2	0.0382
CURVE 7	
315.2	0.0349
333.6	0.0375
354.5	0.0377
373.8	0.0378
401.7	0.0372
CURVE 8	
502.2	0.0218
634.2	0.0214
CURVE 8 (cont.)	
746.2	0.0192
846.2	0.0187
961.2	0.0182
1048.2	0.0175
1172.2	0.0167
1260.2	0.0162
1335.2	0.0159
1409.2	0.0158
1478.2	0.0154
1515.2	0.0150
1552.2	0.0149
1594.2	0.0142
1634.2	0.0141
1652.2	0.0138
CURVE 9	
502.2	0.0179
634.2	0.0187
746.2	0.0171
846.2	0.0172
961.2	0.0165
1048.2	0.0164
1169.2	0.0157
1254.2	0.0155
1329.2	0.0153
1402.2	0.0151
1468.2	0.0149
1501.2	0.0146
1543.2	0.0143
1579.2	0.0142 *
1613.2	0.0136
1629.2	0.0136 *
CURVE 10	
573.2	0.0470
873.2	0.0418
1173.2	0.0383
1473.2	0.0366

* Not shown on plot

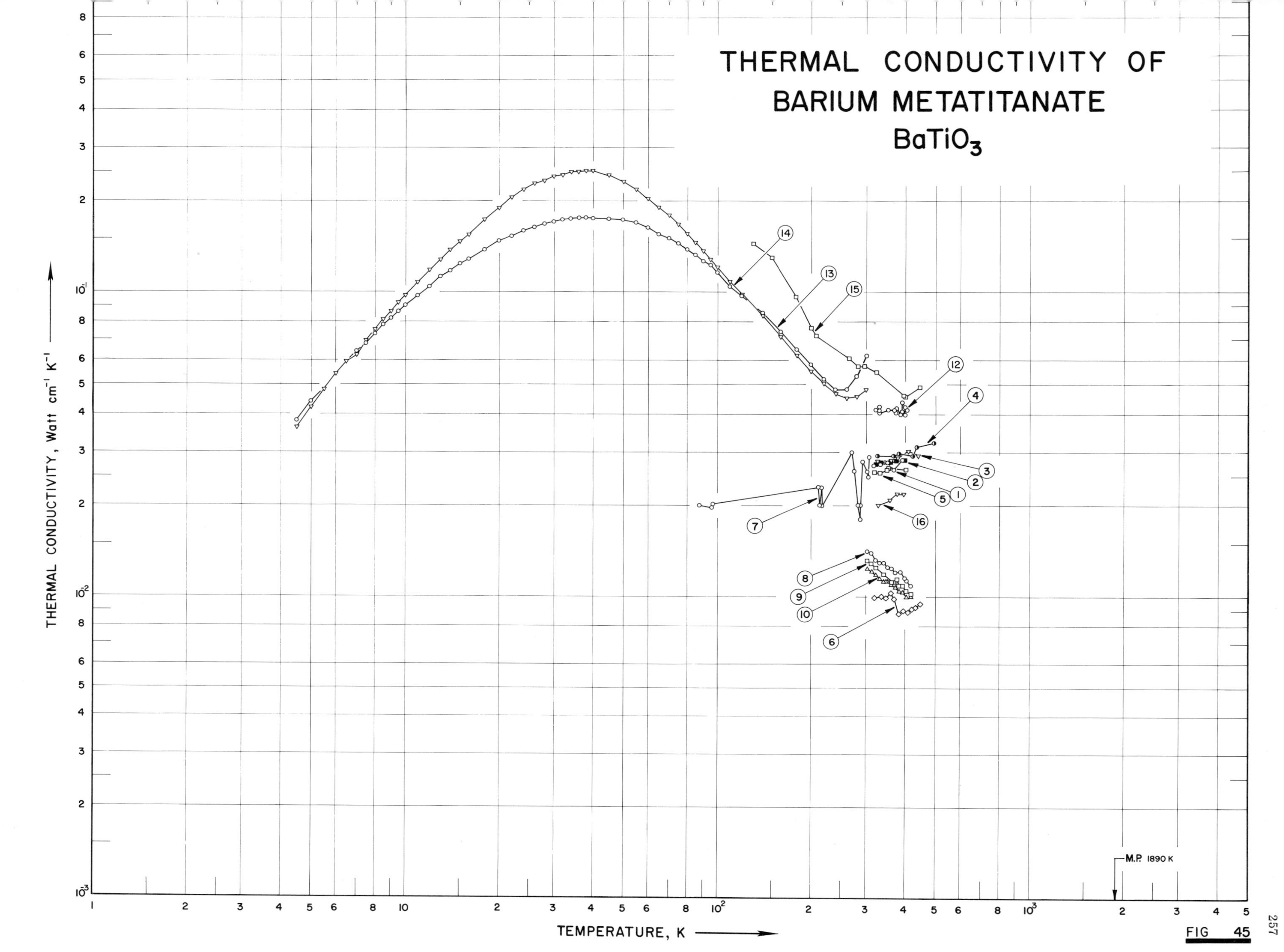
THERMAL CONDUCTIVITY OF
BARIUM METATITANATE
BaTiO3
THERMAL CONDUCTIVITY, Watt cm⁻¹ K⁻¹
TEMPERATURE, K
M.P. 1890 K
1
2
3
4
5
6
7
8
9
10
12
13
14
15
16

FIG 45

SPECIFICATION TABLE NO. 45 THERMAL CONDUCTIVITY OF BARIUM METATITANATE $BaTiO_3$

[For Data Reported in Figure and Table No. 45]

Curve No.	Ref. No.	Method Used	Year	Temp. Range, K	Reported Error, %	Name and Specimen Designation	Composition (weight percent), Specifications and Remarks
1	9	C	1953	317-392		159A-1	Specimen 0.100 in. long, 0.410 in. in dia; bulk density 5.34 g cm^{-3}; copper used as standard.
2	9	C	1953	322-398		159B-1	Specimen 0.100 in. long, 0.410 in. in dia; copper used as standard.
3	9	C	1953	328-440		159C-2	Specimen 0.100 in. long, 0.410 in. in dia; copper used as standard.
4	9	C	1953	327-495		159C-3	Specimen 0.100 in. long, 0.410 in. in dia; copper used as standard.
5	68		1954	318-403	±3	38A2	No other information supplied.
6	201	C	1952	320-450			Specimen 0.12 cm thick, 2.75 cm in dia; copper used as standard.
7	38	L	1960	88-307		No. 2	Specimen 2.06 cm in dia, 1.45 cm thick; made from powdered raw materials of BaO and TiO_2 mixed together with water, pressed into a disc and fired in a siliconite furnace; the product crushed, well pulverized and meshed, then pressed (1.3 tons. cm^{-2}) into disc form and sintered for 6.5 hrs at 1300 C; density 5.03 g cm^{-3}; porosity 16.0%; measurements made in a vacuum better than 10^{-4} mm Hg.
8	202	F	1961	303-418			Pure; specific heat; 0.14 cal g^{-1} deg^{-1}; velocity of sound in specimen 4.5 x 10^5 cm sec^{-1}, mean free path 16.2 Å.
9	202	F	1961	303-418			Similar to above except impurity 0.5 mol % of $Mn_2Nb_2O_7$.
10	202	F	1961	303-418			Similar to above except impurity 1.0 mol % $Mn_2Nb_2O_7$.
11	203	C	1958	323-458			Specimen sintered from $BaCO_3$ and TiO_3 powder of special reagent grade; pure iron used as standard.
12	289	C	1963	323-405	<15		Disc specimen cut from a 1 in. long, 0.75 in. dia cylinder; disc dia to thickness ratio greater than 10; faces of disc lapped parallel.
13	465	L	1965	4.5-300	± 5		0.02 Fe_2O_3 major impurity; single crystal; specimen ~1 x 1 x 7 mm; supplied by Kinsekisha Laboratory; sintered; heat flowed in the direction of <100>; transition temp from rhombohedral to orthorhombic and from orthorhombic to tetragonal at 193 and 273 K, respectively; grown by flux method; thermal conductivity data obtained directly from the author.
14	465	L	1965	4.5-300	± 5		Polycrystalline; specimen 1 x 1 x 7 mm; supplied by Kinsekisha Laboratory; sintered; grown by flux method; thermal conductivity data obtained directly from the author.
15	467	L	1967	132-447			1.5 Sr doped; single crystal; specimen 3 mm in dia and 1 mm thick was grown by floating zone process; supplied by Dr. F. Brown of Williams College, Williamstown, Mass.
16	468	L	1950	329-397		39A	Specimen 0.476 in. dia and 0.488 in. long.

DATA TABLE NO. 45 THERMAL CONDUCTIVITY OF BARIUM METATITANATE $BaTiO_3$

[Temperature, T, K; Thermal Conductivity, k, Watt cm^{-1} K^{-1}]

T	k
CURVE 1	
316.6	0.0271
334.8	0.0274
351.6	0.0277
368.9	0.0263
392.1	0.0283
CURVE 2	
321.8	0.0275
335.9	0.0276
359.8	0.0279
375.7	0.0281
397.6	0.0282
CURVE 3	
327.6	0.0280
360.2	0.0280
380.4	0.0289
410.3	0.0301
440.4	0.0293
CURVE 4	
326.9	0.0293
368.9	0.0293
384.2	0.0297
424.8	0.0293
437.8	0.0314
495.2	0.0322
CURVE 5	
317.8	0.0259
334.7	0.0257
352.4	0.0263
360.7	0.0268
403.3	0.0263
CURVE 6	
320.0	0.00992
338.7	0.0100
348.2	0.00992
364.7	0.0103
CURVE 6 (cont.)	
370.7	0.00983
384.2	0.00879
396.2	0.00900
410.7	0.00891
421.2	0.00912
433.2	0.00925
449.7	0.00950
CURVE 7	
88.2	0.0200
96.2	0.0198
97.2	0.0202
211.2	0.0230
213.2	0.0200
215.2	0.0230
218.2	0.0200
271.2	0.0300
275.2	0.0260
283.2	0.0200
288.2	0.0180
288.2	0.0200
293.2	0.0280
301.2	0.0260
305.2	0.0250
307.2	0.0290
CURVE 8	
303.2	0.0142
313.2	0.0140
323.2	0.0133
333.2	0.0130
343.2	0.0130
353.2	0.0126
363.2	0.0125
373.2	0.0121
383.2	0.0121*
388.2	0.0122
393.2	0.0121*
398.2	0.0116
403.2	0.0113
418.2	0.0109
CURVE 9	
303.2	0.0132
313.2	0.0129
323.2	0.0126
343.2	0.0119
363.2	0.0113
368.2	0.0113
373.2	0.0113
378.2	0.0115
383.2	0.0109
393.2	0.0108
403.2	0.0105
418.2	0.0103
CURVE 10	
303.2	0.0126
313.2	0.0122
323.2	0.0118
333.2	0.0115
343.2	0.0113
348.2	0.0113
353.2	0.0113
358.2	0.0113*
363.2	0.0110
373.2	0.0107
383.2	0.0105
393.2	0.0104
403.2	0.0100
418.2	0.0100
CURVE 11*	
323.2	0.029
328.2	0.032
338.2	0.033
353.2	0.030
355.2	0.031
358.2	0.030
362.2	0.027
367.2	0.028
369.2	0.030
370.2	0.026
378.2	0.029
379.2	0.030
380.2	0.032
CURVE 11 (cont.)*	
383.2	0.032
388.2	0.032
391.2	0.031
392.2	0.031
395.2	0.029
403.2	0.028
418.2	0.027
428.2	0.029
445.2	0.027
451.2	0.026
458.2	0.025
CURVE 12	
322.7	0.0415
331.7	0.0425
331.7	0.0415
333.7	0.0405
355.7	0.0415
369.7	0.0415
375.0	0.0405
375.7	0.0420
387.2	0.0400
391.2	0.0410
392.2	0.0400
395.7	0.0435*
398.2	0.0400
400.2	0.0422
405.2	0.0415
CURVE 13	
4.5	0.038
5.0	0.044
5.5	0.048*
6.0	0.054*
6.5	0.059*
7.0	0.064
7.5	0.068
8.0	0.073
8.5	0.078
9.0	0.082
9.5	0.086
10	0.090
11	0.097
CURVE 13 (cont.)	
12	0.104
13	0.112
14	0.117
15	0.124
16	0.128
18	0.137
20	0.147
22	0.153
24	0.159
26	0.163
28	0.167
30	0.170
32	0.173
34	0.174
36	0.175
38	0.175
40	0.175
45	0.174
50	0.173
55	0.169
60	0.162
65	0.155
70	0.150
75	0.144
80	0.137
85	0.132
90	0.126
95	0.122
100	0.116
110	0.104
120	0.097
140	0.085
160	0.074
180	0.065
200	0.058
220	0.052
240	0.048
260	0.048
280	0.053
300	0.062
CURVE 14	
4.5	0.036
5.0	0.042
5.5	0.048
6.0	0.054
6.5	0.059
7.0	0.062
7.5	0.069
8.0	0.075
8.5	0.081
9.0	0.086
9.5	0.092
10	0.097
11	0.107
12	0.118
13	0.127
14	0.137
15	0.146
16	0.154
18	0.173
20	0.189
22	0.205
24	0.217
26	0.227
28	0.233
30	0.240
32	0.243
34	0.247
36	0.249
38	0.250
40	0.250
45	0.242
50	0.230
55	0.217
60	0.203
65	0.190
70	0.178
75	0.166
80	0.155
85	0.145
90	0.136
95	0.127
100	0.120
110	0.108
120	0.098
140	0.083
CURVE 14 (cont.)	
160	0.071
180	0.062
200	0.055
220	0.050
240	0.0465
260	0.045
280	0.0455
300	0.048
CURVE 15	
131.8	0.144
150.7	0.129
179.5	0.0960
200.0	0.0762
208.4	0.0716
266.7	0.0605
282.5	0.0570
298.5	0.0572
325.8	0.0545
397.2	0.0459
402.7	0.0455
446.7	0.0488
CURVE 16	
329.4	0.0199
360.3	0.0208
378.7	0.0218
397.4	0.0218

*Not shown on plot

SPECIFICATION TABLE NO. 46 THERMAL CONDUCTIVITY OF BARIUM DITITANATE $BaO \cdot 2TiO_2$

Curve No.	Ref. No.	Method Used	Year	Temp. Range, K	Reported Error, %	Name and Specimen Designation	Composition (weight percent), Specifications and Remarks
1	468	L	1950	328-394		38A	Specimen 0.492 in. dia x 0.491 in. long.

DATA TABLE NO. 46 THERMAL CONDUCTIVITY OF BARIUM DITITANATE $BaO \cdot 2TiO_2$

[Temperature, T, K; Thermal Conductivity, k, Watt $cm^{-1} K^{-1}$]

T	k
CURVE 1*	
328.3	0.0302
359.1	0.0304
394.3	0.0304

* No graphical presentation

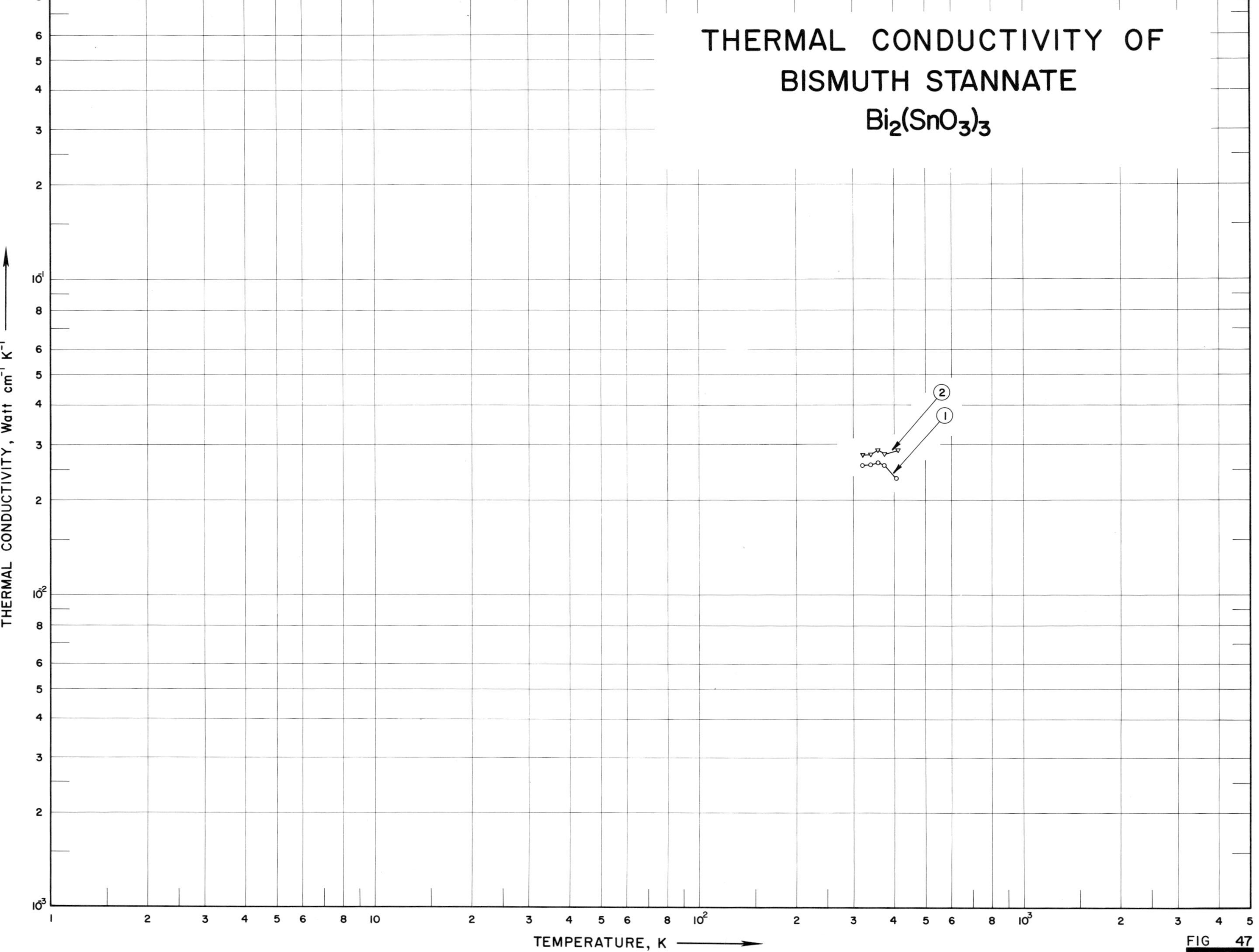
THERMAL CONDUCTIVITY OF
BISMUTH STANNATE
$Bi_2(SnO_3)_3$
THERMAL CONDUCTIVITY, Watt cm⁻¹ K⁻¹
TEMPERATURE, K
1
2

FIG 47

SPECIFICATION TABLE NO. 47 THERMAL CONDUCTIVITY OF BISMUTH STANNATE $Bi_2(SnO_3)_3$

[For Data Reported in Figure and Table No. 47]

Curve No.	Ref. No.	Method Used	Year	Temp. Range, K	Reported Error, %	Name and Specimen Designation	Composition (weight percent), Specifications and Remarks
1	3	L	1953	319-408		169A-1	Specimen supplied by Metal and Thermit Corp.; density 7.64 g cm^{-3} at 25 C; water absorption 0.011%.
2	3	L	1953	319-412		169B-1	Specimen supplied by Metal and Thermit Corp.; density 7.60 g cm^{-3} at 25 C; water absorption 0.011%.

DATA TABLE NO. 47 THERMAL CONDUCTIVITY OF BISMUTH STANNATE $Bi_2(SnO_3)_3$

[Temperature, T, K; Thermal Conductivity, k, Watt cm^{-1} K^{-1}]

T	k
CURVE 1	
319.1	0.0258
339.3	0.0259
357.8	0.0263
374.9	0 0257
408.3	0.0235
CURVE 2	
319.1	0.0277
339.3	0.0278
357.5	0.0287
375.5	0.0280
412.1	0.0287

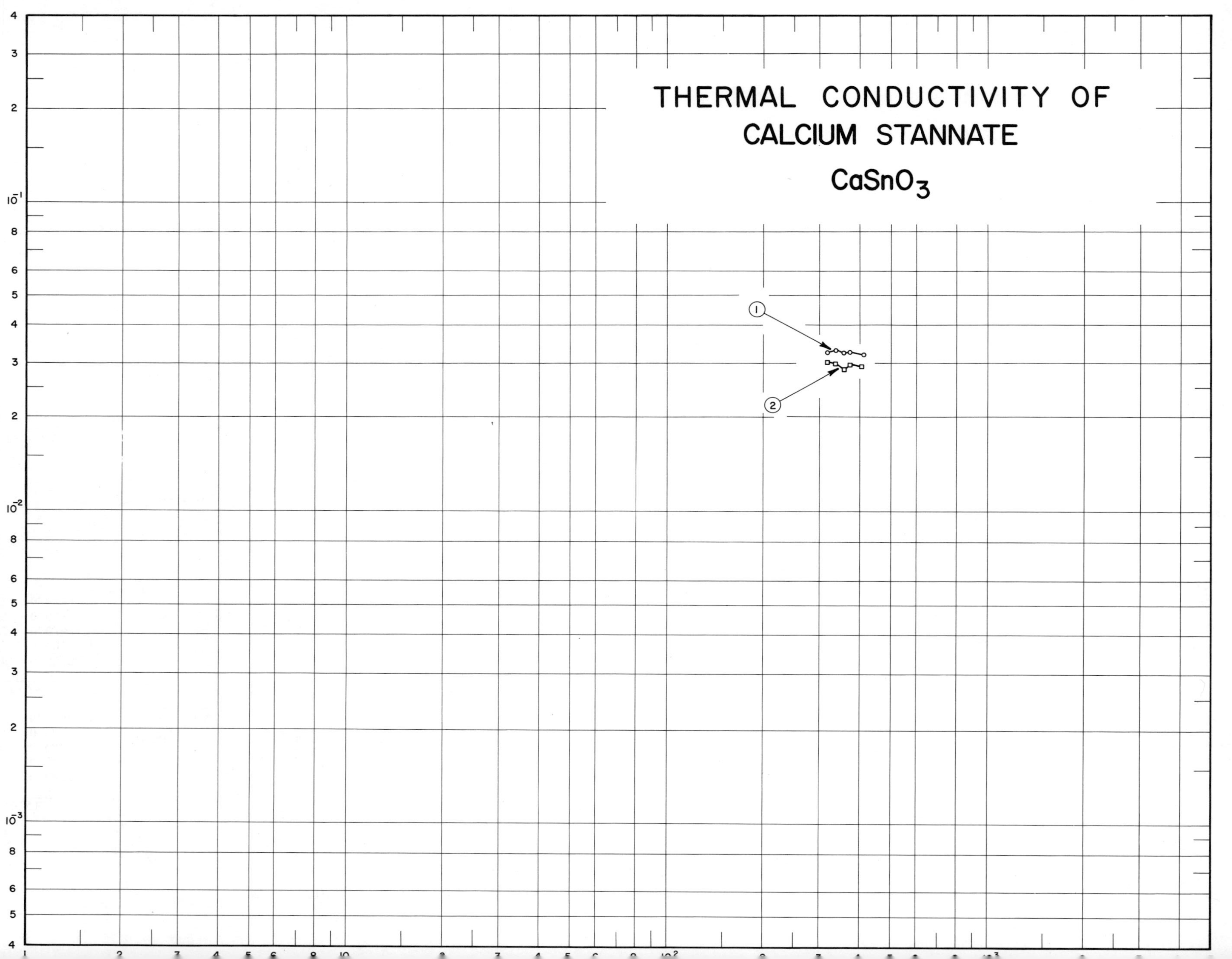
THERMAL CONDUCTIVITY OF
CALCIUM STANNATE
$CaSnO_3$
1
2
THERMAL CONDUCTIVITY, Watt cm^{-1} K^{-1}

SPECIFICATION TABLE NO. 48 THERMAL CONDUCTIVITY OF CALCIUM STANNATE $CaSnO_3$

[For Data Reported in Figure and Table No. 48]

Curve No.	Ref. No.	Method Used	Year	Temp. Range, K	Reported Error, %	Name and Specimen Designation	Composition (weight percent), Specifications and Remarks
1	3	L	1953	319-411		167B-1	Density (25 C) 5.08 g cm^{-3}; water absorption 0.57%.
2	3	L	1953	318-407		167B-2	Separate run of the above specimen.

DATA TABLE NO. 48 THERMAL CONDUCTIVITY OF CALCIUM STANNATE $CaSnO_3$

[Temperature, T, K; Thermal Conductivity, k, Watt $cm^{-1}K^{-1}$]

T	k
CURVE 1	
318.7	0.0327
338.6	0.0333
357.7	0.0327
373.5	0.0328
411.0	0.0323
CURVE 2	
317.9	0.0302
337.4	0.0300
359.3	0.0288
374.0	0.0298
406.8	0.0295

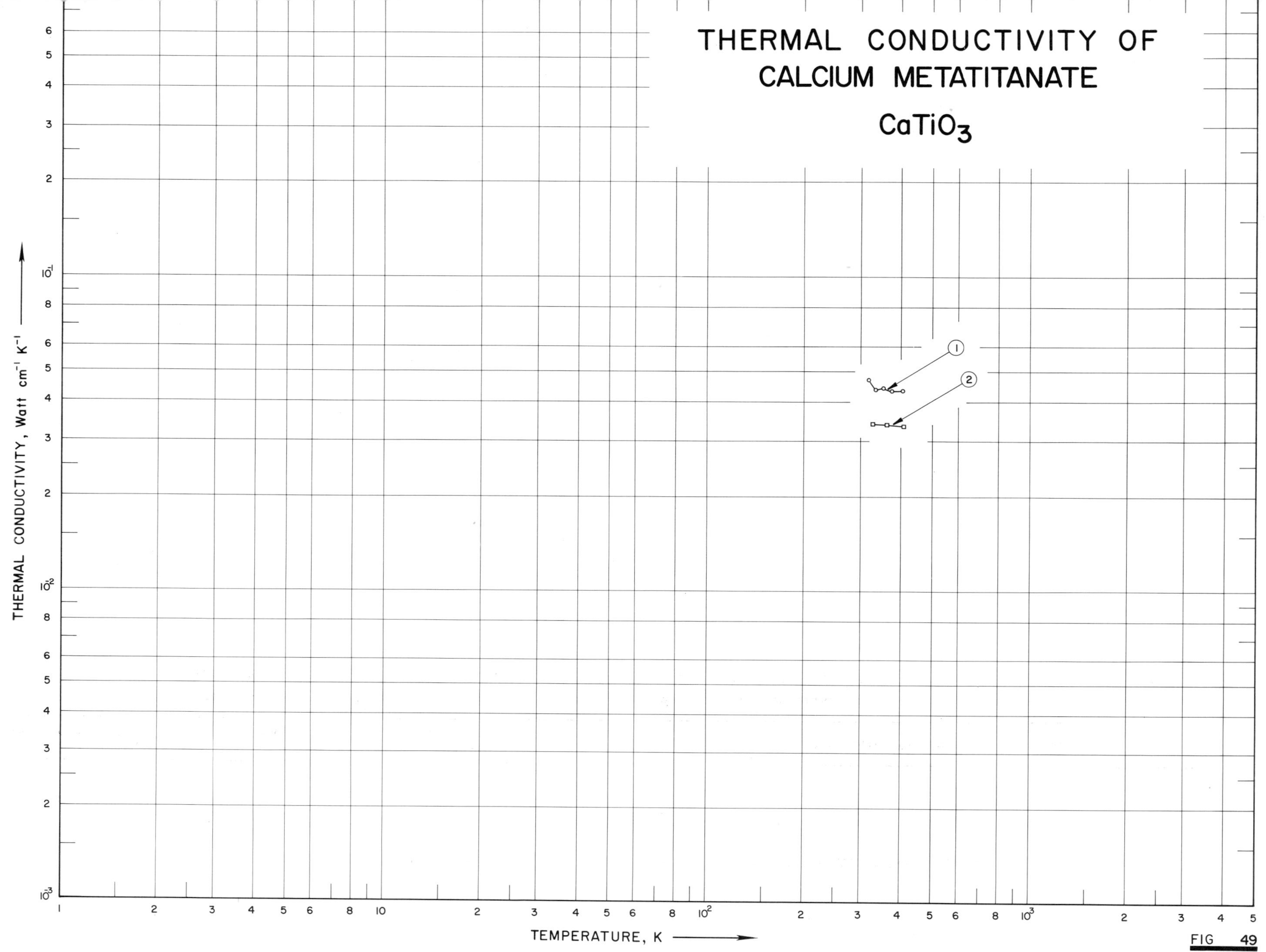
THERMAL CONDUCTIVITY OF
CALCIUM METATITANATE
$CaTiO_3$
1
2
THERMAL CONDUCTIVITY, Watt cm^{-1} K^{-1}
TEMPERATURE, K

FIG 49

SPECIFICATION TABLE NO. 49 THERMAL CONDUCTIVITY OF CALCIUM METATITANATE $CaTiO_3$

[For Data Reported in Figure and Table No. 49]

Curve No.	Ref. No.	Method Used	Year	Temp. Range, K	Reported Error, %	Name and Specimen Designation	Composition (weight percent), Specifications and Remarks
1	68		1954	318-406	±3	35A2	No other details reported.
2	468	L	1950	327-407		42B	0.417 in. dia x 0.513 in. long.

DATA TABLE NO. 49 THERMAL CONDUCTIVITY OF CALCIUM METATITANATE $CaTiO_3$

[Temperature, T, K; Thermal Conductivity, k, Watt $cm^{-1}K^{-1}$]

T	k
CURVE 1	
317.8	0.0473
335.2	0.0439
353.5	0.0444
374.5	0.0435
405.5	0.0435
CURVE 2	
327.4	0.0341
362.3	0.0340
407.4	0.0336

SPECIFICATION TABLE NO. 50 THERMAL CONDUCTIVITY OF CALCIUM TUNGSTATE $CaWO_4$

Curve No.	Ref. No.	Method Used	Year	Temp. Range, K	Reported Error, %	Name and Specimen Designation	Composition (weight percent), Specifications and Remarks
1	584	C	1962	422	15		Copper used as comparative material.

DATA TABLE NO. 50 THERMAL CONDUCTIVITY OF CALCIUM TUNGSTATE $CaWO_4$

[Temperature, T, K; Thermal Conductivity, k, Watt $cm^{-1} K^{-1}$]

T	k
CURVE 1*	
422	0.113

* No graphical presentation

SPECIFICATION TABLE NO. 51 THERMAL CONDUCTIVITY OF TRICOBALT STRONTIUM METATITANATE Co_3SrTiO_3

Curve No.	Ref. No.	Method Used	Year	Temp. Range, K	Reported Error, %	Name and Specimen Designation	Composition (weight percent), Specifications and Remarks
1	283		1959	298.2		No. 3	94 pure $Co_3SrOTiO_2$.

DATA TABLE NO. 51 THERMAL CONDUCTIVITY OF TRICOBALT STRONTIUM METATITANATE Co_3SrTiO_3

[Temperature, T, K; Thermal Conductivity, k, Watt $cm^{-1}K^{-1}$]

T	k
CURVE 1 *	
298.2	0.0724

* No graphical presentation

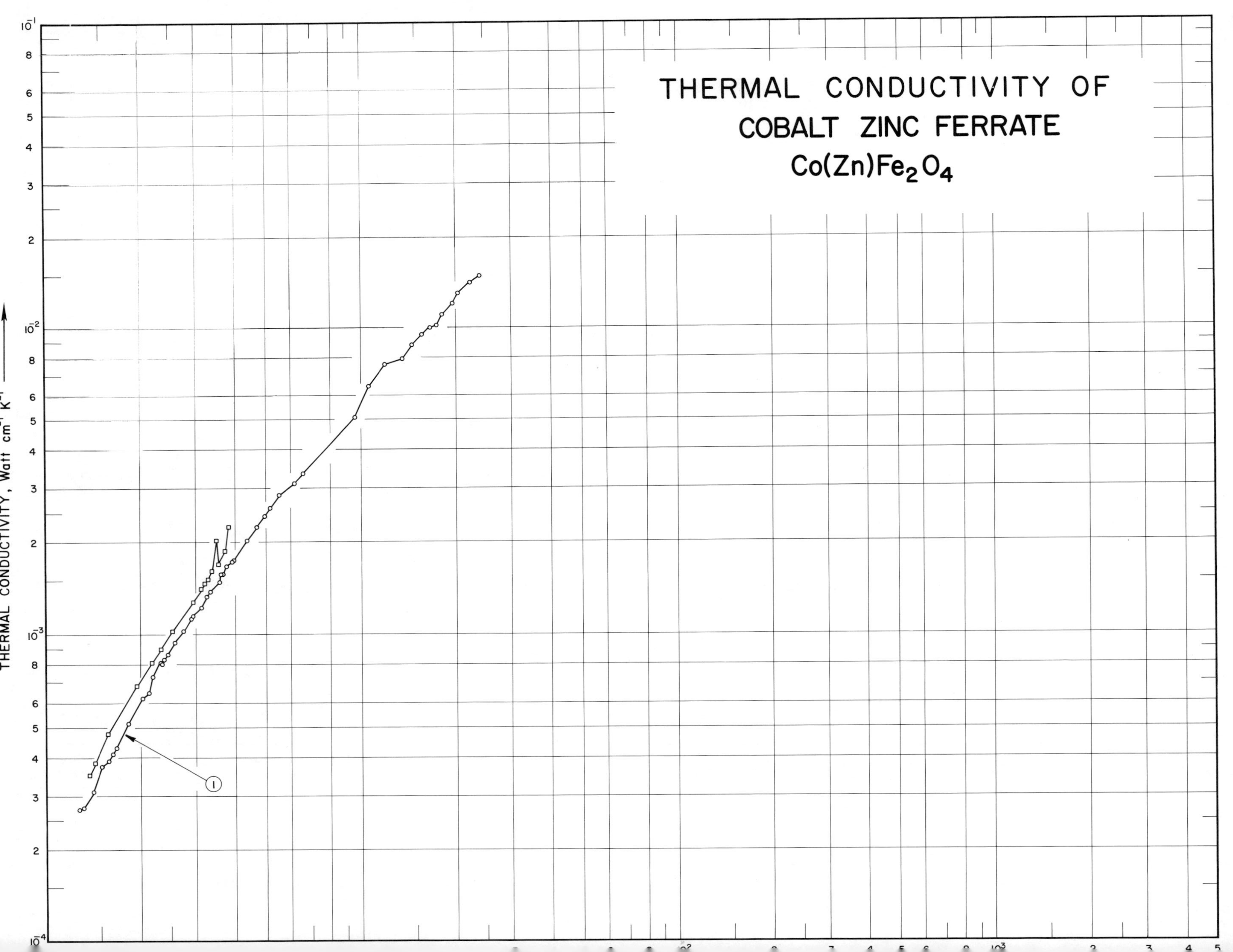
THERMAL CONDUCTIVITY OF
COBALT ZINC FERRATE
$Co(Zn)Fe_2O_4$
THERMAL CONDUCTIVITY, Watt cm^{-1} K^{-1}
1

SPECIFICATION TABLE NO. 52 THERMAL CONDUCTIVITY OF COBALT - ZINC FERRATES $Co(Zn)Fe_2O_4$

[For Data Reported in Figure and Table No. 52]

Curve No.	Ref. No.	Method Used	Year	Temp. Range, K	Reported Error, %	Name and Specimen Designation	Composition (weight percent), Specifications and Remarks
1	466	L	1961	1.3-24			Major metallic components: 55.8 Fe, 11.6 Co, and 6.46 Zn; specimen 4.37 cm in length and 0.329 cm^2 in cross section; single crystal bar with orientation of the long axis [5$\bar{3}$4]; average electrical resistivity 0.2, 53, and $\geq 3 \times 10^{10}$ ohm cm at room temperature, 77 K, and 4.2 K respectively; measured in vacuum.
2	466	L	1961	1.4-3.8			The above specimen measured in a magnetic field of 9400 gauss parallel to the direction of heat flow.

DATA TABLE NO. 52 THERMAL CONDUCTIVITY OF COBALT ZINC FERRATE $Co(Zn)Fe_2O_4$

[Temperature, T, K; Thermal Conductivity, k, Watt cm^{-1} K^{-1}]

T	k
CURVE 1	
1.27	0.000270
1.32	0.000275
1.41	0.000310
1.52	0.000375
1.58	0.000390
1.64	0.000410
1.66	0.000430
1.82	0.000515
2.02	0.000620
2.12	0.000645
2.18	0.000725
2.31	0.000810
2.33	0.000800
2.36	0.000830
2.43	0.000860
2.56	0.000940
2.74	0.00103
2.90	0.00113
2.93	0.00115
3.11	0.00123
3.25	0.00133
3.33	0.00138
3.56	0.00149
3.60	0.00157
3.65	0.00157
3.75	0.00167
3.90	0.00172
3.93	0.00174
4.36	0.00203
4.70	0.00225
4.95	0.00245
5.15	0.00260
5.50	0.00285
6.15	0.00310
6.55	0.00355
9.60	0.00570
9.90	0.00590
10.6	0.00645
12.0	0.00765
13.6	0.00795
14.6	0.00885
15.6	0.00945
16.7	0.0100
17.5	0.0105
18.2	0.0112
CURVE 1 (cont.)	
19.5	0.0121
20.5	0.0131
22.5	0.0143
24.0	0.0150
CURVE 2	
1.36	0.000360
1.43	0.000385
1.57	0.000475
1.67	0.000535
1.95	0.000680
2.18	0.000810
2.32	0.000895
2.53	0.00103
2.92	0.00128
3.12	0.00142
3.20	0.00147
3.29	0.00153
3.39	0.00161
3.46	0.00204
3.51	0.00170
3.72	0.00186
3.76	0.00226

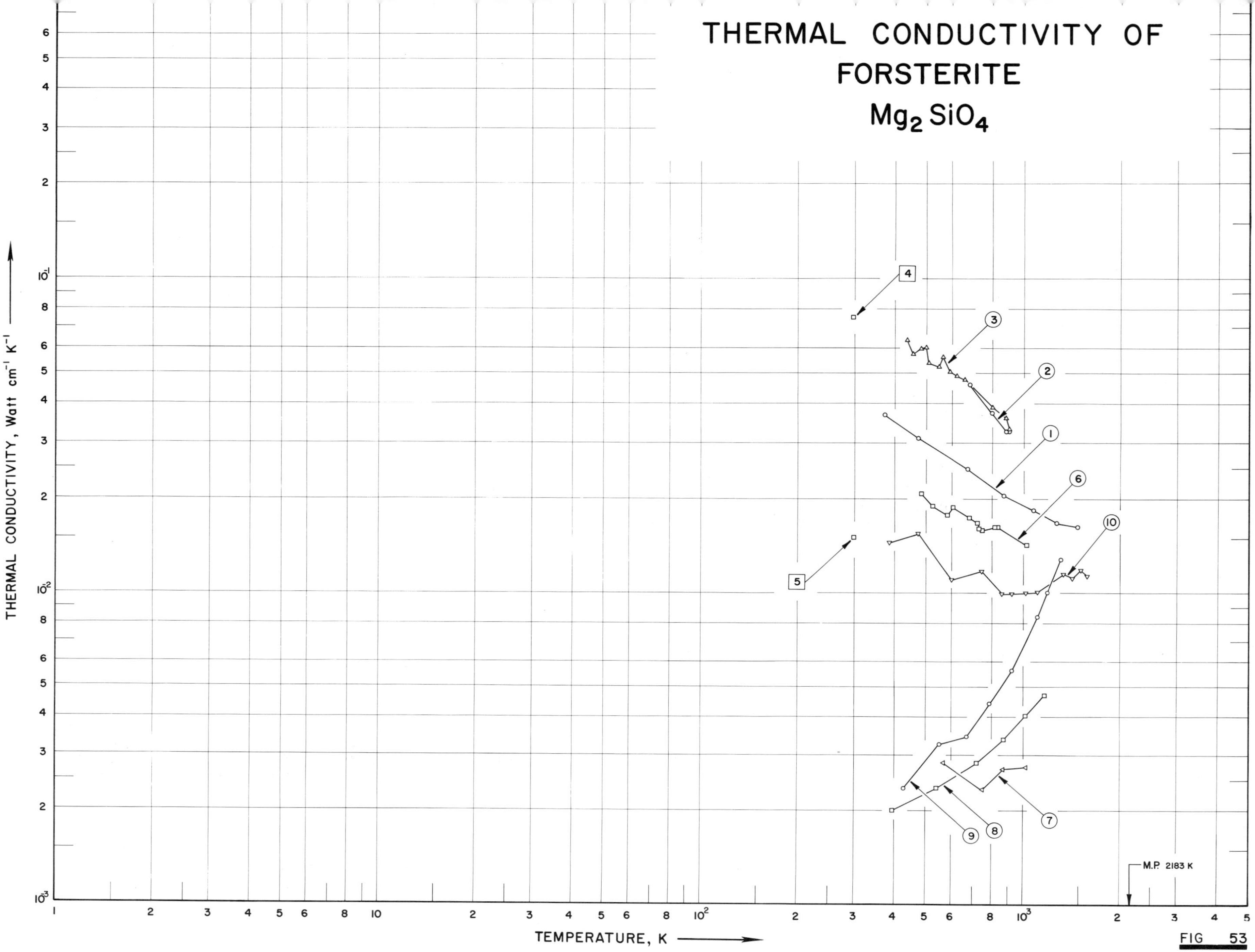

THERMAL CONDUCTIVITY OF FORSTERITE Mg_2SiO_4

FIG 53

SPECIFICATION TABLE NO. 53 THERMAL CONDUCTIVITY OF FORSTERITE Mg_2SiO_4

[For Data Reported in Figure and Table No. 53]

Curve No.	Ref. No.	Method Used	Year	Temp. Range, K	Reported Error, %	Name and Specimen Designation	Composition (weight percent), Specifications and Remarks
1	364	C	1954	373-1473		Forsterite	59.0 MgO, 41.0 SiO_2; polycrystal; prepared from calcined chemically pure magnesium carbonate and silicate acid, ground for 5 hrs in a rubber-lined mill, hydrostatically pressed and fired to 1430 C to form crystalline forsterite, then crushed, ground for 15 hrs and prepared as a casting slip in an ethanol suspension with specific gravity 2.05; fired at 1650 C; bulk density 2.22; porosity 31.1%; dense alumina used as comparative material.
2	65	C	1958	683-901	±2	Forsterite	Milled 45.0 treasure talc, 45.0 $Mg(OH)_2$, 3.6 rex ball clay, and 6.4 $BaCO_3$ in a porcelain mill with porcelain balls for 14 hrs, poured on plaster, dried, pulverized and calcined at 1260 C for 3 hrs, then pulverized again, mixed with equal parts of raw batch, and milled for 12 hrs in a porcelain ball mill, cold-pressed in a steel die at 16000 psi then fired at 1510 C for 8 hrs; bulk density 3.06 g cm^{-3}; porosity 4.4%; alumina (Body Al-300) used as comparative material.
3	65	C	1958	437-901	±2	Forsterite	The above specimen measured with nickelous oxide as standard at temp. below 400 C; measured with another piece of alumina(Body Al-300) as standard at temp. above 400 C.
4	152	L	1958	298.2		Forsterite 243	Specimen 20 mils thick and 0.75 in. in dia; preirradiated with 6 x 10^{19} epithermal neutrons per cm^2 for 480 MWD in the MTR.
5	152	L	1958	298.2		Forsterite 243	The above specimen postirradiated with 6 x 10^{19} epithermal neutrons per cm^2 for megawatt days in the material testing reactor.
6	374	C	1965	483-1023	0-11	Forsterite L (brick)	29.5 SiO_2, 10.9 Al_2O_3, 7.6 Fe_2O_3, 0.7 CaO, 50.3 MgO, 1.0 Cr_2O_3; bulk density 2.6 g cm^{-3}; apparent porosity 21%; specimen prepared to a tolerance of ±0.001 in. in the form of a cylinder 1 in. in dia and 1 in. in length; produced by Harbison-Walker Refractories Co.; alumina AL-300 as reference standard (thermal conductivity determined by J.J. Swica, Alfred University).
7	463,464	R	1961	568-1029			Powder specimen contained in a hollow cylinder of 203 mm long and 91 mm internal dia; grain size < 0.2 mm; bulk density 1.40 g cm^{-3}.
8	463,464	R	1961	396-1167			Similar to the above specimen except grain size 0.2-1 mm and bulk density 0.97 g cm^{-3}.
9	463,464	R	1961	428-1319			Similar to the above specimen except grain size 2-5 mm and bulk density 0.72 g cm^{-3}.
10	463,464	R	1961	384-1581			Prepared from the powder of grain size 0.2-1 mm by pressing and firing at 1650 C; bulk density 1.75 g cm^{-3}.

DATA TABLE NO. 53 THERMAL CONDUCTIVITY OF FORSTERITE Mg_2SiO_4

[Temperature, T, K; Thermal Conductivity, k, Watt cm^{-1} K^{-1}]

T	k
CURVE 1	
373.2	0.0368
473.2	0.0310
673.2	0.0247
873.2	0.0205
1073.2	0.0184
1273.2	0.0167
1473.2	0.0163
CURVE 2	
683.2	0.0460
798.2	0.0372
883.2	0.0326
901.2	0.0326
CURVE 3	
437.2	0.0636
456.2	0.0577
481.2	0.0598
499.2	0.0602
509.2	0.0540
548.2	0.0523
563.2	0.0561
591.2	0.0506
621.2	0.0490
658.2	0.0477
683.2	0.0460*
798.2	0.0389
883.2	0.0360
901.2	0.0331
CURVE 4	
298.2	0.0753
CURVE 5	
298.2	0.0151
CURVE 6	
483.2	0.0208
525.2	0.0189
583.2	0.0178
CURVE 6 (cont.)	
607.2	0.0187
682.2	0.0174
722.2	0.0167
730.2	0.0161
745.2	0.0159
822.2	0.0163
836.2	0.0163
1023.2	0.0143
CURVE 7	
568.2	0.00282
753.2	0.00232
871.2	0.00270
1029	0.00275
CURVE 8	
396.2	0.00199
541.2	0.00235
723.2	0.00282
873.2	0.00335
1019	0.00401
1167	0.00470
CURVE 9	
428.2	0.00235
552.2	0.00324
673.2	0.00343
787.2	0.00442
925.2	0.00561
1108	0.00837
1192	0.0100
1319	0.0129
CURVE 10	
384.2	0.0145
473.2	0.0154
596.2	0.0111
742.2	0.0118
857.2	0.00988
922.2	0.00993
1020	0.00997
CURVE 10 (cont.)	
1110	0.0101
1343	0.0116
1417	0.0112
1510	0.0119
1581	0.0114

*Not shown on plot

SPECIFICATION TABLE NO. 54 THERMAL CONDUCTIVITY OF GARNET $M_3^{II}M_2^{III}(SiO_4)_3$

Curve No.	Ref. No.	Method Used	Year	Temp. Range, K	Reported Error, %	Name and Specimen Designation	Composition (weight percent), Specifications and Remarks
1	68		1954	315-377	± 3	Garnet $[M_3^{II}R_2^{III}(SiO_4)_3]$	Natural single crystal with cubic crystal system.

DATA TABLE NO. 54 THERMAL CONDUCTIVITY OF GARNET $M_3^{II}M_2^{III}(SiO_4)_3$

[Temperature, T, K; Thermal Conductivity, k, Watt $cm^{-1}K^{-1}$]

T	k
CURVE 1*	
315.0	0.0358
334.6	0.0361
357.9	0.0354
376.8	0.0356

* No graphical presentation

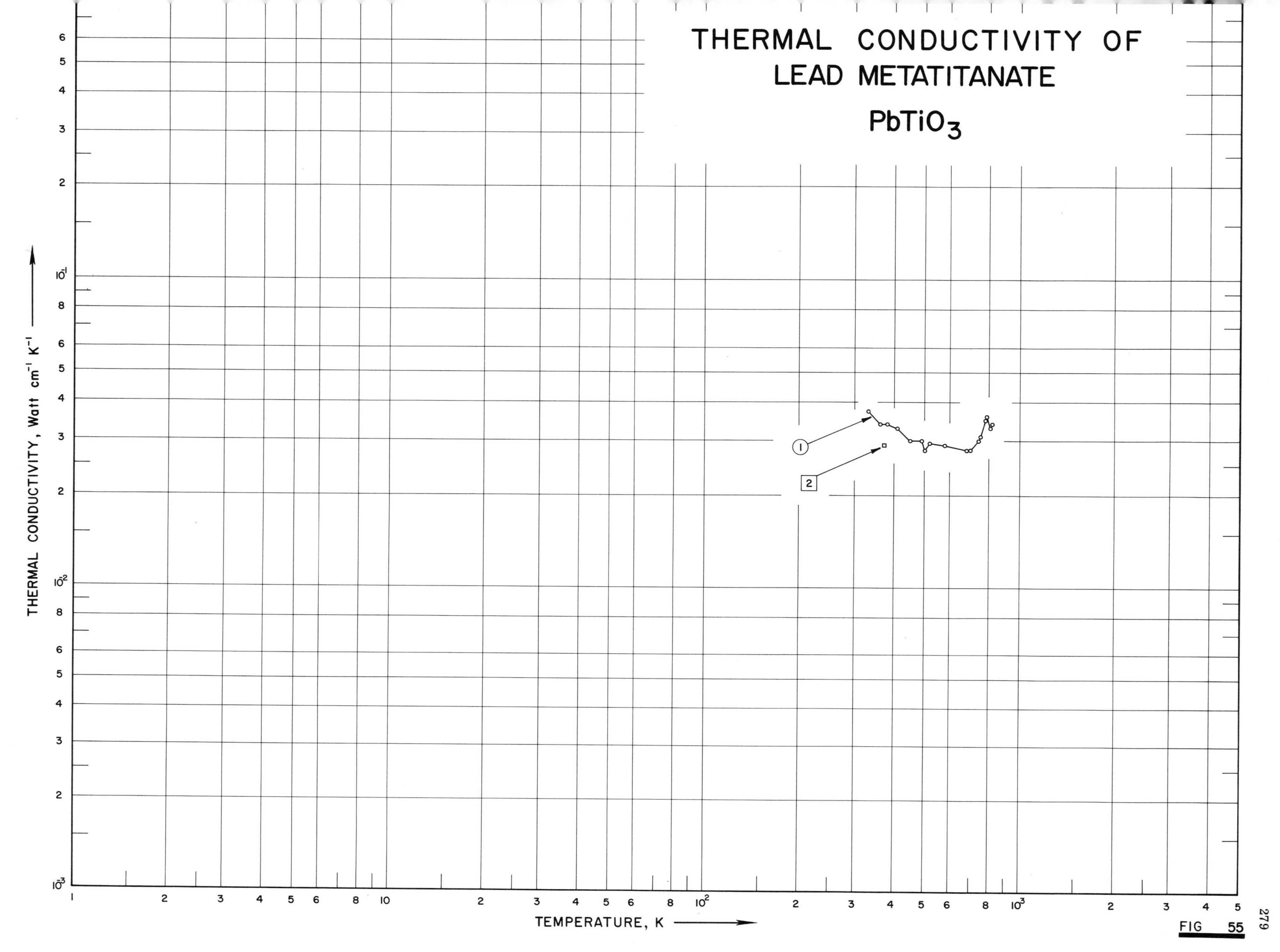

THERMAL CONDUCTIVITY OF
LEAD METATITANATE
PbTiO3
THERMAL CONDUCTIVITY, Watt cm⁻¹ K⁻¹
TEMPERATURE, K
1
2
FIG 55

SPECIFICATION TABLE NO. 55 THERMAL CONDUCTIVITY OF LEAD METATITANATE $PbTiO_3$

[For Data Reported in Figure and Table No. 55]

Curve No.	Ref. No.	Method Used	Year	Temp. Range, K	Reported Error, %	Name and Specimen Designation	Composition (weight percent), Specifications and Remarks
1	38	C	1960	333-823			Powdered raw materials of chemical pure PbO and special grade TiO_2 were mixed together with water, pressed into a disc and fired in a siliconit furnace; the product was crashed, well pulverized and meshed, then pressed (pressure 0.53 cm^{-2}) into a disc, 2.235 cm in dia and 0.395 cm thick; sintered at 1250 C for 2 hrs; density 6.86 g cm^{-3}; porosity 14.6%; measured in vacuum.
2	203	C	1958	373.2			No details reported.

DATA TABLE NO. 55 THERMAL CONDUCTIVITY OF LEAD METATITANATE $PbTiO_3$

[Temperature, T, K; Thermal Conductivity, k, Watt $cm^{-1}K^{-1}$]

T	k
CURVE 1	
333.2	0.0375
363.2	0.034
383.2	0.034
413.2	0.033
453.2	0.030
493.2	0.030
503.2	0.028
523.2	0.0295
583.2	0.029
683.2	0.028
698.2	0.028
743.2	0.030
753.2	0.031
778.2	0.035*
783.2	0.035
788.2	0.036
808.2	0.033
823.2	0.034
CURVE 2	
373.2	0.029

* Not shown on plot

SPECIFICATION TABLE NO. 56 THERMAL CONDUCTIVITY OF LEAD ZIRCONATE $PbZrO_3$

Curve No.	Ref. No.	Method Used	Year	Temp. Range, K	Reported Error, %	Name and Specimen Designation	Composition (weight percent), Specifications and Remarks
1	38	C	1960	345-530			Powdered raw materials of special grade PbO and chemically pure grade ZrO_2 mixed together with water, pressed into a disc and fired in a siliconit furnace; the product crushed, well pulverized and meshed, then pressed (pressure 0.53 ton cm^{-2}) into a disc, with dia 2.12 cm and thickness 0.3274 cm, and sintered at 1200 C for 2 hrs; density 6.05 g cm^{-3}; porosity 25.5%; measured in vacuum.

DATA TABLE NO. 56 THERMAL CONDUCTIVITY OF LEAD ZIRCONATE $PbZrO_3$

[Temperature, T, K; Thermal Conductivity, k, Watt cm^{-1} K^{-1}]

T	k
CURVE 1*	
345.2	0.0136
348.2	0.0136
393.2	0.0130
400.2	0.0130
428.2	0.0130
433.2	0.0130
460.2	0.0138
479.2	0.0145
505.2	0.0145
528.2	0.0145
530.2	0.0150

*No graphical presentation

THERMAL CONDUCTIVITY OF MAGNESIUM ALUMINATE (SPINEL) $MgO \cdot Al_2O_3$

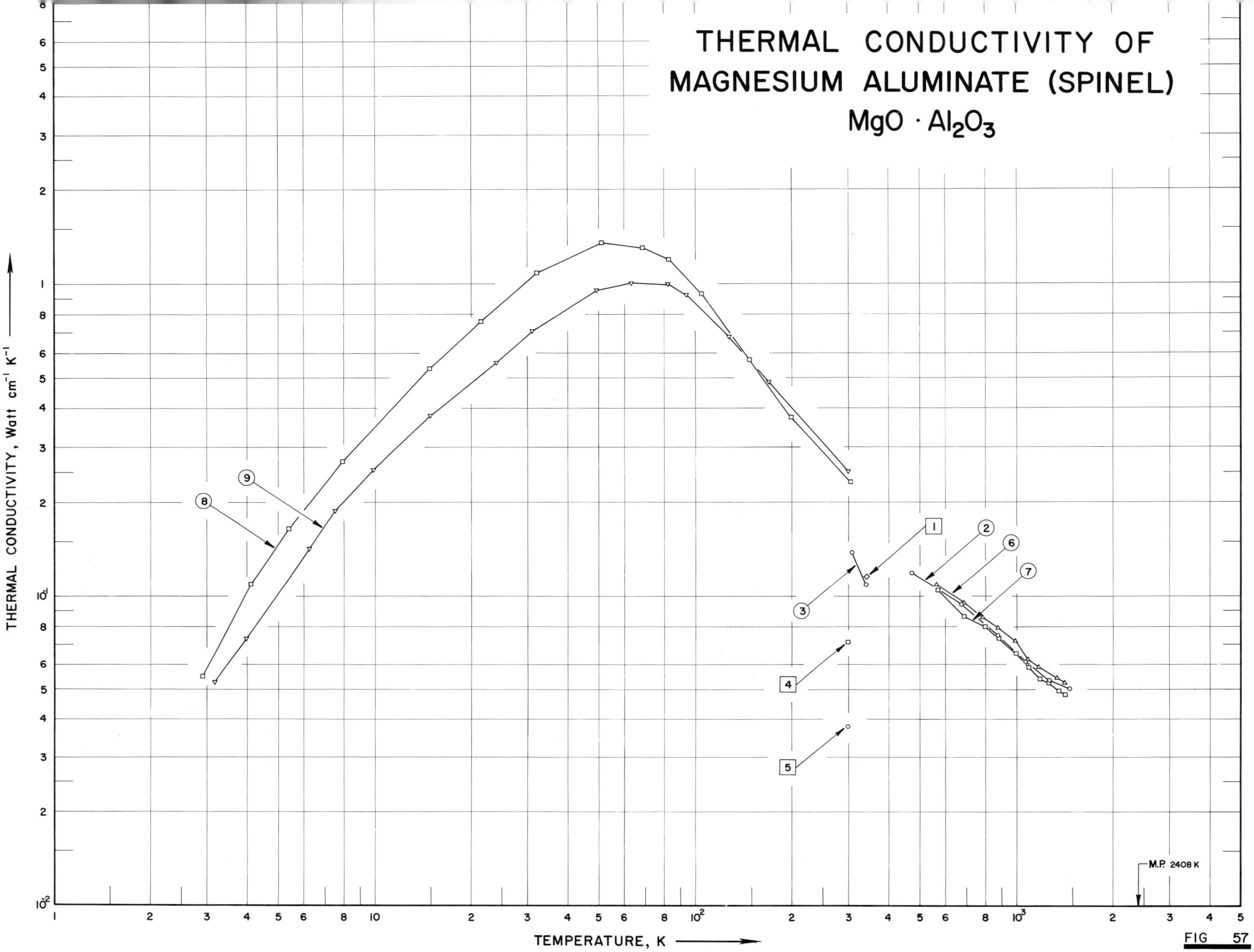

FIG 57

SPECIFICATION TABLE NO. 57 THERMAL CONDUCTIVITY OF MAGNESIUM ALUMINATE $MgO \cdot Al_2O_3$

[For Data Reported in Figure and Table No. 57]

Curve No.	Ref. No.	Method Used	Year	Temp. Range, K	Reported Error, %	Name and Specimen Designation	Composition (weight percent), Specifications and Remarks
1	13	C	1953	343.2	±3.0	Spinel	Single crystal; specimen 1.75 in. long and 0.22 in. in dia; supplied by Linde Air Products Co.; density 3.6 g cm^{-3}; Armco iron used as comparative material.
2	364	C	1954	473-1473		Spinel	71.9 Al_2O_3, 29.0 MgO by chemical analysis; polycrystal; slip-cast from suspension with specific gravity 2.2, pH 3.0; prepared with 1 to 1 molar ratio of MgO to Al_2O_3; bulk density 3.27 g cm^{-3}; porosity 7.65%; dense alumina used as comparative material.
3	71	C	1951	308, 341		Spinel	Cubic isotropic crystal; supplied by Linde Air Products Co. of Tonawanda; z-cut crystalline quartz used as comparative material.
4	152	L	1955	298.2		Spinel	Single crystal; specimen 20 mils thick and 0.75 in. in dia; preirradiated with 7 x 10^{19} epithermal neutrons per cm^2 for 480 Mwd in the MTR.
5	152	L	1955	298.2		Spinel	The above specimen postirradiated with 7 x 10^{19} epithermal neutrons per cm^2 for 480 megawatt days in the material testing reactor.
6	153	R	1951	563-1418			Specimen prepared from pure crystalline spinel formed by calcining a mixture of the oxides, reduced to 20-100 mesh, ground dry for 24 hrs in rubber lined ball mill with spinel balls, then acid-treated, filtered and prepared into a slip and fired at 1200 C, trimmed and fired again to 1840 C for 3 hrs; bulk density 3.27 g cm^{-3}.
7	153	R	1951	568-1423			2nd run of the above specimen.
8	300	L	1962	2.9-305		R-4z; Natural Ruby Spinel	Single natural crystal; impurity (epj in atom cm^{-3}) given in $\log_{10}$ epj: <18.1 Be, <17.7 Ca, 19.2 Cr, 18.9 Fe, 17.7 K, 18.2 Li, 18.1 Mn, 18.3 Na, <17.6 Ni, 18.2 Si, 18.6 Ti, 18.8 V, 20.1 Zn and 17.6 Zr; specimen 0.29 cm long and 0.18 cm avg. dia; lattice constant 8.0866 Å.
9	300	L	1962	3.2-300		R-54 Natural Ruby Spinel	Single natural crystal; impurity (epj in atom cm^{-3}) given in $\log_{10}$ epj: <18.1 Be, <17.7 Ca, 19.3 Cr, 19.2 Fe, 18.3 K, 18.2 Li, 17.9 Mn, 18.3 Na, <17.6 Ni, 18.2 Si, 18.3 Ti, 19.4 V, 19.5 Zn and <17.4 Zr; specimen 0.45 cm long and 0.25 in. avg. dia; lattice constant 8.0866 Å.

DATA TABLE NO. 57 THERMAL CONDUCTIVITY OF MAGNESIUM ALUMINATE $MgO \cdot Al_2O_3$

[Temperature, T, K; Thermal Conductivity, k, Watt $cm^{-1} K^{-1}$]

T	k
CURVE 1	
343.2	0.117
CURVE 2	
473.2	0.119
673.2	0.0941
873.2	0.0749
1073.2	0.0615
1273.2	0.0536
1473.2	0.0502
CURVE 3	
308.2	0.138
341.2	0.109
CURVE 4	
298.2	0.0711
CURVE 5	
298.2	0.0377
CURVE 6	
563.2	0.109
683.2	0.0952
783.2	0.0858
875.2	0.0791
980.2	0.0720
1080.2	0.0624
1173.2	0.0590
1348.2	0.0544
1418.2	0.0528
CURVE 7	
568.2	0.105
688.2	0.0864
798.2	0.0799
883.2	0.0732
998.2	0.0653
1091.2	0.0589
CURVE 7 (cont.)	
1183.2	0.0540
1268.2	0.0528
1358.2	0.0494
1423.2	0.0481
CURVE 8	
2.91	0.055
4.11	0.110
5.39	0.165
7.94	0.271
14.85	0.538
21.50	0.758
32.04	1.08
51.11	1.35
68.56	1.30
82.62	1.19
105.8	0.927
147.51	0.572
199.76	0.375
305.35	0.233
CURVE 9	
3.18	0.053
3.97	0.073
6.25	0.143
7.53	0.188
9.92	0.254
14.91	0.379
24.03	0.560
31.18	0.705
49.42	0.950
63.41	1.00
82.96	0.992
94.40	0.921
128.87	0.678
170.46	0.475
299.64	0.251

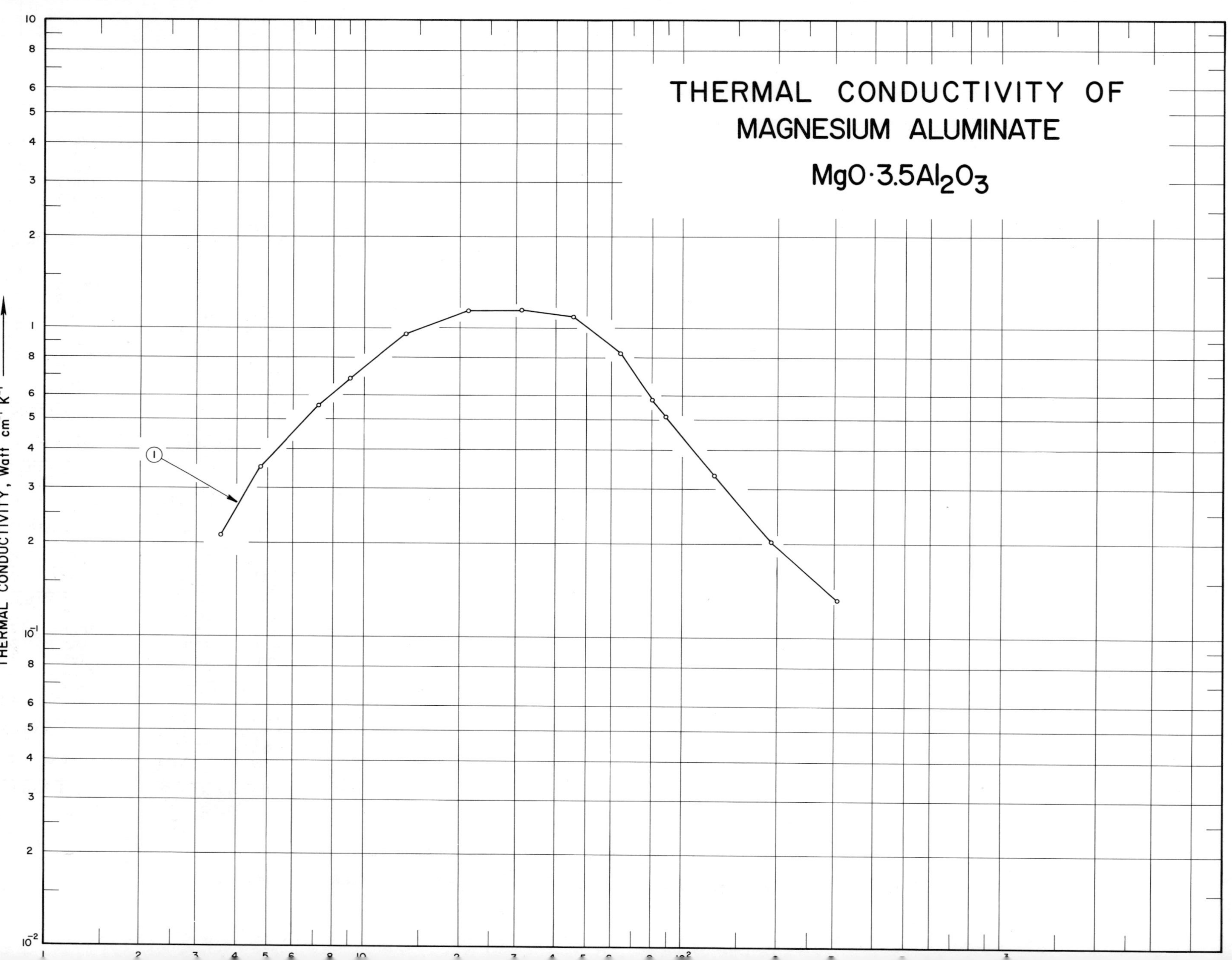
THERMAL CONDUCTIVITY OF
MAGNESIUM ALUMINATE
MgO·3.5Al2O3
THERMAL CONDUCTIVITY, Watt cm-1 K-1
1

SPECIFICATION TABLE NO. 58 THERMAL CONDUCTIVITY OF MAGNESIUM ALUMINATE $MgO \cdot 3.5\ Al_2O_3$

[For Data Reported in Figure and Table No. 58]

Curve No.	Ref. No.	Method Used	Year	Temp. Range, K	Reported Error, %	Name and Specimen Designation	Composition (weight percent), Specifications and Remarks
1	300	L	1962	3.6-301		Synthetic Spinel R-53	Single synthetic crystal; made with the Verneuil process by Linde Air Products Co. N.Y.; impurity (epj in. atoms cm^{-3}) given in $\log_{10}$epj = <18.1 Be, 17.4 Ca, <17.6 Cr, 18.8 Fe, <17.7 K, 18.0 Li, <17.6 Mn, 18.0 Na, <17.6 Ni, 19.1 Si, 18.5 Ti, <17.6 V, <17.8 Zn, <17.4 Zr; specimen 1.12 cm long and 0.40 cm avg dia; lattice constant 7.979Å.

DATA TABLE NO. 58 THERMAL CONDUCTIVITY OF MAGNESIUM ALUMINATE $MgO \cdot 3.5\ Al_2O_3$

[Temperature, T, K; Thermal Conductivity, k, Watt $cm^{-1}K^{-1}$]

T	k
CURVE 1	
3.58	0.212
4.76	0.350
7.21	0.559
9.04	0.685
13.56	0.954
21.44	1.15
31.46	1.16
45.97	1.10
64.38	0.830
80.49	0.585
89.52	0.519
126.47	0.332
190.52	0.204
300.67	0.133

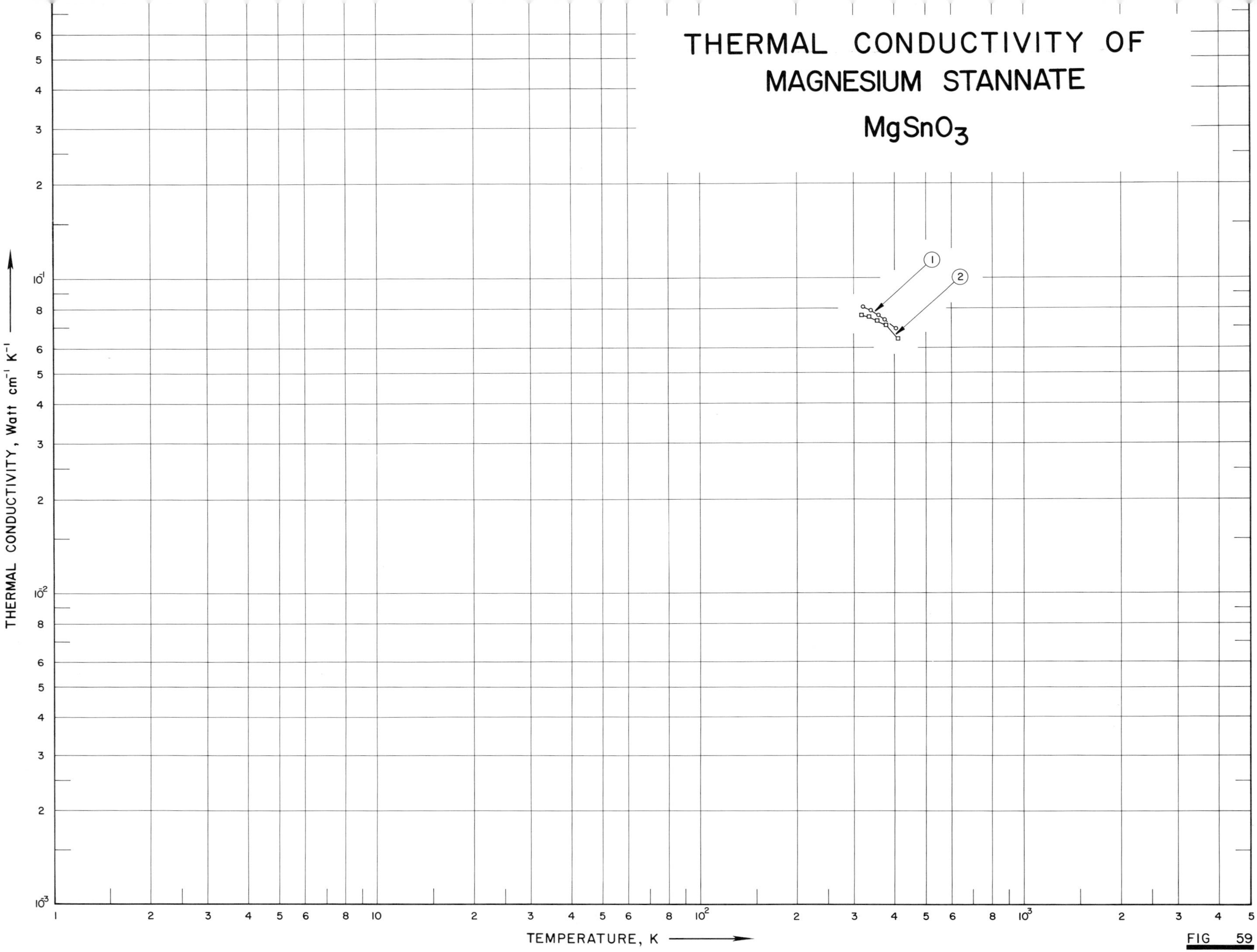

THERMAL CONDUCTIVITY OF MAGNESIUM STANNATE

$MgSnO_3$

FIG 59

SPECIFICATION TABLE NO. 59 THERMAL CONDUCTIVITY OF MAGNESIUM STANNATE $MgSnO_3$

[For Data Reported in Figure and Table No. 59]

Curve No.	Ref. No.	Method Used	Year	Temp. Range, K	Reported Error, %	Name and Specimen Designation	Composition (weight percent), Specifications and Remarks
1	3		1953	322-406		168A-1	Density (25 C) 5.18 g cm^{-3}; water absorption 0.26%.
2	3		1953	317-413		168A-2	Separate run of the above specimen.

DATA TABLE NO. 59 THERMAL CONDUCTIVITY OF MAGNESIUM STANNATE $MgSnO_3$

[Temperature, T, K; Thermal Conductivity, k, Watt $cm^{-1}K^{-1}$]

T	k
CURVE 1	
321.7	0.0808
339.4	0.0787
357.3	0.0761
375.5	0.0736
405.5	0.0690
CURVE 2	
317.3	0.0761
335.6	0.0753
355.6	0.0732
377.1	0.0707
412.9	0.0644

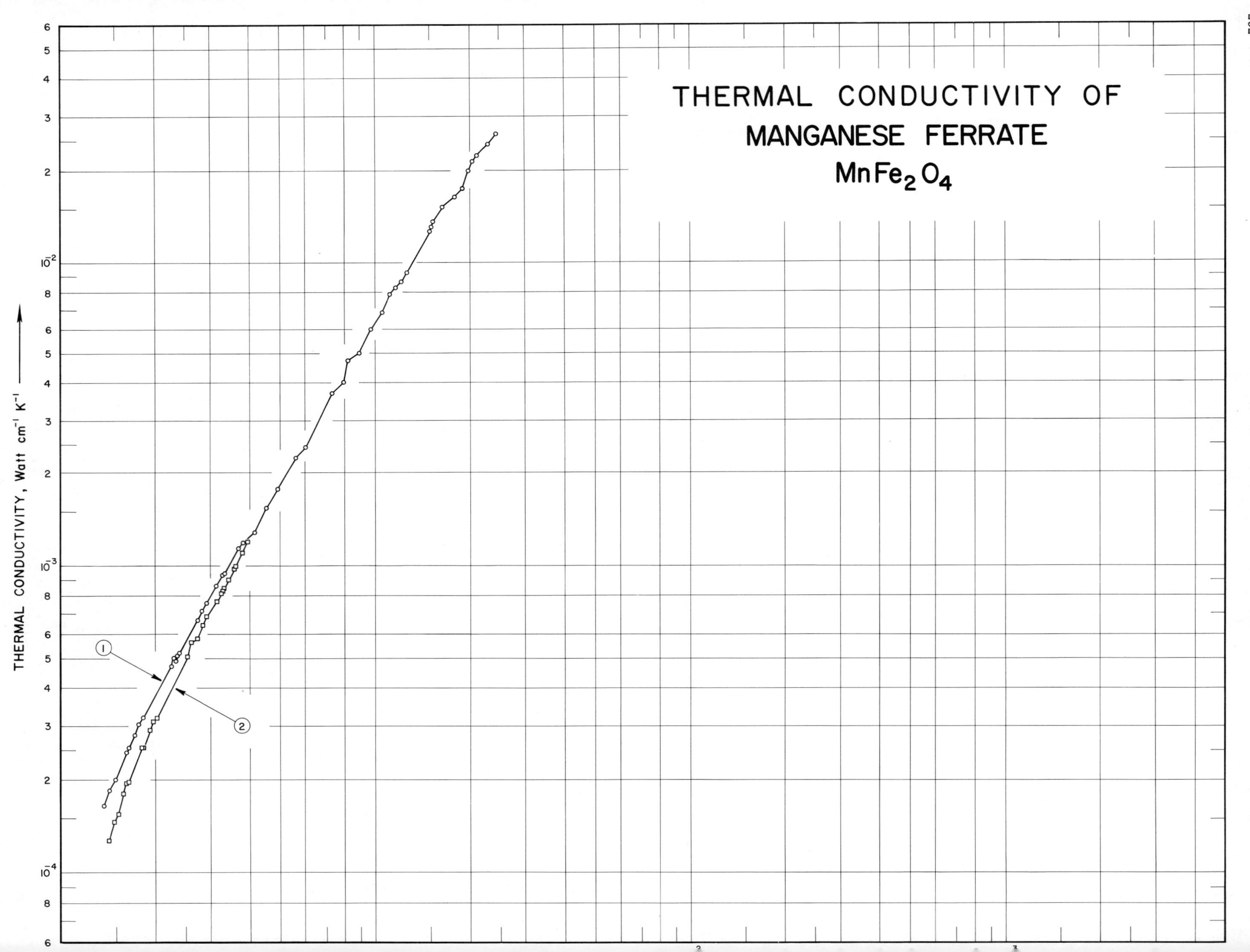

THERMAL CONDUCTIVITY OF
MANGANESE FERRATE
$MnFe_2O_4$
THERMAL CONDUCTIVITY, Watt cm^{-1} K^{-1}
6
5
4
3
2
10^{-2}
8
6
5
4
3
2
10^{-3}
8
6
5
4
3
2
10^{-4}
8
6
1
2

SPECIFICATION TABLE NO. 60 THERMAL CONDUCTIVITY OF MANGANESE FERRATE $MnFe_2O_4$

[For Data Reported in Figure and Table No. 60]

Curve No.	Ref. No.	Method Used	Year	Temp. Range, K	Reported Error, %	Name and Specimen Designation	Composition (weight percent), Specifications and Remarks
1	466	L	1961	1.4-25			Major metallic compositions: 38.4 Fe and 33.7 Mn; specimen 4.92 cm in length and 0.434 cm^2 in cross section; single crystal bar with orientation of the long axis [$\bar{4}10$] ; average electrical resistivity 2 x 10^4, $\sim 10^{12}$, and $\geq 10^2$ ohm cm at room temperature, 77 K, and 4.2 K; measured in vacuum.
2	466	L	1961	1.4-3.9			The above specimen measured in a magnetic field of 9400 gauss parallel to the direction of heat flow.

DATA TABLE NO. 60 THERMAL CONDUCTIVITY OF MANGANESE FERRATE $MnFe_2O_4$

[Temperature, T, K; Thermal Conductivity, k, Watt $cm^{-1} K^{-1}$]

T	k
CURVE 1	
1.38	0.000165
1.43	0.000185
1.50	0.000200
1.62	0.000245
1.65	0.000255
1.72	0.000280
1.77	0.000305
1.83	0.000320
2.25	0.000470
2.29	0.000500
2.32	0.000495
2.35	0.000505
2.37	0.000520
2.72	0.000665
2.80	0.000715
2.90	0.000755
3.13	0.000860
3.27	0.000935
3.32	0.000940
3.68	0.00114
3.80	0.00119
4.15	0.00128
4.50	0.00155
4.90	0.00177
5.60	0.00225
6.00	0.00245
7.30	0.00370
7.95	0.00405
8.20	0.00475
8.90	0.00500
9.70	0.00600
10.6	0.00690
11.3	0.00790
11.7	0.00830
12.3	0.00870
12.7	0.00930
15.0	0.0127
15.2	0.0131
15.4	0.0137
16.5	0.0153
18.0	0.0175
19.0	0.0185
19.8	0.0200
20.5	0.0215
CURVE 1 (cont.)	
21.3	0.0225
23.0	0.0245
24.5	0.0265
CURVE 2	
1.43	0.000127
1.48	0.000145
1.53	0.000155
1.58	0.000180
1.61	0.000195
1.65	0.000197
1.82	0.000255
1.83	0.000255
1.93	0.000290
1.97	0.000310
2.02	0.000317
2.53	0.000505
2.68	0.000565
2.71	0.000580
2.83	0.000640
2.93	0.000683
3.15	0.000765
3.25	0.000815
3.27	0.000827
3.30	0.000845
3.42	0.000900
3.58	0.000972
3.60	0.000990
3.78	0.00110
3.93	0.00120

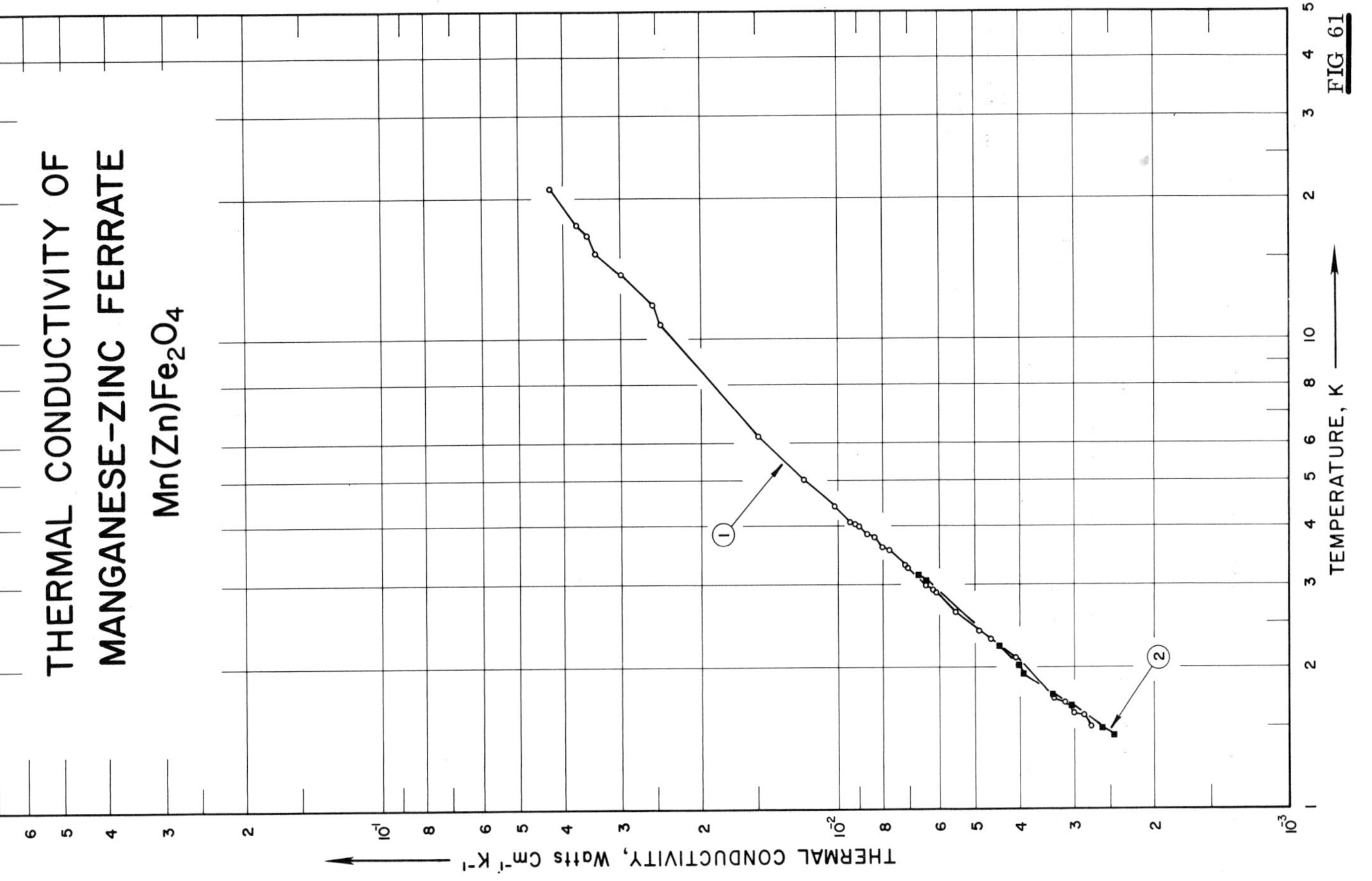
THERMAL CONDUCTIVITY OF
MANGANESE-ZINC FERRATE
$Mn(Zn)Fe_2O_4$
THERMAL CONDUCTIVITY, Watts Cm^{-1} K^{-1}
TEMPERATURE, K
1
2
FIG 61

SPECIFICATION TABLE NO. 61 THERMAL CONDUCTIVITY OF MANGANESE-ZINC FERRATE $Mn(Zn)Fe_2O_4$

[For Data Reported in Figure and Table No. 61]

Curve No.	Ref. No.	Method Used	Year	Temp. Range, K	Reported Error, %	Name and Specimen Designation	Composition (weight percent), Specifications and Remarks
1	466	L	1961	1. 5-21			Major metallic composition: 49. 2 Fe, 20. 95 Mn, and 6. 17 Zn; single crystal bar with long axis oriented [$\bar{2}$ 52]; specimen 4. 29 cm long and 0. 218 cm^2 in cross-section; electrical resistivity 0. 2, 172, $\geq 7 \times 10^9$ ohm cm at room temperature, 77 K and 4. 2 K, respectively; room temperature saturation magnetization 86 c. g. s. units; supplied by Linde Air Products Corp. ; measured in vacuum at zero gauss.
2	466	L	1961	1. 4-4. 4			The above specimen measured in a magnetic field of 9400 gauss parallel to the direction of heat flow.

DATA TABLE NO. 61 THERMAL CONDUCTIVITY OF MANGANESE-ZINC FERRATE $Mn(Zn)Fe_2O_4$

[Temperature, T, K; Thermal Conductivity, k, Watt cm^{-1} K^{-1}]

T	k
CURVE 1	
1.50	0.00275
1.57	0.00285
1.60	0.00300
1.68	0.00315
1.72	0.00333
2.08	0.00410
2.30	0.00463
2.39	0.00490
2.62	0.00550
2.88	0.00610
2.92	0.00620
2.99	0.00645
3.25	0.00705
3.27	0.00715
3.55	0.00775
3.61	0.00800
3.77	0.00835
3.85	0.00865
3.94	0.00895
4.02	0.00920
4.09	0.00940
4.40	0.0102
5.00	0.0120
6.20	0.0150
10.7	0.0245
11.8	0.0265
13.7	0.0300
15.3	0.0340
16.7	0.0355
17.5	0.0375
21.0	0.0430
CURVE 2	
1.44	0.00245
1.48	0.00260
1.65	0.00305
1.75	0.00335
1.94	0.00390
2.03	0.00400
2.06	0.00410*
2.22	0.00445
2.33	0.00475*
2.52	0.00515*
CURVE 2 (cont.)	
2.56	0.00535*
2.65	0.00550
2.79	0.00585*
2.89	0.00600*
3.05	0.00645
3.13	0.00670
3.33	0.00723*
3.43	0.00750*
3.82	0.00850*
4.43	0.0101*

*Not shown on plot

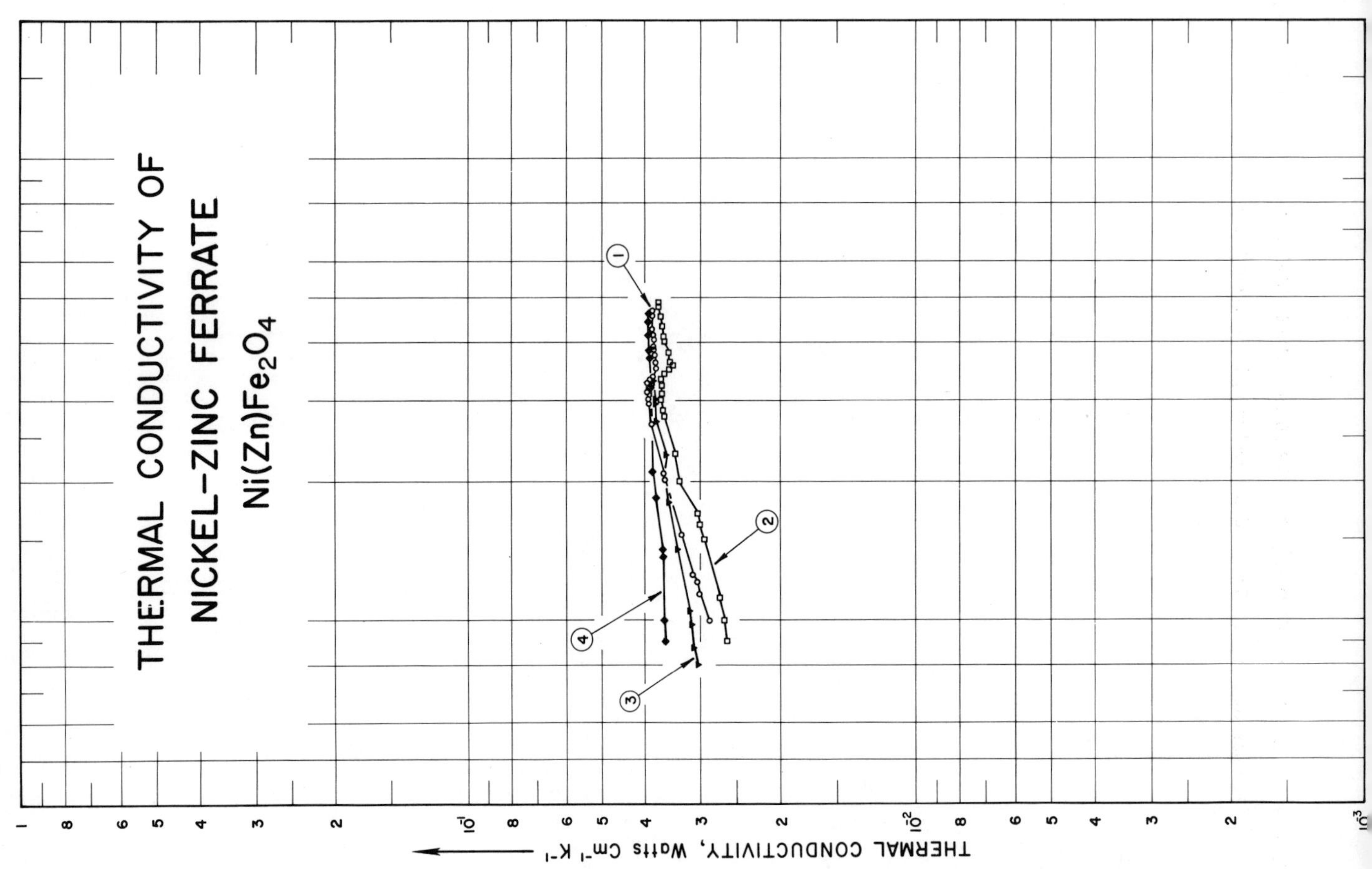
THERMAL CONDUCTIVITY OF
NICKEL-ZINC FERRATE
Ni(Zn)Fe$_2$O$_4$
THERMAL CONDUCTIVITY, Watts Cm^{-1} K^{-1}
1
2
3
4
1 8 6 5 4 3 2 10^{-1} 8 6 5 4 3 2 10^{-2} 8 6 5 4 3 2 10^{-3}

SPECIFICATION TABLE NO. 62 THERMAL CONDUCTIVITY OF NICKEL-ZINC FERRATE $Ni(Zn)Fe_2O_4$

[For Data Reported in Figure and Table No. 62]

Curve No.	Ref. No.	Method Used	Year	Temp. Range, K	Reported Error, %	Name and Specimen Designation	Composition (weight percent), Specifications and Remarks
1	246	L	1963	100-468	1-3	$Ni_{0.3}Zn_{0.7}Fe_2O_4$ No. 3	Initial composition: 66.80 Fe_2O_3, 9.37 NiO and 23.83 ZnO; large grained; specimen 20 x 10 x (2-5) mm; heated at 300-400 C for 3 hrs at a rate of heating and cooling of 50 C per hr; density 5.1 g cm^{-3}; measured in vacuum of 10^{-4} to 10^{-5} mm Hg.
2	246	L	1963	90-487	1-3	$Ni_{0.3}Zn_{0.7}Fe_2O_4$ No. 4	Initial composition: 66.80 Fe_2O_3, 9.37 NiO and 23.83 ZnO; similar to the above specimen except small grained; density 4.3 g cm^{-3}.
3	246	L	1963	80-487	1-3	$Ni_{0.25}Zn_{0.75}Fe_2O_4$ No. 6	Initial composition: 66.52 Fe_2O_3, 4.67 NiO and 28.81 ZnO; similar to the above specimen.
4	246	L	1963	90-463	1-3	$Ni_{0.1}Zn_{0.9}Fe_2O_4$ No. 9	Initial composition: 66.42 Fe_2O_3, 3.11 NiO and 30.47 ZnO; similar to the above specimen.

DATA TABLE NO. 62 THERMAL CONDUCTIVITY OF NICKEL - ZINC FERRATE $Ni(Zn)Fe_2O_4$

[Temperature, T, K; Thermal Conductivity, k, Watts $cm^{-1}K^{-1}$]

T	k
CURVE 1	
100	0.0287
114	0.0301
121	0.0305
126	0.0312
153	0.0331
201	0.0360
209	0.0364
266	0.0387
296	0.0391
302	0.0393
313	0.0395
328	0.0395
332	0.0389
338	0.0383
351	0.0379
364	0.0379
376	0.0379
385	0.0380
392	0.0381
402	0.0381
408	0.0382*
416	0.0383
424	0.0385
458	0.0385
468	0.0385
CURVE 2	
90	0.0264
100	0.0266
112	0.0272
150	0.0295
161	0.0300
171	0.0304
172	0.0308*
200	0.0324
230	0.0341
276	0.0360
285	0.0364
300	0.0366
310	0.0366
322	0.0366
CURVE 2 (cont.)	
332	0.0368
342	0.0363
350	0.0351
355	0.0347
362	0.0351
380	0.0354
400	0.0360
412	0.0362
432	0.0366
451	0.0368
477	0.0374
487	0.0374
CURVE 3	
80	0.0303
87	0.0310
98	0.0314
105	0.0316
142	0.0337
180	0.0351
210	0.0362*
227	0.0356
270	0.0377
295	0.0379
300	0.0379
305	0.0379*
330	0.0381
340	0.0381*
372	0.0381*
392	0.0383*
427	0.0383*
470	0.0383*
487	0.0387*
CURVE 4	
90	0.0360
100	0.0362
137	0.0364
142	0.0364
185	0.0377
CURVE 4 (cont.)	
210	0.0385
260	0.0385*
270	0.0385*
290	0.0385*
294	0.0385*
297	0.0385*
300	0.0387*
310	0.0387*
320	0.0387
350	0.0389*
370	0.0389
385	0.0389
415	0.0391
442	0.0391
463	0.0391

*Not shown on plot

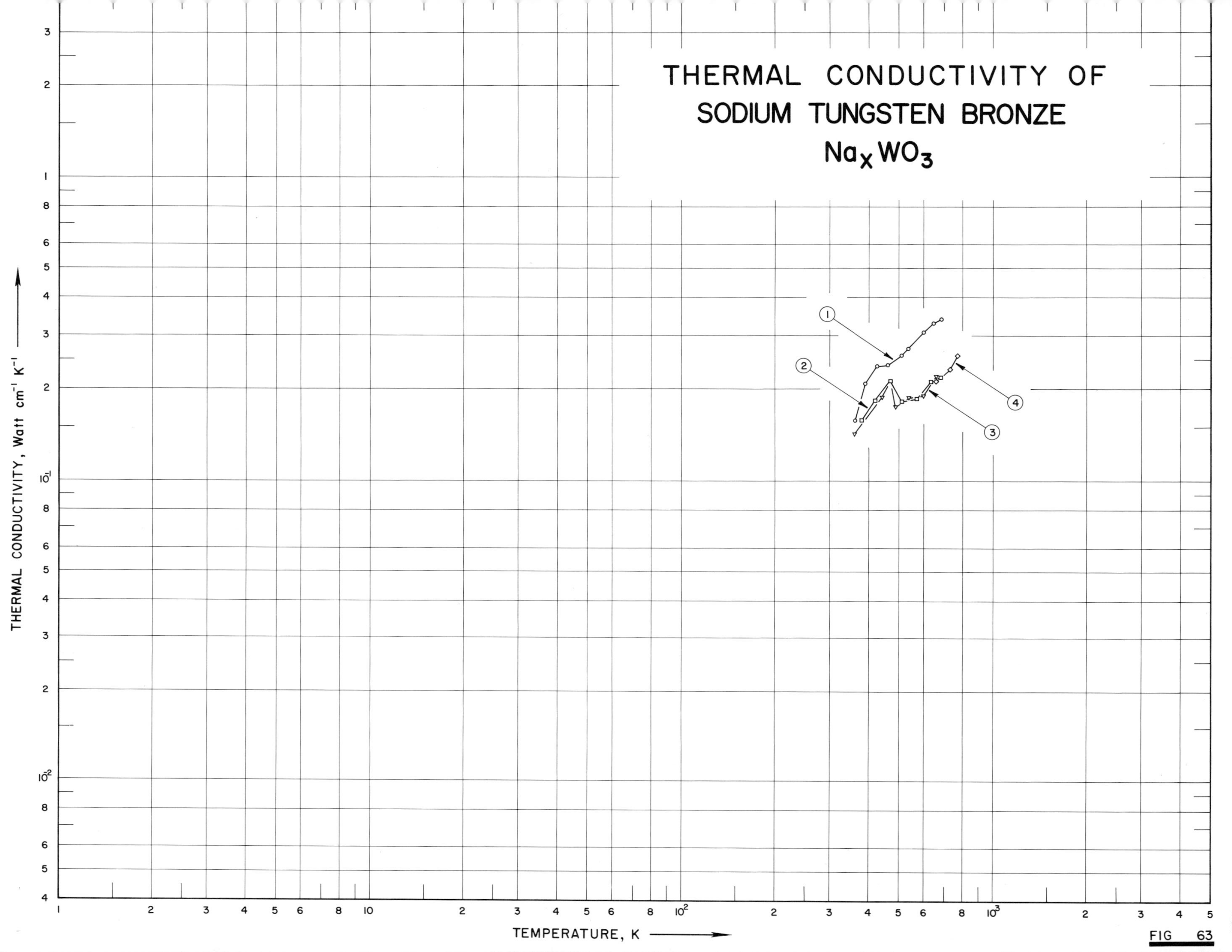
THERMAL CONDUCTIVITY OF
SODIUM TUNGSTEN BRONZE
Na_xWO_3
1
2
3
4
THERMAL CONDUCTIVITY, Watt cm^{-1} K^{-1}
TEMPERATURE, K

FIG 63

SPECIFICATION TABLE NO. 63 THERMAL CONDUCTIVITY OF SODIUM TUNGSTEN BRONZE Na_xWO_3

[For Data Reported in Figure and Table No. 63]

Curve No.	Ref. No.	Method Used	Year	Temp. Range, K	Reported Error, %	Name and Specimen Designation	Composition (weight percent), Specifications and Remarks
1	469	C	1966	362-685		$Na_{0.513}WO_3$	75.5 W, 19.7 O, 4.8 Na; single crystal; 12 mm dia x 7.2 mm thick; prepared by electrolytic reduction of a melt of Na_2WO_4 and WO_3; heat flow in the <100> direction; Battelle Armco iron used as comparative material.
2	469	C	1966	381-680		$Na_{0.804}WO_3$	73.4 W, 19.2 O, 7.4 Na; single crystal; same dimension, fabrication method, and measuring condition as the above specimen.
3	469	C	1966	361-659		$Na_{0.804}WO_3$	2nd run of the above specimen.
4	469	C	1966	656-771		$Na_{0.804}WO_3$	3rd run of the above specimen.

DATA TABLE NO. 63 THERMAL CONDUCTIVITY OF SODIUM TUNGSTEN BRONZE Na_xWO_3

[Temperature, T, K; Thermal Conductivity, k, Watt $cm^{-1} K^{-1}$]

T	k
CURVE 1	
362	0.157
391	0.208
426	0.238
464	0.240
511	0.258
537	0.273
602	0.308
646	0.330
685	0.340
CURVE 2	
381	0.158
420	0.184
473	0.212
512	0.183
570	0.186
634	0.211
680	0.218
CURVE 3	
361	0.143
443	0.188
472	0.212*
486	0.175
537	0.187
596	0.190
659	0.219
CURVE 4	
656	0.212
683	0.219*
730	0.233
771	0.257

*Not shown on plot

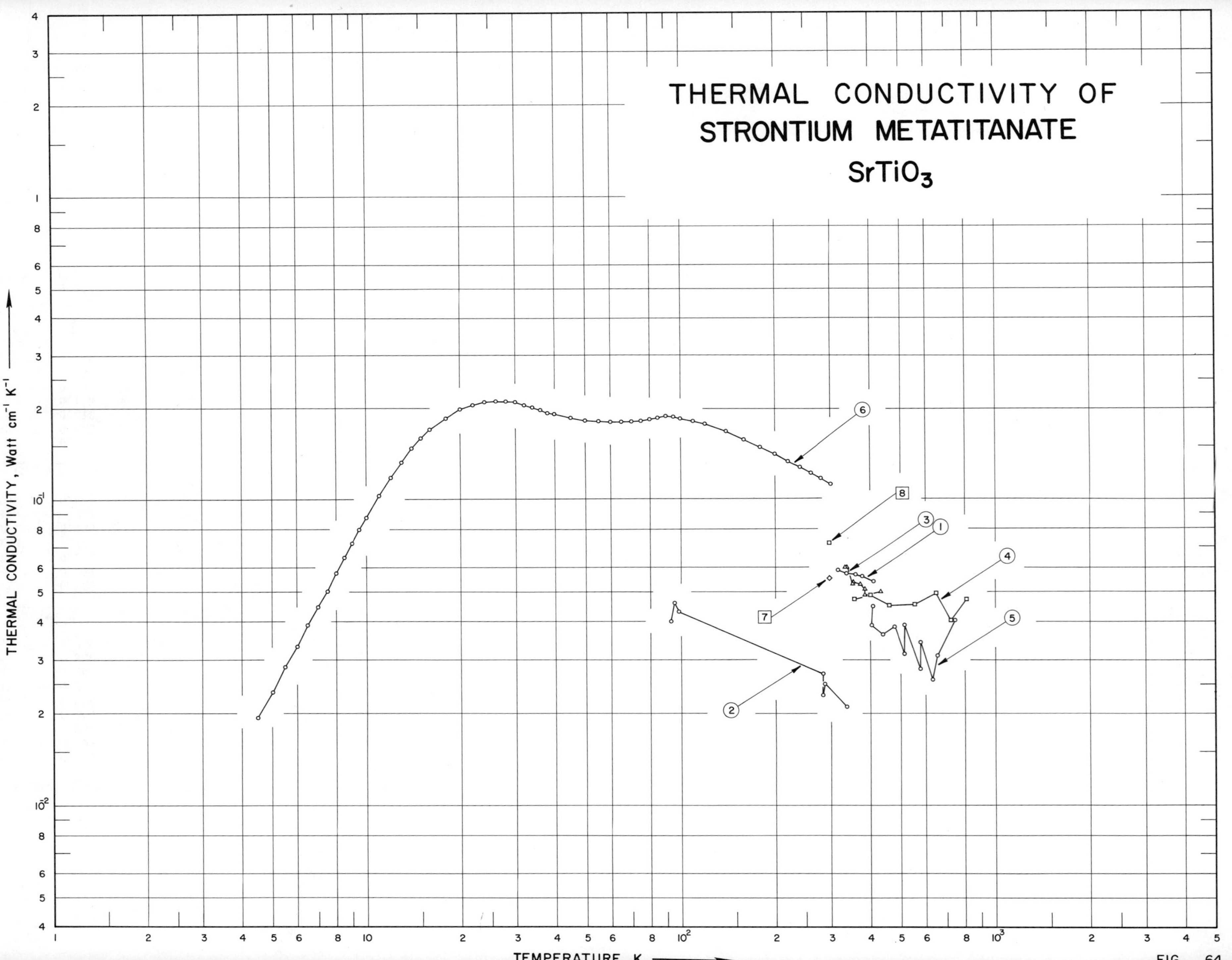
THERMAL CONDUCTIVITY OF
STRONTIUM METATITANATE
$SrTiO_3$
THERMAL CONDUCTIVITY, Watt cm^{-1} K^{-1}
TEMPERATURE, K
FIG. 64
1
2
3
4
5
6
7
8

SPECIFICATION TABLE NO. 64 THERMAL CONDUCTIVITY OF STRONTIUM METATITANATE $SrTiO_3$

[For Data Reported in Figure and Table No. 64]

Curve No.	Ref. No.	Method Used	Year	Temp. Range, K	Reported Error, %	Name and Specimen Designation	Composition (weight percent), Specifications and Remarks
1	68	C	1954	316-410	±3.0	33 B2	Measured in vacuum; copper used as comparative material.
2	38	L	1960	93-337		No. 2	Powdered raw materials of special grade SrO and special grade TiO_2 were mixed together with water, pressed into a disc and fired in a Siliconit furnace; the product was crushed, well pulverized and meshed, then pressed (pressure 0.53 ton cm^{-2}) into a disc; specimen 2.18 cm in dia and 2.47 cm long; sintered at 1400 C for 8 hrs; density 4.01 g cm^{-3}; porosity 19.5%; measured in vacuum less than 10^{-4} mm Hg.
3	203	C	1958	333-433			Specimen sintered from SrO and TiO_2 powder of special reagent grade; pure iron used as comparative material.
4	283		1959	358-808			Pure; 1st run.
5	283		1959	403-743			2nd run of the above specimen.
6	465	L	1965	4.5-300			Single crystal; specimen 1 x 1 x 7 mm; supplied by Fuji Titanium Industry Co.; heat flowed in the direction of <100>; thermal conductivity data obtained directly from the author.
7	470	L	1959	298.2			88.8% theoretical density.
8	470	L	1959	298.2			94% theoretical density.

DATA TABLE NO. 64 THERMAL CONDUCTIVITY OF STRONTIUM METATITANATE $SrTiO_3$

[Temperature, T, K; Thermal Conductivity, k, Watt cm^{-1} K^{-1}]

T	k
CURVE 1	
316.4	0.0586
337.4	0.0573
359.7	0.0569
377.6	0.0561
410.3	0.0540
CURVE 2	
93.2	0.040
95.2	0.046
98.2	0.043
283.2	0.027
283.2	0.023
286.2	0.025
337.2	0.021
CURVE 3	
333.2	0.060
338.2	0.060
353.2	0.053
354.2	0.054
373.2	0.053
385.2	0.051
385.2	0.049
433.2	0.050
CURVE 4	
358.2	0.0473
398.2	0.0490
458.2	0.0452
553.2	0.0456
648.2	0.0494
723.2	0.0406
808.2	0.0473
CURVE 5	
403.2	0.0389
408.2	0.0448
438.2	0.0364
478.2	0.0385
513.2	0.0389

T	k
CURVE 5 (cont.)	
513.2	0.0314
578.2	0.0280
578.2	0.0343
633.2	0.0259
653.2	0.0310
743.2	0.0406
CURVE 6	
4.5	0.0194
5.0	0.0236
5.5	0.0284
6.0	0.0333
6.5	0.0390
7.0	0.0447
7.5	0.0504
8.0	0.0576
8.5	0.0648
9.0	0.0721
9.5	0.0798
10	0.0875
11	0.103
12	0.118
13	0.132
14	0.147
15	0.159
16	0.170
18	0.185
20	0.198
22	0.205
24	0.210
26	0.211
28	0.211
30	0.210
32	0.205
34	0.201
36	0.197
38	0.193
40	0.192
45	0.186
50	0.182
55	0.181
60	0.180
65	0.180

T	k
CURVE 6 (cont.)	
70	0.181
75	0.182
80	0.184
85	0.186
90	0.188
95	0.187
100	0.185
110	0.181
120	0.177
140	0.168
160	0.158
180	0.149
200	0.141
220	0.134
240	0.128
260	0.122
280	0.117
300	0.112
CURVE 7	
298.2	0.0552
CURVE 8	
298.2	0.0724

SPECIFICATION TABLE NO. 65 THERMAL CONDUCTIVITY OF STRONTIUM ZIRCONATE $SrZrO_3$

Curve No.	Ref. No.	Method Used	Year	Temp. Range, K	Reported Error, %	Name and Specimen Designation	Composition (weight percent), Specifications and Remarks
1	470	L	1959	298. 2			88% theoretical density.

DATA TABLE NO. 65 THERMAL CONDUCTIVITY OF STRONTIUM ZIRCONATE $SrZrO_3$

[Temperature, T, K; Thermal Conductivity, k, Watt cm^{-1} K^{-1}]

T	k
CURVE 1*	
298. 2	0. 0226

*No graphical presentation

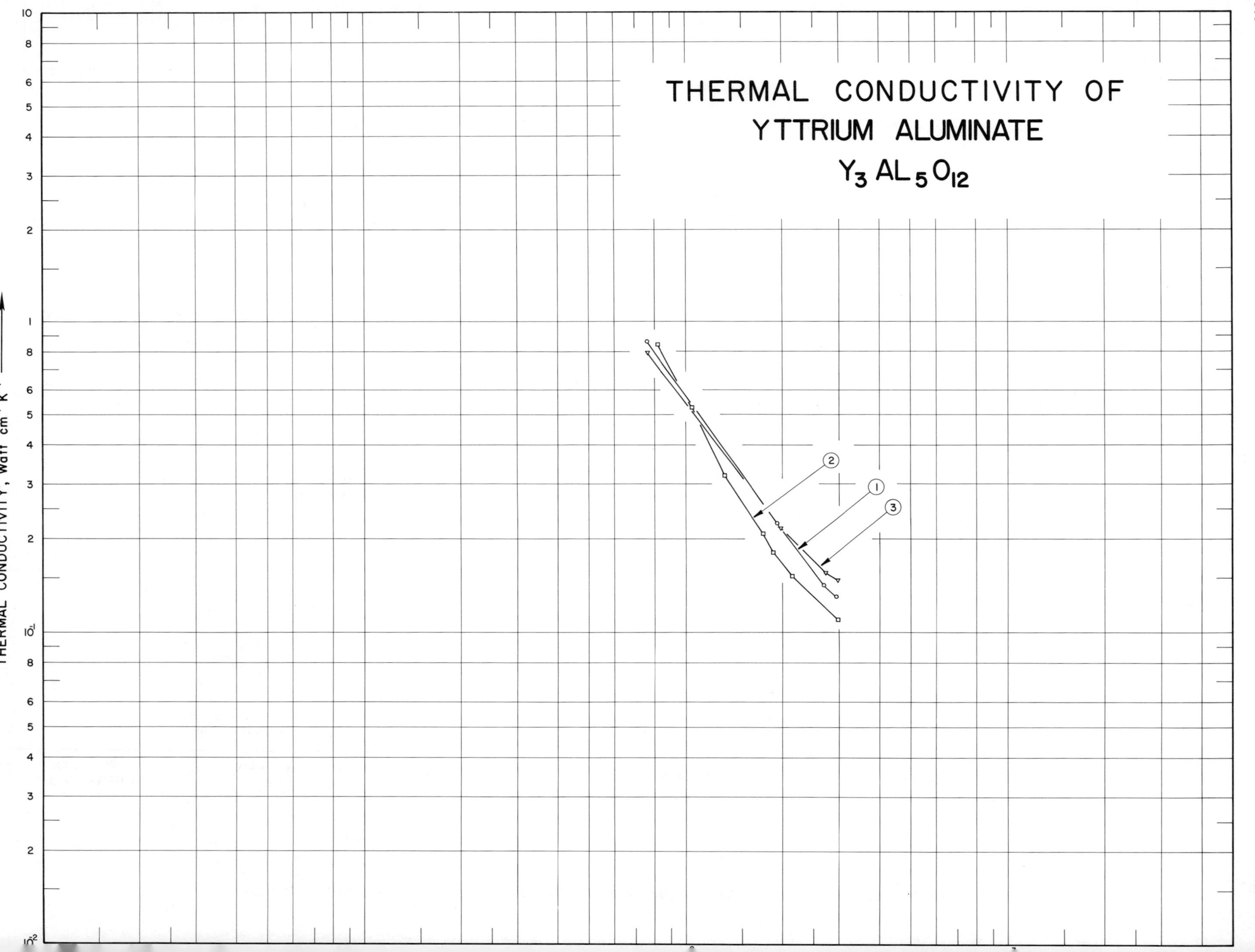
THERMAL CONDUCTIVITY OF
YTTRIUM ALUMINATE
$Y_3 AL_5 O_{12}$
THERMAL CONDUCTIVITY, Watt cm^{-1} K^{-1}
10
8
6
5
4
3
2
1
8
6
5
4
3
2
10^{-1}
8
6
5
4
3
2
10^{-2}
2
1
3

SPECIFICATION TABLE NO. 66 THERMAL CONDUCTIVITY OF YTTRIUM ALUMINATE $Y_3Al_5O_{12}$

[For Data Reported in Figure and Table No. 66]

Curve No.	Ref. No.	Method Used	Year	Temp. Range, K	Reported Error, %	Name and Specimen Designation	Composition (weight percent), Specifications and Remarks
1	461	L	1967	76-294		YAG	Single crystal; prepared from 99.999^+ pure materials by using the Czochralski method; density reported as 4.564, 4.563, 4.563, 4.560, 4.561, 4.558, 4.554, and 4.552 g cm^{-3} at 100, 117, 136, 170, 190, 215, 270, and 297 K, respectively.
2	461	L	1967	82-298		YAG	Single crystal; prepared from 99.999^+ pure materials by using the Czochralski method; Nd^{3+} concentration 4.2×10^{19} cm^{-3}.
3	461	L	1967	76-297		YAG	Single crystal; prepared from 99.999^+ pure materials by using the Czachralski method; Nd^{3+} concentration 1.4×10^{20} cm^{-3}; density reported as 4.556, 4.567, 4.556, 4.563, 4.561, and 4.554 g cm^{-3} at 104, 110, 146, 181, 226, and 294 K, respectively.

DATA TABLE NO. 66 THERMAL CONDUCTIVITY OF YTTRIUM ALUMINATE $Y_3Al_5O_{12}$

[Temperature, T, K; Thermal Conductivity, k, Watt $cm^{-1} K^{-1}$]

T	k
CURVE 1	
76	0.862
193	0.225
269	0.142
294	0.131
CURVE 2	
82	0.840
105	0.529
133	0.319
174	0.208
187	0.181
215	0.152
298	0.110
CURVE 3	
76	0.794
197	0.216
273	0.156
297	0.148

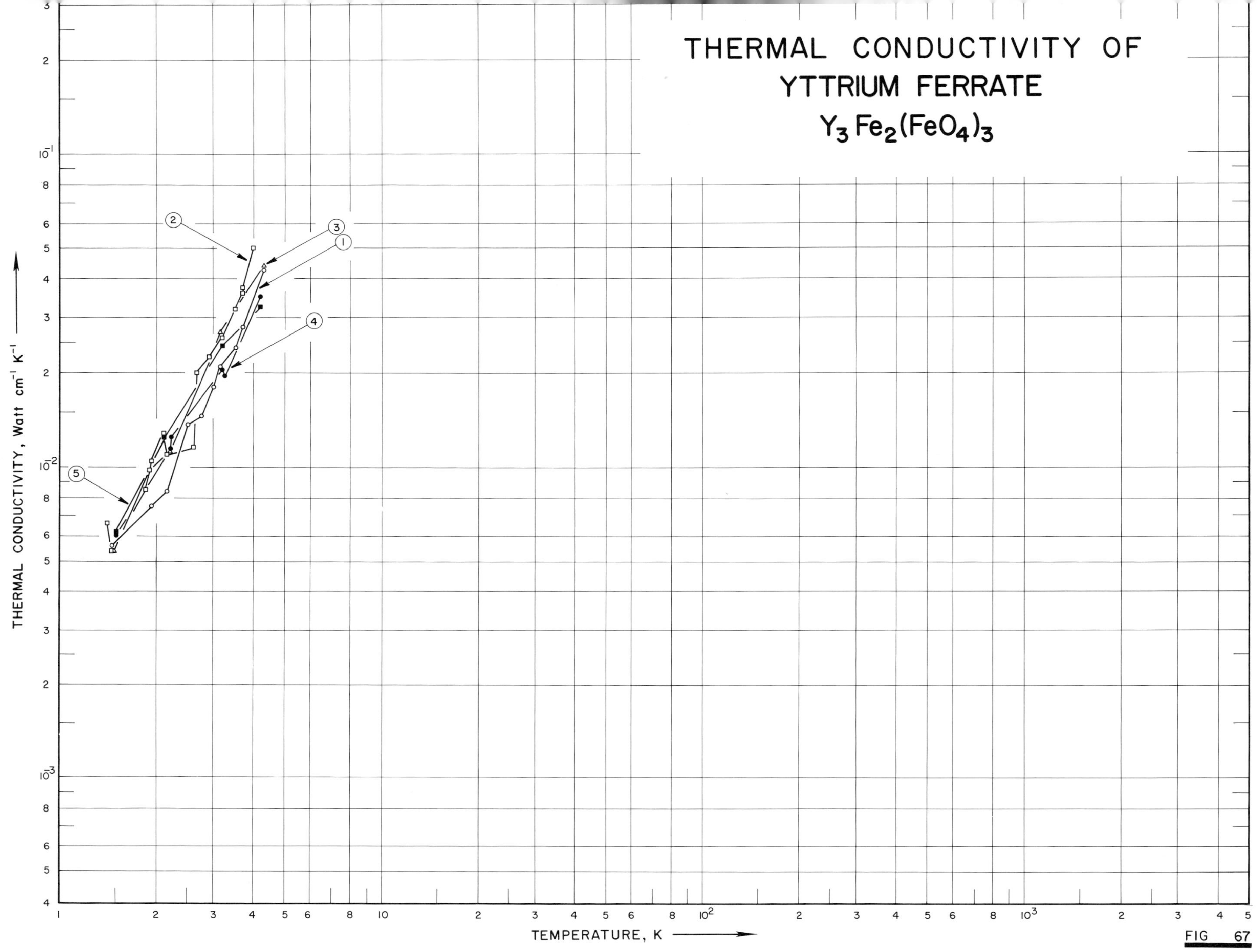
THERMAL CONDUCTIVITY OF
YTTRIUM FERRATE
$Y_3Fe_2(FeO_4)_3$
THERMAL CONDUCTIVITY, Watt cm^{-1} K^{-1}
TEMPERATURE, K
FIG 67

SPECIFICATION TABLE NO. 67 THERMAL CONDUCTIVITY OF YTTRIUM FERRATE $Y_3Fe_2(FeO_4)_3$

[For Data Reported in Figure and Table No. 67]

Curve No.	Ref. No.	Method Used	Year	Temp. Range, K	Reported Error, %	Name and Specimen Designation	Composition (weight percent), Specifications and Remarks
1	517	L	1962	1.5-4.3		YIG I	Single crystal; orientation of axis (321); length between the thermometers 0.81 cm; cross-sectional area 2.31 mm^2; measured in a vacuum of 10^{-7} mm Hg.
2	517	L	1962	1.4-4.2		YIG II	Single crystal; orientation of axis (210); length between the thermometers 0.71 cm; cross-sectional area 1.26 mm^2; measured in a vacuum of 10^{-7} mm Hg.
3	517	L	1962	1.5-4.3		YIG II	The above specimen with a longitudinal field of 300 g; measured in a vacuum of 10^{-7} mm Hg.
4	517	L	1962	1.5-4.2		YIG III	Single crystal; orientation of axis (130); length between the thermometers 0.87 cm; cross-sectional area 0.63 mm^2; measured in a vacuum of 10^{-7} mm Hg.
5	517	L	1962	1.5-4.2		YIG III	The above specimen with a longitudinal field of 300 g; measured in a vacuum of 10^{-7} mm Hg.

DATA TABLE NO. 67 THERMAL CONDUCTIVITY OF YTTRIUM FERRATE $Y_3Fe_2(FeO_4)_3$

[Temperature, T, K; Thermal Conductivity, k, Watt $cm^{-1}K^{-1}$]

T	k
CURVE 1	
1.45	0.0056
1.93	0.0075
2.15	0.0084
2.5	0.0137
2.75	0.0146
3.0	0.018
3.15	0.021
3.5	0.024
3.7	0.028
4.3	0.0425
CURVE 2	
1.4	0.0066
1.45	0.0054
1.85	0.0085
1.90	0.0098
1.93	0.0105
2.1	0.0128
2.15	0.011
2.6	0.016
2.65	0.020
2.9	0.0225
3.2	0.026
3.5	0.032
3.7	0.036
3.7	0.0375
4.2	0.05
CURVE 3	
1.46	0.0054
1.9	0.0098*
2.2	0.0113
3.15	0.027
4.3	0.044
CURVE 4	
1.5	0.006
2.2	0.0115
2.23	0.0125
3.2	0.0205
3.25	0.0195
4.2	0.035
CURVE 5	
1.5	0.0062
2.1	0.0125
3.2	0.0245
4.2	0.0325

*Not shown on plot

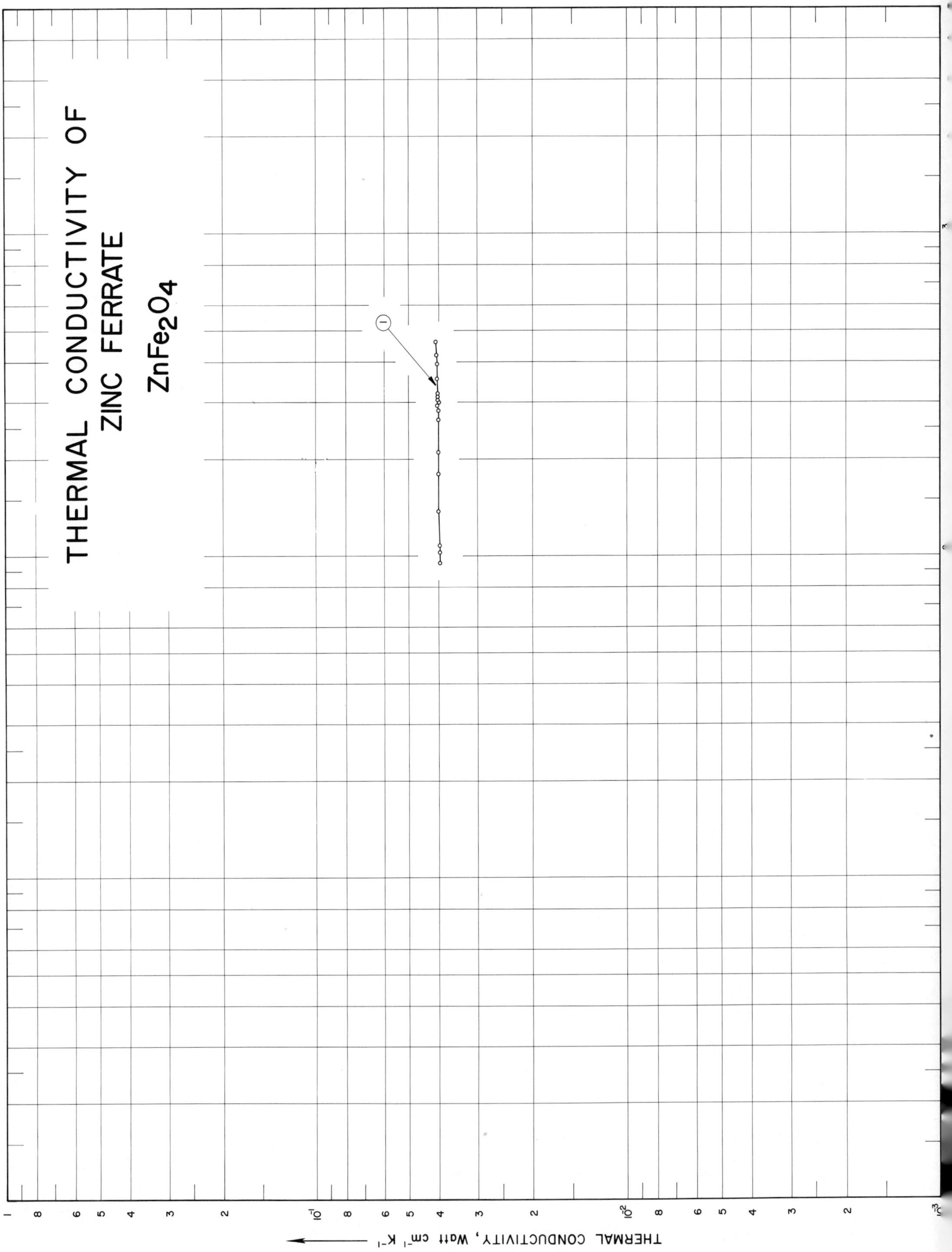

THERMAL CONDUCTIVITY OF ZINC FERRATE $ZnFe_2O_4$

SPECIFICATION TABLE NO. 68 THERMAL CONDUCTIVITY OF ZINC FERRATE $ZnFe_2O_4$

[For Data Reported in Figure and Table No. 68]

Curve No.	Ref. No.	Method Used	Year	Temp. Range, K	Reported Error, %	Name and Specimen Designation	Composition (weight percent), Specifications and Remarks
1	246	L	1963	95-463	1-3	No. 3	Mixture of ground 66.24 analytically pure Fe_2O_3 (CHDA) and 33.76 ZnO; prepared by mixing for 3 days by wet method with raw alcohol as binder and then by special ball mills; powder produced by rubbing the dried charge processed as described, sifted, fired at 700 C, again sifted, and finally pressed at 1500-300 Kg cm^{-2} after polyvinyl alcohol (6-8% weight of the charge) was added; heated at 300-400 C for 3 hrs to remove alcohol and then finally fired at 1300 C for 3 hrs in a furnace with heating and cooling rates at 50 C per hr; specimen 20 x 10 x 2 - 5 mm; measured in a vacuum of 10^{-4} to 10^{-5} mm Hg.

DATA TABLE NO. 68 THERMAL CONDUCTIVITY OF ZINC FERRATE $ZnFe_2O_4$

[Temperature, T, K; Thermal Conductivity, k, Watt $cm^{-1}K^{-1}$]

T	k
CURVE 1	
95	0.0397
102	0.0397
107	0.0397
137	0.0402
180	0.0402
210	0.0402
265	0.0402
282	0.0402
293	0.0406
300	0.0400
305	0.0404
312	0.0404
320	0.0404
355	0.0406
395	0.0406
420	0.0408
463	0.0410

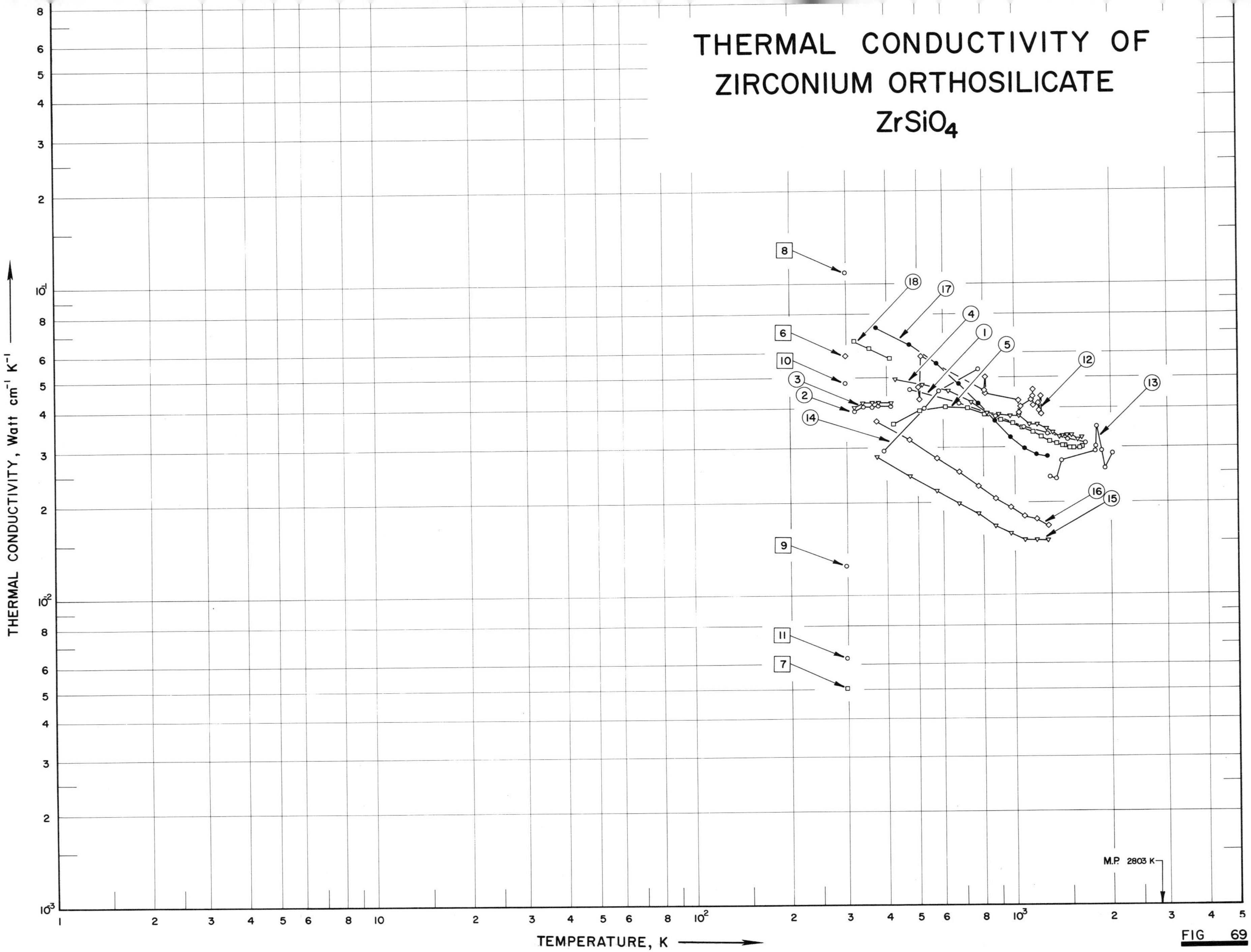

THERMAL CONDUCTIVITY OF ZIRCONIUM ORTHOSILICATE $ZrSiO_4$

FIG 69

SPECIFICATION TABLE NO. 69 THERMAL CONDUCTIVITY OF ZIRCONIUM ORTHOSILICATE $ZrSiO_4$

[For Data Reported in Figure and Table No. 69]

Curve No.	Ref. No.	Method Used	Year	Temp. Range, K	Reported Error, %	Name and Specimen Designation	Composition (weight percent), Specifications and Remarks
1	364	C	1954	473-1673		Zircon	Polycrystalline; supplied by National Lead Co.; prepared from superpax with a mean particle size of 5 microns; as received; acid-treated with 1 N HCl, slip-cast from slip with pH 3.0 and specific gravity 2.4; fired at 1550 C; bulk density 3.69-3.79 g cm^{-3}; porosity 4.5-7.3%.
2	3	L	1953	319-412		254A-1	Single crystal with some impurity; from Zredell County, North Carolina; ground and polished; c (principle) axis parallel to the direction of heat flow.
3	3	L	1953	319-414		254B-1	Similar to the above specimen except c-axis perpendicular to the direction of heat flow.
4	10	R	1952	426-1623		Zircon	Bulk density 3.69 g cm^{-3}; porosity 19.1%.
5	10	R	1952	421-1648		Zircon	Similar to the above specimen.
6	152	L	1955	298.2		Zircon	Specimen 20 mils thick and 0.75 in. in dia; sintered; preirradiated with 3 x 10^{19} epi-thermal neutrons per cm^2 for 480 Mwd in the MTR.
7	152	L	1955	298.2		Zircon	The above specimen postirradiated with 3 x 10^{19} epithermal neutrons per cm^2 for 480 Mwd in the MTR.
8	152	L	1955	298.2		Zircon Tam	Specimen 20 mils thick and 0.75 in. in dia; preirradiated with 3 x 10^{19} epithermal neutrons per cm^2 for 480 Mwd in the MTR.
9	152	L	1955	298.2		Zircon Tam	The above specimen postirradiated with 3 x 10^{19} epithermal neutrons per cm^2 for 480 Mwd in the MTR.
10	152	L	1955	298.2		Zircon 475	Specimen 20 mils thick and 0.75 in. in dia; preirradiated with 5 x 10^{19} epithermal neutrons per cm^2 for 480 Mwd in the MTR.
11	152	L	1955	298.2		Zircon 475	The above specimen postirradiated with 5 x 10^{19} epithermal neutrons per cm^2 for 480 Mwd in the MTR.
12	243	R	1962	503-1219	2-4	Taylor Zircon CZ-5	65-66 ZrO_2, 33-34 SiO_2, 1.0 max Al_2O_3, 0.3 max TiO_2, 0.1 max Fe_2O_3, and 0.2 max others; specimen 0.75 in. long, 0.75 in. O.D. and 0.25 in. I.D.; slip-cast and sintered; max exposure temp 2317 K; denstiy 4.04 g cm^{-3}; SRI run number C51.
13	243	R	1962	1300-2117	2-4	Taylor Zircon CZ-5	Similar to the above specimen but partially melted; SRI run number C58.
14	34	C	1943	391-772		Brazil Zircon	Green single crystal; specimen 1 x 1 x 1 cm; measured normal to c (principal) axis; 18-8 stainless steel used as comparative material.
15	375	C	1960	373-1273		Zircon-149D	Specimen supplied by Titanium Alloy Division, National; Wesgo alumina Al-300 obtained from Western Gold and Platinum Co. as reference standard.
16	375	C	1960	373-1273		Zircon-ZRI-46	Specimen and Wesgo alumina Al-300 reference standard obtained from the above sources.
17	375	C	1960	373-1273		Zircon ZRG-4	Specimen and Wesgo alumina Al-300 reference standard obtained from the above sources.
18	407	L	1949	317-411		B-1	No details reported.

DATA TABLE NO. 69 THERMAL CONDUCTIVITY OF ZIRCONIUM ORTHOSILICATE $ZrSiO_4$

[Temperature, T, K; Thermal Conductivity, k, Watt cm^{-1} K^{-1}]

T	k
CURVE 1	
473.2	0.0460
673.2	0.0418
873.2	0.0377
1073.2	0.0347
1273.2	0.0331
1473.2	0.0318
1673.2	0.0310
CURVE 2	
318.8	0.0392
339.8	0.0405
362.2	0.0403
377.7	0.0408
411.6	0.0407
CURVE 3	
318.7	0.0401
338.3	0.0416
361.2	0.0418
378.6	0.0418
414.2	0.0418
CURVE 4	
425.7	0.0494
518.2	0.0475
623.2	0.0452
733.2	0.0418
823.2	0.0385
893.2	0.0381
970.7	0.0377
1038.2	0.0377
1121.2	0.0354
1188.2	0.0354
1263.2	0.0343
1323.2	0.0333
1378.2	0.0324
1411.2	0.0325
1463.2	0.0326
1503.2	0.0326
1578.2	0.0316
1623.2	0.0322

T	k
CURVE 5	
420.7	0.0356
508.2	0.0393
610.7	0.0402
713.2	0.0400
808.2	0.0383
909.2	0.0366
988.2	0.0358
1063.2	0.0347
1148.2	0.0335
1213.2	0.0324
1293.2	0.0314
1353.2	0.0308
1410.7	0.0303
1433.2	0.0303
1493.2	0.0299
1533.2	0.0299
1600.2	0.0299
1648.2	0.0301
CURVE 6	
298.2	0.0586
CURVE 7	
298.2	0.00502
CURVE 8	
298.2	0.109
CURVE 9	
298.2	0.0126
CURVE 10	
298.2	0.0481
CURVE 11	
298.2	0.00628

T	k
CURVE 12	
502.6	0.0469
507.6	0.0428
512.6	0.0588
805.9	0.0453
812.1	0.0500
812.6	0.0447
1034.3	0.0424
1040.9	0.0384
1044.8	0.0404
1123.2	0.0427
1132.1	0.0433
1138.2	0.0433*
1146.5	0.0456
1148.2	0.0408
1199.8	0.0414
1199.8	0.0392
1209.3	0.0436
1218.7	0.0381
CURVE 13	
1299.8	0.0241
1360.9	0.0239
1377.6	0.0238*
1402.6	0.0273
1794.3	0.0291
1797.1	0.0301
1799.8	0.0313*
1799.8	0.0317*
1805.4	0.0303*
1805.4	0.0348
1872.1	0.0293
1894.3	0.0291*
1916.5	0.0257
2116.5	0.0287
CURVE 14	
391.2	0.0294
525.2	0.0397
582.2	0.0452
772.2	0.0531

T	k
CURVE 15	
373.2	0.0280
473.2	0.0243
573.2	0.0218
673.2	0.0197
773.2	0.0184
873.2	0.0167
973.2	0.0159
1073.2	0.0151
1173.2	0.0151
1273.2	0.0151
CURVE 16	
373.2	0.0364
473.2	0.0318
573.2	0.0276
673.2	0.0251
773.2	0.0226
873.2	0.0205
973.2	0.0192
1073.2	0.0180
1173.2	0.0176
1273.2	0.0167
CURVE 17	
373.2	0.0724
473.2	0.0636
573.2	0.0552
673.2	0.0477
773.2	0.0414
873.2	0.0364
973.2	0.0322
1073.2	0.0297
1173.2	0.0285
1273.2	0.0280
CURVE 18	
317.0	0.0657
354.0	0.0623
410.7	0.0577

*Not shown on plot

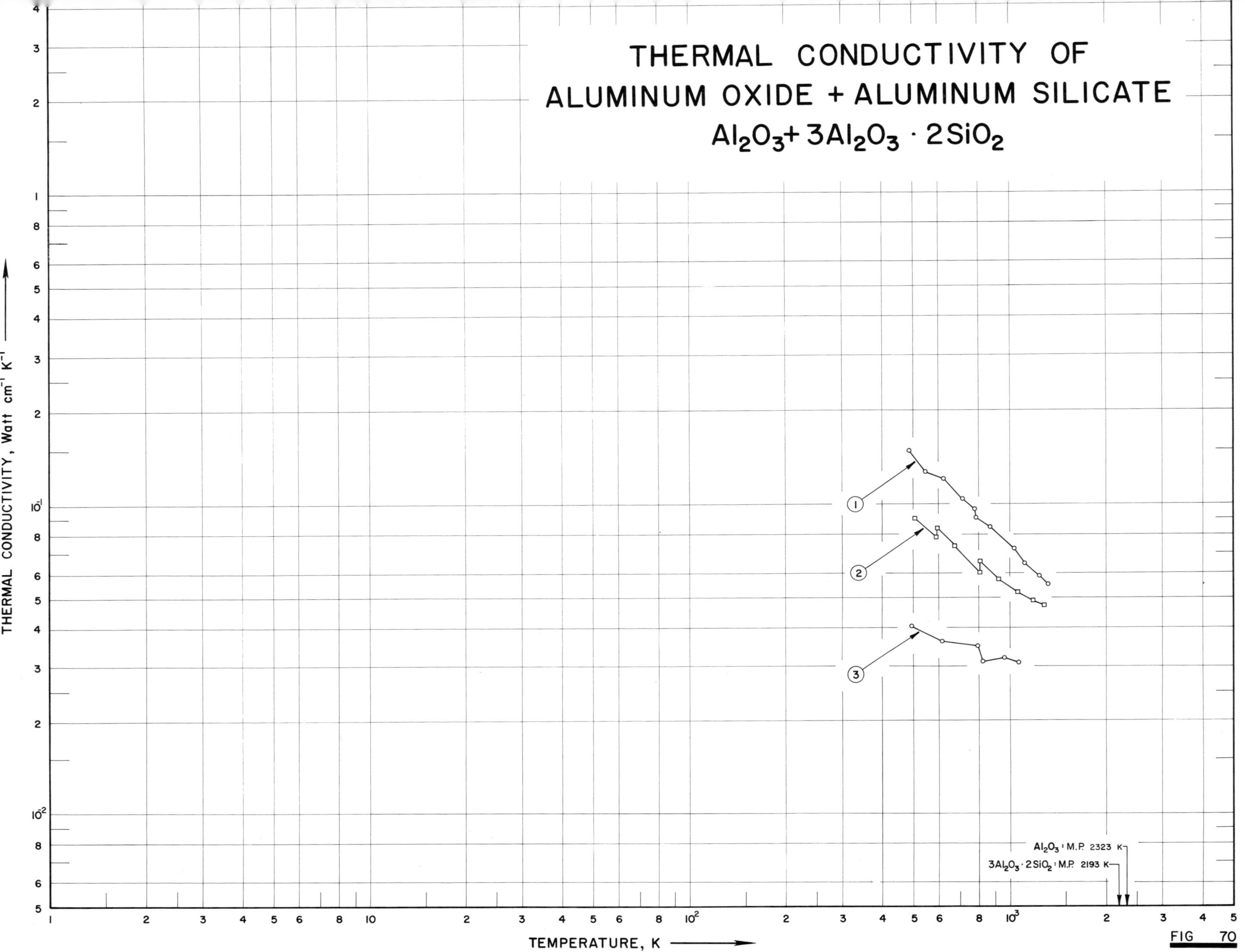
THERMAL CONDUCTIVITY OF
ALUMINUM OXIDE + ALUMINUM SILICATE
Al2O3+3Al2O3 · 2SiO2
Al2O3 : M.P. 2323 K
3Al2O3 · 2SiO2 : M.P. 2193 K
1
2
3
TEMPERATURE, K
THERMAL CONDUCTIVITY, Watt cm-1 K-1

FIG 70

SPECIFICATION TABLE NO. 70 THERMAL CONDUCTIVITY OF [ALUMINUM OXIDE + ALUMINUM SILICATE] $Al_2O_3 + 3Al_2O_3 \cdot 2SiO_2$

[For Data Reported in Figure and Table No. 70]

Curve No.	Ref. No.	Method Used	Year	Temp. Range, K	Reported Error, %	Name and Specimen Designation	Composition (weight percent) Al_2O_3	$3Al_2O_3 \cdot 2SiO_2$	Composition (continued), Specifications and Remarks
1	136	C	1959	486-1323	±4	Alumina + Mullite	90	10	Fired at 1750 C to bulk density 3.20 g cm^{-3}; total porosity 17.5%; data corrected to zero porosity.
2	136	C	1959	505-1288	±4	Alumina + Mullite	75	25	Fired at 1750 C to bulk density 3.08 g cm^{-3}; total porosity 17.4%; data corrected to zero porosity.
3	136	C	1959	493-1070	±4	Alumina + Mullite	50	50	Fired at 1750 C to bulk density 2.68 g cm^{-3}; total porosity 23.8%; data corrected to zero porosity.

DATA TABLE NO. 70 THERMAL CONDUCTIVITY OF [ALUMINUM OXIDE + ALUMINUM SILICATE] $Al_2O_3 + 3Al_2O_3 \cdot 2SiO_2$

[Temperature, T, K; Thermal Conductivity, k, Watt $cm^{-1} K^{-1}$]

T	k
CURVE 1	
486.2	0.149
548.2	0.128
623.2	0.121
713.2	0.104
779.2	0.0962
783.2	0.0905
869.2	0.0845
1038.2	0.0720
1113.2	0.0649
1243.2	0.0589
1323.2	0.0552
CURVE 2	
505.2	0.0900
590.2	0.0787
595.2	0.0837
673.2	0.0736
807.2	0.0602
808.2	0.0657
921.2	0.0573
1063.2	0.0521
1180.2	0.0490
1288.2	0.0473
CURVE 3	
493.2	0.0404
616.2	0.0360
793.2	0.0348
821.2	0.0310
962.2	0.0318
1070.2	0.0308

DATA TABLE NO. 71 THERMAL CONDUCTIVITY OF [ALUMINUM OXIDE + DICHROMIUM TRIOXIDE] $Al_2O_3 + Cr_2O_3$

[Temperature, T, K; Thermal Conductivity, k, Watt $cm^{-1}K^{-1}$]

T	k
CURVE 1	
325.2	0.219
372.2	0.201
395.2	0.190
420.2	0.179
448.2	0.168
481.2	0.154
573.2	0.126
643.2	0.109
756.2	0.0870
896.2	0.0720
1008.2	0.0628
1193.2	0.0531
CURVE 2	
323.2	0.186
356.2	0.176
393.2	0.164
403.2	0.156
432.2	0.147
460.2	0.136
587.2	0.100
690.2	0.0815
773.2	0.0715
873.2	0.0632
973.2	0.0573
1069.2	0.0536
1114.2	0.0519
1148.2	0.0469
1249.2	0.0448

SPECIFICATION TABLE NO. 72 THERMAL CONDUCTIVITY OF [ALUMINUM OXIDE + DIMANGANESE TRIOXIDE] $Al_2O_3 + Mn_2O_3$

Curve No.	Ref. No.	Method Used	Year	Temp. Range, K	Reported Error, %	Name and Specimen Designation	Composition (weight percent) Al_2O_3	Mn_2O_3	Composition (continued), Specifications and Remarks
1	23		1952	313-401		157A	72.1	27.9	Mn_2O_3: 4 Al_2O_3; firing temperature 1811 K; water absorption 0.007%; density 3.65 g cm^{-3}.

DATA TABLE NO. 72 THERMAL CONDUCTIVITY OF [ALUMINUM OXIDE + DIMANGANESE TRIOXIDE] $Al_2O_3 + Mn_2O_3$

[Temperature, T, K; Thermal Conductivity, k, Watt $cm^{-1}K^{-1}$]

T	k
CURVE 1*	
312.7	0.0749
335.7	0.0657
357.8	0.0657
378.0	0.0636
400.9	0.0615

* No graphical presentation

DATA TABLE NO. 73 THERMAL CONDUCTIVITY OF [ALUMINUM OXIDE + SILICON DIOXIDE] $Al_2O_3 + SiO_2$

[Temperature, T, K; Thermal Conductivity, k, Watt cm^{-1} K^{-1}]

T	k
CURVE 1	
640.2	0.0167
675.2	0.0167
811.2	0.0176
823.2	0.0176
1088.2	0.0172
CURVE 2	
1029.2	0.0172
1253.2	0.0176
CURVE 3	
403.4	0.00907
523.0	0.0115
749.4	0.0117
CURVE 4	
347.7	0.0140
620.5	0.0146
723.1	0.0141
CURVE 5	
346.1	0.0156
573.2	0.0151
744.2	0.0147
CURVE 6	
373.2	0.148
473.2	0.126
573.2	0.106
673.2	0.0891
773.2	0.0753
873.2	0.0644
973.2	0.0565
1073.2	0.0502
1173.2	0.0460
1273.2	0.0427

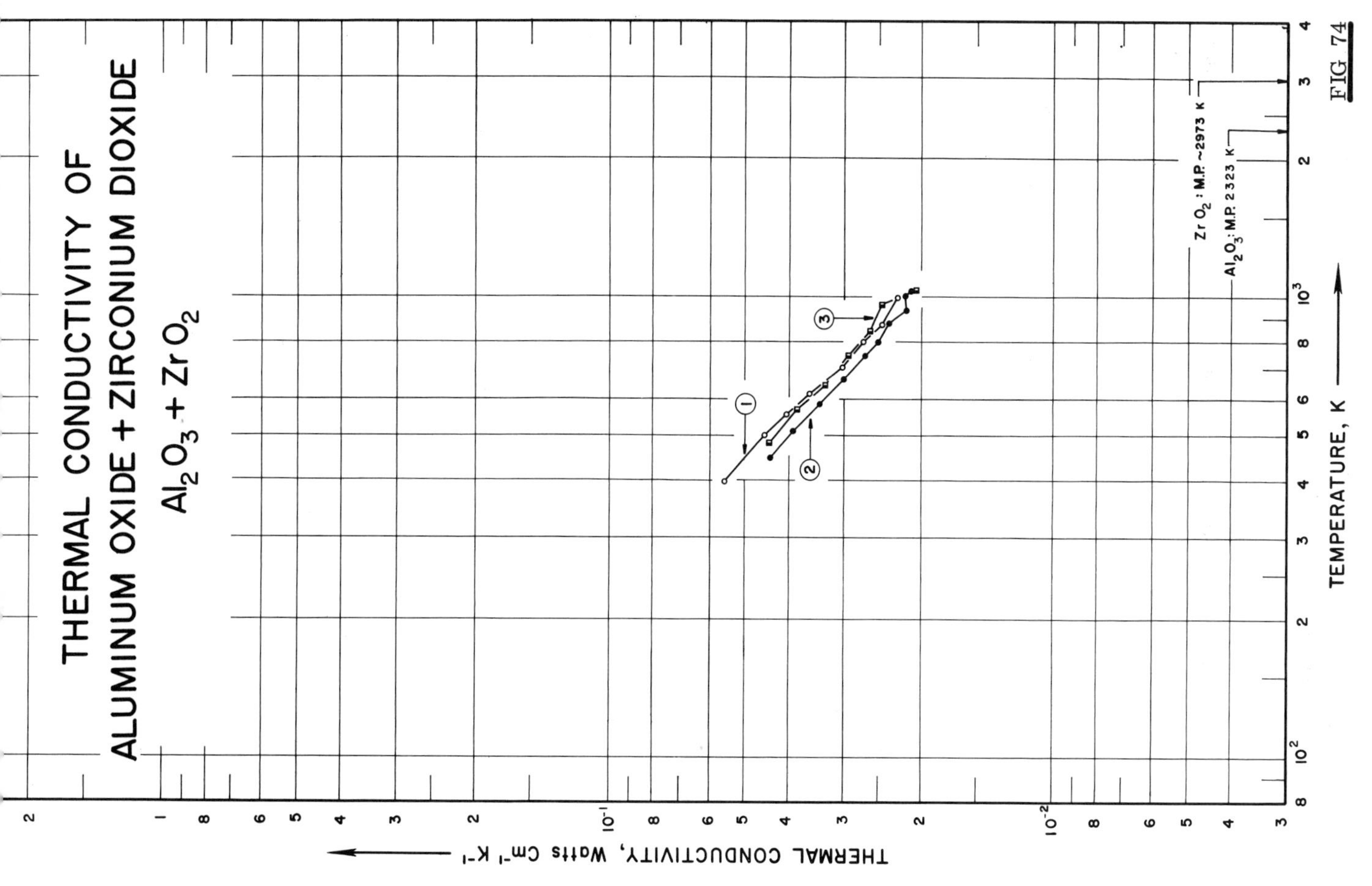
THERMAL CONDUCTIVITY OF
ALUMINUM OXIDE + ZIRCONIUM DIOXIDE
Al2O3 + ZrO2
ZrO2 : M.P. ~2973 K
Al2O3: M.P. 2323 K
TEMPERATURE, K
THERMAL CONDUCTIVITY, Watts Cm-1 K-1
1
2
3

FIG 74

SPECIFICATION TABLE NO. 74 THERMAL CONDUCTIVITY OF [ALUMINUM OXIDE + ZIRCONIUM DIOXIDE] $Al_2O_3 + ZrO_2$

[For Data Reported in Figure and Table No. 74]

Curve No.	Ref. No.	Method Used	Year	Temp. Range, K	Reported Error, %	Name and Specimen Designation	Composition (weight percent) Al_2O_3	ZrO_2	Composition (continued), Specifications and Remarks
1	146	L	1955	400-1008			60	40	Sintered.
2	146	L	1955	451-1078			80	20	Sintered.
3	146	L	1955	488-1093			90	10	Sintered.

DATA TABLE NO. 74 THERMAL CONDUCTIVITY OF [ALUMINUM OXIDE + ZIRCONIUM DIOXIDE] $Al_2O_3 + ZrO_2$

[Temperature, T, K; Thermal Conductivity, k, Watt $cm^{-1}K^{-1}$]

T	k
CURVE 1	
400.2	0.0556
503.2	0.0454
559.2	0.0404
621.2	0.0360
712.5	0.0303
808.7	0.0271
873.2	0.0247
1008.2	0.0228
CURVE 2	
451.2	0.0441
517.2	0.0393
589.2	0.0343
663.2	0.0302
748.2	0.0271
802.2	0.0252
879.2	0.0239
938.2	0.0218
1015.7	0.0220
1078.2	0.0213
CURVE 3	
488.2	0.0431
572.2	0.0389
648.2	0.0333
745.2	0.0295
847.2	0.0262
968.2	0.0247
1093.2	0.0209

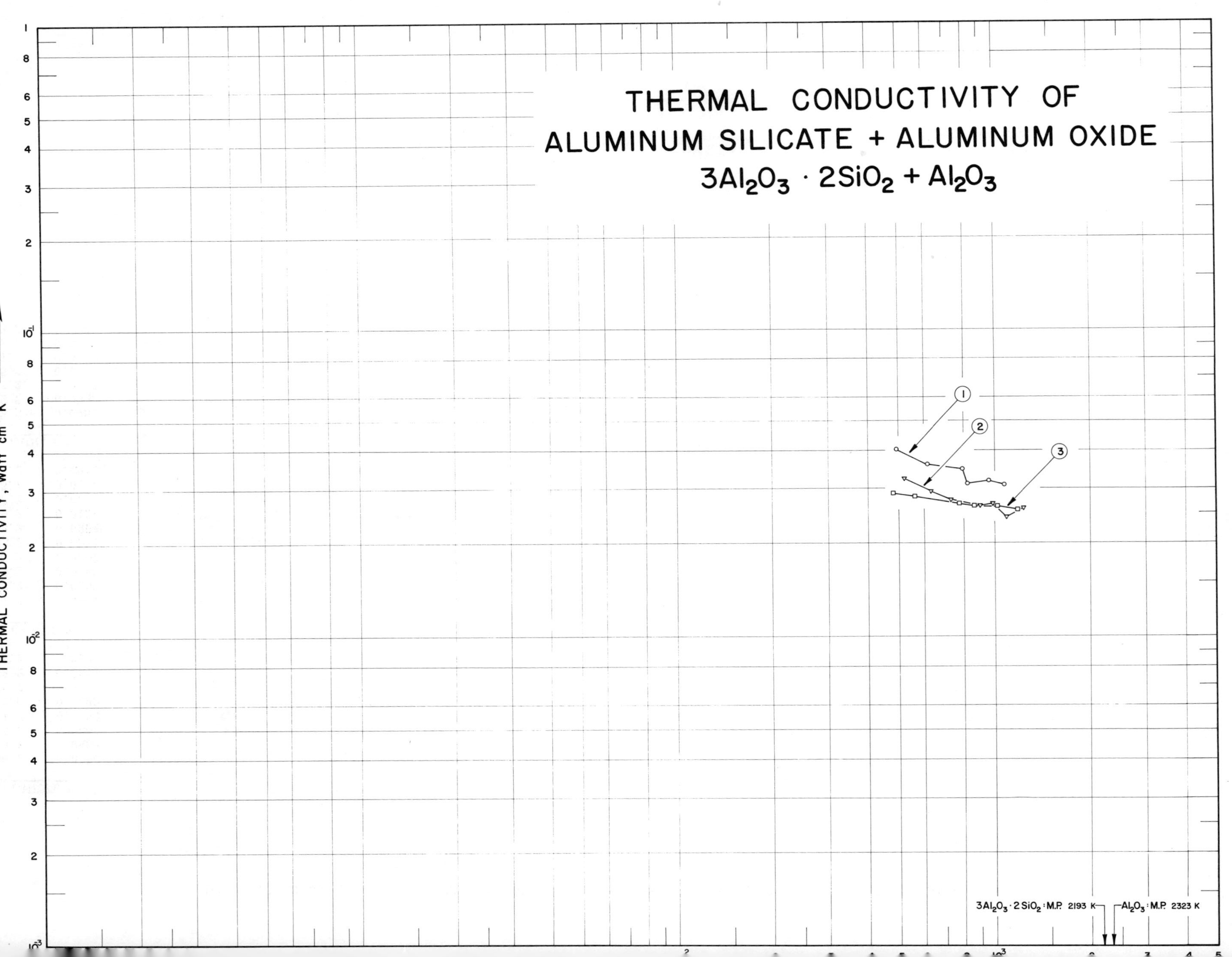
THERMAL CONDUCTIVITY OF
ALUMINUM SILICATE + ALUMINUM OXIDE
3Al2O3 · 2SiO2 + Al2O3
1
2
3
3Al2O3 · 2SiO2 : M.P. 2193 K
Al2O3 : M.P. 2323 K
THERMAL CONDUCTIVITY, Watt cm-1 K-1

SPECIFICATION TABLE NO. 75 THERMAL CONDUCTIVITY OF [ALUMINUM SILICATE + ALUMINUM OXIDE] $3Al_2O_3 \cdot 2SiO_2 + Al_2O_3$

[For Data Reported in Figure and Table No. 75]

Curve No.	Ref. No.	Method Used	Year	Temp. Range, K	Reported Error, %	Name and Specimen Designation	Composition (weight percent) $3Al_2O_3 \cdot 2SiO_2$	Al_2O_3	Composition (continued), Specifications and Remarks
1	136	C	1959	493-1070	±4	Mullite + Alumina	50	50	Fired at 1750 C; bulk density 2.68 g cm^{-3}; porosity 23.8%; data corrected to zero porosity.
2	136	C	1959	523-1241	±4	Mullite + Alumina	74	26	Similar to the above specimen except bulk density 2.35 g cm^{-3}; porosity 29.3%.
3	136	C	1959	481-1188	±4	Mullite + Alumina	90	10	Similar to the above specimen except bulk density 2.42 g cm^{-3}; porosity 25.3%.

DATA TABLE NO. 75 THERMAL CONDUCTIVITY OF [ALUMINUM SILICATE + ALUMINUM OXIDE] $3Al_2O_3 \cdot 2SiO_2 + Al_2O_3$

[Temperature, T, K; Thermal Conductivity, k, Watt $cm^{-1} K^{-1}$]

T	k
CURVE 1	
493.2	0.0404
616.2	0.0360
793.2	0.0348
821.2	0.0310
962.2	0.0318
1070.2	0.0308
CURVE 2	
523.2	0.0322
631.2	0.0293
730.2	0.0276
901.2	0.0264
988.2	0.0267
1094.2	0.0243
1241.2	0.0259
CURVE 3	
481.2	0.0289
560.2	0.0282
775.2	0.0268
865.2	0.0264
1022.2	0.0262
1188.2	0.0255

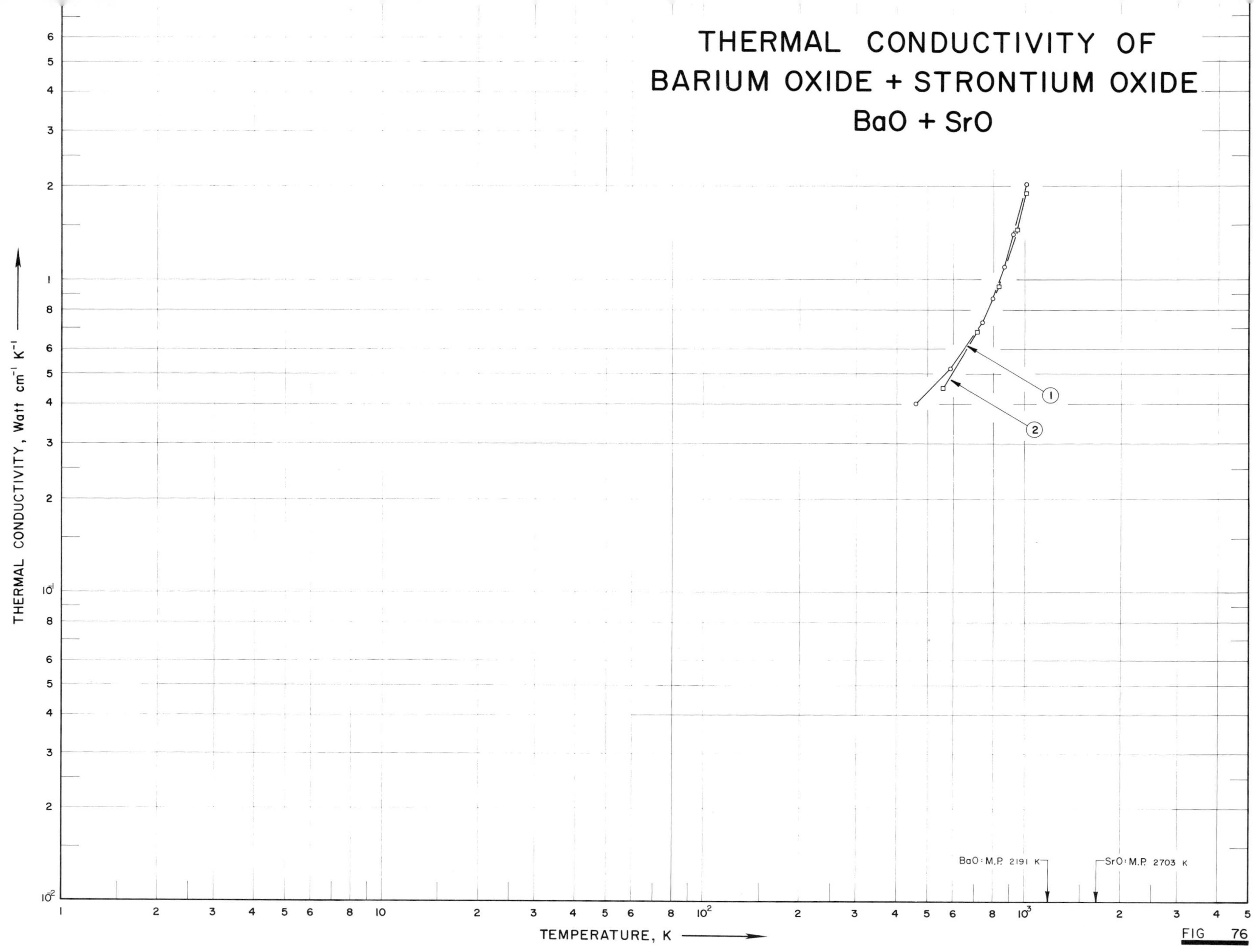
THERMAL CONDUCTIVITY OF
BARIUM OXIDE + STRONTIUM OXIDE
BaO + SrO
1
2
BaO: M.P. 2191 K
SrO: M.P. 2703 K
THERMAL CONDUCTIVITY, Watt cm⁻¹ K⁻¹
TEMPERATURE, K
FIG 76

SPECIFICATION TABLE NO. 76 THERMAL CONDUCTIVITY OF [BARIUM OXIDE + STRONTIUM OXIDE] BaO + SrO

[For Data Reported in Figure and Table No. 76]

Curve No.	Ref. No.	Method Used	Year	Temp. Range, K	Reported Error, %	Name and Specimen Designation	Composition (weight percent) BaO	SrO	Composition (continued), Specifications and Remarks
1	75	L	1955	458-1018	10	Tube No. 7	59.676	40.324	Calculated composition (equimolecular mixture of BaO and SrO); polycrystalline; specimen 15 mm in. dia and 0.9 mm thick; prepared by decomposing $BaCO_3$ and $SrCO_3$ in a vacuum; apparent thermal conductivity (effects of radiation at high temperatures not considered).
2	75	L	1955	558-1013	10	Tube No. 9	59.676	40.324	Similar to the above specimen.

DATA TABLE NO. 76 THERMAL CONDUCTIVITY OF [BARIUM OXIDE + STRONTIUM OXIDE] BaO + SrO

[Temperature, T, K; Thermal Conductivity, k, Watt $cm^{-1} K^{-1}$]

T	k
CURVE 1	
458.2	0.40
588.2	0.52
743.2	0.73
798.2	0.87
868.2	1.10
923.2	1.40
1018.2	2.03
CURVE 2	
558.2	0.45
713.2	0.68
833.2	0.95
948.2	1.45
1013.2	1.90

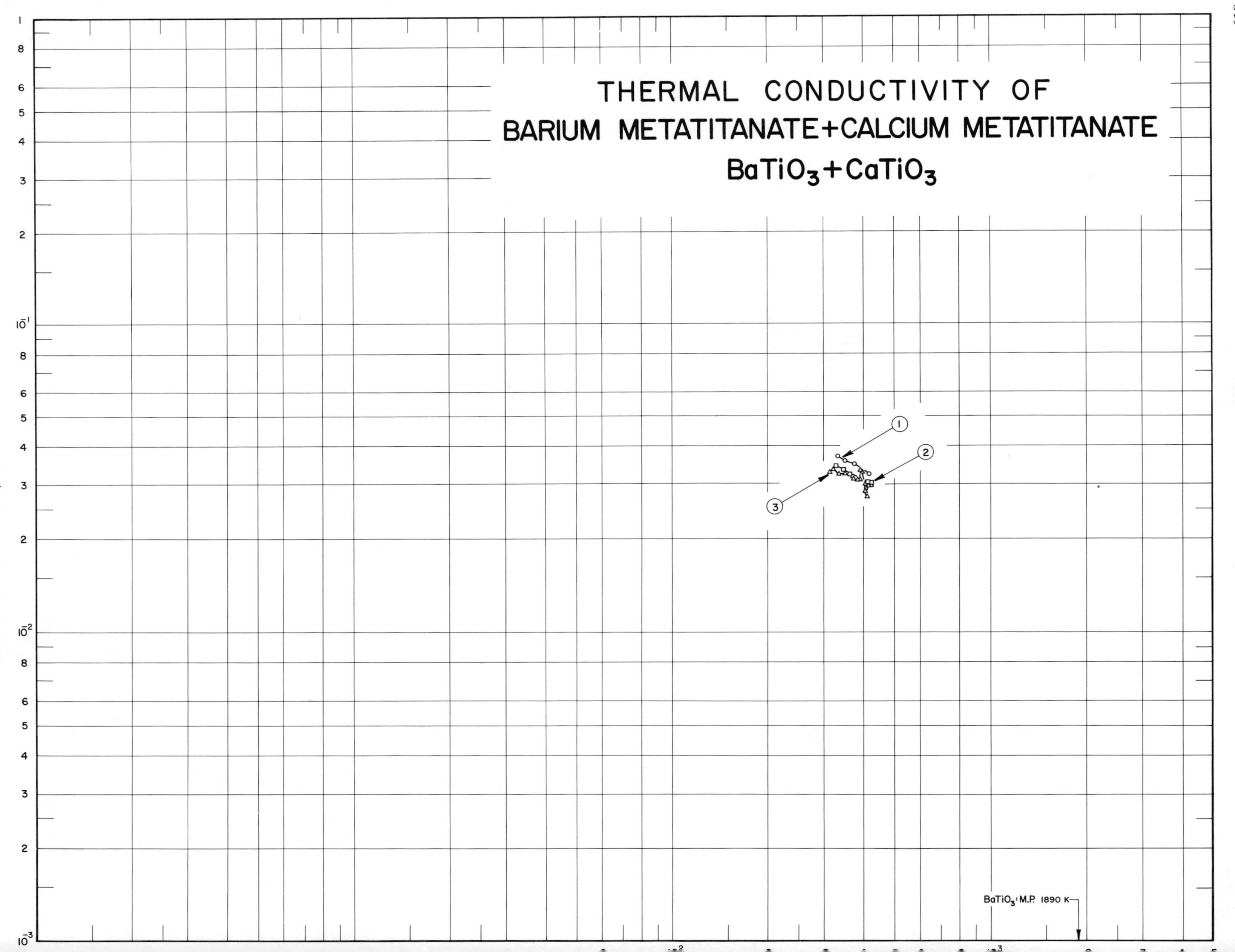

THERMAL CONDUCTIVITY OF
BARIUM METATITANATE+CALCIUM METATITANATE
BaTiO3+CaTiO3
THERMAL CONDUCTIVITY, Watt cm⁻¹ K⁻¹
BaTiO3 : M.P. 1890 K

SPECIFICATION TABLE NO. 77 THERMAL CONDUCTIVITY OF [BARIUM METATITANATE + CALCIUM METATITANATE] $BaTiO_3 + CaTiO_3$

[For Data Reported in Figure and Table No. 77]

Curve No.	Ref. No.	Method Used	Year	Temp. Range, K	Reported Error, %	Name and Specimen Designation	Composition (weight percent)		Composition (continued), Specifications and Remarks
							$BaTiO_3$	$CaTiO_3$	
1	289	C	1962	333-416	<10	$Ca_{0.034}Ba_{0.966}TiO_3$	97.99	2.01	Disk specimen with dia to thickness ratio greater, than 10; density 5.8 g cm^{-3}; Armco iron used as comparative material.
2	289	C	1962	328-423	<10	$Ca_{0.099}Ba_{0.901}TiO_3$	93.98	6.02	Similar to the above specimen except density 5.6 g cm^{-3}.
3	289	C	1962	314-422	<10	$Ca_{0.19}Ba_{0.81}TiO_3$	87.97	12.03	Similar to the above specimen except density 5.28 g cm^{-3}.

DATA TABLE NO. 77 THERMAL CONDUCTIVITY OF [BARIUM METATITANTE + CALCIUM METATITANTE] $BaTiO_3 + CaTiO_3$

[Temperature, T, K; Thermal Conductivity, k, Watt $cm^{-1} K^{-1}$]

T	k
CURVE 1	
333.2	0.037
350.7	0.036
375.7	0.035
416.2	0.0325
CURVE 2	
328.2	0.0345
347.7	0.0335
362.7	0.0325
413.2	0.0307
422.7	0.0305
CURVE 3	
314.2	0.033
320.7	0.034
335.2	0.0325
353.2	0.0325
373.2	0.0315
382.2	0.031
390.2	0.031
392.7	0.0335
395.2	0.033
402.7	0.0287
407.2	0.031
410.2	0.0275
414.7	0.03
418.2	0.03
421.7	0.03

SPECIFICATION TABLE NO. 78 THERMAL CONDUCTIVITY OF [BARIUM METATITANATE + MAGNESIUM ZIRCONATE] $BaTiO_3 + MgZrO_3$

Curve No.	Ref. No.	Method Used	Year	Temp. Range, K	Reported Error, %	Name and Specimen Designation	Composition (weight percent), Specifications and Remarks
1	468	L	1950	331-413		40 B	Specimen 0.456 in. dia x 0.488 in. long.

DATA TABLE NO. 78 THERMAL CONDUCTIVITY OF [BARIUM METATITANATE + MAGNESIUM ZIRCONATE] $BaTiO_3 + MgZrO_3$

[Temperature, T, K; Thermal Conductivity, k, Watt $cm^{-1}K^{-1}$]

T	k
CURVE 1*	
330.6	0.0435
363.9	0.0456
388.1	0.0444
412.5	0.0439

* No graphical presentation

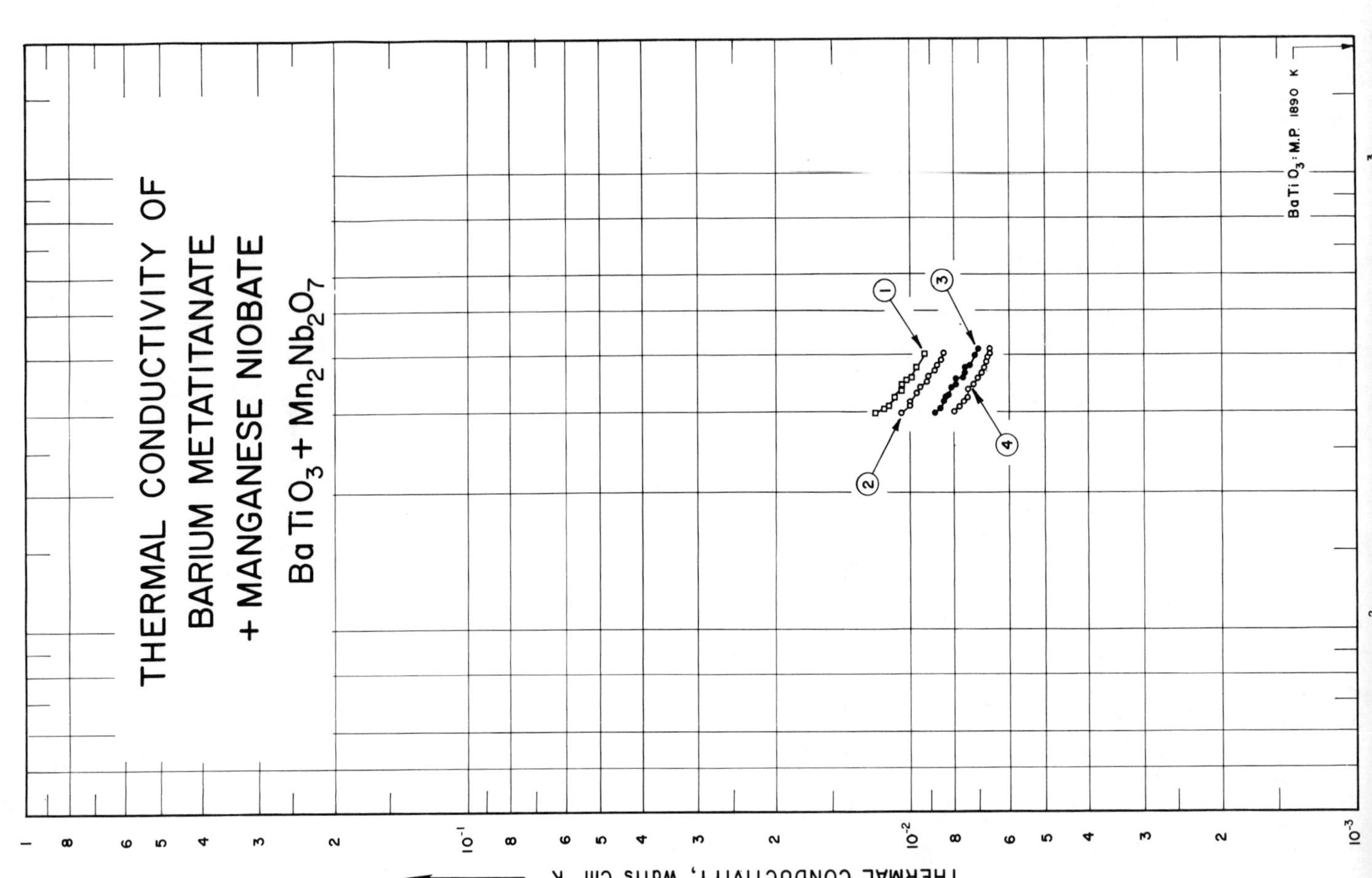
THERMAL CONDUCTIVITY OF
BARIUM METATITANATE
+ MANGANESE NIOBATE
$BaTiO_3 + Mn_2Nb_2O_7$
$BaTiO_3$: M.P. 1890 K
THERMAL CONDUCTIVITY, Watts Cm^{-1} K^{-1}

SPECIFICATION TABLE NO. 79 THERMAL CONDUCTIVITY OF [BARIUM METATITANATE + MANGANESE NIOBATE] $BaTiO_3 + Mn_2Nb_2O_7$

[For Data Reported in Figure and Table No. 79]

Curve No.	Ref. No.	Method Used	Year	Temp. Range, K	Reported Error, %	Name and Specimen Designation	Composition (weight percent) $BaTiO_3$	$MnNb_2O_7$	Composition (continued), Specifications and Remarks
1	202	F	1961	303-403			→	→	2.0 mole % $Mn_2Nb_2O_7$.
2	202	F	1961	303-403			→	→	3.0 mole % $Mn_2Nb_2O_7$.
3	202	F	1961	303-418			→	→	5.0 mole % $Mn_2Nb_2O_7$.
4	202	F	1961	303-418			→	→	7.0 mole % $Mn_2Nb_2O_7$.

DATA TABLE NO. 79 THERMAL CONDUCTIVITY OF [BARIUM METATITANATE + MANGANESE NIOBATE] $BaTiO_3 + Mn_2Nb_2O_7$

[Temperature, T, K; Thermal Conductivity, k, Watt $cm^{-1}K^{-1}$]

T	k	T	k
CURVE 1		CURVE 4	
303.2	0.0121	303.2	0.00795
308.2	0.0115	308.2	0.00774
313.2	0.0112	318.2	0.00761
328.2	0.0109	328.2	0.00745
338.2	0.0105	338.2	0.00745
348.2	0.0105	348.2	0.00720
358.2	0.0102	358.2	0.00703
363.2	0.0100	368.2	0.00690
373.2	0.0100	378.2	0.00686
383.2	0.00971	388.2	0.00678
403.2	0.00929	398.2	0.00674
		403.2	0.00669
CURVE 2		418.2	0.00665
303.2	0.0105		
313.2	0.0100		
319.2	0.0100		
333.2	0.00971		
343.2	0.00954		
353.2	0.00920		
363.2	0.00912		
373.2	0.00883		
383.2	0.00875		
393.2	0.00858		
403.2	0.00849		
CURVE 3			
303.2	0.00879		
308.2	0.00858		
318.2	0.00845		
328.2	0.00837		
333.2	0.00828		
343.2	0.00816		
348.2	0.00795		
358.2	0.00791		
363.2	0.00761		
368.2	0.00761		
378.2	0.00753		
383.2	0.00745		
388.2	0.00745*		
403.2	0.00720		
418.2	0.00703		

*Not shown on plot

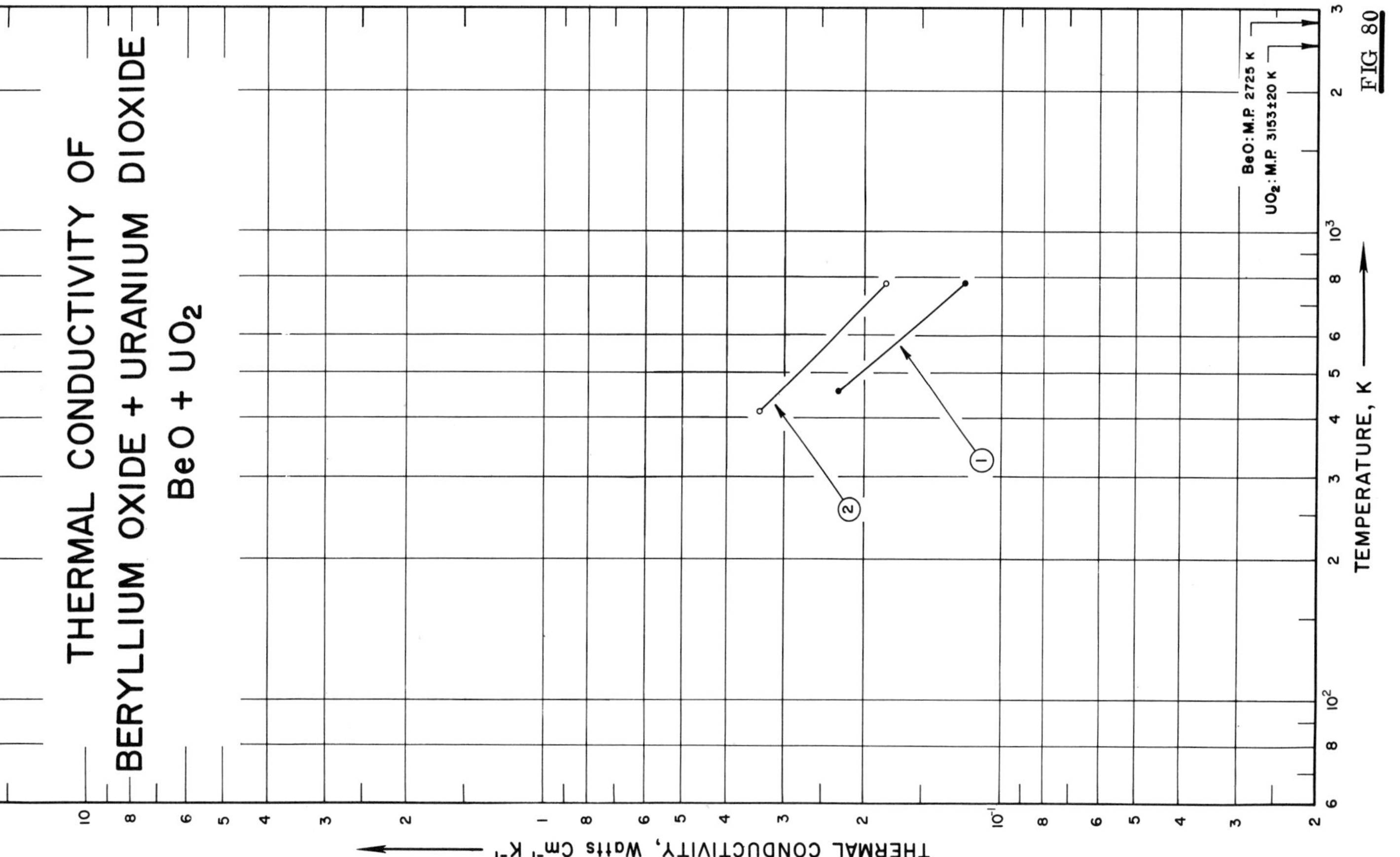
THERMAL CONDUCTIVITY OF
BERYLLIUM OXIDE + URANIUM DIOXIDE
BeO + UO2
THERMAL CONDUCTIVITY, Watts Cm-1 K-1
TEMPERATURE, K
BeO: M.P. 2725 K
UO2: M.P. 3153±20 K
1
2

FIG 80

SPECIFICATION TABLE NO. 80 THERMAL CONDUCTIVITY OF [BERYLLIUM OXIDE + URANIUM DIOXIDE] BeO + UO_2

[For Data Reported in Figure and Table No. 80]

Curve No.	Ref. No.	Method Used	Year	Temp. Range, K	Reported Error, %	Name and Specimen Designation	Composition (weight percent) BeO	UO_2	Composition (continued), Specifications and Remarks
1	76	L	1952	453, 773			53	47	80.0% of theoretical density.
2	76	L	1952	413, 773			53	47	71.2% of theoretical density.

DATA TABLE NO. 80 THERMAL CONDUCTIVITY OF [BERYLLIUM OXIDE + URANIUM DIOXIDE] $BeO + UO_2$

[Temperature, T, K; Thermal Conductivity, k, Watt $cm^{-1}K^{-1}$]

T	k
CURVE 1	
453.2	0.230
773.2	0.120
CURVE 2	
413.2	0.340
773.2	0.180

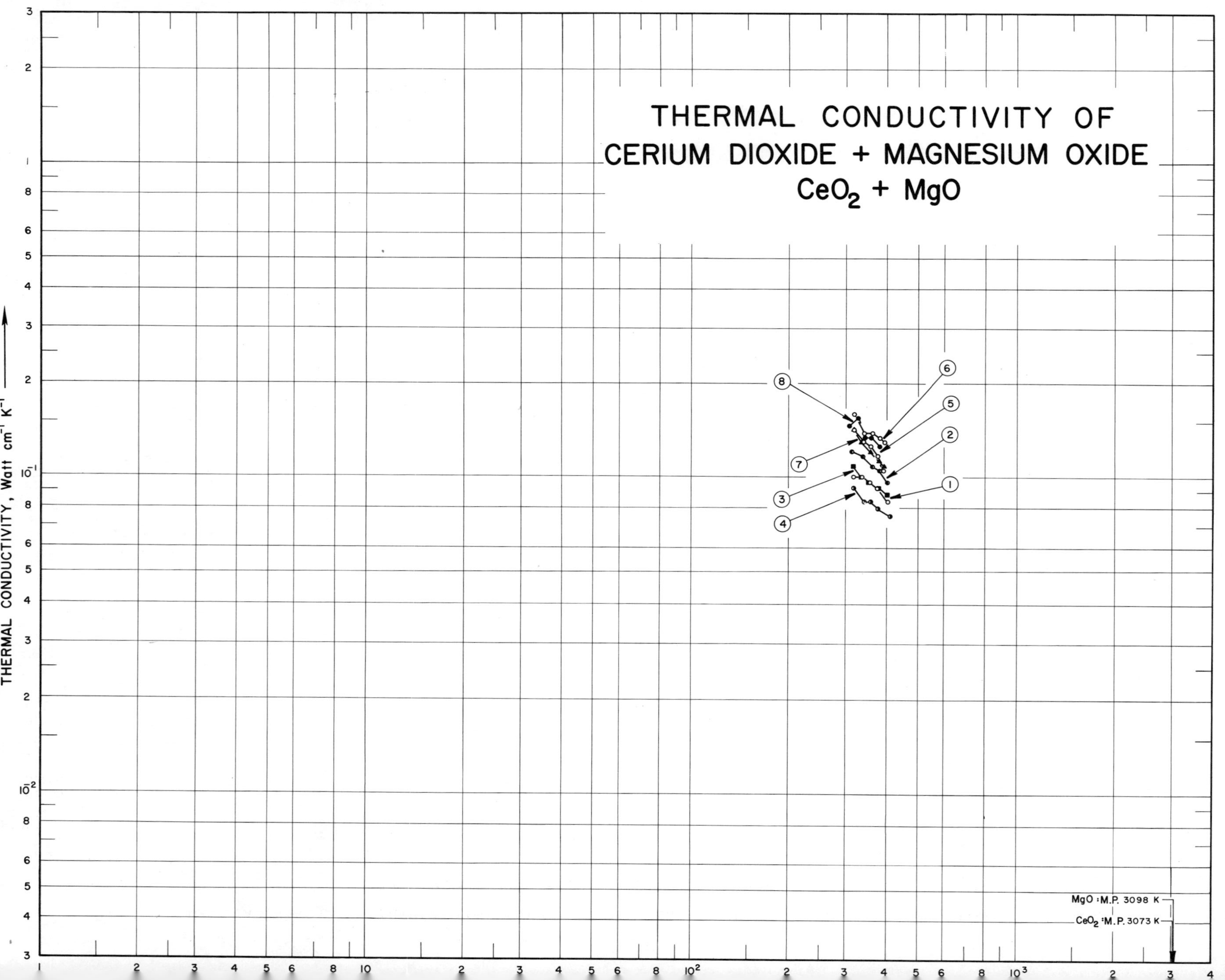

THERMAL CONDUCTIVITY OF
CERIUM DIOXIDE + MAGNESIUM OXIDE
CeO₂ + MgO
THERMAL CONDUCTIVITY, Watt cm⁻¹ K⁻¹
MgO : M.P. 3098 K
CeO₂ : M.P. 3073 K
1
2
3
4
5
6
7
8

SPECIFICATION TABLE NO. 81 THERMAL CONDUCTIVITY OF [CERIUM DIOXIDE + MAGNESIUM OXIDE] CeO_2 + MgO

[For Data Reported in Figure and Table No. 81]

Curve No.	Ref. No.	Method Used	Year	Temp. Range, K	Reported Error, %	Name and Specimen Designation	Composition (weight percent) CeO_2	MgO	Composition (continued), Specifications and Remarks
1	23		1952	316-404		161A-1	96	4	Fire temp 1833 K; water absorption 0.00%; density 6.02 g cm^{-3}; with buff color.
2	23		1952	315-404		186A-1	94	6	Fire temp 1800 K; water absorption 0.95%; density 5.58 g cm^{-3}; with buff color.
3	23		1952	316-403		138A-1	92	8	Fire temp 1711 K; water absorption 0.033%; density 5.63 g cm^{-3}; with buff color.
4	23		1952	319-410		185A-1	92	8	Fire temp 1800 K; water absorption 0.21%; density 5.82 g cm^{-3}; with buff color.
5	23		1952	319-394		106A-1	90	10	Fire temp 1755 K; water absorption 0.34%; density 4.59 g cm^{-3}; with buff color.
6	23		1952	320-396		107A-1	90	10	Fire temp 1769 K; water absorption 0.136%; density 4.68 g cm^{-3}; with buff color.
7	23		1952	319-395		108A-1	80	20	Fire temp 1755 K; water absorption 0.054%; density 4.94 g cm^{-3}; with buff color.
8	23		1952	310-382		109A-1	80	20	Fire temp 1755 K; water absorption 0.003%; density 5.05 g cm^{-3}; with buff color.

DATA TABLE NO. 81 THERMAL CONDUCTIVITY OF [CERIUM DIOXIDE + MAGNESIUM OXIDE] $CeO_2 + MgO$

[Temperature, T, K; Thermal Conductivity, k, Watts $cm^{-1}K^{-1}$]

T	k
CURVE 1	
315.8	0.100
337.1	0.100
356.7	0.096
375.6	0.092
403.5	0.0837
CURVE 2	
314.9	0.121
338.6	0.117
362.1	0.108
380.8	0.105
404.0	0.096
CURVE 3	
316.3	0.108
335.8	0.100
354.9	0.0960
376.1	0.0920
402.7	0.0879
CURVE 4	
319.2	0.0920
341.7	0.0837
358.6	0.0837
376.9	0.0795
410.3	0.0753
CURVE 5	
318.7	0.142
340.3	0.130
359.9	0.126
376.8	0.117
393.5	0.105
CURVE 6	
319.9	0.159
342.9	0.138
364.5	0.138
381.7	0.134
395.8	0.130

T	k
CURVE 7	
318.6	0.142
337.1	0.130
357.4	0.121
380.7	0.113
394.9	0.109
CURVE 8	
309.5	0.146
326.7	0.155
344.8	0.134
360.7	0.134
382.2	0.126

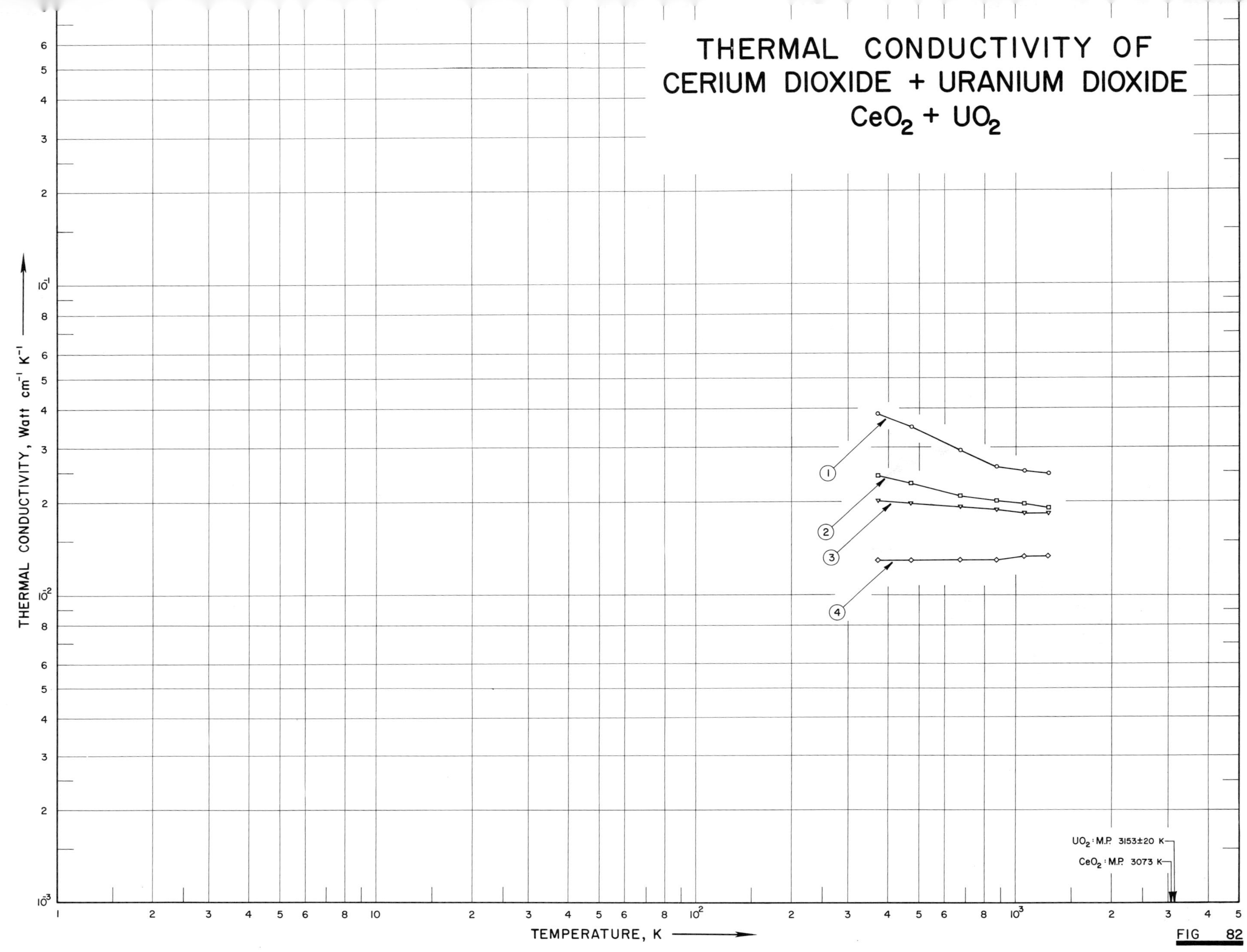

THERMAL CONDUCTIVITY OF
CERIUM DIOXIDE + URANIUM DIOXIDE
CeO2 + UO2
1
2
3
4
UO2 : M.P. 3153±20 K
CeO2 : M.P. 3073 K
THERMAL CONDUCTIVITY, Watt cm-1 K-1
TEMPERATURE, K

FIG 82

SPECIFICATION TABLE NO. 82 THERMAL CONDUCTIVITY OF [CERIUM DIOXIDE + URANIUM DIOXIDE] $CeO_2 + UO_2$

[For Data Reported in Figure and Table No. 82]

Curve No.	Ref. No.	Method Used	Year	Temp. Range, K	Reported Error, %	Name and Specimen Designation	Composition (weight percent) CeO_2	UO_2	Composition (continued), Specifications and Remarks
1	24	C	1958	373-1273	±10	216B	80.2	19.8	15.0 volume % UO_2; bulk density 6.34 g cm^{-3}; specimen dry pressed from U_3O_8 - CeO_2, fired in hydrogen atmosphere to 1600 C for 1 hr, and machined to final size; not oxidized; densed alumina and stabilized zirconia used as comparative material.
2	24	C	1958	373-1273	±10	216B	80.2	19.8	Similar to the above specimen but fired in air; somewhat oxidized.
3	24	C	1958	373-1273	±10	217B	62.5	37.5	30.0 volume % UO_2; bulk density 5.22 g cm^{-3}; specimen dry pressed from U_3O_8 - CeO_2, fired in hydrogen atmosphere to 1600 C for 1 hr, and machined to final size; not oxidized.
4	24	C	1958	373-1273	±10	217B	62.5	37.5	Similar to the above specimen except fired in air; somewhat oxidized.

DATA TABLE NO. 82 THERMAL CONDUCTIVITY OF [CERIUM DIOXIDE + URANIUM DIOXIDE] $CeO_2 + UO_2$

[Temperature, T, K; Thermal Conductivity, k, Watt cm^{-1} K^{-1}]

T	k
CURVE 1	
373.2	0.0385
473.2	0.0347
673.2	0.0293
873.2	0.0259
1073.2	0.0251
1273.2	0.0247
CURVE 2	
373.2	0.0243
473.2	0.0230
673.2	0.0209
873.2	0.0201
1073.2	0.0197
1273.2	0.0192
CURVE 3	
373.2	0.0201
473.2	0.0197
673.2	0.0192
873.2	0.0188
1073.2	0.0184
1273.2	0.0184
CURVE 4	
373.2	0.0130
473.2	0.0130
673.2	0.0130
873.2	0.0130
1073.2	0.0134
1273.2	0.0134

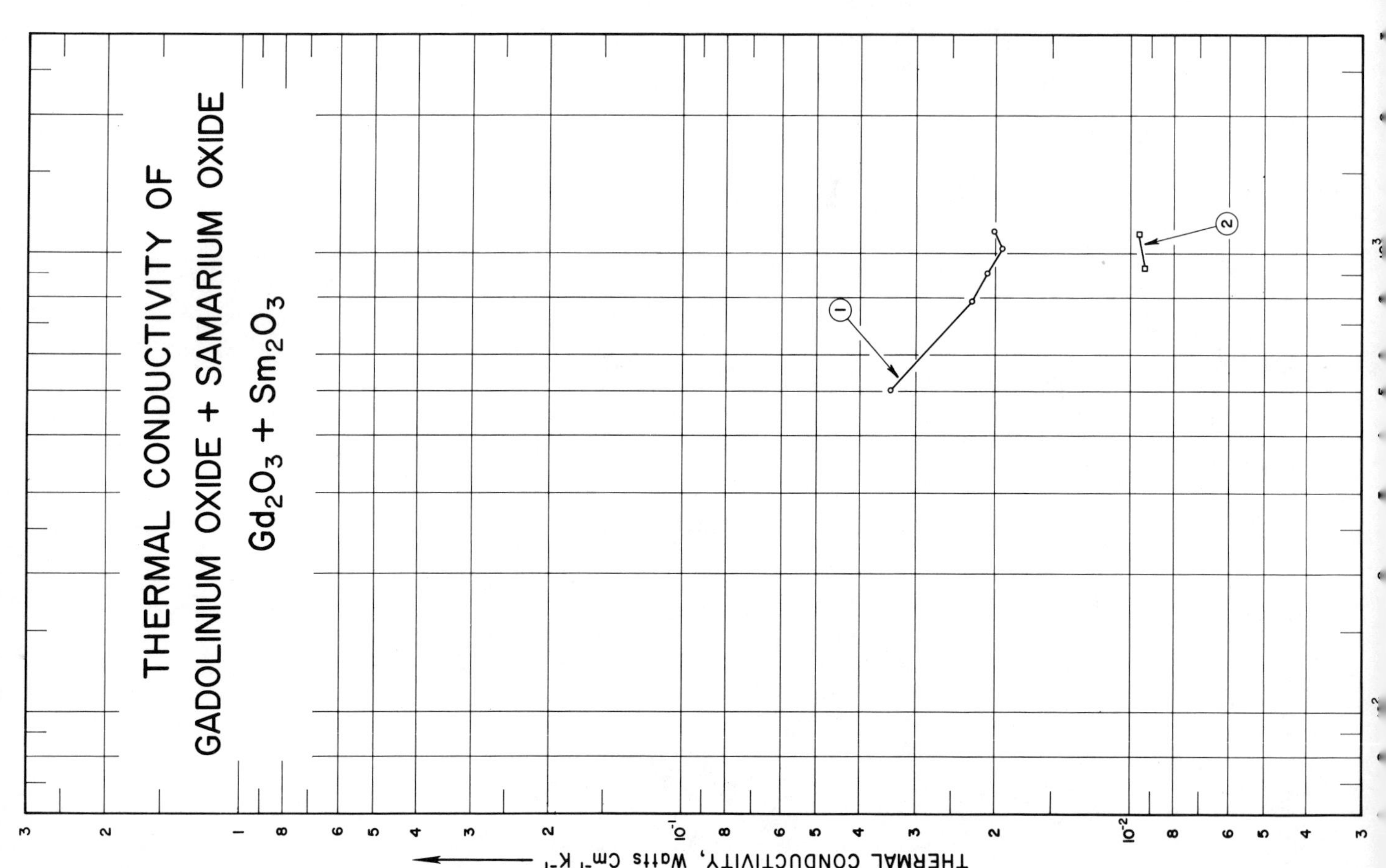
THERMAL CONDUCTIVITY OF
GADOLINIUM OXIDE + SAMARIUM OXIDE
$Gd_2O_3 + Sm_2O_3$
THERMAL CONDUCTIVITY, Watts Cm⁻¹ K⁻¹
1
2

SPECIFICATION TABLE NO. 83 THERMAL CONDUCTIVITY OF [GADOLINIUM OXIDE + SAMARIUM OXIDE] $Gd_2O_3 + Sm_2O_3$

[For Data Reported in Figure and Table No. 83]

Curve No.	Ref. No.	Method Used	Year	Temp. Range, K	Reported Error, %	Name and Specimen Designation	Composition (weight percent)	Composition (continued), Specifications and Remarks
1	145		1955	505-1120				Solid solution; prepared by ORNL; fired to dense condition.
2	145		1955	931, 1110				Solid solution; prepared by ORNL; fired at low temperature; specimen quite porous.

DATA TABLE NO. 83 THERMAL CONDUCTIVITY OF [GADOLINIUM OXIDE + SAMARIUM OXIDE] $Gd_2O_3 + Sm_2O_3$

[Temperature, T, K; Thermal Conductivity, k, Watt $cm^{-1}K^{-1}$]

T	k
CURVE 1	
505.2	0.0341
790.2	0.0225
907.2	0.0209
1023.2	0.0194
1119.7	0.0201
CURVE 2	
930.7	0.00929
1109.7	0.00954

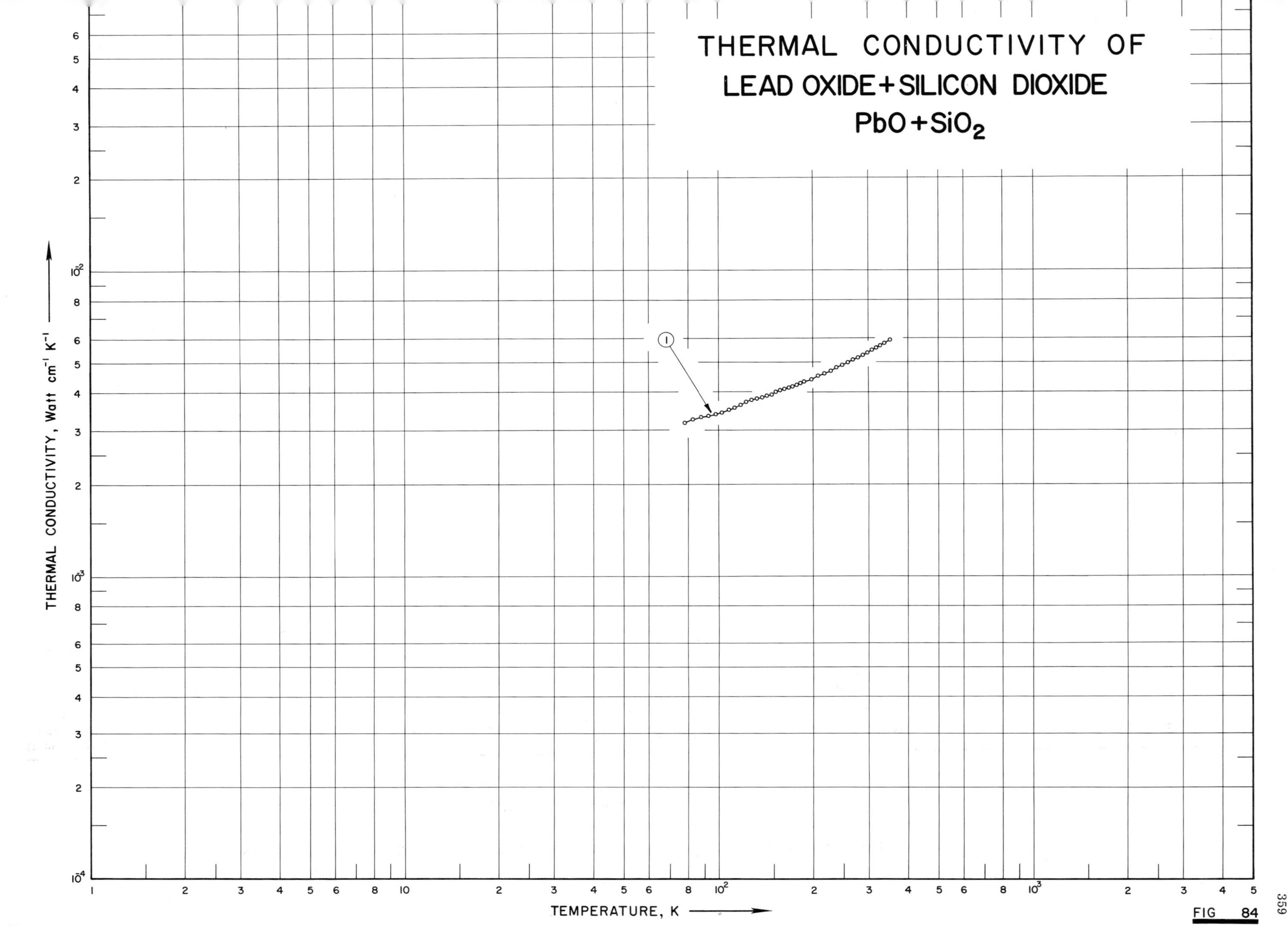
THERMAL CONDUCTIVITY OF
LEAD OXIDE + SILICON DIOXIDE
PbO + SiO2
THERMAL CONDUCTIVITY, Watt cm⁻¹ K⁻¹
TEMPERATURE, K
1

FIG 84

SPECIFICATION TABLE NO. 84 THERMAL CONDUCTIVITY OF [LEAD OXIDE + SILICON DIOXIDE] $PbO + SiO_2$

[For Data Reported in Figure and Table No. 84]

Curve No.	Ref. No.	Method Used	Year	Temp. Range, K	Reported Error, %	Name and Specimen Designation	Composition (weight percent)		Composition (continued), Specifications and Remarks
							PbO	SiO_2	
1	147	L	1960	78-353		L	80.0	20.0	Specimen 3 in. in dia and 0.375 in. thick; density 6.10 g cm^{-3}.

DATA TABLE NO. 84 THERMAL CONDUCTIVITY OF [LEAD OXIDE + SILICON DIOXIDE] $PbO + SiO_2$

[Temperature, T, K; Thermal Conductivity, k, Watt $cm^{-1}K^{-1}$]

T	k	T	k
CURVE 1		CURVE 1 (cont.)	
78.2	0.00318	303.2	0.00540*
83.2	0.00326	308.2	0.00548
88.2	0.00331	313.2	0.00552*
93.2	0.00335	318.2	0.00556
98.2	0.00339	323.2	0.00561*
103.2	0.00343	328.2	0.00569
108.2	0.00351	333.2	0.00573*
113.2	0.00356	338.2	0.00577
118.2	0.00364	343.2	0.00582*
123.2	0.00372	348.2	0.00586*
128.2	0.00377	353.2	0.00594
133.2	0.00381		
138.2	0.00385		
143.2	0.00389		
148.2	0.00393		
153.2	0.00402		
158.2	0.00405		
163.2	0.00410		
168.2	0.00414		
173.2	0.00418		
178.2	0.00423		
183.2	0.00427		
188.2	0.00431		
193.2	0.00435*		
198.2	0.00439		
203.2	0.00444*		
208.2	0.00452		
213.2	0.00456*		
218.2	0.00460		
223.2	0.00464*		
228.2	0.00469		
233.2	0.00473*		
238.2	0.00481		
243.2	0.00485*		
248.2	0.00490		
253.2	0.00494*		
258.2	0.00498		
263.2	0.00502*		
268.2	0.00510		
273.2	0.00515*		
278.2	0.00519		
283.2	0.00523*		
288.2	0.00527		
293.2	0.00531*		
298.2	0.00536		

*Not shown on plot

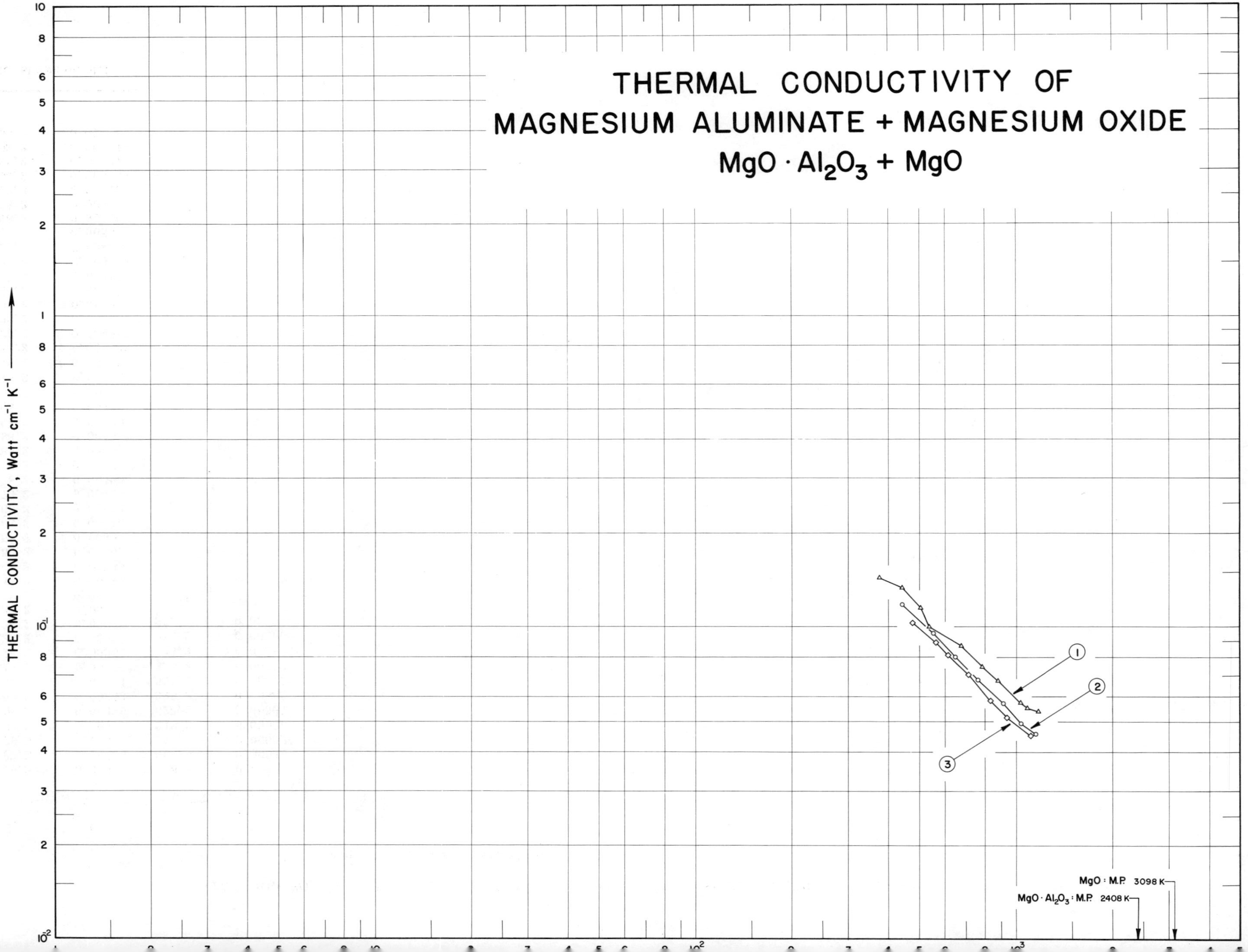
THERMAL CONDUCTIVITY OF
MAGNESIUM ALUMINATE + MAGNESIUM OXIDE
MgO · Al2O3 + MgO
THERMAL CONDUCTIVITY, Watt cm-1 K-1
MgO : M.P. 3098 K
MgO · Al2O3 : M.P. 2408 K
1
2
3

SPECIFICATION TABLE NO. 85 THERMAL CONDUCTIVITY OF [MAGNESIUM ALUMINATE + MAGNESIUM OXIDE] $MgO \cdot Al_2O_3 + MgO$

[For Data Reported in Figure and Table No. 85]

Curve No.	Ref. No.	Method Used	Year	Temp. Range, K	Reported Error, %	Name and Specimen Designation	Composition (weight percent) $MgO \cdot Al_2O_3$	MgO	Composition (continued), Specifications and Remarks
1	136	C	1959	376-1173	±4		50. 1	49. 9	Corresponding to 49. 8 volume % magnesium aluminate; fired at 1800 C to bulk density of 3. 19 g cm^{-3}; porosity 11. 4 volume %; data corrected for porosity; theoretical calculated composition.
2	136	C	1959	443-1166	±4		77. 4	22. 6	Corresponding to 77. 5 volume % magnesium aluminate; fired at 1800 C to bulk density of 3. 04 g cm^{-3}; porosity 14. 8 volume %; data corrected for porosity; theoretical calculated composition.
3	136	C	1959	478-1118	±4		91. 1	8. 9	Corresponding to 91. 2 volume % magnesium aluminate; fired at 1800 C to bulk density of 2. 99 g cm^{-3}; porosity 15. 4 volume %; data corrected for porosity; theoretical calculated composition.

DATA TABLE NO. 85 THERMAL CONDUCTIVITY OF [MAGNESIUM ALUMINATE + MAGNESIUM OXIDE] $MgO \cdot Al_2O_3 + MgO$

[Temperature, T, K; Thermal Conductivity, k, Watt $cm^{-1} K^{-1}$]

T	k
CURVE 1	
376.2	0.144
443.2	0.133
503.2	0.116
537.2	0.100
673.2	0.0872
787.2	0.0749
880.2	0.0678
1033.2	0.0578
1098.2	0.0554
1173.2	0.0540
CURVE 2	
443.2	0.118
553.2	0.0954
650.2	0.0801
765.2	0.0681
914.2	0.0571
1045.2	0.0495
1166.2	0.0457
CURVE 3	
478.2	0.103
563.2	0.0892
615.2	0.0815
716.2	0.0709
833.2	0.0586
935.2	0.0516
1118.2	0.0452

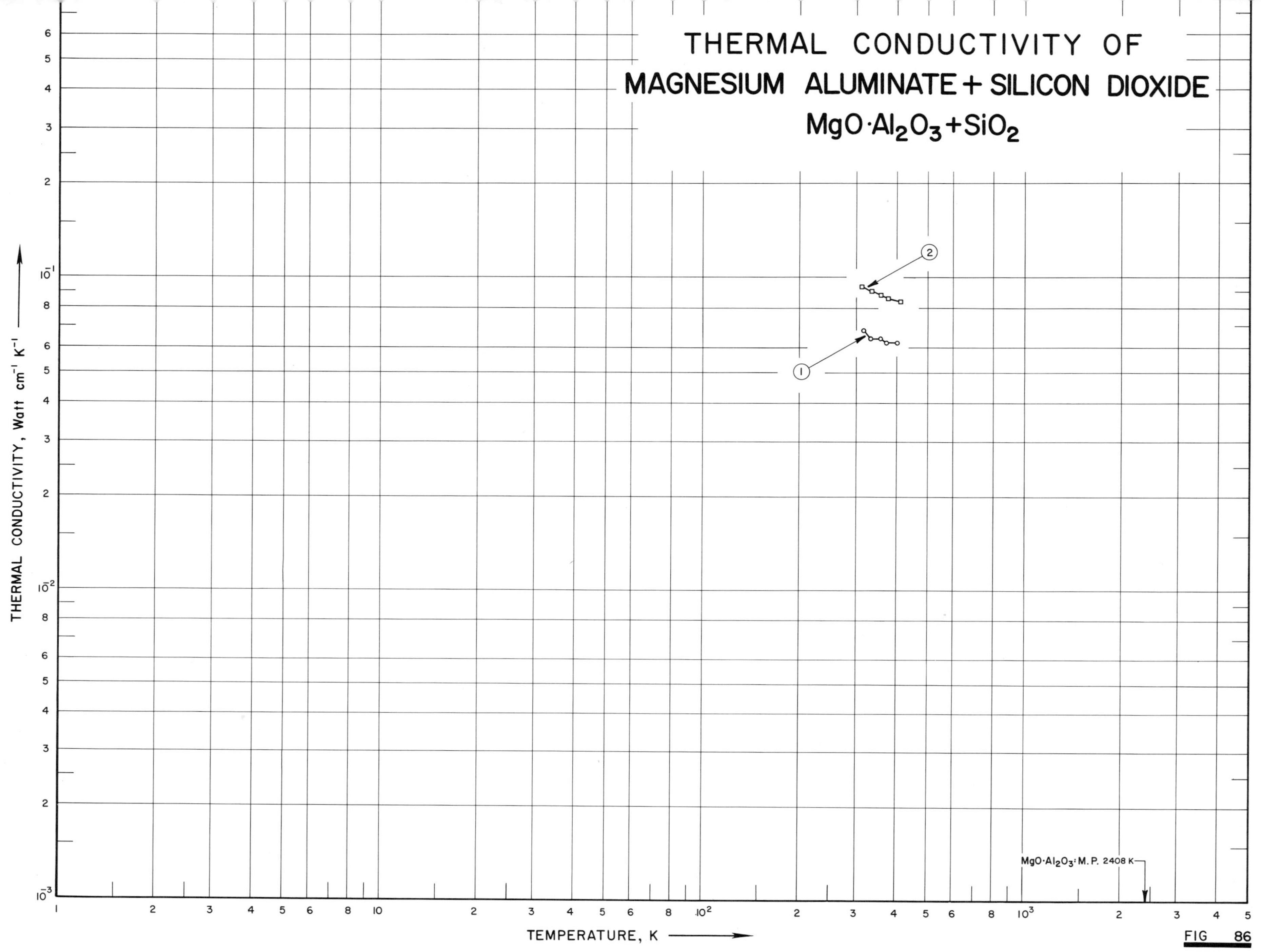
THERMAL CONDUCTIVITY OF
MAGNESIUM ALUMINATE + SILICON DIOXIDE
$MgO \cdot Al_2O_3 + SiO_2$
1
2
$MgO \cdot Al_2O_3$: M.P. 2408 K
THERMAL CONDUCTIVITY, Watt cm^{-1} K^{-1}
TEMPERATURE, K

FIG 86

SPECIFICATION TABLE NO. 86 THERMAL CONDUCTIVITY OF [MAGNESIUM ALUMINATE + SILICON DIOXIDE] $MgO \cdot Al_2O_3 + SiO_2$

[For Data Reported in Figure and Table No. 86]

Curve No.	Ref. No.	Method Used	Year	Temp. Range, K	Reported Error, %	Name and Specimen Designation	Composition (weight percent), Specifications and Remarks
1	68		1954	317-401	±3.0	Spinel; 17A2	20.0 SiO_2.
2	68		1954	315-413	±3.0	Spinel; 27A2	10.0 SiO_2.

DATA TABLE NO. 86 THERMAL CONDUCTIVITY OF [MAGNESIUM ALUMINATE + SILICON DIOXIDE] $MgO \cdot Al_2O_3 + SiO_2$

[Temperature, T, K; Thermal Conductivity, k, Watt $cm^{-1} K^{-1}$]

T	k
CURVE 1	
317.0	0.0682
333.9	0.0644
356.1	0.0644
373.6	0.0628
400.8	0.0628
CURVE 2	
314.7	0.0941
335.4	0.0912
358.8	0.0883
378.5	0.0862
412.5	0.0824

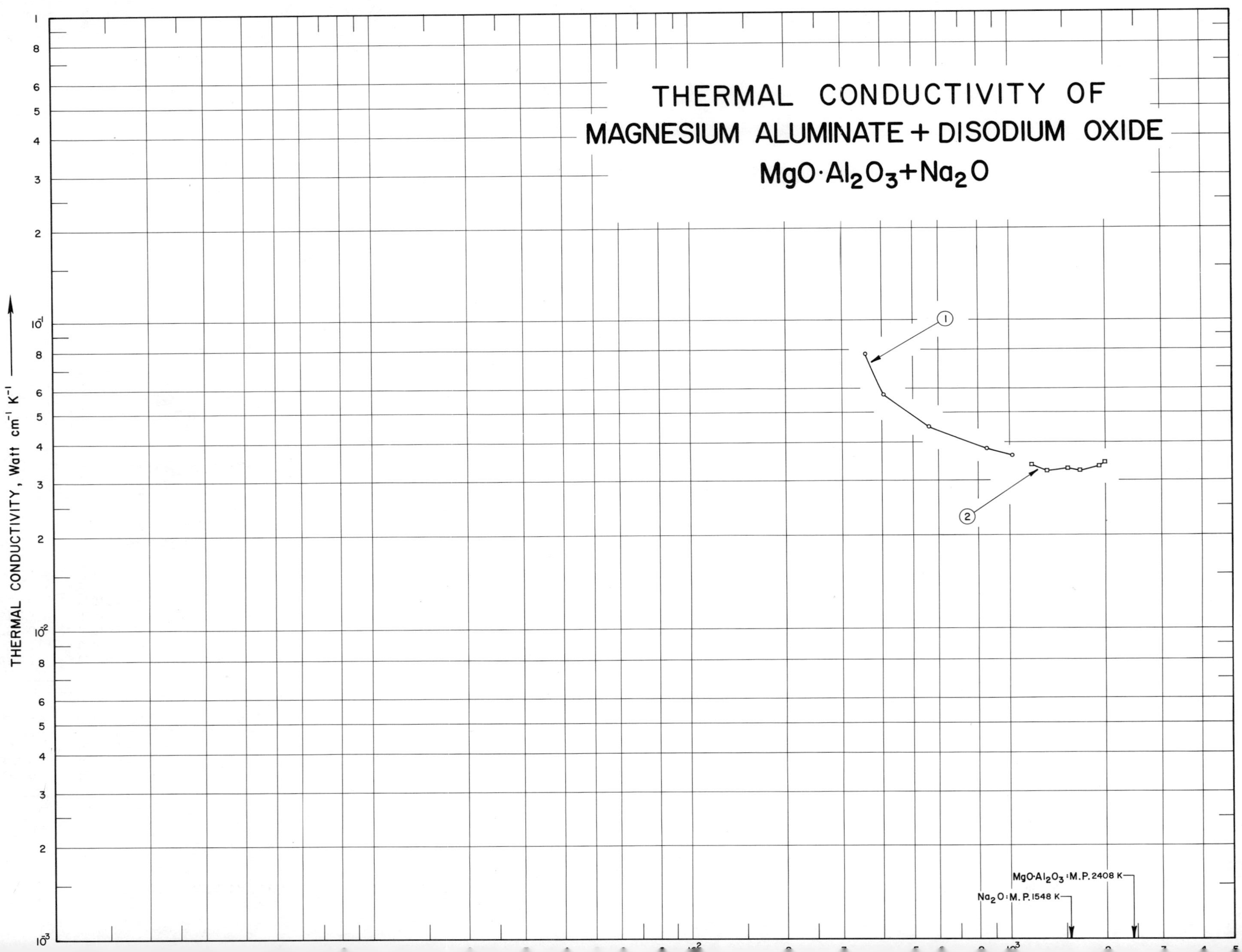

THERMAL CONDUCTIVITY OF MAGNESIUM ALUMINATE + DISODIUM OXIDE

$MgO \cdot Al_2O_3 + Na_2O$

THERMAL CONDUCTIVITY, Watt cm^{-1} K^{-1}

1
8
6
5
4
3
2
10^{-1}
8
6
5
4
3
2
10^{-2}
8
6
5
4
3
2
10^{-3}

10^{2}
10^{3}

1
2

$MgO \cdot Al_2O_3$: M.P. 2408 K

Na_2O : M.P. 1548 K

SPECIFICATION TABLE NO. 87 THERMAL CONDUCTIVITY OF [MAGNESIUM ALUMINATE + DISODIUM OXIDE] $MgO \cdot Al_2O_3 + Na_2O$

[For Data Reported in Figure and Table No. 87]

Curve No.	Ref. No.	Method Used	Year	Temp. Range, K	Reported Error, %	Name and Specimen Designation	Composition (weight percent), Specifications and Remarks
1	251	C	1963	355-1022	±4.0	Spinel	68.26 Al_2O_3, 26.74 MgO, 3.17 Na_2O, 0.33 SiO_2, 0.026 Fe_2O_3, and 1.20 B; specimen 2 in. in dia and 1 in. in thickness; cold pressed; firing temperature 3300 F; both faces of the disc ground flat and parallel; measured in a helium atmosphere; Armco iron used as comparative material.
2	251	P	1963	1174-2000		Spinel	The above specimen measured by another method; thermal conductivity values calculated from measured data of thermal diffusivity, specific heat, and density.

DATA TABLE NO. 87 THERMAL CONDUCTIVITY OF [MAGNISIUM ALUMINATE + DISODIUM OXIDE] $MgO \cdot Al_2O_3 + Na_2O$

[Temperature, T, K; Thermal Conductivity, k, Watt $cm^{-1} K^{-1}$]

T	k
CURVE 1	
355.4	0.0779
505.4	0.0571
660.9	0.0450
849.8	0.0381
1022.1	0.0363
CURVE 2	
1174.8	0.0339
1313.7	0.0322
1527.6	0.0329
1663.7	0.0324
1908.2	0.0334
1999.8	0.0344

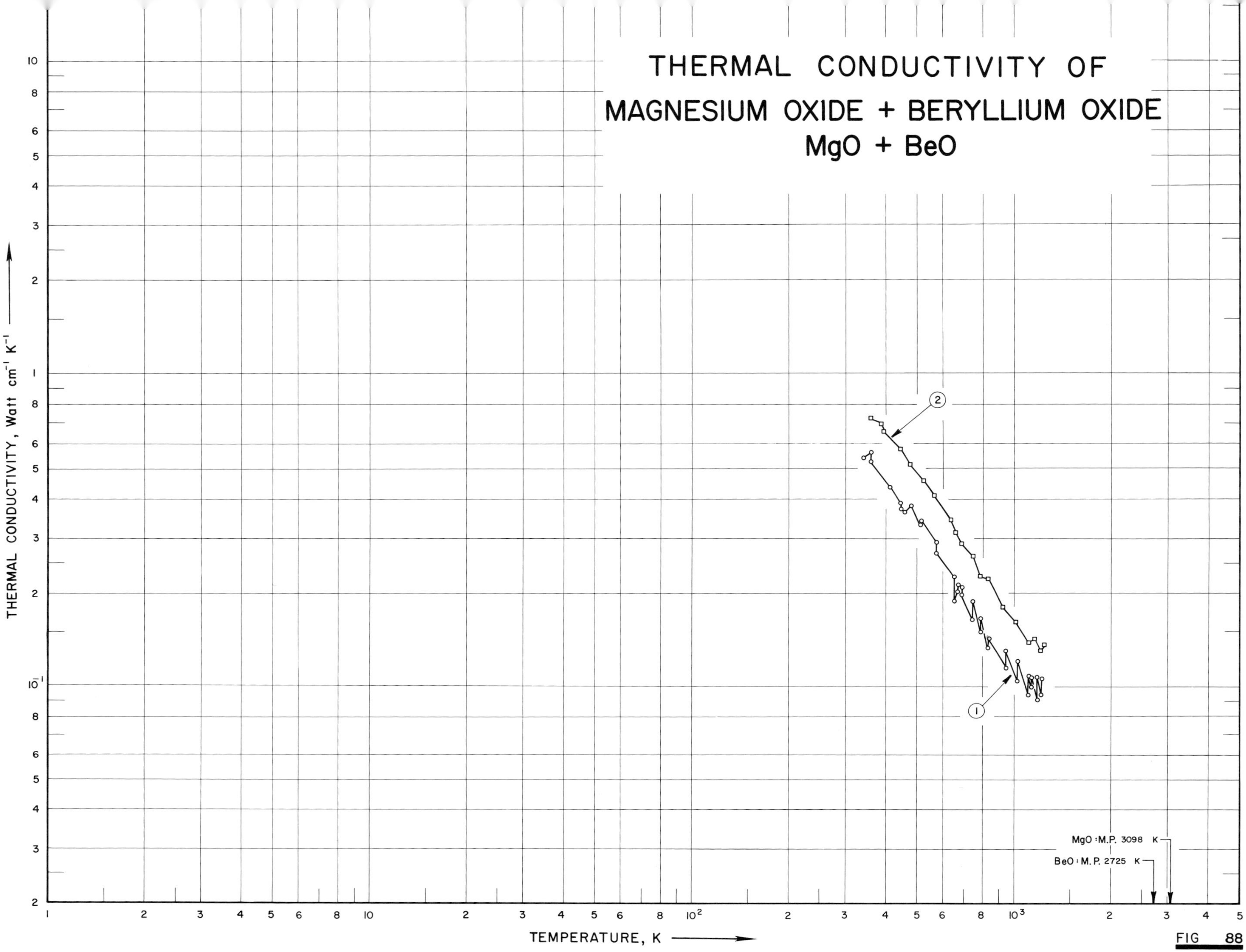

THERMAL CONDUCTIVITY OF MAGNESIUM OXIDE + BERYLLIUM OXIDE $MgO + BeO$

THERMAL CONDUCTIVITY, Watt cm^{-1} K^{-1}

TEMPERATURE, K

MgO : M.P. 3098 K

BeO : M.P. 2725 K

FIG 88

SPECIFICATION TABLE NO. 88 THERMAL CONDUCTIVITY OF [MAGNESIUM OXIDE + BERYLLIUM OXIDE] $MgO + BeO$

[For Data Reported in Figure and Table No. 88]

Curve No.	Ref. No.	Method Used	Year	Temp. Range, K	Reported Error, %	Name and Specimen Designation	Composition (weight percent) MgO	BeO	Composition (continued), Specifications and Remarks
1	232	L	1954	343-1223			61.5	38.5	Specimen prepared by slip casting BeO + MgO suspensions containing 54.3 volume % BeO; fired bulk density 2.43 g cm^{-3}; total porosity 25.7%.
2	136	C	1959	361-1243	±4.0		61.5	38.5	Mixture; specimen prepared by slip casting a composition containing 54.3 volume % BeO, which was sintered at 1800 C; fired bulk density 2.43 g cm^{-3}; total porosity 25.7%; data corrected for porosity.

DATA TABLE NO. 88 THERMAL CONDUCTIVITY OF [MAGNESIUM OXIDE + BERYLLIUM OXIDE] MgO + BeO

[Temperature, T, K; Thermal Conductivity, k, Watt cm^{-1} K^{-1}]

T	k
CURVE 1	
343.2	0.541
363.2	0.563
363.2	0.524
413.2	0.438
448.2	0.390
448.2	0.373
460.2	0.365
480.2	0.381
513.2	0.331
518.2	0.341
578.2	0.293
578.2	0.270
655.2	0.226
655.2	0.189
673.2	0.201
673.2	0.213
693.2	0.209
693.2	0.197
743.2	0.164
748.2	0.188
790.2	0.151
792.2	0.166
835.2	0.133
840.2	0.143
945.2	0.115
945.2	0.130
1027.2	0.105
1029.2	0.121
1105.2	0.0946
1111.2	0.109
1138.2	0.100
1138.2	0.108
1193.2	0.0908
1193.2	0.108
1215.2	0.0941
1223.2	0.107
CURVE 2	
361.2	0.724
389.2	0.695
395.2	0.657
444.2	0.573
474.2	0.519
523.2	0.460
CURVE 2 (cont.)	
566.2	0.410
639.2	0.343
663.2	0.314
688.2	0.289
746.2	0.264
788.2	0.226
835.2	0.222
923.2	0.180
1012.2	0.161
1114.2	0.138
1160.2	0.142
1203.2	0.130
1243.2	0.136

SPECIFICATION TABLE NO. 89 THERMAL CONDUCTIVITY OF [MAGNESIUM OXIDE + CLAY] MgO + Clay

Curve No.	Ref. No.	Method Used	Year	Temp. Range, K	Reported Error, %	Name and Specimen Designation	Composition (weight percent) MgO	Clay	Composition (continued), Specifications and Remarks
1	23		1952	310-421		171 A-1	97.5	2.5	Old mine (No. 4) ball clay; specimen fired at 2800 F and soaked for 0.5 hr; white color; water absorption 0.45%; density 3.031 g cm^{-3}.

DATA TABLE NO. 89 THERMAL CONDUCTIVITY OF [MAGNESIUM OXIDE + CLAY] MgO + Clay

[Temperature, T, K; Thermal Conductivity, k, Watt $cm^{-1} K^{-1}$]

T	k
CURVE 1*	
310.0	0.244
335.4	0.233
360.0	0.217
383.4	0.203
421.0	0.187

* No graphical presentation

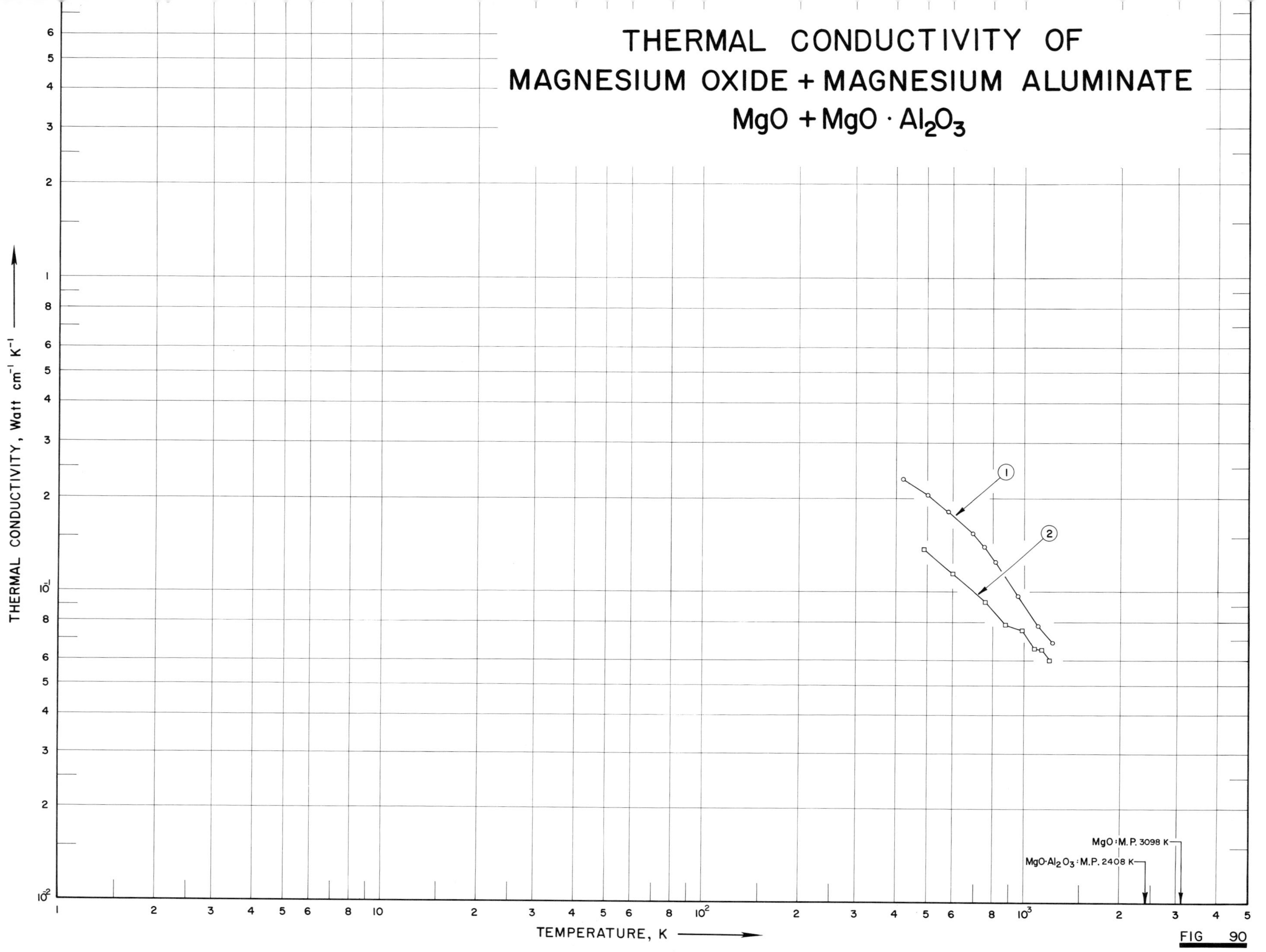

THERMAL CONDUCTIVITY OF
MAGNESIUM OXIDE + MAGNESIUM ALUMINATE
MgO + MgO · Al_2O_3
1
2
MgO : M.P. 3098 K
MgO·Al_2O_3 : M.P. 2408 K
THERMAL CONDUCTIVITY, Watt cm^{-1} K^{-1}
TEMPERATURE, K
FIG 90

SPECIFICATION TABLE NO. 90 THERMAL CONDUCTIVITY OF [MAGNESIUM OXIDE + MAGNESIUM ALUMINATE] $MgO + MgO \cdot Al_2O_3$

[For Data Reported in Figure and Table No. 90]

Curve No.	Ref. No.	Method Used	Year	Temp. Range, K	Reported Error, %	Name and Specimen Designation	Composition (weight percent) MgO	$MgO \cdot Al_2O_3$	Composition (continued), Specifications and Remarks
1	136	C	1959	423-1239	±4		84. 2	15. 8	Corresponding to 14. 4 vol % magnesium aluminate; fired at 1800 C; bulk density 3. 25 g cm^{-3}; porosity 10. 7 vol %; data corrected for porosity; calculated composition.
2	136	C	1959	490-1202	±4		70. 5	29. 5	Corresponding to 28. 7 vol % magnesium aluminate; fired at 1800 C; bulk density 3. 39 g cm^{-3}; porosity 6. 4 vol %; data corrected for porosity; calculated composition.

DATA TABLE NO. 90 THERMAL CONDUCTIVITY OF [MAGNESIUM OXIDE + MAGNESIUM ALUMINATE] $MgO + MgO \cdot Al_2O_3$

[Temperature, T, K; Thermal Conductivity, k, Watt cm^{-1} K^{-1}]

T	k
CURVE 1	
423.2	0.230
502.2	0.205
583.2	0.181
695.2	0.154
753.2	0.140
818.2	0.126
960.2	0.0967
1111.2	0.0776
1239.2	0.0686
CURVE 2	
490.2	0.137
602.2	0.115
760.2	0.0929
877.2	0.0787
989.2	0.0751
1087.2	0.0658
1145.2	0.0651
1202.2	0.0601

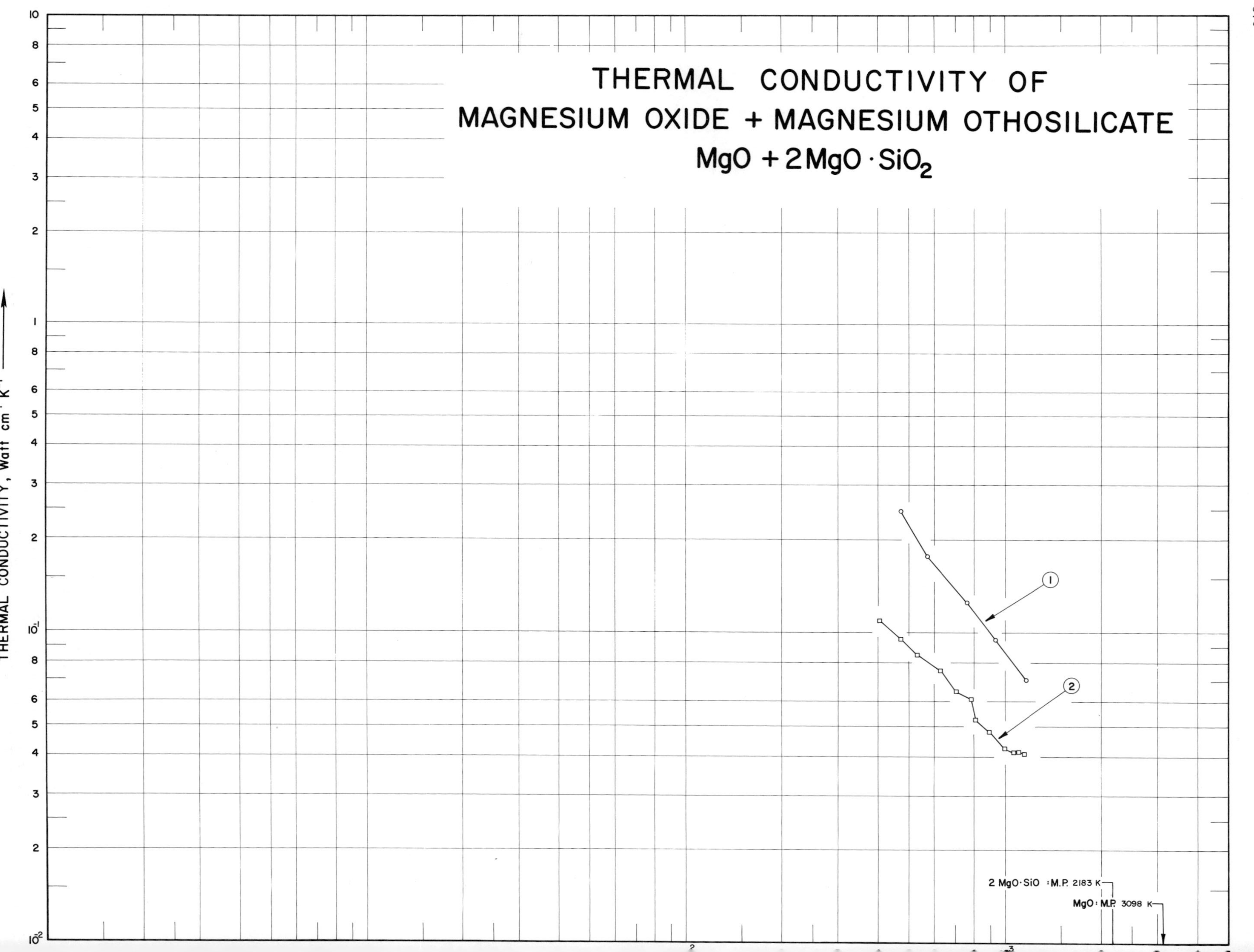

THERMAL CONDUCTIVITY OF
MAGNESIUM OXIDE + MAGNESIUM OTHOSILICATE
MgO + 2MgO · SiO2
THERMAL CONDUCTIVITY, Watt cm⁻¹ K⁻¹
10
8
6
5
4
3
2
1
10⁻¹
10⁻²
1
2
2 MgO·SiO : M.P. 2183 K
MgO : M.P. 3098 K

SPECIFICATION TABLE NO. 91 THERMAL CONDUCTIVITY OF [MAGNESIUM OXIDE + MAGNESIUM ORTHOSILICATE] $MgO + 2MgO \cdot SiO_2$

[For Data Reported in Figure and Table No. 91]

Curve No.	Ref. No.	Method Used	Year	Temp. Range, K	Reported Error, %	Name and Specimen Designation	Composition (weight percent) MgO	$2MgO \cdot SiO_2$	Composition (continued), Specifications and Remarks
1	136	C	1959	473-1166	±4		87.5	12.5	Corresponding to 86.8 vol % MgO; total porosity 11.7%; data corrected to theoretical density; theoretical calculated composition.
2	136	C	1959	406-1153	±4		63.2	36.8	Corresponding to 61.6 vol % MgO; total porosity 13.8%; data corrected to theoretical density; theoretical calculated composition.

DATA TABLE NO. 91 THERMAL CONDUCTIVITY OF [MAGNESIUM OXIDE + MAGNESIUM ORTHOSILICATE] $MgO + 2MgO \cdot SiO_2$

[Temperature, T, K; Thermal Conductivity, k, Watt cm^{-1} K^{-1}]

T	k
CURVE 1	
473.2	0.247
573.2	0.177
763.2	0.126
937.2	0.0950
1166.2	0.0703
CURVE 2	
406.2	0.109
473.2	0.0962
533.2	0.0849
630.2	0.0753
708.2	0.0649
783.2	0.0611
810.2	0.0527
895.2	0.0481
988.2	0.0427
1070.2	0.0416
1111.2	0.0418
1153.2	0.0410

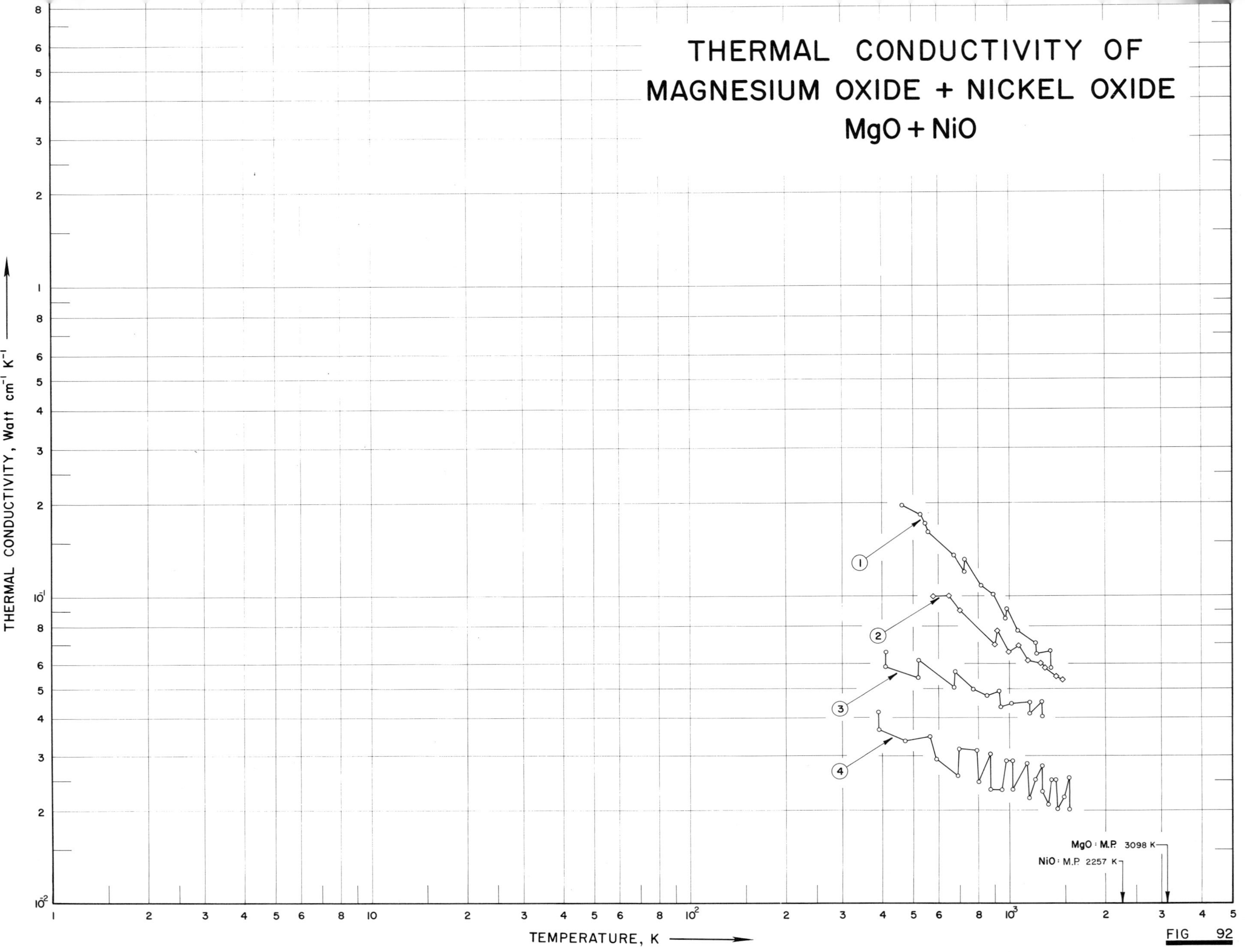
THERMAL CONDUCTIVITY OF
MAGNESIUM OXIDE + NICKEL OXIDE
MgO + NiO
THERMAL CONDUCTIVITY, Watt cm⁻¹ K⁻¹
TEMPERATURE, K
MgO : M.P. 3098 K
NiO : M.P. 2257 K
1
2
3
4
FIG 92

SPECIFICATION TABLE NO. 92 THERMAL CONDUCTIVITY OF [MAGNESIUM OXIDE + NICKEL OXIDE] MgO + NiO

[For Data Reported in Figure and Table No. 92]

Curve No.	Ref. No.	Method Used	Year	Temp. Range, K	Reported Error, %	Name and Specimen Designation	Composition (weight percent) MgO	NiO	Composition (continued), Specifications and Remarks
1	136	C	1959	464-1361	±4.0		→	→	99.5 MgO + NiO, 0.25 SiO_2, 0.1 Al_2O_3; solid solution with 1.0 vol % NiO; prepared by grinding together high-purity fused MgO with analytical reagent-grade NiO in a porcelain ball mill, preparing as a suspension with alcohol, slip-casting, and firing; bulk density 3.12 g cm^{-3}; total porosity 14.7%; data corrected to theoretical density.
2	136	C	1959	579-1468	±4.0		→	→	Similar to the above specimen except having 2.8 vol % NiO; bulk density 3.10 g cm^{-3}; total porosity 18%; data corrected to theoretical density.
3	136	C	1959	413-1272	±4.0		→	→	Similar to the above specimen except having 15 vol % NiO; bulk density 3.34 g cm^{-3}; total porosity 19%; data corrected to theoretical density.
4	136	C	1959	389-1548	±4.0		→	→	Similar to the above specimen except having 34.5 vol % NiO; bulk density 3.26 g cm^{-3}; total porosity 30.5%; data corrected to theoretical density.

DATA TABLE NO. 92 THERMAL CONDUCTIVITY OF [MAGNESIUM OXIDE + NICKEL OXIDE] MgO + NiO

[Temperature, T, K; Thermal Conductivity, k, Watt $cm^{-1} K^{-1}$]

T	k
CURVE 1	
464.2	0.197
530.2	0.184
548.2	0.172
561.2	0.162
676.2	0.136
723.2	0.121
730.2	0.132
823.2	0.108
895.2	0.102
978.2	0.0845
989.2	0.0908
1073.2	0.0774
1213.2	0.0699
1215.2	0.0644
1356.2	0.0657
1361.2	0.0582
CURVE 2	
579.2	0.0996
648.2	0.101
704.2	0.0895
903.2	0.0699
920.2	0.0774
1013.2	0.0657
1078.2	0.0690
1152.2	0.0615
1268.2	0.0602
1300.2	0.0586
1421.2	0.0544
1468.2	0.0536
CURVE 3	
413.2	0.0657
413.2	0.0586
521.2	0.0544
523.2	0.0615
676.2	0.0502
676.2	0.0565
773.2	0.0498
858.2	0.0477
933.2	0.0490
943.2	0.0435

T	k
CURVE 3 (cont.)	
1023.2	0.0448
1162.2	0.0452
1163.2	0.0414
1269.2	0.0456
1272.2	0.0402
CURVE 4	
389.2	0.0418
391.2	0.0364
472.2	0.0335
565.2	0.0347
591.2	0.0293
690.2	0.0259
697.2	0.0318
792.2	0.0314
807.2	0.0247
875.2	0.0305
878.2	0.0234
948.2	0.0234
973.2	0.0289
1035.2	0.0289
1038.2	0.0234
1149.2	0.0285
1153.2	0.0222
1218.2	0.0251
1278.2	0.0276
1278.2	0.0230
1332.2	0.0209
1362.2	0.0251
1417.2	0.0251
1426.2	0.0201
1492.2	0.0222
1546.2	0.0255
1548.2	0.0201

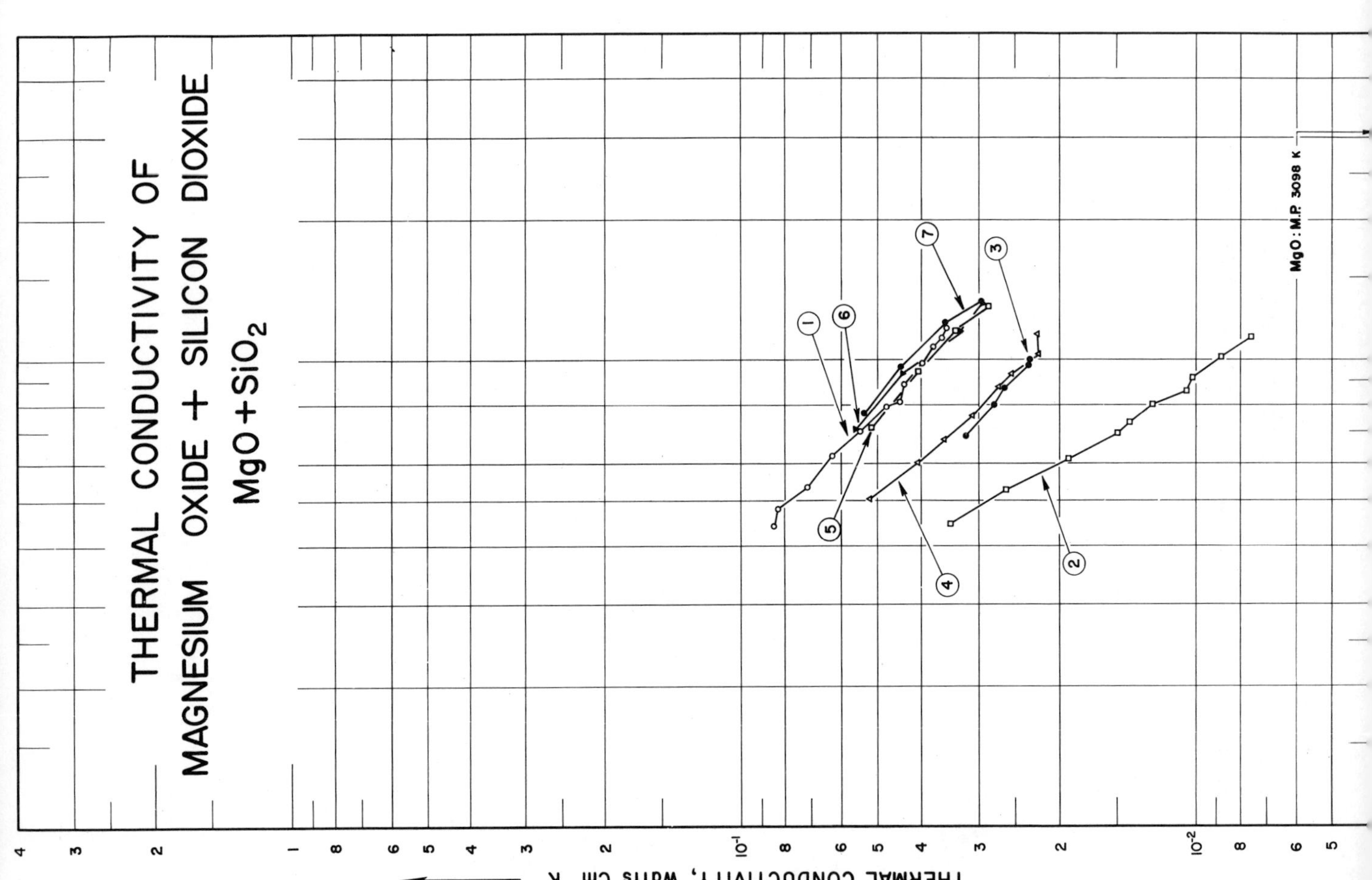

THERMAL CONDUCTIVITY OF
MAGNESIUM OXIDE + SILICON DIOXIDE
MgO + SiO2
THERMAL CONDUCTIVITY, Watts Cm⁻¹ K⁻¹
MgO : M.P. 3098 K
1
2
3
4
5
6
7

SPECIFICATION TABLE NO. 93 THERMAL CONDUCTIVITY OF [MAGNESIUM OXIDE + SILICON DIOXIDE] $MgO + SiO_2$

[For Data Reported in Figure and Table No. 93]

Curve No.	Ref. No.	Method Used	Year	Temp. Range, K	Reported Error, %	Name and Specimen Designation	Composition (weight percent) MgO	SiO_2	Composition (continued), Specifications and Remarks
1	145	R	1955	441-1170			95	5	Sintered.
2	145	R	1955	447-1125			85	15	Sintered.
3	145	R	1955	688-1002			75	25	Sintered.
4	146	L	1955	503-1143			65	35	Sintered.
5	143	L	1955	723-1313		Magnezit; 1	93.88	2.08	0.83 (0.05 TiO_2) Al_2O_3, 1.63 Fe_2O_3, 1.24 CaO, and 0.20 total Ca, Mg, Fe, and Mn; magnesite basic refractory brick; density 2.81 g cm^{-3}; apparent porosity 22.0%; gas permeability 1.34 ml m^{-2} hr^{-1} per mm H_2O.
6	143	L	1955	713-1303		Magnezit; 2			Similar to the above specimen.
7	143	L	1955	773-1348		Magnezit; 3			Similar to the above specimen.

DATA TABLE NO. 93 THERMAL CONDUCTIVITY OF [MAGNESIUM OXIDE + SILICON DIOXIDE] $MgO + SiO_2$

[Temperature, T, K; Thermal Conductivity, k, Watts cm^{-1} K^{-1}]

T	k
CURVE 1	
441.2	0.0845
480.2	0.0828
536.7	0.0741
621.8	0.0628
704.2	0.0544
793.2	0.0478
812.6	0.0447
883.2	0.0435
982.2	0.0397
1069.2	0.0376
1115.5	0.0360
1170.2	0.0351
CURVE 2	
447.2	0.0346
528.4	0.0262
617.2	0.0192
697.2	0.0149
735.5	0.0140
800.7	0.0125
858.2	0.0105
918.2	0.0102
1012.2	0.00879
1125.2	0.00757
CURVE 3	
688.2	0.0318
801.2	0.0277
871.2	0.0262
974.2	0.0233
1001.7	0.0232
CURVE 4	
503.2	0.0520
601.2	0.0404
676.2	0.0356
759.2	0.0309
878.2	0.0270
934.2	0.0254
1027.2	0.0202
1143.2	0.0203

T	k
CURVE 5	
723.2	0.0516
953.2	0.0407
1163.2	0.0337
1313.2	0.0284
CURVE 6	
713.2	0.0558
943.2	0.0437
1163.2	0.0330
1303.2	0.0293
CURVE 7	
773.2	0.0535
973.2	0.0442
1203.2	0.0354
1348.2	0.0294

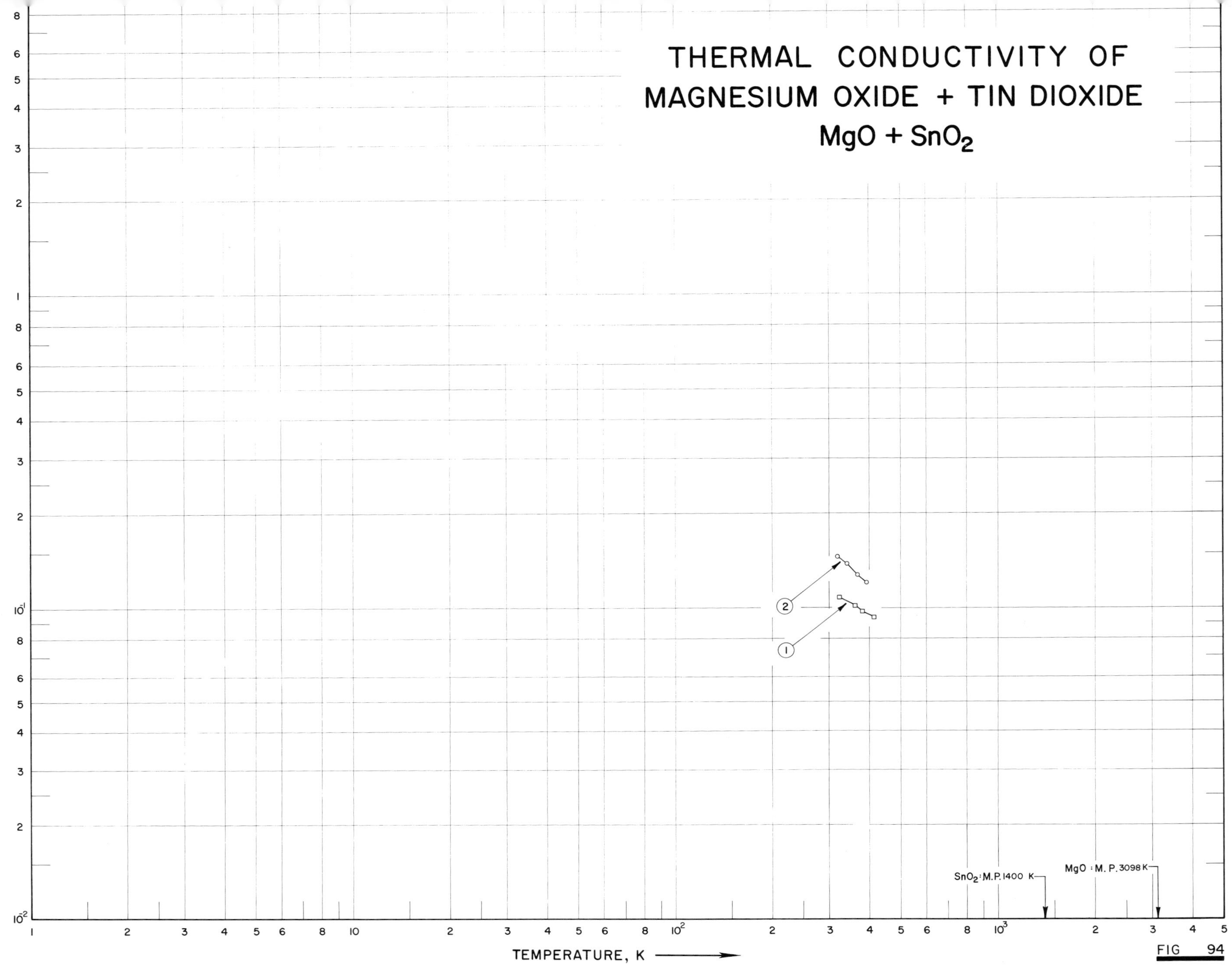
THERMAL CONDUCTIVITY OF
MAGNESIUM OXIDE + TIN DIOXIDE
MgO + SnO2
THERMAL CONDUCTIVITY, Watt cm⁻¹ K⁻¹
TEMPERATURE, K
SnO2: M.P. 1400 K
MgO : M. P. 3098 K
1
2

FIG 94

SPECIFICATION TABLE NO. 94 THERMAL CONDUCTIVITY OF [MAGNESIUM OXIDE + TIN DIOXIDE] $MgO + SnO_2$

[For Data Reported in Figure and Table No. 94]

Curve No.	Ref. No.	Method Used	Year	Temp. Range, K	Reported Error, %	Name and Specimen Designation	Composition (weight percent) MgO	SnO_2	Composition (continued), Specifications and Remarks
1	12		1953	323-415		225A-1	51.7	48.3	4 $MgO + SnO_2$ by mole; prepared by milling pure oxides in water, calcining at 1367 K after drying, then dry-pressing in a 0.5 in. steel die at 15,000 psi; fired at 1755 K and soaked for 1.5 hrs; water absorption 0.028%; density 4.18 g cm^{-3}.
2	12		1953	320-392		226A-1	70.7	29.3	9 $MgO + SnO_2$ by mole; same specimen preparation as the above; fired at 1811 K and soaked for 1.5 hrs; water absorption 0.17%; density 3.84 g cm^{-3}.

DATA TABLE NO. 94 THERMAL CONDUCTIVITY OF [MAGNESIUM OXIDE + TIN DIOXIDE] $MgO + SnO_2$

[Temperature, T, K; Thermal Conductivity, k, Watt cm^{-1} K^{-1}]

T	k
CURVE 1	
322.6	0.108
343.2	0.105*
364.2	0.102
383.6	0.0975
415.0	0.0937
CURVE 2	
319.7	0.146
340.6	0.139
368.1	0.128
391.8	0.121

*Not shown on plot

SPECIFICATION TABLE NO. 95 THERMAL CONDUCTIVITY OF [MAGNESIUM OXIDE + URANIUM DIOXIDE] $MgO + UO_2$

Curve No.	Ref. No.	Method Used	Year	Temp. Range, K	Reported Error, %	Name and Specimen Designation	Composition (weight percent) MgO	UO_2	Composition (continued), Specifications and Remarks
1	76	L	1952	383-793			53	47	87% theorectical density.

DATA TABLE NO. 95 THERMAL CONDUCTIVITY OF [MAGNESIUM OXIDE + URANIUM DIOXIDE] $MgO + UO_2$

[Temperature, T, K; Thermal Conductivity, k, Watt $cm^{-1}K^{-1}$]

T	k
CURVE 1 *	
383.2	0.150
573.2	0.115
793.2	0.065

* No graphical presentation

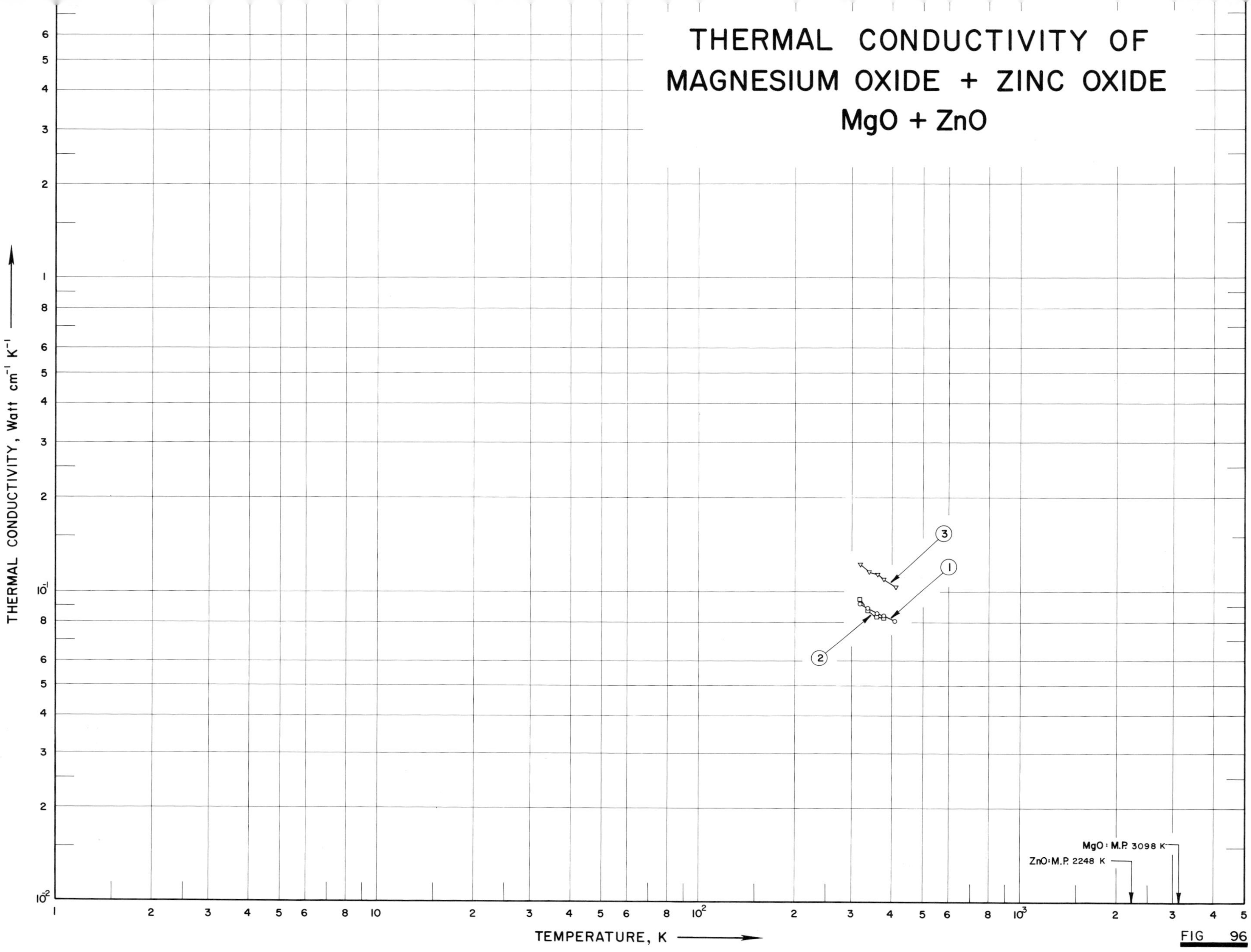
THERMAL CONDUCTIVITY OF
MAGNESIUM OXIDE + ZINC OXIDE
MgO + ZnO
3
1
2
MgO: M.P. 3098 K
ZnO: M.P. 2248 K
THERMAL CONDUCTIVITY, Watt cm⁻¹ K⁻¹
TEMPERATURE, K
FIG 96

SPECIFICATION TABLE NO. 96 THERMAL CONDUCTIVITY OF [MAGNESIUM OXIDE + ZINC OXIDE] $MgO + ZnO$

[For Data Reported in Figure and Table No. 96]

Curve No.	Ref. No.	Method Used	Year	Temp. Range, K	Reported Error, %	Name and Specimen Designation	Composition (weight percent) MgO	ZnO	Composition (continued), Specifications and Remarks
1	12		1953	319-410		264A-1	53.6	46.4	7 MgO + 3 ZnO by mole; prepared from pure oxides, milled in water, dried, calcined at 1367 K, then dry-pressed in 0.5 in. steel die at 15,000 psi; fired at 1700 K and soaked for 2 hrs; water absorption 0.003%; density 5.00 g cm^{-3}.
2	12		1953	318-409		265A-1	66.5	33.5	4 MgO + ZnO by mole; same preparation as that of the above specimen except fired at 1644 K; water absorption 0.015%; density 5.02 g cm^{-3}.
3	12		1953	320-413		266A-1	81.7	18.3	9 MgO + ZnO by mole; same preparation as that of the above specimen; water absorption 0.029%; density 5.22 g cm^{-3}.

DATA TABLE NO. 96 THERMAL CONDUCTIVITY OF [MAGNESIUM OXIDE + ZINC OXIDE] MgO + ZnO

[Temperature, T, K; Thermal Conductivity, k, Watt cm^{-1} K^{-1}]

T	k
CURVE 1	
319.4	0.0916
338.5	0.0887
361.0	0.0854
378.1	0.0841
410.2	0.0808
CURVE 2	
318.2	0.0946
338.1	0.0874
361.8	0.0833
379.7	0.0828
408.8	0.0803*
CURVE 3	
320.4	0.123
340.3	0.116
361.2	0.114
378.2	0.109
412.5	0.103

*Not shown on plot

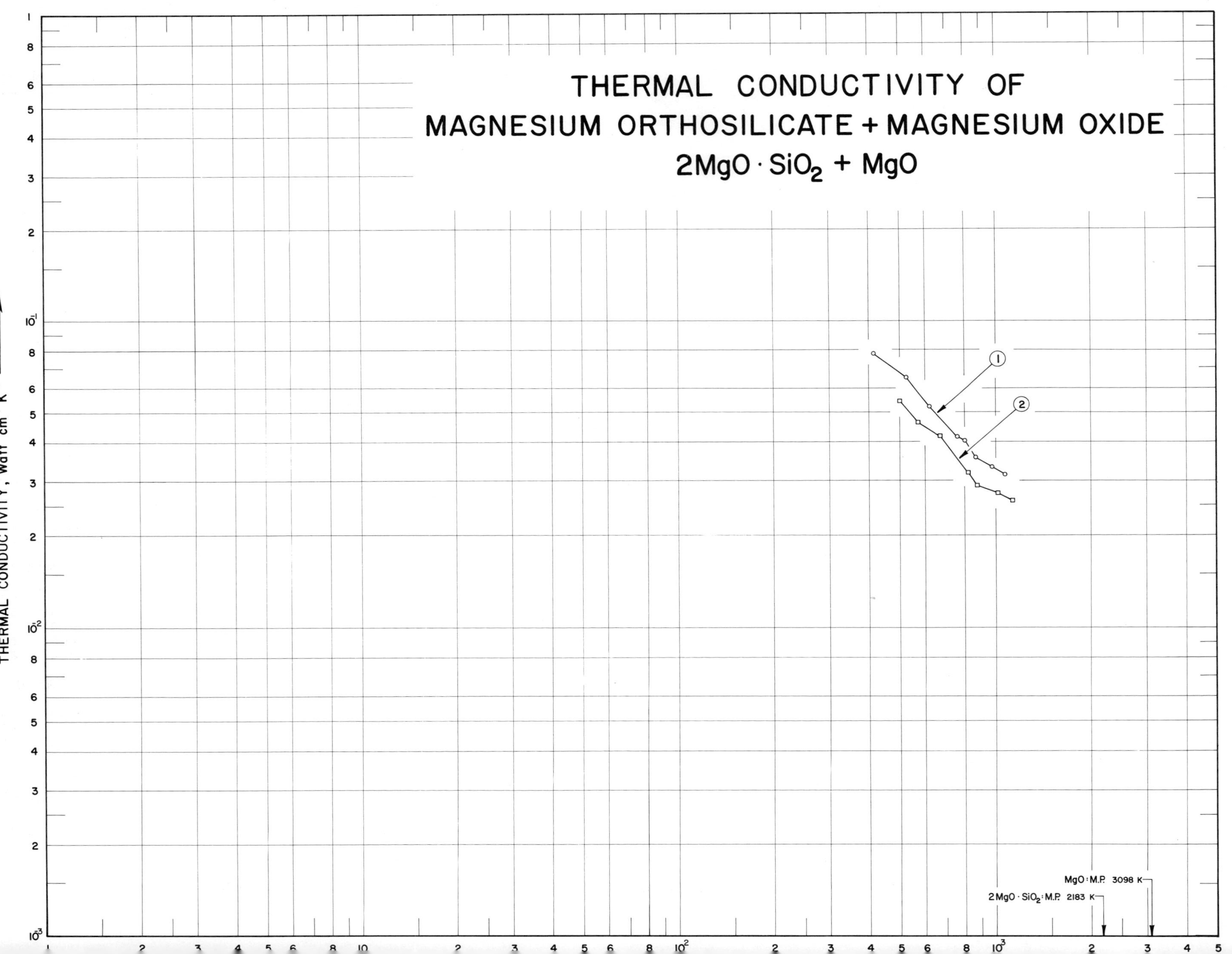

THERMAL CONDUCTIVITY OF
MAGNESIUM ORTHOSILICATE + MAGNESIUM OXIDE
$2MgO \cdot SiO_2 + MgO$
1
2
MgO: M.P. 3098 K
$2MgO \cdot SiO_2$: M.P. 2183 K
THERMAL CONDUCTIVITY, Watt cm^{-1} K^{-1}

SPECIFICATION TABLE NO. 97 THERMAL CONDUCTIVITY OF [MAGNESIUM ORTHOSILICATE + MAGNESIUM OXIDE] $2MgO \cdot SiO_2 + MgO$

[For Data Reported in Figure and Table No. 97]

Curve No.	Ref. No.	Method Used	Year	Temp. Range, K	Reported Error, %	Name and Specimen Designation	Composition (weight percent) $2MgO \cdot SiO_2$	Composition (weight percent) MgO	Composition (continued), Specifications and Remarks
1	136	C	1959	417-1078	±4		60.4	39.6	Corresponding to 38.0 vol % MgO; total porosity 14.3%; data corrected to theoretical density; calculated composition.
2	136	C	1959	502-1144	±4		83.4	16.6	Corresponding to 15.7 vol % MgO; total porosity 16.8%; data corrected to theoretical density; calculated composition.

DATA TABLE NO. 97 THERMAL CONDUCTIVITY OF [MAGNESIUM ORTHOSILICATE + MAGNESIUM OXIDE] $2MgO \cdot SiO_2 + MgO$

[Temperature, T, K; Thermal Conductivity, k, Watt cm^{-1} K^{-1}]

T	k
CURVE 1	
417.2	0.0776
529.2	0.0653
623.2	0.0521
762.2	0.0415
805.2	0.0402
870.2	0.0356
980.2	0.0333
1078.2	0.0314
CURVE 2	
502.2	0.0544
574.2	0.0464
673.0	0.0418
823.2	0.0318
879.2	0.0288
1021.2	0.0273
1144.2	0.0259

SPECIFICATION TABLE NO. 98 THERMAL CONDUCTIVITY OF [DIMANGANESE TRIOXIDE + ALUMINUM OXIDE] $Mn_2O_3 + Al_2O_3$

Curve No.	Ref. No.	Method Used	Year	Temp. Range, K	Reported Error, %	Name and Specimen Designation	Composition (weight percent) Mn_2O_3	Al_2O_3	Composition (continued), Specifications and Remarks
1	23		1952	315-403		156 A	69.89	30.11	$3Mn_2O_3$: $2Al_2O_3$; firing temperature 1811 K; water absorption 0.006%; density 4.13 g cm^{-3}.

DATA TABLE NO. 98 THERMAL CONDUCTIVITY OF [DIMANGANESE TRIOXIDE + ALUMINUM OXIDE] $Mn_2O_3 + Al_2O_3$

[Temperature, T, K; Thermal Conductivity, k, Watt $cm^{-1} K^{-1}$]

T	k
CURVE 1*	
314.6	0.0195
330.7	0.0188
351.9	0.0199
369.9	0.0205
402.7	0.0213

* No graphical presentation

SPECIFICATION TABLE NO. 99 THERMAL CONDUCTIVITY OF [DIMANGANESE TRIOXIDE + MAGNESIUM OXIDE] $Mn_2O_3 + MgO$

Curve No.	Ref. No.	Method Used	Year	Temp. Range, K	Reported Error, %	Name and Specimen Designation	Composition (weight percent) Mn_2O_3	MgO	Composition (continued), Specifications and Remarks
1	23		1952	319-408		177 A	94	6	$4Mn_2O_3$: MgO; fired at 1644 K; density 4.11 g cm^{-3}; water absorption 0.45%.

DATA TABLE NO. 99 THERMAL CONDUCTIVITY OF [DIMANGANESE TRIOXIDE + MAGNESIUM OXIDE] $Mn_2O_3 + MgO$

[Temperature, T, K; Thermal Conductivity, k, Watt $cm^{-1} K^{-1}$]

T	k
CURVE 1*	
319.2	0.0372
334.9	0.0376
355.6	0.0380
375.2	0.0370
408.2	0.0367

* No graphical presentation

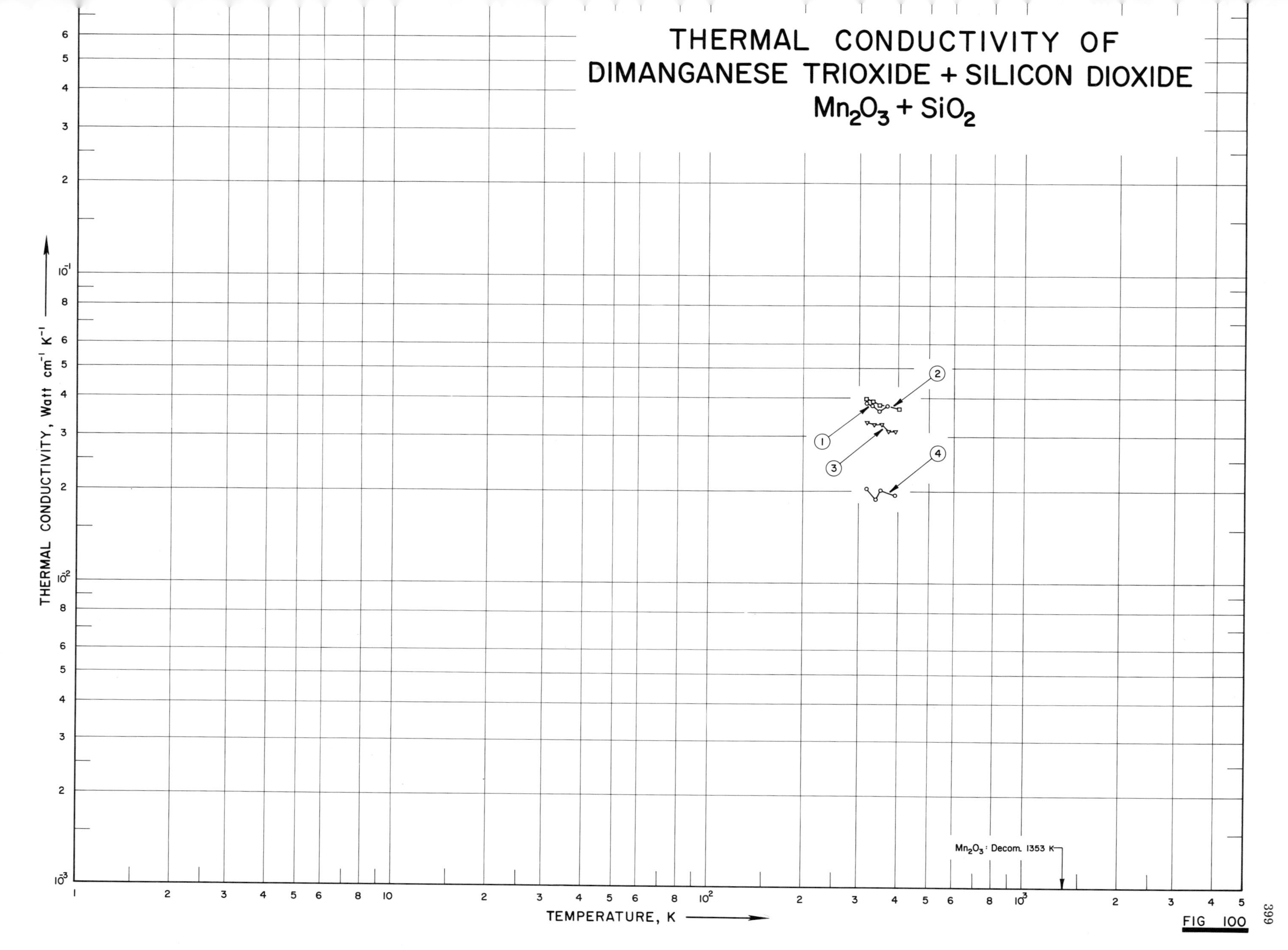

THERMAL CONDUCTIVITY OF
DIMANGANESE TRIOXIDE + SILICON DIOXIDE
$Mn_2O_3 + SiO_2$
THERMAL CONDUCTIVITY, Watt cm^{-1} K^{-1}
TEMPERATURE, K
Mn_2O_3 : Decom. 1353 K
1
2
3
4

FIG 100

SPECIFICATION TABLE NO. 100 THERMAL CONDUCTIVITY OF [DIMANGANESE TRIOXIDE + SILICON DIOXIDE] $Mn_2O_3 + SiO_2$

[For Data Reported in Figure and Table No. 100]

Curve No.	Ref. No.	Method Used	Year	Temp. Range, K	Reported Error, %	Name and Specimen Designation	Composition (weight percent) Mn_2O_3	SiO_2	Composition (continued), Specifications and Remarks
1	23		1952	319-371		172A	91.18	8.82	Composition $4Mn_2O_3 : SiO_2$; firing temperature 1561 K; water absorption 0.01%; density 3.82 g cm^{-3}.
2	23		1952	318-402		172B	91.18	8.82	Composition $4Mn_2O_3 : SiO_2$; firing temperature 1533 K; water absorption 0.012%; density 3.88 g cm^{-3}.
3	23		1952	319-394		173A	79.49	20.51	Composition $3Mn_2O_3 : 2Si_2$; firing temperature 1533 K; water absorption none; density 3.71 g cm^{-3}.
4	23		1952	317-394		174A	63.27	36.73	Composition $2Mn_2O_3 : 3SiO_2$; firing temperature 1478 K; water absorption 0.053%; density 3.53 g cm^{-3}.

DATA TABLE NO. 100 THERMAL CONDUCTIVITY OF [DIMANGANESE TRIOXIDE + SILICON DIOXIDE] $Mn_2O_3 + SiO_2$

[Temperature, T, K; Thermal Conductivity, k, Watt cm^{-1} K^{-1}]

T	k
CURVE 1	
318. 7	0. 0385
332. 8	0. 0378
350. 4	0. 0364
370. 8	0. 0377
CURVE 2	
317. 6	0. 0396
334. 8	0. 0390
351. 6	J. 0380
372. 5	0. 0377*
401. 5	0. 0369
CURVE 3	
318. 8	0. 0333
335. 1	0. 0329
354. 7	0. 0329
373. 5	0. 0312
393. 8	0. 0312
CURVE 4	
317. 3	0. 0205
339. 3	0. 0188
353. 5	0. 0202
393. 5	0. 0195

*Not shown on plot

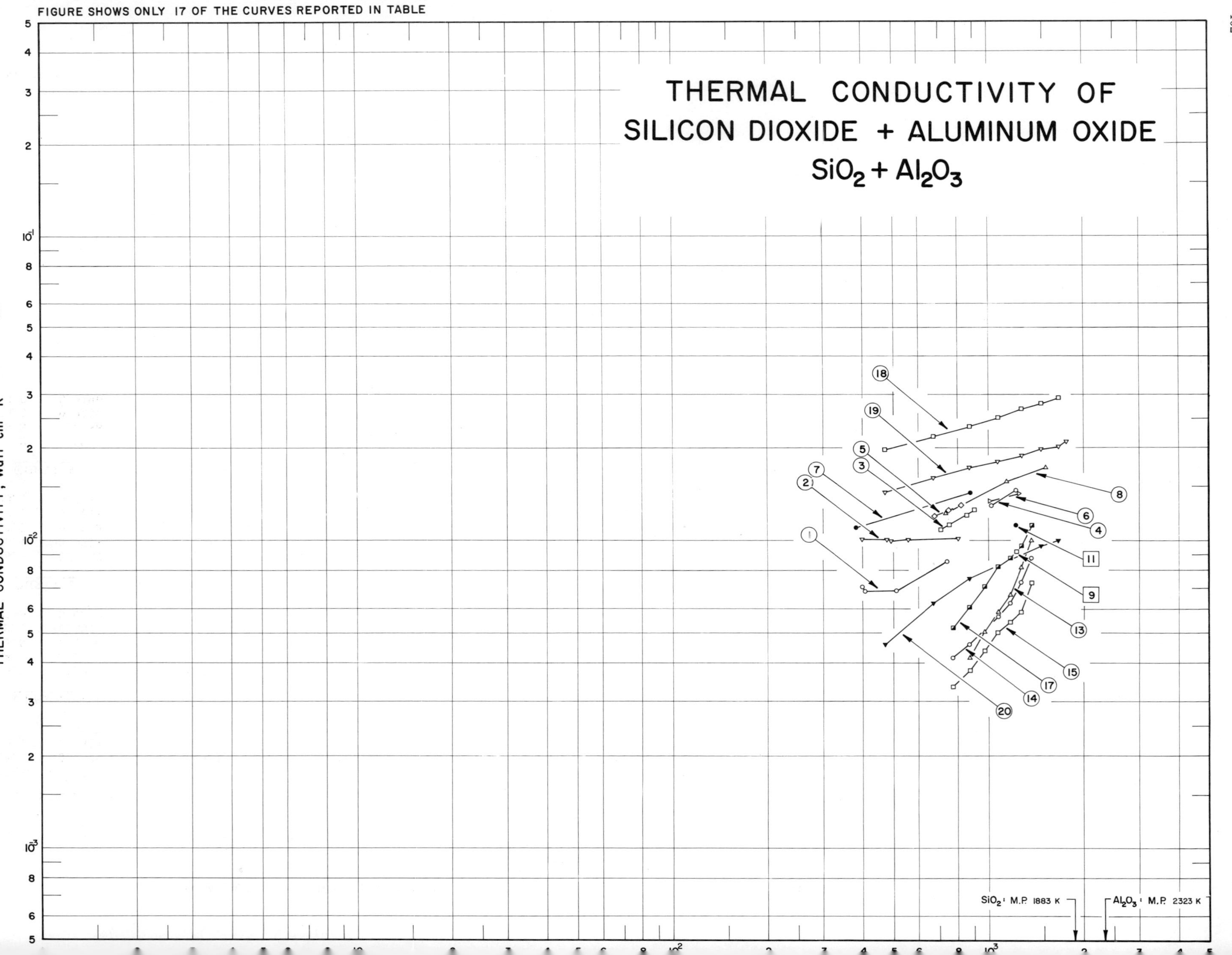

FIGURE SHOWS ONLY 17 OF THE CURVES REPORTED IN TABLE
THERMAL CONDUCTIVITY OF
SILICON DIOXIDE + ALUMINUM OXIDE
$SiO_2 + Al_2O_3$
THERMAL CONDUCTIVITY, Watt cm^{-1} K^{-1}
SiO_2: M.P. 1883 K
Al_2O_3: M.P. 2323 K
18
19
5
3
7
2
1
8
6
4
11
9
13
15
17
14
20

SPECIFICATION TABLE NO. 101 THERMAL CONDUCTIVITY OF [SILICON DIOXIDE + ALUMINUM OXIDE] $SiO_2 + Al_2O_3$

[For Data Reported in Figure and Table No. 101]

Curve No.	Ref. No.	Method Used	Year	Temp. Range, K	Reported Error, %	Name and Specimen Designation	Composition (weight percent) SiO_2	Al_2O_3	Composition (continued), Specifications and Remarks
1	247	L	1932	399-740	1	Fireclay brick 17	53. 56	42. 23	1. 59 Fe_2O_3, 1. 01 CaO, 0. 83 MgO, and 0. 62 TiO_2; density 2. 710 g cm^{-3}, porosity 31. 5%; gas permeability 3. 00 m^3 cm per m^2-hr-mm H_2O.
2	247	L	1932	398-804	1	Fireclay brick 725	<60	>40	Porosity 21. 0%.
3	79, 475	L	1942	709-902		Sillimanite refractory brick; 6	88. 29	9. 24	0. 89 TiO_2, 0. 63 Fe_2O_3, 0. 38 CaO, 0. 10 MgO, and 0. 29 alkali oxides; specimen in the form of a slab measuring 18 in. x 18 in. ; made from natural quartzitic silica sands bonded with clays; fired at 1360 C for 14 hrs; after being fired specimen is of fine texture, containing finely divided cristobalite and considerable residual free quartz in a clayey matrix; porosity 21. 36%; weight lost on ignition 0. 14%.
4	79, 345	C	1942	1023, 1225		Sillimanite refractory brick; 6			Similar to the above specimen but in disc form 8 in. in dia and 1 in. in thickness; steel used as comparative material.
5	79, 345	L	1942	679-822		Sillimanite refractory brick; 7	89. 11	9. 04	0. 75 TiO_2, 0. 53 Fe_2O_3, 0. 13 MgO, 0. 10 CaO, and 0. 17 alkali oxides; specimen in the form of a slab measuring 18 in. x 18 in. ; similar raw material to the above specimen No. 6; fired at 1360 C for 14 hrs; mineralogical constitution of the fired product similar to the above specimen No. 6; porosity 24. 10%; weight lost on ignition 0. 12%.
6	79, 345	C	1942	1006, 1252		Sillimanite refractory			Similar to the above specimen but in disc form 8 in. in dia and 1 in. in thickness; steel used as comparative material.
7	80	L	1934	382, 877		Dense fireclay brick (Mexko-brand)	52. 5	42. 07	2. 0 TiO_2, 1. 6 Fe_2O_3, 0. 5 CaO, trace MgO, and 0. 6 alkali; approximate composition; bulk density 2. 29 g cm^{-3}; porosity 15. 2%.
8	80	L	1934	733-1527		Dense fireclay brick (Mexko-brand)			The above specimen measured with insulating brick placed between the calorimeter and the lower surfaces of the brick.
9	81	L	1924	1223. 2		Pressed fireclay; 2	67. 11	28. 20	1. 61 Fe_2O_3, 1. 60 MgO, 1. 30 TiO_2, 0. 8 CaO, 0. 48 (K_2O + Na_2O), and 0. 27 loss; (consists of 12. 5% sandy fire clay and 37. 5% of plastic fire clay with 50% by volume of 3-mesh to 16-mesh fire clay grog); specimen 8. 5 in. in dia and 2. 04 in. in thickness; porosity 24. 9% calculated from the dry saturated and suspended weight; measured at 740 mm Hg pressure.
10	81	L	1924	1223. 2		Pressed fireclay; 3			Similar to the above specimen except measured at 745. 8 mm Hg pressure.
11	81	L	1924	1223. 2		Light weight fireclay	66. 77	28. 94	2. 0 TiO_2, 1. 52 Fe_2O_3, 0. 95 (K_2O + Na_2O), trace CaO, trace MgO, and 0. 18 loss; (consists of 40% plastic fire clay, 40% lignitic clay, 10% 3-mesh to 16-mesh grog, and 10% 16 F grog) specimen 8. 5 in. in dia and 1. 97 in. thick; porosity 42. 5% calculated from the dry saturated and suspended weight; bulk specific gravity 1. 7; measured at 744. 0 mm Hg pressure.

SPECIFICATION TABLE NO. 101 (continued)

Curve No.	Ref. No.	Method Used	Year	Temp. Range, K	Reported Error, %	Name and Specimen Designation	Composition (weight percent) SiO_2	Al_2O_3	Composition (continued), Specifications and Remarks
12	81	L	1924	1223.2		Light weight fireclay; 2			Similar to the above specimen except porosity 41.2% and measured in 745.8 mm Hg pressure.
13	238	P	1921	873-1373		Firebrick E	57.9	32.96	Very close structure; specimen 9 x 4.5 x 2.5 in.; abundance of fine grained rounded grog and a little larger grained grog; exceptionally good adherence; marked by a fair number of black cores, generally with cavities; faces smooth and edges sharp; porosity 15.9%; heat flow in the length-wise direction.
14	238	P	1921	773-1373		Firebrick F	67.49	27.15	Very open texture; specimen 9 x 4.5 x 2.5 in.; abundance of rounded clay grog of uneven grading, some grains approx to pebbles; unweathered pellets detected; adherence poor - in fact, material is very friable; highly fissured; porosity 24.6%; heat flow in the direction of the length of the specimen.
15	238	P	1921	773-1373		Retort material G	67.1	27.17	Very open texture; specimen 9 x 4.5 x 2.5 in.; very heavily grogged with medium to fine rounded material of uneven grading; abundance of small fissures; adherence of grog very poor; matrix appears to have contracted away from the grog; porosity 24%; heat flow in the direction of the length of the specimen.
16	238	P	1921	773-1373		Retort material H	65.7	28.47	Somewhat closer in texture than G; specimen 9 x 4.5 x 2.5 in.; the above specimen heavily grogged with rounded material of slightly more even grading than G; adherence as a whole fairly good, although some are easily detached; some fissures; very white color with well-defined skin; porosity 28.2%; heat flow in the direction of the length of the specimen.
17	238	P	1921	773-1373		Retort material I	72.46	23.65	Very close in texture; specimen 9 x 4.5 x 2.5 in.; abundant grog which, tending to be rounded, evenly graded, possibly some quartz fragments; black cores present, but scarce; tendency toward layering; fissures, present but scarce, are parallel to outside faces; superficial skin; signs of possible reduction toward end of fire; porosity 24.7%; heat flow in the direction of the length of the specimen.
18	383	L	1927	473-1673	±15	Kaolin firebrick	52.02	45.92	1.51 Fe_2O_3, 0.35 TiO_2, trace of alkalis; specimen 10.8 cm in dia and 22.8 cm long; apparent density 2.66 g cm^{-3}, bulk density 2.36 g cm^{-3}; porosity 10.8%; made of sedimentary kaolin by mixing 65% of 20-mesh prefired grog and 35% of raw clay, and firing to 1575 C for 4 hrs.

SPECIFICATION TABLE NO. 101 (continued)

Curve No.	Ref. No.	Method Used	Year	Temp. Range, K	Reported Error, %	Name and Specimen Designation	Composition (weight percent) SiO_2	Al_2O_3	Composition (continued), Specifications and Remarks
19	383	L	1927	473-1773	±15	Kaolin firebrick	52.02	45.92	1.51 Fe_2O_3, 0.35 TiO_2, trace of alkalis; specimen 10.8 cm in dia and 22.8 cm long; apparent density 2.68 g cm^{-3}, bulk density 2.10 g cm^{-3}; porosity 23.2%; made of sedimentary kaolin by mixing 65% of 4-mesh prefired grog and 35% of raw clay, and firing to 1575 C for 4 hrs.
20	383	L	1927	473-1673	±15	Kaolin firebrick	52.02	45.92	1.51 Fe_2O_3, 0.35 TiO_2, trace of alkalis; specimen 10.8 cm in dia and 22.8 cm long; apparent density 2.50 g cm^{-3}, bulk density 1.27 g cm^{-3}; porosity 49.1%.
21	268	L	1946	573-1573			69.7	27.5	Trace of iron oxide and CaO; density 1.87 g cm^{-3}; porosity 28.5%.
22	458	P	1921	373.2		Red brick	76.32	21.96	1.88 Fe_2O_3, traces of CaO and MgO; commercial brick 4 cm in dia and 8 cm long; density 1.795 g cm^{-3}; thermal conductivity value calculated from measured data of thermal diffusivity, specific heat, and density.
23	458	P	1921	373.2		White Shamotte brick	79.98	19.48	0.40 Fe_2O_3, traces of CaO and MgO; specimen 4 cm in dia and 8 cm long; supplied by Imperial Steel Works; density 1.565 g cm^{-3}; same measuring method as above.
24	458	P	1921	373.2		Red Shamotte brick	71.74	25.56	1.02 Fe_2O_3, 0.82 CaO, and 0.53 MgO; specimen 4 cm in dia and 8 cm long; supplied by Imperical Steel Works; density 1.784 g cm^{-3}; same measuring method as above.

DATA TABLE NO. 101 THERMAL CONDUCTIVITY OF [SILICON DIOXIDE + ALUMINUM OXIDE] $SiO_2 + Al_2O_3$

[Temperature, T, K; Thermal Conductivity, k, Watt $cm^{-1} K^{-1}$]

T	k
CURVE 1	
398.9	0.00709
408.6	0.00687
511.6	0.00688
740.0	0.00853
CURVE 2	
398.2	0.0101
477.9	0.0101
492.0	0.00996
559.6	0.0101
804.0	0.0102
CURVE 3	
709.2	0.0109
754.2	0.0113
853.2	0.0121
902.2	0.0126
CURVE 4	
1023.2	0.0130
1225.2	0.0146
CURVE 5	
679.2	0.0121
748.2	0.0126
822.2	0.0130
CURVE 6	
1006.2	0.0134
1252.2	0.0142
CURVE 7	
381.5	0.0111
876.5	0.0142
CURVE 8	
732.6	0.0123
1145.4	0.0156
1526.5	0.0173
CURVE 9	
1223.2	0.00920
CURVE 10*	
1223.2	0.00782
CURVE 11	
1223.2	0.0113
CURVE 12*	
1223.2	0.00975
CURVE 13	
873.2	0.00418
973.2	0.00502
1073.2	0.00586
1173.2	0.00669
1273.2	0.00816
1373.2	0.0100
CURVE 14	
773.2	0.00418
873.2	0.00460
973.2	0.00502*
1073.2	0.00565
1173.2	0.00628
1273.2	0.00732
1373.2	0.00879
CURVE 15	
773.2	0.00335
873.2	0.00377
973.2	0.00439
1073.2	0.00502
1173.2	0.00544
1273.2	0.00586
1373.2	0.00732
CURVE 16*	
773.2	0.00397
873.2	0.00460
973.2	0.00502
1073.2	0.00586
1173.2	0.00669
1273.2	0.00732
1373.2	0.00879
CURVE 17	
773.2	0.00523
873.2	0.00607
973.2	0.00711
1073.2	0.00816
1173.2	0.00879
1273.2	0.00962
1373.2	0.0113
CURVE 18	
473.2	0.0197
673.2	0.0218
873.2	0.0234
1073.2	0.0251
1273.2	0.0268
1473.2	0.0280
1673.2	0.0293
CURVE 19	
473.2	0.0142
673.2	0.0159
873.2	0.0172
1073.2	0.0180
1273.2	0.0188
1473.2	0.0197
1673.2	0.0201
1773.2	0.0209
CURVE 20*	
473.2	0.00460
673.2	0.00628
873.2	0.00753
1073.2	0.00837*
1273.2	0.00920*
CURVE 20 (cont.)	
1473.2	0.00962
1673.2	0.0100
CURVE 21*	
573.2	0.0108
673.2	0.0115
773.2	0.0122
873.2	0.0129
973.2	0.0135
1073.2	0.0140
1173.2	0.0145
1273.2	0.0150
1373.2	0.0154
1473.2	0.0158
1573.2	0.0161
CURVE 22*	
373.2	0.00674
CURVE 23*	
373.2	0.00523
CURVE 24*	
373.2	0.00661

*Not shown on plot

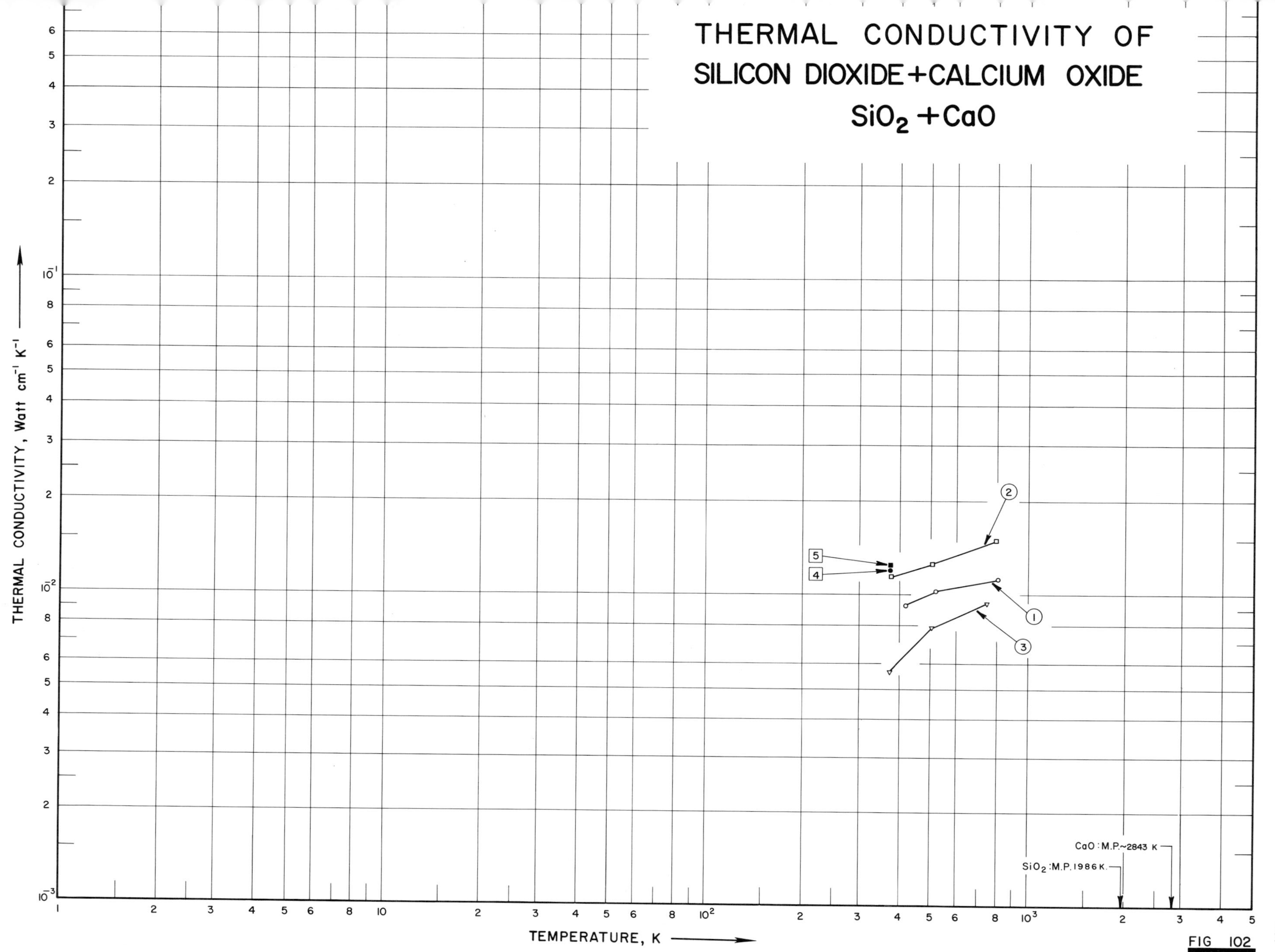
THERMAL CONDUCTIVITY OF
SILICON DIOXIDE + CALCIUM OXIDE
SiO2 + CaO
THERMAL CONDUCTIVITY, Watt cm⁻¹ K⁻¹
TEMPERATURE, K
CaO : M.P. ~2843 K
SiO2 : M.P. 1986 K
1
2
3
4
5
FIG 102

SPECIFICATION TABLE NO. 102 THERMAL CONDUCTIVITY OF [SILICON DIOXIDE + CALCIUM OXIDE] $SiO_2 + CaO$

[For Data Reported in Figure and Table No. 102]

Curve No.	Ref. No.	Method Used	Year	Temp. Range, K	Reported Error, %	Name and Specimen Designation	Composition (weight percent) SiO_2	CaO	Composition (continued), Specifications and Remarks
1	247	L	1932	423-808	1.0	Silica brick 1	95.12	2.37	0.55 Al_2O_3, 0.69 Fe_2O_3, 0.19 MgO, 0.72 TiO_2; density 2.342 g cm^{-3}; porosity 19.0%; gas permeability 0.184 m^3-cm per m^2-hr-mm H_2O.
2	247	L	1932	377-798	1.0	Silica brick 8	94.02	2.98	0.91 Al_2O_3, 0.79 Fe_2O_3, 0.22 MgO, 0.50 TiO_2; density 2.327 g cm^{-3}; porosity 23.1%; gas permeability 0.582 m^3-cm per m^2-hr-mm H_2O.
3	247	L	1932	372-744	1.0	Silica brick 9	92.26	3.26	1.84 Al_2O_3, 0.55 Fe_2O_3, 0.23 MgO, 0.41 TiO_2; density 2.350 g cm^{-3}; porosity 27.6%; gas permeability 1.750 m^3-cm per m^2-hr-mm H_2O.
4	458	R	1921	373.2		Silica brick	94.98	3.06	1.18 Al_2O_3, and 1.13 Fe_2O_3; specimen 10 cm in dia and 4.5 cm thick; density 1.840, 1.836, 1.829, 1.814, and 1.804 g cm^{-3} at 20, 50, 100, 200, and 300 C, respectively.
5	458	P	1921	373.2		Silica brick			Similar to the above specimen; thermal conductivity values calculated from measured data of thermal diffusivity, specific heat, and density.

DATA TABLE NO. 102 THERMAL CONDUCTIVITY OF [SILICON DIOXIDE + CALCIUM OXIDE] $SiO_2 + CaO$

[Temperature, T, K; Thermal Conductivity, k, Watt $cm^{-1} K^{-1}$]

T	k
CURVE 1	
422.9	0.00927
520.9	0.0103
807.7	0.0112
CURVE 2	
377.1	0.0115
504.3	0.0126
798.0	0.0149
CURVE 3	
372.4	0.00564
501.6	0.00787
743.5	0.00939
CURVE 4	
373.2	0.0121
CURVE 5	
373.2	0.0124

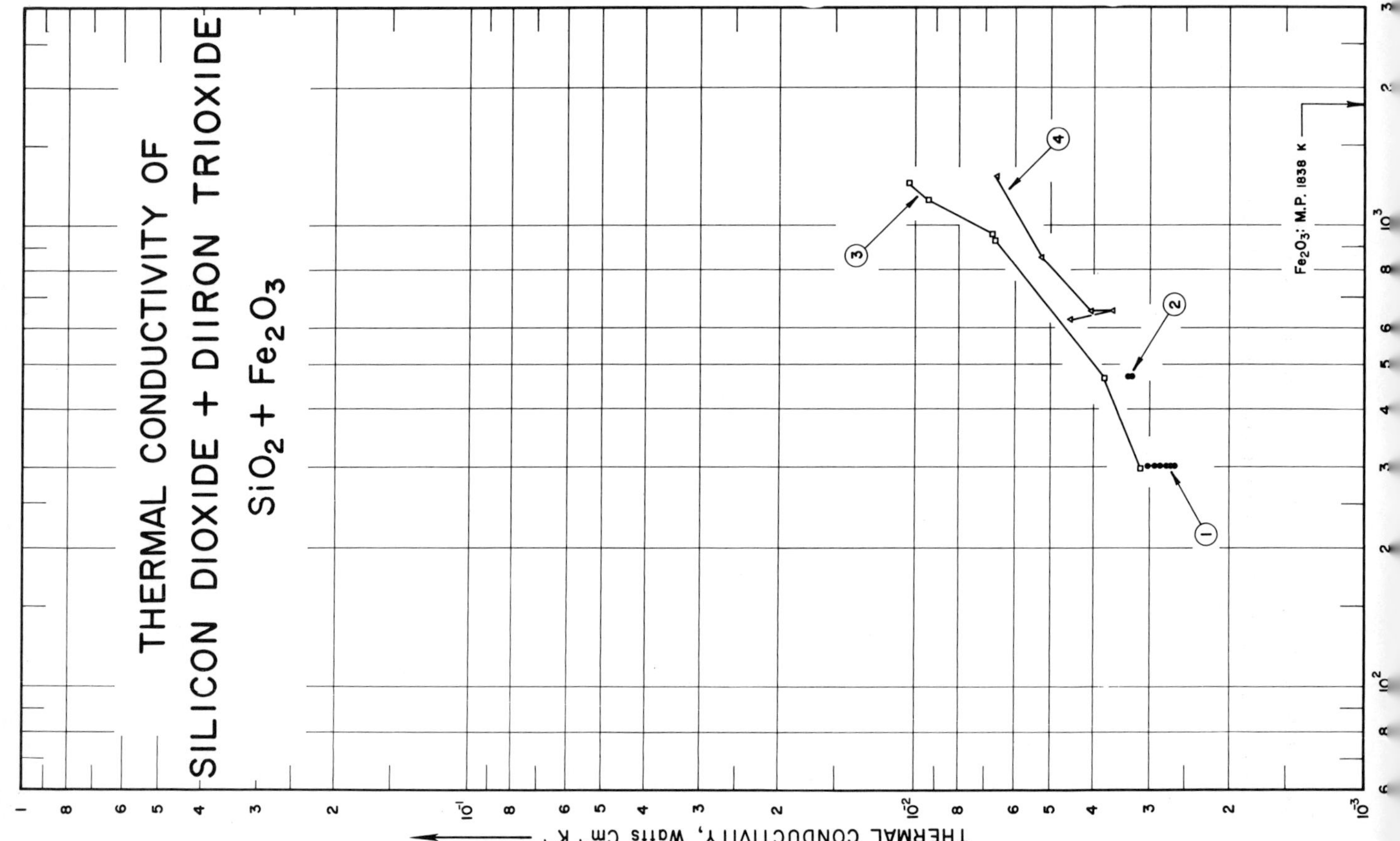

THERMAL CONDUCTIVITY OF
SILICON DIOXIDE + DIIRON TRIOXIDE
$SiO_2 + Fe_2O_3$
THERMAL CONDUCTIVITY, Watts Cm⁻¹ K⁻¹
Fe_2O_3; M.P. 1838 K
1
2
3
4

SPECIFICATION TABLE NO. 103 THERMAL CONDUCTIVITY OF [SILICON DIOXIDE + DIIRON TRIOXIDE] $SiO_2 + Fe_2O_3$

[For Data Reported in Figure and Table No. 103]

Curve No.	Ref. No.	Method Used	Year	Temp. Range, K	Reported Error, %	Name and Specimen Designation	Composition (weight percent) SiO_2	Fe_2O_3	Composition (continued), Specifications and Remarks
1	249	R	1953	303.2		Special No. 4 Silica Sand	96.2	2.15	0.74 CaO, 0.25 Al_2O_3, and trace MgO; trace of ignition loss; grain size 100-120 mesh; measured at constant temperature with increasing apparent specific gravity.
2	249	R	1953	473.2		Special No. 4 Silica Sand	96.2	2.15	0.74 CaO, 0.25 Al_2O_3, and trace MgO; trace of ignition loss; grain size 100-120 mesh; measured at constant temperature with increasing apparent specific gravity.
3	249	R	1953	298-1248		Special No. 4 Silica Sand	96.2	2.15	0.74 CaO, 0.25 Al_2O_3, and trace of ignition loss; grain size 14-20 mesh; apparent specific gravity 1.29.
4	249	R	1953	628-1293		Special No. 4 Silica Sand	96.2	2.15	0.74 CaO, 0.25 Al_2O_3, and trace MgO; trace of ignition loss; grain size 140-200 mesh; apparent specific gravity 1.31.

DATA TABLE NO. 103 THERMAL CONDUCTIVITY OF [SILICON DIOXIDE + DIIRON TRIOXIDE] $SiO_2 + Fe_2O_3$

[Temperature, T, K; Thermal Conductivity, k, Watt cm^{-1} K^{-1}]

Apparent specific gravity	k
CURVE 1 (T = 303.2 K)	
1.220	0.00264
1.228	0.00267
1.228	0.00273
1.296	0.00284
1.296	0.00292
1.296	0.00302
CURVE 2 (T = 473.2 K)	
1.216	0.00329
1.228	0.00335
1.292	0.00336*

T	k
CURVE 3	
298.2	0.00314
468.2	0.00378
928.2	0.00668
963.2	0.00674
1138.2	0.00936
1248.2	0.0104
CURVE 4	
628.2	0.00453
653.2	0.00360
658.2	0.00407
853.2	0.00523
1293.2	0.00662

*Not shown on plot

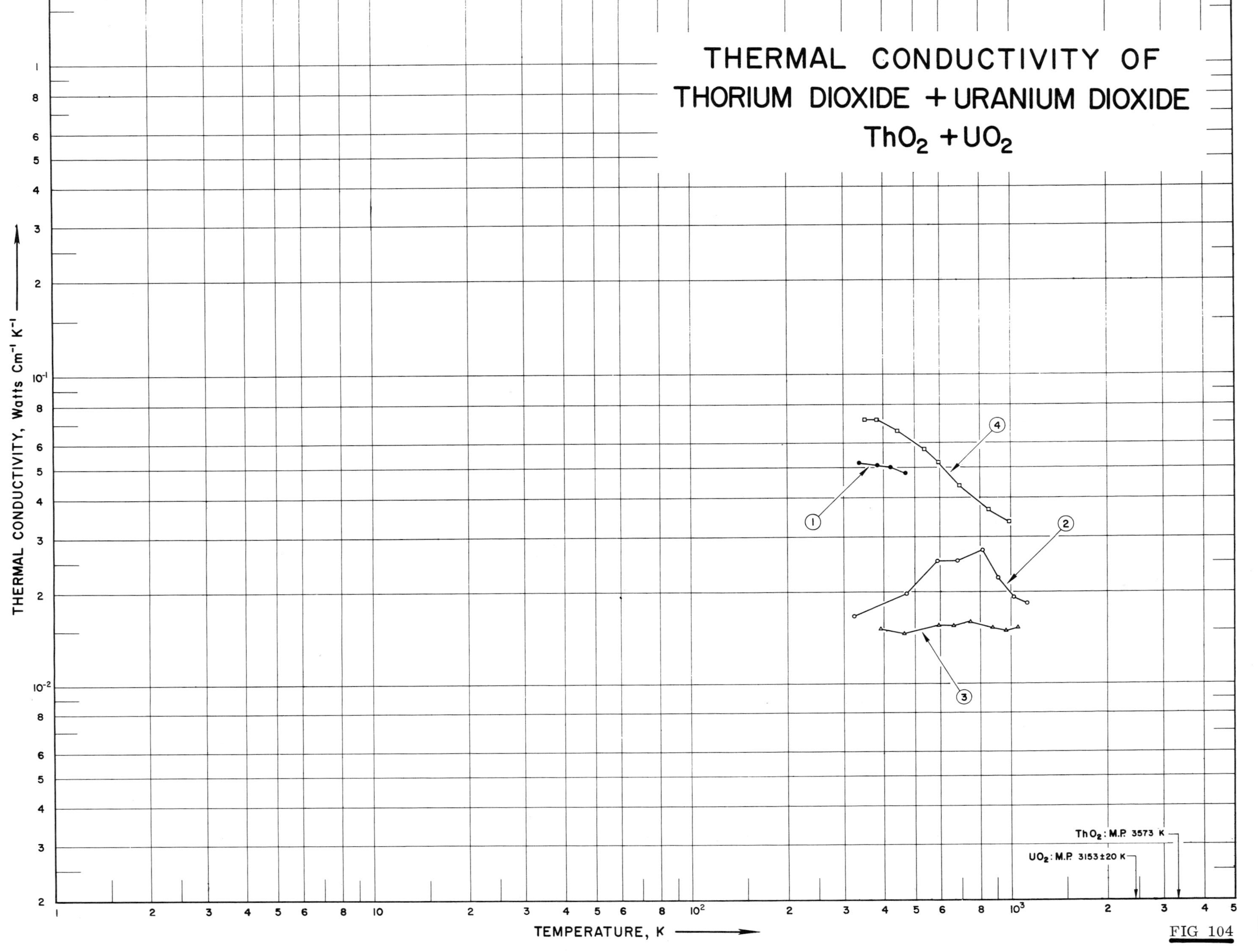

THERMAL CONDUCTIVITY OF
THORIUM DIOXIDE + URANIUM DIOXIDE
$ThO_2 + UO_2$
ThO_2: M.P. 3573 K
UO_2: M.P. 3153±20 K
THERMAL CONDUCTIVITY, Watts Cm^{-1} K^{-1}
TEMPERATURE, K

FIG 104

SPECIFICATION TABLE NO. 104 THERMAL CONDUCTIVITY OF [THORIUM DIOXIDE + URANIUM DIOXIDE] $ThO_2 + UO_2$

[For Data Reported in Figure and Table No. 104]

Curve No.	Ref. No.	Method Used	Year	Temp. Range, K	Reported Error, %	Name and Specimen Designation	Composition (weight percent) ThO_2	UO_2	Composition (continued), Specifications and Remarks
1	136	C	1959	336-471	±4%	$Th_{0.736}U_{0.264}O_2$	73.2	26.8	Corresponding to 26.4 mole % UO_2, 73.6 mole % ThO_2; solid solution.
2	136	C	1959	324-1123	±4%	$Th_{0.736}U_{0.264}O_{2+x}$	73.2	26.8	Corresponding to 26.4 mole % UO_2, 73.6 mole % ThO_2, heated in air above 500 C before testing; having an undetermined oxygen content $0 < x \leq 0.25$; solid solution; density 9.48 g cm^{-3}; porosity 5.0%; data corrected according to theoretical density.
3	136	C	1959	393-1053	±4%	$Th_{0.69}U_{0.31}O_{2+x}$	68.5	31.5	Similar to the above specimen except corresponding to 31 mole % UO_2, 69 mole % ThO_2 and density 8.16 g cm^{-3}, porosity 18.0%.
4	136	C	1959	351-991	±4%	$Th_{0.9}U_{0.1}O_{2+x}$	89.8	10.2	Similar to the above specimen except corresponding to 10 mole % UO_2, 90 mole % ThO_2 and density 8.89 g cm^{-3}, porosity 9.4%.

DATA TABLE NO. 104 THERMAL CONDUCTIVITY OF [THORIUM DIOXIDE + URANIUM DIOXIDE] $ThO_2 + UO_2$

[Temperature, T, K; Thermal Conductivity, k, Watt $cm^{-1} K^{-1}$]

T	k
CURVE 1	
336.2	0.0519
384.2	0.0510
423.2	0.0502
471.2	0.0481
CURVE 2	
324.2	0.0167
472.2	0.0198
591.2	0.0251
682.2	0.0251
818.2	0.0272
916.2	0.0222
1022.2	0.0192
1123.2	0.0184
CURVE 3	
393.2	0.0151
464.2	0.0146
593.2	0.0155
661.2	0.0156
749.2	0.0159
879.2	0.0152
968.2	0.0149
1053.2	0.0153
CURVE 4	
351.2	0.0715
383.2	0.0715
448.2	0.0657
540.2	0.0573
598.2	0.0523
691.2	0.0439
851.2	0.0366
991.2	0.0335

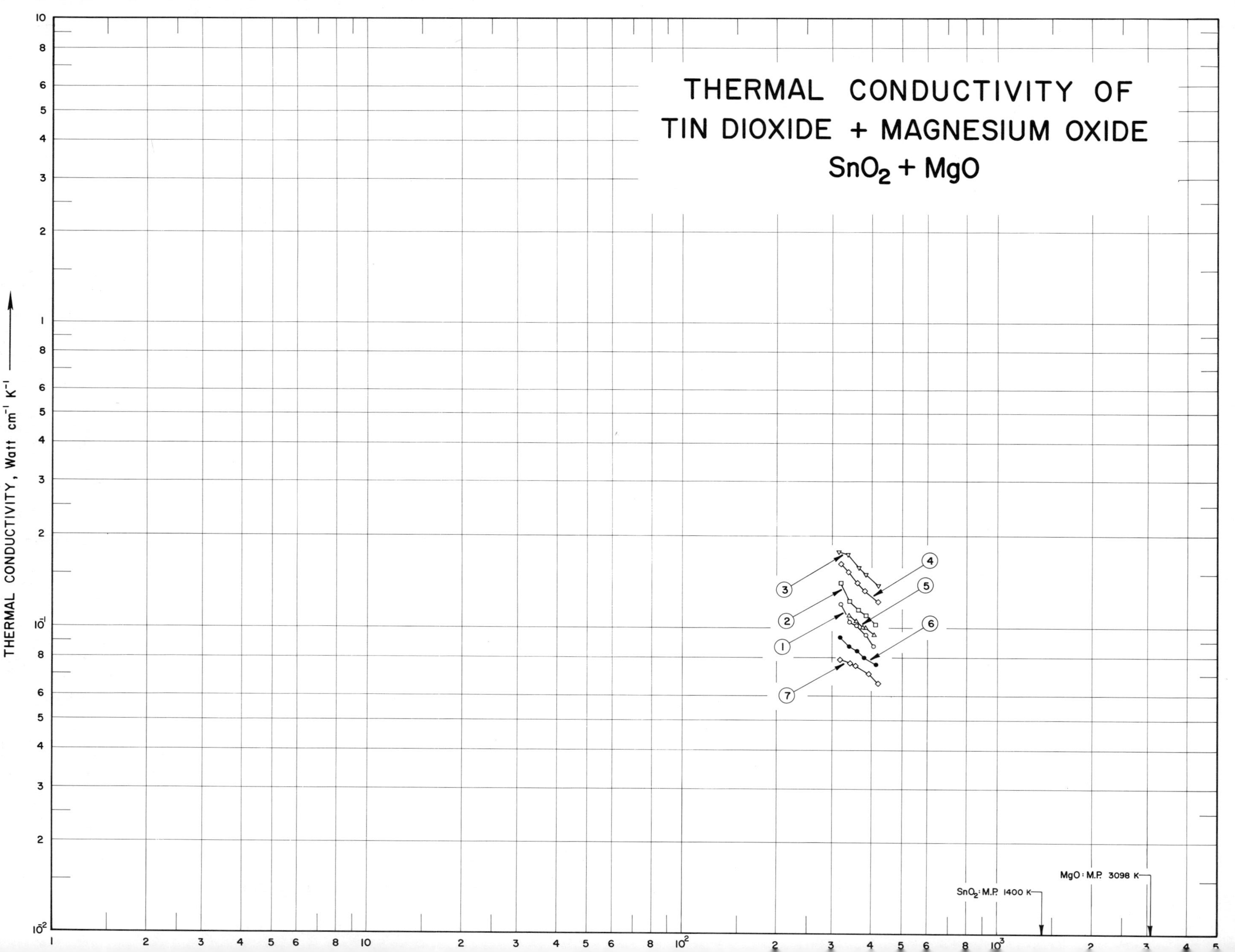
THERMAL CONDUCTIVITY OF
TIN DIOXIDE + MAGNESIUM OXIDE
SnO2 + MgO
THERMAL CONDUCTIVITY, Watt cm-1 K-1
SnO2: M.P. 1400 K
MgO: M.P. 3098 K

SPECIFICATION TABLE NO. 105 THERMAL CONDUCTIVITY OF [TIN DIOXIDE + MAGNESIUM OXIDE] SnO_2 + MgO

[For Data Reported in Figure and Table No. 105]

Curve No.	Ref. No.	Method Used	Year	Temp. Range, K	Reported Error, %	Name and Specimen Designation	Composition (weight percent) SnO_2	MgO	Composition (continued), Specifications and Remarks
1	12		1953	321-408		218A-1	97.1	2.9	1 MgO + 9 SnO_2 by mole; prepared by milling pure oxides in water, calcining at 1367 K after drying and then dry-pressing in a 0.5 in. steel die at 15,000 psi; fired at 1755 K and soaked for 1 hr; water absorption 6.59%; density 4.42 g cm^{-3}.
2	12		1953	321-414		219A-1	93.7	6.3	1 MgO + 4SnO_2 by mole; same specimen preparation as the above; fired at 1755 K and soaked for 1 hr; water absorption 6.86%; density 4.49 g cm^{-3}.
3	12		1953	316-421		220A-1	89.7	10.3	3 MgO + 7 SnO_2 by mole; same specimen preparation as the above; fired at 1783 K and soaked for 1.5 hrs; water absorption 0.70%; density 5.54 g cm^{-3}.
4	12		1953	320-421		221A-1	84.9	15.1	2 MgO + 3 SnO_2 by mole; same specimen preparation as the above; fired at 1783 K and soaked for 1.5 hrs; 0.46% water absorption 0.46%; density 5.45 g cm^{-3}.
5	12		1953	319-409		222A-1	78.9	21.1	1 MgO + SnO_2 by mole; same specimen preparation as the above; fired at 1783 K and soaked for 1.5 hrs; water absorption 0.19%; density 5.20 g cm^{-3}.
6	12		1953	317-415		223A-1	71.4	28.6	3 MgO + 2 SnO_2 by mole; same specimen preparation as the above; fired at 1755 K and soaked for 1.5 hrs; water absorption 0.098%; density 4.81 g cm^{-3}.
7	12		1953	319-422		224A-1	61.6	38.4	7 MgO + 3 SnO_2 by mole; same specimen preparation as the above; fired at 1783 K and soaked for 1.5 hrs; water absorption 0.12%; density 4.26 g cm^{-3}.

DATA TABLE NO. 105 THERMAL CONDUCTIVITY OF [TIN DIOXIDE + MAGNESIUM OXIDE] $SnO_2 + MgO$

[Temperature, T, K; Thermal Conductivity, k, Watt $cm^{-1} K^{-1}$]

T	k
CURVE 1	
320.5	0.120
342.0	0.105
360.2	0.102
385.0	0.0946
407.7	0.0874
CURVE 2	
320.9	0.141
341.0	0.123
364.5	0.115
384.0	0.110
413.9	0.103
CURVE 3	
315.9	0.177
337.4	0.174
365.0	0.158
385.8	0.150
421.4	0.138
CURVE 4	
320.4	0.162
338.8	0.153
362.7	0.141
382.8	0.133
420.6	0.122
CURVE 5	
319.2	0.119*
339.2	0.110
358.2	0.106
376.5	0.101
382.0	0.100
408.6	0.0950
CURVE 6	
317.2	0.0933
340.2	0.0874
361.8	0.0841
CURVE 6 (cont.)	
379.7	0.0799
414.6	0.0761
CURVE 7	
318.7	0.0787
343.1	0.0766
356.5	0.0753
393.8	0.0707
421.9	0.0657

*Not shown on Plot

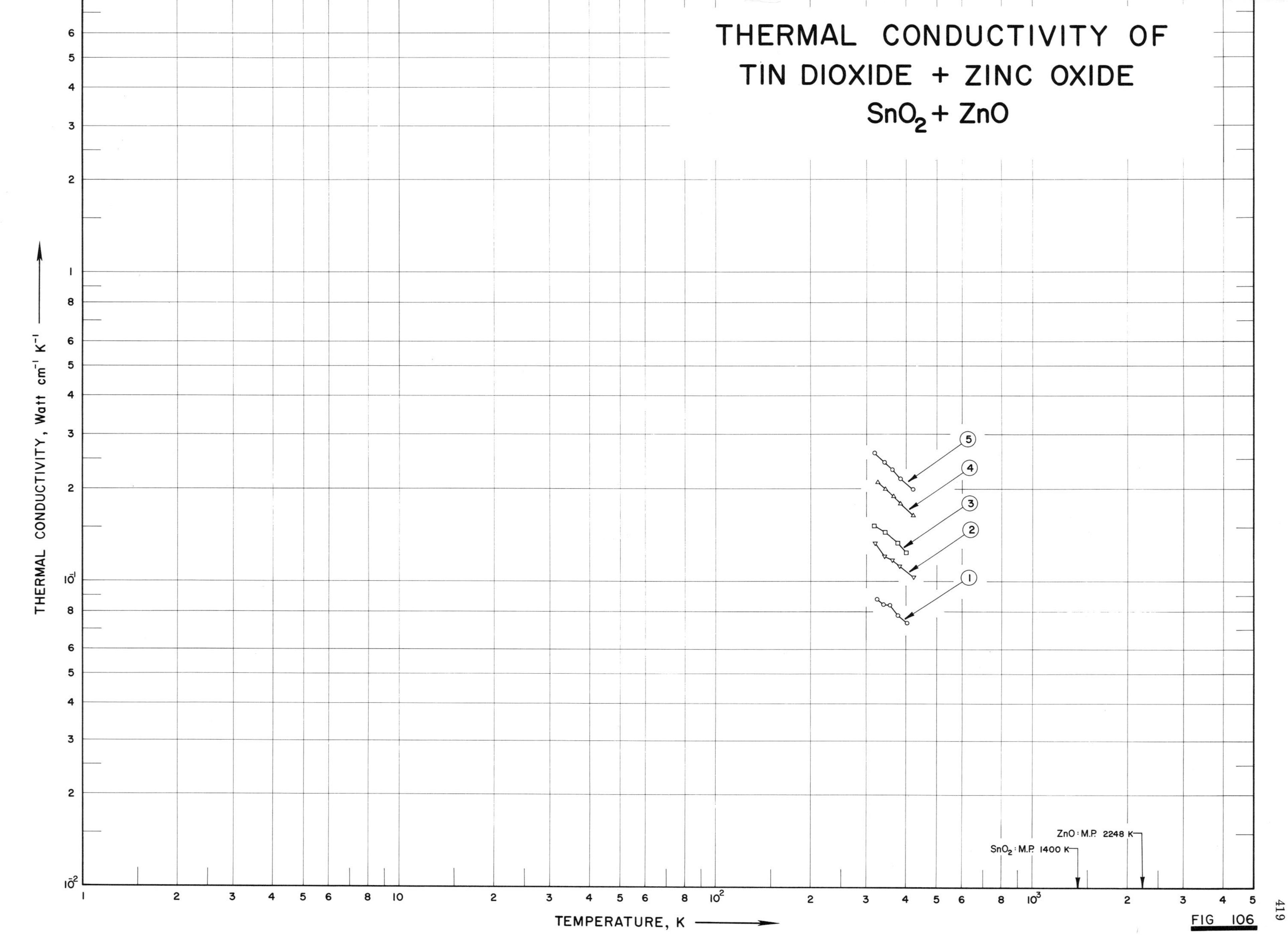
THERMAL CONDUCTIVITY OF
TIN DIOXIDE + ZINC OXIDE
SnO_2 + ZnO
5
4
3
2
1
ZnO : M.P. 2248 K
SnO_2 : M.P. 1400 K
THERMAL CONDUCTIVITY, Watt cm^{-1} K^{-1}
TEMPERATURE, K
FIG 106

SPECIFICATION TABLE NO. 106 THERMAL CONDUCTIVITY OF [TIN DIOXIDE + ZINC OXIDE] SnO_2 + Zno

[For Data Reported in Figure and Table No. 106]

Curve No.	Ref. No.	Method Used	Year	Temp. Range, K	Reported Error, %	Name and Specimen Designation	Composition (weight percent) SnO_2	ZnO	Composition (continued), Specifications and Remarks
1	12		1953	324-403		191A-1	64.9	35.1	1 SnO_2 to 1 ZnO by mole; prepared by milling pure oxides in water, calcining at 1367 K after drying; pressing at 15000 psi; fired at 1644 K and soaked for 1 hr; water absorption 0.224%; density 6.02 g cm^{-3}.
2	12		1953	321-424		192C-1	73.5	26.5	3 SnO_2 to 2ZnO by mole; same preparation as the above specimen; fired at 1700 K and soaked for 2 hrs; water absorption 0.201%; density 6.24 g cm^{-3}.
3	12		1953	317-402		193A-1	81.2	18.8	7 SnO_2 to 3 ZnO by mole; same preparation as the above specimen; fired at 1700 K and soaked for 1 hr; water absorption 0.303%; density 6.16 g cm^{-3}.
4	12		1953	326-421		194A-1	88.1	11.9	4 SnO_2 to 1 ZnO by mole; same preparation as the above specimen; fired at 1728 K and soaked for 2 hrs; density 6.25 g cm^{-3}.
5	12		1953	319-422		195A-1	94.3	5.7	9 SnO_2 to 1 ZnO by mole; same preparation as the above specimen; fired at 1728 K and soaked for 1 hr; water absorption 0.009%; density 6.32 g cm^{-3}.

DATA TABLE NO. 106 THERMAL CONDUCTIVITY OF [TIN DIOXIDE + ZINC OXIDE] $SnO_2 + ZnO$

[Temperature, T, K; Thermal Conductivity, k, Watt $cm^{-1} K^{-1}$]

T	k
CURVE 1	
324.4	0.0879
340.2	0.0845
357.7	0.0841
379.3	0.0778
402.8	0.0736
CURVE 2	
320.6	0.133
342.6	0.121
362.0	0.118
382.6	0.113
424.0	0.103
CURVE 3	
317.2	0.152
344.0	0.145
378.6	0.134
402.0	0.125
CURVE 4	
326.4	0.210
344.9	0.200
365.4	0.189
385.1	0.179
421.0	0.163
CURVE 5	
319.3	0.261
343.2	0.244
363.2	0.231
385.0	0.216
421.5	0.199

SPECIFICATION TABLE NO. 107 THERMAL CONDUCTIVITY OF [TUNGSTEN TRIOXIDE + ZINC OXIDE] $WO_3 + ZnO$

Curve No.	Ref. No.	Method Used	Year	Temp. Range, K	Reported Error, %	Name and Specimen Designation	Composition (weight percent) WO_3	Composition (weight percent) ZnO	Composition (continued), Specifications and Remarks
1	254	E	1955	400-564			94.0	6.0	Coarse crystalline structure; prepared by firing H_2WO_3 and ZnO; considerable porosity.

DATA TABLE NO. 107 THERMAL CONDUCTIVITY OF [TUNGSTEN TRIOXIDE + ZINC OXIDE] $WO_3 + ZnO$

[Temperature, T, K; Thermal Conductivity, k, Watt $cm^{-1} K^{-1}$]

T	k
CURVE 1*	
400.2	0.208
438.2	0.211
473.2	0.218
497.2	0.210
533.2	0.211
564.2	0.218

* No graphical presentation

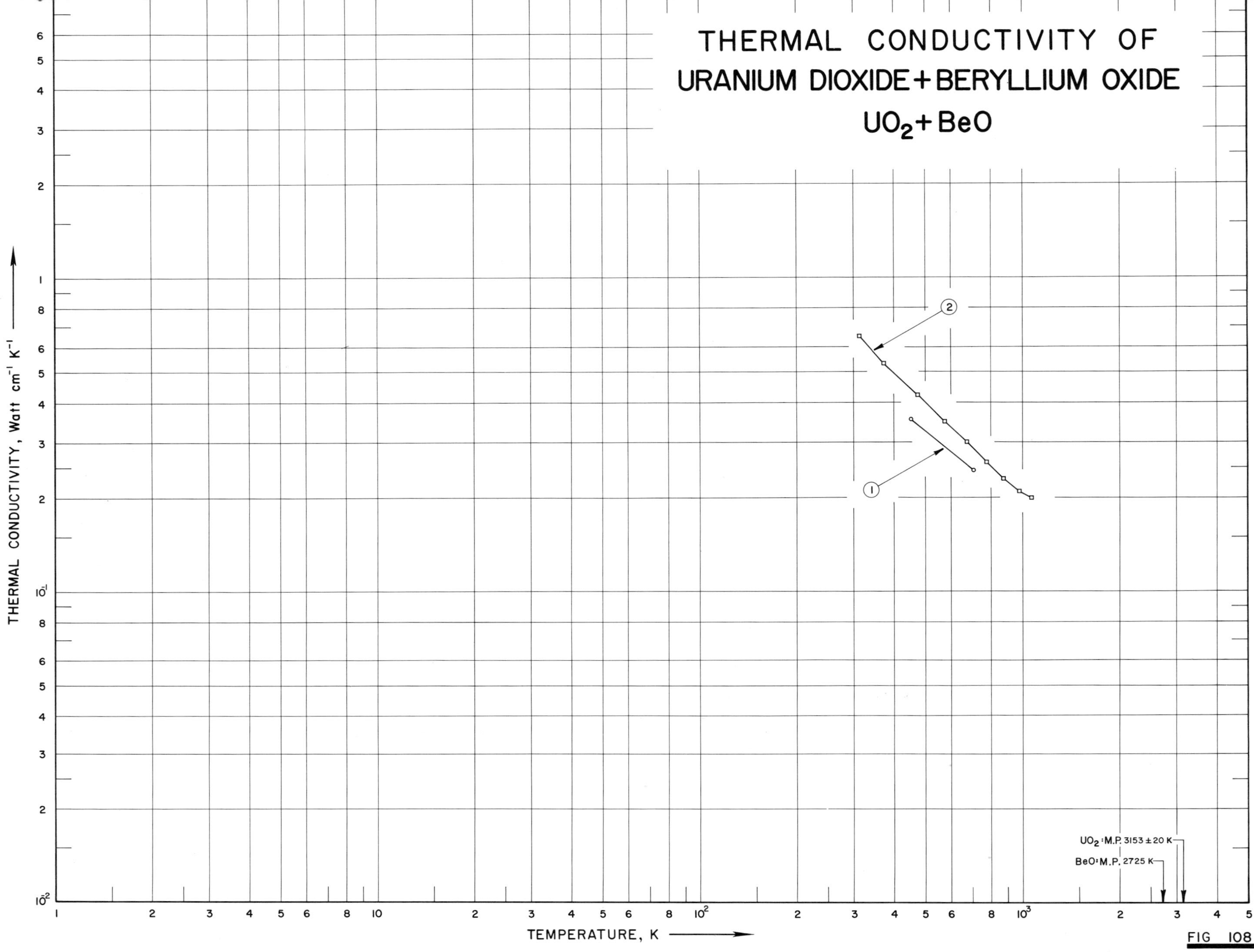
THERMAL CONDUCTIVITY OF
URANIUM DIOXIDE + BERYLLIUM OXIDE
UO_2 + BeO
2
1
UO_2: M.P. 3153 ± 20 K
BeO: M.P. 2725 K
THERMAL CONDUCTIVITY, Watt cm^{-1} K^{-1}
TEMPERATURE, K
FIG 108

SPECIFICATION TABLE NO. 108 THERMAL CONDUCTIVITY OF [URANIUM DIOXIDE + BERYLLIUM OXIDE] UO_2 + BeO

[For Data Reported in Figure and Table No. 108]

Curve No.	Ref. No.	Method Used	Year	Temp. Range, K	Reported Error, %	Name and Specimen Designation	Composition (weight percent) UO_2	BeO	Composition (continued), Specifications and Remarks
1	76	L	1952	453,708			70.9	29.1	79.5% of theoretical density.
2	76	L	1952	313-1073			70.9	29.1	80.5% of theoretical density.

DATA TABLE NO. 108 THERMAL CONDUCTIVITY OF [URANIUM DIOXIDE + BERYLLIUM OXIDE] $UO_2 + BeO$

[Temperature, T, K; Thermal Conductivity, k, Watt $cm^{-1}K^{-1}$]

T	k
CURVE 1	
453.2	0.355
708.2	0.245
CURVE 2	
313.2	0.650
373.2	0.530
473.2	0.425
573.2	0.350
673.2	0.300
773.2	0.260
873.2	0.230
973.2	0.210
1073.2	0.200

SPECIFICATION TABLE NO. 109 THERMAL CONDUCTIVITY OF [URANIUM DIOXIDE + CALCIUM OXIDE] $UO_2 + CaO$

Curve No.	Ref. No.	Method Used	Year	Temp. Range, K	Reported Error, %	Name and Specimen Designation	Composition (weight percent) UO_2	Composition (weight percent) CaO	Composition (continued), Specifications and Remarks
1	518	C	1960	1054, 1079		CA 8-1	98.37	1.63	Prepared by sintering large samples of UO_2 containing CaO for 3 hrs at 1300 C under argon followed by 1 hr in 10%H-90%H at 1300 C; initial oxygen to uranium ratio 2.58, final oxygen to Uranium ratio 1.98; electrical resistivity 58 ohm-cm at 300 K; water absorption 0.02%; density 10.13 g cm^{-3} (97.9% of theoretical value); lattice constant 5.452 Å; measured in vacuum.

DATA TABLE NO. 109 THERMAL CONDUCTIVITY OF [URANIUM DIOXIDE + CALCIUM OXIDE] $UO_2 + CaO$

[Temperature, T, K; Thermal Conductivity, k, Watt $cm^{-1}K^{-1}$]

T	k
CURVE 1 *	
1054.2	0.0326
1079.2	0.0227

* No graphical presentation

SPECIFICATION TABLE NO. 110 THERMAL CONDUCTIVITY OF [URANIUM DIOXIDE + DINIOBIUM PENTOXIDE] $UO_2 + Nb_2O_5$

Curve No.	Ref. No.	Method Used	Year	Temp. Range, K	Reported Error, %	Name and Specimen Designation	Composition (weight percent) UO_2	Composition (weight percent) Nb_2O_5	Composition (continued), Specifications and Remarks
1	518	C	1960	933, 956			96.06	3.94	Specimen 0.810 in. in dia; electrical resistivity 12.090 ohm-cm at 300 K; density 9.56 g cm^{-3}; measured in static argon atmosphere.

DATA TABLE NO. 110 THERMAL CONDUCTIVITY OF [URANIUM DIOXIDE + DINIOBIUM PENTOXIDE] $UO_2 + Nb_2O_5$

[Temperature, T, K; Thermal Conductivity, k, Watt $cm^{-1}K^{-1}$]

T	k
CURVE 1*	
933.2	0.0364
956.2	0.0355

* No graphical presentation

SPECIFICATION TABLE NO. 111 THERMAL CONDUCTIVITY OF [URANIUM DIOXIDE + YTTRIUM OXIDE] $UO_2 + Y_2O_3$

Curve No.	Ref. No.	Method Used	Year	Temp. Range, K	Reported Error, %	Name and Specimen Designation	Composition (weight percent) UO_2	Y_2O_3	Composition (continued), Specifications and Remarks
1	438	C	1959	928-1137			96.63	3.37	Specimen 0.856 ± 0.001 in. in dia and 1.00 ± 0.001 in. height; prepared from ceramic grade low-bulk density UO_2 (supplied by Mallinckroot Chemical Works) and 99.9 pure, 325 mesh Y_2O_3 (supplied by Rare Earths, Inc.); Yttrium added to $UO_{2.36}$ powder in the form of a nitric acid soultion containing 0.11 grames of Yttrium per milliliter; the batch calcined at 600 or 900 C in argon, sintered and reduced in hydrogen at 650 C and re-oxidized in air at 140 C to an U/O 2.5 and sintered again; final density of the stoichiometric specimen 10.05 g cm^{-3} (95.3% of theoretical value).
2	518	C	1960	771-1103		Y-47-1	96.76	3.24	Pellets prepared from calcined UO_2 and $Y(NO_3)_3$; sintered for 3 hrs at 1300 C under argon followed by 1 hr in 90%A-10%H mixture at the same temperature; initial oxygen to uranium ratio 2.5, final oxygen to uranium ratio 2.0; electrical resistivity 69 ohm-cm at 300 K; density 10.01 g cm^{-3} (95.3% of theoretical value); lattice constant 5.455 Å; measured in vacuum.

DATA TABLE NO. 111 THERMAL CONDUCTIVITY OF [URANIUM DIOXIDE + YTTRIUM OXIDE] $UO_2 + Y_2O_3$

[Temperature, T, K; Thermal Conductivity, k, Watt $cm^{-1}K^{-1}$]

T	k	T	k	T	k
CURVE 1*		CURVE 1 (cont.)*		CURVE 2 (cont.)*	
928.2	0.0433	1137.2	0.0471	870.2	0.0359
928.2	0.0518	1137.2	0.0490	876.2	0.0351
1017.2	0.0509			1103.2	0.0322
1017.2	0.0526	CURVE 2*			
1095.2	0.0493				
1095.2	0.0471	771.2	0.0372		
		869.2	0.0355		

* No graphical presentation

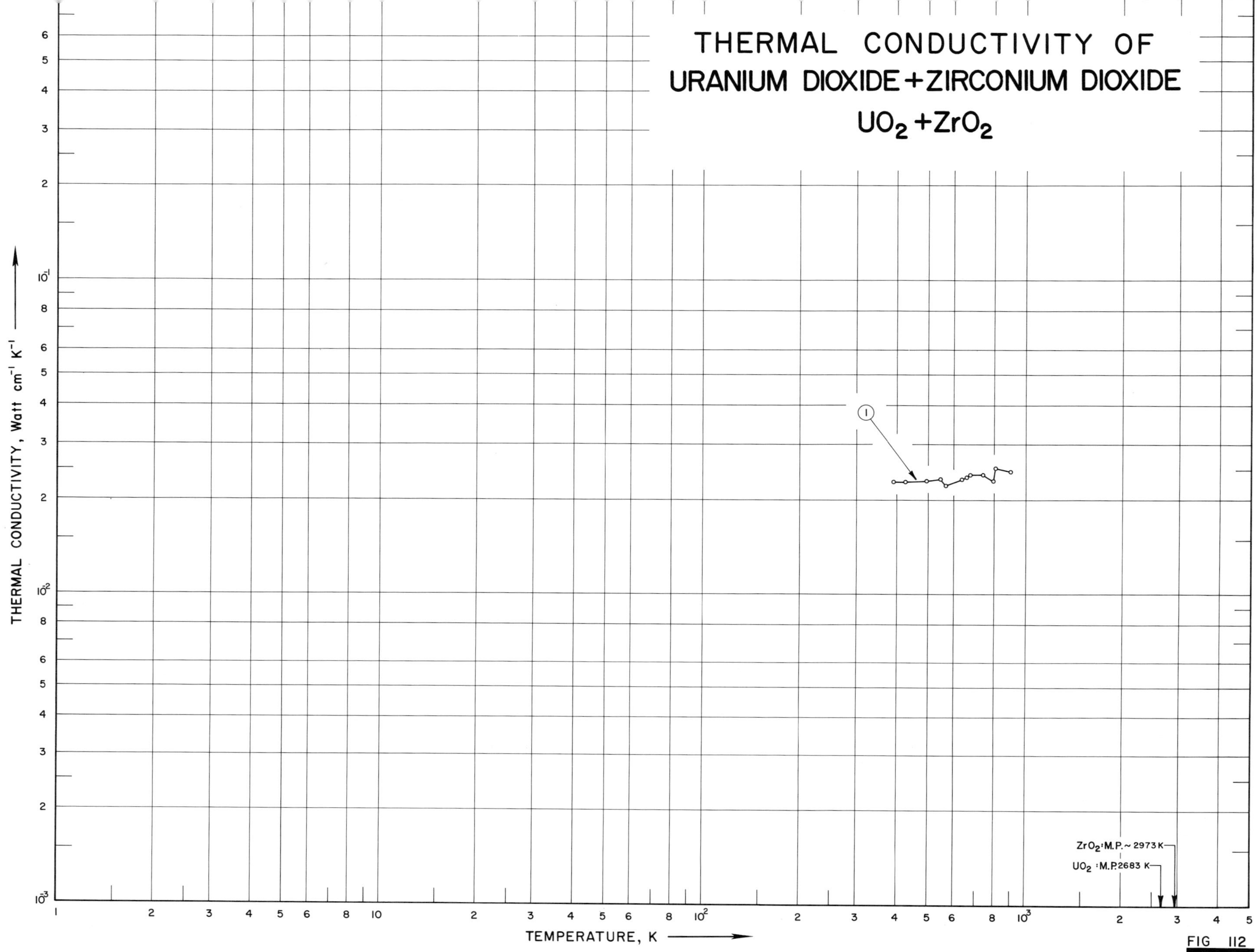

THERMAL CONDUCTIVITY OF
URANIUM DIOXIDE + ZIRCONIUM DIOXIDE
$UO_2 + ZrO_2$
1
ZrO_2: M.P. ~ 2973 K
UO_2: M.P. 2683 K
THERMAL CONDUCTIVITY, Watt cm^{-1} K^{-1}
TEMPERATURE, K

FIG 112

SPECIFICATION TABLE NO. 112 THERMAL CONDUCTIVITY OF [URANIUM DIOXIDE + ZIRCONIUM DIOXIDE] $UO_2 + ZrO_2$

[For Data Reported in Figure and Table No. 112]

Curve No.	Ref. No.	Method Used	Year	Temp. Range, K	Reported Error, %	Name and Specimen Designation	Composition (weight percent)		Composition (continued), Specifications and Remarks
							UO_2	ZrO_2	
1	519	C	1963	390-899	± 5		68.7	31.3	1 UO_2 + 1 ZrO_2 by mole; pressed at 50 tsi and homogenized by sintering for 20 hrs at 1700 C; measured in a vacuum of approximately 2 x 10^{-5} mm Hg.

DATA TABLE NO. 112 THERMAL CONDUCTIVITY OF [URANIUM DIOXIDE + ZIRCONIUM DIOXIDE] $UO_2 + ZrO_2$

[Temperature, T, K; Thermal Conductivity, k, Watt $cm^{-1} K^{-1}$]

T	k
CURVE 1	
390.2	0.0229
424.2	0.0229
493.2	0.0230
545.2	0.0233
567.2	0.0222
636.2	0.0232
657.2	0.0236
671.2	0.0240
739.2	0.0240
792.2	0.0230
807.2	0.0250
899.2	0.0247

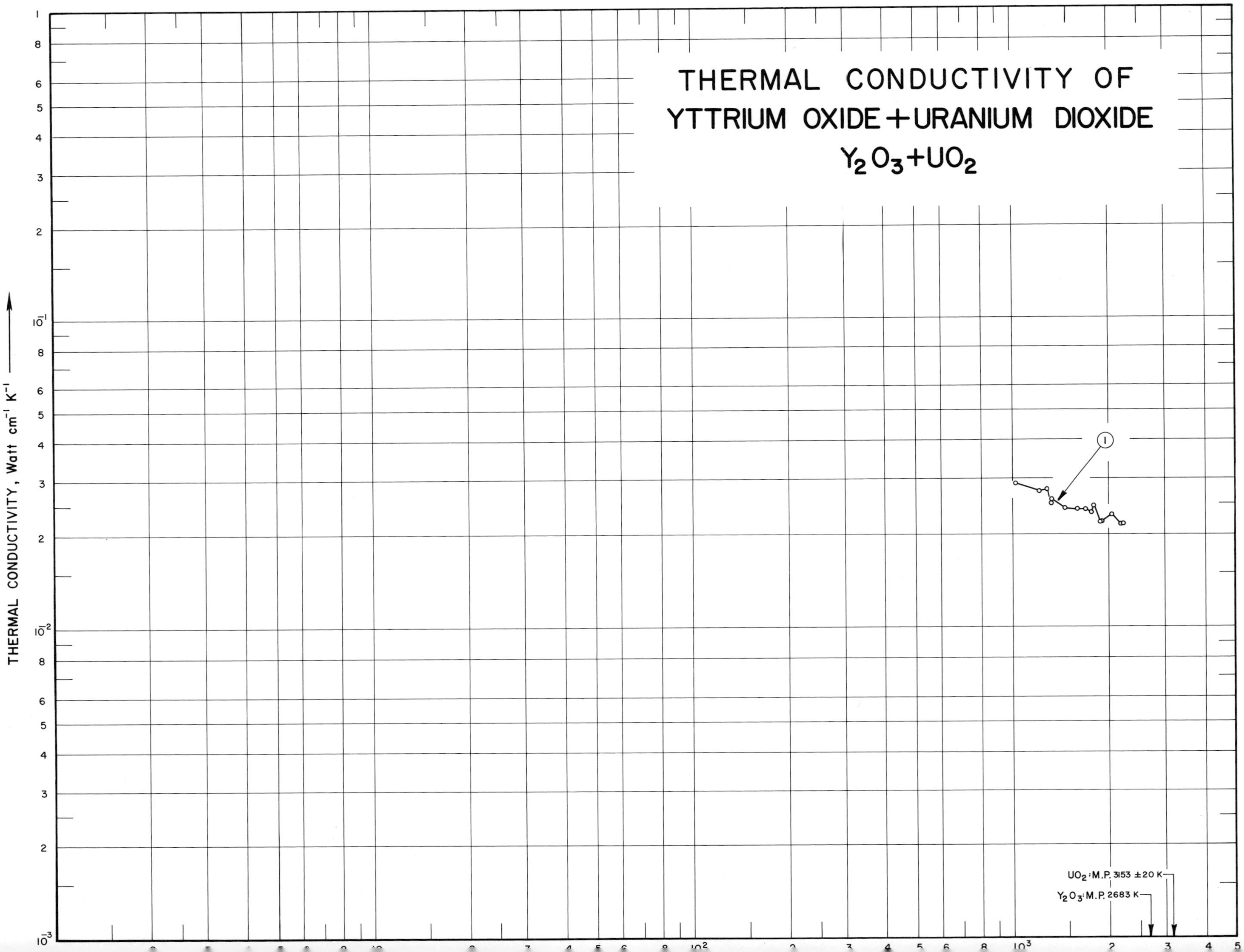

THERMAL CONDUCTIVITY OF YTTRIUM OXIDE + URANIUM DIOXIDE $Y_2O_3 + UO_2$

SPECIFICATION TABLE NO. 113 THERMAL CONDUCTIVITY OF [YTTRIUM OXIDE + URANIUM DIOXIDE] $Y_2O_3 + UO_2$

[For Data Reported in Figure and Table No. 113]

Curve No.	Ref. No.	Method Used	Year	Temp. Range, K	Reported Error, %	Name and Specimen Designation	Composition (weight percent) Y_2O_3	Composition (weight percent) UO_2	Composition (continued), Specifications and Remarks
1	291	R	1964	1038-2233		$3Y_2O_3 + UO_{2.2}$	71.5	28.5	Specimen in disc form of 2 in. O.D. and 0.375 in. I.D.; 99% of theoretical density.

DATA TABLE NO. 113 THERMAL CONDUCTIVITY OF [YTTRIUM OXIDE + URANIUM DIOXIDE] $Y_2O_3 + UO_2$

[Temperature, T, K; Thermal Conductivity, k, Watt $cm^{-1}K^{-1}$]

T	k
CURVE 1	
1038.2	0.029
1223.2	0.0275
1298.2	0.0277
1333.2	0.025
1348.2	0.0259
1473.2	0.0242
1603.2	0.024
1703.2	0.024
1773.2	0.0235
1813.2	0.0246
1893.2	0.0218
1913.2	0.022
2063.2	0.0231
2193.2	0.0215
2233.2	0.0216

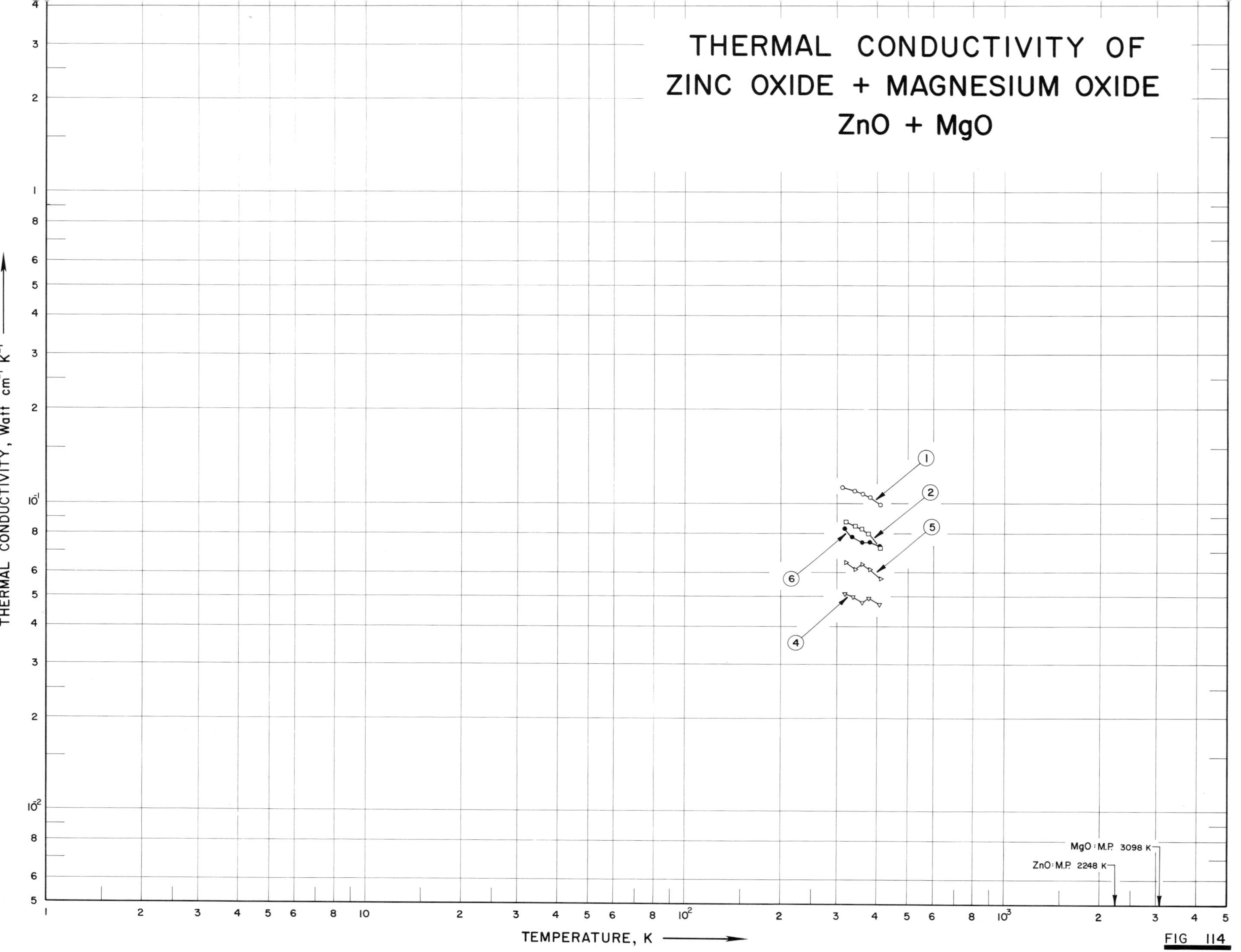
THERMAL CONDUCTIVITY OF
ZINC OXIDE + MAGNESIUM OXIDE
ZnO + MgO
THERMAL CONDUCTIVITY, Watt cm⁻¹ K⁻¹
TEMPERATURE, K
MgO : M.P. 3098 K
ZnO : M.P. 2248 K
1
2
4
5
6

FIG 114

SPECIFICATION TABLE NO. 114 THERMAL CONDUCTIVITY OF [ZINC OXIDE + MAGNESIUM OXIDE] $ZnO + MgO$

[For Data Reported in Figure and Table No. 114]

Curve No.	Ref. No.	Method Used	Year	Temp. Range, K	Reported Error, %	Name and Specimen Designation	Composition (weight percent) ZnO	MgO	Composition (continued), Specifications and Remarks
1	12		1953	313-411		258A-1	94.8	5.2	$9ZnO + MgO$ by mole; prepared from pure oxides, milled in water, dried, calcined at 1367 K; then dry pressed in 0.5 in. steel die at 15000 psi; fired at 1755 K and soaked for 2 hrs; water absorption 0.077%; density 5.04 g cm^{-3}.
2	12		1953	320-412		259A-1	89.0	11.0	$4ZnO + MgO$ by mole; same preparation as that of the above specimen; water absorption 0.002%; density 4.86 g cm^{-3}.
3	12		1953	319-411		260A-1	82.5	17.5	$7ZnO + 3MgO$ by mole; same preparation as that of the above specimen; water absorption 0.020%; density 4.99 g cm^{-3}.
4	12		1953	319-409		261A-1	75.2	24.8	$3ZnO + 2MgO$ by mole; same preparation as that of the above specimen except fired at 1700 K; water absorption 0.006%; density 4.64 g cm^{-3}.
5	12		1953	321-414		262A-1	66.9	33.1	$ZnO + MgO$ by mole; same preparation as that of the above specimen; water absorption 0.040%; density 4.87 g cm^{-3}.
6	12		1953	318-410		363A-1	57.4	42.6	$2ZnO + 3MgO$ by mole; same preparation as that of the above specimen; water absorption 0.010%; density 4.86 g cm^{-3}.

DATA TABLE NO. 114 THERMAL CONDUCTIVITY OF [ZINC OXIDE + MAGNESIUM OXIDE] $ZnO + MgO$

[Temperature, T, K; Thermal Conductivity, k, Watt $cm^{-1} K^{-1}$]

T	k
CURVE 1	
312.7	0.113
341.5	0.110
363.5	0.107
382.5	0.105
411.2	0.0992
CURVE 2	
320.2	0.0874
343.7	0.0845
359.0	0.0828
377.2	0.0799
412.0	0.0720
CURVE 3*	
319.2	0.0904
338.4	0.0858
359.8	0.0841
379.3	0.0812
411.0	0.0778
CURVE 4	
318.9	0.0510
337.7	0.0498
360.8	0.0477
378.2	0.0494
409.0	0.0473
CURVE 5	
320.8	0.0649
342.2	0.0615
359.3	0.0640
381.1	0.0615
413.5	0.0573
CURVE 6	
318.0	0.0833
336.8	0.0782
361.6	0.0753
381.1	0.0753
410.1	0.0724

*Not shown on Plot

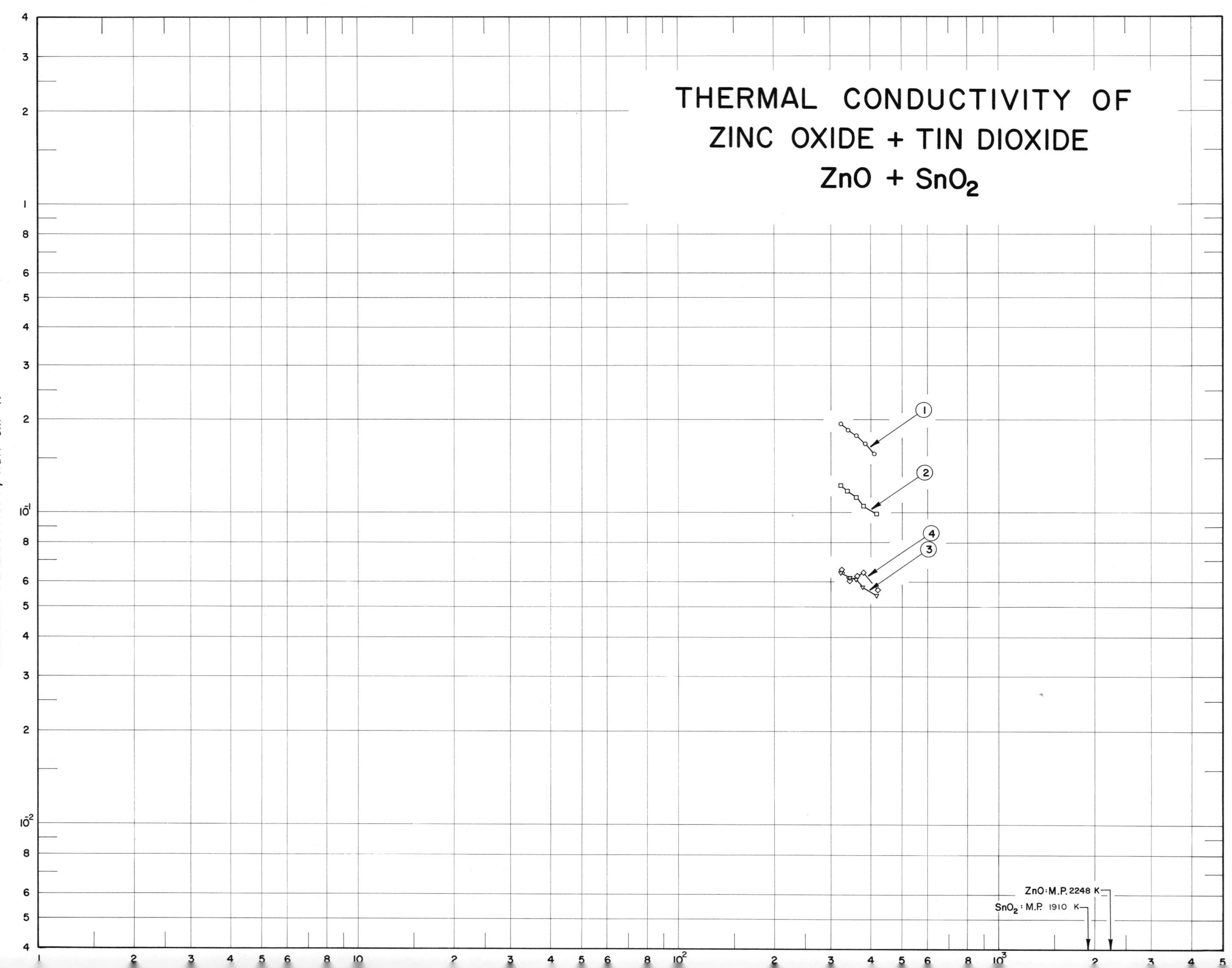
THERMAL CONDUCTIVITY OF
ZINC OXIDE + TIN DIOXIDE
ZnO + SnO2
1
2
3
4
ZnO: M.P. 2248 K
SnO2: M.P. 1910 K
THERMAL CONDUCTIVITY, Watt cm-1 K-1

SPECIFICATION TABLE NO. 115 THERMAL CONDUCTIVITY OF [ZINC OXIDE + TIN DIOXIDE] $ZnO + SnO_2$

[For Data Reported in Figure and Table No. 115]

Curve No.	Ref. No.	Method Used	Year	Temp. Range, K	Reported Error, %	Name and Specimen Designation	Composition (weight percent) ZnO	SnO_2	Composition (continued), Specifications and Remarks
1	12		1953	323-411		187A-1	82.9	17.1	9 ZnO +1 SnO_2 by mole; prepared by milling pure oxides in water, calcining at 1367 K after drying and then dry pressing in a 0.5 in. steel die at 15000 psi; fired at 1645 K and soaked for 1.5 hrs; density 5.73 g cm^{-3}.
2	12		1953	323-418		188A-1	68.4	31.6	4 ZnO + 1 SnO_2 by mole; same preparation as the above specimen; fired at 1645 K and soaked for 1.5 hrs; density 5.90 g cm^{-3}.
3	12		1953	322-417		189A-1	55.8	44.2	7 ZnO + 3 SnO_2 by mole; same preparation as the above specimen; fired at 1645 K and soaked for 1 hr; water absorption 1.062%; density 5.67 g cm^{-3}.
4	12		1953	326-421		190B-1	55.8	44.2	Similar to the above specimen 189A-1 except water absorption 1.57% and density 5.62 g cm^{-3}.

DATA TABLE NO. 115 THERMAL CONDUCTIVITY OF [ZINC OXIDE + TIN DIOXIDE] $ZnO + SnO_2$

[Temperature, T, K; Thermal Conductivity, k, Watt cm^{-1} K^{-1}]

T	k
CURVE 1	
322.9	0.194
340.3	0.186
361.6	0.177
385.0	0.167
411.3	0.156
CURVE 2	
323.1	0.122
338.3	0.117
360.0	0.112
379.9	0.106
418.1	0.0992
CURVE 3	
322.3	0.0640
345.6	0.0619
360.3	0.0611
378.3	0.0577
417.1	0.0544
CURVE 4	
325.5	0.0653
344.7	0.0602
362.3	0.0628
379.0	0.0644
421.1	0.0565

SPECIFICATION TABLE NO. 116 THERMAL CONDUCTIVITY OF [ZIRCONIUM DIOXIDE + ALUMINUM OXIDE] $ZrO_2 + Al_2O_3$

Curve No.	Ref. No.	Method Used	Year	Temp. Range, K	Reported Error, %	Name and Specimen Designation	Composition (weight percent) ZrO_2	Al_2O_3	Composition (continued), Specifications and Remarks
1	522	P	1963	373-823	± 4		91.6	8.4	As received; density 5.13 g cm^{-3} (92.5% of theoretical value); thermal conductivity values calculated from the measured data of thermal diffusivity, specific heat and density.

DATA TABLE NO. 116 THERMAL CONDUCTIVITY OF [ZIRCONIUM DIOXIDE + ALUMINUM OXIDE] $ZrO_2 + Al_2O_3$

[Temperature, T, K; Thermal Conductivity, k, Watt $cm^{-1} K^{-1}$]

T	k
CURVE 1*	
373.2	0.0439
473.2	0.0439
573.2	0.0427
673.2	0.0427
773.2	0.0435
823.2	0.0439

* No graphical presentation

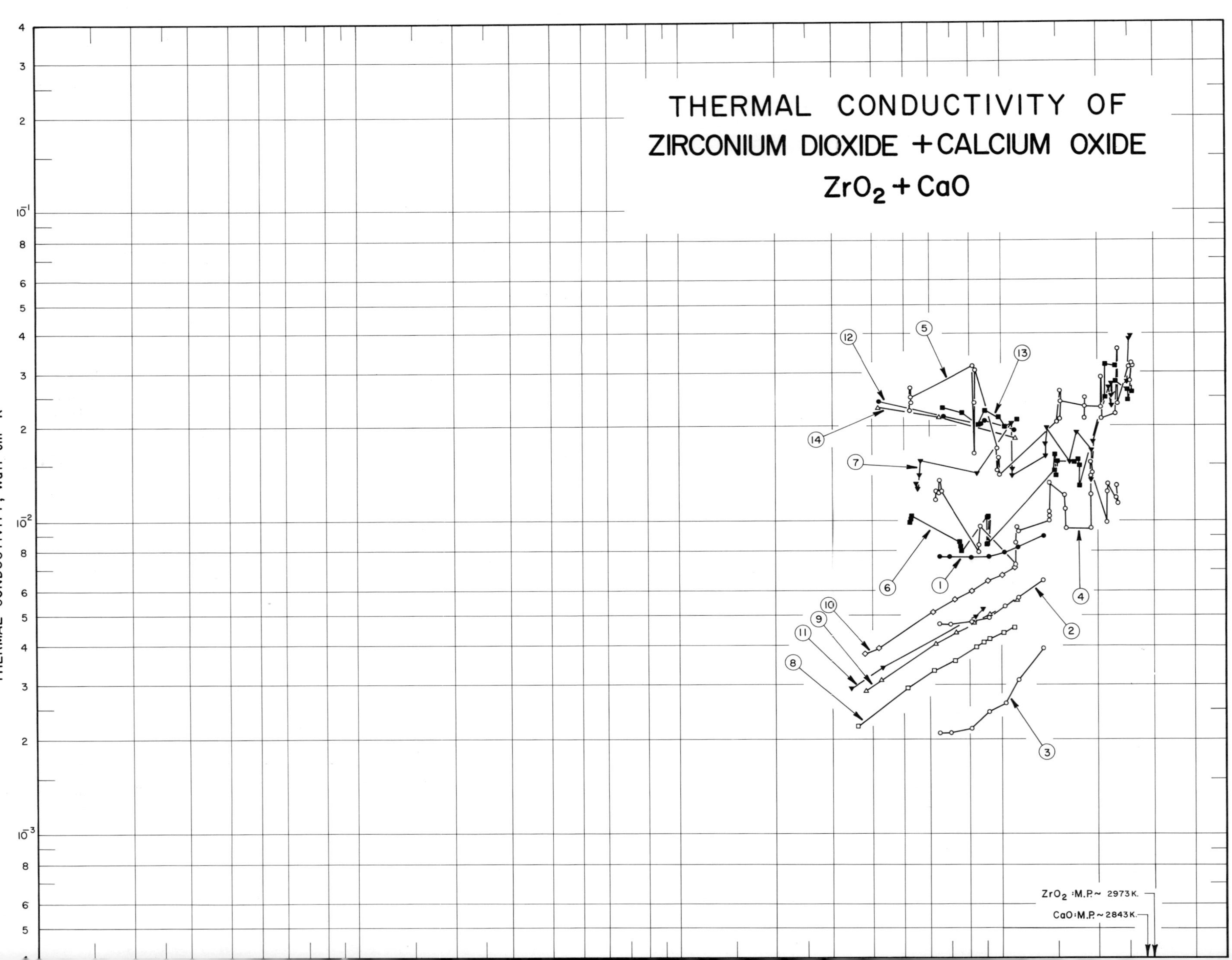
THERMAL CONDUCTIVITY OF
ZIRCONIUM DIOXIDE + CALCIUM OXIDE
$ZrO_2 + CaO$
ZrO_2 : M.P.~ 2973 K.
CaO : M.P.~ 2843 K.
THERMAL CONDUCTIVITY, Watt cm⁻¹ K⁻¹
1
2
3
4
5
6
7
8
9
10
11
12
13
14
4
3
2
10⁻¹
8
6
5
4
3
2
10⁻²
8
6
5
4
3
2
10⁻³
8
6
5

SPECIFICATION TABLE NO. 117 THERMAL CONDUCTIVITY OF [ZIRCONIUM DIOXIDE + CALCIUM OXIDE] $ZrO_2 + CaO$

[For Data Reported in Figure and Table No. 117]

Curve No.	Ref. No.	Method Used	Year	Temp. Range, K	Reported Error, %	Name and Specimen Designation	Composition (weight percent) ZrO_2	CaO	Composition (continued), Specifications and Remarks
1	210		1952	644-1366		Dense brick; A	Bal	4.5/5.0	< 2.0 HfO_2, 0.2-0.5 Fe_2O_3, 0.5-1.0 SiO_2, and 0.4-1.0 TiO_2; stabilized; density (25 C) 4.0 g cm^{-3}; 28 vol % pores.
2	210		1952	644-1366		Insulating brick; B	Bal	4.5/5.0	< 2.0 HfO_2, 0.2-0.5 Fe_2O_3, 0.5-1.0 SiO_2 and 0.4-1.0 TiO_2; stabilized; density (25 C) 2.72 g cm^{-3} and 50 vol % pores.
3	210		1952	644-1366		Insulating brick; C	Bal	4.5/5.0	< 2.0 HfO_2, 0.2-0.5 Fe_2O_3, 0.5-1.0 SiO_2, and 0.4-1.0 TiO_2; stabilized; density (25 C) 1.81 g cm^{-3} and 68 vol % pores.
4	82	R	1962	626-2320			96.0	2.99	0.34 SiO_2, 0.26 $CaSO_4$, 0.21 MgO, and 0.18 SO_4; specimen 1 in. dia and 1 in. long; sintered, stabilized, and molded; fine grain; density 5.92 g cm^{-3}; porosity 0.53%; melting point 2850 K.
5	82	R	1962	526-2565			97.12	2.17	0.267 $CaSO_4$, 0.063 MgO, 0.186 SiO_2, and 0.173 SO_4; specimen 1 in. dia and 1 in. long; sintered, stabilized and molded; coarse grain; formulated by 30% fines (composition: 96 ZrO_2, 2.99 CaO, 0.34 SiO_2, 0.21 MgO, 0.26 $CaSO_4$, 0.18 SO_4) and 70% grog (composition: 97.6 ZrO_2, 1.82 CaO, 0.12 SiO_2, 0.27 $CaSO_4$, 0.17 SO_4); density 4.65 g cm^{-3}; porosity 18.58%; melting point 2816 K.
6	82	R	1962	525-2540					Similar to the above specimen except porosity 25%.
7	82	R	1962	551-2537					Similar to the above specimen except being extruded; density 4.57 g cm^{-3}; porosity 19.4%; melting point 2839 K.
8	400	R	1966	359-1108	± 10	H30F	Bal	3.58	0.01-0.1 Al principal impurity; powder specimen contained in a ~4 in. I.D., 4.5 in. O.D. and ~24 in. long container; supplied by Norton Co.; as-received; mean particle size 292 μ; volume fraction solid 0.58; pore-free density 5.60 g cm^{-3}.
9	400	R	1966	380-1123	± 10	H30F			Similar to the above specimen except volume fraction solid 0.64.
10	400	R	1966	378-1118	± 10	H14F			Similar to the above specimen except mean particle size 1023 μ; volume fraction solid 0.70; pore-free density 5.63 g cm^{-3}.
11	400	P	1966	343-884	± 11	H30F			Similar to the above specimen except mean particle size 292 μ; volume fraction solid 0.64; pore-free density 5.60 g cm^{-3}; thermal conductivity data obtained from the measurement of thermal diffusivity, specific heat and density.
12	374	C	1965	420-1106	0-± 2				c-Type, lime-stabilized zirconia, 93.7 ZrO_2, 3.35 CaO, 1.38 HfO_2, 0.30 SiO_2, 1.07 Al_2O_3, 0.17 Fe_2O_3, 0.03 TiO_2; composed of polygonal grains of anisotropic material with most grains in the range 0.10 to 0.15 mm; prepared to a tolerance of ± 0.001 in. in the form of a cylinder 1 in. in dia and 1 in. high; fabricated by the Zirconium Corp. of America; bulk density 5.4 g cm^{-3}; true density 5.7 g cm^{-3}; true porosity 5% of the total volume; pyroceram 9606 used as comparative material; unknown and standard used for the first time.

SPECIFICATION TABLE NO. 117 (continued)

Curve No.	Ref. No.	Method Used	Year	Temp. Range, K	Reported Error, %	Name and Specimen Designation	Composition (weight percent) ZrO_2	CaO	Composition (continued), Specifications and Remarks
13	374	C	1965	661-1126	0-± 3				Similar to the above specimen; unknown used for the first time, standard used several times up to 1273 K.
14	374	C	1965	417-1116	0-± 4				Similar to the above specimen; unknown used several times up to 1373 K, standard used for the first time.
15	374	C	1965	1313, 1373					Similar to the above specimen except alumina AL-300 used as comparative material.

DATA TABLE NO. 117 THERMAL CONDUCTIVITY OF [ZIRCONIUM DIOXIDE + CALCIUM OXIDE] $ZrO_2 + CaO$

[Temperature, T, K; Thermal Conductivity, k, Watt $cm^{-1}K^{-1}$]

T	k
CURVE 1	
644.2	0.00764
699.2	0.00764
811.2	0.00764
921.2	0.00764
1033.2	0.00793
1144.2	0.00822
1366.2	0.00894
CURVE 2	
644.2	0.00469
699.2	0.00469
811.2	0.00476
921.2	0.00490
1033.2	0.00534
1144.2	0.00570
1366.2	0.00649
CURVE 3	
644.2	0.00209
699.2	0.00209
811.2	0.00216
921.2	0.00245
1033.2	0.00260
1144.2	0.00310
1366.2	0.00392
CURVE 4	
626.0	0.0117
633.2	0.0124
645.4	0.0123
650.4	0.0133
654.3	0.0124
857.1	0.00795
858.7	0.00832
869.3	0.00958
1122.6	0.00723
1126.5	0.00855
1130.4	0.00956
1132.6	0.00929
1423.2	0.0100
CURVE 4 (cont.)	
1427.6	0.0107
1427.6	0.0102
1427.6	0.0131
1599.5	0.0121
1599.8	0.0109
1602.6	0.00946
1602.6	0.00952*
1602.6	0.00950*
1909.8	0.00942
1909.8	0.0122
1909.8	0.0138
1919.3	0.0155
1923.2	0.0142
2147.6	0.00989
2147.6	0.0124
2167.1	0.0130
2167.1	0.0133*
2303.2	0.0128
2303.2	0.0118
2319.8	0.0114
2319.8	0.0118*
CURVE 5	
526	0.0225
528.7	0.0267
529.8	0.0237
529.8	0.0247
822.1	0.0316
830.4	0.0237
832.1	0.0164
836.5	0.0306
983.2	0.0170
986.5	0.0143
988.7	0.0158
989.8	0.0139
1501.5	0.0206
1542.6	0.0262
1544.8	0.0212
1547.1	0.0241
1839.3	0.0233
1840.9	0.0213
1842.1	0.0248
CURVE 5 (cont.)	
1846.5	0.0229*
2062.6	0.0231
2069.3	0.0290
2077.6	0.0213
2083.2	0.0213*
2287.1	0.0219*
2290.4	0.0221
2320.9	0.0358
2320.9	0.0238
2526.5	0.0313
2529.8	0.0284
2562.1	0.0323
2565.4	0.0317
CURVE 6	
524.8	0.00984
526.5	0.0101
531.5	0.0103
747.1	0.00851
750.4	0.00825
756.5	0.00793
904.3	0.0102
909.8	0.00837
913.7	0.0104
920.9	0.00838
1478.7	0.0145
1483.2	0.0162
1486.5	0.0139
1498.2	0.0153
1714.3	0.0153
1742.1	0.0157
1760.9	0.0150
1775.4	0.0129
2083.2	0.0251*
2108.7	0.0250
2134.8	0.0319
2278.7	0.0316
2288.7	0.0286
2305.4	0.0272*
2306.5	0.0268*
2494.3	0.0264
2498.7	0.0255
2520.4	0.0294*
2540.4	0.0261
CURVE 7	
550.9	0.0132
558.2	0.0126
562.1	0.0139
562.6	0.0155
849.8	0.0141
852.1	0.0140*
853.2	0.0143*
1092.1	0.0205
1092.1	0.0146
1093.2	0.0139
1381.5	0.0161
1386.5	0.0176
1405.4	0.0199
1663.2	0.0154
1732.1	0.0191
1935.4	0.0169
1935.4	0.0135
1937.6	0.0180
2185.4	0.0270
2218.7	0.0235
2257.6	0.0276
2265.4	0.0252
2517.6	0.0287*
2523.2	0.0385
2532.1	0.0391
2536.5	0.0385*
CURVE 8	
359.2	0.00219
514.2	0.00292
624.2	0.00332
720.2	0.00357
842.2	0.00394
889.2	0.00411
924.2	0.00421
1023.2	0.00439
1108.2	0.00455
CURVE 9	
380.2	0.00285
425.2	0.00309
628.2	0.00403
726.2	0.00439
825.2	0.00473
922.2	0.00501
1023.2	0.00528*
1123.2	0.00557
CURVE 10	
378.2	0.00376
417.2	0.00392
616.2	0.00512
721.2	0.00560
813.2	0.00596
912.2	0.00643
1016.2	0.00674
1118.2	0.00709
CURVE 11	
343.2	0.00291
430.4	0.00339
835.9	0.00493
883.5	0.00523
CURVE 12	
420.2	0.0240
665.2	0.0216
873.2	0.0204
894.2	0.0209
1106	0.0195
CURVE 13	
661.2	0.0229
662.2	0.0216*
665.1	0.0216*
763.2	0.0221
854.2	0.0203
889.2	0.0224*
CURVE 13 (cont.)	
895.2	0.0225
985.2	0.0214
1035	0.0200
1126.4	0.0210
CURVE 14	
417.2	0.0230
646.2	0.0215
828.2	0.0210*
1078	0.0188*
1116	0.0184
CURVE 15*	
1313	0.0180
1373	0.016

*Not shown on plot

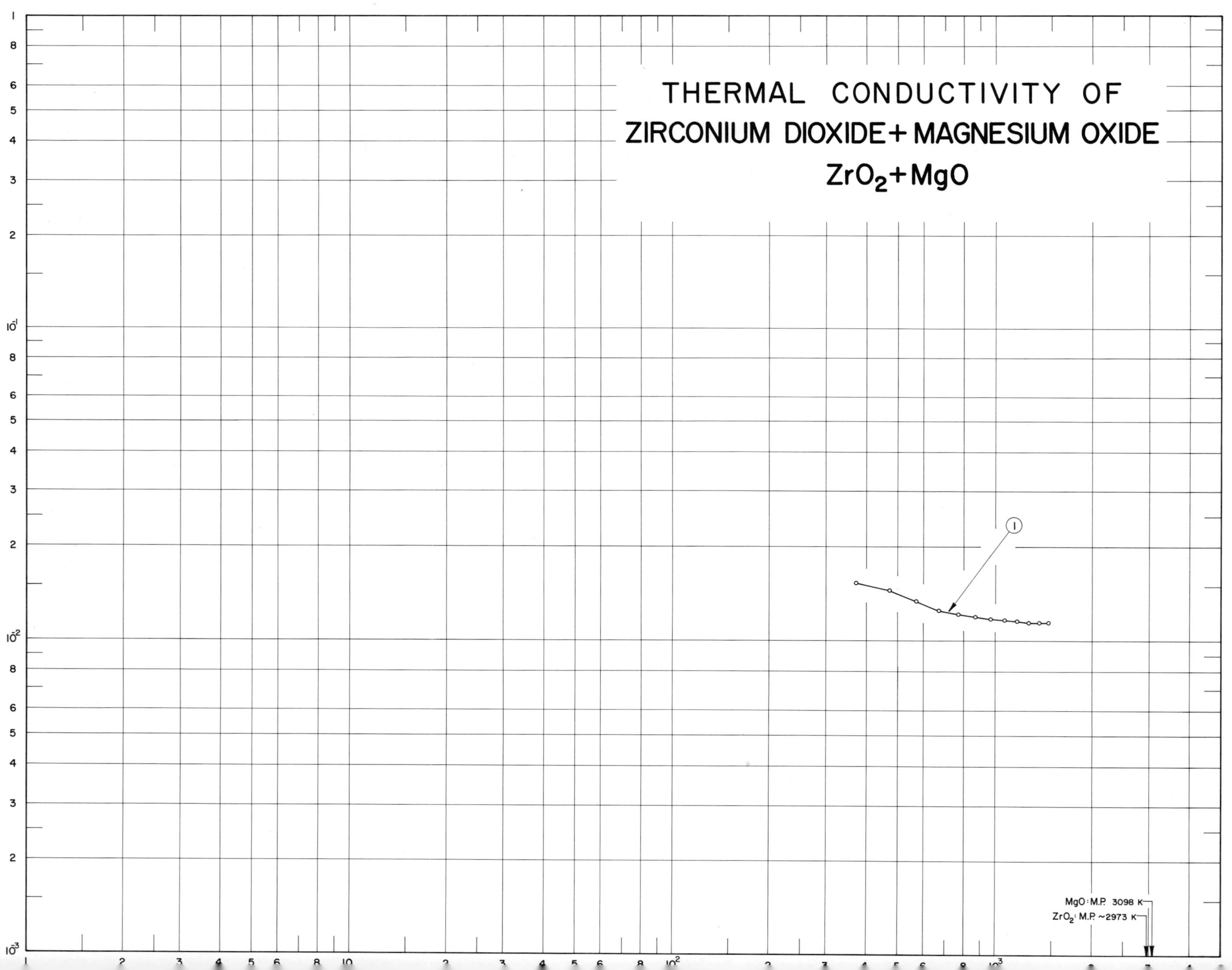
THERMAL CONDUCTIVITY OF
ZIRCONIUM DIOXIDE + MAGNESIUM OXIDE
ZrO_2+MgO
1
MgO: M.P. 3098 K
ZrO_2: M.P. ~2973 K
THERMAL CONDUCTIVITY, Watt cm^{-1} K^{-1}

SPECIFICATION TABLE NO. 118 THERMAL CONDUCTIVITY OF [ZIRCONIUM DIOXIDE + MAGNESIUM OXIDE] $ZrO_2 + MgO$

[For Data Reported in Figure and Table No. 118]

Curve No.	Ref. No.	Method Used	Year	Temp. Range, K	Reported Error, %	Name and Specimen Designation	Composition (weight percent)		Composition (continued), Specifications and Remarks
							ZrO_2	MgO	
1	253	P	1957	373-1473		E-109	95.6	4.1	0.01-0.1 CaO, 0.01-0.1 SiO_2, 0.01-0.1 TiO_2, 0.005-0.05 Al_2O_3, 0.001-0.01 Fe_2O_3, < 0.005 Na_2O, < 0.005 K_2O, < 0.005 Li_2O, < 0.005 BaO; specimen from Corning Glass Works; fired at ~1000 C for 4 hrs; bulk density 3.65 g cm^{-3} at 25 C; porosity 35%.

DATA TABLE NO. 118 THERMAL CONDUCTIVITY OF [ZIRCONIUM DIOXIDE + MAGNESIUM OXIDE] $ZrO_2 + MgO$

[Temperature, T, K; Thermal Conductivity, k, Watt $cm^{-1}K^{-1}$]

T	k
CURVE 1	
373.2	0.0154
473.2	0.0146
573.2	0.0135
673.2	0.0126
773.2	0.0122
873.2	0.0120
973.2	0.0118
1073.2	0.0117
1173.2	0.0116
1273.2	0.0115
1373.2	0.0115
1473.2	0.0115

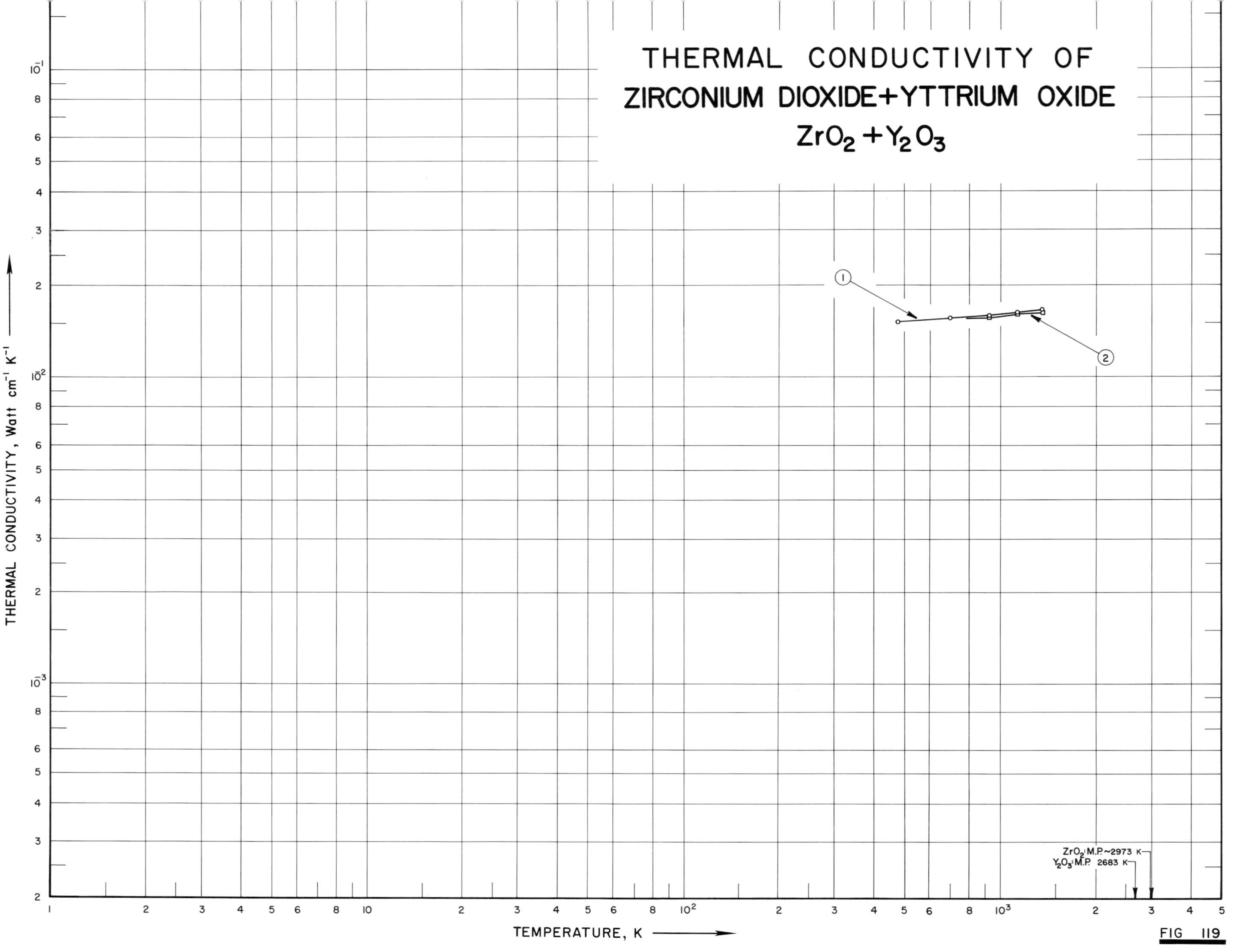
THERMAL CONDUCTIVITY OF
ZIRCONIUM DIOXIDE+YTTRIUM OXIDE
$ZrO_2 + Y_2O_3$
THERMAL CONDUCTIVITY, Watt cm^{-1} K^{-1}
TEMPERATURE, K
ZrO_2: M.P. ~2973 K
Y_2O_3: M.P. 2683 K
1
2
FIG 119

SPECIFICATION TABLE NO. 119 THERMAL CONDUCTIVITY OF [ZIRCONIUM DIOXIDE + YTTRIUM OXIDE] $ZrO_2 + Y_2O_3$

[For Data Reported in Figure and Table No. 119]

Curve No.	Ref. No.	Method Used	Year	Temp. Range, K	Reported Error, %	Name and Specimen Designation	Composition (weight percent) ZrO_2	Y_2O_3	Composition (continued), Specifications and Remarks
1	292	C	1961	478-1367		No. 1	85.0	15.0	Specimen 0.5 x 0.5 x 0.875 in.; prepared from Columbia-National ZrO_2 and Michigan Chemical Y_2O_3 prereacted at 1922 K; fabricated by dry pressing followed by isostatic compaction and then sintered at 1978 K in air for 2 hrs; 86.7% of theoretical density; commercial stabilized ZrO_2 used as comparative material, data corrected to zero porosity.
2	292	C	1961	478-1367		No. 2			Similar to the above specimen.

DATA TABLE NO. 119 THERMAL CONDUCTIVITY OF [ZIRCONIUM DIOXIDE + YTTRIUM OXIDE] $ZrO_2 + Y_2O_3$

[Temperature, T, K; Thermal Conductivity, k, Watt $cm^{-1} K^{-1}$]

T	k
CURVE 1	
477.6	0.0152
699.8	0.0156
922.1	0.0159
1144.2	0.0163
1366.5	0.0166
CURVE 2	
477.6	0.0152*
699.8	0.0156*
922.1	0.0157
1144.3	0.0161
1366.5	0.0163

*Not shown on plot

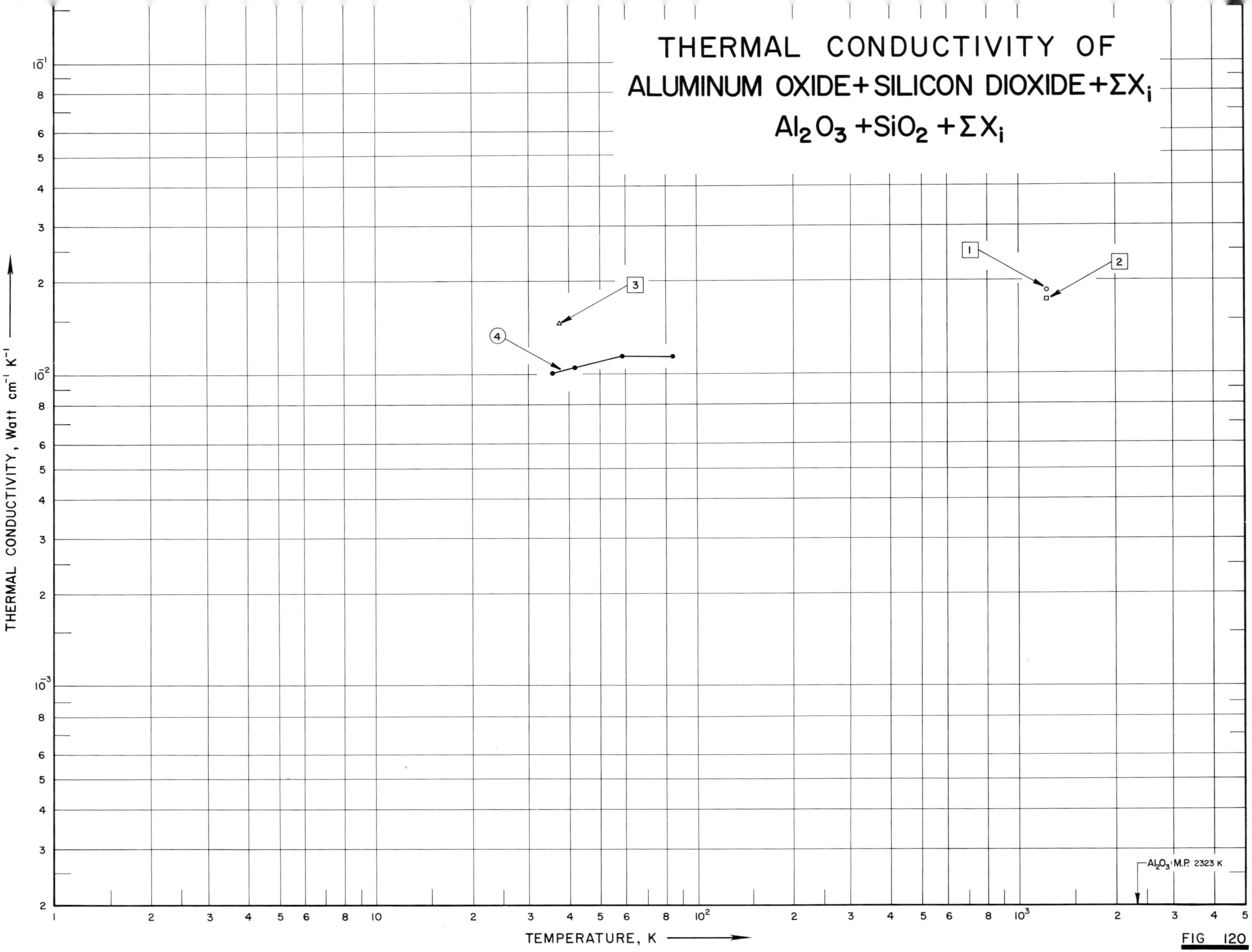

THERMAL CONDUCTIVITY OF ALUMINUM OXIDE + SILICON DIOXIDE + ΣX_i

$Al_2O_3 + SiO_2 + \Sigma X_i$

THERMAL CONDUCTIVITY, Watt cm^{-1} K^{-1}

TEMPERATURE, K

Al_2O_3: M.P. 2323 K

1 2 3 4

FIG 120

SPECIFICATION TABLE NO. 120 THERMAL CONDUCTIVITY OF [ALUMINUM OXIDE + SILICON DIOXIDE + ΣX_i] $Al_2O_3 + SiO_2 + \Sigma X_i$

[For Data Reported in Figure and Table No. 120]

Curve No.	Ref. No.	Method Used	Year	Temp. Range, K	Reported Error, %	Name and Specimen Designation	Composition (weight percent) Al_2O_3	SiO_2	TiO_2	Fe_2O_3	Composition (continued), Specifications and Remarks
1	81	L	1924	1223.2		Sillimanite No. 3	59.75	35.75	2.60	1.92	Specimen 8.5 in dia; 2.053 in. thick; prepared by crushing electric furnace products to pass a No. 14 screen and bonding with fire clay (25.41 Al_2O_3, 59.55 SiO_2, 2.31 Fe_2O_3, 1.33 TiO_2, 1.01 total K_2O and Na_2O, 0.46 CaO, 0.33 MgO and 9.10 loss); fired to cone 16; porosity 15.2%; measured in 748.21 mm Hg pressure.
2	81	L	1924	1223.2		Sillimanite No. 3	59.75	35.75	2.60	1.92	Similar to the above but measured in a pressure of 747.7 mm Hg.
3	458	P	1921	373.2		Chrome brick	30.12	21.80		13.67	19.47 Cr_2O_3, 12.47 MgO, and 0.86 CaO; 4 cm dia x 8 cm long; made of the powder of chrome brick burned at the temperature of Serger's No. 12 cone and pressed into the desired form; density 2.546 g cm^{-3}; thermal conductivity values calculated from measured data of thermal diffusivity, specific heat, and density.
4	247	L	1932	356-840	1.0	Corundum brick	78.82	14.72	2.85	1.35	0.66 MgO, 0.39 CaO, and 0.06 Mn_3O_4; density 3.472 g cm^{-3}; porosity 35.3%; gas permeability 0.068 $m^3 cm\ m^{-2} hr^{-1} (mm\ H_2O)^{-1}$.

DATA TABLE NO. 120 THERMAL CONDUCTIVITY OF [ALUMINUM OXIDE + SILICON DIOXIDE + ΣX_i] $Al_2O_3 + SiO_2 + \Sigma X_i$

[Temperature, T, K; Thermal Conductivity, k, Watt $cm^{-1} K^{-1}$]

T	k
CURVE 1	
1223.2	0.0187
CURVE 2	
1223.2	0.0174
CURVE 3	
373.2	0.0146
CURVE 4	
355.6	0.01012
419.4	0.01051
586.5	0.01147
840.3	0.01146

SPECIFICATION TABLE NO. 121 THERMAL CONDUCTIVITY OF [ALUMINUM OXIDE + TITANIUM DIOXIDE + ΣX_i] $Al_2O_3 + TiO_2 + \Sigma X_i$

Curve No.	Ref. No.	Method Used	Year	Temp. Range, K	Reported Error, %	Name and Specimen Designation	Composition (weight percent) Al_2O_3	TiO_2	SiO_2	Composition (continued), Specifications and Remarks
1	81	L	1924	1223.2		Fused Alundum No. 1	92-96	1.5-4.0	1.0-2.5	0.25-1.0 Fe_2O_3, 0-1.25 ZrO_2; traces of CaO, MgO, and Na_2O; specimen 8.5 in. in dia, 2.54 in. thick; prepared by crushing the electric furnace products to pass a No. 14 screen and by bonding with 10% fire clay (25.41 Al_2O_3, 59.55 SiO_2, 2.31 Fe_2O_3, 1.33 TiO_2, 1.01 total K_2O and Na_2O, 0.46 CaO, 0.33 MgO, and 9.10 loss); fired to cone 16; measured at 752.8 mm Hg pressure; porosity 14.3%.
2	81	L	1924	1223.2		Fused Alundum No. 2	92-96	1.5-4.0	1.0-2.5	0.25-1.0 Fe_2O_3, 0-1.25 ZrO_2; similar to the above specimen but 2.53 in. thick; measured at 736.7 mm Hg pressure; porosity 13.6%.

DATA TABLE NO. 121 THERMAL CONDUCTIVITY OF [ALUMINUM OXIDE + TITANIUM DIOXIDE + ΣX_i] $Al_2O_3 + TiO_2 + \Sigma X_i$

[Temperature, T, K; Thermal Conductivity, k, Watt $cm^{-1} K^{-1}$]

T	k
CURVE 1*	
1223.2	0.0365
CURVE 2*	
1223.2	0.0332

*No graphical presentation

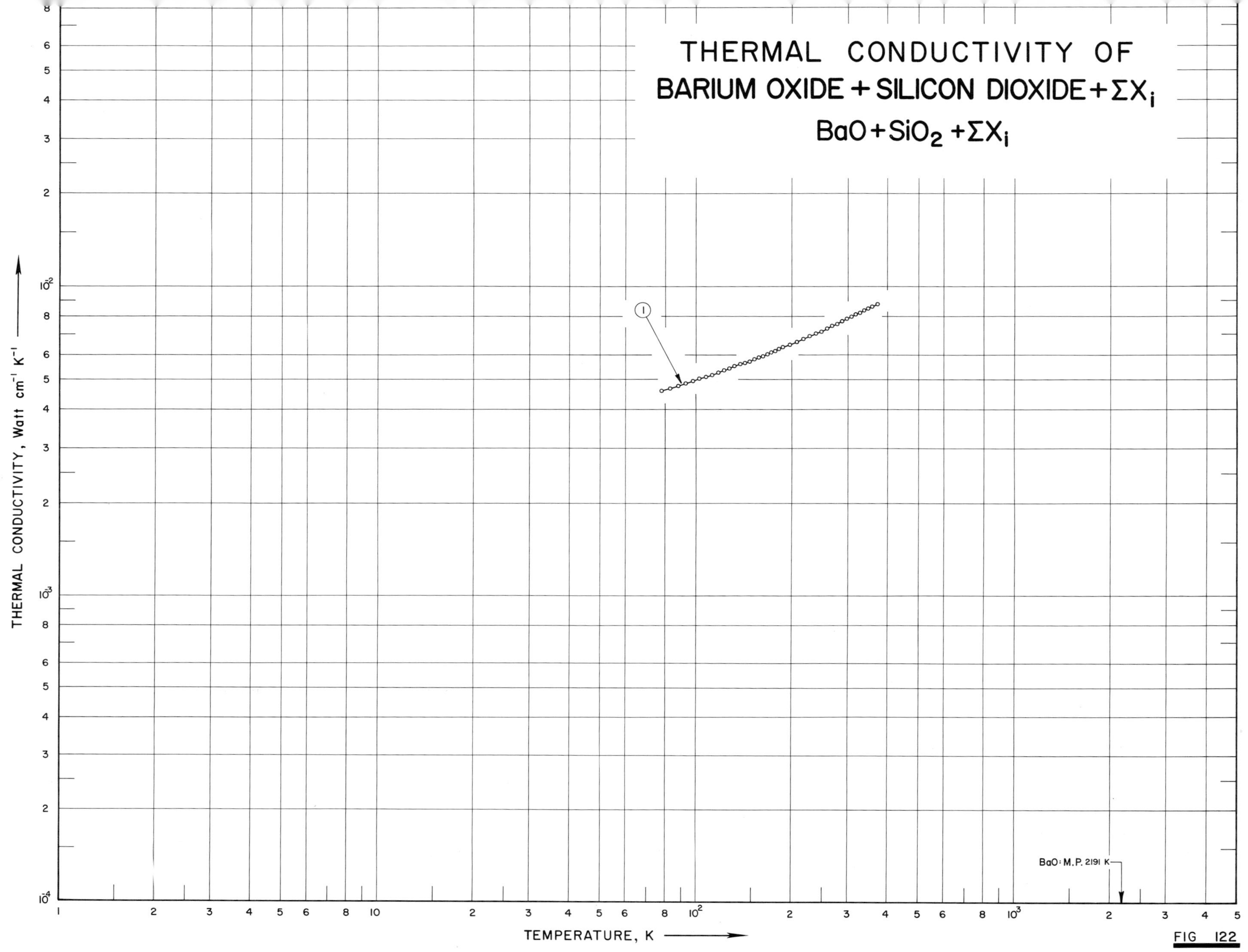
THERMAL CONDUCTIVITY OF
BARIUM OXIDE + SILICON DIOXIDE + ΣXi
BaO+SiO2 +ΣXi
BaO: M.P. 2191 K
1
TEMPERATURE, K
THERMAL CONDUCTIVITY, Watt cm⁻¹ K⁻¹
FIG 122

SPECIFICATION TABLE NO. 122 THERMAL CONDUCTIVITY OF [BARIUM OXIDE + SILICON DIOXIDE + ΣX_i] $BaO + SiO_2 + \Sigma X_i$

[For Data Reported in Figure and Table No. 122]

Curve No.	Ref. No.	Method Used	Year	Temp. Range, K	Reported Error, %	Name and Specimen Designation	Composition (weight percent)				Composition (continued), Specifications and Remarks
							BaO	SiO_2	ZnO	B_2O_3	
1	147	L	1960	78-373		Glass; J	42.9	40.5	7.7	6.5	1.8 Al_2O_3, 0.3 Sb_2O_3, and 0.2 As_2O_3; 3 in. dia x 0.375 in. thick; density 3.56 g cm^{-3}.

DATA TABLE NO. 122 THERMAL CONDUCTIVITY OF [BARIUM OXIDE + SILICON DIOXIDE + ΣX_i] $BaO + SiO_2 + \Sigma X_i$

[Temperature, T, K; Thermal Conductivity, k, Watt $cm^{-1} K^{-1}$]

T	k
CURVE 1	
78.2	0.00460
83.2	0.00469
88.2	0.00477
93.2	0.00485
98.2	0.00494
103.2	0.00502
108.2	0.00510
113.2	0.00519
118.2	0.00527
123.2	0.00536
128.2	0.00544
133.2	0.00552
138.2	0.00561
143.2	0.00565
148.2	0.00573
153.2	0.00582
158.2	0.00590
163.2	0.00594
168.2	0.00602
173.2	0.00611
178.2	0.00619
183.2	0.00628
188.2	0.00636
193.2	0.00640*
198.2	0.00649
203.2	0.00657*
208.2	0.00661
213.2	0.00669*
218.2	0.00678
223.2	0.00682*
228.2	0.00690
233.2	0.00699*
238.2	0.00703
243.2	0.00711*
248.2	0.00715
253.2	0.00724*
258.2	0.00732
263.2	0.00736*
268.2	0.00745
273.2	0.00753*
278.2	0.00757
283.2	0.00766*
288.2	0.00774
293.2	0.00778*
298.2	0.00787
303.2	0.00791*
CURVE 1 (cont.)	
308.2	0.00799
313.2	0.00803*
318.2	0.00812
323.2	0.00820*
328.2	0.00824
333.2	0.00828*
338.2	0.00837
343.2	0.00845*
348.2	0.00849
353.2	0.00854*
358.2	0.00862
363.2	0.00866*
368.2	0.00870*
373.2	0.00879

*Not shown on plot

SPECIFICATION TABLE NO. 123 THERMAL CONDUCTIVITY OF [BARIUM OXIDE + STRONTIUM OXIDE + ΣX_i] BaO + SrO + ΣX_i

Curve No.	Ref. No.	Method Used	Year	Temp. Range, K	Reported Error, %	Name and Specimen Designation	Composition (weight percent), Specifications and Remarks
1	75	L	1955	458-853	<10	Tube No. 6	56.7 BaO, 38.3 SrO, and 5.0 Zr; a mixture of 1 mole BaO, 1 mole SrO and 5 weight percent Zr; BaO and SrO added in the form of their carbonates and heated to 1040 C to drive off the CO_2; apparent thermal conductivity (effects due to radiation at high temperatures not considered).
2	75	L	1955	618-1043		Tube No. 4	An equimolecular mixture of polycrystalline BaO and SrO with 2.5% ZrO added.

DATA TABLE NO. 123 THERMAL CONDUCTIVITY OF [BARIUM OXIDE + STRONTIUM OXIDE + ΣX_i] BaO + SrO + ΣX_i

[Temperature, T, K; Thermal Conductivity, k, Watt $cm^{-1} K^{-1}$]

T	k
CURVE 1*	
458.2	0.000033
583.2	0.000040
663.2	0.000048
773.2	0.000068
813.2	0.000070
853.2	0.000091
CURVE 2*	
618.2	0.0000480
918.2	0.000123
978.2	0.000148
1043.2	0.000178

*No graphical presentation

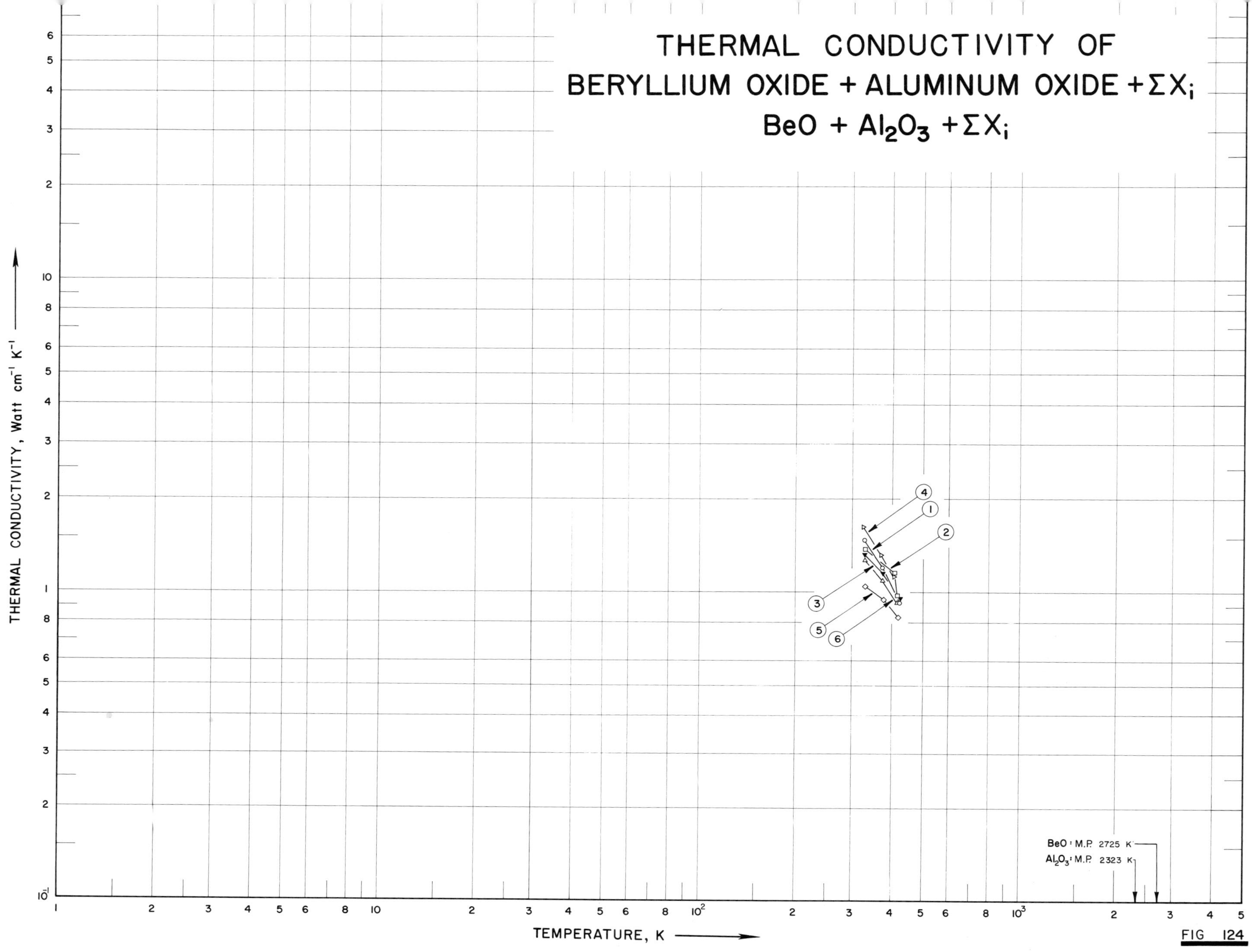
THERMAL CONDUCTIVITY OF
BERYLLIUM OXIDE + ALUMINUM OXIDE + ΣXi
BeO + Al2O3 + ΣXi
THERMAL CONDUCTIVITY, Watt cm-1 K-1
TEMPERATURE, K
BeO: M.P. 2725 K
Al2O3: M.P. 2323 K
1
2
3
4
5
6

FIG 124

SPECIFICATION TABLE NO. 124 THERMAL CONDUCTIVITY OF [BERYLLIUM OXIDE + ALUMINUM OXIDE + ΣX_i] $BeO + Al_2O_3 + \Sigma X_i$

[For Data Reported in Figure and Table No. 124]

Curve No.	Ref. No.	Method Used	Year	Temp. Range, K	Reported Error, %	Name and Specimen Designation	Composition (weight percent) BeO	Al_2O_3	ThO_2	ZrO_2	Composition (continued), Specifications and Remarks
1	141	L	1950	329-423		A	95.0	2.5	2.5		Blended minus 325 mesh or finer size particles of fluorescent grade BeO and special acid-washed Al_2O_3 in the form of a heavy slip, then dried, and pressed (pressure 10,000 lb in^{-2}) with sufficient 2% dextrine solution added to facilitate the pressing; maturing temp 1700 C based on linear firing shrinkage and absorption.
2	141	L	1950	330-415		B	90.0	5.0	5.0		Similar to the above specimen.
3	141	L	1950	330-426		C	80.0	10.0	10.0		Similar to the above specimen.
4	141	L	1950	326-408		N	95.0	2.5		2.5	Prepared by minus 325 mesh or finer sized particles of fluorescent grade BeO, special acid-washed Al_2O_3, and chemically pure ZrO_2; blended in the form of a heavy slip, then dried and pressed (pressure 10,000 lb in^{-2}) with sufficient 2% dextrine solution added to facilitate pressing; maturing temp 1700 C based on linear firing shrinkage and absorption.
5	141	L	1950	330-420		O	90.0	5.0		5.0	Similar to the above specimen.
6	141	L	1950	329-418		P	80.0	10.0		10.0	Similar to the above specimen.

DATA TABLE NO. 124 THERMAL CONDUCTIVITY OF [BERYLLIUM OXIDE + ALUMINUM OXIDE + ΣX_i] $BeO + Al_2O_3 + \Sigma X_i$

[Temperature, T, K; Thermal Conductivity, k, Watt $cm^{-1} K^{-1}$]

T	k
CURVE 1	
329.2	1.477
377.2	1.201
423.2	0.925
CURVE 2	
330.2	1.339
410.2	1.167
415.2	0.975
CURVE 3	
330.2	1.339
379.2	1.163
426.2	0.958
CURVE 4	
326.2	1.62
370.2	1.33
408.2	1.13
CURVE 5	
330.2	1.05
378.2	0.946
420.2	0.833
CURVE 6	
329.2	1.27
375.2	1.09
418.2	0.928

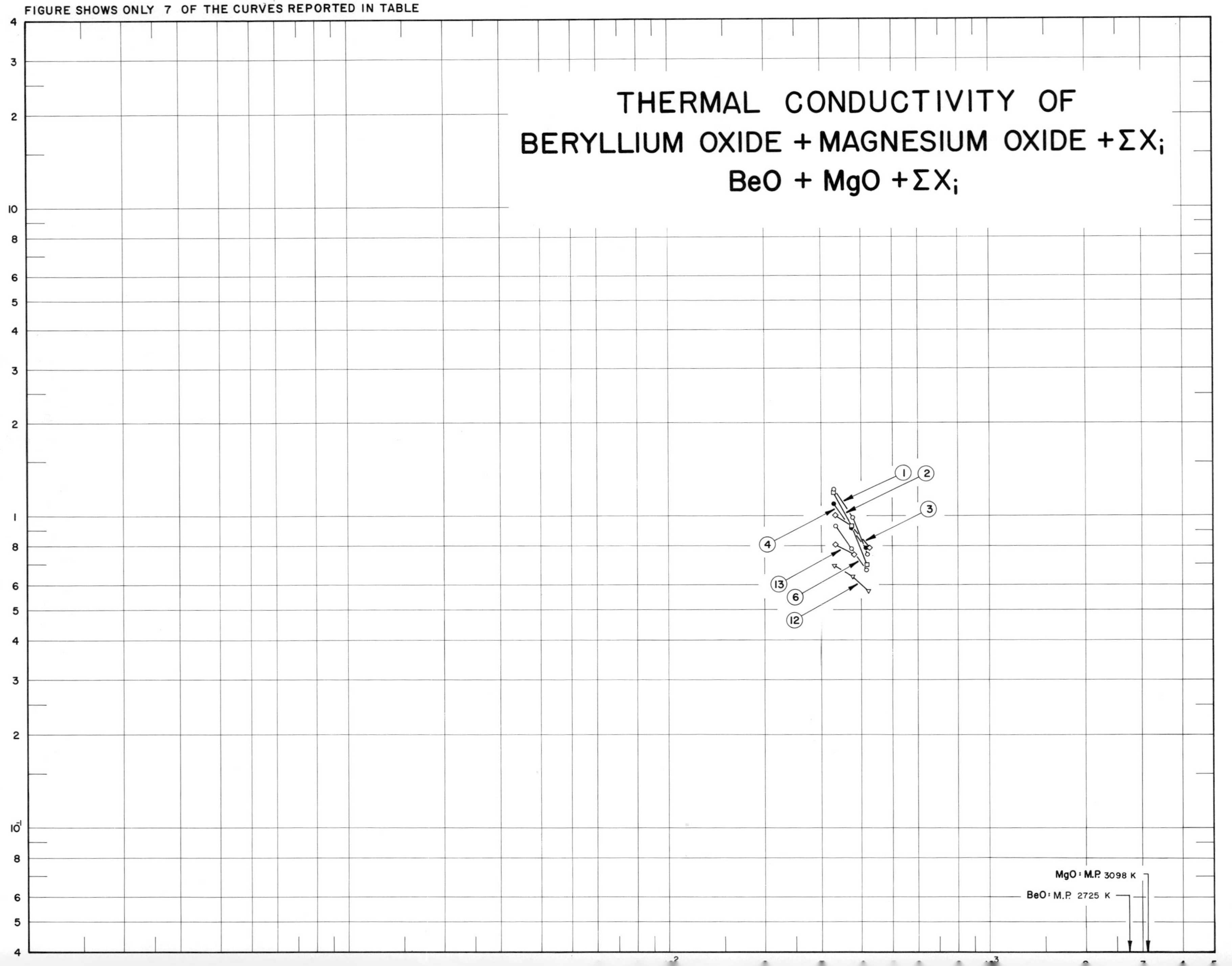

FIGURE SHOWS ONLY 7 OF THE CURVES REPORTED IN TABLE
THERMAL CONDUCTIVITY OF
BERYLLIUM OXIDE + MAGNESIUM OXIDE + ΣXi
BeO + MgO + ΣXi
THERMAL CONDUCTIVITY, Watt cm⁻¹ K⁻¹
MgO: M.P. 3098 K
BeO: M.P. 2725 K
1
2
3
4
6
12
13

SPECIFICATION TABLE NO. 125 THERMAL CONDUCTIVITY OF [BERYLLIUM OXIDE + MAGNESIUM OXIDE + ΣX_i] $BeO + MgO + \Sigma X_i$

[For Data Reported in Figure and Table No. 125]

Curve No.	Ref. No.	Method Used	Year	Temp. Range, K	Reported Error, %	Name and Specimen Designation	Composition (weight percent) BeO	MgO	Al_2O_3	ThO_2	Composition (continued), Specifications and Remarks
1	141	L	1950	329-419		A 4	91.2	4.0	2.4	2.4	Fluorescent grade BeO; special acid-washed Al_2O_3, particle size of all the raw materials was minus 325 mesh or finer; materials were blended first in the proper proportions in the form of a heavy slip, then dried, and then sufficient 2% dextrine solution was added to facilitate pressing; specimen was formed at a pressure of 10,000 lb in^{-2}; maturing temp of 1600 C based on linear firing shrinkage and absorption.
2	141	L	1950	329-418		A 8	87.4	8.0	2.3	2.3	Same description as the above specimen.
3	141	L	1950	331-425		A 15	80.8	15.0	2.1	2.1	Same description as the above specimen.
4	141	L	1950	329-415		B 8	82.8	8.0	4.6	4.6	Same description as the above specimen except maturing temp 1600 C.
5	141	L	1950	327-414		B 15	76.5	15.0	4.25	4.25	Same description as the above specimen.
6	141	L	1950	328-417		C 15	68.0	15.0	8.5	8.5	Same description as the above specimen except maturing temp 1650 C.
7	141	L	1950	327-404		N 4	91.2	4.0	2.4		2.4 ZrO_2; fluorescent grade BeO; special acid-washed Al_2O_3; chemically pure grade ZrO_2; $MgCO_3$ was calcined to 2200 F before being added to the batch; particle size of all the raw materials was minus 325 mesh or finer; materials were blended first in the proper proportions in the form of a heavy slip, then dried, and then sufficient 2% dextrine solution was added to facilitate pressing; specimen was formed at a pressure of 10,000 psi; maturing temp 1500 C based on linear firing shrinkage and absorption.
8	141	L	1950	328-413		N 8	87.4	8.0	2.3		2.3 ZrO_2; same description as the above specimen.
9	141	L	1950	325-409		N 15	80.8	15.0	2.1		2.1 ZrO_2; same description as the above specimen.
10	141	L	1950	329-414		O 8	82.8	8.0	4.6		4.6 ZrO_2; same description as specimen N 4.
11	141	L	1950	328-415		O 15	76.5	15.0	4.25		4.25 ZrO_2; same description as specimen N 4 except maturing temp 1550 C.
12	141	L	1950	331-423		P 15	68.0	15.0	8.5		8.5 ZrO_2; same description as the specimen above.
13	141	L	1950	332-420		Y 8	77.5	8.0	6.5		8.0 ZrO_2; same description as the specimen above.

DATA TABLE NO. 125 THERMAL CONDUCTIVITY OF [BERYLLIUM OXIDE + MAGNESIUM OXIDE + ΣX_i] $BeO + MgO + \Sigma X_i$

[Temperature, T, K; Thermal Conductivity, k, Watt $cm^{-1} K^{-1}$]

T	k	T	k
CURVE 1		CURVE 9*	
329.2	1.218	325.2	1.079
377.2	0.987	367.2	0.908
419.2	0.749	408.7	0.791
CURVE 2		CURVE 10*	
329.2	1.184	329.2	1.008
375.2	0.929	372.2	1.025
418.2	0.695	414.2	0.749
CURVE 3		CURVE 11*	
331.2	1.008	328.2	0.908
374.2	0.925*	375.2	0.795
425.2	0.782	415.2	0.699
CURVE 4		CURVE 12	
329.2	1.092	331.2	0.690
373.2	0.920	378.2	0.636
415.2	0.787	423.2	0.573
CURVE 5*		CURVE 13	
327.2	1.017	332.2	0.808
371.2	0.900	380.2	0.749
414.2	0.778	420.2	0.699*
CURVE 6			
328.2	0.929		
374.2	0.782		
417.2	0.669		
CURVE 7*			
327.2	1.117		
375.2	1.075		
404.2	0.874		
CURVE 8*			
328.2	1.079		
373.2	0.925		
413.2	0.757		

* Not shown on Plot

THERMAL CONDUCTIVITY OF BERYLLIUM OXIDE + THORIUM DIOXIDE $+\Sigma X_i$

$BeO + ThO_2 + \Sigma X_i$

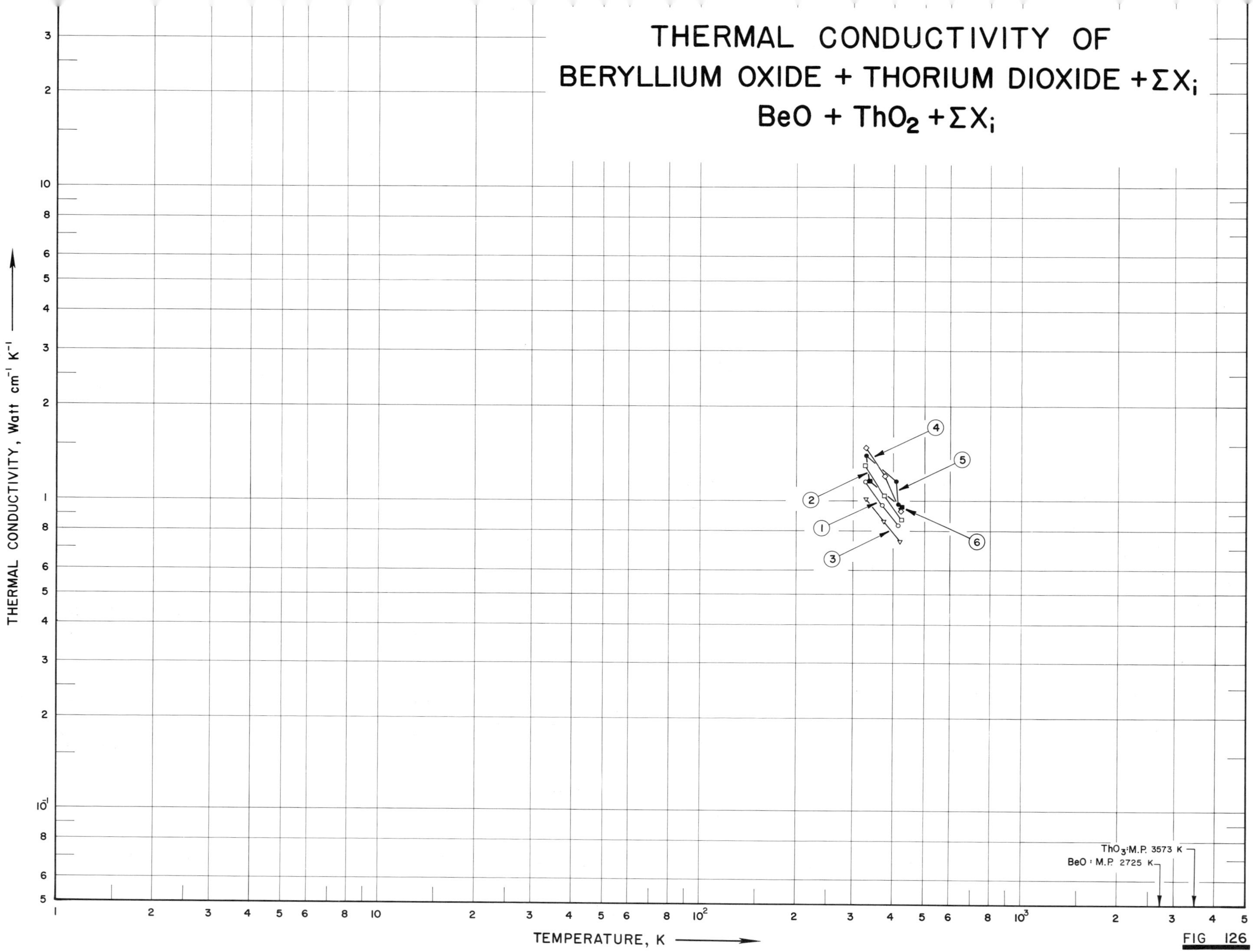

THERMAL CONDUCTIVITY, Watt cm^{-1} K^{-1}

TEMPERATURE, K

ThO_3: M.P. 3573 K

BeO : M.P. 2725 K

1 2 3 4 5 6

FIG 126

SPECIFICATION TABLE NO. 126 THERMAL CONDUCTIVITY OF [BERYLLIUM OXIDE + THORIUM DIOXIDE + ΣX_i] $BeO + ThO_2 + \Sigma X_i$

[For Data Reported in Figure and Table No. 126]

Curve No.	Ref. No.	Method Used	Year	Temp. Range, K	Reported Error, %	Name and Specimen Designation	Composition (weight percent) BeO	ThO_2	Al_2O_3	MgO	Composition (continued), Specifications and Remarks
1	141	L	1950	328-415		B 4	86.4	4.8	4.8	4.0	Prepared from minus 325 mesh fluorescent grade BeO and special acid-washed Al_2O_3 by blending in the form of a slip, drying, and pressing at 10,000 lb in^{-2} with dextrine as binder; fired at 1650 C.
2	141	L	1950	328-425		C 4	76.8	9.6	9.6	4.0	Similar to the above specimen.
3	141	L	1950	329-421		C 8	73.6	9.2	9.2	8.0	Similar to the above specimen.
4	141	L	1950	329-423		A	95.0	2.5	2.5		Blended minus 325 mesh or finer sized particles of fluorescent grade BeO and special acid-washed Al_2O_3 in the form of a heavy slip, then dried, and pressed (pressure 10,000 lb in^{-2}) with sufficient 2% dextrine solution added to facilitate the pressing; maturing temp 1700 C based on linear firing shrinkage and absorption.
5	141	L	1950	330-415		B	90.0	5.0	5.0		Similar to the above specimen.
6	141	L	1950	330-426		C	80.0	10.0	10.0		Similar to the above specimen.

DATA TABLE NO. 126 THERMAL CONDUCTIVITY OF [BERYLLIUM OXIDE + THORIUM DIOXIDE + ΣX_i] $BeO + ThO_2 + \Sigma X_i$

[Temperature, T, K; Thermal Conductivity, k, Watt $cm^{-1} K^{-1}$]

T	k
CURVE 1	
328.2	1.151
369.2	0.967
415.2	0.833
CURVE 2	
328.2	1.297
376.2	1.042
425.2	0.870
CURVE 3	
329.2	1.013*
375.2	0.858
421.2	0.741
CURVE 4	
329.2	1.477
377.2	1.201
423.2	0.925
CURVE 5	
330.2	1.339
410.2	1.167
415.2	0.975
CURVE 6*	
330.2	1.339
379.2	1.163
426.2	0.958

*Not shown on Plot

THERMAL CONDUCTIVITY OF BERYLLIUM OXIDE + ZIRCONIUM DIOXIDE + ΣX_i

$BeO + ZrO_2 + \Sigma X_i$

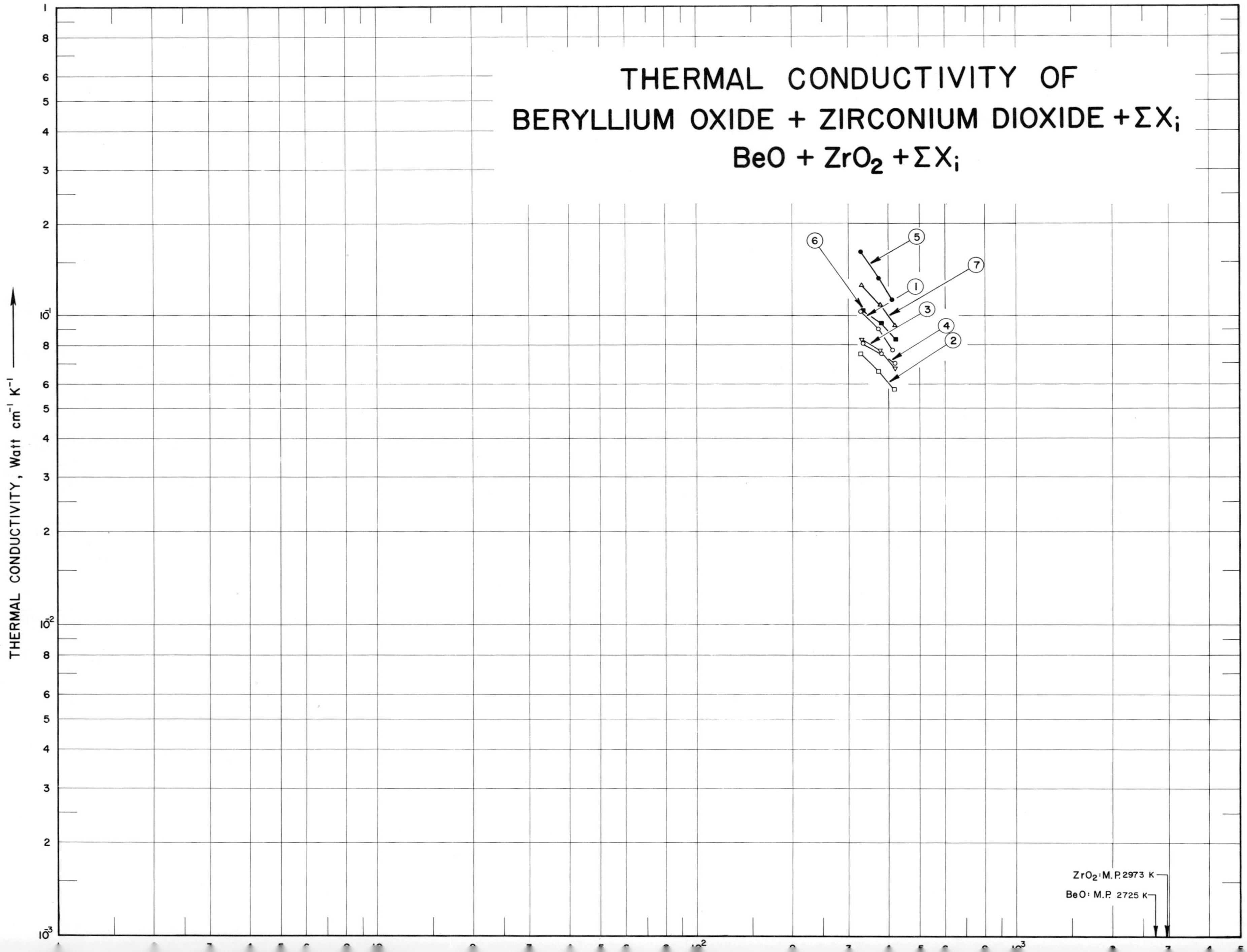

THERMAL CONDUCTIVITY, Watt cm^{-1} K^{-1}

ZrO_2: M.P. 2973 K

BeO: M.P. 2725 K

SPECIFICATION TABLE NO. 127 THERMAL CONDUCTIVITY OF [BERYLLIUM OXIDE + ZIRCONIUM DIOXIDE + ΣX_i] $BeO + ZrO_2 + \Sigma X_i$

[For Data Reported in Figure and Table No. 127]

Curve No.	Ref. No.	Method Used	Year	Temp. Range, K	Reported Error, %	Name and Specimen Designation	Composition (weight percent) BeO	ZrO_2	Al_2O_3	MgO	Composition (continued), Specifications and Remarks
1	141	L	1950	328-414		O 4	86.4	4.8	4.8	4.0	Prepared from -325 mesh fluorescent grade BeO special acid-washed Al_2O_3, chemically pure ZrO_2, and calcined at 2200 F $MgCO_3$ by blending in the form of a slip, drying, and pressing at 10,000 lb in^{-2} with dextrine as binder; fired at 1550 C.
2	141	L	1950	328-417		P 4	76.8	9.6	9.6	4.0	Similar to the above specimen except fired at 1650 C.
3	141	L	1950	330-420		P 8	73.6	9.2	9.2	8.0	Similar to the above specimen except fired at 1550 C.
4	141	L	1950	332-420		Y 8	77.5	8.0	6.5	8.0	Similar to the above specimen.
5	141	L	1950	326-408		N	95.0	2.5	2.5		Prepared by minus 325 mesh or finer sized particles of fluorescent grade BeO, special acid-washed Al_2O_3 and chemcially pure ZrO_2; blended in the form of a heavy slip, then dried and pressed (pressure 10,000 lb in^{-2}) with sufficient 2% dextrine solution added to facilitate pressing; maturing temp 1700 C based on linear firing shrinkage and absorption.
6	141	L	1950	330-420		O	90.0	5.0	5.0		Similar to the above specimen.
7	141	L	1950	329-418		P	80.0	10.0	10.0		Similar to the above specimen.

DATA TABLE NO. 127 THERMAL CONDUCTIVITY OF [BERYLLIUM OXIDE + ZIRCONIUM DIOXIDE + ΣX_i] $BeO + ZrO_2 + \Sigma X_i$

[Temperature, T, K; Thermal Conductivity, k, Watt $cm^{-1} K^{-1}$)

T	k
CURVE 1	
328.2	1.042
373.2	0.900
414.2	0.770
CURVE 2	
328.2	0.745
373.2	0.661
417.2	0.577
CURVE 3	
330.2	0.833
377.2	0.766
420.2	0.674
CURVE 4	
332.2	0.808
380.2	0.749
420.2	0.699
CURVE 5	
326.2	1.62
370.2	1.33
408.2	1.13
CURVE 6	
330.2	1.05
378.2	0.946
420.2	0.833
CURVE 7	
329.2	1.27
375.2	1.09
418.2	0.928

SPECIFICATION TABLE NO. 128 THERMAL CONDUCTIVITY OF [DICHROMIUM TRIOXIDE + MAGNESIUM OXIDE + ΣX_i] $Cr_2O_3 + MgO + \Sigma X_i$

Curve No.	Ref. No.	Method Used	Year	Temp. Range, K	Reported Error, %	Name and Specimen Designation	Composition (weight percent) Cr_2O_3	MgO	Fe_2O_3	Al_2O_3	SiO_2	Mn_2O_4	Composition (continued), Specifications and Remarks
1	247	L	1932	379-794	1.0	Chromite Brick 23	40.09	22.68	13.16	10.48	10.53	2.32	0.68 CaO, trace TiO_2; density 3.988 g cm^{-3}; porosity 27.3%; gas permeability 0.555 m^3-cm per m^2 hr per mm of H_2O.

DATA TABLE NO. 128 THERMAL CONDUCTIVITY OF [DICHROMIUM TRIOXIDE + MAGNESIUM OXIDE + ΣX_i] $Cr_2O_3 + MgO + \Sigma X_i$

[Temperature, T, K; Thermal Conductivity, k, Watt $cm^{-1} K^{-1}$]

T	k
CURVE 1*	
379.4	0.0232
490.5	0.0217
793.9	0.0163

* No graphical presentation

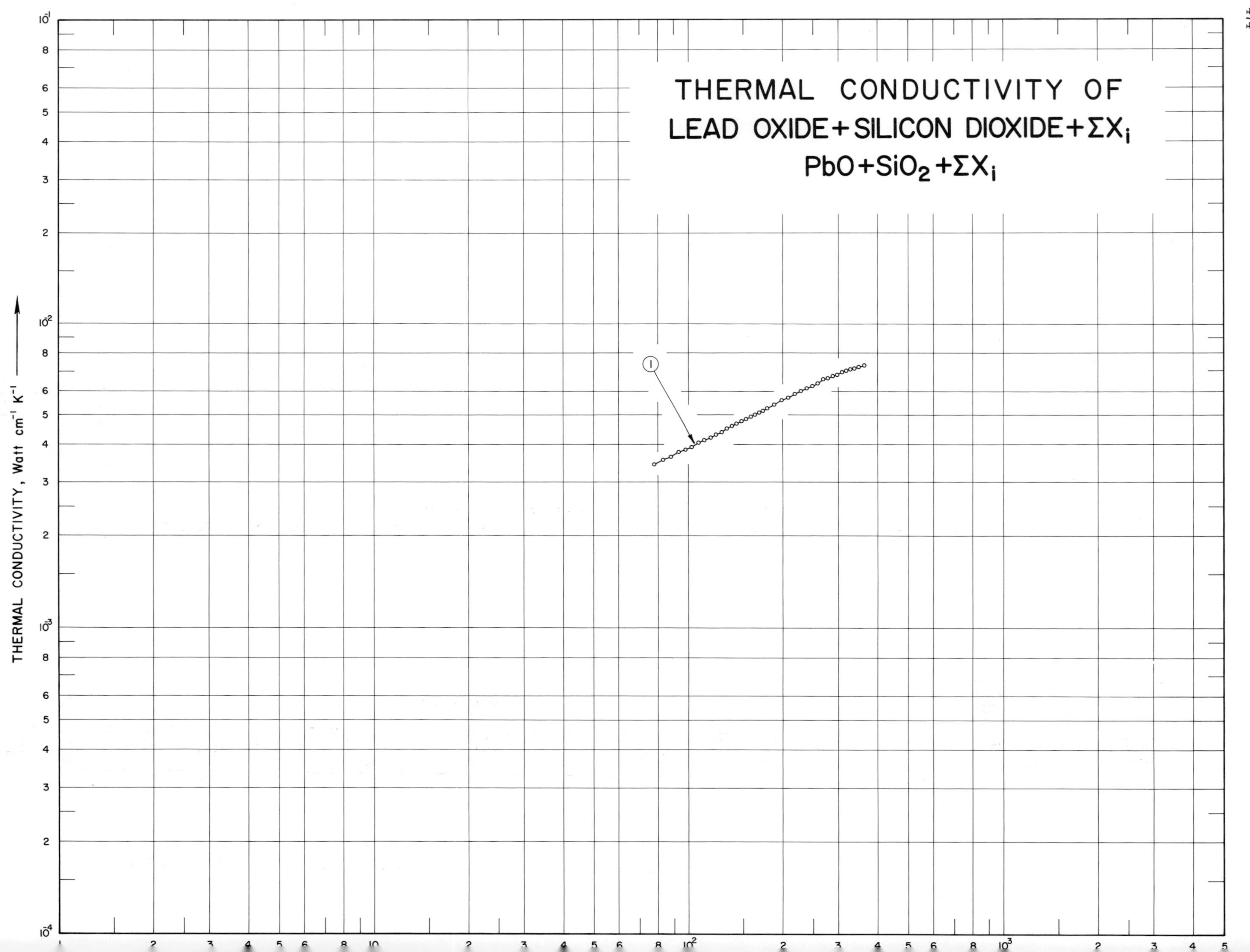
THERMAL CONDUCTIVITY OF
LEAD OXIDE + SILICON DIOXIDE + ΣXi
PbO + SiO2 + ΣXi
1
THERMAL CONDUCTIVITY, Watt cm⁻¹ K⁻¹

SPECIFICATION TABLE NO. 129 THERMAL CONDUCTIVITY OF [LEAD OXIDE + SILICON DIOXIDE + ΣX_i] $PbO + SiO_2 + \Sigma X_i$

[For Data Reported in Figure and Table No. 129]

Curve No.	Ref. No.	Method Used	Year	Temp. Range, K	Reported Error, %	Name and Specimen Designation	Composition (weight percent)				Composition (continued), Specifications and Remarks
							PbO	SiO_2	K_2O	Al_2O_3	
1	147	L	1960	78-363		K	59.7	35.6	4.4	0.2	3 in. dia x 0.375 in. thick; density 4.29 g cm^{-3}.

DATA TABLE NO. 129 THERMAL CONDUCTIVITY OF [LEAD OXIDE + SILICON DIOXIDE + ΣX_i] $PbO + SiO_2 + \Sigma X_i$

[Temperature, T, K; Thermal Conductivity, k, Watt $cm^{-1} K^{-1}$]

T	k	T	k
CURVE 1		CURVE 1 (cont.)	
78.2	0.00343	303.2	0.00690*
83.2	0.00356	308.2	0.00695
88.2	0.00364	313.2	0.00699*
93.2	0.00377	318.2	0.00703
98.2	0.00385	323.2	0.00707*
103.2	0.00393	328.2	0.00711
108.2	0.00406	333.2	0.00715*
113.2	0.00414	338.2	0.00715
118.2	0.00423	343.2	0.00720*
123.2	0.00431	348.2	0.00724
128.2	0.00439	353.2	0.00728*
133.2	0.00452	358.2	0.00732*
138.2	0.00460	363.2	0.00732
143.2	0.00469		
148.2	0.00477		
153.2	0.00485		
158.2	0.00494		
163.2	0.00502		
168.2	0.00510		
173.2	0.00519		
178.2	0.00527		
183.2	0.00536*		
188.2	0.00544		
193.2	0.00552*		
198.2	0.00561		
203.2	0.00569*		
208.2	0.00573		
213.2	0.00582*		
218.2	0.00590		
223.2	0.00598*		
228.2	0.00602		
233.2	0.00611*		
238.2	0.00619		
243.2	0.00623*		
248.2	0.00628		
253.2	0.00636*		
258.2	0.00640		
263.2	0.00649*		
268.2	0.00663		
273.2	0.00661*		
278.2	0.00665		
283.2	0.00669*		
288.2	0.00674		
293.2	0.00678*		
298.2	0.00682		

*Not shown on plot

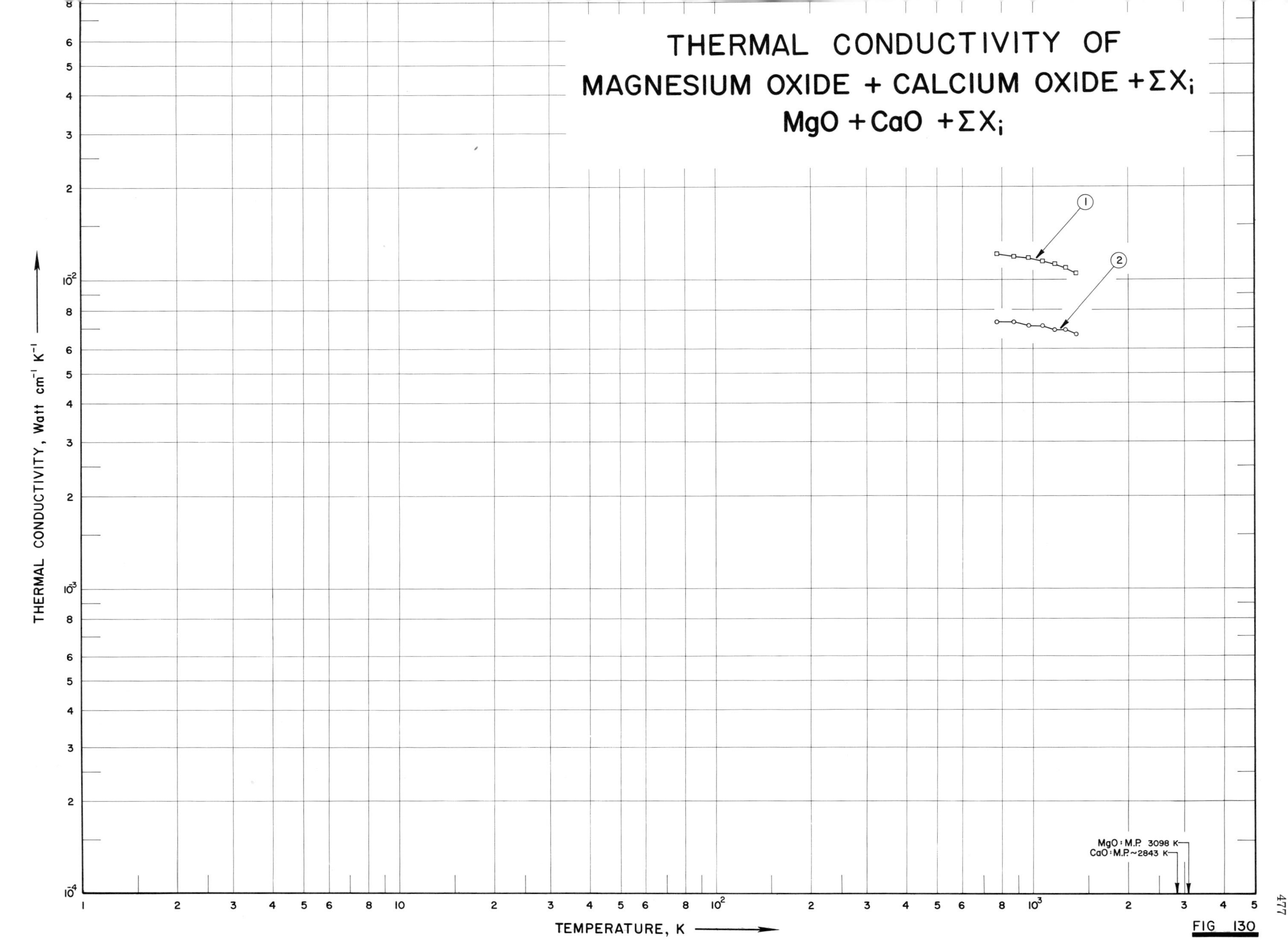
THERMAL CONDUCTIVITY OF
MAGNESIUM OXIDE + CALCIUM OXIDE + ΣXi
MgO + CaO + ΣXi
MgO: M.P. 3098 K
CaO: M.P. ~2843 K
THERMAL CONDUCTIVITY, Watt cm⁻¹ K⁻¹
TEMPERATURE, K
1
2

FIG 130

SPECIFICATION TABLE NO. 130 THERMAL CONDUCTIVITY OF [MAGNESIUM OXIDE + CALCIUM OXIDE + ΣX_i] $MgO + CaO + \Sigma X_i$

[For Data Reported in Figure and Table No. 130]

Curve No.	Ref. No.	Method Used	Year	Temp. Range, K	Reported Error, %	Name and Specimen Designation	Composition (weight percent) MgO	CaO	Fe_2O_3	Composition (continued), Specifications and Remarks
1	238	P	1921	773-1373		Magnesite brick J	81.79	5.24	1.87	Very close texture; brick size: 9 x 4.5 x 2.5 in.; apparent density 2.63 g cm^{-3}; true density 3.29 g cm^{-3} porosity 20.0%; heat flow in the direction of the length of the brick.
2	238	P	1921	773-1373		Magnesite brick L	87.88	4.68	2.56	Texture not so close; porosity 24.5%; apparent density 2.56 g cm^{-3}; true density 3.28 g cm^{-3}.

DATA TABLE NO. 130 THERMAL CONDUCTIVITY OF [MAGNESIUM OXIDE + CALCIUM OXIDE + ΣX_i] $MgO + CaO + \Sigma X_i$

[Temperature, T, K; Thermal Conductivity, k, Watt $cm^{-1}K^{-1}$]

T	k
CURVE 1	
773.2	0.00732
873.2	0.00732
973.2	0.00711
1073.2	0.00711
1173.2	0.00690
1273.2	0.00690
1373.2	0.00690
CURVE 2	
773.2	0.0121
873.2	0.0119
973.2	0.0117
1073.2	0.0115
1173.2	0.0113
1273.2	0.0109
1373.2	0.0105

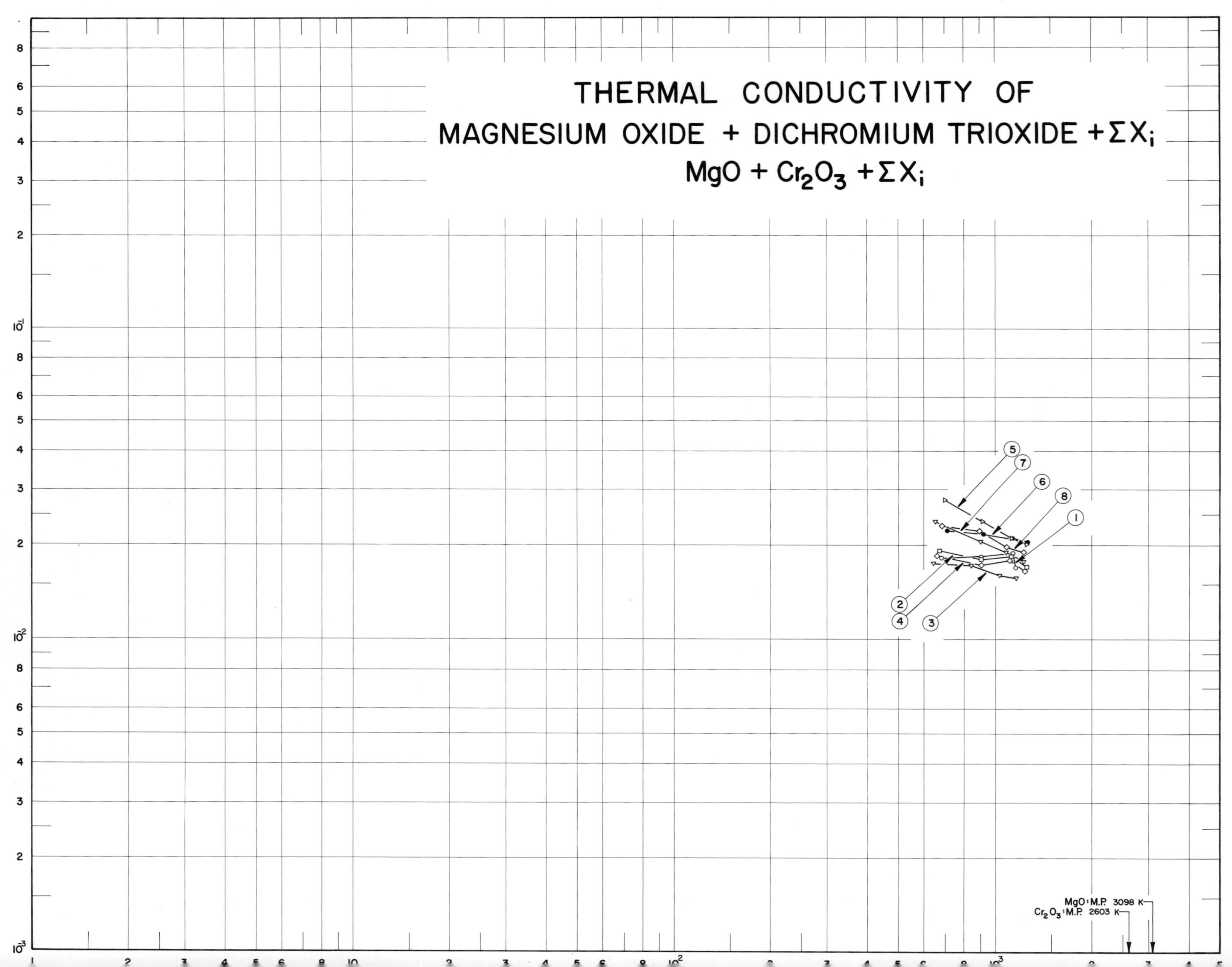
THERMAL CONDUCTIVITY OF
MAGNESIUM OXIDE + DICHROMIUM TRIOXIDE + ΣX_i
$MgO + Cr_2O_3 + \Sigma X_i$
THERMAL CONDUCTIVITY, Watt cm^{-1} K^{-1}
MgO: M.P. 3098 K
Cr_2O_3: M.P. 2603 K
1
2
3
4
5
6
7
8

[For Data Reported in Figure and Table No. 131]

Curve No.	Ref. No.	Method Used	Year	Temp. Range, K	Reported Error, %	Name and Specimen Designation	Composition (weight percent) ~~MgO~~ Al_2O_3	Cr_2O_3	Fe_2O_3	~~Al_2O_3~~ MgO	SiO_2	Composition (continued), Specifications and Remarks
1	143	L	1955	683-1263		Magnezit; 4	12.59	20.48	9.15	49.46	5.24	1.26 CaO, 1.90 FeO, 0.09 total Ca, Mg, Fe, and Mn; chromomagnesite refractory brick; density 2.95 g cm^{-3}; apparent porosity 22.8%; gas permeability 0.455 m^3 x cm $m^{-2}hr^{-1}$ (mm $H_2O)^{-1}$.
2	143	L	1955	673-1253		Magnezit; 5	12.59	20.48	9.15	49.46	5.24	1.26 CaO, 1.90 FeO, 0.09 total Ca, Mg, Fe, and Mn; same as the above specimen.
3	143	L	1955	643-1168		Magnezit; 6	12.59	20.48	9.15	49.46	5.24	1.26 CaO, 1.90 FeO, 0.09 total Ca, Mg, Fe, and Mn; same as the above specimen.
4	143	L	1955	658-1248		K Marksa; 11	12.34	24.35	11.94	42.31	6.14	1.65 CaO, 1.71 FeO, 0.13 total Ca, Mg, Fe, and Mn; chromomagnesite refractory brick; density 3.03 g cm^{-3}; apparent porosity 23.5%; gas permeability 0.303 m^3 x cm $m^{-2}hr^{-1}$ (mm $H_2O)^{-1}$.
5	143	L	1955	698-1258		Magnezit; 7	8.51	12.28	5.80	64.85	4.28	1.44 CaO, 1.67 FeO, 0.21 total Ca, Mg, Fe, and Mn; chromomagnesite heat resistant refractory brick; density 3.04 g cm^{-3}; apparent porosity 19.1%; gas permeability 0.480 m^3 x cm $m^{-2}hr^{-1}$ (mm $H_2O)^{-1}$.
6	143	L	1955	708-1263		Magnezit; 8	8.51	12.28	5.80	64.85	4.28	1.44 CaO, 1.67 FeO, 0.21 total Ca, Mg, Fe, and Mn; chromomagnesite heat resistant refractory brick; similar to the above specimen.
7	143	L	1955	653-1233		Magnezit; 9	8.51	12.28	5.80	64.85	4.28	1.44 CaO, 1.67 FeO, 0.21 total Ca, Mg, Fe, and Mn; chromomagnesite heat resistant refractory brick; similar to the above specimen.
8	143	L	1955	683-1228		Ordzhonikidze; 10	11.46	22.34	13.04	42.87	5.42	2.88 FeO, 1.76 CaO, trace total Ca, Mg, Fe, and Mn; chromomagnesite heat resistant refractory brick; density 2.95 g cm^{-3}; apparent porosity 25.6%; gas permeability 0.598 m^3 x cm $m^{-2}hr^{-1}$ (mm $H_2O)^{-1}$.

DATA TABLE NO. 131 THERMAL CONDUCTIVITY OF [MAGNESIUM OXIDE + DICHROMIUM TRIOXIDE + ΣX_i] $MgO + Cr_2O_3 + \Sigma X_i$

[Temperature, T, K; Thermal Conductivity, k, Watt $cm^{-1}K^{-1}$]

T	k	T	k
CURVE 1		CURVE 7	
683.2	0.0182	653.2	0.0238
903.2	0.0184	898.2	0.0206
1143.2	0.0188	1088.2	0.0189
1263.2	0.0169	1233.2	0.0177
CURVE 2		CURVE 8	
673.2	0.0192	683.2	0.0231
903.2	0.0180	893.2	0.0222
1123.2	0.0184	1083.2	0.0198
1253.2	0.0171	1228.2	0.0189
CURVE 3			
643.2	0.0175		
843.2	0.0173		
1033.2	0.0160		
1168.2	0.0157		
CURVE 4			
658.2	0.0185		
903.2	0.0173		
1113.2	0.0179		
1248.2	0.0166		
CURVE 5			
698.2	0.0277		
913.2	0.0238		
1123.2	0.0211		
1258.2	0.0201		
CURVE 6			
708.2	0.0223		
918.2	0.0217		
1133.2	0.0213		
1263.2	0.0205		

SPECIFICATION TABLE NO. 132 THERMAL CONDUCTIVITY OF [MAGNESIUM OXIDE + DIIRON TRIOXIDE + ΣX_i] $MgO + Fe_2O_3 + \Sigma X_i$

Curve No.	Ref. No.	Method Used	Year	Temp. Range, K	Reported Error, %	Name and Specimen Designation	Composition (weight percent) MgO	Fe_2O_3	CaO	SiO_2	Al_2O_3	Composition (continued), Specifications and Remarks
1	252	L	1933	409-1561		Magnesite brick	86.8	6.3	3.0	2.6	0.8	Vol. density 2.54 g cm^{-3}, true density 3.59 g cm^{-3}, porosity 26-29%.

DATA TABLE NO. 132 THERMAL CONDUCTIVITY OF [MAGNESIUM OXIDE + DIIRON TRIOXIDE + ΣX_i] $MgO + Fe_2O_3 + \Sigma X_i$

[Temperature, T, K; Thermal Conductivity, k, Watt $cm^{-1} K^{-1}$]

T	k
CURVE 1*	
409.3	0.0398
565.4	0.0342
755.4	0.0298
883.2	0.0272
1150	0.0213
1167	0.0220
1467	0.0190
1561	0.0193

*No graphical presentation

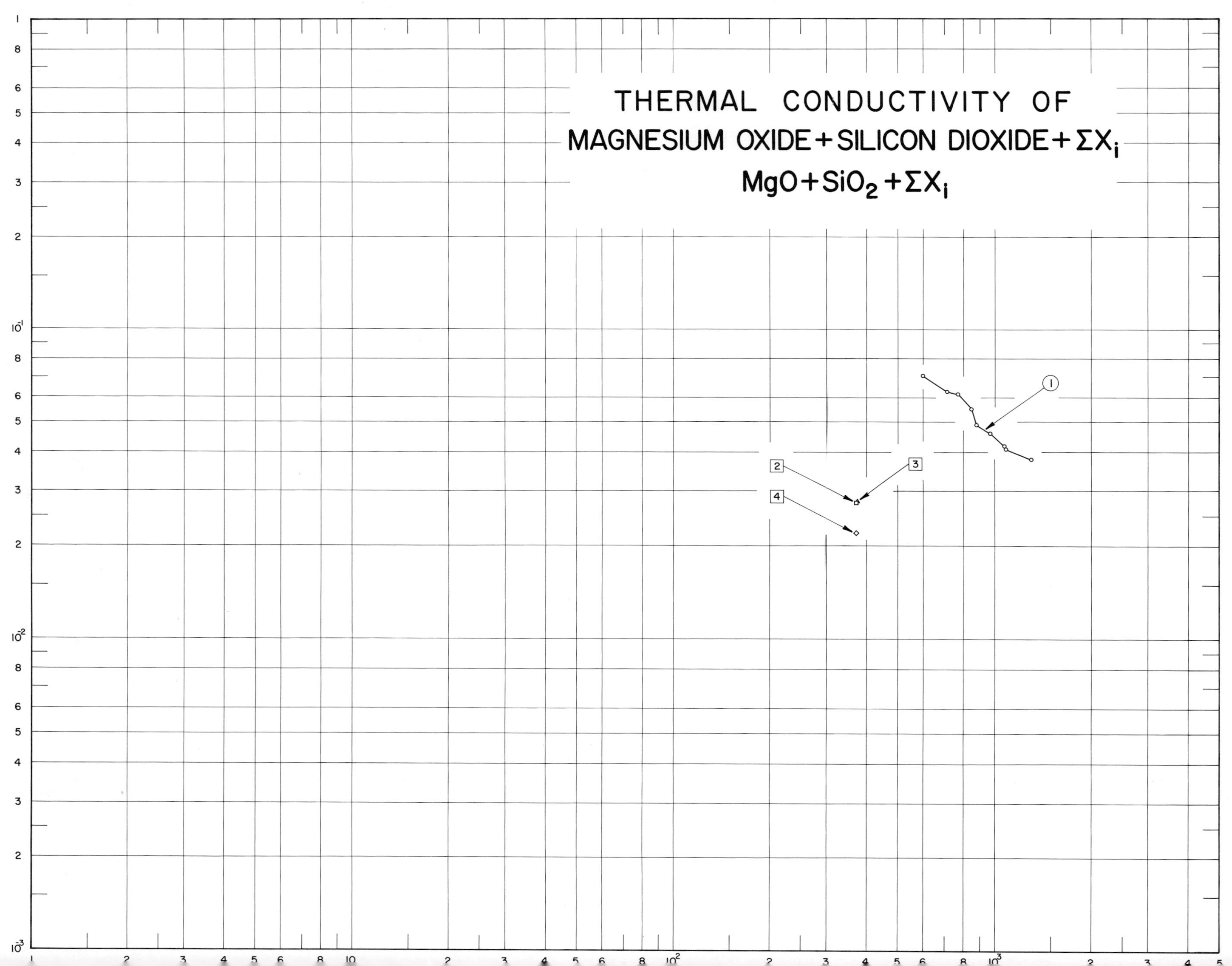
THERMAL CONDUCTIVITY OF
MAGNESIUM OXIDE + SILICON DIOXIDE + ΣX_i
MgO + SiO_2 + ΣX_i
1
2
3
4
THERMAL CONDUCTIVITY, Watt cm^-1 K^-1

SPECIFICATION TABLE NO. 133 THERMAL CONDUCTIVITY OF [MAGNESIUM OXIDE + SILICON OXIDE + ΣX_i] $MgO + SiO_2 + \Sigma X_i$

[For Data Reported in Figure and Table No. 133]

Curve No.	Ref. No.	Method Used	Year	Temp. Range, K	Reported Error, %	Name and Specimen Designation	Composition (weight percent)				Composition (continued), Specifications and Remarks
							MgO	SiO_2	Fe_2O_3	CaO	
1	462	L	1915	598-1303		Magnesia brick (Mabor)	92.1	5.0	1.6	1.7	0.4 Al_2O_3; commercial brand; specimen 2.5 in. thick; fine grained; apparent density 2.40 g cm^{-3}.
2	458	R	1921	373.2		Magnesia brick (Mabor)	53.27	32.46	2.5	4.91	14.78 Al_2O_3; 10 cm dia x 4.5 cm thick; density reported as 2.295, 2.294, 2.291, 2.285, 2.275, and 2.266 g cm^{-3} at 20, 50, 100, 200, 300, and 400 C, respectively.
3	458	P	1921	373.2		Magnesia brick (Mabor)					Similar to the above specimen; thermal conductivity values calculated from measured data of thermal diffusivity, specific heat, and density.
4	458	L	1921	373.2		Magnesia brick (Mabor)	76.43	18.26	0.80	trace	2.56 Al_2O_3, and 0.21 MnO; 4 cm dia x 8 cm long; supplied by Imperial Steel Works; density 2.370 g cm^{-3}; thermal conductivity value calculated from measured data of thermal diffusivity, specific heat, and density.

DATA TABLE NO. 133 THERMAL CONDUCTIVITY OF [MAGNESIUM OXIDE + SILICON DIOXIDE + ΣX_i] $MgO + SiO_2 + \Sigma X_i$

[Temperature, T, K; Thermal Conductivity, k, Watt $cm^{-1}K^{-1}$]

T	k
CURVE 1	
598	0.071
715.7	0.063
773	0.062
848	0.055
883	0.049
973	0.046
1075.7	0.042
1088	0.041
1303	0.038
CURVE 2	
373.2	0.0276
CURVE 3	
373.2	0.0277
CURVE 4	
373.2	0.0221

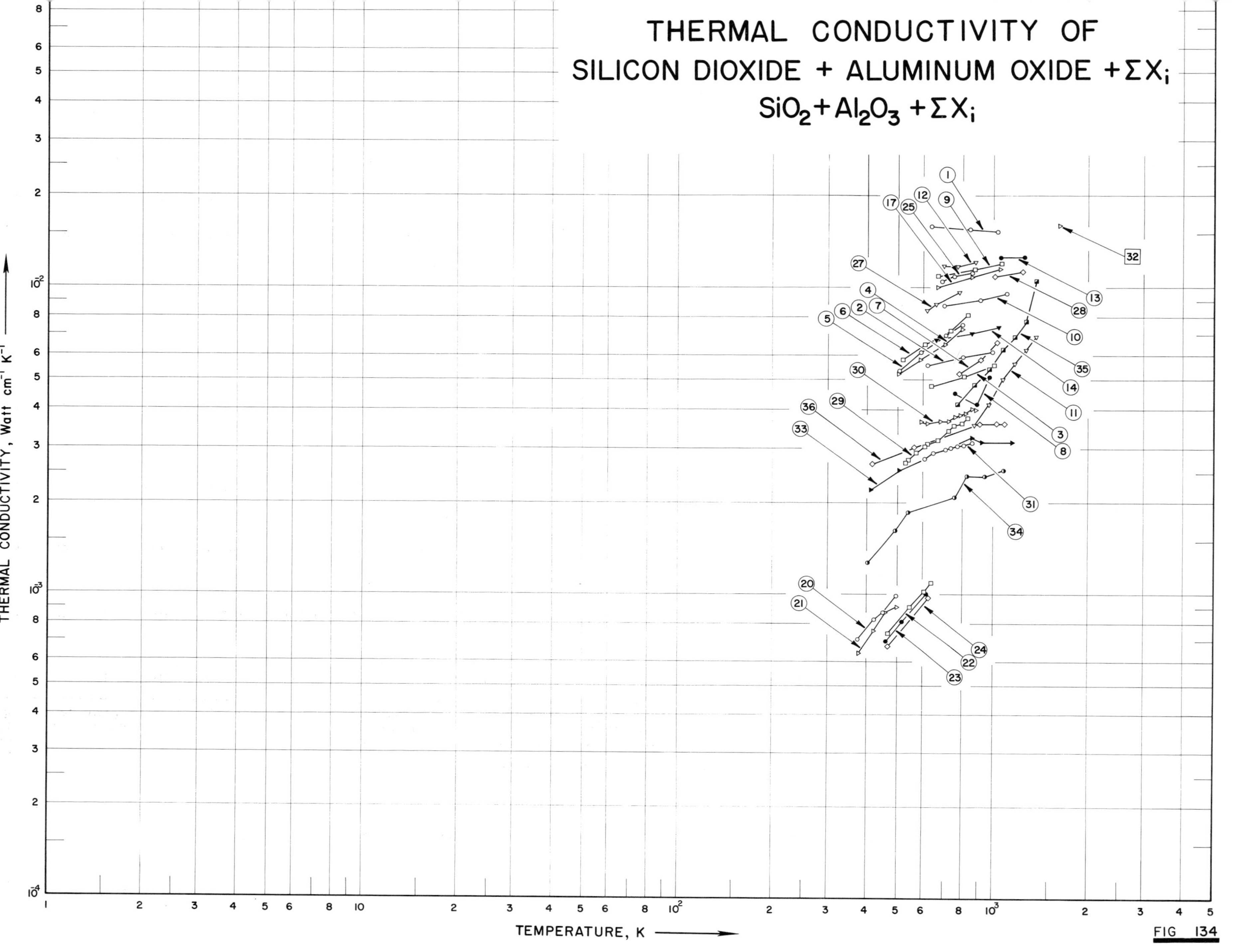
THERMAL CONDUCTIVITY OF
SILICON DIOXIDE + ALUMINUM OXIDE + ΣX_i
$SiO_2 + Al_2O_3 + \Sigma X_i$
THERMAL CONDUCTIVITY, Watt cm^{-1} K^{-1}
TEMPERATURE, K
FIG 134

SPECIFICATION TABLE NO. 134 THERMAL CONDUCTIVITY OF [SILICON DIOXIDE + ALUMINUM OXIDE + ΣX_i] $SiO_2 + Al_2O_3 + \Sigma X_i$

[For Data Reported in Figure and Table No. 134]

Curve No.	Ref. No.	Method Used	Year	Temp. Range, K	Reported Error, %	Name and Specimen Designation	Composition (weight percent) SiO_2	Al_2O_3	Fe_2O_3	TiO_2	Composition (continued), Specifications and Remarks
1	85	L	1949	636-1036	5.0-7.0	High temp. insulating blast furnace brick	56.42	→	1.42	→	40.18 total Al_2O_3 and TiO_2, 0.88 CaO, and 0.36 MgO; semidry pressed; from Semiluksk factory; density 2.16 g cm^{-3} and open porosity 17.8%; no additional shrinkage when exposed at 1350 C for 2 hrs.
2	85	L	1949	623-993	5.0-7.0	Light weight brick	66.08	→	1.04	→	30.48 total Al_2O_3 and TiO_2, 0.92 CaO, and 0.52 MgO; from Shchekinsk factory; density 1.30 g cm^{-3} and open porosity 49.9%; 0.1% additional shrinkage when exposed at 1350 C for 2 hrs.
3	85	L	1949	636-1003	5.0-7.0	Light weight brick	70.98	→	1.38	→	25.08 total Al_2O_3 and TiO_2, 0.96 CaO, and 0.25 MgO; from Snigirevsk factory; density 1.17 g cm^{-3} and open porosity 56.6%; 2.0% additional shrinkage when exposed at 1350 C for 2 hrs.
4	86	C	1948	503-798	3	II	65.72	→	1.93	→	29.02 total TiO_2 and Al_2O_3, 1.08 CaO, and 0.62 MgO; open porosity 68.4%; water absorption 81.0%; gas permeability 211 ml $m^{-2}hr^{-1}$; refractoriness 1630-1650 C; light gray color; pore size up to 1.5 mm in dia; pore distribution even.
5	86	C	1948	505-797	3	III	66.38	→	1.63	→	29.05 total TiO_2 and Al_2O_3, 1.12 CaO, and 0.51 MgO; density 0.86 g cm^{-3}; open porosity 66.3%; water absorption 75.4%; gas permeability 280 ml $m^{-2}hr^{-1}$; refractoriness 1630-1650 C; yellowish red color; pores heterogeneous, a small number of pores of 3-5 mm side by side with pores of 1.5 mm.
6	86	C	1948	518-834	3	IV	67.89	→	1.93	→	27.81 total TiO_2 and Al_2O_3, 0.64 CaO, and 0.52 MgO; density 0.98 g cm^{-3}; open porosity 63%; water absorption 62.6%; gas permeability 46 ml $m^{-2}hr^{-1}$; refractoriness 1630-1650 C; white color; large number of pores with diameters 6-7 mm.
7	86	L	1948	783-1028		IV	67.89	→	1.93	→	27.81 total TiO_2 and Al_2O_3, 0.64 CaO and 0.52 MgO; same as the above specimen.
8	86	L	1948	758-973		II	65.72	→	1.93	→	29.02 total TiO_2 and Al_2O_3, 1.08 CaO, and 0.62 MgO; same as the specimen II.
9	85	L	1949	666-1064	5.0-7.0	Normal brick; 2	63.06	→	1.21	→	33.91 total Al_2O_3 + TiO_2, 1.00 CaO, and 0.67 MgO; semidry pressed; from Semiluksk factory; density 1.86 g cm^{-3}; open porosity 29.1%; no additional shrinkage when exposed at 1350 C for 2 hrs.

SPECIFICATION TABLE NO. 134 (continued)

Curve No.	Ref. No.	Method Used	Year	Temp. Range, K	Reported Error, %	Name and Specimen Designation	Composition (weight percent) SiO_2	Al_2O_3	Fe_2O_3	TiO_2	Composition (continued), Specifications and Remarks
10	85	L	1949	698-1101	5.0-7.0	Normal brick; 5	56.32	→	0.92	→	40.72 Al_2O_3 and TiO_2, 1.02 CaO, and 0.48 MgO; semi-dry pressed; from Borovichsk factory; density 1.92 g cm^{-3}; open porosity 28.1%; 0.5% additional shrinkage when exposed at 1350 C for 2 hrs.
11	238	P	1921	873-1373		Silica brick C	93.36	2.97			2.20 CaO; specimen 9 x 4.5 x 2.5 in.;open texture with large number of fissures of appreciable size; bonding of coarse (angular rock fragments of 0.25 in. and downward) and fine fairly good except fine material easily detached; porosity 20.7%; heat flow in lengthwise direction.
12	79	L	1942	696-880		No.3,aluminous fire clay	52.0	41.3	2.5	2.7	Refractory standard cone 33; 18 in. x 18 in. dimensions.
13	79	C	1942	1062,1254		No.3,aluminous fire clay	52.0	41.3	2.5	2.7	In disc form; steel used as comparative material.
14	85	L	1949	661-1048	5.0-7.0	Light weight brick	54.56	→	2.59	→	40.73 Al_2O_3 + TiO_2, 0.92 CaO and 0.48 MgO; from Borovichsk factory; density 1.24 g cm^{-3}; open porosity 53.0%; 0.5% additional shrinkage when exposed at 1350 C for 2 hrs.
15	86	C	1948	513-827			68.12	→	2.04	→	27.38 Al_2O_3 + TiO_2, 0.88 CaO and 0.59 MgO; density 0.91 g cm^{-3};open porosity 65.1%, water absorption 64.5%; gas permeability 216 ml $m^{-2}hr^{-1}$, and refractoriness 1630 C; light yellow color; numerous pores with size up to 1.5 mm in dia and small number of pores with dia up to 5-6 mm; distribution of pores even.
16	86	L	1948	753-993			68.12	→	2.04	→	27.38 Al_2O_3 + TiO_2, 0.88 CaO and 0.59 MgO; other descriptions same as the above.
17	85	L	1949	668-1053	5.0-7.0	Normal brick; 3	75.70	→	2.59	→	20.41 Al_2O_3 + TiO_2, 0.68 CaO and 0.53 MgO; semi-acid; of sheet form from Latnensk factory; density 1.85 g cm^{-3}; open porosity 30.4%; 0.1% additional shrinkage when exposed at 1350 C for 2 hrs.
18	85	L	1949	668-1053	5.0-7.0	Normal brick; 4	48.40	→	2.24	→	37.36 Al_2O_3 + TiO_2, 0.62 CaO and 0.59 MgO; in sheet form from Borovichsk factory; density 1.919 g cm^{-3}; open porosity 29.0%; 0.3% additional shrinkage when exposed at 1350 C for 2 hrs.
19	262	L	1939	416-1094		β	77.2	15.6	1.9		3.7 alkalis, and 1.6 MgO; density 1.770 g cm^{-3}; porosity 32%.

SPECIFICATION TABLE NO. 134 (continued)

Curve No.	Ref. No.	Method Used	Year	Temp. Range, K	Reported Error, %	Name and Specimen Designation	Composition (weight percent) SiO_2	Al_2O_3	Fe_2O_3	TiO_2	Composition (continued), Specifications and Remarks
20	213	C	1956	376-496		Expanded Vermiculite	35.76	18.7	18.3		7.82 MgO, 3.61 K_2O, 1.40 CaO, 1.02 Na_2O, 0.42 MnO; ignition loss 14.83% (1200 C), 10.10% (1000 C); from Fukushima, Japan; the grain sizes of the vermiculite by sieve analysis as follows : 99 cumulative %(by wt) 100 mesh. (Tyler standard scale, meshes to the in.), 97 cumulative % 60 mesh., 92 cumulative % 28 mesh., 77 cumulative % 12 mesh., 44 cumulative % 7 mesh., 4 cumulative %4 mesh.; produced by firing (with heat rate 2 C min^{-1}) the unsieved vermiculite at 700 C for 5 min (dehydration) then loose-filling into a plate form box (size 19 cm x 19 cm x 2 cm) and pressing it until the volume is 20% less than the original; bulk density (before pressing) 0.25; use diatomaceous earth plate as comparative material.
21	213	C	1956	378-499		Expanded Vermiculite	35.76	18.7	18.3		7.82 MgO, 3.61 K_2O, 1.40 CaO, 1.2 Na_2O, 0.42 MnO; same preparation as the above specimen except firing at 800 C for 2 min; bulk density (before pressing) 0.225; use diatomaceous earth plate as comparative material.
22	213	C	1956	468-640		Expanded Vermiculite	35.76	18.7	18.3		7.82 MgO, 3.61 K_2O, 1.40 CaO, 1.02 Na_2O, 0.42 MnO; same preparation as the above specimen except firing at 900 C for 1 min; bulk density (before pressing) 0.200; use diatomaceous earth plate as comparative material.
23	213	C	1956	463-618		Expanded Vermiculite	35.76	18.7	18.3		7.82 MgO, 3.61 K_2O, 1.40 CaO, 1.02 Na_2O, 0.42 MnO; same preapration as the above specimen except firing at 1000 C for 0.25 min; bulk density (before pressing) 0.175;use diatomaceous earth plate as comparative material.
24	213	C	1956	468, 628		Expanded Vermiculite	35.76	18.7	18.3		7.82 MgO, 3.61 K_2O, 1.40 CaO, 1.02 Na_2O, 0.42 MnO; same preparation as the above specimen except firing at 1100 C for 5 sec.; bulk density (before pressing) 0.15; use diatomaceous earth plate as comparative material.
25	79	L	1942	688-857		Fire clay brick; 1	56.46	36.79	2.58	1.84	1.24 alkali oxides, 0.60 MgO, and 0.38 CaO; refractory test cone 31; in slab form 18 in. x 18 in.
26	79	C	1942	1066, 1301		Fire clay brick; 1	56.46	36.79	2.58	1.84	1.24 alkali oxides, 0.60 MgO, and 0.38 CaO; same as the above specimen but in disc form 8 in. in dia, and 1 in. in thickness; steel used as comparative material.

Curve No.	Ref. No.	Method Used	Year	Temp. Range, K	Reported Error, %	Name and Specimen Designation	Composition (weight percent) SiO_2	Al_2O_3	Fe_2O_3	TiO_2	Composition (continued), Specifications and Remarks
27	79	L	1942	618-784		Fire clay brick; 2	56.46	36.79	2.58	1.34	1.24 alkali oxides 0.60 MgO, and 0.38 CaO; specimen in slab form 18 in. x 18 in.; same firing temp as the above specimen.
28	79	C	1942	1009,1242		Fire clay brick; 2	56.46	36.79	2.58	1.84	1.24 alkali oxides, 0.60 MgO, and 0.38 CaO; specimen 8 in. in dia and 1 in. in thickness; firing temp same as the above specimen; steel used as comparative material.
29	83	L	1956	531-833		Egyptian fire clay brick; A	64.5	26.0	7.0		1.1 CaO and 1.0 MgO; bulk density 1.01 g cm^{-3}; apparent porosity 72.7%.
30	83	L	1956	593-881		Egyptian fire clay brick; B	65.3	29.5	3.5		0.9 MgO and 0.8 CaO; bulk density 1.09 g cm^{-3}; apparent porosity 60.0%.
31	83	L	1956	606-858		Egyptian fire clay brick; C	71.0	24.0	2.5		0.8 CaO and 0.7 MgO; bulk density 0.780 g cm^{-3}; apparent porosity 68.5%.
32	84	L	1925	1623.2	1.0	Fire clay wall; 26	58.50	34.48	3.52	1.80	0.62 MgO, 0.29 CaO, and 0.31 ignition loss; original coarse, fairly open, first quality fire clay; apparent density 2.05 g cm^{-3}; porosity 26.8% calculated by assuming specific gravity 2.60 for fire clay; the wall under test was built with standard 2.5 x 4.5 x 9 in. bricks laid up with cement of the same composition as the brick.
33	262	L	1939	413-1151		α	78.3	17.3	2.2		1.4 alkali oxides, 0.7 CaO, and trace MgO; density 0.805 g cm^{-3}; porosity 68%.
34	262	L	1939	404-1083		γ	61.9	34.1	2.2		1.4 CaO and 0.4 alkali oxides; density 0.710 g cm^{-3}; porosity 73%.
35	238	P	1921	773-1373		Fire-brick D	68.38	26.12	2.50		Close structure; not much clay grog, but large proportion of angular quartz grains; adherence very good; very few fissures; quartz grains very evenly graded; possibly all would pass an 8's lawn; faces of brick not good; very fine black cores; appearance of many pinholes; brick size 9 x 4.5 x 2.5 in.; porosity 17.3%; heat flow is in the direction of the length of the brick.
36	262	L	1939	415-1094			77.2	15.6	1.9		3.7 alkali, 1.6 MgO; density 1.77 g cm^{-3}; porosity 32%.
37	462	L	1915	816-1268		Fire clay brick (Farnley)	66	31	1.2		1.0 alkali, 0.9 MgO, 0.3 CaO; commercial brand; specimen 1.5 in. thick; apparent density 1.95 g cm^{-3}; hard fired to Seger cone 10-11.
38	462	L	1915	1033,1203		Fire clay brick (Farnley)	66	31	1.2		1.0 alkali, 0.9 MgO, 0.3 CaO; similar to the above specimen.

SPECIFICATION TABLE NO. 134 (continued)

Curve No.	Ref. No.	Method Used	Year	Temp. Range, K	Reported Error, %	Name and Specimen Designation	Composition (weight percent) SiO_2	Al_2O_3	Fe_2O_3	TiO_2	Composition (continued), Specifications and Remarks
39	462	L	1915	1278, 1293		Fire clay brick (Farnley)	66	31	1.2		1.0 alkali, 0.9 MgO, 0.3 CaO; similar to the above specimen except apparent density 1.90 g cm^{-3}; soft fired to Seger cone 8-9.
40	462	L	1915	1078		Silicious brick (Farnley)	82.5	16.1	1.2		1.3 alkali, trace CaO and MgO; specimen 3 in. thick; apparent density 1.82 g cm^{-3}; with many silica grains.
41	462	L	1915	1113		Silica brick (Gregory)	95.3	2.0	1.1		1.5 CaO; specimen 2.5 in. thick; coarse grained; apparent density 1.75 g cm^{-3}.
42	462	L	1915	918-1191		Silica brick (Gregory)	95.3	2.0	1.1		1.5 CaO; similar to the above specimen except apparent density 1.74 g cm^{-3}.
43	458	R	1921	373.2		Common brick	76.52	13.67	6.77		1.77 CaO, 0.42 MgO, and 0.27 MnO; 10 cm dia x 4.5 cm thick.
44	458	P	1921	373.2		Common brick					Similar to the above specimen; thermal conductivity values calculated from measured data of thermal diffusivity, specific heat, and density.
45	458	R	1921	373.2		Shamotte brick	60.78	33.95	4.37		0.79 CaO, traces of MgO and MnO; 10 cm dia x 4.5 cm thick; density reported as 1.917, 1.916, 1.913, 1.905, 1.901, 1.896, and 1.892 g cm^{-3} at 20, 50, 100, 200, 300, 400, and 500 C, respectively.
46	458	P	1921	373.2		Shamotte brick					Similar to the above specimen;thermal conductivity values calculated from measured data of thermal diffusivity, specific heat, and density.
47	458	P	1921	373.2		Red brick	76.45	21.13	2.02		Traces of CaO and MgO; commercial brick; 4 cm dia x 8 cm long; density 1.782 g cm^{-3}; thermal conductivity value calculated from measured data of thermal diffusivity, specific heat, and density.
48	383	L	1927	473-1673	±15	Penn. fire brick	54.16	38.84	2.70	2.72	1.14 MgO, 0.20 sulphates, 0.10 CaO; specimen in the form of a cylinder 10.8 cm in dia and 22.8 cm long; apparent density 2.59 g cm^{-3}; bulk density 1.90 g cm^{-3}; porosity 26.7%.
49	383	L	1927	473-1673	±15	Missouri fire brick	53.12	43.3	2.48		0.64 CaO, 0.46 MgO, 0.15 alkalis; specimen in the form of a cylinder 10.8 cm in dia, and 22.8 cm long; apparent density 2.64 g cm^{-3}, bulk density 2.15 g cm^{-3} porosity 18.4%.
50	147	L	1960	83-363		Glass; F	57.9	11.1			9.7 ZnO, 9.4 Na_2O, 4.9 CaO, 2.7 F, 2.3 K_2O, 1.9 B_2O_3, and other oxides < 0.1 each; 3 in. dia x 0.375 in. thick; density 2.55 g cm^{-3}.

DATA TABLE NO. 134 THERMAL CONDUCTIVITY OF [SILICON DIOXIDE + ALUMINUM OXIDE + ΣX_i] $SiO_2 + Al_2O_3 + \Sigma X_i$

[Temperature, T, K; Thermal Conductivity, k, Watt $cm^{-1}K^{-1}$]

T	k
CURVE 1	
636.2	0.0158
843.2	0.0155
1036.2	0.0153
CURVE 2	
623.2	0.00558
803.2	0.00593
993.2	0.00616
CURVE 3	
636.2	0.00477
813.2	0.00511
1003.2	0.00558
CURVE 4	
503.2	0.00523
589.2	0.00581
701.2	0.00651
798.2	0.00732
CURVE 5	
505.2	0.00535
593.2	0.00616
709.2	0.00697
797.2	0.00755
CURVE 6	
518.2	0.00581
608.2	0.00651
735.2	0.00721
834.2	0.00814
CURVE 7	
783.2	0.00523
913.2	0.00581
1028.2	0.00662
CURVE 8	
758.2	0.00453
888.2	0.00418
973.2	0.00511
CURVE 9	
666.2	0.0109
871.2	0.0115
1064.2	0.0120
CURVE 10	
698.2	0.00872
907.2	0.00907
1101.2	0.00953
CURVE 11	
873.2	0.00356
973.2	0.00418
1073.2	0.00502
1173.2	0.00565
1273.2	0.00628
1373.2	0.00690
CURVE 12	
696.2	0.0117
772.2	0.0117
880.2	0.0121
CURVE 13	
1062.2	0.0126
1254.2	0.0126
CURVE 14	
661.2	0.00674
858.2	0.00709
1048.2	0.00744
CURVE 15*	
513.2	0.00558
603.2	0.00628
729.2	0.00709
827.2	0.00779
CURVE 16*	
753.2	0.00488
885.2	0.00558
993.2	0.00616
CURVE 17	
668.2	0.0100
858.2	0.0108
1053.2	0.0115
CURVE 18*	
668.2	0.0102
863.2	0.0109
1053.2	0.0113
CURVE 19	
415.7	0.00267
564.7	0.00302
750.7	0.00360
909.7	0.00360
1032.2	0.00360
1094.2	0.00360
CURVE 20	
376.2	0.000709
423.2	0.000825
450.2	0.000872
496.2	0.000988
CURVE 21	
378.2	0.000639
423.2	0.000755
456.2	0.000872
499.2	0.000907
CURVE 22	
468.2	0.000744
548.2	0.000907
608.2	0.00102
640.2	0.00109
CURVE 23	
463.2	0.000697
518.2	0.000814
618.2	0.00100
CURVE 24	
468.2	0.000674
628.2	0.000976
CURVE 25	
688.2	0.0105
753.2	0.0109
857.2	0.0113
CURVE 26*	
1066.2	0.0121
1301.2	0.0126
CURVE 27	
618.2	0.00837
660.2	0.00879
784.2	0.00962
CURVE 28	
1009.2	0.0109
1242.2	0.0113
CURVE 29	
530.7	0.00269
540.7	0.00276
CURVE 29 (cont.)	
573.2	0.00290
610.2	0.00305
620.2	0.00309
670.2	0.00319
723.2	0.00343
753.2	0.00355
798.2	0.00361
833.2	0.00376
CURVE 30	
593.2	0.00366
620.2	0.00361
681.2	0.00368
725.2	0.00368
759.2	0.00379
788.2	0.00384
818.2	0.00389
860.2	0.00401
881.2	0.00398
CURVE 31	
606.2	0.00278
649.2	0.00290
706.2	0.00297
736.2	0.00301
773.2	0.00306
806.2	0.00307
858.2	0.00313
CURVE 32	
1623.2	0.0159
CURVE 33	
413.2	0.00221
506.2	0.00256
659.7	0.00291*
729.2	0.00302*
864.7	0.00325
924.2	0.00314
1151.2	0.00314
CURVE 34	
404.2	0.00128
492.2	0.00163
540.2	0.00186
758.7	0.00209
829.7	0.00244
942.2	0.00244
1082.7	0.00256
CURVE 35	
773.2	0.00418
873.2	0.00481
973.2	0.00544
1073.2	0.00628
1173.2	0.00690
1273.2	0.00774
1373.2	0.0105
CURVE 36	
415.7	0.00267
564.7	0.00302
750.7	0.00360*
909.7	0.00360
1032.2	0.00360
1094.2	0.00360
CURVE 37*	
815.7	0.012
908	0.012
978	0.015
1268	0.017
CURVE 38*	
1033	0.014
1203	0.016
CURVE 39*	
1278	0.0066
1293	0.0050
CURVE 40*	
1078	0.0104
CURVE 41*	
1113	0.016
CURVE 42*	
918	0.013
1063	0.015
1190.7	0.018
CURVE 43*	
373.2	0.00732
CURVE 44*	
373.2	0.00732
CURVE 45*	
373.2	0.00950
CURVE 46*	
373.2	0.00937
CURVE 47*	
373.2	0.00569
CURVE 48*	
473.2	0.0100
763.2	0.0113
873.2	0.0126
1073	0.0134
1273	0.0142
1473	0.0151
1673	0.0155

*Not shown on plot

DATA TABLE NO. 134 (continued)

T	k
CURVE 49*	
473.2	0.0100
673.2	0.0126
873.2	0.0146
1073.2	0.0155
1273.2	0.0163
1473.2	0.0172
1673.2	0.0176
CURVE 50*	
83.2	0.00653
88.2	0.00674
93.2	0.00690
98.2	0.00711
103.2	0.00732
108.2	0.00749
113.2	0.00766
118.2	0.00778
123.2	0.00795
128.2	0.00812
133.2	0.00828
138.2	0.00841
143.2	0.00854
148.2	0.00870
153.2	0.00883
158.2	0.00895
163.2	0.00908
168.2	0.00916
173.2	0.00929
178.2	0.00941
183.2	0.00950
188.2	0.00962
193.2	0.00975
198.2	0.00983
203.2	0.00992
208.2	0.01004
213.2	0.01013
218.2	0.01021
223.2	0.01029
228.2	0.01038
233.2	0.01046
238.2	0.01054
243.2	0.01063
248.2	0.01071
253.2	0.01079
258.2	0.01084

T	k
CURVE 50 (cont.)*	
263.2	0.01092
268.2	0.01100
273.2	0.01105
278.2	0.01113
283.2	0.01117
288.2	0.01125
293.2	0.01130
298.2	0.01134
303.2	0.01142
308.2	0.01146
313.2	0.01151
318.2	0.01155
323.2	0.01163
328.2	0.01167
333.2	0.01172
338.2	0.01176
343.2	0.01184
348.2	0.01188
353.2	0.01192
358.2	0.01197
363.2	0.01201

*Not shown on plot

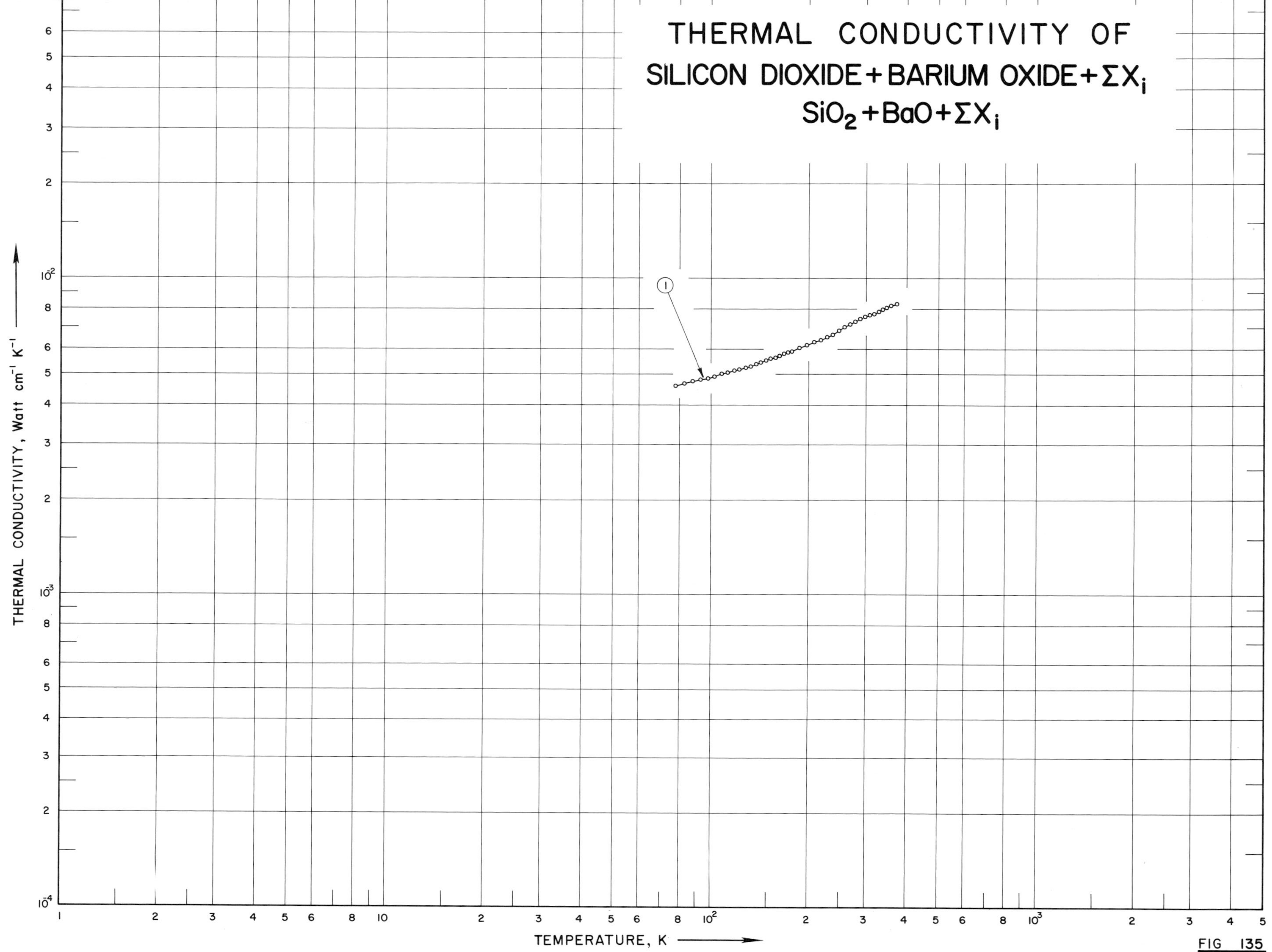
THERMAL CONDUCTIVITY OF
SILICON DIOXIDE + BARIUM OXIDE + ΣXi
SiO2 + BaO + ΣXi
1
THERMAL CONDUCTIVITY, Watt cm-1 K-1
TEMPERATURE, K
FIG 135

SPECIFICATION TABLE NO. 135 THERMAL CONDUCTIVITY OF [SILICON DIOXIDE + BARIUM OXIDE + ΣX_i] $SiO_2 + BaO + \Sigma X_i$

[For Data Reported in Figure and Table No. 135]

Curve No.	Ref. No.	Method Used	Year	Temp. Range, K	Reported Error, %	Name and Specimen Designation	Composition (weight percent)				Composition (continued), Specifications and Remarks
							SiO_2	BaO	K_2O	ZnO	
1	147	L	1960	78-373		Glass; H	49.8	24.2	8.5	7.8	5.9 PbO, 2.9 Na_2O, 0.7 Sb_2O_3, and 0.2 As_2O_3; 3 in. dia x 0.375 in. thick; density 3.18 g cm^{-3}.

DATA TABLE NO. 135 THERMAL CONDUCTIVITY OF [SILICON DIOXIDE + BARIUM OXIDE + ΣX_i] $SiO_2 + BaO + \Sigma X_i$

[Temperature, T, K; Thermal Conductivity, k, Watt $cm^{-1}K^{-1}$]

T	k	T	k
CURVE 1		CURVE 1 (cont.)	
78.2	0.00460	303.2	0.00761*
83.2	0.00469	308.2	0.00766
88.2	0.00477	313.2	0.00700*
93.2	0.00481	318.2	0.00774
98.2	0.00485	323.2	0.00782*
103.2	0.00494	328.2	0.00787
108.2	0.00502	333.2	0.00795*
113.2	0.00506	338.2	0.00799
118.2	0.00515	343.2	0.00803*
123.2	0.00519	348.2	0.00808
128.2	0.00527	353.2	0.00816*
133.2	0.00531	358.2	0.00820
138.2	0.00540	363.2	0.00824*
143.2	0.00548	368.2	0.00828*
148.2	0.00552	373.2	0.00833
153.2	0.00561		
158.2	0.00565		
163.2	0.00573		
168.2	0.00582		
173.2	0.00586		
178.2	0.00590		
183.2	0.00598*		
188.2	0.00607		
193.2	0.00615*		
198.2	0.00619		
203.2	0.00628*		
208.2	0.00632		
213.2	0.00640*		
218.2	0.00644		
223.2	0.00653*		
228.2	0.00657		
233.2	0.00661*		
238.2	0.00669		
243.2	0.00678*		
248.2	0.00686		
253.2	0.00695*		
258.2	0.00703		
263.2	0.00711*		
268.2	0.00720		
273.2	0.00724*		
278.2	0.00732		
283.2	0.00736*		
288.2	0.00745		
293.2	0.00753*		
298.2	0.00757		

*Not shown on plot

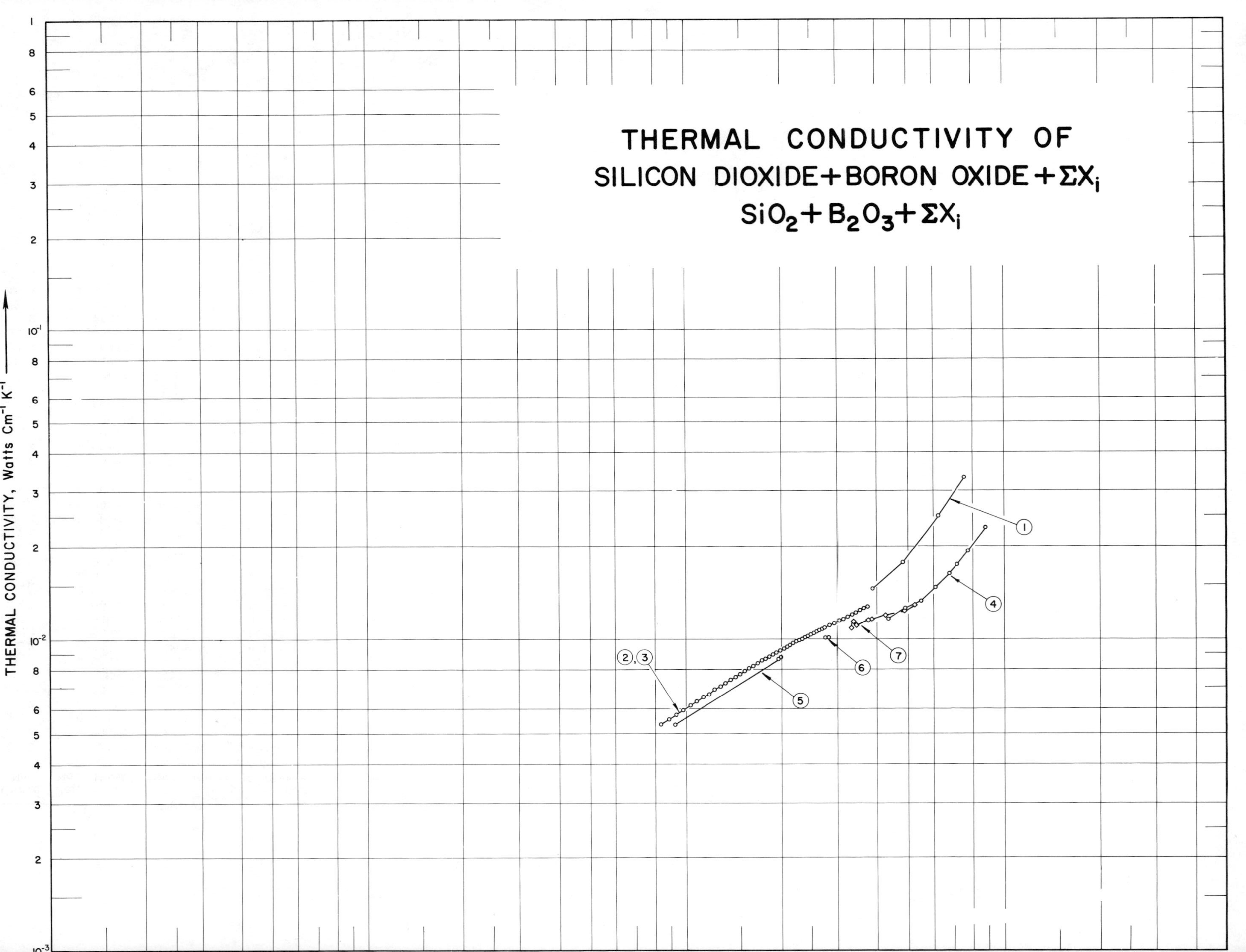

THERMAL CONDUCTIVITY OF
SILICON DIOXIDE+BORON OXIDE+ΣXi
SiO2+B2O3+ΣXi
THERMAL CONDUCTIVITY, Watts Cm⁻¹ K⁻¹
1
2,3
4
5
6
7

SPECIFICATION TABLE NO. 136 THERMAL CONDUCTIVITY OF [SILICON DIOXIDE + BORON OXIDE] $SiO_2 + B_2O_3 + \Sigma X_i$

[For Data Reported in Figure and Table No. 136]

Curve No.	Ref. No.	Method Used	Year	Temp. Range, K	Reported Error, %	Name and Specimen Designation	Composition (weight percent) B_2O_3	SiO_2	Na_2O	Al_2O_3	Composition (continued), Specifications and Remarks
1	34	C	1943	388-751		Pyrex-brand glass	12.9	80.5	3.8	2.2	0.4 PbO, specimen 1 cm cube; ground; 18-8 stainless steel used as standard.
2	147	L	1960	83-373		Glass A	12.8	80.8	4.2	2.2	Oxides < 0.1 omitted; approx composition; specimen measured in a stack of a pair of 3 in. dia discs; density 2.22 g cm^{-3}. (Additional data from the author.)
3	147	L	1960	83-373		Glass B	13.1	79.5	5.3	2.1	Similar to the above specimen except density 2.27 g cm^{-3}. (Additional data from the author.)
4	136	C	1959	430-871	±10.0	Pyrex 7740	12.5	81.0	4.0	2.0	
5	187	R	1932	92-198		Pyrex glass	12.5	80.5	4.0	2.0	A platinum wire threaded through a 60 to 70 cm long pyrex capillary tube of approximately 0.1 cm internal and 1.0 cm external dia, then heated so that the glass was carefully melted around the wire, and finally bent to a form of U-tube; measured in a bath of liquid air and a mixture of solid carbon dioxide; density 2.233 g cm^{-3} at 21 C.
6	187	R	1932	275, 279		Pyrex glass	12.5	80.5	4.0	2.0	Above specimen measured in a bath containing crushed ice and water.
7	187	R	1932	328-523		Pyrex glass	12.5	80.5	4.0	2.0	Above specimen measured in an oil bath.

DATA TABLE NO. 136 THERMAL CONDUCTIVITY OF [SILICON DIOXIDE + BORON OXIDE + ΣX_i] $SiO_2 + B_2O_3 + \Sigma X_i$

[Temperature, T, K; Thermal Conductivity, k, Watt $cm^{-1}K^{-1}$]

T	k
CURVE 1	
388.2	0.0153
482.2	0.0177
625.2	0.0250
751.2	0.0331
CURVE 2	
83.2	0.00536
88.2	0.00552
93.2	0.00573
98.2	0.00594
103.2	0.00615
108.2	0.00632
113.2	0.00653
118.2	0.00669
123.2	0.00690
128.2	0.00707
133.2	0.00724
138.2	0.00741
143.2	0.00757
148.2	0.00774
153.2	0.00791
158.2	0.00808*
163.2	0.00820
168.2	0.00837*
173.2	0.00854
178.2	0.00866*
183.2	0.00879
188.2	0.00895*
193.2	0.00908
198.2	0.00920*
203.2	0.00933
208.2	0.00946*
213.2	0.00958
218.2	0.00971*
223.2	0.00983
228.2	0.00992*
233.2	0.0100
238.2	0.0102*
243.2	0.0103
248.2	0.0104*
253.2	0.0105
258.2	0.0106*
263.2	0.0107
CURVE 2 (cont.)	
268.2	0.0108*
273.2	0.0109
278.2	0.0111*
283.2	0.0111
288.2	0.0112*
293.2	0.0113
298.2	0.0114*
303.2	0.0115
308.2	0.0116*
313.2	0.0116
318.2	0.0117*
323.2	0.0118
328.2	0.0119*
333.2	0.0120
338.2	0.0121*
343.2	0.0122
348.2	0.0123*
353.2	0.0124
358.2	0.0125*
363.2	0.0126
368.2	0.0125*
373.2	0.0127
CURVE 3*	
83.2	0.00536
88.2	0.00552
93.2	0.00573
98.2	0.00594
103.2	0.00615
108.2	0.00632
113.2	0.00653
118.2	0.00669
123.2	0.00690
128.2	0.00707
133.2	0.00724
138.2	0.00741
143.2	0.00757
148.2	0.00774
153.2	0.00791
158.2	0.00808
163.2	0.00820
168.2	0.00837
173.2	0.00854
CURVE 3 (cont.)	
178.2	0.00866
183.2	0.00879
188.2	0.00895
193.2	0.00908
198.2	0.00920
203.2	0.00933
208.2	0.00946
213.2	0.00958
218.2	0.00971
223.2	0.00983
228.2	0.00992
233.2	0.0100
238.2	0.0102
243.2	0.0103
248.2	0.0104
253.2	0.0105
258.2	0.0106
263.2	0.0107
268.2	0.0108
273.2	0.0109
278.2	0.0111
283.2	0.0111
288.2	0.0112
293.2	0.0113
298.2	0.0114
303.2	0.0115
308.2	0.0116
313.2	0.0116
318.2	0.0117
323.2	0.0118
328.2	0.0119
333.2	0.0120
338.2	0.0121
343.2	0.0122
348.2	0.0123
353.2	0.0124
358.2	0.0125
363.2	0.0126
368.2	0.0126
373.2	0.0127
CURVE 4	
430.2	0.0118
CURVE 4 (cont.)	
488.2	0.0128
548.2	0.0135
608.2	0.0150
671.2	0.0166
715.2	0.0178
768.2	0.0197
871.2	0.0234
CURVE 5	
92.2	0.00540
195.5	0.00879
198.3	0.00891
CURVE 6	
275.0	0.0103
279.2	0.0103
CURVE 7	
328.2	0.0110
332.7	0.0110*
338.8	0.0113
343.7	0.0112
372.2	0.0116
378.4	0.0116*
385.2	0.0117
423.2	0.0121
428.2	0.0121*
482.2	0.0126
488.6	0.0127*
523.2	0.0131

*Not shown on plot

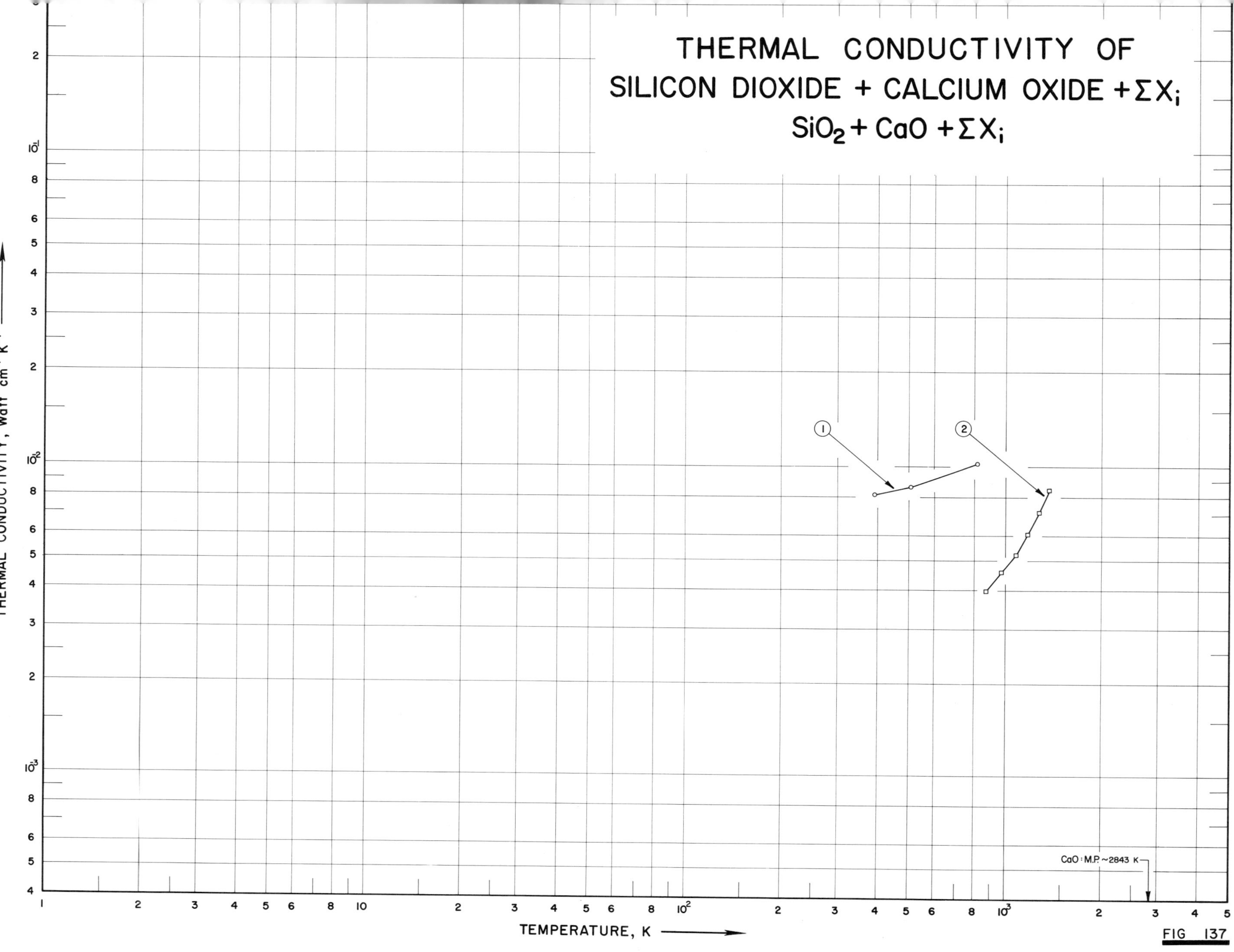
THERMAL CONDUCTIVITY OF
SILICON DIOXIDE + CALCIUM OXIDE + ΣX_i
SiO_2 + CaO + ΣX_i
THERMAL CONDUCTIVITY, Watt cm^{-1} K^{-1}
TEMPERATURE, K
CaO: M.P. ~2843 K
1
2
FIG 137

SPECIFICATION TABLE NO. 137 THERMAL CONDUCTIVITY OF [SILICON DIOXIDE + CALCIUM OXIDE + ΣX_i] $SiO_2 + CaO + \Sigma X_i$

[For Data Reported in Figure and Table No. 137]

Curve No.	Ref. No.	Method Used	Year	Temp. Range, K	Reported Error, %	Name and Specimen Designation	Composition (weight percent) SiO_2	CaO	Al_2O_3	Fe_2O_3	Composition (continued), Specifications and Remarks
1	247	L	1932	394-817	1.0	Silica brick B	92.14	2.95	2.21	1.11	0.34 TiO_2 and 0.23 MgO; density 2.328 g cm^{-3}; porosity 28.1%; gas permeability 2.354 $m^3 \cdot cm\ m^{-2} hr^{-1}$ (mm of $H_2O)^{-1}$.
2	238	P	1921	873-1373		Silica brick B	94.02	2.64	1.78		Exceptionally fine-grained, close, and uniform texture throughout the brick; major portion of material of sand size; with very few fragments of rock of appreciable size; porosity 38.2%, with pores of even size; friable; brick size : 9 x 4.5 x 2.5 in.; heat flow in the lengthwise direction of the brick.

DATA TABLE NO. 137 THERMAL CONDUCTIVITY OF [SILICON DIOXIDE + CALCIUM OXIDE + ΣX_i] $SiO_2 + CaO + \Sigma X_i$

[Temperature, T, K; Thermal Conductivity, k, Watt $cm^{-1}K^{-1}$]

T	k
CURVE 1	
393.5	0.00811
507.3	0.00860
817.3	0.0102
CURVE 2	
873.2	0.00397
973.2	0.00460
1073.2	0.00523
1173.2	0.00607
1273.2	0.00711
1373.2	0.00837

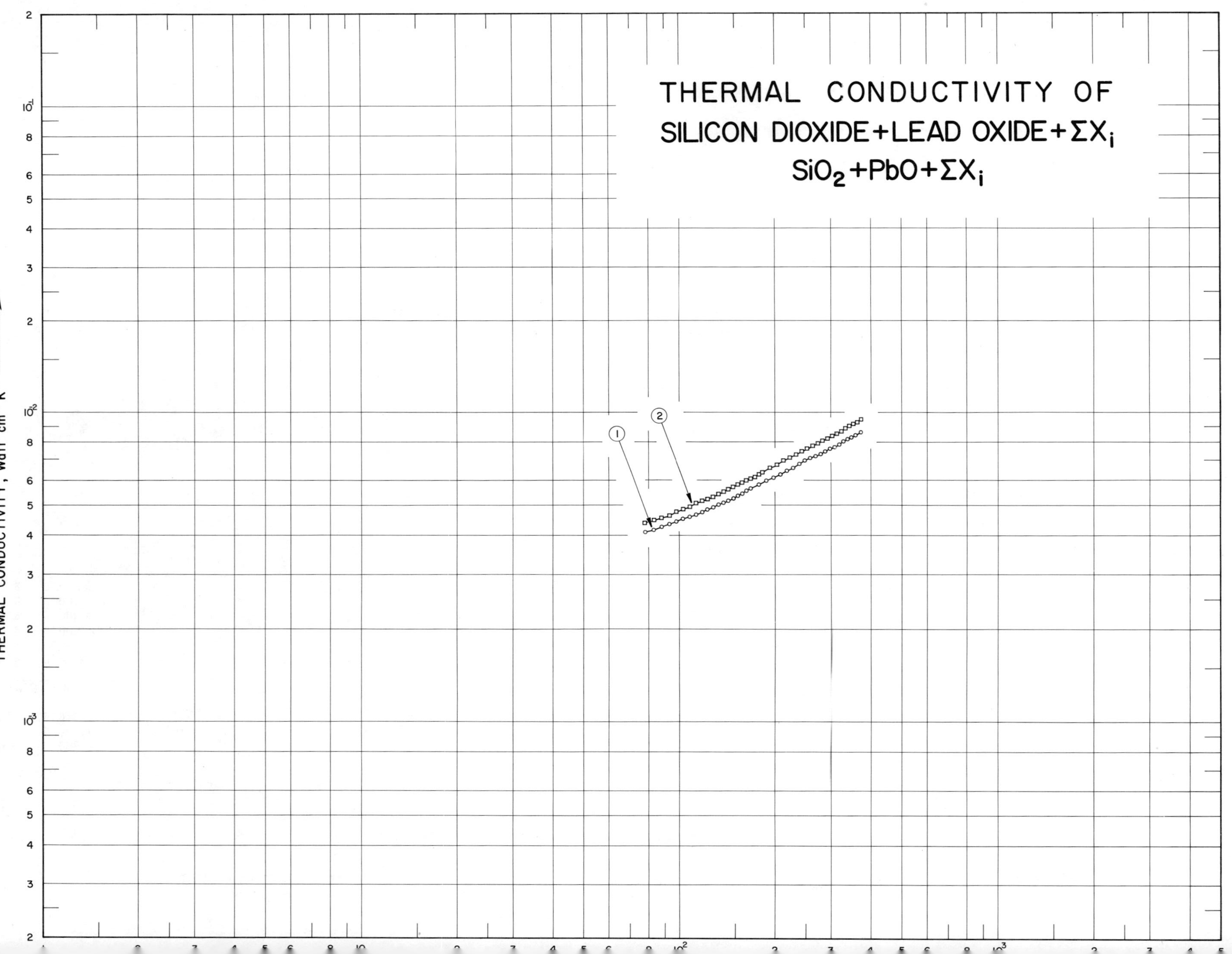
THERMAL CONDUCTIVITY OF
SILICON DIOXIDE + LEAD OXIDE + ΣXi
SiO2 + PbO + ΣXi
THERMAL CONDUCTIVITY, Watt cm⁻¹ K⁻¹
1
2

SPECIFICATION TABLE NO. 138 THERMAL CONDUCTIVITY OF [SILICON DIOXIDE + LEAD OXIDE + ΣX_i] $SiO_2 + PbO + \Sigma X_i$

[For Data Reported in Figure and Table No. 138]

Curve No.	Ref. No.	Method Used	Year	Temp. Range, K	Reported Error, %	Name and Specimen Designation	Composition (weight percent)				Composition (continued), Specifications and Remarks
							SiO_2	PbO	K_2O	Al_2O_3	
1	147	L	1960	78-373		Glass; I	46.0	44.8	9.2	0.2	3 in. dia x 0.375 in. thick; density 3.55 g cm^{-3}.
2	147	L	1960	78-373		Glass; G	54.2	34.2	7.1		2.4 Na_2O, 2.0 Sb_2O_3, and 0.2 As_2O_3, 3 in. dia x 0.375 in. thick; density 3.19 g cm^{-3}.

DATA TABLE NO. 138 THERMAL CONDUCTIVITY OF [SILICON DIOXIDE + LEAD OXIDE + ΣX_i] $SiO_2 + PbO + \Sigma X_i$

[Temperature, T, K; Thermal Conductivity, k, Watt $cm^{-1} K^{-1}$]

T	k	T	k	T	k
CURVE 1		CURVE 1 (cont.)		CURVE 2 (cont.)	
78.2	0.00410	303.2	0.00766*	213.2	0.00695
83.2	0.00418	308.2	0.00770	218.2	0.00703*
88.2	0.00427	313.2	0.00778*	223.2	0.00711
93.2	0.00435	318.2	0.00787	228.2	0.00715*
98.2	0.00444	323.2	0.00795*	233.2	0.00728
103.2	0.00452	328.2	0.00803	238.2	0.00732*
108.2	0.00460	333.2	0.00812*	243.2	0.00745
113.2	0.00469	338.2	0.00820	248.2	0.00753*
118.2	0.00477	343.2	0.00824*	253.0	0.00761
123.2	0.00485	348.2	0.00828	258.2	0.00770*
128.2	0.00494	353.2	0.00837*	263.2	0.00778
133.2	0.00502	358.2	0.00841	268.2	0.00787*
138.2	0.00510	363.2	0.00845*	273.2	0.00791
143.2	0.00519	368.2	0.00854*	278.2	0.00799*
148.2	0.00527	373.2	0.00862	283.2	0.00808
153.2	0.00536			288.2	0.00816*
158.2	0.00544	CURVE 2		293.2	0.00824
163.2	0.00556			298.2	0.00833*
168.2	0.00565	78.2	0.00439	303.2	0.00837
173.2	0.00573*	83.2	0.00448	308.2	0.00845*
178.2	0.00582	88.2	0.00456	313.2	0.00854
183.2	0.00590*	93.2	0.00464	318.2	0.00862*
188.2	0.00598	98.2	0.00477	323.2	0.00870
193.2	0.00602*	103.2	0.00485	328.2	0.00879*
198.2	0.00611	108.2	0.00494	333.2	0.00887
203.2	0.00619*	113.2	0.00506	338.2	0.00895*
208.2	0.00628	118.2	0.00515	343.2	0.00904
213.2	0.00636*	123.2	0.00523	348.2	0.00912*
218.2	0.00644	128.2	0.00531	353.2	0.00916
223.2	0.00649*	133.2	0.00544	358.2	0.00925*
228.2	0.00657	138.2	0.00552	363.2	0.00929
233.2	0.00665*	143.2	0.00561	368.2	0.00937*
238.2	0.00678	148.2	0.00573	373.2	0.00946
243.2	0.00686*	153.2	0.00582		
248.2	0.00695	158.2	0.00590		
253.2	0.00703*	163.2	0.00598		
258.2	0.00707	168.2	0.00607		
263.2	0.00711*	173.2	0.00615		
268.2	0.00720	178.2	0.00628		
273.2	0.00728*	183.2	0.00636		
278.2	0.00732	188.2	0.00644*		
283.2	0.00736*	193.2	0.00657		
288.2	0.00745	198.2	0.00665*		
293.2	0.00753*	203.2	0.00674		
298.2	0.00761	208.2	0.00682*		

*Not shown on plot

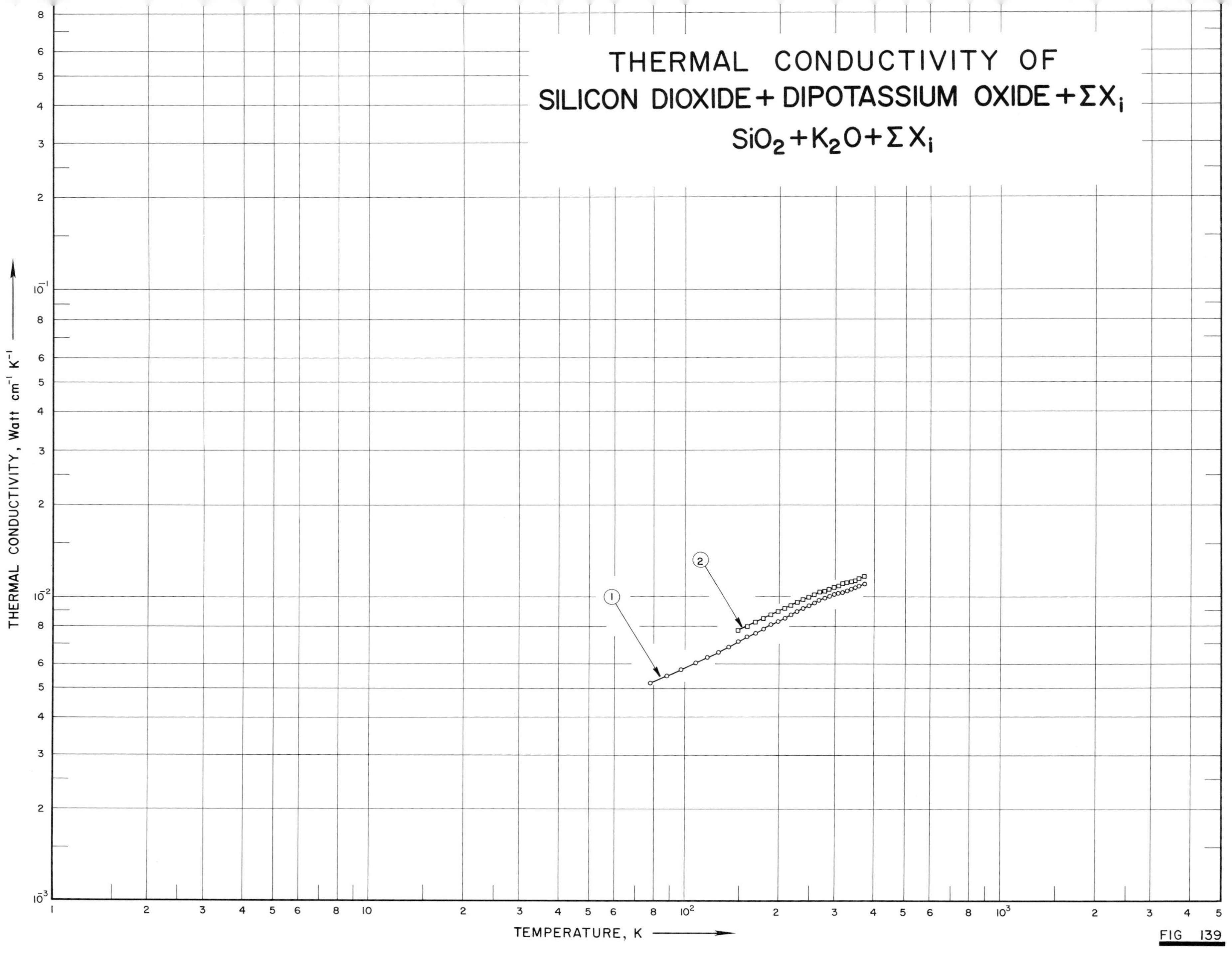

THERMAL CONDUCTIVITY OF
SILICON DIOXIDE + DIPOTASSIUM OXIDE + ΣX_i
$SiO_2 + K_2O + \Sigma X_i$
1
2
THERMAL CONDUCTIVITY, Watt cm^{-1} K^{-1}
TEMPERATURE, K

FIG 139

SPECIFICATION TABLE NO. 139 THERMAL CONDUCTIVITY OF [SILICON DIOXIDE + DIPOTASSIUM OXIDE + ΣX_i] $SiO_2 + K_2O + \Sigma X_i$

[For Data Reported in Figure and Table No. 139]

Curve No.	Ref. No.	Method Used	Year	Temp. Range, K	Reported Error, %	Name and Specimen Designation	Composition (weight percent)				Composition (continued), Specifications and Remarks
							SiO_2	K_2O	Na_2O	CaO	
1	147	L	1960	78-373		Glass; D	71.2	14.5	7.6	5.5	3.0 ZnO, 2.9 B_2O_3, 0.8 Sb_2O_3; 0.1 Al_2O_3, and other oxides < 0.1 each; 3 in. dia x 0.375 in. thick; density 2.52 g cm^{-3}.
2	147	L	1960	148-373		Glass; C	72.7	14.5	4.0		7.7 B_2O_3, 0.4 Al_2O_3, 0.4 Sb_2O_3, 0.4 ZnO, and other oxides < 0.1 each; 3 in. dia x 0.375 in. thick; density 2.45 g cm^{-3}.

DATA TABLE NO. 139 THERMAL CONDUCTIVITY OF [SILICON DIOXIDE + DIPOTASSIUM OXIDE + ΣX_i] $SiO_2 + K_2O + \Sigma X_i$

[Temperature, T, K; Thermal Conductivity, k, Watt $cm^{-1} K^{-1}$]

CURVE 1

T	k
78.2	0.00519
83.2	0.00536*
88.2	0.00548
93.2	0.00561*
98.2	0.00573
103.2	0.00586*
108.2	0.00602
113.2	0.00615*
118.2	0.00628
123.2	0.00640*
128.2	0.00653
133.2	0.00669*
138.2	0.00682
143.2	0.00699*
148.2	0.00711
153.2	0.00724*
158.2	0.00736
163.2	0.00745*
168.2	0.00757
173.2	0.00770*
178.2	0.00782
183.2	0.00795*
188.2	0.00808
193.2	0.00820*
198.2	0.00828
203.2	0.00841*
208.2	0.00849
213.2	0.00862*
218.2	0.00874
223.2	0.00887*
228.2	0.00895
233.2	0.00904*
238.2	0.00912
243.2	0.00925*
248.2	0.00933
253.2	0.00941*
258.2	0.00950
263.2	0.00958*
268.2	0.00971
273.2	0.00979*
278.2	0.00987
283.2	0.00992*
288.2	0.01000
293.2	0.01004*
298.2	0.01013
303.2	0.01017*

CURVE 1 (cont.)

T	k
308.2	0.01025
313.2	0.01029*
318.2	0.01033
323.2	0.01042*
328.2	0.01050
333.2	0.01054*
338.2	0.01063
343.2	0.01071*
348.2	0.01075
353.2	0.01079*
358.2	0.01084
363.2	0.01088*
368.2	0.01096*
373.2	0.01100

CURVE 2

T	k
148.2	0.00774
153.2	0.00787*
158.2	0.00799
163.2	0.00812*
168.2	0.00824
173.2	0.00833*
178.2	0.00845
183.2	0.00858*
188.2	0.00870
193.2	0.00879*
198.2	0.00891
203.2	0.00900*
208.2	0.00912
213.2	0.00920*
218.2	0.00933
223.2	0.00941*
228.2	0.00954
233.2	0.00962*
238.2	0.00975
243.2	0.00983*
248.2	0.00992
253.2	0.01000*
258.2	0.01013
263.2	0.01021*
268.2	0.01030
273.2	0.01038*
278.2	0.01046
283.2	0.01054*
288.2	0.01059

CURVE 2 (cont.)

T	k
293.2	0.01067*
298.2	0.01075
303.2	0.01084*
308.2	0.01088
313.2	0.01096*
318.2	0.01105
323.2	0.01109*
328.2	0.01113
333.2	0.01121*
338.2	0.01125
343.2	0.01130*
348.2	0.01138
353.2	0.01146*
358.2	0.01151
363.2	0.01159*
368.2	0.01163*
373.2	0.01172

*Not shown on plot

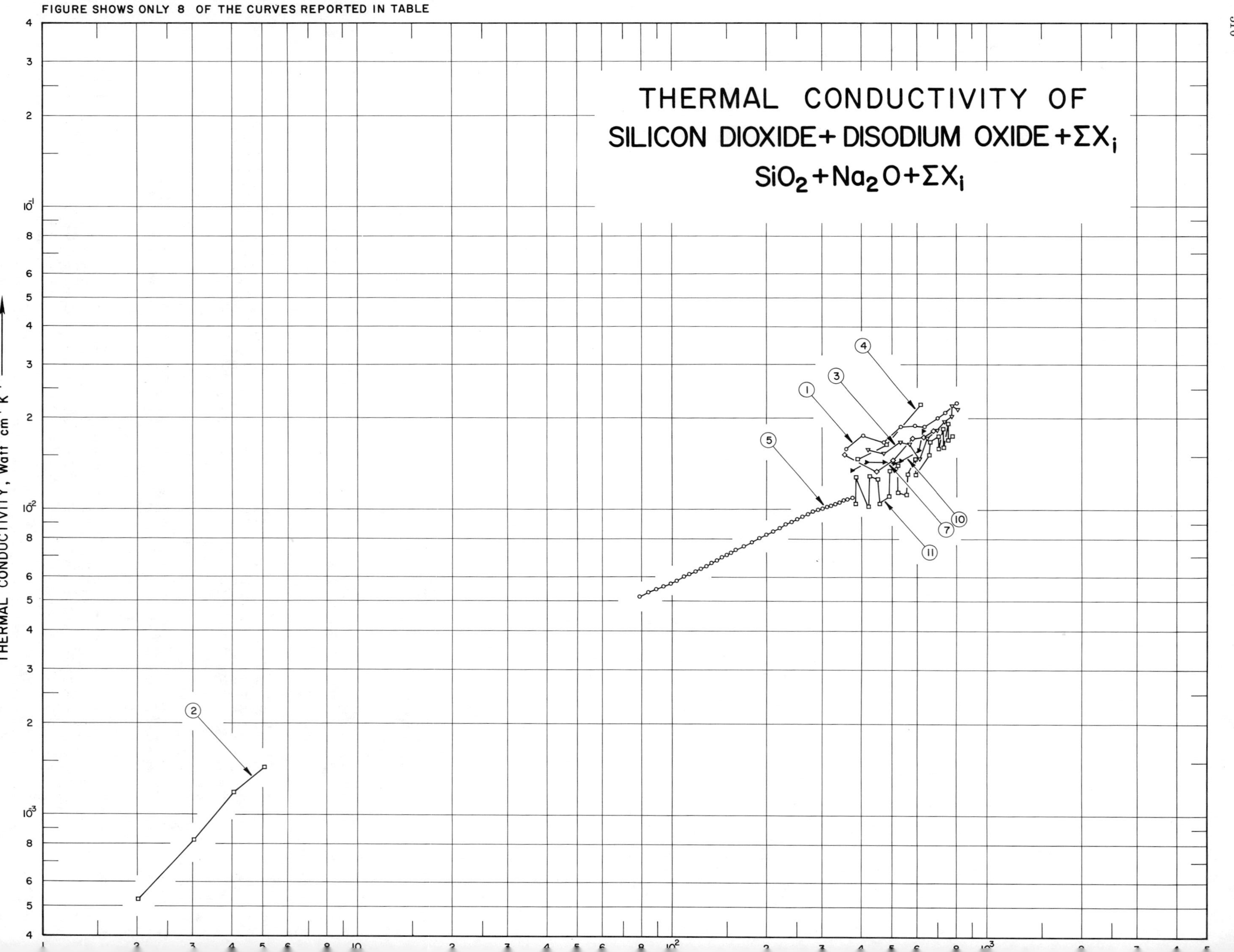
FIGURE SHOWS ONLY 8 OF THE CURVES REPORTED IN TABLE
THERMAL CONDUCTIVITY OF
SILICON DIOXIDE + DISODIUM OXIDE + ΣXi
SiO2 + Na2O + ΣXi
THERMAL CONDUCTIVITY, Watt cm-1 K-1
1
2
3
4
5
7
10
11

SPECIFICATION TABLE NO. 140 THERMAL CONDUCTIVITY OF [SILICON DIOXIDE + DISODIUM OXIDE + ΣX_i] $SiO_2 + Na_2O + \Sigma X_i$

[For Data Reported in Figure and Table No. 140]

Curve No.	Ref. No.	Method Used	Year	Temp. Range, K	Reported Error, %	Name and Specimen Designation	Composition (weight percent) SiO_2	Na_2O	CaO	Al_2O_3	MgO	Composition (continued), Specifications and Remarks
1	136	R	1959	355-795	±10	Silicate glass	69.05	16.38	7.37	3.05	2.80	0.55 Na_2SO_4, 0.48 K_2O, 0.12 NaCl, 0.09 Fe_2O_3, 0.015 CoO and 0.015 NiO.
2	142	L	1955	2.0-5.0		Soft glass	70.12	16.82	5.4	2.58	3.6	0.78 B_2O_3, 0.35 K_2O, and 0.20 SO_3; density 2.50 g cm^{-3}.
3	136	R	1959	418-804	±10	Silicate glass	58.43	19.32	6.0	7.89	3.51	4.25 K_2O, 0.24 Na_2SO_4, 0.12 NaCl, 0.11 Fe_2O_3, 0.013 CaO, and 0.013 NiO.
4	34	C	1943	386-611		Soda-lime glass	69.73	20.96	9.05			0.18B_2O_3, and trace K_2O.
5	147	L	1960	78-373		Glass E	67.7	14.6	5.4	1.8		4.0 B_2O_3, 3.3 BaO, 1.8 K_2O, 1.3 Fe_2O_3, <1.0 As_2O_3, and other oxides <0.1 each; 3 in. dia x 0.375 in. thick; density 2.42 g cm^{-3}.
6	73,136	R	1954	497-826	±10	Soda-lime silica glass; 1	71.25	13.35	11.82	0.26	2.44	0.68 Na_2SO_4, 0.14 Fe_2O_3, and 0.06 NaCl; prepared by Pittsburg Plate Glass Company Laboratory; ellipsoidal specimen.
7	73,136	R	1954	350-724	±10	Soda-lime silica glass; 2	70.84	13.32	11.75	0.22	2.64	0.61 Na_2SO_4, 0.56 Fe_2O_3, and 0.06 NaCl; same source and shape as the above specimen.
8	73,136	R	1954	361-798	±10	Soda-lime silica glass; 3	69.05	16.38	7.37	3.05	2.80	0.55 Na_2O_3, 0.48 K_2O, 0.12 NaCl, 0.09 Fe_2O_3, 0.015 NiO, and 0.015 CoO; same source and shape as the above specimen.
9	73,136	R	1954	421-806	±10	Soda-lime silica glass; 4	58.43	19.32	6.00	7.89	3.51	4.25 K_2O, 0.24 Na_2SO_4, 0.12 NaCl, 0.11 Fe_2O_3, 0.013 NiO, and 0.013 CoO; same source and shape as the above specimen.
10	72,136	C	1955	368-685	±10	Soda-lime silica glass; 1	71.25	13.25	11.82	0.26	2.44	Similar to the above specimen 1 (curve no. 6) except in cubic shape; measured with another method.
11	114,136	C	1954	381-773		Soda-lime silica glass; 2	70.84	13.32	11.75	0.22	2.64	Similar to the above specimen 2 (curve no. 7) except in cubic shape; measured with another method.
12	136	R	1959	393-847	±10	Silica glass, Corning 0080	74.0	16.5	5.0	1.0	3.5	1.0 Al_2O_3; ellipsoidal specimen.
13	136	R	1959	433-888	±10	Silica glass, Corning 0080						Similar to the above specimen.
14	136	R	1959	509-839	±10	Silica glass, Corning 0080						Similar to the above specimen.

DATA TABLE NO. 140 THERMAL CONDUCTIVITY OF [SILICON DIOXIDE + DISODIUM OXIDE + ΣX_i] $SiO_2 + Na_2O + \Sigma X_i$

[Temperature, T, K; Thermal Conductivity, k, Watt $cm^{-1}K^{-1}$]

T	k
CURVE 1	
355.2	0.0159
403.2	0.0176
468.2	0.0167
529.2	0.0188
586.2	0.0189
629.2	0.0188
691.2	0.0201
732.2	0.0209
795.2	0.0226
CURVE 2	
2	0.00053
3	0.00083
4	0.00119
5	0.00144
CURVE 3	
418.2	0.0158
468.2	0.0154
528.2	0.0167
565.2	0.0165
609.2	0.0147
642.2	0.0173
693.2	0.0183
732.2	0.0195
769.2	0.0203
770.2	0.0220
804.2	0.0214
CURVE 4	
386.2	0.0147
478.2	0.0164
611.2	0.0223
CURVE 5*	
78.2	0.00519
83.2	0.00536
88.2	0.00548
93.2	0.00561
98.2	0.00573
103.2	0.00586
CURVE 5 (cont.)	
108.2	0.00602
113.2	0.00615
118.2	0.00628
123.2	0.00640
128.2	0.00653
133.2	0.00670
138.2	0.00682
143.2	0.00699
148.2	0.00711
153.2	0.00724
158.2	0.00736
163.2	0.00745*
168.2	0.00757
173.2	0.00770*
178.2	0.00782
183.2	0.00795*
188.2	0.00808
193.2	0.00820*
198.2	0.00828
203.2	0.00841*
208.2	0.00849
213.2	0.00862*
218.2	0.00874
223.2	0.00887*
228.2	0.00895
233.2	0.00904*
238.2	0.00912
243.2	0.00925*
248.2	0.00933
253.2	0.00941*
258.2	0.00950
263.2	0.00958*
268.2	0.00971
273.2	0.00980*
278.2	0.00987
283.2	0.00992*
288.2	0.01000
293.2	0.01004*
298.2	0.01013
303.2	0.01017*
308.2	0.01025
313.2	0.01029*
318.2	0.01033
323.2	0.01042*
328.2	0.01050
CURVE 5 (cont.)	
333.2	0.01054*
338.2	0.01063
343.2	0.01071*
348.2	0.01075
353.2	0.01079*
358.2	0.01084
363.2	0.01088*
368.2	0.01096*
373.2	0.01100
CURVE 6*	
497.2	0.01736
536.2	0.01866
570.2	0.01632
615.2	0.01778
675.2	0.02038
716.2	0.02075
760.2	0.02163
793.2	0.02042
826.2	0.02159
CURVE 7	
350.2	0.01527
443.2	0.01347
499.2	0.01467
579.2	0.01607
625.2	0.01724
672.2	0.01824
724.2	0.01987*
CURVE 8*	
361.2	0.01527
404.2	0.01707
471.2	0.01648
545.2	0.01833
587.2	0.01862
638.2	0.01879
688.2	0.02029
733.2	0.02084
798.2	0.02264
CURVE 9*	
421.2	0.01540
469.2	0.01510
537.2	0.01695
568.2	0.01632
593.2	0.01527
633.2	0.01736
686.2	0.01820
738.2	0.01979
770.2	0.02042
771.2	0.02192
806.2	0.02167
CURVE 10	
368.2	0.0135
413.2	0.0144
467.2	0.0144
503.2	0.0143
527.2	0.0146
603.2	0.0157
622.2	0.0183
685.2	0.0180*
CURVE 11	
381.2	0.0105
383.2	0.0128
419.2	0.0102
423.2	0.0129
449.2	0.0126
454.2	0.0105
486.2	0.0111
488.2	0.0134
519.2	0.0140
519.7	0.0114
550.2	0.0112
559.2	0.0131
590.2	0.0147
590.2	0.0131
654.2	0.0152
655.2	0.0167
696.2	0.0175
697.2	0.0159
722.7	0.0185
724.2	0.0161
CURVE 11 (cont.)	
746.7	0.0192
748.5	0.0170
771.2	0.0204*
773.2	0.0175
CURVE 12*	
393.2	0.0184
523.2	0.0169
578.2	0.0192
641.2	0.0190
701.2	0.0203
748.2	0.0207
793.2	0.0223
847.2	0.0251
CURVE 13*	
433.2	0.0155
488.2	0.0154
546.2	0.0167
593.2	0.0168
656.2	0.0168
768.2	0.0205
888.2	0.0251
CURVE 14*	
509.2	0.0180
568.2	0.0182
624.2	0.0177
699.2	0.0177
753.2	0.0188
793.2	0.0194
839.2	0.0209

*Not shown on plot

SPECIFICATION TABLE NO. 141 THERMAL CONDUCTIVITY OF [STRONTIUM OXIDE + LITHIUM ALUMINATE + ΣX_i] $SrO + Li_2O \cdot Al_2O_3 + \Sigma X_i$

Curve No.	Ref. No.	Method Used	Year	Temp. Range, K	Reported Error, %	Name and Specimen Designation	Composition (weight percent) SrO	$Li_2O \cdot Al_2O_3$	Al_2O_3	Composition (continued), Specifications and Remarks
1	3	L	1953	324-407		253 A-1	65.0	18.3	16.7	Firing temperature 1589 K; density (25 C) 3.12 g cm^{-3}; water absorption 0%.

DATA TABLE NO. 141 THERMAL CONDUCTIVITY OF [STRONTIUM OXIDE + LITHIUM ALUMINATE + ΣX_i] $SrO + Li_2O \cdot Al_2O_3 + \Sigma X_i$

[Temperature, T, K; Thermal Conductivity, k, Watt cm^{-1} K^{-1}]

T	k
CURVE 1*	
323.7	0.0277
336.7	0.0274
356.0	0.0275
375.2	0.0269
406.5	0.0210

* No graphical presentation

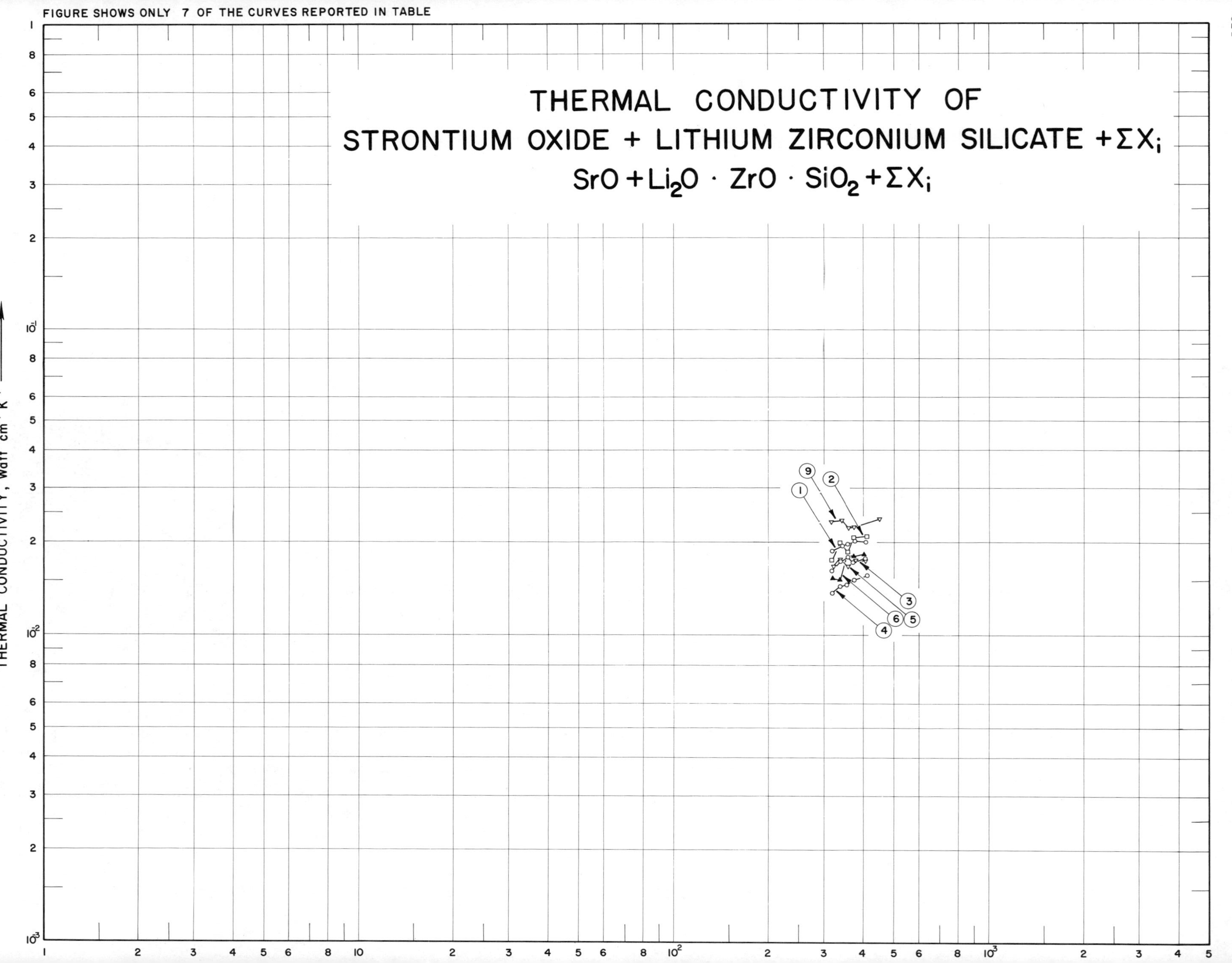
FIGURE SHOWS ONLY 7 OF THE CURVES REPORTED IN TABLE
THERMAL CONDUCTIVITY OF
STRONTIUM OXIDE + LITHIUM ZIRCONIUM SILICATE + ΣX_i
$SrO + Li_2O \cdot ZrO \cdot SiO_2 + \Sigma X_i$
THERMAL CONDUCTIVITY, Watt cm^{-1} K^{-1}
1
2
3
4
5
6
9

SPECIFICATION TABLE NO. 142 THERMAL CONDUCTIVITY OF [STRONTIUM OXIDE + LITHIUM ZIRCONIUM SILICATE + ΣX_i] $SrO + Li_2O \cdot ZrO \cdot SiO_2 + \Sigma X_i$

[For Data Reported in Figure and Table No. 142]

Curve No.	Ref. No.	Method Used	Year	Temp. Range, K	Reported Error, %	Name and Specimen Designation	Composition (weight percent) SrO	$Li_2O \cdot ZrO \cdot SiO_2$	Al_2O_3	Composition (continued), Specifications and Remarks
1	42	L	1953	320-410		232A-1	69.0	14.5	16.5	Density (25 C) 3.21 g cm^{-3}; 0.0% water absorption.
2	42	L	1953	319-412		231A-1	65.0	18.5	16.5	Density (25 C) 3.22 g cm^{-3}; 0.0% water absorption.
3	42	L	1953	320-408		233A-1	68.0	19.5	12.5	Density (25 C) 3.16 g cm^{-3}; 0.0% water absorption.
4	42	L	1953	321-412		234A-1	61.0	26.5	12.5	Density (25 C) 2.49 g cm^{-3}; 0.21% water absorption.
5	42	L	1953	322-406		235A-1	55.5	32.0	12.5	Density (25 C) 2.95 g cm^{-3}; 0.0% water absorption.
6	42	L	1953	319-404		237A-1	53.0	34.5	12.5	Density (25 C) 3.00 g cm^{-3}; 0.0% water absorption.
7	42	L	1953	317-409		238A-1	51.0	36.5	12.5	Density (25 C) 2.75 g cm^{-3}; 0.82% water absorption.
8	3	L	1953	321-411		243A-1	55.7	31.8		12.5 ZnO; density (25 C) 3.27 g cm^{-3}; fired at 1027 C; water absorption 0%.
9	3	L	1953	318-452		244A-1	47.6	27.4		25.0 ZnO; density (25 C) 3.52 g cm^{-3}; fired at 1135 C; water absorption 0.071%.

DATA TABLE NO. 142 THERMAL CONDUCTIVITY OF [STRONTIUM OXIDE + LITHIUM ZIRCONIUM SILICATE + ΣX_i] $SrO + Li_2O \cdot ZrO \cdot SiO_2 + \Sigma X_i$

[Temperature, T, K; Thermal Conductivity, k, Watt cm^{-1} K^{-1}]

T	k	T	k
CURVE 1		CURVE 7*	
320.0	0.0187	316.9	0.0164
343.7	0.0195	336.8	0.0174
357.4	0.0197	355.0	0.0177
377.3	0.0201	374.1	0.0181
409.7	0.0200	408.7	0.0176
CURVE 2		CURVE 8*	
319.0	0.0175	321.0	0.0183
337.6	0.0199	336.6	0.0191
358.0	0.0186	357.5	0.0197
375.6	0.0207	377.7	0.0197
411.7	0.0208	411.2	0.0201
CURVE 3		CURVE 9	
319.9	0.0162	317.5	0.0233
337.3	0.0173	342.5	0.0235
357.3	0.0178	360.0	0.0223
373.3	0.0172	375.9	0.0224
408.3	0.0177	452.3	0.0238
CURVE 4			
320.5	0.0137		
338.3	0.0144		
356.3	0.0146		
376.1	0.0152		
412.3	0.0156		
CURVE 5			
322.1	0.0166		
339.9	0.0171		
359.5	0.0167		
376.7	0.0174		
405.9	0.0175		
CURVE 6			
319.4	0.0153		
336.2	0.0151		
352.8	0.0182*		
373.1	0.0181		
403.5	0.0184		

*Not shown on Plot

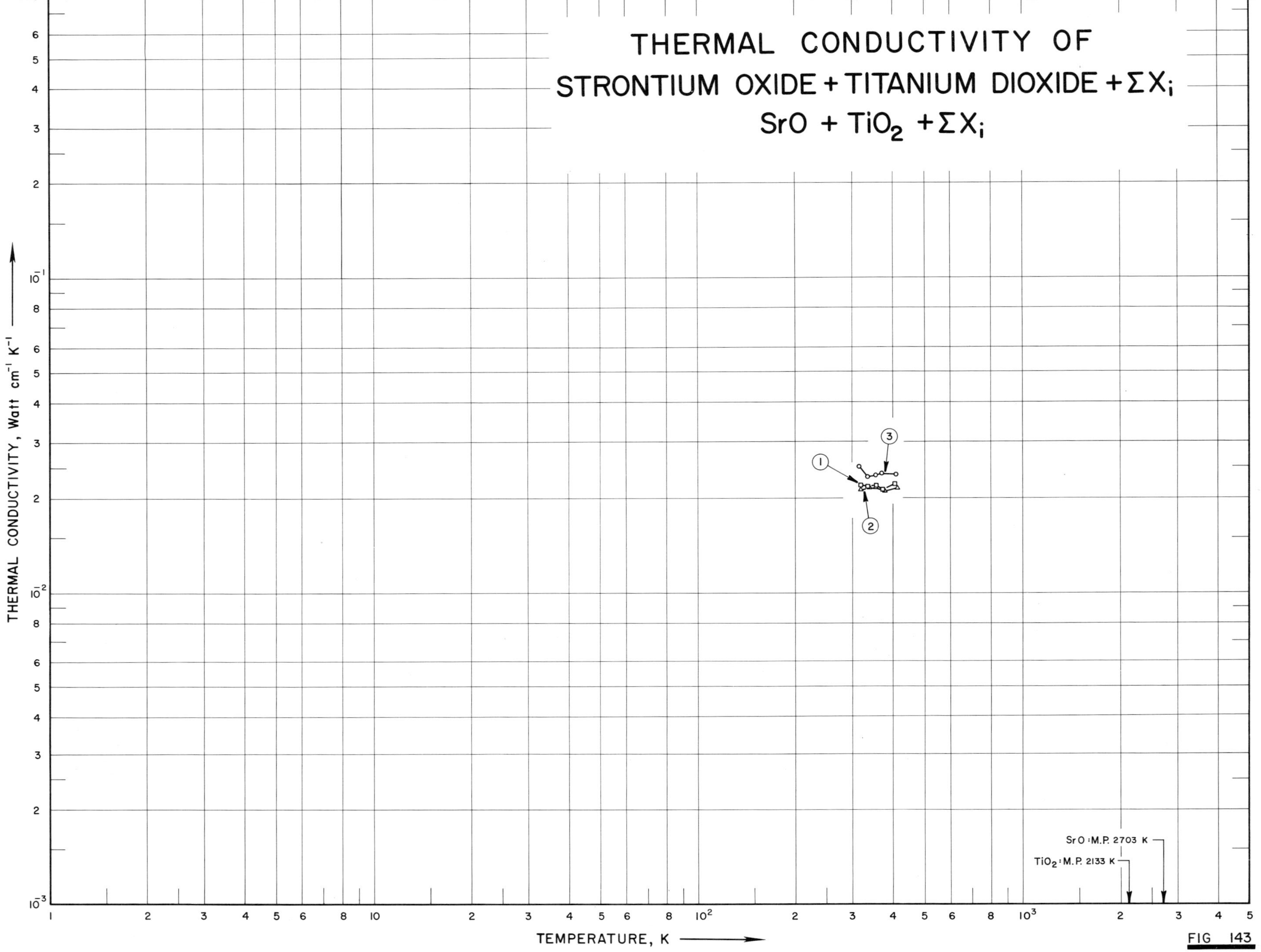
THERMAL CONDUCTIVITY OF
STRONTIUM OXIDE + TITANIUM DIOXIDE + ΣXi
SrO + TiO2 + ΣXi
SrO : M.P. 2703 K
TiO2 : M.P. 2133 K
THERMAL CONDUCTIVITY, Watt cm-1 K-1
TEMPERATURE, K

FIG 143

SPECIFICATION TABLE NO. 143 THERMAL CONDUCTIVITY OF [STRONTIUM OXIDE + TITANIUM DIOXIDE + ΣX_i] $SrO + TiO_2 + \Sigma X_i$

[For Data Reported in Figure and Table No. 143]

Curve No.	Ref. No.	Method Used	Year	Temp. Range, K	Reported Error, %	Name and Specimen Designation	Composition (weight percent) SrO	TiO_2	$Li_2O \cdot ZrO_2 \cdot SiO_2$	Composition (continued), Specifications and Remarks
1	3	L	1953	322-409		240A-1	52.5	25.0	22.5	Density (25 C) 2.80 g cm^{-3}; fired at 1149 C; water absorption 0.024%.
2	3	L	1953	321-415		241A-1	39.8	37.4	22.8	Density (25 C) 2.88 g cm^{-3}; fired at 1149 C; water absorption 0.041%.
3	3	L	1953	317-412		242A-1	43.7	37.5	18.8	Density (25 C) 2.92 g cm^{-3}; fired at 1149 C; water absorption 0.0096%.

DATA TABLE NO. 143 THERMAL CONDUCTIVITY OF [STRONTIUM OXIDE + TITANIUM DIOXIDE + ΣX_i] $SrO + TiO_2 + \Sigma X_i$

[Temperature, T, K; Thermal Conductivity, k, Watt $cm^{-1} K^{-1}$]

T	k
CURVE 1	
321.5	0.0219
336.8	0.0217
355.7	0.0218
375.5	0.0212
409.4	0.0221
CURVE 2	
320.7	0.0212
339.2	0.0215
358.8	0.0216
379.7	0.0210
414.8	0.0215
CURVE 3	
317.0	0.0251
337.0	0.0233
356.9	0.0236
373.4	0.0239
412.2	0.0237

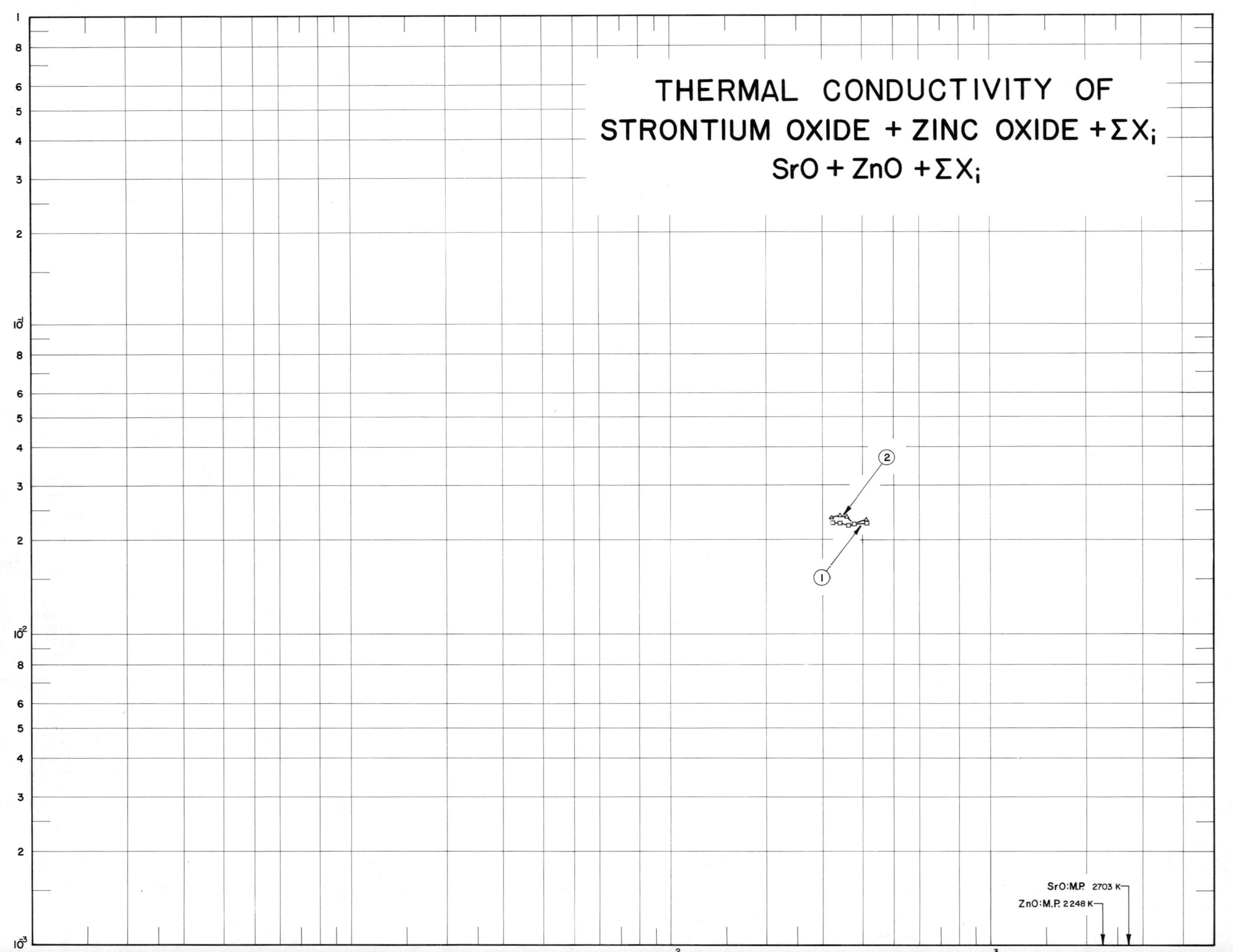
THERMAL CONDUCTIVITY OF
STRONTIUM OXIDE + ZINC OXIDE + ΣX$_i$
SrO + ZnO + ΣX$_i$
SrO: M.P. 2703 K
ZnO: M.P. 2248 K
THERMAL CONDUCTIVITY, Watt cm^{-1} K^{-1}

SPECIFICATION TABLE NO. 144 THERMAL CONDUCTIVITY OF [STRONTIUM OXIDE + ZINC OXIDE + ΣX_i] $SrO + ZnO + \Sigma X_i$

[For Data Reported in Figure and Table No. 144]

Curve No.	Ref. No.	Method Used	Year	Temp. Range, K	Reported Error, %	Name and Specimen Designation	Composition (weight percent) SrO	ZnO	$Li_2O \cdot ZrO_2 \cdot SiO_2$	Composition (continued), Specifications and Remarks
1	3	L	1953	321-415		246A-1	52.5	25.0	22.5	Density (25 C) 3.48 g cm^{-3}; fired at 1232 C; water absorption 0.024%.
2	3	L	1953	320-415		247A-1	43.7	37.5	18.8	Density (25 C) 3.12 g cm^{-3}; fired at 1315 C; water absorption 0%.

DATA TABLE NO. 144 THERMAL CONDUCTIVITY OF [STRONTIUM OXIDE + ZINC OXIDE + ΣX_i] $SrO + ZnO + \Sigma X_i$

[Temperature, T, K; Thermal Conductivity, k, Watt cm^{-1} K^{-1}]

T	k
CURVE 1	
321.0	0.0229
340.2	0.0228
360.3	0.0223
376.3	0.0225
414.9	0.0227
CURVE 2	
319.6	0.0236
339.2	0.0240
356.6	0.0239
375.5	0.0224*
414.6	0.0232

*Not shown on Plot

SPECIFICATION TABLE NO. 145 THERMAL CONDUCTIVITY OF [TIN DIOXIDE + MAGNESIUM OXIDE + ΣX_i] $SnO_2 + MgO + \Sigma X_i$

Curve No.	Ref. No.	Method Used	Year	Temp. Range, K	Reported Error, %	Name and Specimen Designation	Composition (weight percent) SnO_2	MgO	ZnO	Composition (continued), Specifications and Remarks
1	12		1953	323-417		215 A-1	42.7	34.2	23.1	$3MgO + SnO_2 + ZnO$ by mole; fired at 1454 C; water absorption 0.362%, density 4.34 g cm^{-3}.

DATA TABLE NO. 145 THERMAL CONDUCTIVITY OF [TIN DIOXIDE + MAGNESIUM OXIDE + ΣX_i] $SnO_2 + MgO + \Sigma X_i$

[Temperature, T, K; Thermal Conductivity, k, Watt $cm^{-1} K^{-1}$]

T	k
CURVE 1*	
322.8	0.0527
340.9	0.0527
358.7	0.0519
378.8	0.0502
416.7	0.0485

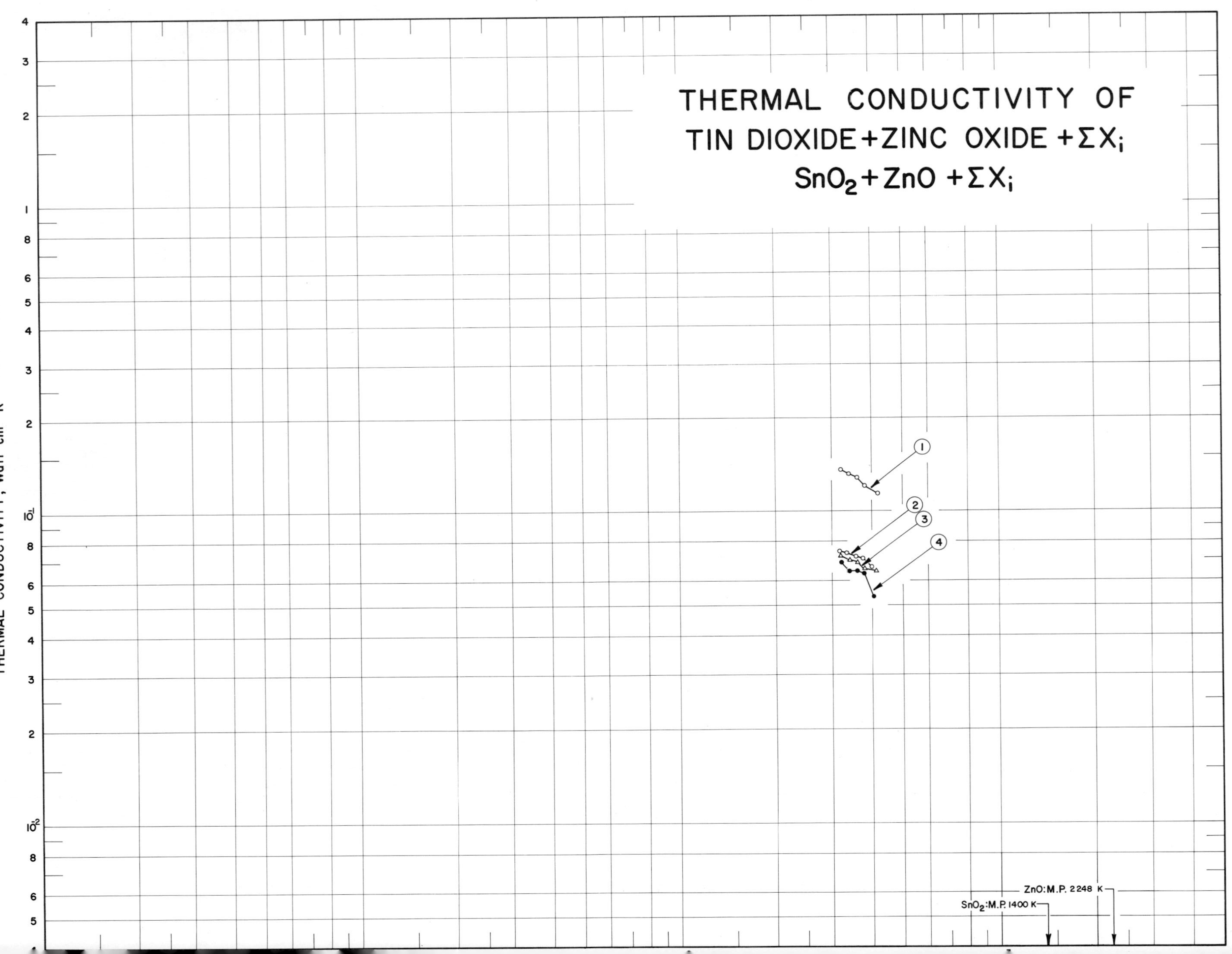
THERMAL CONDUCTIVITY OF
TIN DIOXIDE + ZINC OXIDE + ΣX_i
SnO2 + ZnO + ΣX_i
THERMAL CONDUCTIVITY, Watt cm^-1 K^-1
ZnO: M.P. 2248 K
SnO2: M.P. 1400 K
1
2
3
4

SPECIFICATION TABLE NO. 146 THERMAL CONDUCTIVITY OF [TIN DIOXIDE + ZINC OXIDE + ΣX_i] $SnO_2 + ZnO + \Sigma X_i$

[For Data Reported in Figure and Table No. 146]

Curve No.	Ref. No.	Method Used	Year	Temp. Range, K	Reported Error, %	Name and Specimen Designation	Composition (weight percent) SnO_2	ZnO	MgO	Composition (continued), Specifications and Remarks
1	12		1953	321-420		210A-1	78.9	14.2	6.9	$MgO + 3\ SnO_2 + ZnO$ by mole; prepared by milling pure oxides in water, calcining at 1093 C after drying, then dry pressing in a 0.50 in. steel die at 15000 psi; fired at 1399 C and soaked for 1.50 hr; water absorption 0.171% and density 5.51 g cm^{-3}.
2	12		1953	317-401		211A-1	59.7	32.3	8.0	$MgO + 2\ SnO_2 + 2\ ZnO$ by mole; same preparation as that of the above specimen except fired at 1427 C; water absorption 0.014% and density 5.00 g cm^{-3}.
3	12		1953	319-412		212A-1	65.0	17.6	17.4	$2\ MgO + 2\ SnO_2 + ZnO$ by mole; same preparation as that of the above specimen 210A-1; water absorption 0.097% and density 4.85 g cm^{-3}.
4	12		1953	322-405		216A-1	55.3	29.9	14.8	$MgO + SnO_2 + ZnO$ by mole; same preparation as that of the above specimen except fired at 1482 C; water absorption 0.058% and density 4.50 g cm^{-3}.

DATA TABLE NO. 146 THERMAL CONDUCTIVITY OF [TIN DIOXIDE + ZINC OXIDE + ΣX_i] $SnO_2 + ZnO + \Sigma X_i$

[Temperature, T, K; Thermal Conductivity, k, Watt $cm^{-1}K^{-1}$]

T	k
CURVE 1	
320.8	0.136
340.3	0.131
361.5	0.128
380.1	0.120
420.2	0.114
CURVE 2	
317.0	0.0744
335.2	0.0732
357.3	0.0715
375.8	0.0703
400.9	0.0665
CURVE 3	
319.4	0.0720
341.2	0.0699
360.1	0.0690
378.9	0.0657
411.9	0.0644
CURVE 4	
321.8	0.0686
341.8	0.0644
360.3	0.0644
377.2	0.0632
405.0	0.0536

SPECIFICATION TABLE NO. 147 THERMAL CONDUCTIVITY OF [ZINC OXIDE + STRONTIUM OXIDE + ΣX_i] $ZnO + SrO + \Sigma X_i$

Curve No.	Ref. No.	Method Used	Year	Temp. Range, K	Reported Error, %	Name and Specimen Designation	Composition (weight percent) ZnO	SrO	$Li_2O \cdot ZrO_2 \cdot SiO_2$	Composition (continued), Specifications and Remarks
1	3	L	1953	318-413		245 A-1	50.0	31.8	18.2	Density (25 C) 3.83 g cm^{-3}; fired at 1166 C; water absorption 0.03%.

DATA TABLE NO. 147 THERMAL CONDUCTIVITY OF [ZINC OXIDE + STRONTIUM OXIDE + ΣX_i] $ZnO + SrO + \Sigma X_i$

[Temperature, T, K; Thermal Conductivity, k, Watt $cm^{-1} K^{-1}$]

T	k
CURVE 1*	
318.0	0.0377
338.1	0.0351
356.1	0.0351
377.7	0.0348
412.6	0.0337

* No graphical presentation

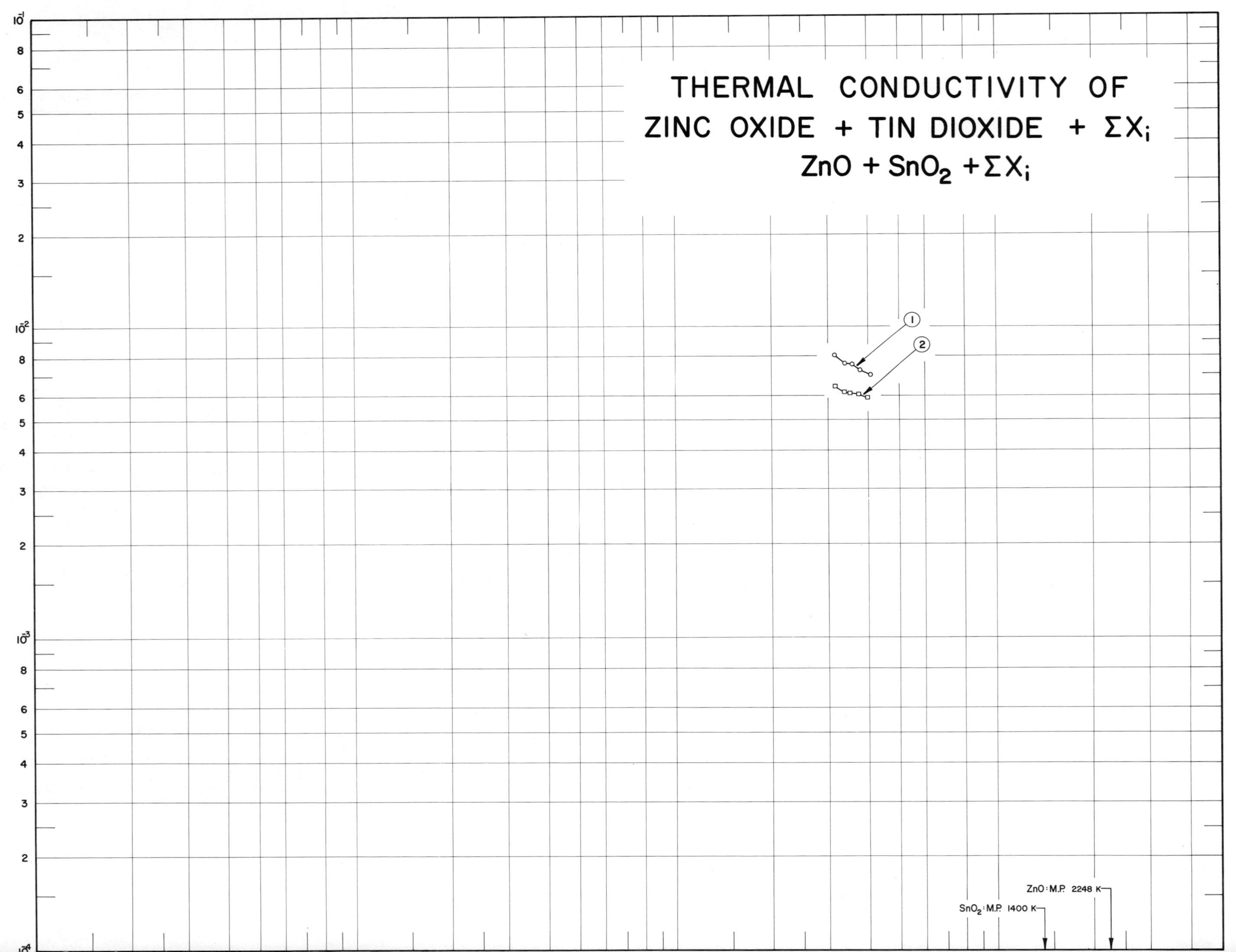
THERMAL CONDUCTIVITY OF
ZINC OXIDE + TIN DIOXIDE + ΣX_i
$ZnO + SnO_2 + \Sigma X_i$
1
2
ZnO: M.P. 2248 K
SnO_2: M.P. 1400 K
10^{-1}
8
6
5
4
3
2
10^{-2}
8
6
5
4
3
2
10^{-3}
8
6
5
4
3
2
10^{-4}
THERMAL CONDUCTIVITY, Watt cm^{-1} K^{-1}

SPECIFICATION TABLE NO. 148 THERMAL CONDUCTIVITY OF [ZINC OXIDE + TIN DIOXIDE + ΣX_i] $ZnO + SnO_2 + \Sigma X_i$

[For Data Reported in Figure and Table No. 148]

Curve No.	Ref. No.	Method Used	Year	Temp. Range, K	Reported Error, %	Name and Specimen Designation	Composition (weight percent) ZnO	SnO_2	MgO	Composition (continued), Specifications and Remarks
1	12		1953	316-410		313A-1	56.1	34.6	9.3	$MgO + SnO_2 + 3\ ZnO$ by mole; prepared by milling pure oxides in water, calcining at 2000 F after drying, then dry pressing in a 0.50 in. steel die at 15000 psi; fired at 2600 F and soaked for 1.50 hr; water absorption 0.082%; density 4.80 g cm^{-3}.
2	12		1953	316-398		214A-1	41.3	38.2	20.5	$2\ MgO + SnO_2 + 2\ ZnO$ by mole; same preparation as that of the above specimen; water absorption 0.006%, density 4.86 g cm^{-3}.

DATA TABLE NO. 148 THERMAL CONDUCTIVITY OF [ZINC OXIDE + TIN DIOXIDE + ΣX_i] $ZnO + SnO_2 + \Sigma X_i$

[Temperature, T, K; Thermal Conductivity, k, Watt $cm^{-1}K^{-1}$]

T	k
CURVE 1	
316.4	0.0803
339.1	0.0757
357.5	0.0753
377.7	0.0724
409.8	0.0699
CURVE 2	
316.4	0.0644
337.5	0.0615
351.9	0.0611
374.1	0.0607
398.2	0.0590

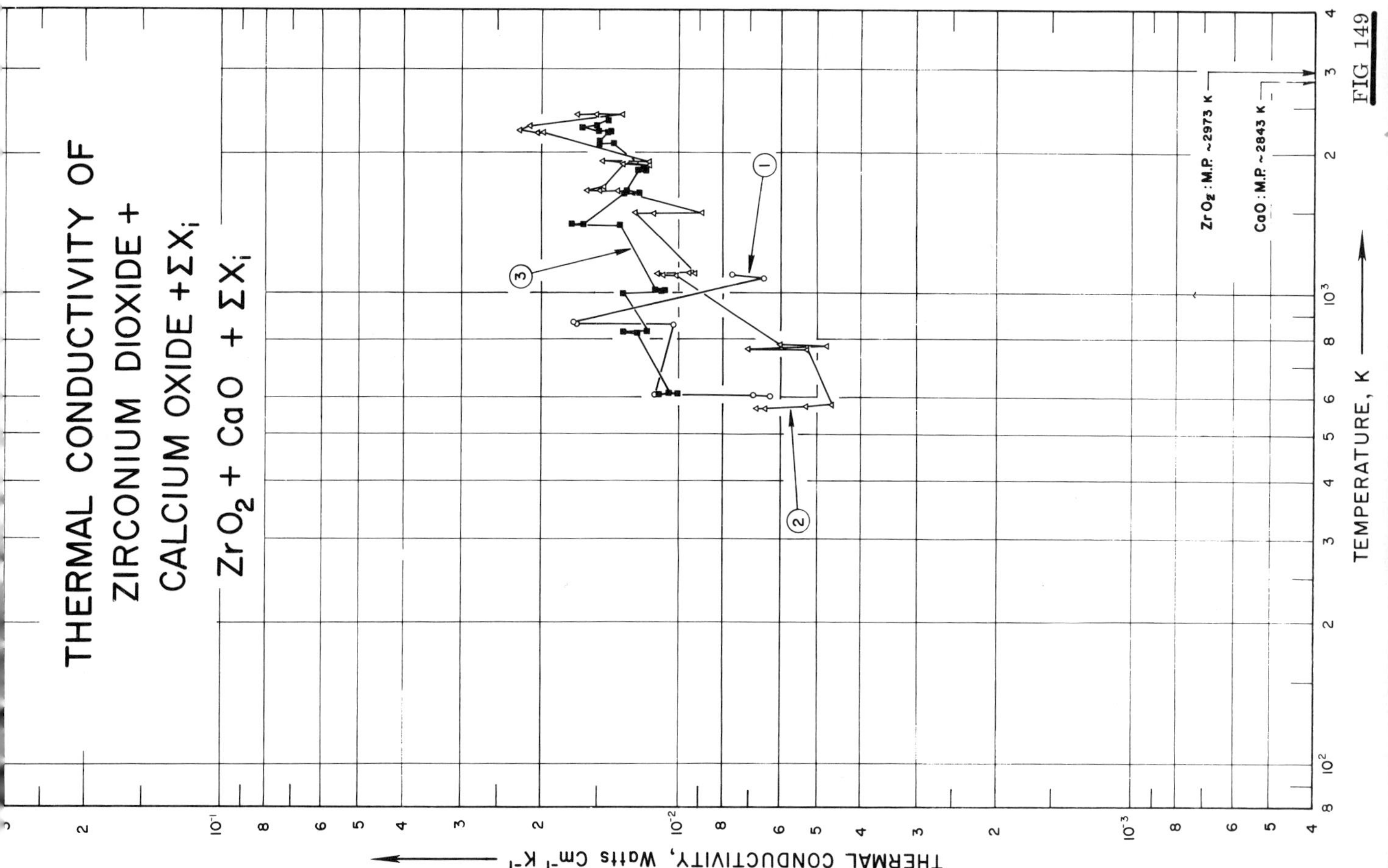
THERMAL CONDUCTIVITY OF
ZIRCONIUM DIOXIDE +
CALCIUM OXIDE + ΣXi
ZrO2 + CaO + ΣXi
ZrO2 : M.P. ~2973 K
CaO : M.P. ~2843 K
TEMPERATURE, K
THERMAL CONDUCTIVITY, Watts Cm-1 K-1

FIG 149

SPECIFICATION TABLE NO. 149 THERMAL CONDUCTIVITY OF [ZIRCONIUM DIOXIDE + CALCIUM OXIDE + ΣX_i] $ZrO_2 + CaO + \Sigma X_i$

[For Data Reported in Figure and Table No. 149]

Curve No.	Ref. No.	Method Used	Year	Temp. Range, K	Reported Error, %	Name and Specimen Designation	Composition (weight percent)			Composition (continued), Specifications and Remarks
							ZrO_2	CaO	CeO_2	
1	82	R	1962	599-1088		Stabilized Zirconia	93.216	2.903	2.67	0.33 SiO_2, 0.204 MgO, 0.252 $CaSO_4$, 0.175 other sulfates and 0.22 other oxides; 1 in. dia x 1 in. long; stabilized zirconia with coarse grains; molded; density 4.153 g cm^{-3}; porosity 31.0%; melting point 2855 K.
2	82	R	1962	568-2387		Stabilized Zirconia	93.216	2.903	2.67	0.33 SiO_2, 0.204 MgO, 0.252 $CaSO_4$, 0.175 other sulfates and 0.22 other oxides; similar to the above specimen.
3	82	R	1962	607-2330		Stabilized Zirconia	93.216	2.903	2.67	0.33 SiO_2, 0.204 MgO, 0.252 $CaSO_4$, 0.175 other sulfates and 0.22 other oxides; similar to the above specimen.

DATA TABLE NO. 149 THERMAL CONDUCTIVITY OF [ZIRCONIUM DIOXIDE + CALCIUM OXIDE + ΣX_i] $ZrO_2 + CaO + \Sigma X_i$

[Temperature, T, K; Thermal Conductivity, k, Watt $cm^{-1}K^{-1}$]

T	k
CURVE 1	
599.3	0.00632
605.4	0.00689
605.4	0.0112
852.6	0.0102
854.3	0.0167
865.4	0.0170
1070.4	0.00653
1088.2	0.00763
CURVE 2	
568.2	0.00649
569.3	0.00674
570.9	0.00528
579.3	0.00462
757.1	0.00528
759.8	0.00704
769.3	0.00476
776.5	0.00601
1080.3	0.0101
1080.9	0.0109
1081.5	0.00922
1081.5	0.0112
1082.6	0.00945
1483.2	0.0124
1485.4	0.0113
1485.4	0.00891
1651.5	0.0136
1651.5	0.0159
1651.5	0.0149
1682.1	0.0146
1874.3	0.0133
1874.3	0.0116
1896.5	0.0147
1896.5	0.0116
2192.1	0.0205
2192.1	0.0198
2203.2	0.0224
2262.1	0.0213
2387.1	0.0133
2387.1	0.0151
2387.1	0.0166

T	k
CURVE 3	
607.1	0.0110
609.3	0.0100
610.4	0.0105
820.4	0.0123
824.8	0.0132
827.1	0.0117
996.5	0.0132
1003.2	0.0109
1008.2	0.0107
1012.6	0.0113
1394.3	0.0134
1398.7	0.0173
1398.7	0.0163
1627.6	0.0122
1635.4	0.0132
1645.9	0.0130
1819.8	0.0123
1819.8	0.0117
1834.8	0.0119
2083.2	0.0139
2083.2	0.0149
2089.8	0.0149
2202.6	0.0143
2206.5	0.0141
2206.5	0.0150
2208.2	0.0150*
2252.6	0.0162
2274.8	0.0151
2330.4	0.0143

*Not shown on Plot

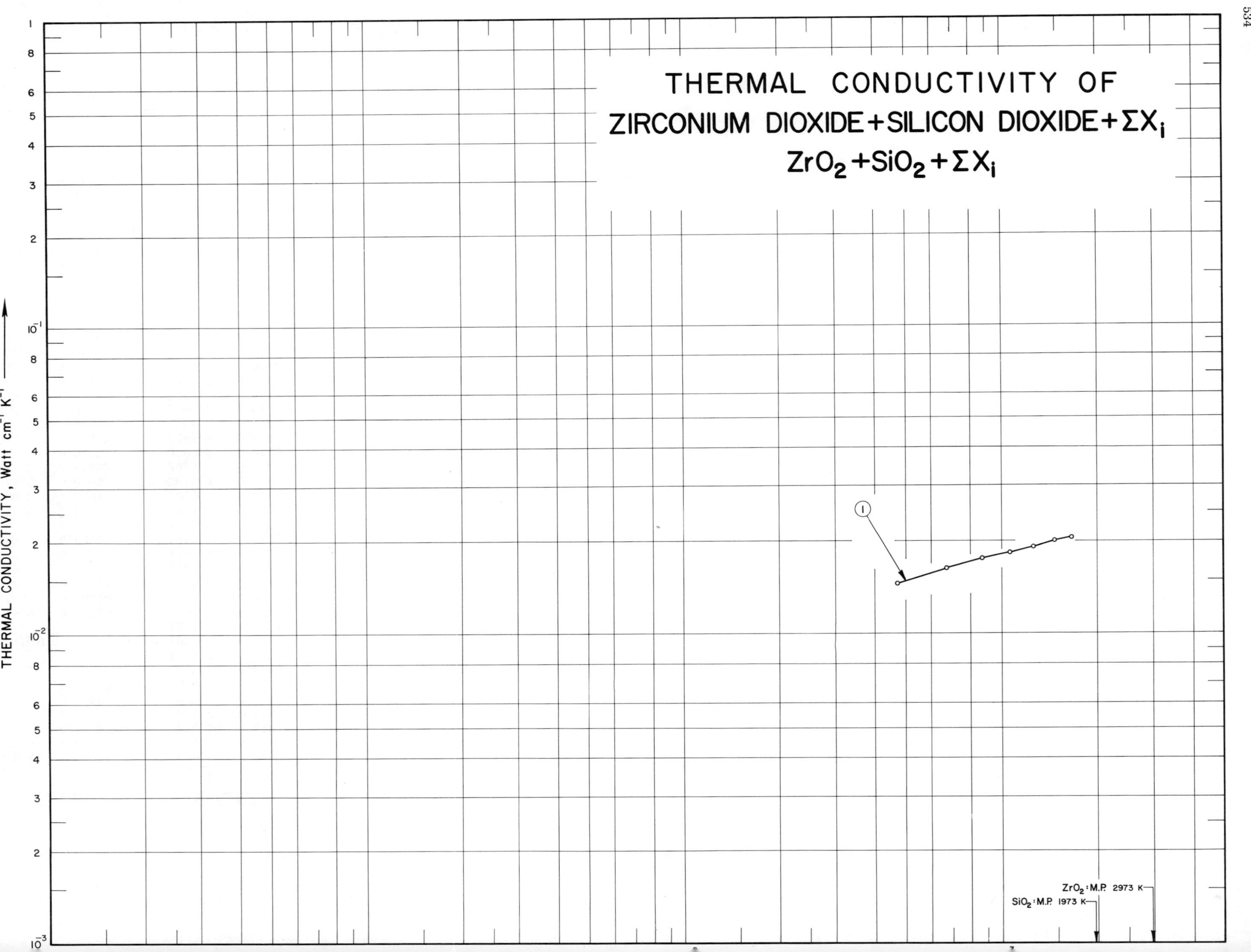
THERMAL CONDUCTIVITY OF
ZIRCONIUM DIOXIDE + SILICON DIOXIDE + ΣXi
ZrO2 + SiO2 + ΣXi
1
ZrO2 : M.P. 2973 K
SiO2 : M.P. 1973 K
THERMAL CONDUCTIVITY, Watt cm-1 K-1
1 8 6 5 4 3 2 10-1 8 6 5 4 3 2 10-2 8 6 5 4 3 2 10-3

SPECIFICATION TABLE NO. 150 THERMAL CONDUCTIVITY OF [ZIRCONIUM DIOXIDE + SILICON DIOXIDE + ΣX_i] $ZrO_2 + SiO_2 + \Sigma X_i$

[For Data Reported in Figure and Table No. 150]

Curve No.	Ref. No.	Method Used	Year	Temp. Range, K	Reported Error, %	Name and Specimen Designation	Composition (weight percent)			Composition (continued), Specifications and Remarks
							ZrO_2	SiO_2	Al_2O_3	
1	383	L	1927	473-1673	±15	Zirconia brick	60.44	27.26	7.75	1.60 Fe_2O_3, 0.04 CaO; specimen 10.8 cm in dia and 22.8 cm long; made of South American baddeleyite ore, calcined and crushed into grog, bonded with some fine ground ore, pressed into bricks, fired at 1923 K; apparent density 4.87 g cm^{-3}; bulk density 3.43 g cm^{-3}; porosity 29.5%.

DATA TABLE NO. 150 THERMAL CONDUCTIVITY OF [ZIRCONIUM DIOXIDE + SILICON DIOXIDE + ΣX_i] $ZrO_2 + SiO_2 + \Sigma X_i$

[Temperature, T, K; Thermal Conductivity, k, Watt $cm^{-1} K^{-1}$]

T	k
CURVE 1	
473.2	0.0146
673.2	0.0163
873.2	0.0176
1073.2	0.0184
1273.2	0.0192
1473.2	0.0201
1673.2	0.0205

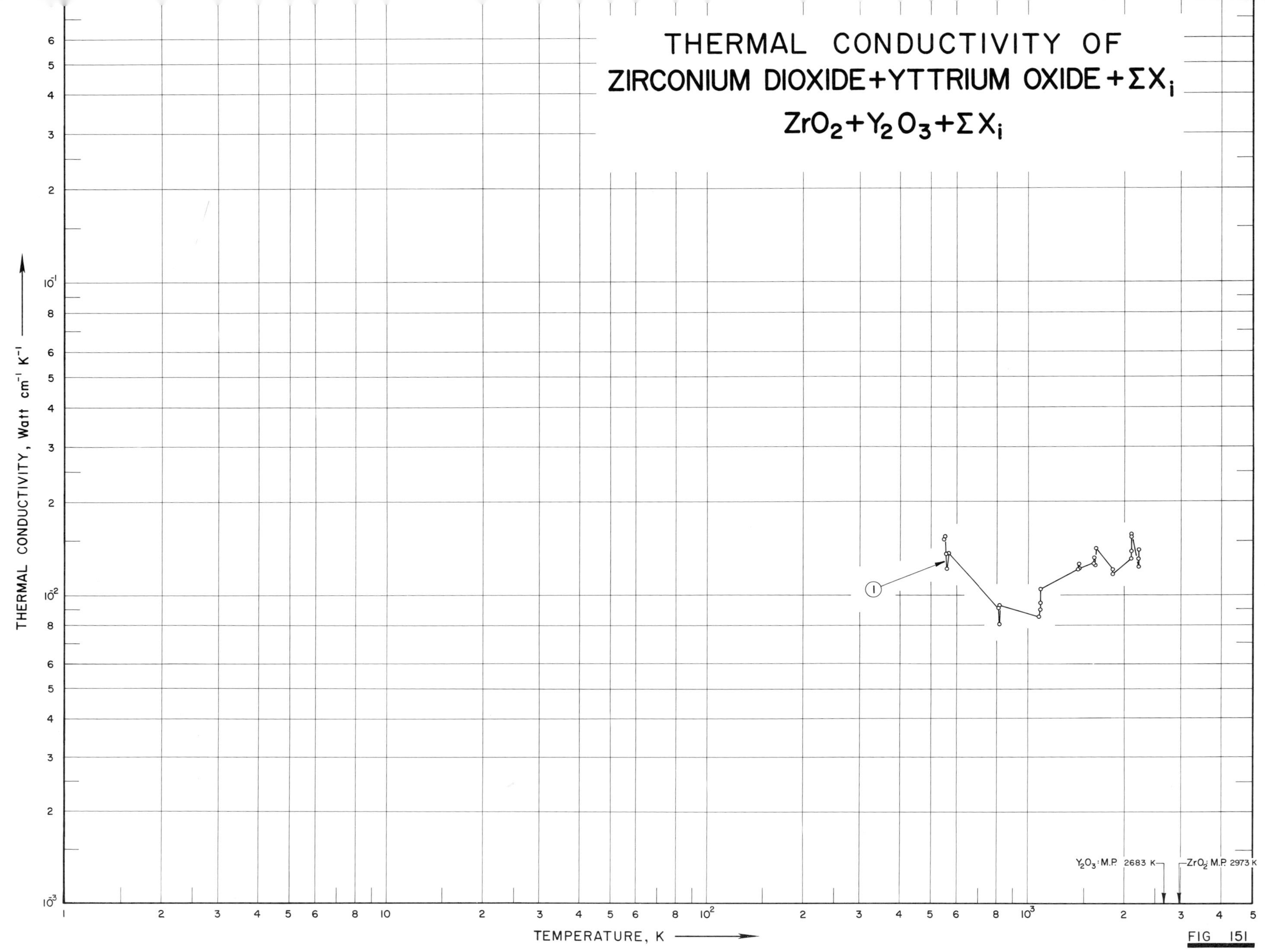
THERMAL CONDUCTIVITY OF
ZIRCONIUM DIOXIDE+YTTRIUM OXIDE+ΣXi
ZrO2+Y2O3+ΣXi
THERMAL CONDUCTIVITY, Watt cm-1 K-1
TEMPERATURE, K
Y2O3: M.P. 2683 K
ZrO2: M.P. 2973 K
1
FIG 151

SPECIFICATION TABLE NO. 151 THERMAL CONDUCTIVITY OF [ZIRCONIUM DIOXIDE + YTTRIUM OXIDE + ΣX_i] $ZrO_2 + Y_2O_3 + \Sigma X_i$

[For Data Reported in Figure and Table No. 151]

Curve No.	Ref. No.	Method Used	Year	Temp. Range, K	Reported Error, %	Name and Specimen Designation	Composition (weight percent)				Composition (continued), Specifications and Remarks
							ZrO_2	Y_2O_3	CeO_2	CaO	
1	82	R	1962	549-2206			88.41	7.09	2.73	0.46	0.25 MgO, 0.25 SiO_2, 0.063 $CaSO_4$, 0.045 SO_4, and 0.702 minor oxides; prepared from 89.3 of (99 ZrO_2, 0.52 CaO, 0.28 MgO, 0.28 SiO, 0.07 $CaSO_4$, and 0.05 SO_4), 7.8 of (90.9 Y_2O_3, 0.80 CeO_2, and 4.27 minor oxides), and 2.9 of (92.0 CeO_2 and 7.6 minor oxides); specimen 1 in. dia x 1 in. long; coarse grain; stabilized; molded; density 4.62 g cm^{-3}; porosity 22.58%; melting point 2905 K.

DATA TABLE NO. 151 THERMAL CONDUCTIVITY OF [ZIRCONIUM DIOXIDE + YTTRIUM OXIDE + ΣX_i] $ZrO_2 + Y_2O_3 + \Sigma X_i$

[Temperature, T, K; Thermal Conductivity, k, Watt $cm^{-1} K^{-1}$]

T	k
CURVE 1	
548.8	0.0151
550.9	0.0154
555.9	0.0136
558.7	0.0122
566.5	0.0137
810.9	0.00900
813.2	0.00799
814.3	0.00922
1084.3	0.00848
1094.3	0.00891
1096.5	0.00939
1098.2	0.0104
1426.5	0.0121
1435.9	0.0126
1450.9	0.0125*
1450.9	0.0122
1602.6	0.0126
1605.4	0.0132
1605.4	0.0124
1614.3	0.0142
1614.3	0.0138*
1833.2	0.0120
1833.2	0.0116
1833.2	0.0115*
1841.5	0.0117*
2083.2	0.0131
2083.2	0.0138
2083.2	0.0153
2083.2	0.0157
2205.9	0.0131
2205.9	0.0123
2205.9	0.0140

*Not shown on plot

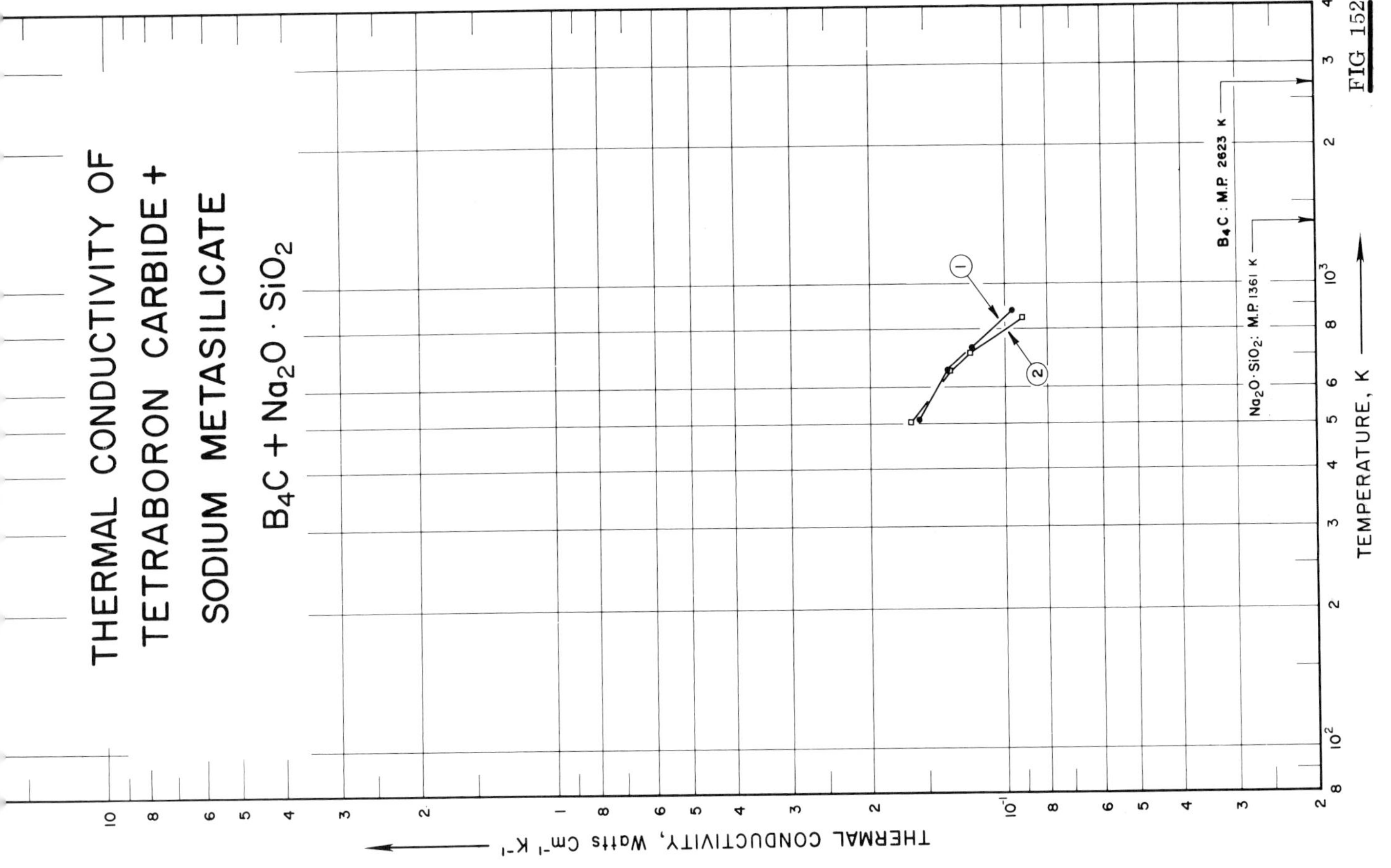
THERMAL CONDUCTIVITY OF
TETRABORON CARBIDE +
SODIUM METASILICATE
$B_4C + Na_2O \cdot SiO_2$
B_4C : M.P. 2623 K
$Na_2O \cdot SiO_2$: M.P. 1361 K
1
2
THERMAL CONDUCTIVITY, Watts Cm^{-1} K^{-1}
TEMPERATURE, K
10
8
6
5
4
3
2
1
10^{-1}
10^{2}
10^{3}

FIG 152

SPECIFICATION TABLE NO. 152 THERMAL CONDUCTIVITY OF [TETRABORON CARBIDE + SODIUM METASILICATE] $B_4C + Na_2O \cdot SiO_2$

[For Data Reported in Figure and Table No. 152]

Curve No.	Ref. No.	Method Used	Year	Temp. Range, K	Reported Error, %	Name and Specimen Designation	Composition (weight percent) B_4C	$Na_2O \cdot SiO_2$	Composition (continued), Specifications and Remarks
1	76	C	1952	511-873		No. 18	97.85	2.15	Rammed; density 2.06 g cm^{-3}.
2	76	C	1952	503-843		No. 19	97.8	2.2	Rammed; density 2.06 g cm^{-3}.

DATA TABLE NO. 152 THERMAL CONDUCTIVITY OF [TETRABORON CARBIDE + SODIUM METASILICATE] $B_4C + Na_2O \cdot SiO_2$

[Temperature, T, K; Thermal Conductivity, k, Watt $cm^{-1}K^{-1}$]

T	k
CURVE 1	
511.2	0.156
653.2	0.134
733.2	0.118
873.2	0.0963
CURVE 2	
503.2	0.163
653.2	0.134
718.2	0.120
843.2	0.0918

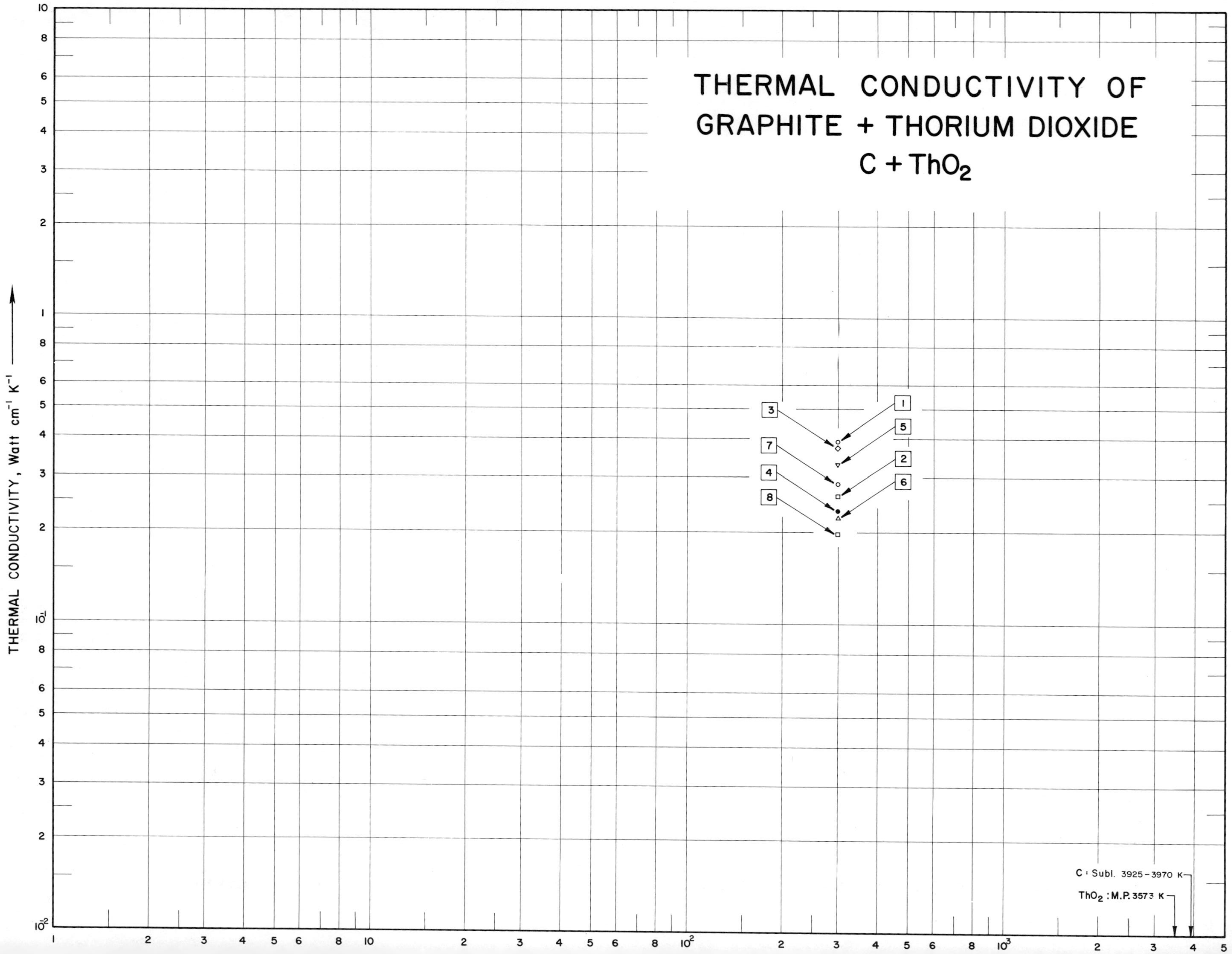

THERMAL CONDUCTIVITY OF GRAPHITE + THORIUM DIOXIDE

$C + ThO_2$

SPECIFICATION TABLE NO. 153 THERMAL CONDUCTIVITY OF [GRAPHITE + THORIUM DIOXIDE] $C + ThO_2$

[For Data Reported in Figure and Table No. 153]

Curve No.	Ref. No.	Method Used	Year	Temp. Range, K	Reported Error, %	Name and Specimen Designation	Composition (weight percent) C	ThO_2	Composition (continued), Specifications and Remarks
1	181		1959	298.2		Fuel-filled Graphite	89.96	10.04	Baked to 1425 C; bulk density 1.827 g cm^{-3}; measured with grain.
2	181		1959	298.2		Fuel-filled Graphite	89.96	10.04	The above specimen measured against grain.
3	181		1959	298.2		Fuel-filled Graphite	80.05	19.95	Baked to 1425 C; bulk density 1.971 g cm^{-3}; measured with grain.
4	181		1959	298.2		Fuel-filled Graphite	80.05	19.95	The above specimen measured against grain.
5	181		1959	298.2		Fuel-filled Graphite	70.03	29.97	Baked to 1425 C; bulk density 2.149 g cm^{-3}; measured with grain.
6	181		1959	298.2		Fuel-filled Graphite	70.03	29.97	The above specimen measured against grain.
7	181		1959	298.2		Fuel-filled Graphite	60.07	39.93	Baked to 1425 C; bulk density 2.369 g cm^{-3}; measured with grain.
8	181		1959	298.2		Fuel-filled Graphite	60.07	39.93	The above specimen measured against grain.

DATA TABLE NO. 153 THERMAL CONDUCTIVITY OF [GRAPHITE + THORIUM DIOXIDE] $C + ThO_2$

[Temperature, T, K; Thermal Conductivity, k, Watt $cm^{-1}K^{-1}$]

T	k
CURVE 1	
298.2	0.391
CURVE 2	
298.2	0.261
CURVE 3	
298.2	0.374
CURVE 4	
298.2	0.234
CURVE 5	
298.2	0.329
CURVE 6	
298.2	0.222
CURVE 7	
298.2	0.286
CURVE 8	
298.2	0.197

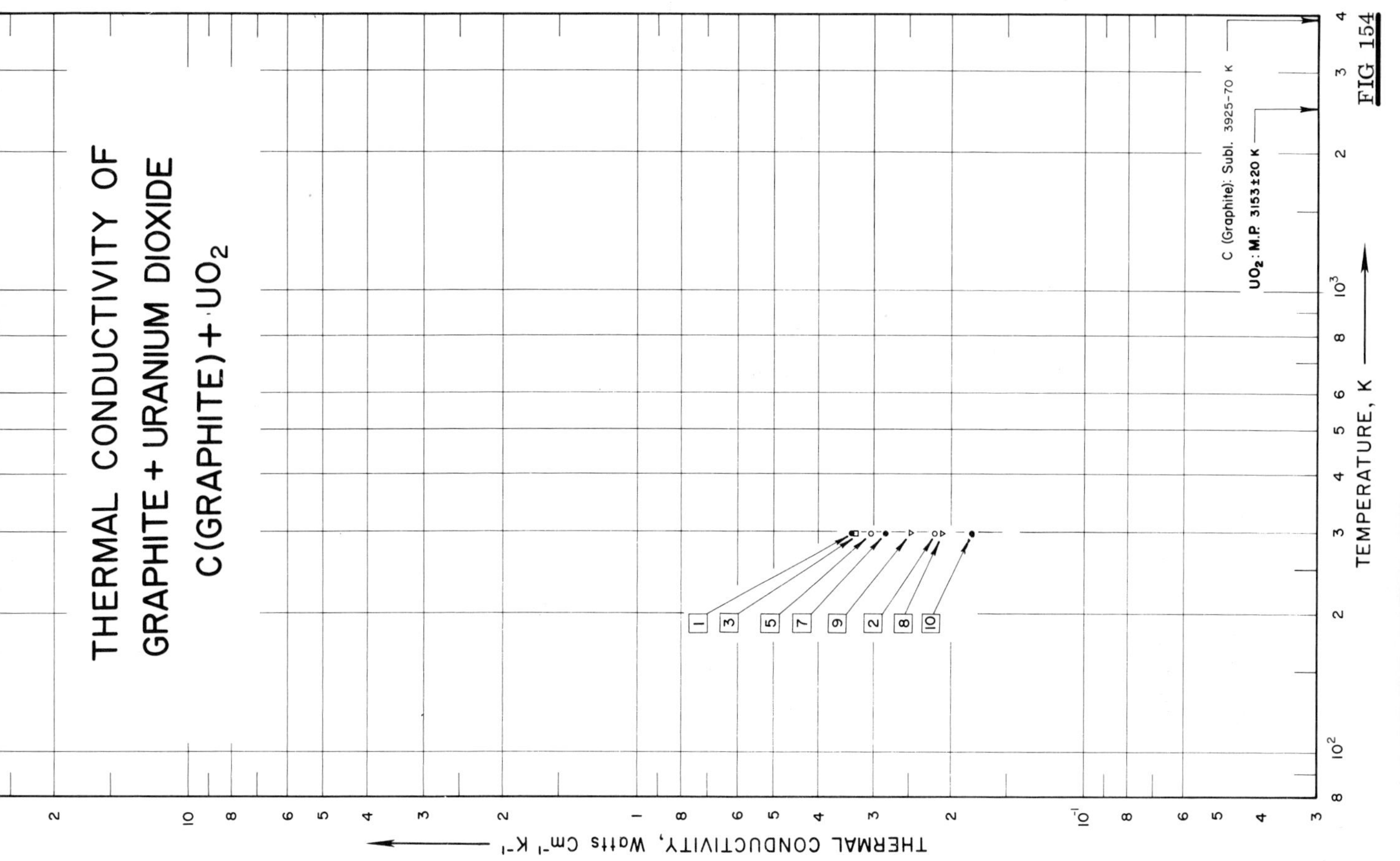
THERMAL CONDUCTIVITY OF
GRAPHITE + URANIUM DIOXIDE
C(GRAPHITE) + UO2
THERMAL CONDUCTIVITY, Watts Cm-1 K-1
TEMPERATURE, K
C (Graphite): Subl. 3925-70 K
UO2: M.P. 3153±20 K
1
3
5
7
9
2
8
10

FIG 154

SPECIFICATION TABLE NO. 154 THERMAL CONDUCTIVITY OF [GRAPHITE + URANIUM DIOXIDE] $C + UO_2$

[For Data Reported in Figure and Table No. 154]

Curve No.	Ref. No.	Method Used	Year	Temp. Range, K	Reported Error, %	Name and Specimen Designation	Composition (weight percent) C	UO_2	Composition (continued), Specifications and Remarks
1	181		1959	298.2		Fuel-filled graphite	96.69	3.31	Baked to 1425 C; bulk density 1.724 g cm^{-3}; measured with grain.
2	181		1959	298.2		Fuel-filled graphite	96.69	3.31	The above specimen measured against grain.
3	181		1959	298.2		Fuel-filled graphite	88.89	11.11	Baked to 1425 C; bulk density 1.832 g cm^{-3}; measured with grain.
4	181		1959	298.2		Fuel-filled graphite	88.89	11.11	The above specimen measured against grain.
5	181		1959	298.2		Fuel-filled graphite	78.47	21.53	Baked to 1425 C; bulk density 2.007 g cm^{-3}; measured with grain.
6	181		1959	298.2		Fuel-filled graphite	78.47	21.53	The above specimen measured against grain.
7	181		1959	298.2		Fuel-filled graphite	67.05	32.95	Baked to 1425 D; bulk density 2.224 g cm^{-3}; measured with grain.
8	181		1959	298.2		Fuel-filled graphite	67.08	32.95	The above specimen measured against grain.
9	181		1959	298.2		Fuel-filled graphite	58.61	41.39	Baked to 1425 C; bulk density 2.441 g cm^{-3}; measured with grain.
10	181		1959	298.2		Fuel-filled graphite	58.61	41.39	The above specimen measured against grain.

DATA TABLE NO. 154 THERMAL CONDUCTIVITY OF [GRAPHITE + URANIUM DIOXIDE] $C + UO_2$

[Temperature, T, K; Thermal Conductivity, k, Watt $cm^{-1}K^{-1}$]

T	k
CURVE 1	
298.2	0.338
CURVE 2	
298.2	0.218
CURVE 3	
298.2	0.334
CURVE 4*	
298.2	0.222
CURVE 5	
298.2	0.306
CURVE 6*	
298.2	0.218
CURVE 7	
298.2	0.282
CURVE 8	
298.2	0.209
CURVE 9	
298.2	0.249
CURVE 10	
298.2	0.180

*Not shown on plot

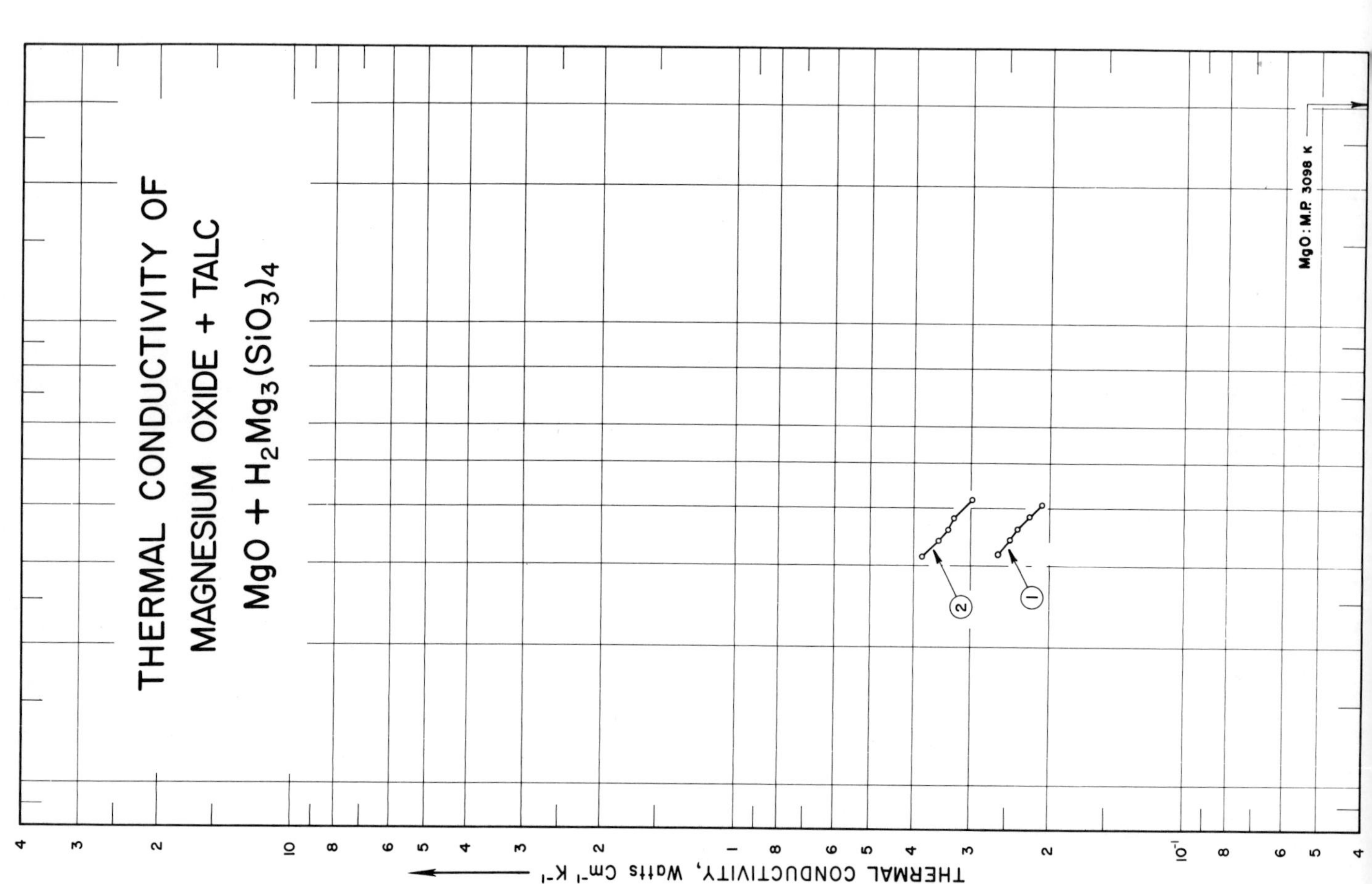
THERMAL CONDUCTIVITY OF
MAGNESIUM OXIDE + TALC
MgO + H$_2$Mg$_3$(SiO$_3$)$_4$
MgO : M.P. 3098 K
THERMAL CONDUCTIVITY, Watts cm^{-1} K^{-1}
1
2

SPECIFICATION TABLE NO. 155 THERMAL CONDUCTIVITY OF [MAGNESIUM OXIDE + TALC] $MgO + H_2Mg_3(SiO_3)_4$

[For Data Reported in Figure and Table No. 155]

Curve No.	Ref. No.	Method Used	Year	Temp. Range, K	Reported Error, %	Name and Specimen Designation	Composition (weight percent) MgO	$H_2Mg_3(SiO_3)_4$	Composition (continued), Specifications and Remarks
1	23		1952	319-408		179A-1	92.5	7.5	White color; fired at 1811 K and soaked for 2 hrs; water absorption 0.029%; density 3.219 g cm^{-3}.
2	12		1953	315-418		181A-1	97.5	2.5	Calcined at 1367 K and dry-pressed at 15000 psi; fired at 1811 K and soaked for 1 hr; density 3.48 g cm^{-3}.

DATA TABLE NO. 155 THERMAL CONDUCTIVITY OF [MAGNESIUM OXIDE + TALC] $MgO + H_2Mg_3(SiO_3)_4$

[Temperature, T, K; Thermal Conductivity, k, Watt $cm^{-1}K^{-1}$]

T	k
CURVE 1	
319.3	0.262
342.7	0.245
360.6	0.236
384.8	0.222
407.7	0.208
CURVE 2	
315.0	0.384
340.8	0.364
360.5	0.346
380.6	0.328
417.6	0.299

SPECIFICATION TABLE NO. 156 THERMAL CONDUCTIVITY OF [SILICON CARBIDE + SILICON DIOXIDE] $SiC + SiO_2$

Curve No.	Ref. No.	Method Used	Year	Temp. Range, K	Reported Error, %	Name and Specimen Designation	Composition (weight percent), Specifications and Remarks
1	84	L	1925	1623.2	1.0	9" carborundum No. 1A Wall	91.51 SiC, 8.02 SiO_2, 0.20 Al_2O_3, 0.65 Fe_2O_3, 0.17 ignition loss; the wall under test was built of standard 2.5 x 4.5 x 9 in. brick laid up with cement of the same composition as the brick; the brick were made of carborundum recrystallized in an electric furnace; apparent density 2.05 g cm^{-3}; porosity 34.1% calculated by assuming SiC specific gravity 3.17; the brick was placed so that the temp gradient was through the 9 in. dimension.
2	84	L	1925	1623.2	1.0	4.5" carborundum No. 1B Wall	93.20 SiC, 4.50 SiO_2, 1.33 Al_2O_3, 1.03 Fe_2O_3; same as the above specimen except having apparent density 2.07 g cm^{-3}, and porosity 33.8%, the brick was placed so that the temp gradient was through the 4.5 in. dimension.
3	84	L	1925	1623.2	1.0	4.5" carborundum No. 1C Wall	93.20 SiC, 4.50 SiO_2, 1.33 Al_2O_3, 1.03 Fe_2O_3; apparent density 2.20 g cm^{-3}; porosity 29.5%; the brick was placed so that the temp gradient was through the 4.5 in. dimension.
4	84	L	1925	1623.2	1.0	4.5" carborundum No. 2 Wall	80.10 SiC, 14.72 SiO_2, 1.47 Al_2O_3, 1.33 Fe_2O_3, 0.88 ignition loss; the wall under test was built of standard 2.5 x 4.5 x 9 in. brick laid up with cement of the same composition as the brick; the brick contained ceramic bonds and were kiln fired at approx 1350 C; apparent density 2.48 g cm^{-3}; porosity 18.4% calculated by assuming 3.17 for SiC specific gravity; the brick was placed so that the temp gradient was through the 4.5 in. dimension.

DATA TABLE NO. 156 THERMAL CONDUCTIVITY OF [SILICON CARBIDE + SILICON DIOXIDE] $SiC + SiO_2$

[Temperature, T, K; Thermal Conductivity, k, Watt $cm^{-1} K^{-1}$]

T	k	T	k
CURVE 1*		CURVE 3*	
1623.2	0.169	1623.2	0.227
CURVE 2*		CURVE 4*	
1623.2	0.180	1623.2	0.160

*No graphical presentation

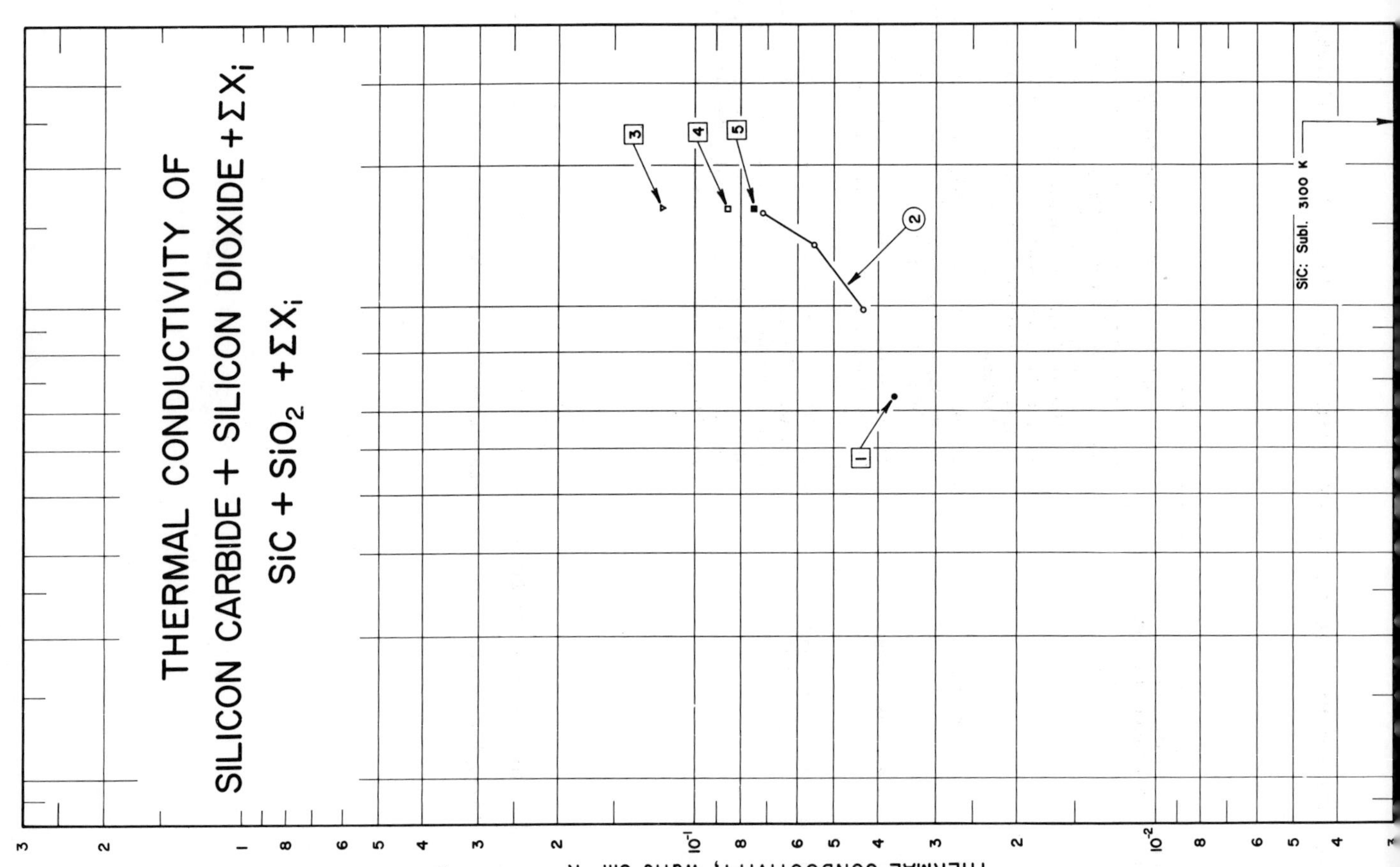
THERMAL CONDUCTIVITY OF
SILICON CARBIDE + SILICON DIOXIDE + ΣXi
SiC + SiO2 + ΣXi
THERMAL CONDUCTIVITY, Watts Cm⁻¹ K⁻¹
SiC: Subl. 3100 K
1
2
3
4
5
3
2
1
8
6
5
4
3
2
10⁻¹
8
6
5
4
3
2
10⁻²
8
6
5
4

SPECIFICATION TABLE NO. 157 THERMAL CONDUCTIVITY OF [SILICON CARBIDE + SILICON DIOXIDE + ΣX_i] $SiC + SiO_2 + \Sigma X_i$

[For Data Reported in Figure and Table No. 157]

Curve No.	Ref. No.	Method Used	Year	Temp. Range, K	Reported Error, %	Name and Specimen Designation	Composition (weight percent) SiC	SiO_2	Al_2O_3	Composition (continued), Specifications and Remarks
1	80	L	1934	647.1		Silicon carbide brick, Norton	86.7	8.2	4.2	0.3 Fe_2O_3, 0.2 CaO, 0.2, TiO_2 and 0.2 Alkali; approximate composition; clay-bonded; bulk density 2.23 g cm^{-3} and porosity 28.3%
2	80	L	1934	986-1598		Silicon carbide brick, Norton	86.7	8.2	4.2	0.3 Fe_2O_3, 0.2 CaO, 0.2 TiO_2, and 0.2 Alkali; the above specimen measured with insulating brick placed between the calorimeter and the lower surface of the specimen.
3	84	L	1925	1623.2	1.0	4.50 in. carborundum No. 3 wall	68.50	23.31	5.58	1.57 Fe_2O_3; made from standard 2.5 x 4.5 x 9 in. brick laid with cement of the same composition; kiln fired at approximately 1350 C; apparent density 2.35 g cm^{-3}; 20.7% porosity calculated by assuming specific gravity of SiC as 3.17; ignition loss 0.24%.
4	84	L	1925	1623.2	1.0	4.50 in. carborundum No. 4 wall	52.60	36.80	8.10	1.97 Fe_2O_3; same as above but apparent density 2.36 g cm^{-3} and porosity 17.7%.
5	84	L	1925	1623.2	1.0	4.50 in. carborundum No. 5 wall	48.35	38.76	11.63	1.97 Fe_2O_3; same as above but apparent density 2.31 g cm^{-3} and porosity 18.9%; ignition loss 0.10%.

DATA TABLE NO. 157 THERMAL CONDUCTIVITY OF [SILICON CARBIDE + SILICON DIOXIDE + ΣX_i] $SiC + SiO_2 + \Sigma X_i$

[Temperature, T, K; Thermal Conductivity, k, Watt $cm^{-1}K^{-1}$]

T	k
CURVE 1	
647.1	0.0369
CURVE 2	
985.9	0.0434
1364.8	0.0552
1597.6	0.0718
CURVE 3	
1623.2	0.118
CURVE 4	
1623.2	0.0854
CURVE 5	
1623.2	0.0745

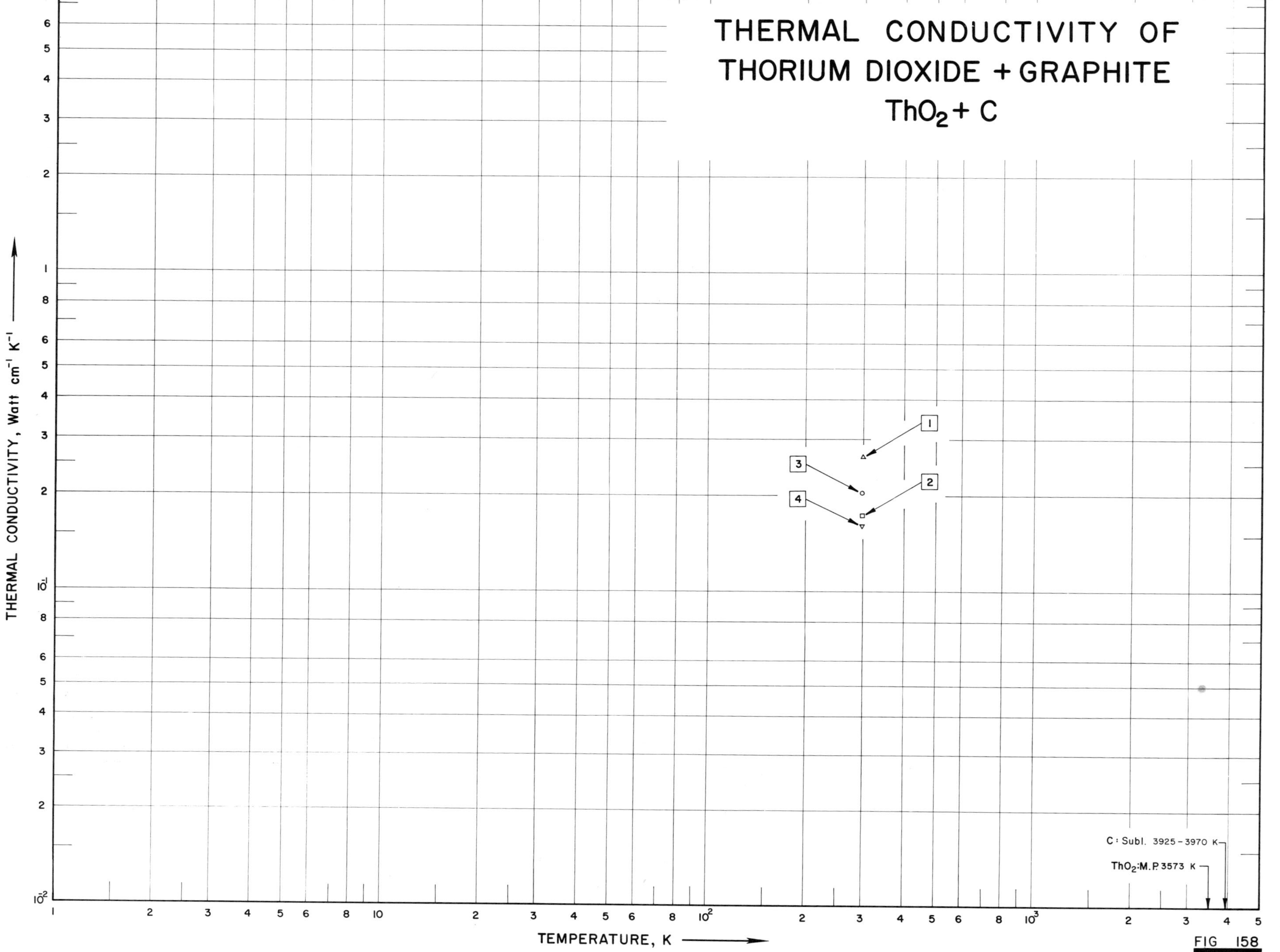
THERMAL CONDUCTIVITY OF
THORIUM DIOXIDE + GRAPHITE
ThO₂ + C
1
2
3
4
C: Subl. 3925 - 3970 K
ThO₂: M.P. 3573 K
THERMAL CONDUCTIVITY, Watt cm⁻¹ K⁻¹
TEMPERATURE, K

FIG 158

SPECIFICATION TABLE NO. 158 THERMAL CONDUCTIVITY OF [THORIUM DIOXIDE + GRAPHITE] $ThO_2 + C$

[For Data Reported in Figure and Table No. 158]

Curve No.	Ref. No.	Method Used	Year	Temp. Range, K	Reported Error, %	Name and Specimen Designation	Composition (weight percent) ThO_2	C	Composition (continued), Specifications and Remarks
1	181		1959	298.2		Fuel-filled graphite	50.06	49.94	Baked to 1425 C; bulk density 2.624 g cm^{-3}; measured with grain.
2	181		1959	298.2		Fuel-filled graphite	50.06	49.94	The above specimen measured against grain.
3	181		1959	298.2		Fuel-filled graphite	60.17	39.83	Baked to 1425 C; bulk density 2.951 g cm^{-3}; measured with grain.
4	181		1959	298.2		Fuel-filled graphite	60.17	39.83	The above specimen measured against grain.

DATA TABLE NO. 158 THERMAL CONDUCTIVITY OF [THORIUM DIOXIDE + GRAPHITE] $ThO_2 + C$

[Temperature, T, K; Thermal Conductivity, k, Watt $cm^{-1}K^{-1}$]

T	k
CURVE 1	
298.2	0.265
CURVE 2	
298.2	0.173
CURVE 3	
298.2	0.204
CURVE 4	
298.2	0.161

SPECIFICATION TABLE NO. 159 THERMAL CONDUCTIVITY OF CESIUM IODIDE CsI

Curve No.	Ref. No.	Method Used	Year	Temp. Range, K	Reported Error, %	Name and Specimen Designation	Composition (weight percent), Specifications and Remarks
1	214	C	1960	227-361			Crystalline specimen supplied by Harshaw Chem. Co. radius 1 cm, thickness 0.5 cm.

DATA TABLE NO. 159 THERMAL CONDUCTIVITY OF CESIUM IODIDE CsI

[Temperature, T, K; Thermal Conductivity, k, Watt $cm^{-1}K^{-1}$]

T	k
CURVE 1*	
227.0	0.0140
242.6	0.0135
265.0	0.0125
277.8	0.0115
296.0	0.0105
316.3	0.0105
346.2	0.00950
360.7	0.00950

*No graphical presentation

SPECIFICATION TABLE NO. 160 THERMAL CONDUCTIVITY OF COPPER IODIDE CuI

Curve No.	Ref. No.	Method Used	Year	Temp. Range, K	Reported Error, %	Name and Specimen Designation	Composition (weight percent), Specifications and Remarks
1	597, 598	P	1963	290			Density 5.63 g cm^{-3}; melting point 605 C; measured by a transient method.

DATA TABLE NO. 160 THERMAL CONDUCTIVITY OF COPPER IODIDE CuI

[Temperature, T, K; Thermal Conductivity, k, Watt $cm^{-1}K^{-1}$]

T	k
CURVE 1*	
290	0.017

* No graphical presentation

SPECIFICATION TABLE NO. 161 THERMAL CONDUCTIVITY OF SILVER IODIDE AgI

Curve No.	Ref. No.	Method Used	Year	Temp. Range, K	Reported Error, %	Name and Specimen Designation	Composition (weight percent), Specifications and Remarks
1	597, 598	P	1963	290			Density 5.67 g cm^{-3}; melting point 557 C; measured by a transient method.

DATA TABLE NO. 161 THERMAL CONDUCTIVITY OF SILVER IODIDE AgI

[Temperature, T, K; Thermal Conductivity, k, Watt $cm^{-1} K^{-1}$]

T	k
CURVE 1*	
290	0.004

* No graphical presentation

SPECIFICATION TABLE NO. 162 THERMAL CONDUCTIVITY OF CESIUM BROMIDE CsBr

Curve No.	Ref. No.	Method Used	Year	Temp. Range, K	Reported Error, %	Name and Specimen Designation	Composition (weight percent), Specifications and Remarks
1	71	C	1951	318, 338			Cubic isotropic crystal.
2	214	C	1960	228–368			Crystalline specimen supplied by Harshaw Chem. Corp. ; 2 cm in dia and 0.5 cm thick.

DATA TABLE NO. 162 THERMAL CONDUCTIVITY OF CESIUM BROMIDE CsBr

[Temperature, T, K; Thermal Conductivity, k, Watt $cm^{-1}K^{-1}$]

T	k
CURVE 1 *	
318.2	0.0092
338.2	0.0109
CURVE 2 *	
228.0	0.0118
248.7	0.0100
269.4	0.00924
295.0	0.00878
312.0	0.00823
337.5	0.00800
367.5	0.00776

* No graphical presentation

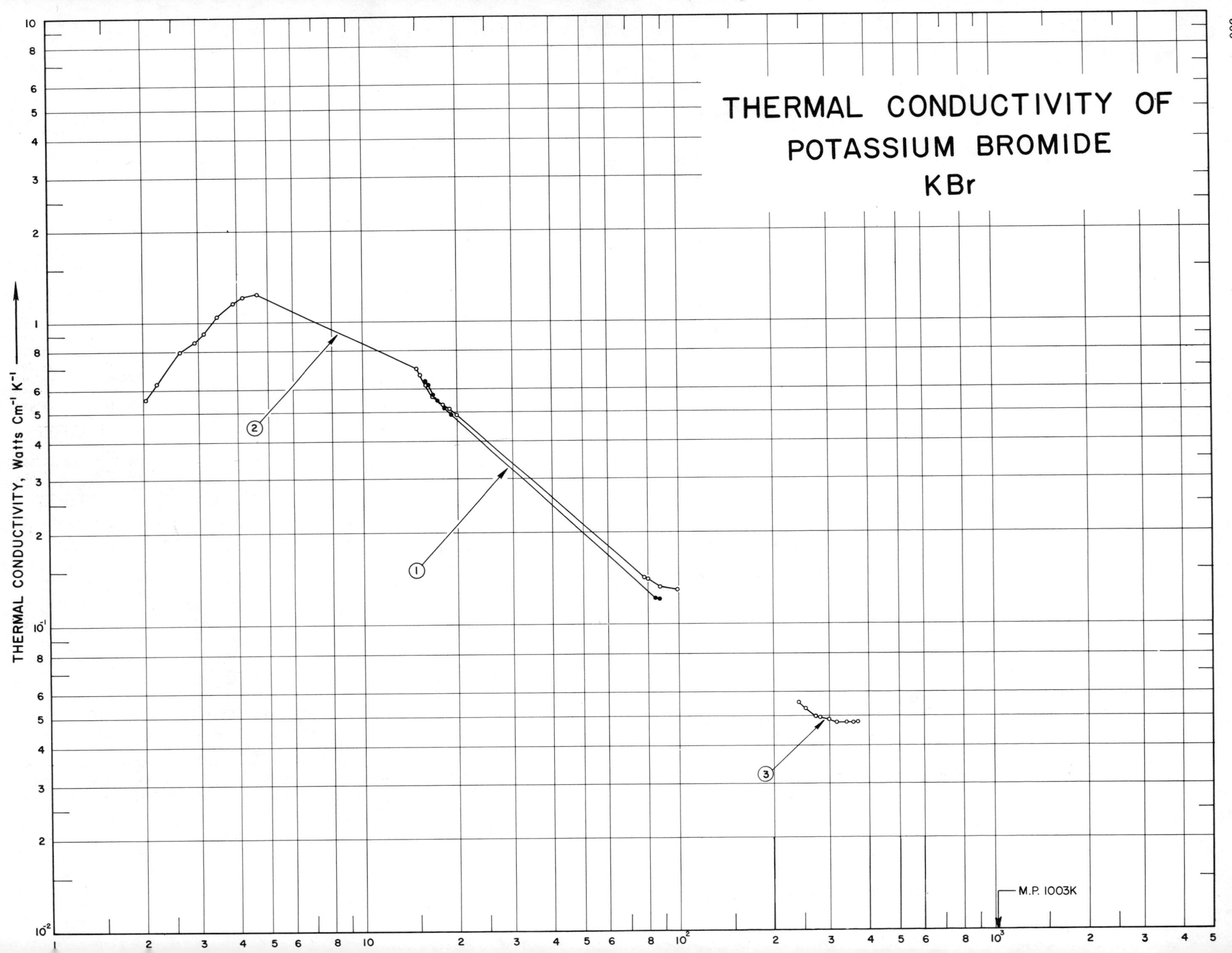
THERMAL CONDUCTIVITY OF
POTASSIUM BROMIDE
KBr
M.P. 1003K
THERMAL CONDUCTIVITY, Watts Cm^{-1} K^{-1}
1
2
3

SPECIFICATION TABLE NO. 163 THERMAL CONDUCTIVITY OF POTASSIUM BROMIDE KBr

[For Data Reported in Figure and Table No. 163]

Curve No.	Ref. No.	Method Used	Year	Temp. Range, K	Reported Error, %	Name and Specimen Designation	Composition (weight percent), Specifications and Remarks
1	32	L	1937	16-88		I	1.938 cm long, 0.294 cm dia; measured using soldered contact.
2	32	L	1937	2.0-90		II	2.437 cm long, 0.310 cm dia; measured using amalgam contact.
3	214	C	1960	241-372			Crystalline sample provided by Harshaw Chem. Comp.; radius 1 cm, thickness 0.5 cm.

DATA TABLE NO. 163 THERMAL CONDUCTIVITY OF POTASSIUM BROMIDE KBr

[Temperature, T, K; Thermal Conductivity, k, Watt $cm^{-1}K^{-1}$]

T	k
CURVE 1	
15.8	0.633
16.3	0.617
16.8	0.571
17.4	0.546
18.3	0.515
19.10	0.490
84.6	0.122
87.7	0.121
CURVE 2	
2.04	0.556
2.21	0.629
2.63	0.794
2.93	0.855
3.13	0.917
3.46	1.05
3.89	1.16
4.18	1.21
4.65	1.24
15.0	0.693
15.3	0.663
15.9	0.615
16.8	0.564
18.0	0.529
18.9	0.512
20.0	0.489
78.2	0.142
80.3	0.140
87.5	0.133
89.5	0.129
CURVE 3	
241.0	0.0555
253.0	0.0531
275.0	0.0500
282.0	0.0498
301.5	0.0487
318.2	0.0480
342.5	0.0480
360.7	0.0480
372.2	0.0480

SPECIFICATION TABLE NO. 164 THERMAL CONDUCTIVITY OF SILVER BROMIDE AgBr

Curve No.	Ref. No.	Method Used	Year	Temp. Range, K	Reported Error, %	Name and Specimen Designation	Composition (weight percent), Specifications and Remarks
1	585	P	1953	308-683			Rod specimen 3 mm in dia; cast.
2	71	C	1951	313, 341			Cubic isotropic crystal.

DATA TABLE NO. 164 THERMAL CONDUCTIVITY OF SILVER BROMIDE AgBr

[Temperature, T, K; Thermal Conductivity, k, Watt $cm^{-1} K^{-1}$]

T	k
CURVE 1*	
308.2	0.00900
353.2	0.00785
413.2	0.00711
473.2	0.00628
485.7	0.00586
545.7	0.00533
580.7	0.00502
615.7	0.00473
630.7	0.00475
648.2	0.00477
673.2	0.00554
683.2	0.00607
CURVE 2*	
313.2	0.0071
341.2	0.0071

* No graphical presentation

SPECIFICATION TABLE NO. 165 THERMAL CONDUCTIVITY OF THALLIUM BROMIDE TlBr

Curve No.	Ref. No.	Method Used	Year	Temp. Range, K	Reported Error, %	Name and Specimen Designation	Composition (weight percent), Specifications and Remarks
1	71	C	1951	316, 343			Cubic isotropic crystals.

DATA TABLE NO. 165 THERMAL CONDUCTIVITY OF THALLIUM BROMIDE TlBr

[Temperature, T, K; Thermal Conductivity, k, Watt $cm^{-1} K^{-1}$]

T	k
CURVE 1*	
316.2	0.00586
343.2	0.00586

* No graphical presentation

SPECIFICATION TABLE NO. 166 THERMAL CONDUCTIVITY OF DIBERYLLIUM CARBIDE Be_2C

Curve No.	Ref. No.	Method Used	Year	Temp. Range, K	Reported Error, %	Name and Specimen Designation	Composition (weight percent), Specifications and Remarks
1	255	R	1950	598-1168			Hollow hot-pressed cylinder; 0. 986 in. O. D. , 0. 609 in. I. D. and 3. 0 in. long.

DATA TABLE NO. 166 THERMAL CONDUCTIVITY OF DIBERYLLIUM CARBIDE Be_2C

[Temperature, T, K; Thermal Conductivity, k, Watt $cm^{-1} K^{-1}$]

T	k	T	k
CURVE 1*		CURVE 1 (cont.)*	
598. 2	0. 0137	923. 2	0. 0175
653. 2	0. 0148	923. 2	0. 0177
696. 2	0. 0146	933. 2	0. 0189
713. 2	0. 0148	963. 2	0. 0184
723. 2	0. 0168	978. 2	0. 0195
740. 2	0. 0158	980. 2	0. 0199
743. 2	0. 0155	1018. 2	0. 0213
758. 2	0. 0163	1033. 2	0. 0211
766. 2	0. 0173	1046. 2	0. 0199
793. 2	0. 0157	1108. 2	0. 0216
803. 2	0. 0173	1113. 2	0. 0218
833. 2	0. 0165	1168. 2	0. 0212
848. 2	0. 0183		
875. 2	0. 0162		
885. 2	0. 0191		
895. 2	0. 0177		

*No graphical presentation

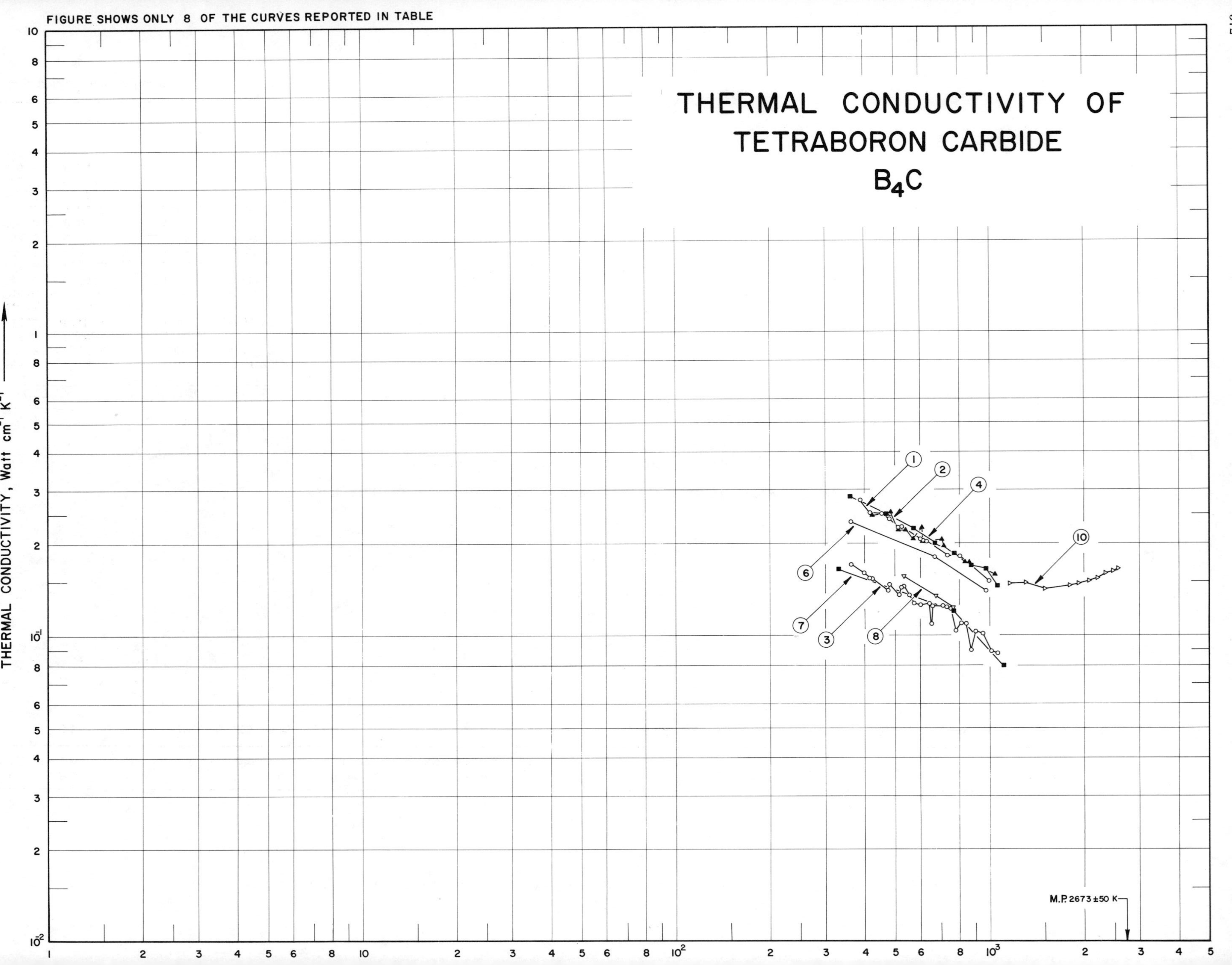

FIGURE SHOWS ONLY 8 OF THE CURVES REPORTED IN TABLE
THERMAL CONDUCTIVITY OF
TETRABORON CARBIDE
B_4C
THERMAL CONDUCTIVITY, Watt cm^{-1} K^{-1}
M.P. 2673±50 K
1
2
4
6
10
7
3
8

SPECIFICATION TABLE NO. 167 THERMAL CONDUCTIVITY OF TETRABORON CARBIDE B_4C

[For Data Reported in Figure and Table No. 167]

Curve No.	Ref. No.	Method Used	Year	Temp. Range, K	Reported Error, %	Name and Specimen Designation	Composition (weight percent), Specifications and Remarks
1	212	C	1951	390-999		No. 5 ($B_{3.85}C$)	77.1 B, 22.2 C chemical composition; specimen 2 cm in dia and 15 cm long; hot-pressed; density 2.5 g cm^{-3}; Armco iron used as comparative material.
2	212	C	1951	425-1033		No. 11 (Norton No. D11, 776-2)	Similar to the above specimen except density 2.33 g cm^{-3}.
3	212	C	1951	364-1066		No. 13 (No. D11, 798-1)	77.1 B, 22.2 C chemical composition; specimen 2 cm in dia and 15 cm long; rammed and sintered; density 1.9 g cm^{-3}; Armco iron used as comparative material.
4	76	C	1952	363-1073			Specimen .75 in. in dia and 9 in. long; hot-pressed; density 2.5 g cm^{-3}; lead used as comparative material.
5	76	C	1952	363-1073			Similar to the above specimen except density 2.33 g cm^{-3}.
6	76	C	1952	363-973			Specimen .75 in. in dia and 9 in. long; rammed and sintered; density 1.9 g cm^{-3}; lead used as comparative material.
7	76	C	1952	333-1103			Similar to the above specimen except density 1.91 g cm^{-3}.
8	76	C	1952	533-763		No. 17	1.85 Sodium silicate; specimen .75 in. in dia and 9 in. long; rammed; density 2.03 g cm^{-3}; lead used as comparative material.
9	583	C	1963	469-1036	±4.0		75.97 B, 21.18 C, 0.07 B_2O_7, 0.27 Fe, 0.40 Si and 0.015 Al_2O_3; specimen 2 in. in dia and 1 in. thick; density 2.5 g cm^{-3}; hot-pressed; measured in helium atmosphere; Armco iron used as comparative material.
10	583	P	1963	1168-2580	±4.0		The above specimen measured by another method; thermal conductivity values calculated from the measurement of thermal diffusivity, specific heat and density.

DATA TABLE NO. 167 THERMAL CONDUCTIVITY OF TETRABORON CARBIDE B_4C

[Temperature, T, K; Thermal Conductivity, k, Watt $cm^{-1}K^{-1}$]

T	k
CURVE 1	
390.2	0.278
418.2	0.252
454.2	0.251
480.2	0.240
513.2	0.227
528.2	0.228
608.2	0.207
618.2	0.206
632.2	0.204
735.2	0.183
806.2	0.182
999.2	0.150
CURVE 2	
425.2	0.248
484.2	0.253
512.2	0.224
545.2	0.224
574.2	0.207
611.2	0.226
611.2	0.204
708.2	0.206
718.2	0.196
838.2	0.172
858.2	0.172
1033.2	0.158
CURVE 3	
364.2	0.171
400.2	0.161
417.2	0.154
425.2	0.154
478.2	0.140
484.2	0.147
519.2	0.136
529.2	0.143
539.2	0.145
558.2	0.136
579.2	0.128
603.2	0.127
644.2	0.128
654.2	0.109
CURVE 3 (cont.)	
659.2	0.127
711.2	0.125
732.2	0.124
751.2	0.123
785.2	0.103
814.2	0.110
842.2	0.108
876.2	0.090
909.2	0.103
950.2	0.102
1021.2	0.089
1066.2	0.088
CURVE 4	
363.2	0.285
473.2	0.250
573.2	0.225
673.2	0.200
773.2	0.185
873.2	0.170
973.2	0.165
1073.2	0.145
CURVE 5*	
363.2	0.287
473.2	0.245
573.2	0.220
673.2	0.195
773.2	0.180
873.2	0.165
973.2	0.155
1073.2	0.145
CURVE 6	
363.2	0.235
673.2	0.180
973.2	0.140
CURVE 7	
333.2	0.165
773.2	0.120
1103.2	0.080
CURVE 8	
533.2	0.156
673.2	0.135
763.2	0.122
CURVE 9*	
469.30	0.247
538.72	0.223
685.94	0.190
844.27	0.170
1035.94	0.156
CURVE 10	
1168.16	0.147
1313.72	0.148
1502.60	0.142
1805.38	0.146
1924.83	0.147
2079.27	0.151
2210.94	0.154
2347.05	0.159
2472.05	0.162
2580.40	0.165

*Not shown on plot

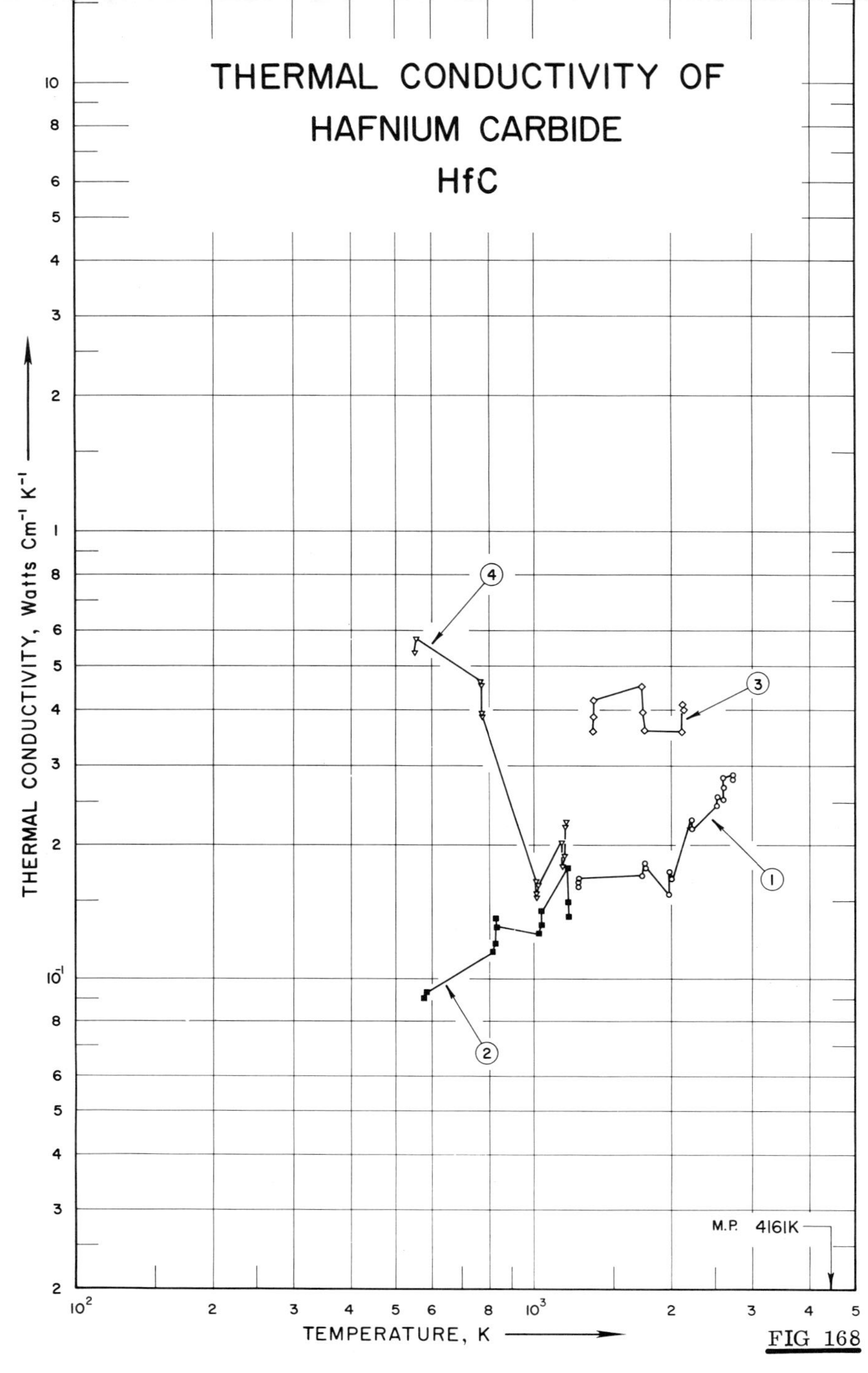
THERMAL CONDUCTIVITY OF
HAFNIUM CARBIDE
HfC
THERMAL CONDUCTIVITY, Watts Cm⁻¹ K⁻¹
TEMPERATURE, K
M.P. 4161K
1
2
3
4

FIG 168

SPECIFICATION TABLE NO. 168 THERMAL CONDUCTIVITY OF HAFNIUM CARBIDE HfC

[For Data Reported in Figure and Table No. 168]

Curve No.	Ref. No.	Method Used	Year	Temp. Range, K	Reported Error, %	Name and Specimen Designation	Composition (weight percent), Specifications and Remarks
1	144	R	1963	1261-2718	5-7	1	Specimen 0.75 in. long, 0.75 in. O.D. and 0.25 in. I.D.; heat-soaked at 2204 C; ground and polished to eliminate all the scratches on specimen surface; specimen found broken on post inspection.
2	243	R	1962	580-1192	2-4		Specimen 0.75 in. O.D., 0.25 in. I.D. and 0.75 in. long; hot-pressed; firing temp near 3593 C and max exposure temp 2882 C; density 10.04 g cm^{-3}.
3	243	R	1962	1351-2122	2-4		The above specimen measured after heat-soaked.
4	243	R	1962	553-1174	2-4		Specimen 0.75 in. O.D., 0.75 in. I.D. and 0.75 in. long; hot-pressed; firing temp near 3593 C and max exposure temp 2882 C; density 10.04 g cm^{-3}.

DATA TABLE NO. 168 THERMAL CONDUCTIVITY OF HAFNIUM CARBIDE HfC

[Temperature, T, K; Thermal Conductivity, k, Watt $cm^{-1}K^{-1}$]

T	k	T	k
CURVE 1		CURVE 3	
1260.9	0.162	1351.0	0.359
1260.9	0.166	1366.5	0.389
1260.9	0.165*	1366.5	0.421
1262.6	0.169	1366.5	0.424*
1747.1	0.171	1738.7	0.454
1755.4	0.183	1749.8	0.398
1758.2	0.179	1752.6	0.401*
1983.2	0.156	1752.6	0.363
1990.9	0.176	2108.2	0.359
1991.5	0.169	2116.5	0.414
2218.7	0.229	2122.1	0.408*
2218.7	0.219	2122.1	0.401
2220.9	0.217*		
2507.6	0.246	CURVE 4	
2509.8	0.258		
2514.8	0.259*	552.6	0.539
2589.3	0.255	555.9	0.578
2595.9	0.270	558.7	0.577*
2598.2	0.284	768.2	0.462
2708.2	0.289	770.9	0.453
2708.7	0.281	772.1	0.395
2718.2	0.285*	773.2	0.385
		1012.1	0.166
CURVE 2		1012.6	0.157
		1013.2	0.153
580.4	0.0909	1014.3	0.164
584.3	0.0937	1167.6	0.203
587.1	0.0923*	1168.7	0.206*
820.4	0.115	1169.8	0.179
823.7	0.115*	1173.2	0.190
827.1	0.120	1173.7	0.221
828.2	0.138	1174.3	0.226
829.3	0.131		
1029.3	0.127		
1040.9	0.133		
1040.9	0.143		
1040.9	0.144*		
1188.2	0.179		
1192.1	0.151		
1192.1	0.138		

*Not shown on plot

SPECIFICATION TABLE NO. 169 THERMAL CONDUCTIVITY OF TRIIRON CARBIDE Fe_3C

Curve No.	Ref. No.	Method Used	Year	Temp. Range, K	Reported Error, %	Name and Specimen Designation	Composition (weight percent), Specifications and Remarks
1	372		1940	298.2		Cementite	No details reported.

DATA TABLE NO. 169 THERMAL CONDUCTIVITY OF TRIIRON CARBIDE Fe_3C

[Temperature, T, K; Thermal Conductivity, k, Watt $cm^{-1}K^{-1}$]

T	k
CURVE 1*	
298.2	0.0711

*No graphical presentation

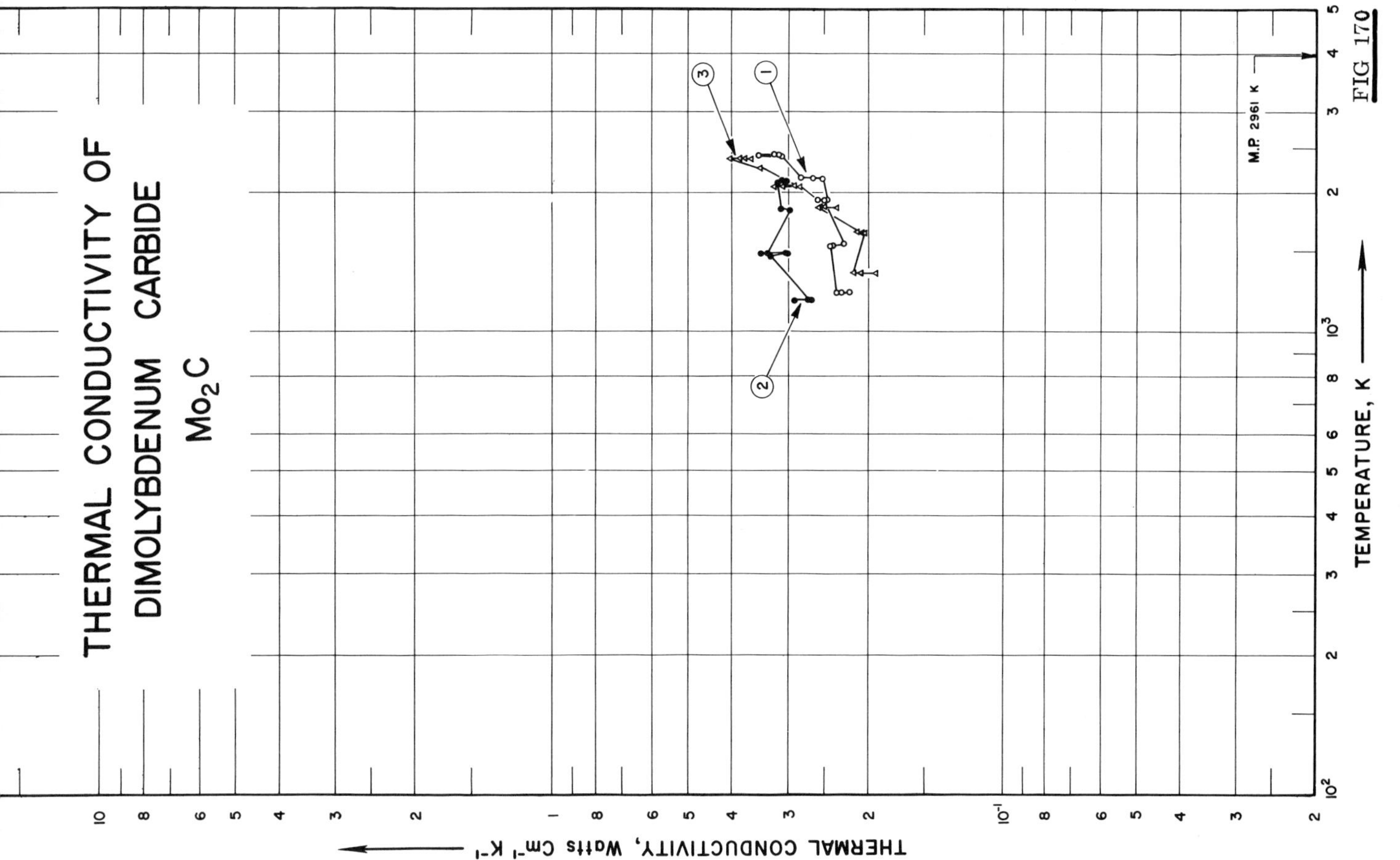
THERMAL CONDUCTIVITY OF
DIMOLYBDENUM CARBIDE
Mo2C
M.P. 2961 K
TEMPERATURE, K
THERMAL CONDUCTIVITY, Watts Cm-1 K-1
1
2
3

FIG 170

SPECIFICATION TABLE NO. 170 THERMAL CONDUCTIVITY OF DIMOLYBDENUM CARBIDE Mo_2C

[For Data Reported in Figure and Table No. 170]

Curve No.	Ref. No.	Method Used	Year	Temp. Range, K	Reported Error, %	Name and Specimen Designation	Composition (weight percent), Specifications and Remarks
1	144	R	1963	1222-2414	5-7	1	Specimen 0.75 in. long, 0.75 in. O.D. and 0.25 in. I.D.; ground and polished to eliminate all the scratches on specimen surface; specimen found broken and partially melted on post inspection.
2	144	R	1963	1172-2114	5-7	2	Similar to the above specimen; specimen found broken on post inspection.
3	144	R	1963	1344-2384	5-7	3	Similar to the above specimen except furthermore heat-soaked at 1538 C; specimen found broken on post inspection.

DATA TABLE NO. 170 THERMAL CONDUCTIVITY OF DIMOLYBDENUM CARBIDE Mo_2C

[Temperature, T, K; Thermal Conductivity, k, Watt $cm^{-1}K^{-1}$]

T	k
CURVE 1	
1222.1	0.219
1222.1	0.229
1222.1	0.235
1223.7	0.233*
1544.8	0.242
1547.6	0.240
1551.5	0.226
1927.1	0.250
1930.9	0.259
1932.1	0.247
2153.2	0.252
2156.5	0.265
2167.1	0.281
2402.1	0.311
2413.7	0.348
2413.7	0.314
2414.3	0.321
CURVE 2	
1172.1	0.268
1172.1	0.291
1172.1	0.270
1469.3	0.328
1472.1	0.301
1473.2	0.305
1475.4	0.345
1475.4	0.334
1842.6	0.299
1844.3	0.311
1847.1	0.310*
2105.4	0.317
2106.5	0.304
2113.7	0.310
CURVE 3	
1344.3	0.207
1344.3	0.215
1344.3	0.192
1344.3	0.216*
1642.1	0.204
1645.9	0.211
1645.9	0.205

T	k
CURVE 3 (cont.)	
1857.1	0.258
1862.6	0.234
1865.4	0.256
1872.6	0.249
2065.4	0.283
2068.2	0.320
2069.8	0.309
2070.4	0.291
2267.1	0.345
2372.6	0.403
2373.2	0.361
2376.5	0.374
2384.3	0.382

* Not shown on plot

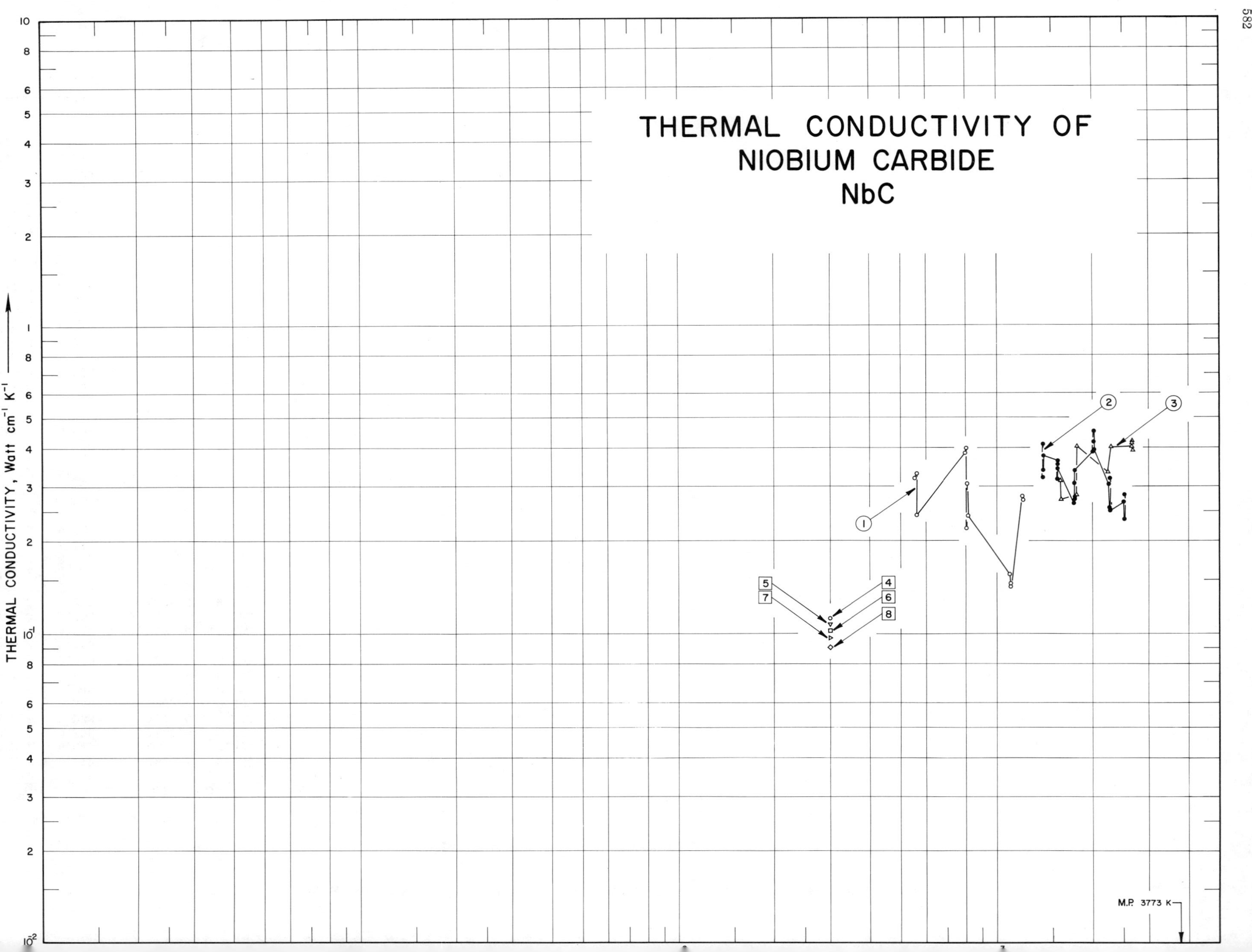

THERMAL CONDUCTIVITY OF
NIOBIUM CARBIDE
NbC
THERMAL CONDUCTIVITY, Watt cm^{-1} K^{-1}
10
8
6
5
4
3
2
1
8
6
5
4
3
2
10^{-1}
8
6
5
4
3
2
10^{-2}
1
2
3
4
5
6
7
8
M.P. 3773 K

SPECIFICATION TABLE NO. 171 THERMAL CONDUCTIVITY OF NIOBIUM CARBIDE NbC

[For Data Reported in Figure and Table No. 171]

Curve No.	Ref. No.	Method Used	Year	Temp. Range, K	Reported Error, %	Name and Specimen Designation	Composition (weight percent), Specifications and Remarks
1	243	R	1962	555-1220	2-4		88.43 Nb, 11.3 C, 0.1 Fe, 0.1 W, 0.07 N, <0.01 each Si, Mn, Mg, Cr, Sn, Ti, Zr and Ni; specimen 0.75 in. O.D., 0.25 in. I.D. and 0.75 in. long; hot-pressed; max exposure temp 2616 C; density 7.62 g cm^{-3}.
2	243	R	1962	1403-2517	2-4		Similar to the above specimen except specimen found cracked and fissured after the measurements.
3	243	R	1962	1586-2694	2-4		Similar to the above specimen.
4	6	L	1965	298.2	±6.0		89.49 Nb, 10.51 C; specimen prepared by pressing and sintering of powder mixtures close to stoichiometric composition of carbide and appropriate metal; sintering was carried out in a TVV-4 vacuum furnace at 10^{-4} to 10^{-5} mm pressure at 2200 to 2400 C; electrical resistivity 89.8 x 10^{-6} ohm cm at room temp.
5	6	L	1965	298.2	±6.0		90.044 Nb, 9.956 C; similar to the above specimen except electrical resistivity 135.2 x 10^{-6} ohm cm at room temp.
6	6	L	1965	298.2	±6.0		90.54 Nb, 9.46 C; similar to the above specimen except electrical resistivity 151.9 x 10^{-6} ohm cm at room temp.
7	6	L	1965	298.2	±6.0		91.065 Nb, 8.935 C; similar to the above specimen except electrical resistivity 150.0 x 10^{-6} ohm cm at room temp.
8	6	L	1965	298.2	6.0		91.6 Nb, 8.4 C; similar to the above specimen except electrical resistivity 171.7 x 10^{-6} ohm cm at room temp.

DATA TABLE NO. 171 THERMAL CONDUCTIVITY OF NIOBIUM CARBIDE NbC

[Temperature, T, K; Thermal Conductivity, k, Watt cm^{-1} K^{-1}]

T	k
CURVE 1	
554.8	0.317
562.1	0.330
563.2	0.244
799.3	0.385
803.2	0.398
809.8	0.220
810.9	0.306
812.6	0.242
1104.8	0.157
1105.4	0.147
1108.7	0.144
1209.3	0.280
1209.3	0.280*
1217.1	0.273
1218.7	0.273*
1220.4	0.271*
CURVE 2	
1402.6	0.320
1405.4	0.412
1405.4	0.399
1405.4	0.378
1408.2	0.410*
1566.5	0.363
1566.5	0.352
1566.5	0.317
1566.5	0.343
1752.6	0.265
1761.0	0.307
1763.7	0.274
1763.7	0.278
1763.7	0.310*
1772.1	0.337
2030.4	0.388
2036.0	0.453
2036.0	0.420
2038.7	0.394
2258.2	0.306
2261.0	0.257
2263.7	0.319
2266.5	0.350
2505.4	0.268
2508.2	0.235
2513.7	0.232*
CURVE 2 (cont.)	
2516.5	0.239*
2516.5	0.284
CURVE 3	
1586.0	0.314
1594.3	0.273
1783.2	0.281
1783.2	0.404
2238.7	0.332
2277.6	0.404
2666.5	0.407
2672.1	0.392
2677.6	0.412
2694.3	0.421
CURVE 4	
298.2	0.112
CURVE 5	
298.2	0.107
CURVE 6	
298.2	0.102
CURVE 7	
298.2	0.097
CURVE 8	
298.2	0.090

*Not shown on plot

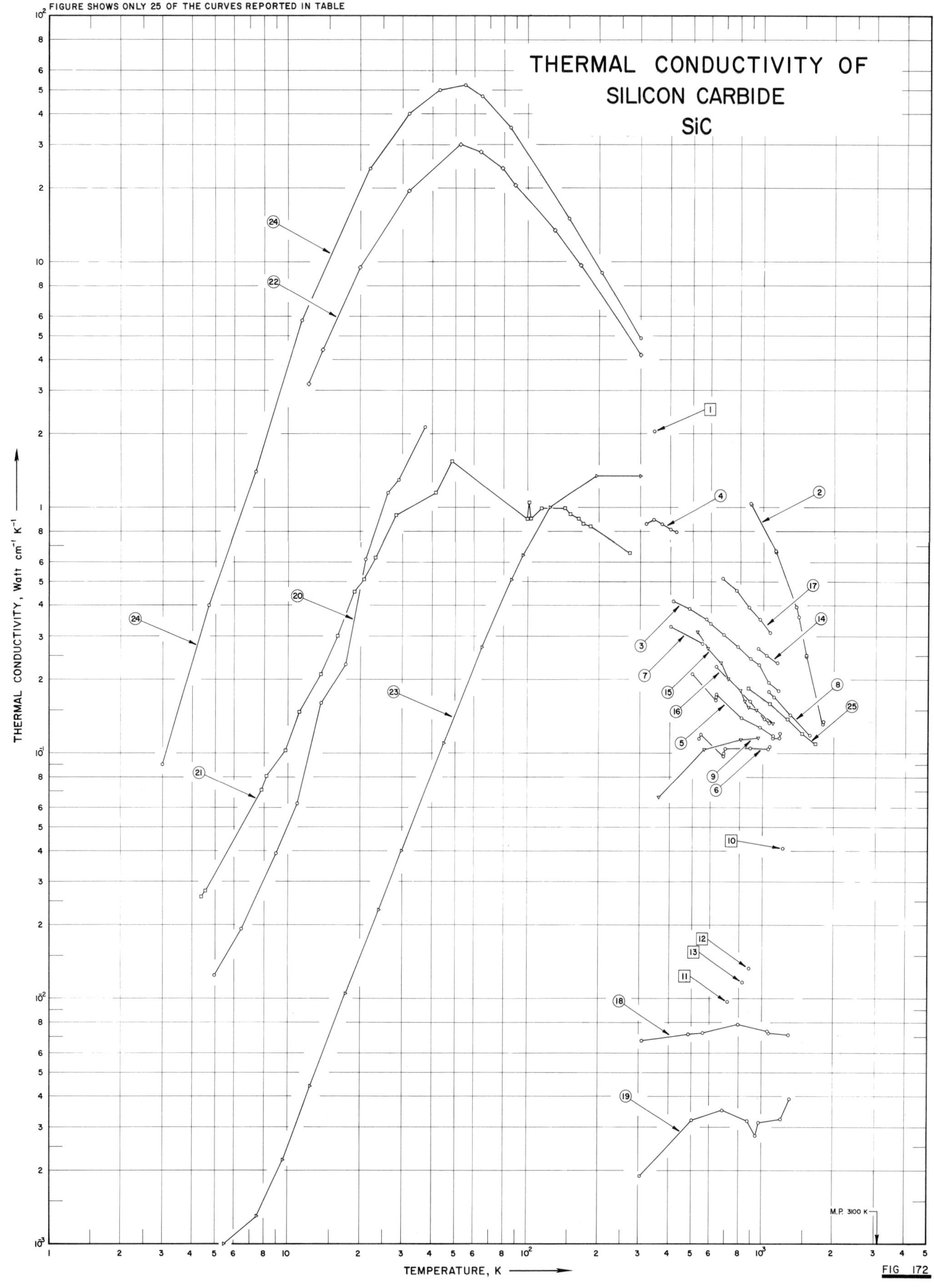
FIGURE SHOWS ONLY 25 OF THE CURVES REPORTED IN TABLE
THERMAL CONDUCTIVITY OF
SILICON CARBIDE
SiC
THERMAL CONDUCTIVITY, Watt cm⁻¹ K⁻¹
TEMPERATURE, K
M.P. 3100 K
FIG 172

SPECIFICATION TABLE NO. 172 THERMAL CONDUCTIVITY OF SILICON CARBIDE SiC

[For Data Reported in Figure and Table No. 172]

Curve No.	Ref. No.	Method Used	Year	Temp. Range, K	Reported Error, %	Name and Specimen Designation	Composition (weight percent), Specifications and Remarks
1	13	C	1953	343.2	±3.0		Polycrystal; specimen 1.75 in. long and 0.22 in. in dia; supplied by Carborundum Co.; density 2.8 g cm^{-3}; Armco iron used as comparative material.
2	37	L	1958	889-1801	5.0		67.46 Si, 28.58 C, 0.73 Al, 0.58 Fe, 0.48 CaO; density 3.1 g cm^{-3}.
3	14	C	1953	413-1168			Cubic specimen; cut from ingot; supplied by Norton Co.; total porosity 21.7%; dense sintered alumina used as comparative material.
4	68	C	1954	318-423	±3.0	288A-1	Single crystal of hexagonal crystal system in α phase.
5	232	L	1954	499-1198		Commercial SiC	Frit-bonded.
6	232	L	1954	538-1063		Commercial SiC	Frit-bonded.
7	80	L	1934	404, 551		SiC brick, Refrax	96.9 SiC, 1.3 Al_2O_3, 1.3 Fe_2O_3 and 0.7 SiO_2; approx composition; recrystallized; supplied by Carborundum Co.; bulk density 2.18 g cm^{-3}; porosity 34.4%.
8	80	L	1934	1063-1582		SiC brick, Refrax	The above specimen measured with insulating brick placed between the calorimeter and the lower surface of the brick.
9	103	L	1937	363-963			> 85 SiC in finished state, containing no clay bond, lime, magnesia, or silicate of soda; specimen ~1 in. thick.
10	81	L	1924	1232.8		Crystolon SiC	98.0 SiC; specimen 8.5 in. in dia and 3.54 in. thick; made by crushing the electric furnace products to pass a No. 14 screen and bonding with fire clay, containing 10% bond clay (composition 25.41 Al_2O_3, 59.55 SiO_2, 2.31 Fe_2O_3, 1.33 TiO_2, 1.01 $K_2O + Na_2O$, 0.46 CaO, 0.33 MgO and 9.10 loss), fired to cone 16; measured at 749.6 mm Hg pressure; porosity 29.14%, calculated from the dry, saturated, and suspended weight.
11	216	R	1951	719.3		No. 8 grain	Specimen measured from the inner half of test annulus.
12	216	R	1951	887.1		No. 8 grain	The above specimen measured from the outer half of the test annulus.
13	216	R	1951	830.4		No. 8 grain	The above specimen measured from the entire annulus.
14	233	R, C	1955	953-1153			Hot-pressed.
15	233	R, C	1955	523-1103		Sample 1	Frit-bonded.
16	233	R, C	1955	633-1063		Sample 2	Frit-bonded.
17	233	R, C	1955	673-1073			Si (~15%) bonded supplied by O.R.N.L.
18	234	R	1959	309-1302			Density 1.59 g cm^{-3}; 320 B mesh powder; measured in helium atmosphere.
19	234	R	1959	305-1320			Density 1.59 g cm^{-3}; 320 B mesh powder; measured in air.
20	381	L	1961	5-38			Polycrystalline specimen of β-type (cubic crystal); dimensions 1x1x10 mm; made by pyrolysis of chlorosilicanes in hydrogen atmosphere.

SPECIFICATION TABLE NO. 172 (continued)

Curve No.	Ref. No.	Method Used	Year	Temp. Range, K	Reported Error, %	Name and Specimen Designation	Composition (weight percent), Specifications and Remarks
21	381	L	1961	4.4-270			Single crystal of α-SiC (hexagonal type); dimensions 6x8x0.6 mm; prepared by Lely procedure from technical-grade polycrystalline SiC; specimen nitrogen doped (10^{18} N atoms cm^{-3}), other impurities $<10^{17}$ atoms cm^{-3} by spectrochemical analysis.
22	382	L	1964	12.3-300	±10	R-43	Single crystal; light green 6H polytype; n-type; 0.14 cm in effective dia and 0.70 cm long; major impurity N, Al $1x10^{19}$ atoms cm^{-3}; electrical conductivity 4.5×10^{-1} $ohm^{-1}cm^{-1}$ at 300 K; heat flow perpendicular to c-axis.
23	382	L	1964	4.0-300	±10	R-52	Single crystal; dark blue and a mixture of 6H and 15R polytypes; p-type; 0.16 cm in effective dia and 0.67 cm long; major impurity AP 4×10^{19} atoms cm^{-3}; electrical conductivity 1.6 $ohm^{-1}cm^{-1}$ at 300 K; heat flow perpendicular to c-axis.
24	382	L	1964	3.0-300	±10	R-66	Single crystal; 6H poly type; colorless; n-type; 0.12 cm in effictive dia and 0.35 cm long; major impurity N 1×10^{17} atoms cm^{-3}, electrical conductivity 9×10^{-2} $ohm^{-1}cm^{-1}$ at 300 K; heat flow perpendicular to c-axis.
25	383	L	1927	873-1673	±25		Commercial recrystallized silicon carbide; specimen 10.8 cm in dia and 22.8 cm long; apparent density 3.19 g cm^{-3}; bulk density 2.06 g cm^{-3}; porosity 35.3%.
26	595	C	1962	589, 882			Foam specimen; Min-k 1301 (Johns Manville Corp.) used as comparative material.
27	472	R	1963	650-1039			Foam specimen contained in a cylindrical annuli of 2.75 in. O.D. and 0.75 in. I.D.; density 0.465 g cm^{-3}.
28	472	R	1963	805-1039			The above specimen measured at decreasing temperatures.

DATA TABLE NO. 172 THERMAL CONDUCTIVITY OF SILICON CARBIDE SiC

[Temperature, T, K; Thermal Conductivity, k, Watt $cm^{-1}K^{-1}$]

T	k
CURVE 1	
343.2	2.05
CURVE 2	
888.8	1.04
889.8	1.03
1130.0	0.670
1133.2	0.666
1388.7	0.395
1389.6	0.395*
1414.6	0.360
1528.3	0.249
1528.6	0.251
1800.0	0.132
1800.5	0.132*
1801.1	0.135
CURVE 3	
413.2	0.418
483.2	0.387
573.2	0.351
598.2	0.339
683.2	0.305
783.2	0.272
883.2	0.243
963.2	0.230
1063.2	0.195
1168.2	0.180
CURVE 4	
317.5	0.862
341.1	0.895
370.1	0.858
400.2	0.816
423.0	0.795
CURVE 5	
499.2	0.211
633.2	0.165
635.2	0.169
637.2	0.174
CURVE 5 (cont.)	
816.2	0.139
973.2	0.128
1105.2	0.118
1107.2	0.117
1186.2	0.117
1198.2	0.121
CURVE 6	
538.2	0.115
543.2	0.119
683.2	0.0979
688.2	0.100
693.2	0.105
888.2	0.106
1062.2	0.105
1063.2	0.107
CURVE 7	
404.1	0.329
550.9	0.280
CURVE 8	
1063.2	0.179
1120.9	0.170
1302.1	0.144
1582.1	0.118
CURVE 9	
362.6	0.0664
566.9	0.103
811.0	0.114
962.7	0.116
CURVE 10	
1232.2	0.0411
CURVE 11	
719.3	0.00974
CURVE 12	
887.1	0.0134
CURVE 13	
830.4	0.0117
CURVE 14	
953.2	0.268
1043.2	0.251
1153.2	0.234
CURVE 15	
523.2	0.314
583.2	0.268
663.2	0.234
713.2	0.201
803.2	0.180
843.2	0.163
873.2	0.155
943.2	0.151
1103.2	0.134
CURVE 16	
633.2	0.226
883.2	0.163
1003.2	0.138
1063.2	0.134
CURVE 17	
673.2	0.519
773.2	0.460
873.2	0.393
973.2	0.351
1073.2	0.310
CURVE 18	
309.3	0.00675
485.2	0.00717
561.0	0.00727
CURVE 18 (cont.)	
791.2	0.00784
1062.2	0.00737
1074.2	0.00727
1301.7	0.00710
CURVE 19	
305.1	0.00190
507.2	0.00320
687.8	0.00351
868.4	0.00318
947.9	0.00277
980.4	0.00313
1204.0	0.00325
1320.1	0.00393
CURVE 20	
4.97	0.0125
6.46	0.0193
8.98	0.0390
11.0	0.0624
13.8	0.160
17.5	0.230
21.2	0.617
26.2	1.15
29.0	1.30
37.5	2.13
CURVE 21	
4.4	0.0260
4.57	0.0275
7.85	0.0707
8.22	0.0808
9.82	0.103
11.2	0.147
13.8	0.210
16.2	0.301
19.0	0.455
20.7	0.510
23.2	0.625
28.3	0.930
41.7	1.15
CURVE 21 (cont.)	
48.5	1.55
100.0	0.895
102.0	1.05
104.0	0.897
115.0	0.987
145.0	0.987
152.0	0.937
165.0	0.897
173.0	0.857
185.0	0.836
270.0	0.655
CURVE 22	
12.3	3.2
14.0	4.4
20.0	9.5
32.0	19.5
52.5	30.0
64.0	28.0
79.0	24.0
89.0	20.5
130.0	13.5
167.0	9.7
300.0	4.2
CURVE 23	
4.0	0.0008
5.5	0.0010
7.5	0.0013
9.6	0.0022
12.5	0.0044
17.5	0.0105
24.0	0.023
30.0	0.040
45.0	0.11
65.0	0.27
86.0	0.51
96.0	0.64
125.0	1.0
195.0	1.35
300.0	1.35
CURVE 24	
3.0	0.09
4.7	0.4
7.4	1.4
11.5	5.8
22.0	24.0
32.0	40.0
43.0	50.0
55.0	52.5
65.0	47.0
85.0	35.0
150.0	15.0
205.0	9.0
300.0	4.9
CURVE 25	
873.2	0.184
1073.2	0.159
1273.2	0.138
1473.2	0.121
1673.2	0.109
CURVE 26*	
588.7	0.00981
882.1	0.0124
CURVE 27*	
649.8	0.00134
760.9	0.00265
855.4	0.00270
944.3	0.00381
1039	0.00462
CURVE 28*	
805.4	0.00322
922.1	0.00402
1039	0.00462

*Not shown on plot

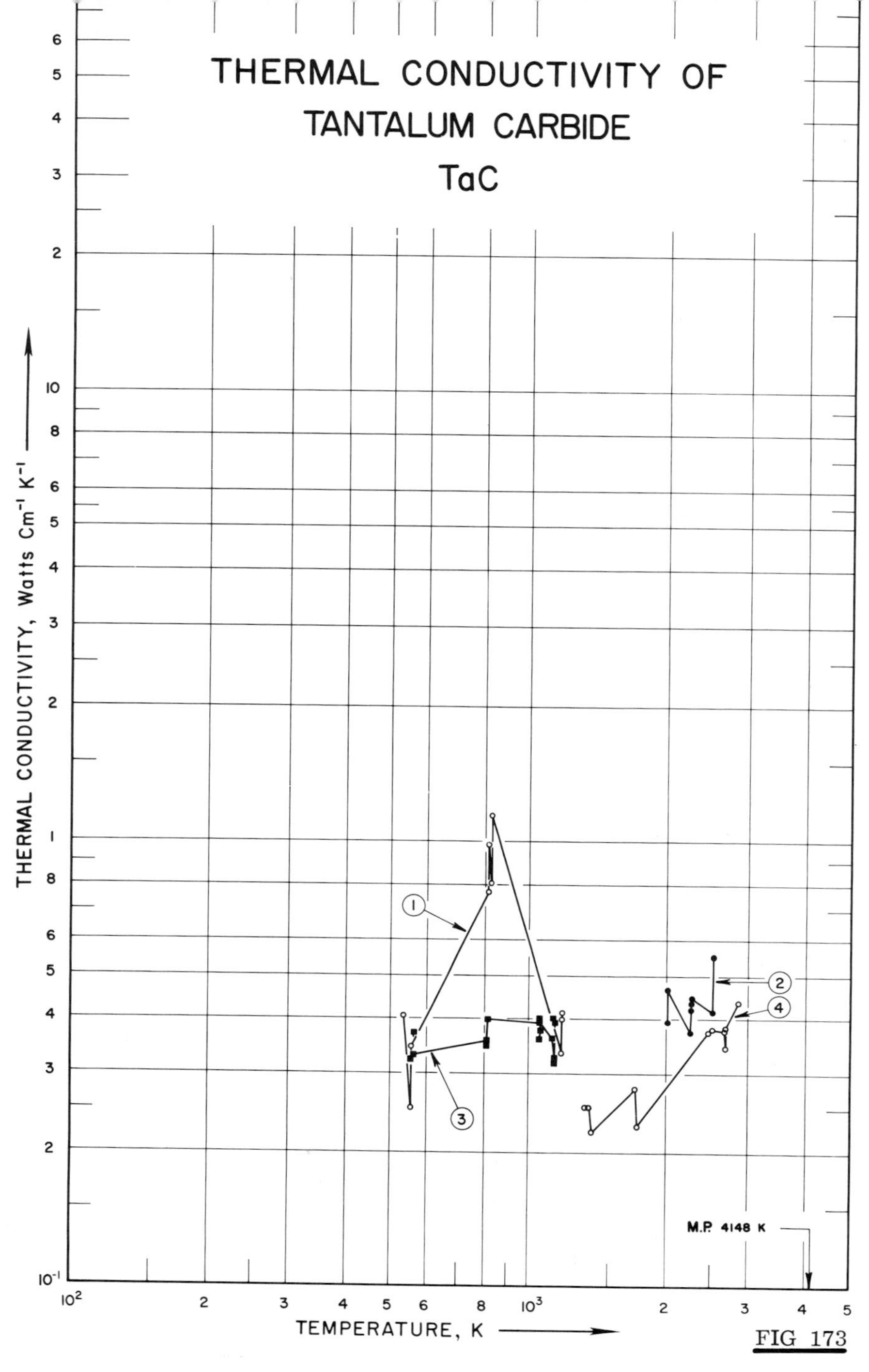
THERMAL CONDUCTIVITY OF
TANTALUM CARBIDE
TaC
THERMAL CONDUCTIVITY, Watts Cm⁻¹ K⁻¹
TEMPERATURE, K
M.P. 4148 K
1
2
3
4

FIG 173

SPECIFICATION TABLE NO. 173 THERMAL CONDUCTIVITY OF TANTALUM CARBIDE TaC

[For Data Reported in Figure and Table No. 173]

Curve No.	Ref. No.	Method Used	Year	Temp. Range, K	Reported Error, %	Name and Specimen Designation	Composition (weight percent), Specifications and Remarks
1	243	R	1962	537-1194			93.75 Ta, 6.14 C, 0.1 W, <0.01 each Si, Mg, Ca, Al, Ti, Nb, Sn, Zr, Fe, Na, Mn, Mg and Ni; specimen 0.75 in. O.D., 0.25 in. I.D. and 0.75 in. long; hot-pressed; max exposure temp 2649 C; density 13.87 g cm^{-3}; SRI run number C72.
2	243	R	1962	2008-2555			Similar to the above but pitting and spalling was found after measurements; SRI run number C83.
3	243	R	1962	552-1151			93.75 Ta, 6.14 C, 0.1 W, <0.01 each Si, Mg, Ca, Al, Ti, Nb, Sn, Zr, Fe, Na, Mn, Mg and Ni; specimen 0.75 in. O.D., 0.25 in. I.D. and 0.75 in. long; hot-pressed; max exposure temp 2649 C; density 13.87 g cm^{-3}; SRI run number C82.
4	243	R	1962	1342-2855			Similar to the above specimen but heat soaked, pitting and spalling found on specimen after measurements; SRI run number C90.

DATA TABLE NO. 173 THERMAL CONDUCTIVITY OF TANTALUM CARBIDE TaC

[Temperature, T, K; Thermal Conductivity, k, Watt $cm^{-1}K^{-1}$]

T	k
CURVE 1	
537.1	0.407
556.5	0.251
559.3	0.343
819.3	0.773
819.8	0.984
829.8	0.808
832.6	1.15
1187.1	0.333
1190.4	0.398
1194.3	0.415
CURVE 2	
2008.2	0.398
2022.1	0.469
2258.2	0.372
2263.7	0.420
2266.5	0.437
2266.5	0.418*
2272.1	0.447
2272.1	0.400*
2536.0	0.417
2555.4	0.551
CURVE 3	
551.5	0.322
554.3	0.320*
558.2	0.342*
565.4	0.371
565.4	0.329
570.9	0.368*
810.4	0.356
810.9	0.345
813.2	0.350*
813.7	0.398
1064.3	0.392
1064.8	0.358
1067.1	0.400
1068.2	0.375
1068.7	0.371*
1133.7	0.361
1139.3	0.400
1147.6	0.325
1149.8	0.317
1150.9	0.392
CURVE 4	
1341.5	0.254
1355.4	0.254
1374.8	0.222
1719.3	0.278
1742.1	0.239
2466.5	0.372
2511.0	0.379
2674.8	0.372
2686.0	0.342
2694.3	0.381
2855.4	0.436

* Not shown on Plot

SPECIFICATION TABLE NO. 174 THERMAL CONDUCTIVITY OF THORIUM CARBIDE ThC

Curve No.	Ref. No.	Method Used	Year	Temp. Range, K	Reported Error, %	Name and Specimen Designation	Composition (weight percent), Specifications and Remarks
1	476	C	1962	438-638			Specimen 20 mm in dia and 15 mm long; measured in vacuum of 10^{-5} mm Hg; density 8.5 g cm^{-3}.
2	476	C	1962	443-624			The above specimen, 2nd run.
3	476	C	1962	457-633			The above specimen, 3rd run.

DATA TABLE NO. 174 THERMAL CONDUCTIVITY OF THORIUM CARBIDE ThC

[Temperature, T, K; Thermal Conductivity, k, Watt $cm^{-1}K^{-1}$]

T	k
CURVE 1*	
438.2	0.0837
478.2	0.0795
531.2	0.0753
563.2	0.0774
638.2	0.0732
CURVE 2*	
443.2	0.0879
483.2	0.0816
535.2	0.0774
566.2	0.0753
624.2	0.0753
CURVE 3*	
457.2	0.0837
523.2	0.0753
568.2	0.0732
593.2	0.0753
633.2	0.0753

*No graphical presentation

SPECIFICATION TABLE NO. 175 THERMAL CONDUCTIVITY OF THORIUM DICARBIDE ThC_2

Curve No.	Ref. No.	Method Used	Year	Temp. Range, K	Reported Error, %	Name and Specimen Designation	Composition (weight percent), Specifications and Remarks
1	476	C	1962	443-627			Specimen 20 mm in dia and 15 mm long; measured in vacuum of the order of 10^{-5} mm Hg; density 6.6 g cm^{-3}.

DATA TABLE NO. 175 THERMAL CONDUCTIVITY OF THORIUM DICARBIDE ThC_2

[Temperature, T, K; Thermal Conductivity, k, Watt $cm^{-1}K^{-1}$]

T	k
CURVE 1*	
443.2	0.241
487.2	0.230
519.2	0.222
546.2	0.213
576.2	0.207
603.2	0.207
627.2	0.205

*No graphical presentation

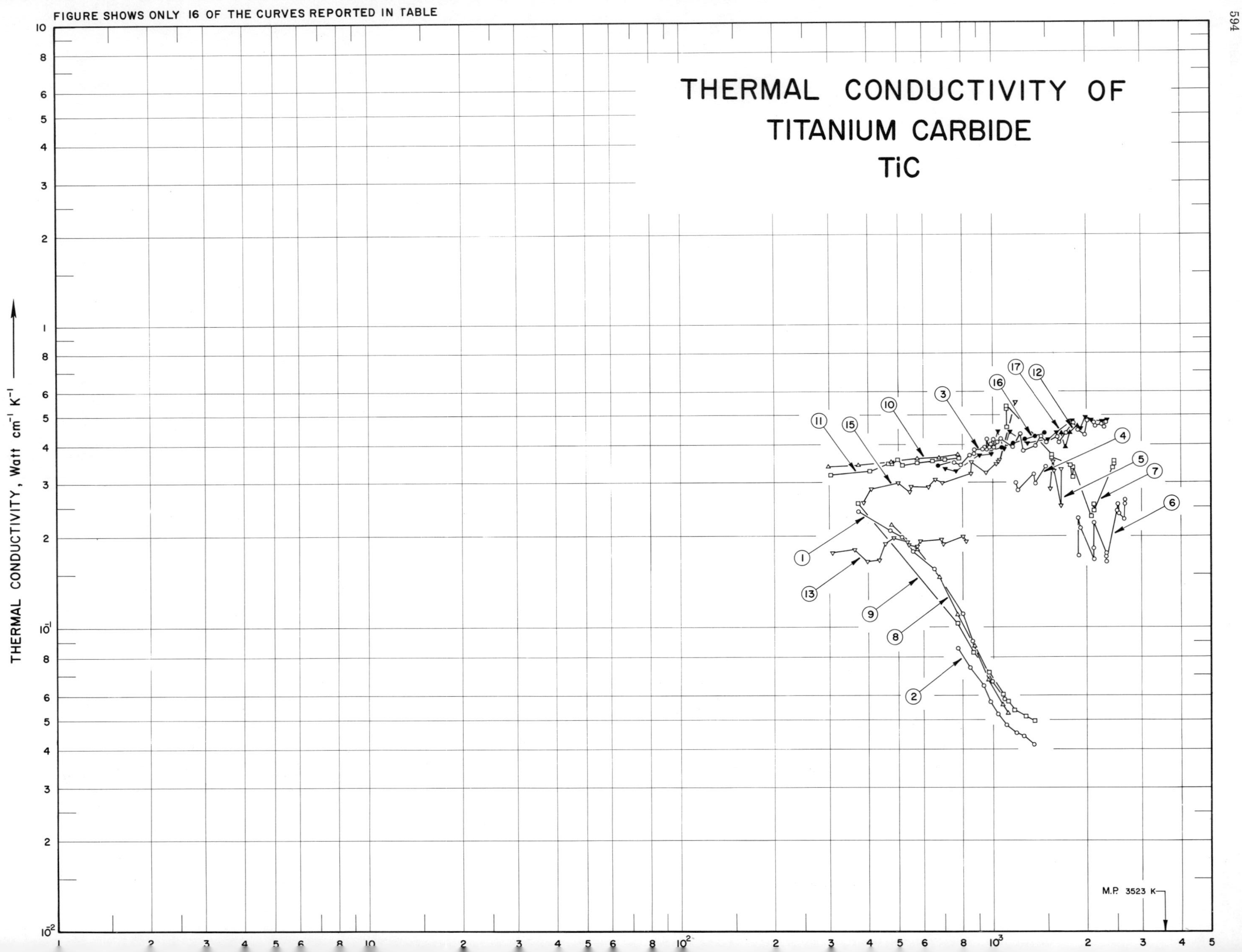
FIGURE SHOWS ONLY 16 OF THE CURVES REPORTED IN TABLE
THERMAL CONDUCTIVITY OF
TITANIUM CARBIDE
TiC
THERMAL CONDUCTIVITY, Watt cm^{-1} K^{-1}
M.P. 3523 K

SPECIFICATION TABLE NO. 176 THERMAL CONDUCTIVITY OF TITANIUM CARBIDE TiC

[For Data Reported in Figure and Table No. 176]

Curve No.	Ref. No.	Method Used	Year	Temp. Range, K	Reported Error, %	Name and Specimen Designation	Composition (weight percent), Specifications and Remarks
1	14	C	1953	373-1093			80.6 Ti, 19.0 C, 0.4 O; 1 in. cubic specimen prepared by hydrostatically pressing and firing to 2000 C; firing shrinkage 15%; total porosity 4.4%; grinding time 100 hrs; dense sintered alumina used as comparative material.
2	14	R	1953	773-1353			80.6 Ti, 19.0 C, 0.4 O; ellipsoidal specimen prepared by casting, hydrostatically pressing and firing to 2000 C; shrinkage in pressing 6%, in firing 5%; grinding time 24 hrs.
3	63	R	1961	758-2313	10.0		< 0.3 metallic impurities by chemical analysis; hot-pressed.
4	144	R	1962	1199-1500	5-7	1	Specimen 0.75 in. long, 0.75 in. O.D. and 0.25 in. I.D.; heat soaked at 2050 K; ground and polished to eliminate all the scratches on specimen's surface.
5	144	R	1962	1543-1860	5-7	1	Similar to the above specimen but cracked during measurement.
6	144	R	1962	1892-2668	5-7	2	Similar to the above specimen except heat soaked at 2200 K; and cracked during experiment.
7	144	R	1962	1121-2460	5-7	3	Similar to the above specimen except heat soaked at 2117 K; specimen cracked.
8	170	C	1953	473-1123			Cubic specimen prepared by hydrostatically pressing at 20,000 psi and firing to 2000 C in vacuo; milled with alcohol in steel for 100 hrs; final bulk density 4.63 g cm^{-3}; lead used as comparative material.
9	170	R	1953	373-1373			Ellipsoidal specimen prepared by slip casting from water suspension, hydrostatically pressing and firing at 2000 C in vacuo; bulk density 3.9 g cm^{-3}.
10	270	P	1962	298-773			No other information reported.
11	277	P	1963	303-778			80.3 ± 0.3 Ti, 19.3 C and < 0.2 metallic impurities; single phase; supplied by the Carborundum Co.; density 4.77 g cm^{-3}; calculated from flash diffusivity.
12	277	P	1963	1673-1873			79.2 Ti, 19.5 C and < 0.2 metallic impurities; single phase; supplied by Norton Co.; density 4.74 g cm^{-3}; calculated from radial diffusivity.
13	277	P	1963	308-823			79.6 ± 1.2 Ti, 17.7 C and 1.4 metallic impurities; single phase; MIT cubic specimen; density 4.56 g cm^{-3}; calculated from flash diffusivity.
14	277	P	1963	1623-1823			80.3 ± 0.3 Ti, 19.3 C and < 0.2 metallic impurities; single phase; supplied by the Carborundum Co.; density 4.77 g cm^{-3}; calculated from radial diffusivity.
15	374	C	1965	386-1185	0.5		79.2 Ti, 20.2 C, 0.02 Fe; specimen in excess of 97% theoretical density; prepared to a tolerance of ± 0.001 in. in the form of a cylinder 1 in. in dia and 1 in. high; fabricated by the Norton Co., supplied by Atomics International; alumina AL-300 as reference standard (thermal conductivity of the standard determined by J.J. Swica, Alfred University).
16	289	R	1961	673-1473			Specimen 2 in. O.D., 0.5 in. I.D. and 3 in. long.

SPECIFICATION TABLE NO. 176 (continued)

Curve No.	Ref. No.	Method Used	Year	Temp. Range, K	Reported Error, %	Name and Specimen Designation	Composition (weight percent), Specifications and Remarks
17	413	R	1963	709-2333	±10	sample 1	80.3 ± 0.3 Ti, 19.3 C, <0.2 metallic impurities; single phase; supplied by Carborundum Co.; specimen 2 in. O.D., 0.5 in. I.D. and 3 in. long; density 4.77 g cm^{-3}; electrical resistivity 72.9, 79.3, 83.2, 87.7 and 92.8 μohm cm at 350, 458, 514, 589 and 686 C, respectively.
18	413	R	1963	770-2181	±10	sample 2	Similar to the above specimen except no electrical resistivity data reported.
19	413	R	1963	760-1075	±10	sample 3	Similar to the above specimen except electrical resistivity 67.9, 74.2, 77.6, 82.0, 86.6 and 47.4 μohm cm at 350, 458, 514, 589, 686 and 21 C, respectively.
20	414	P	1964	296-771	± 4	Carborundum	80.3 ± 0.3 Ti, 19.3 C, and <0.2 metallic impurities; 0.25 in.dia x 0.10 in. thick; supplied by Carborundum Co.; density 4.77 g cm^{-3}; thermal conductivity data calculated from measured thermal diffusivity values (with flash technique) and specific heat data of Naylor, B.F. (J. Am. Ceram. Soc., 68 (3), 370-1, 1946).
21	414	P	1964	311-814	± 4		76.6 ± 1.2 Ti, 17.7 C, and 1.4 metallic impurities; cube specimen obtained from MIT; density 4.56 g cm^{-3}; same measuring method as the above specimen.
22	414	P	1964	1655-1800	± 5		79.2 Ti, 20.2 C, and <0.2 metallic impurities; 0.625 in. dia x 1.375 in. long; supplied by Norton; density 4.74 g cm^{-3}; same measuring method as the above specimen except thermal diffusivity measured with radial technique.
23	414	P	1964	1620-1871	± 5	Carborundum	80.3 ± 0.3 Ti, 19.3 C, and <0.2 metallic impurities; 0.625 in. dia x 1.375 in. long; supplied by Carborundum Co.; density 4.77 g cm^{-3}; same measuring method as the above specimen.

DATA TABLE NO. 176 THERMAL CONDUCTIVITY OF TITANIUM CARBIDE TiC

[Temperature, T, K; Thermal Conductivity, k, Watt cm^{-1}K^{-1}]

CURVE 1

T	k
373.2	0.241
471.2	0.209
513.2	0.198
558.2	0.178
653.2	0.156
803.2	0.111
863.2	0.0900
998.2	0.0669
1093.2	0.0586

CURVE 2

T	k
773.2	0.0858
848.2	0.0741
938.2	0.0649
983.2	0.0573
1048.2	0.0523
1103.2	0.0481
1193.2	0.0452
1263.2	0.0444
1353.2	0.0418

CURVE 3

T	k
758.2	0.349
793.2	0.343
843.2	0.370
872.2	0.374
873.2	0.385
933.2	0.387
943.2	0.391
963.2	0.389
963.2	0.418
978.2	0.391
988.2	0.400
1018.2	0.418
1028.2	0.391
1048.2	0.414
1073.2	0.418
1174.2	0.393
1238.2	0.435
1263.2	0.383
1388.2	0.397
1448.2	0.427

CURVE 3 (cont.)

T	k
1493.2	0.408
1623.2	0.435
1648.2	0.406
1733.2	0.439
1828.2	0.473
1893.2	0.448
1993.2	0.431
2023.2	0.481
2103.2	0.473
2148.2	0.464
2258.2	0.469
2283.2	0.460
2313.2	0.473

CURVE 4

T	k
1198.7	0.299
1212.6	0.284
1362.1	0.320
1385.9	0.299
1499.8	0.340

CURVE 5

T	k
1542.6	0.286
1576.5	0.352
1582.6	0.327
1858.2	0.252
1860.4	0.332

CURVE 6

T	k
1892.1	0.173
1898.7	0.233
1901.5	0.213
2119.8	0.168
2123.7	0.182
2127.1	0.221
2319.3	0.172
2321.5	0.175
2322.6	0.166
2504.8	0.243
2508.2	0.255
2512.6	0.238

CURVE 6 (cont.)

T	k
2648.7	0.228
2666.5	0.257
2667.6	0.263

CURVE 7

T	k
1120.9	0.457
1122.1	0.528
1122.1	0.534
1560.4	0.370
1569.3	0.364
1792.1	0.343
1808.7	0.312
1819.3	0.340
2088.7	0.234
2108.2	0.255
2114.8	0.244
2445.4	0.337
2453.7	0.346
2459.8	0.354

CURVE 8

T	k
473.2	0.219
573.2	0.181
673.2	0.146
773.2	0.110
873.2	0.0872
973.2	0.0674
1073.2	0.0560
1123.2	0.0527

CURVE 9

T	k
373.2	0.256
773.2	0.103
871.2	0.0833
973.2	0.0715
1073.2	0.0606
1123.2	0.0575
1175.7	0.0540
1275.2	0.0513
1373.2	0.0498

CURVE 10

T	k
298.2	0.339
373.2	0.343
473.2	0.352
573.2	0.360
673.2	0.364
773.2	0.372

CURVE 11

T	k
303.2	0.318
403.2	0.326
468.2	0.347
478.2	0.347
498.2	0.358
518.2	0.343
573.2	0.349
643.2	0.354
703.2	0.358
778.2	0.364

CURVE 12

T	k
1673.2	0.439
1728.2	0.393
1773.2	0.439
1873.2	0.460

CURVE 13

T	k
308.2	0.176
363.2	0.180
398.2	0.165
433.2	0.167
453.2	0.188
483.2	0.197
533.2	0.192
538.2	0.186
573.2	0.184
588.2	0.192
688.2	0.195
698.2	0.188
803.2	0.199
823.2	0.192

CURVE 14*

T	k
1623.2	0.439
1668.2	0.464
1673.2	0.418*
1688.2	0.456
1718.2	0.456
1718.2	0.427
1718.2	0.408
1738.2	0.473
1743.2	0.418
1768.2	0.469
1773.2	0.418
1803.2	0.473
1823.2	0.427

CURVE 15

T	k
387.2	0.258
410.2	0.285
500.2	0.298
545.2	0.278
547.2	0.290
625.2	0.288
657.2	0.305
692.2	0.297*
693.2	0.298
857.2	0.320
860.2	0.349
955.2	0.323
1032.2	0.345
1055.2	0.352
1058.2	0.356
1185.2	0.556

CURVE 16

T	k
673.2	0.341
773.2	0.354*
873.2	0.368*
973.2	0.381*
1073.2	0.391
1173.2	0.404
1273.2	0.416
1373.2	0.427
1473.2	0.435

CURVE 17

T	k
709.2	0.331
767.2	0.327
866.2	0.374*
908.2	0.368
928.2	0.390*
981.2	0.400*
993.2	0.372
1003	0.415*
1045	0.444
1188	0.389
1246	0.444
1399	0.402
1446	0.431*
1518	0.414
1602	0.439
1813	0.477
1919	0.452
1994	0.490
2074	0.481
2125	0.477*
2254	0.477
2303	0.473*
2333	0.481

CURVE 18*

T	k
770.2	0.367
841.2	0.339
862.2	0.367
879.2	0.379
966.2	0.396
973.2	0.418
976.2	0.396
1003	0.413
1031	0.400
1069	0.423
1115	0.404
1139	0.427
1620	0.406
1729	0.444
1774	0.418
1982	0.431
2181	0.444

CURVE 19*

T	k
760.2	0.346
852.2	0.363
936.2	0.392
1007	0.410
1075	0.398

CURVE 20*

T	k
296.2	0.322
401.2	0.326
462.2	0.347
474.2	0.347
491.2	0.360
513.2	0.343
568.2	0.351
637.2	0.356
699.2	0.360
771.2	0.364

CURVE 21*

T	k
311.2	0.176
360.2	0.180
392.2	0.167
421.2	0.172
449.2	0.188
478.2	0.197
525.2	0.192
539.2	0.188
568.2	0.188
585.2	0.192
684.2	0.197
696.2	0.192
798.2	0.201
814.2	0.197

CURVE 22*

T	k
1655	0.464
1683	0.460
1716	0.456
1738	0.473
1768	0.469
1800	0.473

CURVE 23*

T	k
1620	0.435
1670	0.439
1670	0.423
1721	0.397
1721	0.410
1721	0.423
1749	0.423
1771	0.418
1771	0.439
1821	0.431
1871	0.460

*Not shown on Plot

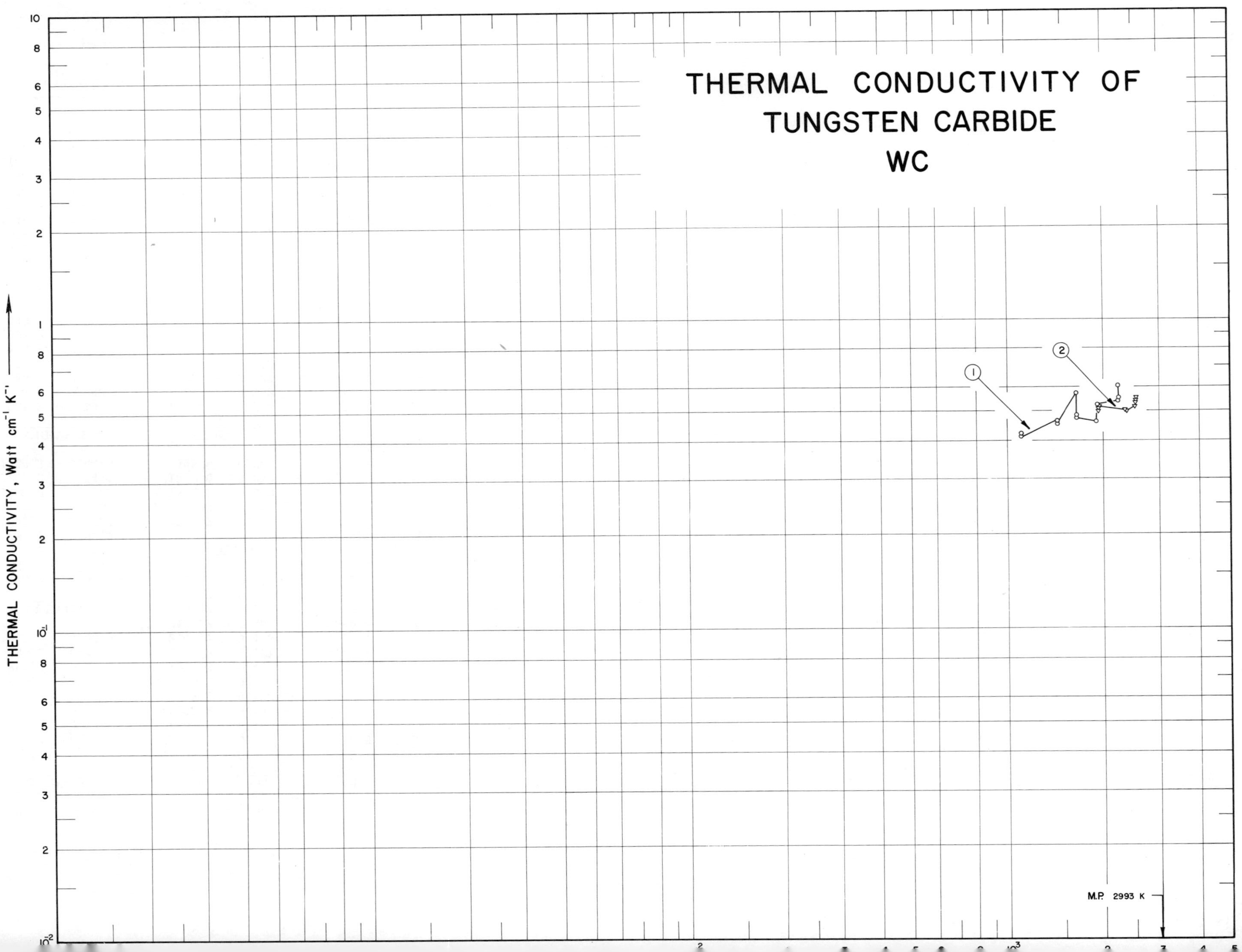

THERMAL CONDUCTIVITY OF TUNGSTEN CARBIDE WC

SPECIFICATION TABLE NO. 177 THERMAL CONDUCTIVITY OF TUNGSTEN CARBIDE WC

[For Data Reported in Figure and Table No. 177]

Curve No.	Ref. No.	Method Used	Year	Temp. Range, K	Reported Error, %	Name and Specimen Designation	Composition (weight percent), Specifications and Remarks
1	144	R	1963	1103-2227	5-7	1	Specimen 0.75 in. long, 0.75 in. O.D. and 0.25 in. I.D.; heat-soaked at 2135 C; ground and polished to eliminate all the scratches on specimen's surface; specimen found cracked on post inspection.
2	144	R	1963	1948-2547	5-7	2	Similar to the above specimen except heat-soaked at 2010-2037 C; specimen found broken on post inspection.

DATA TABLE NO. 177 THERMAL CONDUCTIVITY OF TUNGSTEN CARBIDE WC

[Temperature, T, K; Thermal Conductivity, k, Watt $cm^{-1}K^{-1}$]

T	k
CURVE 1	
1102.6	0.423
1102.6	0.413
1431.5	0.467
1431.5	0.455
1654.8	0.575
1659.3	0.484
1659.3	0.473
1914.8	0.461
1929.3	0.526
2223.2	0.540
2225.4	0.608
2226.5	0.558
CURVE 2	
1947.6	0.497
1948.7	0.510
1953.7	0.518
2357.6	0.504
2375.4	0.498
2376.5	0.497*
2537.1	0.515
2540.4	0.531
2542.1	0.541
2546.5	0.601

* Not shown on Plot

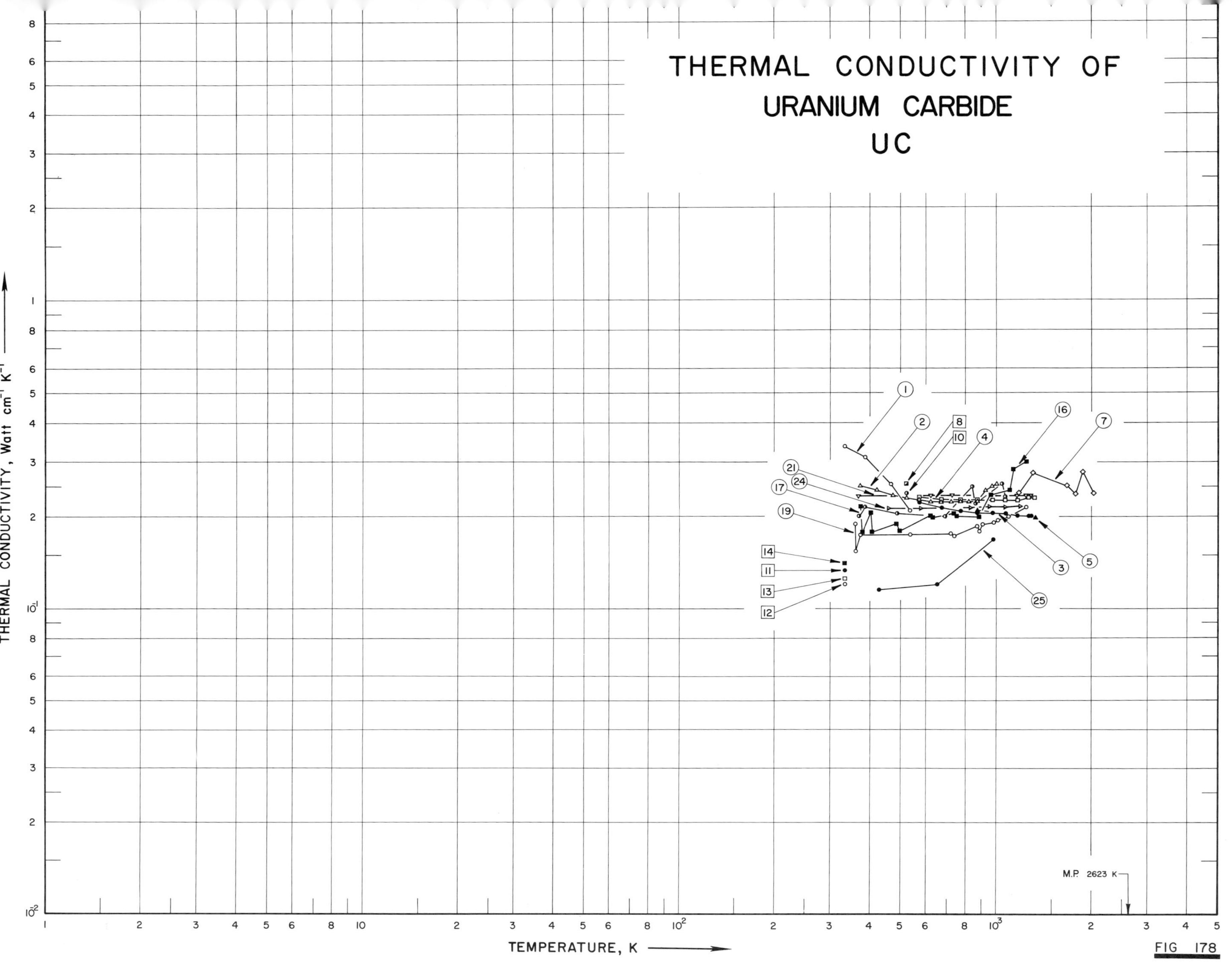
THERMAL CONDUCTIVITY OF
URANIUM CARBIDE
UC
THERMAL CONDUCTIVITY, Watt cm^{-1} K^{-1}
TEMPERATURE, K
M.P. 2623 K
FIG 178

SPECIFICATION TABLE NO. 178 THERMAL CONDUCTIVITY OF URANIUM CARBIDE UC

[For Data Reported in Figure and Table No. 178]

Curve No.	Ref. No.	Method Used	Year	Temp. Range, K	Reported Error, %	Name and Specimen Designation	Composition (weight percent), Specifications and Remarks
1	149	C	1958	333-538	<6		4.815 ± 0.02 total C (0.054 ± 0.02 free carbon), corresponding to 98 mole% UC, 1 mole %U and 1 mole% C; prepared by sintering UC powder itself; density 10.2 ± 0.02 g cm^{-3}; porosity 25%.
2	150	L	1959	373-1008			5.2 total C (0.4 free C); specimen 0.50 in. in dia and 2 in. long; prepared by the drop-casting technique.
3	151	C	1959	473-1293	<5	specimen 100	5.3 total C (0.5 free C); cast; measured in vacuum of ~2 x 10^{-5} mm Hg; Armco iron used as comparative material.
4	151	C	1959	473-1323	<5	specimen 79	4.9 total C (0.1 free C); similar to the above specimen.
5	185		1960	572-1339			4.9 total C (0.1 free C).
6	185		1960	572-1339			5.3 total C (0.5 free C).
7	244	E	1962	1180-2045			100% dense with a composition of 94.7 U, 5.3 total C (determined after measurements) and <0.02 O; specimen 9 cm long and 0.9 cm in dia; supplied by Battelle Memorial Institute; measured in a vacuum of about 10^{-6} torr below 1800 K, and with a cover gas of gettered argon at 100 torr introduced to suppress vapor loss of the specimen above 1800 K.
8	477	C	1962	523.2	± 5		4.5 total C; prepared by arc melting discs of uranium metal with chips of nuclear grade graphite on a copper hearth under argon at 30 mm Hg pressure, then by drop casting.
9	477	C	1962	523.2	± 5		4.8 total C(stoichiometric); similar to the above specimen in preparation.
10	477	C	1962	523.2	± 5		5.1 total C (0.3 free C); similar to the above specimen in preparation.
11	478	C	1958	333.2	± 10	1	Specimen 0.50 in. in dia and 0.50 in. long; prepared at Culcheth by cold compacting of elements and reacting at 1100 C; density 10.3 g cm^{-3}.
12	478	C	1958	333.2	± 10	2	Similar to the above specimen except density 10.14 g cm^{-3}.
13	478	C	1958	333.2	± 10	3	Similar to the above specimen except density 10.03 g cm^{-3}.
14	478	C	1958	333.2	± 15	4	Specimen greater than 0.50 in. in dia and 0.50 in. long; prepared by hot-pressing at A.E.R.E.; density 10.52 g cm^{-3}.
15	478	C	1958	333.2	± 15	5	Similar to the above specimen except density 11.07 g cm^{-3}.
16	480	C	1964	374-1253	<± 5		4.4 analyzed carbon; specimen 0.50 in. in dia and 0.75 in. long; cast; density 13.92 g cm^{-3} (99.8% of theoretical value); stainless steel 347 used as comparative material.
17	480	C	1964	368-1257	<± 5		4.35 analyzed carbon; specimen 0.50 in. in dia and 0.75 in. long; sintered; density 13.72 g cm^{-3} (98.1% of theoretical value); stainless steel 347 used as comparative material.

SPECIFICATION TABLE NO. 178 (continued)

Curve No.	Ref. No.	Method Used	Year	Temp. Range, K	Reported Error, %	Name and Specimen Designation	Composition (weight percent), Specifications and Remarks
18	480	C	1964	379-1259	$< \pm 5$		4.75 analyzed carbon; specimen 0.50 in. in dia and 0.75 in. long; cast; density 13.65 g cm^{-3} (99.8% of theoretical value); stainless steel 347 used as comparative material.
19	480	C	1964	358-1249	$< \pm 5$		4.82 analyzed carbon; specimen 0.50 in. in dia and 0.75 in. long; sintered; density 12.28 g cm^{-3} (90.1% of theoretical value); stainless steel 347 used as comparative material.
20	480	C	1964	383-1260	$< \pm 5$		5.23 analyzed carbon; specimen 0.50 in. in dia and 0.75 in. long; cast; density 13.35 g cm^{-3} (99.0% of theoretical value); stainless steel 347 used as comparative material.
21	481	P	1963	367-1284			4.58 C; polycrystalline disc specimen 0.25 in. in dia and 0.10 in. thick; thermal conductivity data obtained from the smooth curve calculated the measurement of thermal diffusivity, specific heat and density.
22	481	P	1963	362-1286			4.33 C; similar to the above specimen.
23	481	P	1963	369-1185			4.24 C; similar to the above specimen.
24	481	P	1963	458-1184			4.04 C; similar to the above specimen.
25	482, 483		1960	429-978			Prepared from UO_2 and C by heating the mixture in a beryllium oxide crucible in a vacuum furnace with a graphite heater, then sintered; data corrected to zero porosity.

DATA TABLE NO. 178 THERMAL CONDUCTIVITY OF URANIUM CARBIDE UC

[Temperature, T, K; Thermal Conductivity, k, Watt $cm^{-1}K^{-1}$]

T	k
CURVE 1	
333.2	0.335
388.2	0.310
468.2	0.255
538.2	0.209
CURVE 2	
373.2	0.251
473.2	0.243
473.2	0.234
523.2	0.230
573.2	0.226
623.2	0.222
673.2	0.222
723.2	0.222
773.2	0.226
823.2	0.230
873.2	0.238
923.2	0.243
973.2	0.251
1008.2	0.255
CURVE 3*	
473.2	0.234*
573.2	0.222
673.2	0.213
773.2	0.209
873.2	0.205
973.2	0.205
1073.2	0.205
1173.2	0.201
1273.2	0.201
1293.2	0.201
CURVE 4	
473.2	0.234*
573.2	0.230
673.2	0.226
773.2	0.226
873.2	0.226
973.2	0.226
1073.2	0.226
1173.2	0.226
CURVE 4 (cont.)	
1073.2	0.226
1173.2	0.226
1273.2	0.230
1323.2	0.230
CURVE 5	
572.1	0.220*
872.1	0.208
1338.7	0.199
CURVE 6*	
572.1	0.232
872.1	0.225
1338.7	0.232
CURVE 7	
1180	0.239
1310	0.276
1675	0.251
1780	0.236
1870	0.278
2045	0.237
CURVE 8	
523.2	0.255
CURVE 9*	
523.2	0.230
CURVE 10	
523.2	0.238
CURVE 11	
333.2	0.134
CURVE 12	
333.2	0.121
CURVE 13	
333.2	0.126
CURVE 14	
333.2	0.142
CURVE 15*	
333.2	0.138
CURVE 16	
374.2	0.216
377.2	0.178
403.2	0.206
406.2	0.178
485.2	0.188
495.2	0.179
620.2	0.202
630.2	0.198
735.2	0.204
750.2	0.200
768.2	0.210*
783.2	0.199
960.2	0.236
981.2	0.235*
1104.2	0.244
1129.2	0.285
1229.2	0.298*
1253.2	0.300
CURVE 17	
368.2	0.200
372.2	0.211*
386.2	0.215
391.2	0.212*
488.2	0.204
495.2	0.209*
677.2	0.215*
689.2	0.200
843.2	0.251
860.2	0.220
1048.2	0.257
1073.2	0.232
CURVE 17 (cont.)	
1223.2	0.280*
1257.2	0.239*
CURVE 18*	
379.2	0.200
384.2	0.187
435.2	0.203
440.2	0.186
448.2	0.184
457.2	0.181
590.2	0.183
602.2	0.183
694.2	0.188
708.2	0.176
868.2	0.200
890.2	0.208
987.2	0.214
1012.2	0.233
1099.2	0.241
1127.2	0.245
1233.2	0.267
1259.2	0.276
CURVE 19	
358.2	0.189
361.2	0.155
368.2	0.200*
372.2	0.175
537.2	0.175
548.2	0.174*
718.2	0.177
736.2	0.174
871.2	0.187
884.2	0.179
897.2	0.184*
907.2	0.188
982.2	0.192
1010.2	0.195
1095.2	0.200
1127.2	0.204*
1216.2	0.213*
1249.2	0.215
CURVE 20*	
383.2	0.235
391.2	0.245
416.2	0.244
417.2	0.236
514.2	0.214
521.2	0.217
642.2	0.208
651.2	0.215
764.2	0.225
777.2	0.218
841.2	0.220
861.2	0.225
979.2	0.235
1001.2	0.236
1093.2	0.245
1117.2	0.249
1230.2	0.268
1260.2	0.277
CURVE 21	
367.2	0.233
469.2	0.233*
622.2	0.233
725.2	0.233
850.2	0.233*
976.2	0.233*
1072.2	0.233*
1170.2	0.233
1249.2	0.233
1284.2	0.233
CURVE 22*	
362.2	0.221
474.2	0.222
578.2	0.218
673.2	0.223
778.2	0.224
872.2	0.225
974.2	0.226
1071.2	0.226
1180.2	0.227
1286.2	0.227
CURVE 23*	
369.2	0.206
475.2	0.209
583.2	0.210
708.2	0.213
781.2	0.215
872.2	0.216
981.2	0.218
1071.2	0.220
1133.2	0.221
1185.2	0.223
CURVE 24	
458.2	0.212
579.2	0.212
708.2	0.213*
830.2	0.214
953.2	0.215
1070.2	0.215
1184.2	0.216
CURVE 25	
429.2	0.116
653.2	0.121
978.2	0.169

*Not shown on plot

SPECIFICATION TABLE NO. 179 THERMAL CONDUCTIVITY OF URANIUM DICARBIDE UC_2

Curve No.	Ref. No.	Method Used	Year	Temp. Range, K	Reported Error, %	Name and Specimen Designation	Composition (weight percent), Specifications and Remarks
1	159	C	1955	323.2	25		Glycerine coated.
2	244	E	1962	1570-2025	20		8.7 C, 1 Ni and 0.3 O; fairly crystallized with a trace phase of UC present (the last treatment before analysis ended with a rapid cooling from 2150 K); specimen 4.13 cm long and 0.623 cm in dia, with nickel at the cold ends in the grain boundaries while none was observed at the center; hot-pressed in graphite for 15 min at about 1700 C and 6000 psi; a gross measurement of density immediately following fabrication showed 95% of theoretical density; lattice parameter were $a_0 = 3.515 \pm 0.002$ a.u. and $C_0 = 5.976 \pm 0.0002$ a.u.; the specimen was cleaned and polished in a gettered-argon glove box.

DATA TABLE NO. 179 THERMAL CONDUCTIVITY OF URANIUM DICARBIDE UC_2

[Temperature, T, K; Thermal Conductivity, k, Watt $cm^{-1}K^{-1}$]

T	k
CURVE 1*	
323.2	0.335
CURVE 2*	
1570	0.105
1600	0.126
1710	0.126
1770	0.126
1780	0.134
1850	0.151
1940	0.155
1955	0.184
2025	0.157

*No graphical presentation

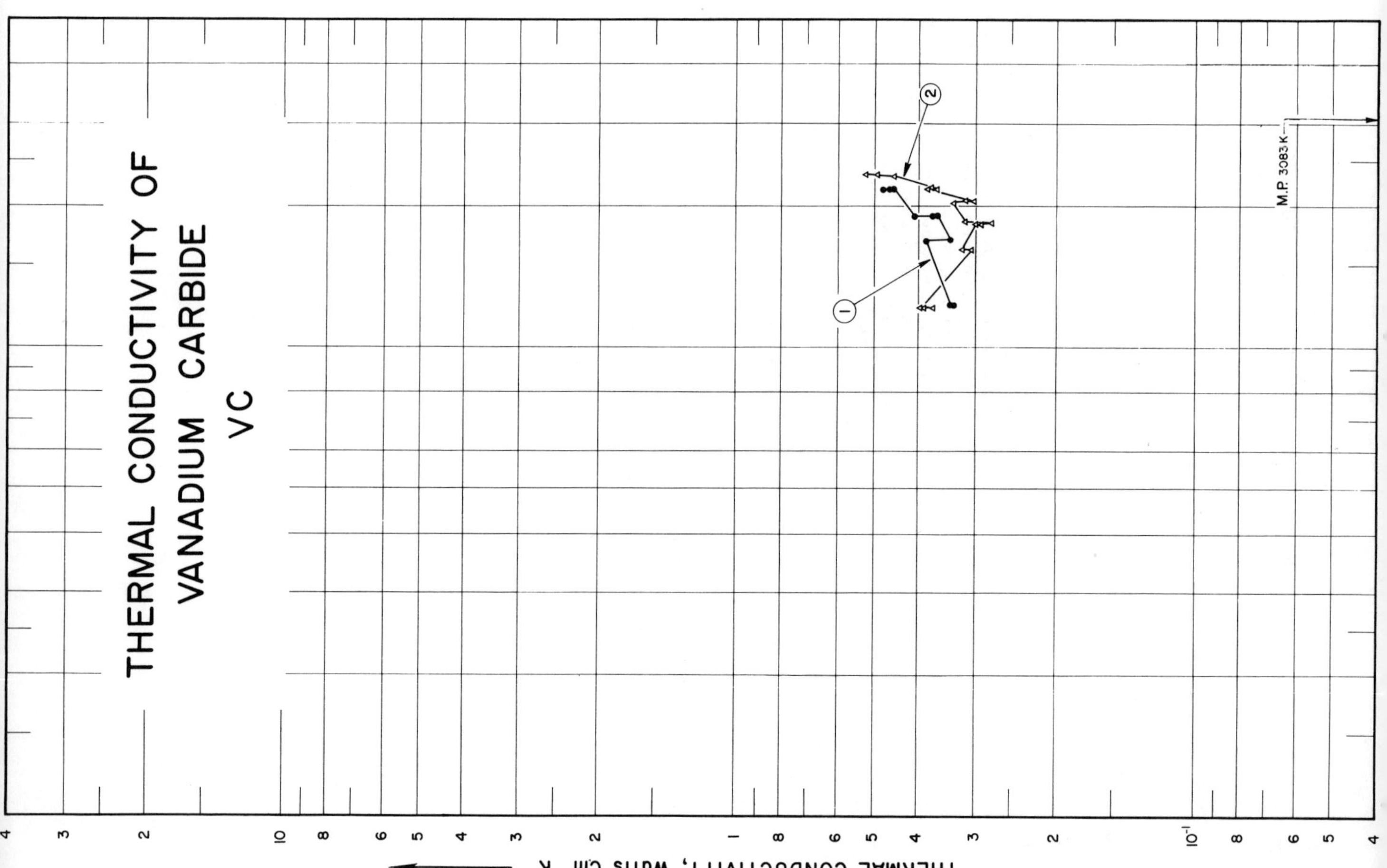
THERMAL CONDUCTIVITY OF
VANADIUM CARBIDE
VC
M.P. 3083 K
1
2
THERMAL CONDUCTIVITY, Watts Cm^{-1} K^{-1}
4 3 2 10 8 6 5 4 3 2 1 8 6 5 4 3 2 10^{-1} 8 6 5 4

SPECIFICATION TABLE NO. 180 THERMAL CONDUCTIVITY OF VANADIUM CARBIDE VC

[For Data Reported in Figure and Table No. 180]

Curve No.	Ref. No.	Method Used	Year	Temp. Range, K	Reported Error, %	Name and Specimen Designation	Composition (weight percent), Specifications and Remarks
1	144	R	1963	1242-2189	5-7	1	Ground and polished to eliminate completely the scratches on the surface of the specimen and then heat-soaked at 2024 C; specimen 0.75 in. long, 0.75 in. O.D., and 0.25 in. I.D.; specimen found cracked on post inspection.
2	144	R	1963	1230-2342	5-7	2	Same as the above specimen except heat-soaked at 1913 C.

DATA TABLE NO. 180 THERMAL CONDUCTIVITY OF VANADIUM CARBIDE VC

[Temperature, T, K; Thermal Conductivity, k, Watt $cm^{-1}K^{-1}$]

T	k
CURVE 1	
1241.5	0.343
1241.5	0.336
1694.3	0.387
1694.3	0.388*
1694.3	0.343
1905.4	0.365
1908.2	0.374
1913.7	0.410
2188.7	0.457
2188.7	0.482
2188.7	0.466
CURVE 2	
1229.8	0.390
1229.8	0.398
1229.8	0.372
1229.8	0.391*
1620.4	0.307
1620.4	0.321
1621.5	0.321*
1844.3	0.301
1848.7	0.293
1852.1	0.278
1853.2	0.319
2047.1	0.335
2051.5	0.303
2065.9	0.316
2065.9	0.313*
2174.3	0.383
2175.9	0.367
2184.3	0.376
2333.2	0.454
2334.3	0.459*
2340.4	0.497
2341.5	0.528

*Not shown on plot

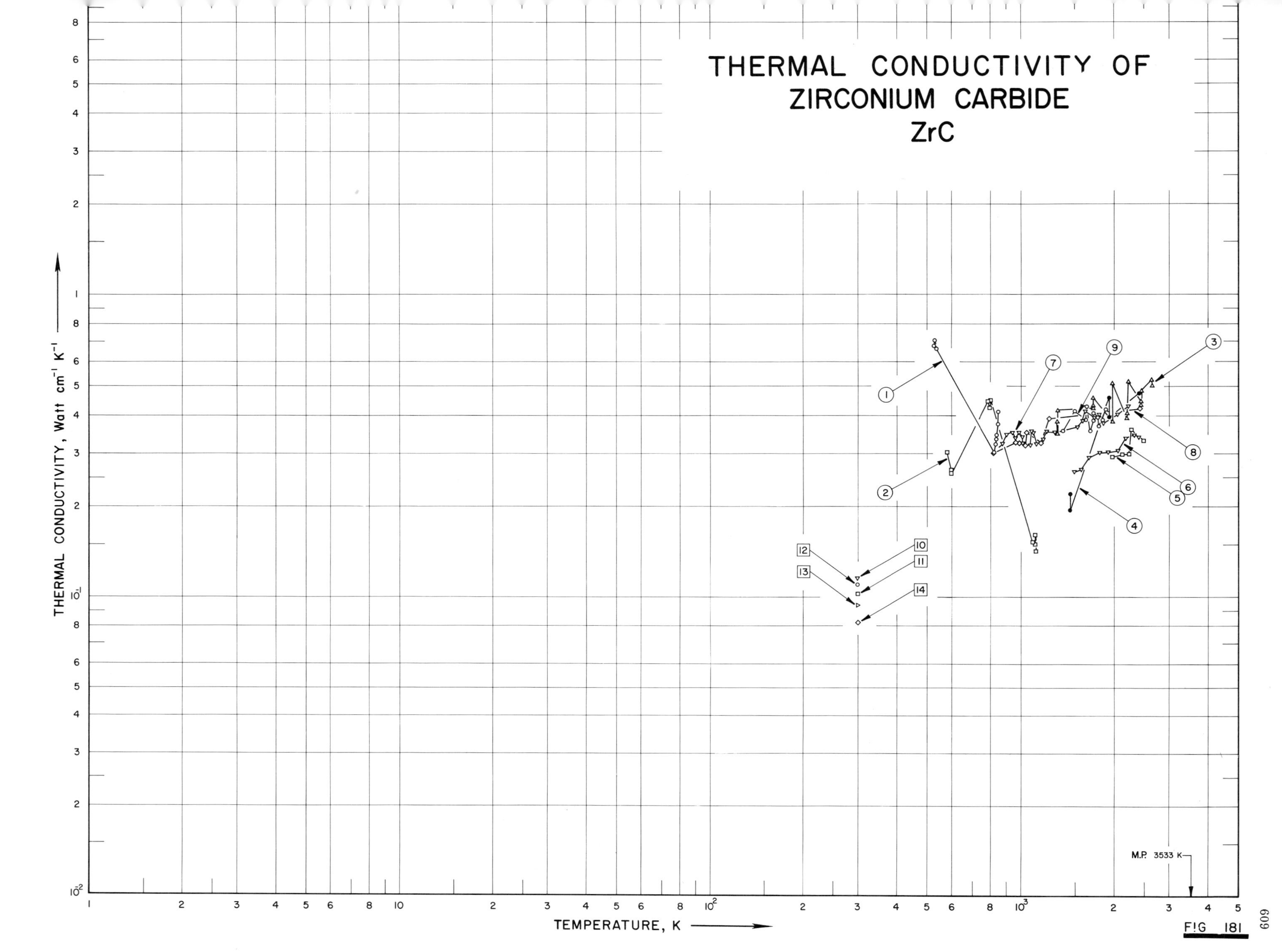
THERMAL CONDUCTIVITY OF
ZIRCONIUM CARBIDE
ZrC
THERMAL CONDUCTIVITY, Watt cm^{-1} K^{-1}
TEMPERATURE, K
M.P. 3533 K
1
2
3
4
5
6
7
8
9
10
11
12
13
14

FIG 181

SPECIFICATION TABLE NO. 181 THERMAL CONDUCTIVITY OF ZIRCONIUM CARBIDE ZrC

[For Data Reported in Figure and Table No. 181]

Curve No.	Ref. No.	Method Used	Year	Temp. Range, K	Reported Error, %	Name and Specimen Designation	Composition (weight percent), Specifications and Remarks
1	243	R	1962	528-847	2-4		Specimen 0.75 in. in dia, 0.75 in. O.D. and 0.25 in. I.D.; supplied by General Electric Co.; pressed and sintered; density 5.94 g cm^{-3}; SRI run No. C-68.
2	243	R	1962	581-1111	2-4		Similar to the above specimen; SRI run No. C-75.
3	243	R	1962	1308-2650	2-4		Similar to the above specimen; SRI run No. C-84.
4	243	R	1962	1450-2425	2-4		Similar to the above specimen but heat-soaked 2755 K before test; specimen deteriorated at above 2866 K; SRI run No. C-94.
5	244	E	1962	1965-2470			11.7 C (including 0.18 free carbon), and <0.002 O_2; specimen 0.607 cm in dia and ~4 cm long; supplied by Wah Chang Co.; hot-pressed from 325 mesh high purity powder in graphite dies at 2300 C for 1 hr at 6000 psi pressure; lattice parameter A_0 4.699 ± 0.002 Å; polished; annealed to 2400 K for 1 hr 90% theoretical density after fabrication and property measurements.
6	244	E	1962	1490-2400			Similar to the above specimen except 11.8 C (including 0.3 free carbon).
7	245, 413, 479, 484	R	1962	813-2208		2	89.8 Zr, 11.0 C and <0.2 metallic impurities; as received; single phase; specimen 2.0 in. O.D., 0.5 in. I.D. and 3.0 in. long; supplied by Carborundum Co.; average grain size 50 μ; hot-pressed; density 6.13 g cm^{-3}.
8	245, 413, 479, 484	R	1962	813-2423		3	87.8 Zr, 12.1 C and <0.6 metallic impurities; similar to the above specimen except density 6.17 g cm^{-3}.
9	245, 413, 479, 484	R	1962	1368-2338		4	89.8 Zr, 11.0 C and <0.2 metallic impurities; similar to the above specimen except density 6.18 g cm^{-3}.
10	6	L	1965	298.2	± 6		89.394 Zr, 10.606 C; specimen prepared by pressing and sintering of powder mixtures close to stoichiometric composition of carbide and appropriate metal; sintering was carried out in a TVV-4 vacuum furnace at 10^{-4} to 10^{-5} mm pressure at 2200 to 2400 C; electrical resistivity 100.0 x 10^{-6} ohm cm at room temp.
11	6	L	1965	298.2	± 6		90.275 Zr, 9.725 C; similar to the above specimen except electrical resistivity 123.6 x 10^{-6} ohm cm at room temp.
12	6	L	1965	298.2	± 6		90.965 Zr, 9.035 C; similar to the above specimen except electrical resistivity 130.2 x 10^{-6} ohm cm at room temp.
13	6	L	1965	298.2	± 6		91.361 Zr, 8.639 C; similar to the above specimen except electrical resistivity 130.0 x 10^{-6} ohm cm at room temp.
14	6	L	1965	298.2	± 6		92.356 Zr, 7.644 C; similar to the above specimen except electrical resistivity 166.2 x 10^{-6} ohm cm at room temp.
15	414	P	1964	1573-1873	± 5		Specimen 0.625 in. dia x 1.375 in. long; thermal conductivity data calculated from measured thermal diffusivity values (with radial diffusivity apparatus) and specific heat value of 0.125 cal g^{-1} C^{-1}.

DATA TABLE NO. 181 THERMAL CONDUCTIVITY OF ZIRCONIUM CARBIDE ZrC

[Temperature, T, K; Thermal Conductivity, k, Watt $cm^{-1}K^{-1}$]

CURVE 1

T	k
528.2	0.678
530.4	0.707
531.5	0.714*
535.4	0.663
818.2	0.303
829.3	0.304*
830.4	0.323
833.2	0.335
833.7	0.346
845.4	0.376
846.5	0.412

CURVE 2

T	k
580.9	0.303
595.9	0.265
597.1	0.258
785.4	0.446
787.1	0.443*
793.2	0.427
793.2	0.450
1097.1	0.153
1101.5	0.151
1105.4	0.162
1107.1	0.143
1108.2	0.150*
1110.9	0.163*

CURVE 3

T	k
1308.2	0.384
1308.2	0.349
1313.7	0.418
1702.6	0.425
1702.6	0.430
1705.4	0.459
1705.4	0.450*
1972.1	0.381
1983.2	0.513
2191.5	0.392
2199.8	0.404
2211.0	0.519
2436.0	0.447
2436.0	0.436
2444.3	0.485
2647.1	0.525
2649.8	0.502

CURVE 4

T	k
1449.8	0.221
1449.8	0.196
1933.2	0.460
1938.7	0.398
2408.2	0.479
2424.8	0.472*
2424.8	0.430*
2424.8	0.434*

CURVE 5

T	k
1965	0.293
2120	0.299
2225	0.299
2265	0.360
2470	0.331

CURVE 6

T	k
1490	0.262
1560	0.266
1655	0.293
1795	0.303
1905	0.305
2055	0.308
2170	0.338
2310	0.346
2400	0.341

CURVE 7

T	k
813.2	0.308*
873.2	0.324
898.2	0.347
938.2	0.354
958.2	0.337
983.2	0.354
998.2	0.351*
1008.2	0.341
1008.2	0.326
1073.2	0.320
1083.2	0.356
1098.2	0 351
1118.2	0.324
1138.2	0.320*
1183.2	0.335
1208.2	0.356
1293.2	0.351
1523.2	0.368
1583.1	0.385
1613.2	0.418
1728.2	0.395
1793.2	0.402
1843.2	0.377
2053.2	0.404
2208.2	0.431

CURVE 8

T	k
813.2	0.301
963.2	0.326
993.2	0.326
1038.2	0.320
1048.2	0.351
1053.2	0.347*
1163.2	0.326
1163.2	0.333*
1238.2	0.393
2423.2	0.423

CURVE 9

T	k
1368.2	0.358
1493.2	0.416
1608.2	0.389
1633.2	0.427
1673.2	0.356
1703.2	0.385
1708.2	0.408
1773.2	0.397
1783.2	0.370
1838.2	0.387
1883.2	0.420
1923.2	0.408*
1973.2	0.389*
2088.2	0.420*
2338.2	0.439*

CURVE 10

T	k
298.2	0.116

CURVE 11

T	k
298.2	0.103

CURVE 12

T	k
298.2	0.110

CURVE 13

T	k
298.2	0.094

CURVE 14

T	k
298.2	0.082

CURVE 15

T	k
1573	0.385
1673	0.406
1773	0.423
1873	0.439

* Not shown on plot

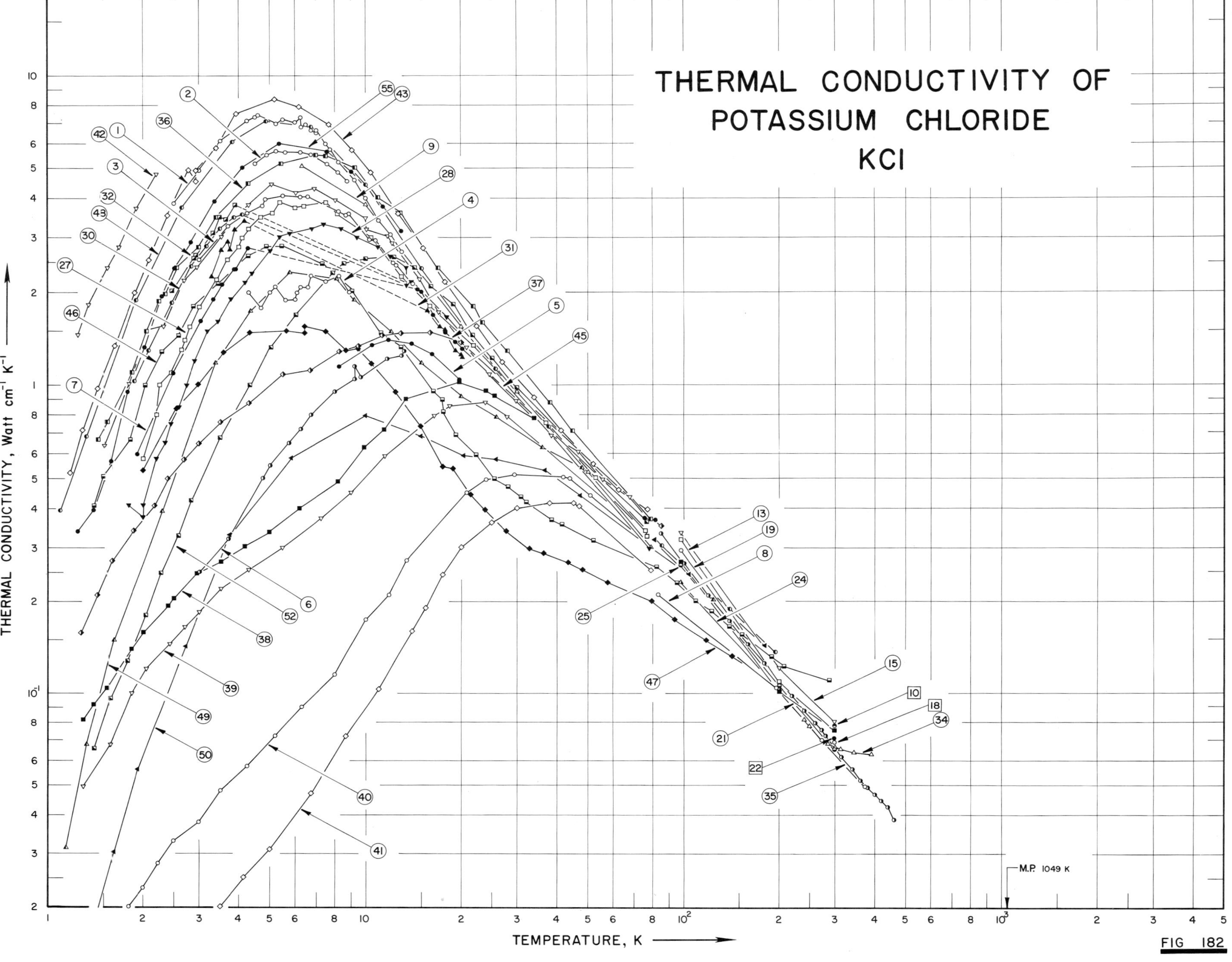
THERMAL CONDUCTIVITY OF
POTASSIUM CHLORIDE
KCl
THERMAL CONDUCTIVITY, Watt cm⁻¹ K⁻¹
TEMPERATURE, K
M.P. 1049 K
FIG 182

SPECIFICATION TABLE NO. 182 THERMAL CONDUCTIVITY OF POTASSIUM CHLORIDE

[For Data Reported in Figure and Table No. 182]

Curve No.	Ref. No.	Method Used	Year	Temp. Range, K	Reported Error, %	Name and Specimen Designation	Composition (weight percent), Specifications and Remarks
1	46	L	1957	2.5-13	± 10	A	Single crystal; supplied by Harshaw Chemical Co.; 40 cm x 0.45 cm^2; annealed in air at 700 C for 0.5 hr, cooled to room temperature at a rate of 30 C hr^{-1}.
2	46	L	1957	4.5-8.8	± 10	K	Cut from the same crystal as the above specimen; 4.0 cm x 0.44 cm^2; remelted and grown in a graphite crucible by using a modified Kyropoules technique at a growth rate of 1.4 cm hr^{-1}, cleaved and ground, annealed in air at 700 C for 0.5 hr, cooled to room temperature at a rate of 30 C hr^{-1}.
3	46	L	1957	3.0-14	± 10	N	Cut from the same crystal as the above specimen; 4.0 cm x 0.41 cm^2; same fabrication method as above except calcium chloride added to the melt during growth (concentrations G = 0.6 x 10^{-4}).
4	46	L	1957	4.3-8.9	± 10	Q	Cut from the same crystal as the above specimen; 4.0 cm x 0.43 cm^2; same fabrication method as above; calcium chloride concentrations G = 1.1 x 10^{-4}.
5	46	L	1957	8.3-20	± 10	R	Cut from the same crystal as the above specimen; 4.0 cm x 0.43 cm^2; same fabrication method as above; G = 1.3 x 10^{-4}.
6	46	L	1957	3.0-13	± 10	S	Cut from the same crystal as the above specimen; 4.0 cm x 0.42 cm^2; same fabrication method as above; G = 2.1 x 10^{-4}.
7	32	L	1937	1.9-82			Red specimen with square cross section of length 3.97 cm and thickness 0.252 cm.
8	22	L	1911	83-373			3.00 cm cubic specimen.
9	28	L	1956	6.3-79			< 0.01 Al, <0.01 Mg, and small concentrations of Fe, Mn, Si, and Na; dielectric crystal supplied by Harshaw Chemical Co.; 5 x 0.5 x 0.5 cm.
10	230	L	1952	298.2		1	Colorless crystal; 20 x 10 x 10 mm; annealed.
11	230	L	1952	298.2		2	Similar to the above specimen.
12	230	L	1952	298.2		3	Similar to the above specimen.
13	230	L	1952	98-298		4	Similar to the above specimen.
14	230	L	1952	298.2		5	Similar to the above specimen.
15	230	L	1952	98-298		8	Similar to the above specimen.
16	230	L	1952	298.2		13	Similar to the above specimen.
17	230	L	1952	298.2		15	Similar to the above specimen.
18	230	L	1952	298.2		10	Similar to the above specimen.
19	230	L	1952	98-298		5	Specimen 5 quenched at 700-710 C; concentration of colored centers n = 0.
20	230	L	1952	298.2		15	Specimen 15 quenched at 700-710 C; n = 0.
21	230	L	1952	98-298		3	Specimen 3 quenched at 700-710 C; n = 0.58 x 10^{18} cm^{-3}.
22	230	L	1952	298.2		6	Similar to the above specimen except n = 1.2 x 10^{18} cm^{-3}.

SPECIFICATION TABLE NO. 182 (continued)

Curve No.	Ref. No.	Method Used	Year	Temp. Range, K	Reported Error, %	Name and Specimen Designation	Composition (weight percent), Specifications and Remarks
23	230	L	1952	298.2		14	Similar to the above specimen except n = 1.7 x 10^{18} cm^{-3}.
24	230	L	1952	98-298		2	Specimen 2 quenched at 700-710 C; n = 1.9 x 10^{18} cm^{-3}.
25	230	L	1952	98-298		9	Similar to the above specimen except n = 4.8 x 10^{18} cm^{-3}.
26	230	L	1952	298.2		10	Specimen 10 quenched at 700-710 C; n = 0.46 x 10^{18} cm^{-3}.
27	62	L	1960	2.0-77			Single crystal; supplied by Harshaw Chemical Co.; 4.0 x 4.5 x 4.0 mm; cleaved and annealed.
28	62	L	1960	1.8-78			The above specimen after the additive coloration n (concentration of colored centers) = 8 x 10^{17} cm^{-3}.
29	225	L	1938	2.3-20			Pure crystalline; 3 3 cm x 0.0626 cm^2.
30	231	L	1938	1.9-20		II	Very pure crystalline square rod specimen of length 3.81 cm and thickness 0.511 cm.
31	231	L	1938	3.3-20		IIA	Cut from the same sample as the above specimen; square rod of length 4.01 cm and thickness 0.383 cm.
32	231	L	1938	2.5-19		III	Cut from the same sample as the above specimen; square rod of length 2.98 cm and thickness 0.763 cm.
33	164	L	1956	5.2-15.0			Optical grade crystal obtained from Harshaw Chemical Co.
34	214	C	1960	240-390			Crystal; supplied by Harshaw Chemical Co ; 1 cm dia x 0.5 cm thick; Z-cut quartz crystal used as comparative material.
35	452, 453	L	1962	85-460	± 3		Single crystal; specimen 8 x 8 x 20 mm; grown from the melt by Kyropoulos method; initial material of chemically pure grade; sample out from a single crystal ingot, annealed in air at 873 K for 6-8 hrs and slowly cooled to room temperature.
36	454	L	1962	1.45-79.0		A	Pure single crystal.
37	454	L	1962	1.27-85.0		B	Single crystal; NKO_2 concentration 9 x 10^{16} cm^{-3}.
38	454	L	1962	1.29-34.0		C	Single crystal; KNO_2 concentration 4 x 10^{17} cm^{-3}.
39	454	L	1962	1.29-79.0		D	Single crystal, KNO_2 concentration 5 x 10^{17} cm^{-3}.
40	454	L	1962	1.48-77.0		E	Single crystal, KNO_2 concentration 1.6 x 10^{18} cm^{-3}.
41	454	L	1962	1.39-79.0		F	Single crystal, KNO_2 concentration 4 x 10^{18} cm^{-3}.
42	455		1967	1.3-2.2		L	High purity crystalline; specimen cross section 12.7 x 13.0 mm^2; zone-refined and seed-pulled.
43	455		1967	1.2-77		2G3	Cut from the above specimen; cross section 6.78 x 5.63 mm^2.
44	440		1964	1.6-46		A	Pure.
45	440		1964	1.5-87		B	Specimen doped with KI with I^- concentration 1.0 x 10^{18} molecules cm^{-3}.
46	440		1964	1.4-288		C	Specimen doped with KI, with I^- concentration 1.25 x 10^{19} molecules cm^{-3}.

SPECIFICATION TABLE NO. 182 (continued)

Curve No.	Ref. No.	Method Used	Year	Temp. Range, K	Reported Error, %	Name and Specimen Designation	Composition (weight percent), Specifications and Remarks
47	440		1964	2.0-203		D	Specimen doped with KI, with I^- concentration 5 x 10^{19} molecules cm^{-3}.
48	440		1964	0.32-195		A	Pure.
49	440		1964	1.1-125		B	Specimen doped with 0.25% in melt of Li Cl.
50	440		1964	0.31-181		C	Specimen doped with 1.0% in melt of Li Cl.
51	456		1965	1.8-194			Rectangular specimen 5 x 5 x 40 mm; grown at Cornell by Kyropoulos technique using high purity argon gas as protective atmosphere, high purity graphite crucible was used; prepared by cleavage then annealed at 650 C for 12 hrs in an atmosphere of distilled chlorine and slowly cooled to remove mechanical strains.
52	456		1965	1.4-77			Similar to the above specimen except compressed 3% in length; dislocation density 2.0 x 10^7 dislocations cm^{-2}.
53	456		1965	1.4-129			Recovery thermal conductivity data of the above specimen.
54	456		1965	1.3-78			The above specimen annealed at 400 C for 15 min.
55	456		1965	1.3-13			The above specimen annealed at 650 K for 15 min.

DATA TABLE NO. 182 THERMAL CONDUCTIVITY OF POTASSIUM CHLORIDE KCl

[Temperature, T, K; Thermal Conductivity, k, Watt $cm^{-1}K^{-1}$]

T	k
CURVE 1	
2.50	3.87
3.00	4.94
3.50	6.16
4.25	7.15
4.50	7.35
4.60	7.45
5.25	7.00
5.50	7.20
6.00	7.08
6.25	7.35
6.30	6.85
6.50	6.95
6.75	6.65
7.00	6.58
7.00	6.68
7.50	6.00
7.75	5.75
8.25	5.25
9.25	4.60
10.0	4.00
11.0	3.40
12.0	3.06
13.0	2.70
CURVE 2	
4.50	5.20
4.90	5.53
5.25	5.70
5.60	5.65
6.25	5.65
6.75	5.55
7.60	5.18
8.20	4.87
8.75	4.56
CURVE 3	
3.00	2.55
3.70	3.26
4.25	3.60
4.75	3.80
4.85	3.98
5.50	4.10
6.25	4.05
6.60	4.05
8.25	3.65
8.70	3.55
8.90	3.56
10.3	2.95
10.7	2.94
12.2	2.48
12.5	2.45
14.1	2.12
CURVE 4	
4.30	2.00
4.70	1.78
5.00	2.00
5.20	2.09
5.60	1.89
6.00	1.90
6.10	2.00
6.25	2.08
6.60	2.08
6.75	2.26
7.50	2.7
8.25	2.26
8.90	2.00
CURVE 5	
8.25	1.15
10.5	1.35
11.8	1.40
14.0	1.36
16.2	1.26
19.7	1.04
CURVE 6	
3.00	0.25
3.70	0.32
4.75	0.50
5.00	0.55
5.75	0.65
6.25	0.72
6.75	0.80
8.00	0.95
9.25	1.05
9.25	1.15
9.65	1.06
11.7	1.22
13.0	1.25
13.2	1.29
CURVE 7	
1.92	0.599
2.49	1.09
3.02	1.61
3.32	1.90
3.55	2.12
3.87	2.38
3.91	2.38
4.28	2.78
14.6	2.04
14.8	1.97
CURVE 7 (cont.)	
16.4	1.69
17.9	1.48
19.2	1.37
20.1	1.31
75.4	0.373
81.8	0.366
CURVE 8	
83.2	0.210
195.2	0.104
273.2	0.0697
373.2	0.0492
CURVE 9	
6.3	5.10
10.0	3.85
20.0	1.55
30.0	0.95
50.0	0.525
79.0	0.300
CURVE 10	
298.2	0.0778
CURVE 11 *	
298.2	0.0774
CURVE 12 *	
298.2	0.0770
CURVE 13	
98.2	0.318
200.2	0.109
298.2	0.0770 *
CURVE 14 *	
298.2	0.0787
CURVE 15	
98.2	0.335
200.2	0.121
298.2	0.0795
CURVE 16*	
298.2	0.0770
CURVE 17*	
298.2	0.0782
CURVE 18	
298.2	0.0686
CURVE 19	
98.2	0.293
200.2	0.109*
298.2	0.0778*
CURVE 20	
298.2	0.0787
CURVE 21	
98.2	0.270
200.2	0.101
298.2	0.0749
CURVE 22	
298.2	0.0703
CURVE 23 *	
298.2	0.0703
CURVE 24	
98.2	0.231
200.2	0.103
298.2	0.0695 *
CURVE 25	
98.2	0.264
200.2	0.106
298.2	0.0649
CURVE 26 *	
298.2	0.0657
CURVE 27 *	
2.0	0.58
2.2	0.80
2.3	1.0
2.5	1.1
2.7	1.3
CURVE 27 (cont.)	
2.7	1.4
2.8	1.6
3.0	1.8
3.4	2.2
3.7	2.6
4.0	2.8
4.1	3.0
4.3	3.2
4.7	3.5
5.1	3.6
5.4	3.9
6.0	3.8
6.4	3.8
7.5	3.9
8.2	3.6
10.5	3.0
13.0	2.3
16.0	1.8
22.0	1.4
37.0	0.75
53.0	0.50
76.0	0.34
77.0	0.325
CURVE 28	
1.8	0.41
2.0	0.37
2.0	0.41
2.2	0.58
2.4	0.65
2.5	0.75
2.6	0.85
2.8	1.0
2.9	1.2
3.2	1.5

* Not shown on plot

DATA TABLE NO. 182 (continued)

T	k
CURVE 28 (cont.)	
3.45	1.6
3.8	1.9
4.2	2.15
4.4	2.3
5.0	2.7
5.4	3.0
5.8	3.1
6.8	3.2
7.4	3.3
8.4	3.2
9.4	3.0
11.0	2.8
12.0	2.6
13.5	2.4
13.5	2.1
14.0	2.15
18.0	1.65
78.0	0.3
CURVE 29*	
2.29	0.829
2.62	1.19
2.87	1.45
2.91	1.52
3.21	1.81
3.41	2.07
3.66	2.13
3.85	2.27
4.06	2.44
16.07	1.79
17.02	1.62
18.06	1.51
19.17	1.42
20.11	1.35
CURVE 30	
1.87	1.03
2.09	1.30
2.46	1.86
2.83	2.43
3.17	2.84
3.50	3.21
3.86	3.50
4.12	3.56
CURVE 30 (cont.)	
14.73	2.01*
15.16	1.96*
15.90	1.80*
16.38	1.68*
17.29	1.57*
18.66	1.39*
19.83	1.32*
CURVE 31	
3.30	2.25
3.52	2.74
3.69	2.92
3.75	2.76
3.88	3.20
4.18	3.40
15.77	1.75
17.05	1.55
17.99	1.50
19.25	1.29
20.12	1.25
20.30	1.24
CURVE 32	
2.47	2.02
2.90	2.58
2.94	2.64
3.31	3.11
3.51	3.50
3.66	3.45
3.90	3.82
15.12	2.03*
16.95	1.64*
18.59	1.44*
19.27	1.39*
CURVE 33*	
5.2	7.14
15.0	5.00
CURVE 34	
240.0	0.0817
249.0	0.0779
CURVE 34 (cont.)	
273.0	0.0705*
286.0	0.0680
302.4	0.0662
314.8	0.0655
344.5	0.0638
390.0	0.0630
CURVE 35	
85	0.335
100	0.268
120	0.209
140	0.172
160	0.144
180	0.126
200	0.111*
220	0.0983
240	0.0879
260	0.0795
273	0.0753
280	0.0720
300	0.0657*
315	0.0619
320	0.0602*
340	0.0561
360	0.0519
380	0.0490
400	0.0464
420	0.0444
440	0.0423
460	0.0385
CURVE 36	
1.45	0.67
1.55	0.76
1.85	1.11
2.05	1.50
2.25	1.88
2.35	1.98
2.55	2.40
3.0	2.80
3.40	3.50
4.30	4.49
5.45	5.20
7.0	5.51
CURVE 36 (cont.)	
7.5	5.50
9.3	5.08
10.9	4.2
12.9	3.6
17.0	2.4
21.8	1.8
23.5	1.6
28.0	1.29
38.0	0.88
45.0	0.71
79.0	0.37
CURVE 37	
1.27	0.158
1.44	0.210
1.60	0.273
1.87	0.340
2.17	0.410
2.40	0.500
2.70	0.575
3.00	0.648
3.50	0.760
4.32	0.875
5.50	1.08
6.72	1.12
8.30	1.28
9.50	1.34
11.3	1.44
13.0	1.47
16.1	1.47
20.0	1.37
85.0	0.34
CURVE 38	
1.29	0.0820
1.40	0.0920
1.55	0.104
1.84	0.140
2.01	0.159
2.40	0.193
2.51	0.210
2.98	0.249
3.50	0.271
4.20	0.304
CURVE 38 (cont.)	
5.00	0.337
6.20	0.400
8.20	0.490
9.9	0.630
11.4	0.720
13.5	0.900
19.7	1.02
24.0	0.960
25.5	0.925
34.0	0.780
CURVE 39	
1.29	0.0495
1.58	0.0775
1.85	0.100
2.05	0.120
2.43	0.146
2.70	0.165
3.00	0.184
3.51	0.220
4.30	0.255
5.49	0.300
7.23	0.374
9.0	0.450
11.5	0.590
16.5	0.795
18.5	0.850
23.7	0.880
28.0	0.790
79.0	0.340*
CURVE 40	
1.48	0.0140*
1.65	0.0180*
1.80	0.0202
2.00	0.0232
2.23	0.0280
2.50	0.0330
2.99	0.0380
3.50	0.0480
4.22	0.0575
5.2	0.0720
6.3	0.0900
8.0	0.126
CURVE 40 (cont.)	
10.0	0.175
11.8	0.210
13.5	0.273
20.7	0.449
24.0	0.495
29.5	0.517
42.0	0.502
44.0	0.500
51.0	0.440
77.0	0.310*
CURVE 41	
1.39	0.0049*
1.41	0.0053*
1.55	0.0067*
1.80	0.0094*
2.00	0.0100*
2.30	0.0120*
2.49	0.0135*
2.90	0.0165*
3.49	0.0202
4.15	0.0252
5.00	0.0310
6.75	0.0470
8.7	0.0720
11.0	0.103
14.0	0.159
15.5	0.190
17.5	0.245
20.5	0.301
24.9	0.360
30.0	0.400
38.0	0.417
45.1	0.417
47.0	0.405
79.0	0.252
CURVE 42	
1.25	1.46
1.35	1.82
1.55	2.40
1.67	2.84
1.91	3.72
2.21	4.80
CURVE 43	
1.17	0.521
1.29	0.713
1.43	0.973
1.63	1.34
1.88	2.00
2.09	2.54
2.39	3.54
2.77	4.96
2.94	4.55
2.94	4.93
3.41	5.88
3.95	7.57
5.21	8.43
6.21	8.00
7.71	6.97
9.10	5.73
10.4	4.83
12.7	3.57
15.2	2.77
17.7	2.16
22.4	1.56
27.4	1.18
34.0	0.910
41.4	0.710
52.1	0.553
62.8	0.459
77.1	0.395
CURVE 44*	
1.61	0.766
1.88	1.12
2.32	1.88
2.58	2.33
2.98	2.75
3.40	3.61
3.81	3.95
4.32	4.57
4.70	4.83
5.38	5.32
6.78	5.61
8.69	5.18
11.3	4.83
15.5	2.88
17.9	2.48
18.5	2.37

*Not shown on plot

DATA TABLE NO. 182 (continued)

T	k
CURVE 44 (cont.)	
22.3	1.80
24.2	1.60
28.3	1.31
38.6	0.867
46.0	0.715
CURVE 45	
1.54	0.638
1.81	1.01
2.08	1.50*
2.33	1.56
2.69	2.17
2.96	2.41
3.53	3.01
4.29	3.83
5.02	4.46
6.05	4.16
6.92	4.31
8.09	3.95
10.0	3.46
10.1	3.19
11.8	3.04
12.2	2.88
14.8	2.20
17.1	1.72
20.7	1.32
24.6	1.07
29.9	0.887
38.6	0.687
46.9	0.604
56.0	0.494
67.8	0.437
86.5	0.355*
CURVE 46	
1.40	0.410
1.50	0.504
1.83	0.766
2.04	0.998
2.29	1.28
2.59	1.45
2.88	1.80
3.45	2.14
4.28	2.63
4.90	2.83
CURVE 46 (cont.)	
5.47	2.83
7.30	2.47
9.08	2.02
11.2	1.47
13.1	1.33
16.5	0.953
17.5	0.895
17.5	0.817
19.2	0.686
22.3	0.593
25.4	0.499
28.1	0.470
30.8	0.433
32.1	0.419
38.6	0.367
41.8	0.356
52.6	0.316
82.0	0.258
95.1	0.228
110.2	0.200
123.6	0.185
140.6	0.166
153.1	0.159
190.5	0.131
207.0	0.123
288.4	0.110
CURVE 47	
1.99	0.536
2.56	0.847
2.98	1.01
3.58	1.28
4.33	1.47
5.65	1.51
6.46	1.47
6.47	1.56
7.50	1.50
8.67	1.29
9.48	1.34
10.5	1.17
12.5	0.953
14.9	0.736
17.5	0.548
18.9	0.541
21.4	0.445
CURVE 47 (cont.)	
23.8	0.391
27.7	0.340
32.7	0.299
36.4	0.288
43.3	0.269
48.1	0.254
57.8	0.231
79.3	0.200
94.0	0.174
117.8	0.148
142.2	0.132
202.8	0.105*
CURVE 48	
0.318	0.00959*
0.391	0.0167*
0.506	0.0352*
0.700	0.0912*
0.897	0.224*
1.09	0.394
1.33	0.681
1.91	1.90
2.66	3.77
3.84	6.12
4.89	7.15
6.76	6.86
10.1	4.46
15.1	2.40
25.7	1.13
37.7	0.726
86.5	0.310
140.6	0.188
194.5	0.137
CURVE 49	
1.14	0.0313
1.33	0.0678
1.62	0.149
2.30	0.493
3.28	1.17
4.36	1.75
5.79	2.33
8.11	2.22
9.20	1.90
CURVE 49 (cont.)	
12.1	1.50
15.2	1.18
20.2	0.920
25.9	0.778
35.7	0.630
47.9	0.543
124.5	0.203
CURVE 50	
0.308	0.0000505*
0.389	0.0000904*
0.432	0.000122*
0.525	0.000171*
0.582	0.000281*
0.684	0.000506*
0.773	0.00114*
0.871	0.00206*
0.977	0.00342*
1.10	0.00630*
1.16	0.00658*
1.32	0.0139*
1.62	0.0306
1.92	0.0561
2.74	0.142
3.76	0.330
5.85	0.676
10.1	0.794
15.1	0.681
20.4	0.592
25.9	0.579
36.8	0.532
47.4	0.441
81.1	0.330
104.7	0.246
181.1	0.143
CURVE 51*	
1.75	0.879
1.97	1.25
2.23	1.61
2.54	2.18
2.81	2.67
3.07	3.29
3.85	4.48
CURVE 51 (cont.)*	
4.33	5.40
6.25	6.10
8.55	5.43
10.5	4.52
12.5	3.46
14.3	2.70
23.4	1.39
28.1	1.11
37.7	0.789
51.4	0.552
67.6	0.420
95.5	0.294
118.3	0.246
159.2	0.197
193.6	0.142
CURVE 52	
1.39	0.0655
1.57	0.0968
1.78	0.127
2.03	0.181
2.28	0.249
2.58	0.327
2.83	0.427
3.49	0.678
4.32	1.01
5.05	1.33
6.03	1.69
7.91	2.32
8.55	2.48
10.1	2.58
12.3	2.60
14.3	2.40
16.3	2.10
18.9	1.82
21.8	1.46
25.0	1.22
29.9	0.989
76.6	0.365
CURVE 53*	
1.37	0.0641
1.55	0.0968
1.77	0.125
CURVE 53 (cont.)*	
1.98	0.178
2.22	0.248
2.52	0.324
2.81	0.426
3.42	0.681
4.22	1.02
4.96	1.32
5.86	1.71
7.89	2.34
8.36	2.52
10.0	2.64
12.1	2.67
14.2	2.46
16.0	2.17
18.6	1.84
21.8	1.47
24.8	1.24
29.7	0.995
80.5	0.346
129.1	0.219
CURVE 54*	
1.30	0.155
1.47	0.233
1.59	0.283
1.79	0.407
2.08	0.522
2.32	0.735
2.58	0.891
2.89	1.13
3.48	1.52
4.18	2.19
4.96	2.70
5.70	3.11
6.67	3.48
8.28	3.74
10.2	3.55
12.4	3.01
16.0	2.40
20.1	1.82
22.5	1.54
77.6	0.370
CURVE 55	
1.25	0.337
1.39	0.496
1.58	0.664
1.78	0.953
2.03	1.35
2.29	1.95
2.54	2.38
2.83	2.91
3.36	3.93
4.11	5.06
5.35	6.07
7.57	5.70
9.04	4.90
11.4	3.78
12.9	3.16

*Not shown on plot

SPECIFICATION TABLE NO. 183 THERMAL CONDUCTIVITY OF SILVER CHLORIDE AgCl

Curve No.	Ref. No.	Method Used	Year	Temp. Range, K	Reported Error, %	Name and Specimen Designation	Composition (weight percent), Specifications and Remarks
1	214	C	1960	221-373			Crystalline; specimen 2 cm in dia and 0.5 cm thick; drawn from a melt (initially chemically pure silver chloride powder).

DATA TABLE NO. 183 THERMAL CONDUCTIVITY OF SILVER CHLORIDE AgCl

[Temperature, T, K; Thermal Conductivity, k, Watt $cm^{-1}K^{-1}$]

T	k
CURVE 1*	
221.3	0.0130
225.0	0.0126
269.8	0.0119
295.0	0.0115
313.0	0.0110
325.0	0.0109
360.4	0.0107
372.5	0.0105

* No graphical presentation

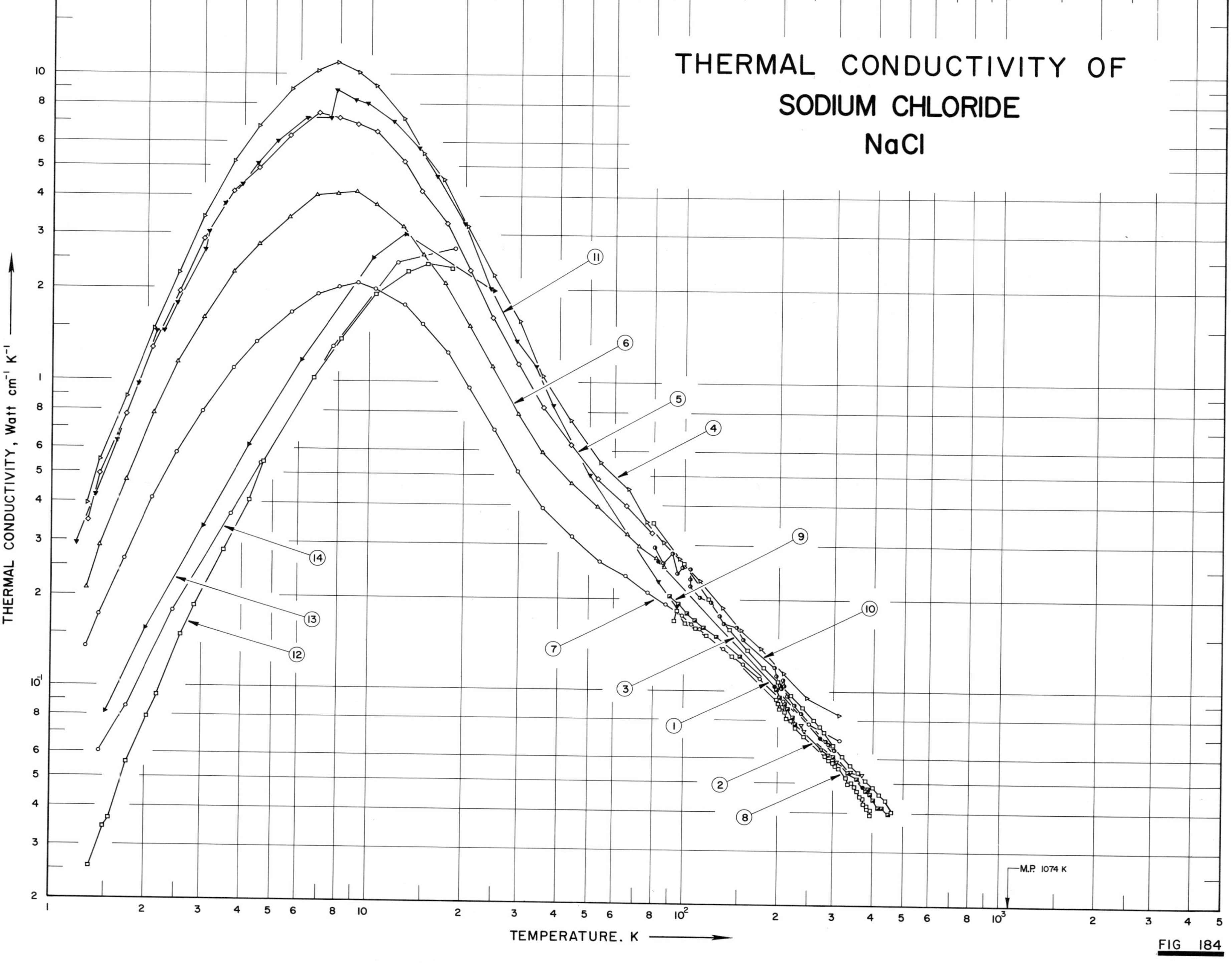
THERMAL CONDUCTIVITY OF
SODIUM CHLORIDE
NaCl
THERMAL CONDUCTIVITY, Watt cm⁻¹ K⁻¹
TEMPERATURE, K
M.P. 1074 K
1
2
3
4
5
6
7
8
9
10
11
12
13
14

FIG 184

SPECIFICATION TABLE NO. 184 THERMAL CONDUCTIVITY OF SODIUM CHLORIDE NaCl

[For Data Reported in Figure and Table No. 184]

Curve No.	Ref. No.	Method Used	Year	Temp. Range, K	Reported Error, %	Name and Specimen Designation	Composition (weight percent), Specifications and Remarks
1	22	L	1911	83-373		Rock salt	Crystalline.
2	214	C	1960	237-370			Crystalline; specimen 1 cm in radius and 0.5 cm in thickness, supplied by Harshaw Chemical Company.
3	453	L	1962	80-460	± 3		Single crystal grown from the melt by Kyropoulos method; initial material, of chemically pure grade; specimen 8 x 8 x 20 mm cut from a single crystal ingot annealed in air at 873 K for 6-8 hrs and slowly cooled to room temperature.
4	496	L	1966	1.3-313	7-8		Pure; treated by bubbling chlorine through the melt.
5	496	L	1966	1.3-79	7-8		0.000077 NaI; prepared from pure NaCl by bubbling chlorine through the melt, then doped with appropriate amount of NaI.
6	496	L	1966	1.3-87	7-8		0.0049 NaI; prepared from pure NaCl by bubbling chlorine through the melt, then doped with appropriate amount of NaI.
7	496	L	1966	1.3-314	7-8		0.019 NaI; prepared from pure NaCl by bubbling chlorine through the melt, then doped with appropriate amount of NaI.
8	497, 498	C	1962	93-393	± 3-± 5		0.4 mole% Ca; single crystal; grown from the melt by the Kyropoulos method using chemically pure materials; pure single crystals of NaCl or KCl used as comparative material.
9	497, 498	C	1962	90-449	± 3-± 5		Similar to the above specimen except with 0.18 mole% Ca.
10	497, 498	C	1962	81-470	± 3-± 5		Similar to the above specimen except the material was pure.
11	456		1965	1.2-95			Rectangular specimen 5 x 5 x 40 mm; prepared by cleavage; grown at Cornell by Kyropoulos technique using high-purity argon gas as protective atmosphere, sintered alumina crucible was used; specimen was cut from a larger rectangular bar compressed along a [100] direction perpendicular to heat flow; annealed at 650 C for 12 hrs in distilled chlorine atmosphere and slowly cooled.
12	456		1965	1.3-18			Similar to the above specimen except plastically deformed 4 % in length; dislocation density 2.5×10^7 dislocations cm^{-2}.
13	456		1965	1.5-25			The above specimen annealed at 355 C for 15 min.
14	456		1965	1.4-19			The above specimen annealed at 400 C for 15 min.
15	456		1965	1.3-16			The above specimen annealed at 450 C for 15 min.

DATA TABLE NO. 184 THERMAL CONDUCTIVITY OF SODIUM CHLORIDE NaCl

[Temperature, T, K; Thermal Conductivity, k, Watt $cm^{-1}K^{-1}$]

T	k
CURVE 1	
83.2	0.266
195.2	0.104
273.2	0.0697
373.2	0.0485
CURVE 2	
237.0	0.0770
242.5	0.0735
276.0	0.0640
289.0	0.0618
308.0	0.0585
342.2	0.0541
369.8	0.0530
CURVE 3	
80	0.351
100	0.261
120	0.199
140	0.159
160	0.136
180	0.119
200	0.107
220	0.0971
240	0.0883
260	0.0799
273	0.0761
280	0.0728
300	0.0661
320	0.0607
340	0.0569
360	0.0540
380	0.0510
400	0.0485
420	0.0460
440	0.0439
460	0.0402
CURVE 4	
1.30	0.396
1.43	0.548
1.72	0.883
2.09	1.47
CURVE 4 (cont.)	
2.51	2.24
3.00	3.40
3.73	5.18
4.46	6.75
5.65	8.92
6.81	10.2
7.87	10.9
9.18	10.1
10.4	9.19
12.7	7.18
14.7	5.50
17.1	4.52
20.4	3.19
24.7	2.23
29.9	1.58
35.5	1.05
43.6	0.754
54.2	0.550
66.7	0.453
76.4	0.355
86.3	0.306
96.8	0.271
113.0	0.230
133.7	0.188
153.1	0.159
176.6	0.138
208.4	0.115
247.7	0.0955
312.6	0.0834
CURVE 5	
1.31	0.347
1.43	0.494
1.72	0.773
2.08	1.28
2.52	1.95
3.00	2.88
3.71	4.13
4.46	4.93
5.59	6.36
6.87	7.45
7.96	7.23
9.14	6.89
10.5	6.52
CURVE 5 (cont.)	
12.8	5.18
14.6	4.15
17.6	3.26
20.8	2.30
24.6	1.62
29.7	1.15
35.7	0.826
43.8	0.628
53.3	0.487
65.9	0.399
79.4	0.327
CURVE 6	
1.30	0.210
1.43	0.289
1.73	0.474
2.10	0.780
2.49	1.15
3.01	1.60
3.73	2.26
4.47	2.77
5.57	3.39
6.78	4.01
7.93	4.08
9.08	4.14
10.5	3.75
12.7	3.17
14.8	2.58
17.4	2.09
20.7	1.53
24.6	1.13
29.8	0.789
35.6	0.593
44.0	0.471
53.3	0.396
66.2	0.324
72.4	0.297
81.7	0.272
86.9	0.255
CURVE 7	
1.30	0.135
1.43	0.172
1.71	0.264
2.08	0.414
2.48	0.581
2.99	0.793
3.72	1.10
4.43	1.34
5.68	1.68
6.86	1.93
8.04	2.03
9.18	2.10
10.4	2.00
12.9	1.78
14.7	1.54
17.7	1.25
20.7	0.958
25.0	0.698
29.7	0.514
35.7	0.390
44.1	0.317
54.2	0.265
65.5	0.237
76.9	0.210
87.9	0.193
98.9	0.177
106.2	0.166
113.2	0.160
134.6	0.137
155.6	0.123
175.0	0.110
207.0	0.0929
251.2	0.0777
314.1	0.0686
CURVE 8	
93	0.169
95	0.184
101	0.167
103	0.170*
109	0.161
118	0.153
143	0.130
197	0.0944
CURVE 8 (cont.)	
200	0.0914
200	0.0885*
202	0.0872
205	0.0898*
207	0.0885
207	0.0863*
210	0.0852
213	0.0809
218	0.0803*
220	0.0795
228	0.0751
241	0.0708
244	0.0701*
275	0.0630*
280	0.0625*
283	0.0614
290	0.0605*
290	0.0592
295	0.0589*
296	0.0596*
299	0.0583
301	0.0590*
303	0.0580*
305	0.0568
308	0.0575*
309	0.0551*
313	0.0558
327	0.0527*
330	0.0523
333	0.0506*
334	0.0497
337	0.0502*
339	0.0493*
341	0.0484*
342	0.0498
350	0.0488
350	0.0476*
352	0.0467*
356	0.0464*
358	0.0473
365	0.0453
365	0.0446*
371	0.0425*
373	0.0446
374	0.0433*
CURVE 8 (cont.)	
374	0.0428
378	0.0424*
384	0.0423
384	0.0413
390	0.0412*
393	0.0409
393	0.0394
CURVE 9	
90	0.204
96	0.193
102	0.180
108	0.171
116	0.162
127	0.151
151	0.130
197	0.102
202	0.0955
208	0.0944*
210	0.0908
214	0.0877
223	0.0816
228	0.0775
241	0.0734*
300	0.0607
306	0.0594*
306	0.0603*
311	0.0593*
326	0.0565*
329	0.0553*
334	0.0541
348	0.0524*
352	0.0515*
358	0.0516
361	0.0507*
368	0.0499*
377	0.0493*
377	0.0485*
380	0.0474
387	0.0466*
388	0.0480
392	0.0461
395	0.0453*
404	0.0443
CURVE 9 (cont.)	
411	0.0437*
418	0.0417
421	0.0422*
426	0.0416
431	0.0411*
439	0.0403*
449	0.0397
CURVE 10	
81	0.295
86	0.263
92	0.283
95	0.243
99	0.254
105	0.251
105	0.232
105	0.221
113	0.202
122	0.196
130	0.177
134	0.167
147	0.162
155	0.148
195	0.120
198	0.113
203	0.105
205	0.102
208	0.109
209	0.104
214	0.0999*
214	0.0969
220	0.0980*
224	0.0922*
226	0.0898
238	0.0842
277	0.0730*
278	0.0702*
284	0.0683
290	0.0669
298	0.0650*
300	0.0664*
302	0.0674*
302	0.0656*
302	0.0648*
CURVE 10 (cont.)	
302	0.0636
306	0.0631*
310	0.0637*
311	0.0627*
317	0.0601*
319	0.0612*
330	0.0588*
347	0.0565*
352	0.0557*
358	0.0547*
359	0.0538*
362	0.0526*
366	0.0523*
368	0.0545*
373	0.0520*
376	0.0531*
379	0.0511*
416	0.0470*
417	0.0459*
419	0.0448*
425	0.0442*
428	0.0448*
431	0.0442*
439	0.0442*
462	0.0421*
463	0.0416*
469	0.0411*
470	0.0418*
CURVE 11	
1.21	0.293
1.38	0.420
1.61	0.630
1.87	0.966
2.14	1.44
2.25	1.45
2.47	1.78
3.02	2.64
3.10	3.03
3.48	3.76
3.95	4.34
4.42	5.07
5.11	6.03
6.34	7.16

*Not shown on plot

DATA TABLE NO. 184 (continued)

T	k	T	k
CURVE 11 (cont.)		CURVE 14	
7.46	7.16	1.43	0.0604
7.82	8.81	1.74	0.0855
8.95	8.19	2.43	0.178
9.75	7.98	3.67	0.367
11.8	6.97	4.56	0.537
14.2	5.73	7.67	1.30
16.2	4.63	12.2	2.43
20.0	3.24	18.6	2.70
24.1	2.00		
29.2	1.36	CURVE 15*	
33.8	1.12		
38.2	0.836	1.33	0.0252
50.0	0.498	1.78	0.0551
66.7	0.322*	2.19	0.0923
83.0	0.227	2.89	0.182
95.1	0.188	4.25	0.403
		6.78	1.01
CURVE 12		10.7	1.88
		15.7	2.37
1.34	0.0255		
1.49	0.0342		
1.55	0.0366		
1.75	0.0556		
2.02	0.0787		
2.18	0.0931		
2.58	0.148		
2.85	0.185		
3.51	0.281		
4.24	0.407		
4.68	0.545		
6.70	1.03		
8.15	1.37		
10.5	1.92		
13.2	2.27		
15.2	2.41		
18.3	2.32		
CURVE 13			
1.50	0.0817		
1.99	0.156		
3.00	0.335		
4.20	0.615		
6.14	1.17		
10.3	2.51		
13.0	2.99		
24.7	1.98		

*Not shown on plot

SPECIFICATION TABLE NO. 185 THERMAL CONDUCTIVITY OF THALLIUM CHLORIDE TlCl

Curve No.	Ref. No.	Method Used	Year	Temp. Range, K	Reported Error, %	Name and Specimen Designation	Composition (weight percent), Specifications and Remarks
1	71	C	1951	311, 345			Cubic isotropic crystal.

DATA TABLE NO. 185 THERMAL CONDUCTIVITY OF THALLIUM CHLORIDE TlCl

[Temperature, T, K; Thermal Conductivity, k, Watt $cm^{-1}K^{-1}$]

T	k
CURVE 1*	
311.2	0.0075
345.2	0.0075

* No graphical presentation

SPECIFICATION TABLE NO. 186 THERMAL CONDUCTIVITY OF ZINC DICHLORIDE $ZnCl_2$

Curve No.	Ref. No.	Method Used	Year	Temp. Range, K	Reported Error, %	Name and Specimen Designation	Composition (weight percent), Specifications and Remarks
1	242	→	1961	563-641	± 3		Specimen of A. R. purity; each data point is the mean value of 4 or 5 different measurements; measured in molten state.
2	242	→	1961	591	± 3		Similar to the above specimen.

DATA TABLE NO. 186 THERMAL CONDUCTIVITY OF ZINC DICHLORIDE $ZnCl_2$

[Temperature, T, K; Thermal Conductivity, k, Watt $cm^{-1} K^{-1}$]

T	k
CURVE 1*	
562.9	0.00313
587.1	0.00296
614.0	0.00294
641.3	0.00290
CURVE 2*	
591	0.00301

* No graphical presentation

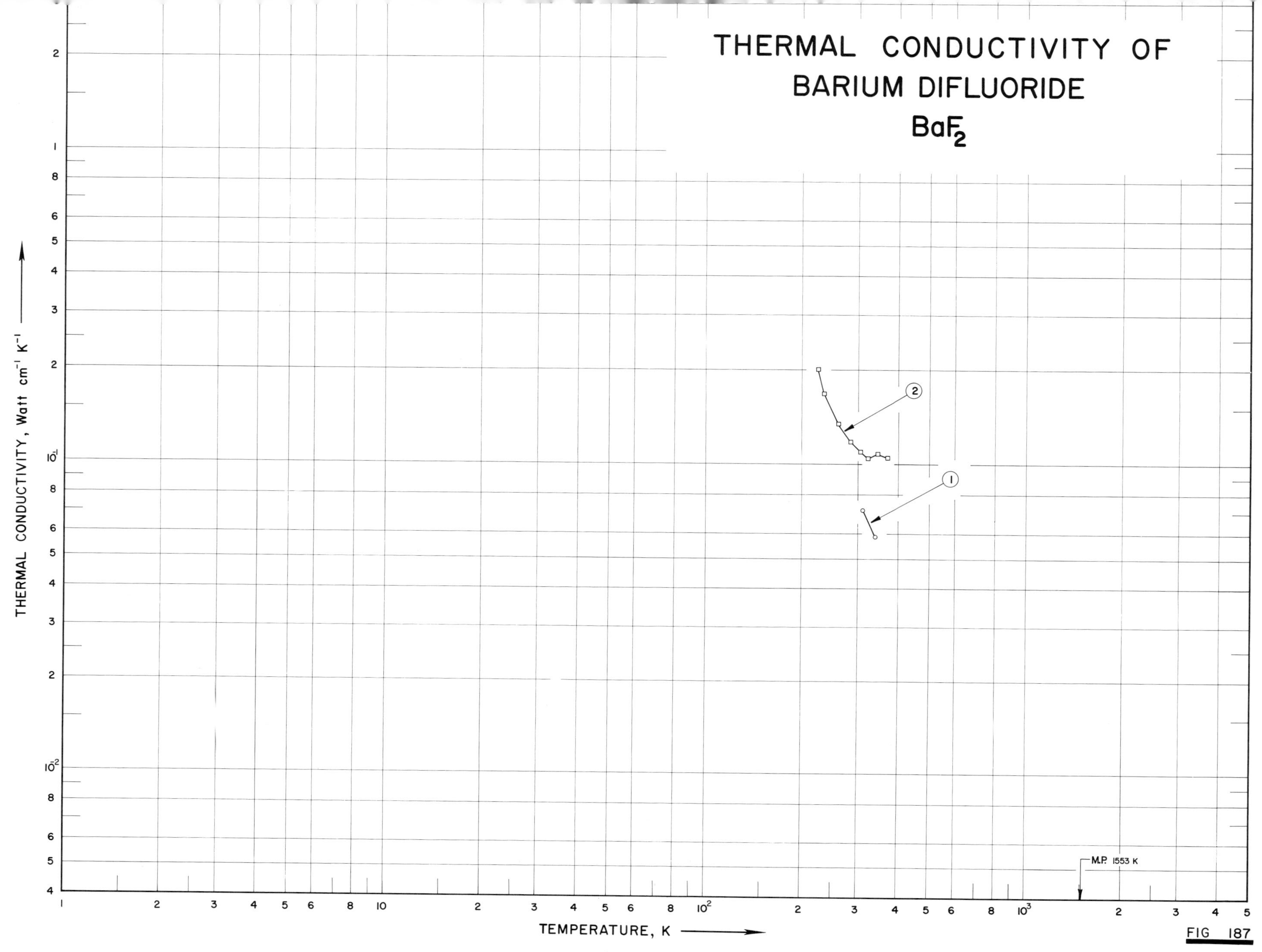
THERMAL CONDUCTIVITY OF
BARIUM DIFLUORIDE
BaF2
THERMAL CONDUCTIVITY, Watt cm⁻¹ K⁻¹
TEMPERATURE, K
M.P. 1553 K
1
2

FIG 187

SPECIFICATION TABLE NO. 187 THERMAL CONDUCTIVITY OF BARIUM DIFLUORIDE BaF_2

[For Data Reported in Figure and Table No. 187]

Curve No.	Ref. No.	Method Used	Year	Temp. Range, K	Reported Error, %	Name and Specimen Designation	Composition (weight percent), Specifications and Remarks
1	71	C	1951	311, 341			Cubic isotropic crystal; specimen supplied by Optovac Co. of Boston; Z-cut crystalline quartz used as standard.
2	214	C	1960	225-370			Crystalline specimen, 1 cm in radius, 0.5 cm thick; specimen supplied by Optovac Co. Z-cut quartz used as standard.

DATA TABLE NO. 187 THERMAL CONDUCTIVITY OF BARIUM DIFLUORIDE BaF_2

[Temperature, T, K; Thermal Conductivity, k, Watt $cm^{-1}K^{-1}$]

T	k
CURVE 1	
311.2	0.0711
341.2	0.0586
CURVE 2	
225.0	0.200
236.6	0.167
260.0	0.134
284.0	0.117
305.0	0.109
323.0	0.104
346.7	0.107
370.0	0.105

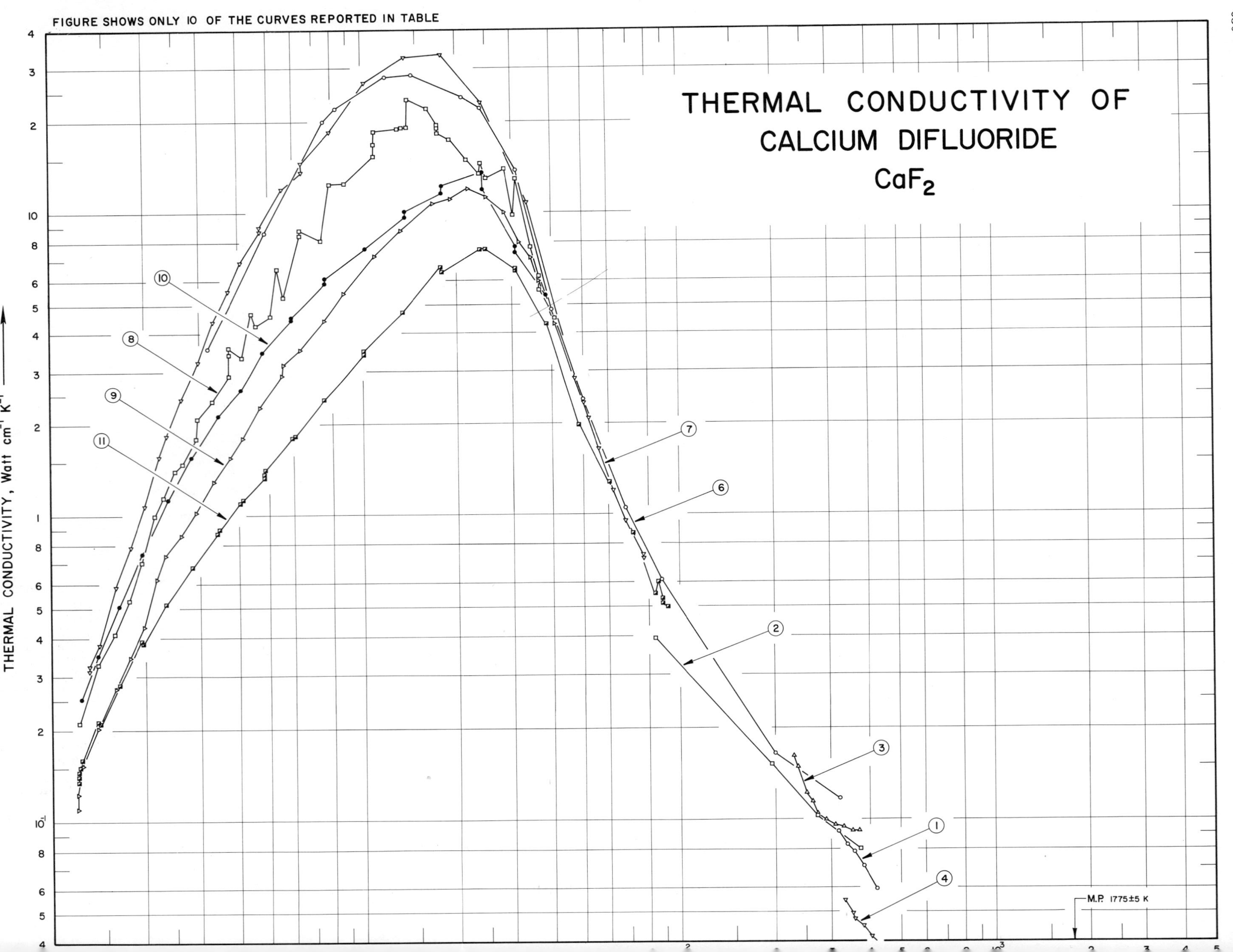

FIGURE SHOWS ONLY 10 OF THE CURVES REPORTED IN TABLE
THERMAL CONDUCTIVITY OF
CALCIUM DIFLUORIDE
CaF_2
THERMAL CONDUCTIVITY, Watt cm^{-1} K^{-1}
M.P. 1775±5 K
1
2
3
4
6
7
8
9
10
11

SPECIFICATION TABLE NO. 188 THERMAL CONDUCTIVITY OF CALCIUM DIFLUORIDE CaF_2

[For Data Reported in Figure and Table No. 188]

Curve No.	Ref. No.	Method Used	Year	Temp. Range, K	Reported Error, %	Name and Specimen Designation	Composition (weight percent), Specifications and Remarks
1	3	L	1953	319-421		255A-1	Single crystal, grown synthetically by Harshaw Chemical Co.
2	22	L	1911	83-373			Single crystal.
3	214	C	1960	229-369			Crystalline, specimen 1 cm in radius, 0.5 cm thick; provided by Optovac Co., Z-cut quartz crystals used as standard.
4	293	C	1957	331-944	±4		Single crystal of optical quality, specimen cube shaped 0.875 x 0.875 x 0.875 in.; raw material privided by Harshaw Chemical Co.; specimen prepared by hydrostatically pressing the powder dry without a binder, and fired; cubic specimen cut from fired slug with faces ground flat and parallel on a diamond lap; polycrystalline alumina used as standard.
5	293	C	1957	440-1031	±4		Polycrystalline of optical quality, specimen cube shaped 0.875 x 0.875 x 0.875 in.; volume porosity 8.17 ~10%, average crystal size 28μ; polycrystalline alumina used as standard.
6	379	L	1960	3.2-320		Sample 30	Debye temperature ~520 K; 0.88 cm dia.
7	499	L	1965	1.3-77		C	Single crystal blank obtained from Harshaw Chemical Company; rectangular rod specimen cut from bulk material with <110> longitudinal axis.
8	499	L	1965	1.2-61		4P	Remainder of the above blank heated to 270 C within 2 hrs, compressed along $[1\bar{2}1]$ axis for 10 min to reach an average deformation of 4.0%, cut a rectangular rod specimen in which edge dislocation was parallel to its longitudinal axis.
9	499	L	1965	1.2-60		4S	Rectangular rod specimen cut from the above blank with edge dislocation perpendicular to the longitudinal axis.
10	499	L	1965	1.2-80		7P	Single crystal blank obtained from Harshaw Chemical Company; heated to 270 C within 2 hrs compressed along $[1\bar{2}1]$ axis for 10 min to reach an average deformation of 7.0%, cut a rectangular rod specimen in which edge dislocation parallel to its longitudinal axis.
11	499	L	1965	1.2-92		7S	Rectangular rod specimen cut from the above blank with edge dislocation perpendicular to the longitudinal axis.

DATA TABLE NO. 188 THERMAL CONDUCTIVITY OF CALCIUM DIFLUORIDE CaF_2

[Temperature, T, K; Thermal Conductivity, k, Watt $cm^{-1}K^{-1}$]

CURVE 1

T	k
318.7	0.0912
338.1	0.0824
356.7	0.0782
382.7	0.0707
420.6	0.0594

CURVE 2

T	k
83.2	0.390
195.2	0.151
273.2	0.103
373.2	0.0799

CURVE 3

T	k
229.1	0.162
236.5	0.149
253.0	0.123
263.5	0.114
273.0	0.104
289.1	0.0985
309.0	0.0960
328.6	0.0941
350.8	0.0918
369.0	0.0918

CURVE 4

T	k
331.0	0.0543
350.0	0.0493
356.0	0.0472
379.0	0.0447
402.0	0.0414
463.0	0.0364*
569.0	0.0322*
651.0	0.0305*
748.0	0.0292*
848.0	0.0284*
944.0	0.0288*

CURVE 5

T	k
440.0	0.0372
528.0	0.0334
661.0	0.0288
810.0	0.0251
937.0	0.0234
1031.0	0.0217

CURVE 6

T	k
3.2	3.55
4.9	8.6
7.6	20.0
8.3	22.0
12.0	28.0
14.6	28.5
21.0	24.0
24.0	22.1
31.0	13.8
40.0	4.80
50.0	2.40
68.0	1.05
88.0	0.61
200.0	0.165
320.0	0.117

CURVE 7

T	k
1.33	0.312
1.33	0.322
1.42	0.380
1.61	0.589
1.80	0.787
1.99	1.07
2.22	1.56
2.36	1.82
2.62	2.41
2.97	3.21
3.33	4.37
3.73	5.55
4.09	6.89
4.72	8.67
4.74	8.95
5.55	11.9
5.57	11.1*
5.60	11.5*
6.40	13.6
6.40	14.5
7.91	18.4
10.2	26.8
13.8	32.4
18.1	33.1
24.1	23.2
33.6	10.7
47.4	2.83
50.4	2.34
52.2	2.07
56.0	1.64
62.2	1.19
68.6	0.951
77.3	0.735
77.3	0.715

CURVE 8

T	k
1.23	0.211
1.41	0.327
1.59	0.413
1.77	0.530
1.95	0.702
2.15	0.995
2.29	1.14
2.50	1.39
2.64	1.47
2.92	1.79
2.95	2.07
3.30	2.38
3.72	2.90
3.72	3.40
3.72	3.58
4.11	3.33
4.43	4.66
4.55	4.25
5.08	4.59
5.35	6.55
5.60	5.30
6.34	8.43
6.34	8.75
7.38	8.09
7.87	12.4
8.79	12.5
10.9	15.4
10.9	16.8
11.0	18.5
13.1	18.9
13.4	19.1
13.9	19.1
14.1	23.6
16.3	22.0
17.5	19.5
17.5	18.9
17.5	18.2
19.1	17.4
21.6	14.9
23.8	13.4
23.9	14.6
24.9	13.0
28.4	13.9
30.2	9.77
30.9	12.9
34.4	7.69
36.6	6.19
36.6	5.55
40.7	4.47
46.1	3.09*
60.5	1.31*

CURVE 9

T	k
1.21	0.110
1.21	0.123
1.25	0.153
1.41	0.202
1.60	0.273
1.78	0.345
1.98	0.435
2.17	0.622
2.32	0.738
2.61	0.859
2.90	1.02
3.33	1.29
3.75	1.55
4.11	1.80
4.66	2.27
5.50	2.90
5.55	3.16
6.31	3.53
7.52	4.41
8.67	5.45
10.9	7.24
13.2	8.75
16.8	10.7
19.1	11.1
21.7	12.0
24.9	11.2
28.2	10.0
31.6	7.98
34.5	7.11
36.3	5.97
40.7	4.29
46.1	2.97*
60.3	1.39*

CURVE 10

T	k
1.24	0.252
1.41	0.350
1.65	0.508
1.96	0.748
1.97	0.776*
2.38	1.13
2.82	1.56
3.44	2.12
4.07	2.61
4.76	3.45
5.92	4.43
5.92	4.51
7.59	5.86
7.59	6.11
10.2	7.62
13.7	9.68
13.7	10.1
18.0	11.6
18.0	12.3
24.3	13.6
24.3	10.9
30.8	7.73
30.8	7.38
38.2	5.35
48.3	2.72*
78.3	0.867*
79.8	0.759*

CURVE 11

T	k
1.22	0.136
1.22	0.141
1.22	0.145
1.23	0.151
1.25	0.159
1.41	0.213
1.42	0.210
1.64	0.281
1.94	0.389
1.96	0.382
2.33	0.515
2.83	0.676
3.40	0.871
3.45	0.900
4.04	1.10
4.13	1.12
4.83	1.32
4.83	1.36
4.85	1.41
5.94	1.80
6.05	1.82
7.48	2.41
10.1	3.39
10.1	3.47
13.4	4.74
17.8	6.61
17.9	6.37
23.8	7.55
24.7	7.55
30.8	6.55
30.8	6.43
38.4	4.27
48.5	1.97
60.3	1.28
71.5	0.867
83.6	0.550
85.1	0.600
89.5	0.530
89.5	0.513
91.6	0.499

*Not shown on plot

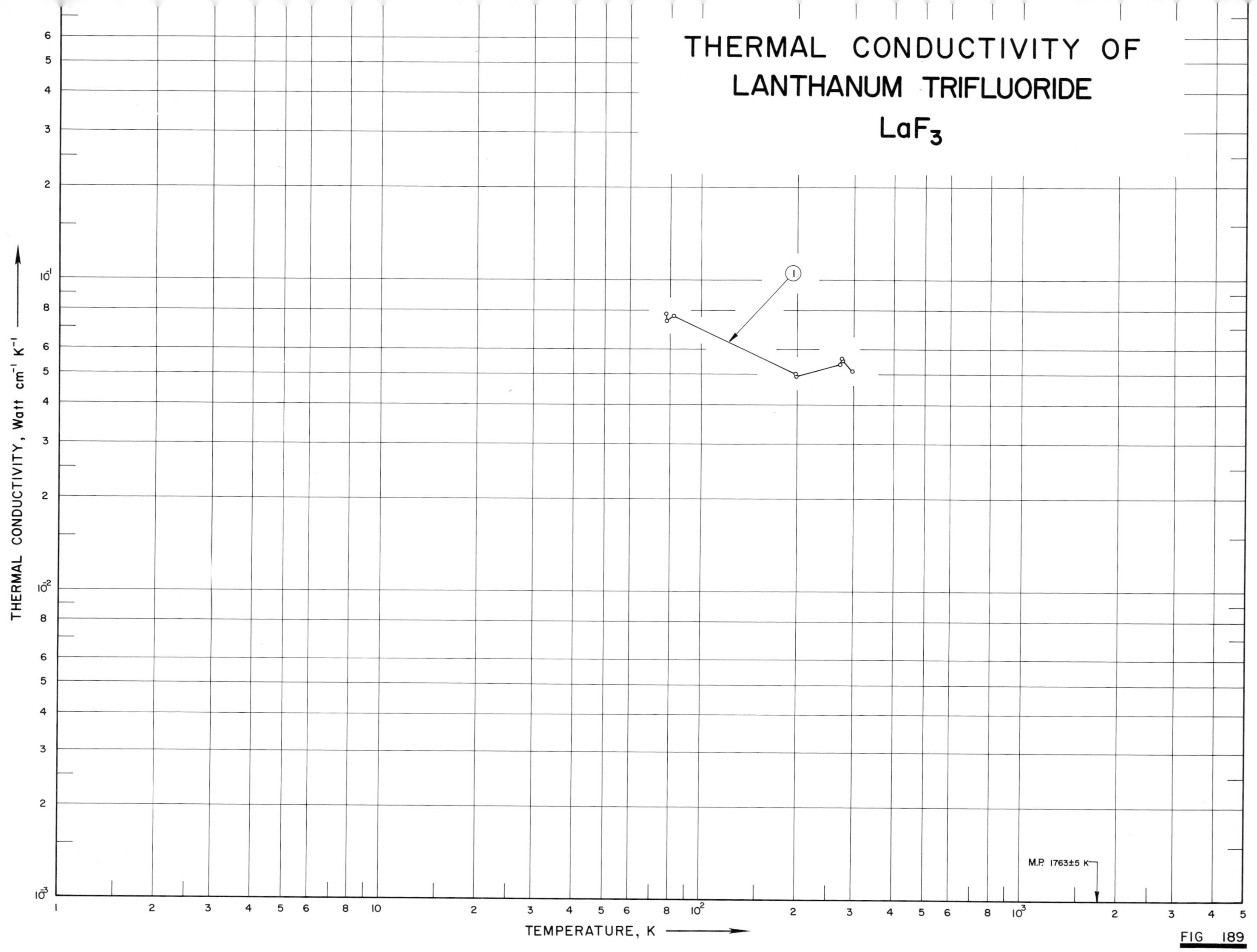

THERMAL CONDUCTIVITY OF
LANTHANUM TRIFLUORIDE
LaF3
THERMAL CONDUCTIVITY, Watt cm⁻¹ K⁻¹
TEMPERATURE, K
M.P. 1763±5 K
1

FIG 189

SPECIFICATION TABLE NO. 189 THERMAL CONDUCTIVITY OF LANTHANUM TRIFLUORIDE LaF_3

[For Data Reported in Figure and Table No. 189]

Curve No.	Ref. No.	Method Used	Year	Temp. Range, K	Reported Error, %	Name and Specimen Designation	Composition (weight percent), Specifications and Remarks
1	461	L	1967	78-299			Single crystal; supplied by Optovac, Inc.; density reported as 5.981, 5.977, 5.972, 5.973, 5.962, 5.959, 5.952, 5.943, 5.941, and 5.930 g cm^{-3} at 111, 126, 144, 176, 208, 231, 254, 271, 299, and 335 K, respectively.

DATA TABLE NO. 189 THERMAL CONDUCTIVITY OF LANTHANUM TRIFLUORIDE LaF_3

[Temperature, T, K; Thermal Conductivity, k, Watt $cm^{-1} K^{-1}$]

T	k
CURVE 1	
77.7	0.0777
78.0	0.0736
82.4	0.0766
197.1	0.0501
197.5	0.0493
273.8	0.0536
273.8	0.0531*
276.6	0.0562
277.0	0.0553
297.2	0.0512
298.7	0.0508*

*Not shown on plot

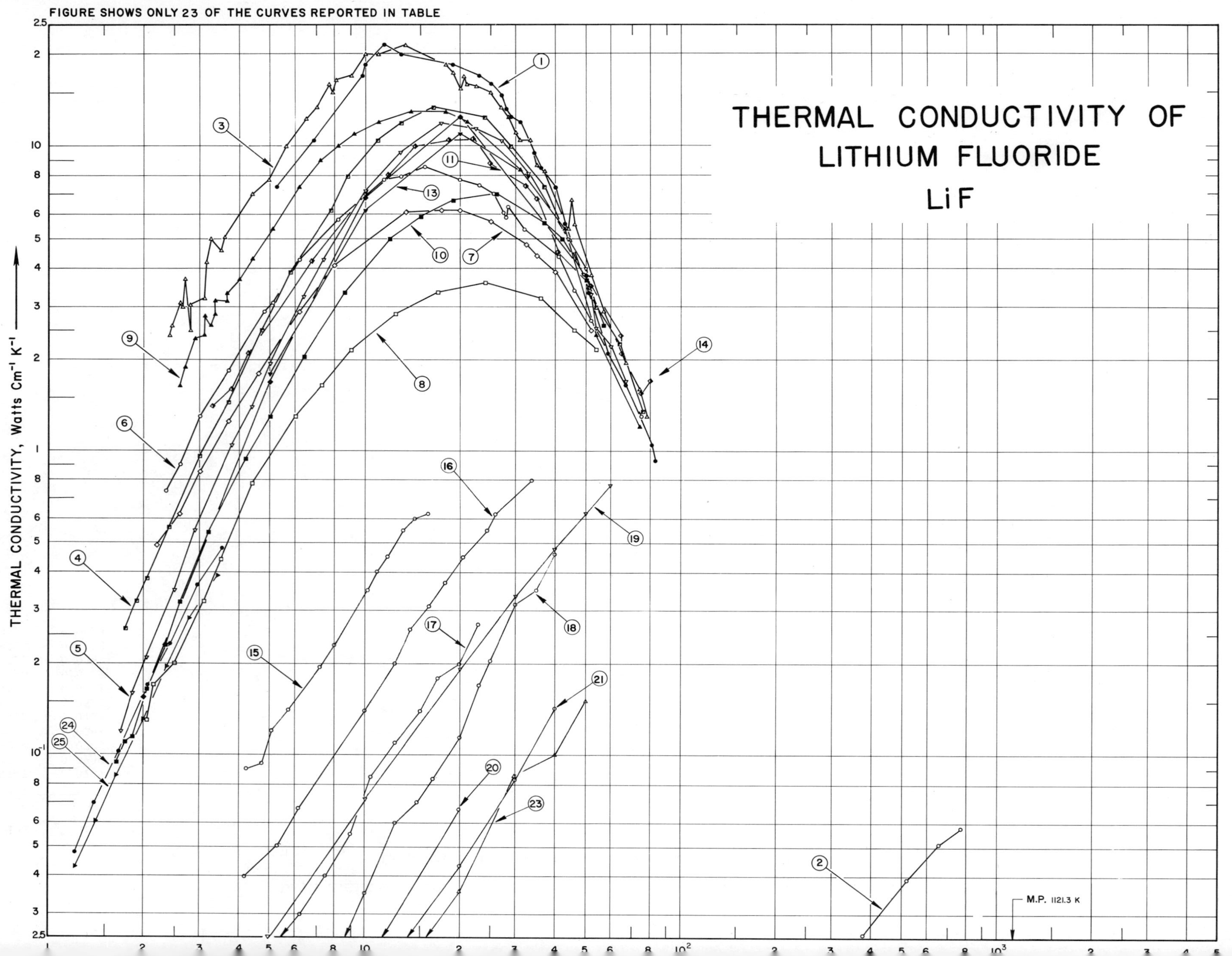

FIGURE SHOWS ONLY 23 OF THE CURVES REPORTED IN TABLE
THERMAL CONDUCTIVITY OF
LITHIUM FLUORIDE
LiF
THERMAL CONDUCTIVITY, Watts Cm⁻¹ K⁻¹
M.P. 1121.3 K

SPECIFICATION TABLE NO. 190 THERMAL CONDUCTIVITY OF LITHIUM FLUORIDE LiF

[For Data Reported in Figure and Table No. 190]

Curve No.	Ref. No.	Method Used	Year	Temp. Range, K	Reported Error, %	Name and Specimen Designation	Composition (weight percent), Specifications and Remarks
1	28	L	1956	5.3-84			Crystalline, with traces of Mg (main inpurity in a concentration of less than 0.001%), Al, Fe, and Si; specimen 5 cm long and of roughly square cross-section of side 5.5 mm, supplied by Harshaw Chemical Co.
2	34	C	1943	378-772			Synthetic, colorless, single crystal, specimen size 1 cm cube; stainless steel rod used as standard.
3	62	L	1960	2.4-78		Harshaw LiF	Single crystal, 6.70 x 7.3 x 40 mm; cleaved specimen; annealed for 3 hrs at 1100 °K (40° below the melting point), cooled at the rate of one degree per min to 500 °K and then more slowly to room temperature.
4	62	L	1960	1.8-76		Harshaw LiF	Single crystal, 0.74 x 0.79 x 40 mm; same treatment as above.
5	62	L	1960	1.7-67		Harshaw LiF	Single crystal 0.74 x 0.79 x 40 mm; first the same treatment as above then sand-blasted and annealed.
6	62	L	1960	2.4-75		Harshaw LiF	Single crystal 6.7 x 7.3 x 40 mm; after cleaving, annealed 3 hrs at 1100 °K (40° below the melting point) and cooled at a rate of one degree per min, to 500 °K and then more slowly to room temperature, then irradiated with x-ray at room temperature to produce F centers (an electrode trapped in a halogen vacancy is an F center), density N_F (number of F centers per unit volume) = 4.2 x 10^{17} cm^{-3}.
7	62	L	1960	2.2-52		Harshaw LiF	The above specimen after successive irradiation of x-rays at room temperature until N_F = 7.1 x 10^{17} cm^{-3}.
8	62	L	1960	2.1-54		Harshaw LiF	The above specimen after successive irradiation with x-rays of 1.5 Mev at room temperature until N_F = 2 x $10^{18}$$cm^{-3}$.
9	62	L	1960	2.6-74		Harshaw LiF	The above specimen (N_F = 2 x 10^{18} cm^{-3}), annealed for 2 hrs at 570 °K, F-band partly bleached but a small band remains which peaks at around 2100 Å wavelength.
10	62	L	1960	1.7-57		Harshaw LiF	Single crystal 0.74 x 0.79 x 40 mm, similarly prepared and treated; irradiated with x-rays at 300 °K; N_F = 2.2 x 10^{17} cm^{-3}.
11	62	L	1960	2.0-50		Harshaw LiF	Single crystal 0.94 x 0.92 x 40 mm, similar to the above but unirradiated.
12	62	L	1960	2-76		Harshaw LiF	The above specimen, irradiated with 60 Kv x-rays at 77 °K; measured without warming up after the irradiation.
13	62	L	1960	5-33		Harshaw LiF	The above specimen irradiated with 60Kv x-rays at 77 °K; measured after a short warmup to 300 °K.
14	229	L	1958	3.3-80			Single crystal, long rod with a cross-section of 0.165 x 0.170 in., grown by Harshaw Chemical Co.; unirradiated.
15	229	L	1958	4.3-16			The above specimen irradiated with a Co^{60} gamma ray dose of 1.936 x 10^8 Roentgens; specimen wrapped in Aluminum foil and temperature of irradiation ~40 °C.

SPECIFICATION TABLE NO. 190 (continued)

Curve No.	Ref. No.	Method Used	Year	Temp. Range, K	Reported Error, %	Name and Specimen Designation	Composition (weight percent), Specifications and Remarks
16	229	L	1958	4.2-35			The above specimen after successive Co^{60} gamma ray dose of 4.035 x 10^8 Roentgens accumulated irradiation.
17	229	L	1958	4.5-23			The above specimen after successive Co^{60} gamma ray dose of 7.575 x 10^8 Roentgens accumulated irradiation.
18	229	L	1958	4.8-40			The above specimen after successive Co^{60} gamma ray dose of 20.885 x 10^8 Roentgens accumulated irradiation.
19	229	L	1958	5-60			Single crystal, long rod with a cross-section of 0.105 x 0.105 in., grown by Harshaw Chemical Co., irradiated with a dose of 4.36 x 10^{15} n cm^{-2} thermal neutrons; the specimen wrapped in 2 S Aluminum foil, and temperature of irradiation $\sim 60\,^{\circ}C$.
20	229	L	1958	7-20			Similar to the above specimen but irradiated at ~60 C with a dose of 2.76 x 10^{16} thermal neutrons cm^{-2}.
21	229	L	1958	8-40			Similar to the above specimen but irradiated with a dose of 5.09 x 10^{16} thermal neutrons cm^{-2}.
22	229	L	1958	6-14			Similar to the above specimen but irradiated with a dose of 1.069 x 10^{17} thermal neutrons cm^{-2}.
23	229	L	1958	5-50			Similar to the above specimen but irradiated with a dose of 5.23 x 10^{17} thermal neutrons cm^{-2}.
24	499	L	1965	1.2-3.5		4.2 P	Single crystal blank deformed at 175 C by compressing along $[0\bar{1}0]$ axis, average deformation 4.2%, rectangular rod specimen cut from the deformed blank with edge dislocation parallel to the longitudinal axis.
25	499	L	1965	1.2-3.4		4.2 S	Rectangular rod specimen cut from the above blank with edge dislocation perpendicular to the longitudinal axis.
26	499	L	1965	7.1-80		4.2 S	Similar to the above specimen.
27	456		1965	1.8-36			Cleaved specimen 0.8 x 1.0 x 40 mm supplied by Harshaw Chemical Co.; as-cleaved condition; dislocation density 5.5 x 10^5 dislocations cm^{-2}.
28	456		1965	1.8-37			Similar to the above specimen except specimen plastically deformed by bending it at room temperature over a rod 8 mm dia (this large piece LiF was annealed and slowly cooled to room temperature), specimen bent immediately after cleaving; average dislocation density 2 x 10^7 dislocations cm^{-2}.

DATA TABLE NO. 190 THERMAL CONDUCTIVITY OF LITHIUM FLUORIDE LiF

[Temperature, T, K; Thermal Conductivity, k, Watt cm^{-1} K^{-1}]

T	k
CURVE 1	
5.25	7.40
6.90	10.5
9.80	17.0
10.0	18.5
11.5	21.5
13.0	20.0
19.0	18.5
23.0	17.0
25.0	16.0
27.0	14.7
28.0	13.3
29.0	12.5
31.0	12.0
34.5	9.5
36.0	8.5
40.0	7.35
43.0	5.60
51.0	3.35
59.0	2.10
67.0	1.65
81.0	1.05
83.5	0.930
CURVE 2	
378.2	0.0256
522.2	0.0390
657.2	0.0510
772.2	0.0577
CURVE 3	
2.4	2.40
2.45	2.60
2.6	3.10
2.65	3.00
2.7	3.70
2.8	2.50
2.8	3.05
3.1	3.20
3.15	4.20
3.35	5.00
3.5	4.60
3.6	5.10
4.4	7.00
4.95	7.80
CURVE 3 (cont.)	
5.6	10.0
6.5	12.4
7.0	13.5
7.7	16.0
7.9	15.0
8.1	16.5
9.0	17.0
10.0	20.0
11.0	20.0
13.5	21.5
18.0	18.5
19.0	17.5
20.0	15.5
20.0	17.0
21.0	16.0
22.5	15.7
25.0	15
27.0	13.5
28.5	12.5
30.0	11.0
31.0	10.5
33.5	10.5
35.0	8.70
37.0	8.30
44.0	5.40
45.0	6.70
46.0	5.60
52.0	3.80
57.0	2.90
58.0	2.65
67.0	1.95
74.0	1.60
78.0	1.30
CURVE 4	
1.75	0.26
1.9	0.32
2.05	0.38
2.4	0.56
3.0	0.96
3.7	1.45
4.7	2.50
5.8	3.90
7.8	6.20
8.8	8.00
CURVE 4 (cont.)	
11	10.5
13	12.0
16.5	13.5
24	12.5
29	10.0
37	7.40
54	3.00
64	2.25
76	1.35
CURVE 5	
1.7	0.12
1.9	0.16
2.1	0.21
2.5	0.35
2.9	0.55
3.8	1.05
4.4	1.40
5.0	1.95
6.4	3.25
7.4	4.30
10.0	7.20
13.0	9.60
17.5	12.0
22.5	11.5
27.0	10.5
33.0	8.20
44.0	5.00
50.0	4.00
52.0	3.20
54.0	2.55
60.0	2.20
67.0	1.70
CURVE 6	
2.35	0.74
2.6	0.90
3.0	1.3
3.7	1.9
4.8	2.9
5.1	3.1
6.2	4.3
8.2	5.8
11.5	7.8
CURVE 6 (cont.)	
13.0	8.0
15.5	8.6
20.0	7.8
23.0	7.5
26.0	7.0
27.5	6.1
28.0	5.9
28.5	6.4
32.0	5.4
41.0	4.4
46.0	3.4
52.0	2.7
75.0	1.3
CURVE 7	
2.2	0.49
2.6	0.62
3.0	0.85
3.7	1.3
4.6	1.8
6.2	2.9
8.0	4.1
13.5	6.1
17.5	6.2
20.0	6.2
25.0	5.7
32.5	4.8
35.0	4.4
40.0	3.9
52.0	2.5
CURVE 8	
2.05	0.13
2.15	0.17
2.5	0.20
3.1	0.32
3.5	0.44
4.4	0.78
6.0	1.30
7.3	1.65
9.0	2.15
12.5	2.85
17.0	3.35
24.0	3.60
CURVE 8 (cont.)	
36.0	3.20
46.0	2.50
54.0	2.15
CURVE 9	
2.6	1.65
2.7	1.90
2.9	2.35
3.1	2.40
3.1	2.80
3.25	2.60
3.35	2.85
3.35	3.15
3.65	3.15
3.65	3.35
4.0	3.70
4.4	4.30
5.1	5.40
6.2	7.40
7.2	9.00
8.2	10.0
9.2	11.0
11.0	12.0
14.0	13.0
18.0	13.0
21.0	12.0
31.0	8.4
43.0	5.4
44.0	5.0*
54.0	2.4
74.0	1.2
CURVE 10	
1.65	0.094
1.75	0.110
1.85	0.115
2.05	0.165
2.35	0.230
2.6	0.320
3.2	0.540
4.2	0.940
5.0	1.30
6.4	2.05
8.6	3.35
CURVE 10 (cont.)	
12	5.00
15	5.90
19	6.70
26	7.00
37	5.60
42	5.00
50	3.65
57	2.60
CURVE 11	
2	0.155
5	1.70
10	6.80
20	12.5
50	3.80
CURVE 12*	
2	0.13
5	1.4
10	4.2
20	7.8
50	3.7
76	1.3
CURVE 13	
5	1.8
10	6.2
20	11.0
33	8.0
CURVE 14	
3.3	1.4
3.8	1.6
4.3	2.1
6.75	4.3
10.0	7.0
12.0	8.0
14.5	10.0
18.5	10.5
22.0	10.5
25.0	8.8
32.5	7.5
CURVE 14 (cont.)	
35.0	6.8
41.0	4.5
52.0	3.5
65.0	2.4
65.0	2.1
75.0	1.6
80.0	1.7
CURVE 15	
4.25	0.090
4.75	0.094
5.10	0.12
5.75	0.14
7.25	0.19
8.0	0.23
10.3	0.35
11.0	0.40
11.8	0.45
13.3	0.55
14.5	0.60
16.0	0.62
CURVE 16	
4.2	0.040
5.25	0.050
6.2	0.077
10.0	0.14
12.5	0.20
14.0	0.26
16.0	0.31
18.0	0.37
20.5	0.45
24.5	0.55
26.0	0.63
34.5	0.80
CURVE 17	
4.50	0.0175*
5.00	0.0210*
5.25	0.0240*
6.25	0.0300
7.50	0.0400
9.00	0.0550
10.5	0.0850

* Not shown on plot

DATA TABLE NO. 190 (continued)

T	k
CURVE 17 (cont.)	
12.5	0.110
15.0	0.140
17.0	0.180
20.0	0.200
23.0	0.270
CURVE 18	
4.75	0.0094*
5.8	0.012*
6.3	0.015*
8.5	0.024*
10.0	0.035
12.5	0.06
14.7	0.07
16.5	0.084
20.0	0.115
23.0	0.17
25.0	0.205
30.0	0.315
35.0	0.35
40.0	0.46
CURVE 19	
5.0	0.0250
10.0	0.0714
20.0	0.192
30.0	0.333
40.0	0.476
50.0	0.625
60.0	0.769
CURVE 20	
7.0	0.0119*
10.0	0.0200*
20.0	0.0667
CURVE 21	
8.0	0.0115*
10.0	0.0159*
20.0	0.0435
30.0	0.0833
40.0	0.143
CURVE 22*	
6.0	0.00625
10.0	0.0114
13.5	0.0167
CURVE 23	
5.0	0.00541*
10.0	0.0133*
20.0	0.0357
30.0	0.0854
40.0	0.100
50.0	0.152
CURVE 24	
1.23	0.0479
1.41	0.0695
1.68	0.102
2.08	0.169
2.42	0.232
2.95	0.363
3.52	0.477
CURVE 25	
1.22	0.0429
1.43	0.0605
1.65	0.0855
2.00	0.131
2.37	0.196
2.78	0.281
3.40	0.389
CURVE 26*	
7.08	1.89
7.08	2.13
8.17	2.37
9.00	2.91
10.5	3.55
11.2	4.41
12.1	4.66
13.6	5.08
15.1	6.11
15.1	6.22
16.9	7.08
CURVE 26 (cont.)*	
18.2	6.11
18.8	6.83
19.1	8.17
20.5	6.03
22.7	7.84
26.9	7.38
29.9	10.6
57.5	3.19
80.2	1.32
CURVE 27*	
1.78	0.159
2.05	0.188
2.43	0.340
2.88	0.547
3.57	0.906
4.14	1.33
4.57	1.55
5.48	2.26
7.03	3.49
8.65	5.02
10.0	6.59
10.5	6.55
12.4	8.79
12.9	7.92
14.1	9.29
15.4	9.06
17.1	9.89
17.1	10.4
18.4	10.1
22.4	10.4
22.4	10.0
26.9	9.84
31.1	8.85
36.1	7.05
CURVE 28*	
1.81	0.0601
2.07	0.0753
2.47	0.129
2.89	0.209
3.65	0.380
4.17	0.560
4.65	0.724
CURVE 28 (cont.)*	
5.48	1.00
7.05	1.69
7.43	1.72
8.77	2.48
9.98	3.54
10.6	3.45
12.5	5.06
13.1	4.84
14.2	5.90
15.2	5.90
15.7	5.90
17.3	6.98
17.3	7.41
18.8	7.10
19.7	7.53
23.8	7.80
29.4	7.62
36.9	6.27

*Not shown on plot

SPECIFICATION TABLE NO. 191 THERMAL CONDUCTIVITY OF [LITHIUM FLUORIDE + POTASSIUM FLUORIDE + ΣX_i] LiF + KF + ΣX_i

Curve No.	Ref. No.	Method Used	Year	Temp. Range, K	Reported Error, %	Name and Specimen Designation	Composition (weight percent) LiF	KF	NaF	UF_4	Composition (continued), Specifications and Remarks
1	257	C	1953	879-973		No. 14	44.5	43.5	10.9	1.1	Armco iron used as comparative material.

DATA TABLE NO. 191 THERMAL CONDUCTIVITY OF [LITHIUM FLUORIDE + POTASSIUM FLUORIDE + ΣX_i] LiF + KF + ΣX_i

[Temperature, T, K; Thermal Conductivity, k, Watt $cm^{-1}K^{-1}$]

T	k
CURVE 1*	
879.2	0.0398
902.2	0.0398
918.2	0.0433
924.2	0.0433
973.2	0.0346

* No graphical presentation

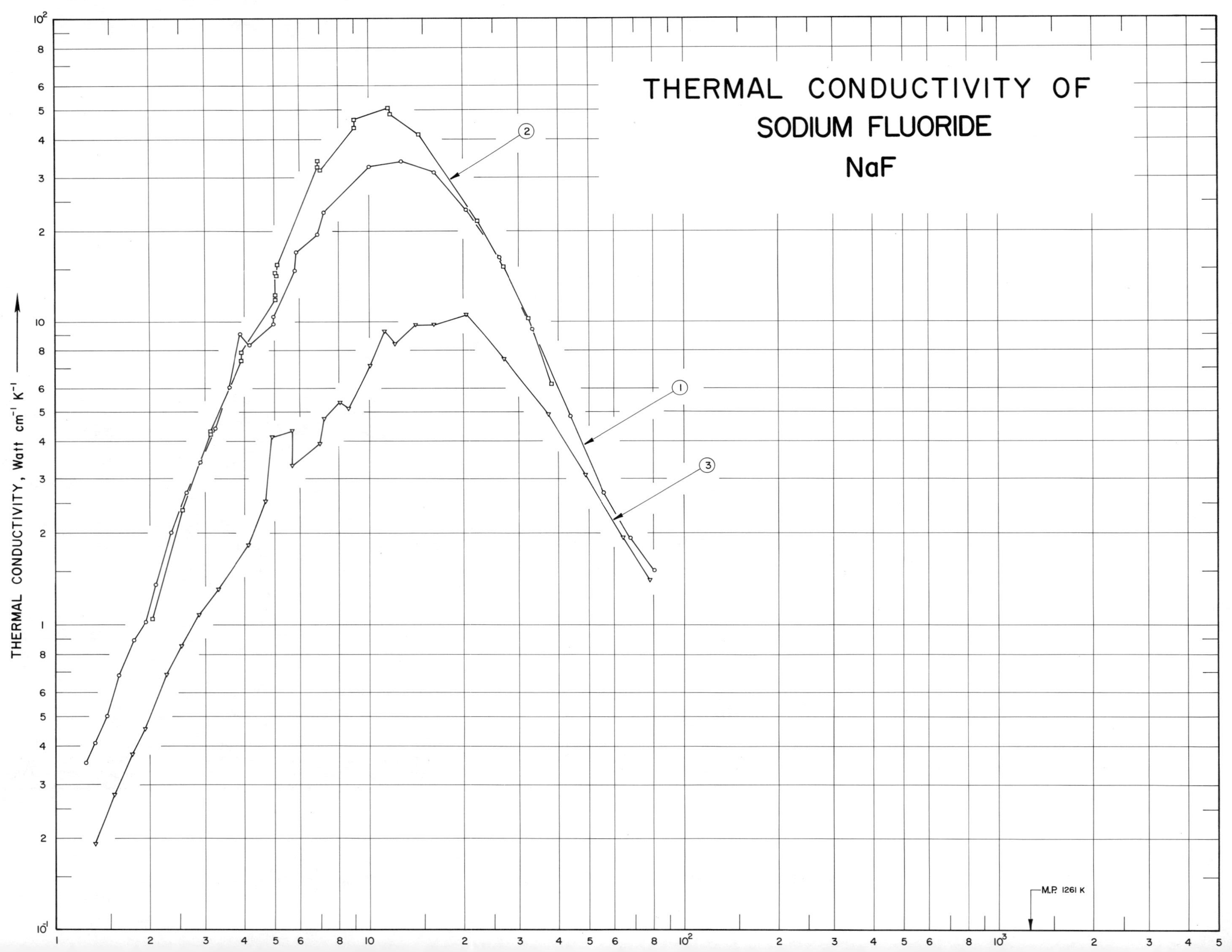
THERMAL CONDUCTIVITY OF
SODIUM FLUORIDE
NaF
M.P. 1261 K
1
2
3
THERMAL CONDUCTIVITY, Watt cm⁻¹ K⁻¹

SPECIFICATION TABLE NO. 192 THERMAL CONDUCTIVITY OF SODIUM FLUORIDE NaF

[For Data Reported in Figure and Table No. 192]

Curve No.	Ref. No.	Method Used	Year	Temp. Range, K	Reported Error, %	Name and Specimen Designation	Composition (weight percent), Specifications and Remarks
1	456		1965	1.3-81			Cleaved specimen supplied by Harshaw Chemical Co.; undeformed crystal; dislocation density 3×10^5 dislocation cm^{-2}.
2	456		1965	2.0-38			Similar to the above specimen except specimen compressed 0.3 % in length parallel to the longest dimensions between the jaws of a toolmaker's vice; dislocation density 1.8×10^6 dislocations cm^{-2}.
3	456		1965	1.3-79			Similar to the above specimen except compressed 0.7 % and dislocation density 8.3×10^6 dislocation cm^{-2}.

DATA TABLE NO. 192 THERMAL CONDUCTIVITY OF SODIUM FLUORIDE NaF

[Temperature, T, K; Thermal Conductivity, k, Watt $cm^{-1} K^{-1}$]

T	k
CURVE 1	
1.25	0.352
1.34	0.410
1.46	0.502
1.59	0.684
1.77	0.891
1.94	1.02
2.09	1.36
2.34	2.01
2.62	2.71
2.90	3.41
3.24	4.42
3.61	6.05
3.90	9.06
4.17	8.32
4.97	9.82
4.97	10.4
5.81	14.8
5.88	17.0
6.87	19.6
7.23	23.1
10.1	32.6
12.7	33.9
16.2	31.2
20.5	23.3
26.1	16.4
33.2	9.40
43.6	4.83
55.6	2.70
67.6	1.93
80.5	1.51
CURVE 2	
2.04	1.05
2.54	2.38
3.13	4.22
3.13	4.32
3.92	7.43
3.92	7.89
5.04	11.9
5.04	12.3
5.04	14.6
5.07	14.3
5.13	15.5
6.89	32.6
CURVE 2 (cont.)	
7.03	31.7
9.00	43.7
9.00	46.5
11.6	50.5
11.7	48.5
14.5	41.6
22.3	21.6
26.8	15.2
32.3	10.2
38.1	6.2
CURVE 3	
1.34	0.191
1.54	0.277
1.76	0.376
1.93	0.455
2.26	0.689
2.52	0.853
2.86	1.08
3.30	1.31
4.14	1.82
4.69	2.54
4.91	4.14
5.69	4.32
5.69	3.30
6.95	3.93
7.16	4.75
8.09	5.38
8.63	5.11
10.1	7.13
11.2	9.25
12.1	8.36
14.1	9.75
16.1	9.75
20.4	10.6
26.9	7.46
37.2	4.90
48.9	3.07
64.1	1.92
78.9	1.40

SPECIFICATION TABLE NO. 193 THERMAL CONDUCTIVITY OF [SODIUM FLUORIDE + BERYLLIUM DIFLUORIDE] $NaF + BeF_2$

Curve No.	Ref. No.	Method Used	Year	Temp. Range, K	Reported Error, %	Name and Specimen Designation	Composition (weight percent) NaF	BeF_2	Composition (continued), Specifications and Remarks
1	263	L	1952	799-801			57	43	No other information reported.

DATA TABLE NO. 193 THERMAL CONDUCTIVITY OF [SODIUM FLUORIDE + BERYLLIUM DIFLUORIDE] $NaF + BeF_2$

[Temperature, T, K; Thermal Conductivity, k, Watt $cm^{-1}K^{-1}$]

T	k
CURVE 1*	
799.2	0.0415
800.2	0.0433
801.2	0.0398

* No graphical presentation

SPECIFICATION TABLE NO. 194 THERMAL CONDUCTIVITY OF [SODIUM FLUORIDE + ZIRCONIUM TETRAFLUORIDE + ΣX_i] $NaF + ZrF_4 + \Sigma X_i$

Curve No.	Ref. No.	Method Used	Year	Temp. Range, K	Reported Error, %	Name and Specimen Designation	Composition (weight percent) NaF	ZrF_4	UF_4	Composition (continued), Specifications and Remarks
1	257	C	1953	787-1049		No. 30	50	46	4	Armco iron used as comparative material.

DATA TABLE NO. 194 THERMAL CONDUCTIVITY OF [SODIUM FLUORIDE + ZIRCONIUM TETRAFLUORIDE + ΣX_i] $NaF + ZrF_4 + \Sigma X_i$

[Temperature, T, K; Thermal Conductivity, k, Watt $cm^{-1} K^{-1}$]

T	k
CURVE 1*	
787.2	0.0294
889.2	0.0242
1049.2	0.0260

* No graphical presentation

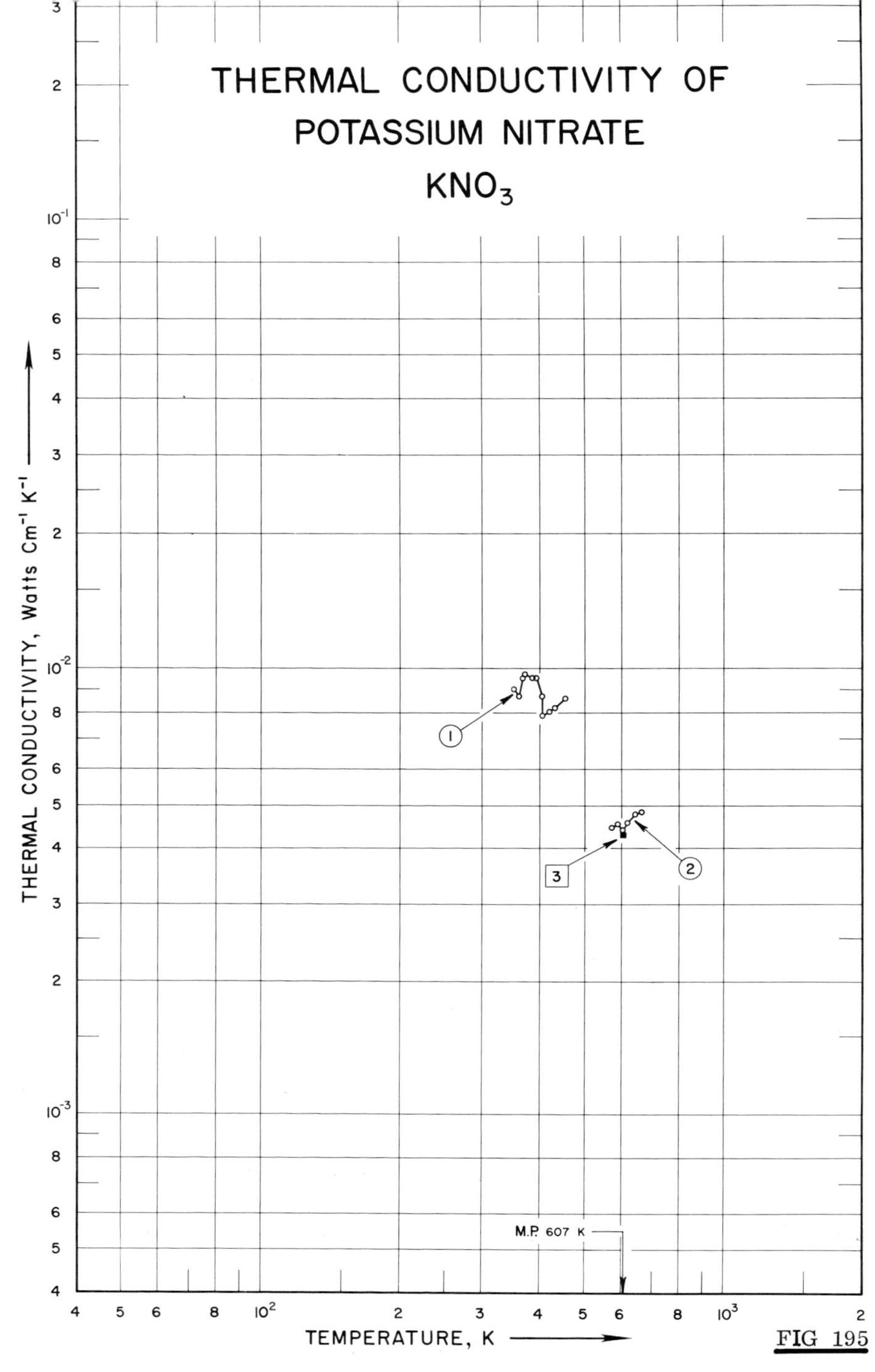
THERMAL CONDUCTIVITY OF
POTASSIUM NITRATE
KNO3
THERMAL CONDUCTIVITY, Watts Cm-1 K-1
TEMPERATURE, K
M.P. 607 K
1
2
3

FIG 195

SPECIFICATION TABLE NO. 195 THERMAL CONDUCTIVITY OF POTASSIUM NITRATE KNO_3

[For Data Reported in Figure and Table No. 195]

Curve No.	Ref. No.	Method Used	Year	Temp. Range, K	Reported Error, %	Name and Specimen Designation	Composition (weight percent), Specifications and Remarks
1	241	C	1960	353-455			Polycrystal obtained by slow cooling of the melt of potassium nitrate of special purity grade.
2	242		1961	573-668	3.0		A.R. purity; data reported as mean of 4 or 5 different measurements; measured in molten state.
3	242		1961	606.0	3.0		Same as the above specimen; measured in molten state.

DATA TABLE NO. 195 THERMAL CONDUCTIVITY OF POTASSIUM NITRATE KNO_3

[Temperature, T, K; Thermal Conductivity, k, Watt $cm^{-1}K^{-1}$]

T	k
CURVE 1	
353.2	0.0090
363.2	0.0087
368.2	0.0095
373.2	0.0097
386.2	0.0095
394.2	0.0095
403.2	0.0087
407.2	0.0079
421.2	0.0080
433.2	0.0082
455.2	0.0086
CURVE 2	
573.0	0.00448
592.0	0.00456
608.4	0.00439
620.2	0.00460
621.8	0.00464*
646.0	0.00479
667.7	0.00485
CURVE 3	
606.0	0.00431

*Not shown on plot

SPECIFICATION TABLE NO. 196 THERMAL CONDUCTIVITY OF SILVER NITRATE $AgNO_3$

Curve No.	Ref. No.	Method Used	Year	Temp. Range, K	Reported Error, %	Name and Specimen Designation	Composition (weight percent), Specifications and Remarks
1	242		1961	469-525	3		Specimen at A.R. purity; each data point is the mean value of 4 or 5 different measurements; measured in molten state.
2	242		1961	484	3		Same as the above specimen; measured in molten state.

DATA TABLE NO. 196 THERMAL CONDUCTIVITY OF SILVER NITRATE $AgNO_3$

[Temperature, T, K; Thermal Conductivity, k, Watt $cm^{-1}K^{-1}$]

T	k
CURVE 1*	
469.3	0.00447
475.4	0.00462
481.4	0.00380
487.6	0.00404
506.0	0.00426
525.4	0.00426
CURVE 2*	
484	0.00377

*No graphical presentation

SPECIFICATION TABLE NO. 197 THERMAL CONDUCTIVITY OF SODIUM NITRATE $NaNO_3$

Curve No.	Ref. No.	Method Used	Year	Temp. Range, K	Reported Error, %	Name and Specimen Designation	Composition (weight percent), Specifications and Remarks
1	242		1961	582	± 3		Specimen of A. R. purity; mean value of 4 or 5 different measurements

DATA TABLE NO. 197 THERMAL CONDUCTIVITY OF SODIUM NITRATE $NaNO_3$

[Temperature, T, K; Thermal Conductivity, k, Watt $cm^{-1}K^{-1}$]

T	k
CURVE 1*	
582	0.00565

* No graphical presentation

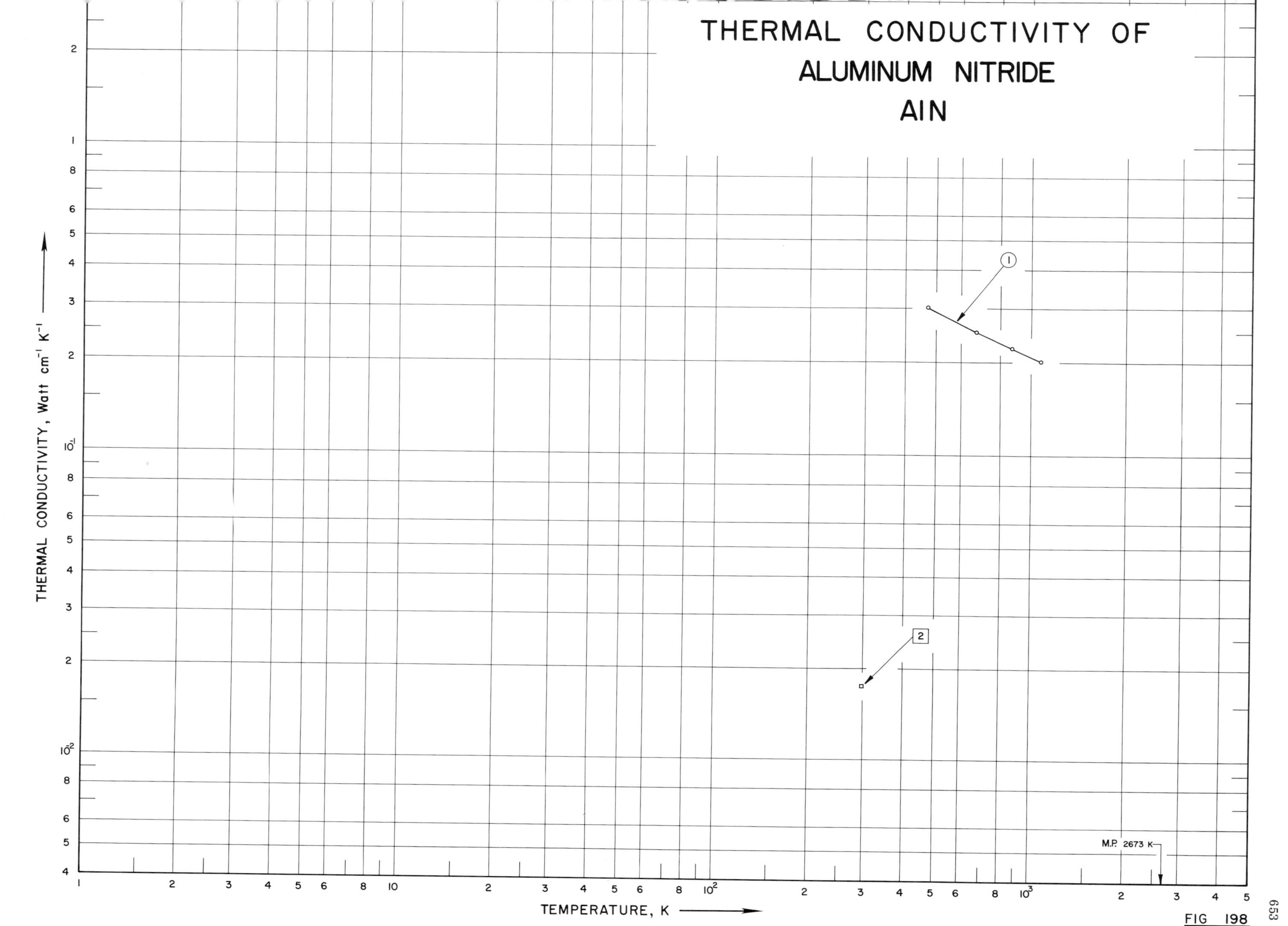

THERMAL CONDUCTIVITY OF
ALUMINUM NITRIDE
AlN
THERMAL CONDUCTIVITY, Watt cm⁻¹ K⁻¹
TEMPERATURE, K
M.P. 2673 K
1
2
FIG 198

SPECIFICATION TABLE NO. 198 THERMAL CONDUCTIVITY OF ALUMINUM NITRIDE AlN

[For Data Reported in Figure and Table No. 198]

Curve No.	Ref. No.	Method Used	Year	Temp. Range, K	Reported Error, %	Name and Specimen Designation	Composition (weight percent), Specifications and Remarks
1	500	C	1960	473-1073			64.8 Al, 32.8 N, 0.2 C, 0.4 Si, 0.1 Fe; fabricated by hot pressing the milled powder at 2273 K in graphite dies, applying pressures of about 5000 psi, test bars 3 x 0.5 x 0.25 in. out from the hot-pressed pieces with a diamond wheel and ground to produce smooth parallel surfaces; bulk density 3.20 g cm^{-3}, (density of the powder 3.23 g cm^{-3}); particle size 0.5 to 25 μ; heat flow parallel to direction of pressing; measured in stagnant nitrogen atmosphere; Inconel used as comparative material.
2	501	P	1959	298.2			High purity; highly sintered.

DATA TABLE NO. 198 THERMAL CONDUCTIVITY OF ALUMINUM NITRIDE AlN

[Temperature, T, K; Thermal Conductivity, k, Watt $cm^{-1} K^{-1}$]

T	k
CURVE 1	
473.2	0.301
673.2	0.251
873.2	0.222
1073.2	0.201
CURVE 2	
298.2	0.0176

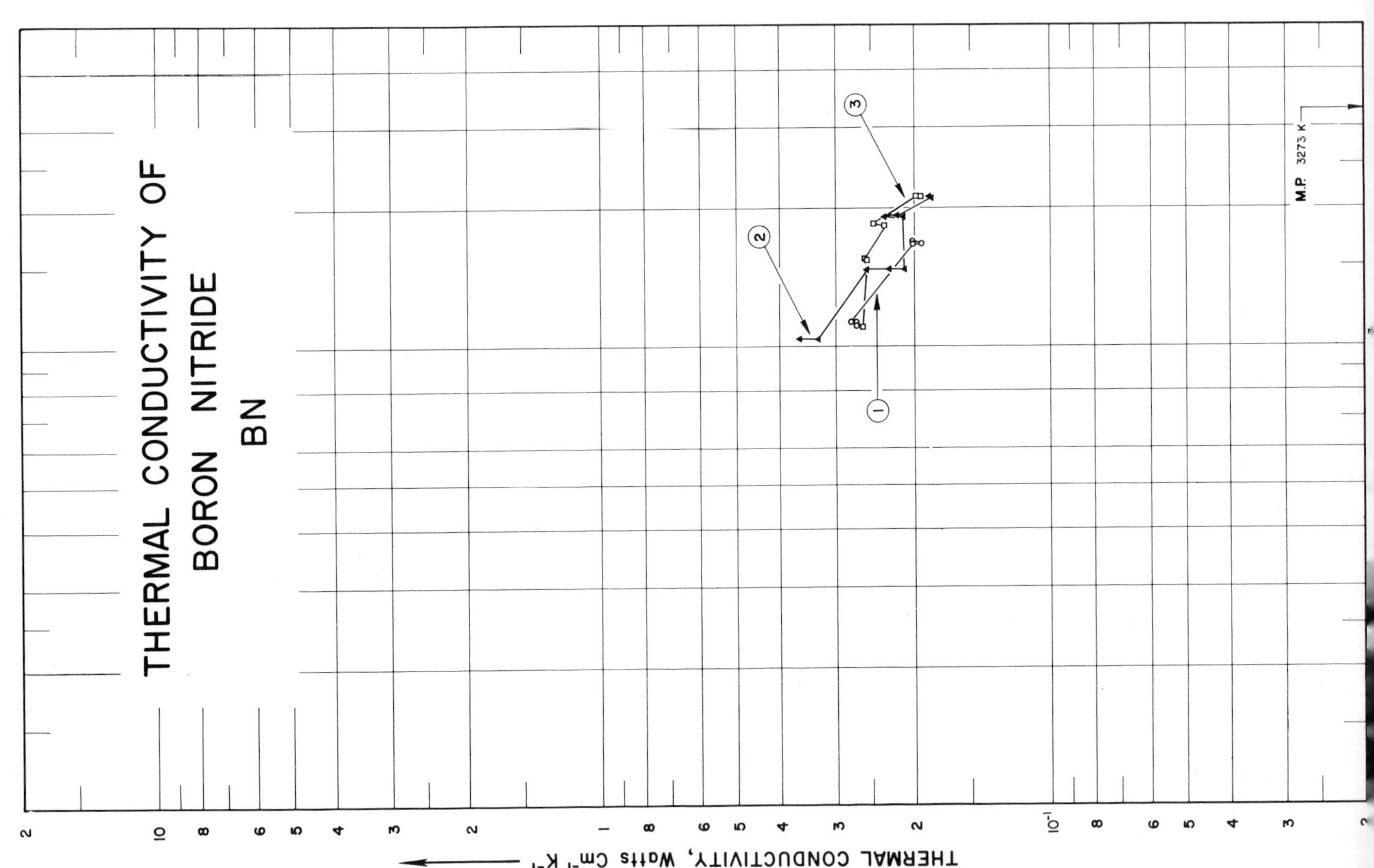
THERMAL CONDUCTIVITY OF
BORON NITRIDE
BN
THERMAL CONDUCTIVITY, Watts Cm^{-1} K^{-1}
M.P. 3273 K
1
2
3

SPECIFICATION TABLE NO. 199 THERMAL CONDUCTIVITY OF BORON NITRIDE BN

[For Data Reported in Figure and Table No. 199]

Curve No.	Ref. No.	Method Used	Year	Temp. Range, K	Reported Error, %	Name and Specimen Designation	Composition (weight percent), Specifications and Remarks
1	144	R	1963	1112-1697		1	Specimen 0.75 in. long, 0.75 in. O.D., and 0.25 in. I.D.; surface scratches eliminated by grinding and polishing.
2	144	R	1963	1047-2114		1	Second run of the above specimen.
3	144	R	1963	1103-2129		2	Similar to the above specimen.

DATA TABLE NO. 199 THERMAL CONDUCTIVITY OF BORON NITRIDE BN

[Temperature, T, K; Thermal Conductivity, k, Watt $cm^{-1}K^{-1}$]

T	k
CURVE 1	
1112.1	0.270
1130.4	0.270
1130.4	0.279
1670.9	0.201
1679.3	0.193
1696.5	0.202
CURVE 2	
1047.1	0.362
1047.1	0.329
1474.8	0.256
1475.4	0.227
1488.7	0.210
1910.4	0.212
1917.1	0.234
1928.2	0.219
1933.2	0.233*
2108.7	0.183
2110.9	0.185
2112.1	0.184*
2114.3	0.185*
CURVE 3	
1102.6	0.262
1102.6	0.260*
1540.4	0.256
1542.6	0.254*
1549.3	0.258
1834.3	0.233
1850.4	0.247
2120.4	0.194
2125.9	0.194*
2129.3	0.198

* Not shown on Plot

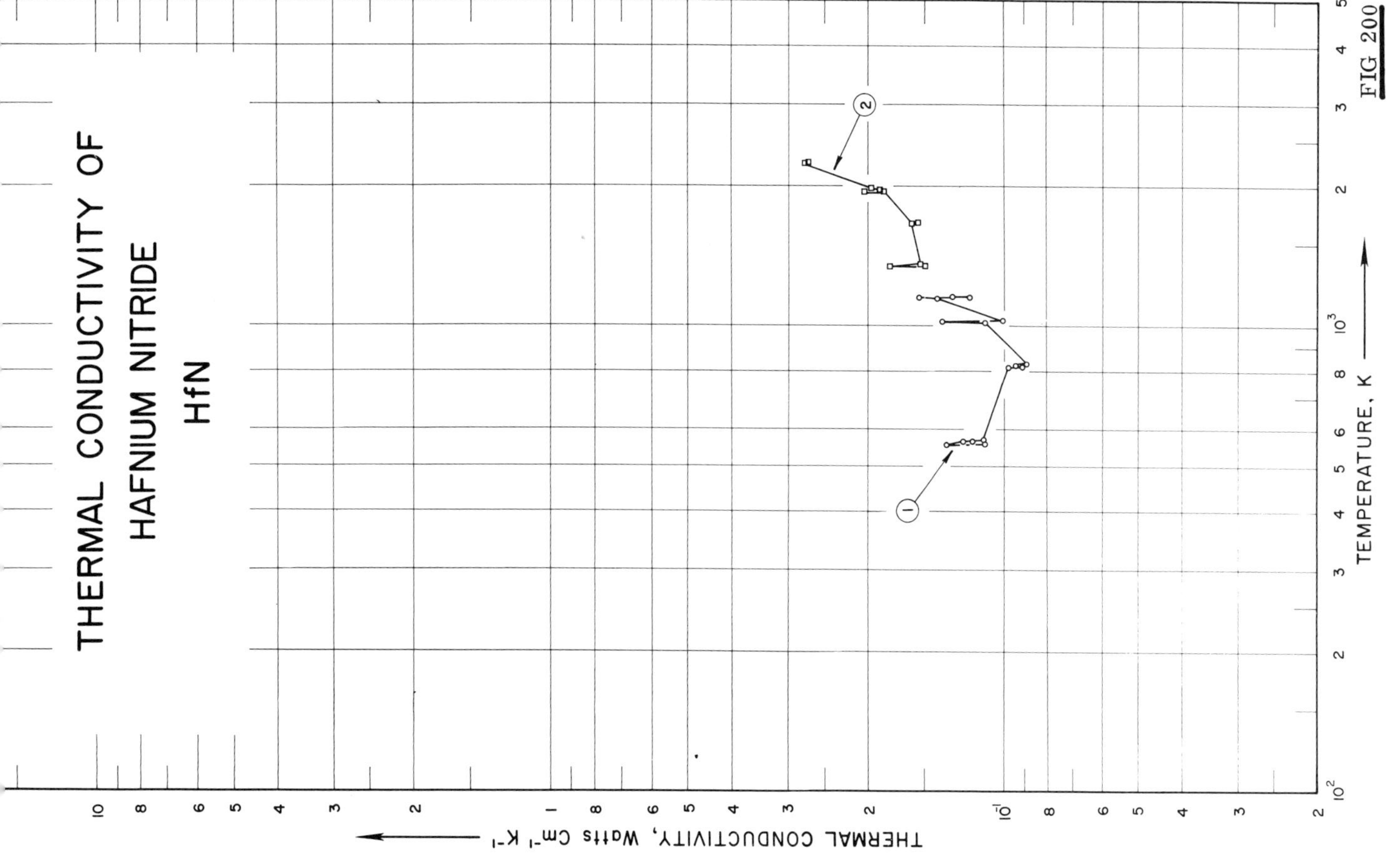
THERMAL CONDUCTIVITY OF
HAFNIUM NITRIDE
HfN
THERMAL CONDUCTIVITY, Watts Cm⁻¹ K⁻¹
TEMPERATURE, K
1
2

FIG 200

SPECIFICATION TABLE NO. 200 THERMAL CONDUCTIVITY OF HAFNIUM NITRIDE HfN

[For Data Reported in Figure and Table No. 200]

Curve No.	Ref. No.	Method Used	Year	Temp. Range, K	Reported Error, %	Name and Specimen Designation	Composition (weight percent), Specifications and Remarks
1	243	R	1962	557-1149	~3		95.4 Hf, 6.61 N_2, and 0.9 O_2 (wet analysis); hot-pressed to 3867 K; supplied by Carborundum Co.; density 10.89 g cm^{-3}.
2	243	R	1962	1336-2232	~3		Second run of the above specimen; fractured during run.

DATA TABLE NO. 200 THERMAL CONDUCTIVITY OF HAFNIUM NITRIDE HfN

[Temperature, T, K; Thermal Conductivity, k, Watt $cm^{-1}K^{-1}$]

T	k
CURVE 1	
557.1	0.111
558.7	0.135
562.6	0.124
562.6	0.118
563.7	0.111
808.7	0.0987
811.5	0.0916
813.2	0.0948
817.1	0.0959
817.1	0.0959*
819.3	0.0920*
821.5	0.0896
1015.4	0.111
1016.5	0.138
1017.1	0.101
1144.8	0.141
1145.4	0.131
1149.3	0.155
1149.3	0.120
CURVE 2	
1335.9	0.150
1349.8	0.180
1349.8	0.153
1661.0	0.160
1661.0	0.156
1666.5	0.160*
1669.3	0.160*
1933.2	0.185
1933.2	0.205
1949.8	0.189
1961.0	0.198
2227.6	0.278
2227.6	0.273
2231.8	0.273*
2231.8	0.273*

* Not shown on Plot

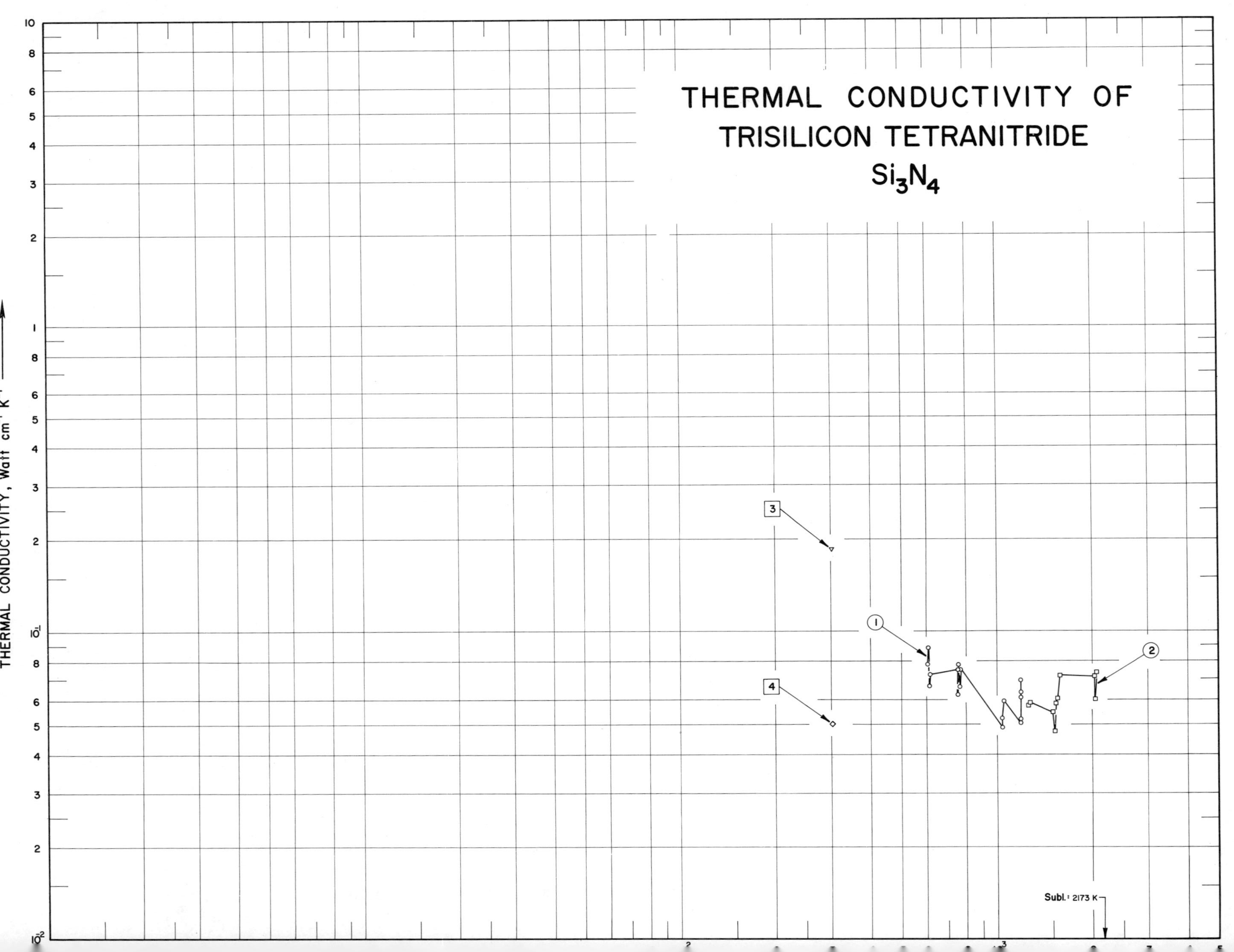
THERMAL CONDUCTIVITY OF
TRISILICON TETRANITRIDE
Si_3N_4
THERMAL CONDUCTIVITY, Watt cm^{-1} K^{-1}
Subl.: 2173 K
1
2
3
4

SPECIFICATION TABLE NO. 201 THERMAL CONDUCTIVITY OF TRISILICON TETRANITRIDE Si_3N_4

[For Data Reported in Figure and Table No. 201]

Curve No.	Ref. No.	Method Used	Year	Temp. Range, K	Reported Error, %	Name and Specimen Designation	Composition (weight percent), Specifications and Remarks
1	243	R	1962	603-1195	2.5		Impurities 0.05 Ca, 0.01 Cu, 0.01 Mg, 0.3 Al, 1.5 Fe, 0.01 Ti and trace Be, Na, and Mn; cast; specimen 0.75 in. O.D., 0.25 in. I.D., 0.75 in. long; density 2.38 g cm^{-3}.
2	243	R	1962	1267-2061	2.5		Second run of the above specimen; melted during run.
3	384	C	1962	303			Density 3.16 g cm^{-3}; data read from calibration of a direct-reading thermal comparator.
4	384	C	1962	303			Data for specimen of density 2.34 g cm^{-3}.

DATA TABLE NO. 201 THERMAL CONDUCTIVITY OF TRISILICON TETRANITRIDE Si_3N_4

[Temperature, T, K; Thermal Conductivity, k, Watt $cm^{-1}K^{-1}$]

T	k
CURVE 1	
603.2	0.0786
607.1	0.0887
611.5	0.0663
613.2	0.0727
753.2	0.0750
755.4	0.0623
757.1	0.0779
763.7	0.0779*
764.3	0.0661
769.3	0.0750
1045.4	0.0490
1054.3	0.0525
1063.2	0.0596
1064.3	0.0594*
1190.4	0.0503
1192.6	0.0522
1193.7	0.0617
1194.3	0.0636
1194.8	0.0698
CURVE 2	
1266.5	0.0575
1272.1	0.0584
1294.3	0.0591*
1511.0	0.0548
1527.6	0.0476
1547.1	0.0584
1555.4	0.0606
1580.6	0.0721
2044.3	0.0715
2047.1	0.0600
2061.0	0.0734
CURVE 3	
303.0	0.185
CURVE 4	
303.0	0.050

* Not shown on Plot

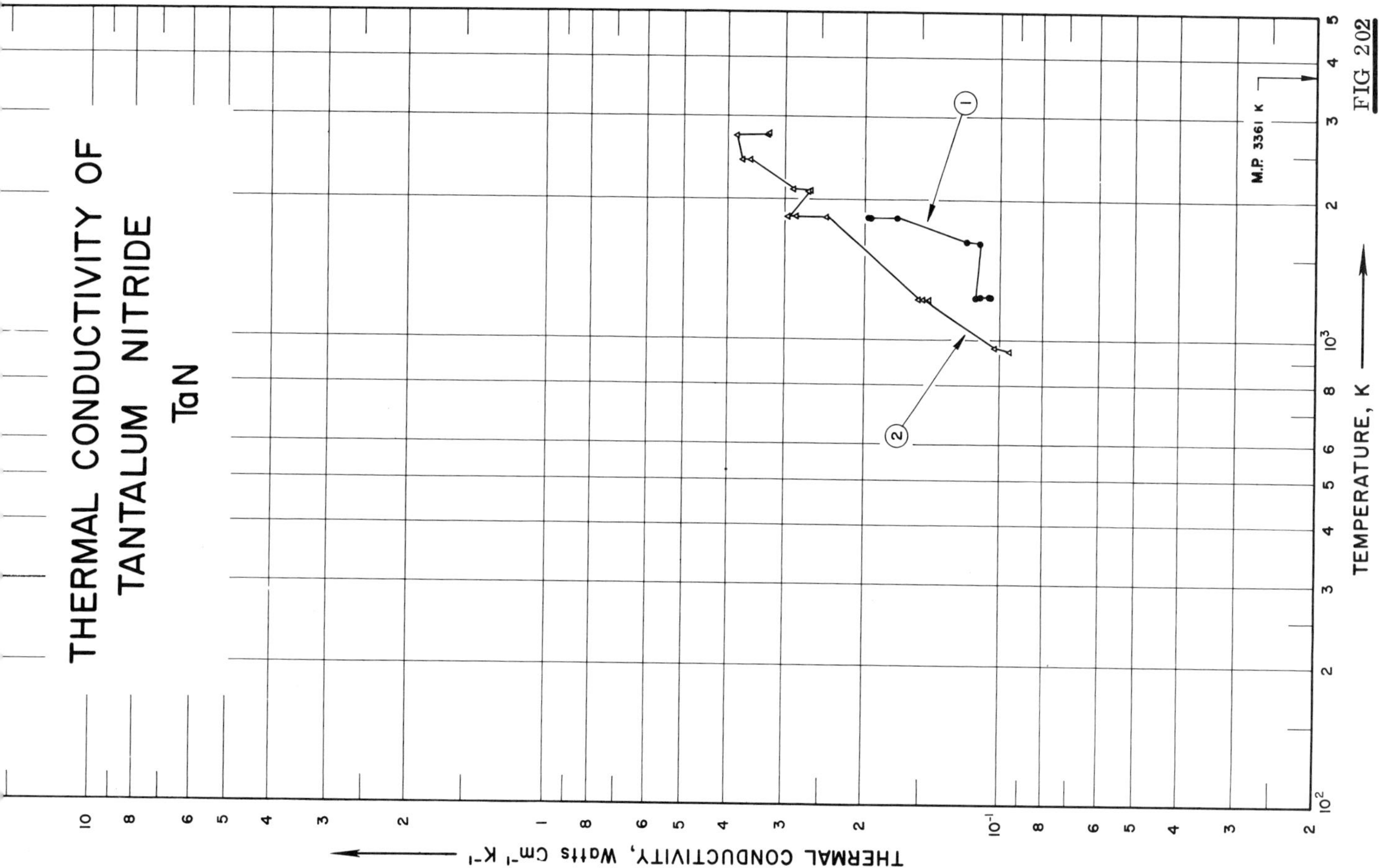
THERMAL CONDUCTIVITY OF
TANTALUM NITRIDE
TaN
M.P. 3361 K
TEMPERATURE, K
THERMAL CONDUCTIVITY, Watts Cm⁻¹ K⁻¹

FIG 202

SPECIFICATION TABLE NO. 202 THERMAL CONDUCTIVITY OF TANTALUM NITRIDE TaN

[For Data Reported in Figure and Table No. 202]

Curve No.	Ref. No.	Method Used	Year	Temp. Range, K	Reported Error, %	Name and Specimen Designation	Composition (weight percent), Specifications and Remarks
1	144	R	1963	1237-1836	~6		Specimen 0.75 in. O.D., 0.25 in. I.D., and 0.75 in. long; ground and polished to eliminate surface scratches; heat soaked at 1422 K; broke during run.
2	144	R	1963	948-2765	~6		Similar to the above specimen but not heat soaked; broke during run.

DATA TABLE NO. 202 THERMAL CONDUCTIVITY OF TANTALUM NITRIDE TaN

[Temperature, T, K; Thermal Conductivity, k, Watt $cm^{-1}K^{-1}$]

T	k
CURVE 1	
1237.1	0.114
1238.7	0.111
1238.7	0.112*
1238.7	0.107
1238.7	0.106
1610.9	0.111
1610.9	0.111*
1612.6	0.119
1820.9	0.170
1832.1	0.195
1835.9	0.196
CURVE 2	
948.2	0.0956
962.6	0.103
1223.7	0.149
1224.8	0.144
1224.8	0.152
1224.8	0.151*
1837.1	0.243
1842.1	0.285
1843.7	0.296
2077.6	0.266
2083.2	0.265
2089.8	0.287
2404.8	0.356
2408.2	0.356*
2414.3	0.372
2739.3	0.382
2750.9	0.329
2765.3	0.328

* Not shown on Plot

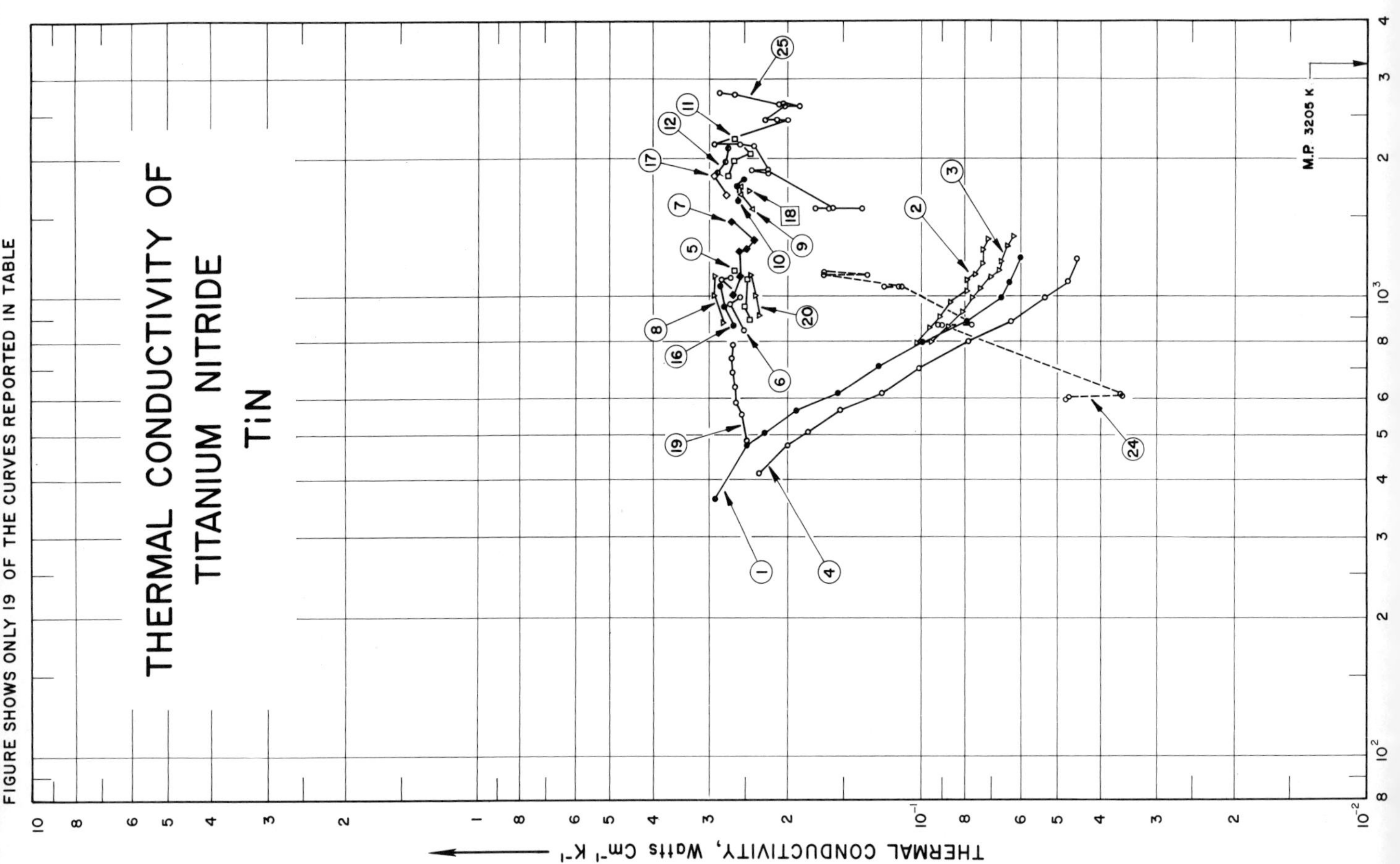
FIGURE SHOWS ONLY 19 OF THE CURVES REPORTED IN TABLE
THERMAL CONDUCTIVITY OF
TITANIUM NITRIDE
TiN
THERMAL CONDUCTIVITY, Watts Cm⁻¹ K⁻¹
M.P. 3205 K

SPECIFICATION TABLE NO. 203 THERMAL CONDUCTIVITY OF TITANIUM NITRIDE TiN

[For Data Reported in Figure and Table No. 203]

Curve No.	Ref. No.	Method Used	Year	Temp. Range, K	Reported Error, %	Name and Specimen Designation	Composition (weight percent), Specifications and Remarks
1	14	C	1953	363-1218			Supplier's analysis: initial composition 77.5 Ti, 18.0 N, 1.0 Ca, 0.2 H, 0.3 C, 0.1 Fe, and 0.1 SiO_2; (after firing) 77.8 Ti, 18.2 N, and 2.6 O; 1 in. cubic specimen, hydrostatically pressed; fired at 2100 C; total porosity 19.0%; firing shrinkage 9%; dense sintered alumina used as comparative material.
2	14	R	1953	793-1343			The above composition; ellipsoidal specimen; slip cast and hydrostatically pressed; fired at 2100 C; total porosity 19.8%; shrinkage: pressing 6%, firing 9%.
3	14	R	1953	798-1353			The second run of the above specimen.
4	45	C	1953	413-1205			Cubic specimen; hydrostatically pressed;fired at 2100 C; measured in vacuo.
5	270,413, 414	R	1962	884-1103		1	76.5 Ti, 17.7 N, <0.5 other metals and carbon; specimen 2 in. O.D.,0.5 in. I.D. and 1.5 in. long; supplied by General Astrometals Corp.; single phase; average grain size 11 μ; density 4.91 g cm^{-3}; porosity 10%; measured in helium atmosphere.
6	270,413, 414	R	1962	840-1100		1	Second run of the above specimen.
7	270,413, 414	R	1962	1010-1455		1	Third run of the above specimen.
8	270,413, 414	R	1962	878-1107		1	Fourth run of the above specimen.
9	270,413, 414	R	1962	1558-1732		1	Fifth run of the above specimen.
10	270,413, 414	R	1962	1611-1800		1	Sixth run of the above specimen.
11	270,413, 414	R	1962	1833-2205		1	Seventh run of the above specimen.
12	270,413, 414	R	1962	1723-2107		1	Eighth run of the above specimen.
13	270,413, 414	R	1962	882-1190		2	77.9 Ti, 17.9 N, <0.9 other metals and carbon; single phase; specimen 2 in. O.D., 0.5 in. I.D. and 1.5 in. long; supplied by General Astrometals Corp; average grain size 11μ; density 4.78 g cm^{-3}; porosity 12%; measured in helium atmosphere.
14	270,413, 414	R	1962	1782.2		2	Second run of the above specimen.
15	270,413, 414	R	1962	964,1127		2	Third run of the above specimen.
16	270,413, 414	R	1962	862-1062		2	Fourth run of the above specimen.

SPECIFICATION TABLE NO. 203 (continued)

Curve No.	Ref. No.	Method Used	Year	Temp. Range, K	Reported Error, %	Name and Specimen Designation	Composition (weight percent), Specifications and Remarks
17	270, 413, 414	R	1962	1671, 1828		2	Fifth run of the above specimen.
18	270, 413, 414	R	1962	1698.2		2	Sixth run of the above specimen.
19	270, 413, 414	R	1962	483-783		2	Seventh run of the above specimen.
20	270, 413, 414	R	1962	904-1115		3	No details reported.
21	270, 413, 414	R	1962	899.2		3	Second run of the above specimen.
22	270, 413, 414	R	1962	892-1130		3	Third run of the above specimen.
23	270, 413, 414	R	1962	888-1052		3	Fourth run of the above specimen.
24	243	R	1962	594-1119	~3		Specimen 0.75 in. O.D., 0.25 in. I.D., and 0.75 in. long; hot pressed; density 4.08 g cm^{-3}.
25	243	R	1962	1553-2769	~3		Second run of the above specimen; melted during measurement.
26	258, 413	P	1963	473-773		2	77.9 Ti, 17.9 N, < 0.9 other metals, remainder probably oxygen tied up as TiO_2 (composition after measurements); single phase; average grain size 11 μ; specimen 0.250 in. dia, 0.100 in. thick; cut from specimen 2, density 4.78 g cm^{-3}; thermal conductivity data calculated from measured diffusivity values (with flash technique) and specific heat data of Naylor, B. F. (J. Am. Ceram. Soc., 68(3), 370-1, 1946)

DATA TABLE NO. 203 THERMAL CONDUCTIVITY OF TITANIUM NITRIDE TiN

[Temperature, T, K; Thermal Conductivity, k, Watts $cm^{-1}K^{-1}$]

T	k
CURVE 1	
363.2	0.291
473.2	0.248
503.2	0.227
563.2	0.192
616.2	0.155
703.2	0.126
796.2	0.100
883.2	0.0795
991.2	0.0669
1075.2	0.0640
1218.2	0.0602
CURVE 2	
793.2	0.103
851.2	0.0962
903.2	0.0912
973.2	0.0866
1033.2	0.0795
1093.2	0.0795
1123.2	0.0761
1181.2	0.0732
1273.2	0.0732
1343.2	0.0711
CURVE 3	
798.2	0.0954
863.2	0.0879
923.2	0.0816
993.2	0.0778
1043.2	0.0741
1103.2	0.0707
1143.2	0.0674
1199.2	0.0669
1291.2	0.0649
1353.2	0.0628
CURVE 4	
413.2	0.234
473.2	0.201
505.2	0.181
565.2	0.154
CURVE 4 (cont.)	
613.2	0.124
697.2	0.102
797.2	0.0791
881.2	0.0636
993.2	0.0536
1078.8	0.0477
1205.2	0.0452
CURVE 5	
884.2	0.224
941.2	0.250
948.2	0.250*
1090.2	0.246
1103.2	0.263
CURVE 6	
840.2	0.251
963.2	0.269
997.2	0.256
1082.2	0.280
1100.2	0.269
CURVE 7	
1010.2	0.263
1118.2	0.254
1251.2	0.256
1270.2	0.248
1337.2	0.238
1455.2	0.267
CURVE 8	
878.2	0.278
1004.2	0.292
1107.2	0.291
CURVE 9	
1558.2	0.241
1654.2	0.255
1732.2	0.255
CURVE 10	
1611.2	0.259
1740.2	0.258
1800.2	0.251
CURVE 11	
1833.2	0.271
1986.2	0.264
2044.2	0.244
2205.2	0.262
CURVE 12	
1723.2	0.259*
1860.2	0.289
1964.2	0.276
1978.2	0.278*
2107.2	0.273
CURVE 13*	
882.2	0.279
983.2	0.257
1076.2	0.267
1190.2	0.273
CURVE 14*	
1782.2	0.254
CURVE 15*	
964.2	0.279
1127.2	0.274
CURVE 16	
862.2	0.264
945.2	0.276
1062.2	0.283
CURVE 17	
1671.2	0.274
1828.2	0.293
CURVE 18	
1698.2	0.243
CURVE 19	
483.2	0.249
553.2	0.255
583.2	0.262
633.2	0.264
683.2	0.266
733.2	0.268
783.2	0.266
CURVE 20	
904.2	0.232
1008.2	0.237
1115.2	0.245
CURVE 21*	
899.2	0.263
CURVE 22*	
892.2	0.242
1034.2	0.257
1130.2	0.253
CURVE 23*	
888.2	0.233
1009.2	0.247
1052.2	0.251
CURVE 24	
594.3	0.0479
600.4	0.0470
604.3	0.0359
612.6	0.0366
874.3	0.0923
874.8	0.0897
874.8	0.0776
1050.9	0.112
1052.1	0.123
CURVE 24	
1052.1	0.113
1117.6	0.167
1117.6	0.134
1119.3	0.169
CURVE 25	
1552.6	0.137
1552.6	0.163
1552.6	0.175
1552.6	0.160
1861.0	0.222
1869.3	0.242
1869.3	0.224
1872.1	0.226*
2127.6	0.238
2141.5	0.257
2144.3	0.265
2149.8	0.294
2416.5	0.200
2427.6	0.212
2436.1	0.226
2597.1	0.202
2613.7	0.186
2622.1	0.213
2622.1	0.209
2763.7	0.264
2769.3	0.284
CURVE 26*	
473.2	0.249
523.2	0.254
573.2	0.259
623.2	0.262
673.2	0.264
723.2	0.266
773.2	0.267

*Not shown on plot

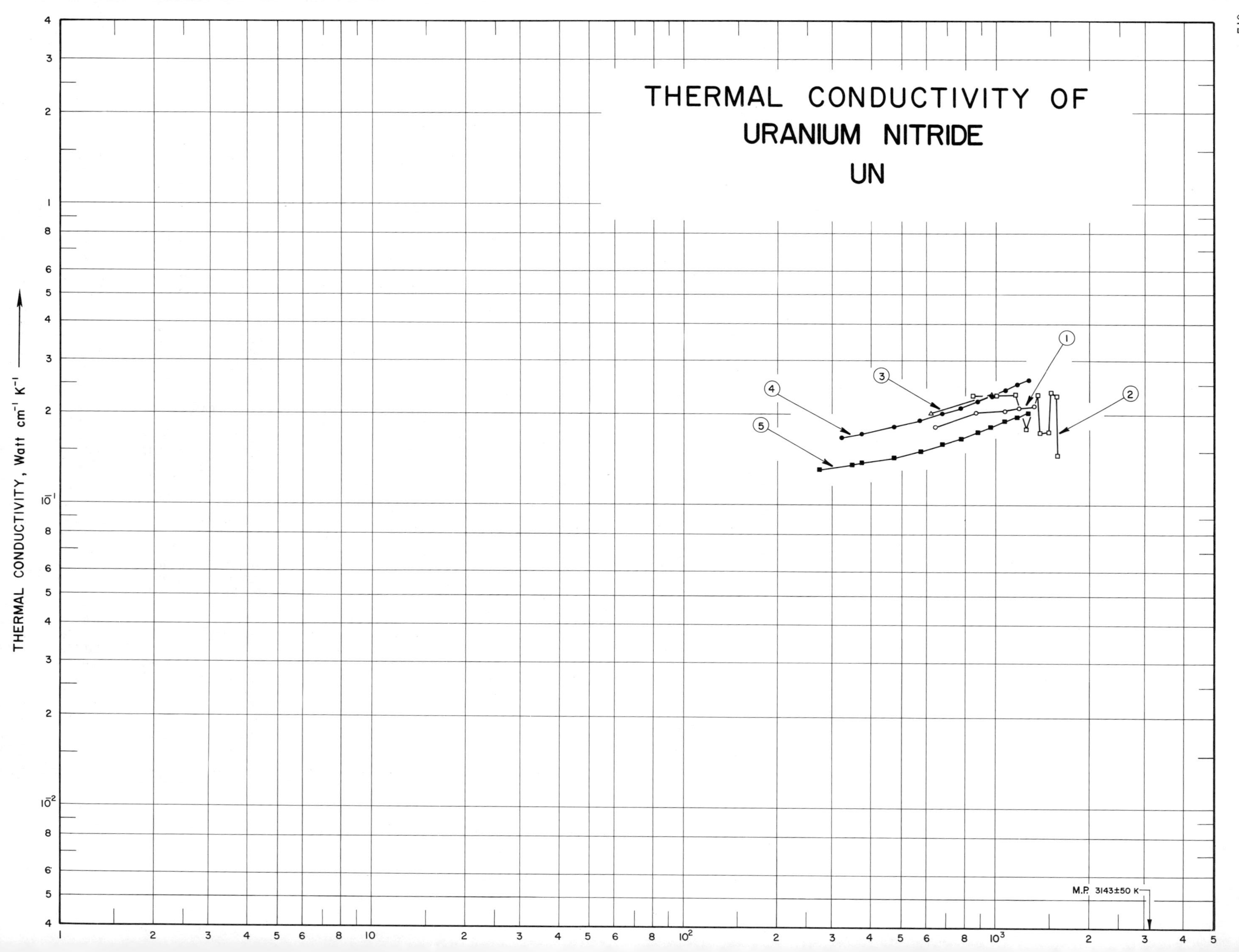
THERMAL CONDUCTIVITY OF
URANIUM NITRIDE
UN
THERMAL CONDUCTIVITY, Watt cm^{-1} K^{-1}
M.P. 3143±50 K
1
2
3
4
5

SPECIFICATION TABLE NO. 204 THERMAL CONDUCTIVITY OF URANIUM NITRIDE UN

[For Data Reported in Figure and Table No. 204]

Curve No.	Ref. No.	Method Used	Year	Temp. Range, K	Reported Error, %	Name and Specimen Designation	Composition (weight percent), Specifications and Remarks
1	502	P	1963	640-1333			95 to 98% dense, prepared by hot pressing.
2	502	P	1963	845-1583			The above specimen, second run.
3	502	P	1963	620, 976			95 to 98% dense, prepared by hot pressing.
4	503	C	1964	323-1273			5.38 N, 0.047 C, 0.0055 O, 0.0025 Si, <0.002 Zn, <0.001 each Al, Fe, and K, <0.006 P, <0.0002 each Ni, Mg, Bi, and Pb, <0.0008 Nb, <0.0005 Sn, Cr, <0.0003 V, Ca, 0.0003 Mo, <0.0001 Mn, 0.0001 each B, Cu, trace Cd and Li; dense, stoichiometric, polycrystalline; prepared by the slow reaction consumable electrode arc-melting method; ground into a 0.5 in. dia cylinder 11/16 in. long and metallographically polished on both ends; measurements made under a vacuum of 4 x 10^{-6} torr; metallographic structure after the measurements free from cracks and contaminations at grain boundaries; stainless steel used as comparative material.
5	504		1964	273-1273			94.5% dense UN; specimen 1 in. in dia and 0.25 in. thick; the thermal conductivity measured values corrected to theoretical density.

DATA TABLE NO. 204 THERMAL CONDUCTIVITY OF URANIUM NITRIDE UN

[Temperature, T, K; Thermal Conductivity, k, Watt $cm^{-1}K^{-1}$]

T	k
CURVE 1	
640.2	0.181
863.2	0.203
1074.2	0.206
1193.2	0.210
1333.2	0.214
CURVE 2*	
845.2	0.232
1001.2	0.232
1165.2	0.234
1251.2	0.179
1251.2	0.179
1369.2	0.233
1394.2	0.174
1492.2	0.175
1510.2	0.237
1573.2	0.230
1583.2	0.147
CURVE 3	
620.2	0.202
976.2	0.233
CURVE 4	
323.2	0.167
373.2	0.172
473.2	0.182
573.2	0.191
673.2	0.201
773.2	0.211
873.2	0.221
973.2	0.231
1073.2	0.241
1173.2	0.251
1273.2	0.260
CURVE 5	
273.2	0.131
348.2	0.136
373.2	0.138
473.2	0.144
CURVE 5 (cont.)	
573.2	0.151
673.2	0.159
773.2	0.166
873.2	0.174
973.2	0.182
1073	0.190
1173	0.196
1273	0.203

*Not shown on plot

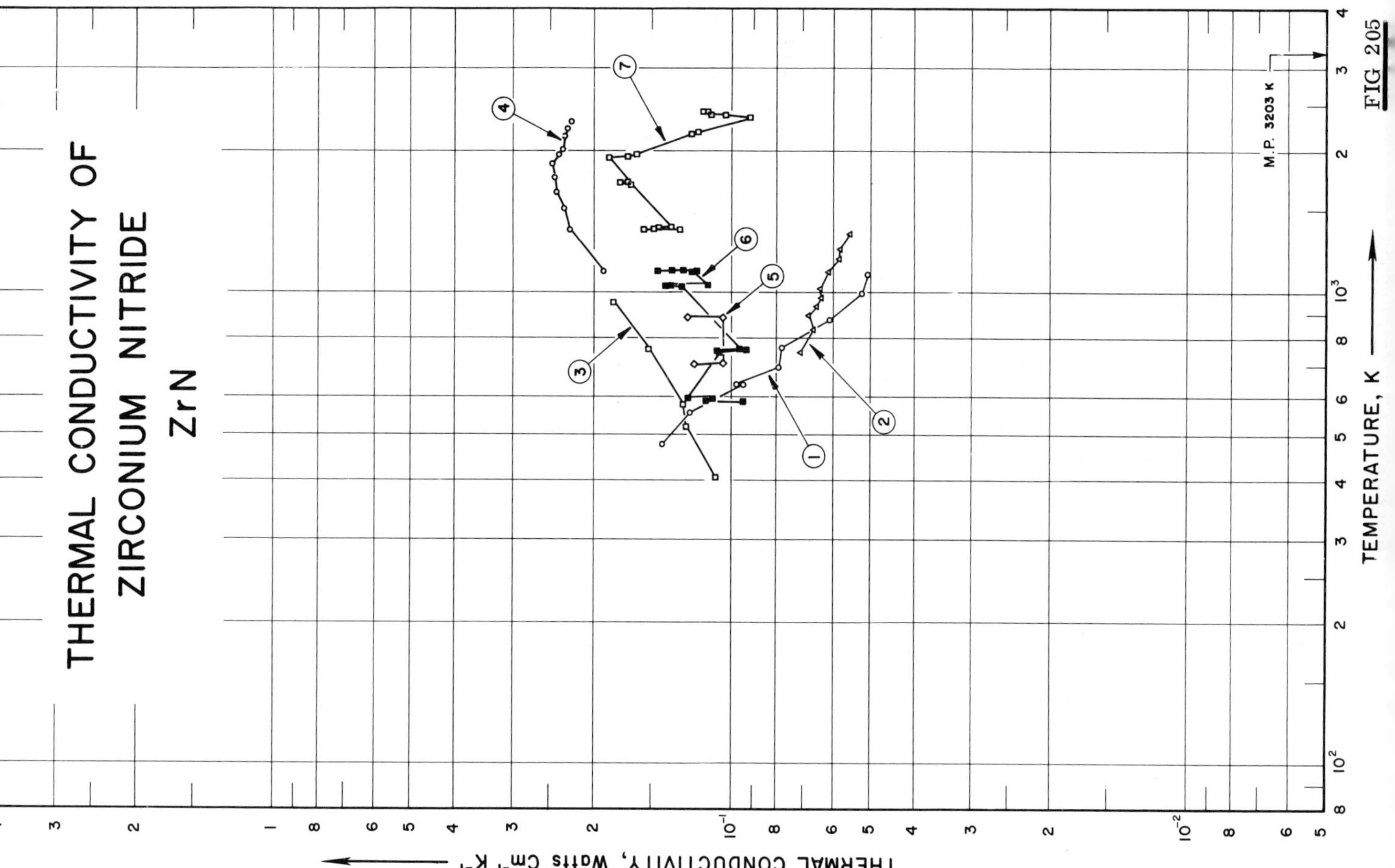
THERMAL CONDUCTIVITY OF
ZIRCONIUM NITRIDE
ZrN
THERMAL CONDUCTIVITY, Watts Cm⁻¹ K⁻¹
TEMPERATURE, K
M.P. 3203 K
1
2
3
4
5
6
7

FIG 205

SPECIFICATION TABLE NO. 205 THERMAL CONDUCTIVITY OF ZIRCONIUM NITRIDE ZrN

[For Data Reported in Figure and Table No. 205]

Curve No.	Ref. No.	Method Used	Year	Temp. Range, K	Reported Error, %	Name and Specimen Designation	Composition (weight percent), Specifications and Remarks
1	14	C	1953	478-1093			Initial composition (supplier's) 9.8 N, 0.12 C, 0.7 SiO_2, and 0.03 H; after firing 81.8 Zr, 8.9 N, and 5.2 O; 1 in cubic specimen; hydrostatically pressed; fired at 2000 C; porosity 19.3% ; total shrinkage 10% ; dense sintered alumina used as comparative material.
2	14	R	1953	748-1341			Initial composition (supplier's) 9.8 N, 0.12 C, 0.7 SiO_2, and 0.03 H; after firing 81.8 Zr, 8.9 N, and 5.2 O; ellipsoidal specimen; slip-cast and hydrostatically pressed; fired at 2000 C; porosity 19.6% ; linear shrinkage: pressing 5% , firing 10% .
3	251	C	1963	408-955	±4		84.6 Zr, 13.5 N, 0.2 Fe, 0.8 H, 0.4 Si, and 0.5 alkali metal oxides; supplied by Norton Co. ; specimen 2 in. dia, 1 in thick; hot-pressed and fired at 2373 K; density 6.50 g cm^{-3}; measured in a helium atmosphere; Armco iron used as comparative material.
4	251	P	1963	1117-2308	±4		The above specimen measured by another method; thermal conductivity values calculated from measured data of thermal diffusivity, specific heat, and density.
5	243	R	1962	704-889	~3		Specimen 0.75 in. O.D., and 0.25 in. I.D., and 0.75 in. long; pressed and sintered; supplied by General Electric Co. ; density 6.84 g cm^{-3}; deteriorated at 2928 K
6	243	R	1962	589-1113	~3		Second run of the above specimen.
7	243	R	1962	1361-2405	~3		Third run of the above specimen; fractured during run

DATA TABLE NO. 205 THERMAL CONDUCTIVITY OF ZIRCONIUM NITRIDE ZrN

[Temperature, T, K; Thermal Conductivity, k, Watt $cm^{-1}K^{-1}$]

T	k
CURVE 1	
478.2	0.142
553.2	0.124
638.2	0.0941
638.2	0.0979
691.2	0.0795
763.2	0.0782
873.2	0.0615
996.2	0.0523
1093.2	0.0510
CURVE 2	
748.2	0.0711
838.2	0.0669
893.2	0.0682
933.2	0.0657
973.2	0.0640
1023.2	0.0644
1110.2	0.0619
1183.2	0.0586
1243.2	0.0582
1341.2	0.0552
CURVE 3	
408.2	0.109
520.9	0.126
579.3	0.128
760.9	0.152
955.4	0.182
CURVE 4	
1116.5	0.192
1372.1	0.227
1502.6	0.232
1638.7	0.242
1758.2	0.244
1883.2	0.246
1966.5	0.239
2019.3	0.235
2149.8	0.232
2238.7	0.228
2307.6	0.225
CURVE 5	
703.7	0.121
709.8	0.105
888.2	0.105
889.3	0.126
CURVE 6	
588.7	0.0946
588.7	0.114
593.2	0.110
594.3	0.125
745.4	0.107
754.8	0.0932
755.4	0.108
755.9	0.0963
1021.5	0.129
1027.1	0.137
1035.4	0.140
1039.3	0.113
1109.8	0.123
1111.5	0.120
1111.5	0.146
1112.1	0.128
1112.6	0.135
CURVE 7	
1360.9	0.156
1363.7	0.148
1363.7	0.130
1366.5	0.144
1372.1	0.136
1699.8	0.166
1705.4	0.176
1705.4	0.169
1938.7	0.186
1944.3	0.169
1952.6	0.162
1955.4	0.163*
2161.0	0.123
2183.2	0.119
2186.0	0.122*
2349.8	0.0914
2388.7	0.104
CURVE 7 (cont.)	
2399.8	0.111
2405.4	0.117
2405.4	0.114

*Not shown on plot

SPECIFICATION TABLE NO. 206 THERMAL CONDUCTIVITY OF AMMONIUM DIHYDROGEN PHOSPHATE $NH_4H_2PO_4$

Curve No.	Ref. No.	Method Used	Year	Temp. Range, K	Reported Error, %	Name and Specimen Designation	Composition (weight percent), Specifications and Remarks
1	71	C	1951	315, 339			Tetragonal crystal; measured with heat flow parallel to the optic axis.
2	71	C	1951	313, 342			As above but heat flow perpendicular to the optic axis.

DATA TABLE NO. 206 THERMAL CONDUCTIVITY OF AMMONIUM DIHYDROGEN PHOSPHATE $NH_4H_2PO_4$

[Temperature, T, K; Thermal Conductivity, k, Watt $cm^{-1} K^{-1}$]

T	k
CURVE 1*	
315. 2	0. 00711
339. 2	0. 00711
CURVE 2*	
313. 2	0. 0126
342. 2	0. 0134

*No graphical presentation

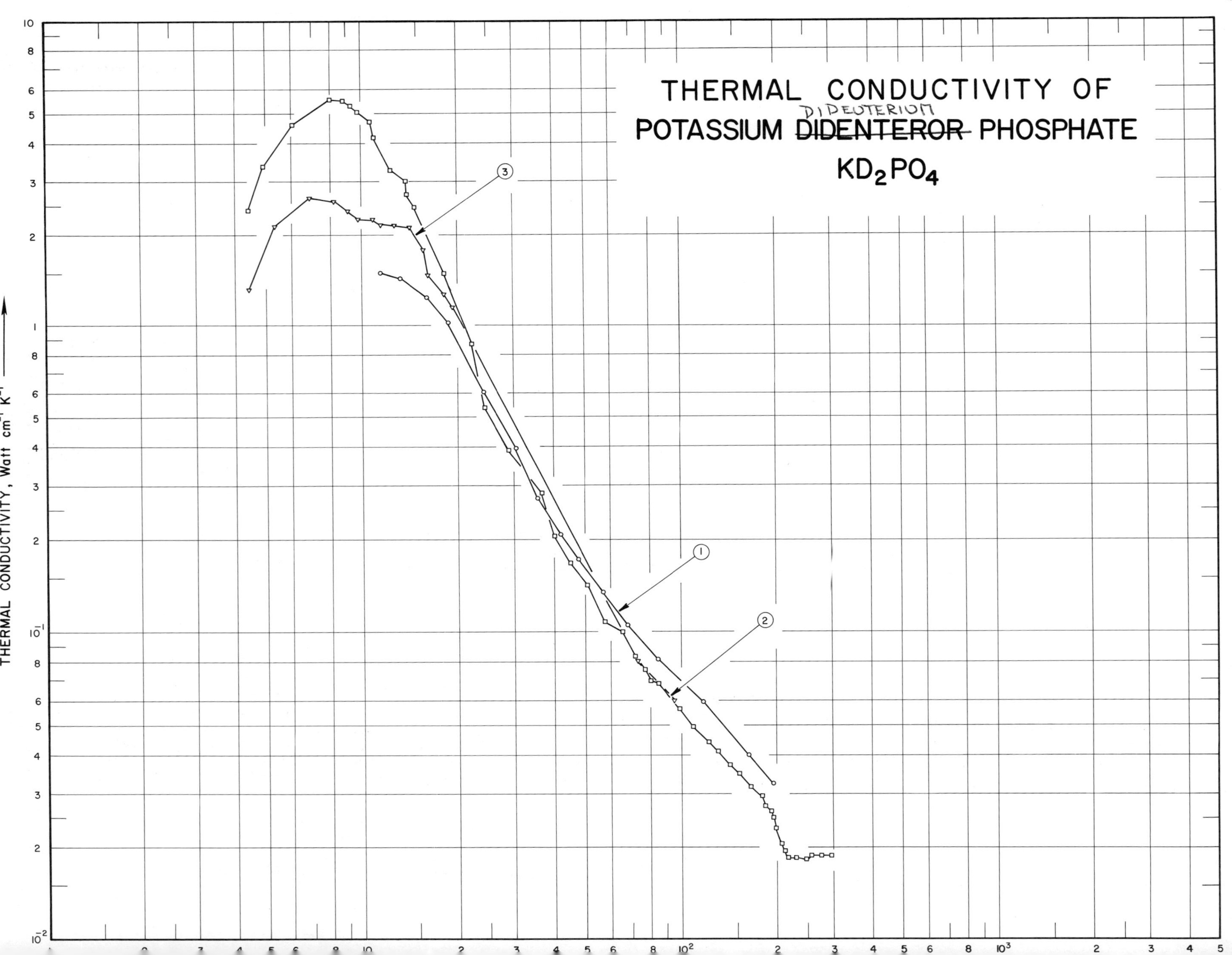
THERMAL CONDUCTIVITY OF
POTASSIUM DIDEUTERIUM PHOSPHATE
KD_2PO_4
THERMAL CONDUCTIVITY, Watt cm^{-1} K^{-1}
1
2
3

SPECIFICATION TABLE NO. 207 THERMAL CONDUCTIVITY OF POTASSIUM DIDEUTERIUM ~~DIDEUTERON~~ PHOSPHATE KD_2PO_4

[For Data Reported in Figure and Table No. 207]

Curve No.	Ref. No.	Method Used	Year	Temp. Range, K	Reported Error, %	Name and Specimen Designation	Composition (weight percent), Specifications and Remarks
1	514	L	1966	12-196		No. 1a	Single crystal with tetragonal symmetry, nominally 92% - deuterated for hydrogen; specimen 2.4 x 2.4 x 6.3 mm, long dimension parallel to a-axis; supplied by Isomet Corporation; cut and formed into rectangular rod using a wet thread and a polishing paper; Curie temperature 213 K; heat flow along a-axis.
2	514	L	1966	4.4-297		No. 2c	Similar to the above specimen except specimen 2.1 x 2.7 x 11.6 mm,long dimension parallel to c-axis; heat flow along c-axis.
3	514,515	L	1966	4.4-95		No. 3c	Similar to the above specimen except specimen 2.2 x 2.2 x 11.5 mm,long dimension parallel to c-axis.

DATA TABLE NO. 207 THERMAL CONDUCTIVITY OF POTASSIUM ~~DIDEUTERON~~ DIDEUTERIUM PHOSPHATE KD_2PO_4

[Temperature, T, K; Thermal Conductivity, k, Watt cm^{-1} K^{-1}]

T	k
CURVE 1	
11.5	1.51
13.3	1.45
16.1	1.25
18.7	1.03
24.3	0.607
30.6	0.398
35.7	0.272
42.2	0.207
47.9	0.172
57.2	0.135
68.1	0.105
84.6	0.0813
118.9	0.0594
164.4	0.0399
196.4	0.0323
CURVE 2	
4.40	2.43
4.93	3.38
6.10	4.61
7.98	5.56
8.79	5.51
9.29	5.33
9.80	5.06
10.7	4.74
11.0	4.18
12.4	3.27
13.8	3.02
13.9	2.73
14.7	2.47
18.3	1.50
22.2	0.867
24.4	0.537
28.9	0.391
36.7	0.283
40.2	0.205
45.0	0.167
50.8	0.143
57.6	0.108
65.9	0.0991
71.8	0.0828
77.1	0.0752
80.2	0.0690
84.7	0.0678
CURVE 2 (cont.)	
99.3	0.0562
109.7	0.0494
122.2	0.0440
131.9	0.0410
143.6	0.0370
152.8	0.0348
166.4	0.0314
180.4	0.0294
184.6	0.0273
193.7	0.0263
196.8	0.0250
199.1	0.0231
208.5	0.0205
213.9	0.0194
217.8	0.0185
230.2	0.0185
247.8	0.0183
258.9	0.0188
276.8	0.0188
297.2	0.0188
CURVE 3	
4.43	1.32
5.35	2.14
6.89	2.65
8.21	2.58
9.06	2.39
9.82	2.26
10.9	2.25
11.6	2.16
12.7	2.15
14.3	2.13
15.7	1.79
16.3	1.47
18.2	1.28
19.4	1.16
73.6	0.0796
95.1	0.0594

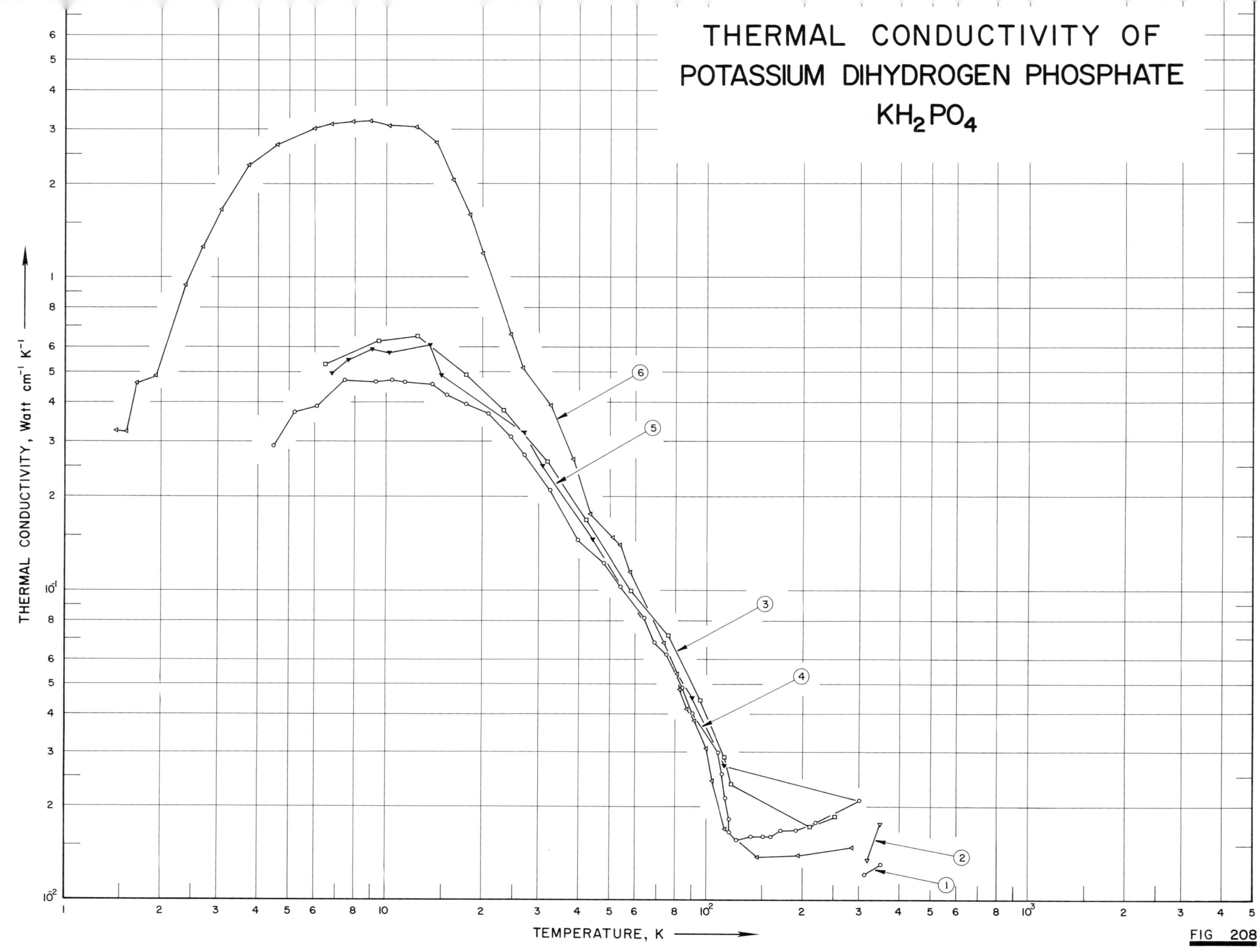
THERMAL CONDUCTIVITY OF
POTASSIUM DIHYDROGEN PHOSPHATE
KH2PO4
THERMAL CONDUCTIVITY, Watt cm-1 K-1
TEMPERATURE, K
1
2
3
4
5
6

FIG 208

SPECIFICATION TABLE NO. 208 THERMAL CONDUCTIVITY OF POTASSIUM DIHYDROGEN PHOSPHATE KH_2PO_4

[For Data Reported in Figure and Table No. 208]

Curve No.	Ref. No.	Method Used	Year	Temp. Range, K	Reported Error, %	Name and Specimen Designation	Composition (weight percent), Specifications and Remarks
1	71	C	1951	312,350			Tetragonal crystal; measured with heat flow parallel to the optic axis.
2	71	C	1951	319,347			Above specimen measured with heat flow perpendicular to the optic axis.
3	514, 515	L	1966	6.5-253		No. 1a	Single crystal with tetragonal symmetry; 2.1 x 2.8 x 10.7 mm,long dimension parallel to a-axis; supplied by Gakushuin University; cut and formed into rectangular rod using a wet thread and a polishing paper; Curie temperature 122 K; heat flow along a-axis.
4	514, 515	L	1966	4.5-300		No. 2c	Similar to the above specimen except specimen 2.3 x 2.4 x 11.7 mm, long dimension parallel to c-axis; heat flow along c-axis.
5	515	L	1967	6.8-301		No. 3c	Similar to the above specimen except specimen 2.4 x 2.6 x 11.9 mm, long dimension parallel to c-axis.
6	440		1964	1.5-286			Ferroelectric - paraelectric transition occurred at 116 K.

DATA TABLE NO. 208 THERMAL CONDUCTIVITY OF POTASSIUM DIHYDROGEN PHOSPHATE KH_2PO_4

[Temperature, T, K; Thermal Conductivity, k, Watt $cm^{-1} K^{-1}$]

T	k
CURVE 1	
312.2	0.0121
350.2	0.0130
CURVE 2	
319.2	0.0134
347.2	0.0176
CURVE 3	
6.50	0.530
9.46	0.628
12.6	0.649
17.8	0.490
23.3	0.378
31.9	0.258
42.4	0.168
58.4	0.0995
76.4	0.0716
95.7	0.0443
114.6	0.0289
120.8	0.0236
211.9	0.0173
253.0	0.0186
CURVE 4	
4.49	0.291
5.20	0.373
6.08	0.389
7.45	0.472
9.27	0.465
10.5	0.474
11.5	0.465
14.0	0.458
15.6	0.423
17.8	0.395
20.9	0.368
24.6	0.311
27.1	0.271
32.6	0.209
39.7	0.145
47.8	0.123
54.0	0.102
64.1	0.0817
CURVE 4 (cont.)	
69.8	0.0684
75.7	0.0622
84.6	0.0485
90.8	0.0401
109.4	0.0299
112.5	0.0254
116.2	0.0213
118.9	0.0182
118.9	0.0166
125.1	0.0156
138.1	0.0160
151.4	0.0160
159.6	0.0160
171.0	0.0167
191.9	0.0167
220.9	0.0177
300.0	0.0209
CURVE 5	
6.84	0.494
7.61	0.545
9.02	0.586
10.3	0.575
13.7	0.607
14.9	0.486
27.1	0.319
30.9	0.249
44.4	0.145
90.0	0.0450
114.0	0.0270
301.4	0.0209*
CURVE 6	
1.45	0.325
1.56	0.323
1.68	0.462
1.92	0.489
2.38	0.944
2.69	1.26
3.08	1.65
3.75	2.31
4.59	2.68
5.96	3.01
CURVE 6 (cont.)	
6.78	3.13
7.87	3.18
8.95	3.17
10.2	3.07
12.5	3.05
14.4	2.72
16.2	2.06
18.2	1.59
20.1	1.19
24.6	0.659
26.8	0.519
32.7	0.391
38.6	0.262
43.6	0.176
51.1	0.147
54.2	0.139
58.9	0.114
74.1	0.0676
81.1	0.0536
83.8	0.0481
87.7	0.0415
92.9	0.0379
100.2	0.0308
104.7	0.0243
115.6	0.0170
145.9	0.0138
194.1	0.0139
286.4	0.0148

*Not shown on plot

SPECIFICATION TABLE NO. 209 THERMAL CONDUCTIVITY OF AMMONIUM HYDROGEN SULFATE NH_4HSO_4

Curve No.	Ref. No.	Method Used	Year	Temp. Range, K	Reported Error, %	Name and Specimen Designation	Composition (weight percent), Specifications and Remarks
1	242	P	1961	413-486	±3		Specimen of A. R. purity; mean value of 4 or 5 different measurements for each data point; measured in molten state by using a thermal conductivity wire probe.
2	242	P	1961	418	±3		Same as above.

DATA TABLE NO. 209 THERMAL CONDUCTIVITY OF AMMONIUM HYDROGEN SULFATE NH_4HSO_4

[Temperature, T, K; Thermal Conductivity, k, Watt $cm^{-1} K^{-1}$]

T	k
CURVE 1*	
413.2	0.00465
417.5	0.00436
436.2	0.00396
461.2	0.00379
486.4	0.00362
CURVE 2*	
418	0.00389

*No graphical presentation

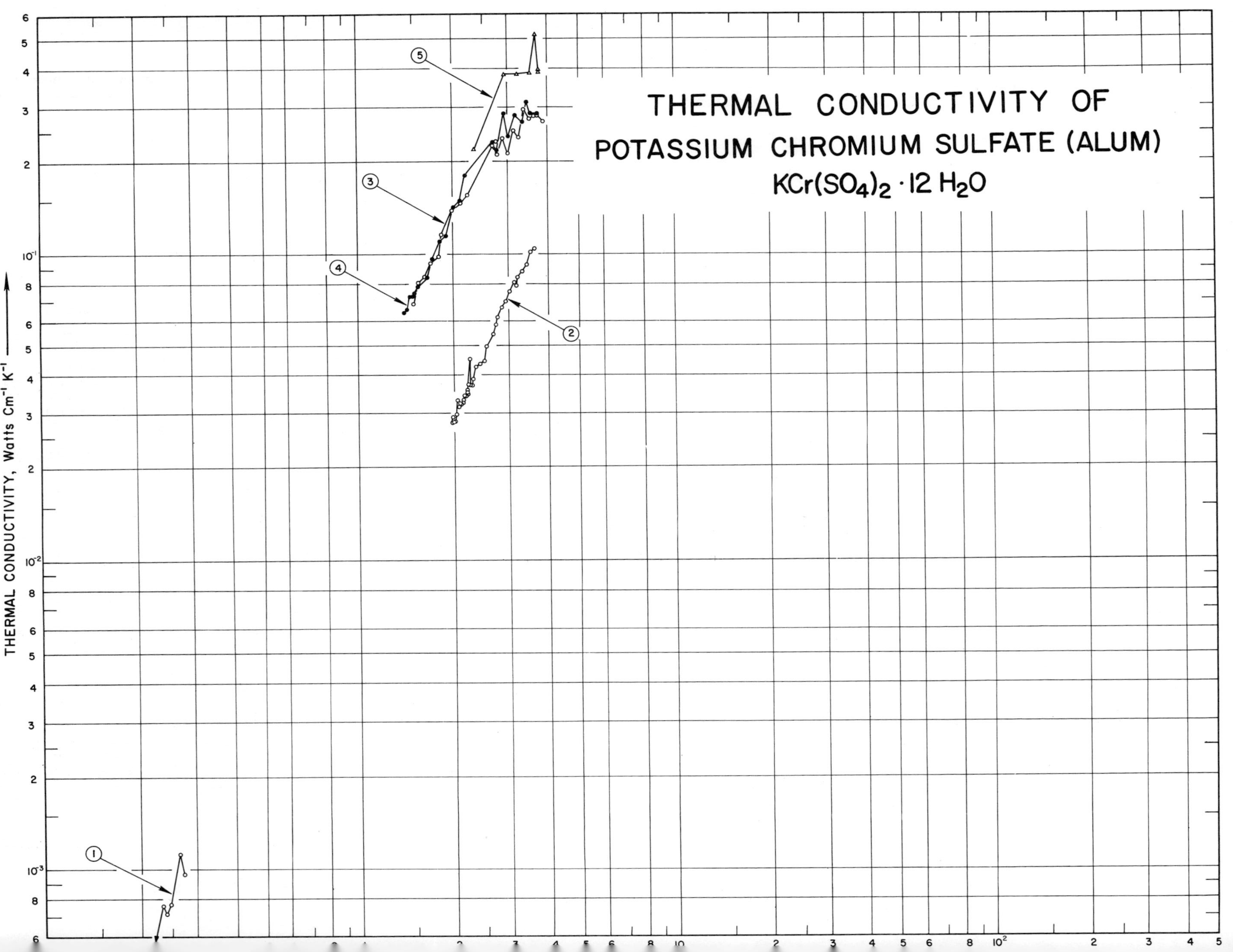
THERMAL CONDUCTIVITY OF
POTASSIUM CHROMIUM SULFATE (ALUM)
$KCr(SO_4)_2 \cdot 12\,H_2O$
THERMAL CONDUCTIVITY, Watts Cm^{-1} K^{-1}
1
2
3
4
5

SPECIFICATION TABLE NO. 210 THERMAL CONDUCTIVITY OF POTASSIUM CHROMIUM SULFATE (ALUM) $KCr(SO_4)_2 \cdot 12H_2O$

[For Data Reported in Figure and Table No. 210]

Curve No.	Ref. No.	Method Used	Year	Temp. Range, K	Reported Error, %	Name and Specimen Designation	Composition (weight percent), Specifications and Remarks
1	77	L	1950	0.15-0.27		Potassium chrome alum salt	Single crystal; 65 mm long, octagonal cross section with 15 mm width; cut from a crystal so that the specimen axis coincided with one of the cubic axes of the crystal.
2	78	L	1948	2.0-3.6		Potassium chrome alum salt; 14-XI	10.46 Cr (theoretically 10.42 Cr), 0.007 Fe; not detectable for contamination with Al and Mn; cut from a large octahedral crystal in the direction of any edge; some turbid spots and some small bubbles shown on rod; cooled as quickly as possible by the help of helium gas passing through glass envelope.
3	78	L	1948	1.5-3.9		Potassium chrome alum salt; 13-XII	Same as the above specimen except cooled slowly by evacuating helium gas from the envelope at temperatures from 70 K to 20 K, but helium gas inserted when cooled from 20 K to 4 K.
4	78	L	1948	1.4-3.7		Potassium chrome alum salt; 19-XII	Same as the above specimen.
5	78	L	1948	2.4-3.8		Potassium chrome alum salt	Same as the above specimen except cooled very slowly with envelope evacuated all the time.

DATA TABLE NO. 210 THERMAL CONDUCTIVITY OF POTASSIUM CHROMIUM SULFATE (ALUM) $KCr(SO_4)_2 \cdot 12H_2O$

[Temperature, T, K; Thermal Conductivity, k, Watt $cm^{-1}K^{-1}$]

T	k
CURVE 1	
0.145	0.000158*
0.149	0.000186*
0.166	0.000222
0.218	0.000552
0.235	0.000756
0.240	0.000715
0.248	0.000768
0.266	0.00112
0.274	0.00096
CURVE 2	
1.97	0.0277
1.98	0.0290
2.00	0.0279
2.01	0.0280
2.04	0.0296
2.05	0.0329
2.07	0.0313
2.09	0.0321
2.11	0.0319
2.13	0.0322
2.15	0.0330
2.16	0.0341
2.18	0.0338
2.20	0.0356
2.20	0.0347
2.20	0.0344
2.22	0.0370
2.25	0.0450
2.25	0.0368
2.28	0.0368
2.30	0.0386
2.35	0.0424
2.42	0.0435
2.50	0.0444
2.58	0.0495
2.66	0.0541
2.72	0.0581
2.75	0.0617
2.84	0.0667
2.93	0.0699
3.02	0.0752
3.11	0.0806
CURVE 2 (cont.)	
3.17	0.0787
3.20	0.0840
3.31	0.0877
3.41	0.0926
3.51	0.101
3.61	0.104
CURVE 3	
1.50	0.0685
1.56	0.0806
1.63	0.0840
1.70	0.0935
1.80	0.0980
1.84	0.116
1.99	0.139
2.11	0.149
2.13	0.147
2.23	0.156
2.67	0.225
2.75	0.232
2.78	0.211
2.88	0.238
2.99	0.213
3.12	0.252
3.24	0.240
3.37	0.296
3.50	0.275
3.62	0.280
3.75	0.283
3.87	0.270
CURVE 5	
1.40	0.0637
1.43	0.0654
1.46	0.0725
1.49	0.0725
1.52	0.0741
1.55	0.0781
1.66	0.0840
1.73	0.0962
1.81	0.110
1.90	0.115
CURVE 4 (cont.)	
2.02	0.143
2.10	0.148*
2.18	0.180
2.67	0.231
2.76	0.219
2.91	0.286
3.00	0.241
3.16	0.282
3.32	0.269
3.45	0.313
3.53	0.286
3.69	0.285
CURVE 5	
2.35	0.218
2.92	0.382
3.22	0.383
3.54	0.386
3.67	0.515
3.74	0.397
3.76	0.389

*Not shown on plot

SPECIFICATION TABLE NO. 211 THERMAL CONDUCTIVITY OF POTASSIUM HYDROGEN SULFATE $KHSO_4$

Curve No.	Ref. No.	Method Used	Year	Temp. Range, K	Reported Error, %	Name and Specimen Designation	Composition (weight percent), Specifications and Remarks
1	242		1961	444-524	± 3		Specimen of A. R. purity; data point is the mean value of 4 or 5 different measurements; measured in molten state.
2	242		1961	479	± 3		Same as the above specimen; measured in molten state.

DATA TABLE NO. 211 THERMAL CONDUCTIVITY OF POTASSIUM HYDROGEN SULFATE $KHSO_4$

[Temperature, T, K; Thermal Conductivity, k, Watt $cm^{-1}K^{-1}$]

T	k
CURVE 1*	
443.8	0.00404
456.2	0.00380
471.7	0.00363
485.2	0.00352
498.2	0.00365
524.2	0.00372
CURVE 2*	
479	0.00339

*No graphical presentation

SPECIFICATION TABLE NO. 212 THERMAL CONDUCTIVITY OF SODIUM HYDROGEN SULFATE $NaHSO_4$

Curve No.	Ref. No.	Method Used	Year	Temp. Range, K	Reported Error, %	Name and Specimen Designation	Composition (weight percent), Specifications and Remarks
1	242		1961	443-518	± 3		Specimen of A.R. purity; data value is the mean value of 4 or 5 different measurements; measured in molten state.
2	242		1961	452	± 3		Same as the above specimen; measured in molten state.

DATA TABLE NO. 212 THERMAL CONDUCTIVITY OF SODIUM HYDROGEN SULFATE $NaHSO_4$

[Temperature, T, K; Thermal Conductivity, k, Watt $cm^{-1}K^{-1}$]

T	k
CURVE 1*	
443.0	0.00626
450.0	0.00604
460.3	0.00480
470.2	0.00502
502.2	0.00509
518.3	0.00515
CURVE 2*	
452	0.00460

*No graphical presentation

SPECIFICATION TABLE NO. 213 THERMAL CONDUCTIVITY OF SODIUM THIOSULFATE $Na_2S_2O_3 \cdot 5H_2O$

Curve No.	Ref. No.	Method Used	Year	Temp. Range, K	Reported Error, %	Name and Specimen Designation	Composition (weight percent), Specifications and Remarks
1	259	R	1932	295.0			The substance was melted and poured into container (5.75 in. x 7 in.) until the depth of liquid was 6 in., when it had solidified and cooled to room temperature the supply to the central heating element was switched on and maintained constant; when temperature conditions had remained steady for two hrs, readings on which the calculations of heat conductivity were based were taken.

DATA TABLE NO. 213 THERMAL CONDUCTIVITY OF SODIUM THIOSULFATE $Na_2S_2O_3 \cdot 5H_2O$

[Temperature, T, K; Thermal Conductivity, k, Watt $cm^{-1}K^{-1}$]

T	k
CURVE 1*	
295.0	0.0136

*No graphical presentation

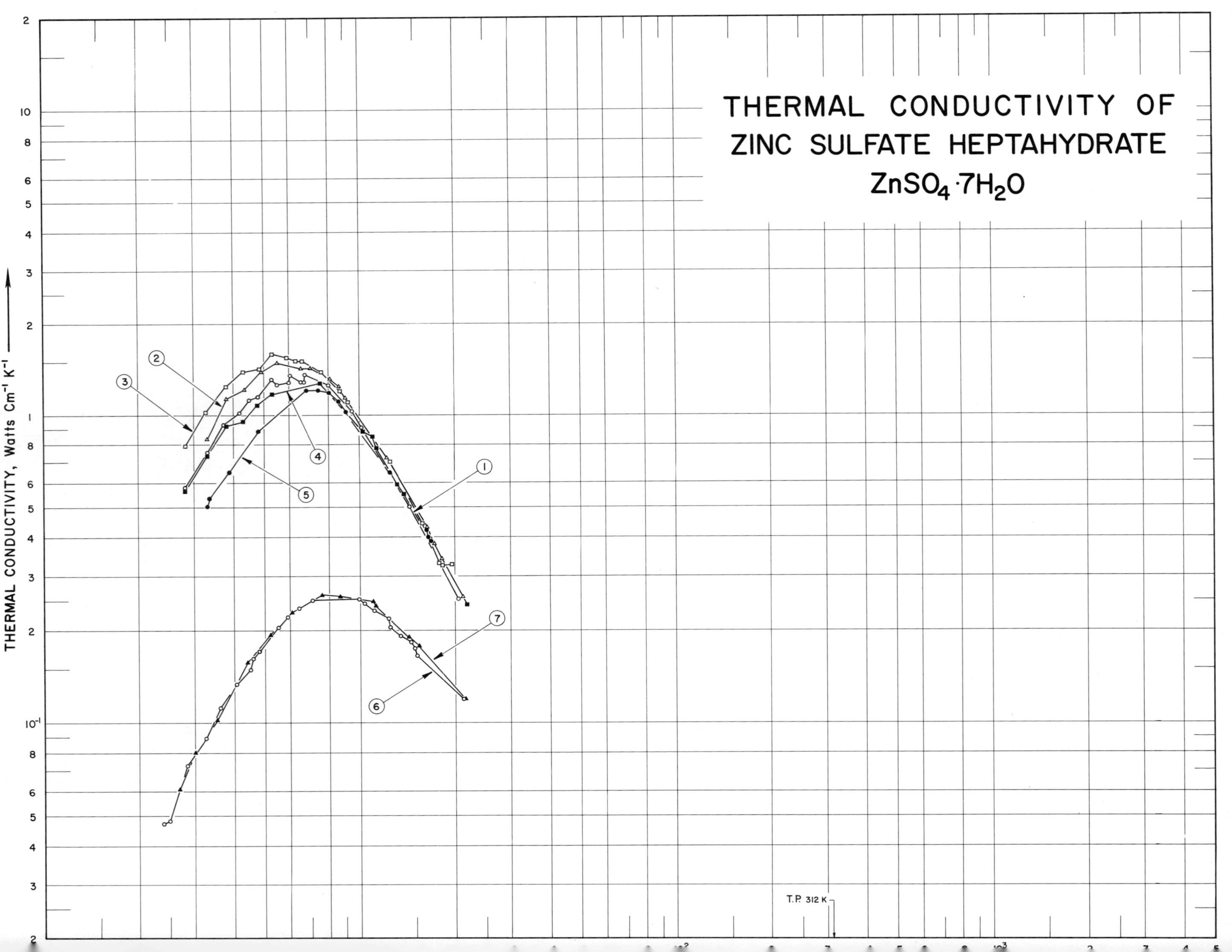
THERMAL CONDUCTIVITY OF
ZINC SULFATE HEPTAHYDRATE
$ZnSO_4 \cdot 7H_2O$
THERMAL CONDUCTIVITY, Watts Cm⁻¹ K⁻¹
2
10
8
6
5
4
3
2
1
8
6
5
4
3
2
10⁻¹
8
6
5
4
3
2
1
2
3
4
5
6
7
T.P. 312 K

SPECIFICATION TABLE NO. 214 THERMAL CONDUCTIVITY OF ZINC SULFATE HEPTAHYDRATE $ZnSO_4 \cdot 7\,H_2O$

[For Data Reported in Figure and Table No. 214]

Curve No.	Ref. No.	Method Used	Year	Temp. Range, K	Reported Error, %	Name and Specimen Designation	Composition (weight per cent) and Remarks
1	148		1961	2.8-21		I	0.25 $FeSO_4 \cdot 7H_2O$; single crystal; the basic material used was of the "Analar" purity grade (max. impurity 0.006%); specimen cut from larger single crystal along the edges to the size of about 3 cm (length) x 0.4 cm x 0.4 cm; the cooling process was that the atmospheric air surrounding the specimen mounted in the apparatus at room temperature was first pumped out after the apparatus had been cooled down to 90 K, then replaced by helium gas at low pressure to continue the slow cooling down process; measurements began at 20 K until 2 K; precautions were taken to avoid dehydrating the specimen.
2	148		1961	3.3-21		I	The second run of the same specimen I; the apparatus was first warmed up to about 90 K; the atmosphere air was let into specimen space to a pressure slightly below one atmosphere; then the temperature was raised to room temperature and pressure to atmospheric pressure before the cooling process.
3	148		1961	2.8-20		I	The third run of the same specimen I; the apparatus was first warmed up to room temperature; then the specimen was wetted with distilled water and dried quickly with absorbent paper. The specimen chamber was resealed and the cooling process started.
4	148		1961	2.8-22		I	The fourth run of the same specimen I; the apparatus was first warmed up to room temperature, then a small amount of acetone was put in the specimen chamber 5 hrs before the cooling down process.
5	148		1961	3.3-22		I	The fifth run of the same specimen I; the apparatus was first warmed up to 200 K and dry air at atmospheric pressure was then admitted into the specimen chamber (the specimen lost some water of crystallization); then the apparatus was warmed up to room temperature and pressure in specimen chamber maintained at atmospheric pressure. Afterwards, the apparatus was cooled down to 90 K in 4 hrs and kept at 90 K for 48 hrs before further cooling down to measurement temperature.
6	148		1961	2.4-21		II	Same as the specimen I except having 1.0 $FeSO_4 \cdot 7H_2O$; process same as that of the first run of the specimen I.
7	148		1961	2.7-22		II	The second run of the same specimen II; the apparatus was first warmed up to and kept at 90 K for about 30 hrs and maintained at high vacuum (order of 10^{-5} mm Hg) in the specimen chamber; then followed the cooling process.

DATA TABLE NO. 214 THERMAL CONDUCTIVITY OF ZINC SULFATE HEPTAHYDRATE $ZnSO_4 \cdot 7H_2O$

[Temperature, T, K; Thermal Conductivity, k, Watt $cm^{-1}K^{-1}$]

CURVE 1

T	k
2.8	0.580
3.3	0.750
3.7	0.931
4.2	1.025
4.5	1.125
4.8	1.150
5.3	1.314
5.5	1.263
6.0	1.268
6.05	1.360
6.55	1.285
6.7	1.286
6.75	1.360
8.0	1.264
9.5	1.032
10.2	0.910
14.4	0.501
20.5	0.252

CURVE 2

T	k
3.3	0.835
3.8	1.113
4.35	1.215
4.9	1.390
5.5	1.490
6.6	1.435
7.0	1.438
8.1	1.315
8.6	1.242
9.1	1.140
12.2	0.720
16.3	0.430
17.2	0.380
18.2	0.340
21.2	0.258

CURVE 3

T	k
2.81	0.79
3.27	1.02
3.8	1.245
4.3	1.39
4.85	1.42
5.3	1.59
5.9	1.55
6.3	1.51
6.6	1.51
7.6	1.39
8.7	1.20
9.2	1.105
12.5	0.700
15.7	0.442
17.8	0.330
18.3	0.325
19.6	0.280

CURVE 4

T	k
2.8	0.565
3.3	0.735
3.8	0.921
4.3	0.955
4.8	1.081
5.3	1.175
7.5	1.278
10.2	0.884
11.0	0.850
11.3	0.770
13.2	0.592
13.8	0.550
21.8	0.242

CURVE 5

T	k
3.30	0.503
3.35	0.535
3.88	0.645
4.79	0.883
6.80	1.207
7.41	1.211
8.0	1.183
8.6	1.114
9.07	1.025
10.29	0.647
16.3	0.421
16.5	0.400
16.75	0.388
21.8	0.242*

CURVE 6

T	k
2.37	0.0471
2.49	0.0481
2.82	0.0728
3.25	0.0890
3.62	0.112
4.05	0.135
4.50	0.150
4.60	0.162
4.80	0.172
5.50	0.205
5.90	0.222
6.41	0.235
7.05	0.250
9.95	0.252
10.32	0.245
11.10	0.232
12.35	0.219
12.50	0.205
13.50	0.192
14.50	0.184
14.94	0.175
15.10	0.166
21.2	0.119

CURVE 7

T	k
2.68	0.0610
3.0	0.0800
3.5	0.102
4.4	0.157
5.2	0.194
6.1	0.229
7.6	0.260
8.65	0.259
11.0	0.248
11.2	0.230
14.25	0.190
15.45	0.178
21.5	0.120

*Not shown on plot

SPECIFICATION TABLE NO. 215 THERMAL CONDUCTIVITY OF CERIUM SULFIDE CeS

Curve No.	Ref. No.	Method Used	Year	Temp. Range, K	Reported Error, %	Name and Specimen Designation	Composition (weight percent), Specifications and Remarks
1	511	T	1962	973-1573			Electrical conductivity 6.3, 6.1, 5.5 and 4.7 mho cm^{-1} at 973, 1203, 1402 and 1573 K, respectively.

DATA TABLE NO. 215 THERMAL CONDUCTIVITY OF CERIUM SULFIDE CeS

[Temperature, T, K; Thermal Conductivity, k, Watt $cm^{-1}K^{-1}$]

T	k
CURVE 1*	
973	0.020
1103	0.018
1203	0.014
1273	0.019
1402	0.011
1523	0.015
1573	0.015

*No graphical presentation

SPECIFICATION TABLE NO. 216 THERMAL CONDUCTIVITY OF DICERIUM TRISULFIDE Ce_2S_3

Curve No.	Ref. No.	Method Used	Year	Temp. Range, K	Reported Error, %	Name and Specimen Designation	Composition (weight percent), Specifications and Remarks
1	505	C	1960	300-1573		A-26	Prepared by resolidifying the molten material; electrical conductivity reported as 8.7, 6.3, 6.37, 6.1, 6.0, 5.5, 5.12, and 4.7 $ohm^{-1}cm^{-1}$ at 300, 973, 1103, 1203, 1273, 1402, 1523, and 1573 K, respectively.

DATA TABLE NO. 216 THERMAL CONDUCTIVITY OF DICERIUM TRISULFIDE Ce_2S_3

[Temperature, T, K; Thermal Conductivity, k, Watt $cm^{-1}K^{-1}$]

T	k
CURVE 1*	
300	0.006
973	0.007
1103	0.011
1203	0.010
1203	0.010
1273	0.011
1402	0.011
1523	0.012
1573	0.011

* No graphical presentation

SPECIFICATION TABLE NO. 217 THERMAL CONDUCTIVITY OF DICOPPER SULFIDE Cu_2S

Curve No.	Ref. No.	Method Used	Year	Temp. Range, K	Reported Error, %	Name and Specimen Designation	Composition (weight percent), Specifications and Remarks
1	256		1955	293.2			Electrical conductivity 3.7 x 10^2 $ohm^{-1}cm^{-1}$ at 20 C.

DATA TABLE NO. 217 THERMAL CONDUCTIVITY OF DICOPPER SULFIDE Cu_2S

[Temperature, T, K; Thermal Conductivity, k, Watt $cm^{-1}K^{-1}$]

T	k
CURVE 1*	
293.2	0.00418

*No graphical presentation

SPECIFICATION TABLE NO. 218 THERMAL CONDUCTIVITY OF [DICOPPER SULFIDE + IRON SULFIDE + TRINICKEL DISULFIDE] $Cu_2S + FeS + Ni_3S_2$

Curve No.	Ref. No.	Method Used	Year	Temp. Range, K	Reported Error, %	Name and Specimen Designation	Composition (weight percent), Specifications and Remarks
1	256		1955	293.2			7.5 mole % FeS.
2	256		1955	293.2			17 mole % FeS.
3	256		1955	293.2			35 mole % FeS.
4	256		1955	293.2			44 mole % FeS.

DATA TABLE NO. 218 THERMAL CONDUCTIVITY OF [DICOPPER SULFIDE + IRON SULFIDE + TRINICKEL DISULFIDE] $Cu_2S + FeS + Ni_3S_2$

[Temperature, T, K; Thermal Conductivity, k, Watt $cm^{-1}K^{-1}$]

T	k
CURVE 1*	
293.2	0.100
CURVE 2*	
293.2	0.0544
CURVE 3*	
293.2	0.0335
CURVE 4*	
293.2	0.0251

*No graphical presentation

SPECIFICATION TABLE NO. 219 THERMAL CONDUCTIVITY OF [DICOPPER SULFIDE + TRINICKEL DISULFIDE] $Cu_2S + Ni_3S_2$

Curve No.	Ref. No.	Method Used	Year	Temp. Range, K	Reported Error, %	Name and Specimen Designation	Composition (weight percent), Specifications and Remarks
1	256		1955	293.2			33 mole % Ni_3S_2.
2	256		1955	293.2			67 mole % Ni_3S_2.

DATA TABLE NO. 219 THERMAL CONDUCTIVITY OF [DICOPPER SULFIDE + TRINICKEL DISULFIDE] $Cu_2S + Ni_3S_2$

[Temperature, T, K; Thermal Conductivity, k, Watt $cm^{-1}K^{-1}$]

T	k
CURVE 1*	
293.2	0.0335
CURVE 2*	
293.2	0.0586

*No graphical presentation

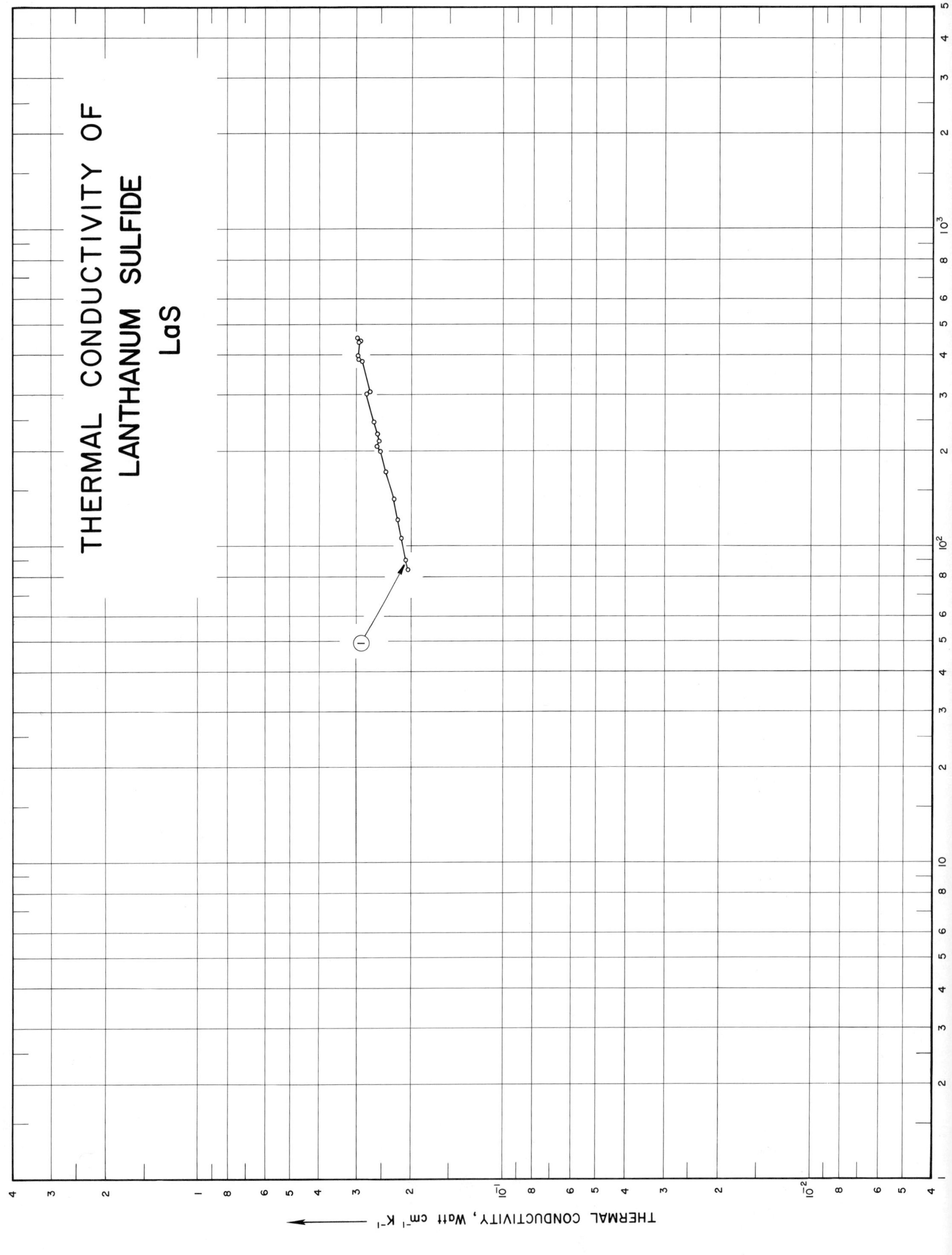

THERMAL CONDUCTIVITY OF LANTHANUM SULFIDE
LaS

THERMAL CONDUCTIVITY, Watt cm^{-1} K^{-1}

4 3 2 1 8 6 5 4 3 2 10^{-1} 8 6 5 4 3 2 10^{-2} 8 6 5 4

1

1 2 3 4 5 6 8 10 2 3 4 5 6 8 10^2 2 3 4 5 6 8 10^3 2 3 4 5

SPECIFICATION TABLE NO. 220 THERMAL CONDUCTIVITY OF LANTHANUM SULFIDE LaS

[For Data Reported in Figure and Table No. 220]

Curve No.	Ref. No.	Method Used	Year	Temp. Range, K	Reported Error, %	Name and Specimen Designation	Composition (weight percent), Specifications and Remarks
1	506, 507	L	1966	84-453	±3-±5		NaCl type compound with ionic-metallic type bonding; prepared by pressing powders of the compound under a pressure of about 8000 kg cm^{-2}, sintering in a vacuum of $\sim 10^{-5}$ Torr for 1 to 2 hrs at 1600 to 1800 C; electrical resistivity reported range from 1.70-5.29 μohm cm at 84-463 K respectively; measured in a vacuum of $10^{-4}\sim 10^{-5}$ mm Hg.

DATA TABLE NO. 220 THERMAL CONDUCTIVITY OF LANTHANUM SULFIDE LaS

[Temperature, T, K; Thermal Conductivity, k, Watt $cm^{-1}K^{-1}$]

T	k
CURVE 1	
84	0.205
90	0.208
106	0.215
122	0.222
141	0.229
172	0.243
199	0.254
206	0.259
215	0.256
227	0.259
246	0.265
303	0.280
307	0.274
381	0.290
389	0.297
398	0.298
441	0.296
446	0.294
453	0.300

SPECIFICATION TABLE NO. 221 THERMAL CONDUCTIVITY OF TRINICKEL DISULFIDE Ni_3S_2

Curve No.	Ref. No.	Method Used	Year	Temp. Range, K	Reported Error, %	Name and Specimen Designation	Composition (weight percent), Specifications and Remarks
1	256		1955	293.2			

DATA TABLE NO. 221 THERMAL CONDUCTIVITY OF TRINICKEL DISULFIDE Ni_3S_2

[Temperature, T, K; Thermal Conductivity, k, Watt $cm^{-1}K^{-1}$]

T	k
CURVE 1*	
293.2	0.0879

*No graphical presentation

SPECIFICATION TABLE NO. 222 THERMAL CONDUCTIVITY OF [ALUMINUM OXIDE + CHROMIUM] CERMETS $Al_2O_3 + Cr$

Curve No.	Ref. No.	Method Used	Year	Temp. Range, K	Reported Error, %	Name and Specimen Designation	Composition (weight percent) Al_2O_3	Composition (weight percent) Cr	Composition (continued), Specifications and Remarks
1	296		1958	293.2		Type LT-1	70	30	Supplied by Haynes Stellite Co.

DATA TABLE NO. 222 THERMAL CONDUCTIVITY OF [ALUMINUM OXIDE + CHROMIUM] CERMETS $Al_2O_3 + Cr$

[Temperature, T, K; Thermal Conductivity, k, Watt cm^{-1} K^{-1}]

T	k
CURVE 1*	
293.2	0.0962

* No graphical presentation

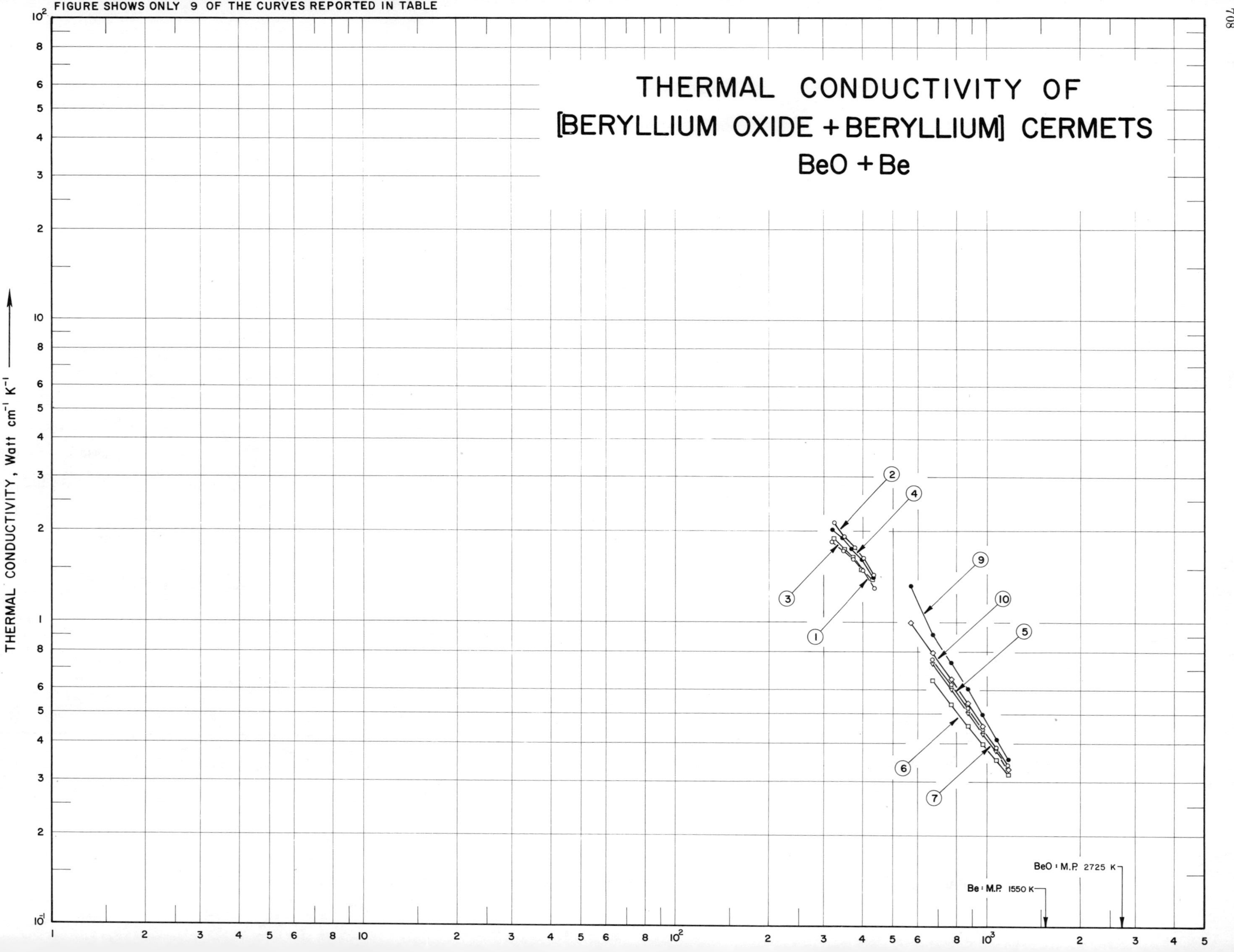
FIGURE SHOWS ONLY 9 OF THE CURVES REPORTED IN TABLE
THERMAL CONDUCTIVITY OF
[BERYLLIUM OXIDE + BERYLLIUM] CERMETS
BeO + Be
THERMAL CONDUCTIVITY, Watt cm⁻¹ K⁻¹
BeO : M.P. 2725 K
Be : M.P. 1550 K

SPECIFICATION TABLE NO. 223 THERMAL CONDUCTIVITY OF [BERYLLIUM OXIDE + BERYLLIUM] CERMETS BeO + Be

[For Data Reported in Figure and Table No. 223]

Curve No.	Ref. No.	Method Used	Year	Temp. Range, K	Reported Error, %	Name and Specimen Designation	Composition (weight percent), Specifications and Remarks
1	12		1953	320–436		267A-1	Metal content 8.24%, metal particle size 2–5μ; density 2.81 g cm^{-3}.
2	12		1953	325–434		268A-1	Metal content 5.65%, metal particle size 2–5μ; density 2.87 g cm^{-3}.
3	12		1953	323–430		269A-1	Metal content 6.06%, metal particle size 15–25μ; density 2.86 g cm^{-3}.
4	12		1953	320–432		270A-1	Metal content 6.06%, metal particle size 15–25μ; density 2.86 g cm^{-3}.
5	378	L	1962	673–1173	±2 - ±4	3A	97 BeO and 3 Be; cylindrical specimen, 1.625 in. in diameter, 1 in. long, bulk density 2.945 g cm^{-3}; > 99% theoretical density; smoothed results; data not corrected to zero porosity.
6	378	L	1962	673–1173	±2 - ±4	4A	94 BeO and 6 Be; cylindrical specimen, 1.625 in. in diameter, 1 in. long; bulk density 2.905 g cm^{-3}; > 99% theoretical density; smoothed results; data not corrected to zero porosity.
7	378	L	1962	673–1173	±2 - ±4	5A	91 BeO and 9 Be; cylindrical specimen, 1.625 in. in diameter, 1 in. long; bulk density 2.834 g cm^{-3}; > 99% theoretical density; smoothed results; data not corrected to zero porosity.
8	378	L	1962	673–1173	±2 - ±4	6A	88 BeO and 12 Be; cylindrical specimen, 1.625 in. in diameter, 1 in. long; bulk density 2.787 g cm^{-3}; > 99% theoretical density; smoothed results; data not corrected to zero porosity.
9	378	L	1962	573–1173	±2 - ±4	7A	93 BeO and 7 Be; cylindrical specimen, 1.625 in. in diameter, 1 in. long; bulk density 2.842 g cm^{-3}; > 98% theoretical density; smoothed results; data not corrected to zero porosity.
10	378	L	1962	573–1173	±2 - ±4	8A	93 BeO and 7 Be; cylindrical specimen, 1.625 in. in diameter, 1 in. long; bulk density 2.917 g cm^{-3}; 100% theoretical density; smoothed results; data not corrected to zero porosity.

DATA TABLE NO. 223 THERMAL CONDUCTIVITY OF [BERYLLIUM OXIDE + BERYLLIUM] CERMETS BeO + Be

[Temperature, T, K; Thermal Conductivity, k, Watt $cm^{-1}K^{-1}$]

T	k
CURVE 1	
319.6	1.85
346.1	1.72
373.8	1.62
400.8	1.48
436.3	1.30
CURVE 2	
324.7	2.13
349.3	1.92
377.8	1.77
402.4	1.63
434.0	1.44
CURVE 3	
323.1	1.89
347.7	1.74
372.0	1.63
396.7	1.49
429.8	1.38
CURVE 4	
319.6	2.01
344.6	1.90
367.6	1.75
398.4	1.62
431.8	1.41
CURVE 5	
673.2	0.756
773.2	0.627
873.2	0.524
973.2	0.448
1073.2	0.388
1173.2	0.340
CURVE 6	
673.2	0.646
773.2	0.538
873.2	0.457
973.2	0.397

T	k
CURVE 6 (cont.)	
1073.2	0.354
1173.2	0.317
CURVE 7	
673.2	0.734
773.2	0.603
873.2	0.503
973.2	0.430
1073.2	0.379
1173.2	0.337*
CURVE 8*	
673.2	0.735
773.2	0.577
873.2	0.482
973.2	0.413
1073.2	0.357
1173.2	0.303
CURVE 9	
573.2	1.132
673.2	0.911
773.2	0.740
873.2	0.605
973.2	0.498
1073.2	0.414
1173.2	0.356
CURVE 10	
573.2	0.995
673.2	0.794
773.2	0.652
873.2	0.542
973.2	0.455
1073.2	0.388*
1173.2	0.329

*Not shown on plot

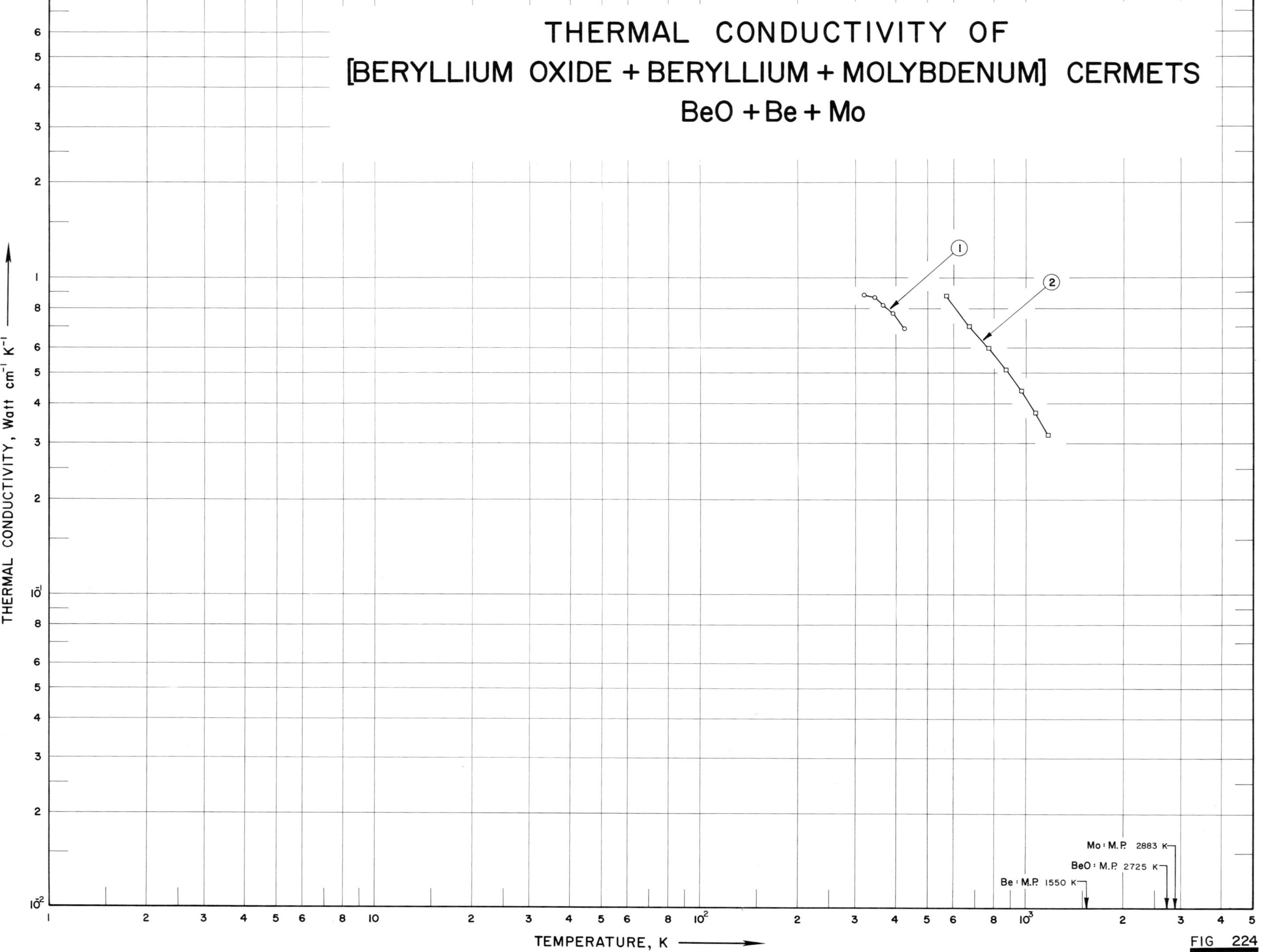
THERMAL CONDUCTIVITY OF
[BERYLLIUM OXIDE + BERYLLIUM + MOLYBDENUM] CERMETS
BeO + Be + Mo
THERMAL CONDUCTIVITY, Watt cm⁻¹ K⁻¹
TEMPERATURE, K
Mo : M.P. 2883 K
BeO : M.P. 2725 K
Be : M.P. 1550 K
1
2

FIG 224

SPECIFICATION TABLE NO. 224 THERMAL CONDUCTIVITY OF [BERYLLIUM OXIDE + BERYLLIUM + MOLYBDENUM] CERMETS BeO + Be + Mo

[For Data Reported in Figure and Table No. 224]

Curve No.	Ref. No.	Method Used	Year	Temp. Range, K	Reported Error, %	Name and Specimen Designation	Composition (weight percent), Specifications and Remarks
1	12		1953	320–428		272A	Apparent porosity 6.2%; density 2.75 g cm^{-3}.
2	378	L	1962	573–1173	± 4	9A	86 BeO, 7 Be and 7 Mo; cylindrical specimen 1.625 in. in diameter, 1 in. long; bulk density 2.975 g cm^{-3}; 98% theoretical density; smoothed data without correction to zero porosity.

DATA TABLE NO. 224 THERMAL CONDUCTIVITY OF [BERYLLIUM OXIDE + BERYLLIUM + MOLYBDENUM] CERMETS BeO + Be + Mo

[Temperature, T, K; Thermal Conductivity, k, Watt $cm^{-1}K^{-1}$]

T	k
CURVE 1	
319.6	0.882
344.2	0.866
366.5	0.816
393.6	0.774
427.6	0.690
CURVE 2	
573.2	0.875
673.2	0.701
773.2	0.597
873.2	0.511
973.2	0.439
1073.2	0.374
1173.2	0.318

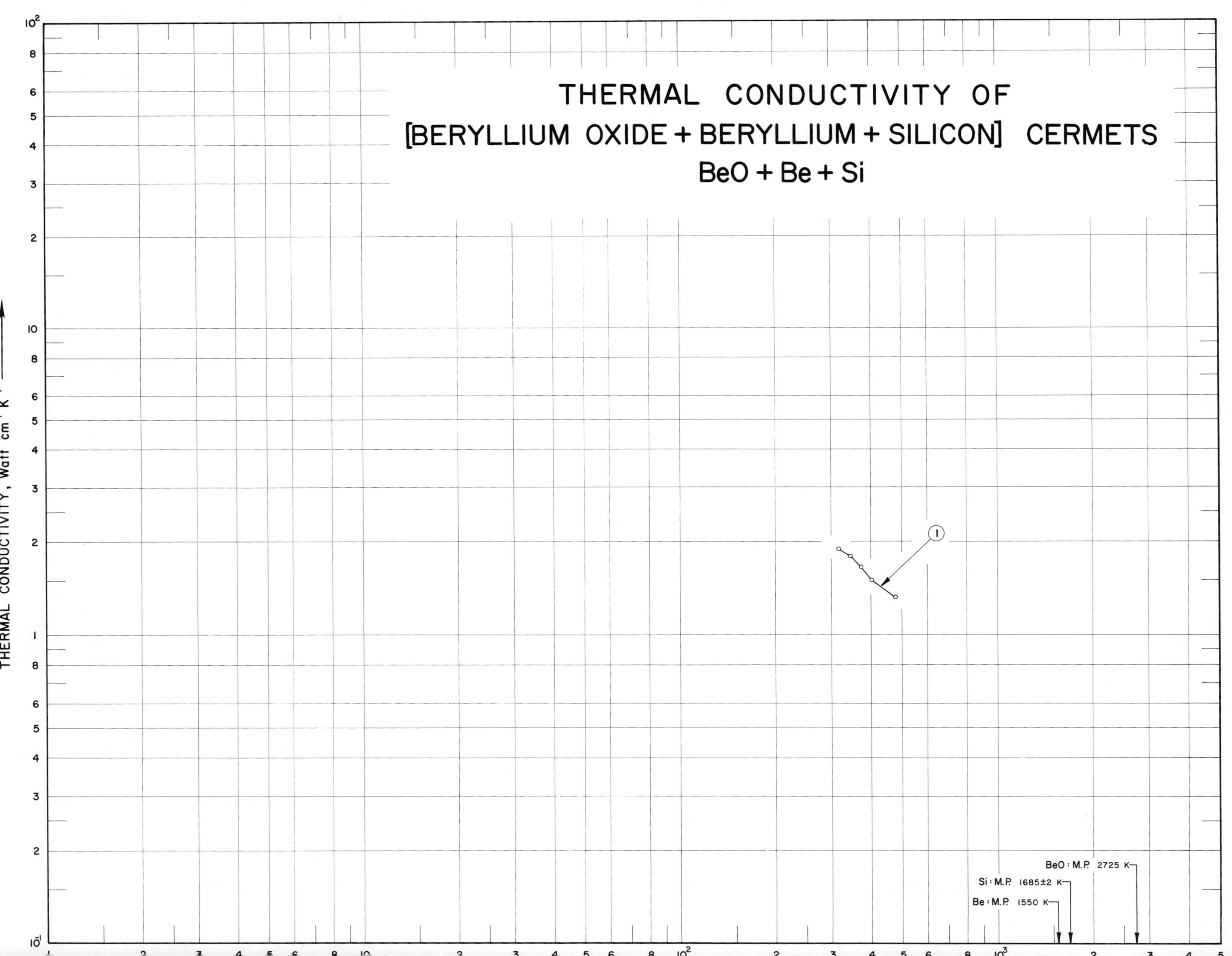

THERMAL CONDUCTIVITY OF
[BERYLLIUM OXIDE + BERYLLIUM + SILICON] CERMETS
BeO + Be + Si
1
BeO : M.P. 2725 K
Si : M.P. 1685±2 K
Be : M.P. 1550 K
THERMAL CONDUCTIVITY, Watt cm⁻¹ K⁻¹

SPECIFICATION TABLE NO. 225 THERMAL CONDUCTIVITY OF [BERYLLIUM OXIDE + BERYLLIUM + SILICON] CERMETS $BeO + Be + Si$

[For Data Reported in Figure and Table No. 225]

Curve No.	Ref. No.	Method Used	Year	Temp. Range, K	Reported Error, %	Name and Specimen Designation	Composition (weight percent), Specifications and Remarks
1	12		1953	317–477		272A-1	Density 2.91 g cm^{-3}.

DATA TABLE NO. 225 THERMAL CONDUCTIVITY OF [BERYLLIUM OXIDE + BERYLLIUM + SILICON] CERMETS BeO + Be + Si

[Temperature, T, K; Thermal Conductivity, k, Watt $cm^{-1}K^{-1}$]

T	k
CURVE 1	
317.2	1.89
345.2	1.79
373.9	1.66
403.4	1.50
476.6	1.33

SPECIFICATION TABLE NO. 226 THERMAL CONDUCTIVITY OF [TETRABORON CARBIDE + ALUMINUM] CERMETS $B_4C + Al$

Curve No.	Ref. No.	Method Used	Year	Temp. Range, K	Reported Error, %	Name and Specimen Designation	Composition (weight percent) B_4C	Composition (weight percent) Al	Composition (continued), Specifications and Remarks
1	473	L	1954	366-533		Boral	50	50	Prepared by adding pre-oxidized B_4C powder (oxidized at 811 C for one hr) to molten Al, mix allowed to stand for 10-15 min, molded, rolled; density 2.53 g cm^{-3}.

DATA TABLE NO. 226 THERMAL CONDUCTIVITY OF [TETRABORON CARBIDE + ALUMINUM] CERMETS $B_4C + Al$

[Temperature, T, K; Thermal Conductivity, k, Watt $cm^{-1}K^{-1}$]

T	k
CURVE 1*	
366	0.433
505	0.332
533	0.329

*No graphical presentation

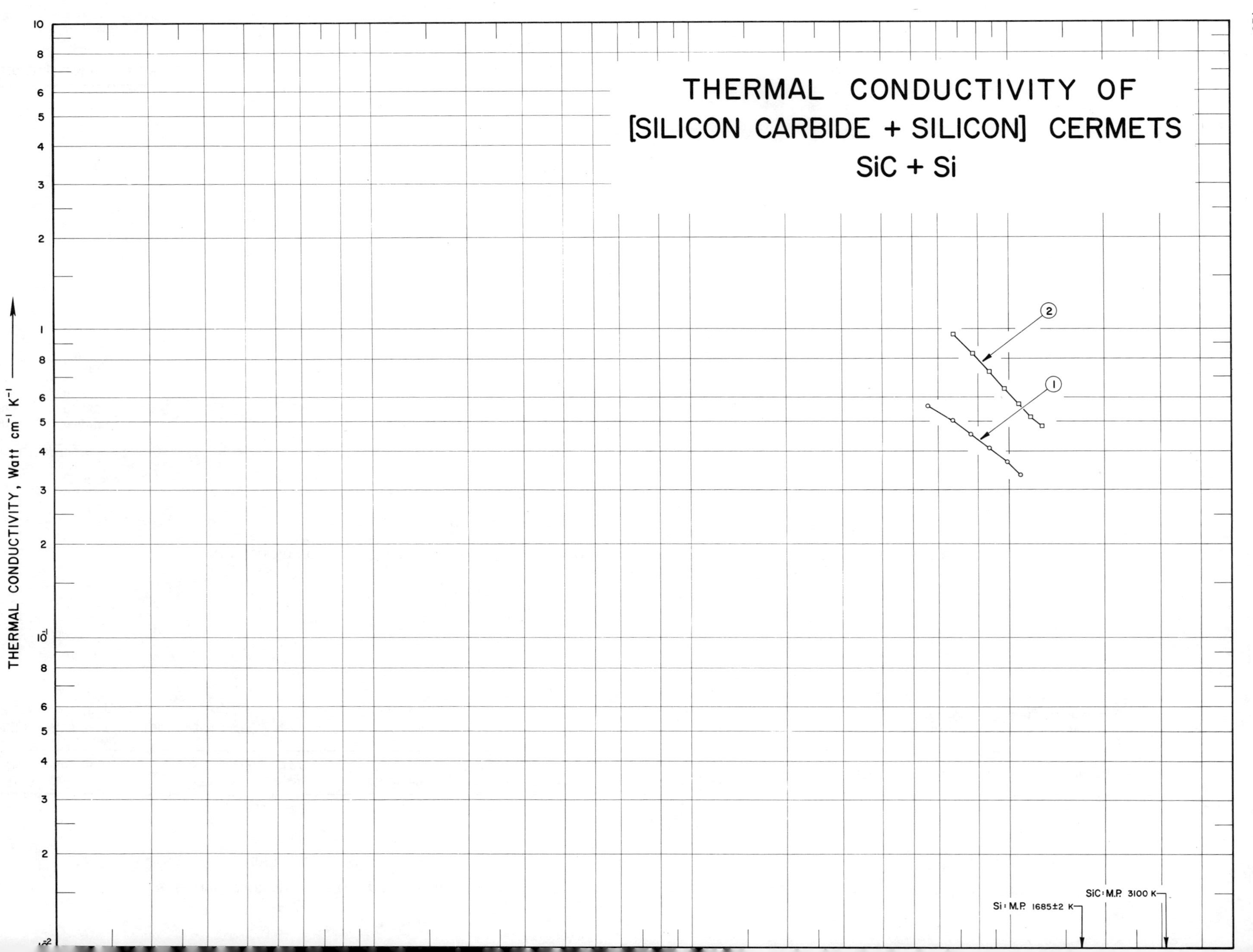

THERMAL CONDUCTIVITY OF
[SILICON CARBIDE + SILICON] CERMETS
SiC + Si
2
1
SiC: M.P. 3100 K
Si: M.P. 1685±2 K
THERMAL CONDUCTIVITY, Watt cm⁻¹ K⁻¹
10
8
6
5
4
3
2
1
8
6
5
4
3
2
10⁻¹
8
6
5
4
3
2
10⁻²

SPECIFICATION TABLE NO. 227 THERMAL CONDUCTIVITY OF [SILICON CARBIDE + SILICON] CERMETS SiC + Si

[For Data Reported in Figure and Table No. 227]

Curve No.	Ref. No.	Method Used	Year	Temp. Range, K	Reported Error, %	Name and Specimen Designation	Composition (weight percent), Specifications and Remarks
1	136	C	1959	556-1089	± 4		β SiC bonded with 30 volume %Si; specimen size 7/8 x 7/8 x 7/8 in. cube, with faces ground flat and parallel on a diamond lap.
2	378	C	1962	673-1273	± 6	10A	96.5 SiC, 2.5 Si, 0.4 C, 0.4 Al, and 0.2 Fe; cylindrical specimen 1.625 in. diameter, 1 in. long; bulk density 3.01 g cm^{-3}; 95% theoretical density; Inconel used as reference material; smoothed data without correction to zero porosity.

DATA TABLE NO. 227 THERMAL CONDUCTIVITY OF [SILICON CARBIDE + SILICON] CERMETS SiC + Si

[Temperature, T, K; Thermal Conductivity, k, Watt $cm^{-1}K^{-1}$]

T	k
CURVE 1	
556.2	0.562
668.2	0.502
760.2	0.455
871.2	0.410
991.2	0.369
1089.2	0.335
CURVE 2	
673.2	0.960
773.2	0.828
873.2	0.725
973.2	0.640
1073.2	0.570
1173.2	0.518
1273.2	0.483

SPECIFICATION TABLE NO. 228 THERMAL CONDUCTIVITY OF [DISODIUM OXIDE + SODIUM] CERMETS $Na_2O + Na$

Curve No.	Ref. No.	Method Used	Year	Temp. Range, K	Reported Error, %	Name and Specimen Designation	Composition (weight percent), Specifications and Remarks
1	399, 398	L	1965	328	<15		59.1 Na_2O, 40.9 Na.
2	399, 398	L	1965	328	<15		50.6 Na_2O, 49.4 Na.

DATA TABLE NO. 228 THERMAL CONDUCTIVITY OF [DISODIUM OXIDE + SODIUM] CERMETS $Na_2O + Na$

[Temperature, T, K; Thermal Conductivity, k, Watt $cm^{-1}K^{-1}$]

T	k
CURVE 1*	
328	0.554
CURVE 2*	
328	0.578

*No graphical presentation

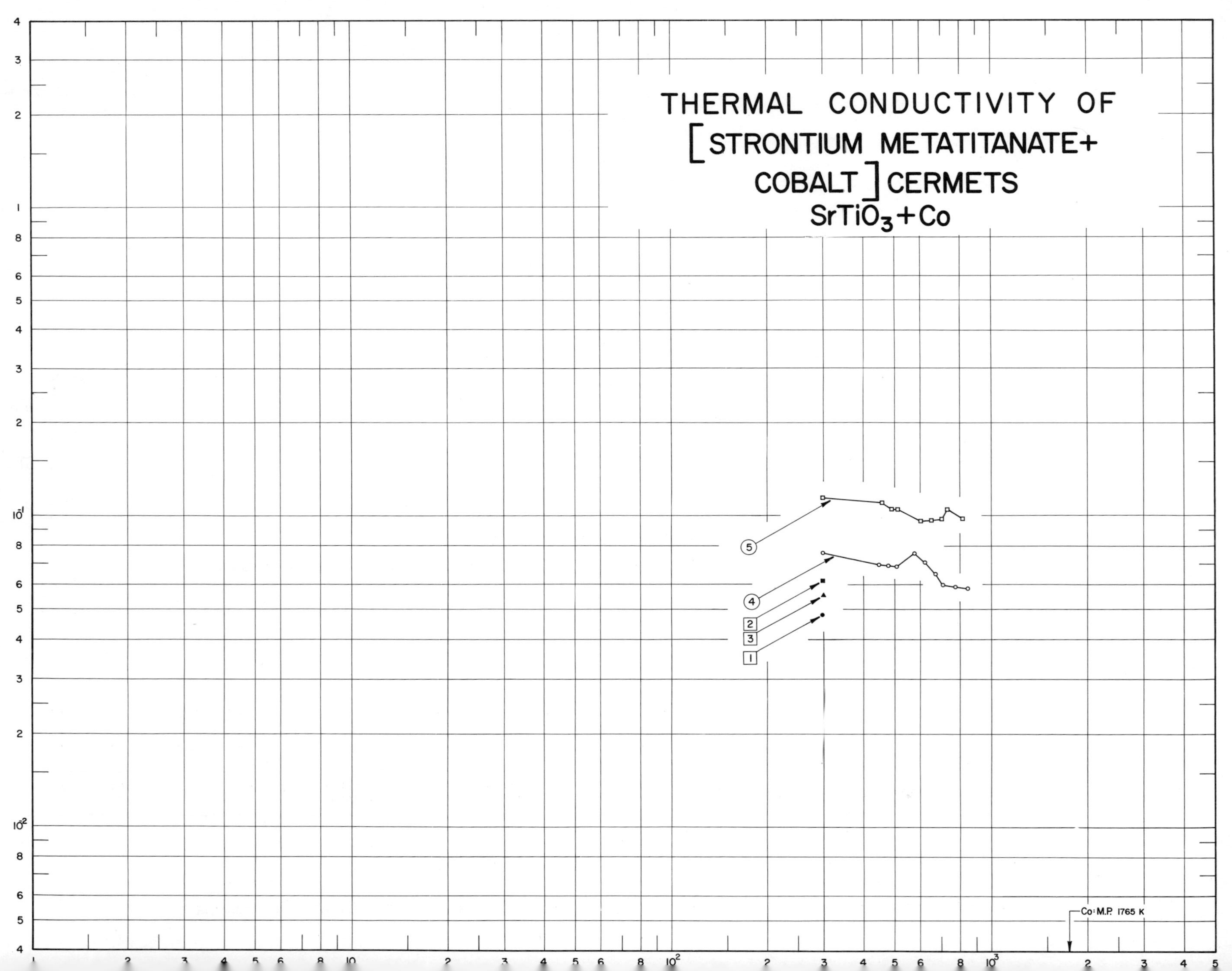
THERMAL CONDUCTIVITY OF
[STRONTIUM METATITANATE +
COBALT] CERMETS
SrTiO$_3$+Co
Co: M.P. 1765 K
THERMAL CONDUCTIVITY, Watt cm^{-1} K^{-1}
5
4
2
3
1

SPECIFICATION TABLE NO. 229 THERMAL CONDUCTIVITY OF [STRONTIUM METATITANATE + COBALT] CERMETS $SrTiO_3 + Co$

[For Data Reported in Figure and Table No. 229]

Curve No.	Ref. No.	Method Used	Year	Temp. Range, K	Reported Error, %	Name and Specimen Designation	Composition (weight percent), Specifications and Remarks
1	283		1959	298.2		Specimen No. 15	10% Co.
2	283		1959	298.2		Specimen No. 4	20% Co.
3	283		1959	298.2		Specimen No. 16	30% Co.
4	283		1959	298-848		Specimen No. 5	30% Co.
5	283		1959	298-818		Specimen No. 12	40% Co.

DATA TABLE NO. 229 THERMAL CONDUCTIVITY OF [STRONTIUM METATITANATE + COBALT] CERMETS $SrTiO_3 + Co$

[Temperature, T, K; Thermal Conductivity, k, Watt $cm^{-1} K^{-1}$]

T	k
CURVE 1	
298.2	0.0481
CURVE 2	
298.2	0.0619
CURVE 3	
298.2	0.0552
CURVE 4	
298.2	0.0761
448.2	0.0695
478.2	0.0690
508.2	0.0686
578.2	0.0757
623.2	0.0707
673.2	0.0649
708.2	0.0598
778.2	0.0590
848.2	0.0582
CURVE 5	
298.2	0.114
458.2	0.110
491.2	0.105
513.2	0.105
601.2	0.0962
653.2	0.0967
701.2	0.0975
733.2	0.105
818.2	0.0979

SPECIFICATION TABLE NO. 230 THERMAL CONDUCTIVITY OF [TITANIUM CARBIDE + COBALT] CERMETS $TiC + Co$

Curve No.	Ref. No.	Method Used	Year	Temp. Range, K	Reported Error, %	Name and Specimen Designation	Composition (weight percent)		Composition (continued), Specifications and Remarks
							TiC	Co	
1	264		1949	298.2			80	20	Density = 5.42 g cm^{-3}.

DATA TABLE NO. 230 THERMAL CONDUCTIVITY OF [TITANIUM CARBIDE + COBALT] CERMETS $TiC + Co$

[Temperature, T, K; Thermal Conductivity, k, Watt $cm^{-1}K^{-1}$]

T	k
CURVE 1*	
298.2	0.356

*No graphical presentation

SPECIFICATION TABLE NO. 231 THERMAL CONDUCTIVITY OF [TITANIUM CARBIDE + COBALT + NIOBIUM CARBIDE] CERMETS $TiC + Co + NbC$

Curve No.	Ref. No.	Method Used	Year	Temp. Range, K	Reported Error, %	Name and Specimen Designation	Composition (weight percent) TiC	Co	NbC	Composition (continued), Specifications and Remarks
1	264		1949	298.2			66.3	18.7	15.0	

DATA TABLE NO. 231 THERMAL CONDUCTIVITY OF [TITANIUM CARBIDE + COBALT + NIOBIUM CARBIDE] CERMETS $TiC + Co + NbC$

[Temperature, T, K; Thermal Conductivity, k, Watt $cm^{-1} K^{-1}$]

T	k
CURVE 1*	
298.2	0.314

* No graphical presentation

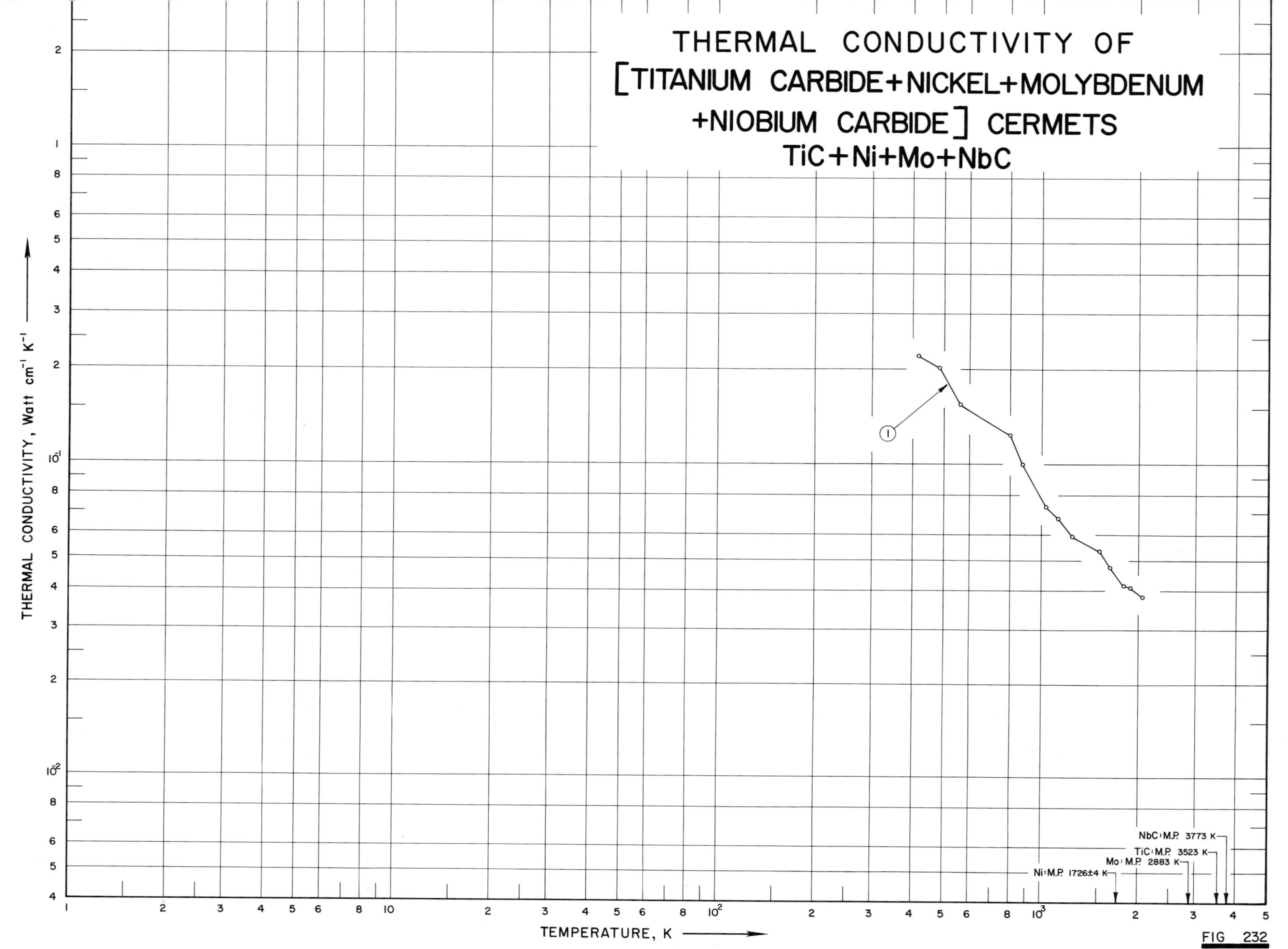

THERMAL CONDUCTIVITY OF
[TITANIUM CARBIDE+NICKEL+MOLYBDENUM
+NIOBIUM CARBIDE] CERMETS
TiC+Ni+Mo+NbC
1
NbC: M.P. 3773 K
TiC: M.P. 3523 K
Mo: M.P. 2883 K
Ni: M.P. 1726±4 K
THERMAL CONDUCTIVITY, Watt cm⁻¹ K⁻¹
TEMPERATURE, K

FIG 232

SPECIFICATION TABLE NO. 232 THERMAL CONDUCTIVITY OF [TITANIUM CARBIDE + NICKEL + MOLYBDENUM + NIOBIUM CARBIDE] CERMETS TiC + Ni + Mo + NbC

[For Data Reported in Figure and Table No. 232]

Curve No.	Ref. No.	Method Used	Year	Temp. Range, K	Reported Error, %	Name and Specimen Designation	Composition (weight percent) TiC	Ni	Mo	NbC	Composition (continued), Specifications and Remarks
1	298	R	1961	420-2062	<5	Kennametals K161B	72.0	16.7	3.3	6.0	Specimen consisted of 5 one-in. disks.

DATA TABLE NO. 232 THERMAL CONDUCTIVITY OF [TITANIUM CARBIDE + NICKEL + MOLYBDENUM + NIOBIUM CARBIDE] CERMETS TiC + Ni + Mo + NbC

[Temperature, T, K; Thermal Conductivity, k, Watt $cm^{-1} K^{-1}$]

T	k
CURVE 1	
419.8	0.222
485.9	0.203
664.8	0.156
802.1	0.124
875.4	0.100
1032.1	0.0736
1134.8	0.0677
1258.7	0.0595
1505.9	0.0537
1649.3	0.0476
1807.6	0.0415
1899.8	0.0412
2061.5	0.0384

SPECIFICATION TABLE NO. 233 THERMAL CONDUCTIVITY OF [TITANIUM CARBIDE + NICKEL + NIOBIUM CARBIDE] CERMETS $TiC + Ni + NbC$

Curve No.	Ref. No.	Method Used	Year	Temp. Range, K	Reported Error, %	Name and Specimen Designation	Composition (weight percent)			Composition (continued), Specifications and Remarks
							TiC	Ni	NbC	
1	296		1958	293.2			70	20	10	Density 5.8 g cm^{-3}.

DATA TABLE NO. 233 THERMAL CONDUCTIVITY OF [TITANIUM CARBIDE + NICKEL + NIOBIUM CARBIDE] CERMETS $TiC + Ni + NbC$

[Temperature, T, K; Thermal Conductivity, k, Watt $cm^{-1}K^{-1}$]

T	k
CURVE 1*	
293.2	0.335

*No graphical presentation

SPECIFICATION TABLE NO. 234 THERMAL CONDUCTIVITY OF [URANIUM CARBIDE + URANIUM] CERMETS UC + U

Curve No.	Ref. No.	Method Used	Year	Temp. Range, K	Reported Error, %	Name and Specimen Designation	Composition (weight percent) UC	U	Composition (continued), Specifications and Remarks
1	482, 483		1960	479-969			80.0	20.0	Prepared by sintering mixture of UC and U in a graphite crucible in a vacuum furnace with a graphite heater; data corrected to zero porosity.

DATA TABLE NO. 234 THERMAL CONDUCTIVITY OF [URANIUM CARBIDE + URANIUM] CERMETS UC + U

[Temperature, T, K; Thermal Conductivity, k, Watt $cm^{-1}K^{-1}$]

T	k
CURVE 1*	
479.2	0.232
642.2	0.200
969.2	0.239

*No graphical presentation

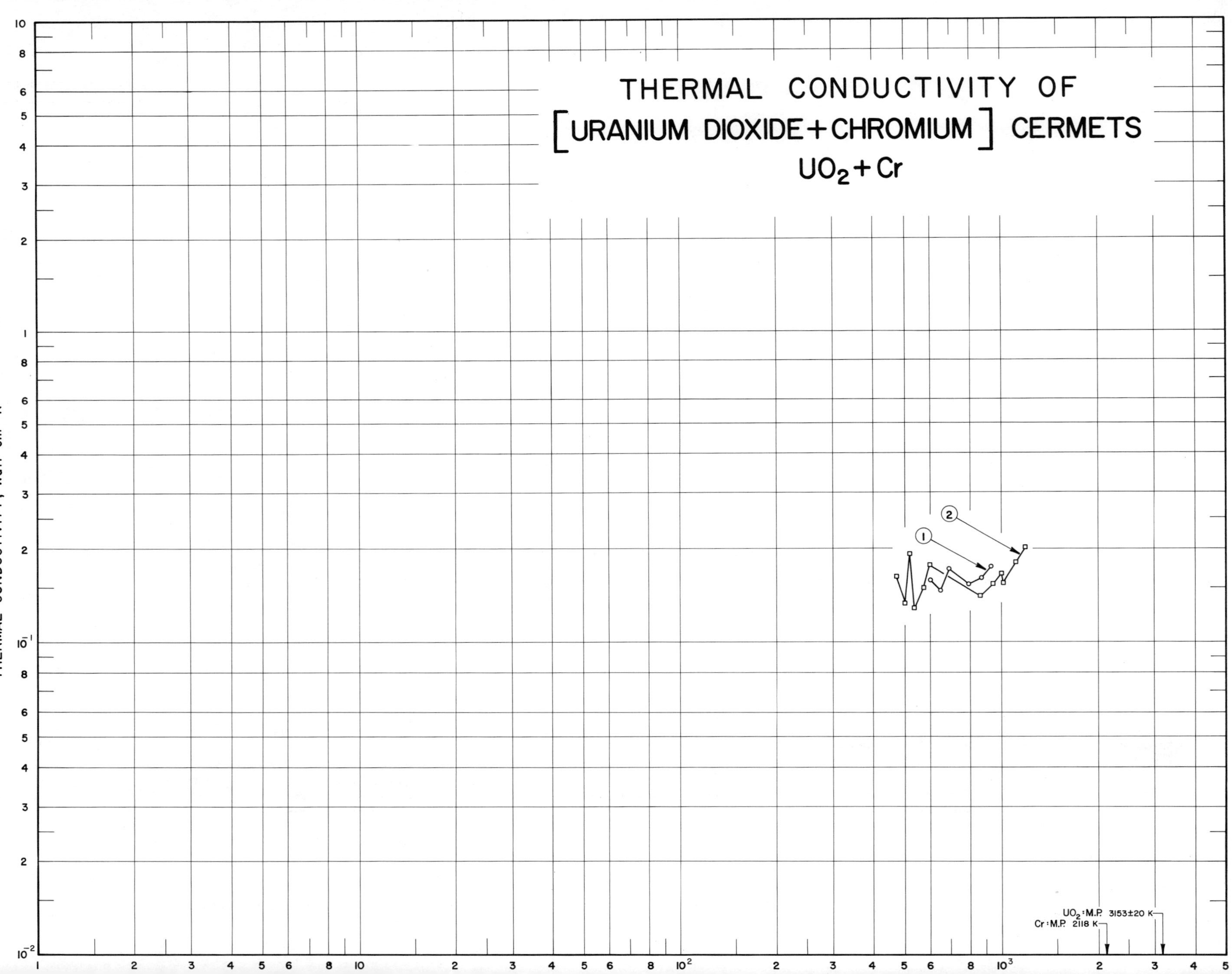

THERMAL CONDUCTIVITY OF
[URANIUM DIOXIDE + CHROMIUM] CERMETS
UO_2 + Cr
THERMAL CONDUCTIVITY, Watt cm^{-1} K^{-1}
UO_2: M.P. 3153±20 K
Cr: M.P. 2118 K

SPECIFICATION TABLE NO. 235 THERMAL CONDUCTIVITY OF [URANIUM DIOXIDE + CHROMIUM] CERMETS $UO_2 + Cr$

[For Data Reported in Figure and Table No. 235]

Curve No.	Ref. No.	Method Used	Year	Temp. Range, K	Reported Error, %	Name and Specimen Designation	Composition (weight percent), Specifications and Remarks
1	385	C	1961	602-927	<±5	TC-104	80 vol% UO_2; 97.1 % of theoretical density; using - 100 + 400 mesh spherical UO_2; gas-pressure bonded for hrs at 2300 F and 10,000 psi of helium gas pressure.
2	385	C	1961	474-1188	<±5	TC-104	Same specimen as above, 2nd run.

DATA TABLE NO. 235 THERMAL CONDUCTIVITY OF [URANIUM DIOXIDE + CHROMIUM] CERMETS $UO_2 + Cr$

[Temperature, T, K; Thermal Conductivity, k, Watt $cm^{-1} K^{-1}$]

T	k
CURVE 1	
602.2	0.158
647.2	0.146
686.2	0.172
794.2	0.154
866.2	0.161
927.2	0.175
CURVE 2	
474.2	0.162
500.2	0.133
522.2	0.192
539.2	0.128
573.2	0.149
599.2	0.176
863.2	0.141
940.2	0.154
1005.2	0.166
1017.2	0.155
1113.2	0.181
1188.2	0.202

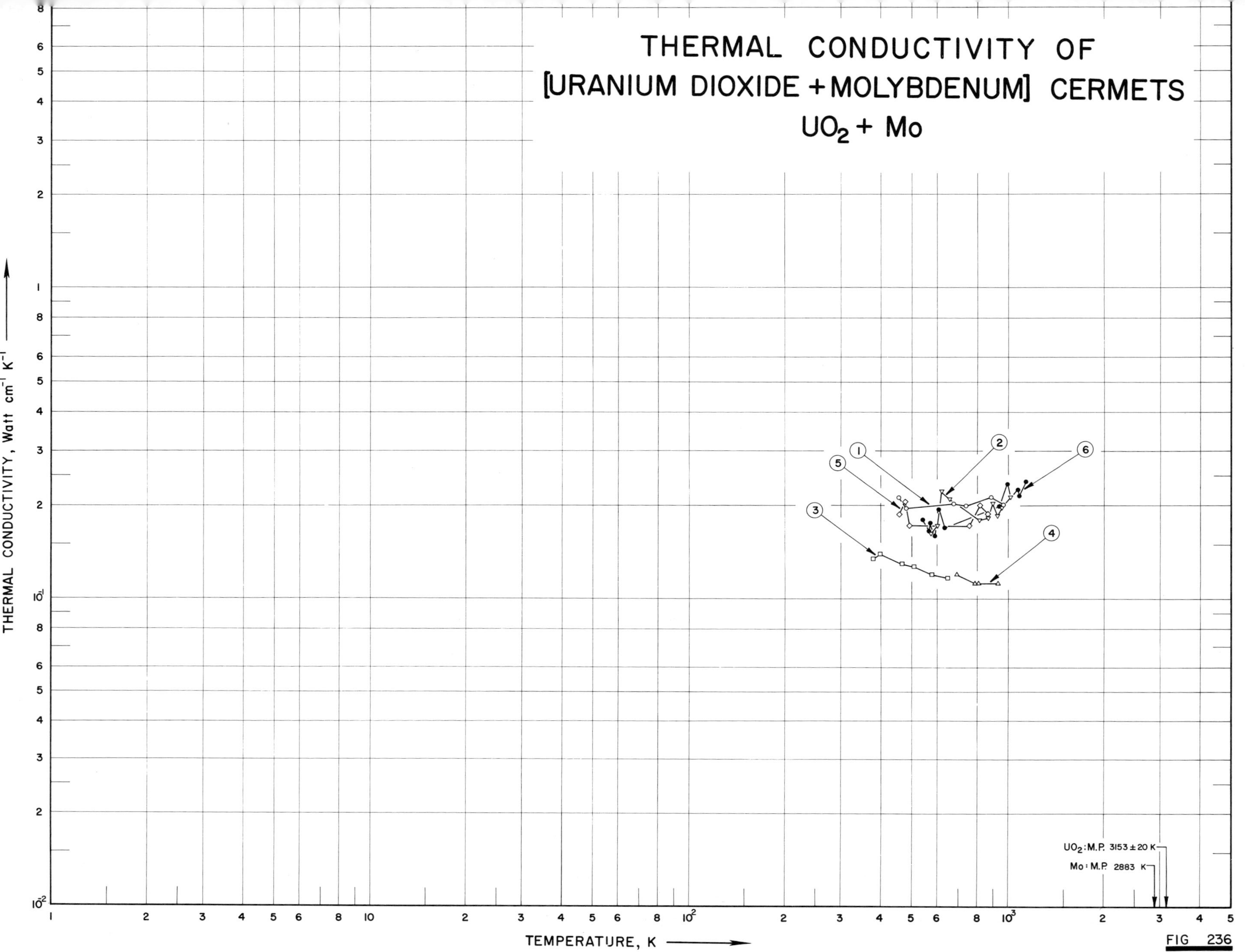
THERMAL CONDUCTIVITY OF
[URANIUM DIOXIDE + MOLYBDENUM] CERMETS
UO2 + Mo
THERMAL CONDUCTIVITY, Watt cm-1 K-1
TEMPERATURE, K
UO2: M.P. 3153 ± 20 K
Mo: M.P. 2883 K
1
2
3
4
5
6

FIG 236

SPECIFICATION TABLE NO. 236 THERMAL CONDUCTIVITY OF [URANIUM DIOXIDE + MOLYBDENUM] CERMETS UO_2 + Mo

[For Data Reported in Figure and Table No. 236]

Curve No.	Ref. No.	Method Used	Year	Temp. Range, K	Reported Error, %	Name and Specimen Designation	Composition (weight percent), Specifications and Remarks
1	385	C	1961	456-972	< ± 5	Tc - 82	70 volume % UO_2, 91.7% of theoretical density, using - 100 + 140 mesh hydrothermal UO_2; specimen prepared in rod form by gas pressure bonding for 3 hours at 1533 K and 10,000 psi of helium gas pressure; Armco iron used as primary standard and Type 347 stainless steel used as secondary standard.
2	385	C	1961	575-1020	< ± 5	Tc - 82	The above specimen, second run.
3	385	C	1961	379-647	< ± 5	Tc - 90	80 volume % UO_2, 91.1% of theoretical density; using - 100 + 140 mesh hydrothermal UO_2; specimen prepared in rod form by gas pressure bonding for 3 hours at 1589 K and 10,000 psi of helium gas pressure; Armco iron used as primary standard and Type 347 stainless steel used as secondary standard.
4	385	C	1961	692-932	< ± 5	Tc - 90	The above specimen, second run.
5	385	C	1961	458-866	< ± 5	Tc - 105	80 volume % UO_2, 94.4% of theoretical density; using - 100 + 140 mesh spherical UO_2; specimen prepared in rod form by gas pressure bonding for 3 hours at 1561 K and 10,000 psi of helium gas pressure; Armco iron used as primary standard and Type 347 stainless steel used as secondary standard.
6	385	C	1961	544-1143	< ± 5	Tc - 105	The above specimen, second run.

DATA TABLE NO. 236 THERMAL CONDUCTIVITY OF [URANIUM DIOXIDE + MOLYBDENUM] CERMETS UO_2 + Mo

[Temperature, T, K; Thermal Conductivity, k, Watt $cm^{-1}K^{-1}$]

T	k
CURVE 1	
456.2	0.212
483.2	0.196
677.2	0.203
740.2	0.199
886.2	0.213
972.2	0.201
CURVE 2	
575.2	0.163
604.2	0.173
623.2	0.222
657.2	0.210
815.2	0.179
866.2	0.183
898.2	0.203
927.2	0.185
956.2	0.197
1020.2	0.212
CURVE 3	
379.2	0.136
398.2	0.140
468.2	0.130
507.2	0.127
579.2	0.120
647.2	0.117
CURVE 4	
692.2	0.120
788.2	0.112
810.2	0.112
932.2	0.112
CURVE 5	
458.2	0.187
476.2	0.206
492.2	0.173
759.2	0.173
816.2	0.200
866.2	0.187
CURVE 6	
544.2	0.180
568.2	0.166
574.2	0.176
591.2	0.160
606.2	0.194
635.2	0.170
944.2	0.199
997.2	0.236
1019.2	0.212*
1076.2	0.225
1084.2	0.214
1143.2	0.239

*Not shown on plot

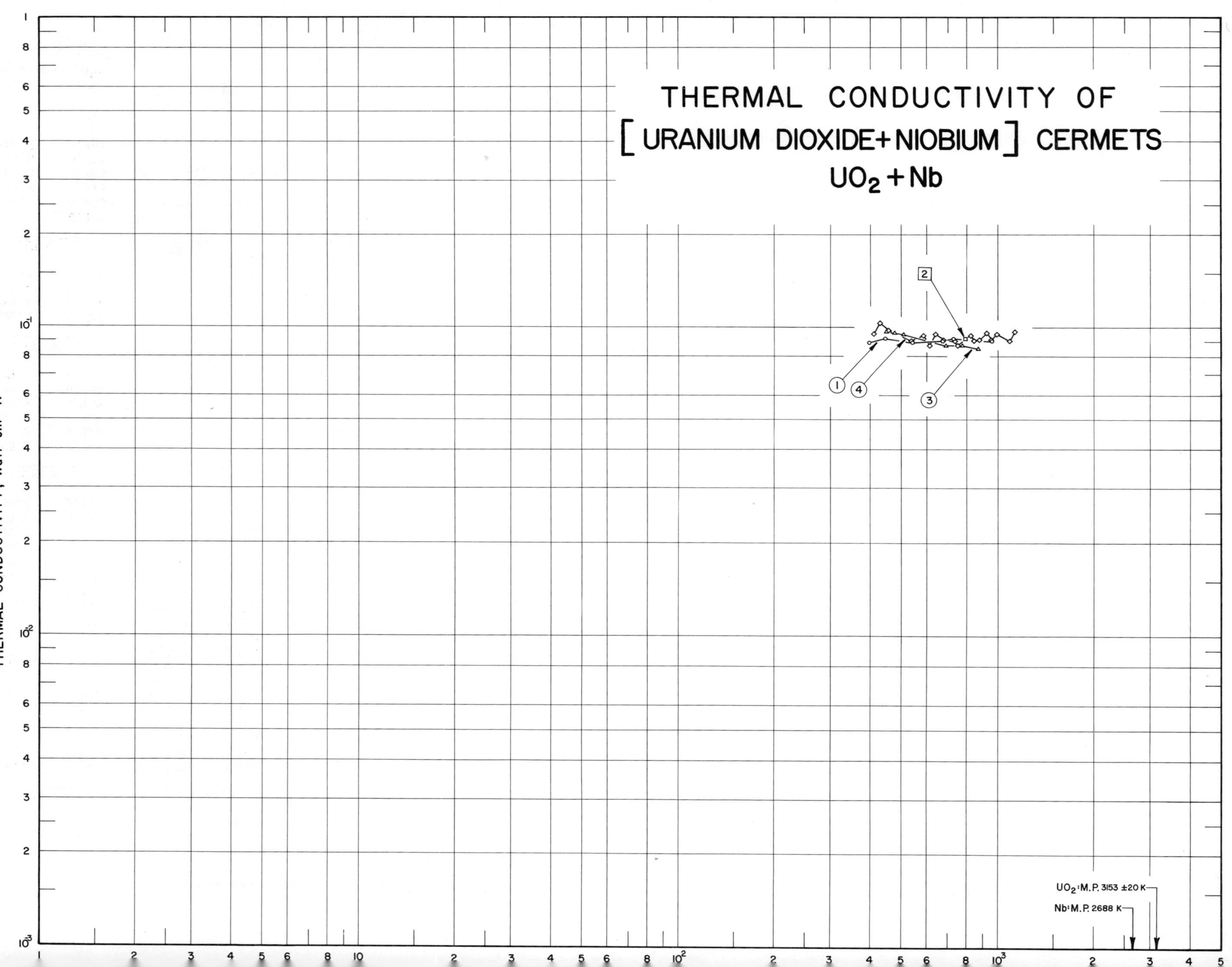
THERMAL CONDUCTIVITY OF
[URANIUM DIOXIDE+NIOBIUM] CERMETS
UO_2 + Nb
THERMAL CONDUCTIVITY, Watt cm^{-1} K^{-1}
UO_2: M.P. 3153 ±20 K
Nb: M.P. 2688 K
1
2
3
4

SPECIFICATION TABLE NO. 237 THERMAL CONDUCTIVITY OF [URANIUM DIOXIDE + NIOBIUM] CERMETS UO_2 + Nb

[For Data Reported in Figure and Table No. 237]

Curve No.	Ref. No.	Method Used	Year	Temp. Range, K	Reported Error, %	Name and Specimen Designation	Composition (weight percent), Specifications and Remarks
1	385	C	1961	397-966	<±5	TC-115	80 vol % UO_2; 85.3 % of theoretical density; niobium coated -100 + 140 mesh spherical UO_2; gas-pressure bonded for 3 hrs at 2100 F and 10,000 psi of helium gas pressure; Armco iron used as primary standard and Type 347 stainless steel used as secondary standard.
2	385	C	1961	795.2	<±5	TC-115	The above specimen, 2nd run.
3	385	C	1961	450-872	<±5	TC-130	80 vol % UO_2; 93.5 % of theoretical density; niobium coated - 140 + 200 mesh hydrothermal UO_2; gas-pressure bonded 3 hrs at 1561 K and 10,000 psi of helium gas pressure; Armco iron used as primary standard and Type 347 stainless steel used as secondary standard.
4	385	C	1961	413-1131	<±5	TC-130	The above specimen, 2nd run.

DATA TABLE NO. 237 THERMAL CONDUCTIVITY OF [URANIUM DIOXIDE + NIOBIUM] CERMETS UO_2 + Nb

[Temperature, T, K; Thermal Conductivity, k, Watt $cm^{-1} K^{-1}$]

T	k
CURVE 1	
397.2	0.089
447.2	0.092
544.2	0.089
677.2	0.090
730.2	0.092
966.2	0.090
CURVE 2	
795.2	0.092
CURVE 3	
450.2	0.097
478.2	0.096
510.2	0.095
690.2	0.087
775.2	0.087
872.2	0.085
CURVE 4	
413.2	0.095
434.2	0.103
457.2	0.098
540.2	0.090
587.2	0.094
614.2	0.087
641.2	0.095
678.2	0.091
736.2	0.090
751.2	0.087
825.2	0.094
843.2	0.090
876.2	0.091
925.2	0.096
958.2	0.091
996.2	0.095
1090.2	0.090
1131.2	0.097

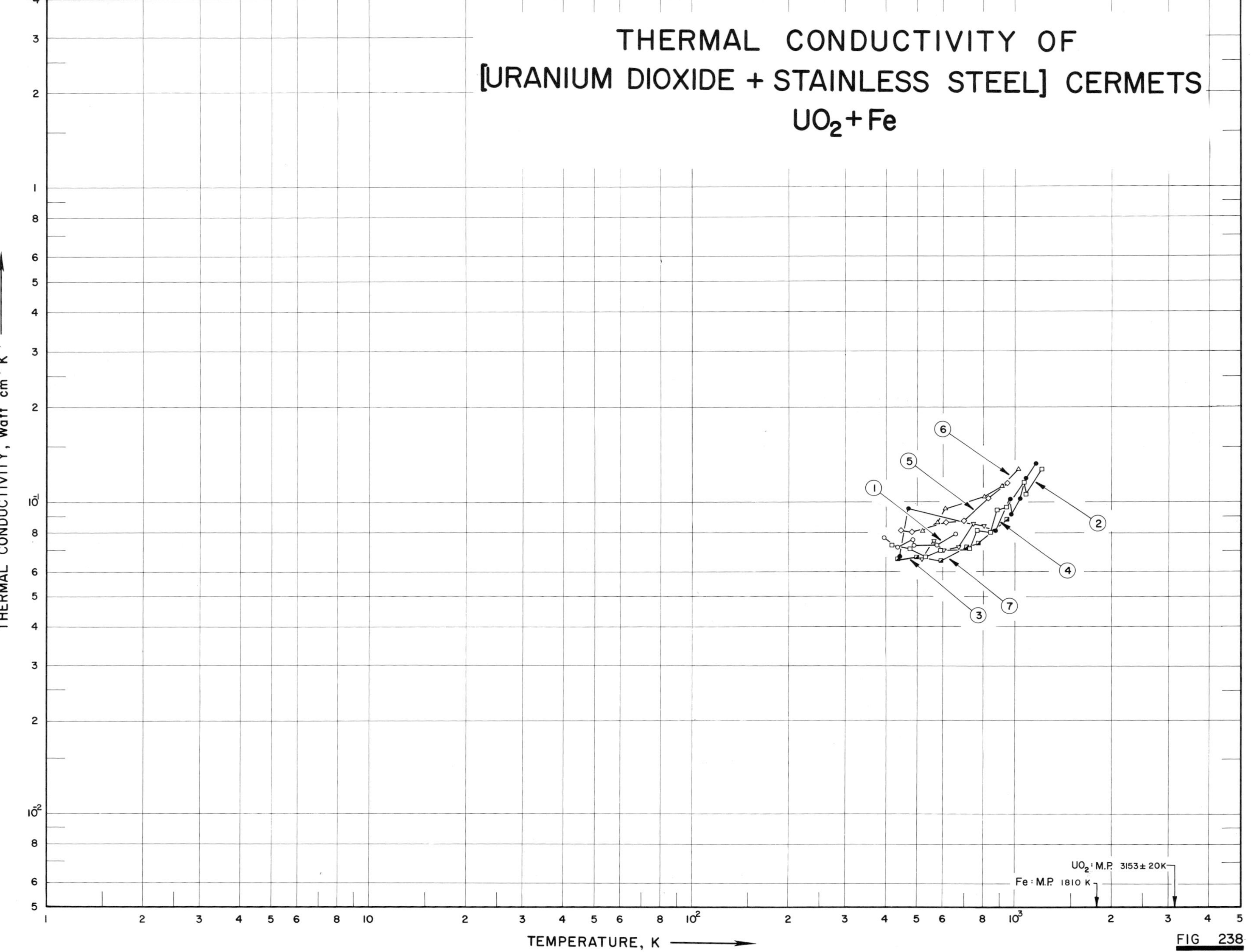

THERMAL CONDUCTIVITY OF
[URANIUM DIOXIDE + STAINLESS STEEL] CERMETS
UO2+Fe
THERMAL CONDUCTIVITY, Watt cm-1 K-1
TEMPERATURE, K
UO2: M.P. 3153± 20K
Fe: M.P. 1810 K
FIG 238

SPECIFICATION TABLE NO. 238 THERMAL CONDUCTIVITY OF [URANIUM DIOXIDE + STAINLESS STEEL ΣX_i] CERMETS $UO_2 + Fe + \Sigma X_i$

[For Data Reported in Figure and Table No. 238]

Curve No.	Ref. No.	Method Used	Year	Temp. Range, K	Reported Error, %	Name and Specimen Designation	Composition (weight percent), Specifications and Remarks
1	385	C	1961	395-656	$< \pm 5$	Tc - 102	80 volume % of UO_2, 98.4% of theoretical density; using - 100 + 140 mesh hydrothermal UO_2 and Type 302B stainless steel; gas pressure bonded for 3 hours at 1533 K and 10,000 psi of helium gas pressure; Armco iron used as primary standard and Type 347 stainless steel as secondary standard.
2	385	C	1961	418-1213	$< \pm 5$	Tc - 102	The above specimen, second run.
3	385	C	1961	518-804	$< \pm 5$	Tc - 103	80 volume % of UO_2, 97.2% of theoretical density; using - 100 + 140 spherical UO_2 and Type 302B stainless steel; gas pressure bonded for 3 hours at 1533 K and 10,000 psi of helium gas pressure; Armco iron used as primary standard and Type 347 stainless steel as secondary standard.
4	385	C	1961	442-1169	$< \pm 5$	Tc - 103	The above specimen, second run.
5	385	C	1961	446-947	$< \pm 5$	Tc - 81	70 volume % of UO_2, 97.0% of theoretical density; using - 100 + 140 mesh hydrothermal UO_2 and Type 302B stainless steel; gas pressure bonded for 3 hours at 1533 K and 10,000 psi of helium gas pressure; Armco iron used as primary standard and Type 347 stainless steel as secondary standard.
6	385	C	1961	521-1031	$< \pm 5$	Tc - 81	The above specimen, second run.
7	385	C	1961	437-945	$< \pm 5$	Tc - 80	80 volume % of UO_2, 95.5% of theoretical density; using - 100 + 140 mesh hydrothermal UO_2 and Type 302B stainless steel; gas pressure bonded for 3 hours at 1533 K and 10,000 psi of helium gas pressure; Armco iron used as primary standard and Type 347 stainless steel as secondary standard.
8	385	C	1961	415-1012	$< \pm 5$	Tc - 80	The above specimen, second run.

DATA TABLE NO. 238 THERMAL CONDUCTIVITY OF [URANIUM DIOXIDE + STAINLESS STEEL] CERMETS $UO_2 + Fe + \Sigma X_i$

[Temperature, T, K; Thermal Conductivity, k, Watt $cm^{-1}K^{-1}$]

T	k	T	k
CURVE 1		CURVE 5	
395.0	0.077	446.2	0.081
433.0	0.072	483.2	0.080
484.0	0.076	614.2	0.086
488.0	0.073	696.2	0.087
575.0	0.073	832.2	0.103
656.0	0.079	947.2	0.115
CURVE 2		CURVE 6	
418.0	0.073	521.2	0.081
476.0	0.071	578.2	0.086
532.0	0.067	714.2	0.095
590.0	0.070	807.2	0.104
727.0	0.071	915.2	0.113
765.0	0.081	1031.2	0.127
844.0	0.080		
884.0	0.094	CURVE 7	
943.0	0.096		
1070.0	0.116	437.2	0.066
1084.0	0.106	499.2	0.067
1213.0	0.127	593.2	0.065
		715.2	0.072
CURVE 3		773.2	0.074
		945.2	0.088
518.0	0.066		
564.0	0.075	CURVE 8*	
604.0	0.070		
674.0	0.072	415.2	0.063
745.0	0.085	465.2	0.071
804.0	0.084	828.2	0.080
		1012.2	0.092
CURVE 4			
442.0	0.067		
470.0	0.095		
873.0	0.081		
970.0	0.102		
978.0	0.091		
1047.0	0.102		
1088.0	0.119		
1169.0	0.133		

*Not shown on plot

SPECIFICATION TABLE NO. 239 THERMAL CONDUCTIVITY OF [URANIUM DIOXIDE + URANIUM] CERMETS $UO_2 + U$

Curve No.	Ref. No.	Method Used	Year	Temp. Range, K	Reported Error, %	Name and Specimen Designation	Composition (weight percent)		Composition (continued), Specifications and Remarks
							UO_2	U	
1	430	→	1966	298.2			55.1	44.9	Spherical uranium powder obtained from National Lead Co. containing impurities; 0.05 Fe, 0.01 Mg, 0.008 Mo, 0.005 Si, <0.005 K, <0.005 P, <0.005 Ti, <0.005 Zn, <0.002 Ca, <0.001 As, <0.001 Na, 0.0005 Ni, <0.0005 Al, <0.0005 Co, <0.0005 Sn, 0.0004 Mn, 0.0002 Cu, 0.0001 Pb, traces of Ag, Bi, Cr, Li, Sb, Be, and B; oxidized to desired percentage by spreading over the bottom of a Petri dish and placed in an oven at 150 C; specimen contained in a 0.75 in. dia x 2 in. long stainless steel cylindrical cell; mesh size -70+80; thermal conductivity measured by using the transient line source method, the heat source was a 36-gauge constantan wire contained in a 0.025 in. O.D. hypodermic tube soldered along the axis of the cylindrical cell, data calculated from measured line temperatures at two certain times; measured in nitrogen at 1 atm.
2	430	→	1966	298.2			55.1	44.9	Similar to the above specimen; measured in nitrogen under pressure in the range $9.12 \times 10^{-3} \sim 5.495 \times 10^{3}$ mm Hg.
3	430	→	1966	298.2			90.0	10.0	Similar to the above specimen; measured in nitrogen at 1 atm.
4	430	→	1966	298.2			60.3	39.7	Same impurities, source, and measuring method as the above specimen; mesh size -230+325; measured in nitrogen at 1 atm.
5	430	→	1966	298.2			54.1	45.9	Similar to the above specimen; measured in nitrogen under pressure in the range $4.90 \times 10^{-3} \sim 4.677 \times 10^{3}$ mm Hg.

DATA TABLE NO. 239 THERMAL CONDUCTIVITY OF [URANIUM DIOXIDE + URANIUM] CERMETS $UO_2 + U$

[Temperature, T, K; Thermal Conductivity, k, Watt $cm^{-1} K^{-1}$]

T	k
CURVE 1*	
298.2	0.00239

p(mm Hg)	k
CURVE 2* T = 298.2	
0.00912	0.0000962
4.27	0.000322
31.3	0.000870
216	0.00146
776	0.00171
1514	0.00181
5495	0.00188

T	k
CURVE 3*	
298.2	0.00254
CURVE 4*	
298.2	0.00205

p(mm Hg)	k
CURVE 5* T = 298.2	
0.00490	0.000130
0.0501	0.000163
0.832	0.000268
14.6	0.000711
135	0.00119
767	0.00157
1738	0.00174
4677	0.00176

*No graphical presentation

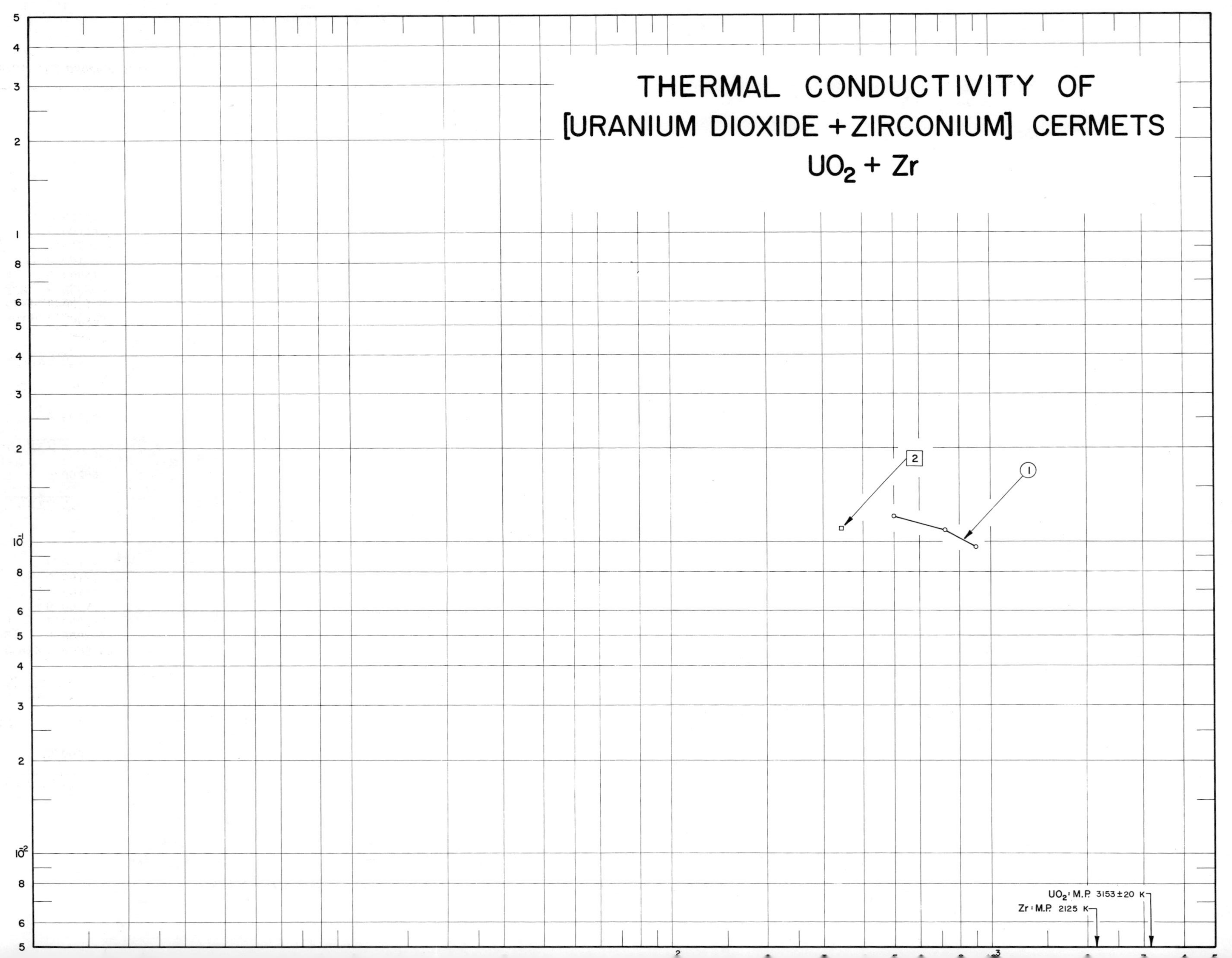

THERMAL CONDUCTIVITY OF
[URANIUM DIOXIDE + ZIRCONIUM] CERMETS
UO2 + Zr
2
1
UO2: M.P. 3153±20 K
Zr: M.P. 2125 K
THERMAL CONDUCTIVITY, Watt cm-1 K-1

SPECIFICATION TABLE NO. 240 THERMAL CONDUCTIVITY OF [URANIUM DIOXIDE + ZIRCONIUM] CERMETS $UO_2 + Zr$

[For Data Reported in Figure and Table No. 240]

Curve No.	Ref. No.	Method Used	Year	Temp. Range, K	Reported Error, %	Name and Specimen Designation	Composition (weight percent), Specifications and Remarks
1	76	C	1952	498-903			43 UO_2, 57 Zr; 59% of theoretical density.
2	386	C	1954	343			80 UO_2, 20 Zr; hot pressed in helium atmosphere at 1750 C and 3500 psi for 30 min; density 8.71 g cm^{-3}.

DATA TABLE NO. 240 THERMAL CONDUCTIVITY OF [URANIUM DIOXIDE + ZIRCONIUM] CERMETS $UO_2 + Zr$

[Temperature, T, K; Thermal Conductivity, k, Watt $cm^{-1}K^{-1}$]

T	k
CURVE 1	
498.2	0.121
723.2	0.109
903.2	0.096
CURVE 2	
343.0	0.111

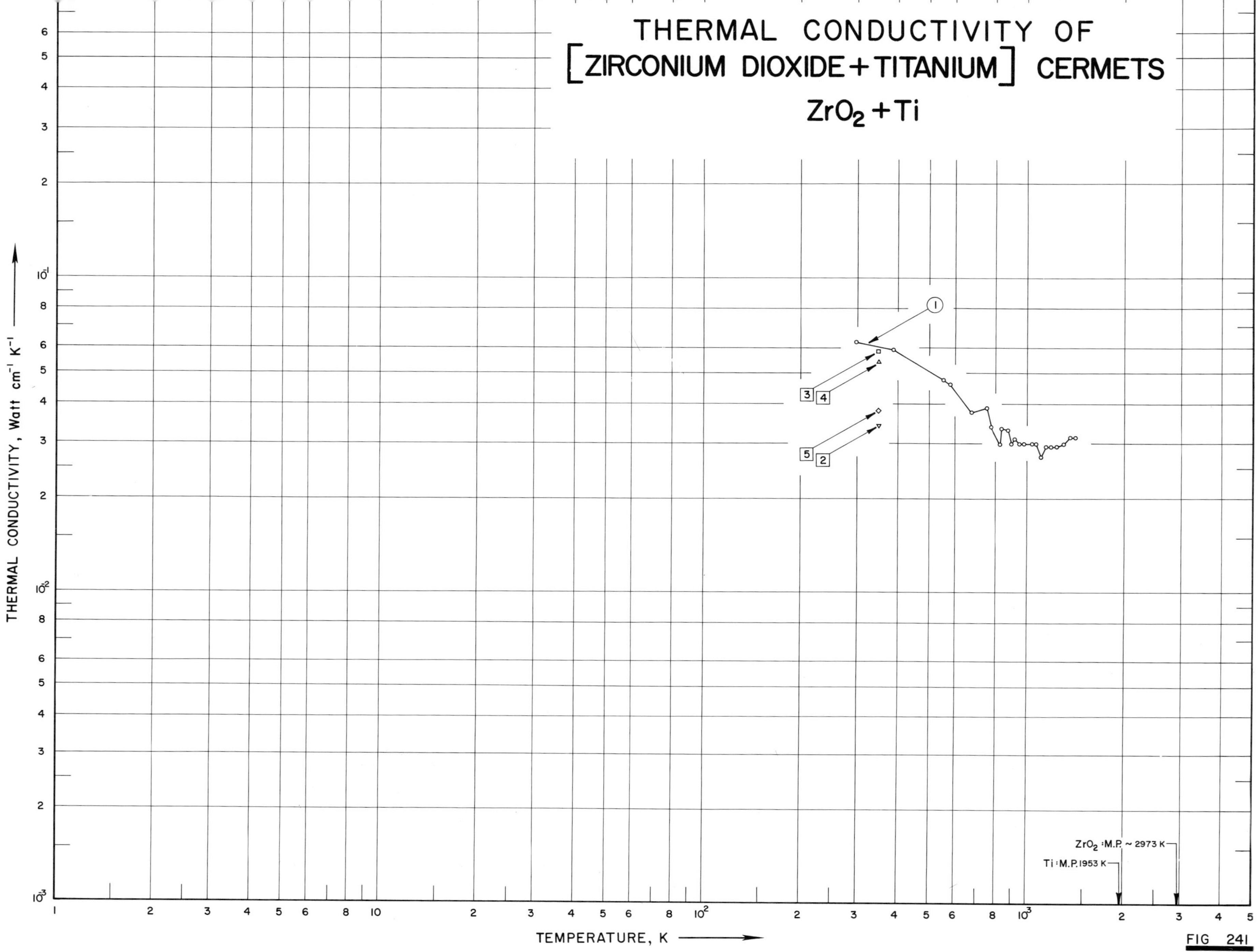
THERMAL CONDUCTIVITY OF
[ZIRCONIUM DIOXIDE + TITANIUM] CERMETS
$ZrO_2 + Ti$
1
3
4
5
2
ZrO_2 : M.P. ~ 2973 K
Ti : M.P. 1953 K
THERMAL CONDUCTIVITY, Watt cm^{-1} K^{-1}
TEMPERATURE, K
FIG 241

SPECIFICATION TABLE NO. 241 THERMAL CONDUCTIVITY OF [ZIRCONIUM DIOXIDE + TITANIUM] CERMETS $ZrO_2 + Ti$

[For Data Reported in Figure and Table No. 241]

Curve No.	Ref. No.	Method Used	Year	Temp. Range, K	Reported Error, %	Name and Specimen Designation	Composition (weight percent) ZrO_2	Composition (weight percent) Ti	Composition (continued), Specifications and Remarks
1	523	R	1964	298-1423			93.58	6.42	The monoclinic zirconia contains 98.8 ZrO_2 (including HfO_2), 0.33 Si, 0.10 TiO_2 and 0.10 CaO, original particle size 0.26 μ; the titanium power contains 98 Ti and 1.1 N, original particle size -325 mesh; specimen was in cylindrical form of 1.375 in. O.D. by 1.000 in. high with a 23/64 in. dia center hole; milled, cold-pressed and then vacuum sintered at 1870 C for 1 hr; density = 5.65 - 5.75 g cm^{-3}.
2	524	C	1963	348.2			98.0	2.0	ZrO_2 has a reported purity of 99.87% for the total oxide including 2% HfO_2; the cermet was prepared by mixing in methyl alcohol , drying, pelletizing, then by pressing into wafer-type specimen, and then was fired in a vacuum resistance furnace for 2 hrs at 1800 C and cooled at a rate of approximately 60 C per min.
3	524	C	1963	348.2			93.6	6.4	Same fabrication as above.
4	524	C	1963	348.2			93.6	6.4	Same fabrication as above,except the firing temperature was 200 C.
5	524	C	1963	348.2			85.7	14.3	Same fabrication as above,except the firing temperature was 1800 C.

DATA TABLE NO. 241 THERMAL CONDUCTIVITY OF [ZIRCONIUM DIOXIDE + TITANIUM] CERMETS $ZrO_2 + Ti$

[Temperature, T, K; Thermal Conductivity, k, Watt $cm^{-1} K^{-1}$]

T	k
CURVE 1	
298.2	0.0628
388.2	0.0594
553.2	0.0477
583.2	0.0464
678.2	0.0377
753.2	0.0389
778.2	0.0339
828.2	0.0297
838.2	0.0335
878.2	0.0331
898.2	0.0297
923.2	0.0310
953.2	0.0297
983.2	0.0297
1048.2	0.0297
1073.2	0.0297
1118.2	0.0272
1153.2	0.0293
1193.2	0.0293
1243.2	0.0293
1303.2	0.0297
1373.2	0.0314
1423.2	0.0314
CURVE 2	
348.2	0.0343
CURVE 3	
348.2	0.0586
CURVE 4	
348.2	0.0544
CURVE 5	
348.2	0.0381

SPECIFICATION TABLE NO. 242 THERMAL CONDUCTIVITY OF [ZIRCONIUM DIOXIDE + ZIRCONIUM] CERMETS $ZrO_2 + Zr$

Curve No.	Ref. No.	Method Used	Year	Temp. Range, K	Reported Error, %	Name and Specimen Designation	Composition (weight percent) ZrO_2	Zr	Composition (continued), Specifications and Remarks
1	430	→	1966	298.2			54.5	45.5	Powder specimen contained in a 0.75 in. dia x 2 in. long stainless steel cylindrical cell; mesh size-70+80; thermal conductivity measured by using the transient line source method, the heat source was a 36-gauge constantan wire contained in a 0.025 in. O.D. hypodermic tube soldered along the axis of the cylindrical cell, data calculated from the measured line temperature at two certain times; measured in nitrogen at 1 atm.
2	430	→	1966	298.2			90.0	10.0	Similar to the above specimen.

DATA TABLE NO. 242 THERMAL CONDUCTIVITY OF [ZIRCONIUM DIOXIDE + ZIRCONIUM] CERMETS $ZrO_2 + Zr$

[Temperature, T, K; Thermal Conductivity, k, Watt $cm^{-1}K^{-1}$]

T	k
CURVE 1*	
298.2	0.00260
CURVE 2*	
298.2	0.00195

*No graphical presentation

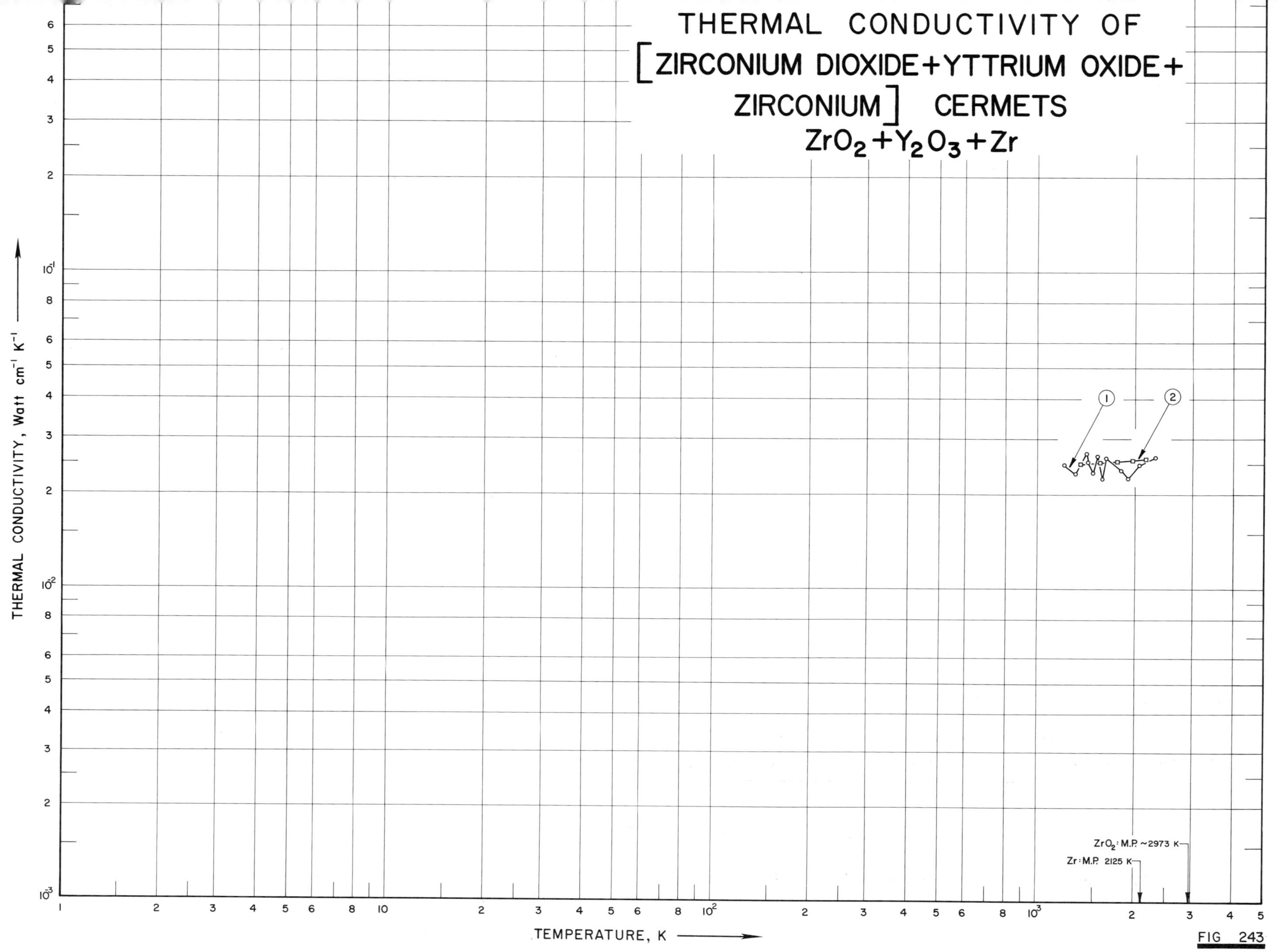

THERMAL CONDUCTIVITY OF
[ZIRCONIUM DIOXIDE + YTTRIUM OXIDE + ZIRCONIUM] CERMETS
$ZrO_2 + Y_2O_3 + Zr$
1
2
ZrO_2: M.P. ~2973 K
Zr: M.P. 2125 K
THERMAL CONDUCTIVITY, Watt cm^{-1} K^{-1}
TEMPERATURE, K

FIG 243

SPECIFICATION TABLE NO. 243 THERMAL CONDUCTIVITY OF [ZIRCONIUM DIOXIDE + YTTRIUM OXIDE + ZIRCONIUM] CERMETS $ZrO_2 + Y_2O_3 + Zr$

[For Data Reported in Figure and Table No. 243]

Curve No.	Ref. No.	Method Used	Year	Temp. Range, K	Reported Error, %	Name and Specimen Designation	Composition (weight percent) ZrO_2	Y_2O_3	Zr	Composition (continued), Specifications and Remarks
1	291	R	1964	1213-2308			80	12	8	Y_2O_3 stabilized ZrO_2 with 8% Zr; prepared by hot-pressing the mixed powders at 1900 C and then machined into disc form with 2 in. O.D. and 0.375 in. I.D.; particles of Zr metal uniformly distributed through the material; density was 97% of theoretical value.
2	291	R	1964	1373-2173						Same specimen as above; data corrected to 100% theoretical value of density.

DATA TABLE NO. 243 THERMAL CONDUCTIVITY OF [ZIRCONIUM DIOXIDE + YTTRIUM OXIDE + ZIRCONIUM] CERMETS $ZrO_2 + Y_2O_3 + Zr$

[Temperature, T, K; Thermal Conductivity, k, Watt $cm^{-1} K^{-1}$]

T	k
CURVE 1	
1213.2	0.0248
1233.2	0.0244*
1318.2	0.0234
1358.2	0.0234*
1423.2	0.027
1433.2	0.0254
1483.2	0.0236
1543.2	0.0266
1588.2	0.0225
1633.2	0.0262
1818.2	0.0239
1833.2	0.0234*
1908.2	0.0226
2073.2	0.0248
2308.2	0.0263
CURVE 2	
1373.2	0.025
1573.2	0.0253
1773.2	0.0255
1973.2	0.0257
2173.2	0.0259

* Not shown on plot

SPECIFICATION TABLE NO. 244 THERMAL CONDUCTIVITY OF AMMONIUM PERCHLORATE NH_4ClO_4

Curve No.	Ref. No.	Method Used	Year	Temp. Range, K	Reported Error, %	Name and Specimen Designation	Composition (weight percent), Specifications and Remarks
1	509	P	1964	323-513	± 5	Reagent grade	Specimen consists of two identical discs 1.125 in. in dia, compressed from powder; particle size range from 43 to 61 μ; density 1.9 g cm^{-3}; porosity 2.3%.

DATA TABLE NO. 244 THERMAL CONDUCTIVITY OF AMMONIUM PERCHLORATE NH_4ClO_4

[Temperature, T, K; Thermal Conductivity, k, Watt $cm^{-1}K^{-1}$]

T	k
CURVE 1*	
323.2	0.00469
373.2	0.00452
423.2	0.00427
473.2	0.00402
513.2	0.00377

*No graphical presentation

SPECIFICATION TABLE NO. 245 THERMAL CONDUCTIVITY OF CADMIUM GERMANIUM PHOSPHIDE $CdGeP_2$

Curve No.	Ref. No.	Method Used	Year	Temp. Range, K	Reported Error, %	Name and Specimen Designation	Composition (weight percent), Specifications and Remarks
1	510	L	1966	298.2		$CdGeP_2$; 6	Stoichiometric n-type polycrystalline; specimen 12 x 4 x 4 mm; prepared from 99.9999 pure Cd, p-type Ge (~50 ohm cm), and 99.9999 pure red phosphorus, materials leaded into a high-purity graphite crucible which was closely fitted in a sealed quartz tube, heated at 10 C hr^{-1} to 580 C, annealed for 12 hrs under an external argon pressure of 60 atm, heated at 10 C hr^{-1} to 980 C, after a period of 30 min, cooled down to 830 C, rotated and vibrated for 30 min, then the Bridgman process performed at a lower rate of 10 mm hr^{-1} till room temperature reached; melting point 800 C; electrical resistivity 4.8 x 10^7 ohm cm at room temperature.

DATA TABLE NO. 245 THERMAL CONDUCTIVITY OF CADMIUM GERMANIUM PHOSPHIDE $CdGeP_2$

[Temperature, T, K; Thermal Conductivity, k, Watt $cm^{-1}K^{-1}$]

T	k
CURVE 1*	
298.2	0.11

*No graphical presentation

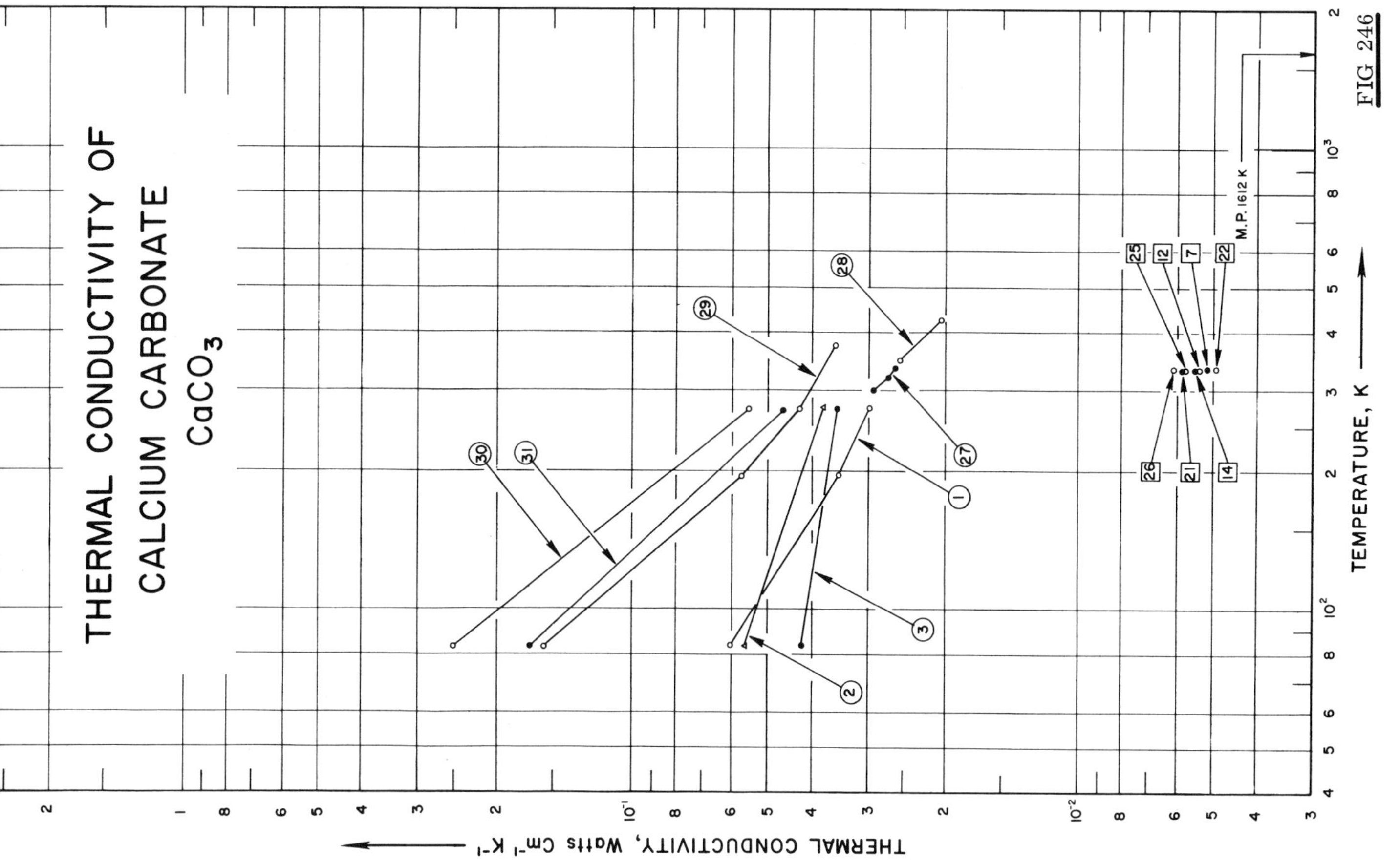
THERMAL CONDUCTIVITY OF
CALCIUM CARBONATE
$CaCO_3$
THERMAL CONDUCTIVITY, Watts Cm^{-1} K^{-1}
TEMPERATURE, K
M.P. 1612 K
1
2
3
7
12
14
21
22
25
26
27
28
29
30
31

FIG 246

SPECIFICATION TABLE NO. 246 THERMAL CONDUCTIVITY OF CALCIUM CARBONATE $CaCO_3$

[For Data Reported in Figure and Table No. 246]

Curve No.	Ref. No.	Method Used	Year	Temp. Range, K	Reported Error, %	Name and Specimen Designation	Composition (weight percent), Specifications and Remarks
1	22	L	1911	83-273		Marble	
2	119	C	1925	83, 273		Marble	Large grain; 103 crystal interrruptions per cm.
3	119	C	1925	83, 273		Marble	Fine grain; 138 crystal interruptions per cm.
4	120	L	1953	331.1		Marble powder	Marble stone powder; apparent density 1.12 g cm^{-3}; porosity 59.3%; particle size -115 + 150 Tyler mesh.
5	120	L	1953	331.2		Marble powder	Marble stone powder; apparent density 1.12 g cm^{-3}; porosity 59.3%; particle size -200 + 325 Tyler mesh.
6	120	L	1953	330.2		Marble powder	Marble stone powder; apparent density 1.13 g cm^{-3}; porosity 58.9%; particle size -150 + 200 Tyler mesh.
7	120	L	1953	333.2		Marble powder	Marble stone powder; apparent density 1.15 g cm^{-3}; porosity 58.2%; particle size -48 + 65 Tyler mesh.
8	120	L	1953	332.2		Marble powder	Marble stone powder; apparent density 1.15 g cm^{-3}; porosity 58.2%; particle size -65 + 100 Tyler mesh.
9	120	L	1953	334.5		Marble powder	Marble stone powder; apparent density 1.18 g cm^{-3}; porosity 57.1%; particle size -28 + 48 Tyler mesh.
10	120	L	1953	330.3		Marble powder	Marble stone powder; apparent density 1.18 g cm^{-3}; porosity 57.1%; particle size -115 + 150 Tyler mesh.
11	120	L	1953	330.4		Marble powder	Marble stone powder; apparent density 1.19 g cm^{-3}; porosity 56.7%; particle size -200 + 325 Tyler mesh.
12	120	L	1953	330.5		Marble powder	Marble stone powder; apparent density 1.22 g cm^{-3}; porosity 55.6%;particle size -48 + 65 Tyler mesh.
13	120	L	1953	331.2		Marble powder	Marble stone powder; apparent density 1.22 g cm^{-3}; porosity 55.6%; particle size -65 + 100 Tyler mesh.
14	120	L	1953	331.1		Marble powder	Marble stone powder; apparent density 1.26 g cm^{-3}; porosity 54.2%; particle size -28 + 48 Tyler mesh.
15	120	L	1953	330.8		Marble powder	Marble stone powder; apparent density 1.26 g cm^{-3}; porosity 54.2%; particle size -115 + 150 Tyler mesh.
16	120	L	1953	330.7		Marble powder	Marble stone powder; apparent density 1.27 g cm^{-3}; porosity 53.8%; particle size -150 + 200 Tyler mesh.
17	120	L	1953	328.8		Marble powder	Marble stone powder; apparent density 1.32 g cm^{-3}; porosity 52.0%; particle size -150 + 200 Tyler mesh.
18	120	L	1953	330.8		Marble powder	Marble stone powder; apparent density 1.32 g cm^{-3}; porosity 52.0%; particle size -200 + 325 Tyler mesh.

Curve No.	Ref. No.	Method Used	Year	Temp. Range, K	Reported Error, %	Name and Specimen Designation	Composition (weight percent), Specifications and Remarks
19	120	L	1953	333.3		Marble powder	Marble stone powder; apparent density 1.32 g cm^{-3}; porosity 52.0%; mixed particle sizes: 22% (-28 + 48) Tyler mesh, 20% (-48 + 65) Tyler mesh, 11% (-65 + 100) Tyler mesh, 5% (-115 + 150) Tyler mesh, 20% (-150 + 200) Tyler mesh, and 22% (-200 + 325) Tyler mesh.
20	120	L	1953	332.1		Marble powder	Marble stone powder; apparent density 1.40 g cm^{-3}; porosity 49.1%; mixed particle sizes as above.
21	120	L	1953	329.5		Marble powder	Marble stone powder; apparent density 1.40 g cm^{-3}; porosity 49.1%; particle size -200 + 325 Tyler mesh.
22	120	L	1953	332.2		Marble powder	Marble stone powder; apparent density 1.07 g cm^{-3}; porosity 61.1%.
23	120	L	1953	331.2		Marble powder	Marble stone powder; apparent density 1.25 g cm^{-3}; porosity 54.6%.
24	120	L	1953	334.0		Marble powder	Marble stone powder; apparent density 1.29 g cm^{-3}; porosity 53.1%.
25	120	L	1953	330.6		Marble powder	Marble stone powder; apparent density 1.36 g cm^{-3}; porosity 50.5%.
26	120	L	1953	332.3		Marble powder	Marble stone powder; apparent density 1.47 g cm^{-3}; porosity 46.5%.
27	117	L	1955	298-337	1.0-2.0	Marble	Specimen thickness 15.2 mm; density 2.68 g cm^{-3}.
28	121	R	1921	348, 423		White Alabama marble	Composed principally of calcium carbonate and a small amount of magnesium carbonate; dried by heating in an oven for 4 hrs at 130 C.
29	22	L	1911	83-374		Calcite	Single crystal; measured perpendicular to the principal axis.
30	119	C	1925	83, 273		Calcite	Single crystal; measured parallel to the principal axis.
31	119	C	1925	83, 273		Calcite	Single crystal; measured perpendicular to the principal axis.
32	508	L	1940	390-633		Brown Marble	98 $CaCO_3$ and organic matter such as oils; specimen 1 in. thick and 8 in. in dia; specimen obtained from St. Marc des Carriers, Que.; density 2.659 g cm^{-3}; coarse-grained.
33	508	L	1940	398-615		White Marble	Specimen 1 in. thick and 8 in. in dia; density 2.755 g cm^{-3}; obtained from Phillisburg, Que.
34	508	L	1940	443.7		White Marble	Similar to the above specimen; measurements done after exposure to high temperature test.
35	508	L	1940	398-608		Black Marble	96 $CaCO_3$ and some organic matter; specimen 1 in. thick and 8 in. in dia; density 2.803 g cm^{-3}; obtained from St. Albert, Ont.
36	508	L	1940	390.4		Black Marble	Similar to the above specimen; measurements done after exposure to high temperature test.

DATA TABLE NO. 246 THERMAL CONDUCTIVITY OF CALCIUM CARBONATE $CaCO_3$

[Temperature, T, K; Thermal Conductivity, k, Watt $cm^{-1}K^{-1}$]

T	k
CURVE 1	
83.2	0.0608
195.2	0.0352
273.2	0.0299
CURVE 2	
83.0	0.0564
273.2	0.0380
CURVE 3	
83.0	0.0425
273.2	0.0352
CURVE 4*	
331.1	0.00502
CURVE 5*	
331.2	0.00505
CURVE 6*	
330.15	0.00502
CURVE 7	
333.2	0.00519
CURVE 8*	
332.2	0.00519
CURVE 9*	
334.45	0.00526
CURVE 10*	
330.3	0.00526
CURVE 11*	
330.4	0.00526
CURVE 12	
330.45	0.00537
CURVE 13*	
331.15	0.00538
CURVE 14	
331.1	0.00550
CURVE 15*	
330.8	0.00543
CURVE 16*	
330.7	0.00550
CURVE 17*	
328.75	0.00561
CURVE 18*	
330.8	0.00562
CURVE 19*	
333.3	0.00562
CURVE 20*	
332.05	0.00587
CURVE 21	
329.45	0.00587
CURVE 22	
332.2	0.00493
CURVE 23*	
331.2	0.00545
CURVE 24*	
334.0	0.00556
CURVE 25	
330.6	0.00578
CURVE 26	
332.3	0.00614
CURVE 27	
298.2	0.0294
318.2	0.0271
337.2	0.0261
CURVE 28	
348.2	0.0257
423.2	0.0206
CURVE 29	
83.2	0.158
195.2	0.0576
273.2	0.0429
374.2	0.0356
CURVE 30	
83.0	0.251
273.2	0.0551
CURVE 31	
83.0	0.170
273.2	0.0466
CURVE 32*	
390.4	0.0167
469.3	0.0151
518.7	0.0138
633.2	0.0114
CURVE 33*	
398.2	0.0144
443.2	0.0144
511.5	0.0150
615.4	0.0138
CURVE 34*	
443.7	0.0131
CURVE 35*	
397.6	0.0156
483.7	0.0151
607.6	0.0137
CURVE 36*	
390.4	0.0131

* Not shown on plot

SPECIFICATION TABLE NO. 247 THERMAL CONDUCTIVITY OF [CALCIUM PHOSPHATE + LITHIUM CARBONATE + MAGNESIUM CARBONATE] $Ca_3(PO_4)_2 + Li_2CO_3 + MgCO_3$

Curve No.	Ref. No.	Method Used	Year	Temp. Range, K	Reported Error, %	Name and Specimen Designation	Composition (weight percent) $Ca_3(PO_4)_2$	Li_2CO_3	$MgCO_3$	Composition (continued), Specifications and Remarks
1	23		1952	312-391		164A	33.33	33.33	33.33	Fired at 1200 K for 1.5 hrs; density (after firing) 2.49 g cm^{-3}; water absorption 0.007%.

DATA TABLE NO. 247 THERMAL CONDUCTIVITY OF [CALCIUM PHOSPHATE + LITHIUM CARBONATE + MAGNESIUM CARBONATE] $Ca_3(PO_4)_2 + Li_2CO_3 + MgCO_3$

[Temperature, T, K; Thermal Conductivity, k, Watt $cm^{-1}K^{-1}$]

T	k
CURVE 1*	
311.7	0.0439
331.2	0.0423
352.2	0.0423
370.2	0.0410
390.7	0.0397

*No graphical presentation

SPECIFICATION TABLE NO. 248 THERMAL CONDUCTIVITY OF [CARBON + OXYGEN] C + O

Curve No.	Ref. No.	Method Used	Year	Temp. Range, K	Reported Error, %	Name and Specimen Designation	Composition (weight percent) C	O	Composition (continued), Specifications and Remarks
1	161	P	1960	1273-3273		Channel Carbon Black	94.59	4.79	0.68 H, and 0.09 ash; heat treated from 1000 to 3000 C in nitrogen atmosphere for a duration of 10 - 30 min; particle dia < 1μ; degree of graphization 9% at the max. exposed temperature ; thermal conductivity data calculated from measurements of diffusivity, bulk weight, and specific heat data; measured in a nitrogen argon atmosphere.

DATA TABLE NO. 248 THERMAL CONDUCTIVITY OF [CARBON + OXYGEN] C + O

[Temperature, T, K, Thermal Conductivity, k, Watt $cm^{-1}K^{-1}$]

T	k
CURVE 1*	
1273.2	0.000256
1573.2	0.000511
1773.2	0.000604
2073.2	0.000814
2273.2	0.000814
2623.2	0.000814
2973.2	0.000814
3273.2	0.000395

*No graphical presentation

SPECIFICATION TABLE NO. 249 THERMAL CONDUCTIVITY OF [CARBON + VOLATILE MATERIALS] C + Volatile Materials

Curve No.	Ref. No.	Method Used	Year	Temp. Range, K	Reported Error, %	Name and Specimen Designation	Composition (weight percent) C	Composition (weight percent) Volatile Materials	Composition (continued), Specifications and Remarks
1	161	P	1960	1273-3073		Petroleum Coke	Bal	5.13	0.08 ash; heat treated 10-30 min before each measurement; particle dia ~0.5 mm; density 1.405 g cm^{-3}; measured in nitrogen + argon atmosphere; thermal conductivity data calculated from measured values of thermal diffusivity and specific heat.

DATA TABLE NO. 249 THERMAL CONDUCTIVITY OF [CARBON + VOLATILE MATERIALS] C + Volatile Materials

[Temperature, T, K; Thermal Conductivity , k, Watt $cm^{-1}K^{-1}$]

T	k
CURVE 1*	
1273.2	0.00214
1673.2	0.00242
2053.2	0.00279
2273.2	0.00316
3073.2	0.00335

*No graphical presentation

SPECIFICATION TABLE NO. 250 THERMAL CONDUCTIVITY OF GALLIUM PHOSPHIDE GaP

Curve No.	Ref. No.	Method Used	Year	Temp. Range, K	Reported Error, %	Name and Specimen Designation	Composition (weight percent), Specifications and Remarks
1	597, 598	P	1963	290			Melting point ~ 1350 C; measured by a transient method.

DATA TABLE NO. 250 THERMAL CONDUCTIVITY OF GALLIUM PHOSPHIDE GaP

[Temperature, T, K; Thermal Conductivity, k, Watt $cm^{-1}K^{-1}$]

T	k
CURVE 1*	
290	1.09

* No graphical presentation

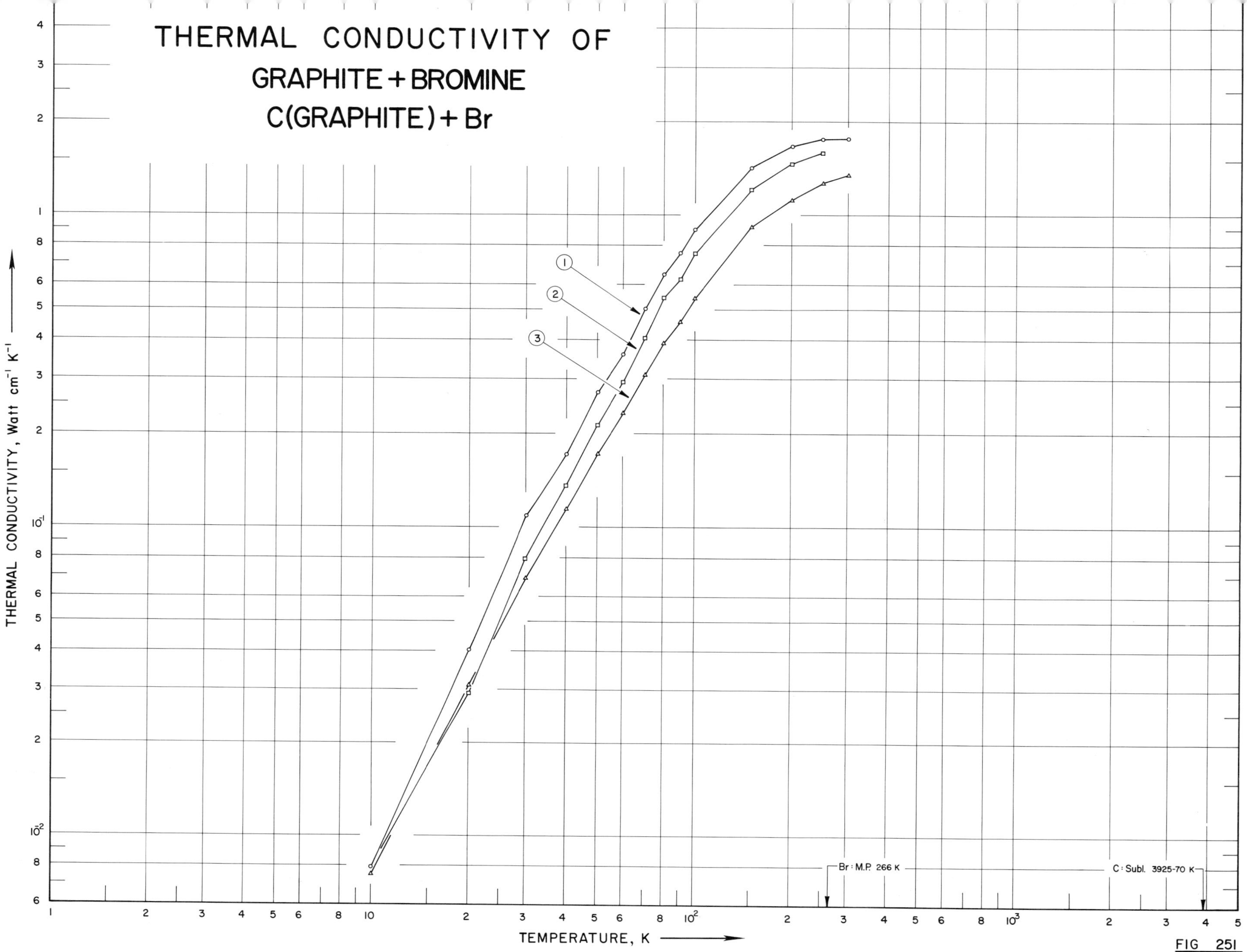

THERMAL CONDUCTIVITY OF GRAPHITE + BROMINE C(GRAPHITE) + Br

THERMAL CONDUCTIVITY, Watt cm^{-1} K^{-1}

TEMPERATURE, K

Br : M.P. 266 K

C : Subl. 3925-70 K

(1) (2) (3)

FIG 251

SPECIFICATION TABLE NO. 251 THERMAL CONDUCTIVITY OF [GRAPHITE + BROMIDE] C + Br

[For Data Reported in Figure and Table No. 251]

Curve No.	Ref. No.	Method Used	Year	Temp. Range, K	Reported Error, %	Name and Specimen Designation	Composition (weight percent)		Composition (continued), Specifications and Remarks
							C	Br	
1	50	E	1956	10-300	±5	Brom-Graphite	97.97	2.03	Calculated composition; brominated AGOT-KC graphite.
2	50	E	1956	10-250	±5	Brom-Graphite	94.43	5.57	Calculated composition; brominated AGOT-KC graphite.
3	50	E	1956	10-300	±5	Brom-Graphite	92.76	7.24	Calculated composition; brominated AGOT-KC graphite.

DATA TABLE NO. 251 THERMAL CONDUCTIVITY OF [GRAPHITE + BROMINE] C + Br

[Temperature, T, K; Thermal Conductivity, k, Watt $cm^{-1} K^{-1}$]

T	k
CURVE 1	
10	0.00795
20	0.0402
30	0.109
40	0.172
50	0.272
60	0.360
70	0.502
80	0.649
90	0.753
100	0.899
150	1.42
200	1.67
250	1.76
300	1.76
CURVE 2	
10	0.00795*
20	0.0293
30	0.0795
40	0.136
50	0.213
60	0.293
70	0.406
80	0.544
90	0.628
100	0.753
150	1.21
200	1.46
250	1.59
CURVE 3	
10	0.00753
20	0.031
30	0.069
40	0.115
50	0.172
60	0.234
70	0.310
80	0.393
90	0.460
100	0.544
150	0.92
200	1.13
250	1.28
300	1.36

* Not shown on plot

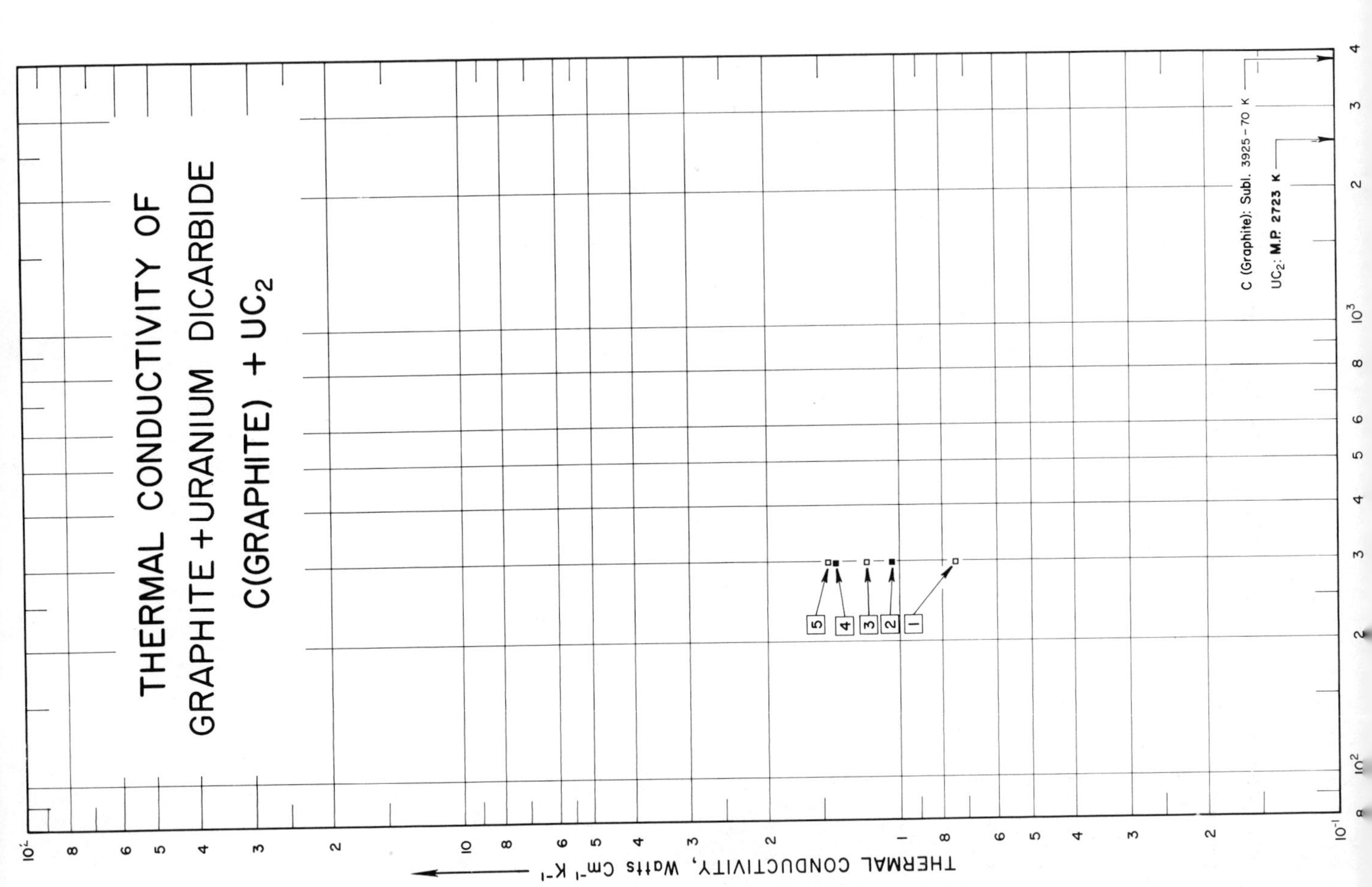

THERMAL CONDUCTIVITY OF
GRAPHITE + URANIUM DICARBIDE
C(GRAPHITE) + UC$_2$
C (Graphite): Subl. 3925 – 70 K
UC$_2$: M.P. 2723 K
THERMAL CONDUCTIVITY, Watts Cm^{-1} K^{-1}
1
2
3
4
5

SPECIFICATION TABLE NO. 252 THERMAL CONDUCTIVITY OF [GRAPHITE + URANIUM DICARBIDE] $C + UC_2$

[For Data Reported in Figure and Table No. 252]

Curve No.	Ref. No.	Method Used	Year	Temp. Range, K	Reported Error, %	Name and Specimen Designation	Composition (weight percent) C	UC_2	Composition (continued), Specifications and Remarks
1	181		1959	298. 2			88. 96	11. 04	Baked to 2800 C; bulk density 1. 866 g cm^{-3}.
2	181		1959	298. 2			78. 07	21. 93	As above but bulk density 1. 986 g cm^{-3}.
3	181		1959	298. 2			72. 52	27. 48	As above but bulk density 2. 067 g cm^{-3}.
4	181		1959	298. 2			65. 92	34. 08	As above but bulk density 2. 157 g cm^{-3}.
5	181		1959	298. 2			56. 13	43. 87	As above but bulk density 2. 333 g cm^{-3}.

DATA TABLE NO. 252 THERMAL CONDUCTIVITY OF [GRAPHITE + URANIUM DICARBIDE] $C + UC_2$

[Temperature, T, K; Thermal Conductivity, k, Watt cm^{-1} K^{-1}]

T	k
CURVE 1	
298.2	0.744
CURVE 2	
298.2	1.04
CURVE 3	
298.2	1.18
CURVE 4	
298.2	1.39
CURVE 5	
298.2	1.45

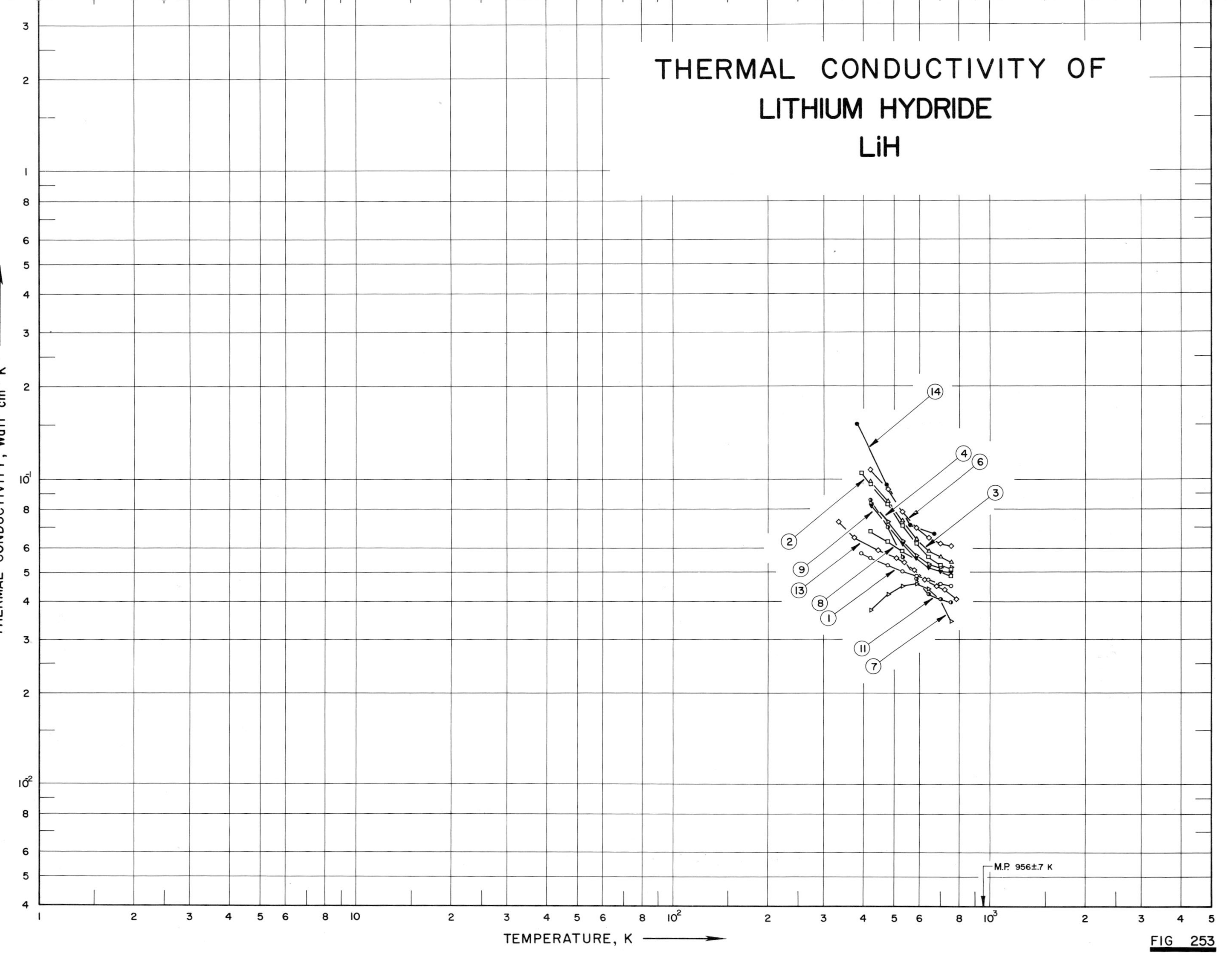

THERMAL CONDUCTIVITY OF
LITHIUM HYDRIDE
LiH
M.P. 956±7 K
TEMPERATURE, K
THERMAL CONDUCTIVITY, Watt cm⁻¹ K⁻¹
FIG 253

SPECIFICATION TABLE NO. 253 THERMAL CONDUCTIVITY OF LITHIUM HYDRIDE LiH

[For Data Reported in Figure and Table No. 253]

Curve No.	Ref. No.	Method Used	Year	Temp. Range, K	Reported Error, %	Name and Specimen Designation	Composition (weight percent), Specifications and Remarks
1	512	R	1962	394-755			Cylindrical specimen 36 in. long and 6 in. dia with 0.03 in. stainless steel walls and a centrally located finned coolant tube, the fin was of a spiral configuration, 1.50 in. in dia, with three spirals per in. of tube, lithium hydride was cast inside the cylinder; thermal conductivity was measured from inside the coolant tube to the outside of the can; vacuum in the voids of the lithium hydride.
2	512	R	1962	394-755			The above specimen measured from inside the coolant tube to the outside of the can; with helium filling the voids in the lithium hydride.
3	512	R	1962	422-755			The above specimen measured from inside the coolant tube to the outside of the can; with hydrogen filling the voids in the lithium hydride.
4	512	R	1962	422-755			The above specimen with vacuum in the voids of the lithium hydride; the fine effective conductivity was measured from inside the finned coolant tube to an intermediate point (radius = 1.25 in.) in the lithium hydride outside the fin area.
5	512	R	1962	422-755			The above specimen with helium in the voids of the lithium hydride; the fin effective conductivity was measured.
6	512	R	1962	422-755			The above specimen with hydrogen in the voids of the lithium hydride; the fin effective conductivity was measured.
7	512	R	1962	422-755			The above specimen with vacuum in the voids of the lithium hydride; the salt (lithium hydride) to the can effective conductivity was measured from the intermediate point (radius = 1.25 in.) to the outside of the can.
8	512	R	1962	422-755			The above specimen with helium in the voids of the lithium hydride; the salt to the can effective conductivity was measured.
9	512	R	1962	422-755			The above specimen with hydrogen in the voids of the lithium hydride; the salt to the can effective conductivity was measured.
10	512	R	1962	422-755			The above specimen with vacuum in the voids of the lithium hydride; the salt effective conductivity was measured between two points in the salt (radius = 1.25 in., 1.70 in.)
11	512	R	1962	422-755			The above specimen with helium in the voids of the lithium hydride; the salt effective conductivity was measured.
12	512	R	1962	422-755			The above specimen with hydrogen in the voids of the lithium hydride; the salt effective conductivity was measured.
13	132	R	1958	335-786			Cold pressed; reinforced with perforated honeycomb oriented with the axis of the cells normal to the direction of heat flow.
14	513		1961	380-670			

DATA TABLE NO. 253 THERMAL CONDUCTIVITY OF LITHIUM HYDRIDE LiH

[Temperature, T, K; Thermal Conductivity, k, Watt $cm^{-1} K^{-1}$]

T	k
CURVE 1	
394.3	0.0576
422.1	0.0557
477.6	0.0528
533.2	0.0502
588.8	0.0485
644.3	0.0472
699.9	0.0459
755.4	0.0450
CURVE 2	
394.3	0.105
422.1	0.0969
477.6	0.0830
533.2	0.0710
588.8	0.0623
644.3	0.0562
699.9	0.0528
755.4	0.0511
CURVE 3	
422.1	0.0981
477.6	0.0848
533.2	0.0736
588.8	0.0644
644.3	0.0588
699.9	0.0562
755.4	0.0540
CURVE 4	
422.1	0.0839
477.6	0.0727
533.2	0.0632
588.8	0.0566
644.3	0.0533
699.9	0.0519
755.4	0.0519
CURVE 5*	
422.1	0.101
477.6	0.0865
533.2	0.0762
CURVE 5 (cont.)*	
588.8	0.0682
644.3	0.0649
699.9	0.0632
755.4	0.0623
CURVE 6	
422.1	0.108
477.6	0.0926
533.2	0.0787
588.8	0.0696
644.3	0.0646
699.9	0.0623
755.4	0.0614
CURVE 7	
422.1	0.0376
477.6	0.0424
533.2	0.0450
588.8	0.0459
644.3	0.0441
699.9	0.0402*
755.4	0.0346
CURVE 8	
422.1	0.0684
477.6	0.0632
533.2	0.0588
588.8	0.0554*
644.3	0.0528*
699.9	0.0502*
755.4	0.0485
CURVE 9	
422.1	0.0822
477.6	0.0710*
533.2	0.0623
588.8	0.0562
644.3	0.0519
699.9	0.0502
755.4	0.0497
CURVE 10*	
422.1	0.0909
477.6	0.0770
533.2	0.0640
588.8	0.0545
644.3	0.0476
699.9	0.0433
755.4	0.0398
CURVE 11	
422.1	0.0857
477.6	0.0696
533.2	0.0562
588.8	0.0476
644.3	0.0424
699.9	0.0407
755.4	0.0398
CURVE 12*	
422.1	0.0813
477.6	0.0718
533.2	0.0640
588.8	0.0573
644.3	0.0533
699.9	0.0502
755.4	0.0481
CURVE 13	
334.8	0.0730
374.3	0.0649
445.4	0.0592
508.7	0.0556
538.7	0.0537
577.1	0.0507
625.4	0.0472
680.9	0.0450
722.1	0.0436
785.9	0.0408
CURVE 14	
380.2	0.151
475.2	0.0962
563.2	0.0711
670.2	0.0669

* Not shown on plot

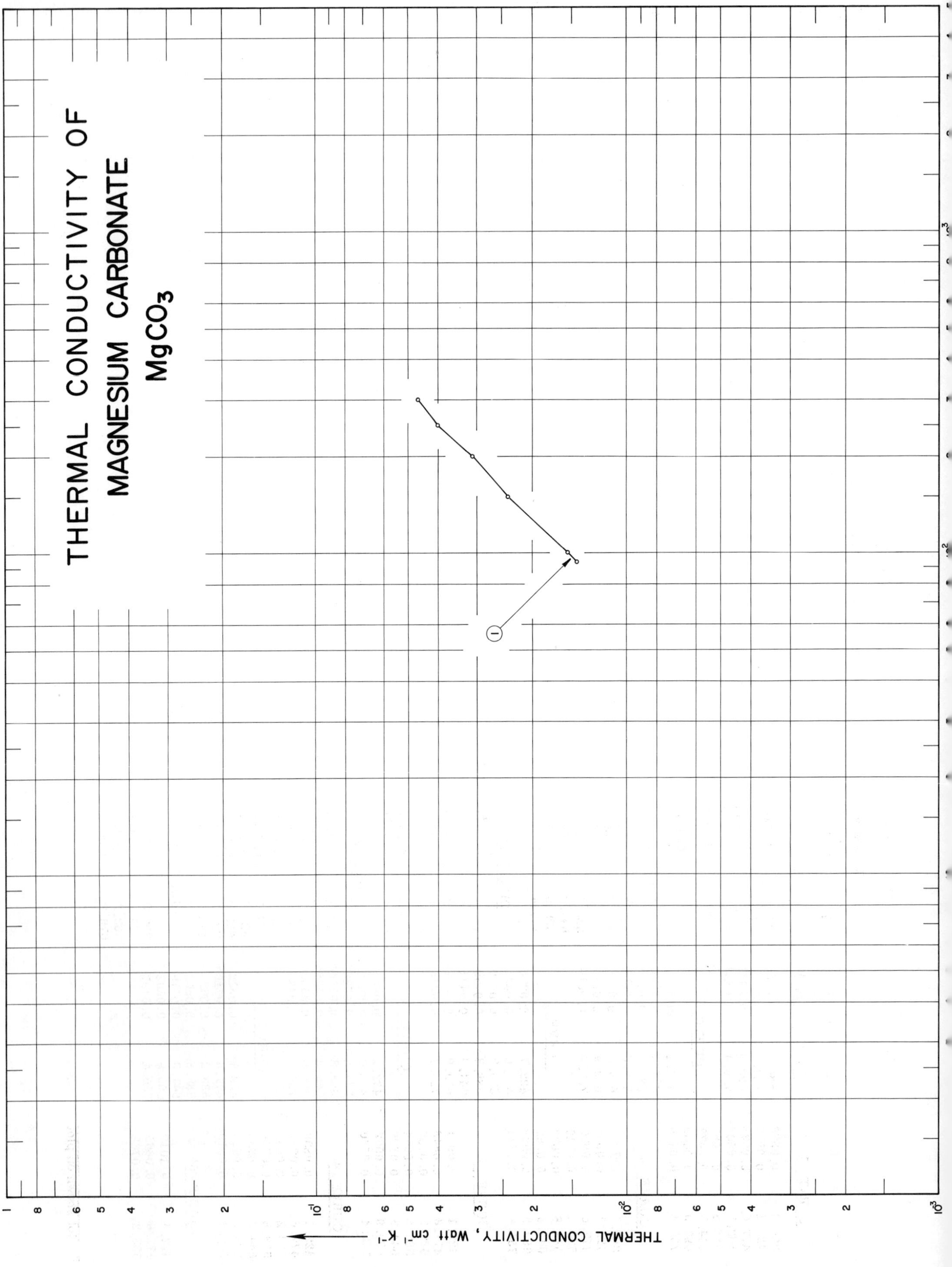

THERMAL CONDUCTIVITY OF
MAGNESIUM CARBONATE
$MgCO_3$
THERMAL CONDUCTIVITY, Watt cm^{-1} K^{-1}
1
8
6
5
4
3
2
10^{-1}
8
6
5
4
3
2
10^{-2}
8
6
5
4
3
2
10^{-3}
1

SPECIFICATION TABLE NO. 254 THERMAL CONDUCTIVITY OF MAGNESIUM CARBONATE $MgCO_3$

[For Data Reported in Figure and Table No. 254]

Curve No.	Ref. No.	Method Used	Year	Temp. Range, K	Reported Error, %	Name and Specimen Designation	Composition (weight percent), Specifications and Remarks
1	139	R	1950	93-300			Powdered; density 0.22 g cm^{-3}.

DATA TABLE NO. 254 THERMAL CONDUCTIVITY OF MAGNESIUM CARBONATE $MgCO_3$

[Temperature, T, K; Thermal Conductivity, k, Watt $cm^{-1}K^{-1}$]

T	k
CURVE 1	
93.2	0.0145
100	0.0155
150	0.0240
200	0.0310
250	0.0400
300	0.0465

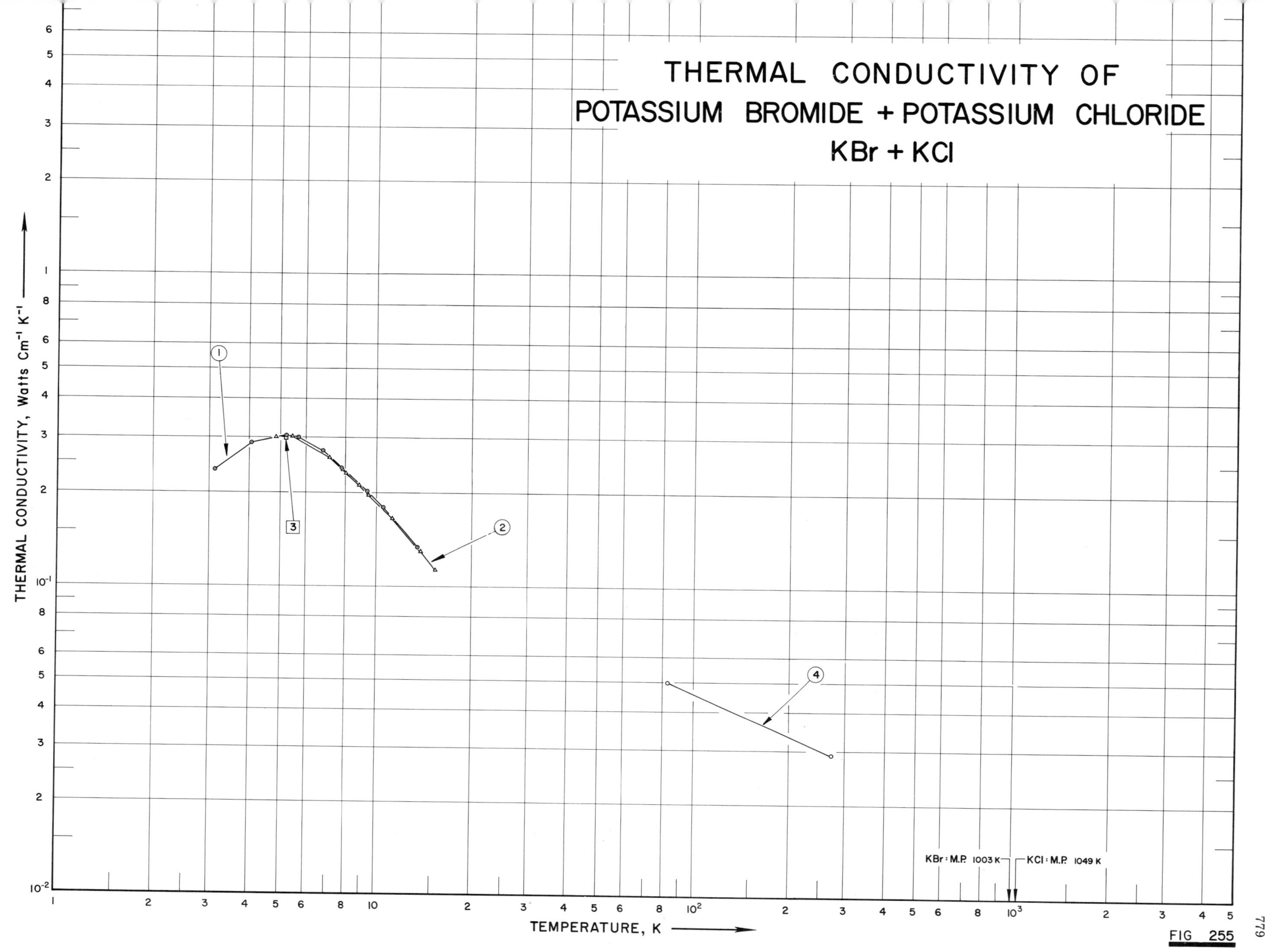
THERMAL CONDUCTIVITY OF
POTASSIUM BROMIDE + POTASSIUM CHLORIDE
KBr + KCl
THERMAL CONDUCTIVITY, Watts Cm⁻¹ K⁻¹
TEMPERATURE, K
KBr : M.P. 1003 K
KCl : M.P. 1049 K
1
2
3
4

FIG 255

SPECIFICATION TABLE NO. 255 THERMAL CONDUCTIVITY OF [POTASSIUM BROMIDE + POTASSIUM CHLORIDE] KBr + KCl

[For Data Reported in Figure and Table No. 255]

Curve No.	Ref. No.	Method Used	Year	Temp. Range, K	Reported Error, %	Name and Specimen Designation	Composition (weight percent) KBr	KCl	Composition (continued), Specifications and Remarks
1	164	L	1956	3.1-14			60.53	39.47	49.0 mole % KBr and 51.0 mole % KCl; single KBr-KCl mixed crystal; specimen obtained in the form of optical grade crystals from Harshaw Chemical Co.; fragments of these melted and single mixed-crystal pulled from the melt; annealed at 75 C below the melting point; helium bath maintained at 2.2 K.
2	164	L	1956	4.9-16			60.53	39.47	As above but the helium bath maintained at 4.2 K.
3	164	L	1956	5.2			61.0	39.0	49.5 mole % KBr and 50.5 mole % KCl; single mixed-crystal.
4	380	L	1956	83, 273			96.15	3.85	90 mole % KBr and 10 mole % KCl; mixed crystals pressed.

DATA TABLE NO. 255 THERMAL CONDUCTIVITY OF [POTASSIUM BROMIDE + POTASSIUM CHLORIDE] KBr + KCl

[Temperature, T, K; Thermal Conductivity, k, Watt cm^{-1} K^{-1}]

T	k
CURVE 1	
3.12	0.236
4.05	0.288
5.21	0.305
5.71	0.300
6.82	0.272
7.82	0.240
9.40	0.201
10.63	0.179
13.61	0.133
CURVE 2	
4.88	0.300
5.46	0.301
7.13	0.258
7.90	0.236
8.03	0.232
8.90	0.211
9.51	0.196
11.35	0.165
13.97	0.129
15.50	0.113
CURVE 3	
5.2	0.302
CURVE 4	
83	0.0496
273	0.0291

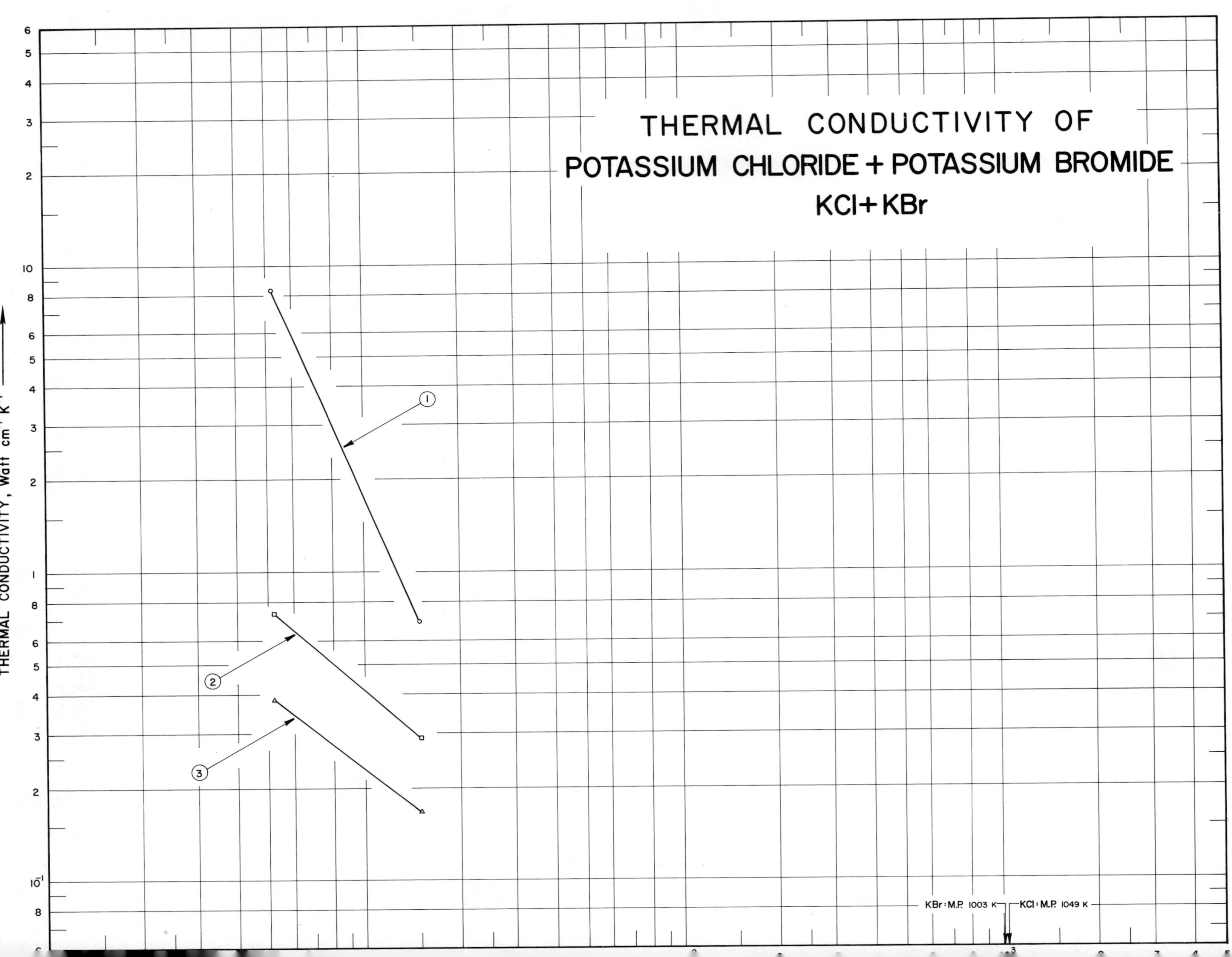
THERMAL CONDUCTIVITY OF
POTASSIUM CHLORIDE + POTASSIUM BROMIDE
KCl+KBr
KBr: M.P. 1003 K
KCl: M.P. 1049 K
1
2
3
THERMAL CONDUCTIVITY, Watt cm^{-1} K^{-1}
6
5
4
3
2
10
8
6
5
4
3
2
1
8
6
5
4
3
2
10^{-1}
8

SPECIFICATION TABLE NO. 256 THERMAL CONDUCTIVITY OF [POTASSIUM CHLORIDE + POTASSIUM BROMIDE] KCl + KBr

[For Data Reported in Figure and Table No. 256]

Curve No.	Ref. No.	Method Used	Year	Temp. Range, K	Reported Error, %	Name and Specimen Designation	Composition (weight percent) KCl	KBr	Composition (continued), Specifications and Remarks
1	164	L	1956	5.2,15			97.63	2.37	98.5 mole % KCl, 1.5 mole % KBr; single mixed crystal.
2	164	L	1956	5.2,15		Single KCl-KBr mixed crystal	85.22	14.78	90.2 mole % KCl, 9,8 mole % KBr; single mixed crystal.
3	164	L	1956	5.2,15		Single KCl-KBr mixed crystal	61.11	38.89	71.5 mole % KCl, 28.5 mole % KBr; single mixed crystal.

DATA TABLE NO. 256 THERMAL CONDUCTIVITY OF [POTASSIUM CHLORIDE + POTASSIUM BROMIDE] KCl + KBr

[Temperature, T, K; Thermal Conductivity, k, Watt $cm^{-1}K^{-1}$]

T	k
CURVE 1	
5.2	1.835
15.0	0.690
CURVE 2	
5.2	0.735
15.0	0.290
CURVE 3	
5.2	0.385
15.0	0.168

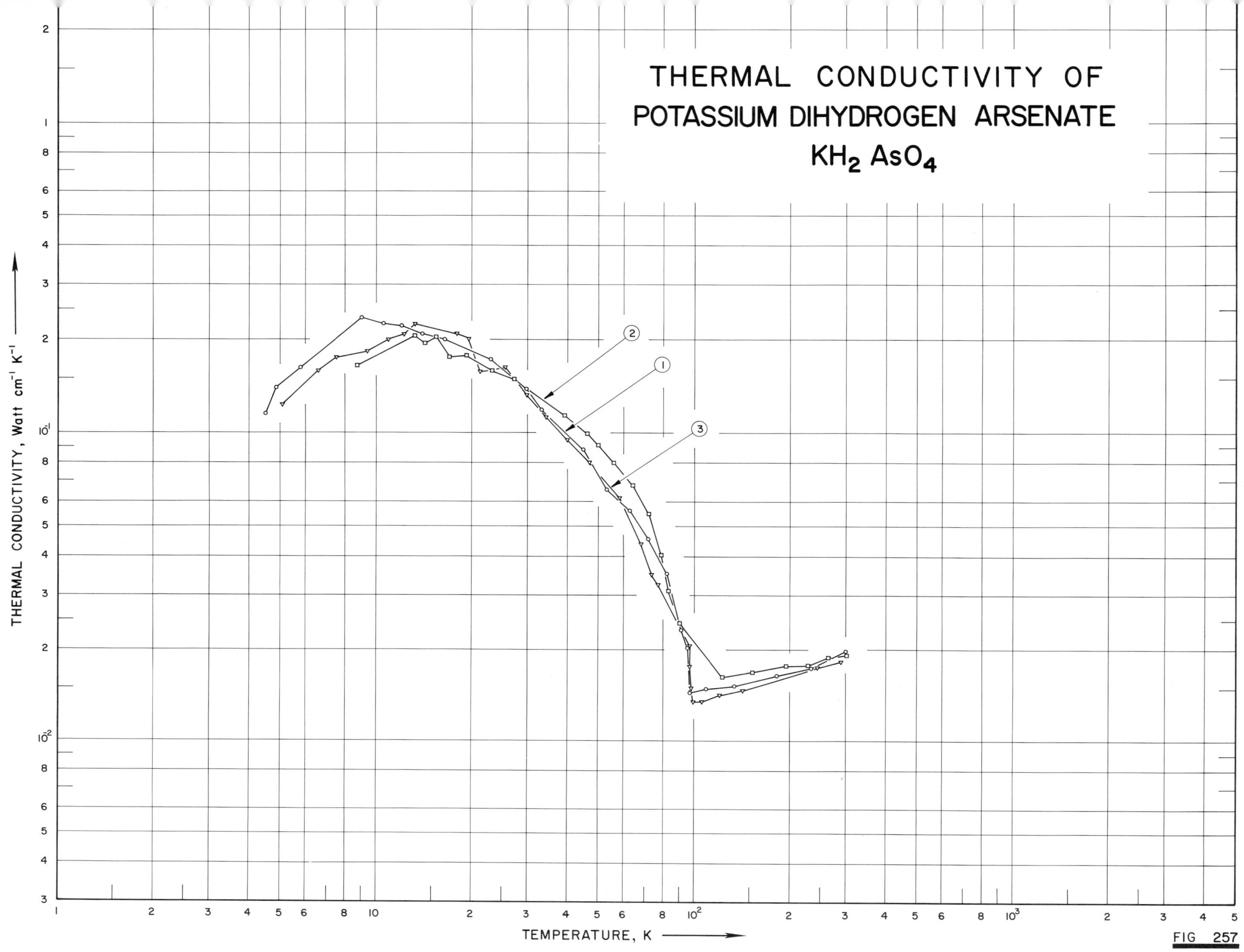
THERMAL CONDUCTIVITY OF
POTASSIUM DIHYDROGEN ARSENATE
KH2 AsO4
1
2
3
THERMAL CONDUCTIVITY, Watt cm-1 K-1
TEMPERATURE, K

FIG 257

SPECIFICATION TABLE NO. 257 THERMAL CONDUCTIVITY OF POTASSIUM DIHYDROGEN ARSENATE KH_2AsO_4

[For Data Reported in Figure and Table No. 257]

Curve No.	Ref. No.	Method Used	Year	Temp. Range, K	Reported Error, %	Name and Specimen Designation	Composition (weight percent), Specifications and Remarks
1	514, 515	L	1966	4.5-300		No. 1c	Single crystal with tetragonal symmetry; specimen 1.8 x 2.2 x 11.0 mm, long dimension parallel to c-axis; prepared by usual recrystallization technique, cut and formed into rectangular rod using a wet thread and a polishing paper; Curie temperature 96 K; heat flow along c-axis.
2	515	L	1967	8.8-302		No. 2c	Similar to the above specimen except specimen 1.9 x 2.2 x 11.5 mm, long dimension parallel to c-axis.
3	515	L	1967	5.1-289		No. 3a	Similar to the above specimen except specimen 1.4 x 2.0 x 6.0 mm, long dimension parallel to a-axis; heat flow along a-axis.

DATA TABLE NO. 257 THERMAL CONDUCTIVITY OF POTASSIUM DIHYDROGEN ARSENATE KH_2AsO_4

[Temperature, T, K; Thermal Conductivity, k, Watt $cm^{-1}K^{-1}$]

T	k
CURVE 1	
4.52	0.116
4.88	0.140
5.81	0.162
9.02	0.235
10.6	0.224
12.1	0.221
14.0	0.208
16.5	0.200
23.0	0.172
29.7	0.139
33.3	0.119
44.8	0.0881
53.1	0.0656
62.8	0.0562
71.6	0.0455
82.3	0.0352
91.4	0.0233
95.3	0.0203
97.1	0.0145
109.9	0.0149
134.0	0.0152
182.0	0.0165
234.0	0.0174
300.0	0.0198
CURVE 2	
8.75	0.165
13.2	0.205
14.2	0.194
15.5	0.202
17.0	0.176
19.3	0.177
23.1	0.159
27.1	0.149
39.2	0.114
46.1	0.0993
49.8	0.0912
55.7	0.0796
64.1	0.0679
72.3	0.0548
78.9	0.0405
83.4	0.0310
90.0	0.0244
122.2	0.0162
CURVE 2 (cont.)	
153.8	0.0169
195.9	0.0177
229.7	0.0177
265.0	0.0188
302.1	0.0191
CURVE 3	
5.06	0.123
6.58	0.158
7.50	0.175
9.40	0.182
10.9	0.199
12.2	0.207
13.2	0.223
18.0	0.207
19.6	0.200
21.2	0.158
25.5	0.163
29.7	0.133
34.3	0.113
39.9	0.0948
46.9	0.0802
58.4	0.0617
68.1	0.0439
73.5	0.0350
77.3	0.0325
96.6	0.0206
96.6	0.0176
97.7	0.0149
99.3	0.0136
106.2	0.0136
121.1	0.0142
142.6	0.0147
245.0	0.0175
289.2	0.0182

SPECIFICATION TABLE NO. 258 THERMAL CONDUCTIVITY OF POTASSIUM THIOCYANATE KSCN

Curve No.	Ref. No.	Method Used	Year	Temp. Range, K	Reported Error, %	Name and Specimen Designation	Composition (weight percent), Specifications and Remarks
1	242		1961	423-486	±3		Specimen of A. R. purity; each data point is the mean value of 4 or 5 different measurements; measured in molten state.
2	242		1961	448	±3		Same as the above specimen; measured in molten state.

DATA TABLE NO. 258 THERMAL CONDUCTIVITY OF POTASSIUM THIOCYANATE KSCN

[Temperature, T, K; Thermal Conductivity, k, Watt $cm^{-1}K^{-1}$]

T	k
CURVE 1*	
423.2	0.00327
440.2	0.00312
454.0	0.00281
465.0	0.00295
486.4	0.00292
CURVE 2*	
448	0.00272

*No graphical presentation

SPECIFICATION TABLE NO. 259 THERMAL CONDUCTIVITY OF [SILICON CARBIDE + GRAPHITE] SiC + C

Curve No.	Ref. No.	Method Used	Year	Temp. Range, K	Reported Error, %	Name and Specimen Designation	Composition (weight percent), Specifications and Remarks
1	216	R	1951	839.3			Mixture of granulated silicon carbide and graphite powder; measured from the inner half of the test annulus.
2	216	R	1951	969.8			Same as the above specimen except measured from the entire annulus.

DATA TABLE NO. 259 THERMAL CONDUCTIVITY OF [SILICON CARBIDE + GRAPHITE] SiC + C

[Temperature, T, K; Thermal Conductivity, k, Watt $cm^{-1}K^{-1}$]

T	k
CURVE 1*	
839.3	0.0134
CURVE 2*	
969.8	0.0159

*No graphical presentation

SPECIFICATION TABLE NO. 260 THERMAL CONDUCTIVITY OF SODIUM HYDROXIDE NaOH

Curve No.	Ref. No.	Method Used	Year	Temp. Range, K	Reported Error, %	Name and Specimen Designation	Composition (weight percent), Specifications and Remarks
1	242		1961	592	±3		Specimen of A. R. purity; mean value of 4 or 5 different measurements; measured in molten state.

DATA TABLE NO. 260 THERMAL CONDUCTIVITY OF SODIUM HYDROXIDE Na OH

[Temperature, T, K; Thermal Conductivity, k, Watt $cm^{-1}K^{-1}$]

T	k
CURVE 1*	
592	0.00920

*No graphical presentation

SPECIFICATION TABLE NO. 261 THERMAL CONDUCTIVITY OF [STRONTIUM DIFLUORIDE + ΣX_i] $SrF_2 + \Sigma X_i$

Curve No.	Ref. No.	Method Used	Year	Temp. Range, K	Reported Error, %	Name and Specimen Designation	Composition (weight percent), Specifications and Remarks
1	470	P	1959	298.2			92% theoretical density.

DATA TABLE NO. 261 THERMAL CONDUCTIVITY OF [STRONTIUM DIFLUORIDE + ΣX_i] $SrF_2 + \Sigma X_i$

[Temperature, T, K; Thermal Conductivity, k, Watt $cm^{-1}K^{-1}$]

T	k
CURVE 1*	
298.2	0.0142

*No graphical presentation

SPECIFICATION TABLE NO. 262 THERMAL CONDUCTIVITY OF ZINC GERMANIUM PHOSPHIDE $ZnGeP_2$

Curve No.	Ref. No.	Method Used	Year	Temp. Range, K	Reported Error, %	Name and Specimen Designation	Composition (weight percent), Specifications and Remarks
1	510	L	1966	298.2		$ZnGeP_2$; 11	Stoichiometric p-type polycrystalline; 12 x 4 x 4 mm; prepared from 99.9998 pure Zn, p-type Ge (~50 ohm cm), and 99.9999 pure red phosphorus, materials loaded into a high-purity graphite crucible which was closely fitted in a sealed quartz tube, heated to 500 C in one hr, then heated to 800 C at 10 C hr^{-1}, then to 1060 C at 20 C hr^{-1}; during the heating process, the tube was subjected to external argon pressure of 7, 13, 45, 80, 100, 125, and 150 atm, respectively in the temperature ranges of room temperature to 400 C, 400-500 C, 500-580 C, 580-630 C, 630-680 C, 680-880 C, and 880-1060 C, rotated and vibrated for 30 min, after another 30 min, the Bridgman process was performed at a lowering rate 10 mm hr^{-1} till reached room temperature; melting point 1025 C.

DATA TABLE NO. 262 THERMAL CONDUCTIVITY OF ZINC GERMANIUM PHOSPHIDE $ZnGeP_2$

[Temperature, T, K; Thermal Conductivity, k, Watt $cm^{-1}K^{-1}$]

T	k
CURVE 1*	
298.2	0.18

*No graphical presentation

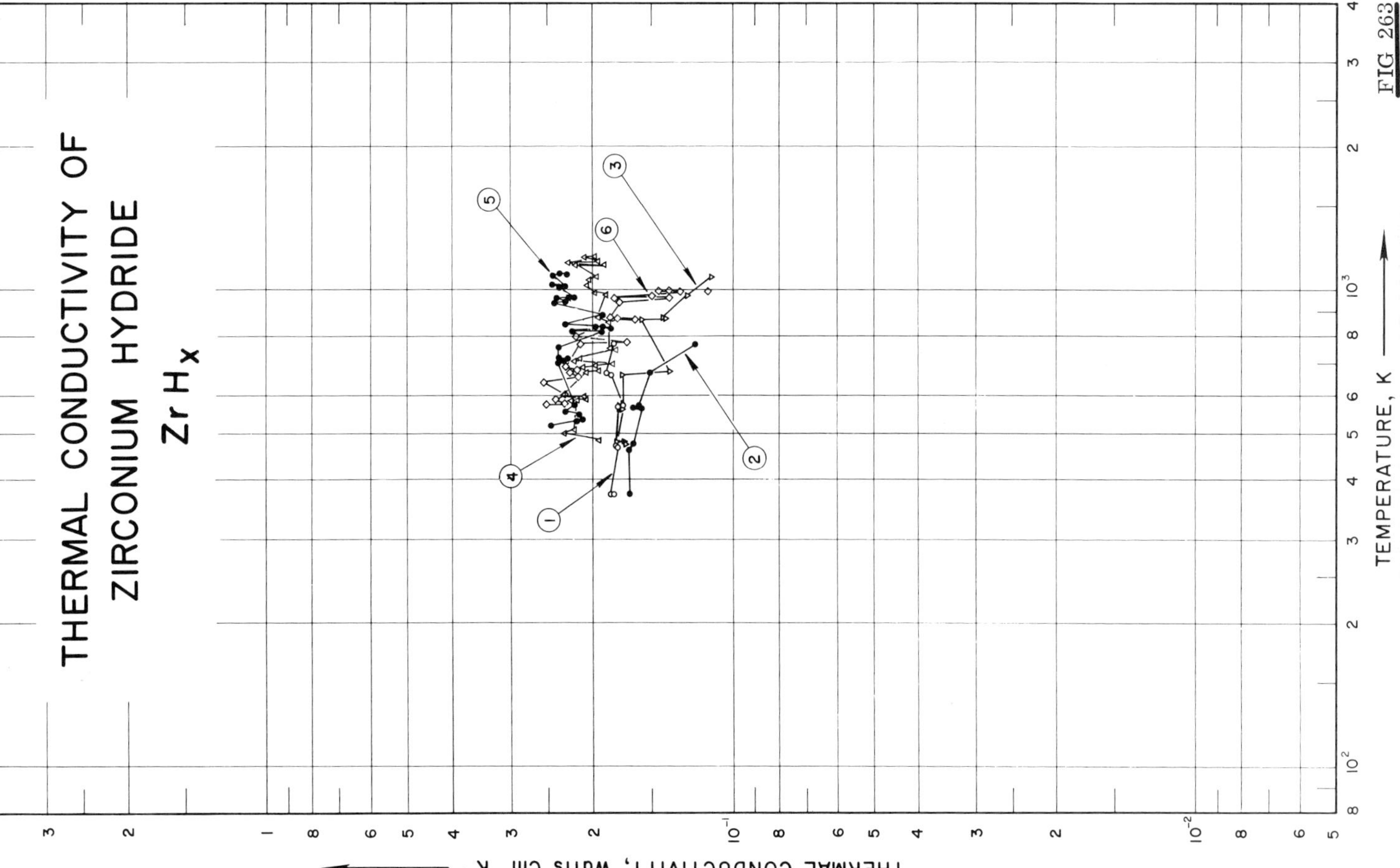

THERMAL CONDUCTIVITY OF
ZIRCONIUM HYDRIDE
Zr Hx
THERMAL CONDUCTIVITY, Watts Cm⁻¹ K⁻¹
TEMPERATURE, K

FIG 263

SPECIFICATION TABLE NO. 263 THERMAL CONDUCTIVITY OF ZIRCONIUM HYDRIDE ZrH_x

[For Data Reported in Figure and Table No. 263]

Curve No.	Ref. No.	Method Used	Year	Temp. Range, K	Reported Error, %	Name and Specimen Designation	Composition (weight percent), Specifications and Remarks
1	36		1956	373-774		$ZrH_{0.536}$ Sample 1	0.59 H (NH = 2.16×10^{22} hydrogen atoms cm^{-3}); specimen 0.5 in. in dia, 4 in. long; density 6.29 g cm^{-3}.
2	36		1956	373-770		$ZrH_{0.536}$ Sample 2-1	As above but heated to 700 C and cooled in a furnace at an undetermined rate; heater being improved to minimize uncertainties of heat transfer in the specimen; 1st run.
3	36		1956	477-1066		$ZrH_{0.536}$ Sample 2-2	2nd run of the above specimen.
4	235	C	1962	490-1181		$ZrH_{0.83}$	45.3 at. % H; prepared in a furnace at 900 C by flowing hydrogen at 1 atm over clean zirconium (the zirconium preheated to 950-1000 C under vacuum); homogenized and slowly cooled; specimen 0.60 in. in dia and 1.56 in. long of delta (fcc) crystal; molybdenum used as comparative material.
5	235	C	1962	526-1101		$ZrH_{1.3}$	Specimen similarly prepared as above with 56.6 at. % H.
6	235	C	1962	581-1001		$ZrH_{1.5}$	Specimen similarly prepared as above with 60 at. % H.

DATA TABLE NO. 263 THERMAL CONDUCTIVITY OF ZIRCONIUM HYDRIDE ZrH_x

[Temperature, T, K; Thermal Conductivity, k, Watts $cm^{-1}K^{-1}$]

T	k
CURVE 1	
373.2	0.180
373.2	0.183
469.2	0.176
472.2	0.178
570.2	0.176
574.2	0.172
666.2	0.183
670.2	0.187
774.2	0.180
CURVE 2	
373.2	0.167
464.2	0.168
468.2	0.169
479.2	0.164
566.2	0.157
567.2	0.164
574.2	0.159
672.2	0.151
770.2	0.121
CURVE 3	
477.2	0.170
482.2	0.172
482.2	0.178
567.2	0.173
568.2	0.174
572.2	0.176
575.2	0.173
665.2	0.173
675.2	0.137
869.2	0.157
873.2	0.140
876.2	0.142
976.2	0.126
1066.2	0.112
CURVE 4	
490.2	0.193
504.2	0.229
516.7	0.218
593.2	0.216
595.2	0.205
597.2	0.220
CURVE 4 (cont.)	
600.9	0.207
605.2	0.230
680.2	0.205
680.7	0.202 *
684.2	0.193
695.7	0.197
698.2	0.209
704.0	0.180
721.2	0.218
725.7	0.211
755.2	0.176
760.2	0.184
869.7	0.184
874.2	0.183 *
883.2	0.194
986.7	0.186
990.2	0.196
1040.7	0.205
1065.5	0.203
1073.2	0.196
1132.2	0.217
1149.6	0.187
1151.1	0.226
1163.2	0.194
1180.4	0.208
1180.8	0.196
CURVE 5	
526.0	0.247
535.7	0.216
537.0	0.210
553.2	0.214
560.3	0.229
578.2	0.217
710.2	0.236
718.2	0.232
723.2	0.226
725.2	0.238
767.4	0.238
823.2	0.190
829.5	0.222
836.2	0.181
840.2	0.195
842.4	0.188
855.2	0.229
892.9	0.189
CURVE 5 (cont.)	
945.2	0.241
956.9	0.228
960.7	0.231 *
962.0	0.239
962.6	0.218
965.6	0.225
1015.3	0.235
1021.1	0.228
1023.2	0.238 *
1024.6	0.231 *
1026.3	0.244
1029.2	0.227 *
1085.7	0.242
1089.7	0.245 *
1095.4	0.226
1095.7	0.234
1100.6	0.226 *
CURVE 6	
581.2	0.251
581.2	0.228
588.2	0.220 *
588.3	0.233 *
598.2	0.240
603.2	0.224 *
606.2	0.226 *
648.2	0.254
668.2	0.213
674.2	0.207 *
678.2	0.222
685.2	0.215
694.2	0.227
779.2	0.211
781.2	0.168
783.7	0.197
784.7	0.200 *
811.2	0.218
859.6	0.192 *
865.2	0.187 *
875.7	0.160
879.2	0.176
883.2	0.183
950.8	0.174
970.8	0.136
971.2	0.178
978.2	0.146 *
CURVE 6 (cont.)	
985.7	0.146
991.2	0.129
991.7	0.136
992.7	0.144
1000.7	0.112

* Not shown on plot

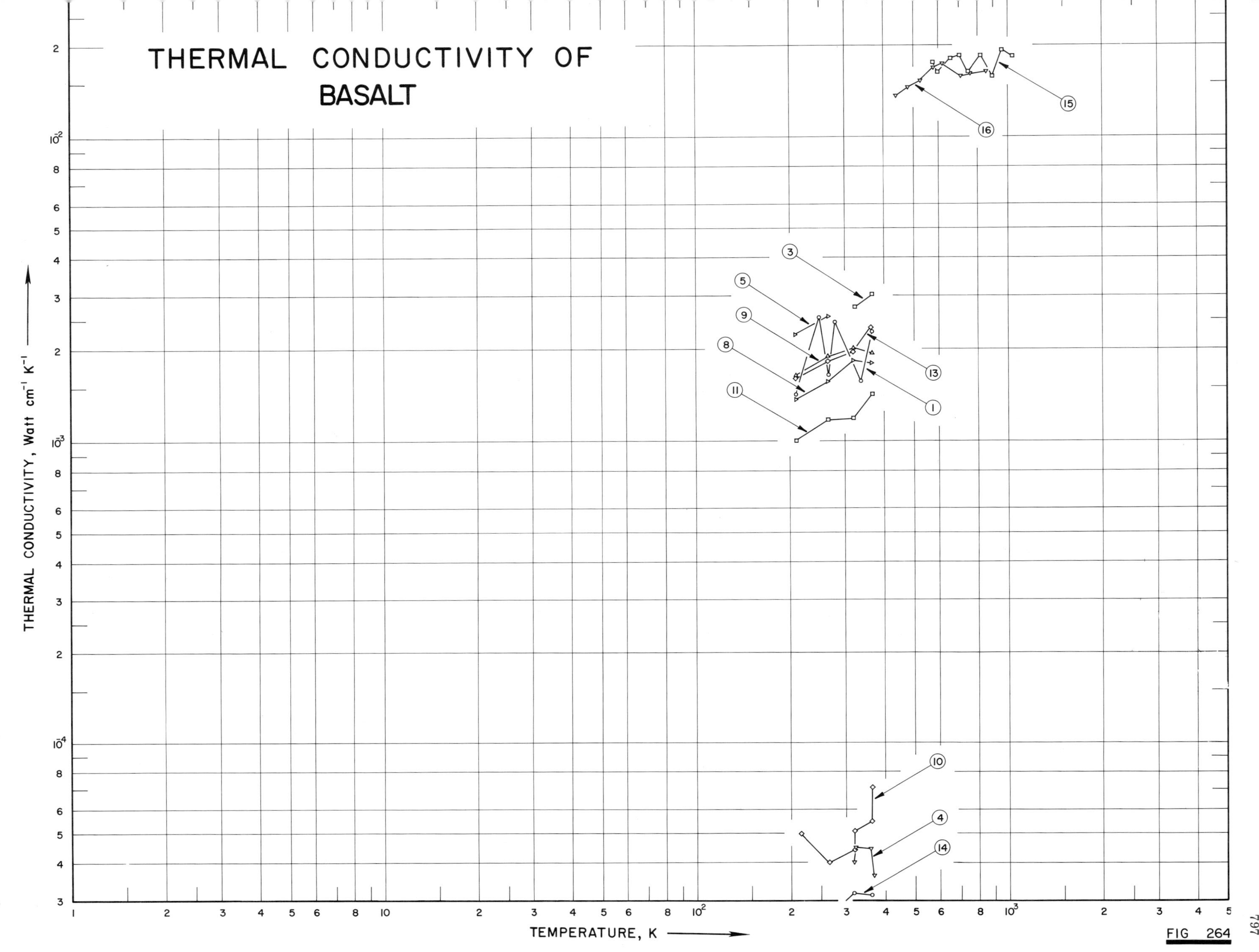
THERMAL CONDUCTIVITY OF BASALT
THERMAL CONDUCTIVITY, Watt cm^{-1} K^{-1}
TEMPERATURE, K
1
2
3
4
5
6
8
10
10^2
10^3
10^{-2}
10^{-3}
10^{-4}
15
16
3
5
9
8
11
13
1
10
4
14

FIG 264

SPECIFICATION TABLE NO. 264 THERMAL CONDUCTIVITY OF BASALT

[For Data Reported in Figure and Table No. 264]

Curve No.	Ref. No.	Method Used	Year	Temp. Range, K	Reported Error, %	Name and Specimen Designation	Composition (weight percent), Specifications and Remarks
1	539	P	1962	209-367		Olivine basalt	Collected from Pisgah Crater, San Bernardino, Calif.; passed through Gates jaw crusher and a stainless steel hammer mill-type pulverizer to reduce to -35 mesh material; density 1.49 g cm^{-3}.
2	539	P	1962	265-363		Olivine basalt	The above specimen measured in a vacuum of 5 x 10^{-6} mm Hg.
3	539	P	1962	325,368		Olivine basalt	Same source and production method as the above specimen; density 1.65 g cm^{-3}.
4	539	P	1962	319-367		Olivine basalt	The above specimen measured in a vacuum of 5 x 10^{-6} mm Hg.
5	539	P	1962	208,266		Olivine basalt	Same source and production method as the above specimen; density 1.95 g cm^{-3}.
6	539	P	1962	242.1		Olivine basalt	Same source and production method as the above specimen; density 1.75 g cm^{-3}; measured in a vacuum of 5 x 10^{-6} mm Hg.
7	539	P	1962	218,267		Olivine basalt	Same source, production method, and measuring condition as the above specimen; density 1.57 g cm^{-3}.
8	539	P	1962	210-367		Olivine basalt	Collected from Pisgah Crater, San Bernardino, Calif.; passed through Gates jaw crusher and a stainless steel hammer mill-type pulverizer to reduce to -35 mesh material, then screened into nominal mesh size -35+48; density 1.36 g cm^{-3}.
9	539	P	1962	210-367		Olivine basalt	Same source and production method as the above specimen; density 1.56 g cm^{-3}.
10	539	P	1962	216-365		Olivine basalt	The above specimen measured in a vacuum of 5 x 10^{-6} mm Hg.
11	539	P	1962	210-367		Olivine basalt	Collected from Pisgah Crater, San Bernardino, Calif.; passed through Gates jaw crusher and a stainless steel hammer mill-type pulverizer to reduce to -35 mesh material then screened into nominal-150 mesh; density 1.14 g cm^{-3}.
12	539	P	1962	319,363		Olivine basalt	The above specimen measured in a vacuum of 5 x 10^{-6} mm Hg.
13	539	P	1962	209-365		Olivine basalt	Same source and production method as the above specimen; density 1.57 g cm^{-3}.
14	539	P	1962	213-362		Olivine basalt	The above specimen measured in a vacuum of 5 x 10^{-6} mm Hg.
15	578	R	1963	576-1048	± 5	NTS Basalt No. 1	(Composition in vol percent); 66 plagioclase, 26 iron minerals (mainly hematite), 6 olivine; fine-grained appearance; specimen 3.5 in. O.D., 0.875 in. I.D. and 18 in. long; obtained from shot No. 12, hole DB-C-4 and shot No. 13, DB-4 of project Buck, Board; density 2.68 g cm^{-3}; specimen does not necessarily represent bulk average properties of its formation.
16	578	R	1963	442-858	± 5	NTS Basalt No. 2	Similar to the above specimen.

DATA TABLE NO. 264 THERMAL CONDUCTIVITY OF BASALT

[Temperature, T, K; Thermal Conductivity, k, Watt $cm^{-1}K^{-1}$]

T	k
CURVE 1	
209.3	0.00142
249.9	0.00254
267.1	0.00165
279.9	0.00246
338.2	0.00157
367.1	0.00228
CURVE 2*	
265.4	0.0000172
319.9	0.0000241
362.6	0.0000294
CURVE 3	
324.9	0.00275
367.6	0.00303
CURVE 4	
318.8	0.0000400
321.5	0.0000447
359.9	0.0000445
367.1	0.0000363
CURVE 5	
207.6	0.00223
266.0	0.00256
CURVE 6*	
242.1	0.0000177
CURVE 7*	
217.6	0.0000157
266.5	0.0000182
CURVE 8	
209.9	0.00137
265.4	0.00157
319.3	0.00182
366.5	0.00180
CURVE 9	
209.9	0.00164
265.4	0.00189
319.3	0.00201
366.5	0.00194
CURVE 10	
216.0	0.0000497
266.5	0.0000400
319.3	0.0000443
319.9	0.0000509
362.6	0.0000547
364.9	0.0000710
CURVE 11	
210.4	0.00100
266.0	0.00117
319.9	0.00118
367.1	0.00143
CURVE 12*	
319.3	0.0000154
363.2	0.0000163
CURVE 13	
208.8	0.00161
266.0	0.00183
318.8	0.00196
364.9	0.00235
CURVE 14	
213.2	0.0000273*
263.8	0.0000265*
318.2	0.0000317
362.1	0.0000313
CURVE 15	
576	0.0176
596	0.0162
658	0.0180
CURVE 15 (cont.)	
704	0.0184
748	0.0163
822	0.0184
895	0.0158
964	0.0192
1048	0.0184
CURVE 16	
442	0.0136
483	0.0145
529	0.0153
584	0.0167
623	0.0172
711	0.0157
762	0.0160
858	0.0163

*Not shown on plot

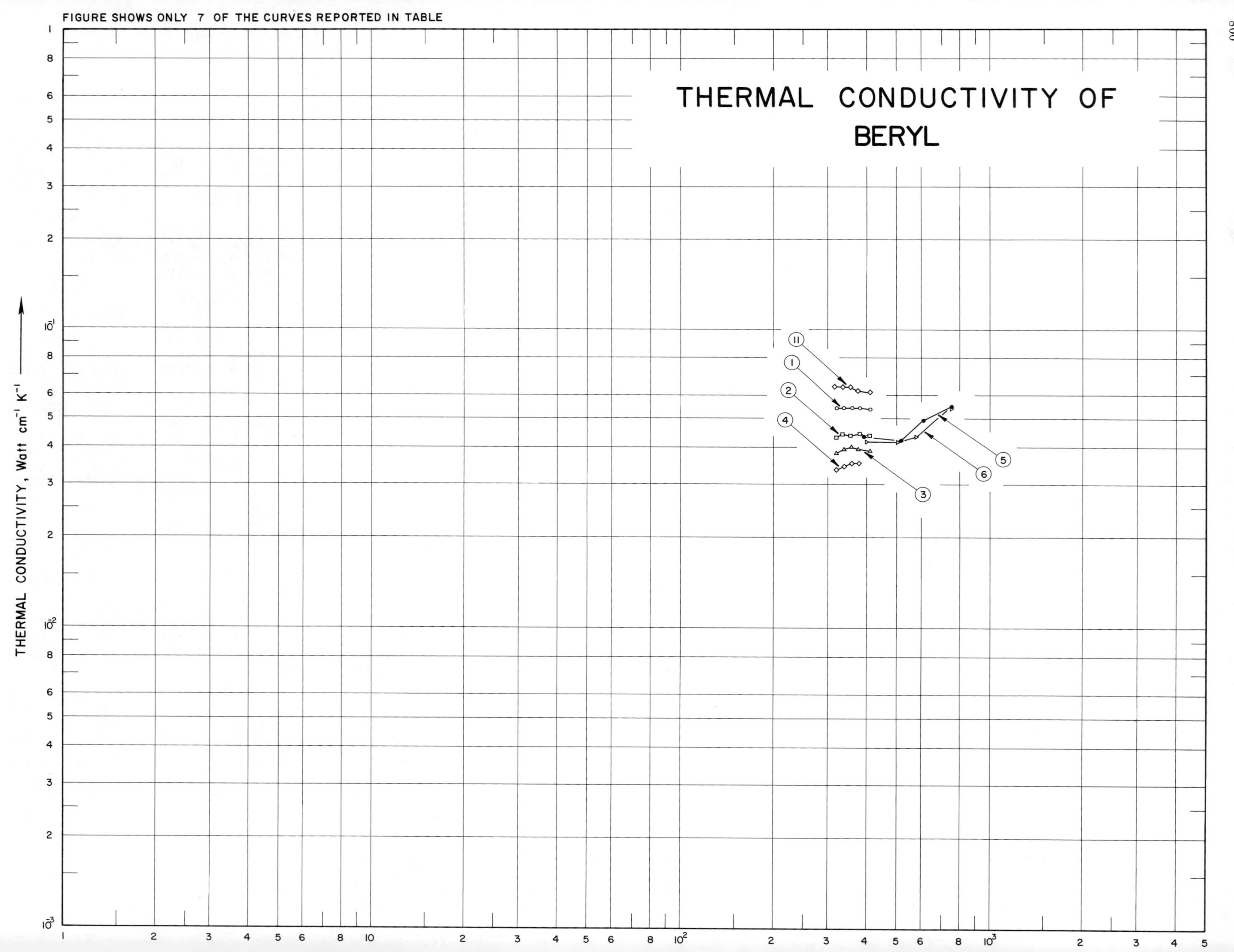
FIGURE SHOWS ONLY 7 OF THE CURVES REPORTED IN TABLE
THERMAL CONDUCTIVITY OF BERYL
THERMAL CONDUCTIVITY, Watt cm⁻¹ K⁻¹
11
1
2
4
3
6
5

SPECIFICATION TABLE NO. 265 THERMAL CONDUCTIVITY OF BERYL

[For Data Reported in Figure and Table No. 265]

Curve No.	Ref. No.	Method Used	Year	Temp. Range, K	Reported Error, %	Name and Specimen Designation	Composition (weight percent), Specifications and Remarks
1	3	L	1953	320-415		256A-1	Single crystal; variation of aquamarine; c-axis parallel to direction of heat flow.
2	3	L	1953	319-408		256B-1	Single crystal; variation of aquamarine; c-axis perpendicular to direction of heat flow.
3	3	L	1953	319-411		257A-1	Single crystal; golden beryl; c-axis parallel to the direction of heat flow.
4	3	L	1953	319-378		257B-1	Single crystal; golden beryl; c-axis perpendicular to the direction of heat flow.
5	34	C	1943	393-754		Brazil $3BeO \cdot Al_2O_3 \cdot 6SiO_2$	Green single crystal; specimen 1 x 1 x 1 cm ; measured parallel to the c-axis; 18-8 stainless steel used as comparative material.
6	34	C	1943	399-756		Brazil $3BeO \cdot Al_2O_3 \cdot 6SiO_2$	The above specimen measured normal to the c-axis; 18-8 stainless steel used as comparative material.
7	68	C	1954	316-380	±3	256C-1, Brazil	Aqua single crystal of hexagonal crystal system; specimen 0.25 in. in dia and 0.25 in. high; measured in the direction of c-axis, and in new "vacuum" (comparative) apparatus.
8	68	C	1954	317-409	±3	256B-2, Brazil	The above specimen measured in the a-axis direction and same apparatus.
9	68	C	1954	316-408	±3	257C-1, Brazil	Golden single crystal of hexagonal crystal system; specimen 0.25 in. in dia and 0.25 in. high; measured in the direction of c-axis; and in new "vacuum" (comparative) apparatus.
10	68	C	1954	314-408	±3	257B-2, Brazil	The above specimen measured in a-axis direction and same apparatus.
11	68	C	1954	315-413	±3	290A-1, India	Aqua single crystal of hexagonal crystal system; specimen 0.25 in. in dia and 0.25 in. high; measured in the direction of c-axis and in new "vacuum" (comparative) apparatus.

DATA TABLE NO. 265 THERMAL CONDUCTIVITY OF BERYL

[Temperature, T, K; Thermal Conductivity, k, Watt $cm^{-1}K^{-1}$]

T	k	T	k
CURVE 1		CURVE 7*	
320.4	0.0544	315.7	0.0552
338.5	0.0544	336.8	0.0548
359.1	0.0544	359.2	0.0548
380.5	0.0544	379.7	0.0548
414.9	0.0536		
CURVE 2		CURVE 8*	
318.6	0.0431	316.5	0.0464
334.1	0.0444	334.6	0.0435
354.5	0.0439	359.5	0.0444
379.8	0.0444	378.5	0.0439
407.6	0.0439	409.4	0.0439
CURVE 3		CURVE 9*	
319.0	0.0384	316.4	0.0389
338.2	0.0395	338.3	0.0388
357.7	0.0401	357.5	0.0401
376.6	0.0396	376.0	0.0403
411.1	0.0390	407.5	0.0403
CURVE 4		CURVE 10*	
319.4	0.0337	314.1	0.0346
338.8	0.0346	332.7	0.0354
358.1	0.0354	353.5	0.0361
378.0	0.0354	376.0	0.0360
378.1	0.0351*	407.5	0.0363
CURVE 5		CURVE 11	
393.2	0.0435	315.3	0.0640
516.2	0.0423	335.1	0.0640
613.2	0.0494	354.2	0.0640
754.2	0.0552	375.3	0.0623
		413.0	0.0615
CURVE 6			
399.2	0.0417		
505.2	0.0418		
577.2	0.0435		
756.2	0.0552		

* Not shown on plot

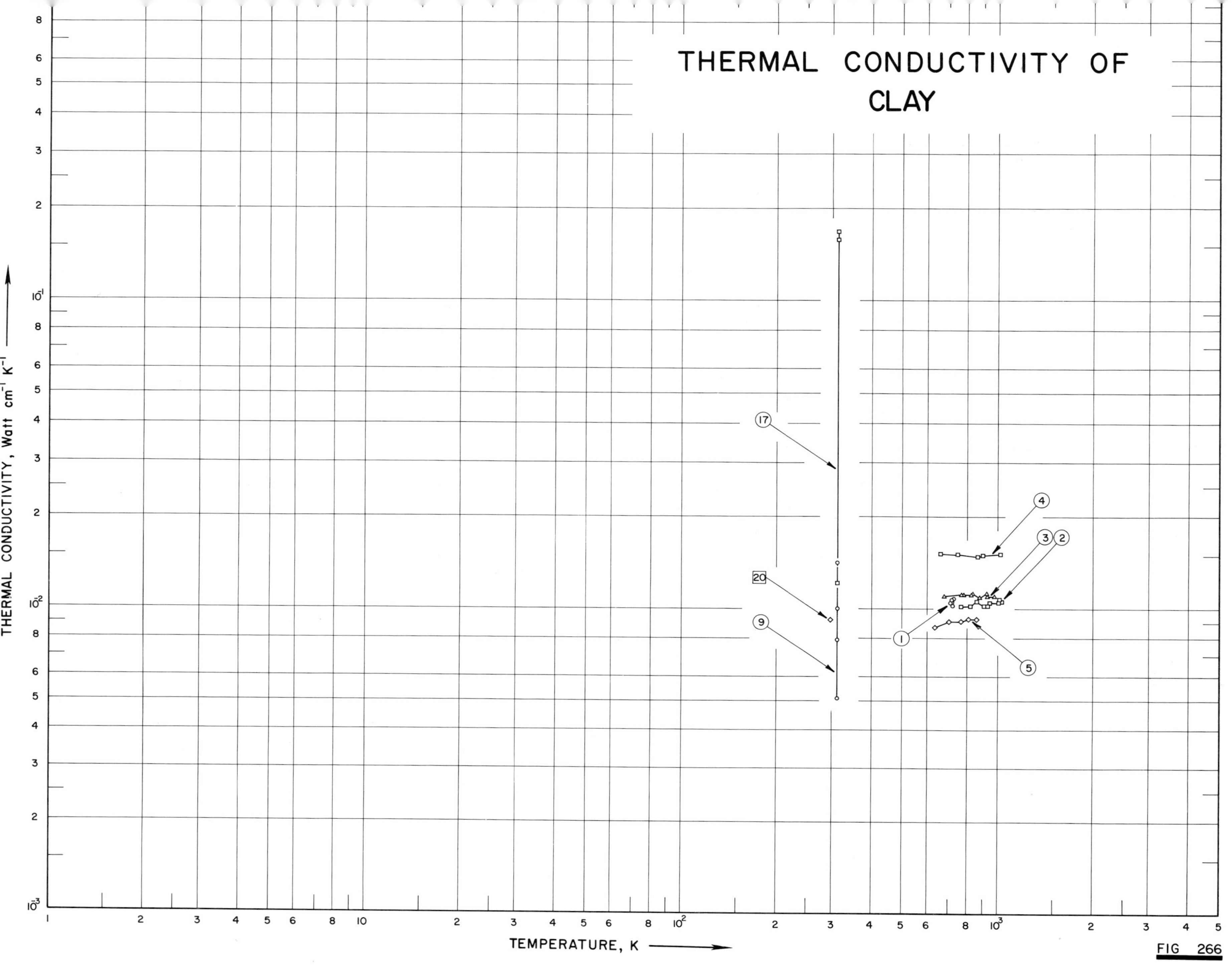

THERMAL CONDUCTIVITY OF
CLAY
THERMAL CONDUCTIVITY, Watt cm⁻¹ K⁻¹
TEMPERATURE, K
1
2
3
4
5
9
17
20

FIG 266

SPECIFICATION TABLE NO. 266 THERMAL CONDUCTIVITY OF CLAY

[For Data Reported in Figure and Table No. 266]

Curve No.	Ref. No.	Method Used	Year	Temp. Range, K	Reported Error, %	Name and Specimen Designation	Composition (weight percent), Specifications and Remarks
1	91	C	1947	713-730		Fire clay; Chamotte	Cylindrical specimen; iron used as comparative material.
2	551	R	1909	769-1030		Fire clay; A	Dark reddish-brown fire clay, containing no gravel, structure similar to sandstone; specimen supplied by the Laclede Christy Clay Products Co., St. Louis, Missouri.
3	551	R	1909	679-978		Fire clay; 3	Almost white, very coarse fire clay, containing a large amount of gravel, specimen supplied by the above-mentioned company.
4	551	R	1909	661-1030		Fire clay; 1	Brown, coarse fire clay, containing very small amount of gravel; specimen supplied by the above-mentioned company.
5	551	R	1909	635-860		Fire clay; B	Reddish-brown fire clay of medium coarse structure, containing very small pieces of white gravel; specimen supplied by the above-mentioned company.
6	556, 557	L	1952	343-393		Kuchin	Specimens 180-190 mm in dia and 25-30 mm thick; measurements done on 5 different specimens with porosities ranging from 14.4 to 21% and different degrees of evacuation (0 to 740 mm Hg).
7	556, 557	L	1952	343-393		Beskhudnikov	Similar to the above specimens except porosity range 13.8 to 17%.
8	556, 557	L	1952	343-393		Ashkhabad	Similar to the above specimens except porosity range 15.7 to 26.6%.
9	556, 557	P	1952	313.2		Beskhudnikov	Specimen 70 mm in dia and 100 mm long; volumetric weight 1230 Kg cm^{-3}; measured over a moisture content range of 1.2 to 26.4% at 0 mm Hg vacuum.
10	556, 557	P	1952	313.2		Beskhudnikov	Similar to the above specimen except moisture content range 1.1 to 26.2%; measured at 300 mm Hg vacuum.
11	556, 557	P	1952	313.2		Beskhudnikov	Similar to the above specimen except moisture content range 1.1 to 26.4%; measured at 500 mm Hg vacuum.
12	556, 557	P	1952	313.2		Beskhudnikov	Similar to the above specimen except measured at 740 mm Hg vacuum.
13	556, 557	P	1952	313.2		Kuchin	Specimen 70 mm in dia and 100 mm long; volumetric weight 1150 Kg cm^{-3}; measured over a moisture content range of 1.2 to 26.8% at 0 mm Hg vacuum.
14	556, 557	P	1952	313.2		Kuchin	Similar to the above specimen except measured at 300 mm Hg vacuum.
15	556, 557	P	1952	313.2		Kuchin	Similar to the above specimen except moisture content range 0.8 to 27.2%; measured at 740 mm Hg vacuum.
16	556, 557	P	1952	313.2		Ashkhabad	Specimen 70 mm in dia and 100 mm long; volumetric weight 1200 Kg cm^{-3}; measured over a moisture content range of 1.2 to 19.8% at 0 mm Hg vacuum.

SPECIFICATION TABLE NO. 266 (continued)

Curve No.	Ref. No.	Method Used	Year	Temp. Range, K	Reported Error, %	Name and Specimen Designation	Composition (weight percent), Specifications and Remarks
17	556, 557	P	1952	313.2		Ashkhabad	Similar to the above specimen except moisture content range 0.8 to 20%; measured at 300 mm Hg vacuum.
18	556, 557	P	1952	313.2		Ashkhabad	Similar to the above specimen except moisture content range 1.2 to 19.2%; measured at 500 mm Hg vacuum.
19	556, 557	P	1952	313.2		Ashkhabad	Similar to the above specimen except moisture content range 1 to 19.3%; measured at 740 mm Hg vacuum.
20	546	P	1924	297.2		Sandy clay	Moisture content 15%; density 1.78 g cm^{-3}

DATA TABLE NO. 266 THERMAL CONDUCTIVITY OF CLAY

[Temperature, T, K; Thermal Conductivity, k, Watt $cm^{-1}K^{-1}$]

T	k
CURVE 1	
713.2	0.0105
715.2	0.0105*
718.2	0.0107
723.2	0.0103*
724.2	0.0103
730.2	0.0108
CURVE 2	
769.4	0.0103
824.8	0.0103
864.9	0.0106
906.2	0.0103
935.3	0.0103
937.9	0.0105*
945.8	0.0105
1011.2	0.0105
1015.8	0.0108
1030.4	0.0106
CURVE 3	
679.7	0.0110
766.4	0.0111
770.9	0.0111
830.7	0.0111
838.6	0.0113
878.9	0.0109
928.4	0.0112
930.3	0.0110
977.8	0.0110
CURVE 4	
661.4	0.0153
747.2	0.0152
869.9	0.0149
900.6	0.0150
1029.7	0.0151
CURVE 5	
635.6	0.00874
703.8	0.00908
CURVE 5 (cont.)	
767.4	0.00908
811.0	0.00925
859.8	0.00925

Total porosity (%)	k
CURVE 6* (T = 343.2-393.2 K)	
14.4	0.00437
15.1	0.00432
17.5	0.00404
19.4	0.00397
21	0.00381
CURVE 7* (T = 343.2-393.2 K)	
13.8	0.00472
14.7	0.00465
15.8	0.00446
16.5	0.00432
17	0.00430
CURVE 8* (T = 343.2-393.2 K)	
15.7	0.00549
16.5	0.00535
18.8	0.00504
22.6	0.00471
26.6	0.00465

Moisture content (%)	k
CURVE 9 (T = 313.2 K)	
1.2	0.00511
6.8	0.00790
13.6	0.0100
26.4	0.0142
CURVE 10* (T = 313.2 K)	
1.1	0.00523
7.2	0.00860
13	0.0105
26.2	0.0151
CURVE 11* (T = 313.2 K)	
1.1	0.00535
7.2	0.00941
13.6	0.0121
26.4	0.0132
CURVE 12* (T = 313.2 K)	
1.1	0.00604
7.8	0.0119
14	0.0144
26.4	0.0172
CURVE 13* (T = 313.2 K)	
1.2	0.00442
6	0.00814
11.4	0.00930
26.8	0.0121
CURVE 14* (T = 313.2 K)	
1.2	0.00465
6.2	0.00883
11.9	0.00976
26.8	0.0130
CURVE 15* (T = 313.2 K)	
0.8	0.00477
6	0.0100
CURVE 15 (cont.)*	
11.9	0.0121
27.2	0.0142
CURVE 16* (T = 313.2 K)	
1.2	0.00465
8	0.0116
14	0.0139
19.8	0.0139
CURVE 17 (T = 313.2 K)	
0.8	0.00511*
7.4	0.0121
13.6	0.158
20	0.167
CURVE 18* (T = 313.2 K)	
1.2	0.00604
8	0.0139
12.8	0.0174
19.2	0.0188
CURVE 19* (T = 313.2 K)	
1	0.00651
7.6	0.0142
13	0.0186
19.3	0.0205

T	k
CURVE 20	
297.2	0.00920

*Not shown on plot

THERMAL CONDUCTIVITY OF COAL

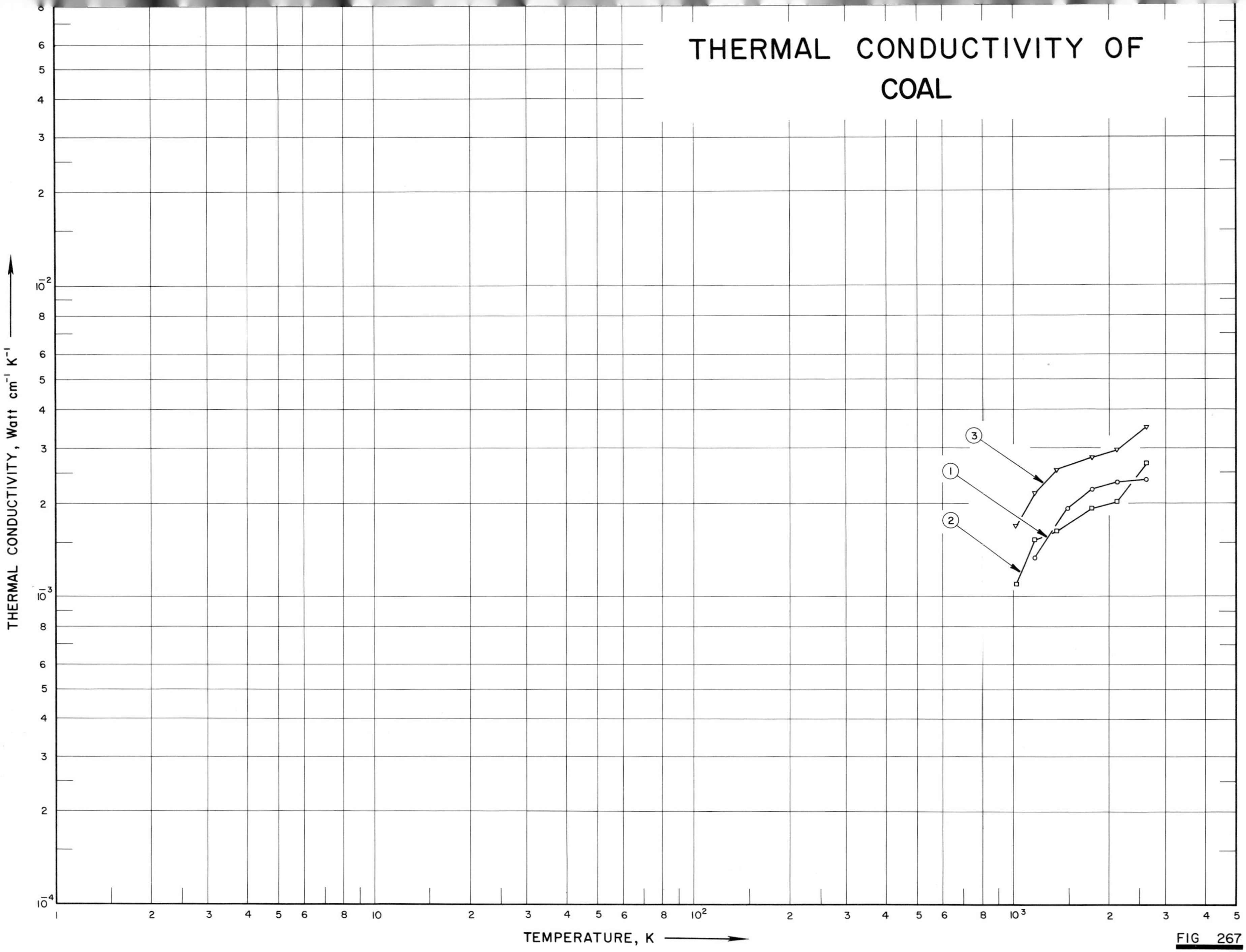

FIG 267

SPECIFICATION TABLE NO. 267 THERMAL CONDUCTIVITY OF COAL

[For Data Reported in Figure and Table No. 267]

Curve No.	Ref. No.	Method Used	Year	Temp. Range, K	Reported Error, %	Name and Specimen Designation	Composition (weight percent), Specifications and Remarks
1	587	P	1960	1173-2623		Angren brown coal	Noncaking Angren brown coal heated to 1100 C at a rate of 8 c min^{-1} and kept at this temperature for 3 hrs; the high-temperature treatment of the specimen carried out in an atmosphere of nitrogen in a furnace with graphite heater, temperature controlled by means of a stepped transformer; total time of heating 45 min; material studied consisting chiefly of carbon (heating to 1100 yields a coke product containing up to 98% carbon, relative to the dry ashless mass); the specimen ground to a particle size of less than 0.5 mm; volume weight between 0.550 and 0.650 g cm^{-3}.
2	587	P	1960	1023-2623		Donets gas coal	Noncaking Donets gas coal; same heat treatment and size as the above specimen.
3	587	P	1960	1023-2623		Donets anthracite	Noncaking Donets anthracite coal; same heat treatment and size as the above specimen.

DATA TABLE NO. 267 THERMAL CONDUCTIVITY OF COAL

[Temperature, T, K; Thermal Conductivity, k, Watt $cm^{-1}K^{-1}$]

T	k
CURVE 1	
1173.2	0.00134
1493.2	0.00192
1773.2	0.00221
2123.2	0.00232
2623.2	0.00238
CURVE 2	
1023.2	0.00110
1173.2	0.00153
1373.2	0.00163
1773.2	0.00192
2123.2	0.00203
2623.2	0.00267
CURVE 3	
1023.2	0.00169
1173.2	0.00215
1373.2	0.00256
1773.2	0.00279
2123.2	0.00296
2623.2	0.00349

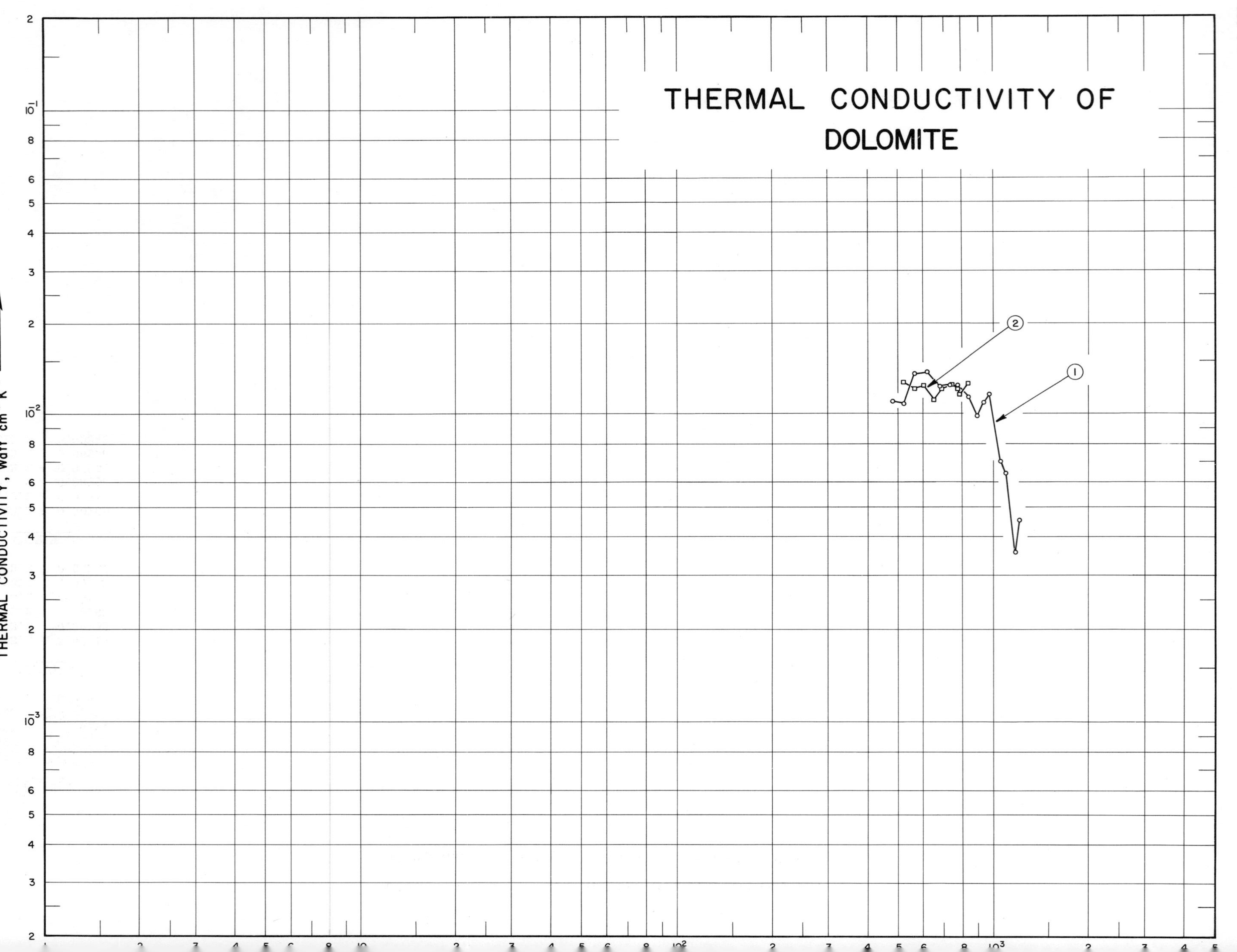
THERMAL CONDUCTIVITY OF
DOLOMITE
THERMAL CONDUCTIVITY, Watt cm^{-1} K^{-1}
1
2

SPECIFICATION TABLE NO. 268 THERMAL CONDUCTIVITY OF DOLOMITE

[For Data Reported in Figure and Table No. 268]

Curve No.	Ref. No.	Method Used	Year	Temp. Range, K	Reported Error, %	Name and Specimen Designation	Composition (weight percent), Specifications and Remarks
1	578	R	1963	484-1208	± 5	NTS dolomite No. 1	(Composition in vol percent); 99 dolomite, 1 others (primarily quartz and muscovite), fine-grained appearance; specimen 2.25 in. O.D., 0.375 in. I.D. and 12 in. long; obtained from exploratory dolomite hole No. 1, dolomite hill at level of 200 ft; density 2.80 g cm^{-3}; specimen does not necessarily represent bulk averages of its formation; data values above 1000 K are in the region of decomposition of the carbonates and are of qualitative value only, due to formation of microcracks and some large fusion in the specimen; other types of carbonate rocks show thermal conductivity values higher by a factor of 2, according to literature.
2	578	R	1963	523-833	± 5	NTS dolomite No. 2	Similar to the above specimen.

DATA TABLE NO. 268 THERMAL CONDUCTIVITY OF DOLOMITE

[Temperature, T, K; Thermal Conductivity, k, Watt $cm^{-1}K^{-1}$]

T	k
CURVE 1	
484	0.0110
521	0.0108
565	0.0136
619	0.0137
678	0.0124
229	0.0125
773	0.0125
835	0.0114
888	0.00983
932	0.0109
970	0.0116
1051	0.00699
1098	0.00640
1173	0.00356
1208	0.00452
CURVE 2	
523	0.0127
565	0.0121
603	0.0124
654	0.0111
687	0.0121
744	0.0125
772	0.0121
774	0.0116
833	0.0126

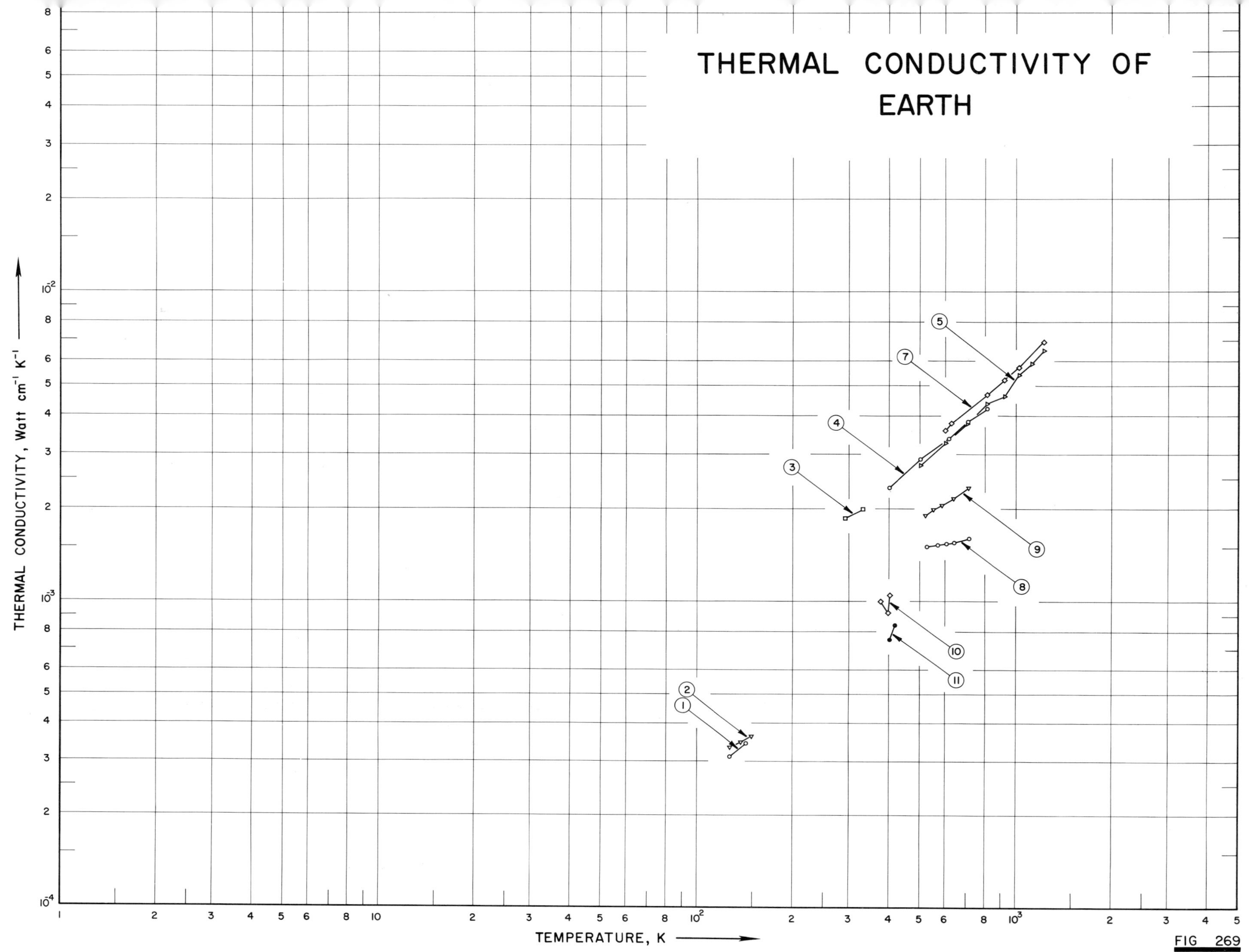

THERMAL CONDUCTIVITY OF EARTH
THERMAL CONDUCTIVITY, Watt cm⁻¹ K⁻¹
TEMPERATURE, K
1
2
3
4
5
7
8
9
10
11

FIG 269

SPECIFICATION TABLE NO. 269 THERMAL CONDUCTIVITY OF EARTH

[For Data Reported in Figure and Table No. 269]

Curve No.	Ref. No.	Method Used	Year	Temp. Range, K	Reported Error, %	Name and Specimen Designation	Composition (weight percent), Specifications and Remarks
1	43	L	1954	128, 144		Diatomaceous Earth	Density 14.0 lb ft^{-3}.
2	43	L	1954	128-150		Diatomaceous Earth	Density 17.0 lb ft^{-3}.
3	273	L	1963	293, 333			Density 0.15 g cm^{-3}; apparatus: NBS guarded hot plate.
4	273	R	1963	403-813	<3.0		Density 0.15 g cm^{-3}; apparatus without end guarding.
5	273	R	1963	503-1223	<3.0		Density 0.15 g cm^{-3}; apparatus with ends guarded.
6	273	R	1963	603-1228	<3.0		Density 0.15 g cm^{-3}; sintered and then pulverized.
7	273	R	1963	603-1228	<3.0		Density 0.51 g cm^{-3}; sintered.
8	100	L	1957	529-716		Kieselguhr (1)	Insulating powders; bulk density 0.388 g cm^{-3}.
9	100	L	1957	523-714		Kieselguhr (2)	Insulating powders; bulk density 0.628 g cm^{-3}.
10	240	R	1919	378-403		Ordinary kieselguhr	Specimen 12.2 mm I.D., 19 mm O.D., and 8 cm long.
11	240	R	1919	401, 419		Ignited kieselguhr	Specimen 12.2 mm I.D., 19 mm O.D., and 8 cm long.

DATA TABLE NO. 269 THERMAL CONDUCTIVITY OF EARTH

[Temperature, T, K; Thermal Conductivity, k, Watt $cm^{-1}K^{-1}$]

T	k
CURVE 1	
127.6	0.000310
144.3	0.000342
CURVE 2	
127.6	0.000332
138.2	0.000346
149.8	0.000361
CURVE 3	
293.2	0.00186
333.2	0.00199
CURVE 4	
403.2	0.00234
503.2	0.00289
618.2	0.00337
708.2	0.00383
813.2	0.00423
CURVE 5	
503.2	0.00276
603.2	0.00326
708.2	0.00377
813.2	0.00437
923.2	0.00462
1023.2	0.00544
1123.2	0.00588
1223.2	0.00649
CURVE 6*	
603.2	0.00347
833.2	0.00444
1023.2	0.00536
1228.2	0.00640
CURVE 7	
603.2	0.00358
628.2	0.00377

T	k
CURVE 7 (cont.)	
813.2	0.00469
923.2	0.00523
1023.2	0.00573
1228.2	0.00690
CURVE 8	
528.7	0.00151
572.6	0.00153
606.5	0.00154
643.2	0.00156
715.9	0.00160
CURVE 9	
523.2	0.00190
552.6	0.00198
587.1	0.00205
637.6	0.00216
713.7	0.00234
CURVE 10	
377.8	0.00100
398.8	0.000920
403.2	0.00105
CURVE 11	
401.2	0.000753
419.0	0.000837

*Not shown on plot

SPECIFICATION TABLE NO. 270 THERMAL CONDUCTIVITY OF GABBRO

Curve No.	Ref. No.	Method Used	Year	Temp. Range, K	Reported Error, %	Name and Specimen Designation	Composition (weight percent), Specifications and Remarks
1	586	L	1933	309, 323	1	Gabbro	Specimen 5 cm in dia and 2 cm long; from Sligachan Skye; density 3.10 g cm^{-3}.

DATA TABLE NO. 270 THERMAL CONDUCTIVITY OF GABBRO

[Temperature, T, K; Thermal Conductivity, k, Watt $cm^{-1} K^{-1}$]

T	k
CURVE 1*	
309.4	0.0255
323.1	0.0247

*No graphical presentation

THERMAL CONDUCTIVITY OF GRANITE

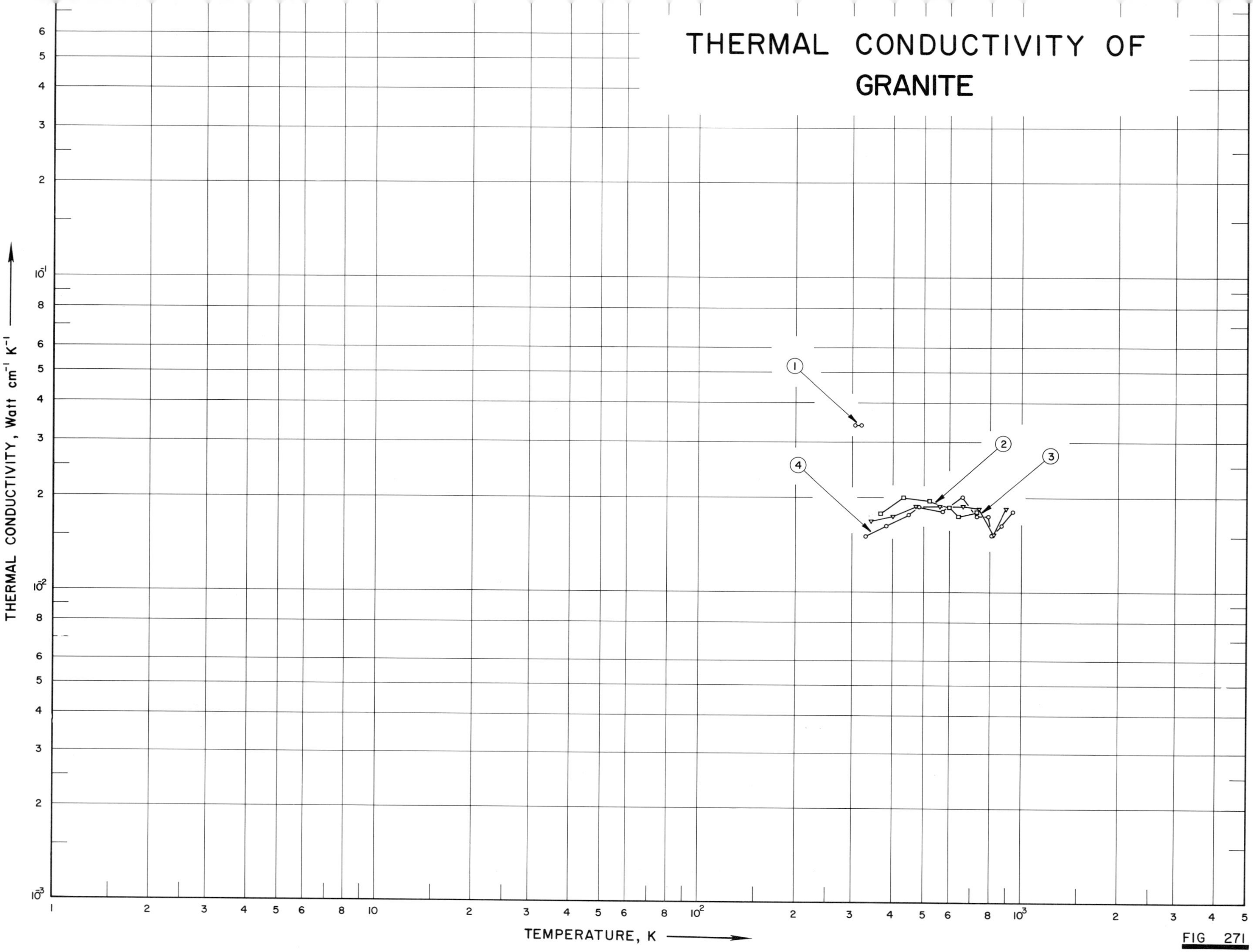

FIG 271

SPECIFICATION TABLE NO. 271 THERMAL CONDUCTIVITY OF GRANITE

[For Data Reported in Figure and Table No. 271]

Curve No.	Ref. No.	Method Used	Year	Temp. Range, K	Reported Error, %	Name and Specimen Designation	Composition (weight percent), Specifications and Remarks
1	586	L	1933	307, 320	1	Granite	Specimen 5 cm in dia and 2 cm long; from New May Quarry, Aberdeenshire; density 2.58 g cm^{-3}.
2	578	R	1963	368-773	±5	NTS Granite No. 1	(Composition in vol percent); 34 plagisclase, 28 ortheoclase, 27 quartz, 9 biotite, 2 others; coarse grained appearance; specimen 3.5 in. O.D., 0.875 in. I.D. and 18 in. long; obtained from U15b exploratory hole area 15, at 1000 ft level; density 2.67 g cm^{-3}; specimen does not necessarily represent bulk average properties of its formation.
3	578	R	1963	343-896	±5	NTS Granite No. 2	Similar to the above specimen.
4	578	R	1963	330-943	±5	NTS Granite No. 3	Similar to the above specimen.

DATA TABLE NO. 271 THERMAL CONDUCTIVITY OF GRANITE

[Temperature, T, K; Thermal Conductivity, k, Watt $cm^{-1}K^{-1}$]

T	k
CURVE 1	
306.9	0.0339
320.2	0.0339
CURVE 2	
368	0.0178
433	0.0200
523	0.0195
600	0.0186
643	0.0174
733	0.0180
CURVE 3	
343	0.0168
400	0.0174
473	0.0187
556	0.0187
662	0.0187
738	0.0184
823	0.0152
896	0.0184
CURVE 4	
330	0.0151
383	0.0163
448	0.0176
483	0.0187
573	0.0176
658	0.0201
728	0.0174
793	0.0174
811	0.0152
873	0.0164
943	0.0180

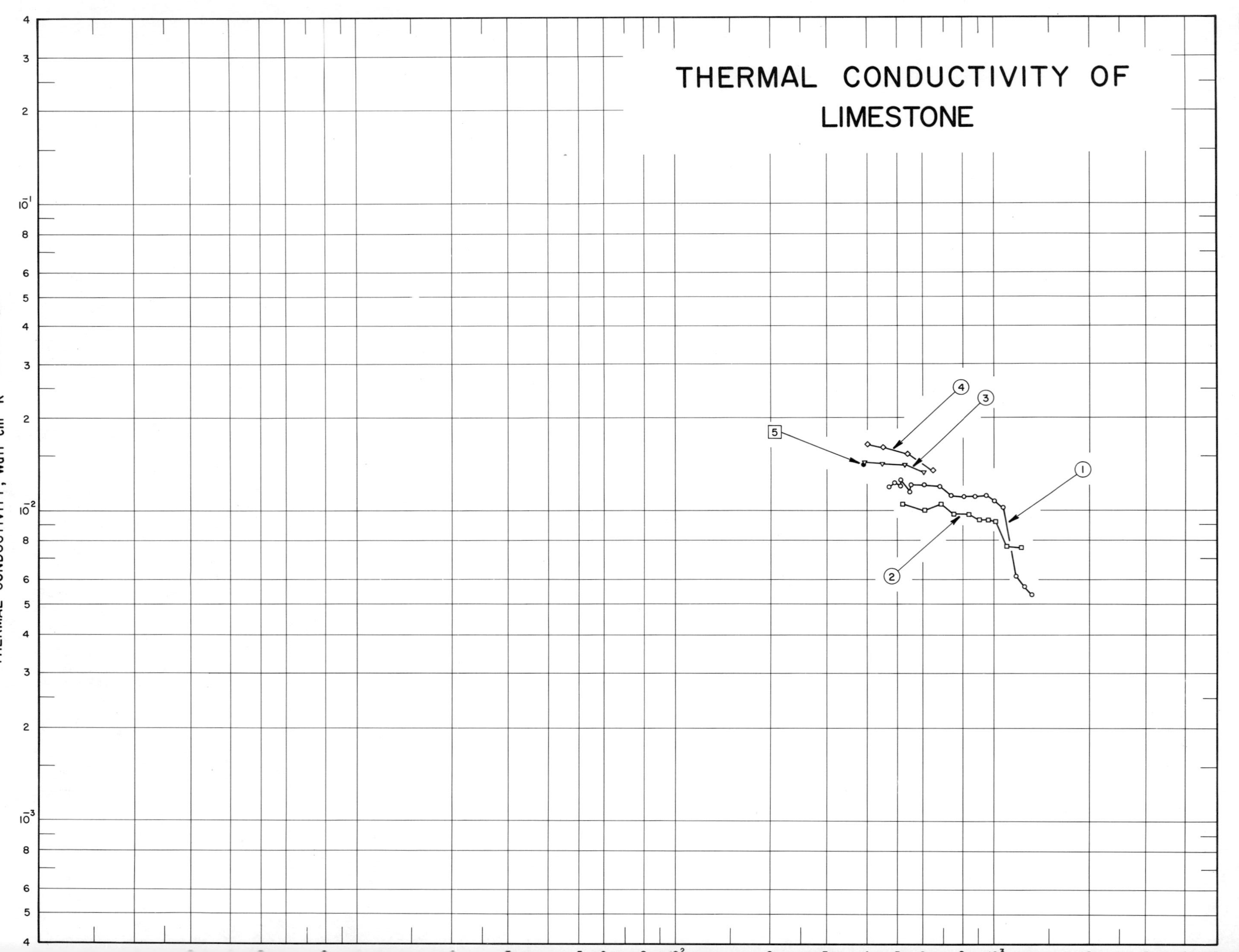
THERMAL CONDUCTIVITY OF
LIMESTONE
THERMAL CONDUCTIVITY, Watt cm^{-1} K^{-1}
4
3
2
10^{-1}
8
6
5
4
3
2
10^{-2}
8
6
5
4
3
2
10^{-3}
8
6
5
4
1
2
3
4
5

SPECIFICATION TABLE NO. 272 THERMAL CONDUCTIVITY OF LIMESTONE

[For Data Reported in Figure and Table No. 272]

Curve No.	Ref. No.	Method Used	Year	Temp. Range, K	Reported Error, %	Name and Specimen Designation	Composition (weight percent), Specifications and Remarks
1	578	R	1963	472-1324	± 6	Indiana Limestone No. 2	(Composition in vol percent); 98.4 calcite, 1.0 quartz, 0.6 hematite; fine grained in appearance; specimen 3.5 in. O.D., 0.875 in. I.D. and 18 in. long; exact origin of specimen unknown; density 2.30 g cm^{-3}; specimen does not necessarily represent the bulk average properties of its formation; data above 1100 K indicate region of decomposition of the carbonates and the thermal conductivity values are qualitative only due to the formation of microcracks and some large fusions in the specimen; other types of carbonate rocks show thermal conductivity values higher by a factor of 2 according to the literature.
2	578	R	1963	520-1228	± 6	Indiana Limestone No. 2	Similar to the above specimen.
3	508	L	1940	396-605		Queenstone grey	Specimen 8 in. in dia and 1 in. thick; a mixture of dolomite and calcite containing 22% $MgCO_3$; density 2.675 g cm^{-3}; obtained from Queenston, Ont.
4	508	L	1940	403-650		Rama	Specimen 8 in. in dia and 1 in. thick; a mixture of dolomite and calcite containing 30% of $MgCO_3$; density 2.563 g cm^{-3}; obtained from Langford Mills, Ont.
5	508	L	1940	392.1		Rama	Similar to the above specimen; measurements done after exposure to high temperature test.

DATA TABLE NO. 272 THERMAL CONDUCTIVITY OF LIMESTONE

[Temperature, T, K; Thermal Conductivity, k, Watt $cm^{-1}K^{-1}$]

T	k	T	k
CURVE 1		CURVE 5	
472	0.0119	392.1	0.0141
493	0.0123		
512	0.0120		
513	0.0126		
546	0.0115		
553	0.0121		
610	0.0121		
683	0.0119		
736	0.0112		
813	0.0111		
878	0.0111		
952	0.0112		
1013	0.0107		
1075	0.0103		
1181	0.00619		
1253	0.00573		
1324	0.00540		
CURVE 2			
520	0.0105		
613	0.0100		
688	0.0105		
754	0.00971		
840	0.00967		
901	0.00933		
965	0.00933		
1028	0.00921		
1101	0.00766		
1228	0.00762		
CURVE 3			
395.9	0.0143		
450.4	0.0141		
527.6	0.0140		
605.4	0.0133		
CURVE 4			
403.2	0.0164		
454.3	0.0160		
540.4	0.0153		
650.4	0.0134		

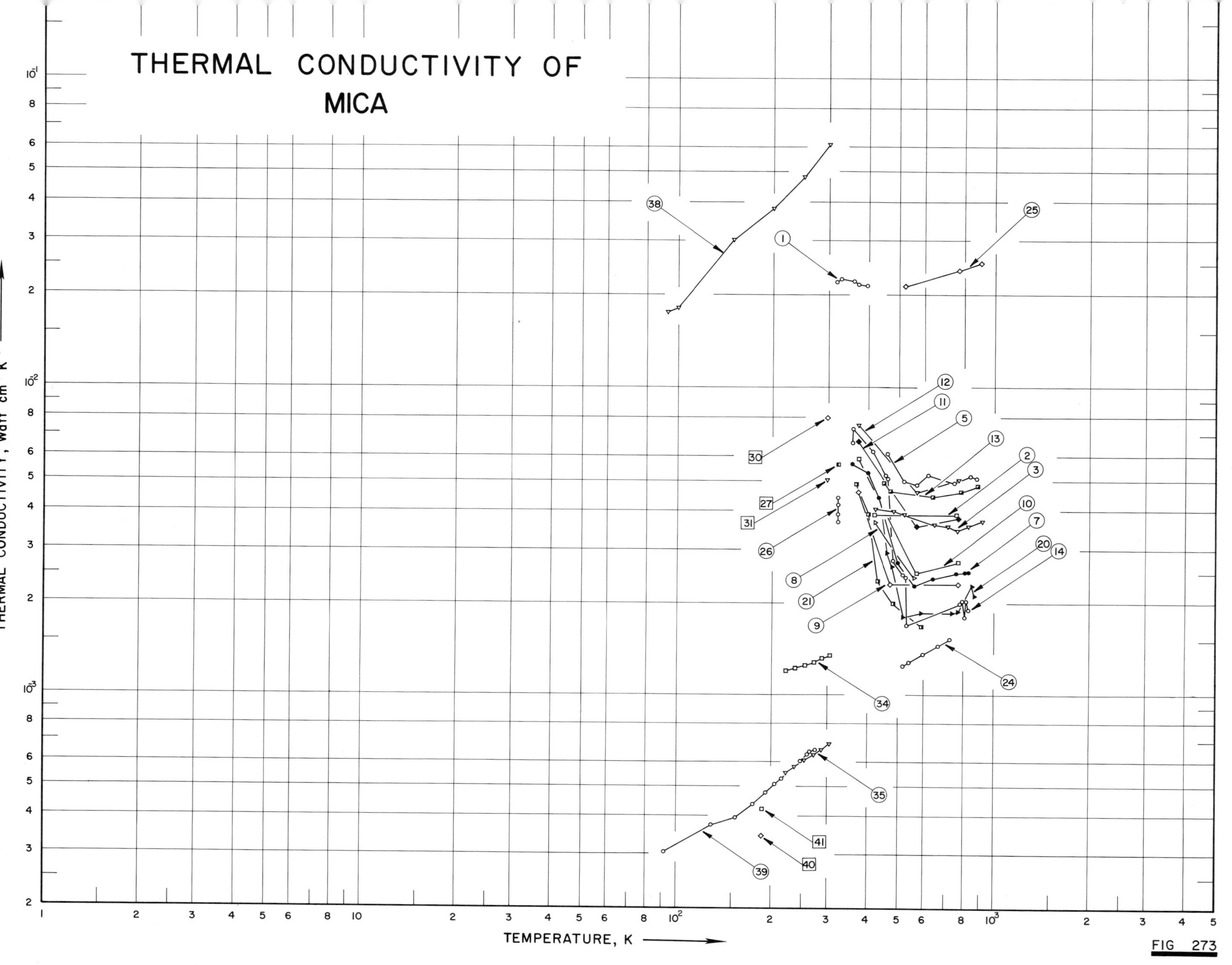
THERMAL CONDUCTIVITY OF
MICA
THERMAL CONDUCTIVITY, Watt cm^{-1} K^{-1}
TEMPERATURE, K

FIG 273

SPECIFICATION TABLE NO. 273 THERMAL CONDUCTIVITY OF MICA

[For Data Reported in Figure and Table No. 273]

Curve No.	Ref. No.	Method Used	Year	Temp. Range, K	Reported Error, %	Name and Specimen Designation	Composition (weight percent), Specifications and Remarks
1	407	L	1949	317-397			Glass bonded.
2	577	L	1937	420,761.2		Madagascan phlogopites; 1	Amber light; measured under 74 psi loading.
3	577	L	1937	424-916		Madagascan phlogopites; 1	The above specimen measured under 176 psi loading; density when loaded 2.66 g cm^{-3}.
4	577	L	1937	426-809		Madagascan phlogopites; 1	The above specimen measured when cooled from 643 C.
5	577	L	1937	460-880		Madagascan phlogopites; 2	Amber dark; measured under 176 psi loading; loaded density 2.90 g cm^{-3}.
6	577	L	1937	496.2		Madagascan phlogopites; 2	The above specimen measured when cooled from 607 C.
7	577	L	1937	357-833		Madagascan phlogopites; 3	Amber dark; measured under 176 psi loading; loaded density 2.87 g cm^{-3}.
8	577	L	1937	423.2,560.2		Madagascan phlogopites; 3	The above specimen measured when cooled from 560 C.
9	577	L	1937	373-773		Madagascan phlogopites; 3-2	Similar to the above specimen 3 but measured under 23 psi loading.
10	577	L	1937	373-773		Madagascan phlogopites; 3-2	The above specimen measured under 48 psi loading.
11	577	L	1937	373-773		Madagascan phlogopites; 3-2	The above specimen measured under 178 psi loading.
12	577	L	1937	373-773		Madagascan phlogopites; 3-2	The above specimen measured under 330 psi loading.
13	577	L	1937	448-891		Canadian phlogopites; 4	Dark; measured under 176 psi loading; loaded density 2.98 g cm^{-3}.
14	577	L	1937	357-833		Canadian phlogopites; 5	Medium; 0.094 cm thickness; measured under 176 psi loading; loaded density 2.85 g cm^{-3}.
15	577	L	1937	364-506		Canadian phlogopites; 5	The above specimen measured when cooled from 250 C.
16	577	L	1937	349-883		Canadian phlogopites; 5	Medium; 0.051 cm thickness; measured under 176 psi loading; loaded density 2.85 g cm^{-3}.
17	577	L	1937	353-541		Canadian phlogopites; 5	The above specimen measured when cooled from 610 C.
18	577	L	1937	346-912		Canadian phlogopites; 6	Medium; measured under 176 psi loading; loaded density 2.82 g cm^{-3}.
19	577	L	1937	419.2,516.2		Canadian phlogopites; 6	The above specimen measured when cooled from 639 C.

Curve No.	Ref. No.	Method Used	Year	Temp. Range, K	Reported Error, %	Name and Specimen Designation	Composition (weight percent), Specifications and Remarks
20	577	L	1937	388-865		Canadian phlogopites; 7	Light; measured under 176 psi loading; loaded density 2.84 g cm^{-3}.
21	577	L	1937	368-593		Canadian phlogopites; 7	The above specimen measured when cooled from 592 C.
22	577	L	1937	383-918		Canadian phlogopites; 8	Light; measured under 176 psi loading; loaded density 2.95 g cm^{-3}.
23	577	L	1937	374.2,413.2		Canadian phlogopites; 8	The above specimen measured when cooled from 645 C.
24	100	L	1957	518-729			Powders; bulk density 0.332 g cm^{-3}.
25	76	C	1952	523-903		Synthetic Mica	98% of theoretical density.
26	125	L	1923	323.2	1		Sample consisting of two large sheets of mica each about 0.7 mm in thickness; measured with increasing pressure.
27	125	L	1923	323.2	1		Specimen 0.013 cm thick.
28	125	L	1923	323.2	1		Specimen 0.013 cm thick.
29	125	L	1923	323.2	1		Various specimens with different thickness from 0.01 cm to 0.025 cm; pressure 120 psi.
30	152	L	1956	298.2		Mica	Unirradiated.
31	152	L	1956	298.2		Mica	The above specimen exposed with 4 x 10^{19} epithermal neutrons per cm^2 for 480 Mwd in the MTR.
32	330	C	1957	298.2		Bonded Mica	Thermal comparator loaded with 100 gram weight applied on the plane lapped surface of the specimen.
33	330	C	1957	298.2		Bonded Mica	Thermal comparator applied to a 3 in. dia and 0.187 in. thick lapped disk specimen.
34	109	L	1947	222-306			Board vermiculite; specimen 1 in. thick; density 0.303 g cm^{-3}; as received.
35	109	L	1947	222-306			Fill vermiculite; expanded; specimen 1 in. thick; density 0.13 g cm^{-3} as received.
36	79	L	1942	565.7			Coarse granules; density 0.152 g cm^{-3}.
37	100	L	1957	532-720			Powder; bulk density 0.268 g cm^{-3}.
38	139	R	1950	93-300			Powdered; density 0.090 g cm^{-3}.
39	561	R	1948	92-275		Granulated vermiculite 2A	Grain size 10-14 mesh; bulk density 0.2163 g cm^{-3}.
40	561	R	1948	187.2		Granulated vermiculite 2B	Grain size 4-10 mesh; bulk density 0.144 g cm^{-3}.
41	561	R	1948	187.2		Granulated vermiculite 2C	Mixed grain size; 7.9% 3-4 mesh, 26.5% 4-10 mesh, and 65.6% 10-14 mesh; bulk density 0.157 g cm^{-3}.

DATA TABLE NO. 273 THERMAL CONDUCTIVITY OF MICA

[Temperature, T, K; Thermal Conductivity, k, Watt $cm^{-1}K^{-1}$]

T	k
CURVE 1	
317.0	0.0220
328.7	0.0225
360.2	0.0221
371.8	0.0216
396.6	0.0214
CURVE 2	
420.2	0.00387
761.2	0.00387
CURVE 3	
424.2	0.00404
482.2	0.00397
521.2	0.00389
646.2	0.00360
715.2	0.00356
765.2	0.00345
830.2	0.00356
916.2	0.00368
CURVE 4*	
426.2	0.00418
586.2	0.00381
719.2	0.00391
809.2	0.00391
CURVE 5	
460.2	0.00611
520.2	0.00498
573.2	0.00485
616.2	0.00523
747.2	0.00494
842.2	0.00519
880.2	0.00510
CURVE 6*	
496.2	0.00481
CURVE 7	
357.2	0.00569
400.2	0.00531
434.2	0.00444
496.2	0.00274
564.2	0.00228
642.2	0.00241
764.2	0.00251
809.2	0.00253
833.2	0.00253
CURVE 8	
423.2	0.00368
560.2	0.00243
CURVE 9	
373.2	0.00460
573.2	0.00230
773.2	0.00230
CURVE 10	
373.2	0.00586
573.2	0.00251
773.2	0.00272
CURVE 11	
373.2	0.00669
573.2	0.00356
773.2	0.00377
CURVE 12	
373.2	0.00753
573.2	0.00460
773.2	0.00502
CURVE 13	
448.2	0.00494
473.2	0.00464
636.2	0.00444
CURVE 13 (cont.)	
785.2	0.00460
891.2	0.00481
CURVE 14	
357.2	0.00661
358.2	0.00732
415.2	0.00623
455.2	0.00523
464.2	0.00506
481.2	0.00276
516.2	0.00247
527.2	0.00243
532.2	0.00169
783.2	0.00199
795.2	0.00203
806.2	0.00180
826.0	0.00203
833.2	0.00190
CURVE 15*	
364.2	0.00590
452.2	0.00408
461.2	0.00410
462.2	0.00448
473.2	0.00272
506.2	0.00247
CURVE 16*	
349.2	0.00674
402.2	0.00619
406.2	0.00653
487.2	0.00234
494.2	0.00234
557.2	0.00188
583.2	0.00167
596.2	0.00192
603.2	0.00190
615.2	0.00169
633.2	0.00176
653.2	0.00192
734.2	0.00169
CURVE 16 (cont.)	
848.2	0.00199
848.2	0.00188
856.2	0.00209
879.2	0.00207
883.2	0.00192
CURVE 17*	
353.2	0.00481
406.2	0.00387
541.2	0.00167
CURVE 18*	
346.2	0.00607
379.2	0.00577
463.2	0.00519
513.2	0.00408
578.2	0.00395
623.2	0.00397
763.2	0.00410
856.2	0.00439
912.2	0.00481
CURVE 19*	
419.2	0.00510
516.2	0.00377
CURVE 20	
388.2	0.00607*
444.2	0.00439*
463.2	0.00293
478.2	0.00264
519.2	0.00180
591.2	0.00186
737.2	0.00186
773.2	0.00188
851.2	0.00228
865.2	0.00211
CURVE 21	
368.2	0.00490
403.2	0.00389
430.2	0.00236
483.2	0.00199
593.2	0.00167
CURVE 22*	
383.2	0.00527
413.2	0.00544
441.2	0.00469
505.2	0.00383
529.2	0.00377
603.2	0.00345
694.2	0.00400
764.2	0.00389
918.2	0.00412
CURVE 23*	
374.2	0.00477
413.2	0.00435
CURVE 24	
517.6	0.00124
541.5	0.00127
599.8	0.00136
669.8	0.00144
728.7	0.00153
CURVE 25	
523.2	0.0212
768.2	0.0240
903.2	0.0252
p(lb in^{-2})	k
CURVE 26 (T = 323.2 K)	
3.75	0.00368
16.25	0.00389
40.50	0.00418
75.00	0.00444
T	k
CURVE 27	
323.2	0.00565
CURVE 28*	
323.2	0.00573
CURVE 29*	
323.2	0.00502
CURVE 30	
298.2	0.00795
CURVE 31	
298.2	0.00502
CURVE 32*	
298.2	0.00828
CURVE 33*	
298.2	0.0075
CURVE 34	
222.0	0.00120
238.7	0.00123
255.4	0.00125
272.2	0.00128
288.9	0.00131
305.5	0.00134
CURVE 35	
222.0	0.000552
238.7	0.000578
255.4	0.000606
272.2	0.000632
288.9	0.000659
305.5	0.000687
CURVE 36*	
565.7	0.00130
CURVE 37*	
532.1	0.00130
548.7	0.00131
611.5	0.00140
686.5	0.00147
720.4	0.00151
CURVE 38*	
93.2	0.0175
100	0.0180
150	0.0300
200	0.0380
250	0.0480
300	0.0610
CURVE 39	
92.2	0.000306
130.2	0.000374
155.2	0.000396
176.2	0.000436
192.2	0.000478
206.2	0.000509
216.4	0.000530
225.2	0.000554*
234.2	0.000571*
248.2	0.000602
260.2	0.000635
265.2	0.000647
275.2	0.000656
CURVE 40	
187.2	0.000344
CURVE 41	
187.5	0.000422

*Not shown on plot

SPECIFICATION TABLE NO. 274 THERMAL CONDUCTIVITY OF PERLITE

Curve No.	Ref. No.	Method Used	Year	Temp. Range, K	Reported Error, %	Name and Specimen Designation	Composition (weight percent), Specifications and Remarks
1	567	L	1955	171-283			Expanded perlite powder; density 0.048 g cm^{-3}.

DATA TABLE NO. 274 THERMAL CONDUCTIVITY OF PERLITE

[Temperature, T, K; Thermal Conductivity, k, Watt cm^{-1} K^{-1}]

T	k
CURVE 1*	
170.7	0.000250
228.2	0.000325
283.2	0.000395

*No graphical presentation

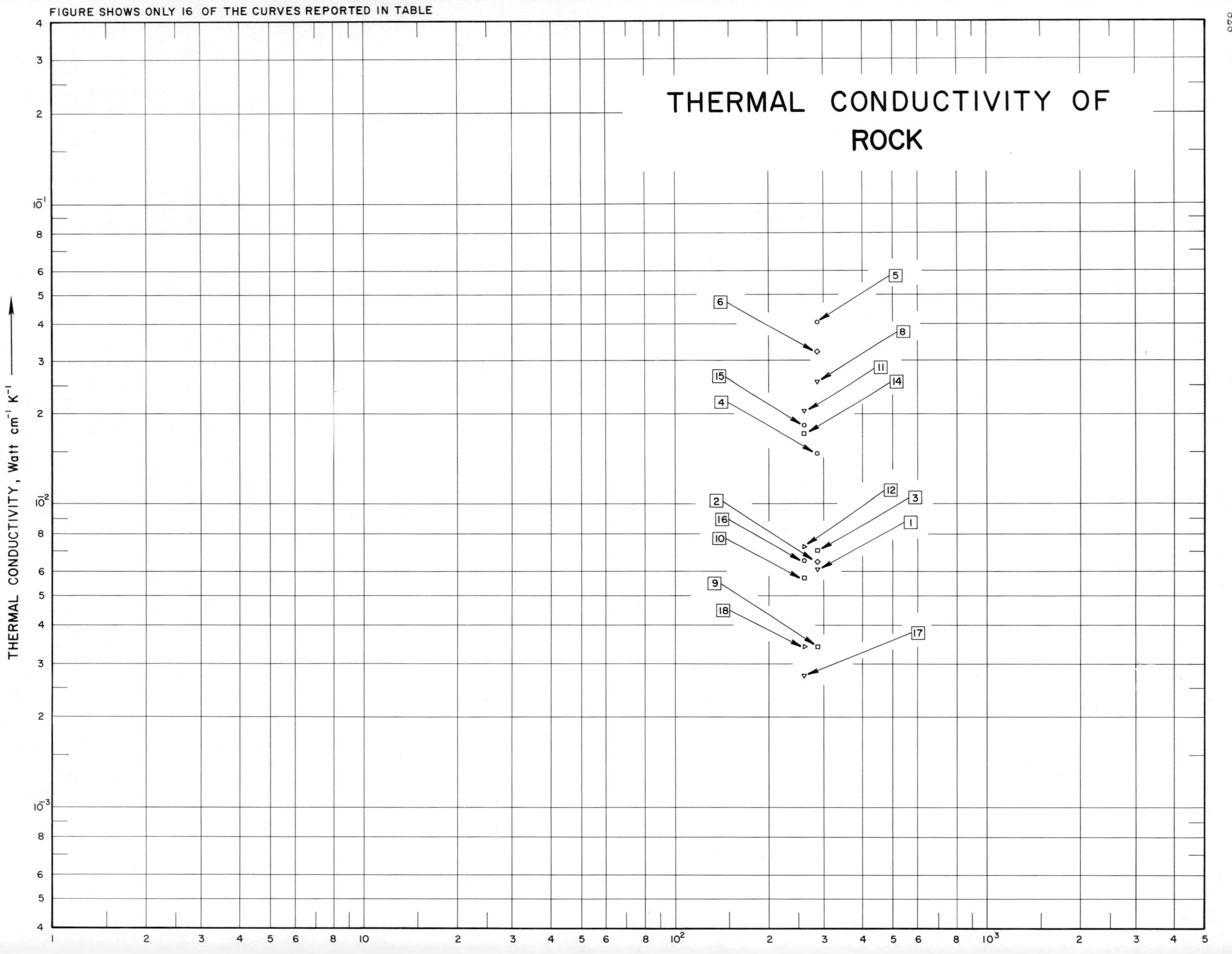
FIGURE SHOWS ONLY 16 OF THE CURVES REPORTED IN TABLE
THERMAL CONDUCTIVITY OF ROCK
THERMAL CONDUCTIVITY, Watt cm^{-1} K^{-1}
5
6
8
15
11
14
4
12
2
3
16
1
10
9
18
17

SPECIFICATION TABLE NO. 275 THERMAL CONDUCTIVITY OF ROCK

[For Data Reported in Figure and Table No. 275]

Curve No.	Ref. No.	Method Used	Year	Temp. Range, K	Reported Error, %	Name and Specimen Designation	Composition (weight percent), Specifications and Remarks
1	545	P	1947	287.3		Winchester crushed trap rock; 3E-1	Consisting of a fine-grained quartz diorite (0.75 in. max size) obtained from quarry at Winchester, Mass.; specimen 5.36 in. in dia and 10.68 in. high. material placed in approximately five equal layers and compacted using an increasing number of blows on each layer to obtain a uniform unit weight through the specimen; unfrozen; immersed in a water bath and brought to a constant temperature of approximately 24 C and removed from this bath and immediately immersed in a second water bath at a temperature of approximately 4.5 C; specific gravity 2.91, unit weight 1.62 g cm^{-3}; unit dry weight 1.59 g cm^{-3}; water content 1.9% dry weight.
2	545	P	1947	287.3		Winchester crushed trap rock; 3E-2	Similar to the above specimen except unit weight 1.636 g cm^{-3}; unit dry weight 1.602 g cm^{-3}; water content 2.1% dry weight.
3	545	P	1947	287.3		Winchester crushed trap rock; 3E-3	Similar to the above specimen except unit weight 1.647 g cm^{-3}; unit dry weight 1.578 g cm^{-3}; water content 4.4% dry weight.
4	545	P	1947	287.3		Winchester crushed trap rock; 3E-4	Similar to the above specimen except unit weight 2.014 g cm^{-3}; unit dry weight 1.578 g cm^{-3}; water content 27.2% dry weight.
5	545	P	1947	287.3		Winchester crushed trap rock; 3E-5	Similar to the above specimen except unit weight 2.04 g cm^{-3}; unit dry weight 1.59 g cm^{-3}; water content 28.4% dry weight; test results are not consistent with results of other tests and time curves are not erratic.
6	545	P	1947	287.3		Winchester crushed trap rock; 3E-6a	Similar to the above specimen except unit weight 2.05 g cm^{-3}; unit dry weight 1.602 g cm^{-3}; water content 27.7% dry weight.
7	545	P	1947	287.3		Winchester crushed trap rock; 3E-7	Similar to specimen 3E-1 except unit weight 1.957 g cm^{-3}; unit dry weight 1.619 g cm^{-3}; water content 20.9% dry weight.
8	545	P	1947	287.3		Winchester crushed trap rock; 3E-8	Similar to the above specimen except unit weight 2.07 g cm^{-3}; unit dry weight 1.63 g cm^{-3}; water content 26.7% dry weight.
9	545	P	1947	287.3		Winchester crushed trap rock; 3E-21	Similar to the above specimen except unit weight 1.804 g cm^{-3}; unit dry weight 1.801 g cm^{-3}; water content 0.2% dry weight.
10	545	P	1947	261.8		Winchester crushed trap rock; 3E-9	Consisting of a fine-grained quartz diorite (0.75 in. max size) obtained from quarry at Winchester, Mass.; specimen 5.36 in. in dia and 10.68 in. high; material placed in approximately five equal layers and compacted using an increasing number of blows on each layer to obtain a uniform unit weight through the specimen; frozen; subjected to a constant freezing temperature of approximately -20 C inside the freezing cabinet until temperature equilibrium is reached, then immersed in a brine bath at a temperature of approximately -2.7 C; specific gravity 2.91; unit weight 1.67 g cm^{-3}; unit dry weight 1.65 g cm^{-3}; water content 1.5% dry weight.

SPECIFICATION TABLE NO. 275 (continued)

Curve No.	Ref. No.	Method Used	Year	Temp. Range, K	Reported Error, %	Name and Specimen Designation	Composition (weight percent), Specifications and Remarks
11	545	P	1947	261. 8		Winchester crushed trap rock; 3E-10	Similar to the above specimen except unit weight 2. 07 g cm^{-3}; unit dry weight 1. 65 g cm^{-3}; water content 25. 8 % dry weight; cover lifted off cylinder due to heaving during test; results slightly affected by leaking of brine into specimen.
12	545	P	1947	261. 8		Winchester crushed trap rock; 3E-11	Similar to specimen 3E-9 except unit weight 1. 74 g cm^{-3}; unit dry weight 1. 071 g cm^{-3}; water content 2. 2% dry weight.
13	545	P	1947	261. 8		Winchester crushed trap rock; 3E-12	Similar to the above specimen except unit weight 1. 68 g cm^{-3}; unit dry weight 1. 66 g cm^{-3}; water content 1. 2%.
14	545	P	1947	261. 8		Winchester crushed trap rock; 3E-13	Similar to the above specimen except unit weight 2. 08 g cm^{-3}; unit dry weight 1. 071 g cm^{-3}; water content 22. 1%, slight leaking.
15	545	P	1947	261. 8		Winchester crushed trap rock; 3E-14	Similar to the above specimen except unit weight 2. 07 g cm^{-3}; unit dry weight 1. 66 g cm^{-3}; water content 25. 0% .
16	545	P	1947	261. 8		Winchester crushed trap rock; 3E-15	Similar to the above specimen except unit weight 1. 71 g cm^{-3}; unit dry weight 1. 68 g cm^{-3}; water content 2. 0%.
17	545	P	1947	261. 8		Winchester crushed trap rock; 3E-17	Similar to the above specimen except unit weight 1. 65 g cm^{-3}; unit dry weight 1. 63 g cm^{-3}; water content 0. 21% dry weight.
18	545	P	1947	261. 8		Winchester crushed trap rock; 3E-18	Similar to the above specimen except unit weight 1. 785 g cm^{-3}; unit dry weight 1. 78 g cm^{-3}; water content 0. 12% dry weight.

DATA TABLE NO. 275 THERMAL CONDUCTIVITY OF ROCK

[Temperature, T, K; Thermal Conductivity, k, Watt $cm^{-1}K^{-1}$]

T	k	T	k
CURVE 1		CURVE 13*	
287.3	0.00606	261.8	0.00578
CURVE 2		CURVE 14	
287.3	0.00642	261.8	0.0171
CURVE 3		CURVE 15	
287.3	0.00697	261.8	0.0183
CURVE 4		CURVE 16	
287.3	0.0147	261.8	0.00649
CURVE 5		CURVE 17	
287.3	0.0402	261.8	0.00272
CURVE 6		CURVE 18	
287.3	0.0320	261.8	0.00339
CURVE 7*			
287.3	0.00642		
CURVE 8			
287.3	0.0256		
CURVE 9			
287.3	0.00339		
CURVE 10			
261.8	0.00568		
CURVE 11			
261.8	0.0204		
CURVE 12			
261.8	0.00722		

*Not shown on plot

SPECIFCATION TABLE NO. 276 THERMAL CONDUCTIVITY OF SALT

Curve No.	Ref. No.	Method Used	Year	Temp. Range, K	Reported Error, %	Name and Specimen Designation	Composition (weight percent), Specifications and Remarks
1	578	R	1963	332-464	±7	Gnome Salt No. 1	Composition (in vol%) 95 NaCl, 3 SiH and clay, 2 polyhalite; grainy in appearance; translucent to red-brown with brown streaks; specimen 3. 5 in. O. D., 0. 875 in. I. D. and 18 in. long; specimen cored from massive blocks from the Gnome event ground zero room (1200 ft level); density 2. 147 g cm^{-3}; the temp dependence of the thermal conductivity for this salt is qualitatively similar to that of the literature values.
2	578	R	1963	370-600	±7	Gnome Salt No. 2	Similar to the above specimen.

DATA TABLE NO. 276 THERMAL CONDUCTIVITY OF SALT

[Temperature, T, K; Thermal Conductivity, k, Watt cm^{-1} K^{-1}]

T	k
CURVE 1*	
332	0. 0291
375	0. 0283
439	0. 0289
464	0. 0278
CURVE 2*	
370	0. 0377
426	0. 0300
496	0. 0260
532	0. 0265
600	0. 0251

*No graphical presentation

THERMAL CONDUCTIVITY OF SAND

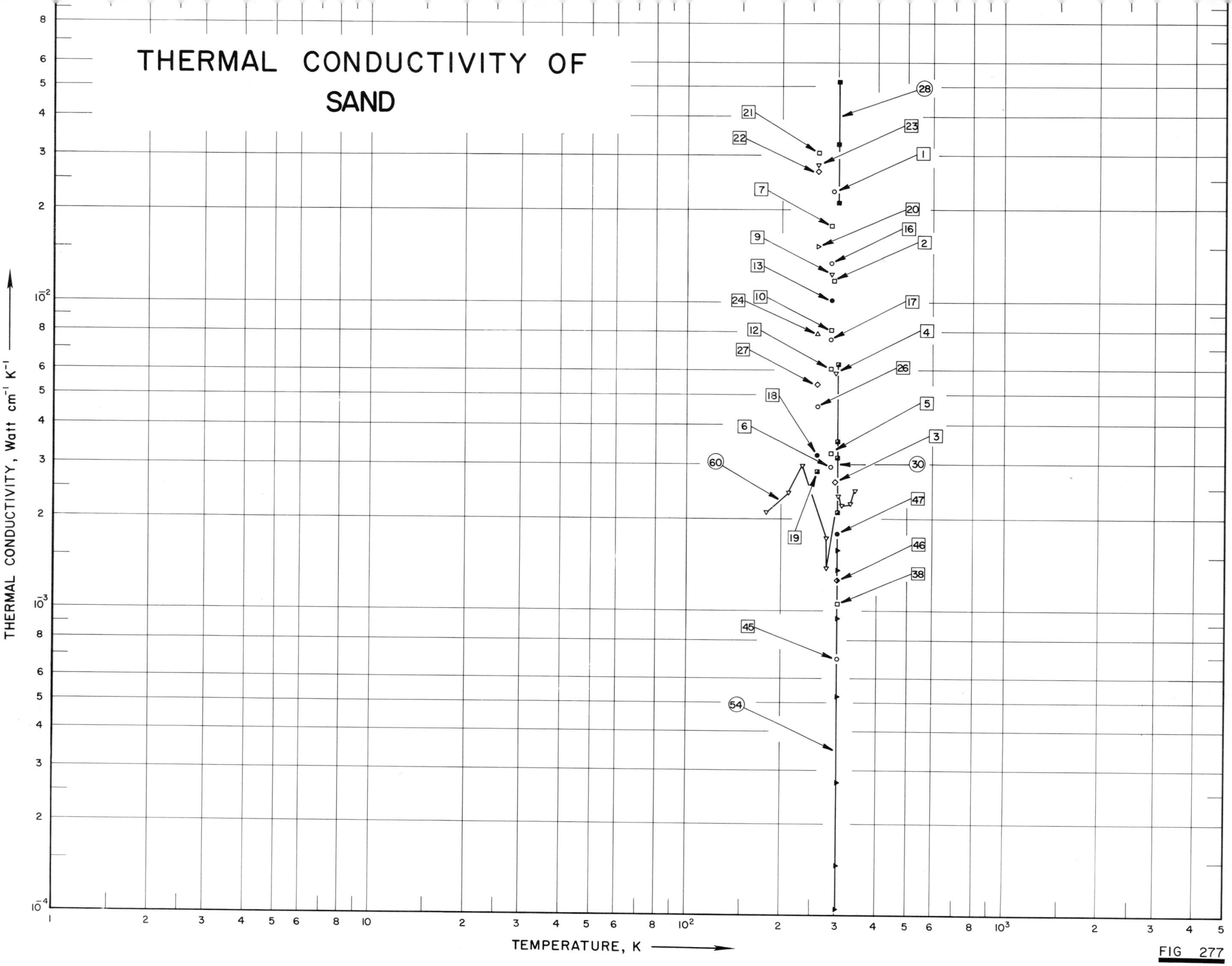

THERMAL CONDUCTIVITY, Watt cm⁻¹ K⁻¹ ⟶

10^{-4} 2 3 4 5 6 8 10^{-3} 2 3 4 5 6 8 10^{-2} 2 3 4 5 6 8

TEMPERATURE, K ⟶

1 2 3 4 5 6 8 10 2 3 4 5 6 8 10^{2} 2 3 4 5 6 8 10^{3} 2 3 4 5

28 21 23 22 1 7 20 16 9 2 13 24 10 17 12 4 27 26 18 5 6 3 60 30 47 19 46 38 45 54

FIG 277

SPECIFICATION TABLE NO. 277 THERMAL CONDUCTIVITY OF SAND

[For Data Reported in Figure and Table No. 277]

Curve No.	Ref. No.	Method Used	Year	Temp. Range, K	Reported Error, %	Name and Specimen Designation	Composition (weight percent), Specifications and Remarks
1	589	F	1958	293.2			Leighton Buzzard sand; particle sizes 0.060-0.085 cm; dry density 1.54 g cm^{-3}; average moisture content 22%.
2	589	F	1958	293.2			Similar to the above specimen except average moisture content 2.3%.
3	546	P	1924	297.2		Quartz sand	Medium fine; dry; density 1.65 g cm^{-3}.
4	546	P	1924	297.2		Quartz sand	Medium fine; density 1.75 g cm^{-3}; moisture content 8.3%.
5	545	P	1947	287.3		Lowell Sand No. 3A-4	Specimen 5.36 in. in dia and 10.68 in. in height; consisting of cohesionless siliceous sand from a glacial outwash deposit at South Lowell, Mass.; the material placed in approx five equal layers and compacted, using an increasing number of blows on each layer to obtain a uniform unit weight throughout the specimen; unfrozen; immersed in a water bath and brought to a constant temp of approx 24 C and then immersed in a second water bath at a temp of approx 4.5 C; specific gravity 2.66; unit dry weight 1.68 g cm^{-3}; water content 0.2% dry weight.
6	545	P	1947	287.3		Lowell Sand No. 3A-5	Similar to the above specimen except specific gravity 2.66; unit weight 1.62 g cm^{-3}; unit dry weight 1.6 g cm^{-3}; specimen not properly sealed letting some water in during test.
7	545	P	1947	287.3		Lowell Sand No. 3A-6	Similar to the above specimen except water content 16.4% dry weight, unit weight 1.987 g cm^{-3}; unit dry weight 1.71 g cm^{-3};specimen not properly sealed letting some water in during test.
8	545	P	1947	287.3		Lowell Sand No. 3A-7	Similar to the above specimen except unit weight 1.96 g cm^{-3}; unit dry weight 1.62 g cm^{-3}; water content 20.9% dry weight.
9	545	P	1947	287.3		Lowell Sand No. 3A-8	Similar to the above specimen except unit weight 1.72 g cm^{-3}; unit dry weight 1.65 g cm^{-3}; water content 4.5% dry weight.
10	545	P	1947	287.3		Lowell Sand No. 3A-9	Similar to the above specimen except unit weight 1.4 g cm^{-3}; unit dry weight 1.34 g cm^{-3}; water content 4.9% dry weight.
11	545	P	1947	287.3		Lowell Sand No. 3A-10	Similar to the above specimen except unit weight 1.38 g cm^{-3}; unit dry weight 1.35 g cm^{-3}; water content 2.3% dry weight.
12	545	P	1947	287.3		Lowell Sand No. 3A-11	Similar to the above specimen except unit weight 1.49 g cm^{-3}; unit dry weight 1.46 g cm^{-3}; water content 1.9% dry weight.
13	545	P	1947	287.3		Lowell Sand No. 3A-12	Similar to the above specimen except unit weight 1.785 g cm^{-3}; unit dry weight 1.75 g cm^{-3}; water content 2.2% dry weight
14	545	P	1947	287.3		Lowell Sand No. 3A-13	Similar to the above specimen except unit weight 1.69 g cm^{-3}; unit dry weight 1.65 g cm^{-3}; water content 2.0% dry weight.
15	545	P	1947	287.3		Lowell Sand No. 3A-15	Similar to the above specimen except unit weight 1.46 g cm^{-3}; unit dry weight 1.43 g cm^{-3}; water content 2.1% dry weight.

SPECIFICATION TABLE NO. 277 (continued)

Curve No.	Ref. No.	Method Used	Year	Temp. Range, K	Reported Error, %	Name and Specimen Designation	Composition (weight percent), Specifications and Remarks
16	545	P	1947	287.3		Lowell Sand No. 3A-16	Similar to the above specimen except unit weight 1.76 g cm^{-3}; unit dry weight 1.68 g cm^{-3}; water content 5.1% dry weight.
17	545	P	1947	287.3		Lowell Sand No. 3A-17	Similar to the above specimen except unit weight 1.48 g cm^{-3}; unit dry weight 1.45 g cm^{-3}; water content 2.1% dry weight.
18	545	P	1947	261.8		Lowell Sand No. 3A-18	Specimen 5.36 in. in dia and 10.68 in. in height; consisting of a cohesionless, siliceous sand from a glacial outwash deposit at South Lowell, Mass.; the material placed in approximately five equal layers and compacted, using an increasing number of blows on each layer to obtain a uniform unit weight throughout the specimen; frozen; subjected to a constant freezing temperature of approximately -20 C inside the freezing cabinet until temperature equilibrium is reached and then immersed in a brine bath at a temperature of approximately -2.77 C; specific gravity 2.66; unit weight 1.70 g cm^{-3}; unit dry weight 1.70 g cm^{-3}; water content 0.17% dry weight.
19	545	P	1947	261.8		Lowell Sand No. 3A-19	Similar to the above specimen except unit weight 1.65 g cm^{-3}; unit dry weight 1.65 g cm^{-3}.
20	545	P	1947	261.8		Lowell Sand No. 3A-20	Similar to the above specimen except unit weight 1.73 g cm^{-3}; unit dry weight 1.65 g cm^{-3}; water content 5.4% dry weight.
21	545	P	1947	261.8		Lowell Sand No. 3A-21	Similar to the above specimen except unit weight 1.98 g cm^{-3}; unit dry weight 1.7 g cm^{-3}; water content 16.5% dry weight.
22	545	P	1947	261.8		Lowell Sand No. 3A-22	Similar to the above specimen except unit weight 1.95 g cm^{-3}; unit dry weight 1.65 g cm^{-3}; water content 18.5% dry weight.
23	545	P	1947	261.8		Lowell Sand No. 3A-23	Similar to the above specimen except unit weight 1.98 g cm^{-3}; unit dry weight 1.65 g cm^{-3}; water content 20.5% dry weight.
24	545	P	1947	261.8		Lowell Sand No. 3A-24	Similar to the above specimen except unit weight 1.71 g cm^{-3}; unit dry weight 1.68 g cm^{-3}; water content 2.2% dry weight.
25	545	P	1947	261.8		Lowell Sand No. 3A-25	Similar to the above specimen except unit weight 1.76 g cm^{-3}; unit dry weight 1.69 g cm^{-3}; water content 4.2% dry weight.
26	545	P	1947	261.8		Lowell Sand No. 3A-26	Similar to the above specimen except unit weight 1.79 g cm^{-3}; unit dry weight 1.79 g cm^{-3}; water content 0.66% dry weight.
27	545	P	1947	261.8		Lowell Sand No. 3A-27	Similar to the above specimen except unit weight 1.8 g cm^{-3}; unit dry weight 1.78 g cm^{-3}; water content 0.98% dry weight.
28	590	P	1961	303.2		Quartz sand	Unconsolidated sand; specimen 5 in. in dia and 7 in long; measured with water saturated (at 1 atm) and various porosities.
29	590	P	1961	303.2		Quartz sand	Unconsolidated sand; specimen 5 in. in dia and 7 in. long; measured with saturated oil (i.e. n-heptane, at 1 atm) and various porosities.

SPECIFICATION TABLE NO. 277 (continued)

Curve No.	Ref. No.	Method Used	Year	Temp. Range, K	Reported Error, %	Name and Specimen Designation	Composition (weight percent), Specifications and Remarks
30	590	P	1961	303.2		Quartz sand	Unconsolidated sand; 5 in. in dia and 7 in. long; measured with saturated air (at 1 atm) and various porosities.
31	590	P	1961	303.2		Quartz sand	Unconsolidated sand; 20/30 mesh Ottawa sand and 140/200 Wassau sand; specimen 5 in. in dia and 7 in. long; measured with saturated Freon-12; porosity = 19.4%; K_f is saturant thermal conductivity.
32	590	P	1961	303.2		Quartz sand	Similar to the above specimen except with saturated argon.
33	590	P	1961	303.2		Quartz sand	Similar to the above specimen except with saturated air.
34	590	P	1961	303.2		Quartz sand	Similar to the above specimen except with saturated n-heptane.
35	590	P	1961	303.2		Quartz sand	Similar to the above specimen except with saturated helium.
36	590	P	1961	303.2		Quartz sand	Similar to the above specimen except with saturated hydrogen.
37	590	P	1961	303.2		Quartz sand	Similar to the above specimen except with saturated water.
38	590	P	1961	303.2		Quartz sand	Unconsolidated sand; 20/30 mesh Ottawa sand; specimen 5 in. in dia and 7 in. long; measured with saturated Freon-12; porosity 36.1%; K_f is saturant thermal conductivity.
39	590	P	1961	303.2		Quartz sand	Similar to the above specimen except with saturated argon.
40	590	P	1961	303.2		Quartz sand	Similar to the above specimen except with saturated air.
41	590	P	1961	303.2		Quartz sand	Similar to the above specimen except with saturated n-heptane.
42	590	P	1961	303.2		Quartz sand	Similar to the above specimen except with saturated helium.
43	590	P	1961	303.2		Quartz sand	Similar to the above specimen except with saturated hydrogen.
44	590	P	1961	303.2		Quartz sand	Similar to the above specimen except with saturated water.
45	590	P	1961	303.2		Quartz sand	Unconsolidated sand; 40/200 mesh Wassau sand; specimen 5 in. in dia and 7 in. long; measured with saturated Freon-12; porosity 59%; K_f is saturant thermal conductivity.
46	590	P	1961	303.2		Quartz sand	Similar to the above specimen except with saturated argon.
47	590	P	1961	303.2		Quartz sand	Similar to the above specimen except with saturated air.
48	590	P	1961	303.2		Quartz sand	Similar to the above specimen except with saturated n-heptane.
49	590	P	1961	303.2		Quartz sand	Similar to the above specimen except with saturated helium.
50	590	P	1961	303.2		Quartz sand	Similar to the above specimen except with saturated hydrogen.
51	590	P	1961	303.2		Quartz sand	Similar to the above specimen except with saturated water.
52	590	P	1961	303.2		Quartz sand	Unconsolidated sand; 20/30 mesh Ottawa and 140/200 mesh Wassau sand; specimen 5 in. in dia and 7 in. long; porosity = 18%; measured with various air pressure. (interstitial air pressure).

SPECIFICATION TABLE NO. 277 (continued)

Curve No.	Ref. No.	Method Used	Year	Temp. Range, K	Reported Error, %	Name and Specimen Designation	Composition (weight percent), Specifications and Remarks
53	590	P	1961	303.2		Quartz sand	Unconsolidated sand; 20/30 mesh Ottawa sand; specimen 5 in. in dia and 7 in. long; porosity 33%; measured with various interstitial air pressure.
54	590	P	1961	303.2		Quartz sand	Unconsolidated sand; 140/200 mesh Wassau sand; specimen 5 in. in dia and 7 in. long; porosity 59%; measured with various interstitial air pressure.
55	590	P	1961	303.2		Quartz sand	Unconsolidated quartz sand; specimen 5 in. in dia and 7 in. long; porosity = 19%; measured with various interstitial hydrogen gas pressure.
56	590	P	1961	303.2		Quartz sand	Similar to the above specimen except measured with various interstitial helium gas pressure.
57	590	P	1961	303.2		Quartz sand	Similar to the above specimen except measured with various interstitial air pressure.
58	590	P	1961	303.2		Quartz sand	Similar to the above specimen except measured with various interstitial argon gas pressure.
59	590	P	1961	303.2		Quartz sand	Similar to the above specimen except measured with various interstitial Freon-12 pressure.
60	539	P	1962	180-343		Silica sand	99 silica, 70 mesh commercial grade foundry sand; moisture content < 0.03; density 1.60 g cm^{-3}; specific heat data obtained from Goldsmith, A., Waterman, T. E., and Hirschhorn, H. J., "Handbook of Thermophysical Properties of Solid Materials", and from Smithsonian Institute Physical Tables.
61	539	P	1962	319-368		Silica sand	The above specimen measured in a vacuum of 5×10^{-6} mm Hg.

DATA TABLE NO. 277 THERMAL CONDUCTIVITY OF SAND

[Temperature, T, K; Thermal Conductivity, k, Watt $cm^{-1}K^{-1}$]

T	k
CURVE 1	
293.2	0.0230
CURVE 2	
293.2	0.0117
CURVE 3	
297.2	0.00263
CURVE 4	
297.2	0.00586
CURVE 5	
287.3	0.00325
CURVE 6	
287.3	0.00292
CURVE 7	
287.3	0.0177
CURVE 8*	
287.3	0.0173
CURVE 9	
287.3	0.0124
CURVE 10	
287.3	0.00812
CURVE 11*	
287.3	0.00580

T	k
CURVE 12	
287.3	0.00609
CURVE 13	
287.3	0.0101
CURVE 14*	
287.3	0.00824
CURVE 15*	
287.3	0.00801
CURVE 16	
287.3	0.0134
CURVE 17	
287.3	0.00756
CURVE 18	
261.8	0.00320
CURVE 19	
261.8	0.00284
CURVE 20	
261.8	0.0153
CURVE 21	
261.8	0.0304
CURVE 22	
261.8	0.0267

T	k
CURVE 23	
261.8	0.0279
CURVE 24	
261.8	0.00796
CURVE 25*	
261.8	0.0158
CURVE 26	
261.8	0.00459
CURVE 27	
261.8	0.00543
porosity (%)	**k**
CURVE 28 (T = 303.2 K)	
19.5	0.0519
36	0.0324
52	0.0210
CURVE 29* (T = 303.2 K)	
19.5	0.0210
36	0.0142
52	0.00690
CURVE 30 (T = 303.2 K)	
18	0.00628
33	0.00356
36	0.00314
59.5	0.00209

K_f(W/cm K)	k
CURVE 31* (T = 303.2 K)	
0.000100	0.00209
CURVE 32* (T = 303.2 K)	
0.000178	0.00418
CURVE 33* (T = 303.2 K)	
0.000251	0.00649
CURVE 34* (T = 303.2 K)	
0.00134	0.0220
CURVE 35* (T = 303.2 K)	
0.00146	0.0251
CURVE 36* (T = 303.2 K)	
0.00178	0.0272
CURVE 37* (T = 303.2 K)	
0.00649	0.0523
CURVE 38 (T = 303.2 K)	
0.000100	0.00105
CURVE 39* (T = 303.2 K)	
0.000178	0.00209

K_f(W/cm K)	k
CURVE 40* (T = 303.2 K)	
0.000251	0.00335
CURVE 41* (T = 303.2 K)	
0.00134	0.0146
CURVE 42* (T = 303.2 K)	
0.00146	0.0167
CURVE 43* (T = 303.2 K)	
0.00178	0.0188
CURVE 44* (T = 303.2 K)	
0.00649	0.0314
CURVE 45 (T = 303.2 K)	
0.000100	0.000690
CURVE 46 (T = 303.2 K)	
0.000178	0.00126
CURVE 47 (T = 303.2 K)	
0.000251	0.00178
CURVE 48* (T = 303.2 K)	
0.00134	0.00690

K_f(W/cm K)	k
CURVE 49* (T = 303.2 K)	
0.00146	0.00732
CURVE 50* (T = 303.2 K)	
0.00178	0.00837
CURVE 51* (T = 303.2 K)	
0.00649	0.0209
P_{air}(mm Hg)	**k**
CURVE 52* (T = 303.2 K)	
0.16	0.000314
0.85	0.000690
8.0	0.00176
12	0.00211
40	0.00377
75	0.00429
215	0.00481
400	0.00544
700	0.00569
1100	0.00575
1700	0.00586
CURVE 53* (T = 303.2 K)	
0.02	0.000209
0.045	0.000209
0.13	0.000314
0.5	0.000523
2	0.00100
5	0.00178
10	0.00209
30	0.00262
75	0.00328

P_{air}(mm Hg)	k
CURVE 53 (cont.)*	
180	0.00377
400	0.00418
700	0.00431
1000	0.00431
CURVE 54 (T = 303.2 K)	
0.01	0.000105
0.03	0.000105*
0.06	0.000105*
0.11	0.000105*
0.55	0.000146
3	0.000272
10	0.000523
40	0.000941
110	0.00136
280	0.00157
500	0.00161*
750	0.00178*
900	0.00180*
CURVE 55* (T = 303.2 K)	
11	0.00377
55	0.0138
200	0.0203
750	0.0268
1600	0.0285
CURVE 56* (T = 303.2 K)	
12	0.00335
20	0.00460
90	0.0119
190	0.0163
430	0.0201
750	0.0220
1050	0.0236
1550	0.0251

*Not shown on plot

P_{air}(mm Hg)	k
CURVE 57* (T = 303.2 K)	
10	0.00188
52.5	0.00418
200	0.00565
750	0.00649
1600	0.00669
CURVE 58* (T = 303.2 K)	
15	0.00126
110	0.00314
400	0.00397
700	0.00418
850	0.00418
CURVE 59* (T = 303.2 K)	
110	0.00209
750	0.00209
1700	0.00230

T	k
CURVE 60	
180.4	0.00209
212.6	0.00243
235.4	0.00296
279.3	0.00172
279.9	0.00138
303.8	0.00236
312.1	0.00220
332.1	0.00222
342.6	0.00246
CURVE 61*	
319.3	0.0000369
320.4	0.0000396
329.9	0.0000438
344.3	0.0000417
366.0	0.0000500
368.2	0.0000346

* Not shown on plot

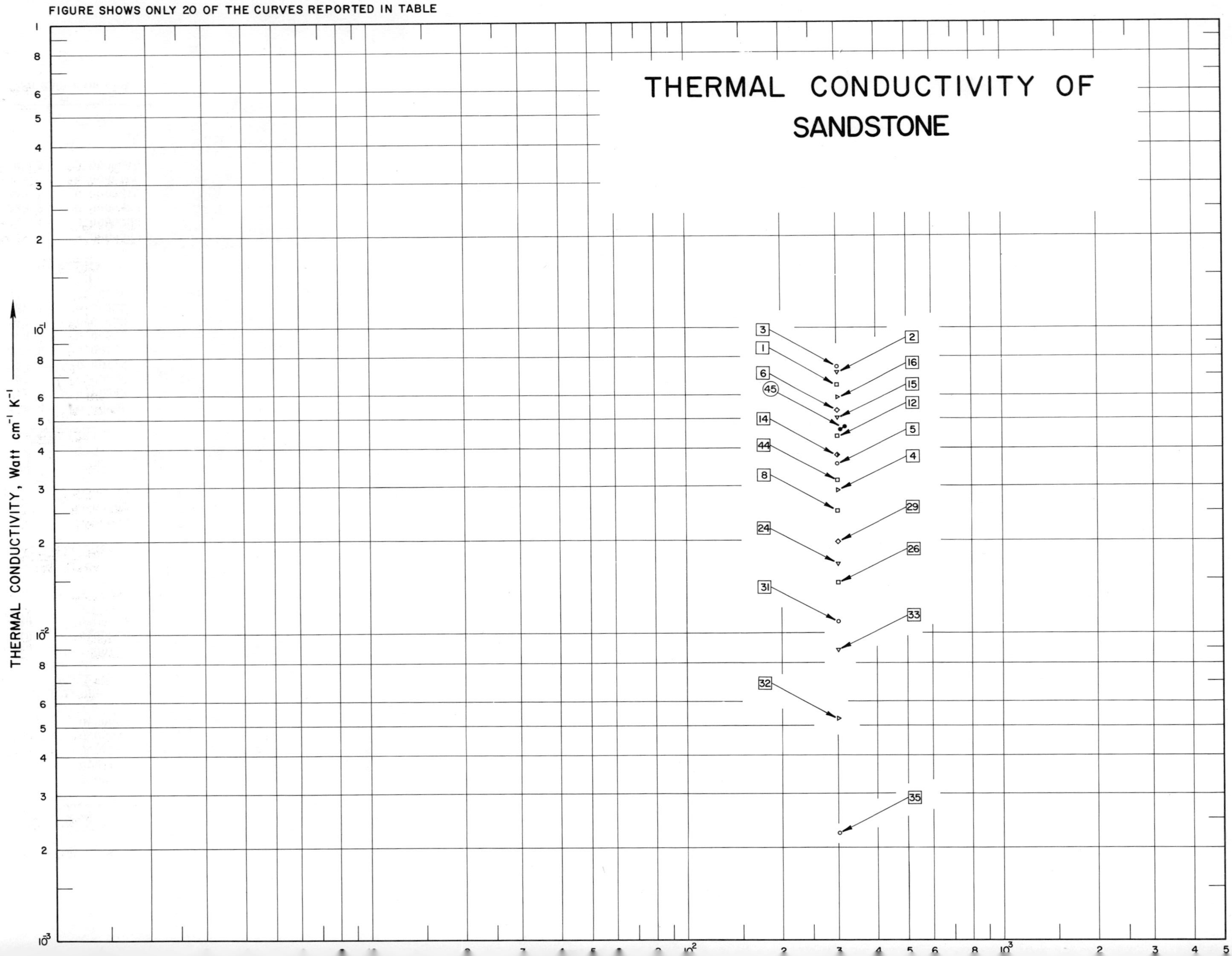

FIGURE SHOWS ONLY 20 OF THE CURVES REPORTED IN TABLE

THERMAL CONDUCTIVITY OF SANDSTONE

THERMAL CONDUCTIVITY, Watt cm^{-1} K^{-1}

3 1 6 45 14 44 8 24 31 32

2 16 15 12 5 4 29 26 33 35

SPECIFICATION TABLE NO. 278 THERMAL CONDUCTIVITY OF SANDSTONE

[For Data Reported in Figure and Table No. 278]

Curve No.	Ref. No.	Method Used	Year	Temp. Range, K	Reported Error, %	Name and Specimen Designation	Composition (weight percent), Specifications and Remarks
1	591	P	1961	303.2		Berkeley	Consolidated sandstone; (98-99 quartz, 1-2 kaolinite); with saturant (air) at atmospheric pressure; specimen 3 in. in dia and 6.5 in. long; porosity 3.0%; permeability < 0.1 md; K_f is saturant conductivity.
2	591	P	1961	303.2		Berkeley	The above specimen with n-heptane saturated at atmospheric pressure.
3	591	P	1961	303.2		Berkeley	The above specimen with water saturated at atmospheric pressure.
4	591	P	1961	303.2		Berkeley	The above specimen with vacuo.
5	591	P	1961	303.2		St. Peters	Consolidated santstone; (98-99 quartz, 1-2 kaolinite); with saturant (air) at atmospheric pressure; specimen 3 in. in dia and 6.5 in. long; porosity 11%; permeability 3.4 md; K_f is saturant conductivity.
6	591	P	1961	303.2		St. Peters	The above specimen with n-heptane saturated at atmospheric pressure.
7	591	P	1961	303.2		St. Peters	The above specimen with water saturated at atmospheric pressure.
8	591	P	1961	303.2		St. Peters	The above specimen with vacuo.
9	591	P	1961	303.2		Tensleep	Consolidated sandstone; (90-95 quartz, 5-10 amorphous silica); with saturant (freon) at atmospheric pressure; specimen 3 in. in dia and 6.5 in. long; porosity 15.5%; permeability 220 md; K_f is saturant conductivity.
10	591	P	1961	303.2		Tensleep	The above specimen with saturated argon at atmospheric pressure.
11	591	P	1961	303.2		Tensleep	The above specimen with saturated air at atmospheric pressure.
12	591	P	1961	303.2		Tensleep	The above specimen with saturated n-heptane at atmospheric pressure.
13	591	P	1961	303.2		Tensleep	The above specimen with saturated helium at atmospheric pressure.
14	591	P	1961	303.2		Tensleep	The above specimen with saturated hydrogen at atmospheric pressure.
15	591	P	1961	303.2		Tensleep	The above specimen with saturated 60% ethanol at atmospheric pressure.
16	591	P	1961	303.2		Tensleep	The above specimen with saturated water at atmospheric pressure.
17	591	P	1961	303.2		Tensleep	The above specimen with vacuo.
18	591	P	1961	303.2		Berea	Consolidated sandstone; (88-89 quartz, 9-10 kaolinite, 2 illite); with saturant (N_2O) at atmospheric pressure; specimen 3 in. in dia and 6.5 in. long; porosity 22%; permeability 480 md; K_f is saturant conductivity.
19	591	P	1961	303.2		Berea	The above specimen with saturated air at atmospheric pressure.
20	591	P	1961	303.2		Berea	The above specimen with saturated n-heptane at atmospheric pressure.
21	591	P	1961	303.2		Berea	The above specimen with saturated helium at atmospheric pressure.
22	591	P	1961	303.2		Berea	The above specimen with saturated hydrogen at atmospheric pressure.
23	591	P	1961	303.2		Berea	The above specimen with saturated water at atmospheric pressure.

SPECIFICATION TABLE NO. 278 (continued)

Curve No.	Ref. No.	Method Used	Year	Temp. Range, K	Reported Error, %	Name and Specimen Designation	Composition (weight percent), Specifications and Remarks
24	591	P	1961	303.2		Berea	The above specimen with vacuo.
25	591	P	1961	303.2		Teapot	Consolidated sandstone; (88 quartz, 7 kaolinite, 5 illite); with saturant (freon) at atmospheric pressure; specimen 3 in. in dia and 6.5 in. long; porosity 29%; permeability 1960 md; K_f is saturant conductivity.
26	591	P	1961	303.2		Teapot	The above specimen with saturated argon at atmospheric pressure.
27	591	P	1961	303.2		Teapot	The above specimen with saturated air at atmospheric pressure.
28	591	P	1961	303.2		Teapot	The above specimen with saturated n-heptane at atmospheric pressure.
29	591	P	1961	303.2		Teapot	The above specimen with saturated helium at atmospheric pressure.
30	591	P	1961	303.2		Teapot	The above specimen with saturated water at atmospheric pressure.
31	591	P	1961	303.2		Teapot	The above specimen with vacuo.
32	591	P	1961	303.2		Tripolite	Consolidated sandstone; (85-90 quartz, 10-15 amorphous silica); with saturant (air) at atmospheric pressure; specimen 3 in. in dia and 6.5 in. long; porosity 59%; permeability 650 md; K_f is saturant conductivity.
33	591	P	1961	303.2		Tripolite	The above specimen with saturated n-heptane at atmospheric pressure.
34	591	P	1961	303.2		Tripolite	The above specimen with saturated water at atmospheric pressure.
35	591	P	1961	303.2		Tripolite	The above specimen with vacuo.
36	591	P	1961	303.2		Berkeley	Similar to the above Berkeley specimen with various air pressures (dry air values).
37	591	P	1961	303.2		Berkeley	Similar to the above Berkeley specimen with various air pressures (60% RH values).
38	591	P	1961	303.2		Tensleep	Similar to the above Tensleep specimen with various air pressures (dry air values).
39	591	P	1961	303.2		Berea	Similar to the above Berea specimen with various internal nitrogen gas pressures and zero net overburden pressure.
40	591	P	1961	303.2		Berea	Similar to the above Berea specimen with various internal nitrogen gas pressures and 4000 psi net overburden pressure.
41	591	P	1961	303.2		Berea	Similar to the above Berea specimen (except evacuated) measured with increasing overburden pressure P_{ob}.
42	591	P	1961	303.2		Berea	Similar to the above Berea specimen (except evacuated) measured with decreasing overburden pressure P_{ob}.
43	591	P	1961	303.2		Berea	Similar to the above Berea specimen (air saturated) measured with increasing overburden pressure P_{ob}.
44	591	P	1961	303.2		Berea	Similar to the above Berea specimen (air saturated) measured with increasing overburden pressure P_{ob}.

SPECIFICATION TABLE NO. 278 (continued)

Curve No.	Ref. No.	Method Used	Year	Temp. Range, K	Reported Error, %	Name and Specimen Designation	Composition (weight percent), Specifications and Remarks
45	586	L	1933	310, 320	1		Recrystallized; specimen 5 cm in dia and 2 cm long; density 2.40 g cm^{-3}; obtained from Lower Permian, The Old Quarry, Penrith, Cumberland.

DATA TABLE NO. 278 THERMAL CONDUCTIVITY OF SANDSTONE

[Temperature, T, K; Thermal Conductivity, k, Watt $cm^{-1}K^{-1}$]

K_f(W/cmK)	k
CURVE 1 (T = 303.2 K)	
0.000264	0.0649
CURVE 2 (T = 303.2 K)	
0.00128	0.0711
CURVE 3 (T = 303.2 K)	
0.00628	0.0741
CURVE 4 (T = 303.2 K)	
0.00	0.0290
CURVE 5 (T = 303.2 K)	
0.000264	0.0356
CURVE 6 (T = 303.2 K)	
0.00128	0.0534
CURVE 7* (T = 303.2 K)	
0.00628	0.0636
CURVE 8 (T = 303.2 K)	
0.00	0.0249
CURVE 9* (T = 303.2 K)	
0.0000979	0.0300

K_f(W/cmK)	k
CURVE 10* (T = 303.2 K)	
0.000177	0.0287
CURVE 11* (T = 303.2 K)	
0.000264	0.0304
CURVE 12 (T = 303.2 K)	
0.00128	0.0437
CURVE 13* (T = 303.2 K)	
0.00147	0.0348
CURVE 14 (T = 303.2 K)	
0.00179	0.0379
CURVE 15 (T = 303.2 K)	
0.00305	0.0502
CURVE 16 (T = 303.2 K)	
0.00628	0.0586
CURVE 17* (T = 303.2 K)	
0.00	0.0262
CURVE 18* (T = 303.2 K)	
0.000167	0.0241

K_f(W/cmK)	k
CURVE 19* (T = 303.2 K)	
0.000264	0.0239
CURVE 20* (T = 303.2 K)	
0.00128	0.0374
CURVE 21* (T = 303.2 K)	
0.00147	0.0292
CURVE 22* (T = 303.2 K)	
0.00179	0.0300
CURVE 23* (T = 303.2 K)	
0.00628	0.0448
CURVE 24 (T = 303.2 K)	
0.00	0.0168
CURVE 25* (T = 303.2 K)	
0.0000979	0.0146
CURVE 26 (T = 303.2 K)	
0.000177	0.0142
CURVE 27* (T = 303.2 K)	
0.000264	0.0154

K_f(W/cmK)	k
CURVE 28* (T = 303.2 K)	
0.00128	0.0265
CURVE 29 (T = 303.2 K)	
0.00147	0.0198
CURVE 30* (T = 303.2 K)	
0.00628	0.0405
CURVE 31 (T = 303.2 K)	
0.00	0.0109
CURVE 32 (T = 303.2 K)	
0.000264	0.00527
CURVE 33 (T = 303.2 K)	
0.00128	0.00879
CURVE 34* (T = 303.2 K)	
0.00628	0.0203
CURVE 35 (T = 303.2 K)	
0.00	0.00222

P(mm Hg)	k
CURVE 36* (T = 303.2 K)	
0.011	0.0290
0.019	0.0299
0.035	0.0328
0.3	0.0408
0.7	0.0418
4	0.0433
8.5	0.0452
105	0.0519
750	0.0649
CURVE 37* (T = 303.2 K)	
0.06	0.0364
0.33	0.0450
750	0.0680
CURVE 38* (T = 303.2 K)	
0.01	0.0262
0.02	0.0264
0.05	0.0264
0.6	0.0264
10.05	0.0270
55	0.0274
190	0.0282
300	0.0285
750	0.0301
1600	0.0324
P_{N_2}(PSIA)	**k**
CURVE 39* (T = 303.2 K)	
0.16	0.0167
0.57	0.0178
1.5	0.0184
3	0.0192
5.5	0.0213

P(mm Hg)	k
CURVE 39 (cont.)*	
14	0.0236
77	0.0243
150	0.0262
300	0.0274
550	0.0276
1000	0.0282
1200	0.0282
1700	0.0291
2100	0.0301
CURVE 40* (T = 303.2 K)	
0.045	0.0324
0.38	0.0337
1.6	0.0337
6	0.0337
15	0.0339
57.5	0.0345
100	0.0356
170	0.0351
300	0.0349
550	0.0348
1300	0.0358
1800	0.0354
2300	0.0356
P_{ob}(psi)	**k**
CURVE 41* (T = 303.2 K)	
525	0.0234
1140	0.0262
1830	0.028
2480	0.0296
3190	0.0304
3965	0.0305
CURVE 42* (T = 303.2 K)	
0	0.0173
225	0.0213
2000	0.030

P_{ob}(psi)	k
CURVE 43* (T = 303.2 K)	
0	0.0235
410	0.0269
999	0.0286
1575	0.0298
2125	0.0307
2780	0.0314
3600	0.0320
CURVE 44 (T = 303.2 K)	
2000	0.0312
T	**k**
CURVE 45	
309.7	0.0460
319.8	0.0470

*Not shown on plot

SPECIFICATION TABLE NO. 279 THERMAL CONDUCTIVITY OF SILLIMANITE

Curve No.	Ref. No.	Method Used	Year	Temp. Range, K	Reported Error, %	Name and Specimen Designation	Composition (weight percent), Specifications and Remarks
1	3		1953	333.2			Max water absorption 0.05%; flexural strength 18,500 psi; coefficient of expansion 4.4×10^6 at 25 to 700 C.

DATA TABLE NO. 279 THERMAL CONDUCTIVITY OF SILLIMANITE

[Temperature, T, K; Thermal Conductivity, k, Watt $cm^{-1} K^{-1}$]

T	k
CURVE 1*	
333.2	0.0260

*No graphical presentation

SPECIFICATION TABLE NO. 280 THERMAL CONDUCTIVITY OF SLATE

Curve No.	Ref. No.	Method Used	Year	Temp. Range, K	Reported Error, %	Name and Specimen Designation	Composition (weight percent), Specifications and Remarks
1	508	L	1940	393-578		Slate	Specimen 8 in. in dia and 1 in. thick; density 2.947 g cm^{-3}; obtained from Madoc, Ont.
2	508	L	1940	395-577		Slate	Similar to the above specimen; measurements done after exposure to high temp test.

DATA TABLE NO. 280 THERMAL CONDUCTIVITY OF SLATE

[Temperature, T, K; Thermal Conductivity, k, Watt cm^{-1} K^{-1}]

T	k
CURVE 1*	
393.2	0.0153
461.5	0.0164
577.6	0.0146
CURVE 2*	
394.8	0.0149
483.2	0.0147
576.5	0.0143

*No graphical presentation

SPECIFICATION TABLE NO. 281 THERMAL CONDUCTIVITY OF SOIL

Curve No.	Ref. No.	Method Used	Year	Temp. Range, K	Reported Error, %	Name and Specimen Designation	Composition (weight percent), Specifications and Remarks
1	31	R	1958	293.2			Mineral; density 2.65 g cm^{-3}.
2	31	R	1958	293.2			Organic; density 1.3 g cm^{-3}.
3	31	R	1958	293.2			Mineral; dry; density 1.5 g cm^{-3}.
4	31	R	1958	293.2			Mineral; saturated; density 1.93 g cm^{-3}.
5	31	R	1958	293.2			Organic; dry; density 0.13 g cm^{-3}.
6	31	R	1958	293.2			Organic; saturated; density 1.03 g cm^{-3}.

DATA TABLE NO. 281 THERMAL CONDUCTIVITY OF SOIL

[Temperature, T, K; Thermal Conductivity, k, Watt $cm^{-1} K^{-1}$]

T	k
CURVE 1*	
293.2	0.0293
CURVE 2*	
293.2	0.0251
CURVE 3*	
293.2	0.0109
CURVE 4*	
293.2	0.0209
CURVE 5*	
293.2	0.000335

T	k
CURVE 6*	
293.2	0.00502

*No graphical presentation

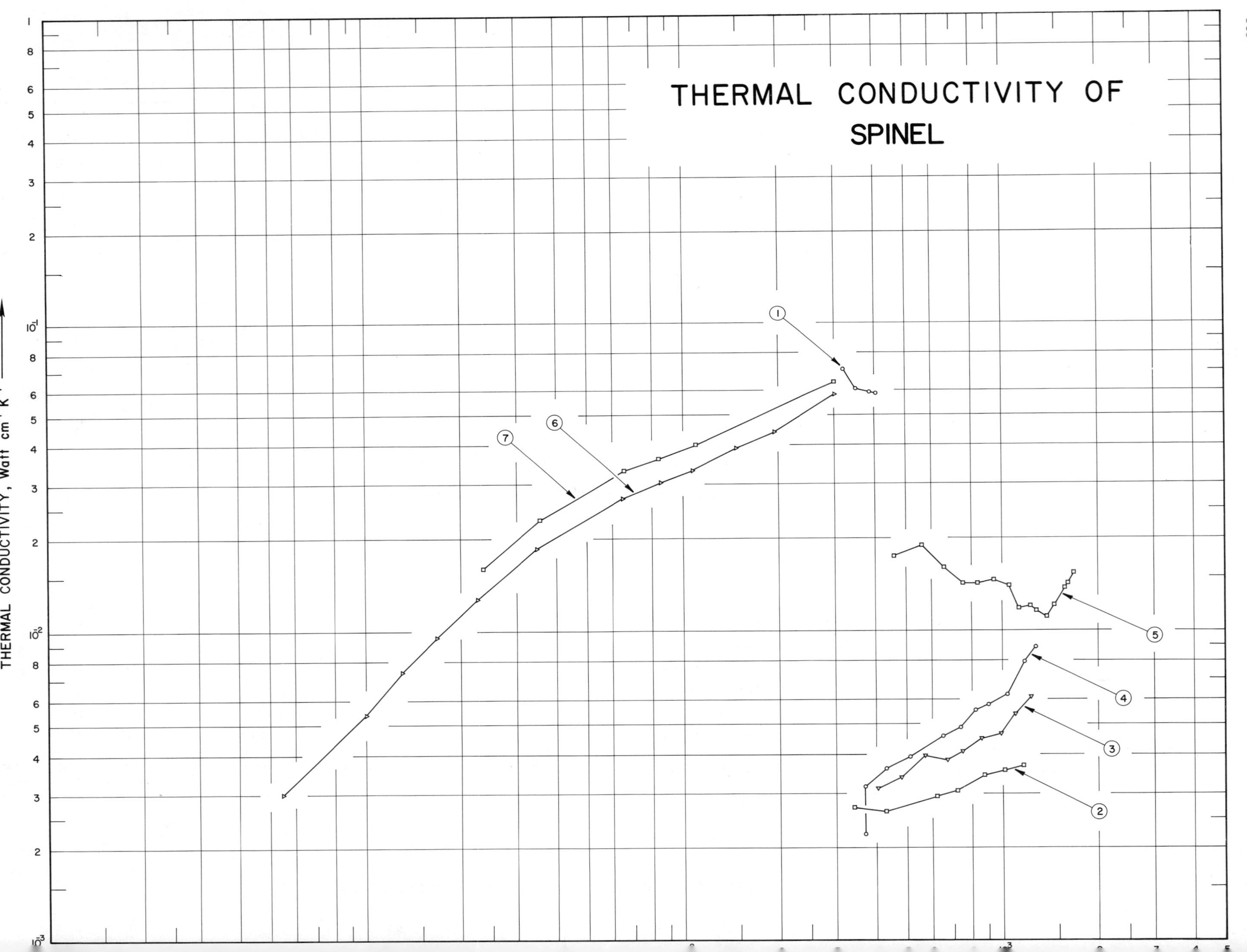
THERMAL CONDUCTIVITY OF SPINEL
THERMAL CONDUCTIVITY, Watt cm⁻¹ K⁻¹
1
2
3
4
5
6
7

SPECIFICATION TABLE NO. 282 THERMAL CONDUCTIVITY OF SPINEL

[For Data Reported in Figure and Table No. 282]

Curve No.	Ref. No.	Method Used	Year	Temp. Range, K	Reported Error, %	Name and Specimen Designation	Composition (weight percent), Specifications and Remarks
1	407	L	1949	322-406			No details reported.
2	463	R	1961	341-1169			Powder specimen contained in a hollow cylinder of 91 mm internal dia and 203 mm long; grain size < 0.2 mm; bulk density 1.51 g cm^{-3}.
3	463	R	1961	405-1232			Similar to the above specimen except grain size 0.2 ~1 mm and bulk density 1.2 g cm^{-3}.
4	463	R	1961	369-1273			Similar to the above specimen except grain size 2~5 mm and bulk density 0.92 g cm^{-3}.
5	463	R	1961	459-1670			Prepared from the powder of grain size 0.2 ~1 mm by pressing and firing at 1650 C; bulk density 1.79 g cm^{-3}.
6	300	L	1962	5.5 300		R-56 ($Mg_{0.75}Fe_{0.41}Al_{1.85}O_4$)	Single natural crystal of pleonaste spinel from Queensland; chemical analysis indicated a formula of $(Ti_{0.02}Mg_{0.75}Fe_{0.41}Al_{1.85})O_4$; other impurities low Cr, Mn, Ni, V, Zn, and trace Si; not detected for Be, Ca, K, Li, Na and Zr; free of second phase inclusions; specimen 0.34 cm average dia and 1.12 cm long; opaque and water worn; lattice constant 8.128 Å.
7	300	L	1962	24-300		R-62 ($Mg_{0.73}Fe_{0.33}Al_{1.93}O_4$)	Similar to the above specimen except chemical analysis indicated a formula of $(Ti_{0.01}Mg_{0.73}Fe_{0.33}Al_{1.93})O_4$; specimen 0.24 cm average dia and 1.00 cm long; lattice constant 8.117 Å.

DATA TABLE NO. 282 THERMAL CONDUCTIVITY OF SPINEL

[Temperature, T, K; Thermal Conductivity, k, Watt $cm^{-1}K^{-1}$]

T	k
CURVE 1	
321.5	0.0707
351.2	0.0607
388.7	0.0594
405.6	0.0586
CURVE 2	
341.2	0.00270
430.2	0.00263
621.2	0.00293
717.2	0.00306
875.2	0.00342
1017	0.00356
1169	0.00368
CURVE 3	
405.2	0.00311
482.2	0.00337
573.2	0.00396
673.2	0.00384
749.2	0.00407
856.2	0.00449
987.2	0.00465
1100	0.00538
1232	0.00608
CURVE 4	
369.2	0.00221
369.2	0.00316
430.2	0.00360
511.2	0.00392
651.2	0.00456
736.2	0.00487
823.2	0.00554
903.2	0.00576
1041	0.00622
1177	0.00794
1273	0.00884
CURVE 5	
459.2	0.0175
561.2	0.0189
656.2	0.0160
751.2	0.0142
839.2	0.0142
942.2	0.0146
1058	0.0140
1138	0.0118
1233	0.0120
1287	0.0116
1381	0.0111
1462	0.0121
1577	0.0138
1609	0.0142
1670	0.0154
CURVE 6	
5.47	0.0030
10.06	0.0054
13.09	0.0074
16.86	0.0096
22.65	0.0128
34.75	0.0186
64.78	0.027
85.19	0.0303
107.52	0.033
147.8	0.039
195.8	0.044
300	0.058
CURVE 7	
23.6	0.016
35.6	0.023
65.5	0.033
84.1	0.036
110.0	0.040
300.0	0.064

SPECIFICATION TABLE NO. 283 THERMAL CONDUCTIVITY OF SPODUMENE

Curve No.	Ref. No.	Method Used	Year	Temp. Range, K	Reported Error, %	Name and Specimen Designation	Composition (weight percent), Specifications and Remarks
1	282	R	1955	398-1133			β- Spodumene; 17. 7 chemically pure Li_2CO_3, 24. 5 Al_2O_3 and AlO, 57. 8 pottery flint; corresponding to the compound LiO · Al_2O_3 · 4 SiO_2 (there is a discrepancy between the LiO_2 term of this formula and the above-mentioned composition of Li_2CO_3); specimen made up of eleven rings of 0. 524 cm O. D. , 2. 74 cm I. D. , and 0. 5 in. thick each, stacked to form a cylinder 5. 5 in. high but measured only over the centrally-placed rings; 7% binder (500 g carbowax, 10 g methocel and 1000 cc water) added; hydraulically pressed at 4400 psi; set on aluminum and fired at a rate of 120 C per hr and held at a peak of 1345 C for 5 hrs; specific gravity 2. 13; apparent porosity 10. 0% and shrinkage 7. 2%.

DATA TABLE NO. 283 THERMAL CONDUCTIVITY OF SPODUMENE

[Temperature, T, K; Thermal Conductivity, k, Watt $cm^{-1} K^{-1}$]

T	k
CURVE 1*	
398. 2	0. 0108
511. 2	0. 0113
623. 2	0. 0120
708. 2	0. 0119
783. 2	0. 0130
856. 2	0. 0134
923. 2	0. 0138
978. 2	0. 0138
1028. 2	0. 0138
1078. 2	0. 0140
1133. 2	0. 0141

*No graphical presentation

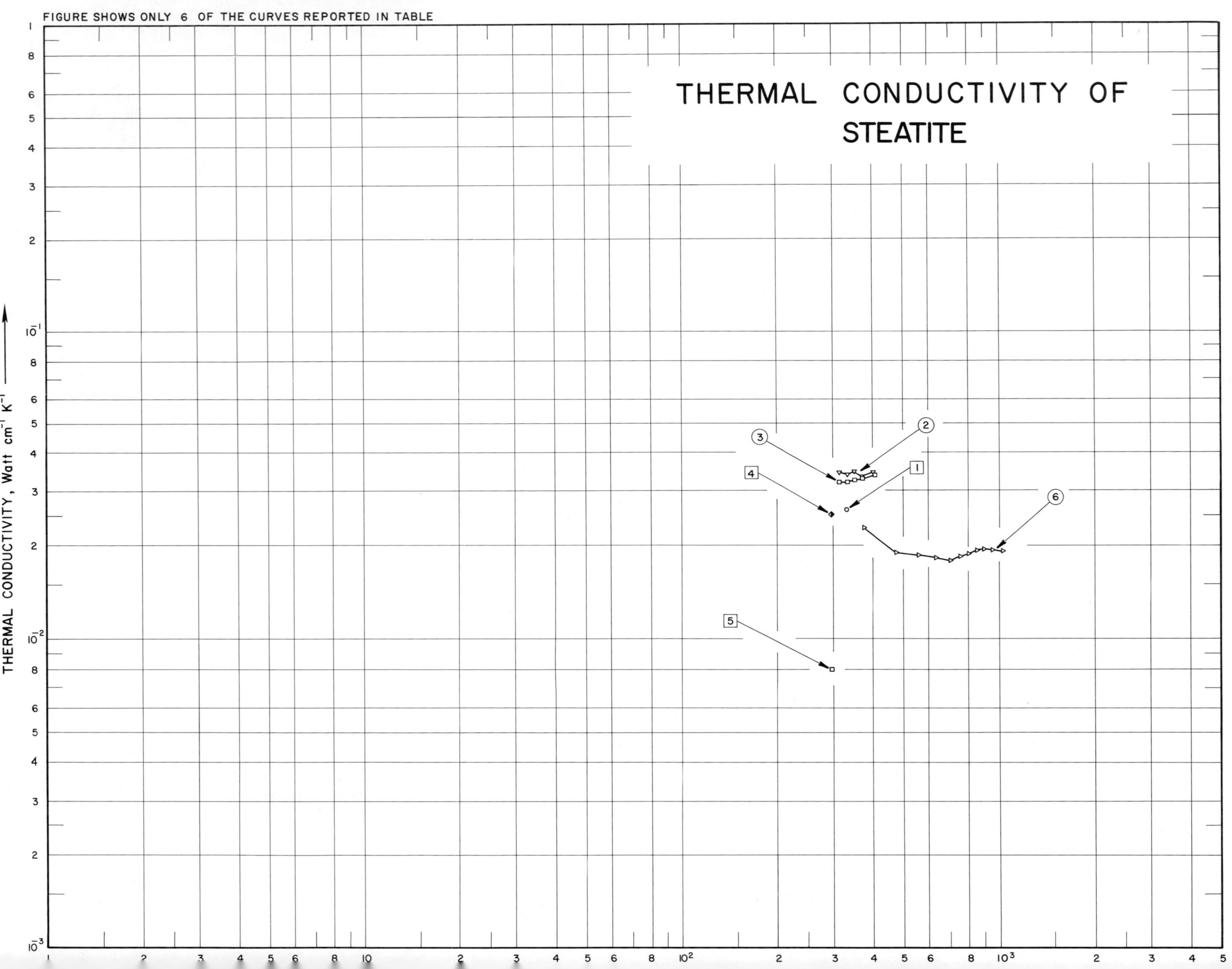
FIGURE SHOWS ONLY 6 OF THE CURVES REPORTED IN TABLE
THERMAL CONDUCTIVITY OF STEATITE
THERMAL CONDUCTIVITY, Watt cm⁻¹ K⁻¹
1
2
3
4
5
6

SPECIFICATION TABLE NO. 284 THERMAL CONDUCTIVITY OF STEATITE

[For Data Reported in Figure and Table No. 284]

Curve No.	Ref. No.	Method Used	Year	Temp. Range, K	Reported Error, %	Name and Specimen Designation	Composition (weight percent), Specifications and Remarks
1	3		1953	333.2			Maximum water absorption 0.05%; flexural strength 18,000 psi; coefficient of expansion (25-700 C) 8.3 x 10^{-6}.
2	68		1954	317-404	±3.0	10B2	
3	68	C	1954	316-409	±3.0	12C2	Commercial steatite.
4	152	L	1956	298.2		Steatite 228	Specimen in form of wafers 0.75 in. in dia and 20 mils thick.
5	152	L	1956	298.2		Steatite 228	The above specimen exposed with 7 x 10^{19} epithermal neutrons per cm^2 for 480 Mwd in the MTR.
6	282	R	1955	378-1030			82.0 manchurian talc, 8.0 Main feldspar, and 10.0 Tennessee ball clay; specimen made up of eleven rings 5.31 cm O.D., 2.79 cm I.D. and 0.5 in. thick each, stacked to form a cylinder 5.5 in. high, but measured only over the centrally-placed rings; milled in distilled water for five hrs; 10% binder (500 g carbowax, 10 g methocel, and 1000 cc water) added; fired on pottery flint to 1350 C at a rate of 60 C per hr; sp gr 2.56; mod. rupture 8,890 psi; apparent porosity 0.2%; shrinkage 8.4%; coefficient of expansion (25-1000 C) 82.0 x 10^{-7}.
7	108	L	1920	368.2		Soapstone	Specimen 9 in. in dia and 0.715 in. thick; specific gravity 2.87.

DATA TABLE NO. 284 THERMAL CONDUCTIVITY OF STEATITE

[Temperature, T, K; Thermal Conductivity, k, Watt $cm^{-1}K^{-1}$]

T	k
CURVE 1	
333.2	0.0260
CURVE 2	
316.8	0.0342
335.0	0.0338
353.8	0.0346
372.8	0.0333
404.1	0.0344
CURVE 3	
316.3	0.0319
335.6	0.0320
352.8	0.0325
374.4	0.0327
408.6	0.0337
CURVE 4	
298.2	0.0251
CURVE 5	
298.2	0.00795
CURVE 6	
378.2	0.0228
478.2	0.0189
560.2	0.0186
635.2	0.0181
701.2	0.0178
758.2	0.0184
805.2	0.0187
853.2	0.0191
898.2	0.0194
958.2	0.0192
1030.2	0.0191
CURVE 7*	
368.2	0.0335

*Not shown on plot

SPECIFICATION TABLE NO. 285 THERMAL CONDUCTIVITY OF TOURMALINE

Curve No.	Ref. No.	Method Used	Year	Temp. Range, K	Reported Error, %	Name and Specimen Designation	Composition (weight percent), Specifications and Remarks
1	34	C	1943	398-723		Brazil	Green; specimen 1 x 1 x 1 cm; cut from a single crystal with its c- (principle) axis normal to the two opposite surfaces and parallel to other surfaces; heat flow parallel to c-axis; 18-8 stainless steel used as comparative material.
2	34	C	1943	398-729		Brazil	The above specimen measured with heat flow perpendicular to c-axis; 18-8 stainless steel used as comparative material.

DATA TABLE NO. 285 THERMAL CONDUCTIVITY OF TOURMALINE

[Temperature, T, K; Thermal Conductivity, k, Watt $cm^{-1} K^{-1}$]

T	k
CURVE 1*	
398.2	0.0291
540.2	0.0320
613.2	0.0322
723.2	0.0351
CURVE 2*	
393.2	0.0296
492.2	0.0346
591.2	0.0383
729.2	0.0417

*No graphical presentation

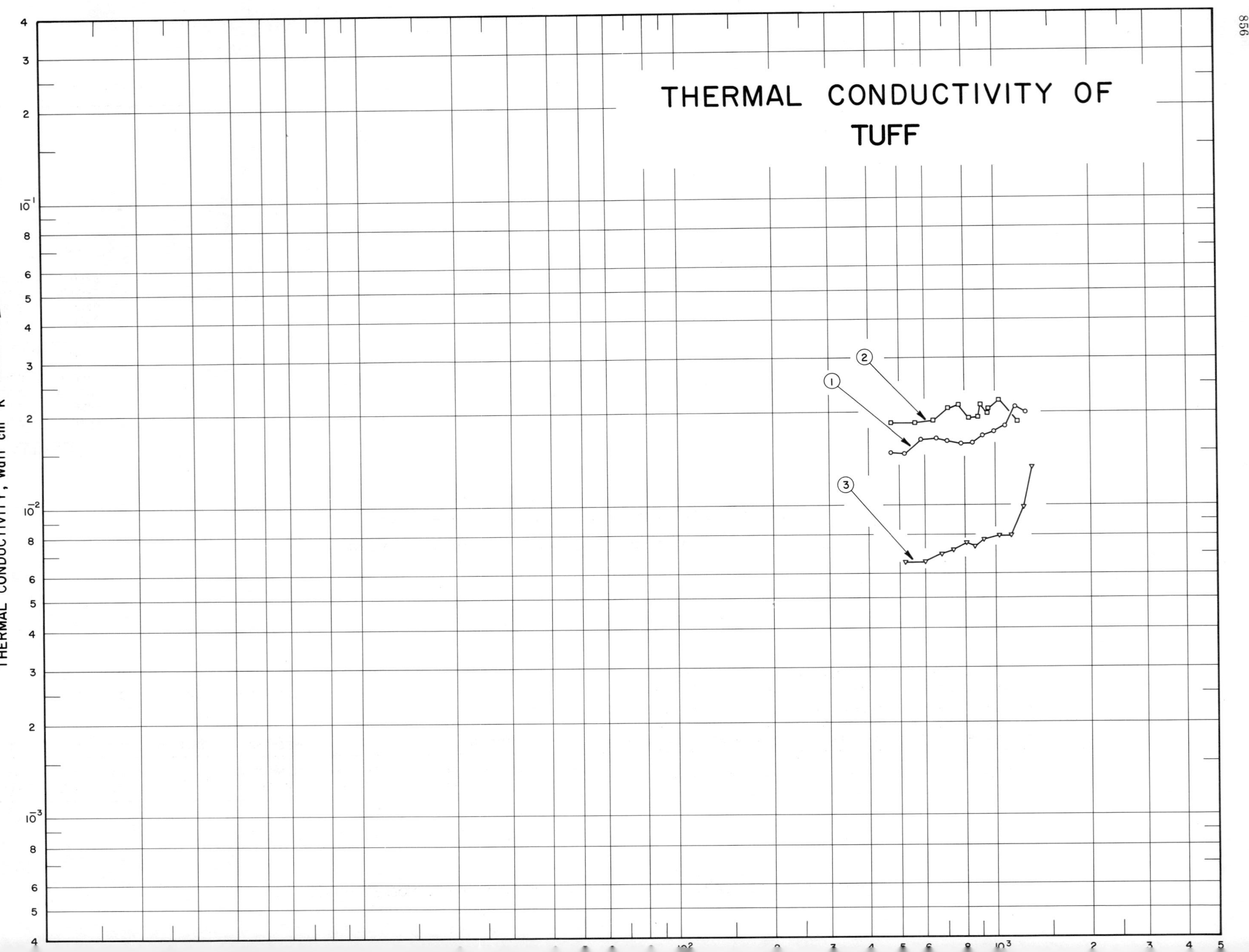
THERMAL CONDUCTIVITY OF
TUFF
THERMAL CONDUCTIVITY, Watt cm⁻¹ K⁻¹
1
2
3

SPECIFICATION TABLE NO. 286 THERMAL CONDUCTIVITY OF TUFF

[For Data Reported in Figure and Table No. 286]

Curve No.	Ref. No.	Method Used	Year	Temp. Range, K	Reported Error, %	Name and Specimen Designation	Composition (weight percent), Specifications and Remarks
1	578	R	1963	471-1253	± 9	TOS-4	(Composition in vol percent) 72 matrix (clay, chalcedony, zeolite, etc.), 14 chalcedony segregations, 8 quartz phenocrysts, 6 feldspar phenocrysts; moderately grained appearance; specimen 3.5 in. O.D., 0.875 in. I.D. and 18 in. long; obtained from U12b07 at a drill depth of 75-78 ft; density 2.24 g cm^{-3} (dry); specimen fairly representative of most of the coherent NTS Tuffs (TOS-1 to 5, and basal part of 7).
2	578	R	1963	471-1173	± 9	TOS-4	Similar to the above specimen.
3	578	R	1963	523-1307	± 9	TOS-7	(Composition in vol percent) 86 glassy matrix, 12 quartz phenocrysts, 1 feldspar phenocrysts, 1 rock fragments; grainly friable in appearance; specimen 3.5 in. O.D., 0.875 in. I.D. and 18 in. long; extraction location unknown; sintered above 1173 K; density 1.27 g cm^{-3}; specimen representative of friable materials.

DATA TABLE NO. 286 THERMAL CONDUCTIVITY OF TUFF

[Temperature, T, K; Thermal Conductivity, k, Watt $cm^{-1}K^{-1}$]

T	k
CURVE 1	
471	0.0147
523	0.0146
588	0.0163
658	0.0164
711	0.0161
785	0.0158
850	0.0159
918	0.0167
998	0.0172
1083	0.0180
1165	0.0207
1253	0.0199
CURVE 2	
471	0.0184
563	0.0184
643	0.0187
715	0.0205
771	0.0210
831	0.0191
893	0.0192
903	0.0210
951	0.0198
959	0.0204
1033	0.0217
1173	0.0186
CURVE 3	
523	0.00653
601	0.00653
680	0.00690
744	0.00711
813	0.00749
863	0.00728
923	0.00766
1031	0.00791
1134	0.00791
1240	0.00983
1307	0.0132

SPECIFICATION TABLE NO. 287 THERMAL CONDUCTIVITY OF WOLLASTONITE

Curve No.	Ref. No.	Method Used	Year	Temp. Range, K	Reported Error, %	Name and Specimen Designation	Composition (weight percent), Specifications and Remarks
1	3		1953	333. 2			Max water absorption 0. 05%; flexural strength 22, 000 psi; coefficient of expansion 5. 9 x 10^{-6} at 25-700 C.
2	23		1952	317-397		W-34	60 wollastonite, 30 Kentucky Old Mine No. 4 ball clay, 5 barium carbonate, 5 litharge.

DATA TABLE NO. 287 THERMAL CONDUCTIVITY OF WOLLASTONITE

[Temperature, T, K; Thermal Conductivity, k, Watt $cm^{-1} K^{-1}$]

T	k
CURVE 1*	
333. 2	0. 0260
CURVE 2*	
317. 2	0. 0271
334. 2	0. 0268
354. 7	0. 0272
373. 2	0. 0259
397. 2	0. 0262

*No graphical presentation

SPECIFICATION TABLE NO. 288 THERMAL CONDUCTIVITY OF CEMENT

Curve No.	Ref. No.	Method Used	Year	Temp. Range, K	Reported Error, %	Name and Specimen Designation	Composition (weight percent), Specifications and Remarks
1	125	L	1923	328.2	1		Portland burnt cement; thickness = 0.42 cm; pressure = 21 lb in.$^{-2}$.
2	580	L	1955	307.3		Portland cement	100% under binding; dry density 2.010 g cm^{-3}.
3	580	L	1955	311.5		Slag-Portland cement	Dry density 1.730 g cm^{-3}.
4	580	L	1955	311.0		Slag cement	12-15% under binding; dry density 1.760 g cm^{-3}.

DATA TABLE NO. 288 THERMAL CONDUCTIVITY OF CEMENT

[Temperature, T, K; Thermal Conductivity, k, Watt cm^{-1} K^{-1}]

T	k
CURVE 1*	
328.2	0.0126
CURVE 2*	
307.3	0.00675
CURVE 3*	
311.5	0.00530
CURVE 4*	
311.0	0.00416

*No graphical presentation

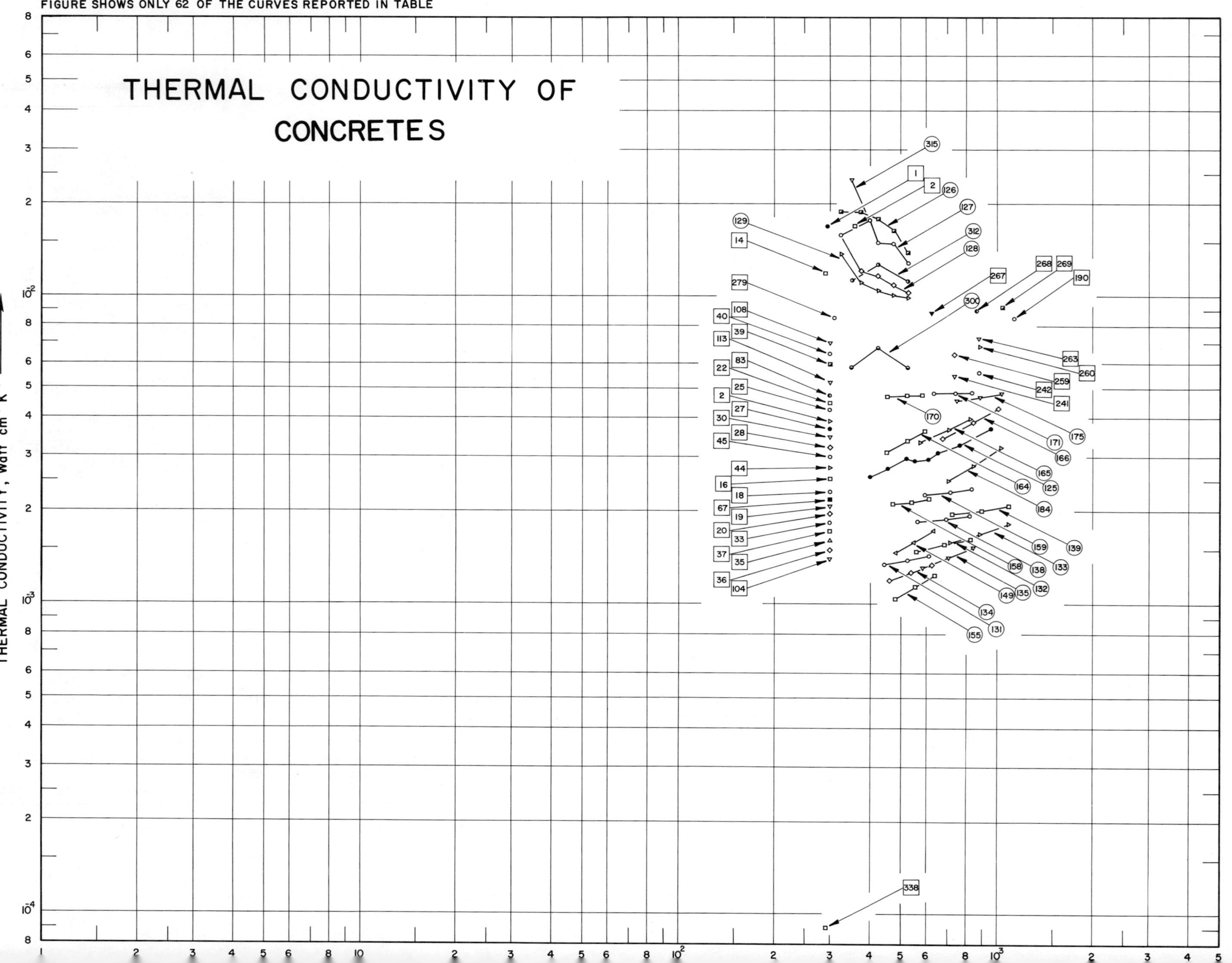

FIGURE SHOWS ONLY 62 OF THE CURVES REPORTED IN TABLE
THERMAL CONDUCTIVITY OF CONCRETES
THERMAL CONDUCTIVITY, Watt cm⁻¹ K⁻¹

SPECIFICATION TABLE NO. 289 THERMAL CONDUCTIVITY OF CONCRETES

[For Data Reported in Figure and Table No. 289]

Curve No.	Ref. No.	Method Used	Year	Temp. Range, K	Reported Error, %	Name and Specimen Designation	Composition (weight percent), Specifications and Remarks
1	88	L	1926	292.8	1.0	Bitumin concrete	Bitumin used as conglutinant; density 2.25 g cm^{-3}.
2	88	L	1926	356.1	1.0	Bitumin concrete	Similar to the above specimen.
3	88	L	1926	356.2	1.0	Bitumin concrete	Similar to the above specimen.
4	88	L	1926	292.8	1.0	Bitumin concrete	Similar to the above specimen.
5	88	L	1926	376.5	1.0	Bitumin concrete	Similar to the above specimen.
6	88	L	1926	376.6	1.0	Bitumin concrete	Similar to the above specimen.
7	88	L	1926	292.8	1.0	Paraffin concrete	Paraffin used as conglutinant; density 2.25 g cm^{-3}.
8	88	L	1926	292.9	1.0	Paraffin concrete	Similar to the above specimen.
9	88	L	1926	292.7	1.0	Paraffin concrete	Similar to the above specimen.
10	88	L	1926	293.3	1.0	Paraffin concrete	Similar to the above specimen.
11	88	L	1926	293.0	1.0	Paraffin concrete	Similar to the above specimen.
12	88	L	1926	293.6	1.0	Paraffin concrete	Similar to the above specimen.
13	545	P	1947	287.3		Blended bituminous concrete aggregate; 3G-1	Prepared from locally processed aggregates of sand and partially crushed gravel obtained from bins at plant; specimen contained in brass hollow cylinder of 5.36 in. I.D. and 10.68 in. long; materials compacted by placing into the cylinder in 5 approx equal layers, applying increasing number of blows on each layer; density 2.14 g cm^{-3}; water content zero; thermal conductivity values calculated from measured transient surface temp change and specific heat data.
14	545	P	1947	287.3		Asphaltic bituminous concrete; 3F-1	Prepared from the blended bituminous concrete aggregate and 4.5% bitumin; same dimensions, compacting procedure, and measuring method as the above specimen; density 2.40 g cm^{-3}.
15	545	P	1947	287.3		Asphaltic bituminous concrete; 3F-2	Similar to the above specimen.
16	552	L	1960	298.2		Metallurgical pumice concrete	20 x 20 x 4 cm; mixed by hand, filled into prescribed form and tamped, prestored for 24 hrs, submitted to heat treatment for about 2 hrs at 60 C and for 4 hrs at 85 C; grain size 0~7 mm; density 1.23 g cm^{-3}.
17	552	L	1960	298.2		Metallurgical pumice concrete	Similar to the above specimen except density 1.17 g cm^{-3}.
18	552	L	1960	298.2		Metallurgical pumice concrete	Similar to the above specimen except density 1.13 g cm^{-3}.
19	552	L	1960	298.2		Metallurgical pumice concrete	Similar to the above specimen.

SPECIFICATION TABLE NO. 289 (continued)

Curve No.	Ref. No.	Method Used	Year	Temp. Range, K	Reported Error, %	Name and Specimen Designation	Composition (weight percent), Specifications and Remarks
20	552	L	1960	298.2		Metallurgical pumice concrete	Similar to the above specimen except density 1.08 g cm^{-3}.
21	552	L	1960	298.2		Direct process slag concrete	Same dimensions and fabrication method as the above specimen; grain size 0-7 mm; density 1.57 g cm^{-3}.
22	552	L	1960	298.2		Direct process slag concrete	Similar to the above specimen.
23	552	L	1960	298.2		Direct process slag concrete	Similar to the above specimen except density 1.52 g cm^{-3}.
24	552	L	1960	298.2		Direct process slag concrete	Similar to the above specimen.
25	552	L	1960	298.2		Direct process slag concrete	Similar to the above specimen except density 1.51 g cm^{-3}.
26	552	L	1960	298.2		Direct process slag concrete	Similar to the above specimen except density 1.48 g cm^{-3}.
27	552	L	1960	298.2		Leuna slag concrete	Same dimensions and fabrication method as the above specimen; density 1.80 g cm^{-3}; grain size 0~15 mm.
28	552	L	1960	298.2		Leuna slag concrete	Similar to the above specimen except density 1.73 g cm^{-3}.
29	552	L	1960	298.2		Leuna slag concrete	Similar to the above specimen except density 1.69 g cm^{-3}.
30	552	L	1960	298.2		Leuna slag concrete	Similar to the above specimen except density 1.71 g cm^{-3}.
31	552	L	1960	298.2		Leuna slag concrete	Similar to the above specimen except density 1.63 g cm^{-3}.
32	552	L	1960	298.2		Leuna slag concrete	Similar to the above specimen except density 1.61 g cm^{-3}.
33	552	L	1960	298.2		Slag concrete	Same dimensions and fabrication method as the above specimen; density 0.87 g cm^{-3}; grain size 0~15 mm.
34	552	L	1960	298.2		Slag concrete	Similar to the above specimen except density 0.85 g cm^{-3}.
35	552	L	1960	298.2		Slag concrete	Similar to the above specimen except density 0.86 g cm^{-3}.
36	552	L	1960	298.2		Slag concrete	Similar to the above specimen except density 0.83 g cm^{-3}.
37	552	L	1960	298.2		Slag concrete	Similar to the above specimen except density 0.81 g cm^{-3}.
38	552	L	1960	298.2		Slag concrete	Similar to the above specimen except density 0.80 g cm^{-3}.
39	552	L	1960	298.2		Limestone gravel concrete	Same dimensions and fabrication method as the above specimen; density 1.89 g cm^{-3}; grain size 0~15 mm.
40	552	L	1960	298.2		Limestone gravel concrete	Similar to the above specimen except density 1.87 g cm^{-3}.

SPECIFICATION TABLE NO. 289 (continued)

Curve No.	Ref. No.	Method Used	Year	Temp. Range, K	Reported Error, %	Name and Specimen Designation	Composition (weight percent), Specifications and Remarks
41	552	L	1960	298.2		Limestone gravel concrete	Similar to the above specimen.
42	552	L	1960	298.2		Limestone gravel concrete	Similar to the above specimen except density 1.82 g cm^{-3}.
43	552	L	1960	298.2		Limestone gravel concrete	Similar to the above specimen except density 1.81 g cm^{-3}.
44	552	L	1960	298.2		K 1	100% blast furnace slag; cement content in concrete 280 K gm^{-3}; 20 x 20 x 4 cm; mixed by hand, filled in prescribed forms and tamped, prestored for 24 hrs, submitted to a heat treatment for about 2 hrs at 60 C and 4 hrs at 85 C; grain size 0-7 mm; density (dry) 1.66 g cm^{-3}.
45	552	L	1960	298.2		K 2	80% blast furnace slag (grain size 0~7 mm) and 20% metallurgical pumice (grain size 3~15 mm); cement content in concrete 283 K gm^{-3}; size 20 x 20 x 4 cm; same treatment as the above specimen; density (dry) 1.57 g cm^{-3}.
46	552	L	1960	298.2		K 4	60% blast furnace slag (grain size 0~7 mm) and 40% metallurgical pumice (grain size 3~15 mm); cement content in concrete 264 K gm^{-3}; size 20 x 20 x 4 cm; same treatment as the above specimen; density (dry) 1.61 g cm^{-3}.
47	552	L	1960	298.2		K 6	40% blast furnace slag (grain size 0~7mm) and 60% metallurgical pumice (grain size 3~15 mm); cement content in concrete 272 K gm^{-3}; size 20 x 20 x 4 cm; same treatment as the above specimen; density (dry) 1.46 g cm^{-3}.
48	552	L	1960	298.2		K 3	80% blast furnace slag (grain size 0~7 mm) and 20% hirschfelder slag (grain size 0~15 mm); cement content in concrete 275 K gm^{-3}; size 20 x 20 x 4 cm; same treatment as the above specimen; density (dry) 1.46 g cm^{-3}.
49	552	L	1960	298.2		K 5	60% blast furnace slag (grain size 0~7 mm) and 40% hirschfelder slag (grain size 0~15 mm); cement content in concrete 275 K gm^{-3}; size 20 x 20 x 4 cm; same treatment as the above specimen; density (dry) 1.37 g cm^{-3}.
50	552	L	1960	298.2		K 7	40% blast furnace slag (grain size 0~7 mm) and 60% hirschfelder slag (grain size 0~15 mm); cement content in concrete 273 K gm^{-3}; size 20 x 20 x 4 cm; same treatment as the above specimen; density (dry) 1.25 g cm^{-3}.
51	552	L	1960	298.2		K 22	40% blast furnace slag and 60% Calbe blast furnace slag (grain size 0~7 mm); cement content in concrete 275 K gm^{-3}; compression strength 42.7 K cm^{-2}; size 20 x 20 x 4 cm; same treatment as the above specimen; density (dry) 1.69 g cm^{-3}.
52	552	L	1960	298.2		K 28	40% blast furnace slag (grain size 0~7 mm) and 60% Lenna coal slag (grain size 0~15 mm); cement content in concrete 275 K gm^{-3}; size 20 x 20 x 4 cm; same treatment as the above specimen; density (dry) 1.72 g cm^{-3}.
53	552	L	1960	298.2		K 17	100% Calbe blast furnace slag (grain size 0~7 mm); cement content in concrete 275 K gm^{-3}; size 20 x 20 x 4 cm; same treatment as the above specimen; density (dry) 1.73 g cm^{-3}.

SPECIFICATION TABLE NO. 289 (continued)

Curve No.	Ref. No.	Method Used	Year	Temp. Range, K	Reported Error, %	Name and Specimen Designation	Composition (weight percent), Specifications and Remarks
54	552	L	1960	298.2		K 18	80% Calbe blast furnace slag (grain size 0~7 mm) and 20% metallurgical pumice (grain size 3~15 mm); cement content in concrete 275 K gm^{-3}; size 20 x 20 x 4 cm; same treatement as the above specimen; density (dry) 1.68 g cm^{-3}.
55	552	L	1960	298.2		K 20	60% Calbe blast furnace slag (grain size 0~7 mm) and 40% metallurgical pumice (grain size 3~15 mm); cement content in concrete 275 K gm^{-3}; compression strength 61.5 K cm^{-3}; size 20 x 20 x 4 cm; same treatment as the above specimen; density (dry) 1.60 g cm^{-3}.
56	552	L	1960	298.2		K 19	80% Calbe blast furnace slag (grain size 0~7 mm) and 20% hirschfelder slag (grain size 0~15 mm); cement content in concrete 283 K gm^{-3}; compression strength 41.1 Kp cm^{-2}; size 20 x 20 x 4 cm; same treatment as the above specimen; density (dry) 1.53 g cm^{-3}.
57	552	L	1960	298.2		K 21	60% Calbe blast furnace slag (grain size 0~7 mm) and 40% hirschfelder slag (grain size 0~15 mm); cement content in concrete 275 K gm^{-3}; compression strength 37.1 Kp cm^{-2}; size 20 x 20 x 4 cm; same treatment as the above specimen; density (dry) 1.33 g cm^{-3}.
58	552	L	1960	298.2		K 46	80% Stalinstadt blast furnace slag (grain size 0~7 mm) and 20% sand (grain size 0~7 mm); cement content in concrete not measured; size 20 x 20 x 4 cm; same treatment as the above specimen; density (dry) 1.79 g cm^{-3}.
59	552	L	1960	298.2		K 34	100% Stalinstadt dump slag; cement content in concrete 275 K gm^{-3}; size 20 x 20 x 4 cm; same treatment as the above specimen; grain size 0-15 mm; density (dry) 1.22 g cm^{-3}.
60	552	L	1960	298.2		K 38	80% Stalinstadt dump slag (grain size 0~15 mm) and 20% metallurgical pumice (grain size 3~15 mm); cement content in concrete 277 K gm^{-3}; size 20 x 20 x 4 cm; same treatment as the above specimen; density (dry) 1.24 g cm^{-3}.
61	552	L	1960	298.2		K 39	60% Stalinstadt dump slag (grain size 0~15 mm) and 40% metallurgical pumice (grain size 0~15 mm); cement content in concrete 277 K gm^{-3}; size 20 x 20 x 4 cm; same treatment as the above specimen; density (dry) 1.28 g cm^{-3}.
62	552	L	1960	298.2		K 40	40% Stalinstadt dump slag (grain size 0~15 mm) and 60% metallurgical pumice (grain size 3~15 mm); cement content in concrete 286 K gm^{-3}; size 20 x 20 x 4 cm; same treatment as the above specimen; density (dry) 1.38 g cm^{-3}.
63	552	L	1960	298.2		K 41	80% Stalinstadt dump slag (grain size 0~15 mm) and 20% Unterwellenborn direct process slag (grain size 0~7 mm); cement content in concrete 275 K cm^{-3}; size 20 x 20 x 4 cm; same treatment as the above specimen; density (dry) 1.25 g cm^{-3}.
64	552	L	1960	298.2		K 42	60% Stalinstadt dump slag (grain size 0~15 mm) and 40% Unterwellenborn direct process slag (grain size 0~7 mm); cement content 275 K gm^{-3}; size 20 x 20 x 4 cm; same treatment as the above specimen; density (dry) 1.35 g cm^{-3}.
65	552	L	1960	298.2		K 43	40% Stalinstadt dump slag (grain size 0~15 mm) and 60% Unterwellenborn direct process slag (grain size 0~7 mm); cement content 275 K gm^{-3}; size 20 x 20 x 4 cm; same treatment as the above specimen; density (dry) 1.42 g cm^{-3}.

Curve No.	Ref. No.	Method Used	Year	Temp. Range, K	Reported Error, %	Name and Specimen Designation	Composition (weight percent), Specifications and Remarks
66	552	L	1960	298.2		K 35	80% Stalinstadt dump slag (grain size 0~15 mm) and 20% blast furnace slag (gray)(grain size 0~7 mm); cement content 275 K gm^{-3}; size 20 x 20 x 4 cm; same treatment as the above specimen; density (dry) 1.30 g cm^{-3}.
67	552	L	1960	298.2		K 36	60% Stalinstadt dump slag (grain size 0~15 mm) and 40% blast furnace slag (gray) (grain size 0~7 mm); cement content 275 K gm^{-3}; size 20 x 20 x 4 cm; same treatment as the above specimen; density (dry) 1.45 g cm^{-3}.
68	552	L	1960	298.2		K 37	40% Stalinstadt dump slag (grain size 0~15 mm) and 60% blast furnace slag (gray) (grain size 0~7 mm); cement content 274 K gm^{-3}; 20 x 20 x 4 cm; same treatment as the above specimen; density (dry) 1.47 g cm^{-3}.
69	552	L	1960	298.2		K 44	60% Stalinstadt dump slag (grain size 0~15 mm) and 40% limestone gravel (grain size 0~7 mm); cement content 275 K gm^{-3}; size 20 x 20 x 4 cm; same treatment as the above specimen; density (dry) 1.36 g cm^{-3}.
70	552	L	1960	298.2		K 45	40% Stalinstadt dump slag (grain size 0~7 mm); cement content 280 K gm^{-3}; compression strength 30.0 Kp cm^{-2}; size 20 x 20 x 4 cm; same treatment as the above specimen; density (dry) 1.43 g cm^{-3}.
71	552	L	1960	298.2		K 10	100% Stalinstadt metallurgical pumice (grain size 0-7 mm); cement content 275 K gm^{-3}; size 20 x 20 x 4 cm; same treatment as the above specimen; density (dry) 1.23 g cm^{-3}.
72	552	L	1960	298.2		K 8	80% Stalinstadt metallurgical pumice (grain size 0~7 mm) and 20% sand (grain size 0~3 mm); cement content 273 K gm^{-3}; size 20 x 20 x 4 cm; same treatment as the above specimen; density (dry) 1.38 g cm^{-3}.
73	552	L	1960	298.2		K 12	100% Unterwellenborn direct process slag (grain size 0-7 mm); cement content 275 K gm^{-3}; size 20 x 20 x 4 cm; same treatment as the above specimen; density (dry) 1.56 g cm^{-3}.
74	552	L	1960	298.2		K 13	80% Unterwellenborn direct process slag (grain size 0~7 mm) and 20% metallurgical pumice (grain size 0~15 mm); cement content 275 K gm^{-3}; compression strength 66.8 Kp cm^{-2}; size 20 x 20 x 4 cm; same treatment as the above specimen; density (dry) 1.49 g cm^{-3}.
75	552	L	1960	298.2		K 15	60% Unterwellenborn direct process slag (grain size 0~7 mm) and 40% metallurgical pumice (grain size 0~15 mm); cement content 275 K gm^{-3}; size 20 x 20 x 4 cm; same treatment as the above specimen; density (dry) 1.48 g cm^{-3}.
76	552	L	1960	298.2		K 14	80% Unterwellenborn direct process slag (grain size 0~7 mm) and 20% hirschfelder slag (grain size 0~15 mm); cement content 27 K gm^{-3}; size 20 x 20 x 4 cm; same treatment as the above specimen; density (dry) 1.42 g cm^{-3}.
77	552	L	1960	298.2		K 16	60% Unterwellenborn direct process slag (grain size 0~7 mm) and 40% hirschfelder slag (grain 0~15 mm); cement content 272 K gm^{-3}; size 20 x 20 x 4 cm; same treatment as the above specimen; density (dry) 1.27 g cm^{-3}.
78	552	L	1960	298.2		K 23	100% Leuna coal slag (grain size 0-15 mm) cement content 275 K gm^{-3}; size 20 x 20 x 4 cm; same treatment as the above specimen; density (dry) 1.74 g cm^{-3}.

SPECIFICATION TABLE NO. 289 (continued)

Curve No.	Ref. No.	Method Used	Year	Temp. Range, K	Reported Error, %	Name and Specimen Designation	Composition (weight percent), Specifications and Remarks
79	552	L	1960	298. 2		K 24	80% Leuna coal slag (grain size 0~15 mm) and 20% metallurgical pumice (grain size 3-15 mm); cement content 275 K gm^{-3}; size 20 x 20 x 4 cm; same treatment as the above specimen; density (dry) 1. 64 g cm^{-3}.
80	552	L	1960	298. 2		K 27	60% Leuna coal slag (grain size0~15 mm) and 40% metallurgical pumice (grain size 3-15 mm); cement content 275 K gm^{-3}; size 20 x 20 x 4 cm; same treatment as the above specimen; density (dry) 1. 63 g cm^{-3}.
81	552	L	1960	298. 2		K 25	80% Leuna coal slag (grain size 0~15 mm) and 20% hirschfelder slag (grain size 0~7 mm); cement content 275 K gm^{-3}; size 20 x 20 x 4 cm; same treatment as the above specimen; density 1. 51 g cm^{-3}.
82	552	L	1960	298. 2		K 26	60% Leuna coal slag (grain size 0~15 mm) and 40% hirschfelder slag (grain size 0~7 mm); cement content 272 K gm^{-3}; size 20 x 20 x 4 cm; same treatment as the above specimen; density (dry) 1. 33 g cm^{-3}.
83	552	L	1960	298. 2		K 29	100% Limestone gravel; grain size 0-15 mm; cement content 275 K gm^{-3}; size 20 x 20 x 4 cm; same treatment as the above specimen; density (dry) 1. 63 g cm^{-3}.
84	552	L	1960	298. 2		K 30	80% Limestone gravel (grain size 0-15 mm) and 20% metallurgical pumice (grain size 3-15 mm); cement content 275 K gm^{-3}; size 20 x 20 x 4 cm; same treatment as the above specimen; density (dry) 1. 58 g cm^{-3}.
85	552	L	1960	298. 2		K 31	60% Limestone gravel (grain size 0-15 mm) and 40% metallurgical pumice (grain size 3-15 mm); cement content 275 K gm^{-3}; size 20 x 20 x 4 cm; same treatment as the above specimen; density (dry) 1. 40 g cm^{-3}.
86	552	L	1960	298. 2		K 32	80% Limestone gravel (grain size 0~15 mm) and 20% hirschfelder slag (grain size 0-15 mm); cement content 275 K gm^{-3}; size 20 x 20 x 4 cm; same treatment as the above specimen; density (dry) 1. 47 g cm^{-3}.
87	552	L	1960	298. 2		K 33	60% Limestone gravel (grain size 0~15 mm) and 40% hirschfelder slag (grain size 0-15 mm); cement content 275 K gm^{-3}; size 20 x 20 x 4 cm; same treatment as the above specimen; density (dry) 1. 27 g cm^{-3}.
88	552	L	1960	298. 2		K 11	100% hirschfelder slag [from coal (liquid) slag]; grain size 0-15 mm; cement content 277 K gm^{-3}; size 20 x 20 x 4 cm; same treatment as the above specimen; density (dry) 1. 00 g cm^{-3}.
89	552	L	1960	298. 2		K 9	80% hirschfelder slag [from coal (liquid) slag] (grain size 0-15 mm) and 20% sand (grain size 0-3 mm); cement content 275 K gm^{-3}; size 20 x 20 x 4 cm; same treatment as the above specimen; density (dry) 1. 13 g cm^{-3}.
90	553	L	1937	297. 1		Sand and gravel aggregate; 1	Mix proportions (by volume): cement 1, fine aggregate 0 to No. 4 2. 00, and coarse aggregate No. 4 to 0. 5 in. 2. 75; 24 x 24 x 2 in. ; density 2. 406 g cm^{-3}.

SPECIFICATION TABLE NO. 289 (continued)

Curve No.	Ref. No.	Method Used	Year	Temp. Range, K	Reported Error, %	Name and Specimen Designation	Composition (weight percent), Specifications and Remarks
91	553	L	1937	297.1		Sand and gravel aggregate; 2	Mix proportions (by volume): cement 1, fine aggregate 0 to No. 4 2.75, coarse aggregate No. 4 to 0.5 in. 4.50; 24 x 24 x 2 in.; density 2.404 g cm^{-3}.
92	553	L	1937	297.1		Sand and gravel aggregate; 3	Mix proportions (by volume): cement 1, fine aggregate 0 to No. 4 3.50, and coarse aggregate No. 4 to 0.5 in. 5.50; 24 x 24 x 2 in.; density 2.372 g cm^{-3}.
93	553	L	1937	297.1		Sand and gravel aggregate; 4	Similar to specimen 1 but with density 2.374 g cm^{-3} and slump inches 5.
94	553	L	1937	297.1		Sand and gravel aggregate; 4	2nd run of the above specimen.
95	553	L	1937	297.1		Sand and gravel aggregate; 5	Similar to specimen 2 but with density 2.345 g cm^{-3} and slump inches 5.
96	553	L	1937	297.1		Sand and gravel aggregate; 5	2nd run of the above specimen.
97	553	L	1937	297.1		Sand and gravel aggregate; 6	Similar to specimen 3 but with density 2.318 g cm^{-3} and slump inches 5.
98	553	L	1937	297.1		Sand and gravel aggregate; 6	2nd run of the above specimen.
99	553	L	1937	297.1		Limestone aggregate; 1	Mix proportions (by volume): cement 1, fine aggregate 0 to No. 4 2.00, and coarse aggregate No. 4 to 0.5 in. 2.75; 24 x 24 x 2 in.; density 2.259 g cm^{-3}.
100	553	L	1937	297.1		Limestone aggregate; 2	Mix proportions (by volume): cement 1, fine aggregate 0 to No. 4 2.75, and coarse aggregate No. 4 to 0.5 in. 4.50; 24 x 24 x 2 in.; density 2.295 g cm^{-3}.
101	553	L	1937	297.1		Limestone aggregate; 3	Mix proportions (by volume): cement 1, fine aggregate 0 to No. 4 3.50, and coarse aggregate No. 4 to 0.5 in. 5.50; 24 x 24 x 2 in.; density 2.270 g cm^{-3}.
102	553	L	1937	297.1		Limestone aggregate; 4	Similar to the specimen 1 but with density 2.183 g cm^{-3} and slump inches = 3.
103	553	L	1937	297.1		Limestone aggregate; 5	Similar to the specimen 2 but with density 2.124 g cm^{-3} and slump inches = 3.
104	553	L	1937	297.1		Limestone aggregate; 6	Similar to the specimen 3 but with density 2.135 g cm^{-3} and slump inches = 3.
105	553	L	1937	297.1		Cinders aggregate; 1	Mix proportions (by volume): cement 1, fine aggregate 0 to No. 4 2.00, and coarse aggregate No. 4 to 0.5 in. 2.75; 24 x 24 x 2 in.; density 1.762 g cm^{-3}.
106	553	L	1937	297.1		Cinders aggregate; 2	Mix proportions (by volume): cement 1, fine aggregate 0 to No. 4 2.75, and coarse aggregate No. 4 to 0.5 in. 4.50; 24 x 24 x 2 in.; density 1.664 g cm^{-3}.
107	553	L	1937	297.1		Cinders aggregate; 3	Mix proportions (by volume): cement 1, fine aggregate 0 to No. 4 3.50, and coarse aggregate No. 4 to 0.5 in. 5.50; 24 x 24 x 2 in.; density 1.554 g cm^{-3}.

SPECIFICATION TABLE NO. 289 (continued)

Curve No.	Ref. No.	Method Used	Year	Temp. Range, K	Reported Error, %	Name and Specimen Designation	Composition (weight percent), Specifications and Remarks
108	553	L	1937	297.1		Cinders aggregate; 4	Similar to the specimen 1 but with slump inches = 3.
109	553	L	1937	297.1		Cinders aggregate; 5	Similar to the specimen 2 but with density 1.632 g cm^{-3} and slump inches = 3.
110	553	L	1937	297.1		Cinders aggregate; 6	Similar to the specimen 3 but with density 1.626 g cm^{-3} and slump inches = 3.
111	553	L	1937	297.1		Haydite aggregate; 1	Mix proportions (by volume): cement 1, fine aggregate 0 to No. 4 2.00, and coarse aggregate No. 4 to 0.5 in. 2.75; 24 x 24 x 2 in.; density 1.405 g cm^{-3}.
112	553	L	1937	297.1		Haydite aggregate; 2	Mix proportions (by volume): cement 1, fine aggregate 0 to No. 4 2.75, and coarse aggregate No. 4 to 0.5 in. 4.50; 24 x 24 x 2 in.; density 1.301 g cm^{-3}.
113	553	L	1937	297.1		Haydite aggregate; 3	Mix proportions (by volume): cement 1, fine aggregate 0 to No. 4 3.50, and coarse aggregate No. 4 to 0.5 in. 5.50; 24 x 24 x 2 in.; density 1.198 g cm^{-3}.
114	553	L	1937	297.1		Haydite aggregate; 4	Similar to specimen 1 but with density 1.424 g cm^{-3} and slump inches = 4.
115	553	L	1937	297.1		Haydite aggregate; 5	Similar to specimen 2 but with density 1.293 g cm^{-3} and slump inches = 4.
116	553	L	1937	297.1		Haydite aggregate; 5	2nd run of above specimen.
117	553	L	1937	297.1		Haydite aggregate; 6	Similar to specimen 3 but with density 1.286 g cm^{-3} and slump inches = 4.
118	553	L	1937	297.1		Haydite aggregate; 7	Mix proportions (by volume): cement 1 and fine aggregate 0 to No. 4 8.50; 24 x 24 x 2 in.; density 1.129 g cm^{-3}.
119	553	L	1937	297.1		Haydite aggregate; 7	2nd run of above specimen.
120	553	L	1937	297.1		Expanded burned clay aggregate	Mix proportions (by volume): cement 1 and fine aggregate 0 to No. 4 8.00; 24 x 24 x 2 in.; density 0.960 g cm^{-3}.
121	553	L	1937	297.1		Treated limestone slag aggregate	Mix proportions (by volume): cement 1 and fine aggregate 0 to No. 4 7.00; 24 x 24 x 2 in.; density 1.241 g cm^{-3}.
122	553	L	1937	297.1			By-product of manufacturing phosphate used as aggregate; mix proportions (by volume): cement 1 and fine aggregate 0 to No. 4 8.00; 24 x 24 x 2 in.; density 1.426 g cm^{-3}.
123	553	L	1937	297.1			Similar to the above specimen but with density 1.496 g cm^{-3}.
124	553	L	1937	297.1			Similar to the above specimen but with density 1.108 g cm^{-3} and pumice as the aggregate.
125	186	L	1928	395-946		Sil-o-cel C-3 concrete	Cured in moist air for 28 days, oven dried at 394 K and fired slowly to 700 K; density 0.923 g cm^{-3}; porosity 65.2%.

SPECIFICATION TABLE NO. 289 (continued)

Curve No.	Ref. No.	Method Used	Year	Temp. Range, K	Reported Error, %	Name and Specimen Designation	Composition (weight percent), Specifications and Remarks
126	583	R	1953	324-525		Barytes concrete	45.15 sweetwater barytes (1 in.), 39.23 sweetwater barytes (3/8 in.), 9.37 Portland cement (type I or II) and 6.25 water; density (25 C) 3.5 g cm^{-3}.
127	583	R	1953	324-525		Barytes concrete	5.7% water; density (25 C) 3.5 g cm^{-3}.
128	583	R	1953	324-525		Barytes concrete	1.4% water; density (25 C) 3.5 g cm^{-3}.
129	583	R	1953	324-525		Barytes concrete	0.75% water; density (25 C) 3.5 g cm^{-3}.
130	583	R	1953	366.2		Portland cement concrete	Density (25 C) 2.3 g cm^{-3}.
131	582		1953	443-607		Lummite cement concrete	1 lummite to 4 vermiculite by dry loose volume; 13.5 x 9.0 x 2.5 in.; prefired at 1366 K on one side for 24 hrs; fired density 0.614 g cm^{-3}; measured at 755 K furnace temp.
132	582		1953	556-824		Lummite cement concrete	The above specimen measured at 1033 K furnace temp.
133	582		1953	707-1088		Lummite cement concrete	The above specimen measured at 1366 K furnace temp.
134	582		1953	456-624		Lummite cement concrete	Same composition and dimensions as the above specimen; fired at 1366 K on all sides for 24 hrs; fired density 0.620 g cm^{-3}; measured at 755 K furnace temp.
135	582		1953	583-844		Lummite cement concrete	The above specimen measured at 1033 K furnace temp.
136	582		1953	750-1121		Lummite cement concrete	The above specimen measured at 1366 K furnace temp.
137	582		1953	447-602		Commercial castable	13.5 x 9.0 x 2.5 in.; prefired at 1255 K on one side for 24 hrs; fired density 1.016 g cm^{-3}; measured at 755 K furnace temp.
138	582		1953	563-818		Commercial castable	The above specimen measured at 1033 K furnace temp.
139	582		1953	719-1082		Commercial castable	The above specimen measured at 1366 K furnace temp.
140	582		1953	461-612		Commercial castable	Same dimensions as the above specimen; prefired at 1255 K on all sides for 24 hrs; fired density 1.016 g cm^{-3}; measured at 755 K furnace temp.
141	582		1953	586-832		Commercial castable	The above specimen measured at 1033 K furnace temp.
142	582		1953	743-1100		Commercial castable	The above specimen measured at 1366 K furnace temp.
143	582		1953	463-615			1 lumnite to 4 vermiculite by dry loose volume; 9.0 x 4.5 x 2.5 in.; cast on end in steel mold so as to obtain smooth faces, cured for 24 hrs in the molds in a moist cabinet or in laboratory air, dried for 18 to 24 hrs at 378 K after removing molds, then prefired for 24 to 72 hrs at 1144 to 1366 K on all sides; dried density 0.708 g cm^{-3}; measured at 755 K furnace temp.
144	582		1953	591-835			The above specimen measured at 1033 K furnace temp.
145	582		1953	743-1099			The above specimen measured at 1366 K furnace temp.
146	582		1953	454-627			1 lumnite to 6 vermiculte by dry loose volume; same preparation procedures and dimensions as the above specimen; dried density 0.586 g cm^{-3} and fired density 0.517 g cm^{-3}; measured at 755 K furnace temp.
147	582		1953	568-848			The above specimen measured at 1033 K furnace temp.

SPECIFICATION TABLE NO. 289 (continued)

Curve No.	Ref. No.	Method Used	Year	Temp. Range, K	Reported Error, %	Name and Specimen Designation	Composition (weight percent), Specifications and Remarks
148	582		1953	711-1094			The above specimen measured at 1366 K furnace temp.
149	582		1953	480-626			1 lumnite to 4 perlite by dry loose volume; same preparation procedures and dimensions as the above specimen; dried density 0.726 g cm^{-3} and fired density 0.673 g cm^{-3}; measured at 755 K furnace temp.
150	582		1953	616-851			The above specimen measured at 1033 K furnace temp.
151	582		1953	728-1035			The above specimen measured at 1255 K furnace temp.
152	582		1953	464-624			1 lumnite to 6 perlite by dry loose volume; same preparation procedures and dimensions as the above specimen; fired density 0.583 g cm^{-3}; measured at 755 K furnace temp.
153	582		1953	590-846			The above specimen measured at 1033 K furnace temp.
154	582		1953	691-1027			The above specimen measured at 1255 K furnace temp.
155	582		1953	478-636			1 lumnite to 8 perlite by dry loose volume; same preparation procedures and dimensions as the above specimen; dired density 0.394 g cm^{-3} and fired density 0.372 g cm^{-3}; measured at 755 K furnace temp.
156	582		1953	615-866			The above specimen measured at 1033 K furnace temp.
157	582		1953	730-1057			The above specimen measured at 1255 K furnace temp.
158	582		1953	467-608			1 lumnite to 4 calcined diatomaceous earth by dry loose volume; same preparation procedures and dimenions as the above specimen; dried density 0.916 g cm^{-3} and fired density 0.847 g cm^{-3}; measured at 755 K furnace temp.
159	582		1953	594-827			The above specimen measured at 1033 K furnace temp.
160	582		1953	699-1006			The above specimen measured at 1255 K furnace temp.
161	582		1953	455-610			1 lumnite to 6 calcined diatomaceous earth by dry loose volume; same preparation procedures and dimensions as the above specimen; dried density 0.795 g cm^{-3} and fired density 0.753 g cm^{-3}; measured at 755 K furnace temp.
162	582		1953	572-828			The above specimen measured at 1033 K furnace temp.
163	582		1953	680-1016			The above specimen measured at 1255 K furnace temp.
164	582		1953	447-595			1 lumnite to 4 pumice by dry loose volume; same preparation procedures and dimensions as the above specimen; dried density 1.245 g cm^{-3} and fired density 1.141 g cm^{-3}; measured at 755 K furnace temp.
165	582		1953	575-819			The above specimen measured at 1033 K furnace temp.
166	582		1953	673-1002			The above specimen measured at 1255 K furnace temp.
167	582		1953	419-588			1 lumnite to 6 pumice by dry loose volume; same preparation procedures and dimensions as the above specimen; dried density 1.147 g cm^{-3} and fired density 1.105 g cm^{-3}; measured at 755 K furnace temp.

SPECIFICATION TABLE NO. 289 (continued)

Curve No.	Ref. No.	Method Used	Year	Temp. Range, K	Reported Error, %	Name and Specimen Designation	Composition (weight percent), Specifications and Remarks
168	582		1953	533-810			The above specimen measured at 1033 K furnace temp.
169	582		1953	607-980			The above specimen measured at 1255 K furnace temp.
170	582		1953	453-580			1 lumnite to 4 haydite by dry loose volume; same preparation procedures and dimensions as the above specimen; dried density 1.555 g cm^{-3} and fired density 1.463 g cm^{-3}; measured at 755 K furnace temp.
171	582		1953	633-833			The above specimen measured at 1033 K furnace temp.
172	582		1953	751-1018			The above specimen measured at 1255 K furnace temp.
173	582		1953	464-597			1 lumnite to 6 haydite by dry loose volume; same preparation procedures and dimensions as the above specimen; dried density 1.440 g cm^{-3} and fired density 1.381 g cm^{-3}; measured at 755 K furnace temp.
174	582		1953	633-846			The above specimen measured at 1033 K furnace temp.
175	582		1953	745-1032			The above specimen measured at 1255 K furnace temp.
176	582		1953	463-596			1 lumnite to 1 vermiculite to 5 haydite by dry loose volume; same preparation procedures and dimensions as the above specimen; dried density 1.376 g cm^{-3} and fired density 1.314 g cm^{-3}; measured at 755 K furnace temp.
177	582		1953	614-839			The above specimen measured at 1033 K furnace temp.
178	582		1953	712-1014			The above specimen measured at 1255 K furnace temp.
179	582		1953	462-601			1 lumnite to 2 vermiculite to 4 haydite by dry loose volume; same preparation procedures and dimensions as the above specimen; dried density 1.256 g cm^{-3} and fired density 1.203 g cm^{-3}; measured at 755 K furnace temp.
180	582		1953	604-835			The above specimen measured at 1033 K furnace temp.
181	582		1953	708-1018			The above specimen measured at 1255 K furnace temp.
182	582		1953	461-609			1 lumnite to 3 vermiculite to 3 haydite by dry loose volume; same preparation procedures and dimensions as the above specimen; dried density 1.044 g cm^{-3} and fired density 0.995 g cm^{-3}; measured at 755 K furnace temp.
183	582		1953	600-841			The above specimen measured at 1033 K furnace temp.
184	582		1953	700-1023			The above specimen measured at 1255 K furnace temp.
185	582		1953	459-610			1 lumnite to 4 vermiculite to 2 haydite by dry loose volume; same preparation procedures and dimensions as the above specimen; dried density 0.938 g cm^{-3} and fired density 0.883 g cm^{-3}; measured at 755 K furnace temp.
186	582		1953	591-841			The above specimen measured at 1033 K furnace temp.
187	582		1953	694-1024			The above specimen measured at 1255 K furnace temp.

SPECIFICATION TABLE NO. 289 (continued)

Curve No.	Ref. No.	Method Used	Year	Temp. Range, K	Reported Error, %	Name and Specimen Designation	Composition (weight percent), Specifications and Remarks
188	582		1953	621.0			1 lumnite to 4 crushed firebrick by dry loose volume; some preparation procedures and dimensions as the above specimen; dried density 1.911 g cm^{-3} and fired density 1.834 g cm^{-3}; measured at 755 K furnace temp.
189	582		1953	859.3			The above specimen measured at 1033 K furnace temp.
190	582		1953	1139.7			The above specimen measured at 1366 K furnace temp.
191	582		1953	624.9			1 lumnite to 6 crushed firebrick by dry loose volume; same preparation procedures and dimensions as the above specimen; dried density 1.876 g cm^{-3} and fired density 1.804 g cm^{-3}; measured at 755 K furnace temp.
192	582		1953	860.5			The above specimen measured at 1033 K furnace temp.
193	582		1953	1140.7			The above specimen measured at 1366 K furnace temp.
194	195	L	1958	327-422			Contained 50 crushed insulating firebrick and 50 aluminous cement; cut from cement-gunned slabs of a cast-and-heat-at-811 K cement having a fired density 1.04 to 1.11 g cm^{-3}, cured at 297 K for 24 hrs with a timed intermittent water spray on the exposed surface during the last 18 hrs, followed by drying in air for a day and 18 hrs at 378 K, then heated for 18 hrs at 533 K; measured in air at 1 atm.
195	195	L	1958	328-427			The above specimen measured in helium at 1 atm.
196	195	L	1958	328-422			The above specimen measured in helium at 200 psi g.
197	195	L	1958	367, 644			The above specimen measured in pure helium.
198	564	L	1944	297-325			Thickness of the specimen 2 in.; density 1.679 g cm^{-3}.
199	564	L	1944	297-325			Thickness of the specimen 2 in.; density 1.922 g cm^{-3}.
200	79	L	1942	428-803		Diatomaceous aggregate concrete	5 parts of diatomaceous aggregate to one part of Portland cement; soaked with water and rammed.
201	79	L	1942	503-753		Sand cement concrete	One part of ordinary sand to three parts of cement.
202	100	L	1957	513-682		Light weight concrete	Bulk density 0.870 g cm^{-3}.
203	100	L	1957	514-682		Light weight concrete	Bulk density 1.012 g cm^{-3}.
204	581		1958	525.5			1 lumnite to 6 vermiculite by dry loose volume; 9.0 x 4.5 x 2.0 in.; placed with a dry gun; cured for 24 hrs in the molds in a moist cabinet or in laboratory air, dried for 18 to 24 hrs at 378 K after removing molds, and then prefired for 24 to 72 hrs at 1144 to 1366 K on all sides; dried density 0.684 g cm^{-3} and fired density 0.605 g cm^{-3}; measured at 755 K furnace temp.
205	581		1958	689.3			The above specimen measured at 1033 K furnace temp.
206	581		1958	820.9			The above specimen measured at 1255 K furnace temp.

SPECIFICATION TABLE NO. 289 (continued)

Curve No.	Ref. No.	Method Used	Year	Temp. Range, K	Reported Error, %	Name and Specimen Designation	Composition (weight percent), Specifications and Remarks
207	581		1958	540.5			1 lumnite to 4 vermiculite by dry loose volume; placed with dry gun; same preparation procedures and dimensions as the above specimen; dried density 0.843 g cm^{-3} and fired density 0.735 g cm^{-3}; measured at 755 K furnace temp.
208	581		1958	709.4			The above specimen measured at 1033 K furnace temp.
209	581		1958	847.2			The above specimen measured at 1255 K furnace temp.
210	581		1958	527.5		Commercial castable	Specimen cast on end in steel mold so as to obtain homogeneous surface; same preparation procedures and dimensions as the above specimen; fired density 0.755 g cm^{-3}; measured at 755 K furnace temp.
211	581		1958	693.8		Commercial castable	The above specimen measured at 1033 K furnace temp.
212	581		1958	823.5		Commercial castable	The above specimen measured at 1255 K furnace temp.
213	581		1958	528.0			1 lumnite to 6 perlite by dry loose volume; cast on end in steel mold so as to obtain homogeneous surface; same preparation procedures and dimensions as the above specimen; dried density 0.865 g cm^{-3} and fired density 0.823 g cm^{-3}; measured at 755 K furnace temp.
214	581		1958	699.3			The above specimen measured at 1033 K furnace temp.
215	581		1958	838.0			The above specimen measured at 1255 K furnace temp.
216	581		1958	542.9			The specimen was placed with dry gun; same preparation procedures and dimensions as the above specimen; dried density 1.008 g cm^{-3} and fired density 0.879 g cm^{-3}; measured at 755 K furnace temp.
217	581		1958	717.7			The above specimen measured at 1033 K furnace temp.
218	581		1958	857.0			The above specimen measured at 1255 K furnace temp.
219	581		1958	538.9			1 lumnite to 4 perlite by dry loose volume; cast on end in steel mold so as to obtain homogeneous surface; same preparation procedures and dimensions as the above specimen; dried density 0.980 g cm^{-3}, fired density 0.907 g cm^{-3}; measured at 755 K furnace temp.
220	581		1958	718.5			The above specimen measured at 1033 K furnace temp.
221	581		1958	865.7			The above specimen measured at 1255 K furnace temp.
222	581		1958	538.7			The specimen was cast on end in steel mold so as to obtain homogeneous surface; same preparation procedures and dimensions as the above specimens; fired density 1.012 g cm^{-3}; measured at 755 K furnace temp.
223	581		1958	712.2			The above specimen measured at 1033 K furnace temp.
224	581		1958	845.6			The above specimen measured at 1255 K furnace temp.

SPECIFICATION TABLE NO. 289 (continued)

Curve No.	Ref. No.	Method Used	Year	Temp. Range, K	Reported Error, %	Name and Specimen Designation	Composition (weight percent), Specifications and Remarks
225	581		1958	604.0			1 lumnite to 1 topaz to 1 CFB to 4 calcined diatomaceous earth by weight; cast on end in steel mold so as to obtain homogeneous surface; same preparation procedures and dimensions as the above specimen; fired density 1.177 g cm^{-3}; measured at 755 K furnace temp.
226	581		1958	824.9			The above specimen measured at 1033 K furnace temp.
227	581		1958	1082.1			The above specimen measured at 1255 K furnace temp.
228	581		1958	543.3			Aerated 1 lumnite to 4 expanded shale; cast on end in steel mold so as to obtain homogeneous surface; same preparation procedures and dimensions as the above specimen; dried density 1.265 g cm^{-3} and fired density 1.201 g cm^{-3}; measured at 755 K furnace temp.
299	581		1958	733.7			The above specimen measured at 1033 K furnace temp.
230	581		1958	872.9			The above specimen measured at 1255 K furnace temp.
231	581		1958	551.4		Commercial castable	Specimen placed with dry gun; same preparation procedures and dimensions as the above specimen; dried density 1.309 g cm^{-3} and fired density 1.224 g cm^{-3}; measured at 755 K furnace temp.
232	581		1958	739.2		Commercial castable	The above specimen measured at 1033 K furnace temp.
233	581		1958	878.8		Commercial castable	The above specimen measured at 1255 K furnace temp.
234	581		1958	541.2			1 lumnite to 3 vermiculite to 2 expanded shale; placed with wet gun; same preparation procedures and dimensions as the above specimen; dried density 1.322 g cm^{-3} and fired density 1.233 g cm^{-3}; measured at 755 K furnace temp.
235	581		1958	723.2			The above specimen measured at 1033 K furnace temp.
236	581		1958	862.9			The above specimen measured at 1255 K furnace temp.
237	581		1958	548.7			1 lumnite to 4 expanded shale (-8 M to Dust); cast on end in steel mold so as to obtain homogeneous surface; same preparation procedures and dimensions as the above specimen; fired density 1.374 g cm^{-3}; measured at 755 K furnace temp.
238	581		1958	744.4			The above specimen measured at 1033 K furnace temp.
239	581		1958	889.9			The above specimen measured at 1255 K furnace temp.
240	581		1958	535.9			1 lumnite to 2.4 "A" expanded shale to 3.0 "C" expanded shale; same preparation procedures and dimensions as the above specimen; fired density 1.480 g cm^{-3}; measured at 755 K furnace temp.
241	581		1958	732.6			The above specimen measured at 1033 K furnace temp.
242	581		1958	875.8			The above specimen measured at 1255 K furnace temp.
243	581		1958	535.7			Aerated 1 lumnite to 4 CFB; same preparation procedures and dimensions as the above specimen; dried density 1.567 g cm^{-3} and fired density 1.509 g cm^{-3}; measured at 755K furnace temp.
244	581		1958	723.0			The above specimen measured at 1033 K furnace temp.

Curve No.	Ref. No.	Method Used	Year	Temp. Range, K	Reported Error, %	Name and Specimen Designation	Composition (weight percent), Specifications and Remarks
245	581		1958	876.8			The above specimen measured at 1255 K furnace temp.
246	581		1958	555.1			1 lumnite to 4 expanded shale; placed with dry gun; same preparation procedures and dimensions as the above specimen; dried density 1.591 g cm^{-3} and fired density 1.525 g cm^{-3}; measured at 755 K furnace temp.
247	581		1958	741.4			The above specimen measured at 1033 K furnace temp.
248	581		1958	883.8			The above specimen measured at 1255 K furnace temp.
249	581		1958	618.1			1 lumnite to 3 expanded shale; placed with dry gun; same preparation procedures and dimensions as the above specimen; fired density 1.535 g cm^{-3}; measured at 755 K furnace temp.
250	581		1958	822.4			The above specimen measured at 1033 K furnace temp.
251	581		1958	1013.1			The above specimen measured at 1255 K furnace temp.
252	581		1958	533.0			1 lumnite to 1 CFB to 3 expanded shale; cast on end in steel mold so as to obtain homogeneous surface; same preparation procedures and dimensions as the above specimen; fired density 1.569 g cm^{-3}; measured at 755 K furnace temp.
253	581		1958	728.9			The above specimen measured at 1033 K furnace temp.
254	581		1958	871.9			The above specimen measured at 1255 K furnace temp.
255	581		1958	538.6			Vibrated 1 lumnite to 4 expanded shale; cast on end in steel mold so as to obtain homogeneous surface; same preparation procedures and dimensions as the above specimen; dried density 1.655 g cm^{-3}; fired density 1.570 g cm^{-3}; measured at 755 K furnace temp.
256	581		1958	741.9			The above specimen measured at 1033 K furnace temp.
257	581		1958	888.5			The above specimen measured at 1255 K furnace temp.
258	581		1958	537.3			1 lumnite to 2 CFR to 2 expanded shale; cast on end in steel mold so as to obtain homogeneous surface; same preparation procedures and dimensions as the above specimen; fired density 1.656 g cm^{-3}; measured at 755 K furnace temp.
259	581		1958	731.9			The above specimen measured at 1033 K furnace temp.
260	581		1958	878.4			The above specimen measured at 1255 K furnace temp.
261	581		1958	531.1			1 lumnite to 3 CFB to 1 expanded shale; cast on end in steel mold so as to obtain homogeneous surface; same preparation procedures and dimensions as the above specimen; fired density 1.762 g cm^{-3}; measured at 755 K furnace temp.
262	581		1958	725.1			The above specimen measured at 1033 K furnace temp.
263	581		1958	871.1			The above specimen measured at 1255 K furnace temp.
264	581		1958	635.5		Commercial castable	The specimen was cast on end in steel mold so as to obtain homogeneous surface; same preparation procedures and dimensions as the above specimen; fired density 1.788 g cm^{-3}; measured at 755 K furnace temperature.

SPECIFICATION TABLE NO. 289 (continued)

Curve No.	Ref. No.	Method Used	Year	Temp. Range, K	Reported Error, %	Name and Specimen Designation	Composition (weight percent), Specifications and Remarks
265	581		1958	868.6		Commercial castable	The above specimen measured at 1033 K furnace temp.
266	581		1958	1143.2		Commercial castable	The above specimen measured at 1255 K furnace temp.
267	581		1958	623.5			1 lumnite to 4 CFB (stiff mud); cast on end in steel mold so as to obtain homogeneous surface; same preparation procedures and dimensions as the above specimen; dried density 2.002 g cm^{-3}; fired density 1.922 g cm^{-3}; measured at 755 K furnace temp.
268	581		1958	860.2			The above specimen measured at 1033 K furnace temp.
269	581		1958	1145.6			The above specimen measured at 1255 K furnace temp.
270	581		1958	630.0			Vibrated 1 lumnite to 4 CFB (stiff mud); cast on end in steel mold so as to obtain homogeneous surface; same preparation procedures and dimensions as the above specimen; dried density 2.046 g cm^{-3}; fired density 1.972 g cm^{-3}; measured at 755 K furnace temp.
271	581		1958	862.9			The above specimen measured at 1033 K furnace temp.
272	581		1958	1146.7			The above specimen measured at 1255 K furnace temp.
273	580		1955	296.5			Made from granulated blast furnace slag and slag cement; dry density 1.750 g cm^{-3}.
274	580		1955	303.9			Made from granulated blast furnace slag and Portland cement; water binding ratio 55%; dry density 1.840 g cm^{-3}.
275	580		1955	311.7			Similar to the above specimen except water binding ratio 63%; dry density 1.920 g cm^{-3}.
276	580		1955	310.0			Made from boiler slag and slag cement; dry density 1.720 g cm^{-3}.
277	580		1955	312.4			Slag concrete from clay coagulation binding; dry density 1.850 g cm^{-3}.
278	580		1955	309.1			Slag concrete from tripolite clay binding; dry density 1.850 g cm^{-3}.
279	580		1955	308.2			Activated blast-furnace slag mixed with crushed grainite; density 2.270 g cm^{-3}.
280	580		1955	303, 313			Activated blast-furnace slag mixed with crushed chips from solid waste of blast-furnace slag; density from 2.170-2.360 g cm^{-3}.
281	580		1955	308.2			Activated blast-furnace slag mixed with powdered rubble from blast-furnace slag; density 1.820 g cm^{-3}.
282	580		1955	305.2			Similar to the above specimen except density 1.520 g cm^{-3}.
283	579		1958	298.2		Expanded slag concrete; A	0.375 in. to 0. aggregate and 4.08 sacks · yd^{-3} cement factor; moist-cured for 28 days, oven-dried and ground both sides, then re-dried in the oven prior to test; air content 19.5%; oven dry weight 1.394 g cm^{-3}.
284	579		1958	298.2		Expanded slag concrete; A	0.375 in. to 0 aggregate and 5.51 sacks · yd^{-3} cement factor; same curing and drying process as the above specimen; air content 23.0%; oven dry weight 1.408 g cm^{-3}.

SPECIFICATION TABLE NO. 289 (continued)

Curve No.	Ref. No.	Method Used	Year	Temp. Range, K	Reported Error, %	Name and Specimen Designation	Composition (weight percent), Specifications and Remarks
285	579		1958	298.2		Expanded slag concrete; A	0.5 in. aggregate and 7.14 sacks · yd^{-3} cement factor; same curing and drying process as the above specimen; air content 13.1%; oven dry weight 1.575 g cm^{-3}.
286	579		1958	298.2		Expanded slag concrete; A	0.75 in. aggregate and 7.36 sacks · yd^{-3} cement factor; same curing and drying process and the above specimen; air content 15.8%; oven dry weight 1.525 g cm^{-3}.
287	579		1958	298.2		Expanded slag concrete; A	0.375 in. to 0 aggregate and 7.41 sacks · yd^{-3} cement factor; same curing and drying process as the above specimen; air content 14.3%; oven dry weight 1.543 g cm^{-3}.
288	579		1958	298.2		Expanded slag concrete; A	0.375 in. to 0 aggregate and 8.69 sacks · yd^{-3} cement factor; same curing and drying process as the above specimen; air content 8.7%; oven dry weight 1.656 g cm^{-3}.
289	579		1958	298.2		Expanded slag concrete; B	0.375 in. to 0 aggregate and 7.01 sacks · yd^{-3} cement factor; same curing and drying process as the above specimen; air content 14.5%; oven dry weight 1.373 g cm^{-3}.
290	579		1958	298.2		Expanded slag concrete; B	0.375 in. to 0 aggregate and 7.00 sacks · yd^{-3} cement factor; 10% natural sand by dry loose volume of fine aggregate; same curing and drying process as the above specimen; air content 13.5%; oven dry weight 1.410 g cm^{-3}.
291	579		1958	298.2		Expanded slag concrete; B	0.375 in. to 0 aggregate and 7.22 sacks · yd^{-3} cement factor; 20% natural sand by dry loose volume of fine aggregate; same curing and drying process as the above specimen; air content 7.4%; oven dry weight 1.538 g cm^{-3}.
292	579		1958	298.2		Expanded slag concrete; B	0.375 in. to 0 aggregate and 7.15 sacks · yd^{-3} cement factor; 30% natural sand by dry loose volume of fine aggregate; same curing and drying process as the above specimen; air content 8.7%; oven dry weight 1.583 g cm^{-3}.
293	579		1958	298.2		Expanded slag concrete; C	0.375 in. to 0 aggregate and 3.92 sacks · yd^{-3} cement factor; same curing and drying process as the above specimen; air content 25.8%; oven dry weight 0.961 g cm^{-3}.
294	579		1958	298.2		Expanded slag concrete; C	0.375 in. to 0 aggregate and 5.43 sacks · yd^{-3} cement factor; same curing and drying process as the above specimen; air content 21.4%; oven dry weight 1.089 g cm^{-3}.
295	579		1958	298.2		Expanded slag concrete; C	0.375 in. to 0 aggregate and 7.10 sacks · yd^{-3} cement factor; same curing and drying process as the above specimen; air content 10.5%; oven dry weight 1.278 g cm^{-3}.
296	579		1958	298.2		Expanded slag concrete; C	0.375 in. to 0 aggregate and 8.41 sacks · yd^{-3} cement factor; same curing and drying process as the above specimen; air content 16.4%; oven dry weight 1.171 g cm^{-3}.
297	579		1958	298.2		Expanded slag concrete; D	0.375 in. to 0 aggregate and 5.86 sacks · yd^{-3} cement factor; same curing and drying process as the above specimen; air content 26.4%; oven dry weight 1.280 g cm^{-3}.
298	579		1958	298.2		Expanded slag concrete; D	0.375 in. to 0 aggregate and 7.50 sacks · yd^{-3} cement factor; same curing and drying process as the above specimen; air content 25%; oven dry weight 1.378 g cm^{-3}.
299	579		1958	298.2		Expanded slag concrete; D	0.375 in. to 0 aggregate and 8.85 sacks · yd^{-3} cement factor; same curing and drying process as the above specimen; air content 18.0%; oven dry weight 1.563 g cm^{-3}.

SPECIFICATION TABLE NO. 289 (continued)

Curve No.	Ref. No.	Method Used	Year	Temp. Range, K	Reported Error, %	Name and Specimen Designation	Composition (weight percent), Specifications and Remarks
300	121	R	1921	348-523			Made with 1 ft^3 "Universal" Portland cement and 0.384 ft^3 of water; the concrete was set for 24 hrs, stored in damp sand for two weeks and then thoroughly dried in a dry room.
301	121	R	1921	348-523			Ratio of Portland cement: sand : gravel = 1:1.2:1.1 (in vol); 100% water content (normal); same preparation procedures as above.
302	121	R	1921	348-523			Similar to the above specimen except 110% water content.
303	121	R	1921	373, 473			Similar to the above specimen except 120% water content.
304	121	R	1921	348-523			Ratio of cement: sand : gravel = 1:1.9:1.7; 100% water content; same preparation procedures as above.
305	121	R	1921	348, 423			Similar to the above specimen except 110% water content.
306	121	R	1921	348-523			Similar to the above specimen except 120% water content.
307	121	R	1921	373, 473			Ratio of cement: sand : gravel = 1:2.4:2.3; 100% water content; same preparation procedures as above.
308	121	R	1921	348-523			Similar to the above specimen except 110% water content.
309	121	R	1921	348, 423			Similar to the above specimen except 120% water content.
310	121	R	1921	373, 473			Ratio of cement: sand : gravel = 1:3.1:3.0; 100% water content; same preparation procedures as above.
311	121	R	1921	348-523			Similar to the above specimen except 110% water content.
312	121	R	1921	348-523			Similar to the above specimen except 120% water content.
313	121	R	1921	348, 423			Ratio of cement: sand : gravel = 1:4.3:4.0; 110% water content; same preparation procedures as above.
314	121	R	1921	373, 473			Similar to the above specimen except 120% water content.
315	121	R	1921	348, 423			Ratio of cement: sand : gravel = 1:5.6:5.1; 110% water content; same preparation procedures as above.
316	121	R	1921	373, 473			Similar to the above specimen except 120% water content.
317	552	L	1960	298.2		Slag concrete	20 x 20 x 4 cm; mixed by hand, filled in prescribed form, tempted, then normal heated with seven days of damp storage and subsequent air storaged.
318	552	L	1960	298.2		Slag concrete	Similar to the above specimen except density 1.49 g cm^{-3}.
319	552	L	1960	298.2		Slag concrete	Similar to the above specimen except density 1.52 g cm^{-3}.
320	552	L	1960	298.2		Slag concrete	Similar to the above specimen except density 1.70 g cm^{-3}.
321	552	L	1960	298.2		Slag concrete	Similar to the above specimen except density 1.79 g cm^{-3}.

SPECIFICATION TABLE NO. 289 (continued)

Curve No.	Ref. No.	Method Used	Year	Temp. Range, K	Reported Error, %	Name and Specimen Designation	Composition (weight percent), Specifications and Remarks
322	552	L	1960	298.2		Slag concrete	Similar to the above specimen except density 2.02 g cm^{-3}.
323	552	L	1960	298.2		Slag concrete	Similar to the above specimen except density 1.58 g cm^{-3}.
324	552	L	1960	298.2		Slag concrete	Similar to the above specimen except density 1.59 g cm^{-3}.
325	552	L	1960	298.2		Slag concrete	Similar to the above specimen except density 1.68 g cm^{-3}.
326	552	L	1960	298.2		Slag concrete	Similar to the above specimen except density 1.80 g cm^{-3}.
327	552	L	1960	298.2		Slag concrete	Similar to the above specimen except density 1.66 g cm^{-3}.
328	552	L	1960	298.2		Slag concrete	Similar to the above specimen except density 1.73 g cm^{-3}.
329	552	L	1960	298.2		Slag concrete	Similar to the above specimen except density 1.30 g cm^{-3}.
330	552	L	1960	298.2		Slag concrete	Similar to the above specimen except density 1.22 g cm^{-3}.
331	552	L	1960	298.2		Slag concrete	Similar to the above specimen except density 2.00 g cm^{-3}.
332	552	L	1960	298.2		Slag concrete	Similar to the above specimen except density 1.83 g cm^{-3}.
333	552	L	1960	298.2		Slag concrete	Similar to the above specimen except density 1.80 g cm^{-3}.
334	552	L	1960	298.2		Slag concrete	Similar to the above specimen except density 1.77 g cm^{-3}.
335	552	L	1960	298.2		Slag concrete	Similar to the above specimen except density 1.70 g cm^{-3}.
336	552	L	1960	298.2		Slag concrete	Similar to the above specimen except density 1.70 g cm^{-3}.
337	529	P	1956	291.5		Foamed lightweight concrete	12 in. cube; water content 12.0%; density 0.384 g cm^{-3}; thermal conductivity value calculated from measured transient temp change.
338	529	P	1956	291.5		Foamed lightweight concrete	The above specimen dried; density 0.336 g cm^{-3}.

DATA TABLE NO. 289 THERMAL CONDUCTIVITY OF CONCRETE

[Temperature, T,K; Thermal Conductivity, k, Watt $cm^{-1}K^{-1}$]

T	k	T	k	T	k	T	k	T	k	T	k	T	k	T	k
CURVE 1		CURVE 12*		CURVE 23*		CURVE 34*		CURVE 45		CURVE 56*		CURVE 67		CURVE 78*	
292.8	0.0171	293.6	0.0175	298.2	0.00395	298.2	0.00186	298.2	0.00302	298.2	0.00256	298.2	0.00221	298.2	0.00267
CURVE 2		CURVE 13*		CURVE 24*		CURVE 35		CURVE 46*		CURVE 57*		CURVE 68*		CURVE 79*	
356.1	0.0171	287.3	0.00542	298.2	0.00407	298.2	0.00163	298.2	0.00325	298.2	0.00209	298.2	0.00291	298.2	0.00267
CURVE 3*		CURVE 14		CURVE 25		CURVE 36		CURVE 47*		CURVE 58*		CURVE 69*		CURVE 80*	
356.2	0.0171	287.3	0.0120	298.2	0.00430	298.2	0.00151	298.2	0.00256	298.2	0.00314	298.2	0.00267	298.2	0.00337
CURVE 4*		CURVE 15*		CURVE 26*		CURVE 37		CURVE 48*		CURVE 59*		CURVE 70*		CURVE 81*	
292.8	0.0172	287.3	0.0119	298.2	0.00395	298.2	0.00174	298.2	0.00244	298.2	0.00232	298.2	0.00349	298.2	0.00256
CURVE 5*		CURVE 16		CURVE 27		CURVE 38*		CURVE 49*		CURVE 60*		CURVE 71*		CURVE 82*	
376.5	0.0173	298.2	0.00256	298.2	0.00372	298.2	0.00174	298.2	0.00232	298.2	0.00209	298.2	0.00291	298.2	0.00209
CURVE 6*		CURVE 17*		CURVE 28		CURVE 39		CURVE 50*		CURVE 61*		CURVE 72*		CURVE 83	
376.6	0.0174	298.2	0.00244	298.2	0.00325	298.2	0.00604	298.2	0.00267	298.2	0.00232	298.2	0.00279	298.2	0.00477
CURVE 7*		CURVE 18		CURVE 29*		CURVE 40		CURVE 51*		CURVE 62*		CURVE 73*		CURVE 84*	
292.8	0.0175	298.2	0.00232	298.2	0.00325	298.2	0.00651	298.2	0.00325	298.2	0.00314	298.2	0.00430	298.2	0.00395
CURVE 8*		CURVE 19		CURVE 30		CURVE 41*		CURVE 52*		CURVE 63*		CURVE 74*		CURVE 85*	
292.9	0.0173	298.2	0.00209	298.2	0.00349	298.2	0.00581	298.2	0.00256	298.2	0.00198	298.2	0.00407	298.2	0.00407
CURVE 9*		CURVE 20		CURVE 31*		CURVE 42*		CURVE 53*		CURVE 64*		CURVE 75*		CURVE 86*	
292.7	0.0174	298.2	0.00198	298.2	0.00267	298.2	0.00616	298.2	0.00291	298.2	0.00291	298.2	0.00349	298.2	0.00395
CURVE 10*		CURVE 21*		CURVE 32*		CURVE 43*		CURVE 54*		CURVE 65*		CURVE 76*		CURVE 87*	
293.3	0.0174	298.2	0.00395	298.2	0.00267	298.2	0.00593	298.2	0.00349	298.2	0.00232	298.2	0.00349	298.2	0.00325
CURVE 11*		CURVE 22		CURVE 33		CURVE 44		CURVE 55*		CURVE 66*		CURVE 77*		CURVE 88*	
293.0	0.0175	298.2	0.00453	298.2	0.00186	298.2	0.00279	298.2	0.00314	298.2	0.00232	298.2	0.00314	298.2	0.00198

*Not shown on plot

DATA TABLE NO. 289 (continued)

T	k
CURVE 89*	
298.2	0.00244
CURVE 90*	
297.1	0.0189
CURVE 91*	
297.1	0.0186
CURVE 92*	
297.1	0.0190
CURVE 93*	
297.1	0.0175
CURVE 94*	
297.1	0.0179
CURVE 95*	
297.1	0.0179
CURVE 96*	
297.1	0.0175
CURVE 97*	
297.1	0.0185
CURVE 98*	
297.1	0.0180
CURVE 99*	
297.1	0.0162
CURVE 100*	
297.1	0.0173
CURVE 101*	
297.1	0.0166
CURVE 102*	
297.1	0.0151
CURVE 103*	
297.1	0.0144
CURVE 104	
297.1	0.0141
CURVE 105*	
297.1	0.00668
CURVE 106*	
297.1	0.00620
CURVE 107*	
297.1	0.00538
CURVE 108	
297.1	0.00705
CURVE 109*	
297.1	0.00632
CURVE 110*	
297.1	0.00612
CURVE 111*	
297.1	0.00599
CURVE 112*	
297.1	0.00545
CURVE 113	
297.1	0.00529
CURVE 114*	
297.1	0.00632
CURVE 115*	
297.1	0.00561
CURVE 116*	
297.1	0.00557
CURVE 117*	
297.1	0.00577
CURVE 118*	
297.1	0.00417
CURVE 119*	
297.1	0.00407
CURVE 120*	
297.1	0.00329
CURVE 121*	
297.1	0.00327
CURVE 122*	
297.1	0.00460
CURVE 123*	
297.1	0.00493
CURVE 124*	
297.1	0.00349
CURVE 125	
395	0.00260
451	0.00278
456	0.00277*
515	0.00299
547	0.00294
605	0.00297
648	0.00313
758	0.00330
946	0.00372
CURVE 126	
323.6	0.0190
373.2	0.0189
424.4	0.0180
474.8	0.0164
525.2	0.0140
CURVE 127	
323.6	0.0160
373.2	0.0173*
399.2	0.0177
424.4	0.0151
474.8	0.0150
525.2	0.0129
CURVE 128	
323.6	0.0160*
373.2	0.0123
424.4	0.0118
474.8	0.0110
525.2	0.0104
CURVE 129	
323.6	0.0139
373.2	0.0112
424.4	0.0106
474.8	0.0102
525.2	0.0100
CURVE 130*	
366.2	0.01731
CURVE 131	
442.5	0.00136
521.8	0.00140
607.1	0.00145
CURVE 132	
555.7	0.00149
683.3	0.00157
823.7	0.00164
CURVE 133	
706.6	0.00160
883.4	0.00171
1087.7	0.00184
CURVE 134	
456.0	0.00121
535.0	0.00128
623.6	0.00136
CURVE 135	
583.1	0.00133
704.0	0.00142
844.2	0.00154
CURVE 136*	
750.4	0.00147
919.1	0.00160
1120.9	0.00176
CURVE 137*	
446.9	0.00175
522.7	0.00179
601.8	0.00183
CURVE 138	
563.1	0.00187
687.5	0.00191
817.6	0.00196
CURVE 139	
718.7	0.00197
894.2	0.00203
1081.5	0.00210
CURVE 140*	
460.6	0.00166
532.2	0.00175
612.1	0.00185
CURVE 141*	
585.9	0.00177
702.3	0.00186
832.1	0.00197
CURVE 142*	
742.5	0.00187
909.3	0.00199
1100.3	0.00214
CURVE 143*	
462.5	0.00137
531.5	0.00152
614.5	0.00168
CURVE 144*	
590.6	0.00152
702.8	0.00168
835.4	0.00184
CURVE 145*	
742.8	0.00173
904.6	0.00189
1099.1	0.00210
CURVE 146*	
453.7	0.00112
531.9	0.00123
626.8	0.00135
CURVE 147*	
568.2	0.00122
694.7	0.00134
847.6	0.00148
CURVE 148*	
711.0	0.00137
871.2	0.00160
1094.3	0.00192
CURVE 149	
480.4	0.00148
545.4	0.00160
626.0	0.00176
CURVE 150*	
616.1	0.00163
719.2	0.00179
850.5	0.00199
CURVE 151*	
728.3	0.00175
860.8	0.00195
1035.0	0.00221
CURVE 152*	
463.8	0.00131
539.4	0.00142
624.0	0.00154
CURVE 153*	
589.6	0.00147
707.9	0.00162
846.2	0.00180
CURVE 154*	
691.3	0.00157
841.8	0.00178
1027.1	0.00203
CURVE 155	
477.5	0.00105
552.6	0.00115
635.6	0.00126
CURVE 156*	
615.1	0.00127
729.9	0.00144
866.2	0.00163
CURVE 157*	
730.2	0.00145
873.0	0.00170
1056.5	0.00202
CURVE 158	
467.4	0.00212
537.1	0.00216
607.5	0.00221
CURVE 159	
593.9	0.00228
710.1	0.00232
827.0	0.00237
CURVE 160*	
698.8	0.00234
849.3	0.00242
1005.6	0.00250
CURVE 161*	
454.6	0.00188
527.7	0.00196
610.3	0.00207
CURVE 162*	
571.5	0.00202
692.5	0.00211
828.3	0.00220

* Not shown on plot

DATA TABLE NO. 289 (continued)

T	k
CURVE 163*	
679.6	0.00211
837.1	0.00222
1016.1	0.00233
CURVE 164	
447.1	0.00314
524.3	0.00341
594.6	0.00365
CURVE 165	
574.9	0.00338
701.3	0.00371
819.3	0.00402
CURVE 166	
672.9	0.00348
839.2	0.00392
1002.0	0.00435
CURVE 167*	
419.3	0.00283
529.2	0.00304
588.0	0.00316
CURVE 168*	
532.8	0.00307
711.0	0.00338
809.5	0.00355
CURVE 169*	
606.9	0.00322
841.4	0.00368
979.7	0.00395
CURVE 170	
452.9	0.00477
518.8	0.00479
580.1	0.00481
CURVE 171	
633.0	0.00487
736.5	0.00489
832.8	0.00492
CURVE 172*	
751.1	0.00491
887.3	0.00502
1018.3	0.00512
CURVE 173*	
463.7	0.00437
529.3	0.00442
596.8	0.00449
CURVE 174*	
632.7	0.00451
737.9	0.00458
846.3	0.00464
CURVE 175	
744.7	0.00460
883.8	0.00475
1031.6	0.00489
CURVE 176*	
463.2	0.00377
526.1	0.00397
596.2	0.00420
CURVE 177*	
614.0	0.00384
718.7	0.00411
839.0	0.00441
CURVE 178*	
711.9	0.00402
853.3	0.00427
1013.8	0.00456
CURVE 179*	
462.0	0.00294
530.2	0.00301
601.3	0.00307
CURVE 180*	
603.6	0.00305
713.4	0.00321
835.1	0.00338
CURVE 181*	
708.3	0.00316
852.2	0.00337
1017.9	0.00363
CURVE 182*	
461.4	0.00225
528.8	0.00247
609.1	0.00275
CURVE 183*	
599.5	0.00244
705.4	0.00272
840.7	0.00306
CURVE 184	
699.9	0.00253
839.6	0.00284
1023.3	0.00326
CURVE 185*	
458.7	0.00192
528.6	0.00209
610.4	0.00229
CURVE 186*	
591.4	0.00204
704.8	0.00224
840.8	0.00250
CURVE 187*	
694.3	0.00213
841.7	0.00237
1023.8	0.00268
CURVE 188*	
621.0	0.00788
CURVE 189*	
859.3	0.00824
CURVE 190	
1139.7	0.00856
CURVE 191*	
624.9	0.00783
CURVE 192*	
860.5	0.00821
CURVE 193*	
1140.7	0.00859
CURVE 194*	
326.5	0.00277
363.7	0.00280
422.1	0.00326
CURVE 195*	
327.6	0.00306
370.4	0.00320
427.1	0.00326
CURVE 196*	
327.6	0.00453
370.4	0.00506
422.1	0.00513
CURVE 197*	
366.5	0.00483
644.3	0.00620
CURVE 198*	
297.3	0.00541
311.3	0.00550
324.8	0.00547
CURVE 199*	
297.3	0.00702
311.4	0.00711
311.7	0.00711
325.0	0.00711
CURVE 200*	
428.2	0.00188
553.2	0.00201
678.2	0.00213
803.2	0.00242
CURVE 201*	
503.2	0.00105
753.2	0.00130
CURVE 202*	
512.6	0.00199
558.2	0.00202
609.3	0.00205
682.1	0.00208
CURVE 203*	
514.3	0.00218
568.7	0.00219
624.3	0.00221
682.1	0.00222
CURVE 204*	
525.5	0.00149
CURVE 205*	
689.3	0.00159
CURVE 206*	
820.9	0.00167
CURVE 207*	
540.5	0.00175
CURVE 208*	
709.4	0.00180
CURVE 209*	
847.2	0.00190
CURVE 210*	
527.5	0.00167
CURVE 211*	
693.8	0.00186
CURVE 212*	
823.5	0.00200
CURVE 213*	
528	0.00238
CURVE 214*	
699.3	0.00262
CURVE 215*	
838.0	0.00284
CURVE 216*	
542.9	0.00224
CURVE 217*	
717.7	0.00239
CURVE 218*	
857.0	0.00258
CURVE 219*	
538.9	0.00265
CURVE 220*	
718.5	0.00299
CURVE 221*	
865.7	0.00323
CURVE 222*	
538.7	0.00228
CURVE 223*	
712.2	0.00255
CURVE 224*	
845.6	0.00262
CURVE 225*	
604.0	0.00394
CURVE 226*	
824.9	0.00418
CURVE 227*	
1082.1	0.00453
CURVE 228*	
543.3	0.00349
CURVE 229*	
733.7	0.00368
CURVE 230*	
872.9	0.00388
CURVE 231*	
551.4	0.00385
CURVE 232*	
739.2	0.00392
CURVE 233*	
878.8	0.00412
CURVE 234*	
541.2	0.00327
CURVE 235*	
723.2	0.00333
CURVE 236*	
862.9	0.00339
CURVE 237*	
548.7	0.00391
CURVE 238*	
744.4	0.00412
CURVE 239*	
889.9	0.00427
CURVE 240*	
535.9	0.00531

* Not shown on plot

DATA TABLE NO. 289 (continued)

T	k
CURVE 241	
732.6	0.00552
CURVE 242	
875.8	0.00568
CURVE 243*	
535.7	0.00529
CURVE 244*	
723.0	0.00551
CURVE 245*	
876.8	0.00570
CURVE 246*	
555.1	0.00531
CURVE 247*	
741.4	0.00548
CURVE 248*	
883.8	0.00555
CURVE 249*	
618.1	0.00453
CURVE 250*	
822.4	0.00467
CURVE 251*	
1013.1	0.00482
CURVE 252*	
533.0	0.00557
CURVE 253*	
728.9	0.00574
CURVE 254*	
871.9	0.00584
CURVE 255*	
538.6	0.00573
CURVE 256*	
741.9	0.00571
CURVE 257*	
888.5	0.00580
CURVE 258*	
537.3	0.00633
CURVE 259	
731.9	0.00648
CURVE 260	
878.4	0.00668
CURVE 261*	
531.1	0.00679
CURVE 262*	
725.1	0.00708
CURVE 263	
871.1	0.00734
CURVE 264*	
635.5	0.00787
CURVE 265*	
868.6	0.00806
CURVE 266*	
1143.2	0.00845
CURVE 267	
623.5	0.00890
CURVE 268	
860.2	0.00907
CURVE 269	
1145.6	0.00933
CURVE 270*	
630.0	0.00909
CURVE 271*	
862.9	0.00929
CURVE 272*	
1146.7	0.00955
CURVE 273*	
296.5	0.00442
CURVE 274*	
303.9	0.00581
CURVE 275*	
311.7	0.00477
CURVE 276*	
310.0	0.00674
CURVE 277*	
312.4	0.00616
CURVE 278*	
309.1	0.00477
CURVE 279	
308.2	0.00860
CURVE 280*	
303.2	0.00721
313.2	0.00686
CURVE 281*	
308.2	0.00546
CURVE 282*	
305.2	0.00407
CURVE 283*	
298.2	0.00342
CURVE 284*	
298.2	0.00348
CURVE 285*	
298.2	0.00420
CURVE 286*	
298.2	0.00408
CURVE 287*	
298.2	0.00395
CURVE 288*	
298.2	0.00460
CURVE 289*	
298.2	0.00346
CURVE 290*	
298.2	0.00376
CURVE 291*	
298.2	0.00444
CURVE 292*	
298.2	0.00436
CURVE 293*	
298.2	0.00216
CURVE 294*	
298.2	0.00251
CURVE 295*	
298.2	0.00325
CURVE 296*	
298.2	0.00313
CURVE 297*	
298.2	0.00306
CURVE 298*	
298.2	0.00352
CURVE 299*	
298.2	0.00408
CURVE 300	
348.2	0.00586
423.2	0.00682
523.2	0.00586
CURVE 301*	
348.2	0.0153
423.2	0.0135
523.2	0.0134
CURVE 302*	
348.2	0.0126
423.2	0.0139
523.2	0.0130
CURVE 303*	
373.2	0.0133
473.2	0.0133
CURVE 304*	
348.2	0.0145
423.2	0.0153
523.2	0.0142
CURVE 305*	
348.2	0.0144
423.2	0.0164
CURVE 306*	
348.2	0.0148
423.2	0.0144
523.2	0.0130
CURVE 307*	
373.2	0.0149
473.2	0.0149
CURVE 308*	
348.2	0.0174
423.2	0.0156
523.2	0.0135
CURVE 309*	
348.2	0.0172
423.2	0.0132
CURVE 310*	
373.2	0.0148
473.2	0.0148
CURVE 311*	
348.2	0.0159
423.2	0.0159
523.2	0.0159
CURVE 312	
348.2	0.0114
423.2	0.0128
523.2	0.0113
CURVE 313*	
348.2	0.0168
423.2	0.0162
CURVE 314*	
373.2	0.0126
473.2	0.0126
CURVE 315	
348.2	0.0240
423.2	0.0150*
CURVE 316*	
373.2	0.0114
473.2	0.0114
CURVE 317*	
298.2	0.00291
CURVE 318*	
298.2	0.00349
CURVE 319*	
298.2	0.00360
CURVE 320*	
298.2	0.00267
CURVE 321*	
298.2	0.00314
CURVE 322*	
298.2	0.00407
CURVE 323*	
298.2	0.00232
CURVE 324*	
298.2	0.00232
CURVE 325*	
298.2	0.00267
CURVE 326*	
298.2	0.00279
CURVE 327*	
298.2	0.00279
CURVE 328*	
298.2	0.00291
CURVE 329*	
298.2	0.00232
CURVE 330*	
298.2	0.00232
CURVE 331*	
298.2	0.00384

*Not shown on plot

DATA TABLE NO. 289 (continued)

T	k
CURVE 332*	
298.2	0.00325
CURVE 333*	
298.2	0.00372
CURVE 334*	
298.2	0.00325
CURVE 335*	
298.2	0.00349
CURVE 336*	
298.2	0.00325
CURVE 337*	
291.5	0.00175
CURVE 338	
291.5	0.000909

* Not shown on plot

SPECIFICATION TABLE NO. 290 THERMAL CONDUCTIVITY OF PLASTER

Curve No.	Ref. No.	Method Used	Year	Temp. Range, K	Reported Error, %	Name and Specimen Designation	Composition (weight percent), Specifications and Remarks
1	576		1957	298. 2		Plaster of Paris	In powdered form; density (25 C) = 1. 13 g cm^{-3}.

DATA TABLE NO. 290 THERMAL CONDUCTIVITY OF PLASTER

[Temperature, T, K; Thermal Conductivity, k, Watt cm^{-1} K^{-1}]

T	k
CURVE 1*	
298. 2	0. 00134

*No graphical presentation

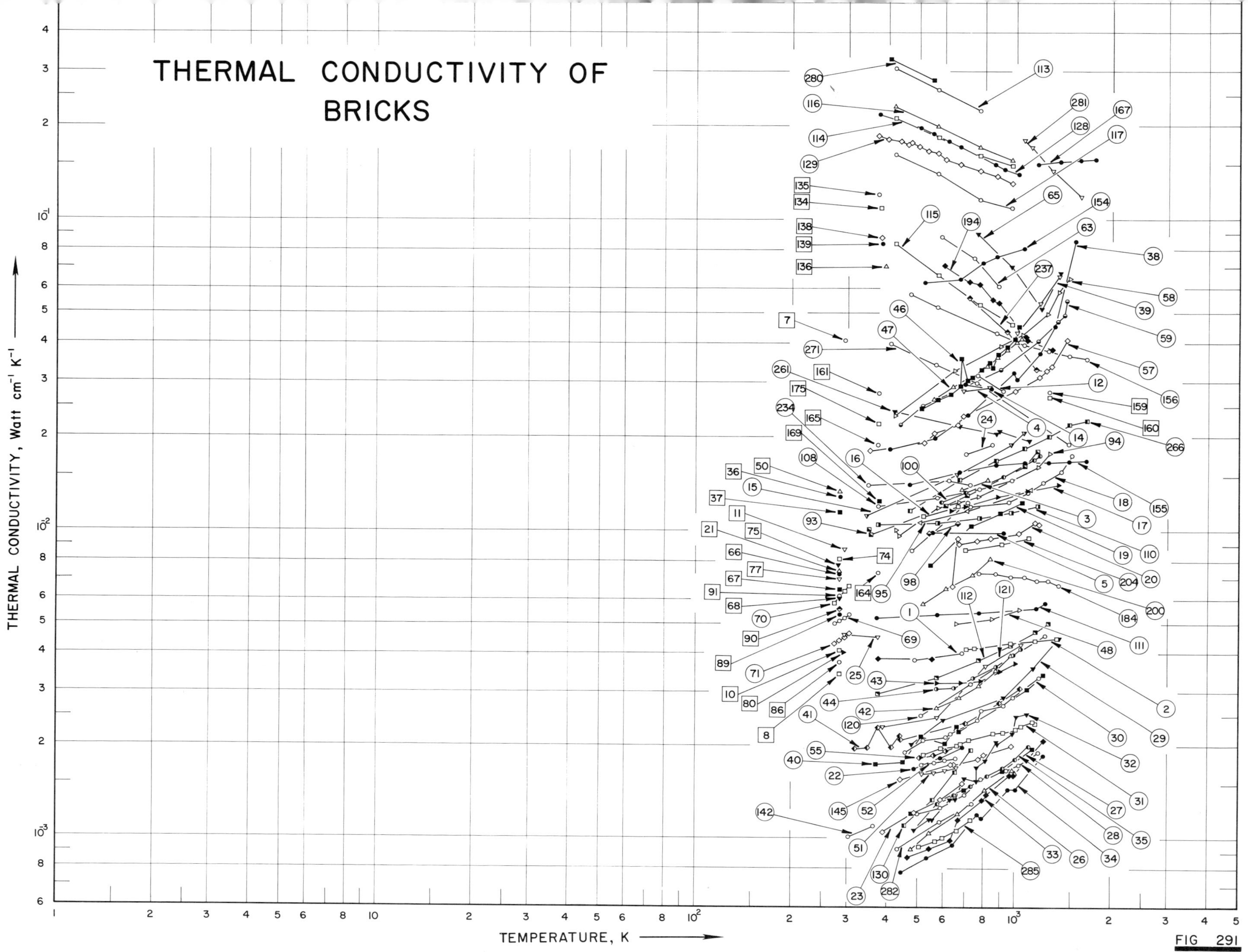

THERMAL CONDUCTIVITY OF BRICKS
THERMAL CONDUCTIVITY, Watt cm⁻¹ K⁻¹
TEMPERATURE, K

FIG 291

SPECIFICATION TABLE NO. 291 THERMAL CONDUCTIVITY OF BRICKS

[For Data Reported in Figure and Table No. 291]

Curve No.	Ref. No.	Method Used	Year	Temp. Range, K	Reported Error, %	Name and Specimen Designation	Composition (weight percent), Specifications and Remarks
1	90	L	1943	489, 685	<1.5	B & W No. 80 Kaolin 9-in. straight	8.93 in. length, 4.48 in. width, 2.44 in. thickness; bulk density 2.21 g cm^{-3}; condition of brick stability obtained by repeated heating.
2	90	L	1943	706-1358	<1.5	B & W No. 80 Kaolin 9-in. straight	Same as above but measured with the backing-up of 0.50 in. of group 20 insulating fire brick.
3	91	C	1947	681, 824	1.7	Chamotte	20% carborundum; cylindrical specimen; iron used as comparative material.
4	91	C	1947	688, 830		Chamotte	60% carborundum; cylindrical specimen; iron used as comparative material.
5	92	F	1950	553, 923		Chamotte (fire brick)	Normal; density 1.85 g cm^{-3}.
6	92	F	1950	373, 553		Chamotte (fire brick)	Porous; density 1.23 g cm^{-3}.
7	93		1957	293.2		Carbon brick (plenia)	Density 1.5 g cm^{-3}; porosity 30%.
8	94	L	1942	283.2		Cement porous brick	Measured at wall; density 0.725 g cm^{-3}; moisture content 5.0 vol. %.
9	94	L	1942	283.2		Cement porous brick	Measured at laboratory; density 0.720 g cm^{-3}; moisture content 4.1 vol. %.
10	94	L	1942	283.2		Cement porous brick	Measured at wall; density 0.722 g cm^{-3}; moisture content 6.1 vol. %.
11	93		1957	293.2		Ceramic brick K60	Specimen prepared by factory Hoganas-Billesholm AB, Hoganas; density 2.1 g cm^{-3}; porosity 17%.
12	91	C	1947	763, 912		Chrome magnesite brick	Cylindrical specimens from batches of different compositions; iron used as comparative material.
13	91	C	1947	743, 893		Chrome magnesite brick	Similar to the above specimens.
14	91	C	1947	696, 846		Chrome magnesite brick	Similar to the above specimens.
15	95	L	1935	347-1070	10.0	Mexko; dense fire clay brick	0.989 in. in thickness and 3.97 in. in dia; density 2.34 g cm^{-3}.
16	90	L	1943	516, 662	<1.5	Mexko; superduty fire-clay 9-in. straight	9.00 in. in length, 4.42 in. in width, 2.48 in. in thickness; bulk density 2.27 g cm^{-3}; condition of brick stability obtained by repeated heating.
17	90	L	1943	632-1369	<1.5	Mexko; superduty fire-clay 9-in. straight	Same as the above specimen; measured with 0.50 in. of group 29 backing-up insulation.
18	96	L	1934	478-1505		Mexko; fire brick	Specimen consisted of 8 standard-size bricks forming a section 18 x 18 x 2.5 in. in size.
19	96	L	1934	544-1055		Mexko; fire brick	Similar to the above specimen but measured by using a different apparatus.
20	97	C	1958	638-1183	±6.0	Superduty fire-clay brick	From "Ordzhonikidze" factory; 55 mm dia x 56 mm long; porosity 20.2%; carborundum used as comparative material.
21	94	L	1942	283.2			Measured at wall; density 1.67 g cm^{-3}; moisture content 1.55 vol. %.
22	79	L	1942	488-688		Diatomaceous	Main constituents: hydrated amorphous silica and diatomaceous earth; crushing strength 850 lb $in.^{-2}$ and density 0.72 g cm^{-3}.

SPECIFICATION TABLE NO. 291 (continued)

Curve No.	Ref. No.	Method Used	Year	Temp. Range, K	Reported Error, %	Name and Specimen Designation	Composition (weight percent), Specifications and Remarks
23	79	L	1942	388-688		Diatomaceous	Same as the above specimen but with a crushing strength 250 lb in.$^{-2}$ and density 0.42 g cm^{-3}.
24	91	C	1947	700, 848		Dinas	Specimen prepared by "Firey Ural Dinas" factory; iron used as comparative material.
25	44	F	1914	293, 373		Fire brick	Density 1.73 g cm^{-3}.
26	98	L	1942	478-1005		Fire brick; A	Density 0.316 g cm^{-3}.
27	98	L	1942	478-1144		Fire brick; B	Density 0.442 g cm^{-3}.
28	98	L	1942	478-1189		Fire brick; C	Density 0.493 g cm^{-3}.
29	98	L	1942	478-1367		Fire brick; D	Density 0.639 g cm^{-3}.
30	98	L	1942	455-1194		Fire brick; E	Density 0.638 g cm^{-3}.
31	98	L	1942	522-1167		Fire brick; F	Density 0,606 g cm^{-3}.
32	98	L	1942	489-1100		Fire brick; A	With coarse pores; 1.04 lb per 9-in. straight; density 0.285 g cm^{-3}.
33	98	L	1942	467-1233		Fire brick; B	Intermediate pores, 1.01 lb per 9-in. straight; density 0.277 g cm^{-3}.
34	98	L	1942	444-1233		Fire brick; C	Fine pores, 1.01 lb per 9-in. straight; density 0.277 g cm^{-3}.
35	99	L	1952	489-1122		Fire brick	No details reported.
36	94	L	1942	283.2		Hand-burned face brick	Measured at wall; density 1.885 g cm^{-3}; moisture content 1.7 vol. %.
37	94	L	1942	283.2		Hand-burned face brick	Measured in the laboratory; density 1.952 g cm^{-3}; moisture content 1.4 vol. %.
38	96	L	1934	405-1533		High temp. insulating brick; 2	Specimen consisted of 8 standard-size bricks forming a section 18 x 18 x 2.5 in. in size.
39	96	L	1934	1009-1378		High temp. insulating brick; 2	The above specimen with cracks opening on the hot surface.
40	95	L	1935	369-1225	2.0	Lightweight brick; A	1.021 in. thick and 3.98 in. in dia; measured in direct contact with heater; density 0.509 g cm^{-3}.
41	95	L	1935	318-1048	2.0	Lightweight brick; B	1.013 in. thick and 4.0 in. in dia; measured by placing the specimen between heater and the specimen A; density 0.513 g cm^{-3}.
42	4	R	1951	573-973	5	B & W K-28 insulating fire brick	No details reported.
43	7	R	1949	571-1014		B & W K-28 insulating fire brick	Weight 2.5 lb per standard brick; supplied by Babcock and Wilcox Co.
44	1		1950	573-1046		B & W K-28 insulating fire brick	Specimen supplied by Babcock and Wilcox Co.

SPECIFICATION TABLE NO. 291 (continued)

Curve No.	Ref. No.	Method Used	Year	Temp. Range, K	Reported Error, %	Name and Specimen Designation	Composition (weight percent), Specifications and Remarks
45	1		1950	576-1082		B & W K-28 insulating brick	Specimen supplied by Babcock and Wilcox Co.
46	2	R	1951	510-1021		B & W K-28 insulating brick	Specimen supplied by Babcock and Wilcox Co.
47	2	R	1951	553-1040		B & W K-28 insulating brick	2nd run of the above specimen.
48	85	L	1949	655-1038	5.0-7.0	Lightweight brick; 7	Specimen prepared by "Borovichsk" factory; density 1.00 g cm^{-3}; open porosity 61.2%; additional shrinkage 0.8% when exposed at 1350 C for 2 hrs.
49	94	L	1942	283.2		Lime sand brick	Measured at wall; density 1.760 g cm^{-3}; moisture content 8.3 vol. %.
50	94	L	1942	283.2		Lime sand brick	Measured in the laboratory; density 1.884 g cm^{-3}; moisture content 4.9 vol. %.
51	100	L	1957	515-659		Mica brick	White; bulk density 0.703 g cm^{-3}.
52	100	L	1957	512-656		Mica brick	Red; bulk density 0.705 g cm^{-3}.
53	100	L	1957	523-644		Mica brick	White; fired at 800 C; bulk density 0.684 g cm^{-3}.
54	100	L	1957	531-684		Mica brick	White; fired at 900 C; bulk density 0.660 g cm^{-3}.
55	100	L	1957	505-673		Mica brick	White; fired at 1000 C; bulk density 0.666 g cm^{-3}.
56	100	L	1957	507-653		Mica brick	White; fired at 1100 C; bulk density 0.666 g cm^{-3}.
57	96	L	1934	353-1450		Refractory insulating brick	Specimen consisted of 8 standard-size bricks forming a section 18 x 18 x 2.5 in. in size.
58	96	L	1934	422-1478		Refractory insulating brick	Similar to the above specimen.
59	96	L	1934	439-1450		Refractory insulating brick	Similar to the above specimen.
60	96	L	1934	561-1339		Johns-Manville C-22 refractory insulating brick	Similar to the above specimen.
61	86	C	1948	841.2	3	Refractory insulating common chamotte brick	Specimen prepared by "Semiluksk" factory; density 1.98 g cm^{-3}; iron used as comparative material.
62	86	L	1948	738.2		Refractory insulating chamotte brick	Similar to the above specimen but measured in a different apparatus.
63	91	C	1947	587-883		Magnesite brick	Specimen prepared by "Magnezit" factory; iron used as comparative material.
64	92	F	1950	813-1023		Magnesite brick	No details reported.
65	97	C	1958	763-1373	±6.0	Magnesite brick	Porosity 22.5%; carborundum used as comparative material.
66	94	L	1942	284.1		Metallurgical brick; HS150	Measured at wall; density 1.910 g cm^{-3}; moisture content 3.99 vol. %; aged for 83 days.
67	94	L	1942	284.3		Metallurgical brick; HS150	Measured at wall; density 1.910 g cm^{-3}; moisture content 2.91 vol. %; aged for 198 days.

SPECIFICATION TABLE NO. 291 (continued)

Curve No.	Ref. No.	Method Used	Year	Temp. Range, K	Reported Error, %	Name and Specimen Designation	Composition (weight percent), Specifications and Remarks
68	94	L	1942	284.1		Metallurgical brick; HS150	Measured at wall; density 1.910 g cm^{-3}; moisture content 2.32 vol.%; aged for 402 days.
69	94	L	1942	273-303		Metallurgical brick; HS150	Measured in the laboratory; density 2.115 g cm^{-3}; moisture content 2.8 vol.%.
70	94	L	1942	273-303		Metallurgical brick; HS100	Measured in the laboratory; density 1.786 g cm^{-3}; moisture content 15.5 vol.%.
71	94	L	1942	273-303		Metallurgical brick; HS100	Measured in the laboratory; density 1.786 g cm^{-3}; moisture content 1.7 vol.%.
72	94	L	1942	285.4		Metallurgical brick; HS150	Measured at wall; density 2.030 g cm^{-3}; moisture content 4.3 vol.% (brick), 4.8 vol.% (wall).
73	94	L	1942	284.1		Metallurgical brick; HS150	Laboratory measurement; density 1.910 g cm^{-3}; moisture content 2.2 vol.% (brick), 2.3 vol.% (wall).
74	94	L	1942	282.0		Metallurgical brick; HS150	Measured at wall; density 1.975 g cm^{-3}; moisture content 17.2 vol.% (brick), 13.1 vol.% (wall).
75	94	L	1942	280.7		Metallurgical brick; HS100	Measured at wall; density 1.570 g cm^{-3}; moisture content 18.1 vol.% (brick), 14.2 vol.% (wall).
76	94	L	1942	282.4		Metallurgical brick; HS100	Measured at wall; density 1.620 g cm^{-3}; moisture content 15.1 vol.% (brick), 11.5 vol.% (wall).
77	94	L	1942	283.8		Metallurgical brick; HS100	Measured at wall; density 1.740 g cm^{-3}; moisture content 16.2 vol.% (brick), 12.2 vol.% (wall).
78	94	L	1942	287.3		Metallurgical brick; HS100	Measured at wall; density 1.690 g cm^{-3}; moisture content 14.8 vol.% (brick), 10.4 vol.% (wall).
79	94	L	1942	283.0		Metallurgical brick; HS100	Measured at wall; density 1.900 g cm^{-3}; moisture content 15.8 vol.% (brick), 11.1 vol.% (wall).
80	94	L	1942	292.1		Metallurgical porous brick	Measured at wall; density 0.955 g cm^{-3}; moisture content 6.8 vol.% (brick), 5.3 vol.% (wall).
81	94	L	1942	288.2		Metallurgical porous brick	Measured at wall; density 1.035 g cm^{-3}; moisture content 7.0 vol.% (brick), 5.2 vol.% (wall).
82	94	L	1942	294.9		Metallurgical porous brick	Measured at wall; density 1.000 g cm^{-3}; moisture content 10.9 vol.% (brick), 10.1 vol.% (wall).
83	94	L	1942	283.6		Metallurgical porous brick	Measured at wall; density 0.950 g cm^{-3}; moisture content 15.6 vol.% (brick), 13.6 vol.% (wall).
84	94	L	1942	300.3		Metallurgical porous brick	Measured at wall; density 0.955 g cm^{-3}; moisture content 6.6 vol.% (brick), 5.5 vol.% (wall).
85	94	L	1942	283.2		Metallurgical porous brick	Measured at wall; density 0.985 g cm^{-3}; moisture content 8.5 vol.%.
86	94	L	1942	283.2		Metallurgical porous brick (special)	Measured at wall; density 0.955 g cm^{-3}; moisture content 5.5 vol.%.

SPECIFICATION TABLE NO. 291 (continued)

Curve No.	Ref. No.	Method Used	Year	Temp. Range, K	Reported Error, %	Name and Specimen Designation	Composition (weight percent), Specifications and Remarks
87	94	L	1942	283.2		Porous brick	Measured at wall; density 1.230 g cm^{-3}; moisture content 1.1 vol %.
88	94	L	1942	283.2		Porous brick	Measured in laboratory; density 1.279 g cm^{-3}; moisture content 1.0 vol %.
89	94	L	1942	283.2		Porous concrete brick	Measured at wall; density 1.010 g cm^{-3}; moisture content 7.5 vol %.
90	94	L	1942	283.2		Slag brick	Measured at wall; density 1.147 g cm^{-3}; moisture content 4.6 vol %.
91	94	L	1942	283.2		Slag brick	Measured at wall; density 1.315 g cm^{-3}; moisture content 6.0 vol %.
92	94	L	1942	283.2		Slag brick	Laboratory measurement; density 1.357 g cm^{-3}; moisture content 3.3 vol %.
93	95	L	1935	350-1173	10.0	Silica fire brick (Star brand); I	1.009 in. thick and 4.02 in. in dia; measured with specimen II (the following specimen) in between the heater and this specimen; density 1.69 g cm^{-3}.
94	95	L	1935	425-1277	10.0	Silica fire brick (Star brand); II	0.990 in. thick and 4.02 in. in dia; measured in direct contact with the heater and with the above specimen I on top; density 1.69 g cm^{-3}.
95	95	L	1935	503-1121	10.0	Silica fire brick (Star brand); I	Same as the above specimen I except measured in direct contact with the heater and with the above specimen II on top.
96	95	L	1935	457-1030	10.0	Silica fire brick (Star brand); II	Same as the above specimen II except measured in contact with the above specimen I which is in direct contact with the heater.
97	95	L	1935	482-1078	10.0	Silica fire brick (Star brand); II	Same as the above specimen II except measured alone.
98	90	L	1943	544, 663	< 1.5	Commercial 9 in. straight silica brick (Star)	8.99 in. in length, 4.49 in. in width, 2.47 in. in thickness; bulk density 1.68 g cm^{-3}; condition of brick stability obtained by repeated heating.
99	90	L	1943	731-1367	< 1.5	Commercial 9 in. straight silica brick (Star)	Same as above but measured with 0.50 in. of group 29 insulating fire brick backing-up.
100	101	L	1935	589-1190	± 5.0	Silica brick; 1	Supplied by Wilkes, G. B.
101	101	L	1935	506-1190	± 5.0	Silica brick; 2	Same supplier as above; used as a guard brick in the measurement of the above specimen.
102	101	L	1935	456-1190	± 5.0	Silica brick; 1	Second run of specimen 1, after the brick and the thermocouples removed and then restored.
103	101	L	1935	489-1190	± 5.0	Silica brick; 3	Supplied by Birch, R. E. of Harbison-Walker Refractories Co.
104	101	L	1935	556-1190	± 5.0	Silica brick; 3	Second run of the above specimen, after the brick and the thermocouples removed and then restored.
105	101	L	1935	500-1123	± 5.0	Silica brick	Porosity 32%; specimen thickness 2.5 in.
106	100	L	1957	520-659		Vermiculite brick	Bulk density 0.484 g cm^{-3}.
107	92	F	1950	413, 733		Tripolite (tripoli earth) brick	No details reported.
108	102	L	1957	373-1173		Shamotte	Obtained from Harbison-Walker Refractories Co.; 30 mm dia x 20 mm high.
109	102	L	1957	373-1173		Silica	Obtained from General Refractories Co.; 30 mm dia x 20 mm high.

Curve No.	Ref. No.	Method Used	Year	Temp. Range, K	Reported Error, %	Name and Specimen Designation	Composition (weight percent), Specifications and Remarks
110	102	L	1957	373-1173		Zirconia	Obtained from Norton Co.; 30 mm dia x 30 mm long; dense; stabilized.
111	102	L	1957	373-1248		Zirconia insulating	Obtained from Norton Co.; 30 mm dia x 30 mm long.
112	102	L	1957	373-1273		Fire-brick HW28	Obtained from Harbison-Walker Refractories Co.; 30 mm dia x 30 mm long.
113	102	L	1957	423-773		Carsiat carborundum	Obtained from Didier Co.; 50 x 22 x 22 mm.
114	102	L	1957	423-973		Carbofrax carborundum	50 x 22 x 22 mm.
115	102	L	1957	423-973		Magnesite brick	30 mm dia x 20 mm high.
116	102	L	1957	423-973		SiC brick; 1	Manufactured in Japan; 30 mm dia x 30 mm long.
117	102	L	1957	423-973		SiC brick; 2	Similar to the above specimen.
118	103	L	1937	393, 855	<10	Silica-fire brick	Flat circular disk specimen.
119	103	L	1937	534, 692	<10	Silica-fire brick	Same sample as above but measured in different apparatus.
120	103	L	1937	513-1244	<10	Kaolin insulating refractory brick; I	0.958 in. in thickness, 4.44 in. in dia; density 0.641 g cm^{-3}; specimen in direct contact with center heater.
121	103	L	1937	386-1046	<10	Kaolin insulating refractory brick; II	0.959 in. in thickness, 4.50 in. in dia; density 0.639 g cm^{-3}; in measurement this specimen placed on top of the above specimen I.
122	103	L	1937	606-1215	<10	Kaolin insulating refractory brick; II	The above specimen II placed in direct contact with center heater.
123	103	L	1937	491-1033	<10	Kaolin insulating refractory brick; I	Specimen I placed on top of the above specimen II.
124	103	L	1937	424-1321	<10	Kaolin insulating refractory brick; I	The above specimen I in direct contact with center heater in a different apparatus.
125	103	L	1937	362-1173	<10	Kaolin insulating refractory brick; II	Specimen II placed on top of the above specimen I.
126	103	L	1937	485-1420	<10	Kaolin insulating refractory brick; II	Specimen II in direct contact with center heater in the 2nd apparatus.
127	103	L	1937	420-1250	<10	Kaolin insulating refractory brick; I	Specimen I placed on top of the above specimen II.
128	104	L	1956	376-1023		Carbofrax	Produced by U.S. Carborundum Co.; SiC 89%; specimen 20 mm in dia and 30 mm long; porosity 16.52%; clay guard ring is used.
129	104	L	1956	373-973		SiC brick; domestic (Japan)	Specimen 22 x 22 x 50 mm; porosity 15.34%; clay guard ring is used in measurement.
130	105	R	1958	454-732		Porous fire-brick (Italy)	Density 0.520 g cm^{-3}.
131	105	R	1958	382-567		Porous fire-brick (Italy)	Different run of the above specimen.

SPECIFICATION TABLE NO. 291 (continued)

Curve No.	Ref. No.	Method Used	Year	Temp. Range, K	Reported Error, %	Name and Specimen Designation	Composition (weight percent), Specifications and Remarks
132	81	L	1924	1223. 2	< 2. 0	Star silica brick; 2	8. 5 in. dia x 3 in. thick; porosity 27. 7%.
133	81	L	1924	1223. 2	< 2. 0	Star silica brick; 3	8. 5 in. dia x 3 in. thick; porosity 27. 7%.
134	106	C	1956	377. 7	3	Carbon brick; A1	80 mm dia x 125 mm long; apparent density 1. 594 g cm^{-3}; electrical resistivity 0. 00433 ohm cm; used for lining of Al electrolysis; copper used as comparative material.
135	106	C	1956	372. 2	3	Carbon brick; A2	80 mm dia x 125 mm long; apparent density 1. 586 g cm^{-3}; electrical resistivity 0. 00446 ohm cm; used for lining of Al electrolysis; same comparative material as above.
136	106	C	1956	392. 7	3	Carbon brick; B1	80 mm dia x 125 mm long; apparent density 1. 548 g cm^{-3}; electrical resistivity 0. 00505 ohm cm; used for lining of Al electrolysis; same comparative material as above.
137	106	C	1956	388. 2	3	Carbon brick; B2	80 mm dia x 125 mm long; apparent density 1. 556 g cm^{-3}; electrical resistivity 0. 00477 ohm cm; used for lining of Al electrolysis; same comparative material as above.
138	106	C	1956	381. 2	3	Carbon brick; C1	80 mm dia x 125 mm long; apparent density 1. 654 g cm^{-3}; electrical resistivity 0. 00418 ohm cm; used for lining of Al electrolysis; same comparative material as above.
139	106	C	1956	383. 2	3	Carbon brick; C2	80 mm dia x 125 mm long; apparent density 1. 674 g cm^{-3}; electrical resistivity 0. 00412 ohm cm; used for lining of Al electrolysis; same comparative material as above.
140	107	L	1924	413–498		Hytex hydraulic pressed building brick	12 in. x 12 in. x 2 in. thick; heat flow approximately perpendicular to surface of the brick.
141	107	L	1924	518–1183		Georgia fire brick (G–3)	Same as the above specimen.
142	108	L	1920	303, 363		Sil–O–Cel	Specimen 9 in. in dia, 0. 977 in. thick; density 0. 495; heat flow in transverse direction.
143	186	L	1928	547–1093	±2. 5	Sil–O–Cel, No. 2	Density 0. 641 g cm^{-3}; porosity 75%.
144	186	L	1928	573–1081	±2. 5	Sil–O–Cel C–22, No. 3	Density 0. 577 g cm^{-3}; porosity 75. 7%.
145	186	L	1928	443–979	±2. 5	Special Sil–O–Cel, No. 4	Density 0. 541 g cm^{-3}; porosity 76. 7%.
146	186	L	1928	492–1025	±2. 5	Sil–O–Cel super, No. 5	Density 0. 703 g cm^{-3}; porosity 72. 6%.
147	186	L	1928	600–1120	±2. 5	Sil–O–Cel super, No. 6	Density 0. 652 g cm^{-3}; porosity 76. 7%.
148	186	L	1928	472–1043	±2. 5	Sil–O–Cel calcined, No. 8	Density 0. 450 g cm^{-3}; porosity 83. 9%.
149	186	L	1928	426–961	±2. 5	Sil–O–Cel calcined, No. 9	Density 0. 509 g cm^{-3}; porosity 84. 0%.
150	186	L	1928	412–903	±2. 5	Sil–O–Cel natural, No. 10	Density 0. 437 g cm^{-3}; porosity 82. 0%.
151	186	L	1928	410–977	±2. 5	Sil–O–Cel natural, No. 11	Density 0. 450 g cm^{-3}; porosity 85. 6%.
152	186	L	1928	598–1220	±2. 5	Fire–clay brick, 1st quality Missouri	Density 1. 93 g cm^{-3}; porosity 26. 5%.
153	186	L	1928	623–1187	±2. 5	Red brick, hard burned	Density 2. 10 g cm^{-3}; porosity 23. 7%.
154	186	L	1928	522–1069	±2. 5	Red brick, soft burned	Density 1. 75 g cm^{-3}; pososity 38. 3%.

SPECIFICATION TABLE NO. 291 (continued)

Curve No.	Ref. No.	Method Used	Year	Temp. Range, K	Reported Error, %	Name and Specimen Designation	Composition (weight percent), Specifications and Remarks
155	383	L	1927	473-1673	±15	Alumina fused brick	Commercial brick; specimen 10.8 cm dia x 22.8 cm long; apparent density 3.67 g cm^{-3}; bulk density 2.90 g cm^{-3}; porosity 27.3%; made of fused alumina grog and bonded with fire clay.
156	383	L	1927	473-1673	±20	Magnesite fire brick	Commercial brick; 10.8 cm dia x 22.8 cm long.
157	383	L	1927	473-1673	±15	Chrome fire brick	Commercial brick; 10.8 cm dia x 22.8 cm long; apparent density 3.94 g cm^{-3}; bulk density 2.74 g cm^{-3}; porosity 30.5%.
158	457	C	1954	293.2			Cork used as comparative material.
159	81	L	1924	1223	< 2.0	Magnesia brick; 1	92.0-95.0 MgO; electrically sintered, fired to cone 16; 0.5 in. dia x 1.983 in. thick; prepared by crushing the electric furnace products to pass a No. 14 screen and bonding with fire clay; specimen contains 3% bond clay of 25.41 Al_2O_3, 59.55 SiO_2, 2.31 Fe_2O_3, 1.33 TiO_2, 1.01 ($K_2O + Na_2O$), 0.46 CaO, 0.33 MgO, and 9.10 loss; measured at 744.0 mm Hg pressure; porosity 23.4% calculated from the dry saturated suspended weight.
160	81	L	1924	1223	< 2.0	Magnesia brick; 2	Same as the above specimen except measured at 741.3 mm Hg pressure.
161	458	R	1921	373.2		Magnesia brick	53.27 MgO (original composition from author was 53.27 MnO that could be mistyped), 32.46 SiO_2, 14.78 Al_2O_3, 4.91 CaO, 2.5 Fe_2O_3; specimen 10 cm in dia and 45 mm in height; density reported as 2.295, 2.294, 2.291, 2.285, 2.275, and 2.266 g cm^{-3} at 20.50, 100, 200, 300, and 400 C, respectively.
162	458	P	1921	373.2		Magnesia brick	Similar to the above specimen except measured with different method, thermal conductivity data calculated from diffusibility, specific heat and specific gravity.
163	458	R	1921	373.2		Common brick	76.52 SiO_2, 13.67 Al_2O_3, 6.77 Fe_2O_3, 1.77 CaO, 0.42 MgO, and 0.27 MnO; 10 cm dia x 4.5 cm thick.
164	458	P	1921	373.2		Common brick	Similar to the above specimen; thermal conductivity values calculated from measured data of thermal diffusivity, specific heat, and density.
165	458	R	1921	373.2		Chrome	31.89 Cr_2O_3, 24.86 Al_2O_3, 16.48 MgO, 16.91 Fe_2O_3, 8.30 SiO_2, 0.26 CaO; specimen 10 cm in dia and 45 mm in height; density reported as 2.982, 2.981, 2.978, and 2.970 g cm^{-3} at 20, 50, 100, and 200 C, respectively.
166	458	P	1921	373.2		Chrome	Similar to the above specimen except measured in different method; thermal conductivity data calculated from the measurement of diffusibility, specific heat and specific gravity.
167	84	L	1925	1173-1773		Carbofrax	Test wall built of standard 9 x 4.5 x 2.5 in. bricks laid up with cement of the same composition as the bricks; heat flow through the 4.5 in. dimension; measured at decreasing temperatures.
168	458	R	1921	373.2		Silica brick	94.98 SiO_2, 3.06 CaO, 1.18 Al_2O_3, and 1.13 Fe_2O_3; 10 cm dia x 4.5 cm thick; density reported as 1.840, 1.836, 1.829, 1.814, and 1.804 g cm^{-3} at 20, 50, 100, 200, and 300 C, respectively.

SPECIFICATION TABLE NO. 291 (continued)

Curve No.	Ref. No.	Method Used	Year	Temp. Range, K	Reported Error, %	Name and Specimen Designation	Composition (weight percent), Specifications and Remarks
169	458	P	1921	373.2		Silica brick	Similar to the above specimen; thermal conductivity values calculated from measured data of thermal diffusivity, specific heat, and density.
170	458	R	1921	373.2		Shamotte brick	60.78 SiO_2, 33.95 Al_2O_3, 4.37 Fe_2O_3, 0.79 CaO, traces of MgO and MnO; 10 cm dia x 4.5 cm thick; density reported as 1.917, 1.916, 1.913, 1.905, 1.901, 1.896, and 1.892 g cm^{-3} at 20, 50, 100, 200, 300, 400, and 500 C, respectively.
171	458	P	1921	373.2		Shamotte brick	Similar to the above specimen; thermal conductivity values calculated from measured data of thermal diffusivity, specific heat, and density.
172	458	R	1921	373.2		Slag brick	43.27 CaO, 26.78 SiO_2, 15.21 Al_2O_3, 12.60 Cr_2O_3, 6.75 Fe_2O_3, and 2.20 MnO; 10 cm dia x 4.5 cm thick.
173	458	P	1921	373.2		Slag brick	Similar to the above specimen; thermal conductivity values calculated from measured data of thermal diffusivity, specific heat, and density.
174	458	P	1921	373.2		Chrome brick	36.42 Cr_2O_3, 19.51 Al_2O_3, 14.56 MgO, 1.88 SiO_2, 1.843 Fe_2O_3, and 0.94 CaO; 4 cm dia x 8 cm long; density 3.028 g cm^{-3}; thermal conductivity values calculated from measured data of thermal diffusivity, specific heat, and density.
175	458	P	1921	373.2		Magnesia brick	76.43 MgO, 18.26 SiO_2, 2.56 Al_2O_3, 0.80 Fe_2O_3, 0.21 MnO, and trace of CaO; 4 cm dia x 8 cm long; supplied by Imperial Steel Works; density 2.370 g cm^{-3}; same measuring method as above.
176	458	P	1921	373.2		Silica brick	91.20 SiO_2, 4.01 Fe_2O_3, 2.80 CaO, 1.27 Al_2O_3, and 0.46 MnO; 4 cm dia x 8 cm long; prepared by Imperial Steel Works; density 1.891 g cm^{-3}; same measuring method as above.
177	458	P	1921	373.2		Red brick	76.32 SiO_2, 21.96 Al_2O_3, 1.88 Fe_2O_3, traces of CaO and MgO; commercial brick; 4 cm dia x 8 cm long; density 1.795 g cm^{-3}; same measuring method as above.
178	458	P	1921	373.2		Red brick	76.45 SiO_2, 21.13 Al_2O_3, 2.02 Fe_2O_3, traces of CaO and MgO; same source, dimensions, and measuring method as the above specimen; density 1.782 g cm^{-3}.
179	458	P	1921	373.2		Shamotte brick	79.98 SiO_2, 19.48 Al_2O_3, 0.40 Fe_2O_3, traces of CaO and MgO; 4 cm dia x 8 cm long; supplied by Imperial Steel Works; density 1.565 g cm^{-3}; same measuring method as above.
180	458	P	1921	373.2		Shamotte brick	71.74 SiO_2, 25.56 Al_2O_3, 1.02 Fe_2O_3, 0.82 CaO, and 0.53 MgO; 4 cm dia x 8 cm long; supplied by Imperial Steel Works; density 1.784 g cm^{-3}; same measuring method as above.
181	458	P	1921	373.2		Slag brick	40.39 CaO, 26.34 SiO_2, 12.90 Al_2O_3, 1.71 MnO, 0.34 MgO, and 0.26 Fe_2O_3; 4 cm dia x 8 cm long; density 1.572 g cm^{-3}; same measuring method as above.
182	458	P	1921	373.2		Slag brick	41.32 CaO, 25.64 SiO_2, 13.72 Al_2O_3, 2.09 MnO, 0.4 MgO, and 0.28 Fe_2O_3; 4 cm dia x 8 cm long; density 1.585 g cm^{-3}; same measuring method as above.

SPECIFICATION TABLE NO. 291 (continued)

Curve No.	Ref. No.	Method Used	Year	Temp. Range, K	Reported Error, %	Name and Specimen Designation	Composition (weight percent), Specifications and Remarks
183	238	P	1921	773-1373		Magnesite brick	81.79 MgO, 5.24 CaO, and 1.87 Fe_2O_3; very close texture; 9 x 4.5 x 2.5 in.; apparent density 2.63 g cm^{-3}, true density 3.29 g cm^{-3}; porosity 20.0%; heat flow along the length of the specimen.
184	238	P	1921	773-1373		Magnesite brick	87.88 MgO, 4.68 CaO, and 2.56 Fe_2O_3; texture not so close; apparent density 2.56 g cm^{-3}, true density 3.28 g cm^{-3}; porosity 24.5%.
185	247	L	1932	379-794	1.0	Chromite brick	40.09 Cr_2O_3, 22.68 MgO, 13.16 Fe_2O_3, 10.48 Al_2O_3, 10.53 SiO_2, 2.32 Mn_3O_4, 0.68 CaO, and trace of TiO_2; density 3.988 g cm^{-3}; porosity 27.3%.
186	143	L	1955	683-1263		Magnezit; 4	49.46 MgO, 20.48 Cr_2O_3, 12.59 Al_2O_3, 9.15 Fe_2O_3, 5.24 SiO_2, 1.90 FeO, 1.26 CaO, 0.09 total Ca, Mg, Fe, and Mn; chromomagnesite refractory brick; density 2.95 g cm^{-3}; apparent porosity 22.8%; gas permeability 0.455 m^3 x cm m^{-2} hr^{-1} (mm of H_2O)$^{-1}$.
187	143	L	1955	673-1253		Magnezit; 5	Similar to the above specimen.
188	143	L	1955	643-1168		Magnezit; 6	Similar to the above specimen.
189	143	L	1955	658-1248		K Marksa; 11	42.31 MgO, 24.35 Cr_2O_3, 12.34 Al_2O_3, 11.94 Fe_2O_3, 6.14 SiO_2, 1.71 FeO, 1.65 CaO, 0.13 total Ca, Mg, Fe, and Mn; chromomagnesite refractory brick; density 3.03 g cm^{-3}; apparent porosity 23.5%; gas permeability 0.303 m^3 x cm m^{-2} hr^{-1} (mm H_2O)$^{-1}$.
190	143	L	1955	698-1258		Magnezit; 7	64.85 MgO, 12.28 Cr_2O_3, 8.51 Al_2O_3, 5.80 Fe_2O_3, 4.28 SiO_2, 1.67 FeO, 1.44 CaO, 0.21 total Ca, Mg, Fe, and Mn; chromomagnesite heat resistant refractory brick; density 3.04 g cm^{-3}; apparent porosity 19.1%; gas permeability 0.480 m^3 x cm m^{-2} hr^{-1} (mm H_2O)$^{-1}$.
191	143	L	1955	708-1263		Magnezit; 8	Similar to the above specimen.
192	143	L	1955	653-1233		Magnezit; 9	Similar to the above specimen.
193	143	L	1955	683-1228		Ordzhonikidze; 10	42.87 MgO, 22.34 Cr_2O_3, 13.04 Fe_2O_3, 11.46 Al_2O_3, 5.42 SiO_2, 2.88 FeO, 1.76 CaO, trace total Ca, Mg, Fe, and Mn; chromomagnesite heat resistant refractory brick; density 2.95 g cm^{-3}; apparent porosity 25.6%; gas permeability 0.598 m^3 x cm m^{-2} hr^{-1} (mm H_2O)$^{-1}$.
194	462	L	1915	598-1303		Magnesia brick	92.1 MgO, 5.0 SiO_2, 1.7 CaO, 1.6 Fe_2O_3, and 0.4 Al_2O_3; commercial brand; 2.5 in. thick; fine grained; apparent density 2.40 g cm^{-3}.
195	85	L	1949	636-1036	5-7	High temp insulating blast furnace brick	56.42 SiO_2, 1.42 Fe_2O_3, 40.18 total Al_2O_3 and TiO_2, 0.88 CaO, and 0.36 MgO; semi-dry pressed; from Semiluksk factory; density 2.16 g cm^{-3} and open porosity 17.8%; no additional shrinkage when exposed at 1350 C for 2 hrs.
196	85	L	1949	623-993	5-7	Light weight brick	66.08 SiO_2, 1.04 Fe_2O_3, 30.48 total Al_2O_3 and TiO_2, 0.92 CaO, and 0.52 MgO; from Shchekinsk factory; density 1.30 g cm^{-3} and open porosity 49.9%; 0.1% additional shrinkage when exposed at 1350 C for 2 hrs.
197	85	L	1949	636-1003	5-7	Light weight brick	70.98 SiO_2, 1.38 Fe_2O_3, 25.08 total Al_2O_3 and TiO_2, 0.96 CaO, and 0.25 MgO; from Snigirevsk factory; density 1.17 g cm^{-3} and open porosity 56.6%; 2.0% additional shrinkage when exposed at 1350 C for 2 hrs.

SPECIFICATION TABLE NO. 291 (continued)

Curve No.	Ref. No.	Method Used	Year	Temp. Range, K	Reported Error, %	Name and Specimen Designation	Composition (weight percent), Specifications and Remarks
198	86	C	1948	503-798	3	II	65. 72 SiO_2, 1. 93 Fe_2O_3, 29. 02 total TiO_2 and Al_2O_3, 1. 08 CaO, and 0. 62 MgO; open porosity 0. 86 %; water absorption 81. 0%; gas permeability 211 ml m^{-2} hr^{-1}; refractoriness 1630-1650 C; light gray color; pore size up to 1. 5 mm in dia; pore distribution even.
199	86	C	1948	505-797	3	III	66. 38 SiO_2, 1. 63 Fe_2O_3, 29. 05 total TiO_2 and Al_2O_3, 1. 12 CaO, and 0. 51 MgO; density 0. 86 g cm^{-3}; open porosity 66. 3%; water absorption 75. 4%; gas permeability 280 ml m^{-2} hr^{-1}; refractoriness 1630-1650 C; yellowish red color; pores heterogeneous, a small number of pores of 3-5 mm side by side with pores of 1. 5 mm.
200	86	C	1948	518-834	3	IV	67. 89 SiO_2, 1. 93 Fe_2O_3, 27. 81 total TiO_2 and Al_2O_3, 0. 64 CaO, and 0. 52 MgO; density 0. 98 g cm^{-3}; open porosity 63%; water absorption 62. 6%; gas permeability 46 ml m^{-2} hr^{-1}; refractoriness 1630-1650 C; white color; large number of pores with diameters 6-7 mm.
201	86	L	1948	783-1028		IV	67. 89 SiO_2, 27. 81 total TiO_2 and Al_2O_3, 1. 93 Fe_2O_3, 0. 64 CaO, and 0. 52 MgO; same as the above specimen.
202	86	L	1948	758-973		II	65. 72 SiO_2, 29. 02 total TiO_2 and Al_2O_3, 1. 93 Fe_2O_3, 1. 08 CaO, and 0. 52 MgO; same as the above specimen.
203	85	L	1949	666-1064	5-7	Normal brick; 2	63. 06 SiO_2, 33. 91 total Al_2O_3 + TiO_2, 1. 21 Fe_2O_3, 1. 00 CaO, and 0. 67 MgO; semidry pressed; from Semiluksk factory; density 1. 86 g cm^{-3}; open porosity 29. 1%; no additional shrinkage when exposed at 1350 C for 2 hrs.
204	85	L	1949	698-1101	5-7	Normal brick; 5	56. 32 SiO_2, 40. 72 Al_2O_3 and TiO_2, 1. 02 CaO, 0. 92 Fe_2O_3, and 0. 48 MgO; semidry pressed; from Borovichsk factory; density 1. 92 g cm^{-3}; open porosity 28. 1%; 0. 5% additional shrinkage when exposed at 1350 C for 2 hrs.
205	238	P	1921	873-1373		Silica brick C	93. 36 SiO_2, 2. 97 Al_2O_3, and 2. 20 CaO; specimen 9 x 4. 5 x 2. 5 in. ; open texture with large number of fissures of appreciable size; bonding of coarse (angular rock fragments of 0. 25 in. and downward) and fine fairly good except fine material easily detached; porosity 20. 7%; heat flow in lengthwise direction.
206	79	L	1942	696-880		No. 3, aluminous fire clay	52. 0 SiO_2, 41. 3 Al_2O_3, 2. 7 TiO_2, and 2. 5 Fe_2O_3; fired at the temperature of refractory standard cone 33; 18 in. x 18 in. dimensions.
207	79	C	1942	1062, 1254		No. 3, aluminous fire clay	Similar to the above specimen except in disc form; steel used as comparative material.
208	85	L	1949	661-1048	5-7	Light weight brick	54. 56 SiO_2, 40. 73 Al_2O_3 + TiO_2, 2. 59 Fe_2O_3, 0. 92 CaO, and 0. 48 MgO; from Borovichsk factory; density 1. 24 g cm^{-3}; open porosity 53. 0%; 0. 5% additional shrinkage when exposed at 1350 C for 2 hrs.
209	86	C	1948	513-827			68. 12 SiO_2, 27. 38 Al_2O_3 + TiO_2, 2. 04 Fe_2O_3, 0. 88 CaO, and 0. 59 MgO; density 0. 91 g cm^{-3}; open porosity 65. 1%, water absorption 64. 5%; gas permeability 216 ml m^{-2} hr^{-1}, and refractoriness 1630 C; light yellow color; numerous pores with size up to 1. 5 mm in dia and small number of pores with dia up to 5-6 mm; distribution of pores even.

SPECIFICATION TABLE NO. 291 (continued)

Curve No.	Ref. No.	Method Used	Year	Temp. Range, K	Reported Error, %	Name and Specimen Designation	Composition (weight percent), Specifications and Remarks
210	86	L	1948	753-993			68.12 SiO_2, 27.38 Al_2O_3 + TiO_2, 2.04 Fe_2O_3, 0.88 CaO, and 0.59 MgO; other descriptions same as the above.
211	85	L	1949	668-1053	5-7	Normal brick; 3	75.70 SiO_2, 20.41 Al_2O_3 + TiO_2, 2.59 Fe_2O_3, 0.68 CaO, and 0.53 MgO; semi-acid; of sheet form from Latnensk factory; density 1.85 g cm^{-1}; open porosity 30.4%; 0.1% additional shrinkage when exposed at 1350 C for 2 hrs.
212	85	L	1949	668-1053	5-7	Normal brick; 4	48.40 SiO_2, 37.36 Al_2O_3 + TiO_2, 2.24 Fe_2O_3, 0.62 CaO, and 0.59 MgO; in sheet form from Borovichsk factory; density 1.919 g cm^{-3}; open porosity 29.0%; 0.3% additional shrinkage when exposed at 1350 C for 2 hrs.
213	247	L	1932	399-851	1	Bauxite brick; 21	73.04 Al_2O_3, 19.60 SiO_2, 2.86 TiO_2, 2.30 Fe_2O_3, 1.42 CaO, and 0.65 MgO; density 3.310 g cm^{-3}; porosity 27.2%; gas permeability 0.013 m^3-cm per m^2-hr-mm H_2.
214	79	L	1942	688-857		Fire clay brick; 1	56.46 SiO_2, 36.79 Al_2O_3, 2.58 Fe_2O_3, 1.84 TiO_2, 1.24 alkali oxides, 0.60 MgO, and 0.38 CaO; fired at the temp of the refractory test cone 31; in slab form 18 in. x 18 in.
215	79	C	1942	1066, 1301		Fire clay brick; 1	Similar to the above specimen except in disk form of 8 in. in dia and 1 in. in thickness; steel used as comparative material.
216	79	L	1942	618-784		Fire clay brick; 2	Similar to the above specimen except in slab form 18 in. x 18 in.
217	79	C	1942	1009, 1242		Fire clay brick; 2	Similar to the above specimen except in disk form 8 in. in dia and 1 in. in thickness; steel used as comparative material.
218	83	C	1956	531-833		Egyptian fire clay brick; A	64.5 SiO_2, 26.0 Al_2O_3, 7.0 Fe_2O_3, 1.1 CaO, and 1.0 MgO; bulk density 1.01 g cm^{-3}; apparent porosity 72.7%.
219	83	L	1956	593-881		Egyptian fire clay brick; B	65.3 SiO_2, 29.5 Al_2O_3, 3.5 Fe_2O_3, 0.9 MgO, and 0.8 CaO; bulk density 1.09 g cm^{-3}; apparent porosity 60.0%.
220	83	L	1956	606-858		Egyptian fire clay brick; C	71.0 SiO_2, 24.0 Al_2O_3, 2.5 Fe_2O_3, 0.8 CaO, and 0.7 MgO; bulk density 0,780 g cm^{-3}; apparent porosity 68.5%.
221	84	L	1925	1623	1	Fire clay wall; 26	58.50 SiO_2, 34.48 Al_2O_3, 3.52 Fe_2O_3, 1.80 TiO_2, 0.62 MgO, 0.29 CaO, and 0.31 ignition loss; original coarse, fairly open, first quality fire clay; apparent density 2.05 g cm^{-3}; porosity 26.8% calculated by assuming specific gravity 2.60 for fire clay; the wall under test was built with standard 2.5 x 4.5 x 9 in. bricks laid up with cement of the same composition as the brick.
222	462	L	1915	816-1268		Fire clay brick (Farnley)	66 SiO_2, 31 Al_2O_3, 1.2 TiO_2, 1.0 alkalies, 0.9 MgO, 0.3 CaO; commercial brand; specimen 1.5 in. thick; apparent density 1.95 g cm^{-3}; hard fired to Seger Cone 10-11.
223	462	L	1915	1033-1203		Fire clay brick (Farnley)	Similar to the above specimen.
224	462	L	1915	1278-1293		Fire clay brick (Farnley)	Similar to the above specimen except apparent density 1.90 g cm^{-3}; soft fired to Seger Cone 8-9.

SPECIFICATION TABLE NO. 291 (continued)

Curve No.	Ref. No.	Method Used	Year	Temp. Range, K	Reported Error, %	Name and Specimen Designation	Composition (weight percent), Specifications and Remarks
225	462	L	1915	1078		Silicious brick (Farnley)	82.5 SiO_2, 16.1 Al_2O_3, 1.2 TiO_2, 1.3 alkalies, trace CaO and MgO; specimen 3 in. thick; apparent density 1.82 g cm^{-3}; with many silica grains.
226	462	L	1915	1113		Silica brick (Gregory)	95.3 SiO_2, 2.0 Al_2O_3, 2.0 TiO_2, 1.5 CaO; specimen 2.5 in. thick; coarse grain; apparent density 1.75 g cm^{-3}.
227	462	L	1915	918-1191		Silica brick (Gregory)	Similar to the above specimen except apparent density 1.74 g cm^{-3}.
228	238	P	1921	773-1373		Fire brick; D	68.38 SiO_2, 26.12 Al_2O_3, and 2.50 Fe_2O_3; close structure; not much clay grog, but large proportion of angular quartz grains; adherence very good; very few fissures; quartz grains very evenly graded; faces of brick not good; very fine black cores; appearance of many pinholes; brick size 9 x 4.5 x 2.5 in.; porosity 17.3%; heat flow along the length of the specimen.
229	247	L	1932	394-817	1	Silica brick; 8	92.14 SiO_2, 2.59 CaO, 2.21 Al_2O_3, 1.11 Fe_2O_3, 0.34 TiO_2, and 0.23 MgO; density 2.328 g cm^{-3}; porosity 28.1%; gas permeability 2.354 $m^3 \cdot cm\ m^{-1}\ hr^{-1}$ (mm of $H_2O)^{-1}$.
230	238	P	1921	873-1373		Silica brick; B	94.02 SiO_2, 2.64 CaO, and 1.78 Al_2O_3; exceptionally fine-grained, close, and uniform textrue throughout the brick; major portion of material of sand size; with very few fragments of rock of appreciable size; porosity 38.2%, with pores of even size; friable; brick size: 9 x 4.5 x 2.5 in.; heat flow in the lengthwise direction of the brick.
231	79	L	1942	640-1088		Sillimanite refractory brick; 4	58.08 Al_2O_3, 39.6 SiO_2, 1.49 TiO_2, 0.53 Fe_2O_3, 0.39 CaO, 0.15 MgO, and 0.25 alkali oxides; specimen in the form of a slab measuring 18 in. x 18 in.; prepared from well calcined and graded sillimanite bonded with highly refractory plastic clays; fired at 1410 to 1420 C for 60 hrs; fired material showing considerable mullite development; porosity 23.16%; weight lost on ignition 0.08%.
232	79	C	1942	1029, 1253		Sillimanite refractory brick; 4	Similar to the above specimen but in disc form 8 in. in dia and 1 in. thick; steel used as comparative material.
233	247	L	1932	403-749	1	Bauxite brick 20	59.89 Al_2O_3, 36.62 SiO_2, 1.12 CaO, 0.87 Fe_2O_3, 0.66 MgO, and 0.62 TiO_2; density 3.029 g cm^{-3}; porosity 35.9%; gas permeability 0.010 m^3-cm per m^2-hr-mm H_2O.
234	247	L	1932	348-723	1	Sillimanite brick E	59.70 Al_2O_3, 36.88 SiO_2, 1.44 TiO_2, 1.01 Fe_2O_3, 0.50 CaO, 0.07 MgO, and 0.06 Mn_3O_4; density 2.989 g cm^{-3}; porosity 20.7%; gas permeability 0.100 m^3-cm per m^2-hr-mm H_2O.
235	247	L	1932	346-744	1	Sillimanite brick G	59.47 Al_2O_3, 37.00 SiO_2, 1.54 TiO_2, 0.91 Fe_2O_3, 0.44 CaO, 0.09 MgO, and 0.06 Mn_3O_4; density 3.053 g cm^{-3}; porosity 20.6%; gas permeability 0.073 m^3-cm per m^2-mm H_2O.
236	143	L	1955	723-1313		Magnezit; 1	93.88 MgO, 2.08 SiO_2, 0.83 (0.05 TiO_2) Al_2O_3, 1.63 Fe_2O_3, 1.24 CaO, and 0.20 total Ca, Mg, Fe, and Mn; magnesite basic refractory brick; density 2.81 g cm^{-3}; apparent porosity 22.0%; gas permeability 1.34 ml $m^{-2}\ hr^{-1}$ per mm H_2O.
237	143	L	1955	713-1303		Magnezit; 2	Similar to the above specimen.
238	143	L	1955	773-1348		Magnezit; 3	Similar to the above specimen.

Curve No.	Ref. No.	Method Used	Year	Temp. Range, K	Reported Error, %	Name and Specimen Designation	Composition (weight percent), Specifications and Remarks
239	247	L	1932	399-740	1	Fire clay brick 17	53.56 SiO_2, 42.23 Al_2O_3, 1.59 Fe_2O_3, 1.01 CaO, 0.83 MgO, and 0.62 TiO_2; density 2.710 g cm^{-3}, porosity 31.5%; gas permeability 3.00 m^3 cm per m^2-hr-mm H_2O.
240	247	L	1932	598-804	1	Fire clay brick 725	<60 SiO_2, >40 Al_2O_3; porosity 21.0%.
241	79	L	1942	709-902		Sillimanite refractory brick; 6	88.29 SiO_2, 9.24 Al_2O_3, 0.89 TiO_2, 0.63 Fe_2O_3, 0.38 CaO, 0.10 MgO, and 0.29 alkali oxides; specimen in the form of a slab measuring 18 in. x 18 in.; made from natural quartzitic silica sands bonded with clays; fired at 1360 C for 14 hrs; after being fired specimen is of fine texture, containing finely divided cristobalite and considerable residual free quartz in a clayey matrix; porosity 21.36%; weight lost on ignition 0.14%.
242	79	C	1942	1023, 1225		Sillimanite refractory brick; 6	Similar to the above specimen but in disc form 8 in. in dia and 1 in. in thickness; steel used as comparative material.
243	79	L	1942	679-822		Sillimanite refractory brick; 7	89.11 SiO_2, 9.04 Al_2O_3, 0.75 TiO_2, 0.53 Fe_2O_3, 0.13 MgO, 0.10 CaO, and 0.17 alkali oxides; specimen in the form of a slab measuring 18 in. x 18 in.; similar raw material to the above specimen No. 6; fired at 1360 C for 14 hrs; mineralogical constitution of the fired product similar to the above specimen No. 6; porosity 24.10%; weight lost on ignition 0.12%.
244	79	C	1942	1006, 1252		Sillimanite refractory brick; 7	Similar to the above specimen but in disc form 8 in. in dia and 1 in. in thickness; steel used as comparative material.
245	80	L	1934	382, 877		Dense fire clay brick (Mexko-brand)	52.5 SiO_2, 42.07 Al_2O_3, 2.0 TiO_2, 1.6 Fe_2O_3, 0.5 CaO, trace MgO, and 0.6 alkali; approx composition; bulk density 143 lb ft^{-3}; porosity 15.2%.
246	80	L	1934	733-1527		Dense fire clay brick (Mexko-brand)	The above specimen measured with insulating brick placed between the calorimeter and the lower surfaces of the brick.
247	238	P	1921	873-1373		Fire brick E	57.9 SiO_2, 32.96 Al_2O_3; very close structure; abundance of fine grained rounded grog and a little larger grained grog; exceptionally good adherence; marked by a fair number of black cores, generally with cavities; faces smooth and edges sharp; brick size 9 x 4.5 x 2.5 in.; porosity 15.9%; heat flow in the length-wise direction.
248	238	P	1921	773-1373		Fire brick F	67.49 SiO_2, 27.15 Al_2O_3; very open texture; abundance of rounded clay grog of uneven grading, some grains approximating to pebbles; unweathered pellets detected; adherence poor - in fact, material is very friable; highly fissured; brick size 9 x 4.5 x 2.5 in.; porosity 24.6%; heat flow in the direction of the length of the brick.
249	238	P	1921	773-1373		Retort material G	67.1 SiO_2, 27.17 Al_2O_3, very open texture; very heavily grogged with medium to fine rounded material of uneven grading; abundance of small fissures; adherence of grog very poor; matrix appears to have contracted away from the grog; brick size 9 x 4.5 x 2.5 in.; porosity 24%; heat flow in the direction of the length of the brick.

SPECIFICATION TABLE NO. 291 (continued)

Curve No.	Ref. No.	Method Used	Year	Temp. Range, K	Reported Error, %	Name and Specimen Designation	Composition (weight percent), Specifications and Remarks
250	238	P	1921	773-1373		Retort material H	65.7 SiO_2, 28.47 Al_2O_3; somewhat closer in texture than G; heavily grogged with rounded material of slightly more even grading than G; adherence as a whole fairly good, although some are easily detached; some fissures; very white color with well-defined skin; brick size 9 x 4.5 x 2.5 in.; porosity 28.2%; heat flow in the direction of the length of the brick.
251	238	P	1921	773-1373		Retort material I	72.46 SiO_2, 23.65 Al_2O_3; very close in texture; abundant grog which, tending to be rounded, is evenly graded, possibly some quartz fragments; black cores present, but scarce; tendency toward layering; fissures, present but scarce, are parallel to outside faces; superficial skin; signs of possible reduction toward end of fire; brick size 9 x 4.5 x 2.5 in.; porosity 24.7%; heat flow in the direction of the length of the brick.
252	383	L	1927	473-1623	±15	Kaolin firebrick	52.02 SiO_2, 45.92 Al_2O_3, 1.51 Fe_2O_3, 0.35 TiO_2, and traces of alkalis; specimen 10.8 cm in dia and 22.8 cm long; apparent density 2.66 g cm^{-3}, bulk density 2.36 g cm^{-3}; porosity 10.8%; made of sedimentary kaolin by mixing 65% of 20-mesh prefired grog and 35% of raw clay, and firing to 1575 C for 4 hrs.
253	383	L	1927	473-1773	±15	Kaolin firebrick	52.02 SiO_2, 45.92 Al_2O_3, 1.51 Fe_2O_3, 0.35 TiO_2, and traces of alkalis; specimen 10.8 cm in dia and 22.8 cm long; apparent density 2.68 g cm^{-3}, bulk density density 2.10 g cm^{-3}; porosity 23.2%; made of sedimentary kaolin by mixing 65% of 4-mesh prefired grog and 35% of raw clay, and firing to 1575 C for 4 hrs.
254	383	L	1927	473-1673	±15	Kaolin firebrick	52.02 SiO_2, 45.92 Al_2O_3, 1.51 Fe_2O_3, 0.35 TiO_2, and traces of alkalis; specimen 10.8 cm in dia and 22.8 cm long; apparent density 2.50 g cm^{-3}, bulk density 1.27 g cm^{-3}; porosity 49.1%.
255	247	L	1932	423-808	1	Silica brick 1	95.12 SiO_2, 2.37 CaO, 0.55 Al_2O_3, 0.69 Fe_2O_3, 0.19 MgO, and 0.72 TiO_2; density 2.342 g cm^{-3}; porosity 19.0%; gas permeability 0.184 m^3-cm per m^2-hr-mm H_2O.
256	247	L	1932	377-798	1	Silica brick 8	94.02 SiO_2, 2.98 CaO, 0.91 Al_2O_3, 0.79 Fe_2O_3, 0.22 MgO, and 0.50 TiO_2; density 2.327 g cm^{-3}; porosity 23.1%; gas permeability 0.582 m^3-cm per m^2-hr-mm H_2O.
257	247	L	1932	372-744	1	Silica brick 9	92.26 SiO_2, 3.26 CaO, 1.84 Al_2O_3, 0.55 Fe_2O_3, 0.23 MgO, and 0.41 TiO_2; density 2.350 g cm^{-3}; porosity 27.6%; gas permeability 1.750 m^3-cm per m^2-hr-mm H_2O.
258	210		1952	644-1366		Dense brick; A	4.5/5.0 CaO, $<$2.0 HfO_2, 0.2-0.5 Fe_2O_3, 0.5-1.0 SiO_2, 0.4-1.0 TiO_2, and balance of ZrO_2; stabilized; density (25 C) 4.0 g cm^{-3}; 28 vol % pores.
259	210		1952	644-1366		Insulating brick; B	4.5/5.0 CaO, $<$2.0 HfO_2, 0.2-0.5 Fe_2O_3, 0.5-1.0 SiO_2, 0.4-1.0 TiO_2, and balance of ZrO_2; stabilized; density (25 C) 2.72 g cm^{-3} and 50 vol % pores.
260	210		1952	644-1366		Insulating brick; C	4.5/5.0 CaO, $<$2.0 HfO_2, 0.2-0.5 Fe_2O_3, 0.5-1.0 SiO_2, 0.4-1.0 TiO_2, and balance of ZrO_2; stabilized; density (25 C) 1.81 g cm^{-3} and 68 vol % pores.

SPECIFICATION TABLE NO. 291 (continued)

Curve No.	Ref. No.	Method Used	Year	Temp. Range, K	Reported Error, %	Name and Specimen Designation	Composition (weight percent), Specifications and Remarks
261	374	C	1965	420-1106	0-±2		Type C, lime-stabilized zirconia, 93.7 ZrO_2, 3.35 CaO, 1.38 HfO_2, 0.30 SiO_2, 1.07 Al_2O_3, 0.17 Fe_2O_3, and 0.03 TiO_2; composed of polygonal grains of anisotropic material with most grains in the range 0.10 to 0.15 mm; prepared to a tolerance of ±0.001 in. in the form of a cylinder 1 in. in dia and 1 in. high; fabricated by the Zirconium Corp. of America; bulk density 5.4 g cm^{-3}; true density 5.7 g cm^{-3}; true porosity 5% of the total volume; pyroceram 9606 as comparative material; unknown and standard used for the first time.
262	374	C	1965	661-1126	0-±3		Similar to the above specimen; unknown used first time, standard used several times up to 1273 K.
263	374	C	1965	417-1116	0-±4		Similar to the above specimen; unknown used several times up to 1373 K, standard used for the first time.
264	374	C	1965	1313, 1373			Similar to the above specimen except alumina AL-300 as reference standard.
265	247	L	1932	356-840	1.0	Corundum brick	78.82 Al_2O_3, 14.72 SiO_2, 2.85 TiO_2, 1.35 Fe_2O_3, 0.66 MgO, 0.39 CaO, and 0.06 Mn_3O_4; density 3.472 g cm^{-3}; porosity 35.3%; gas permeability 0.068 m^3-cm m^{-2} hr^{-1} (mm $H_2O)^{-1}$.
266	383	L	1927	473-1673	±15	Silica firebrick	97 SiO_2; 10.8 cm dia x 22.8 cm long; apparent density 2.34 g cm^{-3}; bulk density 1.64 g cm^{-3}; porosity 30.4%.
267	383	L	1927	473-1673	±15	Penn. firebrick	54.16 SiO_2, 38.84 Al_2O_3, 2.72 TiO_2, 2.70 Fe_2O_3, 1.14 MgO, 0.20 sulphates and 0.10 CaO; 10.8 cm dia x 22.8 cm long; apparent density 2.59 g cm^{-3}; bulk density 1.90 g cm^{-3}; porosity 26.7%.
268	383	L	1927	473-1673	±15	Missouri firebrick	53.12 SiO_2, 43.3 Al_2O_3, 2.48 Fe_2O_3, 0.64 CaO, 0.46 MgO, and 0.15 alkalis; 10.8 cm dia x 22.8 cm long; apparent density 2.64 g cm^{-3}; bulk density 2.15 g cm^{-3}; porosity 18.4%.
269	383	L	1927	473-1673	±15	Zirconia brick	60.44 ZrO_2, 27.26 SiO_2, 7.75 Al_2O_3, 1.60 Fe_2O_3, and 0.04 CaO; 10.8 cm dia x 22.8 cm long; made of South American baddeleyite ore, calcined and crushed into grog, bonded with some fine ground ore, pressed into brick, fired at 1923 K; apparent density 4.87 g cm^{-3}; bulk density 3.43 g cm^{-3}; porosity 29.5%.
270	383	L	1927	473-1673	±15	Spinel firebrick	65 Al_2O_3 and 26 MgO; specimen 10.8 cm in dia and 22.8 cm long; made up by grinding together Grecian magnesite and Dutch Guiana bauxite heated to 2048 K until the mixture became a dense spinel biscuit, then crushed into grog and bonded together with raw magnesite and bauxite in the same proportions to form a brick, then fired to 2023 K; apparent density 3.51 g cm^{-3}; bulk density 2.23 g cm^{-3}; porosity 36.3%.
271	252	L	1933	409-1561		Magnesite brick	86.8 MgO, 6.3 Fe_2O_3, 3.0 CaO, 2.6 SiO_2, and 0.8 Al_2O_3; bulk density 2.54 g cm^{-3}; true density 3.59 g cm^{-3}; porosity 26-29%.

SPECIFICATION TABLE NO. 291 (continued)

Curve No.	Ref. No.	Method Used	Year	Temp. Range, K	Reported Error, %	Name and Specimen Designation	Composition (weight percent), Specifications and Remarks
272	79	L	1942	582-752		Silica refractory brick; No. 5	95.16 SiO_2, 1.96 CaO, 1.57 TiO_2, 1.46 Al_2O_3, 0.85 Fe_2O_3, 0.08 MgO, and 0.21 alkali oxides; slab specimen 18 in. x 18 in.; prepared from ganister-type quartzite rocks; fired at 1410-1420 C for 60 hrs; density 1.81 g cm^{-3}; porosity 22.77%; weight lost on ignition 0.14%.
273	79	C	1942	1091, 1223		Silica refractory	Similar to the above specimen but in disc form of dimensions 8 in. dia x 1 in. thick; steel used as comparative material.
274	80	L	1934	393-1101		Star-brand brick	95.9 SiO_2, 2.0 CaO, 1.0 Al_2O_3, 1.0 Fe_2O_3, 0.1 MgO, and 0.1 alkalis; supplied by Harbison-Walker Refractories Co.; bulk density 1.52 g cm^{-3}; porosity 28.0%.
275	80	L	1934	1569		Star-brand brick	The above specimen measured with insulating brick placed between the calorimeter and the lower surface of the specimen.
276	238	P	1921	974-1256		Silica brick; $1A_1$	95.4 SiO_2, 1.68 CaO, and 0.90 Al_2O_3; 9 x 4.5 x 2.5 in.; texture very open and many large and sub-angular rock fragments; bonding of coarse and fine fairly good, although adherence of some of the grains is only fair; abundant large fissures; apparent density 1.75 g cm^{-3}; porosity 24.0%; heat flow in the direction of the length of brick with thermocouple at a distance of 4.0 cm from the hot face; thermal conductivity values calculated from author's measured thermal diffusivity data and the specific heat data of Bradshaw and Emergy (Trans., <u>19</u>, 84, 1919).
277	238	P	1921	940-1208		Silica brick; $1A_2$	The above specimen measured with thermocouple at a distance of 5.4 cm from the hot face.
278	238	P	1921	968-1229		Silica brick; $2A_3$	Similar to the above specimen except apparent density 1.80 g cm^{-3} and porosity 22.3%; heat flow in the direction of the length of the brick with thermocouple at a distance of 4.3 cm from the hot face.
279	238	P	1921	830-1159		Silica brick; $2A_4$	The above specimen measured with thermocouple at a distance of 6.4 cm from the hot face.
280	80	L	1934	404, 551		SiC brick, refrax	96.9 SiC, 1.3 Al_2O_3, 1.3 Fe_2O_3, and 0.7 SiO_2; recrystallized; supplied by Carborundum Co.; bulk density 2.18 g cm^{-3}; porosity 34.4%.
281	80	L	1934	1063-1582		SiC brick, refrax	The above specimen measured with insulating brick placed between the calorimeter and the lower surface of the brick.
282	547	L	1938	434-739		Diatomaceous insulating brick	9 x 9 x 3 in.
283	547	L	1938	433-751		Diatomaceous insulating brick	5 x 3 x 3 in.
284	547	L	1938	446-754		Diatomaceous insulating brick	2 in. dia x 6.8 cm thick

SPECIFICATION TABLE NO. 291 (continued)

Curve No.	Ref. No.	Method Used	Year	Temp. Range, K	Reported Error, %	Name and Specimen Designation	Composition (weight percent), Specifications and Remarks
285	547	L	1938	505-732		Diatomaceous insulating brick	2 in. dia x 6. 8 cm thick.
286	547	L	1938	527-732	3. 2	Diatomaceous insulating brick	9 x 9 x 3 in.
287	547	L	1938	396-705	3. 2	Diatomaceous insulating brick	9 x 9 x 3 in.
288	547	L	1938	468-723	3. 2	Diatomaceous insulating brick	9 x 9 x 3 in.
289	547	L	1938	501-719	3. 2	Diatomaceous insulating brick	9 x 9 x 3 in.
290	547	L	1938	413-702	3. 2	Diatomaceous insulating brick	9 x 9 x 3 in.
291	547	L	1938	462-722	3. 2	Diatomaceous insulating brick	9 x 9 x 3 in.

DATA TABLE NO. 291 THERMAL CONDUCTIVITY OF BRICKS

[Temperature, T, K; Thermal Conductivity, k, Watt $cm^{-1} K^{-1}$]

T	k
CURVE 1	
488.8	0.00379
684.9	0.00400
CURVE 2	
706.0	0.00414
747.1	0.00415
971.0	0.00434
973.2	0.00434*
1169.4	0.00440
1358.3	0.00449
CURVE 3	
681.2	0.0136
824.2	0.0146
CURVE 4	
688.2	0.0281
830.2	0.0289
CURVE 5	
553.2	0.00988
923.2	0.00988
CURVE 6*	
373.2	0.00383
553.2	0.00383
CURVE 7	
293.2	0.0407
CURVE 8	
283.2	0.00344
CURVE 9*	
283.2	0.00341
CURVE 10	
283.2	0.00409
CURVE 11	
293.2	0.00872
CURVE 12	
763.2	0.0315
912.2	0.0281
CURVE 13*	
743.2	0.0253
892.7	0.0238
CURVE 14	
696.2	0.0296
846.2	0.0286
CURVE 15	
347.3	0.0112
657.6	0.0152
974.8	0.0188
1070.2	0.0206
CURVE 16	
516.0	0.0112
662.1	0.0120
CURVE 17	
632.1	0.0116
697.1	0.0120
882.1	0.0129
1097.1	0.0135
1368.8	0.0142
CURVE 18	
477.6	0.00865
710.9	0.0123
724.8	0.0120
955.4	0.0124
1099.8	0.0134
1227.6	0.0143
1394.3	0.0156
1505.4	0.0175
CURVE 19	
544.3	0.00779
727.3	0.0104
899.8	0.0114
1055.3	0.0123
CURVE 20	
638.2	0.00662
663.2	0.00947
668.2	0.00907
763.2	0.00933
838.2	0.00947
888.2	0.00970*
1026.2	0.00982
1158.2	0.0107
1183.2	0.0105
CURVE 21	
283.2	0.00744
CURVE 22	
488.2	0.00168
588.2	0.00182
688.2	0.00197
CURVE 23	
388.2	0.00105
488.2	0.00122
588.2	0.00134
688.2	0.00151
CURVE 24	
700.2	0.0177
847.7	0.0189
CURVE 25	
293.2	0.00460
373.2	0.00452
CURVE 26	
477.6	0.000923
544.3	0.00104
666.5	0.00121
810.9	0.00144
988.7	0.00167
1005.4	0.00164
CURVE 27	
477.6	0.00121
699.8	0.00144
922.1	0.00167
1144.3	0.00196
CURVE 28	
477.6	0.00121*
499.8	0.00121
588.7	0.00127
699.8	0.00138
788.7	0.00156
944.3	0.00167
1033.2	0.00173
1188.7	0.00190
CURVE 29	
477.6	0.00202
699.8	0.00242
922.1	0.00288
1144.3	0.00358
1366.5	0.00448
CURVE 30	
455.4	0.00190
522.1	0.00208
610.9	0.00213
633.2	0.00219
766.5	0.00242
788.7	0.00260
922.1	0.00271
988.7	0.00288
1194.3	0.00335
CURVE 31	
522.1	0.00186
588.7	0.00196
660.9	0.00199
677.6	0.00202
699.8	0.00203*
722.1	0.00208
855.4	0.00219
922.1	0.00221
1010.9	0.00226
1055.4	0.00231
1144.3	0.00239
1166.5	0.00237
CURVE 32	
488.7	0.00107
544.3	0.00115
555.4	0.00115
633.2	0.00133
655.4	0.00133
699.8	0.00156
766.5	0.00153
766.5	0.00170
810.9	0.00179
877.6	0.00205
888.7	0.00202
988.7	0.00219
1010.9	0.00248
1099.8	0.00254
CURVE 33	
466.5	0.000865
544.3	0.000923
633.2	0.000981
666.5	0.00115
794.3	0.00133
816.5	0.00138
966.5	0.00162
999.8	0.00162
1233.2	0.00208
CURVE 34	
444.3	0.000779
533.2	0.000865
633.2	0.000981*
644.3	0.000952
766.5	0.00120
794.3	0.00117
955.4	0.00144
1010.9	0.00144
1233.2	0.00187
CURVE 35	
488.7	0.00120
577.6	0.00130
649.8	0.00138
733.2	0.00148
822.1	0.00159
916.5	0.00169
1005.4	0.00182
1122.1	0.00199
CURVE 36	
283.2	0.0129
CURVE 37	
283.2	0.0115
CURVE 38	
405.4	0.0183
560.9	0.0198
710.9	0.0235
988.7	0.0322
1016.5	0.0507
1183.2	0.0371
1327.6	0.0454
1533.2	0.0855
CURVE 39	
1009.3	0.0433
1199.8	0.0537
1377.6	0.0655
CURVE 40	
368.7	0.00174
449.7	0.00177
510.2	0.00214
609.5	0.00204
664.1	0.00230
673.5	0.00222
921.2	0.00270*
1098.5	0.00305
1225.3	0.00342
CURVE 41	
317.8	0.00195
347.5	0.00196
375.8	0.00229
414.2	0.00197
438.6	0.00215
441.9	0.00209
708.0	0.00236
898.9	0.00277
1047.6	0.00308

* Not shown on plot

DATA TABLE NO. 291 (continued)

T	k
CURVE 42	
573.2	0.00266
673.2	0.00287
773.2	0.00315
873.2	0.00351
973.2	0.00394
CURVE 43	
571.2	0.00322
678.2	0.00322
783.2	0.00326
896.2	0.00351
1014.2	0.00372
CURVE 44	
573.2	0.00307
648.2	0.00310
740.2	0.00333
873.2	0.00362
986.2	0.00395
1046.2	0.00415
CURVE 45*	
576.2	0.00302
648.2	0.00312
741.2	0.00333
870.2	0.00366
982.2	0.00395
1042.2	0.00415
1082.2	0.00432
CURVE 46	
510.2	0.0246
574.2	0.0263
626.2	0.0273
674.2	0.0292
677.2	0.0357
705.2	0.0303
731.2	0.0311
783.2	0.0328
830.2	0.0346
848.2	0.0333
883.2	0.0369
941.2	0.0388
996.2	0.0414
1021.2	0.0454
CURVE 47	
553.2	0.0264
636.2	0.0290
729.2	0.0306*
810.2	0.0339
877.2	0.0361
942.2	0.0382
1008.2	0.0403
1040.2	0.0414
CURVE 48	
655.2	0.00500
843.2	0.00523
1038.2	0.00558
CURVE 49*	
283.2	0.0116
CURVE 50	
283.2	0.0134
CURVE 51	
515.4	0.00162
562.1	0.00163
609.3	0.00166
658.7	0.00169
CURVE 52	
512.1	0.00173
564.8	0.00176
605.9	0.00179
655.9	0.00182
CURVE 53*	
523.2	0.00162
568.2	0.00163
598.7	0.00164
644.3	0.00166
CURVE 54*	
530.9	0.00175
595.4	0.00176
643.2	0.00177
CURVE 54(cont.)*	
684.3	0.00179
CURVE 55	
505.4	0.00183
560.9	0.00187
611.5	0.00193
673.2	0.00199*
CURVE 56*	
506.5	0.00190
556.5	0.00193
600.4	0.00196
652.6	0.00198
CURVE 57	
352.6	0.0180
519.3	0.0192
558.2	0.0206
660.9	0.0219
688.7	0.0234
847.1	0.0262*
999.8	0.0281
1183.2	0.0312
1235.9	0.0327
1294.3	0.0335
1433.2	0.0405*
1449.8	0.0410
CURVE 58	
422.1	0.0234
649.8	0.0327
899.8	0.0392
1055.4	0.0427
1260.9	0.0498
1372.1	0.0587
1477.6	0.0648
CURVE 59	
438.7	0.0219
513.7	0.0252
649.8	0.0284*
899.8	0.0329
1177.6	0.0408
1355.4	0.0470
CURVE 59(cont.)	
1422.1	0.0493
1449.8	0.0548
CURVE 60*	
560.9	0.0226
616.5	0.0242
727.6	0.0248
838.7	0.0267
1116.5	0.0293
1299.8	0.0342
1338.7	0.0355
CURVE 61*	
841.2	0.0184
CURVE 62*	
738.2	0.0124
CURVE 63	
586.7	0.0883
741.7	0.0753
883.2	0.0607
CURVE 64*	
813.2	0.0251
1023.2	0.0251
CURVE 65	
763.2	0.0898
973.2	0.0703
1208.2	0.0517
1373.2	0.0668
CURVE 66	
284.1	0.00726
CURVE 67	
284.3	0.00649
CURVE 68	
284.1	0.00604
CURVE 69	
273.2	0.00500
283.2	0.00511
293.2	0.00523
303.2	0.00535
CURVE 70	
273.2	0.00581
283.2	0.00604*
293.2	0.00639
303.2	0.00662
CURVE 71	
273.2	0.00430
283.2	0.00442
293.2	0.00453
303.2	0.00465
CURVE 72*	
285.4	0.00604
CURVE 73*	
284.1	0.00604
CURVE 74	
282.0	0.00814
CURVE 75	
280.7	0.00779
CURVE 76*	
282.4	0.00721
CURVE 77	
283.8	0.00697
CURVE 78*	
287.3	0.00662
CURVE 79*	
283.0	0.00639
CURVE 80	
292.1	0.00401
CURVE 81*	
288.2	0.00430
CURVE 82*	
294.9	0.00386
CURVE 83*	
283.6	0.00401
CURVE 84*	
300.3	0.00389
CURVE 85*	
283.2	0.00397
CURVE 86	
283.2	0.00374
CURVE 87*	
283.2	0.00511
CURVE 88*	
283.2	0.00511
CURVE 89	
283.2	0.00535
CURVE 90	
283.2	0.00558
CURVE 91	
283.2	0.00616
CURVE 92*	
283.2	0.00633
CURVE 93	
349.9	0.0102
354.8	0.00972
512.8	0.0112*
605.8	0.0121
708.7	0.0131
836.3	0.0141*
1072.8	0.0163
1172.8	0.0181
CURVE 94	
425.3	0.0102
434.6	0.00966
575.6	0.0118
665.8	0.0124
766.9	0.0129
916.6	0.0137*
1176.4	0.0161
1277.1	0.0178
CURVE 95	
502.8	0.0107
709.1	0.0117
1120.7	0.0136
CURVE 96*	
457.1	0.00988
648.5	0.0123
1029.7	0.0142

* Not shown on plot

DATA TABLE NO. 291 (continued)

T	k
CURVE 97*	
481.8	0.0110
749.2	0.0142
774.1	0.0137
803.5	0.0143
1078.2	0.0159
CURVE 98	
543.8	0.00984
663.2	0.0107
CURVE 99*	
731.0	0.0119
862.7	0.0128
1053.3	0.0138
1367.2	0.0155
CURVE 100	
589.2	0.0123
700.4	0.0132
811.6	0.0141*
922.8	0.0150
1034.0	0.0160*
1145.2	0.0170
1189.7	0.0175
CURVE 101*	
505.7	0.0112
589.2	0.0120
700.4	0.0130
811.6	0.0141
922.8	0.0150
1034.0	0.0160
1145.2	0.0170
1189.7	0.0175
CURVE 102*	
455.8	0.00959
478.0	0.00981
589.2	0.0112
700.4	0.0123
811.6	0.0133
922.8	0.0144
1034.0	0.0153
CURVE 102 (cont.)*	
1145.2	0.0164
1189.7	0.0168
CURVE 103*	
489.1	0.0105
589.2	0.0117
700.4	0.0127
811.6	0.0138
922.8	0.0147
1034.0	0.0157
1145.2	0.0167
1189.7	0.0177
CURVE 104*	
555.8	0.0112
589.2	0.0115
700.4	0.0125
811.6	0.0134
922.8	0.0144
1034.0	0.0153
1145.2	0.0164
1189.7	0.0168
CURVE 105*	
500.2	0.0110
567.0	0.0113
605.9	0.0116
711.5	0.0123
728.2	0.0123
800.5	0.0130
883.9	0.0135
1034.0	0.0146
1123.0	0.0152
CURVE 106*	
520.4	0.00172
567.6	0.00176
617.1	0.00182
659.3	0.00186
CURVE 107*	
413.2	0.00198
733.2	0.00198
CURVE 108	
373.2	0.0120
573.2	0.0128
773.2	0.0137
973.2	0.0146
1173.2	0.0160*
CURVE 109*	
373.2	0.0100
573.2	0.0116
773.2	0.0130
973.2	0.0146
1173.2	0.0153
CURVE 110	
373.2	0.0105
573.2	0.0107
773.2	0.0110
973.2	0.0115
1173.2	0.0121
CURVE 111	
373.2	0.00523
573.2	0.00536
773.2	0.00544
973.2	0.00552*
1173.2	0.00565
1248.2	0.00586
CURVE 112	
373.2	0.00297
573.2	0.00335
773.2	0.00381
973.2	0.00427
1173.2	0.00473
1273.2	0.00502
CURVE 113	
423.2	0.307
573.2	0.262
773.2	0.224
CURVE 114	
423.2	0.212
573.2	0.184
773.2	0.161
973.2	0.149
CURVE 115	
423.2	0.0837
573.2	0.0661
773.2	0.0531
973.2	0.0460
CURVE 116	
423.2	0.231
573.2	0.199
773.2	0.171
973.2	0.156
CURVE 117	
423.2	0.162
573.2	0.141
773.2	0.119
973.2	0.109
CURVE 118*	
393.3	0.0089
855.2	0.0127
CURVE 119*	
533.7	0.0105
692.3	0.0118
CURVE 120	
512.5	0.00251
726.2	0.00320
976.1	0.00392*
1243.8	0.00459
CURVE 121	
385.8	0.00229
571.9	0.00297
806.6	0.00363
CURVE 121 (cont.)	
1046.1	0.00425
CURVE 122*	
605.7	0.00297
852.9	0.00358
1024.6	0.00417
1215.1	0.00443
CURVE 123*	
491.4	0.00264
709.2	0.00320
866.4	0.00384
1032.8	0.00404
CURVE 124*	
423.7	0.00216
617.4	0.00291
864.4	0.00313
1015.7	0.00339
1098.7	0.00389
1236.8	0.00430
1320.9	0.00482
CURVE 125*	
361.9	0.00218
509.4	0.00290
717.4	0.00303
810.3	0.00323
978.6	0.00366
1103.6	0.00418
1173.2	0.00462
CURVE 126*	
484.6	0.00271
662.9	0.00297
905.9	0.00349
1033.3	0.00372
1249.7	0.00446
1324.4	0.00470
1420.4	0.00525
CURVE 127*	
420.1	0.00254
546.8	0.00270
751.6	0.00317
851.2	0.00337
1092.4	0.00423
1174.9	0.00443
1249.8	0.00462
CURVE 128	
376.2	0.218
428.2	0.209*
503.2	0.197
553.2	0.188
618.2	0.178
673.2	0.172
773.2	0.159*
863.2	0.151
923.2	0.146
1023.2	0.140
CURVE 129	
373.2	0.186
398.2	0.182
438.2	0.178
463.2	0.174
473.2	0.176
498.2	0.172
533.2	0.167
573.2	0.163
603.2	0.157
673.2	0.151
773.2	0.144
873.2	0.138
973.2	0.132
CURVE 130	
454.2	0.001103
555.2	0.001334
651.2	0.001647
732.2	0.001932
CURVE 131*	
381.8	0.001013
444.9	0.001106
CURVE 131 (cont.)*	
506.7	0.001237
566.8	0.001404
CURVE 132*	
1223.2	0.00372
CURVE 133*	
1223.2	0.00456
CURVE 134	
377.7	0.109
CURVE 135	
372.2	0.121
CURVE 136	
392.7	0.0711
CURVE 137*	
388.2	0.0711
CURVE 138	
381.2	0.0879
CURVE 139	
383.2	0.0837
CURVE 140*	
413.2	0.0121
428.2	0.0117
498.2	0.0117
CURVE 141*	
518.2	0.00916
643.2	0.00900
1183.2	0.0129
CURVE 142	
303.2	0.00101
363.2	0.00110

* Not shown on plot

DATA TABLE NO. 291 (continued)

T	k
CURVE 143*	
547.1	0.00248
668.2	0.00267
690.9	0.00267
812.1	0.00288
875.4	0.00287
1092.6	0.00310
CURVE 144*	
573.2	0.00234
687.1	0.00245
695.4	0.00255
852.6	0.00283
861.5	0.00267
1080.9	0.00291
CURVE 145	
442.6	0.00156
523.7	0.00162
529.8	0.00163
607.1	0.00166
632.6	0.00173
647.6	0.00173
771.0	0.00182
800.4	0.00187
978.7	0.00199
CURVE 146*	
492.1	0.00255
544.3	0.00264
584.8	0.00273
669.8	0.00280
686.5	0.00287
809.3	0.00297
825.4	0.00293
1024.8	0.00320
CURVE 147*	
599.8	0.00235
705.4	0.00241
730.4	0.00260
881.5	0.00270
902.6	0.00291
1120.4	0.00309
CURVE 148*	
472.1	0.000995
583.7	0.00107
605.9	0.00107
767.1	0.00116
790.9	0.00116
1043.2	0.00127
CURVE 149*	
425.9	0.00105
488.7	0.00108
520.9	0.00115
567.6	0.00113
618.2	0.00119
630.9	0.00126
739.8	0.00128
775.9	0.00134
960.9	0.00148
CURVE 150*	
411.5	0.000940
485.9	0.000949
506.5	0.00101
534.8	0.000966
615.9	0.00108
623.2	0.00105
698.7	0.00108
790.9	0.00118
902.6	0.00122
CURVE 151*	
409.8	0.000848
473.7	0.000922
505.9	0.000900
547.1	0.000969
612.1	0.000956
617.1	0.00100
741.5	0.00107
784.3	0.00110
976.5	0.00119
CURVE 152*	
598.2	0.00988
688.7	0.0102
786.5	0.0103
888.2	0.0111
992.1	0.0111
1062.6	0.0119
1135.9	0.0117
1220.4	0.0123
CURVE 153*	
623.2	0.0109
741.5	0.0116
903.7	0.0139
1186.5	0.0165
CURVE 154	
522.1	0.0626
670.4	0.0642
789.8	0.0727
872.6	0.0762
1068.7	0.0809
CURVE 155	
473.2	0.0142
673.2	0.0155
873.2	0.0163
1073.2	0.0167
1273	0.0167
1473	0.0168*
1673	0.0168
CURVE 156	
473.2	0.0573
673.2	0.0523
873.2	0.0431
1073	0.0397
1273	0.0377
1473	0.0364
1673	0.0356
CURVE 157*	
473.2	0.0259
673.2	0.0310
873.2	0.0343
1073	0.0372
1273	0.0397
1473	0.0423
1673	0.0439
CURVE 158*	
293.2	0.0050
CURVE 159	
1223	0.0279
CURVE 160	
1223	0.0269
CURVE 161	
373.2	0.0276
CURVE 162*	
373.2	0.0277
CURVE 163*	
373.2	0.00732
CURVE 164	
373.2	0.00732
CURVE 165	
373.2	0.0188
CURVE 166*	
373.2	0.0188
CURVE 167	
1773	0.156
1573	0.155
1373	0.153
1173	0.151
CURVE 168*	
373.2	0.0121
CURVE 169	
373.2	0.0124
CURVE 170*	
373.2	0.00950
CURVE 171*	
373.2	0.00937
CURVE 172*	
373.2	0.00448
CURVE 173*	
373.2	0.00435
CURVE 174*	
373.2	0.0164
CURVE 175	
373.2	0.0221
CURVE 176*	
373.2	0.0101
CURVE 177*	
373.2	0.00674
CURVE 178*	
373.2	0.00569
CURVE 179*	
373.2	0.00523
CURVE 180*	
373.2	0.00661
CURVE 181*	
373.2	0.00364
CURVE 182*	
373.2	0.00435
CURVE 183*	
773.2	0.0121
873.2	0.0119
973.2	0.0117
1073	0.0115
1173	0.0113
1273	0.0109
1373	0.0105
CURVE 184	
773.2	0.00732
873.2	0.00732
973.2	0.00711
1073	0.00711
1173	0.00690
1273	0.00690
1373	0.00669
CURVE 185*	
379.4	0.0232
490.5	0.0217
793.9	0.0163
CURVE 186*	
683.2	0.0182
903.2	0.0184
1143	0.0188
1263	0.0169
CURVE 187*	
673.2	0.0192
903.2	0.0180
1123	0.0184
1253	0.0171
CURVE 188*	
643.2	0.0175
843.2	0.0173
1033	0.0160
1168	0.0157
CURVE 189*	
658.2	0.0185
903.2	0.0173
1113	0.0179
1248	0.0166
CURVE 190*	
698.2	0.0277
913.2	0.0238
1123	0.0211
1258	0.0201
CURVE 191*	
708.2	0.0223
918.2	0.0217
1133	0.0213
1263	0.0205
CURVE 192*	
653.2	0.0238
898.2	0.0206
1088	0.0189
1233	0.0177
CURVE 193*	
683.2	0.0231
893.2	0.0222
1083	0.0198
1228	0.0189

* Not shown on plot

DATA TABLE NO. 291 (continued)

T	k
CURVE 194	
598	0.071
715.7	0.063
773	0.062
848	0.055
883	0.049
973	0.046*
1076	0.042
1088	0.041
1303	0.038
CURVE 195*	
636.2	0.0158
843.2	0.0155
1036	0.0153
CURVE 196*	
623.2	0.00558
803.2	0.00593
993.2	0.00616
CURVE 197*	
636.2	0.00477
813.2	0.00511
1003	0.00558
CURVE 198*	
503.2	0.00523
589.2	0.00581
701.2	0.00651
798.2	0.00732
CURVE 199*	
505.2	0.00535
593.2	0.00616
709.2	0.00697
797.2	0.00755
CURVE 200	
518.2	0.00681
608.2	0.00651

T	k
CURVE 200 (cont.)	
735.2	0.00721
834.2	0.00814
CURVE 201*	
783.2	0.00523
913.2	0.00581
1028	0.00662
CURVE 202*	
758.2	0.00453
888.2	0.00418
973.2	0.00511
CURVE 203*	
666.2	0.0109
871.2	0.0115
1064	0.0120
CURVE 204	
698.2	0.00872
907.2	0.00907
1101	0.00953
CURVE 205*	
873.2	0.00351
973.2	0.00418
1073	0.00502
1173	0.00565
1273	0.00628
1373	0.00690
CURVE 206*	
696.2	0.0117
772.2	0.0117
880.2	0.0121
CURVE 207*	
1062	0.0126
1254	0.0126

T	k
CURVE 208*	
661.2	0.00674
858.2	0.00709
1048	0.00744
CURVE 209*	
513.2	0.00558
603.2	0.00628
729.2	0.00709
827.2	0.00779
CURVE 210*	
753.2	0.00488
885.2	0.00558
993.2	0.00616
CURVE 211*	
668.2	0.0100
858.2	0.0108
1053	0.0115
CURVE 212*	
668.2	0.0102
863.2	0.0109
1053	0.0113
CURVE 213*	
399.2	0.0156
507.5	0.0149
851.3	0.0132
CURVE 214*	
688.2	0.0105
753.2	0.0109
857.2	0.0113
CURVE 215*	
1066.	0.0121
1301	0.0126

T	k
CURVE 216*	
618.2	0.00837
660.2	0.00879
784.2	0.00962
CURVE 217*	
1009	0.0109
1242	0.0113
CURVE 218*	
530.7	0.00269
540.7	0.00276
573.2	0.00290
610.2	0.00305
620.2	0.00309
670.2	0.00319
723.2	0.00343
753.2	0.00355
798.2	0.00361
833.2	0.00376
CURVE 219*	
593.2	0.00366
620.2	0.00361
681.2	0.00368
725.2	0.00368
759.2	0.00379
788.2	0.00384
818.2	0.00389
860.2	0.00401
881.2	0.00398
CURVE 220*	
606.2	0.00278
649.2	0.00290
706.2	0.00297
736.2	0.00301
773.2	0.00306
806.2	0.00307
858.2	0.00313

T	k
CURVE 221*	
1623	0.0159
CURVE 222*	
815.7	0.012
908	0.012
978	0.015
1268	0.017
CURVE 223*	
1033	0.014
1203	0.016
CURVE 224*	
1278	0.0066
1293	0.0050
CURVE 225*	
1078	0.0104
CURVE 226*	
1113	0.016
CURVE 227*	
918	0.013
1063	0.015
1191	0.018
CURVE 228*	
773.2	0.00418
873.2	0.00481
973.2	0.00544
1073	0.00628
1173	0.00690
1273	0.00774
1373	0.0105

T	k
CURVE 229*	
393.5	0.00811
507.3	0.00860
817.3	0.0102
CURVE 230*	
873.2	0.00356
973.2	0.00418
1073	0.00502
1173	0.00565
1273	0.00628
1373	0.00690
CURVE 231*	
640.2	0.0167
675.2	0.0167
811.2	0.0176
823.2	0.0176
1088	0.0172
CURVE 232*	
1029	0.0172
1253	0.0176
CURVE 233*	
403.4	0.00907
523.0	0.0115
749.4	0.0117
CURVE 234*	
347.7	0.0140
620.5	0.0146
723.1	0.0141
CURVE 235*	
346.1	0.0156
573.2	0.0151
744.2	0.0147

T	k
CURVE 236*	
723.2	0.0516
953.2	0.0407
1163.2	0.0337
1313.2	0.0284
CURVE 237	
713.2	0.0558
943.2	0.0437
1163	0.0330
1303	0.0293*
CURVE 238*	
773.2	0.0535
973.2	0.0442
1203	0.0354
1348	0.0294
CURVE 239*	
398.9	0.00709
408.6	0.00687
511.6	0.00688
740.0	0.00853
CURVE 240*	
598.2	0.0101
477.9	0.0101
402.0	0.00996
559.6	0.0101
804.0	0.0102
CURVE 241*	
709.2	0.0109
754.2	0.0113
853.2	0.0121
902.2	0.0126
CURVE 242*	
1023	0.0130
1225	0.0146

* Not shown on plot

DATA TABLE NO. 291 (continued)

T	k
CURVE 243*	
679.2	0.0121
748.2	0.0126
822.2	0.0130
CURVE 244*	
1006	0.0134
1252	0.0142
CURVE 245*	
381.5	0.0111
876.5	0.0142
CURVE 246*	
732.6	0.0123
1145	0.0156
1527	0.0173
CURVE 247*	
873.2	0.00418
973.2	0.00502
1073	0.00586
1173	0.00669
1273	0.00816
1373	0.0100
CURVE 248*	
773.2	0.00418
873.2	0.00460
973.2	0.00502
1073	0.00565
1173	0.00628
1273	0.00732
1373	0.00879
CURVE 249*	
773.2	0.00335
873.2	0.00377
973.2	0.00439
1073	0.00502
1173	0.00544
CURVE 249 (cont.)*	
1273	0.00586
1373	0.00732
CURVE 250*	
773.2	0.00397
873.2	0.00460
973.2	0.00502
1073	0.00586
1173	0.00669
1273	0.00732
1373	0.00879
CURVE 251*	
773.2	0.00523
873.2	0.00607
973.2	0.00711
1073	0.00816
1173	0.00879
1273	0.00962
1373	0.0113
CURVE 252*	
473.2	0.0197
673.2	0.0218
873.2	0.0234
1073	0.0251
1273	0.0268
1473	0.0280
1673	0.0293
CURVE 253*	
473.2	0.0142
673.2	0.0159
873.2	0.0172
1073	0.0180
1273	0.0188
1473	0.0197
1673	0.0201
1773	0.0209
CURVE 254*	
473.2	0.00460
673.2	0.00628
873.2	0.00753
1073	0.00837
1273	0.00920
1473	0.00962
1673	0.0100
CURVE 255*	
422.9	0.00927
520.9	0.0103
807.7	0.0112
CURVE 256*	
377.1	0.0115
504.3	0.0126
798.0	0.0149
CURVE 257*	
372.4	0.00564
501.6	0.00787
743.5	0.00939
CURVE 258*	
644.2	0.00764
699.2	0.00764
811.2	0.00764
921.2	0.00764
1033	0.00793
1144	0.00822
1366	0.00894
CURVE 259*	
644.2	0.00469
699.2	0.00469
811.2	0.00476
921.2	0.00490
1033	0.00534
1144	0.00570
1366	0.00649
CURVE 260*	
644.2	0.00209
699.2	0.00209
811.2	0.00216
921.2	0.00245
1033	0.00260
1144	0.00310
1366	0.00392
CURVE 261	
420.2	0.0240
665.2	0.0216
873.2	0.0204
894.2	0.0209
1106	0.0195
CURVE 262*	
661.2	0.0229
662.2	0.0216
665.1	0.0216
763.2	0.0221
854.2	0.0203
889.2	0.0224
895.2	0.0225
985.2	0.0214
1035	0.0200
1126	0.0210
CURVE 263*	
417.2	0.0230
646.2	0.0215
828.2	0.0210
1078	0.0188
1116	0.0184
CURVE 264*	
1313	0.0180
1373	0.016
CURVE 265*	
355.6	0.01012
419.4	0.01051
CURVE 265 (cont.)*	
586.5	0.01147
840.3	0.01146
CURVE 266	
473.2	0.0117
673.2	0.0146
873.2	0.0167
1073	0.0184
1273	0.0201
1473	0.0218
1673	0.0226
CURVE 267*	
473.2	0.0100
673.2	0.0113
873.2	0.0126
1073	0.0134
1273	0.0142
1473	0.0151
1673	0.0155
CURVE 268*	
473.2	0.0100
673.2	0.0126
873.2	0.0146
1073	0.0155
1273	0.0163
1473	0.0172
1673	0.0176
CURVE 269*	
473.2	0.0146
673.2	0.0163
873.2	0.0176
1073	0.0184
1273	0.0192
1473	0.0201
1673	0.0205
CURVE 270*	
473.2	0.0151
673.2	0.0167
CURVE 270 (cont.)*	
873.2	0.0176
1073	0.0188
1273	0.0197
1473	0.0205
1673	0.0213
CURVE 271	
409.3	0.0398
565.4	0.0342
755.4	0.0298
883.2	0.0272
1150	0.0213*
1167	0.0220*
1467	0.0190
1561	0.0193*
CURVE 272*	
582.2	0.0151
683.2	0.0163
752.2	0.0167
CURVE 273*	
1091	0.0172
1223	0.0184
CURVE 274*	
393.0	0.00796
610.4	0.0117
742.1	0.0133
939.3	0.0155
1101	0.0176
CURVE 275*	
1569	0.0223
CURVE 276*	
974.2	0.00360
1061	0.00406
1156	0.00448
1256	0.00515
CURVE 277*	
940.2	0.00377
1020	0.00402
1114	0.00448
1208	0.00519
CURVE 278*	
967.7	0.00368
1045	0.00444
1134	0.00481
1229	0.00590
CURVE 279*	
829.7	0.00393
996.7	0.00460
1080	0.00498
1159	0.00540
CURVE 280	
404.1	0.329
550.9	0.280
CURVE 281	
1063	0.179
1121	0.170
1302	0.144
1582	0.118
CURVE 282	
434.2	0.000920
586.2	0.00113
739.2	0.00130
CURVE 283*	
433.2	0.00109
586.2	0.00120
751.2	0.00132
CURVE 284*	
446.2	0.000920
530.2	0.00103

*Not shown on plot

DATA TABLE NO. 291 (continued)

T	k
CURVE 284 (cont.)*	
614.2	0.00110
754.2	0.00123
CURVE 285	
505.2	0.000937
568.2	0.00979
596.2	0.00100
664.2	0.00107
732.2	0.00115
CURVE 286*	
527.2	0.00111
598.2	0.00121
732.2	0.00132
CURVE 287*	
396.2	0.000996
619.2	0.00121
705.2	0.00128
CURVE 288*	
468.2	0.00104
618.2	0.00120
723.2	0.00128
CURVE 289*	
501.2	0.00112
606.2	0.00120
719.2	0.00132
CURVE 290*	
413.2	0.00100
599.2	0.00115
702.2	0.00126
CURVE 291*	
462.2	0.00104
604.2	0.00118
722.2	0.00129

*Not shown on plot

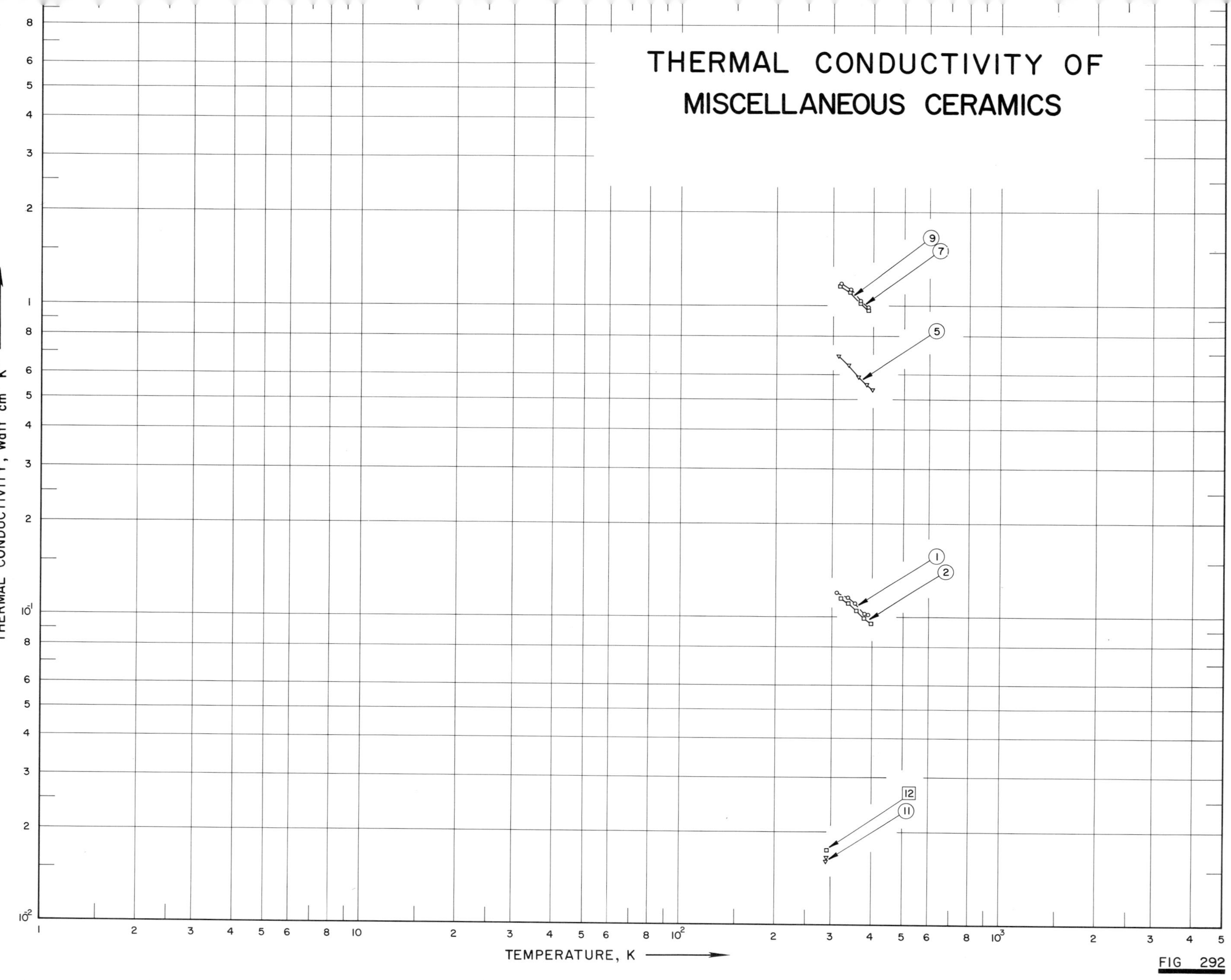

THERMAL CONDUCTIVITY OF
MISCELLANEOUS CERAMICS
9
7
5
1
2
12
11
THERMAL CONDUCTIVITY, Watt cm⁻¹ K⁻¹
TEMPERATURE, K
FIG 292

SPECIFICATION TABLE NO. 292 THERMAL CONDUCTIVITY OF MISCELLANEOUS CERAMICS

[For Data Reported in Figure and Table No. 292]

Curve No.	Ref. No.	Method Used	Year	Temp. Range, K	Reported Error, %	Name and Specimen Designation	Composition (weight percent), Specifications and Remarks
1	87	C	1954	311-389	± 3.0	98 A-1	0.500 in. dia x 0.500 in. long; measured in a vacuum of < 1 μ; copper used as comparative material.
2	87	C	1954	320-397	± 3.0	98 A-2	Second run of the above specimen.
3	87	C	1954	318-414	± 3.0	139 A-1	Similar to the above specimen.
4	87	C	1954	316-408	± 3.0	140 A-1	Similar to the above specimen.
5	87	C	1954	312-399	± 3.0	140 B-1	Similar to the above specimen.
6	87	C	1954	315-444	± 3.0	14 C-1	Similar to the above specimen.
7	87	C	1954	318-386	± 3.0	78 A-1	Similar to the above specimen.
8	87	C	1954	317-395	± 3.0	78 A-2	Second run of the above specimen.
9	87	C	1954	316-389	± 3.0	78 B-1	Similar to the above specimen.
10	88	L	1926	292.0, 292.4			Specimen thickness 12.6 mm; density 2.26 g cm^{-3}.
11	88	L	1926	290.6, 290.7			Specimen thickness 4.7 mm; density 2.26 g cm^{-3}.
12	88	L	1926	293.6	1.0		Specimen dia 100 mm, thickness 14 mm; density 2.25 g cm^{-3}.

DATA TABLE NO. 292 THERMAL CONDUCTIVITY OF MISCELLANEOUS CERAMICS

[Temperature, T, K; Thermal Conductivity, k, Watt $cm^{-1}K^{-1}$]

T	k
CURVE 1	
311 .3	0.120
337.8	0.116
355.3	0.111
379.6	0.103
389.2	0.102
CURVE 2	
320.3	0.115
338.2	0.111
358.5	0.105
378.8	0.0992
397.4	0.0954
CURVE 3*	
317.6	0.120
344.3	0.112
359.6	0.106
376.8	0.103
414.2	0.0941
CURVE 4*	
316.2	0.121
335.8	0.115
357.6	0.106
376.8	0.102
407.7	0.0941
CURVE 5	
312.3	0.686
335.1	0.644
361.9	0.590
382.4	0.556
399.1	0.536
CURVE 6*	
315.2	0.657
338.1	0.649
365.1	0.598
384.9	0.565
443.6	0.540

T	k
CURVE 7	
318.3	1.180
340.8	1.134
366.4	1.042
386.1	0.987
CURVE 8*	
316.5	1.180
338.0	1.138
362.4	1.038
395.0	0.962
CURVE 9	
316.4	1.167
339.6	1.117
366.9	1.029
388.8	0.967
CURVE 10*	
292.0	0.0165
292.4	0.0164
CURVE 11	
290.6	0.0161
290.7	0.0165
CURVE 12	
293.64	0.0175

*Not shown on plot

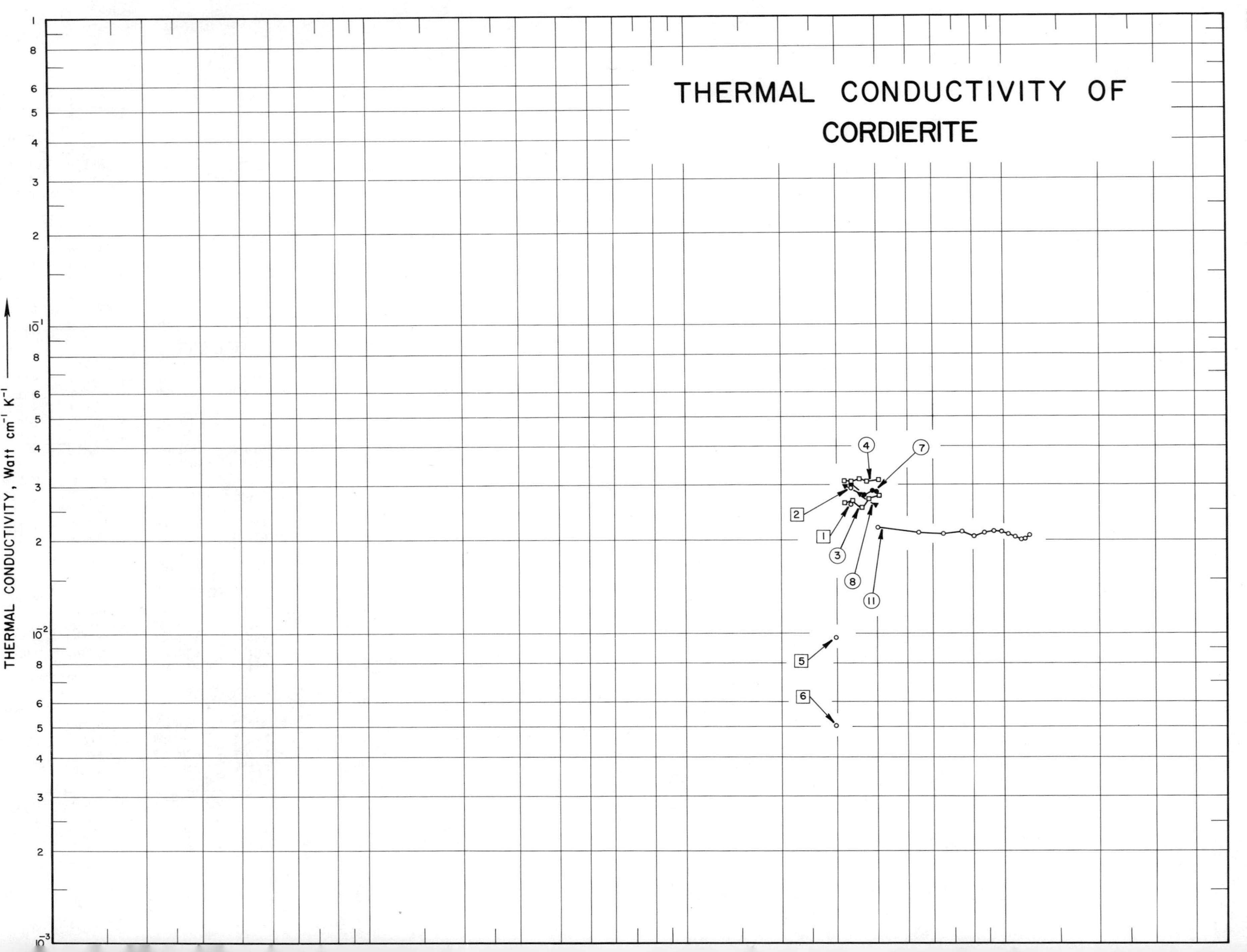
THERMAL CONDUCTIVITY OF
CORDIERITE
THERMAL CONDUCTIVITY, Watt cm⁻¹ K⁻¹
1
8
6
5
4
3
2
10⁻¹
10⁻²
10⁻³
1
2
3
4
5
6
7
8
11

SPECIFICATION TABLE NO. 293 THERMAL CONDUCTIVITY OF CORDIERITE $4(Mg, Fe)O \cdot 4Al_2O_3 \cdot 10SiO_2 \cdot H_2O$

[For Data Reported in Figure and Table No. 293]

Curve No.	Ref. No.	Method Used	Year	Temp. Range, K	Reported Error, %	Name and Specimen Designation	Composition (weight percent), Specifications and Remarks
1	3		1953	333.2			Max. water absorption 0.05%; flexural strength 11,200 psi; coefficient of expansion (25 - 700 C) 2.9 x 10^{-6}.
2	3		1953	333.2		Rutgers	Max. water absorption 0.05%; flexural strength 15,000 psi; coefficient of expansion (25 - 700 C) 2.3 x 10^{-6}.
3	68	C	1954	320-408		274A-1; G-10-N	35.95 calcine 23 D [27.1 hi-grade talc, 20.05 magnesium oxide, 44.9 Edgar plastic Kaolin, 7.59 N. Carolina Kaolin (sparks), and 14.92 $Ba_2P_2O_7$ (Monsanto)], 10.0 old mine No. 4, 29.57 Edgar Plastic Kaolin, and 29.57 N. Carolina Kaolin; fabricated by using 0.5 in. extrusion die; measured in vacuum; copper used as comparative standard.
4	68	C	1954	317-406		43A2; G-10-A	35.95 calcine 23 D (same as above), 10.0 old mine No. 4, and 59.15 Edgar Plastic Kaolin; similar to the above specimen.
5	152	L	1956	298.2		Cordierite 202	Preirradiated with 6 x 10^{19} epithermal neutrons per cm^2 for 480 Mwd in the MTR; weight change -0.6%; specimen 20 mils thick and 0.75 in. in dia.
6	152	L	1956	298.2		Cordierite 202	The above specimen postirradiated with 6 x 10^{19} epithermal neutrons per cm^2 for 480 Mwd in The MTR; weight change -0.6%.
7	468	L	1950	336-399		G-10A	No details reported.
8	468, 407	L	1950	321-398		M-244-A	Commercial cordierite.
9	468, 407	L	1950	328-392		Commercial steatite	No details reported.
10	468, 407	L	1950	320-383		Steatite	No details reported.
11	282	R	1955	403-1208			93% cordierte (composition: 52.1 Florida kaolin [EPK], 34.6 Champion and Challenger ball clay, 7.3 MgO [Westvaco FN-722], 6.0 Sierra talc) was mixed in a ball mill with distilled water and flind pebbles, and after drying, 7% binder (composition: 500 g carbowax, 10 g methocel and 1000 cc water) was added; specimens (ring type) were set on alumina powder, fired at a rate of 120° per hr, and held at a peak of 1400 C for 5 hrs; specimen 5.08 cm O.D., 2.67 cm I.D. and 0.5 in. thickness; eleven rings are stacked to form a cylinder 5.5 in. high but measured only over the centrally-placed rings; density 2.12 g cm^{-3}; 5,000 (estimated) psi mod. rupture; 0.3% apparent porosity; 11.2% shrinkage; 26.2 x 10^{-7} coeff. of exp. (25 - 1000 C).

DATA TABLE NO. 293 THERMAL CONDUCTIVITY OF CORDIERITE $4(Mg, Fe)O \cdot 4Al_2O_3 \cdot 10SiO_2 \cdot H_2O$

[Temperature, T, K; Thermal Conductivity, k, Watt $cm^{-1} K^{-1}$]

T	k
CURVE 1	
333.2	0.0260
CURVE 2	
333.2	0.0294
CURVE 3	
320.0	0.0263
337.9	0.0266
360.6	0.0254
378.1	0.0269
408.4	0.0277
CURVE 4	
317.0	0.0309
335.2	0.0308
354.7	0.0315
372.7	0.0308
406.3	0.0312
CURVE 5	
298.2	0.00962
CURVE 6	
298.2	0.00502
CURVE 7	
336.4	0.0304
365.2	0.0279
387.1	0.0287
399.0	0.0285
CURVE 8	
320.8	0.0297
355.2	0.0280
398.1	0.0259
CURVE 9*	
327.7	0.0311
348.2	0.0309
369.4	0.0306
391.7	0.0294
CURVE 10*	
319.9	0.0245
337.4	0.0266
366.7	0.0278
383.1	0.0263
CURVE 11	
403.2	0.0220
543.2	0.0212
648.2	0.0209
743.2	0.0213
808.2	0.0206
873.2	0.0211
938.2	0.0214
988.2	0.0213
1038.2	0.0209
1088.2	0.0205
1138.2	0.0201
1173.2	0.0202
1208.2	0.0207

*Not shown on plot

SPECIFICATION TABLE NO. 294 THERMAL CONDUCTIVITY OF ENAMEL

Curve No.	Ref. No.	Method Used	Year	Temp. Range, K	Reported Error, %	Name and Specimen Designation	Composition (weight percent), Specifications and Remarks
1	575	R	1953	293.2		Silicon enamel; Heyden L28	From the company VVB (Z) Alcid Chemische Fabrick von Heyden, Radebeul-Dresden; heated at 400 C for about 4 min.

DATA TABLE NO. 294 THERMAL CONDUCTIVITY OF ENAMEL

[Temperature, T, K; Thermal Conductivity, k, Watt $cm^{-1} K^{-1}$]

T	k
CURVE 1*	
293.2	0.00470

*No graphical presentation

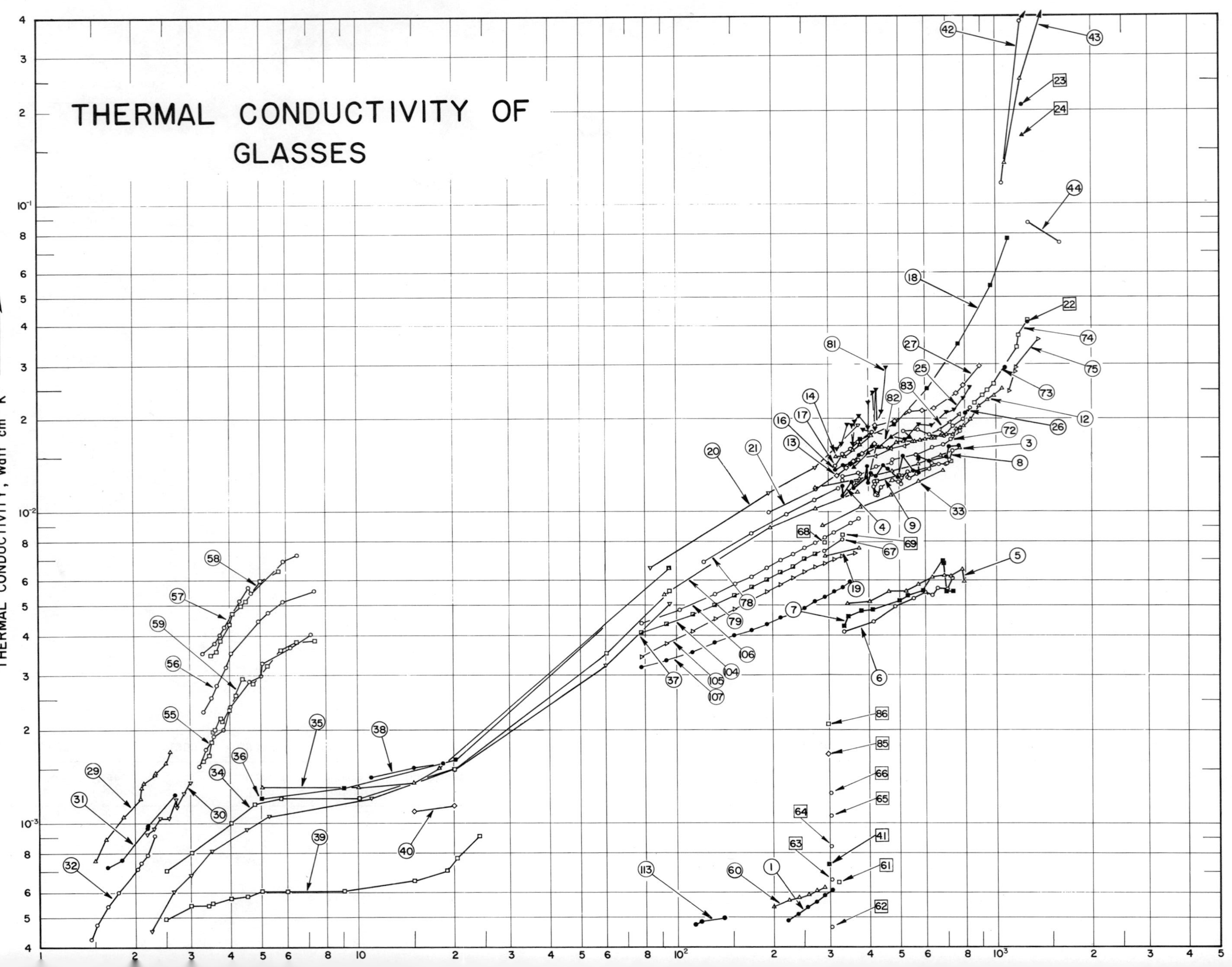
THERMAL CONDUCTIVITY OF GLASSES
THERMAL CONDUCTIVITY, Watt cm^{-1} K^{-1}

SPECIFICATION TABLE NO. 295 THERMAL CONDUCTIVITY OF GLASSES

[For Data Reported in Figure and Table No. 295]

Curve No.	Ref. No.	Method Used	Year	Temp. Range, K	Reported Error, %	Name and Specimen Designation	Composition (weight percent), Specifications and Remarks
1	109	L	1947	222-305		Cellular Glass	Cellular glass block; density 0.165 g cm^{-3}; 1 in. in thickness; as received.
2	110	R	1945	1.3		Thuringian glass	Two tubes of O.D. 0.95 cm, thickness 0.091 cm, and different lengths viz. 5.8 and 2 cm.
3	330	C	1957	298.2		Borosilicate glass	3 in. dia x 0.20 in. thick lapped disk specimen; measured under 100 g load; thermal comparator No. 4 used.
4	111	L	1952	331-768		Pyrex glass 774	Clear chemical glass supplied by Cincinnati Gasket and Packing Co.; density reported as 2.2213 and 2.2205 g cm^{-3} at 0 and 50 C, respectively.
5	112	L	1958	344-800		Aluminate silicate glass 723	14 x 14 x 0.250 in.; supplied by Corning Glass Works.
6	112	L	1958	335-733		Borosilicate glass 3235	14 x 14 x 0.250 in.; supplied by Pittsburgh Plate Glass Co.
7	112	L	1958	337-736		Soda-lime silica plate glass 9330	14 x 14 x 0.250 in.; supplied by Libbey-Owen-Ford Glass Co.
8	33	C	1954	419-736		White plate glass	Disk specimen; supplied by Pittsburgh Plate Glass Co.; measured under a load of ~23 psi; Armco iron used as comparative material.
9	33	C	1954	426-728		White plate glass	The above specimen measured with aluminum foil of 0.00025 in. thick adjacent to specimen surface; same comparative material.
10	33	C	1954	413-714		Solex 2808 plate glass	Similar to above but without the aluminum foil.
11	33	C	1954	421-731		Solex "S" plate glass	Similar to above.
12	33	C	1954	442-1059			Similar to above but supplied by Corning Glass Works.
13	9	L	1953	319-420	±1	Window glass; 248 AI-1	0.497 in. in length, 0.410 in. in dia; density 2.49 g cm^{-3}.
14	9	L	1953	319-410	±1	Borosilicate glass; 249 AI-1	0.496 in. in length, 0.411 in. in dia; density 2.35 g cm^{-3}.
15	9	L	1953	317-407	±1	Lead glass; 250 AI-1	0.494 in. in length, 0.409 in. in dia; density 3.04 g cm^{-3}.
16	9	L	1953	316-411	±1	Plate glass; 251 AI-1	0.496 in. in length, 0.410 in. in dia; density 2.50 g cm^{-3}.
17	9	L	1953	316-412	±1	Pyrex glass; 252 AI-1	0.497 in. in length, 0.410 in. in dia; density 2.22 g cm^{-3}.
18	34	C	1943	377-1092		Silica glass	Pure; fused; 1 cm cubic specimen; 18-8 stainless steel used as comparative material.
19	44	F	1914	293, 373		Soda glass	0.332 cm dia x 6.0 cm long; density 2.59 g cm^{-3}.
20	22	L	1911	83-373		Quartz glass	No details reported.
21	22	L	1911	195-373		Borosilicate crown glass	No details reported.
22	113	L	1954	1268.2		Green glass	0.6 Fe_2O_3.

SPECIFICATION TABLE NO. 295 (continued)

Curve No.	Ref. No.	Method Used	Year	Temp. Range, K	Reported Error, %	Name and Specimen Designation	Composition (weight percent), Specifications and Remarks
23	113	L	1954	1206.2		Colorless glass	0.1 Fe_2O_3.
24	113	L	1954	1216.2		Amber glass	0.2 Fe_2O_3.
25	45	R	1953	387-840		Soda-lime silica glass; 0080	Ellipsoidal specimen supplied by Corning Glass Works.
26	45	R	1953	517-834		Soda-lime silica glass; 0080	Similar to the above specimen.
27	45	R	1953	422-892		Soda-lime silica glass; 0080	Similar to the above specimen.
28	114		1954	489-858		Pyrex glass	No details reported.
29	78	L	1948	1.5-2.6		Jena Gerate 20	Cylindrical specimen of dia 0.78 cm.
30	78	L	1948	2.2-3.0		Jena Gerate 16	Similar to the above specimen.
31	78	L	1948	1.6-2.7		Thuringian glass	Similar to the above specimen.
32	78	L	1948	1.5-2.3		Monax	Similar to the above specimen.
33	302	P	1962	289-685		Pyrex glass 7740	Approx composition: 80.4 SiO_2, 13.3 B_2O_3, 4.4 Na_2O, and 2.0 Al_2O_3; thermal conductivity values calculated from measured thermal diffusivity data and literature data of specific heat and density.
34	69	L	1951	2.5-95		Quartz glass	Specimen dia 6.1 mm, length 2.3 cm.
35	69	L	1951	5.0-95		Quartz glass	Specimen dia 7.7 mm, length 2.25 cm.
36	69	L	1951	5.0-95		Quartz glass	Specimen dia 7.4 mm, length 4.6 cm.
37	69	L	1951	2.3-95		Phoenix glass	A borosilicate glass; specimen dia 6.0 mm, length 2.15 cm.
38	69	L	1951	11-20		Phoenix glass	A borosilicate glass; specimen dia 9.4 mm, length 3.05 cm.
39	330	C	1957	298.2		Borosilicate glass	3 in. dia x 0.25 in. thick lapped disk specimen; thermal comparator No. 4 used.
40	116	L	1949	15, 20		Phoenix glass	A borosilicate glass; specimen dia 6.1 mm, length 27 mm.
41	117	P	1955	297.2		Foamglass	Specimen thickness 14.8 mm; density 0.15 g cm^{-3}; thermal conductivity values calculated from measured transient temp changes.
42	118	→	1952	1050-1800		Window glass	Thermal conductivity values calculated from measured data of spectral absorption constant of the specimen.
43	118	→	1952	1073-1540		Commercial glass	Similar to above.
44	118	→	1952	1273, 1600		X-ray protection glass	Similar to above.
45	152	L	1956	298.2		Plate glass	Wafer specimen 0.75 in. dia x 0.020 in. thick.

SPECIFICATION TABLE NO. 295 (continued)

Curve No.	Ref. No.	Method Used	Year	Temp. Range, K	Reported Error, %	Name and Specimen Designation	Composition (weight percent), Specifications and Remarks
46	152	L	1956	298. 2		Plate glass	The above specimen exposed in 3 x 10^{19} epithermal neutrons per cm^2 for 480 MWD in the MTR.
47	152	L	1956	298. 2		Silica glass	No details reported.
48	152	L	1956	298. 2		Silica glass	The above specimen exposed in 7 x 10^{19} epithermal neutrons per cm^2 for 480 MWD in the MTR.
49	39	C	1960	367-700		White plate glass	Specimen 3 in. in dia and 0. 25 in. in thickness prepared by Pittsburgh Plate Glass Co. ; density 2. 52 g cm^{-3}; Armco iron used as comparative material.
50	39	C	1960	367-700		Pyrex glass 774	Specimen 3 in. in dia and 0. 25 in. in thickness; prepared by Cincinnati Gasket and Packing Co. ; density 2. 22 g cm^{-3} at 273 K; same comparative material as above.
51	39	C	1960	273-573		Pyrex glass 774	Same as the above specimen.
52	39	C	1960	367-700		Solex 2808X	Specimen 3 in. in dia and 0. 25 in. in thickness; prepared by Pittsburgh Plate Glass Co. ; density 2. 53 g cm^{-3} at 273 K; same comparative material as above.
53	39	C	1960	367-700		Solex "S"	Similar to the above specimen except density 2. 52 g cm^{-3} at 273 K.
54	39	C	1960	423-1073		Fused silica	Specimen 3 in. in dia and 0. 25 in. in thickness; density 2. 20 g cm^{-3} at 273 K; same comparative material as above.
55	188	L	1958	3. 2-7. 2		Silica glass	High purity fused silica square rod of 19. 8 mm^2 cross-section supplied by Corning Glass Works; density 2. 2002 g cm^{-3}.
56	188	L	1958	3. 3-7. 4		Silica glass	The above specimen irradiated with 1. 71 x 10^{19} neutrons cm^{-2}; density 2. 24 g cm^{-3}.
57	188	L	1958	3. 3-6. 5		Silica glass	The above specimen again irradiated with 4. 13 x 10^{19} neutrons (total irradiation 5. 84 x 10^{19} n cm^{-2}); density 2. 26 g cm^{-3}.
58	188	L	1958	3. 5-5. 7		Silica glass	The above specimen again irradiated with 3. 5 x 10^{19} n cm^{-2}; density 2. 26 g cm^{-3}.
59	188	L	1958	3. 3-7. 4		Silica glass	The above specimen annealed at 925 C for 9 hrs; density 2. 2045 g cm^{-3}.
60	189		1950	200-288		Foam glass	Density 0. 160-0. 176 g cm^{-3}; negligible water absorption (in vol % per 24 hrs).
61	100	L	1957	310. 9		Foam glass	Bulk density 0. 171 g cm^{-3}.
62	190		1954	303. 2		Foam glass	Density 0. 1 g cm^{-3}.
63	190		1954	303. 2		Foam glass	Density 0. 2 g cm^{-3}.
64	190		1954	303. 2		Foam glass	Density 0. 3 g cm^{-3}.
65	190		1954	303. 2		Foam glass	Density 0. 4 g cm^{-3}.
66	190		1954	303. 2		Foam glass	Density 0. 5 g cm^{-3}.
67	108	L	1920	293, 333		Plate glass	Dia 9 in. , thickness 0. 252 in. ; specific gravity 2. 49; transverse heat flow.
68	108	L	1920	293. 2		Plate glass	Dia 9 in. , thickness 0. 289 in. ; specific gravity 2. 60; transverse heat flow.

SPECIFICATION TABLE NO. 295 (continued)

Curve No.	Ref. No.	Method Used	Year	Temp. Range, K	Reported Error, %	Name and Specimen Designation	Composition (weight percent), Specifications and Remarks
69	108	L	1920	333.2		Plate glass	Dia 9 in., thickness 0.289 in.; transverse heat flow.
70	34	C	1943	388-751		Pyrex glass	80.5 SiO_2, 12.9 B_2O_3, 3.8 Na_2O, 2.2 Al_2O_3, and 0.4 PbO; 1 x 1 x 1 cm; ground; 18-8 stainless steel used as comparative material.
71	34	C	1943	386-611		Soda-lime glass	69.73 SiO_2, 20.96 Na_2O, 9.05 CaO, 0.18 B_2O_3, and trace of K_2O; 1 x 1 x 1 cm; 18-8 stainless steel used as comparative material.
72	303	L	1940	333-773		Pyrex glass 774	Disk specimen obtained from Corning Glass Works; density 2.229 g cm^{-3}.
73	295	C	1960	513-1083		Vycor-brand glass, V-1	Obtained in the leached condition, then heat treated, cut, and polished; sintered; density 2.18 g cm^{-3}; zero porosity; polycrystalline Al_2O_3 and ZrO_2 used as comparative material.
74	295	C	1960	493-1273		Vycor-brand glass, V-2	Obtained in the leached condition, then heat treated, cut, and polished; sintered; density 1.62 g cm^{-3}; porosity 2.57%, pore size 0.005 μ; polycrystalline Al_2O_3 and ZrO_2 used as comparative material.
75	295	C	1960	553-1363		Vycor-brand glass, V-2	Obtained in the leached condition, then heat treated, cut, and polished; sintered; density 2.12 g cm^{-3}, microscopic porosity 4.1%; gravimetric porosity 2.6%; pore size 14-20 μ; polycrystalline Al_2O_3 and ZrO_2 used as comparative material.
76	301	L	1963	273-573		Pyrex glass 7740	Density 2.2258 g cm^{-3} at 24.3 C; refractive index 1.47257 ±0.00003 at 23 C for the D lines of sodium (5893 Å); specimen was somewhat strained; after the thermal conductivity measurement the glass was annealed and the index of refraction decreased to 1.47211 ±0.00003; data reported are the deduced probable values.
77	136	C	1959	343-869	±4	Pyrex glass 7740	Approx composition: 81.0 SiO_2, 12.5 B_2O_3, 4.5 Na_2O, and 2.0 Al_2O_3.
78	147	L	1960	123-373		A	Approx composition: 80.8 SiO_2, 12.8 B_2O_3, 4.2 Na_2O, and 2.2 Al_2O_3; 3 in. dia x 0.375 in. thick; density 2.22 g cm^{-3}.
79	187	R	1932	92-523			80.5 SiO_2, 12.5 B_2O_3, 4.0 Na_2O, and 2.0 Al_2O_3; density 2.233 g cm^{-3} at 21 C.
80	304	L	1961	273-773		Pyrex glass 7740	Cylindrical specimen of dia 2.540 cm; tentative data.
81	321	P	1961	297-458	±5	Pyrex glass 7740	Tubing; resistance 8.5-21.7 ohm; value of (ρ ck) measured by platinum film gage deposited parallel to axis of the tube; value of specific heat C taken from calculated data by Kelly, K. K., U. S. Dept. of the Interior, Bur. of Mines Bull. No. 476, 1949.
82	321	P	1961	294-544		Soda-lime plate glass	Tubing; resistance 12.3 ohm; value of (ρ ck) measured by platinum film gage deposited parallel to axis of the tube; value of specific heat C taken from calculated data by Kelly, K. K., U. S. Dept. of the Interior, Bur. of Mines Bull. No. 476, 1949.
83	303	L	1940	323-773		Silica glass	1.500 in. dia and 0.250 in. thick; obtained from Thermal Syndicate, Ltd.; density 2.199 g cm^{-3}.

Curve No.	Ref. No.	Method Used	Year	Temp. Range, K	Reported Error, %	Name and Specimen Designation	Composition (weight percent), Specifications and Remarks
84	430	P	1966	298.2		Pyrex	Spherical beads supplied by Minnesota Mining and Mfg. Co.; specimen contained in a 0.75 in. dia x 2 in. long cylindrical cell; mesh size -140 + 200; thermal conductivity measured by using the transient line source method; the heat source was a 36-gauge constantan wire contained in a 0.025 in. O.D. hypodermic tube soldered along the axis of the cylindrical cell; data calculated from measured line temp at two certain times; measured in Freon-12 under a pressure of ~100 psig.
85	430	P	1966	298.2		Pyrex	Similar to the above specimen; measured in argon under a pressure of ~100 psig.
86	430	P	1966	298.2		Pyrex	Similar to the above specimen; measured in nitrogen under a pressure of ~100 psig.
87	430	P	1966	298.2		Pyrex	Similar to the above specimen; measured in methane under a pressure of ~100 psig.
88	430	P	1966	298.2		Pyrex	Similar to the above specimen; measured in helium under a pressure of ~100 psig.
89	430	P	1966	298.2		Pyrex	Similar to the above specimen; measured in hydrogen under a pressure of ~100 psig.
90	430	P	1966	298.2		Pyrex	Similar to the above specimen with water filled in the container.
91	147	L	1960	78-373		E	67.7 SiO_2, 14.6 Na_2O, 5.4 CaO, 4.0 B_2O_3, 3.3 BaO, 1.8 Al_2O_3, 1.8 K_2O, 1.3 Fe_2O_3, and <1.0 As_2O_3; 3 in. dia x 9.375 in. thick; density 2.42 g cm^{-3}.
92	147	L	1960	83-363		F	57.9 SiO_2, 11.1 Al_2O_3, 9.7 ZnO, 9.4 Na_2O, 4.9 CaO, 3.0 ZnO, 2.9 B_2O_3, 0.8 Sb_2O_3, and 0.1 Al_2O_3; 3 in. dia x 0.375 in. thick; density 2.52 g cm^{-3}.
93	147	L	1960	78-373		D	71.2 SiO_2, 14.5 K_2O, 7.6 Na_2O, 5.5 CaO, 3.0 ZnO, 2.9 B_2O_3, 0.8 Sb_2O_3, and 0.1 Al_2O_3; 3 in. dia x 0.375 in. thick; density 2.52 g cm^{-3}.
94	147	L	1960	148-373		C	72.7 SiO_2, 14.5 K_2O, 7.7 B_2O_3, 4.0 Na_2O, 0.4 Al_2O_3, 0.4 Sb_2O_3, and 0.4 ZnO; 3 in. dia x 0.375 in. thick; density 2.45 g cm^{-3}.
95	73, 136	R	1954	497-826	±10	Soda-lime silica glass; 1	71.25 SiO_2, 13.35 Na_2O, 11.82 CaO, 244 MgO, 0.68 Na_2SO_4, 0.26 Al_2O_3, 0.14 Fe_2O_3, and 0.06 NaCl; prepared by Pittsburgh Plate Glass Co. Lab.; ellipsoidal specimen.
96	73, 136	R	1954	350-724	±10	Soda-lime silica glass; 2	70.84 SiO_2, 13.32 Na_2O, 11.75 CaO, 2.64 MgO, 0.61 Na_2SO_4, 0.56 Fe_2O_3, 0.22 Al_2O_3, and 0.06 CaCl; same source and shape as the above specimen.
97	73, 136	R	1954	361-798	±10	Soda-lime silica glass; 3	69.05 SiO_2, 16.38 Na_2O, 7.37 CaO, 3.05 Al_2O_3, 2.80 MgO, 0.55 Na_2SO_4, 0.48 K_2O, 0.12 NaCl, 0.09 Fe_2O_3, 0.015 NiO, and 0.015 CoO; same source and shape as the above specimen.
98	73, 136	R	1954	421-806	±10	Soda-lime silica glass; 4	58.43 SiO_2, 19.32 Na_2O, 7.89 Al_2O_3, 6.00 CaO, 4.25 K_2O, 3.51 MgO, 0.24 Na_2SO_4, 0.12 CaCl, 0.11 Fe_2O_3, 0.013 NiO, and 0.013 CoO; same source and shape as the above specimen.
99	72, 136	C	1955	368-685	±10	Soda-lime silica glass; 1	Similar to the above specimen 1 except in cubic shape; measured with another method.
100	114, 72, 136	C	1954	381-773	±10	Soda-lime silica glass;2	Similar to the above specimen 2 except in cubic shape; measured with another method.

SPECIFICATION TABLE NO. 295 (continued)

Curve No.	Ref. No.	Method Used	Year	Temp. Range, K	Reported Error, %	Name and Specimen Designation	Composition (weight percent), Specifications and Remarks
101	136	R	1959	393-847		Silicate glass, Corning 0080	74.0 SiO_2, 16.5 Na_2O, 5.0 CaO, 3.5 MgO, and 1.0 Al_2O_3; ellipsoidal specimen.
102	136	R	1959	433-888		Silicate glass, Corning 0080	Similar to the above specimen.
103	136	R	1959	509-839		Silicate glass, Corning 0080	Similar to the above specimen.
104	147	L	1960	78-373		I	46.0 SiO_2, 44.8 PbO, 9.2 K_2O, and 0.1 Al_2O_3; 3 in. dia x 0.375 in. thick; density 3.55 g cm^{-3}.
105	147	L	1960	78-363		K	59.7 PbO, 35.6 SiO_2, 4.4 K_2O, and 0.2 Al_2O_3; 3 in. dia x 0.375 in. thick; density 4.29 g cm^{-3}.
106	147	L	1960	78-373		G	54.2 SiO_2, 34.2 PbO, 7.1 K_2O, 2.4 Na_2O, 2.0 Sb_2O_3, and 0.2 As_2O_3; 3 in. dia x 0.375 in. thick; density 3.19 g cm^{-3}.
107	147	L	1960	78-353		L	80.0 PbO and 20.0 SiO_2; 3 in. dia x 0.375 in. thick; density 6.10 g cm^{-3}.
108	147	L	1960	78-373		J	42.9 BaO, 40.5 SiO_2, 7.7 ZnO, 6.5 B_2O_3, 1.8 Al_2O_3, 0.3 Sb_2O_3, and 0.2 As_2O_3; 3 in. dia x 0.375 in. thick; density 3.56 g cm^{-3}.
109	147	L	1960	78-373		H	49.8 SiO_2, 24.2 BaO, 8.5 K_2O, 7.8 ZnO, 5.9 PbO, 2.9 Na_2O, 0.7 Sb_2O_3, and 0.2 As_2O_3; 3 in. dia x 0.375 in. thick; density 3.18 g cm^{-3}.
110	128	L	1946	325.4			2.05 in. thick; expanded; density 0.170 g cm^{-3}.
111	128	L	1946	194,243			Similar to above except 1.07 in. thick.
112	563	L	1945	222-305			Cellular block; thickness of sample 1 in.; density (25 C) 0.165 g cm^{-3}.
113	43	L	1954	115-141			Cellular block; density 0.147 g cm^{-3}.

DATA TABLE NO. 295 THERMAL CONDUCTIVITY OF GLASSES

[Temperature, T, K; Thermal Conductivity, k, Watt $cm^{-1} K^{-1}$]

T	k
CURVE 1	
222.0	0.000486
238.7	0.000509
255.4	0.000535
272.2	0.000558
288.9	0.000583
305.5	0.000609
CURVE 2*	
1.3	0.000335
CURVE 3*	
298.2	0.0121
CURVE 4	
331.3	0.0122
331.4	0.0113
357.1	0.0125
360.6	0.0119
397.3	0.0129
398.0	0.0141
398.8	0.0123
407.8	0.0133
423.0	0.0130
445.6	0.0142
460.7	0.0138
494.8	0.0129
516.8	0.0152
553.5	0.0136
577.0	0.0134
578.3	0.0149
578.7	0.0151
624.5	0.0147
702.7	0.0152
705.0	0.0150
717.0	0.0164
768.0	0.0165
CURVE 5	
344.3	0.00508
404.2	0.00516
CURVE 5 (cont.)	
464.7	0.00554
526.3	0.00557
575.0	0.00584
576.2	0.00581
632.2	0.00619
689.8	0.00626
723.4	0.00622
782.2	0.00655
800.2	0.00596
CURVE 6	
335.4	0.00414
414.9	0.00444
486.0	0.00498
554.9	0.00529
610.3	0.00552
633.8	0.00541
659.0	0.00570
704.4	0.00565
733.0	0.00614
CURVE 7	
336.5	0.00430
346.6	0.00462
381.3	0.00480
383.5	0.00480*
411.5	0.00487
501.1	0.00519
501.7	0.00518*
531.9	0.00538
594.1	0.00559
680.9	0.00695
683.7	0.00681
702.7	0.00554
736.3	0.00557
CURVE 8	
419.2	0.0126
419.2	0.0121
420.2	0.0123
421.2	0.0114
CURVE 8 (cont.)	
424.2	0.0115
426.2	0.0114
439.2	0.0120
478.2	0.0127
496.2	0.0126
507.2	0.0123
509.2	0.0128
531.2	0.0135
532.2	0.0129
536.2	0.0128
569.2	0.0133
579.2	0.0134*
622.2	0.0138
634.2	0.0144
666.2	0.0143
668.2	0.0144*
700.2	0.0143
709.2	0.0144
718.2	0.0148
736.2	0.0158
CURVE 9	
426.2	0.0126
500.2	0.0131
533.2	0.0129*
728.2	0.0147
CURVE 10*	
413.2	0.0125
454.2	0.0126
486.2	0.0129
513.2	0.0123
527.2	0.0128
586.2	0.0128
644.2	0.0131
714.2	0.0133
CURVE 11*	
421.2	0.0120
439.2	0.0123
482.2	0.0126
CURVE 11 (cont.)*	
529.2	0.0127
535.2	0.0123
562.2	0.0133
639.2	0.0137
731.2	0.0143
CURVE 12	
442.2	0.0163
464.2	0.0162
491.2	0.0168
519.2	0.0170
551.2	0.0170
560.2	0.0170
582.2	0.0172
604.2	0.0173
633.2	0.0174
646.2	0.0174
683.2	0.0180
703.2	0.0178
739.2	0.0179
753.2	0.0186
775.2	0.0187
795.2	0.0191
833.2	0.0200
890.2	0.0221
948.2	0.0232
991.2	0.0240
1059.2	0.0254
CURVE 13	
319.3	0.0132
341.4	0.0139
360.6	0.0147
382.9	0.0155
420.4	0.0167
CURVE 14	
318.5	0.0151
338.6	0.0152
359.5	0.0159
377.1	0.0166
409.9	0.0178
CURVE 15*	
316.9	0.0121
336.1	0.0121
354.2	0.0124
374.3	0.0133
406.6	0.0141
CURVE 16	
316.3	0.0137
334.0	0.0142
352.6	0.0144
375.1	0.0154
411.4	0.0164
CURVE 17	
315.8	0.0142
334.6	0.0155
354.8	0.0161
374.2	0.0168
412.2	0.0182
CURVE 18	
377.2	0.0172
487.2	0.0192
611.2	0.0252
761.2	0.0349
964.2	0.0548
1092.2	0.0774
CURVE 19	
293.2	0.00720
373.2	0.00761
CURVE 20	
83.2	0.00661
195.2	0.0116
273.2	0.0139
373.2	0.0191
CURVE 21	
195.2	0.0100
273.2	0.0118
373.2	0.0148
CURVE 22	
1268.2	0.0418
CURVE 23	
1206.2	0.209
CURVE 24	
1216.2	0.167
CURVE 25	
387.2	0.0185
525.2	0.0172
579.2	0.0192
634.4	0.0191
703.6	0.0211
745.2	0.0215
795.2	0.0235
839.6	0.0257
CURVE 26	
517.2	0.0184
573.6	0.0185
624.0	0.0179
694.0	0.0178
739.2	0.0190
790.8	0.0200
834.4	0.0218
CURVE 27	
422.4	0.0191
487.2	0.0197
541.2	0.0213
592.4	0.0214
645.2	0.0217
CURVE 27 (cont.)	
754.0	0.0244
795.2	0.0259
891.6	0.0297
CURVE 28*	
489.2	0.0134
491.2	0.0124
543.2	0.0144
546.2	0.0134
594.2	0.0151
595.2	0.0140
650.2	0.0160
654.7	0.0169
662.2	0.0156
666.2	0.0164
717.2	0.0180
719.2	0.0174
764.2	0.0206
766.2	0.0197
854.2	0.0223
858.2	0.0234
CURVE 29	
1.49	0.000752
1.62	0.000885
1.84	0.00105
2.083	0.00120
2.084	0.00130
2.14	0.00135
2.31	0.00143
2.32	0.00145
2.50	0.00156
2.59	0.00169
CURVE 30	
2.19	0.000917
2.30	0.000952
2.40	0.00103
2.56	0.00103
2.69	0.00119
2.70	0.00112
CURVE 30 (cont.)	
2.84	0.00125
2.98	0.00135
CURVE 31	
1.64	0.000719
1.81	0.000758
2.18	0.00962
2.19	0.000971
2.66	0.00123
CURVE 32	
1.45	0.000422
1.52	0.000469
1.64	0.000538
1.77	0.000595
2.04	0.000709
2.09	0.000741
2.18	0.000787
2.31	0.000909
CURVE 33	
289.2	0.00954
378.2	0.0104
473.2	0.0114
577.2	0.0125
685.2	0.0136
CURVE 34	
2.5	0.00070
3.0	0.00080
4.0	0.00100
4.75	0.00115
5.75	0.00120
10.15	0.00120
20.25	0.00150
60.0	0.00350
95.0	0.00550

*Not shown on plot

DATA TABLE NO. 295 (continued)

T	k
CURVE 35	
5.0	0.00130
10.0	0.00130
15.0	0.00135
18.0	0.00150
95.0	0.00650
CURVE 36	
5.0	0.00120
9.0	0.00130
20.25	0.00160
95.0	0.00650
CURVE 37	
2.25	0.00045
2.65	0.00060
3.0	0.00068
3.5	0.00081
4.5	0.00095
5.25	0.00105
11.0	0.00120
20.25	0.00150*
60.0	0.0032
95.0	0.0050
CURVE 38	
11.0	0.00140
15.0	0.00150
18.5	0.00155
20.25	0.00160*
CURVE 39*	
298.2	0.0117
CURVE 40	
15.0	0.00109
20.0	0.00113
CURVE 41	
297.2	0.000731
CURVE 42	
1050	0.116
1200	0.384
1400	0.480*
1600	0.465*
1800	0.436*
CURVE 43	
1073	0.136
1200	0.252
1540	0.558*
CURVE 44	
1273	0.0872
1600	0.0755
CURVE 45*	
298.2	0.0126
CURVE 46*	
298.2	0.0126
CURVE 47*	
298.2	0.0146
CURVE 48*	
298.2	0.0146
CURVE 49*	
366.5	0.0114
477.6	0.0126
588.8	0.0135
699.8	0.0144
CURVE 50*	
366.5	0.0125
477.6	0.0135
588.8	0.0145
699.8	0.0157
CURVE 51*	
273.2	0.0119
373.2	0.0128
473.2	0.0137
573.2	0.0144
CURVE 52*	
366.5	0.0123
477.6	0.0126
588.8	0.0130
699.8	0.0132
CURVE 53*	
366.5	0.0118
477.6	0.0126
588.8	0.0133
699.8	0.0138
CURVE 54*	
423.2	0.0189
473.2	0.0192
573.2	0.0201
673.2	0.0213
773.2	0.0222
873.2	0.0230
973.2	0.0239
1073.2	0.0247
CURVE 55	
3.2	0.00115
3.35	0.00175
3.55	0.00193
3.8	0.00202
4.0	0.0024
4.6	0.0029
5.0	0.003
5.05	0.0033
6.2	0.0037
7.2	0.00408
CURVE 56	
3.3	0.0023
3.5	0.00255
3.65	0.0028
3.9	0.0032
4.05	0.00355
4.95	0.0045
5.3	0.0048
5.9	0.0052
7.4	0.0056
CURVE 57	
3.3	0.00355
3.6	0.0038
3.75	0.00405
3.85	0.0043
4.3	0.0052
4.6	0.00575
4.7	0.0055
5.9	0.0070
6.5	0.0073
CURVE 58	
3.5	0.0035
3.65	0.0036
3.75	0.0039
4.0	0.0044
4.1	0.00475
4.35	0.0050
4.5	0.0052
5.0	0.0060
5.7	0.0065
CURVE 59	
3.3	0.0016
3.45	0.00167
3.5	0.00186
3.55	0.00198
3.6	0.00203
3.75	0.0022
3.8	0.00215
4.0	0.00235
4.2	0.0026
4.4	0.00295
CURVE 59 (cont.)	
4.75	0.00285
5.25	0.00325
5.8	0.00365
6.5	0.00385
7.4	0.0039
CURVE 60	
200.0	0.000548
222.2	0.000562
239.2	0.000577
255.5	0.000591
272.1	0.000606
288.1	0.000620
CURVE 61	
310.9	0.000649
CURVE 62	
303.2	0.000465
CURVE 63	
303.2	0.000651
CURVE 64	
303.2	0.000848
CURVE 65	
303.2	0.00106
CURVE 66	
303.2	0.00126
CURVE 67	
293.2	0.00747
333.2	0.00814
CURVE 68	
293.2	0.00797
CURVE 69	
333.2	0.00843
CURVE 70*	
388.2	0.0153
482.2	0.0177
625.2	0.0250
751.2	0.0331
CURVE 71*	
386.2	0.0147
478.2	0.0164
611.2	0.0223
CURVE 72	
333.2	0.0126
333.2	0.0130
373.2	0.0134
378.2	0.0133
405.2	0.0136*
423.2	0.0140
473.2	0.0144
478.2	0.0146
568.2	0.0153
633.2	0.0161
683.2	0.0165
723.2	0.0172
773.2	0.0182
CURVE 73	
513.2	0.0165*
533.2	0.0167*
565.2	0.0172*
585.2	0.0172*
686.2	0.0184*
776.2	0.0201*
803.2	0.0209
1083.2	0.0293
CURVE 74	
493.2	0.0162*
635.2	0.0173*
733.2	0.0202*
833.2	0.0218*
863.2	0.0226
913.2	0.0239
948.2	0.0249
993.2	0.0260
1173.2	0.0339
1193.2	0.0372
1273.2	0.0415*
CURVE 75	
553.2	0.0174*
613.2	0.0179*
703.2	0.0188*
705.2	0.0183*
791.2	0.0192*
813.2	0.0213*
903.2	0.0222*
933.2	0.0251*
1103.2	0.0247
1153.2	0.0284
1163.2	0.0292
1363.2	0.0360
CURVE 76*	
273.2	0.0111
323.2	0.0116
373.2	0.0122
423.2	0.0127
473.2	0.0133
523.2	0.0138
573.2	0.0143
CURVE 77*	
343.2	0.01088
433.2	0.01167
493.2	0.01272
553.2	0.01339
613.2	0.01473
671.2	0.01653
718.2	0.01795
CURVE 77 (cont.)*	
768.2	0.01975
869.2	0.02343
CURVE 78	
123.2	0.0069
173.2	0.0085
223.2	0.0098
273.2	0.0109
323.2	0.0118
373.2	0.0127*
CURVE 79	
92	0.0054
198	0.0089
275	0.0103
372	0.0116
423	0.0121*
482	0.0126*
523	0.0131*
CURVE 80*	
273.2	0.0106
373.2	0.0119
473.2	0.0133
573.2	0.0146
673.2	0.0160
773.2	0.0173
CURVE 81	
297	0.0135*
300	0.0134*
306	0.0139*
319	0.0142*
322	0.0159
322	0.0145*
329	0.0143*
330	0.0156*
332	0.0166
336	0.0168*
345	0.0191
356	0.0190
360	0.0167

*Not shown on plot

DATA TABLE NO. 295 (continued)

T	k
CURVE 81 (cont.)	
362	0.0159*
366	0.0195
373	0.0164*
374	0.0201*
374	0.0162*
376	0.0204
394	0.0212*
395	0.0214*
400	0.0187
401	0.0226
402	0.0182*
414	0.0246
416	0.0219*
422	0.0186
424	0.0249
427	0.0192
441	0.0211*
458	0.0293
CURVE 82	
294	0.0120*
315	0.0148*
362	0.0138
400	0.0156
431	0.0162
476	0.0174
518	0.0185*
544	0.0174
CURVE 83	
323.2	0.0143*
383.2	0.0149*
423.2	0.0152
473.2	0.0159
523.2	0.0163
573.2	0.0172*
623.2	0.0176*
673.2	0.0184
723.2	0.0194
773.2	0.0207
CURVE 84*	
298.2	0.00107
CURVE 85	
298.2	0.00167
CURVE 86	
298.2	0.00209
CURVE 87*	
298.2	0.00293
CURVE 88*	
298.2	0.00439
CURVE 89*	
298.2	0.00565
CURVE 90*	
298.2	0.00628
CURVE 91*	
78.2	0.00519
93.2	0.00561
113.2	0.00615
133.2	0.00670
153.2	0.00724
173.2	0.00770
193.2	0.00820
213.2	0.00862
233.2	0.00904
253.2	0.00941
273.2	0.00980
293.2	0.01004
313.2	0.01029
333.2	0.01054
353.2	0.01079
373.2	0.01100
CURVE 92*	
83.2	0.00653
93.2	0.00690
113.2	0.00766
CURVE 92 (cont.)*	
133.2	0.00828
153.2	0.00883
173.2	0.00929
193.2	0.00975
213.2	0.01013
233.2	0.01046
253.2	0.01079
273.2	0.01105
293.2	0.01130
313.2	0.01151
333.2	0.01172
353.2	0.01192
363.2	0.01201
CURVE 93*	
78.2	0.00519
93.2	0.00561
113.2	0.00615
133.2	0.00669
153.2	0.00724
173.2	0.00770
193.2	0.00820
213.2	0.00862
233.2	0.00904
253.2	0.00941
273.2	0.00979
293.2	0.01004
313.2	0.01029
333.2	0.01054
353.2	0.01079
373.2	0.01100
CURVE 94*	
148.2	0.00774
173.2	0.00833
193.2	0.00879
213.2	0.00920
233.2	0.00962
253.2	0.01000
273.2	0.01038
293.2	0.01067
313.2	0.01096
333.2	0.01121
353.2	0.01146
373.2	0.01172
CURVE 95*	
497.2	0.01736
536.2	0.01866
570.2	0.01632
615.2	0.01778
675.2	0.02038
716.2	0.02075
760.2	0.02163
793.2	0.02042
826.2	0.02159
CURVE 96*	
350.2	0.01527
443.2	0.01347
499.2	0.01469
579.2	0.01607
625.2	0.01724
672.2	0.01824
724.2	0.01987
CURVE 97*	
361.2	0.01527
404.2	0.01707
471.2	0.01648
545.2	0.01833
587.2	0.01862
638.2	0.01879
688.2	0.02029
733.2	0.02084
798.2	0.02264
CURVE 98*	
421.2	0.01540
469.2	0.01510
537.2	0.01695
568.2	0.01632
593.2	0.01527
633.2	0.01736
686.2	0.01820
738.2	0.01979
770.2	0.02042
771.2	0.02192
806.2	0.02167
CURVE 99*	
368.2	0.0135
413.2	0.0144
467.2	0.0144
503.2	0.0143
527.2	0.0146
603.2	0.0159
622.2	0.0183
685.2	0.0180
CURVE 100*	
381.2	0.0105
383.2	0.0138
419.2	0.0102
423.2	0.0129
449.2	0.0126
454.2	0.0105
486.2	0.0111
488.2	0.0134
519.2	0.0140
519.7	0.0114
550.2	0.0112
559.2	0.0131
590.2	0.0147
590.2	0.0131
654.2	0.0152
655.2	0.0167
696.2	0.0175
697.2	0.0159
722.7	0.0185
724.2	0.0161
746.7	0.0192
748.5	0.0170
771.2	0.0204
773.2	0.0175
CURVE 101*	
393.2	0.0184
523.2	0.0169
578.2	0.0192
641.2	0.0190
701.2	0.0203
748.2	0.0207
793.2	0.0223
847.2	0.0251
CURVE 102*	
433.2	0.0155
488.2	0.0154
546.2	0.0167
593.2	0.0168
656.2	0.0168
768.2	0.0205
888.2	0.0251
CURVE 103*	
509.2	0.0180
568.2	0.0182
624.2	0.0177
699.2	0.0177
753.2	0.0188
793.2	0.0194
839.2	0.0209
CURVE 104	
78.2	0.00410
93.2	0.00435
113.2	0.00469
133.2	0.00502
153.2	0.00536
173.2	0.00573
193.2	0.00602
213.2	0.00636
233.2	0.00665
253.2	0.00703
273.2	0.00728
293.2	0.00753*
313.2	0.00778*
333.2	0.00812*
353.2	0.00837*
373.2	0.00862*
CURVE 105	
78.2	0.00343
93.2	0.00377
113.2	0.00414
133.2	0.00452
153.2	0.00485
173.2	0.00519*
193.2	0.00552
CURVE 105 (cont.)	
213.2	0.00582
233.2	0.00611
253.2	0.00636
273.2	0.00661
293.2	0.00678
313.2	0.00699
333.2	0.00715
363.2	0.00732
CURVE 106	
78.2	0.00439
103.2	0.00485
133.2	0.00544
153.2	0.00582
173.2	0.00615
193.2	0.00657
213.2	0.00695
233.2	0.00728
253.2	0.00761
273.2	0.00791
293.2	0.00824
313.2	0.00854
333.2	0.00887
353.2	0.00916
373.2	0.00946
CURVE 107	
78.2	0.00318
93.2	0.00335
113.2	0.00356
133.2	0.00381
153.2	0.00402
173.2	0.00418
193.2	0.00435
213.2	0.00456*
233.2	0.00473
253.2	0.00494
273.2	0.00515
293.2	0.00531
313.2	0.00552
333.2	0.00573
353.2	0.00594
CURVE 108*	
78.2	0.00460
93.2	0.00485
113.2	0.00519
133.2	0.00552
153.2	0.00582
173.2	0.00611
193.2	0.00640
213.2	0.00669
233.2	0.00699
253.2	0.00724
273.2	0.00753
293.2	0.00778
313.2	0.00803
333.2	0.00828
353.2	0.00854
373.2	0.00879
CURVE 109*	
78.2	0.00460
93.2	0.00481
113.2	0.00506
133.2	0.00531
153.2	0.00561
173.2	0.00586
193.2	0.00615
213.2	0.00640
233.2	0.00661
253.2	0.00695
273.2	0.00724
293.2	0.00753
313.2	0.00770
333.2	0.00795
353.2	0.00816
373.2	0.00833
CURVE 110	
325.4	0.000149
CURVE 111	
194.3	0.000131
243.2	0.000141

*Not shown on plot

DATA TABLE NO. 295 (continued)

T	k
CURVE 112	
222.2	0.000383
239.2	0.000509
255.5	0.000539
272.1	0.000558
288.8	0.000583
305.3	0.000609
CURVE 113	
115	0.000476
120	0.000483
141	0.000498

FIGURE AND TABLE NO. 295R RECOMMENDED THERMAL CONDUCTIVITY OF CORNING CODE 7740 GLASS

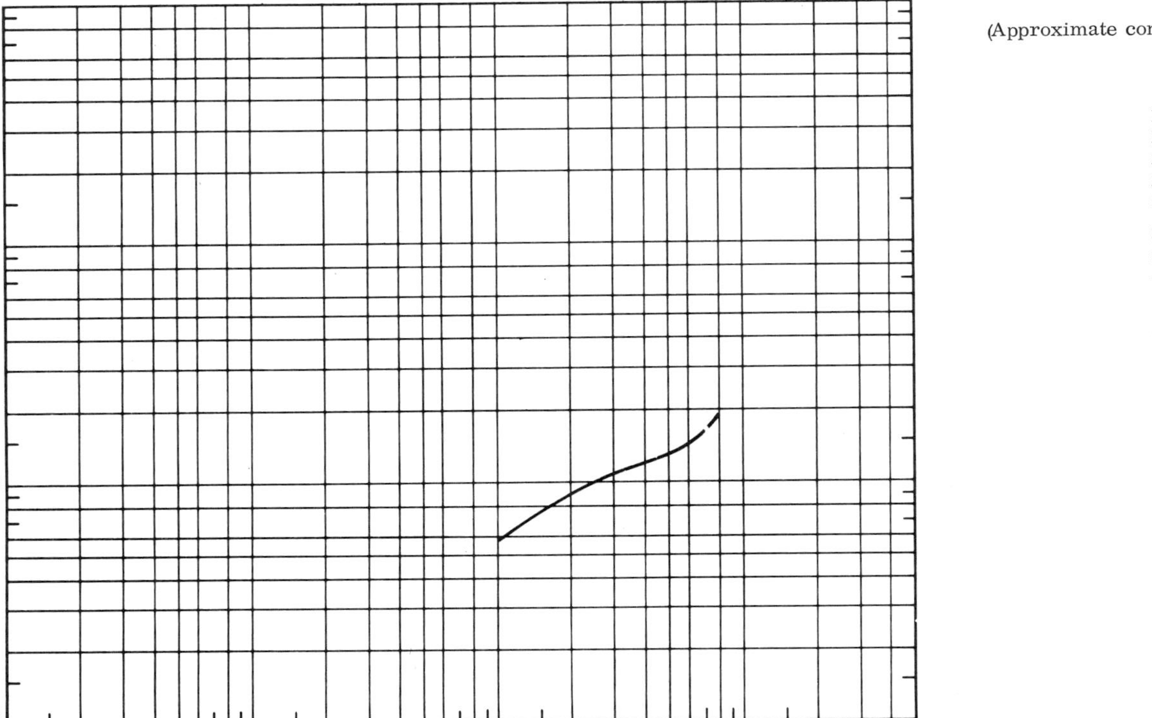

RECOMMENDED VALUES*
(Approximate composition: 80.6% SiO_2, 13% B_2O_3, 4.3% Na_2O, and 2.1% Al_2O_3)

T_1	k_1	k_2	T_2
100	0.0058	0.335	-279.7
150	0.0076	0.439	-189.7
200	0.0090	0.520	- 99.7
250	0.0101	0.584	- 9.7
273.2	0.0106	0.612	32.0
300	0.0110	0.636	80.3
350	0.0117	0.676	170.3
400	0.0124	0.716	260.3
500	0.0136	0.786	440.3
600	0.0149	0.861	620.3
700	0.0165	0.953	800.3
800	(0.0189) ‡	(1.09)	980.3

REMARKS

The recommended values are thought to be accurate to within 5% of the true values near room temperature and 5 to 10% at other temperatures.

*T_1 in K, k_1 in Watt cm^{-1}K^{-1}, T_2 in F, and k_2 in Btu hr^{-1}ft^{-1}F^{-1}.

‡ Values in parentheses are extrapolated.

SPECIFICATION TABLE NO. 296 THERMAL CONDUCTIVITY OF MULLITE

Curve No.	Ref. No.	Method Used	Year	Temp. Range, K	Reported Error, %	Name and Specimen Designation	Composition (weight percent), Specifications and Remarks
1	407	L	1949	327-446		C-1	No details reported.

DATA TABLE NO. 296 THERMAL CONDUCTIVITY OF MULLITE

[Temperature, T, K; Thermal Conductivity, k, Watt $cm^{-1} K^{-1}$]

T	k
CURVE 1*	
326.6	0.0323
362.3	0.0301
445.6	0.0287

*No graphical presentation

SPECIFICATION TABLE NO. 297 THERMAL CONDUCTIVITY OF PETALITE

Curve No.	Ref. No.	Method Used	Year	Temp. Range, K	Reported Error, %	Name and Specimen Designation	Composition (weight percent), Specifications and Remarks
1	468	L	1950	329-392			No details reported.

DATA TABLE NO. 297 THERMAL CONDUCTIVITY OF PETALITE

[Temperature, T, K; Thermal Conductivity, k, Watt $cm^{-1} K^{-1}$]

T	k
CURVE 1*	
328.6	0.0266
357.8	0.0268
391.5	0.0277

*No graphical presentation

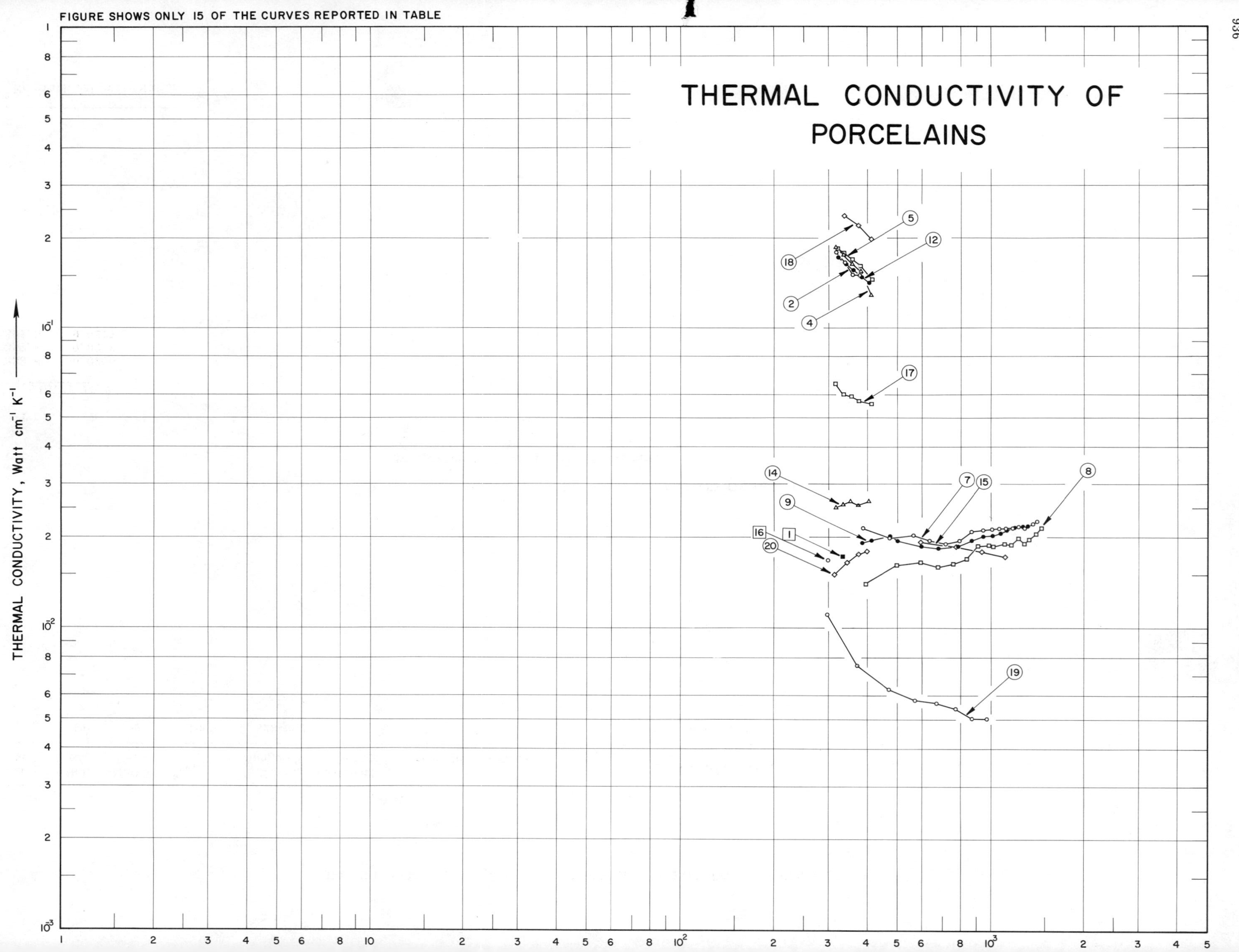
FIGURE SHOWS ONLY 15 OF THE CURVES REPORTED IN TABLE
THERMAL CONDUCTIVITY OF
PORCELAINS
THERMAL CONDUCTIVITY, Watt cm⁻¹ K⁻¹
5
12
18
2
4
17
14
7
15
8
9
16
1
20
19

[For Data Reported in Figure and Table No. 298]

Curve No.	Ref. No.	Method Used	Year	Temp. Range, K	Reported Error, %	Name and Specimen Designation	Composition (weight percent), Specifications and Remarks
1	3		1953	333.2			Maximum water absorption 0.05%.
2	42	L	1953	319-377	±3	6 M-1	High Al_2O_3; 0.349 in. in dia and 0.500 in. in length.
3	42	L	1953	316-410	±3	6 P-1	High Al_2O_3; 0.450 in. in dia and 0.496 in. in length.
4	42	L	1953	318-412	±3	6 J-1	High Al_2O_3; 0.250 in. in dia and 0.250 in. in length.
5	42	L	1953	320-416	±3	6 K-1	High Al_2O_3; 0.249 in. in dia and 0.500 in. in length.
6	42	L	1953	313-387	±3	6 L-1	High Al_2O_3; 0.300 in. in dia and 0.501 in. in length.
7	89	R	1952	388-1418		Electrical porcelain; A	Starting material consisting of 19.0 flint, 37.0 feldspar, 7.0 Edgar plastic kaolin, 22.0 Edgar Nocarb clay, and 15.0 Kentucky old mine No. 4 ball clay, ball-milled for 15 hrs, slip-cast, and fired to 1250 C; 0.25% open pores; bulk density 2.5 g cm^{-3}.
8	89	R	1952	395-1456		Electrical porcelain; B	Same as the above specimen.
9	89	R	1952	385-1396		Electrical porcelain; 3	Same as the above specimen.
10	42, 87	C	1953	322-425	±3	6 R-1	High Al_2O_3; 0.500 in. in dia and 0.497 in. in length; copper used as comparative material.
11	42, 87	C	1953	382-410	±3	6 R-2	Second run of the above specimen.
12	42, 87	C	1953	323-406	±3	6 Q-1	High Al_2O_3; 0.501 in. in dia and 0.500 in. in length.
13	42, 87	C	1953	319-384	±3	6 N-1	High Al_2O_3; 0.410 in. in dia and 0.500 in. in length.
14	68	C	1954	317-404	±3	Wet process; 7A2	Copper used as comparative material.
15	76	C	1952	593-1113		$MgTiO_3$ porcelain	0.75 in. dia x 9 in. long; density 2.87 g cm^{-3}.
16	152	L	1956	298.2		Porcelain 576	Exposed with 6 x 10^{19} epithermal neutrons per cm^2 for 480 MWD in the MTR.
17	9	C	1953	315-411		High Zircon; 283A-1	62.5 zircon G, 25.0 calcium zirconium silicate, and 12.5 old mine No. 4 ball clay; bulk density 3.91 g cm^{-3}; water absorption 0.03%.
18	136	C	1959	336-413	±4	Alumina porcelain	91 vol % α- alumina bonded with a continuous glassy matrix.
19	550	P	1963	298-973		Coors; type AB-2	High alumina; cylindrical specimen from Coors Porcelain Co.; density 3.42 g cm^{-3}; thermal conductivity values calculated from measured data of thermal diffusivity and specific heat.
20	468	L	1950	314-400		Wet process	No details reported.

DATA TABLE NO. 298 THERMAL CONDUCTIVITY OF PORCELAINS

[Temperature, T, K; Thermal Conductivity, k, Watt $cm^{-1} K^{-1}$]

T	k
CURVE 1	
333.2	0.0173
CURVE 2	
318.7	0.179
339.7	0.166
358.0	0.151
377.3	0.151
CURVE 3*	
315.5	0.178
334.1	0.172
358.1	0.156
374.2	0.149
409.6	0.138
CURVE 4	
317.7	0.187
335.6	0.176
358.2	0.164
381.4	0.155
411.8	0.129
CURVE 5	
320.0	0.185
337.3	0.178
356.0	0.170
378.1	0.161
416.2	0.146
CURVE 6*	
312.8	0.177
332.0	0.176
353.4	0.161
372.7	0.153
386.7	0.150
CURVE 7	
388.2	0.0215
473.2	0.0199
567.2	0.0203
636.2	0.0195
715.2	0.0190
793.2	0.0195
868.2	0.0209
943.2	0.0211
1006.2	0.0213
1066.2	0.0215
1123.2	0.0215
1178.2	0.0215
1238.2	0.0218
1288.2	0.0215
1368.2	0.0222
1418.2	0.0227
CURVE 8	
395.2	0.0140
496.2	0.0161
593.2	0.0165
676.2	0.0159
756.2	0.0163
836.2	0.0169
910.2	0.0187
988.2	0.0188
1046.2	0.0186
1106.2	0.0190
1166.2	0.0188
1230.2	0.0199
1280.2	0.0190
1330.2	0.0197
1400.2	0.0205
1456.2	0.0215
CURVE 9	
385.2	0.0191
412.2	0.0195
473.2	0.0202
500.2	0.0195
596.2	0.0187
678.2	0.0184
CURVE 9 (cont.)	
784.2	0.0186
866.2	0.0195
943.2	0.0201
1016.2	0.0203
1076.2	0.0207
1128.2	0.0211
1198.2	0.0216
1273.2	0.0218
1323.2	0.0219
1396.2	0.0224
CURVE 10*	
321.5	0.169
341.2	0.162
362.2	0.154
424.9	0.146
CURVE 11*	
381.7	0.162
404.1	0.141
410.1	0.135
CURVE 12	
322.5	0.173
342.5	0.164
360.5	0.157
383.6	0.148
405.5	0.143
CURVE 13*	
318.5	0.182
338.5	0.166
359.4	0.160
383.6	0.150
CURVE 14	
316.8	0.0251
334.4	0.0257
353.7	0.0262
CURVE 14 (cont.)	
372.0	0.0256
404.1	0.0264
CURVE 15	
593.2	0.0193
773.2	0.0186
933.2	0.0179
1113.2	0.0172
CURVE 16	
298.2	0.0167
CURVE 17	
314.7	0.0653
334.1	0.0598
355.5	0.0594
375.7	0.0573
411.0	0.0556
CURVE 18	
336.2	0.237
373.2	0.220
413.2	0.198
CURVE 19	
298.2	0.0111
373.2	0.00753
473.2	0.00628
573.2	0.00577
673.2	0.00565
773.2	0.00544
873.2	0.00502
973.2	0.00502
CURVE 20	
314.0	0.0150
345.5	0.0164
375.6	0.0176
399.5	0.0179

*Not shown on plot

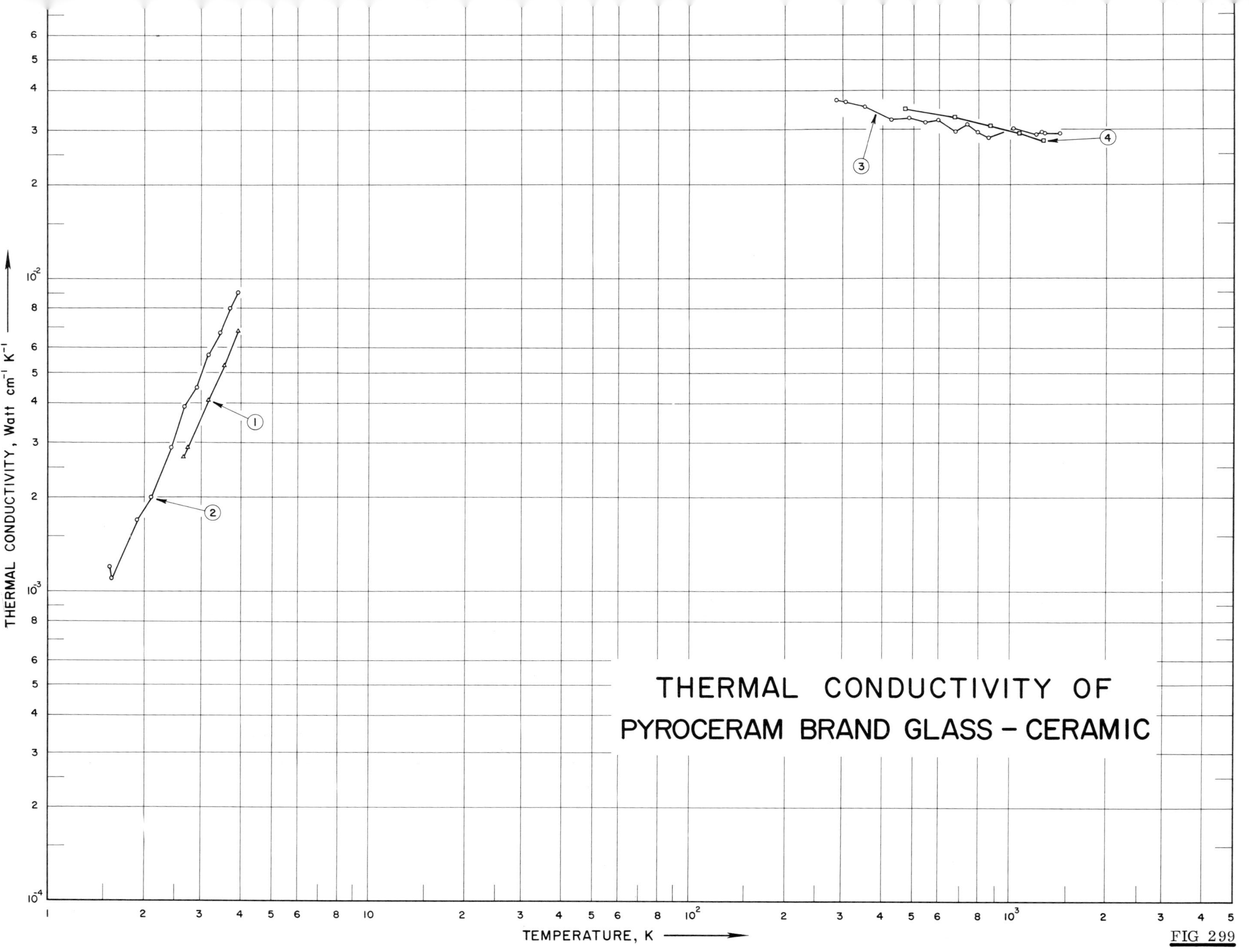
THERMAL CONDUCTIVITY OF
PYROCERAM BRAND GLASS - CERAMIC
THERMAL CONDUCTIVITY, Watt cm⁻¹ K⁻¹
TEMPERATURE, K
1
2
3
4

FIG 299

SPECIFICATION TABLE NO. 299 THERMAL CONDUCTIVITY OF PYROCERAM BRAND GLASS-CERAMIC

[For Data Reported in Figure and Table No. 299]

Curve No.	Ref. No.	Method Used	Year	Temp. Range, K	Reported Error, %	Name and Specimen Designation	Composition (weight percent), Specifications and Remarks
1	306	L	1962	2.7-3.9		Pyroceram 9606 (a)	From Corning Glass Works; specimen cross section 0.034 cm x 1.666 cm.
2	306	L	1962	1.6-3.9		Pyroceram 9606 (b)	From Corning Glass Works; specimen cross section 0.130 cm x 1.607 cm.
3	275	P	1963	288-1433		Pyroceram 9606	From Corning Glass Works; calculated from thermal diffusivity data.
4	305	L	1962	473-1273		Pyroceram 9606	Pyroceram 9606 (a microcrystalline glass), product of Corning Glass Works; specimen 2.540 cm in dia and 1.269 cm in length; density 2.601 g cm^{-3}; data obtained before and after the specimen held at 1000 C for about 275 hrs agree with each other.

DATA TABLE NO. 299 THERMAL CONDUCTIVITY OF PYROCERAM BRAND GLASS-CERAMIC

[Temperature, T, K; Thermal Conductivity, k, Watt $cm^{-1} K^{-1}$]

T	k	T	k
CURVE 1		CURVE 4	
2.65	0.0027	473.2	0.0350
2.74	0.0029	673.2	0.0327
3.16	0.0041	873.2	0.0308
3.56	0.0053	1073.2	0.0290
3.94	0.0068	1273.2	0.0275
CURVE 2			
1.56	0.0012		
1.59	0.0011		
1.90	0.0017		
2.10	0.0020		
2.43	0.0029		
2.67	0.0039		
2.91	0.0045		
3.18	0.0057		
3.45	0.0067		
3.70	0.0080		
3.91	0.0090		
CURVE 3			
288.2	0.0372		
308.2	0.0368		
353.2	0.0354		
428.2	0.0324		
483.2	0.0328		
543.2	0.0317		
598.2	0.0321		
673.2	0.0295		
735.2	0.0312		
791.2	0.0294		
858.2	0.0281		
1021.2	0.0301		
1201.2	0.0289		
1253.2	0.0294		
1263.2	0.0290		
1433.2	0.029		

FIGURE AND TABLE NO. 299R RECOMMENDED THERMAL CONDUCTIVITY OF PYROCERAM BRAND GLASS-CERAMIC CODE 9606

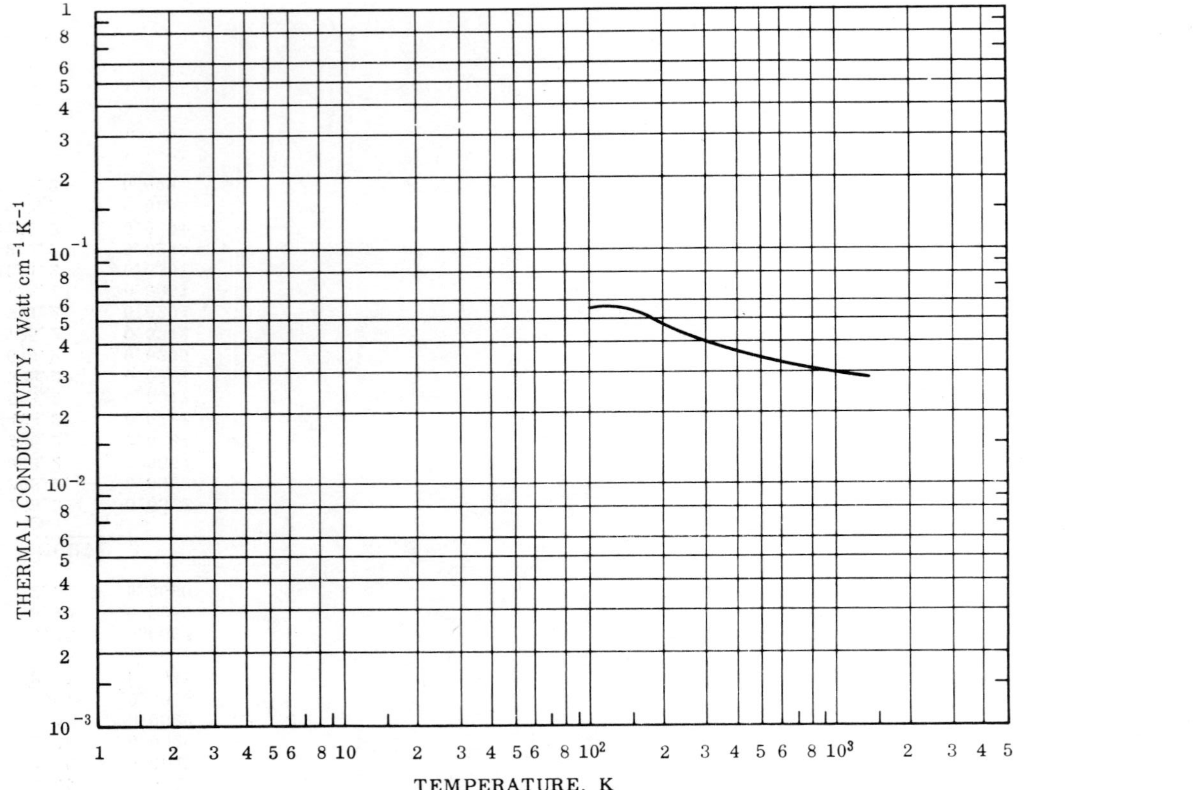

RECOMMENDED VALUES*

T_1	k_1	k_2	T_2
100	0.0542	3.13	-279.7
150	0.0550	3.18	-189.7
200	0.0474	2.74	- 99.7
250	0.0428	2.47	- 9.7
273.2	0.0413	2.39	32.0
300	0.0399	2.31	80.3
350	0.0379	2.19	170.3
400	0.0365	2.11	260.3
500	0.0345	1.99	440.3
600	0.0331	1.91	620.3
700	0.0319	1.84	800.3
800	0.0310	1.79	980.3
900	0.0303	1.75	1160
1000	0.0297	1.72	1340
1100	0.0291	1.68	1520
1200	0.0287	1.66	1700
1300	0.0284	1.64	1880
1400	0.0282	1.63	2060

REMARKS

The recommended values are thought to be accurate to within 5% of the true values near room temperature and 5 to 10% at other temperatures.

* T_1 in K, k_1 in Watt $cm^{-1}K^{-1}$, T_2 in F, and k_2 in Btu $hr^{-1}ft^{-1}F^{-1}$.

SPECIFICATION TABLE NO. 300 THERMAL CONDUCTIVITY OF COPOLY(CHLOROETHYLENE-VINYL ACETATE)

Curve No.	Ref. No.	Method Used	Year	Temp. Range, K	Reported Error, %	Name and Specimen Designation	Composition (weight percent), Specifications and Remarks
1	388	L	1960	289-375		PVC III	Degree of polymerization: 1500.
2	388	L	1960	289-365		PVC IV	Degree of polymerization: 800.

DATA TABLE NO. 300 THERMAL CONDUCTIVITY OF COPOLY(CHLOROETHYLENE-VINYL ACETATE)

[Temperature, T, K; Thermal Conductivity, k, Watt $cm^{-1} K^{-1}$]

T	k
	CURVE 1*
289	0.00126
292	0.00134
298	0.00138
312	0.00138
325	0.00146
347	0.00146
361	0.00172
375	0.00218
	CURVE 2*
289	0.000921
298	0.00100
308	0.000921
313	0.00100
322	0.00109
332	0.00126
345	0.00134
354	0.00151
365	0.00159

*No graphical presentation

SPECIFICATION TABLE NO. 301 THERMAL CONDUCTIVITY OF COPOLY(FORMALDEHYDE-UREA), MIPORA

Curve No.	Ref. No.	Method Used	Year	Temp. Range, K	Reported Error, %	Name and Specimen Designation	Composition (weight percent), Specifications and Remarks
1	525		1951	293.2			Foamed; density 0.018-0.022 g cm^{-3}; porosity 97-98%.
2	525		1951	178.2			The above specimen measured at one atm.
3	525		1951	178.2			The above specimen measured at 0.01 mm Hg.
4	525		1951	178.2			The above specimen measured by filling H_2 in the apparatus and then evacuating by forepump.

DATA TABLE NO. 301 THERMAL CONDUCTIVITY OF COPOLY(FORMALDEHYDE-UREA), MIPORA

[Temperature, T, K; Thermal Conductivity, k, Watt cm^{-1} K^{-1}]

T	k
CURVE 1*	
293.2	0.000430
CURVE 2*	
178.2	0.000256
CURVE 3*	
178.2	0.0000291
CURVE 4*	
178.2	0.0000360

*No graphical presentation

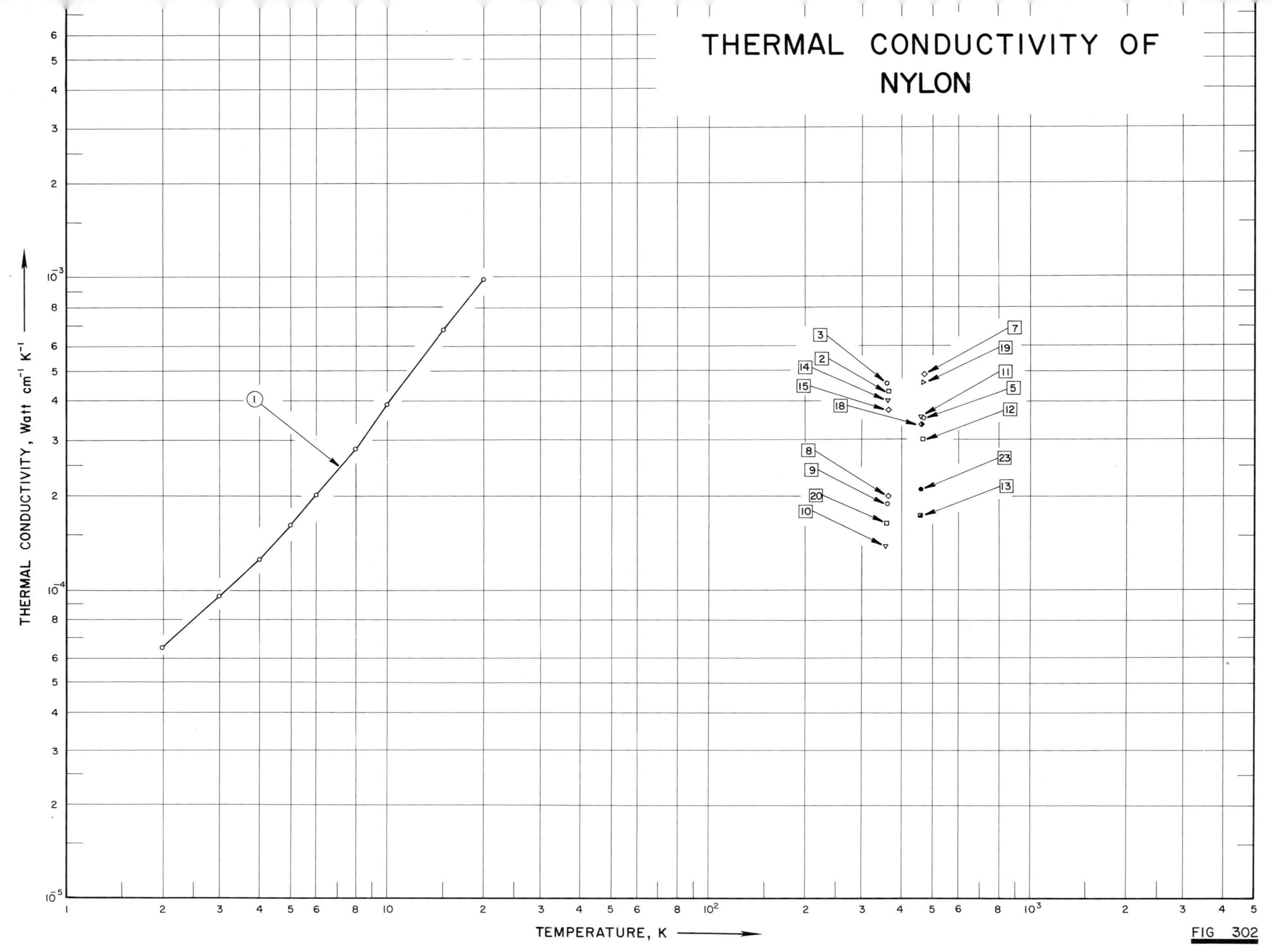
THERMAL CONDUCTIVITY OF
NYLON
THERMAL CONDUCTIVITY, Watt cm⁻¹ K⁻¹
TEMPERATURE, K
FIG 302
1
2
3
5
7
8
9
10
11
12
13
14
15
18
19
20
23

SPECIFICATION TABLE NO. 302 THERMAL CONDUCTIVITY OF NYLON

[For Data Reported in Figure and Table No. 302]

Curve No.	Ref. No.	Method Used	Year	Temp. Range, K	Reported Error, %	Name and Specimen Designation	Composition (weight percent), Specifications and Remarks
1	142	L	1955	2.0-20			A drawn monofilament, 2 mm in dia; supplied by Imperial Chemical Industries, Ltd.
2	526	P	1962	366.4	±7	IN	Plain weave, 64 ends in.$^{-1}$, 68 picks in.$^{-1}$ nylon; 2.10 oz yd^{-2}; average thickness 0.00531 in.; thickness under load 0.00458 in.; measured under 16.7 psi pressure, 35.6 ppi tensile stress; and 11.8 psi compressive stress.
3	526	P	1962	361.6	±7	IN	Similar to the above specimen except under 15.3 psi pressure.
4	526	P	1962	363.7	±7	IN	Similar to the above specimen except under 16.3 psi pressure, 39.9 ppi tensile stress and 13.6 compressive stress.
5	526	P	1962	469.8	±7	IN	Similar to the above specimen except under 18.2 psi pressure, 35.6 ppi tensile stress and 11.8 compressive stress.
6	526	P	1962	469.5	±7	IN	Similar to the above specimen except under 17.2 psi pressure.
7	526	P	1962	471.2	±7	IN	Similar to the above specimen except under 18.2 psi pressure, 38.9 ppi tensile stress and 13.0 compressive stress.
8	526	P	1962	364.8	±7	IN	Similar to the above specimen except in 12.7 mm Hg vacuum, under 35.6 ppi tensile stress and 11.8 psi compressive stress.
9	526	P	1962	362.1	±7	IN	Similar to the above specimen except in 12.7 mm Hg vacuum.
10	526	P	1962	357.9	±7	IN	Similar to the above specimen except in 1.0 mm Hg vacuum, under 41.0 ppi tensile stress and 13.7 psi compressive stress.
11	526	P	1962	466.0	±7	IN	Similar to the above specimen except in 38.0 mm Hg vacuum, under 35.6 ppi tensile stress and 11.8 psi compressive stress.
12	526	P	1962	467.1	±7	IN	Similar to the above specimen except in 25.4 mm Hg vacuum, under 35.6 ppi tensile stress and 12.3 psi compressive stress.
13	526	P	1962	459.0	±7	IN	Similar to the above specimen except in 2.0 mm Hg vacuum, under 38.6 ppi tensile stress and 12.9 psi compressive stress.
14	526	P	1962	362.6	±7	IIN	Twill (2 x 1) type; 64 ends in.$^{-1}$, 68 picks in.$^{-1}$ nylon; 2.06 oz yd^{-2}; average thickness 0.00656 in.; thickness under load 0.00573 in.; measured under 15.7 psi pressure, 35.6 ppi tensile stress and 11.8 psi compressive stress.
15	526	P	1962	365.2	±7	IIN	Similar to the above specimen except 0.00573 in. thickness under load, 16.1 psi pressure, 35.6 ppi tensile stress and 11.8 psi compressive stress.
16	526	P	1962	364.1	±7	IIN	Similar to the above specimen except 0.00570 in. thickness under load, 16.3 psi pressure, 39.5 ppi tensile stress and 13.2 psi compressive stress.
17	526	P	1962	464.5	±7	IIN	Similar to the above specimen except 0.00573 in. thickness under load, 20.2 psi pressure, 35.6 ppi tensile stress and 11.9 psi compressive stress.
18	526	P	1962	462.0	±7	IIN	Similar to the above specimen except 17.5 psi pressure, 35.6 ppi tensile stress and 11.8 psi compressive stress.

SPECIFICATION TABLE NO. 302 (continued)

Curve No.	Ref. No.	Method Used	Year	Temp. Range, K	Reported Error, %	Name and Specimen Designation	Composition (weight percent), Specifications and Remarks
19	526	P	1962	467.7	±7	IIN	Similar to the above specimen except 0.00571 in. thickness under load, 16.0 psi pressure, 38.4 ppi tensile stress and 12.8 psi compressive stress.
20	526	P	1962	359.0	±7	IIN	Similar to the above specimen except 0.00572 in. thickness under load, 12.7 mm Hg vacuum, 35.6 ppi tensile stress and 12.3 psi compressive stress.
21	526	P	1962	357.6	±7	IIN	Similar to the above specimen except 0.00573 in. thickness under load, 12.7 mm Hg vacuum, 35.6 ppi tensile stress and 11.8 psi compressive stress.
22	526	P	1962	359.0	±7	IIN	Similar to the above specimen except 0.00570 in. thickness under load, 0.4 mm Hg vacuum, 39.9 ppi tensile stress and 13.3 psi compressive stress.
23	526	P	1962	462.6	±7	IIN	Similar to the above specimen except 0.00573 in. thickness under load, 12.7 mm Hg vacuum, 35.6 ppi tensile stress and 11.8 psi compressive stress.
24	526	P	1962	460.4	±7	IIN	Similar to the above specimen except 12.7 mm Hg vacuum, 35.6 ppi tensile stress and 11.8 psi compressive stress.
25	526	P	1962	458.9	±7	IIN	Similar to the above specimen except 0.00571 in. thickness under load, 2.0 mm Hg vacuum, 37.8 ppi tensile stress and 12.6 psi compressive stress.

DATA TABLE NO. 302 THERMAL CONDUCTIVITY OF NYLON

[Temperature, T, K; Thermal Conductivity, k, Watt $cm^{-1} K^{-1}$]

T	k	T	k	T	k
CURVE 1		CURVE 10		CURVE 21*	
2	0.000065	357.9	0.000138	357.6	0.000168
3	0.000095				
4	0.000125	CURVE 11		CURVE 22*	
5	0.000160				
6	0.000200	466.0	0.000358	359.0	0.000137
8	0.000280				
10	0.000390	CURVE 12		CURVE 23	
15	0.000680				
20	0.000980	467.1	0.000301	462.6	0.000209
CURVE 2		CURVE 13		CURVE 24*	
366.4	0.000431	459.0	0.000173	460.4	0.000208
CURVE 3		CURVE 14		CURVE 25*	
361.6	0.000457	362.6	0.000403	458.9	0.000171
CURVE 4*		CURVE 15			
363.7	0.000436	365.2	0.000376		
CURVE 5		CURVE 16*			
469.8	0.000355	364.1	0.000412		
CURVE 6*		CURVE 17*			
469.5	0.000358	464.5	0.000294		
CURVE 7		CURVE 18			
471.2	0.000488	462.0	0.000337		
CURVE 8		CURVE 19			
364.8	0.000199	467.7	0.000460		
CURVE 9		CURVE 20			
362.1	0.000187	359.0	0.000163		

*Not shown on plot

SPECIFICATION TABLE NO. 303 THERMAL CONDUCTIVITY OF PHENOLIC RESIN

Curve No.	Ref. No.	Method Used	Year	Temp. Range, K	Reported Error, %	Name and Specimen Designation	Composition (weight percent), Specifications and Remarks
1	527	L	1956	293.5	1.2		0.1 amino acid; cured.
2	527	L	1956	310.4	1.2		0.2 amino acid; cured.
3	527	L	1956	310.2	1.2		0.3 amino acid; cured.
4	527	L	1956	296.6	1.2		Cured.

DATA TABLE NO. 303 THERMAL CONDUCTIVITY OF PHENOLIC RESIN

[Temperature, T, K; Thermal Conductivity, k, Watt $cm^{-1} K^{-1}$]

T	k
CURVE 1*	
293.5	0.00328
CURVE 2*	
310.4	0.00323
CURVE 3*	
310.2	0.00294
CURVE 4*	
296.6	0.00455

*No graphical presentation

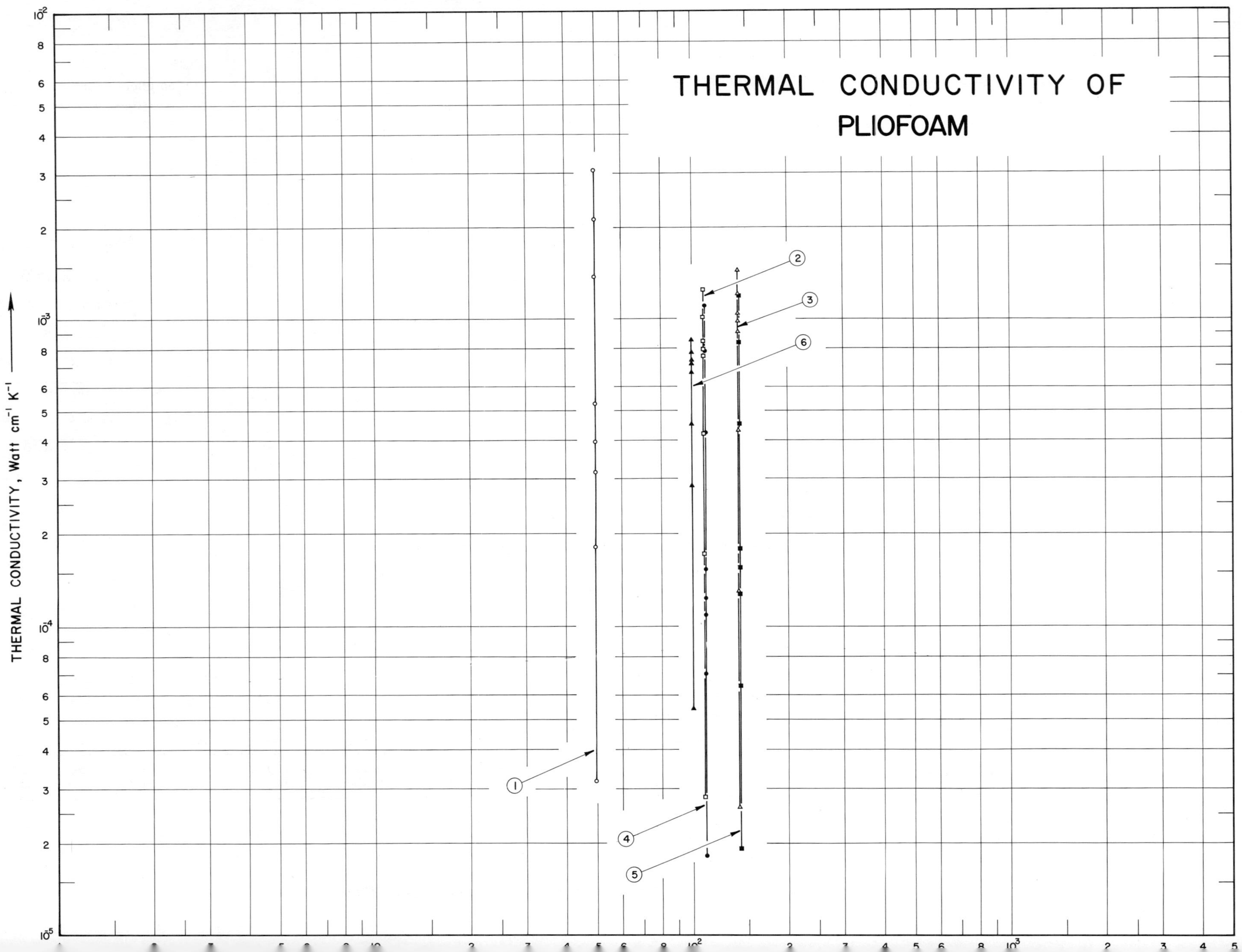

THERMAL CONDUCTIVITY OF PLIOFOAM

SPECIFICATION TABLE NO. 304 THERMAL CONDUCTIVITY OF PLIOFOAM

[For Data Reported in Figure and Table No. 304]

Curve No.	Ref. No.	Method Used	Year	Temp. Range, K	Reported Error, %	Name and Specimen Designation	Composition (weight percent), Specifications and Remarks
1	122	R	1953	49.4			Foam specimen supplied by Goodyear Co.; packing density 0.014 g cm^{-3}; filled with hydrogen; measured at various hydrogen pressures ranging from 0.0059 to 595 mm Hg.
2	122	R	1953	109.8			Similar to the above specimen; measured at various hydrogen pressures ranging from 0.0050 to 607.5 mm Hg.
3	122	R	1953	140.9			Similar to the above specimen; measured at various hydrogen pressures ranging from 0.0049 to 601.0 mm Hg.
4	122	R	1953	110.1			Similar to the above specimen but filled with nitrogen; measured at various nitrogen pressures ranging from 0.0055 to 591.5 mm Hg.
5	122	R	1953	141.0			Similar to the above specimen; measured at various pressures nitrogen ranging from 0.0056 to 585.5 mm Hg.
6	122	R	1953	98.5			Foam specimen supplied by U.S. Rubber Co.; packing density 0.024 g cm^{-3}; filled with hydrogen; measured at various hydrogen pressures ranging from 0.0221 to 625.3 mm Hg.

DATA TABLE NO. 304 THERMAL CONDUCTIVITY OF PLIOFOAM

[Temperature, T, K; Thermal Conductivity, k, Watt cm^{-1} K^{-1}]

p(mm Hg)	k
CURVE 1 T = 49.4K	
0.0059	0.0000318
0.181	0.000181
1.17	0.000315
16.0	0.000395
50.0	0.000528
224.0	0.00139
405.0	0.00214
595.0	0.00307
CURVE 2 T = 109.8K	
0.005	0.0000281
0.112	0.000173
0.946	0.000420
17.0	0.000757
50.5	0.000792
209.0	0.000847
406.5	0.00102
607.5	0.00126
CURVE 3 T = 140.9K	
0.0049	0.0000262
0.103	0.000131
0.956	0.000433
17.0	0.000910
54.5	0.000987
205.0	0.00105
417.5	0.00123
601.0	0.00146
CURVE 4 T = 110.1K	
0.0055	0.0000183
0.181	0.0000706
1.37	0.000109
15.5	0.000124
50.5	0.000154
195.5	0.000425

p(mm Hg)	k
CURVE 4 (cont.)	
390.0	0.000782
591.5	0.00111
CURVE 5 T = 141 K	
0.0056	0.0000193
0.108	0.0000643
0.122	0.000128
15.0	0.000156
48.0	0.000179
196.5	0.000456
384.5	0.000833
585.5	0.00120
CURVE 6 T = 98.5K	
0.0221	0.0000542
0.507	0.000285
1.77	0.000452
15.5	0.000671
69.9	0.000719
230.2	0.000738
441.3	0.000778
625.3	0.000853

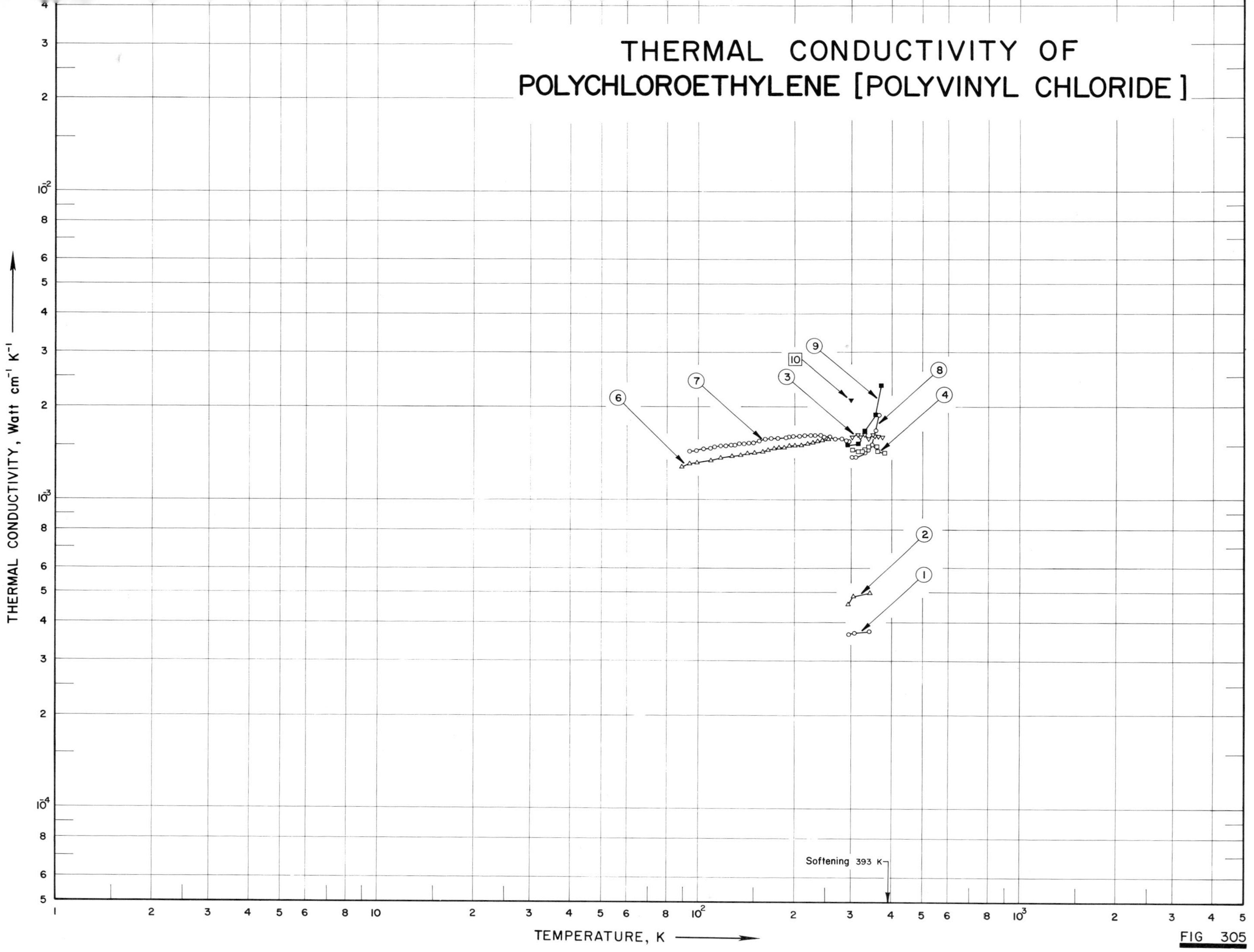
THERMAL CONDUCTIVITY OF
POLYCHLOROETHYLENE [POLYVINYL CHLORIDE]
THERMAL CONDUCTIVITY, Watt cm⁻¹ K⁻¹
TEMPERATURE, K
Softening 393 K
1
2
3
4
6
7
8
9
10
FIG 305

SPECIFICATION TABLE NO. 305 THERMAL CONDUCTIVITY OF POLYCHLOROETHYLENE (POLYVINYL CHLORIDE)

[For Data Reported in Figure and Table No. 305]

Curve No.	Ref. No.	Method Used	Year	Temp. Range, K	Reported Error, %	Name and Specimen Designation	Composition (weight percent), Specifications and Remarks
1	58	L	1953	296-343		No. 1	Specimen from Gen. Tire and Rubber Co.
2	58	L	1953	295-342		No. 2	Similar to the above specimen.
3	59	L	1953	303-377			Low plasticizer content; approx 0.05 in. thick placed under a pressure of about 2 psi.
4	59	L	1953	303-380			As above but with high plasticizer content.
5	60	P	1960	323.2			Specimen 20 mm in dia, 150 mm long; the rod was initially cooled to 0 C and put in boiling water; thermal conductivity values calculated from measured transient temperature changes.
6	387	P	1962	89-363	±4		10% plasticizer content.
7	387	P	1962	94-362	±4		40% plasticizer content.
8	388	L	1960	303-368		PVC - I	Molecular weight 1000.
9	388	L	1960	293-373		PVC - II	Molecular weight 1300.
10	330	C	1957	298.2		Plasticized PVC	Specimen .25 in. thick and 3 in. dia lapped disk; measured by thermal comparator No. 4.

DATA TABLE NO. 305 THERMAL CONDUCTIVITY OF POLYCHLOROETHYLENE (POLYVINYL CHLORIDE)

[Temperature, T, K; Thermal Conductivity, k, Watt $cm^{-1}K^{-1}$]

T	k
CURVE 1	
296.3	0.000366
308.8	0.000369
342.7	0.000374
CURVE 2	
294.8	0.000459
305.8	0.000485
342.1	0.000498
CURVE 3	
303.2	0.00159
316.2	0.00163
320.2	0.00159
331.2	0.00163
340.2	0.00157
351.2	0.00163
355.7	0.00161
364.2	0.00161
374.2	0.00160*
376.7	0.00159
CURVE 4	
303.2	0.00146
315.2	0.00144
325.2	0.00144
331.2	0.00146
339.2	0.00149
341.2	0.00149*
360.2	0.00149
362.2	0.00144
380.2	0.00142
CURVE 5*	
323.2	0.00164
CURVE 6	
89.1	0.00128
94.6	0.00131
99.0	0.00133
CURVE 6 (cont.)	
103.4	0.00134*
109.1	0.00135
113.7	0.00136*
118.6	0.00137
123.2	0.00138*
127.0	0.00139
131.3	0.00140*
136.1	0.00140
139.4	0.00141*
143.1	0.00142
146.7	0.00142*
150.8	0.00143
154.4	0.00144*
159.0	0.00144
161.8	0.00145*
166.2	0.00146
170.1	0.00146*
173.1	0.00147
176.0	0.00148*
179.9	0.00148
183.3	0.00149*
186.9	0.00148
190.3	0.00149*
193.2	0.00150
196.7	0.00149*
200.1	0.00150
203.2	0.00151*
206.0	0.00151*
209.3	0.00151
212.1	0.00152*
215.5	0.00152*
219.5	0.00153
221.9	0.00153*
224.5	0.00153*
227.6	0.00154
229.9	0.00154*
232.7	0.00155*
236.5	0.00155
239.1	0.00155*
241.4	0.00156*
244.2	0.00156
246.6	0.00156*
250.3	0.00156*
252.9	0.00157*
CURVE 6 (cont.)	
255.5	0.00157
257.9	0.00158*
260.1	0.00157*
263.4	0.00158*
265.6	0.00158*
273.1	0.00159*
276.2	0.00159*
283.3	0.00160*
286.7	0.00160*
287.7	0.00159*
291.8	0.00160*
295.7	0.00160*
303.4	0.00161*
308.0	0.00161*
311.1	0.00161*
313.2	0.00161*
317.3	0.00161*
319.5	0.00160*
322.2	0.00160*
324.1	0.00160*
326.3	0.00160*
329.1	0.00160*
331.3	0.00161*
333.8	0.00160*
336.3	0.00161*
338.8	0.00160*
341.8	0.00160*
344.2	0.00159*
348.7	0.00159*
350.3	0.00160*
352.6	0.00160*
354.6	0.00159*
356.2	0.00159*
359.1	0.00158*
362.6	0.00158*
CURVE 7	
94.3	0.00144
99.5	0.00145
104.2	0.00147
109.1	0.00147
113.4	0.00149
CURVE 7 (cont.)	
118.6	0.00150
123.2	0.00150
127.3	0.00151
131.8	0.00151
136.1	0.00153
139.6	0.00153
144.2	0.00154
148.1	0.00154
151.5	0.00155*
156.3	0.00156
158.9	0.00156*
162.6	0.00157
166.6	0.00157*
169.6	0.00158
173.2	0.00158*
177.0	0.00158
181.0	0.00158*
183.6	0.00159*
188.0	0.00159
190.7	0.00159*
193.7	0.00160
197.5	0.00161
201.2	0.00161*
203.7	0.00161*
207.2	0.00161
209.8	0.00161*
213.4	0.00162*
216.3	0.00162*
219.4	0.00162*
223.1	0.00162*
225.6	0.00162
228.7	0.00162*
231.2	0.00162*
233.4	0.00162
236.9	0.00162*
239.9	0.00161*
242.7	0.00162
245.3	0.00161*
247.9	0.00160
252.8	0.00160*
255.7	0.00160*
258.8	0.00160
260.9	0.00159*
263.7	0.00159*
CURVE 7 (cont.)	
266.6	0.00159*
268.8	0.00158
271.5	0.00157*
272.5	0.00158*
276.4	0.00157*
278.7	0.00157*
281.7	0.00157
283.9	0.00156*
286.6	0.00156*
289.1	0.00156
291.3	0.00155*
294.2	0.00155*
296.8	0.00155
298.8	0.00155*
301.4	0.00155*
304.2	0.00154*
306.1	0.00154*
308.4	0.00154*
310.7	0.00153*
312.7	0.00153*
316.7	0.00153*
317.3	0.00153*
320.7	0.00152*
322.7	0.00151*
324.8	0.00152*
326.8	0.00151*
329.2	0.00151*
332.7	0.00151*
335.1	0.00150*
336.8	0.00151*
339.4	0.00150*
341.7	0.00150*
345.1	0.00150*
349.1	0.00150*
350.6	0.00149*
352.7	0.00149*
354.5	0.00149*
356.7	0.00149*
359.4	0.00148*
362.1	0.00148*
CURVE 8	
303	0.00138
310	0.00138
315	0.00138*
328	0.00142*
333	0.00142
339	0.00146
349	0.00151
357	0.00167
368	0.00188
CURVE 9	
293	0.00151
316	0.00153
331	0.00167
339	0.00163*
345	0.00159*
359	0.00188
373	0.00234
CURVE 10	
298.2	0.0021

*Not shown on Plot

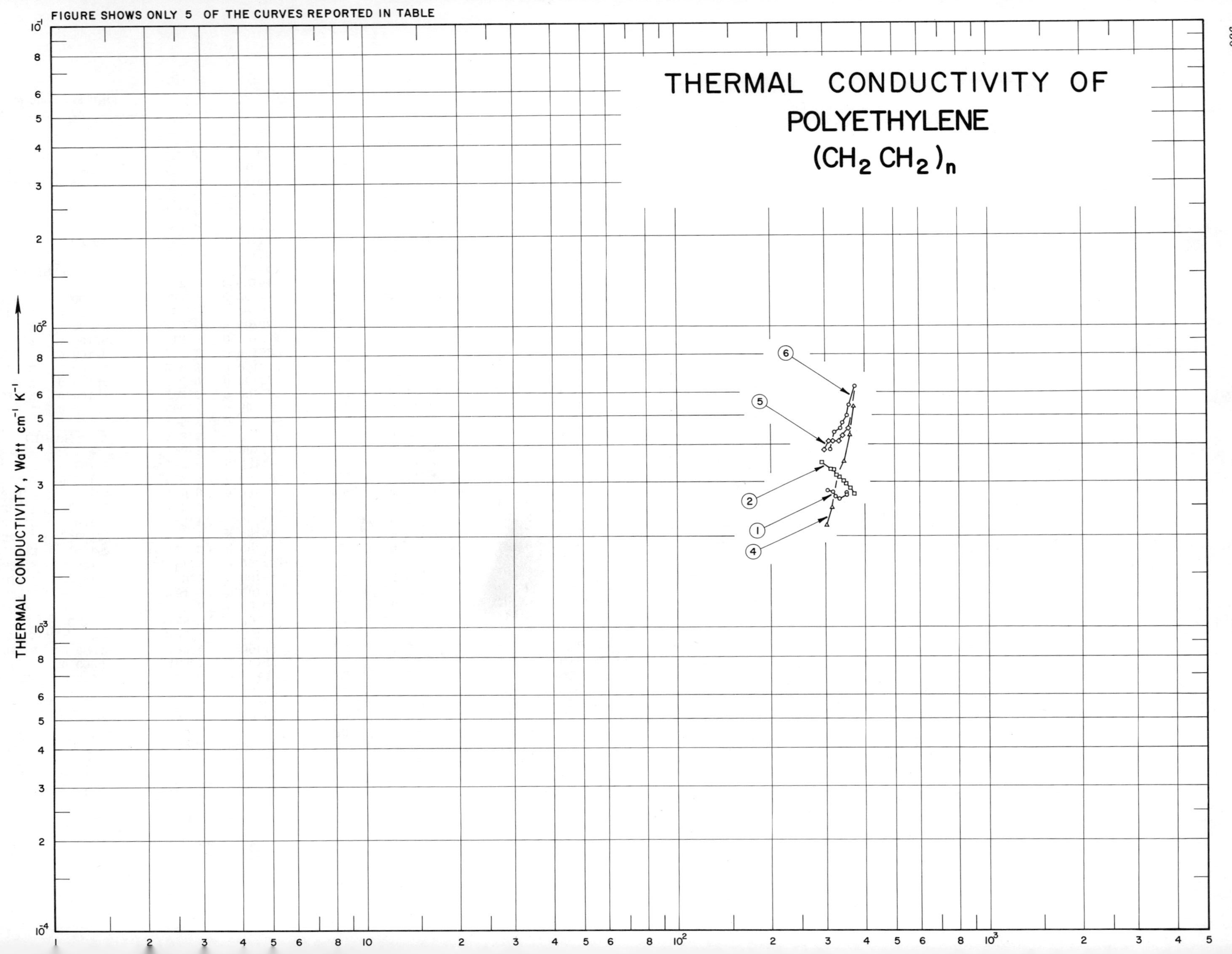

FIGURE SHOWS ONLY 5 OF THE CURVES REPORTED IN TABLE
THERMAL CONDUCTIVITY OF
POLYETHYLENE
$(CH_2 CH_2)_n$
THERMAL CONDUCTIVITY, Watt cm^{-1} K^{-1}
6
5
2
1
4

SPECIFICATION TABLE NO. 306 THERMAL CONDUCTIVITY OF POLYETHYLENE $(CH_2CH_2)_n$

[For Data Reported in Figure and Table No. 306]

Curve No.	Ref. No.	Method Used	Year	Temp. Range, K	Reported Error, %	Name and Specimen Designation	Composition (weight percent), Specifications and Remarks
1	59	L	1953	305-353			Dry surface; approximately 0.05 in. thick placed under a slight pressure estimated at 2 lb in.$^{-2}$
2	59	L	1953	293-371			Glycerine on surfaces; approximately 0.05 in. thick placed under a slight pressure estimated at 2 lb in.$^{-2}$
3	330	C	1957	298.2			0.2188 in. thick and 3 in. dia lapped disk specimen; thermal comparator No. 4 used.
4	388	L	1960	302-369		PE I	M.W. 21,000.
5	388	L	1960	297-372		PE II	M.W. 70,000-80,000.
6	388	L	1960	311-373		PE III	M.W. 70,000-80,000.
7	388	L	1960	308-393		PE IV	M.W. 70,000-80,000.

DATA TABLE NO. 306 THERMAL CONDUCTIVITY OF POLYETHYLENE $(CH_2CH_2)_n$

[Temperature, T, K; Thermal Conductivity, k, Watt cm^{-1} K^{-1}]

T	k
CURVE 1	
305.2	0.00280
317.2	0.00278
322.2	0.00268
333.7	0.00264
333.7	0.00268*
352.7	0.00270
352.7	0.00274
CURVE 2	
293.2	0.00349
313.2	0.00331
319.2	0.00320
326.2	0.00316
328.2	0.00314*
328.7	0.00316*
333.2	0.00310
344.7	0.00301
347.2	0.00299*
349.7	0.00295
360.2	0.00285
371.2	0.00272
CURVE 3*	
298.2	0.00335
CURVE 4	
302	0.00218
314	0.00247
327	0.00318*
344	0.00352
359	0.00431
369	0.00536
CURVE 5	
297	0.00385
308	0.00410
316	0.00410
331	0.00410
340	0.00427
CURVE 5 (cont.)	
356	0.00452
372	0.00636*
CURVE 6	
311	0.00385
321	0.00439
335	0.00452
340	0.00473
352	0.00498
357	0.00544
373	0.00628
CURVE 7*	
308	0.00335
319	0.00385
324	0.00300
335	0.00402
348	0.00402
374	0.00494
393	0.00561

*Not shown on plot

SPECIFICATION TABLE NO. 307 THERMAL CONDUCTIVITY OF POLYHEXAHYDRO-2H-AZEPIN-2-ONE, SILON

Curve No.	Ref. No.	Method Used	Year	Temp. Range, K	Reported Error, %	Name and Specimen Designation	Composition (weight percent), Specifications and Remarks
1	538	L	1955	333, 373			Rod 41/71 x 210 mm; density 1.15 g cm^3.

DATA TABLE NO. 307 THERMAL CONDUCTIVITY OF POLYHEXAHYDRO-2H-AZEPIN-2-ONE, SILON

[Temperature, T, K; Thermal Conductivity, k, Watt cm^{-1} K^{-1}]

T	k
CURVE 1*	
333.2	0.00259
373.2	0.00267

*No graphical presentation

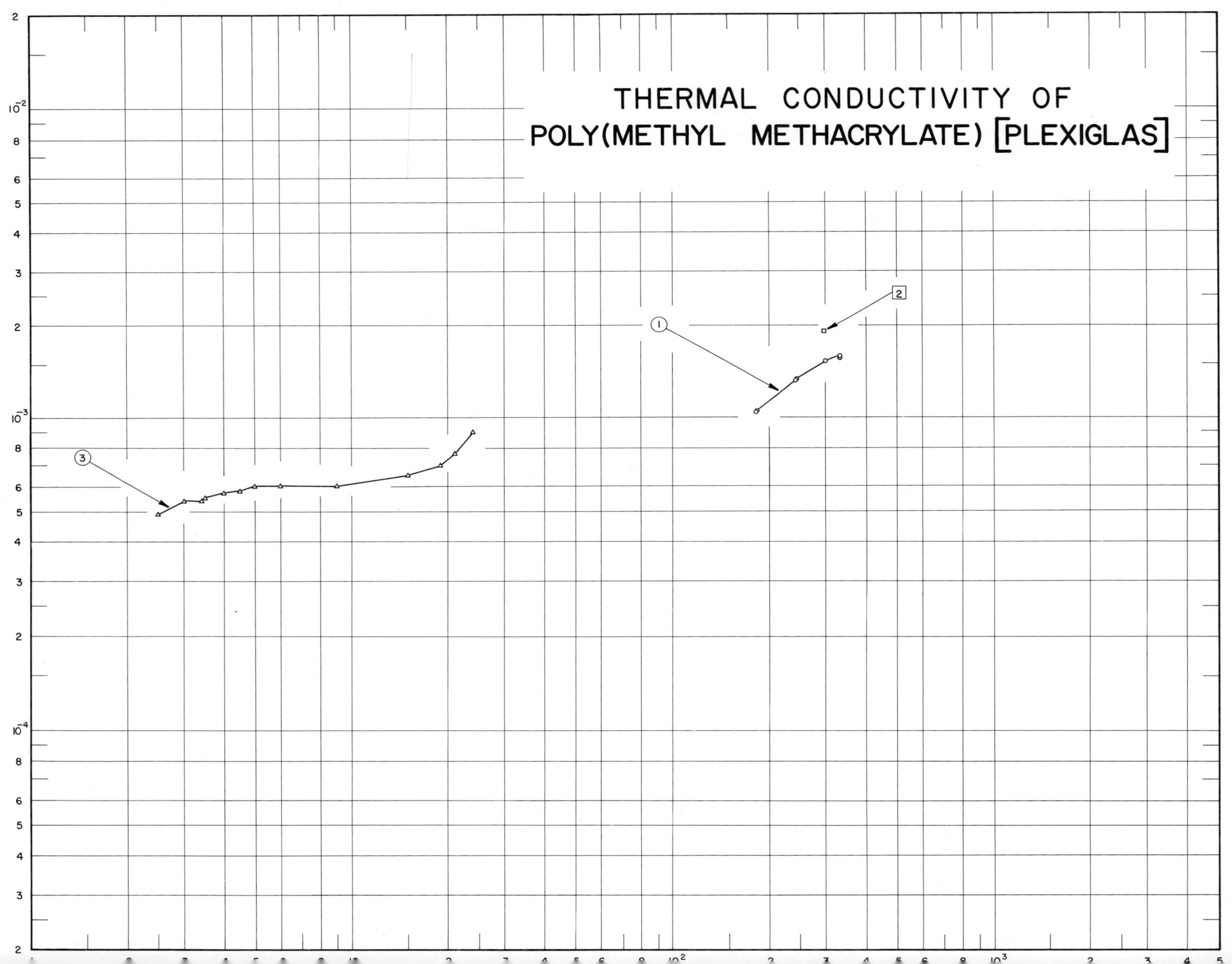
THERMAL CONDUCTIVITY OF
POLY(METHYL METHACRYLATE) [PLEXIGLAS]
THERMAL CONDUCTIVITY, Watt cm⁻¹ K⁻¹
1
2
3
2
10⁻²
8
6
5
4
3
2
10⁻³
8
6
5
4
3
2
10⁻⁴
8
6
5
4
3
2
10²
10³

SPECIFICATION TABLE NO. 308 THERMAL CONDUCTIVITY OF POLY(METHYL METHACRYLATE) [PLEXIGLAS] $(CH_2CCH_3COOCH_3)_n$

[For Data Reported in Figure and Table No. 308]

Curve No.	Ref. No.	Method Used	Year	Temp. Range, K	Reported Error, %	Name and Specimen Designation	Composition (weight percent), Specifications and Remarks
1	111	L	1952	183-332		AN-P-44A	Aircraft quality; supplied by Rohm and Haas Chemical Co.
2	117	P	1955	298. 2	1-2		Specimen thickness 10. 0 mm; density 1. 18 g cm^{-3}.
3	69	L	1951	2. 5-24		Perspex	1. 27 cm dia x 3. 05 cm long.

DATA TABLE NO. 308 THERMAL CONDUCTIVITY OF POLY(METHYL METHACRYLATE) [PLEXIGLAS]

[Temperature, T, K; Thermal Conductivity, k, Watt cm^{-1} K^{-1}]

T	k
CURVE 1	
183.3	0.00105
184.4	0.00106
244.8	0.00134
245.6	0.00135
300.3	0.00154
332.1	0.00158
332.2	0.00157*
CURVE 2	
298.2	0.00192
CURVE 3	
2.5	0.00049
3.0	0.00054
3.4	0.00054
3.5	0.00055
4.0	0.00057
4.5	0.00058
5.0	0.00060
6.0	0.00060
9.0	0.00060
15.0	0.00065
19.0	0.00070
21.0	0.00077
24.0	0.00090

*Not shown on plot

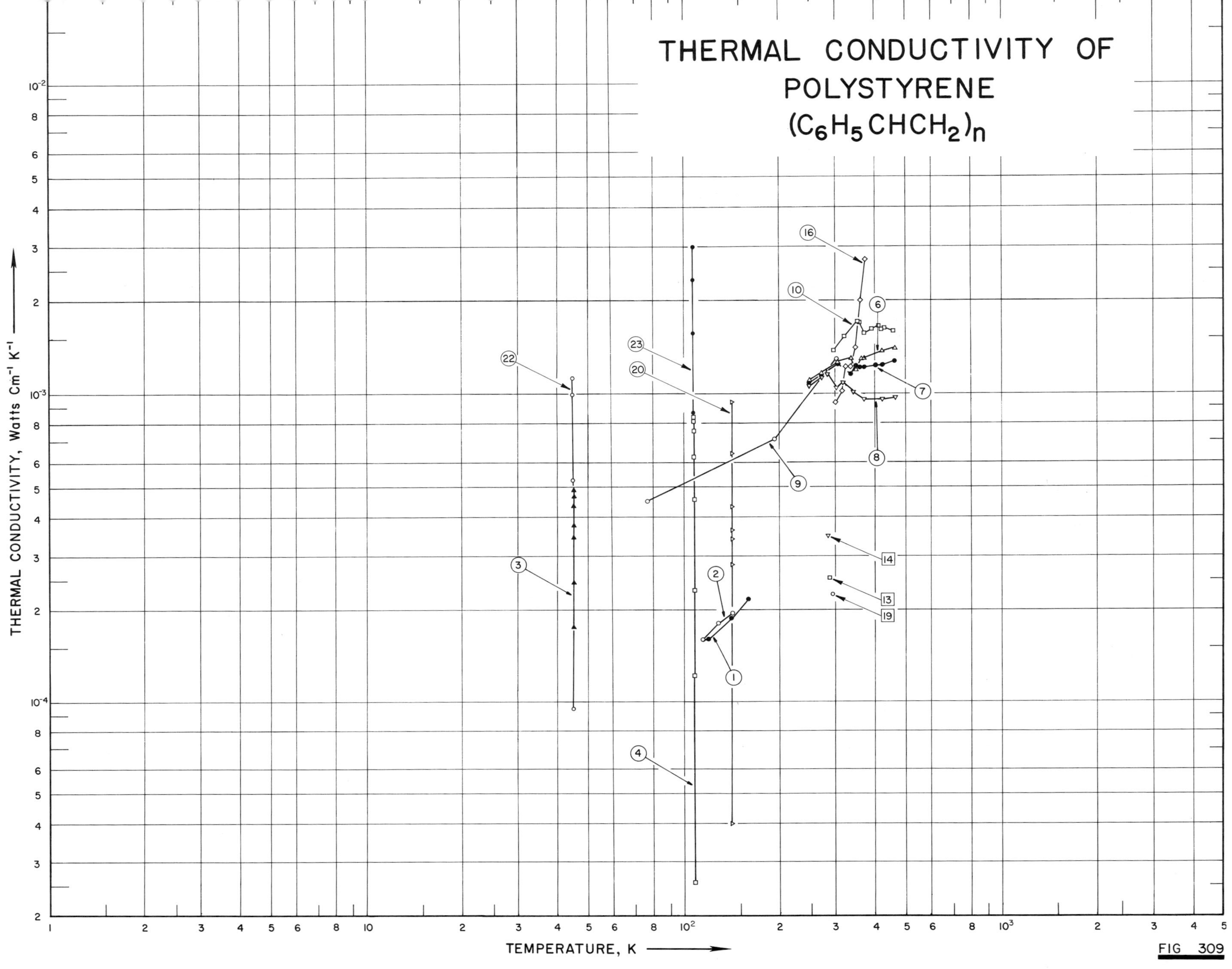

THERMAL CONDUCTIVITY OF
POLYSTYRENE
$(C_6H_5CHCH_2)_n$
THERMAL CONDUCTIVITY, Watts Cm^{-1} K^{-1}
TEMPERATURE, K
FIG 309

SPECIFICATION TABLE NO. 309 THERMAL CONDUCTIVITY OF POLYSTYRENE $(C_6H_5CHCH_2)_n$

[For Data Reported in Figure and Table No. 309]

Curve No.	Ref. No.	Method Used	Year	Temp. Range, K	Reported Error, %	Name and Specimen Designation	Composition (weight percent), Specifications and Remarks
1	43	L	1954	120-160			Expanded board specimen; density 0.0240 g cm^{-3}.
2	43	L	1954	115-143			Expanded board specimen; flameproofed; density 0.0288 g cm^{-3}.
3	122	R	1953	45.1			Colloidal aggregate; packing density 0.282 g cm^{-3}; filled with hydrogen; measured at various hydrogen pressures ranging from 0.495 to 621.6 mm Hg.
4	122	R	1953	108.8			Similar to the above specimen; measured at various hydrogen pressures ranging from 0.0251 to 632.6 mm Hg.
5	122	R	1953	109.1			Similar to the above specimen but filled with nitrogen; measured at various nitrogen pressures ranging from 0.0241 to 590.5 mm Hg.
6	123	P	1953	249-466			Molecular weight 3650; density reported as 1.099, 1.096, 1.097, 1.090, 1.087, 1.045, 1.024, and 0.9775 g cm^{-3} at -20, -10, 0, 10, 20, 50, 100, and 185 C, respectively; thermal conductivity values calculated from measured data of thermal diffusivity, specific heat, and density.
7	123	P	1953	249-466			Molecular weight 2300; density reported as 1.095, 1.093, 1.091, 1.088, 1.052, 1.045, 1.023, and 0.9807 g cm^{-3} at -20, -10, 0, 10, 20, 50, 100, and 180 C, respectively; thermal conductivity values calculated from measured data of thermal diffusivity, specific heat, and density.
8	123	P	1953	249-466			Molecular weight 860; density reported as 1.071, 1.068, 1.065, 1.098, 1.095, 1.044, 1.014, and 0.0694 g cm^{-3} at -20, -10, 0, 10, 20, 50, 100, and 185 C, respectively thermal conductivity values calculated from measured data of thermal diffusivity, specific heat, and density.
9	124	L	1960	78-305	2		Density 1.03 g cm^{-3}.
10	528	P	1953	298-460			Prepared by polymerizing monostyrene with 5 mole % of very pure p-divinylbenzene at 200 C without a catalyst; thermal conductivity values calculated from measured data of thermal diffusivity and specific heat.
11	528	P	1953	301-458			Similar to the above specimen except prepared with 9 mole % of very pure p-divinylbenzene.
12	528	P	1953	302-458			Similar to the above specimen except prepared with 15 mole % of very pu e p-divinylbenzene.
13	529	→	1956	289.8			Expanded; density 0.0441 g cm^{-3}; thermal conductivity data calculated from measured transient rise of the specimen temperature.
14	529	→	1956	286.5			Similar to the above specimen except density 0.0641 g cm^{-3}.
15	388	L	1960	298-368		PS-II	Degree of polymerization: M.W. 1000.
16	388	L	1960	303-375		PS-I	Degree of polymerization: M.W. 700.

SPECIFICATION TABLE NO. 309 (continued)

Curve No.	Ref. No.	Method Used	Year	Temp. Range, K	Reported Error, %	Name and Specimen Designation	Composition (weight percent), Specifications and Remarks
17	530	L	1956	296.6			Foam specimen; 20 cm dia x 2.5 cm thick; density 0.030 g cm^{-3}; measured in air.
18	530	L	1956	297.8			The above specimen measured in carbon dioxide.
19	530	L	1956	296.7			The above specimen measured in chloromethane.
20	122	R	1953	142.0		Styrofoam	Cellular specimen, filled with hydrogen; packing density 0.0258 g cm^{-3}; measured at various hydrogen pressures ranging from 0.0061 to 602.0 mm Hg.
21	122	R	1953	109.0		Styrofoam	Finely ground specimen filled with nitrogen; packing density 0.031 to 0.0586 g cm^{-3}; measured at various nitrogen pressures ranging ffom 0.032 to 603.9 mm Hg.
22	122	R	1953	45.2		Styrofoam	Similar to the above specimen but filled with hydrogen; measured at various hydrogen pressures ranging from 0.0194 to 645.4 mm Hg.
23	122	R	1953	109.2		Styrofoam	Similar to the above specimen; measured at various hydrogen pressures ranging from 0.056 to 601.5 mm Hg.
24	520	L	1964	288-386	±2		Specimen 0.5 in. thick; unoriented.
25	520	L	1964	301-361	±2		Specimen 0.5 in. thick; biaxially oriented; provided by the Plax Company; perfectly clear and bubble free.
26	554	R	1949	45.1		Colloidal aggregate	Packing density 0.282 g cm^{-3}; measured in the presence of hydrogen gas with liquid hydrogen in the cryostat, hydrogen pressure ranged from 0.495 to 621.6 mm Hg.
27	554	R	1949	109.1			Same as above except measured with liquid nitrogen in the cryostat, nitrogen pressure ranged from 0.0241 to 590.5 mm Hg.
28	555	R	1949	109.2			Same as above except measured with liquid hydrogen in the cryostat, hydrogen pressure ranged from 0.025 to 632.6 mm Hg.

DATA TABLE NO. 309 THERMAL CONDUCTIVITY OF POLYSTYRENE $(C_6H_5CHCH_2)_n$

[Temperature, T, K; Thermal Conductivity, k, Watt $cm^{-1}K^{-1}$]

CURVE 1

T	k
120	0.000159
141	0.000187
160	0.000216

CURVE 2

T	k
115	0.000159
129	0.000180
143	0.000195

CURVE 3
T = 45.1 K

p(mm Hg)	k
0.495	0.000174
1.73	0.000243
18.9	0.000343
49.3	0.000376
203.6	0.000432
421.4	0.000465
621.6	0.000489

CURVE 4
T = 108.8 K

p(mm Hg)	k
0.0251	0.0000256
0.264	0.000121
0.778	0.000230
7.68	0.000455
60.5	0.000622
242.6	0.000753
452.9	0.000807
632.6	0.000836

CURVE 5*
T = 109.1 K

p(mm Hg)	k
0.0241	0.0000157
0.264	0.0000652
1.23	0.000126
17.9	0.000208
62.8	0.000237
191.0	0.000257

CURVE 5 (cont.)*

p(mm Hg)	k
396.3	0.000276
590.5	0.000284

CURVE 6

T	k
249.2	0.00109
273.2	0.00116
306.2	0.00126
339.2	0.00129
350.2	0.00118
363.2	0.00128
373.2	0.00129
423.2	0.00136
466.2	0.00139

CURVE 7

T	k
249.2	0.00108
273.2	0.00115
323.2	0.00126
339.2	0.00115
349.2	0.00121
358.2	0.00120
373.2	0.00121
403.2	0.00123
423.2	0.00123
466.2	0.00126

CURVE 8

T	k
249.2	0.00106
273.2	0.00113
285.2	0.00115
303.2	0.00103
319.2	0.00107
343.2	0.00100
373.2	0.00095
423.2	0.00095
466.2	0.00096

CURVE 9

T	k
77.6	0.00045
194.3	0.00071
305.4	0.00128

CURVE 10

T	k
298.2	0.00136
323.2	0.00151
355.2	0.00169
361.2	0.00167
373.2	0.00155
394.2	0.00159
413.2	0.00163
423.2	0.00159
433.2	0.00160
460.2	0.00157

CURVE 11*

T	k
301.2	0.00135
323.2	0.00146
342.2	0.00155
352.2	0.00157
373.2	0.00158
378.2	0.00159
388.2	0.00161
406.2	0.00156
419.2	0.00162
458.2	0.00159

CURVE 12*

T	k
302.2	0.00132
307.2	0.00134
313.2	0.00137
323.2	0.00141
332.2	0.00146
347.2	0.00148
367.2	0.00149
381.2	0.00156
389.2	0.00151
398.2	0.00157
415.2	0.00156
423.2	0.00156
458.2	0.00159

CURVE 13

T	k
289.8	0.000252

CURVE 14

T	k
286.5	0.000346

CURVE 15*

T	k
298	0.000921
308	0.00113
315	0.00117
326	0.00121
333	0.00130
348	0.00134
353	0.00134
358	0.00176
368	0.00243

CURVE 16

T	k
303	0.000921
317	0.00100
324	0.00121
338	0.00121
350	0.00138
353	0.00167*
363	0.00197
375	0.00268

CURVE 17*

T	k
296.6	0.000337

CURVE 18*

T	k
297.8	0.000268

CURVE 19

T	k
296.7	0.000223

CURVE 20
T = 142 K

p(mm Hg)	k
0.0061	0.0000397
0.175	0.000184*
1.78	0.000279
16.0	0.000338
50.0	0.000360
204.5	0.000429
398.0	0.000635
602.0	0.000927

CURVE 21*
T = 109 K

p(mm Hg)	k
0.032	0.0000843
0.322	0.000104
2.43	0.000111
14.1	0.000170
52.8	0.000348
196.6	0.000749
395.8	0.00114
603.9	0.00146

CURVE 22
T = 45.2 K

p(mm Hg)	k
0.0194	0.0000944
0.238	0.000271
2.65	0.000351
17.7	0.000368
61.3	0.000380
197.8	0.000523
416.9	0.000983
645.4	0.00111

CURVE 23
T = 109.2 K

p(mm Hg)	k
0.056	0.000174*
0.265	0.000557*
2.33	0.000717*
15.4	0.000828*
52.5	0.000857
195.8	0.00155
394.8	0.00231
601.5	0.00294

CURVE 24*

T	k
287.7	0.00153
299.1	0.00155
313.4	0.00157
318.3	0.00158
327.7	0.00158
339.4	0.00161
347.6	0.00162
349.5	0.00162
352.4	0.00163
352.7	0.00161
355.6	0.00163
357.7	0.00164
359.3	0.00162
361.4	0.00163
362.4	0.00162
364.8	0.00164
366.5	0.00165
368.9	0.00163
369.6	0.00162
371.4	0.00163
375.8	0.00164
385.7	0.00163

CURVE 25*

T	k
301.3	0.00149
311.1	0.00148
312.0	0.00151
320.9	0.00153
331.7	0.00155
340.4	0.00155
345.9	0.00157
350.6	0.00158
356.3	0.00160
360.6	0.00161

CURVE 26*
T = 45.09 K

Pressure of hydrogen (mm Hg)	k
0.495	0.000190
1.732	0.000243
18.95	0.000343
49.3	0.000376
203.65	0.000432
421.45	0.000465
621.8	0.000489

CURVE 27*
T = 109.09 K

Pressure of nitrogen (mm Hg)	k
0.0241	0.0000157
0.264	0.0000652
1.23	0.000126
17.9	0.000208
62.8	0.000236
191.0	0.000257
590.5	0.000284

CURVE 28*
T = 109.16 K

Pressure of hydrogen (mm Hg)	k
0.0251	0.0000256
0.264	0.000121
0.778	0.000230
7.68	0.000455
60.5	0.000622
242.6	0.000753
452.9	0.000807
632.6	0.000833

*Not shown on plot

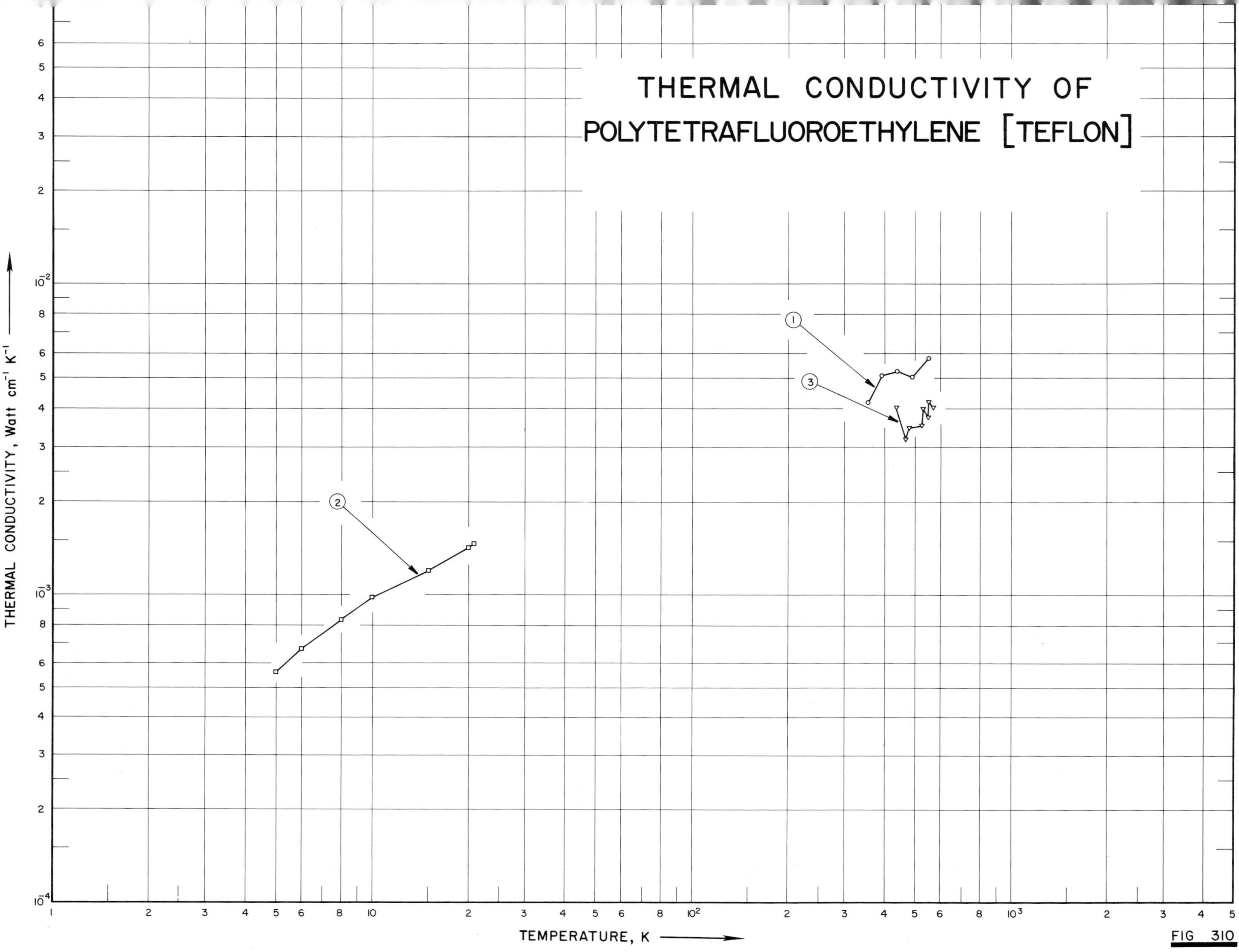
THERMAL CONDUCTIVITY OF
POLYTETRAFLUOROETHYLENE [TEFLON]
1
2
3
THERMAL CONDUCTIVITY, Watt cm⁻¹ K⁻¹
TEMPERATURE, K
FIG 310

SPECIFICATION TABLE NO. 310 THERMAL CONDUCTIVITY OF POLYTETRAFLUOROETHYLENE (TEFLON)

[For Data Reported in Figure and Table No. 310]

Curve No.	Ref. No.	Method Used	Year	Temp. Range, K	Reported Error, %	Name and Specimen Designation	Composition (weight percent), Specifications and Remarks
1	531	C	1958	359-554	<5	Duroid 5600	Fiber reinforced; annealed and heat treated at 583 K for 4 hrs; density 1.99 g cm^{-3}; alumina used as comparative material.
2	532	L	1957	5.0-21	10		2.54 cm dia x 20.5 cm long; extruded; density 2.218 g cm^{-3}.
3	531	C	1958	439-572	<5		Crystalline; annealed and heat treated at 583 K for 4 hrs, cooled; density 2.17 g cm^{-3}; melting point 600 K, alumina used as comparative material.

DATA TABLE NO. 310 THERMAL CONDUCTIVITY OF POLYTETRAFLUOROETHYLENE (TEFLON)

[Temperature, T, K; Thermal Conductivity, k, Watt $cm^{-1} K^{-1}$]

T	k
CURVE 1	
358.7	0.00420
394.3	0.00511
442.1	0.00529
494.3	0.00505
553.7	0.00584
CURVE 2	
5	0.00056
6	0.00067
8	0.00083
10	0.00098
15	0.0012
20	0.00142
20.8	0.00146
CURVE 3	
439.3	0.00401
470.9	0.00317
481.5	0.00346
485.4	0.00346*
527.6	0.00352
532.1	0.00397
553.2	0.00375
553.2	0.00420
572.1	0.00404

*Not shown on plot

SPECIFICATION TABLE NO. 311 THERMAL CONDUCTIVITY OF POLYTRIFLUOROCHLOROETHYLENE

Curve No.	Ref. No.	Method Used	Year	Temp. Range, K	Reported Error, %	Name and Specimen Designation	Composition (weight percent), Specifications and Remarks
1	531	C	1958	318-465	< 5	Kel-F	Annealed and heat treated at 472 K for 24 hrs, cooled slowly; density 2.15 g cm^{-3}; alumina used as comparative material.

DATA TABLE NO. 311 THERMAL CONDUCTIVITY OF POLYTRIFLUOROCHLOROETHYLENE

[Temperature, T, K; Thermal Conductivity, k, Watt cm^{-1} K^{-1}]

T	k
CURVE 1*	
317.6	0.00146
350.7	0.00185
370.9	0.00211
383.2	0.00245
390.9	0.00232
402.6	0.00237
408.2	0.00249
435.9	0.00235
437.6	0.00251
460.9	0.00267
464.6	0.00248

*No graphical presentation

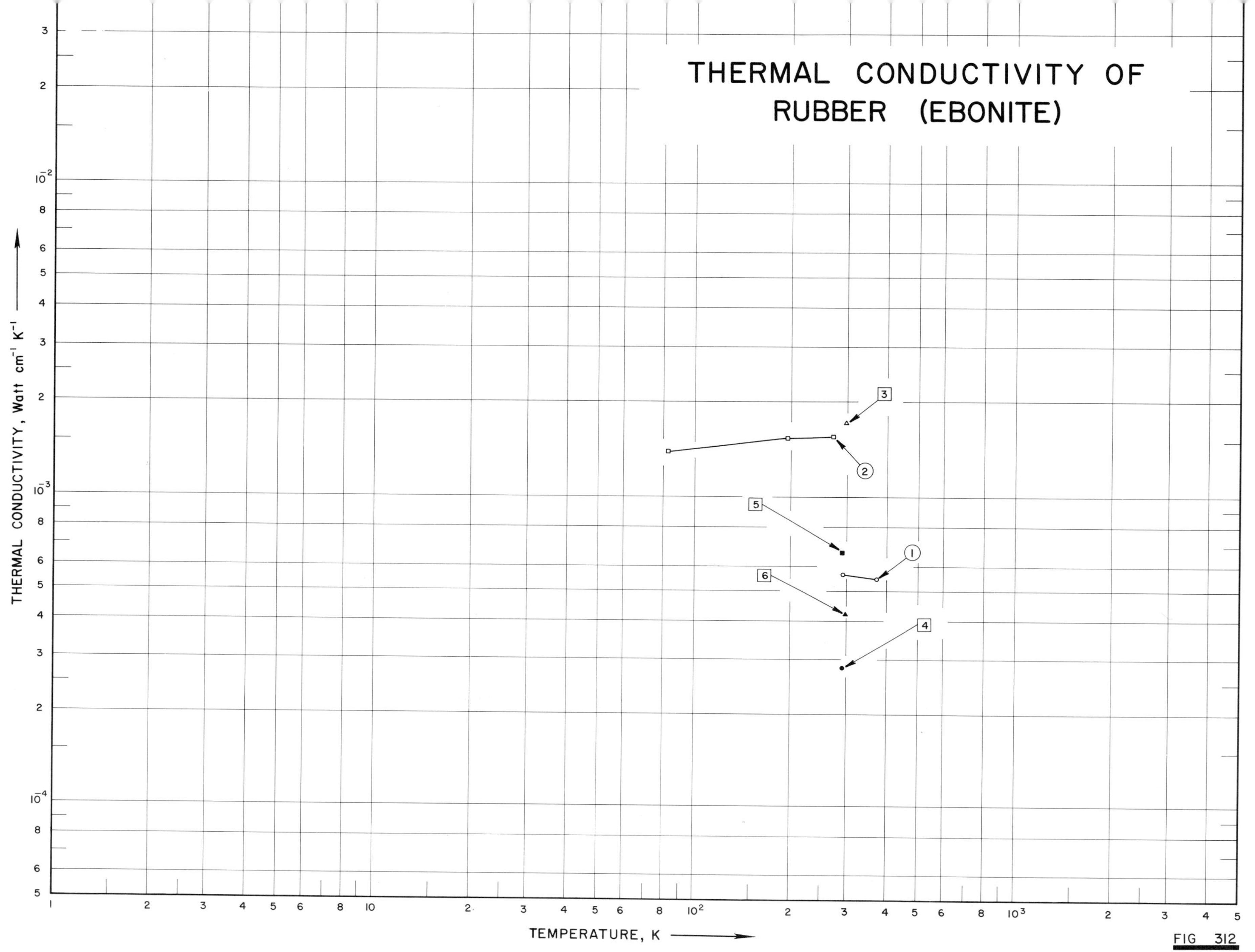
THERMAL CONDUCTIVITY OF
RUBBER (EBONITE)
THERMAL CONDUCTIVITY, Watt cm⁻¹ K⁻¹
TEMPERATURE, K
1
2
3
4
5
6

FIG 312

SPECIFICATION TABLE NO. 312 THERMAL CONDUCTIVITY OF RUBBER (EBONITE)

[For Data Reported in Figure and Table No. 312]

Curve No.	Ref. No.	Method Used	Year	Temp. Range, K	Reported Error, %	Name and Specimen Designation	Composition (weight percent), Specifications and Remarks
1	44	F	1914	293,373		Hard Rubber	Radius 0.327 cm, length 6.3 cm; density 1.19 g cm^{-3}.
2	22	L	1911	83-273		Hard Rubber	No details reported.
3	125	L	1923	298.2	1	Hard Rubber	Average value of measurements for various specimens with different thicknesses ranging from 0.035 cm to 0.15 cm; measured under pressure 21 lb $in.^{-2}$
4	529	P	1956	290.4		Hard Rubber	New sample; expanded; density 0.069 g cm^{-3}; thermal conductivity values calculated from measured transient temperature changes.
5	529	P	1956	290.4		Hard Rubber	Old weathered sample, expanded; same measuring method as above.
6	330	C	1957	298.2		Hard Rubber	Expanded; 0.3125 in. thick and 3 in. dia lapped disk specimen; thermal comparator No. 4 used.

DATA TABLE NO. 312 THERMAL CONDUCTIVITY OF RUBBER (EBONITE)

[Temperature, T, K; Thermal Conductivity, k, Watt $cm^{-1}K^{-1}$]

T	k
CURVE 1	
293.2	0.000569
373.2	0.000548
CURVE 2	
83.2	0.00138
195.2	0.00153
273.2	0.00155
CURVE 3	
298.2	0.00172
CURVE 4	
290.4	0.000281
CURVE 5	
290.4	0.000663
CURVE 6	
298.2	0.00042

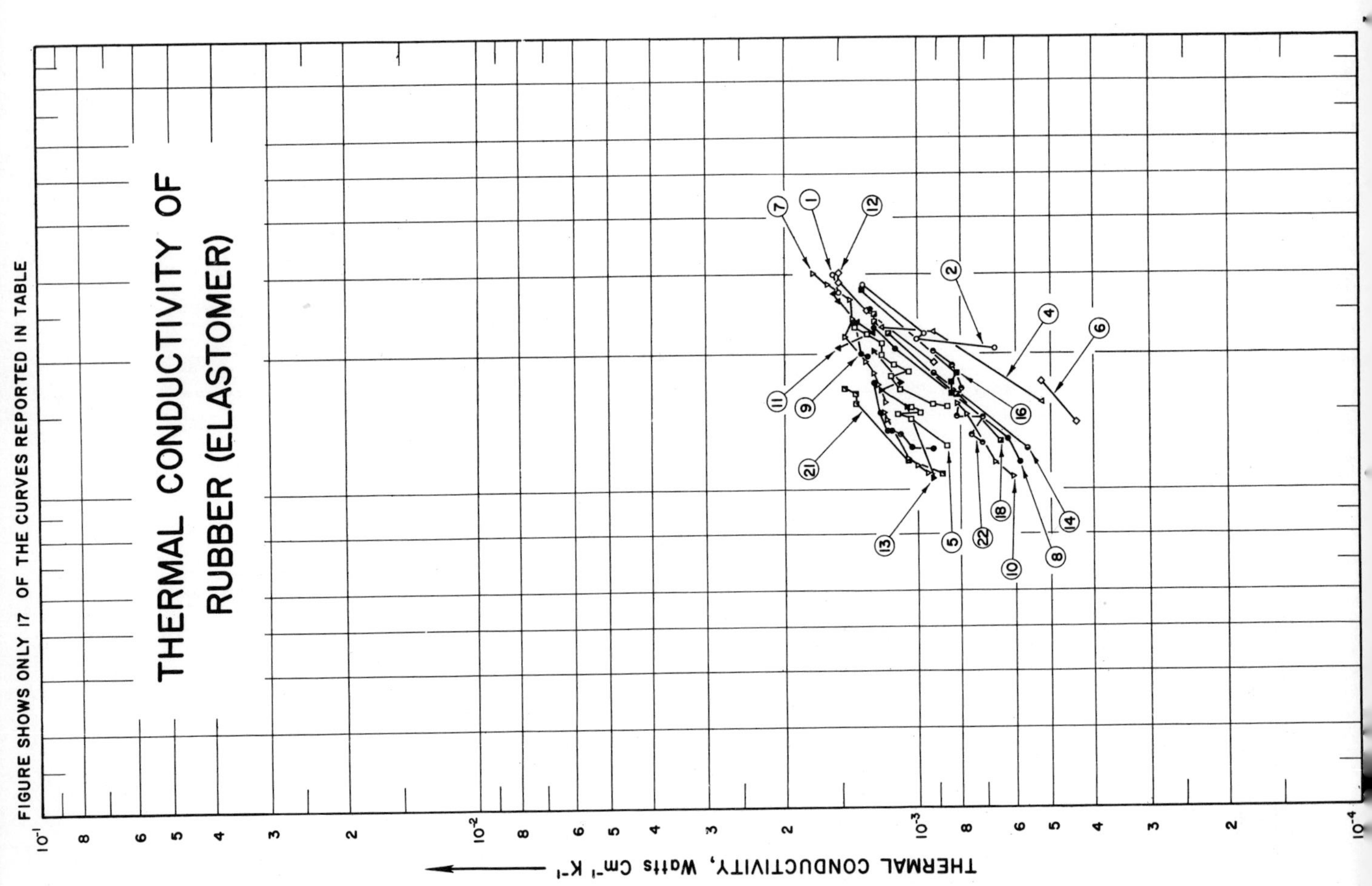
FIGURE SHOWS ONLY 17 OF THE CURVES REPORTED IN TABLE
THERMAL CONDUCTIVITY OF
RUBBER (ELASTOMER)
THERMAL CONDUCTIVITY, Watts Cm⁻¹ K⁻¹
10⁻¹
8
6
5
4
3
2
10⁻²
8
6
5
4
3
2
10⁻³
8
6
5
4
3
2
10⁻⁴
1
2
4
5
6
7
8
9
10
11
12
13
14
16
18
21
22

SPECIFICATION TABLE NO. 313 THERMAL CONDUCTIVITY OF RUBBER (ELASTOMER)

[For Data Reported in Figure and Table No. 313]

Curve No.	Ref. No.	Method Used	Year	Temp. Range, K	Reported Error, %	Name and Specimen Designation	Composition (weight percent), Specifications and Remarks
1	191	L	1950	275, 301	4.0	Run No. 1	87.64 pure latex, 2.61 zinc oxide, 2.4 Sulphur, 0.79 captax, 3.5 stearic acid, 1.75 pine tar, 1.31 antioxidant; cured 120 min at 274 F; no stretch; measured with increasing temperature.
2	191	L	1950	205-285	4.0	Run No. 1	The above specimen measured with decreasing temperature.
3	191	L	1950	151-253	4.0	Run No. 2	Same as the above specimen; no stretch; measured with increasing temperature.
4	191	L	1950	157-229	4.0	Run No. 2	The above specimen measured with decreasing temperature.
5	191	L	1950	125-229	4.0	Run No. 3	Same as the above specimen; no stretch; measured with increasing temperature.
6	191	L	1950	141, 173	4.0	Run No. 3	The above specimen measured with decreasing temperature.
7	191	L	1950	109-301	4.0	Run No. 4	Same as the above specimen; 50% stretch; measured with increasing temperature.
8	191	L	1950	115-251	4.0	Run No. 4	The above specimen measured with decreasing temperature.
9	191	L	1950	123-227	4.0	Run No. 5	Same as the above specimen; 50% stretch; measured with increasing temperature.
10	191	L	1950	107-155	4.0	Run No. 5	The above specimen measured with decreasing temperature.
11	191	L	1950	203-273	4.0	Run No. 6	Same as the above specimen; 100% stretch; measured with increasing temperature.
12	191	L	1950	191-301	4.0	Run No. 6	The above specimen measured with decreasing temperature.
13	191	L	1950	107-203	4.0	Run No. 7	Same as the above specimen; 100% stretch; measured with increasing temperature.
14	191	L	1950	123, 181	4.0	Run No. 7	The above specimen; measured with decreasing temperature.
15	191	L	1950	183-257	4.0	Run No. 8	Same as the above specimen; 100% stretch; measured with increasing temperature.
16	191	L	1950	163-278	4.0	Run No. 8	The above specimen measured with decreasing temperature.
17	191	L	1950	118-204	4.0	Run No. 9	Same as the above specimen; 100% stretch; measured with increasing temperature.
18	191	L	1950	128-246	4.0	Run No. 9	The above specimen measured with decreasing temperature.
19	191	L	1950	169-268	4.0	Run No. 10	Same as the above specimen; 50% stretch; measured with increasing temperature.
20	191	L	1950	195, 209	4.0	Run No. 10	The above specimen measured with decreasing temperature.
21	191	L	1950	108-268	4.0	Run No. 11	Same as the above specimen; no stretch; measured with increasing temperature.
22	191	L	1950	127-273	4.0	Run No. 11	The above specimen measured with decreasing temperature.

DATA TABLE NO. 313 THERMAL CONDUCTIVITY OF RUBBER (ELASTOMER)

[Temperature, T, K; Thermal Conductivity, k, Watt $cm^{-1}K^{-1}$]

T	k
CURVE 1	
275.2	0.00151
301.2	0.00155
CURVE 2	
205.2	0.000669
216.2	0.00100
221.2	0.000962
285.2	0.00134
CURVE 3 *	
151.2	0.00117
163.2	0.00105
163.2	0.00111
193.2	0.00119
195.2	0.00128
205.2	0.00117
243.2	0.00138
253.2	0.00138
CURVE 4	
157.2	0.000523
223.2	0.000920
229.2	0.00121
CURVE 5	
125.2	0.000858
143.2	0.00103
147.2	0.00111
147.2	0.000983
153.2	0.00103
153.2	0.000858
155.2	0.000920
163.2	0.00107
167.2	0.00109
179.2	0.00115
183.2	0.00105
189.2	0.00113
199.2	0.00121
211.2	0.00121
221.2	0.00130
229.2	0.00138
CURVE 6	
141.2	0.000439
173.2	0.000523
CURVE 7	
109.2	0.000941
113.2	0.00100
117.2	0.00105
143.2	0.00117
149.2	0.00119
156.2	0.00118
157.2	0.00118 *
171.2	0.00123
181.2	0.00126
193.2	0.00130
217.2	0.00146
239.2	0.00140
263.2	0.00142
285.2	0.00159
301.2	0.00172
CURVE 8	
115.2	0.000586
129.2	0.000628
205.2	0.00113
229.2	0.00126
251.2	0.00128
CURVE 9	
123.2	0.000920
123.2	0.00103
133.2	0.00109
136.2	0.00115
136.2	0.00117
148.2	0.00121
173.2	0.00126
197.2	0.00130
200.2	0.00134
227.2	0.00138 *
CURVE 10	
107.2	0.000607
115.2	0.000669
CURVE 10 (cont.)	
147.2	0.000774
155.2	0.000816
CURVE 11	
203.2	0.00134 *
207.2	0.00151
211.2	0.00151 *
222.2	0.00126
226.2	0.00134 *
235.2	0.00138
237.2	0.00140 *
261.2	0.00151
273.2	0.00155
CURVE 12	
191.2	0.000920
249.2	0.00130
288.2	0.00151
301.2	0.00151
CURVE 13	
107.2	0.000920
153.2	0.00105
165.2	0.00121
173.2	0.00109
203.2	0.00126
CURVE 14	
123.2	0.000565
181.2	0.000920
CURVE 15 *	
183.2	0.00126
213.2	0.00130
227.2	0.00140
243.2	0.00142
257.2	0.00146
CURVE 16	
163.2	0.000837
173.2	0.000837
181.2	0.000816
278.2	0.00134
CURVE 17 *	
118.2	0.00105
129.2	0.00100
141.2	0.00109
143.2	0.00107
161.2	0.00111
173.2	0.00119
178.2	0.00113
193.2	0.00119
204.2	0.00121
CURVE 18	
128.2	0.000649
163.2	0.000816
223.2	0.00117
235.2	0.00126
246.2	0.00126
CURVE 19 *	
169.2	0.00113
187.2	0.00124
203.2	0.00142
207.2	0.00136
223.2	0.00140
223.2	0.00144
233.2	0.00132
233.2	0.00136
235.2	0.00132
245.2	0.00140
249.2	0.00138
258.2	0.00142
260.2	0.00151
268.2	0.00153
CURVE 20 *	
195.2	0.00100
209.2	0.00111
CURVE 21	
108.2	0.000879
118.2	0.00105 *
255.2	0.00138
263.2	0.00138
268.2	0.00146
CURVE 22	
127.2	0.000711
133.2	0.000753
145.2	0.000711
146.2	0.000816
168.2	0.000795
188.2	0.000837
203.2	0.000920
223.7	0.00111 *
236.2	0.00115 *
248.2	0.00119 *
258.2	0.00126 *
273.2	0.00121 *

* Not shown on plot

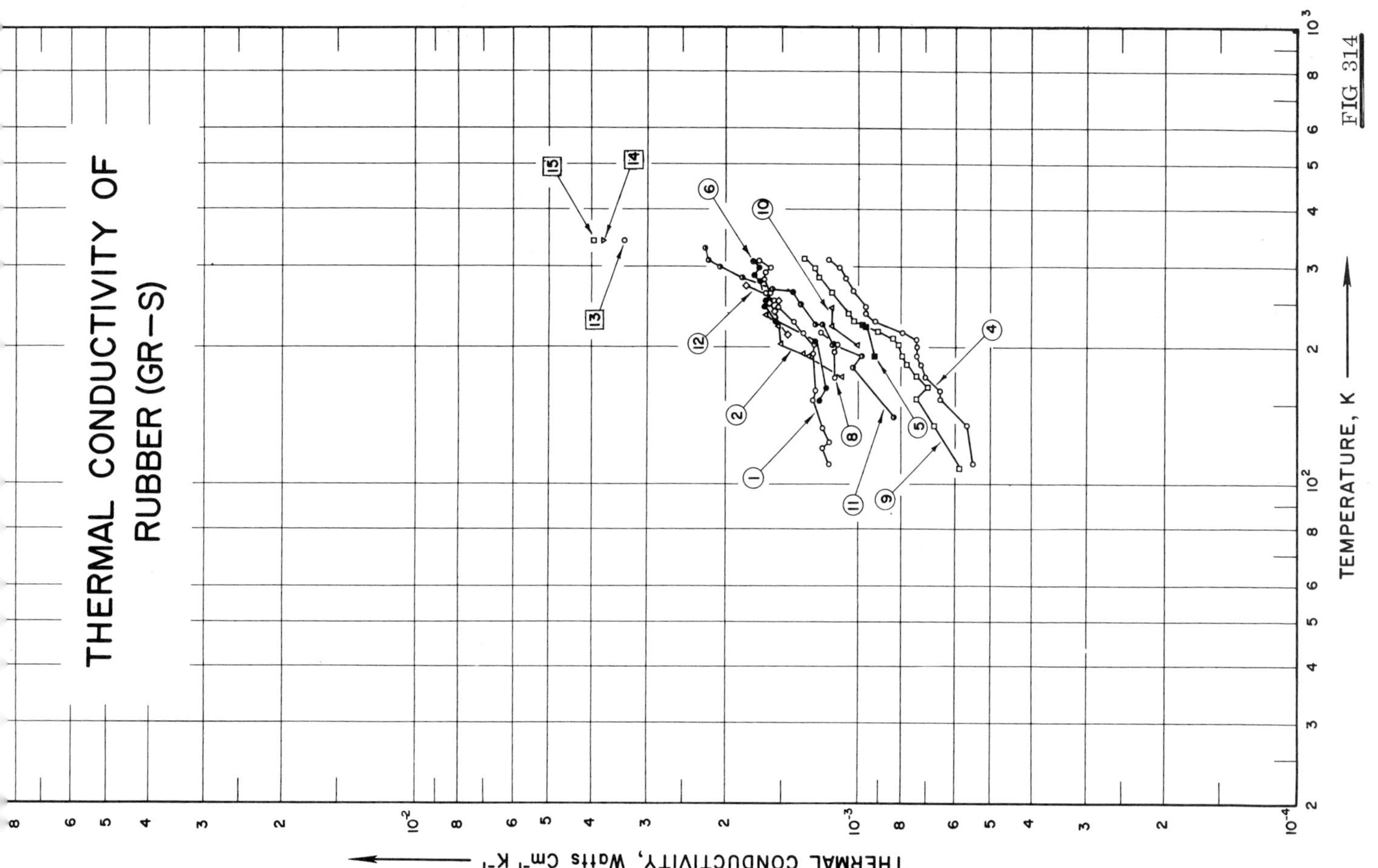
THERMAL CONDUCTIVITY OF
RUBBER (GR–S)
THERMAL CONDUCTIVITY, Watts Cm^{-1} K^{-1}
TEMPERATURE, K

FIG 314

SPECIFICATION TABLE NO. 314 THERMAL CONDUCTIVITY OF RUBBER (GOVERNMENT RUBBER-STYRENE)

[For Data Reported in Figure and Table No. 314]

Curve No.	Ref. No.	Method Used	Year	Temp. Range, K	Reported Error, %	Name and Specimen Designation	Composition (weight percent), Specifications and Remarks
1	191	L	1950	110-309	< 8.5	GR-S	88.1 GR-S, 4.4 zinc oxide, 1.8 sulphur, 1.3 captax, 4.4 bardol B; optimum-cured 60 min at 421 K, specific gravity 0.99; no stretch; measured at increasing temperatures.
2	191	L	1950	173-236	< 8.5	GR-S	The above specimen measured at decreasing temperatures.
3	191	L	1950	173-215	< 8.5	GR-S	Similar to the above specimen except measured at increasing temperatures.
4	191	L	1950	111-311	< 8.5	GR-S	Similar to the above specimen except 100% stretched and measured at increasing temperatures.
5	191	L	1950	191-225	< 8.5	GR-S	The above specimen measured at decreasing temperatures.
6	191	L	1950	153-309	< 8.5	GR-S	Same chemical composition and specific gravity as the above specimen; over-cured 120 min at 421 K; unstretched; measured at increasing temperatures.
7	191	L	1950	191-233	< 8.5	GR-S	The above specimen measured atdecreasing temperatures.
8	191	L	1950	173-215	< 8.5	GR-S	The above specimen measured at increasing temperatures.
9	191	L	1950	109-313	< 8.5	GR-S	Similar to the above specimen except 100% stretched and measured at increasing temperatures.
10	191	L	1950	203-245	< 8.5	GR-S	The above specimen measured at decreasing temperatures.
11	191	L	1950	141-329	< 8.5	GR-S	Kept for two years at room temperature, constantly exposed to air and light.
12	191	L	1950	203-274	< 8.5	GR-S	The above specimen measured at decreasing temperatures.
13	192	L	1945	343.2	1.3	GR-S tread type	58 carbon black (by weight base on 100 parts of GR-S).
14	192	L	1945	343.2	1.3	GR-S tread type	25 aluminum powder, 33 carbon black (by weight based on 100 parts of GR-S).
15	192	L	1945	343.2	1.3	GR-S tread type	50 aluminum powder, 8 carbon black (by weight based on 100 parts of GR-S).

DATA TABLE NO. 314 THERMAL CONDUCTIVITY OF RUBBER (GOVERNMENT RUBBER-STYRENE)

[Temperature, T, K; Thermal Conductivity, k, Watt $cm^{-1}K^{-1}$]

T	k
CURVE 1	
110.2	0.00117
120.2	0.00121
124.2	0.00117
134.2	0.00121 *
134.2	0.00120
153.2	0.00128
161.2	0.00126
193.2	0.00128
203.2	0.00127
215.2	0.00134
227.2	0.00140
245.2	0.00153
253.2	0.00157
263.2	0.00163
263.2	0.00159
283.2	0.00165
291.2	0.00163
295.2	0.00159 *
299.2	0.00159
309.2	0.00169
CURVE 2	
173.2	0.00109
191.2	0.00130
193.2	0.00134
203.2	0.00151
205.2	0.00155 *
223.2	0.00153
236.2	0.00163
236.2	0.00161 *
CURVE 3 *	
173.2	0.00109
197.2	0.00117
204.2	0.00117
215.2	0.00126
CURVE 4	
111.2	0.000544
135.2	0.000565
155.2	0.000649
161.2	0.000649
173.2	0.000699
CURVE 4 (cont.)	
183.2	0.000720
191.2	0.000732
201.2	0.000732
209.2	0.000736
215.2	0.000795
227.2	0.000920
237.2	0.000962
245.2	0.000962
265.2	0.00103
283.2	0.00107
299.2	0.00111
311.2	0.00117
CURVE 5	
191.2	0.000920
223.2	0.000962
225.2	0.000983
CURVE 6	
153.2	0.00123
163.2	0.00119
205.2	0.00126
217.2	0.00134 *
217.2	0.00138 *
227.2	0.00155
245.2	0.00165
253.2	0.00163
265.2	0.00165 *
279.2	0.00167
289.2	0.00174
299.2	0.00167
309.2	0.00174
CURVE 7 *	
191.2	0.00128
193.2	0.00126
193.2	0.00130
205.2	0.00149
205.2	0.00159
223.2	0.00163
233.2	0.00165
CURVE 8	
173.2	0.00113
195.2	0.00113
203.2	0.00111
215.2	0.00123
CURVE 9	
109.2	0.000586
135.2	0.000669
155.2	0.000732
163.2	0.000690
173.2	0.000732
183.2	0.000774
191.2	0.000795
201.2	0.000816
209.2	0.000837
216.2	0.000900
227.2	0.00103
237.2	0.00105
263.2	0.00115
283.2	0.00123
299.2	0.00126
313.2	0.00134
CURVE 10	
203.2	0.00100
223.2	0.00115
225.2	0.00115 *
245.2	0.00115
CURVE 11	
141.2	0.000837
181.2	0.00103
191.2	0.000983
203.2	0.00115
225.2	0.00121
225.2	0.00126
226.2	0.00128 *
249.2	0.00136
265.2	0.00142
267.2	0.00159
285.2	0.00184
286.2	0.00180 *
287.2	0.00186 *
CURVE 11 (cont.)	
300.2	0.00205
310.2	0.00218
329.2	0.00222
CURVE 12	
203.2	0.00115 *
213.2	0.00146
229.2	0.00157 *
241.2	0.00155
251.2	0.00161
253.2	0.00153
274.2	0.00180
CURVE 13	
343.2	0.00339
CURVE 14	
345.2	0.00377
CURVE 15	
343.2	0.00397

* Not shown on plot

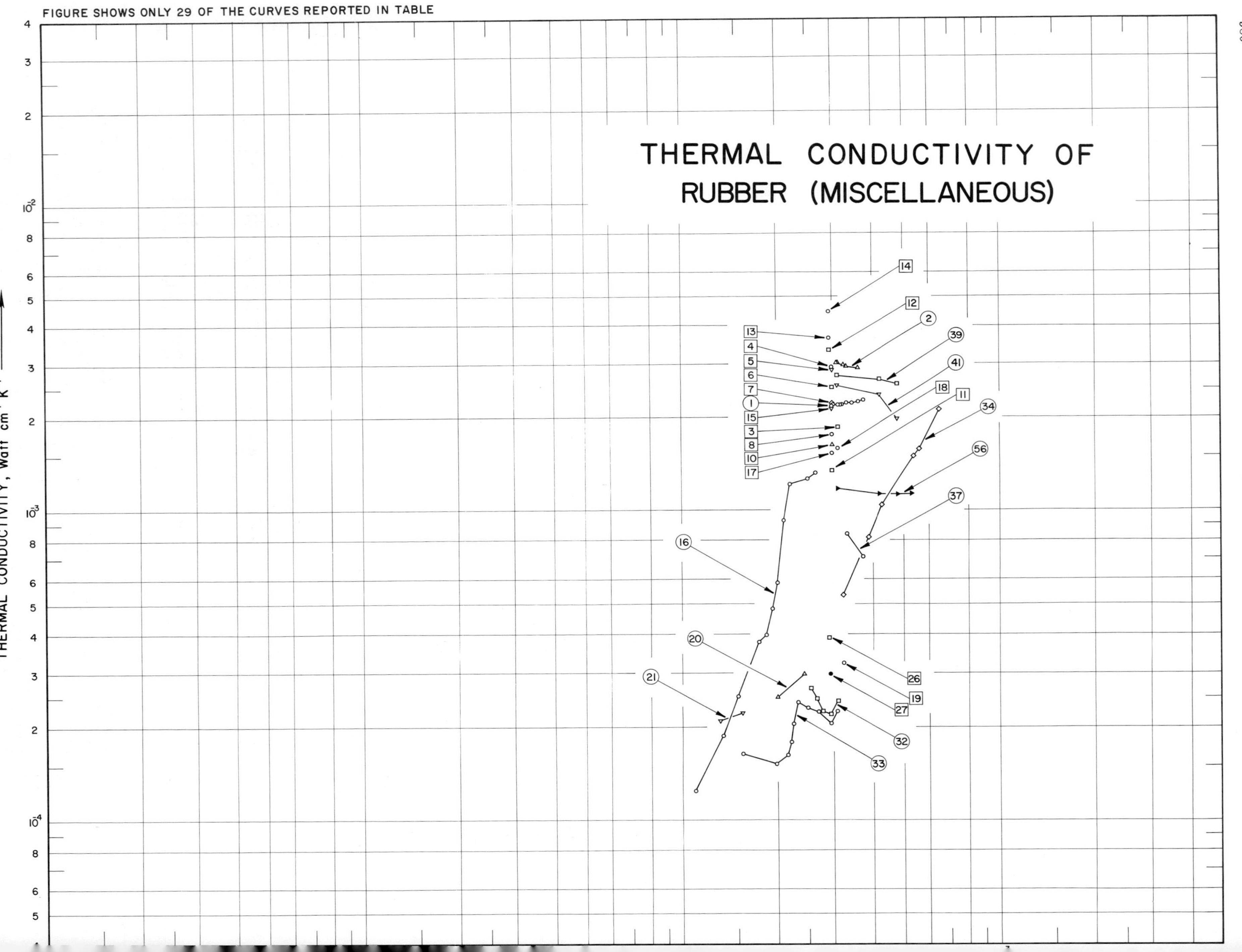
FIGURE SHOWS ONLY 29 OF THE CURVES REPORTED IN TABLE
THERMAL CONDUCTIVITY OF
RUBBER (MISCELLANEOUS)
THERMAL CONDUCTIVITY, Watt cm⁻¹ K⁻¹

SPECIFICATION TABLE NO. 315 THERMAL CONDUCTIVITY OF RUBBER (Miscellaneous)

[For Data Reported in Figure and Table No. 315]

Curve No.	Ref. No.	Method Used	Year	Temp. Range, K	Reported Error, %	Name and Specimen Designation	Composition (weight percent), Specifications and Remarks
1	59	L	1953	299-375			Typical dielectric mix; approximately 0.05 in. thick placed under a slight pressure estimated at 2 lb in.$^{-2}$.
2	59	L	1953	308-360			Designed for high conductivity; approximately 0.05 in. thick placed under a slight pressure estimated at 2 lb in.$^{-2}$.
3	125	L	1923	311.2	1	Commercial rubber	Cross-sectional area 15.59 cm^{-2}; thickness 0.286 cm; under pressure 8 lb in.$^{-2}$.
4	125	L	1923	298.2	1	Commercial rubber	Vulcanized; 38% rubber; thickness 0.385 cm; under pressure 8.4 lb in.$^{-2}$.
5	125	L	1923	298.2	1	Commercial rubber	Vulcanized; 40% rubber; thickness 0.381 cm; under pressure 8.4 lb in.$^{-2}$.
6	125	L	1923	298.2	1	Commercial rubber	Vulcanized; 44% rubber; thickness 0.287 cm; under pressure 8.4 lb in.$^{-2}$.
7	125	L	1923	298.2	1	Commercial rubber	Vulcanized; 50% rubber; thickness 0.387 cm; under pressure 8.4 lb in.$^{-2}$.
8	125	L	1923	298.2	1	Commercial rubber	Vulcanized; 67% rubber; thickness 0.286 cm; under pressure 8.4 lb in.$^{-2}$.
9	125	L	1923	298.2	1	Commercial rubber	Vulcanized; 83% rubber; thickness 0.412 cm; under pressure 8.4 lb in.$^{-2}$.
10	125	L	1923	298.2	1	Commercial rubber	Vulcanized; 92% rubber; thickness 0.324 cm; under pressure 8.4 lb in.$^{-2}$.
11	125	L	1923	298.2	1	Commercial rubber	Vulcanized; 100% rubber; plantation rubber crepe; thickness 0.103 cm; under pressure 34 lb in.$^{-3}$
12	125	L	1923	293.2	1	X-ray protective rubber	Lead equivalent = 0.35%; thickness 0.173 cm; under pressure 2.5 lb in.$^{-2}$.
13	125	L	1923	293.2	1	X-ray protective rubber	Lead equivalent = 0.41%; thickness 0.260 cm; under pressure 2.5 lb in.$^{-2}$.
14	125	L	1923	293.2	1	X-ray protective rubber	Lead equivalent = 0.43%; thickness 0.463 cm; under pressure 2.5 lb in.$^{-2}$.
15	117	L	1955	297.2	1.0-2.0		Thickness 15.1 mm; density 1.08 g cm^{-3}; thermal conductivity values calculated from measured data of thermal diffusivity and specific heat.
16	126	L	1941	110-265			Vulcanized flat ring specimen of 30 mm O.D. and 10 mm I.D.
17	127	F	1950	298.2			Thickness 1.025 cm; density 1.10 g cm^{-3}; thermal conductivity values calculated from measured transient temperature changes.
18	108	L	1920	310.7		Hard rubber	9 in. in dia, 0.380 in. thick; specific gravity 1.19; heat flow in transverse direction.
19	128	L	1946	323.7			Expanded rubber board 2.01 in. in thickness; density 0.078 g cm^{-3}.
20	128	L	1946	200,243			Expanded rubber board 1.00 in. in thickness; density 0.078 g cm^{-3}.
21	43	L	1954	132,155			Expanded rubber board density 0.072 g cm^{-3}.
22	536	P	1953	301.2			10 x 10 x 0.42 cm; density 1.60 g cm^{-3}.
23	536	P	1953	315.2			5 mm thick plate; 13 x 15 cm^2 surface; density 1.96 g cm^{-3}.
24	189		1950	200-289		Rubatex	Density 0.13-0.14 g cm^{-3}; water absorption (24 hrs) 0.2 vol %.
25	533	L	1962	300-334		Rubatex R203-H	Buna-N foam rubber; specimen size 0.25 in. x 12 in. x 12 in.

SPECIFICATION TABLE NO. 315 (continued)

Curve No.	Ref. No.	Method Used	Year	Temp. Range, K	Reported Error, %	Name and Specimen Designation	Composition (weight percent), Specifications and Remarks
26	534	L	1961	291.8		Adiprene rubber	Flexible polyurethane foam rubber; 8 x 8 x 0.469 in.; density 0.0320 g cm^{-3}.
27	535	L	1962	294.3		Adiprene rubber; 2	Polyurethane foam rubber; crushed and re-expanded with CCl_3F in pure air, aged at 333 K for 28 days; density 0.0481 g cm^{-3}.
28	535	L	1962	294.3		Adiprene rubber; 2	The cells of the above specimen curshed and then re-expanded with CCl_3F in pure air the second time, aged at 333 K for 30 days.
29	535	L	1962	294.3		Adiprene rubber; 2	The cells of the above specimen crushed and then re-expanded with CCl_3F in pure air the third time, aged at 333 K for 33 days.
30	535	L	1962	294.3		Adiprene rubber; 2	The cells of the above specimen crushed and then re-expanded with CCl_3F in pure air the fourth time, aged at 333 K for 36 days.
31	535	L	1962	294.3		Adiprene rubber; 3	Cut from the same sample as the above specimen; aged at 333 K continuously.
32	535	L	1962	255-310		Adiprene rubber; A	Rigid polyurethane foam rubber; prepared by using CCl_3F as captive blowing agent.
33	535	L	1962	157-309		Adiprene rubber; B	Similar to the above specimen.
34	488, 489	L	1961	323-647		SRI 3	Resin type R-7002 silicone foam rubber; 30 x 12 x 0.5 in.; average thickness 0.688 in.; manufactured by Dow Corning Corp.; fabricated by heating at 472 K for 1 hr, 497 K for 1 hr, and 522 K for 24-48 hrs, cooled at 497 K for 1 hr, 472 K for 1 hr, and at 339 K for 1 hr; density 0.186 g cm^{-3}.
35	192	L	1945	333, 373			Rubber sponge; initially uncured.
36	192	L	1945	333, 373			Rubber sponge; cured, undried.
37	192	L	1945	333, 373			Rubber sponge; cured, dried.
38	537	L	1960	311, 422		Acrylate Rubber	83.7 Vyram N 5400, 11.6 MgO, 2.9 PbO_2, and 1.8 $CH_3(CH_2)_{16}COOH$ (stearic acid); 6 x 6 x 0.250 in.; optimum-cured at 439 K for 60 min.
39	537	L	1960	311-478		Carboxy Nitrile	66.9 Hycar 1072, 26.8 FEF black, 3.3 ZnO, 2.3 Tuad, and 1.7 $CH_3(CH_2)_{16}COOH$; 6 x 6 x 0.250 in.; vulcanized; optimum-cured at 428 K for 30 min.
40	537	L	1960	311-478		Acrylic Rubber	64.4 Acrylon EA-5, 32.2 HAF black, 1.3 Tetrone A, 1.3 triethylene tetramine peroxide, and 0.8 $CH_3(CH_2)_{16}COOH$; 6 x 6 x 0.250 in.; volcanized; optimum-cured at 428 K for 20 min.
41	537	L	1960	311-478		Carboxy Rubber	Firestone proprietary compound butaprene T; 6 x 6 x 0.250 in.; vulcanized.
42	537	L	1960	311-478		Nitrile Rubber	68.3 Butaprene NL, 20.5 HAF black, 6.8 dibutyl phthalate, 2.0 ZnO, 1.7 S, and 0.7 altax; 6 x 6 x 0.250 in.; vulcanized.
43	537	L	1960	310.9		Thiokel ST	63.3 Thiokel ST, 31.7 HAF black, 2.8 GMF (30%), 1.9 $CH_3(CH_2)_{16}COOH$, and 0.3 ZnO; 6 x 6 x 0.250 in.; vulcanized; optimum-cured at 433 K for 40 min.
44	537	L	1960	311, 422		Butaprene E	Firestone proprietary compound; 6 x 6 x 0.250 in.; vulcanized; optimum-cured at 411 K for 20 min.

SPECIFICATION TABLE NO. 315 (continued)

Curve No.	Ref. No.	Method Used	Year	Temp. Range, K	Reported Error, %	Name and Specimen Designation	Composition (weight percent), Specifications and Remarks
45	537	L	1960	311-478		Hevea	61.2 smoked sheet, 30.6 HAF black, 1.8 pine tar, 1.8 $CH_3(CH_2)_{16}COOH$, 1.8 ZnO, 1.6 S, 0.7 BLE, 0.31 santo cure, and 0.2 ajone LX; 6 x 6 x 0.250 in.; vulcanized; optimum-cured at 411 K for 60 min.
46	537	L	1960	311-422		Hypalon S2	65.2 Hypalon S2, 19.5 MgO, 13.0 HAF black, 1.6 staybelite resin, and 0.7 Tetrone A; 6 x 6 x 0.250 in.; vulcanized; optimum-cured at 411 K for 45 min.
47	537	L	1960	311-533		Chloroprene Rubber	61.6 Neoprene GRT, 30.8 MPC black, 3.1 ZnO, 2.5 MgO, 1.2 agerite stalite, 0.5 altex, and 0.3 $CH_3(CH_2)$ 16 COOH; 6 x 6 x 0.250 in.; vulcanized; optimum-cured at 411 K for 45 min.
48	537	L	1960	311-533		Viton Rubber	72.6 Viton A, 14.5 MT black, 10.9 MgO, 1.3 copper inhibitor 65, and 0.7 HMDAC; 6 x 6 x 0.250 in.; vulcanized optimum-cured at 422 K for 60 min; post-cured at 373, 394, 422, and 450 K for 1 hr each, and at 478 K for 8 hrs.
49	537	L	1960	311,422		LTP	60.0 FR-S 1500, 30.0 HAF black, 4.8 med. proc. oil, 1.8 ZnO, 1.2 $CH_3(CH_2)_{16}COOH$, 1.2 S, 0.7 santo cure, and 0.3 PBNA; 6 x 6 x 0.250 in.; vulcanized; optimum-cured at 411 K for 60 min.
50	537	L	1960	311,422		Methacrylate Rubber	45.2 butadiene, 19.35 methyl methacrylate, 19.35 HAF black, 9.7 $Ba(OH)_2 \cdot 8H_2O$, 3.2 pine tar, 1.9 Di cup 40 C, and 1.3 Monsanto 4010 antioxidant; 6 x 6 x 0.250 in.; vulcanized; optimum-cured at 433 K for 45 min.
51	537	L	1960	311-533		Resin-cured Butyl	53.5 Enjay butyl 325, 32.8 HAF black, 6.6 amberol ST-137, 6.6 brominated butyl, and 0.5 $CH_3(CH_2)_{16}COOH$; 6 x 6 x 0.250 in.; vulcanized; optimum-cured at 433 K for 90 min.
52	537	L	1960	311-533		Poly(ethyl acrylate)	64.8 Poly(ethyl acrylate), 19.4 HAF black, 6.5 $Ca(OH)_2$, 3.2 CH_2OHCH_2OH (ethylene glycol), 3.2 pine tar, 1.3 PBNA, 1.3 $CH_3(CH_2)_{16}COOH$, and 0.3 copper inhibitor 65; 6 x 6 x 0.250 in.; vulcanized; optimum-cured at 433 K for 90 min.
53	537	L	1960	311-533		Silicone Rubber	96.1 silastic 916 U, 3.4 silastic S-2084 (20% di-t-butyl peroxide), and 0.5 $CH_3(CH_2)_{16}COOH$; 6 x 6 x 0.250 in.; vulcanized; optimum-cured at 433 K for 20 min; post-cured at 422 and 450 K for 1 hr each, at 478 K for 4 hrs, and at 505 and 522 K for 8 hrs each.
54	537	L	1960	311-478		Tellurac-cured Butyl	52.5 Enjay butyl 325, 26.2 MPC black, 13.1 ZnO, 2.6 PbO_2, 2.6 vistanex B-80, 1.6 ethyl tellurac, 1.1 S, and 0.3 polyac; 6 x 6 x 0.250 in.; vulcanized; optimum-cured at 433 K for 30 min.
55	537	L	1960	311-478		Dibenzo GMF-cured Butyl	51.4 Enjay butyl 325, 25.7 MPC black, 10.3 ZnO, 5.1 Pb_3O_4, 3.3 dibenzo GMF, 2.6 vistanex B-80, 1.0 S, 0.3 polyac, and 0.3 $CH_3(CH_2)_{16}COOH$; 6 x 6 x 0.250 in.; vulcanized; optimum-cured at 433 K for 30 min.
56	537	L	1960	311-533		Kel-F 3700	82.4 Kel-F 3700, 8.2 dyphos, 8.2 ZnO, and 1.2 benzoyl peroxide; 6 x 6 x 0.250 in.; vulcanized; optimum-cured at 394 K for 30 min, post-cured at 394 K for 1 hr and at 422 K for 16 hrs.

DATA TABLE NO. 315 THERMAL CONDUCTIVITY OF RUBBER (Miscellaneous)

[Temperature, T, K; Thermal Conductivity, k, Watt $cm^{-1}K^{-1}$]

T	k
CURVE 1	
299.2	0.00218
302.2	0.00220
313.2	0.00220
317.2	0.00220
321.2	0.00222
332.2	0.00224
345.2	0.00224
361.2	0.00226
375.2	0.00228
CURVE 2	
308.2	0.00301
310.2	0.00301
325.2	0.00297
332.2	0.00295
360.2	0.00291
CURVE 3	
311.2	0.00186
CURVE 4	
298.2	0.00293
CURVE 5	
298.2	0.00285
CURVE 6	
298.2	0.00251
CURVE 7	
298.2	0.00222
CURVE 8	
298.2	0.00176
CURVE 9*	
298.2	0.00176
CURVE 10	
298.2	0.00163
CURVE 11	
298.2	0.00134
CURVE 12	
293.2	0.00331
CURVE 13	
293.2	0.00364
CURVE 14	
293.2	0.00448
CURVE 15	
297.2	0.00214
CURVE 16	
110	0.000125
135	0.000188
150	0.000251
175	0.000377
185	0.000397
193	0.000481
200	0.000586
210	0.000928
220	0.00121
250	0.00126
265	0.00132
CURVE 17	
298.2	0.00153
CURVE 18	
310.7	0.00159
CURVE 19	
323.7	0.000322
CURVE 20	
199.8	0.000250
243.2	0.000297
CURVE 21	
132	0.000209
155	0.000222
CURVE 22*	
301.2	0.00209
CURVE 23*	
315.2	0.00207
CURVE 24*	
200.0	0.000245
222.2	0.000288
239.2	0.000296
255.5	0.000303
272.1	0.000303
288.8	0.000317
CURVE 25*	
300	0.000620
306	0.000649
312	0.000649
313	0.000692
313	0.000591
315	0.000663
334	0.000663
CURVE 26	
291.8	0.000389
CURVE 27	
294.3	0.000297
CURVE 28*	
294.3	0.000329
CURVE 29*	
294.3	0.000335
CURVE 30*	
294.3	0.000336
Aging Time (days)	k
CURVE 31* (T = 294.3K)	
1	0.000209
16	0.000221
28	0.000219
31	0.000219
35	0.000219
T	k
CURVE 32	
255.4	0.000268
266.5	0.000248
277.6	0.000226
294.3	0.000221
309.8	0.000242
CURVE 33	
156.5	0.000164
197.6	0.000153
216.5	0.000102
221.5	0.000179
225.4	0.000205
233.2	0.000241
249.8	0.000231
269.3	0.000225
294.3	0.000206
308.7	0.000226
CURVE 34	
323.2	0.000535
387.1	0.000819
428.2	0.00104
538.7	0.00149
562.6	0.00158
646.5	0.00213
CURVE 35*	
333.2	0.00172
373.2	0.00105
CURVE 36*	
333.2	0.00121
373.2	0.00105
CURVE 37	
333.2	0.00084
373.2	0.00071
CURVE 38*	
310.9	0.00280
422.1	0.00280
CURVE 39	
310.9	0.00276
422.1	0.00268
477.6	0.00259
CURVE 40*	
310.9	0.00268
422.1	0.00268
477.6	0.00268
CURVE 41	
310.9	0.00255
422.1	0.00238
477.6	0.00197
CURVE 42*	
310.9	0.00255
422.1	0.00264
477.6	0.00264
CURVE 43*	
310.9	0.00268
CURVE 44*	
310.9	0.00255
422.1	0.00251
CURVE 45*	
310.9	0.00247
422.1	0.00230
477.6	0.00213
CURVE 46*	
310.9	0.00247
422.1	0.00243
CURVE 47*	
310.9	0.00243
422.1	0.00230
533.2	0.00209
CURVE 48*	
310.9	0.00230
422.1	0.00205
533.2	0.00180
CURVE 49*	
310.9	0.00226
422.1	0.00230
CURVE 50*	
310.9	0.00226
422.1	0.00230
CURVE 51*	
310.9	0.00213
422.1	0.00230
533.2	0.00213
CURVE 52*	
310.9	0.00213
422.1	0.00226
477.6	0.00234
533.2	0.00226
CURVE 53*	
310.9	0.00197
422.1	0.00192
477.6	0.00192
533.2	0.00172
CURVE 54*	
310.9	0.00180
422.1	0.00184
477.6	0.00188
CURVE 55*	
310.9	0.00172
422.1	0.00176
477.6	0.00172
CURVE 56	
310.9	0.00117
422.1	0.00113
477.6	0.00113
533.2	0.00113

*Not shown on plot

SPECIFICATION TABLE NO. 316 THERMAL CONDUCTIVITY OF ANTHRACENE $C_6H_4(CH)_2C_6H_4$

Curve No.	Ref. No.	Method Used	Year	Temp. Range, K	Reported Error, %	Name and Specimen Designation	Composition (weight percent), Specifications and Remarks
1	67	P	1950	298-361			Specimen in long cylindrical form; distilled and recrystallized; melting point 490 K; mean error = $\pm 0.5 \times 10^{-3}$ cal $cm^{-1}sec^{-1}C^{-1}$.

DATA TABLE NO. 316 THERMAL CONDUCTIVITY OF ANTHRACENE $C_6H_4(CH)_2C_6H_4$

[Temperature, T, K; Thermal Conductivity, k, Watt $cm^{-1}K^{-1}$]

T	k
CURVE 1*	
297.9	0.00339
313.6	0.00337
333.2	0.00335
343.2	0.00333
352.8	0.00331
360.6	0.00328

*No graphical presentation

SPECIFICATION TABLE NO. 317 THERMAL CONDUCTIVITY OF BENZENE, P-DIBROMO $C_6H_4Br_2$

Curve No.	Ref. No.	Method Used	Year	Temp. Range, K	Reported Error, %	Name and Specimen Designation	Composition (weight percent), Specifications and Remarks
1	67	P	1950	293-357			Specimen in long cylindrical form; distilled and recrystallized; melting point 360 K; mean error = $\pm 0.5 \times 10^{-3}$ cal $cm^{-1} sec^{-1} C^{-1}$.

DATA TABLE NO. 317 THERMAL CONDUCTIVITY OF BENZENE, P-DIBROMO $C_6H_4Br_2$

[Temperature, T, K; Thermal Conductivity, k, Watt $cm^{-1}K^{-1}$]

T	k
CURVE 1*	
293.0	0.00215
309.6	0.00218
324.0	0.00220
338.4	0.00218
357.1	0.00209

*No graphical presentation

SPECIFICATION TABLE NO. 318 THERMAL CONDUCTIVITY OF BENZENE, P-DICHLORO $C_6H_4Cl_2$

Curve No.	Ref. No.	Method Used	Year	Temp. Range, K	Reported Error, %	Name and Specimen Designation	Composition (weight percent), Specifications and Remarks
1	67	P	1950	293-323			Specimen in long cylindrical form; distilled and recrystallized; melting point 326 K; mean error = 0.5 x 10^{-3} cal $cm^{-1}sec^{-1}C^{-1}$.

DATA TABLE NO. 318 THERMAL CONDUCTIVITY OF BENZENE, P-DICHLORO $C_6H_4Cl_2$

[Temperature, T, K; Thermal Conductivity, k, Watt $cm^{-1}K^{-1}$]

T	k
CURVE 1*	
293.4	0.00247
299.9	0.00228
306.4	0.00201
313.0	0.00176
322.7	0.00134

*No graphical presentation

SPECIFICATION TABLE NO. 319 THERMAL CONDUCTIVITY OF BENZENE, P-DIIODO $C_6H_4I_2$

Curve No.	Ref. No.	Method Used	Year	Temp. Range, K	Reported Error, %	Name and Specimen Designation	Composition (weight percent), Specifications and Remarks
1	67	P	1950	293-324			Specimen in long cylindrical form; distilled and recrystillized; melting point 402 K; mean error = 0.5 x 10^{-3} cal $cm^{-1}sec^{-1}C^{-1}$.

DATA TABLE NO. 319 THERMAL CONDUCTIVITY OF BENZENE, P-DIIODO $C_6H_4I_2$

[Temperature, T, K; Thermal Conductivity, k, Watt $cm^{-1}K^{-1}$]

T	k
CURVE 1*	
293.1	0.00184
305.5	0.00184
323.8	0.00180

*No graphical presentation

SPECIFICATION TABLE NO. 320 THERMAL CONDUCTIVITY OF DIPHENYL $C_6H_5C_6H_5$

Curve No.	Ref. No.	Method Used	Year	Temp. Range, K	Reported Error, %	Name and Specimen Designation	Composition (weight percent), Specifications and Remarks
1	67	P	1950	292-331			Specimen in long cylindrical form; destilled and recrystallized; mean error = $\pm 0.5 \times 10^{-3}$ (cal $cm^{-1}sec^{-1}C^{-1}$).
2	260	R	1961	365-582	1.5		Supply from Hopkin and Williams Ltd. with crystallizing point 68.7 C min and distillation range 2.5 C max; measured in liquid state.

DATA TABLE NO. 320 THERMAL CONDUCTIVITY OF DIPHENYL $C_6H_5C_6H_5$

[Temperature, T, K; Thermal Conductivity, k, Watt $cm^{-1}K^{-1}$]

T	k
CURVE 1*	
291.6	0.00251
301.8	0.00297
315.6	0.00301
331.3	0.00301
CURVE 2*	
365.2	0.00136
402.0	0.00131
437.5	0.00124
472.7	0.00119
484.4	0.00117
517.4	0.00113
578.9	0.00105
581.6	0.00105

*No graphical presentation

SPECIFICATION TABLE NO. 321 THERMAL CONDUCTIVITY OF DIPHENYL OXIDE $(C_6H_5)_2O$

Curve No.	Ref. No.	Method Used	Year	Temp. Range, K	Reported Error, %	Name and Specimen Designation	Composition (weight percent), Specifications and Remarks
1	260	R	1961	330-511	1.5		B.D.H. Laboratory reagent; M.P./F.P. 26-27 C; very slightly tinted crystalline solid; in liquid state.

DATA TABLE NO. 321 THERMAL CONDUCTIVITY OF DIPHENYL OXIDE $(C_6H_5)_2O$

[Temperature, T, K; Thermal Conductivity, k, Watt $cm^{-1}K^{-1}$]

T	k
CURVE 1*	
329.9	0.00138
352.2	0.00136
402.6	0.00128
461.7	0.00118
493.0	0.00114
511.2	0.00111

*No graphical presentation

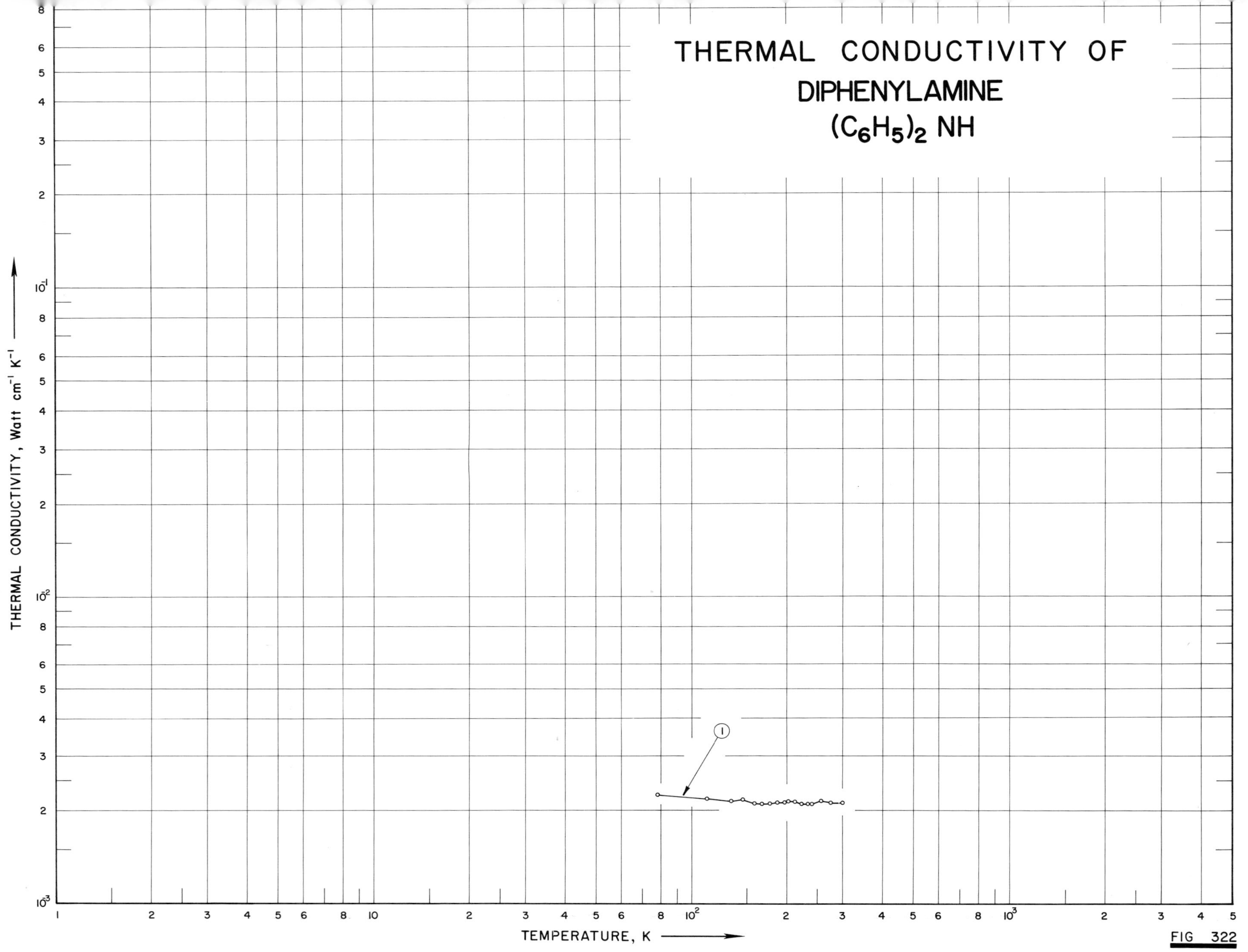
THERMAL CONDUCTIVITY OF
DIPHENYLAMINE
$(C_6H_5)_2$ NH
1
THERMAL CONDUCTIVITY, Watt cm^{-1} K^{-1}
TEMPERATURE, K
FIG 322

SPECIFICATION TABLE NO. 322 THERMAL CONDUCTIVITY OF DIPHENYLAMINE $(C_6H_5)_2NH$

[For Data Reported in Figure and Table No. 322]

Curve No.	Ref. No.	Method Used	Year	Temp. Range, K	Reported Error, %	Name and Specimen Designation	Composition (weight percent), Specifications and Remarks
1	66	R	1905	78-299			Melted over a water bath and frozen slowly from beneath.

DATA TABLE NO. 322 THERMAL CONDUCTIVITY OF DIPHENYLAMINE $(C_6H_5)_2NH$

[Temperature, T, K; Thermal Conductivity, k, Watt cm^{-1} K^{-1}]

T	k
CURVE 1	
78.2	0.00225
112.2	0.00218
134.2	0.00215
146.2	0.00217
158.2	0.00211
167.2	0.00211
177.2	0.00211
187.2	0.00213
197.2	0.00213
202.2	0.00215
212.2	0.00214
215.2	0.00211*
223.2	0.00210
228.5	0.00210*
234.2	0.00210
240.2	0.00210
246.2	0.00213*
250.2	0.00214*
256.2	0.00215
266.2	0.00213*
275.2	0.00212
299.2	0.00212

*Not shown on plot

SPECIFICATION TABLE NO. 323 THERMAL CONDUCTIVITY OF [DIPHENYLMETHANE + NAPHTHALENE] $(C_6H_5)_2CH_2 + C_{10}H_8$

Curve No.	Ref. No.	Method Used	Year	Temp. Range, K	Reported Error, %	Name and Specimen Designation	Composition (weight percent) $(C_6H_5)_2CH_2$	$C_{10}H_8$	Composition (continued), Specifications and Remarks
1	261	E	1955	293.2	±0.5		Bal	2.6	

DATA TABLE NO. 323 THERMAL CONDUCTIVITY OF [DIPHENYLMETHANE + NAPHTHALENE] $(C_6H_5)_2CH_2 + C_{10}H_8$

[Temperature, T, K; Thermal Conductivity, k, Watt $cm^{-1}K^{-1}$]

T	k
CURVE 1*	
293.2	0.00349

*No graphical presentation

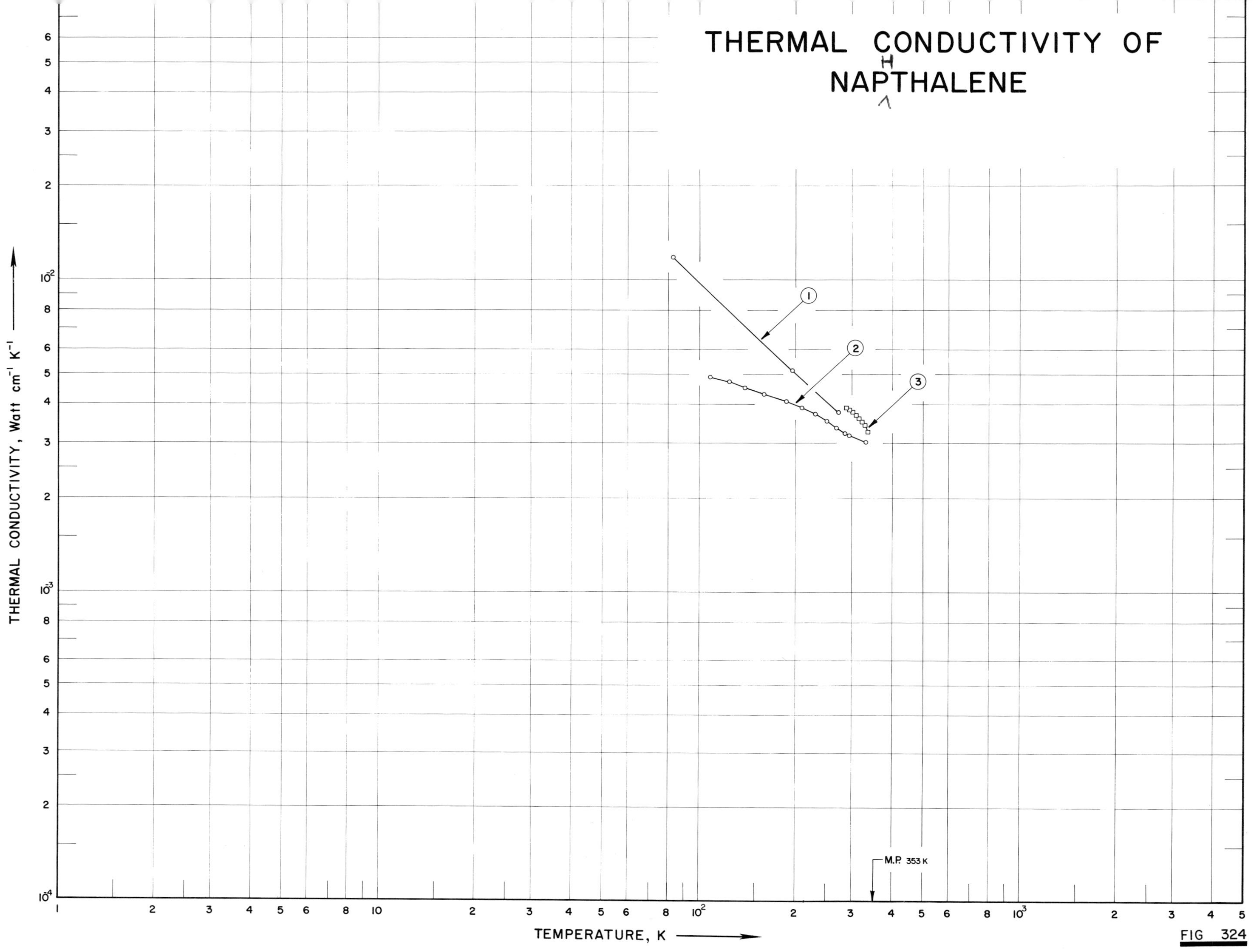
THERMAL CONDUCTIVITY OF NAPTHALENE
THERMAL CONDUCTIVITY, Watt cm⁻¹ K⁻¹
TEMPERATURE, K
M.P. 353 K
1
2
3

FIG 324

SPECIFICATION TABLE NO. 324 THERMAL CONDUCTIVITY OF NAPHTHALENE $C_{10}H_8$

[For Data Reported in Figure and Table No. 324]

Curve No.	Ref. No.	Method Used	Year	Temp. Range, K	Reported Error, %	Name and Specimen Designation	Composition (weight percent), Specifications and Remarks
1	22	L	1911	83-273			No other details reported.
2	66	R	1905	108-332			Melting point 79 C.
3	67	P	1950	290-339			Purified by distillation and recrystallization; melting point 353 K; data mean error $\pm 0.5 \times 10^{-3}$ cal cm^{-1} sec^{-1} C^{-1}.

DATA TABLE NO. 324 THERMAL CONDUCTIVITY OF NAPHTHALENE $C_{10}H_8$

[Temperature, T, K; Thermal Conductivity, k, Watt $cm^{-1} K^{-1}$]

T	k
CURVE 1	
83.2	0.0118
195.2	0.00512
273.2	0.00377
CURVE 2	
108.2	0.00490
125.2	0.00473
139.2	0.00452
160.2	0.00431
187.2	0.00410
210.2	0.00390
231.2	0.00374
251.2	0.00354
269.2	0.00336
286.2	0.00323
295.2	0.00318
332.2	0.00302
CURVE 3	
289.5	0.00387
296.5	0.00383
303.5	0.00377
310.6	0.00368
317.7	0.00360
324.8	0.00351
331.8	0.00343
338.9	0.00326

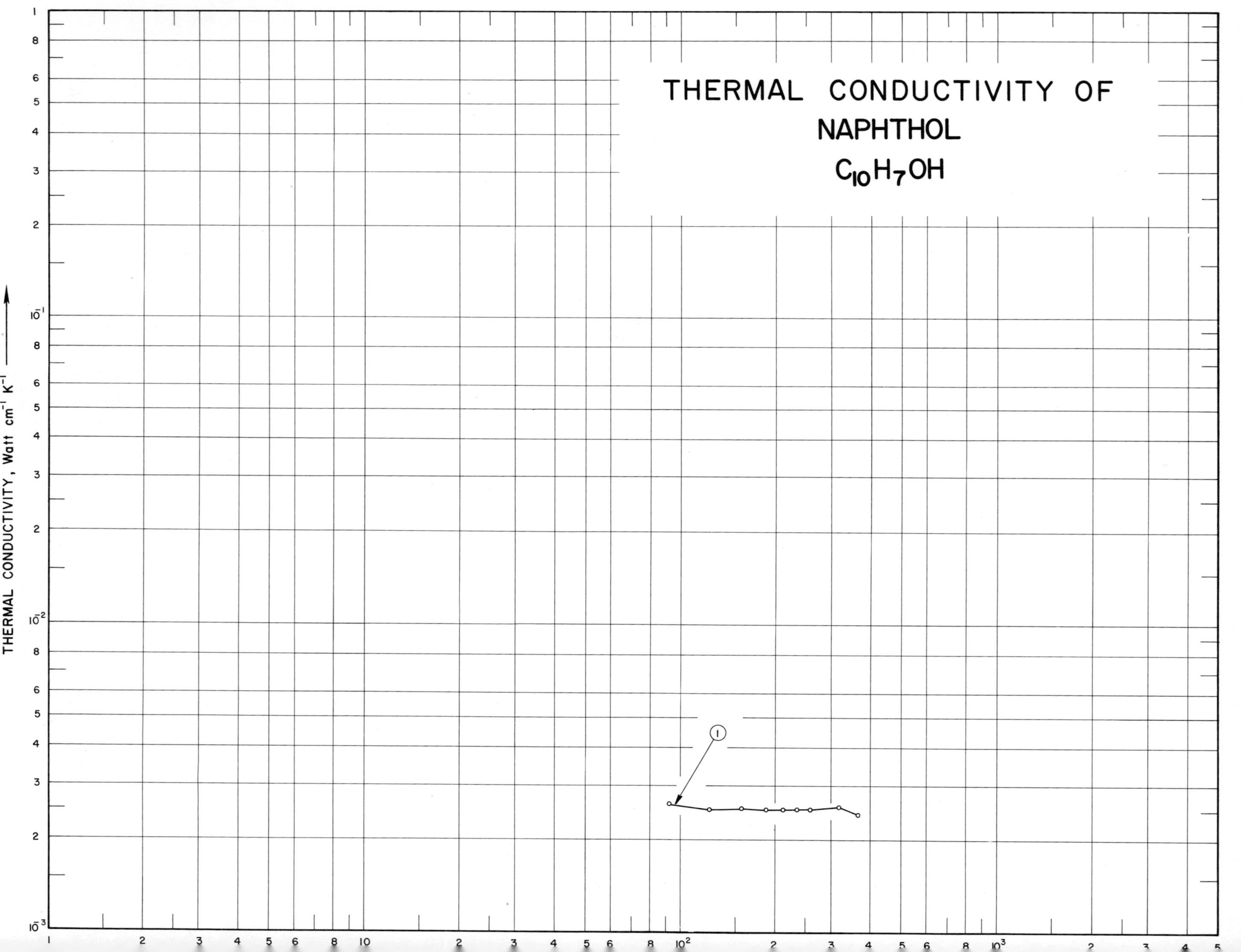
THERMAL CONDUCTIVITY OF
NAPHTHOL
$C_{10}H_7OH$
THERMAL CONDUCTIVITY, Watt cm^{-1} K^{-1}
1

SPECIFICATION TABLE NO. 325 THERMAL CONDUCTIVITY OF NAPHTHOL $C_{10}H_7OH$

[For Data Reported in Figure and Table No. 325]

Curve No.	Ref. No.	Method Used	Year	Temp. Range, K	Reported Error, %	Name and Specimen Designation	Composition (weight percent), Specifications and Remarks
1	66	R	1905	92-365			Heated in an oil bath and cooled slowly underneath.

DATA TABLE NO. 325 THERMAL CONDUCTIVITY OF NAPHTHOL $C_{10}H_7OH$

[Temperature, T, K; Thermal Conductivity, k, Watt cm^{-1} K^{-1}]

T	k
CURVE 1	
92.2	0.00261
123.2	0.00250
156.2	0.00253
187.2	0.00251
212.2	0.00250
235.2	0.00251
258.2	0.00251
318.2	0.00256
365.2	0.00242

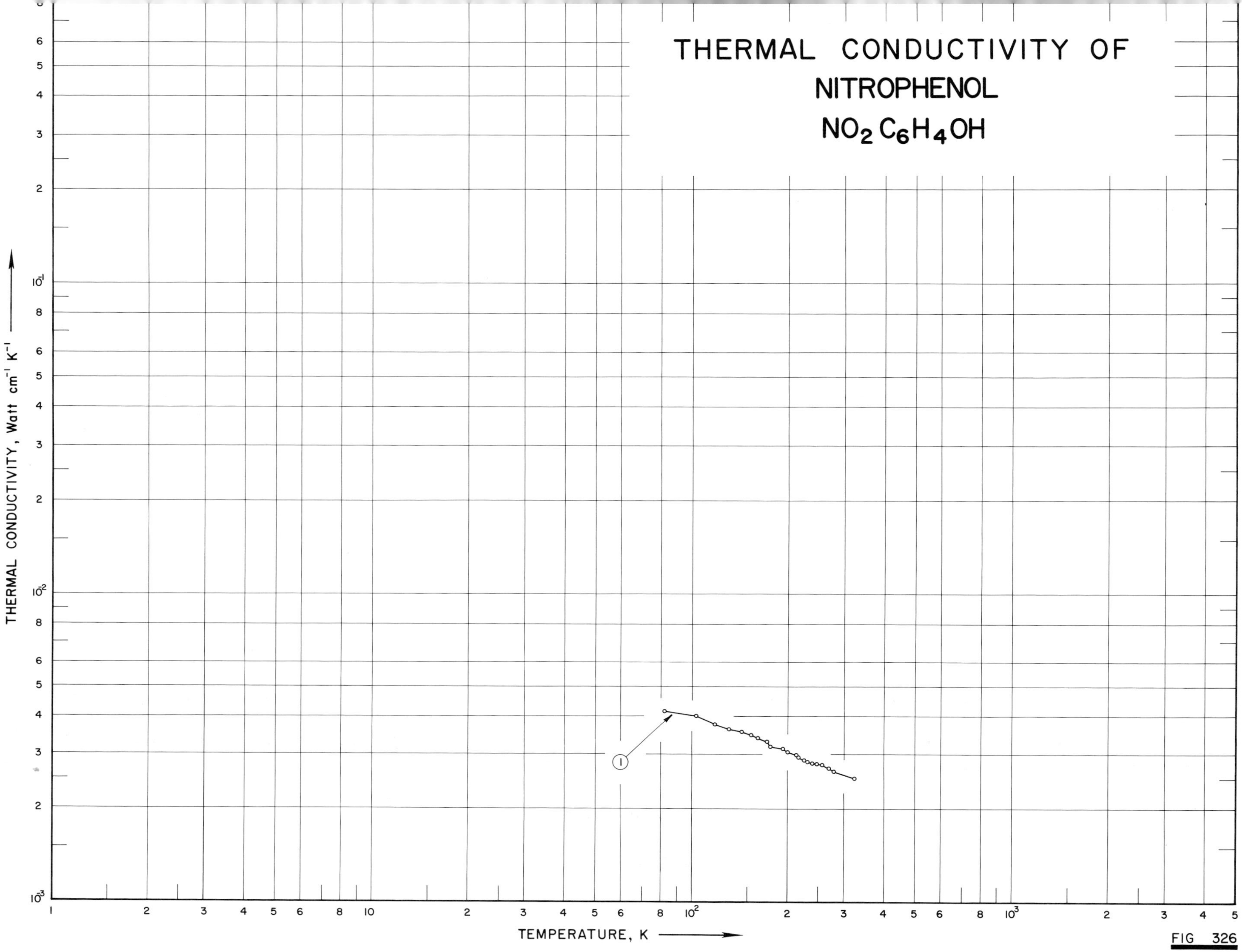

THERMAL CONDUCTIVITY OF NITROPHENOL $NO_2C_6H_4OH$

THERMAL CONDUCTIVITY, Watt cm^{-1} K^{-1}

TEMPERATURE, K

FIG 326

SPECIFICATION TABLE NO. 326 THERMAL CONDUCTIVITY OF NITROPHENOL $NO_2C_6H_4OH$

[For Data Reported in Figure and Table No. 326]

Curve No.	Ref. No.	Method Used	Year	Temp. Range, K	Reported Error, %	Name and Specimen Designation	Composition (weight percent), Specifications and Remarks
1	66	R	1905	82-323			Pure; heated in an oil bath and cooled slowly beneath.

DATA TABLE NO. 326 THERMAL CONDUCTIVITY OF NITROPHENOL $NO_2C_6H_4OH$

[Temperature, T, K; Thermal Conductivity, k, Watt $cm^{-1} K^{-1}$]

T	k
CURVE 1	
82.2	0.00418
103.2	0.00401
118.2	0.00377
131.2	0.00363
143.2	0.00356
153.2	0.00347
161.2	0.00338
172.2	0.00330
177.2	0.00318
194.2	0.00312
200.2	0.00305
213.2	0.00298
216.2	0.00293
226.2	0.00287
231.2	0.00284
239.2	0.00280
247.2	0.00278
257.2	0.00277
265.2	0.00272*
269.2	0.00269
273.2	0.00266*
279.2	0.00264
280.2	0.00264*
320.2	0.00251*
323.2	0.00250

*Not shown on plot

SPECIFICATION TABLE NO. 327 THERMAL CONDUCTIVITY OF PHENANTHREN $C_{14}H_{10}$

Curve No.	Ref. No.	Method Used	Year	Temp. Range, K	Reported Error, %	Name and Specimen Designation	Composition (weight percent), Specifications and Remarks
1	67	P	1950	291-325			Specimen in long cylindrical form; destilled and recrystallized; mean error = $\pm 0.5 \times 10^{-3}$ cal $cm^{-1} sec^{-1} C^{-1}$.

DATA TABLE NO. 327 THERMAL CONDUCTIVITY OF PHENANTHREN $C_{14}H_{10}$

[Temperature, T, K; Thermal Conductivity, k, Watt $cm^{-1}K^{-1}$]

T	k
CURVE 1*	
290.9	0.00205
305.9	0.00203
313.3	0.00201
324.5	0.00192

*No graphical presentation

SPECIFICATION TABLE NO. 328 THERMAL CONDUCTIVITY OF SANTOWAX R

Curve No.	Ref. No.	Method Used	Year	Temp. Range, K	Reported Error, %	Name and Specimen Designation	Composition (weight percent), Specifications and Remarks
1	260	R	1961	429-665	1.5		52 meta-terphenyl, 39 para-terphenyl, 8 ortho-terphenyl, 1 diphenyl, and higher polyphenyls; in liquid state.

DATA TABLE NO. 328 THERMAL CONDUCTIVITY OF SANTOWAX R

[Temperature, T, K; Thermal Conductivity, k, Watt $cm^{-1}K^{-1}$]

T	k
CURVE 1*	
428.6	0.00132
428.8	0.00132
462.2	0.00130
481.8	0.00127
508.6	0.00126
546.8	0.00120
549.2	0.00120
570.3	0.00119
615.0	0.00115
618.4	0.00114
664.7	0.00110

*No graphical presentation

SPECIFICATION TABLE NO. 329 THERMAL CONDUCTIVITY OF SODIUM ACETATE $NaC_2H_3O_2 \cdot 3H_2O$

Curve No.	Ref. No.	Method Used	Year	Temp. Range, K	Reported Error, %	Name and Specimen Designation	Composition (weight percent), Specifications and Remarks
1	259	R	1932	298.8			The substance was melted and poured into container (5.75 in. dia x 7 in.) until the depth of liquid was 6 in., when it had solidified and cooled to room temperature the supply to the central heating element was switched on and maintained constant; when temperature conditions had remained steady for two hrs, readings on which the calculations of heat conductivity were based were taken.

DATA TABLE NO. 329 THERMAL CONDUCTIVITY OF SODIUM ACETATE $NaC_2H_3O_2 \cdot 3H_2O$

[Temperature, T, K; Thermal Conductivity, k, Watt $cm^{-1}K^{-1}$]

T	k
CURVE 1*	
298.8	0.00115

*No graphical presentation

SPECIFICATION TABLE NO. 330 THERMAL CONDUCTIVITY OF TRINITROTOLUENE $CH_3C_6H_2(NO_2)_3$

Curve No.	Ref. No.	Method Used	Year	Temp. Range, K	Reported Error, %	Name and Specimen Designation	Composition (weight percent), Specifications and Remarks
1	521	R	1948	293-393			Grade I; contained almost entirely of 2:4:6-trinitrotoluene; of setting-point 80.4 C.

DATA TABLE NO. 330 THERMAL CONDUCTIVITY OF TRINITROTOLUENE $CH_3C_6H_2(NO_2)_3$

[Temperature, T, K; Thermal Conductivity, k, Watt $cm^{-1}K^{-1}$]

T	k	T	k
CURVE 1*		CURVE 1 (cont.)*	
293.2	0.00151	383.7	0.000586
307.8	0.00147	393.2	0.000552
317.5	0.00145	393.2	0.000628
338.0	0.00140		
342.0	0.00139		
344.2	0.00139		
346.2	0.00131		
348.2	0.00136		
350.2	0.000732		
352.0	0.000142		
353.2	0.000360		
355.0	0.000251		
356.8	0.000293		
357.2	0.000473		
357.9	0.000418		
360.7	0.000506		
362.6	0.000490		
363.2	0.000439		
368.9	0.000506		
372.7	0.000515		
373.7	0.000607		
377.2	0.000556		
382.9	0.000552		

*No graphical presentation

SPECIFICATION TABLE NO. 331 THERMAL CONDUCTIVITY OF APPLIED COATINGS (NONMETALLIC)

Curve No.	Ref. No.	Method Used	Year	Temp. Range, K	Reported Error, %	Name and Specimen Designation	Composition Base	Composition Coating	Composition (continued), Specifications and Remarks
1	485	L	1963	323-374		No. 9	Glass laminate	Paint	7 sheets of 184 Volan (glass cloth with a Volan finish) impregnated with 37-9X phenyl silane resin, plus one coat of plastic primer and two coats of SAF paint (alkyd resin base); specimen 7 x 7 x 0.172 in.; pressed at 100 psi at 450 K for 1 hr, post-cured at 450 K for 16 hrs; density 1.925 g cm^{-3}.
2	485	L	1963	337-351		No. 10	Glass laminate	Paint	Similar to the above specimen but applied two coats of TiC paint (water base) instead of SAF paint.

DATA TABLE NO. 331 THERMAL CONDUCTIVITY OF APPLIED COATINGS (NONMETALLIC)

[Temperature, T, K; Thermal Conductivity, k, Watt $cm^{-1} K^{-1}$]

T	k
CURVE 1*	
323.2	0.00177
343.2	0.00160
348.7	0.00226
356.5	0.00192
374.3	0.00190
CURVE 2*	
336.5	0.00182
338.7	0.00156
350.9	0.00180

*No graphical presentation

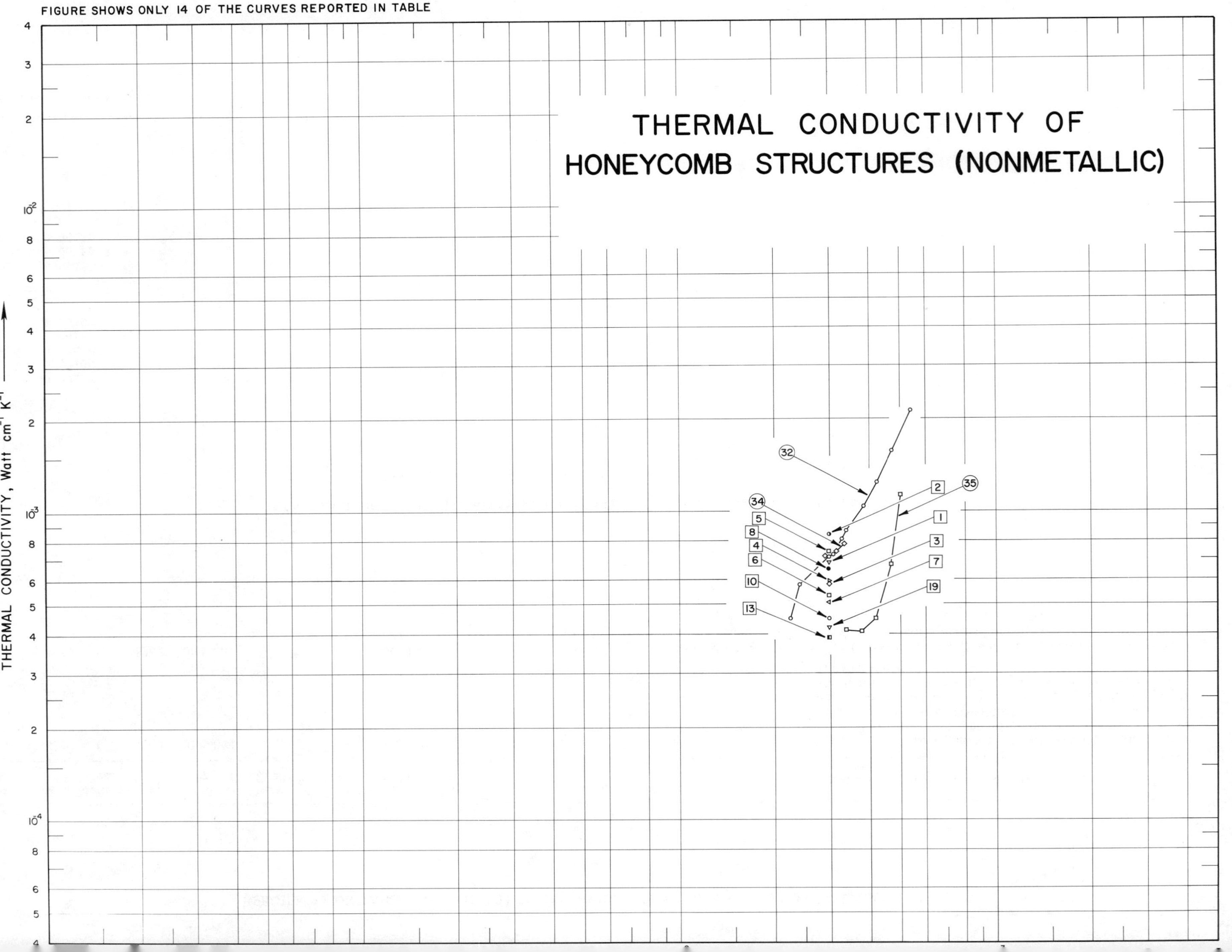
FIGURE SHOWS ONLY 14 OF THE CURVES REPORTED IN TABLE
THERMAL CONDUCTIVITY OF
HONEYCOMB STRUCTURES (NONMETALLIC)
THERMAL CONDUCTIVITY, Watt cm^{-1} K^{-1}
4
3
2
10^{-2}
8
6
5
4
3
2
10^{-3}
8
6
5
4
3
2
10^{-4}
8
6
5
4
32
2
35
34
5
8
4
6
10
13
1
3
7
19

[For Data Reported in Figure and Table No. 332]

Curve No.	Ref. No.	Method Used	Year	Temp. Range, K	Reported Error, %	Name and Specimen Designation	Composition: Core	Composition: Facing	Composition (continued), Specifications and Remarks
1	205	L	1953	297.2			Paper	Air	Paper honeycomb with hexagonal cell section; core description: corrugated sheets laid perpendicular to the faces and parallel to each other; weight of paper 50 lb per 3000 ft^2 and density of core 0.0471 g cm^{-3}.
2	205	L	1953	298.7			Paper	Air	Similar to the above specimen except core density 0.0557 g cm^{-3}.
3	205	L	1953	298.2			Paper	Foamed Resin A	Similar to the above specimen except space filled with foamed resin A and core weight 0.0854 g cm^{-3}.
4	205	L	1953	298.9			Paper	Foamed Resin B	Similar to the above specimen except space filled with foamed resin B and core weight 0.139 g cm^{-3}.
5	205	L	1953	298.8			Paper	Foamed Resin C	Similar to the above specimen except space filled with foamed resin C and core weight 0.169 g cm^{-3}.
6	205	L	1953	298.4			Paper	Urea Foam	Similar to the above specimen except space filled with shredded urea foam and core weight 0.0756 g cm^{-3}.
7	205	L	1953	298.5			Paper	Silica Aerogel	Similar to the above specimen except space filled with silica aerogel and core weight 0.103 g cm^{-3}.
8	205	L	1953	298.8			Paper	Siliceous Rock	Similar to the above specimen except space filled with puffed siliceous rock; core weight 0.171 g cm^{-3}.
9	205	L	1953	298.3			Paper	Air	Paper honeycomb with core description: corrugated paper laid perpendicular to the faces and parallel to each other with treated flat paper insertion; weight of paper 50 lb per 3000 ft^2 and weight of core 0.0879 g cm^{-3}.
10	205	L	1953	297.4			Paper	Air	Paper honeycomb with core description: corrugated paper laid parallel to the faces and parallel to each other with treated flat paper insertions; weight of paper 50 lb per 3000 ft^2 and weight of core 0.0681 g cm^{-3}.
11	205	L	1953	298.1			Paper	Air	Paper honeycomb with core description: corrugated sheets laid perpendicular to the faces and parallel to each other with treated flat paper insertion; weight of paper 30 lb per 3000 ft^2 and weight of core 0.0537 g cm^{-3}.
12	205	L	1953	297.1			Paper	Air	Paper honeycomb with core description: corrugated sheets were laid parallel to the faces and parallel to each other with treated flat paper insertion; weight of paper 50 lb per 3000 ft^2 and weight of core 0.0750 g cm^{-3}.
13	205	L	1953	298.5			Paper	Silica Aerogel	Similar to the above specimen except space filled with silica aerogel and core weight 0.136 g cm^{-3}.
14	205	L	1953	299.1			Paper	Air	Paper honeycomb with core description: the flutes of adjacent corrugated sheets, which were perpendicular to the faces, were at right angles; paper weight 50 lb per 3000 ft^2 and core weight 0.0441 g cm^{-3}.

SPECIFICATION TABLE NO. 332 (continued)

Curve No.	Ref. No.	Method Used	Year	Temp. Range, K	Reported Error, %	Name and Specimen Designation	Composition: Core		Composition: Facing	Composition (continued), Specifications and Remarks
15	205	L	1953	297.1			Paper	Air		Paper honeycomb with core description: the flutes of adjacent corrugated sheets, which were laid parallel to the faces, were at right angles; paper weight 50 lb per 3000 ft^2 and core weight 0.0441 g cm^{-3}.
16	205	L	1953	298.1			Paper	Air		Paper honeycomb with core description: the flutes of adjacent corrugated sheets, which were laid perpendicular to the faces, were at right angles and inclining 45° to the faces; paper weight 50 lb per 3000 ft^2 and core weight 0.0431 g cm^{-3}.
17	205	L	1953	297.8			Paper	Urea Foam		Paper honeycomb with core description: the flutes of adjacent corrugated sheets, which were laid perpendicular to the faces, were at right angles; space filled with shredded urea foam; paper weight 50 lb per 3000 ft^2 and core weight 0.0654 g cm^{-3}.
18	205	L	1953	297.8			Paper	Air		Paper honeycomb with core description: the flutes of adjacent corrugated sheets, which were laid perpendicular to the faces, were at right angles with flat sheets laid in between; paper weight 50 lb per 3000 ft^2 and core weight 0.0849 g cm^{-3}.
19	205	L	1953	297.4			Paper	Air		Paper honeycomb with core description: the flutes of adjacent corrugated sheets, which were laid parallel to the faces, were at right angles with flat sheets laid in between; paper weight 50 lb per 3000 ft^2 and core weight 0.0750 g cm^{-3}.
20	205	L	1953	297.7			Paper	Air		Paper honeycomb with circular cell sections; paper weight 50 lb per 3000 ft^2 and core weight 0.0463 g cm^{-3}.
21	205	L	1953	298.5			Paper	Foamed Resin A		Paper honeycomb with space filled with foamed resin A; paper weight 50 lb per 3000 ft^2 and core weight 0.0881 g cm^{-3}.
22	205	L	1953	298.6			Paper	Foamed Resin A		Paper honeycomb of circular cell sections with space filled with foamed resin A; large loops 1.25 in. in dia and made with 20 lb kraft paper; core weight 0.0301 g cm^{-3}.
23	206	L	1956	298.2			Paper	Air		Paper honeycomb with 15% water-soluble phenolic resin treated paper; core made of corrugated sheets assembled with principal flute directions of adjacent sheet at right angles; paper weight 50 lb per 3000 ft^2; core weight 0.0441 g cm^{-3} and thickness 1 in.
24	206	L	1956	298.2			Paper	Air		Paper honeycomb with resin treated paper; core made of corrugated sheets assembled parallel to each other and bonded at the crest; paper weight 30 lb per 3000 ft^2; core weight 0.0471 g cm^{-3} and thickness 1 in.
25	206	L	1956	298.2			Paper	Air		Similar to the above specimen except with core weight 0.0876 g cm^{-3}.

SPECIFICATION TABLE NO. 332 (continued)

Curve No.	Ref. No.	Method Used	Year	Temp. Range, K	Reported Error, %	Name and Specimen Designation	Composition			Composition (continued), Specifications and Remarks
							Core		Facing	
26	206	L	1956	298.2			Paper	Air		Paper honeycomb with resin treated paper; core made of corrugated sheets of paper assembled parallel to each other and separated by a single treated and uncorrugated sheet; paper weight 30 lb per 3000 ft^2; core weight 0.0537 g cm^{-3} and thickness 1 in.
27	206	L	1956	298.2			Paper	Air		Similar to the above specimen except with core weight 0.0881 g cm^{-3}.
28	206	L	1956	298.2			Paper	Air		Paper honeycomb with resin treated paper; core consisted of sheets of paper looped and bonded to form circular cells; core weight 0.0463 g cm^{-3} and thickness 1 in.
29	206	L	1956	298.2			Paper	Air		Paper honeycomb with resin treated paper; core filled with foamed resin and consisted of corrugated sheets assembled parallel to each other and bonded at crest; paper weight 30 lb per 3000 ft^2; core weight 0.0859 g cm^{-3} and thickness 1 in.
30	206	L	1956	298.2			Paper	Insulation		Similar to the above specimen except core weight 0.0756 g cm^{-3} and filled with insulation in the core.
31	206	L	1956	298.2			Paper	Air		Paper honeycomb with resin treated paper; core consisted of sheets of paper looped and bonded to form circular cells; core weight 0.0301 g cm^{-3} and thickness 1 in.
32	193	L	1957	225-545		BBC1	Polyester	Air	Polyester	Polyester honeycomb panel (0.25 in. cell size, 0.144 g cm^{-3} density) with polyester glass fabric faces (181 volan A); face resin No.: P-43; face catalyst: 2% A. T. C.; core 0.625 in. thick.
33	208	L	1957	289-339		A	Phenol	Air	Glass	0.034 in. thick skins made of combinations of 181 and 112 glass fabric with Garan finish, impregnated with polyester resin (BRS 142); 0.0641 g cm^{-3} nylon phenolic honeycomb core 0.309 in. thick; cell size honeycomb 0.25 in.; core bonded to the skins with epoxy resin.
34	208	L	1957	290-333		B	Phenol	Air		0.034 in. thick skins made of combination of 181 and 112 glass fabric with Garan finish, impregnated with polyester resin (BRS 142); 0.0961 g cm^{-3} CTL (glass fabric) phenolic honeycomb core 0.311 in. thick; cell size honeycomb 0.25 in.; core bonded to the skins with epoxy resin.
35	196	L	1955	337-504						Heat-resistant sandwich panel supplied by Goodyear Co.; 0.362 in. thick; fabricated using a phenolic-honeycomb core and laminate faces of 181 glass fabric and TAC-polyester resin 135; density 0.159 g cm^{-3}.

DATA TABLE NO. 332 THERMAL CONDUCTIVITY OF HONEYCOMB STRUCTURES (NONMETALLIC)

[Temperature, T, K; Thermal Conductivity, k, Watt $cm^{-1}K^{-1}$]

T	k
CURVE 1	
297.2	0.000678
CURVE 2	
298.7	0.000837
CURVE 3	
298.2	0.000577
CURVE 4	
298.9	0.000591
CURVE 5	
298.8	0.000736
CURVE 6	
298.4	0.000534
CURVE 7	
298.5	0.000505
CURVE 8	
298.8	0.000649
CURVE 9*	
298.3	0.000851
CURVE 10	
297.4	0.000447
CURVE 11*	
298.1	0.000678
CURVE 12*	
297.1	0.000447
CURVE 13	
298.5	0.000389
CURVE 14*	
299.1	0.000649
CURVE 15*	
297.1	0.000519
CURVE 16*	
298.1	0.000692
CURVE 17*	
297.8	0.000505
CURVE 18*	
297.8	0.000736
CURVE 19	
297.4	0.000418
CURVE 20*	
297.7	0.000764
CURVE 21*	
298.5	0.000548
CURVE 22*	
298.6	0.000447
CURVE 23*	
298.2	0.000649
CURVE 24*	
298.2	0.000663
CURVE 25 *	
298.2	0.000837
CURVE 26*	
298.2	0.000678
CURVE 27*	
298.2	0.000851
CURVE 28*	
298.2	0.000764
CURVE 29*	
298.2	0.000577
CURVE 30*	
298.2	0.000447
CURVE 31*	
298.2	0.000534
CURVE 32*	
225.2	0.000447
241.6	0.000577
299.3	0.000707
307.2	0.000721
327.6	0.000808
338.8	0.000865
382.9	0.00104
CURVE 32 (cont.)	
424.6	0.00124
474.7	0.00157
544.5	0.00213
CURVE 33*	
289.3	0.000668
318.8	0.000701
338.8	0.000734
CURVE 34	
290.4	0.000711
314.9	0.000738
333.2	0.000776
CURVE 35	
337.1	0.000411
377.4	0.000408
419.8	0.000446
468.5	0.000668
503.9	0.00112

* Not shown on plot

THERMAL CONDUCTIVITY OF HONEYCOMB STRUCTURES (METALLIC–NONMETALLIC)

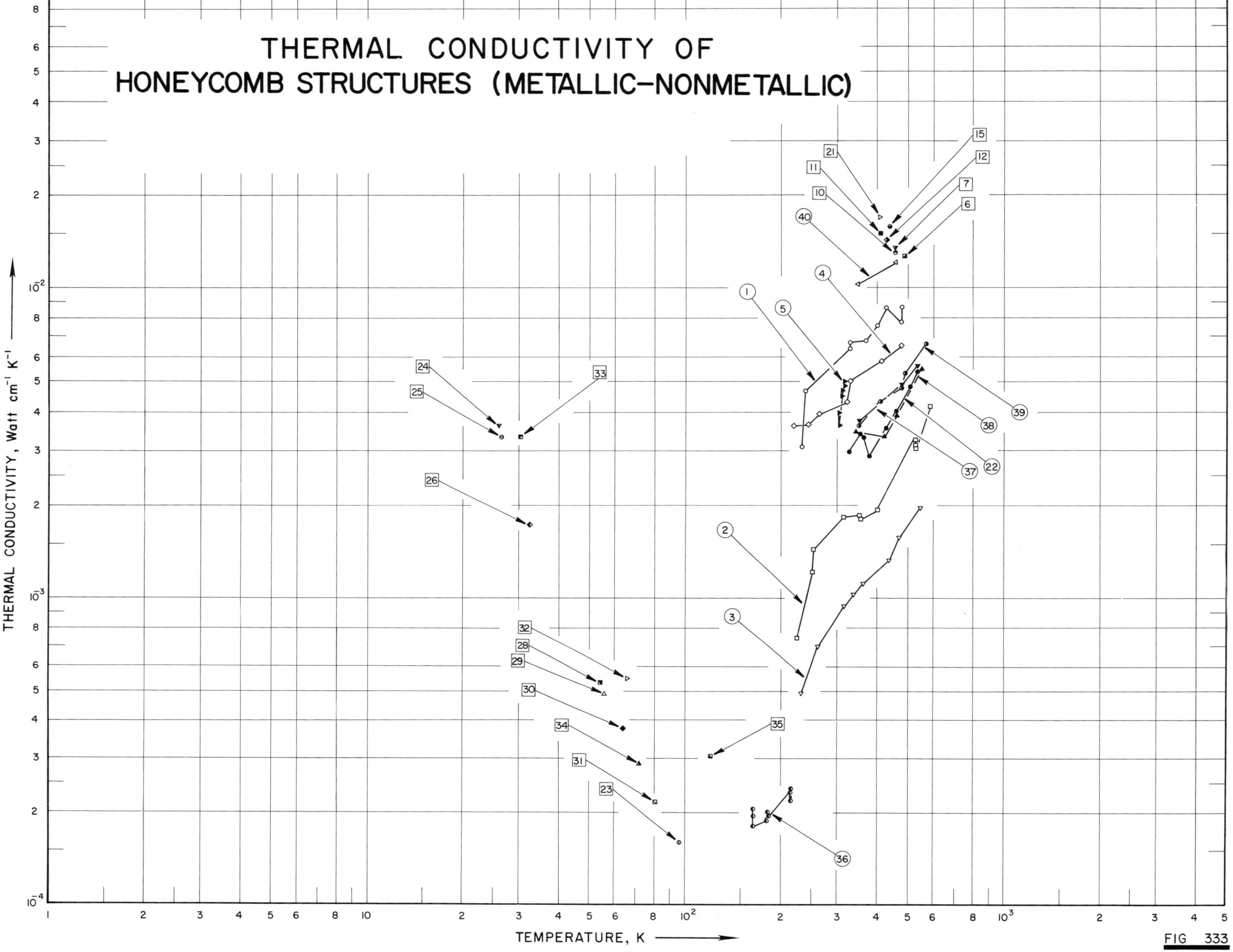

FIG 333

SPECIFICATION TABLE NO. 333 THERMAL CONDUCTIVITY OF HONEYCOMB STRUCTURES (METALLIC-NONMETALLIC)

[For Data Reported in Figure and Table No. 333]

Curve No.	Ref. No.	Method Used	Year	Temp. Range, K	Reported Error, %	Name and Specimen Designation	Composition: Core		Composition: Facing	Composition (continued), Specifications and Remarks
1	193	L	1957	232-480		BBC3	Aluminum Alloy	Air	Aluminum Alloy	Aluminum alloy honeycomb panel (0.25 in. cell size, 0.0689 g cm^{-3} density) with 0.02 in. thick aluminum alloy (2024-T3) faces; core resin 0.002 in. thick and core 0.625 in thick.
2	193	L	1957	224-587		GLM9	Stainless Steel	Air	Stainless Steel	Stainless steel honeycomb panel (0.375 in. cell size, 0.0015 in. thick foil) with stainless steel faces (0.032 in. thick); core and face of 17-7 PH alloys; face brazed to core; core 0.625 in thick.
3	193	L	1957	232-546		BBC4	Phenol	Air	Alumina Alloy	Phenolic honeycomb panel (0.25 in. cell size, 0.144 g cm^{-3} density) with 0.02 in. thick aluminum alloy (2024-T3) faces (catalyst: agent A) core 0.625 in. thick.
4	193	L	1957	219-480		BBC5	Aluminum Alloy	Air	Polyester	Aluminum alloy honeycomb panel (0.25 in. cell size, 0.0689 g cm^{-3} density) with polyester glass fabric faces (181 volan A); face resin No.: P-43 and face catalyst 2% A. T. C.; core resin 0.002 in. aluminum alloy 2024-T3; core 0.625 in. thick.
5	207	L	1956	304-318			Aluminum	Air	Aluminum	Two 0.008 in. thick 24ST aluminum face sheets bonded with elastomer-resin adhesive to a 0.25 in. cellular aluminum core of 0.003 in.; 3S-H19 aluminum core density 0.0689 g cm^{-3}.
6	209	L	1962	490.4			Stainless Steel	Air	Stainless Steel	A 12 in. x 12 in. x 0.532 in. brazed PH5-7Mo stainless steel honeycomb panel with nominal 0.016 in. PH15-7Mo stainless steel faces; honeycomb foil 0.0015 in. nominal thickness and 0.1875 in. cell spacing; brazing foil a 72% silver, 7.3% copper, and 0.2% lithium alloy with thickness 0.002 in. prior to brazing; hot surface temp controlled by ignition temp controller; run No. 1.
7	209	L	1962	457.0			Stainless Steel	Air	Stainless Steel	Run No. 2 of the above specimen.
8	209	L	1962	490.4			Stainless Steel	Air	Stainless Steel	Run No. 3 of the above specimen.
9	209	L	1962	490.4			Stainless Steel	Air	Stainless Steel	Run No. 4 of the above specimen.
10	209	L	1962	457.0			Stainless Steel	Air	Stainless Steel	Run No. 5 of the above specimen.
11	209	L	1962	414.3			Stainless Steel	Air	Stainless Steel	Run No. 6 of the above specimen.
12	209	L	1962	431.5			Stainless Steel	Air	Stainless Steel	Run No. 7 of the above specimen.
13	209	L	1962	414.3			Stainless Steel	Air	Stainless Steel	Run No. 8 of the above specimen.

SPECIFICATION TABLE NO. 333 (continued)

Curve No.	Ref. No.	Method Used	Year	Temp. Range, K	Reported Error, %	Name and Specimen Designation	Composition: Core	Composition:	Composition: Facing	Composition (continued), Specifications and Remarks
14	209	L	1962	478.1			Stainless Steel	Air	Stainless Steel	The above specimen, hot surface temperature controlled by multichannel recorder; run No. 1.
15	209	L	1962	444.3			Stainless Steel	Air	Stainless Steel	Run No. 2 of the above specimen.
16	209	L	1962	476.5			Stainless Steel	Air	Stainless Steel	Run No. 3 of the above specimen.
17	209	L	1962	476.5			Stainless Steel	Air	Stainless Steel	Run No. 4 of the above specimen.
18	209	L	1962	451.5			Stainless Steel	Air	Stainless Steel	Run No. 5 of the above specimen.
19	209	L	1962	411.5			Stainless Steel	Air	Stainless Steel	Run No. 6 of the above specimen.
20	209	L	1962	424.3			Stainless Steel	Air	Stainless Steel	Run No. 7 of the above specimen.
21	209	L	1962	409.3			Stainless Steel	Air	Stainless Steel	Run No. 8 of the above specimen.
22	279	L	1959	329-537			Stainless Steel	Air		17-7 PH stainless steel panel with shell adhesive No. 422 as bonding agent; overall height 0.622 in. with 0.052 in. face thickness; solidity of core 0.0136 in. and cell 0.25 in. square; core density 0.104 g cm^{-3}.
23	280	C	1962	96.5			Glass + Phenolic	Vacuum	CRES Metal	3 in. thick flat glass fiber/phenolic honeycomb bonded between 0.010 thick CRES skins and closed with 0.030 thick CRES channels of the same depth as the honeycomb core at the panel edges to form gas-tight boxes; sine wave glass-fiber/phenolic core with numerous 0.125 in dia drilled holes; cell size 1.125 in.; core density 0.0160 g cm^{-3}, bonded to CRES skins with Bloomingdale Rubber Co.'s HT424 adhesives, and filled by 0.0320 g cm^{-3} density polyurethane foam which was bonded with Hexcel's No. 1252 adhesives; core manufactured by the Narmco Materials Div.; measured with vacuum of 5 - 10 x 10^{-4} cm Hg in the core; Stafoam 604 used as comparative material.
24	280	C	1962	25.9			Glass + Phenolic	Helium	CRES Metal	Same description as the above specimen except HRP (heat resistant phenolic)/glass-fiber 0.0352 g cm^{-3} density core with cell size 0.375 in., manufactured by Hexcel Products Inc.; measured with helium filled core at 1 atm pressure.

SPECIFICATION TABLE NO. 333 (continued)

Curve No.	Ref. No.	Method Used	Year	Temp. Range, K	Reported Error, %	Name and Specimen Designation	Composition Core		Facing	Composition (continued), Specifications and Remarks
25	280	C	1962	26.5			Glass + Phenolic	Helium	CRES Metal	The above specimen measured with helium in the core at 0.50 atm pressure.
26	280	C	1962	32.6			Glass + Phenolic	Helium	CRES Metal	The above specimen measured with helium in the core at 0.20 atm pressure.
27	280	C	1962	25.9			Glass + Phenolic	Hydrogen	CRES Metal	The above specimen measured with hydrogen in the core at 1 atm pressure.
28	280	C	1962	54.3			Glass + Phenolic	Hydrogen	CRES Metal	The above specimen measured with hydrogen in the core at 35 x 10^{-4} cm Hg pressure.
29	280	C	1962	55.9			Glass + Phenolic	Hydrogen	CRES Metal	The above specimen measured with hydrogen in the core at 10 x 10^{-4} cm Hg pressure.
30	280	C	1962	64.3			Glass + Phenolic	Hydrogen	CRES Metal	The above specimen measured with hydrogen in the core at 2 x 10^{-4} cm Hg pressure.
31	280	C	1962	80.9			Glass + Phenolic	Helium	CRES Metal	The above specimen measured with helium (probably) in the core at 10^{-4} cm Hg pressure.
32	280	C	1962	65.9			Glass + Phenolic	Helium	CRES Metal	Similar to the above specimen except 5 in. in thickness; measured with helium in the core at 40 x 10^{-4} cm Hg.
33	280	C	1962	30.4			Glass + Phenolic	Helium	CRES Metal	The above specimen measured with helium in the core at 1 atm.
34	280	C	1962	72.0			Glass + Phenolic	Vacuum	CRES Metal	Same descriptions as the above specimen except HRP (heat resistant phenolic)/glass-fiber 0.0381 g cm^{-3} density core, manufactured by Hexcel Products Inc; measured at a pressure of 11 x 10^{-4} cm Hg self cryopumped.
35	280	C	1962	121.5			Glass + Phenolic	Vacuum	CRES Metal	The above specimen measured at a pressure of 5 x 10^{-4} cm Hg (pump on).
36	281	L	1963	165-217			Glass	Air		Glass fiber with syntactic foam; specimen dia 100 in. and mean thickness 0.68 in.
37	486	L	1961	353-539			Steel	Air	Steel	Core made from 0.002 in. thick 17-7 PH steel; face sheets PH 15-7 Mo steel 0.020 in. thick, brazed to core with Cu 1700 (75.0 Cu and 25.0 Mn); specimen 9 in. in dia; core section 0.5 in. thick manufactured by Tool Research Co.; prepared by brazing at 1228 K, austenite conditioning in argon 1228 ±14 K for 15 min, air cooling to room temp, transforming at 200 ±6 K for 8 hrs, and aging at 783 ±6 K for 1 hr; a pressure of 80 psi applied to the specimen during testing.

SPECIFICATION TABLE NO. 333 (continued)

Curve No.	Ref. No.	Method Used	Year	Temp. Range, K	Reported Error, %	Name and Specimen Designation	Composition			Composition (continued), Specifications and Remarks
							Core		Facing	
38	486	L	1961	344-553			Steel	Air	Steel	Similar to the above specimen except brazing alloy Lithobraze 501 (50. 0 Ni, 46. 4 Ag, 3. 5 Cu, and 0. 1 Li).
39	486	L	1961	350-572			Steel	Air	Steel	Similar to the above specimen except brazing alloy Silvalloy T-50 (62. 4 Ag, 32. 07 Cu, and 4. 43 Ni), and austenite conditioning at 1172 K for 15 min.
40	486	L	1961	350, 458			Steel	Air	Steel	Similar to the above specimen except brazing alloy Lithobraze 925 (92. 8 Ag, 7. 0 Cu, and 0. 2 Li), and heat treatment was austenite conditioning in argon at 1033 ±14 K for 90 min; air cooling to 289 K, and aged at 839 ±14 K for 90 min.

DATA TABLE NO. 333 THERMAL CONDUCTIVITY OF HONEYCOMB STRUCTURES (METALLIC-NONMETALLIC)

[Temperature, T, K; Thermal Conductivity, k, Watt $cm^{-1}K^{-1}$]

T	k
CURVE 1	
232.1	0.00309
239.6	0.00469
330.2	0.00637
330.2	0.00668
369.0	0.00678
403.9	0.00757
430.7	0.00868
478.4	0.00776
479.1	0.00767*
479.6	0.00762*
480.1	0.00865
CURVE 2	
224.2	0.000736
251.2	0.00121
252.1	0.00144
315.4	0.00183
353.5	0.00185
356.2	0.00180
403.8	0.00193
526.8	0.00325
528.8	0.00313
529.0	0.00306
587.3	0.00418
CURVE 3	
232.4	0.000490
261.6	0.000692
316.3	0.000937
338.8	0.00102
363.5	0.00111
438.6	0.00133
470.7	0.00157
546.3	0.00196
CURVE 4	
219.1	0.00361
244.3	0.00365
265.6	0.00394
323.7	0.00431
CURVE 4 (cont.)	
331.3	0.00502
415.5	0.00581
479.8	0.00653
CURVE 5	
304.3	0.00398
305.4	0.00363
311.9	0.0045
313.9	0.00467
318.3	0.00485
318.4	0.00502
CURVE 6	
490.4	0.0127
CURVE 7	
457.0	0.0135
CURVE 8*	
490.4	0.0133
CURVE 9*	
490.4	0.0129
CURVE 10	
457.0	0.0130
CURVE 11	
414.3	0.0150
CURVE 12	
431.5	0.0142
CURVE 13*	
414.3	0.0152
CURVE 14*	
478.1	0.0133
CURVE 15	
444.3	0.0157
CURVE 16*	
476.5	0.0143
CURVE 17*	
476.5	0.0138
CURVE 18*	
451.5	0.0132
CURVE 19*	
411.5	0.0157
CURVE 20*	
424.3	0.0155
CURVE 21	
409.3	0.0169
CURVE 22	
328.7	0.00298
355.4	0.00339
365.2	0.00331
379.8	0.00288
428.7	0.00355
463.7	0.00402
511.2	0.00483
537.3	0.00537
CURVE 23	
96.5	0.000159
CURVE 24	
25.9	0.00361
CURVE 25	
26.5	0.00332
CURVE 26	
32.6	0.00173
CURVE 27*	
25.9	0.00375
CURVE 28	
54.3	0.000534
CURVE 29	
55.9	0.000490
CURVE 30	
64.3	0.000375
CURVE 31	
80.9	0.000216
CURVE 32	
65.9	0.000548
CURVE 33	
30.4	0.00332
CURVE 34	
72.0	0.000288
CURVE 35	
121.5	0.000303
CURVE 36	
164.8	0.000194
164.8	0.000204
164.8	0.000197*
164.8	0.000180
181.5	0.000187
182.0	0.000200
183.2	0.000194
214.3	0.000231
214.8	0.000217
215.9	0.000237
216.5	0.000232*
CURVE 37	
352.6	0.00375
483.2	0.00490
538.7	0.00562
CURVE 38	
344.3	0.00346
422.1	0.00332
463.7	0.00389
552.6	0.00548
CURVE 39	
349.8	0.00361
410.9	0.00433
477.6	0.00476
494.3	0.00534
572.1	0.00663
CURVE 40	
349.8	0.0103
458.2	0.0121

* Not shown on plot

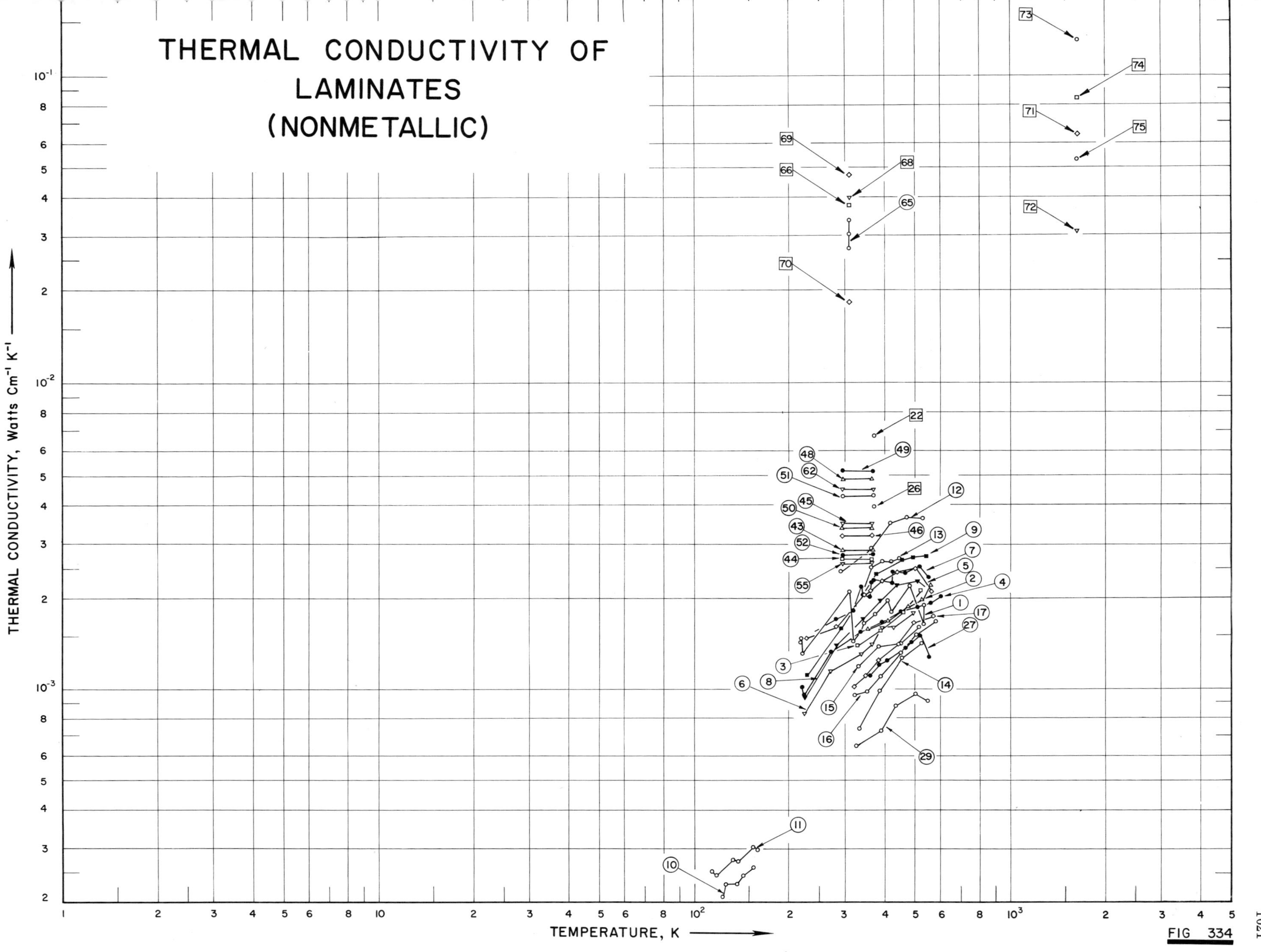
THERMAL CONDUCTIVITY OF
LAMINATES
(NONMETALLIC)
THERMAL CONDUCTIVITY, Watts Cm⁻¹ K⁻¹
TEMPERATURE, K
73
74
71
75
69
66
68
65
72
70
22
48
49
62
51
26
12
45
50
46
13
9
43
52
7
44
5
55
2
4
1
17
27
3
8
6
14
15
16
29
11
10

FIG 334

SPECIFICATION TABLE NO. 334 THERMAL CONDUCTIVITY OF LAMINATES (NONMETALLIC)

[For Data Reported in Figure and Table No. 334]

Curve No.	Ref. No.	Method Used	Year	Temp. Range, K	Reported Error, %	Name and Specimen Designation	Base Material	Binder	Specifications and Remarks
1	389	L	1957	218-535		BBC-10	Glass Fabric (No. 181)	Epon-828 Resin	Specimen 12 x 12 in. cut from laminate of Epon-828 resin with glass fabric (class No. 181 with Volan A finish); manufactured by Brunswick-Balke Collender Co.
2	389	L	1957	358-566		DC-1	Glass Fabric (No. 181)	Silicone (Dow Corning 2104)	Similar to the above.
3	389	L	1957	331-523		DC-2	Glass Fabric (No. 181)	Silicone (Dow Corning 2106)	Similar to the above.
4	194	L	1958	222-609		Asbestos-Phenolic 9526D	Raybestos (Manhattan 9526D)	Phenolic Resin	Specimen of 0.225 in. nominal thickness; 3 plies; cured at 300 F for 0.50 hr at 400 psi, post cured at 300 F for 4 hrs and 350 F for 3 hrs; laminate density 78 lb ft^{-3} (1.25 g cm^{-3}); fabricated by Brunswick-Balke Collender Co.
5	194	L	1958	229-565		Shell X-131	Glass Fabric (No. 181)	Epoxy (Shell X-131) Resin	0.50 in. nominal thickness; 12 plies; glass fabric of Volan A finish and the catalyst consisted of 13 parts (by weight) pyromellitic dianhydride and 19 parts maleic anhydride; the laminate cured at 200 F for 2 hrs at 25 psi and 300 F for 4 hrs; density 117 lb ft^{-3} (1.87 g cm^{-3}).
6	194	L	1958	225-611		CTL 37-9X	Glass Fabric (No. 181)	Phenolic (CTL 37-9X) Resin	0.225 in. nominal thickness; 12 plies; modified phenolic and glass fabric of A-1100 finish; laminate cured at 280 F for 0.50 hr at 175 psi, post cured at 250 F for 24 hrs, 300 F for 24 hrs, 350 F for 24 hrs and 400 F for 24 hrs; calculated density 103 lb ft^{-3} (1.65 g cm^{-3}).
7	194	L	1958	227-560		X-1068	Glass Fabric (No. 181)	Tac-Polyester (Vibrin X-1068) Resin	0.225 in. nominal thickness; 13 plies; glass fabric of Garan Finish and the catalyst composed of 1.5% Benzoyl Peroxide; laminate cured at 200 F for 12 hrs, 250 F for 3 hrs in vacuum, post cured at 300 F for 1 hr, 350 F for 5 hrs and 400 F for 1 hr; density 112 lb ft^{-3} (1.79 g cm^{-3}).
8	194	L	1958	227-569		Selectron 5016 Fluted Core	Glass Fabric (No. 181)	Polyester (Selectron 5016) Resin	Fluted core sandwich, flute size: 0.143 in. thick, 0.375 in. wide and 0.0075 in. wall thickness; 15 plies on both sides of the sandwich; glass fabric of Volan A finish, 2% Benzoyl Peroxide used as catalyst; cured at 250 F for 5 hrs; density 105 lb ft^{-3} (1.68 g cm^{-3}).

Curve No.	Ref. No.	Method Used	Year	Temp. Range, K	Reported Error, %	Name and Specimen Designation	Base Material	Binder	Specifications and Remarks
9	194	L	1958	230-547		Shell X-131 Fluted Core	Glass Fabric (No. 181)	Epoxy (Shell X-131) Resin	Fluted core sandwich, fluted size: 0.300 in. wide, 0.350 in. thick and 0.0075 in. wall thickness, 15 plies on one side and 11 plies on the other side of the sandwich; glass fabric of Volan A finish; 13 parts by weight of maleic anhydride used as catalyst; cured at 200 F for 2 hrs under vacuum and post cured at 300 F for 4 hrs; density 124 lb ft^{-3}(1.99 g cm^{-3}).
10	43	L	1954	124-155		Resin-bonded mineral wool board	Mineral Wool	Resin	Resin-bonded mineral wool board; density 16.4 lb ft^{-3} (0.26 g cm^{-3}).
11	43	L	1954	115-160		Asphalt-bonded mineral wool board	Mineral Wool	Asphalt Resin	Asphalt-bonded mineral wool board; density 16.6 lb ft^{-3} (0.27 g cm^{-3}).
12	390	C	1961	294-533			Asbestos	Phenolic Resin	Laminate 0.225 in thick; Transite used as comparative material.
13	390	C	1961	339-450		Lamicoid No. 6045			Transite used as comparative material.
14	196	L	1955	336-527		B-1	Glass Fabric (No. 181)	Phenolic (BV 17085) Resin	Laminate 0.119 in. thick; 14 plies; manufactured by Bakelite Co.; resin content 39.3% by weight; Hexasol used as catalyst.
15	196	L	1955	334-519		N-1	Glass Fabric (No. 181)	Phenolic (Conolon 506) Resin	Laminate 0.115 in. thick; 12 plies; manufactured by Narmco Inc.; glass fabric of Volan A finish; resin content 33% by weight; no catalyst.
16	196	L	1955	325-585		DC-2	Glass Fabric (No. 181)	Silicone (Dow Corning 2106) Resin	Laminate 0.105 in. thick; 12 plies; manufactured by Dow Corning Corp.; glass fabric of No. 112 finish; XY-15 used as catalyst.
17	196	L	1955	324-575		NK-1	Glass Fabric (No. 181	Tac-Polyester (Vibrin 135) Resin	Laminate 0.143 in. thick; glass fabric of No. 301 finish; resin content 39.3% by weight; Benzoyl Peroxide used as catalyst; manufactured by Naugatuck Chem. Co.
18	197	P	1952	373.2		Insurok (XXX-T-640)	Paper	Phenolic Resin	Specific gravity ~1.35; measured in the x-direction (perpendicular to lamination).
19	197	P	1952	373.2		Insurok (XXX-T-640)	Paper	Phenolic Resin	The above specimen measured in the y-direction (parallel to lamination).
20	197	P	1952	373.2		Insurok (XXX-T-640)	Paper	Phenolic Resin	The above specimen measured in the z-direction (parallel to lamination).
21	197	P	1952	373.2		Insurok (C-T-601)	Cloth	Phenolic Resin	Laminate of 1 in. nominal thickness; specific gravity ~1.34; measured in the x-direction.

SPECIFICATION TABLE NO. 334 (continued)

Curve No.	Ref. No.	Method Used	Year	Temp. Range, K	Reported Error, %	Name and Specimen Designation	Base Material	Binder	Specifications and Remarks
22	197	P	1952	373.2		Insurok (C-T-601)	Cloth	Phenolic Resin	The above specimen measured in the y-direction.
23	197	P	1952	373.2		Insurok (C-T-601)	Cloth	Phenolic Resin	The above specimen measured in the z-direction.
24	197	P	1952	373.2		Lamicoid (C-6030)	Cloth	Phenolic Resin	Laminate 0.50 in. thick; specific gravity 1.31; measured in the x-direction.
25	197	P	1952	373.2		Lamicoid (C-6030)	Cloth	Phenolic Resin	The above specimen measured in the y-direction.
26	197	P	1952	373.2		Lamicoid (C-6030)	Cloth	Phenolic Resin	The above specimen measured in the z-direction.
27	198	L	1955	361-555		GY-F	Glass Fabric (No. 181)	Selectron 5003 Resin	Laminate of 11-12 plies; glass fabric of Volan A finish; resin content 37.5% by weight; manufactured by Goodyear Co.; density 114.3 lb ft^{-3} (1.83 g cm^{-3}); laminate placed in a cabinet of constant temp at 73 F and 50% RH for at least 1 week prior to test.
28	198	L	1955	332-552		GY-G	Glass Fabric (No. 181)	Selectron 5003 Resin	Laminate of 12 plies with 42.8% resin content; similarly treated as above; density 113.51 lb ft^{-3} (1.82 g cm^{-3}).
29	198	L	1955	327-550		GY-H	Glass Fabric (No. 181)	Selectron 5003 Resin	4 plies; 24.5% resin content; similarly treated as above; density 118.99 lb ft^{-3} (1.90 g cm^{-3})
30	198	L	1955	328-560		GY-I	Glass Fabric (No. 181)	Selectron 5003 Resin	11 plies; 34.1% resin content; similarly treated as above; density 101.15lb ft^{-3} (1.62 g cm^{-3}).
31	198	L	1955	362-564		CTL-A	Glass Fabric (No. 181)	Phenolic (CTL-91-LD) Resin	14 plies; 28.3% resin content; manufactured by Cincinnati Testing and Research Lab; similarly treated as above; density 119.18 lb ft^{-3}(1.75 g cm^{-3}).
32	198	L	1955	336-584		CTL-B	Glass Fabric (No. 181)	Phenolic (CTL-91-LD)	Similarly treated as above; density 94.22 lb ft^{-3} (1.51 g cm^{-3}).
33	198	L	1955	366-552		CTL-C	Glass Fabric (No. 181)	Phenolic (CTL-91-LD)	Similarly treated as above; density 102.34 lb ft^{-3} (1.64 g cm^{-3}).
34	198	L	1955	314-507		S	Glass Fabric (No. 181)	33 E pon + 67 Plyophen (1001/5023)	24.5% resin content; density 108.58 lb ft^{-3} (1.74 g cm^{-3}); laminate from Shell Co.
35	198	L	1955	361-613		GY-F	Glass Fabric (No. 181)	Selectron 5003 Resin	Another run of the specimen in curve No. 27.

SPECIFICATION TABLE NO. 334 (continued)

Curve No.	Ref. No.	Method Used	Year	Temp. Range, K	Reported Error, %	Name and Specimen Designation	Base Material	Binder	Specifications and Remarks
36	198	L	1955	330-557		GY-A	Glass Fabric (No. 116)	Selectron 5003 Resin	Laminates of 30-32 plies; 39.5% resin content, glass fabric of Garan finish; laminate placed in a cabinet of constant temp at 73 F and 50% RH for at least 1 week prior to test; density 110.12 lb ft^{-3} (1.76 g cm^{-3}); specimen from Goodyear Co.
37	198	L	1955	375-553		GY-C	Glass Fabric (No. 181)	Selectron 5003 Resin	12 ply-laminate with glass fabric of Volan A finish and 40.8% resin content; similarly treated as above; density 114.07 lb ft^{-3} (1.82 g cm^{-3}).
38	198	L	1955	331-523		GY-D	Glass Fabric (No. 143)	Selectron 5003 Resin	11 ply-laminate; 43.5% resin content; similarly treated as above; density 107.08 lb ft^{-3} (1.61 g cm^{-3}).
39	199	L	1954	344-556		DC-1	Glass Fabric (No. 181)	Silicone (No. 2104) Resin	14 ply-laminate; glass fabric of OC 112 finish, 32-35% resin content; laminated by Dow Corning Co.; specimen 0.1415 in. thick and 64 in.^{2}; density 101.03 lb ft^{-3} (1.62 g cm^{-3}); similarly treated before test.
40	199	L	1954	335-551		GY-E-1	Glass Fabric (No. 181)	Selectron 5003 Resin	11 ply-laminate; glass fabric of Volan A finish, 29.2% resin content; laminated by Goodyear Co.; specimen 0.1055 in. thick, 64 in.^{2} area; similarly treated as above before test; density 119.75 lb ft^{-3} (1.92 g cm^{-3})
41	199	L	1954	340-537		GY-B-1	Glass Fabric (No. 128)	Selectron 5003 Resin	18 ply-laminate; glass fabric of Volan A finish, 39.2% resin content; laminated by Goodyear Co.; specimen 0.113 in. thick and 64 in.^{2} area; similarly treated as above before test; density 115.32 lb ft^{-3} (1.85 g cm^{-3}).
42	199	L	1954	320-575		AC-3 & 4	Glass Fabric (No. 181)	Laminac (PDL-7-669) Resin	12 ply-laminate; glass fabric of OC 136 finish, 33.9% resin content; LuperCo ATC-1% used as catalyst; laminated by American Cynamid Co.; specimen 0.1165 in. thiek 64 in.^{2} area; similarly treated before test; density 116.38 lb ft^{-3} (1.86 g cm^{-3}).
43	200	P	1951	297, 367		PBE-A	Paper	Phenolic Resin	Specimen 5 x 5 x 0.50 in.; resin content 50%; density 83.6 lb ft^{-3} (1.34 g cm^{-3}).

SPECIFICATION TABLE NO. 334 (continued)

Curve No.	Ref. No.	Method Used	Year	Temp. Range, K	Reported Error, %	Name and Specimen Designation	Base Material	Binder	Specifications and Remarks
44	200	P	1951	297,367		PBE-B	Paper	Phenolic Resin	As above but resin content 57.5%, density 57.5 lb ft^{-3} (0.92 g cm^{-3}).
45	200	P	1951	297,367		FBG-A	Cotton Fabric	Phenolic Resin	As above but resin content 50%, density 82.9 lb ft^{-3} (1.33 g cm^{-3}).
46	200	P	1951	297,367		FBG-B	Cotton Fabric	Phenolic Resin	As above but resin content 57.5%, density 82.1 lb ft^{-3} (1.31 g cm^{-3}).
47	200	P	1951	297,367		GBE-B	Glass mat	Phenol Aniline Resin	As above but resin content 59.5%, density 92.8 lb ft^{-3} (1.49 g cm^{-3}).
48	200	P	1951	297,367		GMG-A	Fiberglass	Melamine Resin	As above but resin content 40%, density 117.5 lb ft^{-3} (1.88 g cm^{-3}).
49	200	P	1951	297,367		GMG-C	Fiberglass	Melamine Resin	As above but resin content 45%, density 118.9 lb ft^{-3} (1.90 g cm^{-3}).
50	200	P	1951	297,367		Poly-D	Fiberglass	Polyester (low-pressure)	As above but resin content 31%, density 112.2 lb ft^{-3} (1.80 g cm^{-3}).
51	200	P	1951	297,367		LPP-D	Fiberglass	Phenolic (low-pressure)	As above but resin content 28.4%, density 115.9 lb ft^{-3} (1.86 g cm^{-3}).
52	200	P	1951	297,367		GSG-E	Glass	Silicone Resin	As above but resin content 52%, density 101.6 lb ft^{-3} (1.63 g cm^{-3}).
53	200	P	1951	297,367		GSG-F	Glass	Silicone Resin	As above but resin content 40%,density 106.2 lb ft^{-3} (1.70 g cm^{-3}).
54	200	P	1951	297,367		PBE-A	Paper	Phenolic Resin	Specimen 5 x 5 x 0.225 in.; resin content 50%, density 86.9 lb ft^{-3} (1.38 g cm^{-3}).
55	200	P	1951	297,367		PBE-B	Paper	Phenolic Resin	Specimen 5 x 5 x 0.225 in.; resin content 57.5%, density 82.5 lb ft^{-3} (1.32 g cm^{-3}).
56	200	P	1951	297,367		FBG-A	Cotton Fabric	Phenolic Resin	Specimen 5 x 5 x 0.225 in.; resin content 50%, density 81.8 lb ft^{-3} (1.31 g cm^{-3}).
57	200	P	1951	297,367		FBG-B	Cotton Fabric	Phenolic Resin	Specimen 5 x 5 x 0.225 in.; resin content 57.5%, density 81.5 lb ft^{-3} (1.31 g cm^{-3}).
58	200	P	1951	297,367		GBE-B	Glass mat	Phenol Aniline Resin	Specimen 5 x 5 x 0.225 in.; resin content 59.5%; density 91.1 lb ft^{-3} (1.46 g cm^{-3}).
59	200	P	1951	297,367		GMG-B	Fiberglass	Melamine Resin	Specimen 5 x 5 x 0.225 in.; resin content 40%; density 118.2 lb ft^{-3} (1.89 g cm^{-3}).
60	200	P	1951	297,367		GMG-C	Fiberglass	Melamine Resin	Specimen 5 x 5 x 0.225 in.; resin content 45%, density 118.5 lb ft^{-3} (1.90 g cm^{-3}).

SPECIFCATION TABLE NO. 334 (continued)

Curve No.	Ref. No.	Method Used	Year	Temp. Range, %	Reported Error, %	Name and Specimen Designation	Base Material	Binder	Specifications and Remarks
61	200	P	1951	297, 367		Poly-D	Fiberglass	Polyester (low pressure)	Specimen 5 x 5 x 0. 225 in.; resin content 31. 0%, density 113. 1 lb ft^{-3} (1. 81 g cm^{-3}).
62	200	P	1951	297, 367		LPP-D	Fiberglass	Phenolic (low pressure)	Specimen 5 x 5 x 0. 225 in.; resin content 28. 4%, density 124. 8 lb ft^{-3} (2. 00 g cm^{-3}).
63	200	P	1951	297, 367		GSG-E	Glass	Silicone Resin	Specimen 5 x 5 x 0. 225 in.; resin content 52%, density 100. 4 lb ft^{-3} (1. 61 g cm^{-3}).
64	200	P	1951	297, 367		GSG-F	Glass	Silicone Resin	Specimen 5 x 5 x 0. 225 in.; resin content 40%, density 106. 4 lb ft^{-3} (1. 70 g cm^{-3}).
65	487	L	1957	308. 2			Glass	Polyester Resin	Thermal conductivities were measured on pairs of discs 3 in. in dia each about 0. 25 in. thick; bulk density range from 1. 7 to 1. 84 g cm^{-3} (105. 9 to 115. 0 lb ft^{-3}).
66	487	L	1957	308. 2			Glass	Phenolic Resin	Similar to the above specimen except bulk density 1. 86 g cm^{-3} (116. 1 lb ft^{-3}).
67	487	L	1957	308. 2			Glass	Silicone Resin	Similar to the above specimen except bulk density 1. 90 g cm^{-3} (118. 9 lb ft^{-3}).
68	487	L	1957	308. 2			Glass	Epoxy-phenolic Resin	Similar to the above specimen except bulk density 1. 89 g cm^{-3} (118. 0 lb ft^{-3}).
69	487	L	1957	308. 2			Glass	Melamine Resin	Similar to the above specimen except bulk density 1. 97 g cm^{-3} (124. 0 lb ft^{-3}).
70	487	L	1957	308. 2			Glass	Unfilled polyester Resin	Similar to the above specimen except bulk density 1. 21 g cm^{-3} (75. 5 lb ft^{-3}).
71	84	L	1925	1623			Brick	Fire clay	9 in. carborundum No. 1A wall was insulated with 4. 5 in. fire-clay No. 26 brick wall with a thin layer of fire clay in between; the wall under test was built of standard 2. 5 x 4. 5 x 9 in. bricks laid up with cement of the same composition as the brick; carborundum No. 1A was made of carborundum recrystallized in an electric furnace containing 91. 51 SiC, 8. 02 SiO_2, 0. 20 Al_2O_3, and 0. 65 Fe_2O_3 (0. 17 ignition loss) with apparent density 2. 05 g cm^{-3} and 34. 1% porosity calculated by assuming specific gravity 3. 17 for SiC; fire clay No. 26 was coarse, fairly open, first quality fire clay containing 58. 50 SiO_2, 34. 48 Al_2O_3, 3. 52 Fe_2O_3, 1. 80 TiO_2, 0. 29 CaO, and 0. 62 MgO (0. 31 ignition loss) with apparent density 2. 05 g cm^{-3} and 26. 8% porosity calculated by assuming specific gravity 2. 60 for fire clay; the bricks were placed so that the temp gradient was through the 4. 5 in. dimension.

SPECIFICATION TABLE NO. 334 (continued)

Curve No.	Ref. No.	Method Used	Year	Temp. Range, K	Reported Error, %	Name and Specimen Designation	Base Material	Binder	Specifications and Remarks
72	84	L	1925	1623			Brick	Fire clay	Similar to the above specimen except the 9 in. carborundum No. 1A wall was insulated with 4.5 in. fire clay No. 75 brick wall; fire clay No. 75 containing 56.90 SiO_2, 37.70 Al_2O_3, 2.37 Fe_2O_3, 1.74 TiO_2, 0.82 CaO, and 0.20 MgO with 0.67 g cm^{-3} apparent density, and 74.2% porosity calculated by assuming 2.60 specific gravity for fire clay, were fired at 1350 C.
73	84	L	1925	1623			Brick	Fire clay	Similar to the above specimen except 4.5 in. carborundum No. 1C wall was insulated with 4.5 in. fire clay No. 26 brick wall; the carborundum No. 1C brick containing 93.20 SiC, 4.50 SiO_2, 1.33 Al_2O_3, and 1.03 Fe_2O_3 with 2.20 g cm^{-3} apparent density, and 29.5% porosity calculated with 3.17 specific gravity for SiC were made of carborundum recrystallized in an electric furnace.
74	84	L	1925	1623			Brick	Fire clay	Similar to the above specimen except 4.5 in. carborundum No. 2 wall was insulated with 4.5 in. fire clay No. 26 brick wall; the carborundum No. 2 brick containing 80.10 SiC, 14.72 SiO_2, 1.47 Al_2O_3, and 1.33 Fe_2O_3 (0.88 ignition loss) with 2.48 g cm^{-3} apparent density and 18.4% porosity calculated by assuming 3.17 specific gravity for SiC, contained ceramic bonds and were kiln fired at approx. 1350 C.
75	84	L	1925	1623			Brick	Fire clay	Similar to the above specimen except 4.5 in. carborundum No. 2 wall was insulated with 4.5 in. fire clay No. 75 brick wall; see above specimen for composition, specifications and remarks about carborundum No. 2 brick.
76	84	L	1925	1623			Brick	Fire clay	Similar to the above specimen except 4.5 in. carborundum No. 3 wall was insulated with 4.5 in. fire clay No. 26 brick wall; the carborundum No. 3 brick containing 5.58 Al_2O_3, 68.50 SiC, 23.31 SiO_2, and 1.57 Fe_2O_3 (0.24 ignition loss) with 2.35 g cm^{-3} apparent density and 20.7% porosity calculated by assuming 3.17 for SiC specific gravity, contained ceramic bonds and were kiln fired at approx 1350 C.

SPECIFICATION TABLE NO. 334 (continued)

Curve No.	Ref. No.	Method Used	Year	Temp. Range, K	Reported Error, %	Name and Specimen Designation	Base Material	Binder	Specifications and Remarks
77	84	L	1925	1623			Brick	Fire clay	Similar to the above specimen except 4.5 in. carborundum No. 4 wall was insulated with 4.5 in. fire clay No. 26 brick wall; the carborundum No. 4 brick containing 8.10 Al_2O_3, 52.60 SiC, 36.80 SiO_2, and 1.97 Fe_2O_3 (0.02 ignition loss) with 2.36 g cm^{-3} apparent density and 17.7% porosity calculated by assuming 3.17 for SiC specific gravity, contained ceramic bonds and were kiln fired at approx 1350 C.
78	84	L	1925	1623			Brick	Fire clay	Similar to the above specimen except 4.5 in. carborundum No. 5 wall was insulated with 4.5 in. fire clay No. 26 brick wall; carborundum No. 5 containing 11.63 Al_2O_3, 48.35 SiC, 38.76 SiO_2, and 1.97 Fe_2O_3 (0.10 ignition loss) with 2.31 g cm^{-3} apparent density, and 18.9% porosity calculated by assuming 3.17 for SiC specific gravity contained ceramic bonds and were kiln fired at approx 1350 C.
79	488, 489	L	1961	225-548		Scotchply; SRI 1-1	Plastic	3MXP-175 Epoxy Resin	30 ±1.5 resin; 60 N roving reinforcing; unidirectional laminate; specimen 18 x 12 x 0.125 in.; average thickness 0.125 in.; manufactured by Minnesota Mining and Mfg. Co.; pressed at 408-411 K under 90-95 psi for 40 min; post-cured at 380 K for 16 hrs; 20 plies; density 1.841 g cm^{-3}.
80	488, 489	L	1961	230-548		Scotchply; SRI 1-2	Plastic	3MXP-175 Epoxy Resin	30 ±1.5 resin; 60 N roving reinforcing; isotropic laminate; specimen 18 x 12 x 0.125 in.; average thickness 0.122 in.; manufactured by Minnesota Mining and Mfg. Co.; pressed at 403-405 K under 90 psi for 40 min; post-cured at 380 K for 16 hrs; 21 plies; density 1.841 g cm^{-3}.
81	488, 489	L	1961	443, 548		Scotchply; SRI 1-2	Plastic	3MXP-175 Epoxy Resin	The above specimen measured at decreasing temp.
82	488, 489	L	1961	222-661		Astrolite; SRI 7-1	Plastic	Phenolic resin SC-1008	30-35 resin; refrasil 184 weave reinforcing; parallel layup; specimen 18 x 18 x 0.125 in.; average thickness 0.126 in.; manufactured by H. I. Thompson Fiber Glass Co.; curing at 422 K for 2 hrs; 6 plies; density 1.442 g cm^{-3}.
83	488, 489	L	1961	514-661		Astrolite; SRI 7-1			The above specimen measured at decreasing temp.

SPECIFICATION TABLE NO. 334 (continued)

Curve No.	Ref. No.	Method Used	Year	Temp. Range, K	Reported Error, %	Name and Specimen Designation	Base Material	Binder	Specifications and Remarks
84	488, 489	L	1961	219-623		Astrolite; SRI 7-2	Plastic	Phenolic resin SC-1008	30-35 resin; refrasil 184 weave reinforcing; specimen 12 x 8 x 0.125 in.; edgewise layup with thickness in warp direction; average thickness 0.125 in.; manufactured by H.I. Thompson Fiber Glass Co.; cured at 366 K for 1 hr, at 394 K for 1 hr, and at 422 K for 2 hrs; 50 plies per in.; density 1.442 g cm^{-3}.
85	488, 489	L	1961	514-623		Astrolite; SRI 7-2			The above specimen measured at decreasing temp.
86	488, 489	L	1961	224-661		SRI 8-1	Plastic	CTL 37-9X resin	25.7 resin; 181 "E" glass fabric reinforcing; parallel layup; specimen 18 x 18 x 0.125 in.; average thickness 0.132 in.; manufactured by Cincinnati Testing and Research Lab.; molded with part pressure 100 psi, cured at 411-416 K for 0.5 hr, post-cured at 394 K for 4 days; 14 plies; density 1.866 g cm^{-3}.
87	488, 489	L	1961	515-661		SRI 8-1			The above specimen measured at decreasing temp.
88	488, 489	L	1961	220-645		SRI 8-2	Plastic	CTL 37-9X resin	31.272 resin; 181 "E" glass fabric reinforcing; specimen 7 x 4.75 x 0.7 in.; edgewire layup with thickness in warp direction; average thickness 0.699 in.; manufactured by Cincinnati Testing and Research Lab; molded with part pressure 100 psi, cured at 411-416 K for 4 hrs, post-cured at 394 K for 4 days; 91 plies per in.; density 1.778 g cm^{-3}.
89	488, 489	L	1961	465-645		SRI 8-2			The above specimen measured at decreasing temp.
90	488, 489	L	1961	223-645		SRI 8-3	Plastic	CTL 37-9X resin	24.312 resin; 181 "E" glass fabric reinforcing; specimen 10 x 10 x 0.7 in.; edgewise layup with thickness at 45° of warp direction; average thickness 0.708 in.; manufactured by Cincinnati Testing and Research Lab; augmented bag molded with part pressure 100 psi, cured at 411-416 K for 4 hrs; post-cured at 394 K for 4 days; density 1.789 g cm^{-3}.
91	488, 489	L	1961	475-645		SRI 8-3			The above specimen measured at decreasing temp.

SPECIFICATION TABLE NO. 334 (continued)

Curve. No.	Ref. No.	Method Used	Year	Temp. Range, K	Reported Error, %	Name and Specimen Designation	Base Material	Binder	Specifications and Remarks
92	488, 489	L	1961	218-662		YN 25; SRI 9	Plastic	CTL-91 LD Phenolic resin	42 resin; heat set and scoured SN-19 nylon reinforcing; chopped fabric construction; specimen 6 in. dia x 0.25 in. thick; average thickness 0.265 in.; manufactured by U.S. Polymeric Chemicals, Inc; molded at 408 K for 1 hr; density 1.153 g cm^{-3}.
93	488, 489	L	1961	457-662		YN 25; SRI 9			The above specimen measured at decreasing temp.
94	488, 489	L	1961	219-656		SRI 10	Plastic	DC-2106 Silicon resin	40 RPD asbestos reinforcing; specimen 12 x 12 x 0.125 in.; average thickness 0.121 in.; manufactured by U.S. Polymeric Chemicals, Inc; fabricated by loading at 439 K, applying full pressure and bumping after 20 sec, cured at 439 K and 850 psi for 1.5 hrs, removed hot from press; 37 plies; density 1.820 g cm^{-3}.
95	488, 489	L	1961	444-656		SRI 10			The above specimen measured at decreasing temp.
96	488, 489	L	1961	234-581		SRI 11	Plastic	CTL 91-LD Phenolic resin	31 resin; 38-40 resin in prepreg. prior to molding; glass type YM-31A weave style 181 reinforcing; specimen 18 x 18 x 0.125 in.; average thickness 0.116 in.; manufactured by Wright Air Development Div, USAF; pressed at 345 psi; cured at 422 K for 20 min, post-cured at 422 K for 24 hrs, at 450 K for 24 hrs, and at 478 K for 24 hrs; density 2.098 g cm^{-3}.
97	488, 489	L	1961	461-581		SRI 11			The above specimen measured at decreasing temp.
98	488, 489	L	1961	227-588		SRI 12	Plastic	Epon 1031 Epoxy resin	22 resin; 38-40 resin in prepreg. prior to molding; glass type YM-31A weave style 181 reinforcing; specimen 18 x 18 x 0.125 in.; average thickness 0.110 in.; manufactured by Wright Air Development div, USAF; pressed at 245 psi, cured at 444 K for 30 min, post-cured at 422 K for 24 hrs; density 2.203 g cm^{-3}.
99	488, 489	L	1961	463-558		SRI 12			The above specimen measured at decreasing temp.
100	488, 489	L	1961	227-648		SRI 13	Plastic	R/M High heat resistant phenolic resin	25-30 resin; R/M style 42 RPD asbestos reinforcing; specimen 18 x 18 x 0.125 in.; average thickness 0.147 in.; manufactured by Raybestos-Manhatten, Inc; 40 plies; density 1.874 g cm^{-3}.

SPECIFICATION TABLE NO. 334 (continued)

Curve No.	Ref. No.	Method Used	Year	Temp. Range, K	Reported Error, %	Name and Specimen Designation	Base Material	Binder	Specifications and Remarks
101	488, 489	L	1961	481-648		SRI 13			The above specimen measured at decreasing temp.
102	485	L	1963	315-434		No. 1	Graphite	101 Phenolic Resin	6 sheets of graphite mat impregnated with 101 phenolic resin; specimen 7 x 7 x 0.141 in.; materials supplied by U.S. Polymeric Chemicals Co., Inc; pressed at 100 psi at 450 K for 1 hr; post-cured at 450 K for 16 hrs; density 1.064 g cm^{-3} at 60 C.
103	485	L	1963	312-409		No. 2	Graphite	101 Phenolic Resin	13 sheets of WC-001 graphite cloth impregnated with 101 phenolic resin; specimen 7 x 7 x 0.163 in.; pressed at 100 lb $in.^{-2}$ at 450 K for 1 hr, removed from the press, and post-cured for 16 hrs at 450 K; density (60 C) 1.275 g cm^{-3}.
104	485	L	1963	330-428		No. 3	Glass	37-9X Phenyl silane resin	7 sheets of 184 Volan (glass cloth with a Volan finish) impregnated with 37-9X phenyl silane resin; specimen 7 x 7 x 0.168 in.; pressed at 100 lb $in.^{-2}$ at 450 K for 1 hr, removed from the press, and post-cured for 16 hrs at 450 K; density (60 C) 1.938 g cm^{-3}.
105	485	L	1963	336-432		No. 5	Glass	37-9X Phenyl silane resin	Similar to the above specimen except specimen thickness 0.170 in.
106	485	L	1963	337-407		No. 8	Refrasil	91 LD Phenolic Resin	Refrasil cloth impregnated with 91 LD phenolic resin; specimen 7 x 7 x 0.170 in.; pressed at 100 lb $in.^{-2}$ at 450 K for 1 hr, removed from the press, and post-cured for 16 hrs at 450 K; density (60 C) 1.554 g cm^{-3}.
107	588	C	1956	302-697		G-7	Glass	Silicone resin	Glass cloth impregnated with silicone resin; 6 in. square plate specimen; supplied by Cadillac Plastic Co.; foam glass used as comparative material.
108	588	C	1956	301-728		Armalon 410L	Glass	Teflon resin	Glass cloth impregnated with teflon resin; 6 in. square plate specimen; supplied by E.I. DuPont deNemours and Co.; same comparative material as above.
109	588	C	1956	318-775		DC 301	Glass	Silicone resin	Glass fiber molding compound impregnated with silicone resin; 6 in. square plate specimen; supplied by Dow Corning Corp; same comparative material as above.

DATA TABLE NO. 334 THERMAL CONDUCTIVITY OF LAMINATES (NONMETALLIC)

[Temperature, T, K; Thermal Conductivity, k, Watt $cm^{-1}K^{-1}$]

T	k
CURVE 1	
218.1	0.00143
219.7	0.00147
220.7	0.00131
313.2	0.00209
320.3	0.00143
345.8	0.00166
377.4	0.00177
409.8	0.00195
423.3	0.00179
485.4	0.00218
533.1	0.00163
534.8	0.00189
CURVE 2	
358.4	0.00157
413.9	0.00169
480.2	0.00186
529.9	0.00196
565.6	0.00219
CURVE 3	
330.9	0.00138
390.4	0.00156
463.2	0.00180
523.0	0.00211
CURVE 4	
222.1	0.00102
223.7	0.000959
273.2	0.00133
334.8	0.00154
395.4	0.00167
450.9	0.00179
507.1	0.00186
563.2	0.00193
564.3	0.00193*
608.7	0.00202
CURVE 5	
229.3	0.00147
284.5	0.00160
347.1	0.00205
347.6	0.00203*
364.8	0.00211
395.9	0.00226
398.6	0.00228*
443.7	0.00244
502.1	0.00251
565.4	0.00208
CURVE 6	
224.8	0.000829
272.2	0.00114
338.7	0.00130
367.1	0.00140
391.5	0.00162*
393.7	0.00160*
430.4	0.00160
495.9	0.00177
552.6	0.00195*
610.9	0.00205*
CURVE 7	
227.1	0.00150*
258.7	0.00151*
283.2	0.00169
320.4	0.00182
339.3	0.00218
340.4	0.00219*
343.2	0.00203*
346.5	0.00202*
346.5	0.00206*
361.5	0.00203
367.1	0.00225
374.8	0.00229
426.5	0.00224
426.5	0.00224*
428.2	0.00244
429.3	0.00237*
467.1	0.00242
519.8	0.00254
CURVE 7 (cont.)	
522.1	0.00251*
555.4	0.00234
559.8	0.00231*
CURVE 8	
226.5	0.000937
284.3	0.00137
346.5	0.00170
389.2	0.00195
441.5	0.00219
516.5	0.00228
569.3	0.00209*
CURVE 9	
230.4	0.00112
293.7	0.00159
378.2	0.00239
455.4	0.00267
499.3	0.00271
547.1	0.00274
CURVE 10	
124	0.000209
127	0.000231
138	0.000231
143	0.000245
155	0.000260
CURVE 11	
115	0.000252
118	0.000245
134	0.000275
138	0.000274
155	0.000301
160	0.000296
CURVE 12	
294.3	0.00243
366.5	0.00290
CURVE 12 (cont.)	
422.1	0.00348
477.6	0.00363
533.2	0.00362
CURVE 13	
338.7	0.00206*
366.5	0.00251
394.3	0.00263
422.1	0.00263
449.8	0.00269
CURVE 14	
336.1	0.000737
389.3	0.000981
460.6	0.00127
526.8	0.00151
CURVE 15	
334.3	0.00119
385.6	0.00138
453.2	0.00142
518.5	0.00162
CURVE 16	
325.2	0.000952
353.7	0.000979
388.9	0.00110
454.4	0.00131
507.0	0.00151
585.3	0.00167
CURVE 17	
323.7	0.00102
351.0	0.00111
385.0	0.00124
448.2	0.00140
499.3	0.00165
574.8	0.00174
CURVE 18*	
373.2	0.00230
CURVE 19*	
373.2	0.0148
CURVE 20*	
373.2	0.0148
CURVE 21*	
373.2	0.00330
CURVE 22	
373.2	0.00674
CURVE 23*	
373.2	0.00674
CURVE 24*	
373.2	0.00313
CURVE 25*	
373.2	0.0138
CURVE 26	
373.2	0.00395
CURVE 27	
360.5	0.00112
385.3	0.00120
408.9	0.00124
469.9	0.00136
489.6	0.00143
519.4	0.00151
554.5	0.00127
CURVE 28*	
331.5	0.00139
376.6	0.00153
419.8	0.00163
465.4	0.00175
519.1	0.00184
552.1	0.00151
CURVE 29	
326.8	0.000646
391.1	0.000723
436.3	0.000875
501.8	0.000961
550.0	0.000912
CURVE 30*	
327.7	0.00119
367.0	0.00128
407.7	0.00134
467.9	0.00140
531.5	0.00153
559.5	0.00119
CURVE 31*	
362.0	0.00126
401.7	0.00140
439.0	0.00153
491.7	0.00164
564.4	0.00152
CURVE 32*	
336.2	0.00153
392.2	0.00157
438.7	0.00150
497.6	0.00149
583.8	0.00147
CURVE 33*	
365.8	0.00114
417.2	0.00139
CURVE 33 (cont.)*	
450.4	0.00145
482.9	0.00156
551.9	0.00162
CURVE 34*	
314.2	0.00111
359.2	0.00129
390.3	0.00140
430.1	0.00151
456.3	0.00162
475.5	0.00152
507.2	0.00166
CURVE 35*	
361.1	0.00155
393.9	0.00170
475.9	0.00170
521.5	0.00164
613.0	0.00164
CURVE 36*	
329.9	0.00109
368.3	0.00131
426.8	0.00146
510.1	0.00153
557.1	0.00140
CURVE 37*	
374.8	0.00175
445.2	0.00182
507.5	0.00187
553.1	0.00119
CURVE 38*	
331.0	0.00126
376.4	0.00146
426.2	0.00155
469.8	0.00160
509.3	0.00155
523.2	0.00137

* Not shown on plot

DATA TABLE NO. 334 (continued)

CURVE 39*

T	k
344.2	0.00137
384.6	0.00158
424.5	0.00172
474.2	0.00182
533.3	0.00196
555.5	0.00204

CURVE 40*

T	k
335.2	0.00170
378.3	0.00182
424.5	0.00196
458.4	0.00203
501.7	0.00207
540.4	0.00167
550.6	0.00139

CURVE 41*

T	k
339.6	0.00130
388.4	0.00151
438.9	0.00161
499.6	0.00163
536.7	0.00133

CURVE 42*

T	k
319.5	0.00161
353.0	0.00166
399.8	0.00179
458.0	0.00193
490.0	0.00206
541.6	0.00216
575.0	0.00202

CURVE 43

T	k
297.1	0.00287
366.5	0.00287

CURVE 44

T	k
297.1	0.00268
366.5	0.00268

CURVE 45

T	k
297.1	0.00348
366.5	0.00348

CURVE 46

T	k
297.1	0.00320
366.5	0.00320

CURVE 47*

T	k
297.1	0.00312
366.5	0.00312

CURVE 48

T	k
297.1	0.00486
366.5	0.00486

CURVE 49

T	k
297.1	0.00507
366.5	0.00507

CURVE 50

T	k
297.1	0.00339
366.5	0.00339

CURVE 51

T	k
297.1	0.00427
366.5	0.00427

CURVE 52

T	k
297.1	0.00279
366.5	0.00279

CURVE 53*

T	k
297.1	0.00294
366.5	0.00294

CURVE 54*

T	k
297.1	0.00289
366.5	0.00289

CURVE 55

T	k
297.1	0.00260
366.5	0.00260

CURVE 56*

T	k
297.1	0.00322
366.5	0.00322

CURVE 57*

T	k
297.1	0.00325
366.5	0.00325

CURVE 58*

T	k
297.1	0.00280
366.5	0.00280

CURVE 59*

T	k
297.1	0.00495
366.5	0.00495

CURVE 60*

T	k
297.1	0.00479
366.5	0.00479

CURVE 61*

T	k
297.1	0.00348
366.5	0.00348

CURVE 62

T	k
297.1	0.00447
366.5	0.00447

CURVE 63*

T	k
297.1	0.00272
366.5	0.00272

CURVE 64*

T	k
297.1	0.00292
366.5	0.00292

CURVE 65
T = 308.2

Density ($g\ cm^{-3}$)	k
1.70	0.0303
1.70	0.0335
1.79	0.0273
1.84	0.0296*

CURVE 66

T	k
308.2	0.0374

CURVE 67*

T	k
308.2	0.0342

CURVE 68

T	k
308.2	0.0396

CURVE 69

T	k
308.2	0.0472

CURVE 70

T	k
308.2	0.0183

CURVE 71

T	k
1623	0.0640

CURVE 72

T	k
1623	0.0310

CURVE 73

T	k
1623	0.131

CURVE 74

T	k
1623	0.0841

CURVE 75

T	k
1623	0.0531

CURVE 76*

T	k
1623	0.0770

CURVE 77*

T	k
1623	0.0782

CURVE 78*

T	k
1623	0.0736

CURVE 79*

T	k
224.8	0.00103
307.6	0.00131
364.3	0.00160
411.5	0.00179
453.7	0.00193
497.6	0.00173
548.2	0.00143

CURVE 80*

T	k
230.4	0.00103
315.4	0.00126
321.5	0.00149
364.8	0.00158
374.3	0.00163
422.1	0.00196
502.6	0.00194
548.2	0.00184

CURVE 81*

T	k
442.6	0.00172
548.2	0.00184

CURVE 82*

T	k
221.5	0.00103
317.1	0.00154
362.6	0.00171
429.3	0.00185
503.7	0.00183
660.9	0.00207

CURVE 83*

T	k
513.7	0.00180
619.8	0.00197
660.9	0.00207

CURVE 84*

T	k
219.3	0.00105
313.7	0.00167
368.7	0.00196
426.5	0.00199
477.6	0.00218
544.8	0.00238
622.6	0.00225

CURVE 85*

T	k
513.7	0.00199
602.6	0.00211
622.6	0.00225

CURVE 86*

T	k
223.7	0.00104
312.6	0.00156
357.1	0.00174
431.5	0.00196
505.4	0.00210
580.9	0.00223
660.9	0.00216

CURVE 87*

T	k
515.4	0.00187
596.5	0.00201
660.9	0.00216

CURVE 88*

T	k
219.8	0.00255
312.1	0.00311
374.8	0.00389
430.9	0.00443
510.4	0.00496
563.2	0.00499
644.8	0.00510

CURVE 89*

T	k
464.8	0.00435
567.1	0.00475
644.8	0.00510

CURVE 90*

T	k
223.2	0.00210
319.8	0.00235
379.3	0.00266
438.7	0.00324
509.8	0.00400
573.2	0.00414
645.4	0.00434

CURVE 91*

T	k
475.4	0.00369
559.3	0.00400
645.4	0.00434

CURVE 92*

T	k
217.6	0.00115
311.5	0.00187
367.1	0.00204
420.9	0.00196
505.9	0.00188
564.3	0.00189
662.1	0.00167

CURVE 93*

T	k
456.5	0.00932
565.9	0.00114
662.1	0.00167

CURVE 94*

T	k
219.3	0.00108
312.1	0.00190
362.6	0.00207
420.9	0.00222
498.7	0.00223
571.5	0.00237
655.9	0.00229

CURVE 95*

T	k
444.3	0.00199
559.3	0.00213
655.9	0.00229

CURVE 96*

T	k
223.7	0.00119
313.2	0.00196
378.7	0.00223
435.9	0.00244
524.8	0.00275
580.9	0.00257

CURVE 97*

T	k
460.9	0.00231
525.9	0.00237
580.9	0.00257

CURVE 98*

T	k
227.0	0.00142
312.6	0.00209
383.7	0.00235
448.2	0.00263
538.7	0.00250
588.2	0.00219

CURVE 99*

T	k
463.2	0.00193
521.5	0.00210
558.2	0.00219

CURVE 100*

T	k
227.1	0.00123
319.3	0.00195
376.5	0.00219
447.1	0.00239
520.4	0.00271
593.2	0.00285
647.6	0.00281

* Not shown on plot

DATA TABLE NO. 334 (continued)

T	k
CURVE 101*	
480.9	0.00249
558.7	0.00260
647.6	0.00281
CURVE 102*	
315.4	0.00179
332.6	0.00172
367.6	0.00180
380.4	0.00208
394.3	0.00229
411.5	0.00262
418.2	0.00238
434.	0.00274
CURVE 103*	
311.5	0.00469
324.	0.00459
345.4	0.00427
357.6	0.00394
382.6	0.00430
408.7	0.00415
CURVE 104*	
329.8	0.000952
340.4	0.00192
344.8	0.00190
358.7	0.00157
385.9	0.00146
397.1	0.00163
420.9	0.00164
427.6	0.00185
CURVE 105*	
335.9	0.00128
358.7	0.00150
377.6	0.00141
404.3	0.00146
431.5	0.00153

T	k
CURVE 106*	
337.1	0.00185
360.4	0.00193
384.3	0.00173
389.8	0.00176
399.8	0.00169
407.1	0.00185
CURVE 107*	
302.0	0.00120
302.5	0.00144
315.9	0.00132
316.3	0.00133
340.0	0.00140
340.0	0.00139
364.2	0.00151
364.3	0.00155
393.6	0.00155
394.4	0.00156
424.8	0.00160
424.8	0.00161
458.9	0.00164
464.8	0.00164
533.6	0.00170
533.6	0.00171
566.1	0.00177
568.1	0.00179
606.5	0.00184
606.8	0.00187
640.9	0.00199
696.8	0.00197
CURVE 108*	
300.9	0.00128
306.2	0.00126
329.6	0.00153
329.6	0.00152
367.7	0.00155
368.0	0.00156
410.2	0.00156
411.8	0.00156
459.2	0.00160
459.8	0.00160
508.8	0.00161
509.7	0.00159

T	k
CURVE 108 (cont.)*	
557.9	0.00156
558.7	0.00157
610.5	0.00150
611.5	0.00151
657.7	0.00151
661.1	0.00122
661.2	0.00142
662.3	0.00119
723.9	0.00132
728.1	0.00126
CURVE 109*	
317.6	0.00186
317.9	0.00212
362.3	0.00219
362.9	0.00218
429.8	0.00234
429.9	0.00229
494.1	0.00247
494.4	0.00243
563.6	0.00251
564.7	0.00253
647.5	0.00268
647.8	0.00266
726.7	0.00309
730.2	0.00310
769.6	0.00363
774.9	0.00363

* Not shown on plot

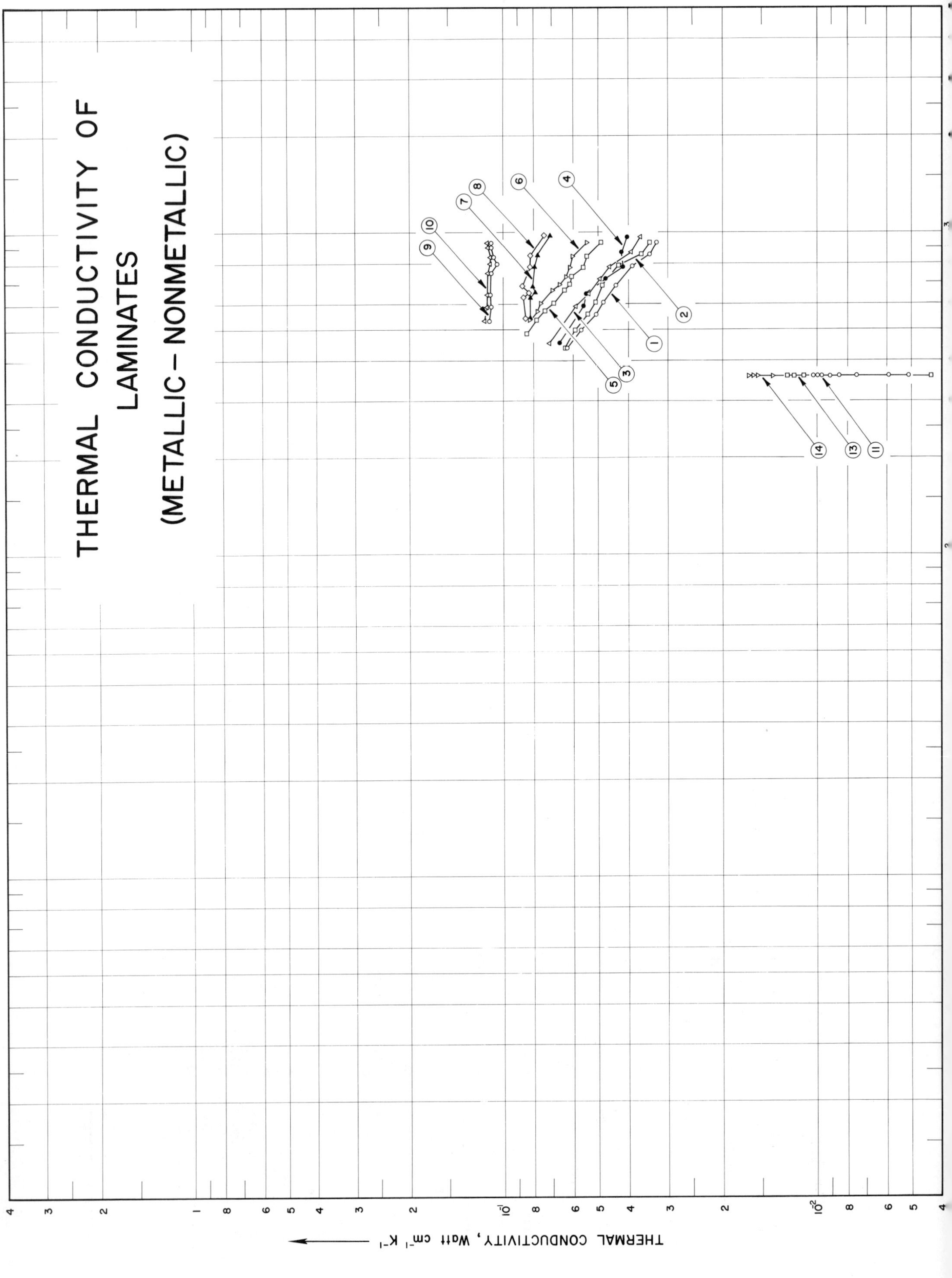

THERMAL CONDUCTIVITY OF LAMINATES (METALLIC – NONMETALLIC)

SPECIFICATION TABLE NO. 335 THERMAL CONDUCTIVITY OF LAMINATES (METALLIC-NONMETALLIC)

[For Data Reported in Figure and Table No. 335]

Curve No.	Ref. No.	Method Used	Year	Temp. Range, K	Reported Error, %	Name and Specimen Designation	Composition: Composite Materials		Composition: Binder	Composition (continued), Specifications and Remarks
1	65	C	1958	438-933	±2		Forsterite	Steel	Cu-Ag alloy	85% forsterite, 15% stainless steel (430); the ceramic and the metal brazed together by braze alloy (72 Cu, 18 Ag, , M. P. 780 C); measured perpendicular to lamination using alumina (97. 6 Al_2O_3, designated as Body Al -300) from Western Gold and Platinum Co. as standard.
2	65	C	1958	438-933	±2		Forsterite	Steel	Cu-Ag alloy	Data of the above specimen obtained by comparing to another alumina standard.
3	65	C	1958	454-966	±2		Forsterite	Steel	Cu-Ag alloy	67% forsterite, 33% stainless steel (430); similarly prepared and tested as above.
4	65	C	1958	454-966	±2		Forsterite	Steel	Cu-Ag alloy	Data of the above specimen obtained by comparing to another alumina standard.
5	65	C	1958	485-933	±2		Forsterite	Steel	Cu-Ag alloy	50% forsterite, 50% stainless steel (430); similarly prepared and tested as above.
6	65	C	1958	485-933	±2		Forsterite	Steel	Cu-Ag alloy	Data of the above specimen obtained by comparing to another alumina standard.
7	65	C	1958	543-983	±2		Forsterite	Steel	Cu-Ag alloy	85% forsterite, 15% stainless steel (430); thermal conductivity parallel to lamination; temp gradient from thermocouples inserted below the ceramic.
8	65	C	1958	543-983	±2		Forsterite	Steel	Cu-Ag alloy	Data of the above specimen from thermocouples below the metal.
9	65	C	1958	533-926	±2		Forsterite	Steel	Cu-Ag alloy	67% forsterite, 33% stainless steel (430); thermal conductivity parallel to lamination; temp gradient from thermocouples inserted below the ceramic.
10	65	C	1958	533-926	±2		Forsterite	Steel	Cu-Ag alloy	Data of the above specimen from thermocouples below the metal.
11	490	L	1920	358. 2			Steel	Air		30 varnished silicon steel foils each of thickness 0. 014 in.; density 7. 36 g cm^{-3}; measured under pressures in the range 0 to 132 psi.
12	490	L	1920	358. 2			Steel	Air		30 oxide coated silicon steel foils each of thickness 0. 014 in.; density 7. 54 g cm^{-3}; measured under pressures in the range 0 to 136 psi.
13	490	L	1920	358. 2			Steel	Air		30 varnished silicon steel foils each of thickness 0. 0172 in.; density 7. 51 g cm^{-3}; measured under pressures in the range 0 to 128 psi.

SPECIFICATION TABLE NO. 335 (continued)

Curve No.	Ref. No.	Method Used	Year	Temp. Range, K	Reported Error, %	Name and Specimen Designation	Composition			Composition (continued), Specifications and Remarks
							Composite Materials		Binder	
14	490	L	1920	358.2			Steel	Air		30 silicon steel foils each of thickness 0.0172 in.; density 7.79 g cm^{-3}; measured under pressures in the range 0 to 125 psi.

DATA TABLE NO. 335 THERMAL CONDUCTIVITY OF LAMINATES (METALLIC-NONMETALLIC)

[Temperature, T, K; Thermal Conductivity, k, Watt $cm^{-1}K^{-1}$]

T	k
CURVE 1	
438.2	0.0628
499.2	0.0569
559.2	0.0506
608.2	0.0481
685.2	0.0439
789.2	0.0389
855.2	0.0343
933.2	0.0326
CURVE 2	
438.2	0.0636
499.2	0.0594
559.2	0.0540
608.2	0.0510
685.2	0.0485
789.2	0.0431
855.2	0.0364
933.2	0.0343
CURVE 3	
454.2	0.0715
591.2	0.0590
645.2	0.0536
716.2	0.0494
783.2	0.0460
868.2	0.0393
966.2	0.0368
CURVE 4	
454.2	0.0661
591.2	0.0556
645.2	0.0544
716.2	0.0473
783.2	0.0418
868.2	0.0423
966.2	0.0402
CURVE 5	
485.2	0.0841
535.2	0.0782

T	k
CURVE 5 (cont.)	
570.2	0.0736
602.2	0.0690
663.2	0.0640
687.2	0.0615
729.2	0.0607
776.2	0.0556
843.2	0.0544
933.2	0.0490
CURVE 6	
485.2	0.0841*
535.2	0.0824
570.2	0.0778
602.2	0.0757
663.2	0.0695
687.2	0.0661
729.2	0.0632
776.2	0.0615
843.2	0.0604
933.2	0.0544
CURVE 7	
543.2	0.0820
629.2	0.0820
653.2	0.0791
681.2	0.0803
783.2	0.0795
848.2	0.0778
983.2	0.0711
CURVE 8	
543.2	0.0849
629.2	0.0862
653.2	0.0828
681.2	0.0870
783.2	0.0824
848.2	0.0820
983.2	0.0741

T	k
CURVE 9	
533.2	0.111
589.2	0.109
641.2	0.110
746.2	0.109
796.2	0.105
838.2	0.107
898.2	0.109
926.2	0.109
CURVE 10	
533.2	0.115
589.2	0.113
641.2	0.113
746.2	0.113
796.2	0.111
838.2	0.110
898.2	0.112
926.2	0.113

p(psi)	k
CURVE 11 (T = 358.2K)	
0	0.00512
20	0.00748
40	0.00846
60	0.00906
80	0.00925*
100	0.00965
120	0.00992
132	0.0102
120	0.0100*
100	0.0996*
80	0.00984*
60	0.00945*
40	0.00906*
20	0.00846*
0	0.00591

p(psi)	k
CURVE 12* (T = 358.2K)	
0	0.003937
20	0.00591
40	0.00689
60	0.00748
80	0.00827
100	0.00866
120	0.00906
136	0.00945
120	0.00925
100	0.00906
80	0.00886
60	0.00866
40	0.00787
20	0.00709
0	0.00512
CURVE 13 (T = 358.2K)	
0	0.00433
20	0.00807*
40	0.00965*
60	0.0104*
80	0.0110
100	0.0118
120	0.0124
128	0.0126*
120	0.0126*
100	0.0122*
80	0.0118*
60	0.0114*
40	0.0110*
20	0.00984*
0	0.00630*

p(psi)	k
CURVE 14 (T = 358.2K)	
0	0.00496*
10	0.00748*
22.5	0.00945*

p(psi)	k
CURVE 14 (cont.)	
125	0.0165
100	0.0159
80	0.0154
47	0.0138
20	0.0114*
0	0.00709*

*Not shown on plot

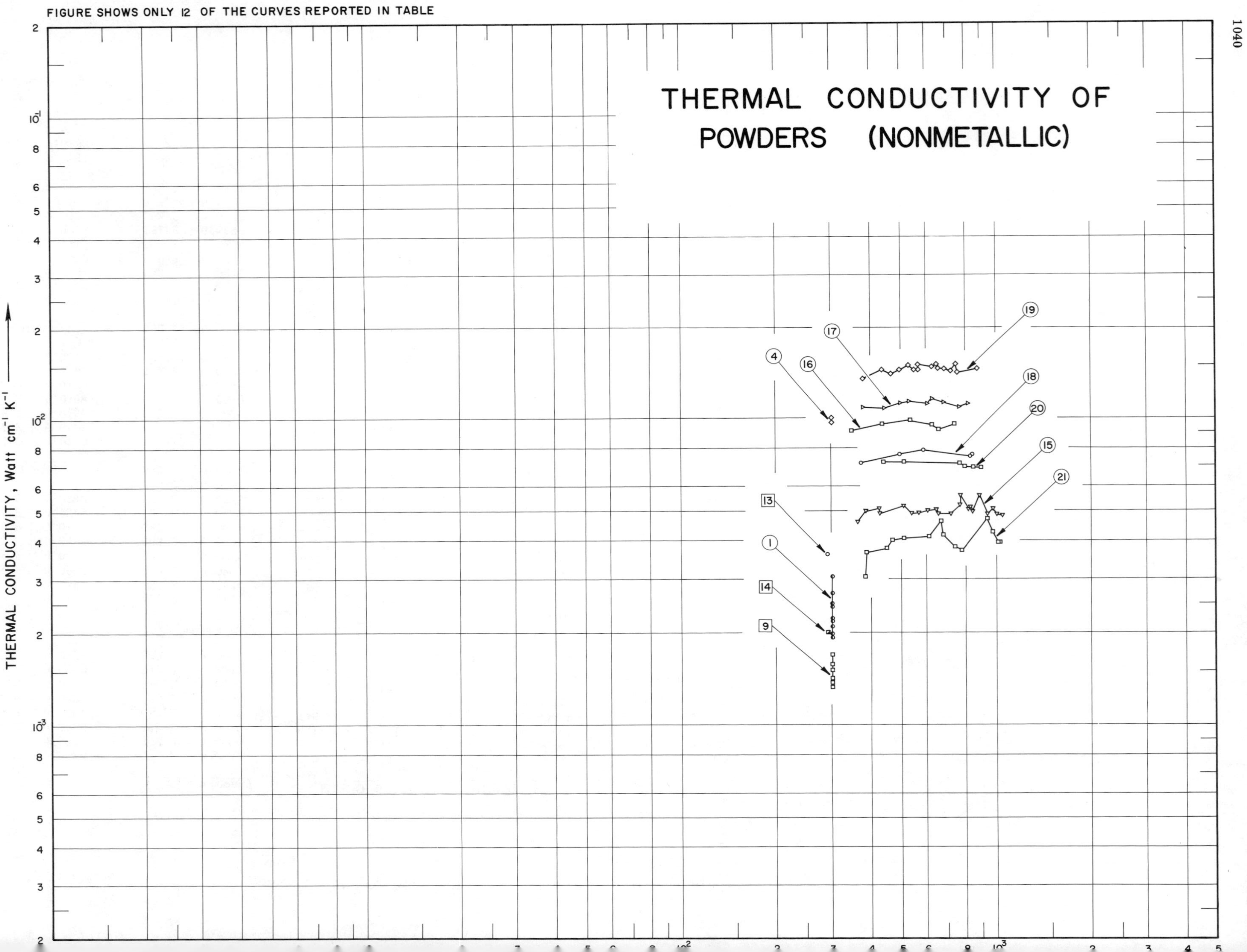

FIGURE SHOWS ONLY 12 OF THE CURVES REPORTED IN TABLE
THERMAL CONDUCTIVITY OF POWDERS (NONMETALLIC)
THERMAL CONDUCTIVITY, Watt cm^{-1} K^{-1}
19
17
4
16
18
20
15
21
13
1
14
9
2
10^{-1}
8
6
5
4
3
2
10^{-2}
8
6
5
4
3
2
10^{-3}
8
6
5
4
3
2
10^{2}
2
3
4
5
6
8
10^{3}
2
3
4
5

SPECIFICATION TABLE NO. 336 THERMAL CONDUCTIVITY OF POWDERS (NONMETALLIC)

[For Data Reported in Figure and Table No. 336]

Curve No.	Ref. No.	Method Used	Year	Temp. Range, K	Reported Error, %	Name and Specimen Designation	Composition: Solid Particles	Composition: Environment	Composition (continued), Specifications and Remarks
1	471	R	1951	303.2			Aluminum Oxide	Carbon Dioxide	Alumina grains supplied by Norton Co. contained in a cylindrical cell of 2.210 cm O.D., 0.9476 cm I.D., and total volume 110.06 cm^3; carbon dioxide filled the cell; alumina grain size 1.088 mm; measured as CO_2 pressure varied from 1.4 to 63.3 atm.
2	471	R	1951	303.2			Aluminum Oxide	Nitrogen	Similar to the above specimen except nitrogen filled the cell; measured as N_2 pressure varied from 1.5 to 84.1 atm.
3	471	R	1951	303.2			Aluminum Oxide	Helium	Similar to the above specimen except helium filled the cell; measured as He_2 pressure varied from 2.5 to 83.6 atm.
4	471	R	1951	303.2			Aluminum Oxide	Hydrogen	Similar to the above specimen except hydrogen filled the cell; measured as H_2 pressure varied from 3.8 to 83.0 atm.
5	471	R	1951	303.2			Aluminum Oxide	Carbon Dioxide	Similar to the above specimen except alumina grain size 0.4797 mm and carbon dioxide filled the cell; measured as CO_2 pressure varied from 1.4 to 64.5 atm.
6	471	R	1951	303.2			Aluminum Oxide	Nitrogen	Similar to the above specimen except nitrogen filled the cell; measured as N_2 pressure varied from 1.6 to 83.7 atm.
7	471	R	1951	303.2			Aluminum Oxide	Helium	Similar to the above specimen except helium filled the cell; measured as He_2 pressure varied from 3.2 to 82.8 atm.
8	471	R	1951	303.2			Aluminum Oxide	Hydrogen	Similar to the above specimen except hydrogen filled the cell; measured as H_2 pressure varied from 1.0 to 84.3 atm.
9	471	R	1951	303.2			Borosilicate	Carbon Dioxide	Borosilicate glass grains contained in a cylindrical cell of 2.210 cm O.D., 0.9476 cm I.D., and total volume 110.06 cm^3; carbon dioxide filled the cell; average glass grain size 0.5404 mm; measured as CO_2 pressure varied from 1.3 to 62.5 atm.
10	471	R	1951	303.2			Borosilicate	Nitrogen	Similar to the above specimen except nitrogen filled the cell; measured as N_2 pressure varied from 1.2 to 80.0 atm.
11	471	R	1951	303.2			Borosilicate	Helium	Similar to the above specimen except helium filled the cell; measured as He_2 pressure varied from 1.0 to 80.2 atm.
12	471	R	1951	303.2			Borosilicate	Hydrogen	Similar to the above specimen except hydrogen filled the cell; measured as H_2 pressure varied from 1.1 to 83.8 atm.

SPECIFICATION TABLE NO. 336 (continued)

Curve No.	Ref. No.	Method Used	Year	Temp. Range, K	Reported Error, %	Name and Specimen Designation	Composition: Solid Particles	Composition: Environment	Composition (continued), Specifications and Remarks
13	491	R	1958	292.7			Charcoal	Helium	Sutcliffe Speakman 208 C. 8-14 mesh charcoal obtained from I. C. I. contained in a cylindrical cell of I. D. 1.5 cm, O. D. 3.9 cm, and length 5 cm; measured in helium.
14	491	R	1958	293.3			Charcoal	Helium	Similar to the above specimen but measured in air.
15	492	R	1954	365-1058			Uranium Dioxide	Argon	Cylindrical cell of 0.375 in. I. D., 1.75 in. O. D., and 16.5 in. long filled with UO_2 powder, -100 + 150 mesh size 59% by weight; -325 mesh size 41% by weight packed to 58% of theoretical density of UO_2, void volume of 42% filled with argon gas; UO_2 powder supplied by KAPL, General Electric Co.
16	492	R	1954	349-631			Uranium Dioxide	Argon & Helium	Similar to the above specimen but the 42% void volume filled with 51.2 vol % argon and 48.8 vol % helium.
17	492	R	1954	380-819			Uranium Dioxide	Argon & Helium	Similar to the above specimen but the 42% void volume filled with 35 vol % argon and 65 vol % helium.
18	492	R	1954	375-845			Uranium Dioxide	Argon & Helium	Similar to the above specimen but the 42% void volume filled with 75 vol % argon and 25 vol % helium.
19	492	R	1954	381-880			Uranium Dioxide	Helium	Similar to the above specimen but the 42% void volume filled with helium.
20	492	R	1954	444-900			Uranium Dioxide	Nitrogen	Similar to the above specimen but the 42% void volume filled with nitrogen.
21	492	R	1954	386-1033			Uranium Dioxide	Xenon & Krypton	Similar to the above specimen but the 42% void volume filled with mixture of xenon and krypton with Xe to Kr ratio = 4.898.

DATA TABLE NO. 336 THERMAL CONDUCTIVITY OF POWDERS (NONMETALLIC)

[Temperature, T, K; Thermal Conductivity, k, Watt $cm^{-1}K^{-1}$]

p(atm)	k
CURVE 1 (T = 303.2 K)	
1.4	0.001928
4.6	0.001974
11.1	0.002008*
14.7	0.002016*
20.5	0.002038*
27.9	0.002096
38.0	0.002180
42.4	0.002230
49.5	0.002418
52.6	0.002494
57.4	0.002690
63.3	0.003059
CURVE 2* (T = 303.2 K)	
1.5	0.002690
20.9	0.002761
32.1	0.002807
41.7	0.002837
46.7	0.002841
47.5	0.002849
66.2	0.002916
80.9	0.002958
84.1	0.002971
CURVE 3* (T = 303.2 K)	
2.5	0.00831
3.8	0.00839
5.4	0.00855
7.0	0.00864
11.0	0.00866
20.4	0.00867
23.4	0.00868
27.1	0.00869
35.3	0.00870
41.4	0.00872
49.2	0.00874
62.0	0.00875
71.8	0.00878
81.7	0.00880
83.6	0.00876

p(atm)	k
CURVE 4 (T = 303.2 K)	
3.8	0.00969
5.9	0.00975*
7.5	0.00980*
21.0	0.00980*
40.0	0.00992*
41.5	0.00987*
51.6	0.00996*
67.0	0.00998*
80.7	0.01003*
83.0	0.01004
CURVE 5* (T = 303.2 K)	
1.4	0.001757
4.8	0.001799
8.3	0.001812
14.8	0.001874
21.7	0.001954
30.0	0.001992
36.3	0.002054
38.7	0.002096
45.7	0.002209
52.4	0.002356
56.3	0.002464
60.8	0.002644
64.5	0.002816
CURVE 6* (T = 303.2 K)	
1.6	0.002548
3.2	0.002615
6.6	0.002628
9.3	0.002661
12.1	0.002678
15.2	0.002686
21.5	0.002707
28.9	0.002715
31.7	0.002724
43.0	0.002766
48.7	0.002795
56.4	0.002820
65.3	0.002841

p(atm)	k
CURVE 6 (cont.)*	
76.4	0.002908
83.7	0.002916
CURVE 7* (T = 303.2 K)	
3.2	0.00849
8.3	0.00869
18.0	0.00874
20.0	0.00875
33.5	0.00873
41.7	0.00878
48.3	0.00879
55.4	0.00882
66.9	0.00884
82.8	0.00886
CURVE 8* (T = 303.2 K)	
1.0	0.00916
7.3	0.00967
11.3	0.00967
17.5	0.00970
27.1	0.00972
47.9	0.00975
56.6	0.00982
65.6	0.00982
69.2	0.00982
84.3	0.00992
CURVE 9 (T = 303.2 K)	
1.3	0.001331
6.7	0.001351*
15.1	0.001372
22.6	0.001389*
28.6	0.001418
35.5	0.001448*
42.2	0.001498
47.6	0.001561
51.4	0.001611*
55.0	0.001695
62.5	0.001996*

p(atm)	k
CURVE 10* (T = 303.2 K)	
1.2	0.001849
10.5	0.001858
17.6	0.001866
29.4	0.001883
59.0	0.001925
80.0	0.001958
CURVE 11* (T = 303.2 K)	
1.0	0.003908
9.7	0.003996
21.5	0.003996
35.0	0.004004
45.6	0.004029
54.4	0.004012
63.3	0.004029
70.4	0.004058
80.2	0.004050
CURVE 12* (T = 303.2 K)	
1.1	0.004217
9.4	0.004310
15.0	0.004310
29.3	0.004293
32.0	0.004305
44.8	0.004322
47.0	0.004310
62.2	0.004326
67.5	0.004314
83.8	0.004335
T	**k**
CURVE 13	
292.7	0.0036
CURVE 14	
293.3	0.0020

T	k
CURVE 15	
365.4	0.00460
387.1	0.00497
428.7	0.00509
430.4	0.00490
513.7	0.00517
544.3	0.00490
574.3	0.00493
610.9	0.00500
649.3	0.00504
652.6	0.00497*
659.8	0.00490
724.8	0.00488
770.4	0.00523
774.8	0.00559
819.3	0.00504
833.2	0.00498*
833.7	0.00511
845.4	0.00495
891.5	0.00557
941.5	0.00486
984.3	0.00507
1008.2	0.00486
1058.2	0.00483
CURVE 16	
349.3	0.00914
437.6	0.00954
537.6	0.00981
663.2	0.00917
741.5	0.00954
631.5	0.00947
CURVE 17	
380.4	0.0109
445.4	0.0108
497.6	0.0112
532.6	0.0114
609.3	0.0112
630.4	0.0116
685.4	0.0113
768.7	0.0109
818.7	0.0112

T	k
CURVE 18	
374.8	0.00715
496.5	0.00763
594.3	0.00789
834.3	0.00753
845.4	0.00763
CURVE 19	
380.9	0.0136
385.4	0.0135*
438.7	0.0145
468.2	0.0141
472.1	0.0140*
497.1	0.0145
533.2	0.0150
552.6	0.0145
572.1	0.0145
572.6	0.0151
578.7	0.0150*
632.1	0.0148
652.1	0.0151
664.8	0.0146
691.5	0.0146
713.2	0.0147*
723.7	0.0144
747.1	0.0151
760.9	0.0142
879.8	0.0146
CURVE 20	
443.7	0.00720
514.3	0.00720
769.3	0.00711
798.7	0.00694
847.6	0.00692
900.4	0.00691
CURVE 21	
385.9	0.00305
388.2	0.00365
450.4	0.00377
470.9	0.00400
513.2	0.00407

T	k
CURVE 21 (cont.)	
615.4	0.00412
674.8	0.00464
683.2	0.00417
742.6	0.00381
783.7	0.00369
943.2	0.00471
979.8	0.00426
1019.8	0.00395
1033.2	0.00395

* Not shown on plot

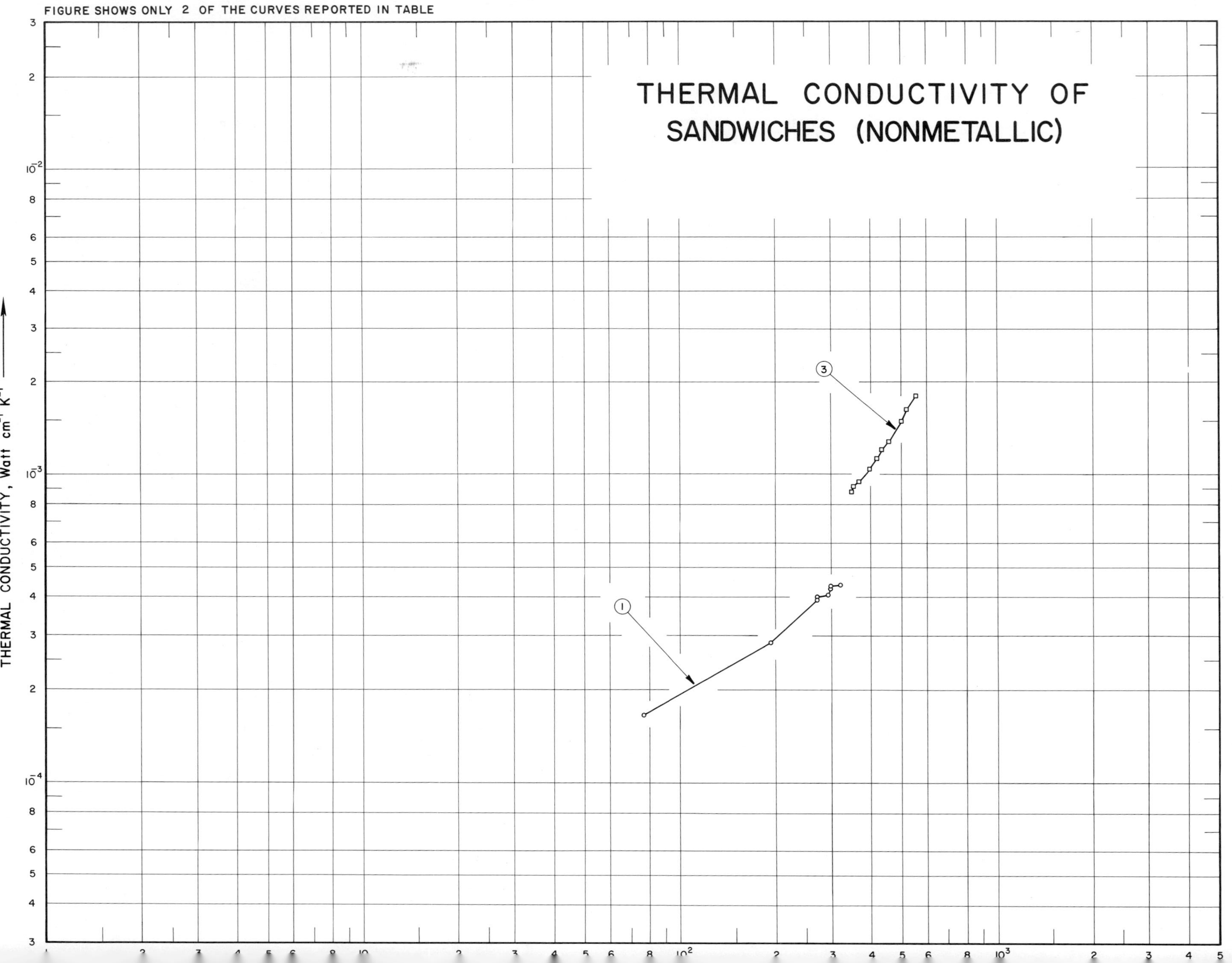
FIGURE SHOWS ONLY 2 OF THE CURVES REPORTED IN TABLE
THERMAL CONDUCTIVITY OF
SANDWICHES (NONMETALLIC)
THERMAL CONDUCTIVITY, Watt cm^{-1} K^{-1}
1
3

SPECIFICATION TABLE NO. 337 THERMAL CONDUCTIVITY OF SANDWICHES (NONMETALLIC)

[For Data Reported in Figure and Table No. 337]

Curve No.	Ref. No.	Method Used	Year	Temp. Range, K	Reported Error, %	Name and Specimen Designation	Composition: Composite Materials		Composition: Binder	Composition (continued), Specifications and Remarks
1	124	L	1960	77-322	< 2		Cork	Rubber		Specimen 0.249 in. in thickness; density 0.168 g cm^{-3}.
2	124	L	1960	297.6	< 2		Cork	Rubber		Specimen 0.141 in. in thickness; density 0.168 g cm^{-3}.
3	196	L	1955	349-559		SRI GP-2	Plastic	Glass	Polyester Resin	One foamed-in-place sandwich panel; 1/3 in. thick; fabricated using alkyd-isocyanate plastic core and laminate faces of 181 glass fabric and TAC-polyester resin.

DATA TABLE NO. 337 THERMAL CONDUCTIVITY OF SANDWICHES (NONMETALLIC)

[Temperature, T, K; Thermal Conductivity, k, Watt $cm^{-1}K^{-1}$]

T	k
CURVE 1	
77.1	0.000166
194.3	0.000284
273.2	0.000391
273.2	0.000398
294.3	0.000405
299.8	0.000424
300.4	0.000434
322.1	0.000438
CURVE 2*	
297.6	0.000421
CURVE 3	
349.2	0.000873
356.8	0.000906
369.9	0.000943
398.2	0.00104
423.3	0.00112
436.6	0.00120
457.8	0.00127
461.3	0.00127*
502.9	0.00149
524.1	0.00162
558.9	0.00180

* Not shown on plot

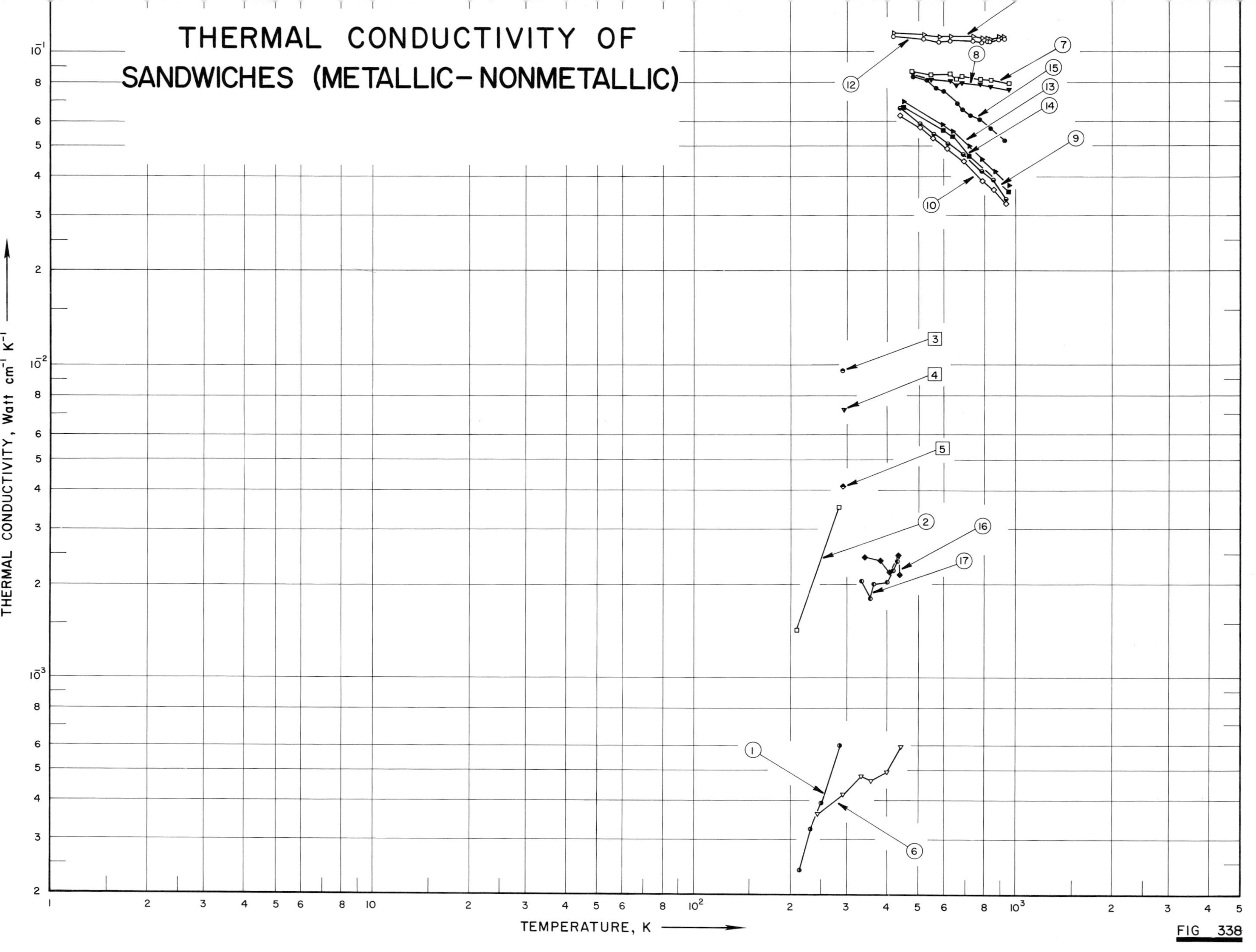
THERMAL CONDUCTIVITY OF
SANDWICHES (METALLIC-NONMETALLIC)
THERMAL CONDUCTIVITY, Watt cm^{-1} K^{-1}
TEMPERATURE, K
FIG 338
1
2
3
4
5
6
7
8
9
10
12
13
14
15
16
17

SPECIFICATION TABLE NO. 338 THERMAL CONDUCTIVITY OF SANDWICHES (METALLIC-NONMETALLIC)

[For Data Reported in Figure and Table No. 338]

Curve No.	Ref. No.	Method Used	Year	Temp. Range, K	Reported Error, %	Name and Specimen Designation	Composition: Composite Materials		Composition: Binder	Composition (continued), Specifications and Remarks
1	493	R	1949	215-286			Aluminum foil	Asbestine strips + Air		46 mm dia x 305 mm long; consisted of 3 sheets of 0.2 mm thick aluminum foil lapped in three layers about a 12 mm dia cylindrical heater, asbestine strips of about 4 mm thick and 4 mm wide used as separators between successive metal sheet layers, with air of 1 atm filling the vacant spaces.
2	493	R	1949	210, 285			Aluminum foil	Asbestine strips + Helium		Similar to the above specimen, but with helium of 1 atm filling the vacant spaces between the metal layers.
3	491	R	1958	293.2			Charcoal + Helium	Copper gauze		Sutcliffe Speakman 208 C. 8-14 mesh charcoal obtained from I. C. I. contained in a cylindrical cell of I. D. 1.5 cm, O. D. 3.9 cm, and length 5 cm, eleven copper gauze disk made from 28 gauge wire 20 holes $in.^{-1}$ placed parallel to heat flow in the charcoal bed spaced at ~0.5 cm intervals; measured in helium.
4	491	R	1958	295.1			Charcoal + Helium	Copper gauze		Similar to the above specimen except six gauze discs placed in the bed at ~1.0 cm intervals.
5	491	R	1958	293.4			Charcoal + Air	Copper gauze		Similar to the above specimen but measured in air.
6	389	L	1957	244-445		BBC 2	Foam	Aluminum alloy		Specimen consisted of core of foam with alkyd isocyanate resin manufactured by Brunswick-Balke Collender Co., and faces of aluminum alloy; core thickness 0.625 in., face thickness 0.25 in.; foam density 0.160 g cm^{-3}.
7	494	C	1960	483-963	±6		Forsterite	Stainless steel		85 vol % forsterite, 15 vol % stainless steel 430; 1 in. cube specimen fabricated using 4 layers of forsterite and 3 layers of stainless steel; heat flow parallel to the laminas; MIT alumina standard used as comparative material.
8	494	C	1960	483-963	±6					Second run of the above specimen.
9	494	C	1960	441-936	±3					Similar to the above specimen except heat flow perpendicular to the laminas.
10	494	C	1960	440-936	±3					Second run of the above specimen.
11	494	C	1960	420-924	±6		Forsterite	Steel		67 vol % forsterite, 15 vol % stainless steel 430; 1 in. cube specimen fabricated using 2 layers of forsterite and 1 layer of stainless steel; heat flow parallel to the laminas; same comparative material as above.
12	494	C	1960	420-928	±6					Second run of the above specimen.

SPECIFICATION TABLE NO. 338 (continued)

Curve No.	Ref. No.	Method Used	Year	Temp. Range, K	Reported Error, %	Name and Specimen Designation	Composition: Composite Materials		Binder	Composition (continued), Specifications and Remarks
13	494	C	1960	453-956	±3					Similar to the above specimen except heat flow perpendicular to the laminas.
14	494	C	1960	451-958	±3					Second run of the above specimen.
15	494	C	1960	485-931			Forsterite	Stainless steel		50 vol % forsterite, 50 vol % stainless steel 430; 1 in. cube specimen fabricated using 2 layers of forsterite and 1 layer of stainless steel; heat flow perpendicular to the laminas; MIT standard alumina used as comparative material.
16	485	L	1963	343-438			Asbestos	Silver	91 LD Phenolic resin	Asbestos cloth with one sheet of silver foil in center impregnated with 91 LD phenolic resin; 0.254 in. x 0.34 ft^2; materials supplied by U.S. Polymeric Chemicals Co., Inc.; pressed at 100 psi at 450 K for 1 hr, post-cured at 450 K for 16 hrs; density 1.810 g cm^{-3} at 60 C.
17	485	L	1963	334-433			Asbestos	Aluminum	91 LD Phenolic resin	Asbestos cloth with one sheet of aluminum foil in center impregnated with 91 LD phenolic resin; 0.281 in. x 0.34 ft^2; same supplier and fabrication method as the above specimen; density 1.727 g cm^{-3} at 60 C.

DATA TABLE NO. 338 THERMAL CONDUCTIVITY OF SANDWICHES (METALLIC-NONMETALLIC)

[Temperature, T, K; Thermal Conductivity, k, Watt $cm^{-1}K^{-1}$]

T	k
CURVE 1	
214.7	0.000238
232.7	0.000323
251.7	0.000391
286.2	0.000598
CURVE 2	
210.2	0.00143
284.7	0.00351
CURVE 3	
293.2	0.0096
CURVE 4	
295.1	0.0072
CURVE 5	
293.4	0.0041
CURVE 6	
244.3	0.000361
283.6	0.000418
332.9	0.000476
357.0	0.000462
358.4	0.000462*
400.6	0.000490
444.6	0.000591
CURVE 7	
483.2	0.0868
548.2	0.0849
632.2	0.0851
659.2	0.0824
685.2	0.0837
783.2	0.0823
845.2	0.0816
963.2	0.0795
CURVE 8	
483.2	0.0847
548.2	0.0824
632.2	0.0812
659.2	0.0787
685.2	0.0796
783.2	0.0791
843.2	0.0778
963.2	0.0761
CURVE 9	
441.2	0.0665
506.2	0.0592
563.2	0.0548
620.2	0.0511
693.2	0.0471
786.2	0.0418
856.2	0.0390
936.2	0.0339
CURVE 10	
440.2	0.0630
509.2	0.0579
559.2	0.0531
615.2	0.0492
695.2	0.0448
791.2	0.0387
861.2	0.0364
936.2	0.0328
CURVE 11	
420.2	0.115
524.2	0.114
583.2	0.113
631.2	0.113
742.7	0.113
793.2	0.111
817.2	0.111
837.2	0.110
893.2	0.112
924.2	0.113
CURVE 12	
420.2	0.113
523.2	0.110
583.2	0.108
632.2	0.109
743.2	0.109
794.2	0.107
820.2	0.108
841.2	0.108
893.2	0.110
928.2	0.110
CURVE 13	
453.2	0.0698
595.2	0.0590
644.7	0.0561
723.2	0.0500
789.2	0.0455
868.2	0.0418
956.2	0.0376
CURVE 14	
451.2	0.0667
598.2	0.0561
644.0	0.0536
722.2	0.0467
790.2	0.0418*
868.2	0.0389*
958.2	0.0357
CURVE 15	
485.2	0.0837
535.2	0.0816
571.5	0.0769
603.2	0.0750
665.2	0.0686
689.2	0.0657
728.5	0.0630
778.2	0.0613
843.2	0.0573
931.2	0.0523
CURVE 16	
242.6	0.00244
383.7	0.00238
409.8	0.00219
435.9	0.00247
438.2	0.00215
CURVE 17	
334.3	0.00205
356.5	0.00179
365.4	0.00200
401.5	0.00203
417.7	0.00221
432.6	0.00237

* Not shown on plot

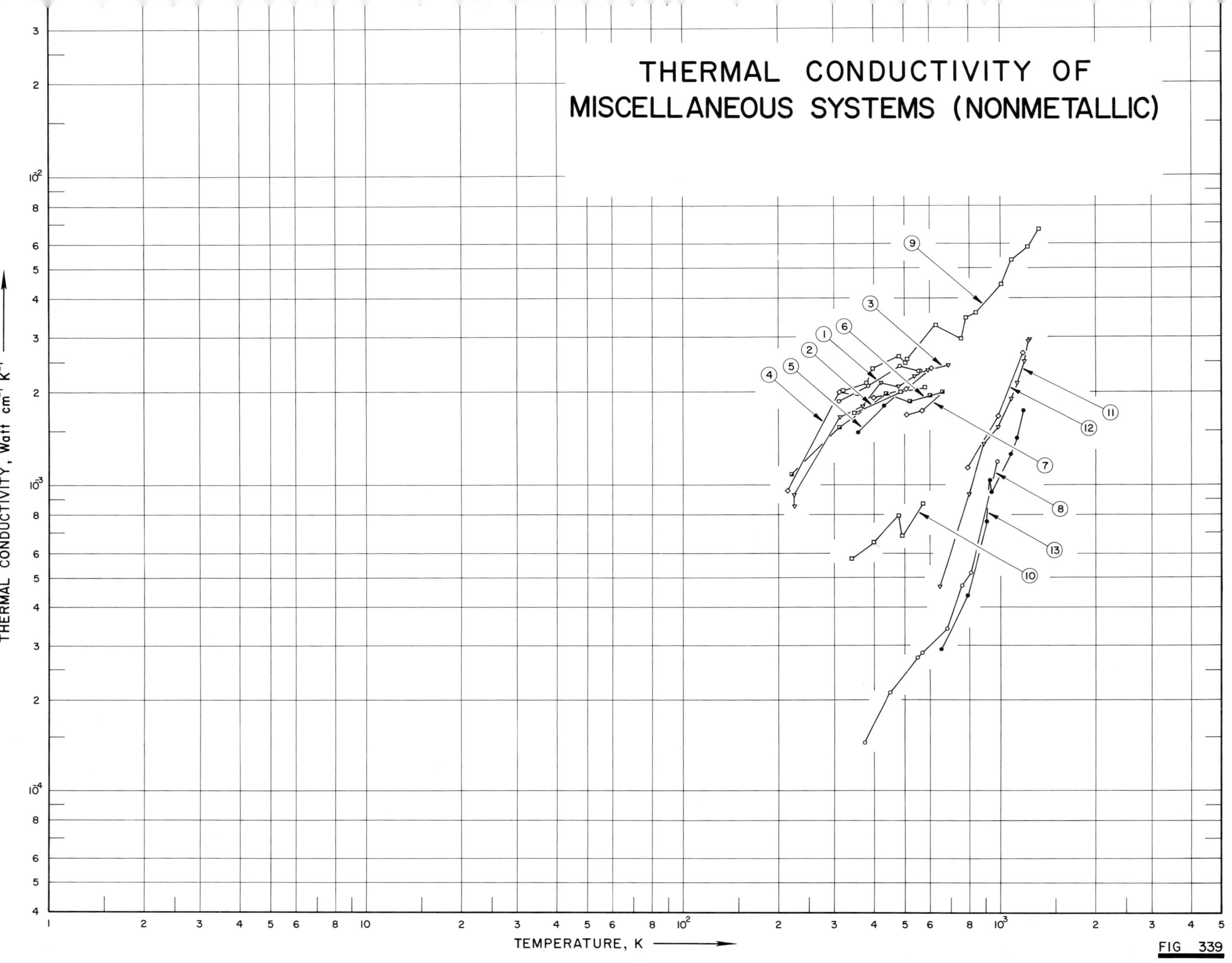

THERMAL CONDUCTIVITY OF
MISCELLANEOUS SYSTEMS (NONMETALLIC)
THERMAL CONDUCTIVITY, Watt cm⁻¹ K⁻¹
TEMPERATURE, K
FIG 339

SPECIFICATION TABLE NO. 339 THERMAL CONDUCTIVITY OF MISCELLANEOUS SYSTEMS (NONMETALLIC)

[For Data Reported in Figure and Table No. 339]

Curve No.	Ref. No.	Method Used	Year	Temp. Range, K	Reported Error, %	Name and Specimen Designation	Composition: Composite Materials		Composition: Binder	Composition (continued), Specifications and Remarks
1	488, 489	L	1961	312-563		SRI 2	Plastic	Glass	CTL-91 LD Phenolic resin	35% ±5% resin; 181 chopped glass reinforcing; 8 x 8 x 0.25 in.; average thickness 0.281 in.; manufactured by Reihold Engineering and Plastics Co.; molded at 408 K for 1 hr, post-cured at 408 K for 24 hrs; density 1.829 g cm^{-3}.
2	488, 489	L	1961	348-585		SRI 2				The above specimen measured at decreasing temp.
3	488, 489	L	1961	225-689		Fiberite 4030-190; SRI 4	Plastic	Glass	Phenolic resin R181	48-53 resin; glass roving No. F846 reinforcing; 8 x 8 x 0.25 in.; average thickness 0.230 in.; prepared by curing hydraulic press at 2000 psi, cured at 433 K for 20 min; density 1.728 g cm^{-3}.
4	488, 489	L	1961	215-607		Coast F130R; SRI 5	Plastic	Glass	Silicone resin 4P020, F130R DC-2106	35% ±3% resin; chopped glass 1B603 reinforcing; 10 x 6.5 x 0.25 in.; average thickness 0.273 in.; manufactured by Coast Manufacturing and Supply Co.; heated at 366 K for 16 hrs, at 400 K for 2 hrs, at 422 K for 2 hrs, at 450 K for 2 hrs, at 478 K for 2 hrs, at 500 K for 2 hrs, and at 523 K for 12 hrs, then cooled to 366 K before removing from oven; density 1.698 to 1.757 g cm^{-3}.
5	488, 489	L	1961	357-607		Coast F130R; SRI 5				The above specimen measured at decreasing temp.
6	488, 489	L	1961	222-658		Astrolite; SRI 6	Plastic	Cloth	Phenolic resin SC-1008	30-35 resin; 0.5 x 0.5 in. squares of 1201 V cloth reinforcing; 16 x 6 x 0.125 in.; manufactured by H. I. Thompson Fiber Glass Co.; average thickness 0.132 in.; molded, cured at 422 K for 2 hrs; density 1.442 g cm^{-3}.
7	488, 489	L	1961	508-658		Astrolite; SRI 6				The above specimen measured at decreasing temp.
8	495	R	1964	375-981			Carbon	Vacuum		Type VDF carbon felt supplied by National Carbon Co.; specimen prepared by wrapping carbon felt on a 0.875 in. dia iron tube; specimen 1.25 in. in O.D. and 10 in. in length; density of carbon felt 0.083 g cm^{-3}; measured in a vacuum of about 200 μ.
9	495	R	1964	319-1332			Carbon	Helium		Similar to the above specimen but measured in helium at 1 atm.
10	495	R	1964	340-571			Carbon	Air		Similar to the above specimen but measured in air at 1 atm.

SPECIFICATION TABLE NO. 339 (continued)

Curve No.	Ref. No.	Method Used	Year	Temp. Range, K	Reported Error, %	Name and Specimen Designation	Composition: Composite Materials		Composition: Binder	Composition (continued), Specifications and Remarks
11	472	R	1963	650-1250	8		Aluminum Oxide	Graphite		Density of Al_2O_3 foam 0.474 g cm^{-3}, graphite fibers 0.087 g cm^{-3}; solid section of Al_2O_3 (foam brick) were formed into cylinders, which filled the annulus between the measuring apparatus and a tantalum shield 2 in. in dia, two layers of graphite fiber (14 μ average dia, type WDF) mats were placed in the annulus between 2 in. tantalum shield and outer container; measured in a vacuum of 4 x 10^{-7} to 2 x 10^{-4} mm Hg.
12	472	R	1963	794-1189	7		Aluminum Oxide	Graphite		The above specimen measured at decreasing temp.
13	472	R	1963	655-1189			Aluminum Oxide	Graphite		Density of graphite fibers 0.082 g cm^{-3}, alumina 1.057 g cm^{-3}; alumina bubbles (250-350 μ dia) placed in annulus formed by tantalum shield (2 in. dia) and the measuring apparatus (0.75 in. dia); graphite fibers (14 μ average dia, 2 layers of matting) in annulus formed by 2 in. dia shield and outer tantalum container (2.75 in. dia); measured in a vacuum of 0.7 x 10^{-6} to 1.2 x 10^{-4} mm Hg.
14	472	R	1963	911-1144	7		Aluminum Oxide	Graphite		The above specimen measured at decreasing temp.

DATA TABLE NO. 339 THERMAL CONDUCTIVITY OF MISCELLANEOUS SYSTEMS (NONMETALLIC)

[Temperature, T, K; Thermal Conductivity, k, Watt $cm^{-1}K^{-1}$]

T	k
CURVE 1	
311.5	0.00185
384.3	0.00207
485.4	0.00240
555.9	0.00231
563.2	0.00231
CURVE 2	
348.2	0.00169
489.3	0.00198
584.8	0.00205
CURVE 3	
225.4	0.000844
225.4	0.000917
314.8	0.00164
369.8	0.00178
374.3	0.00178*
423.7	0.00211
481.5	0.00206
539.8	0.00223
595.9	0.00233
688.7	0.00241
CURVE 4	
214.8	0.000948
313.2	0.00198
400.9	0.00189
507.6	0.00202
607.1	0.00236
CURVE 5	
357.1	0.00147
433.2	0.00179
607.1	0.00236*
CURVE 6	
221.5	0.00107
313.7	0.00153
376.5	0.00176*
CURVE 6 (cont.)	
440.4	0.00195
522.6	0.00184
603.7	0.00192
658.2	0.00197
CURVE 7	
507.6	0.00168
568.2	0.00172
658.2	0.00197*
CURVE 8	
375	0.000142
447	0.000209
550	0.000270
569	0.000280
681	0.000337
758	0.000465
808	0.000513
981	0.00118
CURVE 9	
319	0.00199*
321	0.00200
397	0.00236
481	0.00258
505	0.00246
511	0.00253
755	0.00294
782	0.00350*
783	0.00343
1008	0.00440
1092	0.00527
379	0.00211
844	0.00356
1232	0.00580
1332	0.00662
626	0.00324
CURVE 10	
340	0.000571
400	0.000645
480	0.000786
491	0.000677
571	0.000860
CURVE 11	
649.8	0.000462
799.8	0.000923
894.3	0.00134
988.7	0.00153
1088.7	0.00187
1133.2	0.00211*
1141.5	0.00211
1194.3	0.00247
1238.7	0.00288
1249.8	0.00291
CURVE 12	
794.3	0.00112
988.7	0.00166
1188.7	0.00264
CURVE 13	
655.4	0.000288
794.3	0.000433
910.9	0.000750
933.2	0.00102
944.3	0.000937
1088.7	0.00125
1133.2	0.00141
1188.7	0.00173
CURVE 14*	
910.9	0.000721
1099.8	0.00118
1144.3	0.00151

* Not shown on plot

THERMAL CONDUCTIVITY OF MISCELLANEOUS SYSTEMS (METALLIC–NONMETALLIC)

Continued on Figure 340-2

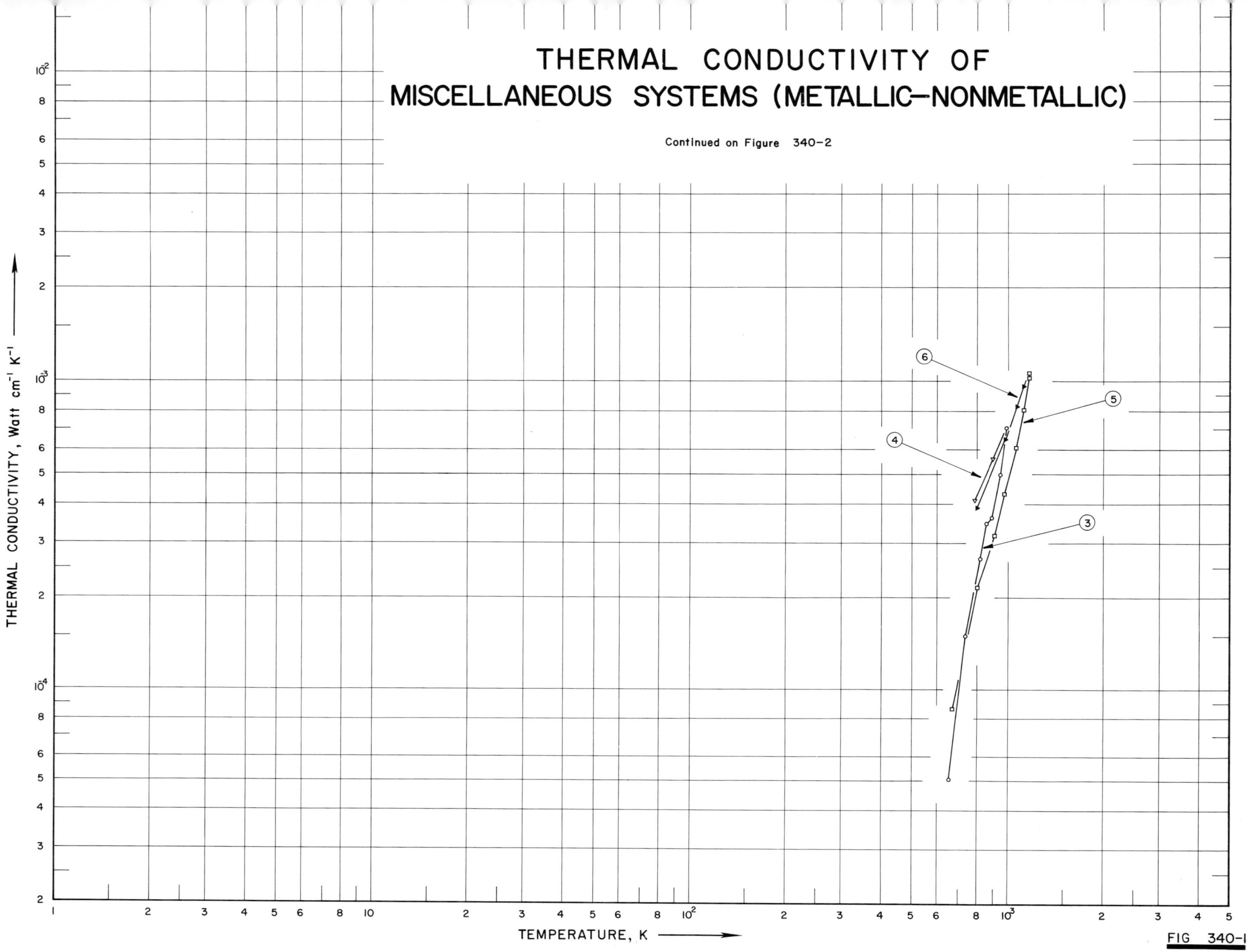

FIG 340-1

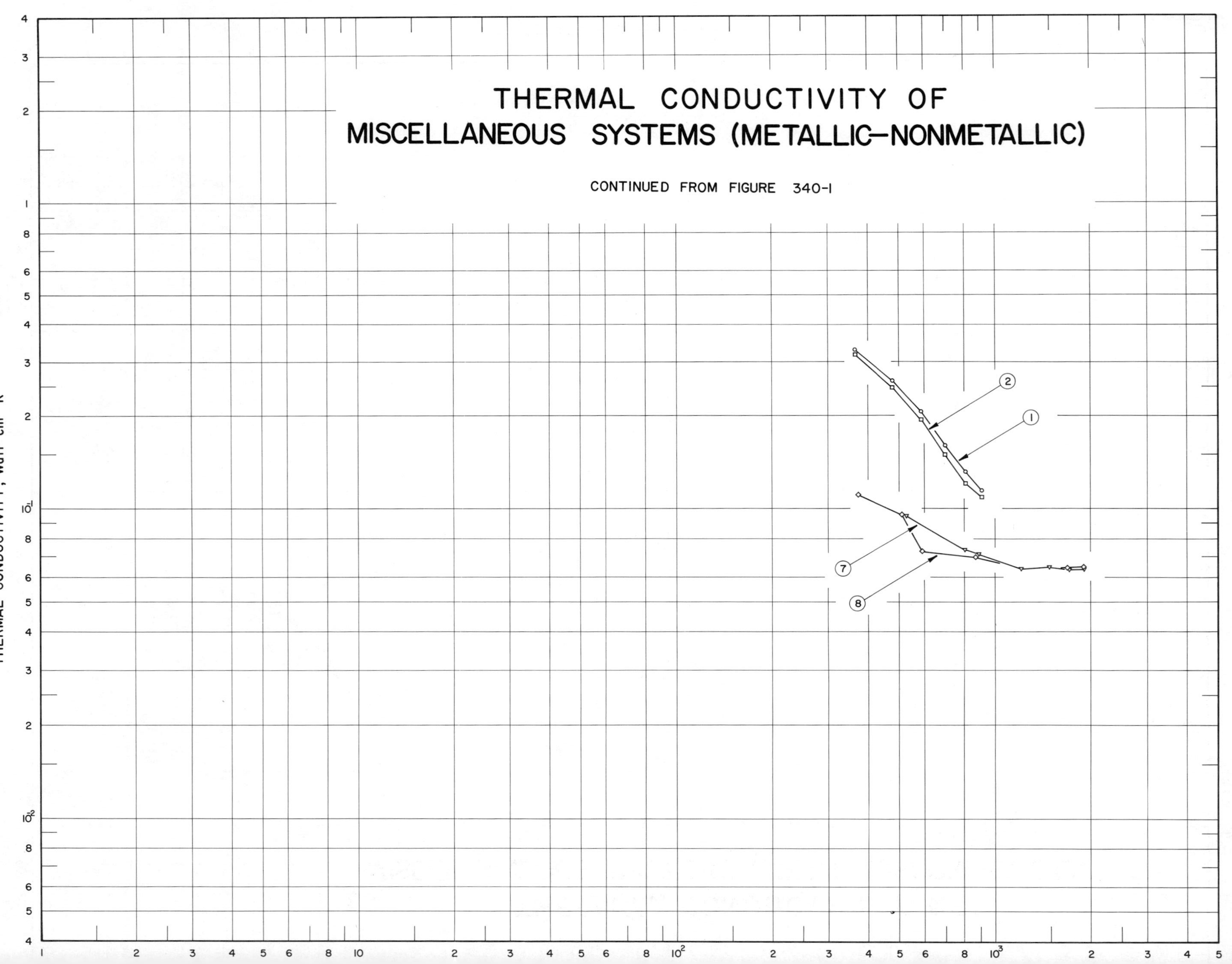
THERMAL CONDUCTIVITY OF
MISCELLANEOUS SYSTEMS (METALLIC—NONMETALLIC)
CONTINUED FROM FIGURE 340-I
THERMAL CONDUCTIVITY, Watt cm⁻¹ K⁻¹
1
2
7
8

SPECIFICATION TABLE NO. 340 THERMAL CONDUCTIVITY OF MISCELLANEOUS SYSTEMS (METALLIC-NONMETALLIC)

[For Data Reported in Figure and Table No. 340]

Curve No.	Ref. No.	Method Used	Year	Temp. Range, K	Reported Error, %	Name and Specimen Designation	Composition: Composite Materials	Composition: Binder	Composition (continued), Specifications and Remarks
1	286	C	1958	367-914			Aluminum Oxide Molybdenum		90.0 Al_2O_3 and 10.0 Mo; 1 x 1 x 1 in.; prepared from Norton 38-900 alumina of grain size 5-9 μ, and from 0.002 in. dia unannealed molybdenum wire obtained from Fansteel Metallurgical Corp and cut to 0.125 in. long by Rayon Processing Co.; batches thoroughly mixed, charged into a graphite mold, hot pressed at 1650 C under 3000 psi, cooled to room temp in about 4 hrs; heat flow parallel to the Mo fiber orientation; alumina used as comparative material.
2	286	C	1958	367-911			Aluminum Oxide Molybdenum		Similar to the above specimen but heat flow perpendicular to the Mo fiber orientation.
3	472	R	1963	655-994	6-9	Test 11	Paper, Aluminum, Stainless Steel, Graphite		Multifoil insulation formed by wrapping Dexiglas paper and 0.002 in. aluminum foil alternately around the measuring apparatus, the foil was slit to reduce conduction, thirteen layers of paper and ten layers of foil were used, the insulation was surrounded by a 1.25 in. dia Type 304 stainless steel tube, four layers of graphite fiber (14 μ average dia, type WDF) mats filled the annulus between the stainless steel tube and the outer tantalum container; total heating time 91 hrs; pressure range 1.4 x 10^{-7} to 2.0 x 10^{-5} mm Hg; density of graphite fiber 0.088 g cm^{-3}; measured in increasing temp.
4	472	R	1963	789-994	6-9				The above specimen measured at decreasing temp.
5	472	R	1963	672-1161	6-9		Paper, Tantalum, Graphite		Multifoil insulation of alternate layers of "Fiberfrax" paper and tantalum foil (0.0005 in.) was formed around the measuring apparatus, there were ten layers of paper, nine foil layers, foils were slit to reduce conduction, an outer layer of tantalum (0.005 in.) was used for the last foil, three layers of graphite fibers (10 μ average dia, type WDF) were wrapped around the multifoil insulation, tantalum foils were also used to separate the graphite layers, the entire insulation was enclosed by a tantalum cyclinder (2.75 in. dia); total heating time 122 hrs; pressure range 1 x 10^{-6} to 6 x 10^{-4} mm Hg; density of graphite 0.083 g cm^{-3}; measured in increasing temp.
6	472	R	1963	800-1161	6-9				The above specimen measured at decreasing temp.
7	204	R	1957	529-1913			Thorium Dioxide, Molybdenum		10 Mo; 0.5 CaF_2; reinforcing Mo fibers 0.002 in. in dia and 0.125 in. in nominal length; specimen made up of two 1 in. thick discs of 0.50 in. I.D. and 2.5 in. O.D.; hot pressed at 1500 $\pm$50 C and at pressure of 100 psi for 30 min; average bulk density 9.26 g cm^{-3}.
8	236	R	1962	373-1903			Thorium Dioxide, Molybdenum		10 Mo; reinforcing Mo fibers 0.005 cm in dia and 0.5 cm in length.

DATA TABLE NO. 340 THERMAL CONDUCTIVITY OF MISCELLANEOUS (METALLIC-NONMETALLIC)

[Temperature, T, K; Thermal Conductivity, k, Watt $cm^{-1}K^{-1}$]

T	k
CURVE 1	
366.5	0.331
477.6	0.262
588.7	0.207
699.8	0.161
810.9	0.132
913.7	0.115
CURVE 2	
366.5	0.318
477.6	0.249
588.7	0.195
699.8	0.151
810.9	0.121
910.9	0.109
CURVE 3	
655.4	0.0000505
738.7	0.000151
816.5	0.000267
855.4	0.000346
894.3	0.000361
949.8	0.000498
994.3	0.000707
CURVE 4	
788.7	0.000411
899.8	0.000562
994.3	0.000707*
CURVE 5	
672.1	0.0000865
802.6	0.000216
910.9	0.000317
977.6	0.000433
1066.5	0.000606
1124.8	0.000808
1160.9	0.00103
1160.9	0.00106
CURVE 6	
799.8	0.000389
983.2	0.000649
1069.3	0.000829
1119.3	0.000966
1160.9	0.00106*
CURVE 7	
529.3	0.0950
802.4	0.0736
892.7	0.0715
1209.9	0.0640
1490.2	0.0649
1705.9	0.0636
1912.8	0.0644
CURVE 8	
373.2	0.111
513.2	0.0962
593.2	0.0732
873.2	0.0699
1203.2	0.0644*
1483.2	0.0644*
1698.2	0.0644
1903.2	0.0649

* Not shown on plot

SPECIFICATION TABLE NO. 341 THERMAL CONDUCTIVITY OF ASH

Curve No.	Ref. No.	Method Used	Year	Temp. Range, K	Reported Error, %	Name and Specimen Designation	Composition (weight percent), Specifications and Remarks
1	125	L	1923	293,298	1	3	3.63 cm dia x 0.091 cm thick; sp. gravity = 0.74; moisture content 14.3%; heat flow parallel to the grain, measured under a load of 21 psi.
2	125	L	1923	293,298	1	1	3.63 cm dia x 0.100 cm thick; sp. gravity = 0.74; moisture content 15.5%; heat flow perpendicular to the grain, radial to annual rings; measured under a load of 21 psi.
3	125	L	1923	293,298	1	2	3.63 cm dia x 0.102 cm thick; sp. gravity = 0.74; moisture content 15%; heat flow perpendicular to the grain, tangential to annual rings; measured under a load of 21 psi.
4	125	L	1923	293,298	1	6	3.63 cm dia x 0.081 cm thick; sp. gravity = 0.74; moisture content 14.9%; same measuring conditions as the above specimen 3.
5	125	L	1923	293,298	1	4	3.63 cm dia x 0.100 cm thick; sp. gravity = 0.74; moisture content 16.0%; same measuring conditions as the above specimen 1.
6	125	L	1923	293,298	1	5	3.63 cm dia x 0.100 cm thick; sp. gravity = 0.74; moisture content 16.6%; same measuring conditions as the above specimen 2.

DATA TABLE NO. 341 THERMAL CONDUCTIVITY OF ASH

[Temperature, T, K; Thermal Conductivity, k, Watt $cm^{-1}K^{-1}$]

T	k	T	k	T	k
CURVE 1*		CURVE 3*		CURVE 5*	
293.2	0.00299	293.2	0.00154	293.2	0.00194
298.2	0.00303	298.2	0.00155	298.2	0.00196
CURVE 2*		CURVE 4*		CURVE 6*	
293.2	0.00153	293.2	0.00315	293.2	0.00174
298.2	0.00159	298.2	0.00323	298.2	0.00176

* No graphical presentation

SPECIFICATION TABLE NO. 342 THERMAL CONDUCTIVITY OF BALSA

Curve No.	Ref. No.	Method Used	Year	Temp. Range, K	Reported Error, %	Name and Specimen Designation	Composition (weight percent), Specifications and Remarks
1	125	L	1923	293.2	1		Sp. gravity 0.1; moisture content 13%; heat flor perpendicular to grain and tangential to annual rings; measured under a load of 21 psi.
2	125	L	1923	293.2	1	Waterproofed balsa	Sp. gravity 0.1; moisture content 13%; heat flow perpendicular to grain and tangential to annual rings; measured under a load of 21 psi.
3	125	L	1923	293.2	1	Pseudo balsa	Sp. gravity 0.25; moisture content 13%; heat flow parallel to grain; measured under a load of 21 psi.
4	125	L	1923	293.2	1	Pseudo balsa	Sp. gravity 0.25; moisture content 13%; heat flow perpendicular to grain; measured under a load of 21 psi.

DATA TABLE NO. 342 THERMAL CONDUCTIVITY OF BALSA

[Temperature, T,K; Thermal Conductivity, k, Watt $cm^{-1}K^{-1}$]

T	k
CURVE 1*	
293.2	0.000460
CURVE 2*	
293.2	0.000544
CURVE 3*	
293.2	0.00121
CURVE 4*	
293.2	0.000669

*No graphical presentation

SPECIFICATION TABLE NO. 343 THERMAL CONDUCTIVITY OF BOXWOOD

Curve No.	Ref. No.	Method Used	Year	Temp. Range, K	Reported Error, %	Name and Specimen Designation	Composition (weight percent), Specifications and Remarks
1	44	F	1914	293, 373			0.654 cm dia x 5.7 cm long; density 0.901 g cm^{-3}.

DATA TABLE NO. 343 THERMAL CONDUCTIVITY OF BOXWOOD

[Temperature, T, K; Thermal Conductivity, k, Watt $cm^{-1}K^{-1}$]

T	k
CURVE 1*	
293.2	0.00149
373.2	0.00173

* No graphical presentation

SPECIFICATION TABLE NO. 344 THERMAL CONDUCTIVITY OF CEDAR

Curve No.	Ref. No.	Method Used	Year	Temp. Range, K	Reported Error, %	Name and Specimen Designation	Composition (weight percent), Specifications and Remarks
1	125	L	1923	293.2	1		0.230 in. thick; specific gravity 0.49, moisture content 13.4%; measured under a load of 21 psi; heat flow perpendicular to grain and tangent to the annual rings.
2	125	L	1923	293.2	1		Similar to above but 0.200 in. thick, specific gravity 0.48, and moisture content 12.7%.
3	125	L	1923	293.2	1		Similar to above but 0.197 in. thick, moisture content 10.6%.

DATA TABLE NO. 344 THERMAL CONDUCTIVITY OF CEDAR

[Temperature, T, K; Thermal Conductivity, k, Watt $cm^{-1}K^{-1}$]

T	k
CURVE 1*	
293.2	0.00121
CURVE 2*	
293.2	0.00105
CURVE 3*	
293.2	0.00116

* No graphical presentation

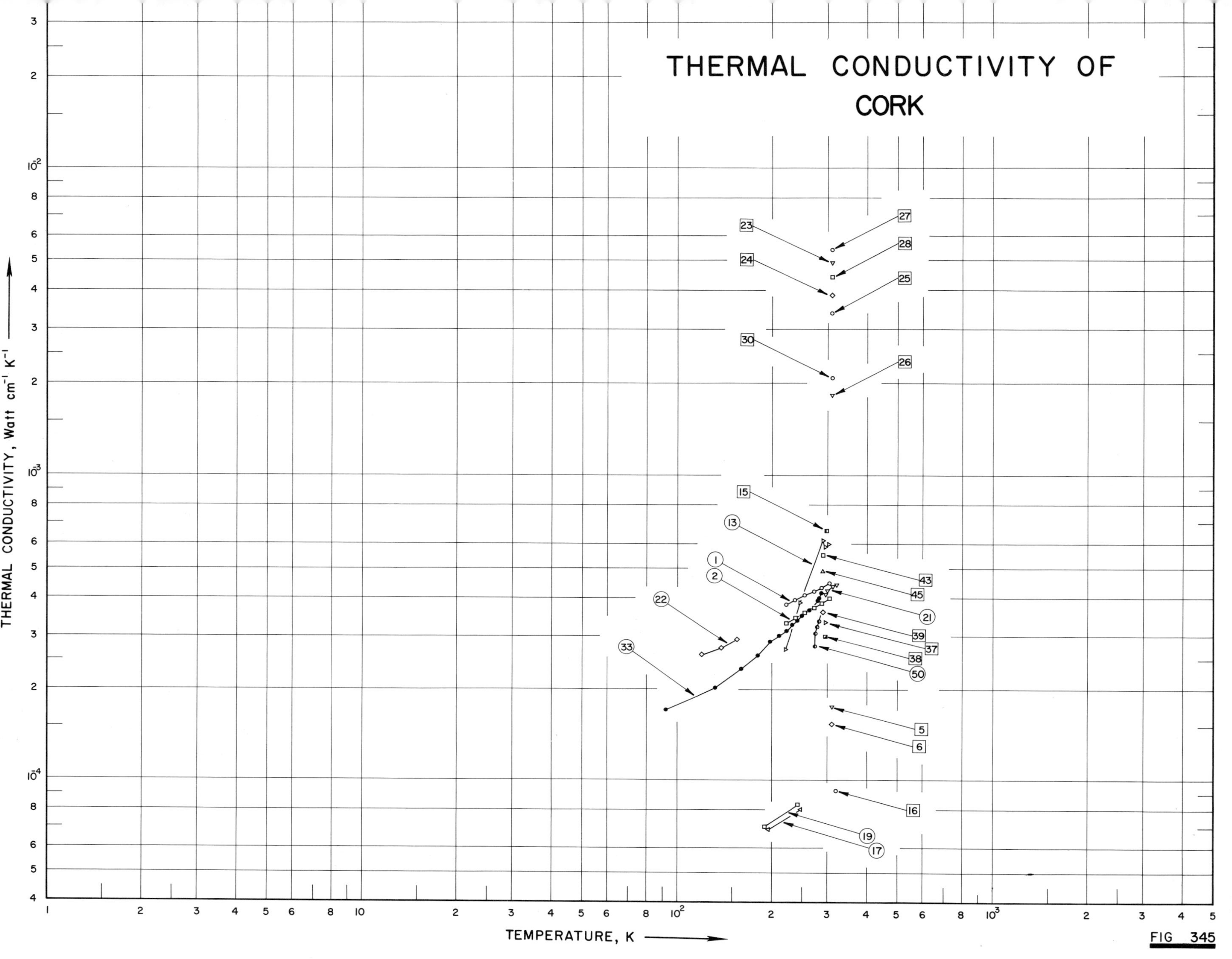

THERMAL CONDUCTIVITY OF
CORK
THERMAL CONDUCTIVITY, Watt cm⁻¹ K⁻¹
TEMPERATURE, K
27
28
25
23
24
30
26
15
13
1
2
22
33
43
45
21
39
37
38
50
5
6
16
19
17

FIG 345

SPECIFICATION TABLE NO. 345 THERMAL CONDUCTIVITY OF CORK

[For Data Reported in Figure and Table No. 345]

Curve No.	Ref. No.	Method Used	Year	Temp. Range, K	Reported Error, %	Name and Specimen Designation	Composition (weight percent), Specifications and Remarks
1	109	L	1947	222-306			Thickness of specimen 1 in.; oven dried; density 0.195 g cm^{-3}.
2	109	L	1947	222-306			Thickness of specimen 1 in.; oven dried; density 0.128 g cm^{-3}.
3	109	L	1947	222-306			Thickness of specimen 1 in.; oven dried; density 0.104 g cm^{-3}.
4	558	L	1951	310.9			Thickness of specimen 1 in.; density 0.107 g cm^{-3} at 25 C.
5	558	L	1951	310.9			Thickness of specimen 1 in.; density 0.107 g cm^{-3} at 25 C; measured in a vacuum of 0.08 μ Hg.
6	558	L	1951	310.9			Thickness of specimen 1 in.; density 0.107 g cm^{-3} at 25 C; measured in a vacuum of 0.06 μ Hg.
7	558	L	1951	310.9			Thickness of specimen 1 in.; density 0.199 g cm^{-3} at 25 C.
8	558	L	1951	310.9			Thickness of specimen 1 in.; density 0.199 g cm^{-3} at 25 C; measured in a vacuum of 0.13 μ Hg.
9	558	L	1951	222-305			Thickness of specimen 1 in.; density 0.195 g cm^{-3} at 25 C.
10	558	L	1951	222-305			Thickness of specimen 1 in.; density 0.128 g cm^{-3} at 25 C.
11	558	L	1951	222-305			Thickness of specimen 1 in.; density 0.104 g cm^{-3} at 25 C.
12	189		1950	200-289			Density 0.120-0.160 g cm^{-3}; water absorption (24 hrs submerged) 5.2 vol %.
13	493	R	1949	222-302		A	Uniformly carbonized; bulk density 0.208 g cm^{-3}.
14	493	C	1949	220, 296		B	20 x 20 x 8 cm; uniformly carbonized; bulk density 0.208 g cm^{-3}; cork board A used as comparative material.
15	493	C	1949	295.2		B	20 x 20 x 8 cm; over-carbonized; bulk density 0.138 g cm^{-3}; same comparative material as above.
16	128	L	1946	319.3		A	2.02 in. thick; dry; density 0.113 g cm^{-3}.
17	128	L	1946	195, 248		A	Similar to above except 1.12 in. thick.
18	128	L	1946	322.6		B	1.99 in. thick; dry; density 0.107 g cm^{-3}.
19	128	L	1946	192, 244		B	Similar to above except 1.04 in. thick.
20	559	L	1957	289-313			Density 0.123 g cm^{-3}.
21	559	L	1957	296-320			Density 0.138 g cm^{-3}.
22	43	L	1954	120-155			Density 0.151 g cm^{-3}.
23	560	L	1954	310.9			Specimen 1 in. thick; density 0.107 g cm^{-3}; measured in air.
24	560	L	1954	310.9			The above specimen measured in CO_2.
25	560	L	1954	310.9			The above specimen measured in F-12.

SPECIFICATION TABLE NO. 345 (continued)

Curve No.	Ref. No.	Method Used	Year	Temp. Range, K	Reported Error, %	Name and Specimen Designation	Composition (weight percent), Specifications and Remarks
26	560	L	1954	310.9			The above specimen measured in vacuum.
27	560	L	1954	310.9			Specimen 1 in. thick; density 0.199 g cm^{-3}; measured in air.
28	560	L	1954	310.9			The above specimen measured in CO_2.
29	560	L	1954	310.9			The above specimen measured in F-12.
30	560	L	1954	310.9			The above specimen measured in vacuum.
31	100	L	1957	323, 332			Bulk density 0.284 g cm^{-3}.
32	534	L	1961	271.5			9 x 9 x 1 in.; density 0.128 g cm^{-3}.
33	561	R	1948	92-288		1A	Granulated slab specimen packed into a hollow cylinder of 2 in. I.D. and 10 in. O.D.; baked; grain size 49.5% 10-20 mesh and 50.5% through 20 mesh; bulk density 0.101-0.103 g cm^{-3}.
34	561	R	1948	190.5		1A	The above specimen repacked; bulk density 0.103 g cm^{-3}.
35	561	R	1948	189.8		1B	Similar to the above specimen but grain size 4-10 mesh and bulk density 0.0865 g cm^{-3}.
36	530	L	1956	297.1			Expanded; 20 cm dia x 2.5 cm long; density 0.160 g cm^{-3}; measured in air.
37	530	L	1956	296.6			The above specimen measured in CO_2.
38	530	L	1956	296.2			The above specimen measured in CH_3Cl.
39	529	P	1956	290.4			Moisture content 3.0%; thermal conductivity data calculated from measured transient temp change.
40	529	P	1956	290.4			Similar to the above specimen but moisture content 10.0%.
41	529	P	1956	290.4			Density 0.208 g cm^{-3}; same measuring method as above.
42	529	P	1956	290.4			Density 0.112 g cm^{-3}; moisture content 2.5%; same measuring method as above.
43	529	P	1956	290.9			Density 0.336 g cm^{-3}; moisture content 2.0%; same measuring method as above.
44	529	P	1956	290.9			Density 0.128 g cm^{-3}; moisture content 3.0%; same measuring method as above.
45	529	P	1956	290.9			Similar to the above specimen but moisture content 8.0%.
46	529	P	1956	290.4			Granulated; baked; grain size 2-3 mesh; density 0.0961 g cm^{-3}; moisture content 8.0%; same measuring method as above.
47	529	P	1956	290.4			The above specimen dried for 36 hrs.
48	529	P	1956	290.4			Granulated; baked; grain size 4-6 mesh; water content 3.0%; density 0.144 g cm^{-3}; same measuring method as above.

SPECIFICATION TABLE NO. 345 (continued)

Curve No.	Ref. No.	Method Used	Year	Temp. Range, K	Reported Error, %	Name and Specimen Designation	Composition (weight percent), Specifications and Remarks
49	529	P	1956	290.4			The above specimen dried for 36 hrs.
50	529	P	1956	274-290			Slab cork 3 in. x 3 in. x 6 in.; wrapped in "polythene"; density 0.128 g cm^{-3}; moisture content 2.3%; same measuring method as above.
51	529	P	1956	278-290			The above specimen after drying out for 48 hrs and cooling for 72 hrs.

DATA TABLE NO. 345 THERMAL CONDUCTIVITY OF CORK

[Temperature, T,K; Thermal Conductivity, k, Watt $cm^{-1}K^{-1}$]

Curve	T	k
CURVE 1	222.0	0.000381
	238.7	0.000394
	255.4	0.000407
	272.2	0.000421
	288.9	0.000433
	305.5	0.000446
CURVE 2	222.0	0.000333
	238.7	0.000346
	255.4	0.000359
	272.2	0.000372
	288.9	0.000385
	305.5	0.000398
CURVE 3*	222.0	0.000320
	238.7	0.000336
	255.4	0.000352
	272.2	0.000368
	288.9	0.000384
	305.5	0.000400
CURVE 4*	310.9	0.000408
CURVE 5	310.9	0.000175
CURVE 6	310.9	0.000153
CURVE 7*	310.9	0.000453
CURVE 8*	310.9	0.000173
CURVE 9*	222.2	0.000381
	239.2	0.000394
	255.5	0.000407
	272.1	0.000421
	288.8	0.000433
	305.3	0.000446
CURVE 10*	222.2	0.000331
	239.2	0.000346
	255.5	0.000359
	272.1	0.000372
	288.8	0.000385
	305.3	0.000398
CURVE 11*	222.2	0.000320
	239.2	0.000336
	255.5	0.000352
	272.1	0.000368
	288.8	0.000384
	305.3	0.000400
CURVE 12*	200.0	0.000317
	222.2	0.000332
	239.2	0.000346
	255.5	0.000361
	272.1	0.000375
	288.8	0.000389
CURVE 13	221.7	0.000271
	246.7	0.000388
	290.2	0.000615
	294.2	0.000586
	302.2	0.000598
CURVE 14*	219.7	0.000258
	295.7	0.000594
CURVE 15	295.2	0.000661
CURVE 16	319.3	0.0000924
CURVE 17	195.4	0.0000693
	247.6	0.0000803
CURVE 18*	322.6	0.0000910
CURVE 19	191.5	0.0000703
	243.7	0.0000834
CURVE 20*	288.8	0.000397
	289	0.000397
	289.1	0.000399
	289.3	0.000395
	289.6	0.000397
	290	0.000396
	291.1	0.000398
	309.9	0.000418
	310.2	0.000416
	310.6	0.000415
	311.1	0.000418
	311.1	0.000420
	312.9	0.000419
CURVE 21*	296	0.000407
	298.8	0.000421*
	299.1	0.000421
	310.2	0.000435*
	312.2	0.000435
	319.0	0.000443*
	320.2	0.000443
CURVE 22	120	0.000260
	138	0.000274
	155	0.000293
CURVE 23	310.9	0.00490
CURVE 24	310.9	0.00384
CURVE 25	310.9	0.00337
CURVE 26	310.9	0.00183
CURVE 27	310.9	0.00543
CURVE 28	310.9	0.00443
CURVE 29*	310.9	0.00377
CURVE 30	310.9	0.00208
CURVE 31*	322.6	0.000433
	331.5	0.000476
CURVE 32*	271.5	0.000490
CURVE 33	92.2	0.000171
	132.2	0.000202
	160.2	0.000234
	181.7	0.000258
	198.2	0.000287
	211.2	0.000301
	222.6	0.000312
	232.4	0.000327
	241.2	0.000339
	248.9	0.000351
	256.0	0.000357*
	262.4	0.000365
	268.4	0.000372*
	273.9	0.000377*
	279.0	0.000393
	283.5	0.000398
	287.7	0.000417
CURVE 34*	190.5	0.000280
CURVE 35*	189.8	0.000291
CURVE 36*	297.1	0.000415
CURVE 37	296.6	0.000334
CURVE 38	296.2	0.000300
CURVE 39	290.4	0.000361
CURVE 40*	290.4	0.000447
CURVE 41*	290.4	0.000462
CURVE 42*	290.4	0.000340
CURVE 43	290.9	0.000554
CURVE 44*	290.9	0.000397
CURVE 45	290.9	0.000490
CURVE 46*	290.4	0.000325
CURVE 47*	290.4	0.000313
CURVE 48*	290.4	0.000430
CURVE 49*	290.4	0.000348
CURVE 50	274.2	0.000277
	275.9	0.000307
	277.6	0.000313*
	278.7	0.000323
	282.6	0.000337
	284.8	0.000346*
	287.1	0.000368*
	288.7	0.000397*
	289.8	0.000415*
	290.4	0.000433*
CURVE 51*	278.2	0.000309
	280.4	0.000312
	283.2	0.000309
	287.1	0.000352
	288.7	0.000372
	289.8	0.000387

* Not shown on plot

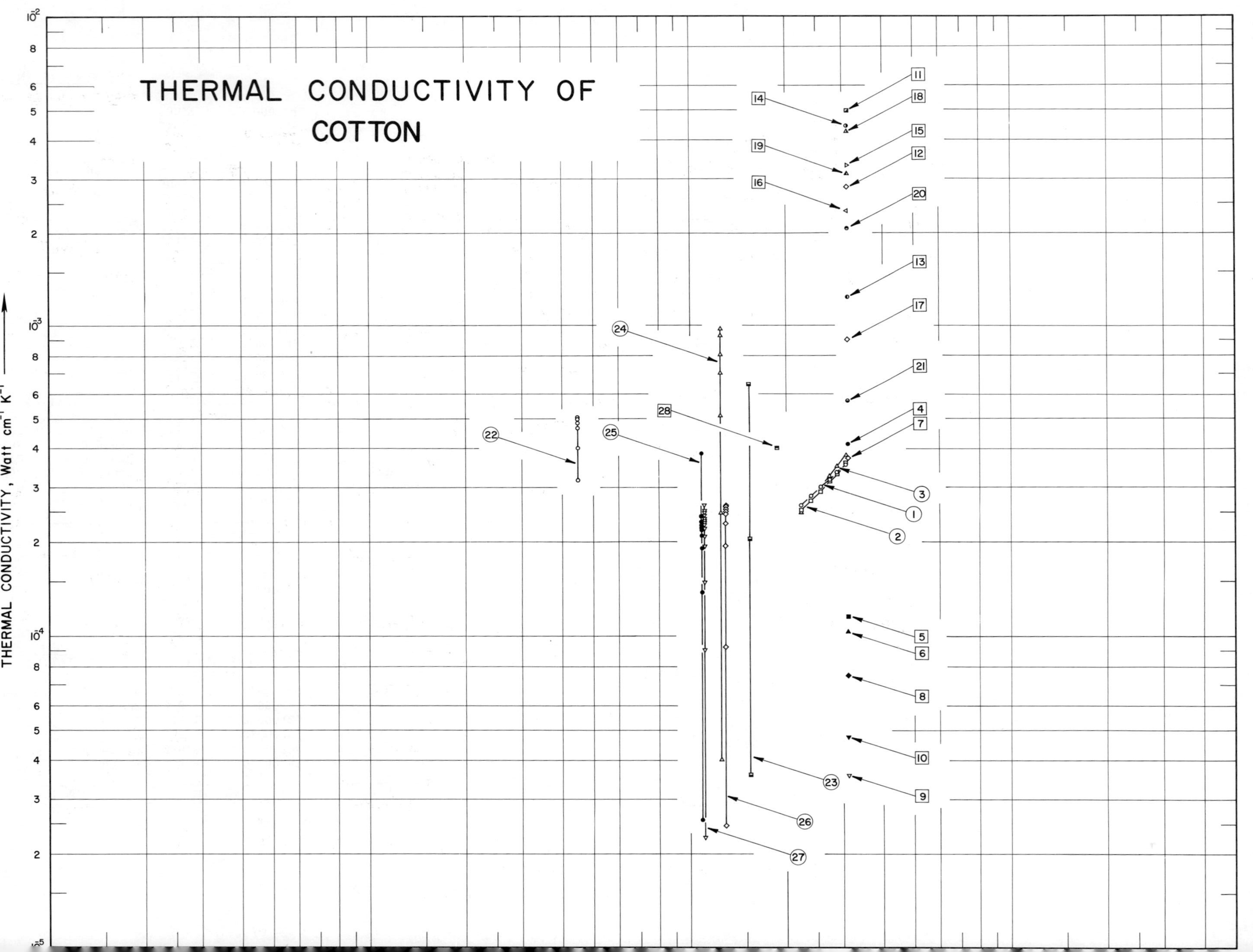
THERMAL CONDUCTIVITY OF COTTON
THERMAL CONDUCTIVITY, Watt cm^{-1} K^{-1}

SPECIFICATION TABLE NO. 346 THERMAL CONDUCTIVITY OF COTTON

[For Data Reported in Figure and Table No. 346]

Curve No.	Ref. No.	Method Used	Year	Temp. Range, K	Reported Error, %	Name and Specimen Designation	Composition (weight percent), Specifications and Remarks
1	109	L	1947	222-305			Specimen 1 in. thick; density 0.0424 g cm^{-3}; as received.
2	109	L	1947	222-305			Specimen 1 in. thick; density 0.0245 g cm^{-3}; as received.
3	109	L	1947	222-305			Specimen 1 in. thick; density 0.0152 g cm^{-3}; as received.
4	558	L	1951	310.9			8 x 8 x 1 in.; density 0.0125 g cm^{-3} at 25 C.
5	558	L	1951	310.9			The above specimen measured in a vacuum of 0.11 μ Hg.
6	558	L	1951	310.9			The above specimen measured in a vacuum of 0.07 μ Hg.
7	558	L	1951	310.9			8 x 8 x 1 in.; density 0.0245 g cm^{-3} at 25 C.
8	558	L	1951	310.9			The above specimen measured in a vacuum of 0.12 μ Hg.
9	558	L	1951	310.9			8 x 8 x 1 in.; density 0.0420 g cm^{-3} at 25 C.
10	558	L	1951	310.9			The above specimen measured in a vacuum of 0.14 μ Hg.
11	560	L	1954	310.9			Specimen 1 in. thick; density 0.0125 g cm^{-3}; measured in F-12.
12	560	L	1954	310.9			Specimen 1 in. thick; density 0.0245 g cm^{-3}; measured in CO_2.
13	560	L	1954	310.9			The above specimen measured in F-12.
14	560	L	1954	310.9			Specimen 1 in. thick; density 0.0420 g cm^{-3}; measured in CO_2.
15	560	L	1954	310.9			The above specimen measured in F-12.
16	122	R	1953	49.4		Medical cotton	Filled with hydrogen; measured under various hydrogen pressures ranging from 0.322 to 322.0 mm Hg.
17	122	R	1953	154.8		Medical cotton	Filled with hydrogen; measured under various hydrogen pressures ranging from 0.0089 to 0.791 mm Hg.
18	122	R	1953	125.1		Medical cotton	Filled with hydrogen; measured under various hydrogen pressures ranging from 0.0104 to 501.4 mm Hg.
19	122	R	1953	109.2		Medical cotton	Filled with nitrogen; measured under various nitrogen pressures ranging from 0.0090 to 831.8 mm Hg.
20	122	R	1953	128.5		Medical cotton	Filled with nitrogen; measured under various nitrogen pressures ranging from 0.0084 to 800.0 mm Hg.
21	122	R	1953	111.0		Medical cotton	Measured in air under various atmospheric pressures ranging from 0.0080 to 708.0 mm Hg.
22	122	R	1953	49.4		Medical cotton	Specimen contained between a solid copper ball and a concentric spherical shell of dia 2 in. and 5 in., respectively; measured in hydrogen at pressures ranging from 0.322 to 322.0 mm Hg.

SPECIFICATION TABLE NO. 346 (continued)

Curve No.	Ref. No.	Method Used	Year	Temp. Range, K	Reported Error, %	Name and Specimen Designation	Composition (weight percent), Specifications and Remarks
23	122	R	1953	154.8		Medical cotton	Similar to above except measured in hydrogen at pressures ranging from 0.0089 to 0.791 mm Hg.
24	122	R	1953	125.1		Medical cotton	Similar to above except measured in hydrogen at pressures ranging from 0.0104 to 501.4 mm Hg.
25	122	R	1953	109.2		Medical cotton	Similar to above except measured in nitrogen at pressures ranging from 0.0090 to 831.8 mm Hg.
26	122	R	1953	128.5		Medical cotton	Similar to above except measured in nitrogen at pressures ranging from 0.0084 to 800.0 mm Hg.
27	122	R	1953	111.0		Medical cotton	Similar to above except measured in air at pressures ranging from 0.0080 to 986.3 mm Hg.
28	561	R	1948	186.5		Cotton waste	Tangled thread form; packed in a hollow cylinder of 2 in. I.D. and 10 in. O.D.; bulk density 0.131 g cm^{-3}.

DATA TABLE NO. 346 THERMAL CONDUCTIVITY OF COTTON

[Temperature, T, K; Thermal Conductivity, k, Watt $cm^{-1} K^{-1}$]

T	k
CURVE 1	
222.0	0.000264
238.7	0.000281
255.4	0.000300
272.2	0.000319
288.9	0.000336
305.5	0.000355
CURVE 2	
222.0	0.000254
238.7	0.000271
255.4	0.000291
272.2	0.000312
288.9	0.000332
305.5	0.000356
CURVE 3	
222.0	0.000250
238.7	0.000275
255.4	0.000300*
272.2	0.000327
288.9	0.000353
305.5	0.000384
CURVE 4	
310.9	0.000415
CURVE 5	
310.9	0.000116
CURVE 6	
310.9	0.000104
CURVE 7	
310.9	0.00371
CURVE 8	
310.9	0.0000750

T	k
CURVE 9	
310.9	0.0000358
CURVE 10	
310.9	0.0000476
CURVE 11	
310.9	0.00498
CURVE 12	
310.9	0.00284
CURVE 13	
310.9	0.00125
CURVE 14	
310.9	0.00445
CURVE 15	
310.9	0.00332
CURVE 16	
310.9	0.00236
CURVE 17	
310.9	0.000900
CURVE 18	
310.9	0.00429
CURVE 19	
310.9	0.00312
CURVE 20	
310.9	0.00208

T	k
CURVE 21	
310.9	0.000571
p(mm Hg)	**k**
CURVE 22 T = 49.4 K	
0.322	0.000318
1.03	0.000401
7.00	0.000469
28.0	0.000486
62.9	0.000491*
195.9	0.000495*
396.3	0.000500
322.0	0.000506
CURVE 23 T = 154.8K	
0.0089	0.0000360
0.0847	0.000205
0.791	0.000653
CURVE 24 T = 125.1K	
0.0104	0.0000405
0.243	0.000250
0.625	0.000515
2.01	0.000711
4.00	0.000812
35.0	0.000933
303.4	0.000983
501.4	0.000987
CURVE 25 T = 109.2K	
0.0090	0.0000258
0.266	0.000138
1.48	0.000192
5.00	0.000210
20.0	0.000220
50.0	0.000224
200.0	0.000227

p(mm Hg)	k
CURVE 25 (cont.)	
398.0	0.000229*
603.6	0.000233
798.0	0.000243
831.8	0.000388
CURVE 26 T = 128.5K	
0.0084	0.0000246
0.0837	0.0000927
0.838	0.000195
4.00	0.000230
20.0	0.000248
50.0	0.000252
200.0	0.000256
402.7	0.000256*
602.7	0.000262
800.0	0.000264
CURVE 27 T = 111 K	
0.0080	0.0000225
0.0721	0.0000903
0.330	0.000149
1.61	0.000194
4.00	0.000208
20.0	0.000222
50.0	0.000226
200.9	0.000229*
394.5	0.000231
394.5	0.000231*
597.0	0.000235
625.2	0.000235*
746.4	0.000239
918.2	0.000249
201.8	0.000229*
400.9	0.000230*
602.6	0.000235*
743.0	0.000239*
986.3	0.000264
760.3	0.000252
708.0	0.000242

T	k
CURVE 28	
186.5	0.000402

* Not shown on plot

SPECIFICATION TABLE NO. 347 THERMAL CONDUCTIVITY OF FAT

Curve No.	Ref. No.	Method Used	Year	Temp. Range, K	Reported Error, %	Name and Specimen Designation	Composition (weight percent), Specifications and Remarks
1	562	R	1954	293,333		Beef fat	No details reported.
2	562	R	1954	293.2		Bone fat	No details reported.
3	562	R	1954	293.2		Pig fat	No details reported.

DATA TABLE NO. 347 THERMAL CONDUCTIVITY OF FAT

[Temperature, T, K; Thermal Conductivity, k, Watt $cm^{-1}K^{-1}$]

T	k
CURVE 1*	
293.2	0.00354
333.2	0.00175
CURVE 2*	
293.2	0.00186
CURVE 3*	
293.2	0.00238

* No graphical presentation

SPECIFICATION TABLE NO. 348 THERMAL CONDUCTIVITY OF FIR

Curve No.	Ref. No.	Method Used	Year	Temp. Range, K	Reported Error, %	Name and Specimen Designation	Composition (weight percent), Specifications and Remarks
1	125	L	1923	293.2			Specific gravity 0.6; moisture content 15%; measured under a load of 21 psi; heat flow perpendicular to the grain.

DATA TABLE NO. 348 THERMAL CONDUCTIVITY OF FIR

[Temperature, T, K; Thermal Conductivity, k, Watt $cm^{-1}K^{-1}$]

T	k
CURVE 1*	
293.2	0.00117

* No graphical presentation

SPECIFICATION TABLE NO. 349 THERMAL CONDUCTIVITY OF GREENHEART

Curve No.	Ref. No.	Method Used	Year	Temp. Range, K	Reported Error, %	Name and Specimen Designation	Composition (weight percent), Specifications and Remarks
1	44	F	1914	293,373			0.668 cm dia x 7.5 cm long; density 1.08 g cm^{-3}.

DATA TABLE NO. 349 THERMAL CONDUCTIVITY OF GREENHEART

[Temperature, T, K; Thermal Conductivity, k, Watt $cm^{-1}K^{-1}$]

T	k
CURVE 1*	
293.2	0.00469
373.2	0.00460

* No graphical presentation

SPECIFICATION TABLE NO. 350 THERMAL CONDUCTIVITY OF HARDWOOD

Curve No.	Ref. No.	Method Used	Year	Temp. Range, K	Reported Error, %	Name and Specimen Designation	Composition (weight percent), Specifications and Remarks
1	596	L	1951	292-325			No details reported.
2	596	C	1951	328.8			3 in. dia x 0.25 in. thick; Armco iron used as comparative material.

DATA TABLE NO. 350 THERMAL CONDUCTIVITY OF HARDWOOD

[Temperature, T, K; Thermal Conductivity, k, Watt $cm^{-1}K^{-1}$]

T	k
CURVE 1*	
292.4	0.00158
316.3	0.00155
317.8	0.00154
324.7	0.00154
CURVE 2*	
328.8	0.00152

* No graphical presentation

SPECIFICATION TABLE NO. 351 THERMAL CONDUCTIVITY OF IVORY

Curve No.	Ref. No.	Method Used	Year	Temp. Range, K	Reported Error, %	Name and Specimen Designation	Composition (weight percent), Specifications and Remarks
1	125	L	1923	353.2	1	African ivory	Measured under a load of 113 psi; heat flow perpendicular to the axis.
2	125	L	1923	353.2	1	African ivory	Similar to the above specimen.
3	125	L	1923	353.2	1	African ivory	Similar to the above specimen except thickness 0.178 cm; heat flow parallel to the axis.

DATA TABLE NO. 351 THERMAL CONDUCTIVITY OF IVORY

[Temperature, T, K; Thermal Conductivity, k, Watt $cm^{-1}K^{-1}$]

T	k
CURVE 1*	
353.2	0.00452
CURVE 2*	
353.2	0.00523
CURVE 3*	
353.2	0.00573

* No graphical presentation

SPECIFICATION TABLE NO. 352 THERMAL CONDUCTIVITY OF KAPOK

Curve No.	Ref. No.	Method Used	Year	Temp. Range, K	Reported Error, %	Name and Specimen Designation	Composition (weight percent), Specifications and Remarks
1	558	L	1951	310.9			1 in. thick; density 0.00416 g cm^{-3} at 25 C; measured in air at various air pressures ranging from 12.7 to 743.0 mm Hg.
2	558	L	1951	310.9			1 in. thick; density 0.0159 g cm^{-3} at 25 C; measured in air at 1 atm.
3	558	L	1951	310.9			The above specimen measured in a vacuum of 0.1 μHg.
4	558	L	1951	310.9			1 in. thick; density 0.0634 g cm^{-3} at 25 C; measured in air at 1 atm.
5	558	L	1951	310.9			The above specimen measured in a vacuum of 0.12 μ Hg.
6	560	L	1954	310.9			1 in. thick; density 0.00416 g cm^{-3}; measured in air.
7	560	L	1954	310.9			The above specimen measured in CO_2.
8	560	L	1954	310.9			The above specimen measured in F-12.
9	560	L	1954	310.9			The above specimen measured in vacuum.
10	560	L	1954	310.9			1 in. thick; density 0.0159 g cm^{-3}; measured in CO_2.
11	560	L	1954	310.9			The above specimen measured in F-12.
12	560	L	1954	310.9			1 in. thick; density 0.0634 g cm^{-3}; measured in CO_2.
13	560	L	1954	310.9			The above specimen measured in F-12.
14	560	L	1954	310.9			1 in. thick; density 0.00416 g cm^{-3}; measured in air with various % content of CO_2.
15	560	L	1954	310.9			The above specimen measured in air with various % content of F-12.

DATA TABLE NO. 352 THERMAL CONDUCTIVITY OF KAPOK

[Temperature, T, K; Thermal Conductivity, k, Watt $cm^{-1}K^{-1}$]

p(mm Hg)	k
CURVE 1* (T = 310.9K)	
12.7	0.000513
236.2	0.000516
490.2	0.000516
743.0	0.000516

T	k
CURVE 2*	
310.9	0.000353
CURVE 3*	
310.9	0.0000541
CURVE 4*	
310.9	0.000352
CURVE 5*	
310.9	0.0000252
CURVE 6*	
310.9	0.00633
CURVE 7*	
310.9	0.00509
CURVE 8*	
310.9	0.00403
CURVE 9*	
310.9	0.00157
CURVE 10*	
310.9	0.00305
CURVE 11*	
310.9	0.00214
CURVE 12*	
310.9	0.00292
CURVE 13*	
310.9	0.00199

% of CO_2 in air	k
CURVE 14* (T = 310.9K)	
0	0.00625
25	0.00599
50	0.00561
75	0.00535
100	0.00505

% of F-12 in air	k
CURVE 15* (T = 310.9K)	
0	0.00625
25	0.00531
50	0.00467
75	0.00427
100	0.00395

* No graphical presentation

SPECIFICATION TABLE NO. 353 THERMAL CONDUCTIVITY OF LIGNUM VITAE

Curve No.	Ref. No.	Method Used	Year	Temp. Range, K	Reported Error, %	Name and Specimen Designation	Composition (weight percent), Specifications and Remarks
1	44	F	1914	293, 373			0.668 cm dia x 6.8 cm long; density 1.16 g cm^{-3}.

DATA TABLE NO. 353 THERMAL CONDUCTIVITY OF LIGNUM VITAE

[Temperature, T, K; Thermal Conductivity, k, Watt $cm^{-1}K^{-1}$]

T	k
CURVE 1*	
293.2	0.00253
373.2	0.00300

* No graphical presentation

SPECIFICATION TABLE NO. 354 THERMAL CONDUCTIVITY OF MAHOGANY

Curve No.	Ref. No.	Method Used	Year	Temp. Range, K	Reported Error, %	Name and Specimen Designation	Composition (weight percent), Specifications and Remarks
1	44	F	1914	293,373			0.65 cm dia x 5.5 cm long; density 0.55 g cm^{-3}.
2	125	L	1923	293.2	1		Sp. gravity 0.70; moisture content 15%; heat flow parallel to grain.
3	125	L	1923	293.2	1		Sp. gravity 0.70; moisture content 15%; heat flow perpendicular to grain, radial to annual rings.
4	125	L	1923	293.2	1		Sp. gravity 0.70; moisture content 15%; heat flow perpendicular to grain, tangential to annual rings.

DATA TABLE NO. 354 THERMAL CONDUCTIVITY OF MAHOGANY

[Temperature, T, K; Thermal Conductivity, k, Watt $cm^{-1}K^{-1}$]

T	k
CURVE 1*	
293.2	0.00213
373.2	0.00253
CURVE 2*	
293.2	0.00310
CURVE 3*	
293.2	0.00167
CURVE 4*	
293.2	0.00155

* No graphical presentation

SPECIFICATION TABLE NO. 355 THERMAL CONDUCTIVITY OF MAPLE

Curve No.	Ref. No.	Method Used	Year	Temp. Range, K	Reported Error, %	Name and Specimen Designation	Composition (weight percent), Specifications and Remarks
1	108	L	1920	293, 323			9 in. dia x 0.733 in. thick; specific gravity 0.72; heat flow along grain.
2	108	L	1920	323.2			Similar to the above specimen except thickness 0.508 in.

DATA TABLE NO. 355 THERMAL CONDUCTIVITY OF MAPLE

[Temperature, T, K; Thermal Conductivity, k, Watt $cm^{-1}K^{-1}$]

T	k
CURVE 1*	
293.2	0.00425
323.2	0.00434
CURVE 2*	
323.2	0.00182

* No graphical presentation

SPECIFICATION TABLE NO. 356 THERMAL CONDUCTIVITY OF OAK

Curve No.	Ref. No.	Method Used	Year	Temp. Range, K	Reported Error, %	Name and Specimen Designation	Composition (weight percent), Specifications and Remarks
1	108	L	1920	323.2		White oak	9 in. dia x 0.516 in. thick; specific gravity 0.60; heat flow across grain.
2	108	L	1920	328.2		White oak	Similar to the above specimen but thickness 0.754 in.
3	44	F	1914	293,373			0.65 in. dia x 6.3 cm long; density 0.65 g cm^{-3}.
4	125	L	1923	293.2	1		Specific gravity 0.60; moisture content 14%; heat flow perpendicular to grain and tangent to annual rings.

DATA TABLE NO. 356 THERMAL CONDUCTIVITY OF OAK

[Temperature, T, K; Thermal Conductivity, k, Watt $cm^{-1}K^{-1}$]

T	k
CURVE 1*	
323.2	0.00190
CURVE 2*	
328.2	0.00395
CURVE 3*	
293.2	0.00244
373.2	0.00254
CURVE 4*	
293.2	0.00117

* No graphical presentation

SPECIFICATION TABLE NO. 357 THERMAL CONDUCTIVITY OF PINES

Curve No.	Ref. No.	Method Used	Year	Temp. Range, K	Reported Error, %	Name and Specimen Designation	Composition (weight percent), Specifications and Remarks
1	108	L	1920	343.2		White pine	9 in. dia by 0.519 in. thick; specific gravity 0.45; heat flow in longitudinal direction.
2	108	L	1920	328.2		White pine	Similar to the above specimen but thickness 0.732 in.
3	125	L	1923	293.2		Pitch pine	Moisture content 15%; measured under a load of 21 psi.
4	109	L	1947	222-306			1.25 in. thick; density 0.386 g cm^{-3}; as received.

DATA TABLE NO. 357 THERMAL CONDUCTIVITY OF PINES

[Temperature, T, K; Thermal Conductivity, k, Watt cm^{-1} K^{-1}]

T	k
CURVE 1*	
343.2	0.00107
CURVE 2*	
328.2	0.00256
CURVE 3*	
293.2	0.00138
CURVE 4*	
222.0	0.000886
238.7	0.000913
255.4	0.000939
272.2	0.000966
288.9	0.000994
305.5	0.00102

*No graphical presentation

SPECIFICATION TABLE NO. 358 THERMAL CONDUCTIVITY OF REDWOOD

Curve No.	Ref. No.	Method Used	Year	Temp. Range, K	Reported Error, %	Name and Specimen Designation	Composition (weight percent), Specifications and Remarks
1	563	L	1945	222-305		Redwood bark	Thickness of specimen 1 in.; density 0.0641 g cm^{-3}; as received.
2	128	L	1946	318.7		Redwood bark	Shredded bark; 2.00 in. thick; density 0.0625 g cm^{-3}.
3	128	L	1946	189,242		Redwood bark	Shredded bark; 1.00 in. thick; density 0.0641 g cm^{-3}.

DATA TABLE NO. 358 THERMAL CONDUCTIVITY OF REDWOOD

[Temperature, T, K; Thermal Conductivity, k, Watt $cm^{-1}K^{-1}$]

T	k
CURVE 1*	
222.2	0.000286
239.2	0.000307
255.5	0.000330
272.1	0.000356
288.8	0.000379
305.3	0.000407
CURVE 2*	
318.7	0.000107
CURVE 3*	
189.3	0.0000593
241.5	0.0000717

* No graphical presentation

SPECIFICATION TABLE NO. 359 THERMAL CONDUCTIVITY OF SAWDUST

Curve No.	Ref. No.	Method Used	Year	Temp. Range, K	Reported Error, %	Name and Specimen Designation	Composition (weight percent), Specifications and Remarks
1	529	P	1956	290.4			Dry mixed timbers; density 0.128 g cm^{-3}; thermal conductivity data calculated from measured transient temperature changes.
2	529	P	1956	290.4			Similar to the above specimen except density 0.224 g cm^{-3}.

DATA TABLE NO. 359 THERMAL CONDUCTIVITY OF SAWDUST

[Temperature, T, K; Thermal Conductivity, k, Watt $cm^{-1}K^{-1}$]

T	k
CURVE 1*	
290.4	0.000505
CURVE 2*	
290.4	0.000699

* No graphical presentation

SPECIFICATION TABLE NO. 360 THERMAL CONDUCTIVITY OF SPRUCE

Curve No.	Ref. No.	Method Used	Year	Temp. Range, K	Reported Error, %	Name and Specimen Designation	Composition (weight percent), Specifications and Remarks
1	125	L	1923	373.2	1		Specimen gradually dried out in an electric oven maintained at 100 C; measurements taken with various reducing moisture contents; heat flow across the grain and radial to the annual rings.
2	125	L	1923	293.2	1		Specific gravity 0.41; moisture content 16%; heat flow along grain.
3	125	L	1923	293.2	1		Similar to the above specimen except heat flow perpendicular to the grain and radial to the annual rings.
4	125	L	1923	293.2	1		Similar to the above specimen except heat flow perpendicular to the grain and tangent to the annual rings.

DATA TABLE NO. 360 THERMAL CONDUCTIVITY OF SPRUCE

[Temperature, T, K; Thermal Conductivity, k, Watt $cm^{-1}K^{-1}$]

Moisture content(%)	k
CURVE 1* (T = 373.2K)	
3.40	0.00122
5.80	0.00126
7.70	0.00129
9.95	0.00133
17.00	0.00142

T	k
CURVE 2*	
293.2	0.00222
CURVE 3*	
293.2	0.00121
CURVE 4*	
293.2	0.00105

* No graphical presentation

SPECIFICATION TABLE NO. 361 THERMAL CONDUCTIVITY OF TEAK

Curve No.	Ref. No.	Method Used	Year	Temp. Range, K	Reported Error, %	Name and Specimen Designation	Composition (weight percent), Specifications and Remarks
1	125	L	1923	293.2	1		Specific gravity 0.72; moisture content 10%; measured under a load of 21 psi; heat flow perpendicular to grain and tangent to annual rings.

DATA TABLE NO. 361 THERMAL CONDUCTIVITY OF TEAK

[Temperature, T, K; Thermal Conductivity, k, Watt $cm^{-1}K^{-1}$]

T	k
CURVE 1*	
293.2	0.00138

* No graphical presentation

SPECIFICATION TABLE NO. 362 THERMAL CONDUCTIVITY OF VULCANIZED FIBER

Curve No.	Ref. No.	Method Used	Year	Temp. Range, K	Reported Error, %	Name and Specimen Designation	Composition (weight percent), Specifications and Remarks
1	125	L	1923	323.2	1		Measured under a load of 21 psi.
2	125	L	1923	323.2	1		Similar to above.

DATA TABLE NO. 362 THERMAL CONDUCTIVITY OF VULCANIZED FIBER

[Temperature, T, K; Thermal Conductivity, k, Watt $cm^{-1}K^{-1}$]

T	k
CURVE 1*	
323.2	0.00209
CURVE 2*	
323.2	0.00335

* No graphical presentation

SPECIFICATION TABLE NO. 363 THERMAL CONDUCTIVITY OF WALNUT

Curve No.	Ref. No.	Method Used	Year	Temp. Range, K	Reported Error, %	Name and Specimen Designation	Composition (weight percent), Specifications and Remarks
1	44	F	1914	293,373		Satin walnut	Density 0.5 g cm^{-3}; measured along grain.
2	44	F	1914	293,373		Satin walnut	Similar to the above specimen.
3	125	L	1923	293.2	1		Heat flow across the grain and tangent to the annual rings; measured under various load ranging from 33.2 to 61.1 psi.
4	125	L	1923	293.2	1		The above specimen measured at decreasing loads ranging from 61.1 to 23.2 psi.
5	125	L	1923	293.8	1		0.0593 cm thick; measured under a load of 21 psi.
6	125	L	1923	294.4	1		Similar to above except thickness 0.0913 cm.
7	125	L	1923	291.4	1		Similar to above except thickness 0.1503 cm.
8	125	L	1923	293.2	1		Specific gravity 0.65; moisture content 11.8%; measured under a load of 21 psi; heat flow along grain.
9	125	L	1923	293.2	1		Specific gravity 0.65; moisture content 12.1%; measured under a load of 21 psi; heat flow perpendicular to grain and radial to annual rings.
10	125	L	1923	293.2	1		Specific gravity 0.65; moisture content 11.3%; measured under a load of 21 psi; heat flow perpendicular to grain and tangent to annual rings.

DATA TABLE NO. 363 THERMAL CONDUCTIVITY OF WALNUT

[Temperature, T, K; Thermal Conductivity, k, Watt $cm^{-1}K^{-1}$]

T	k	p(psi)	k	p(psi)	k	T	k	T	k	T	k
CURVE 1*		CURVE 3* (T = 293.2K)		CURVE 4* (T = 293.2K)		CURVE 5*		CURVE 7*		CURVE 9*	
293.2	0.00182	33.2	0.00137	23.2	0.00138	293.8	0.00136	291.4	0.00136	293.2	0.00145
373.2	0.00227	46.1	0.00138	33.2	0.00136						
CURVE 2*		61.1	0.00137	37.7	0.00137	CURVE 6*		CURVE 8*		CURVE 10*	
293.2	0.000669			46.1	0.00138	294.4	0.00137	293.2	0.00332	293.2	0.00136
373.2	0.00104			52.0	0.00137						
				61.1	0.00137						

* No graphical presentation

SPECIFICATION TABLE NO. 364 THERMAL CONDUCTIVITY OF WHITE WOOD

Curve No.	Ref. No.	Method Used	Year	Temp. Range, K	Reported Error, %	Name and Specimen Designation	Composition (weight percent), Specifications and Remarks
1	44	F	1914	293,373		American white wood	0.65 cm dia x 6.5 cm long; density 0.575 g cm^{-3}.

DATA TABLE NO. 364 THERMAL CONDUCTIVITY OF WHITE WOOD

[Temperature, T, K; Thermal Conductivity, k, Watt $cm^{-1}K^{-1}$]

T	k
CURVE 1*	
293.2	0.00170
373.2	0.00187

* No graphical presentation

SPECIFICATION TABLE NO. 365 THERMAL CONDUCTIVITY OF WOOD FIBERS

Curve No.	Ref. No.	Method Used	Year	Temp. Range, K	Reported Error, %	Name and Specimen Designation	Composition (weight percent), Specifications and Remarks
1	558	L	1951	310.9			Fiber; 1 in. thick; density 0.056 g cm^{-3} at 25 C; measured in air.
2	558	L	1951	310.9			The above specimen measured in a vacuum of 0.08 μ Hg.
3	558	L	1951	310.9			Fiber; 1 in. thick; density 0.111 g cm^{-3} at 25 C; measured in air.
4	558	L	1951	310.9			The above specimen measured in a vacuum of 0.09 μ Hg.
5	44	F	1914	293, 373		Red fiber	Fiber; density 2.59 g cm^{-3}.
6	564	L	1944	297-325			Fiber; 1 in. thick; density 0.0641 g cm^{-3}.
7	564	L	1944	297-325			1 in. thick; density 0.128 g cm^{-3}.
8	564	L	1944	296-325			1 in. thick; density 0.0961 g cm^{-3}.
9	560	L	1954	310.9			1 in. thick; density 0.056 g cm^{-3}; measured in F-12.
10	560	L	1954	310.9			1 in. thick; density 0.111 g cm^{-3}; measured in CO_2.
11	560	L	1954	310.9			The above specimen measured in F-12.

DATA TABLE NO. 365 THERMAL CONDUCTIVITY OF WOOD FIBERS

[Temperature, T, K; Thermal Conductivity, k, Watt $cm^{-1} K^{-1}$]

T	k	T	k	T	k	T	k	T	k
CURVE 1*		CURVE 4*		CURVE 6*		CURVE 8*		CURVE 10*	
310.9	0.000392	310.9	0.0000476	297.1	0.000387	296.2	0.000389	310.9	0.00350
				311	0.000410	311	0.000411		
CURVE 2*		CURVE 5*		324.6	0.000436	324.6	0.000436	CURVE 11*	
310.9	0.0000750	293.2	0.00469	CURVE 7*		CURVE 9*		310.9	0.00246
		373.2	0.00498						
CURVE 3*				297.1	0.000404	310.9	0.00247		
				310.7	0.000418				
310.9	0.000400			324.8	0.000441				

*No graphical presentation

SPECIFICATION TABLE NO. 366 THERMAL CONDUCTIVITY OF WOOL

Curve No.	Ref. No.	Method Used	Year	Temp. Range, K	Reported Error, %	Name and Specimen Designation	Composition (weight percent), Specifications and Remarks
1	565	L	1950	293.2		Angora wool	Density 0.19 g cm^{-3}.
2	565	L	1950	293.2		Sheep wool	Density 0.13 g cm^{-3}.
3	565	L	1950	293.2		Sheep wool	Density 0.16 g cm^{-3}.

DATA TABLE NO. 366 THERMAL CONDUCTIVITY OF WOOL

[Temperature, T, K; Thermal Conductivity, k, Watt $cm^{-1}K^{-1}$]

T	k
CURVE 1*	
293.2	0.000464
CURVE 2*	
293.2	0.000477
CURVE 3*	
293.2	0.000565

* No graphical presentation

THERMAL CONDUCTIVITY OF COTTON FABRIC

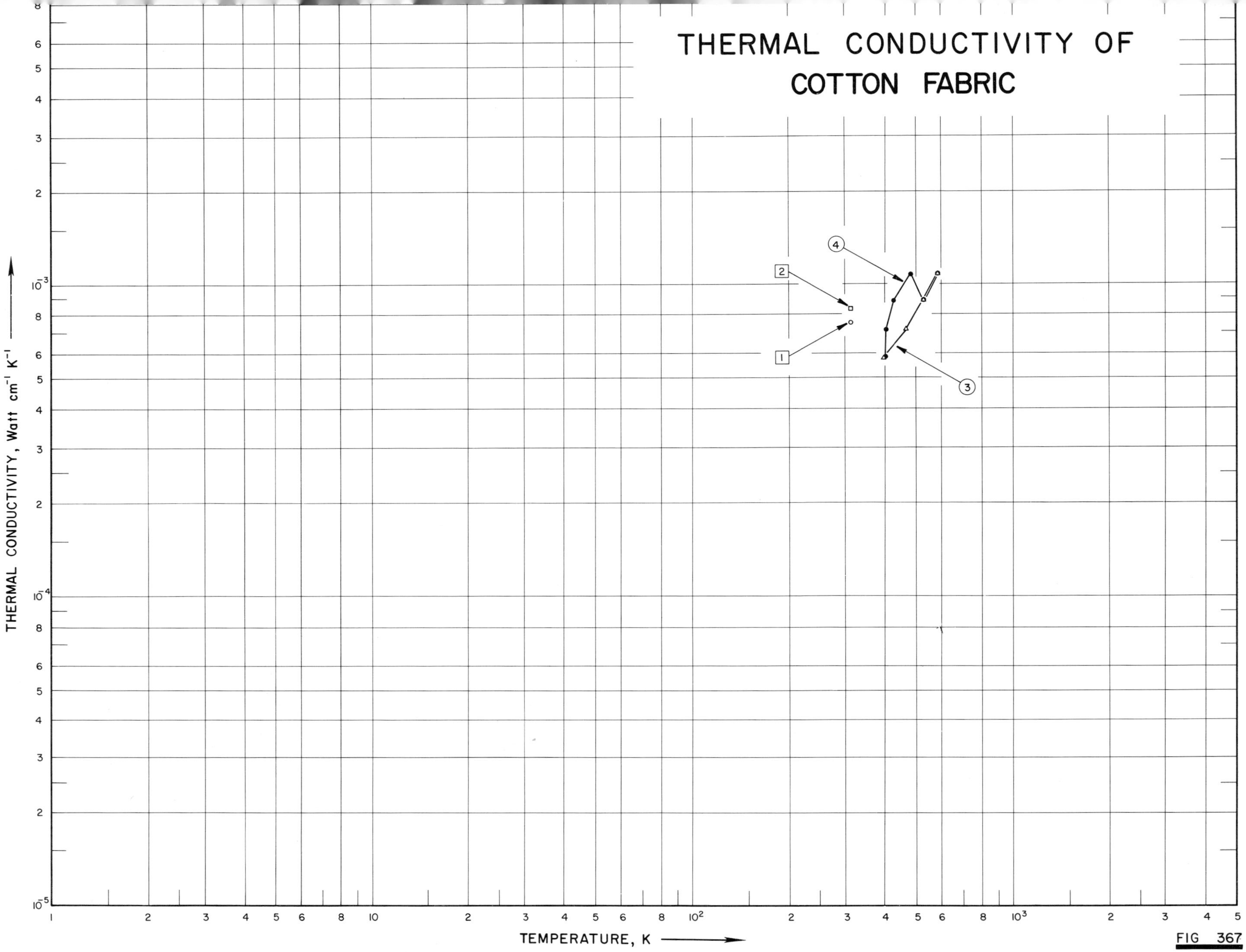

THERMAL CONDUCTIVITY, Watt cm^{-1} K^{-1}

TEMPERATURE, K

1
2
3
4

FIG 367

SPECIFICATION TABLE NO. 367 THERMAL CONDUCTIVITY OF COTTON FABRICS

[For Data Reported in Figure and Table No. 367]

Curve No.	Ref. No.	Method Used	Year	Temp. Range, K	Reported Error, %	Name and Specimen Designation	Composition (weight percent), Specifications and Remarks
1	125	L	1923	313. 2	1	Cotton fabric	3. 63 cm dia disk specimen; measured under a load of 21 psi.
2	125	L	1923	313. 2	1	Cotton fabric	Another run of the above specimen.
3	105	R	1958	398-584		Cotton silicate felt	Density 0. 185 g cm^{-3}.
4	105	R	1958	401-582		Cotton silicate felt	Second run of the above specimen.

DATA TABLE NO. 367 THERMAL CONDUCTIVITY OF COTTON FABRIC

[Temperature, T, K; Thermal Conductivity, k, Watt cm^{-1} K^{-1}]

T	k
CURVE 1	
313.2	0.000753
CURVE 2	
313.2	0.000837
CURVE 3	
398.2	0.000580
467.2	0.000717
528.7	0.000890
583.9	0.001080
CURVE 4	
400.9	0.000582
402.7	0.000715
527.2	0.000885
582.2	0.001074

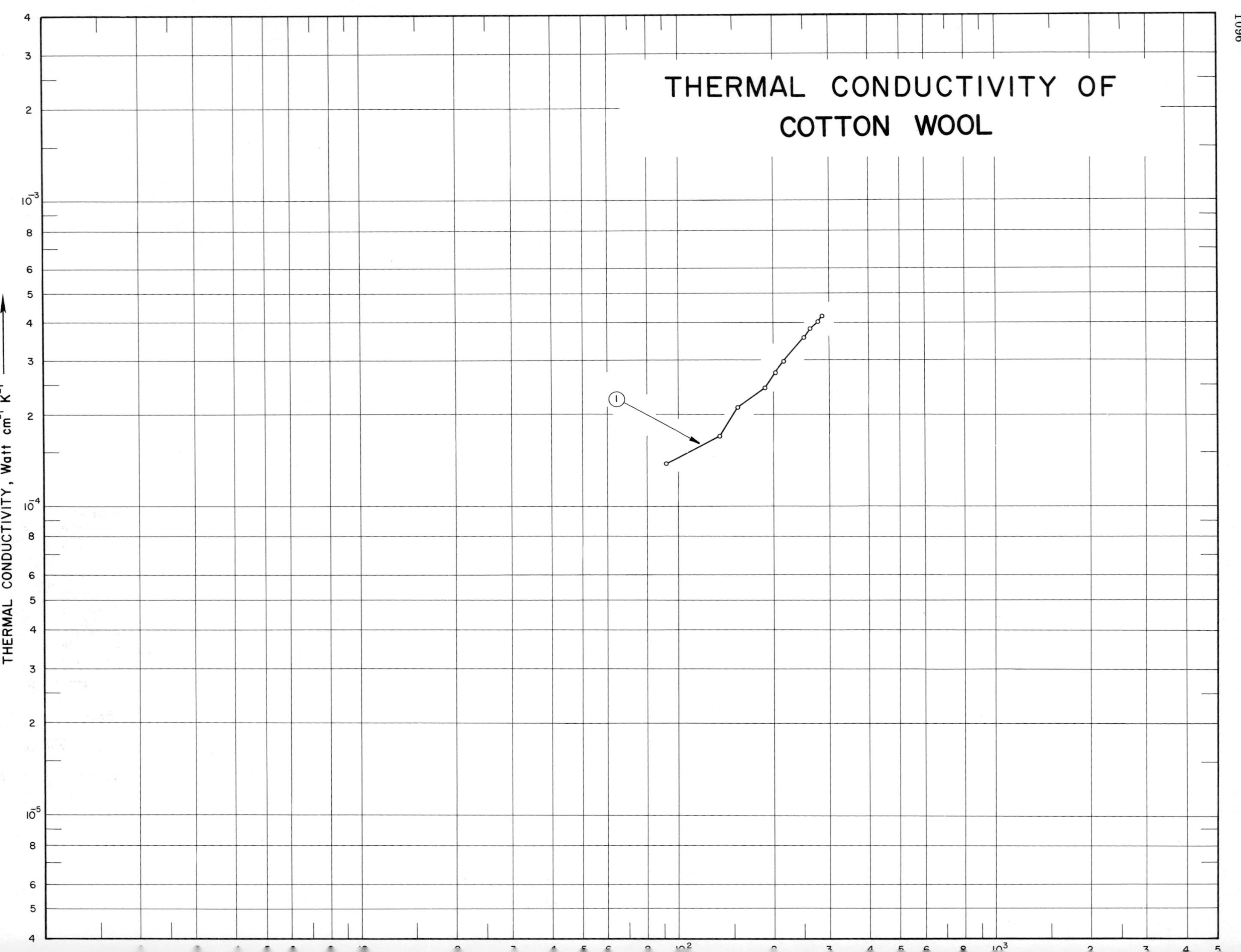
THERMAL CONDUCTIVITY OF
COTTON WOOL
1
THERMAL CONDUCTIVITY, Watt cm^{-1} K^{-1}

SPECIFICATION TABLE NO. 368 THERMAL CONDUCTIVITY OF COTTON WOOL

[For Data Reported in Figure and Table No. 368]

Curve No.	Ref. No.	Method Used	Year	Temp. Range, K	Reported Error, %	Name and Specimen Designation	Composition (weight percent), Specifications and Remarks
1	561	R	1948	92-286			Wadding specimen packed in a hollow cylinder of 2 in. I.D. and 10 in. O.D.; cruded; bulk density 0.0416 g cm^{-3}.

DATA TABLE NO. 368 THERMAL CONDUCTIVITY OF COTTON WOOL

[Temperature, T, K; Thermal Conductivity, k, Watt $cm^{-1} K^{-1}$]

T	k
CURVE 1	
92.2	0.000138
136.0	0.000170
155.9	0.000211
188.0	0.000246
204.4	0.000275
217.2	0.000298
251.2	0.000360
263.5	0.000384
278.0	0.000406
286.2	0.000422

SPECIFICATION TABLE NO. 369 THERMAL CONDUCTIVITY OF HAIR FELT

Curve No.	Ref. No.	Method Used	Year	Temp. Range, K	Reported Error, %	Name and Specimen Designation	Composition (weight percent), Specifications and Remarks
1	109	L	1947	222-306			Thickness 0.75 in.; density 0.175 g cm^{-3}; as received.
2	558	L	1951	310.9			Thickness 0.75 in.; density 0.178 g cm^{-3} at 25 C.
3	558	L	1951	310.9			The above specimen measured in a vacuum of 0.1 μ Hg.
4	560	L	1954	310.9			0.75 in. thick; density 0.178 g cm^{-3}; measured in air.
5	560	L	1954	310.9			The above specimen measured in CO_2.
6	560	L	1954	310.9			The above specimen measured in F-12.
7	560	L	1954	310.9			The above specimen measured in vacuum.

DATA TABLE NO. 369 THERMAL CONDUCTIVITY OF HAIR FELT

[Temperature, T, K; Thermal Conductivity, k, Watt $cm^{-1} K^{-1}$]

T	k
CURVE 1*	
222.0	0.000287
238.7	0.000303
255.4	0.000322
272.2	0.000340
288.9	0.000361
305.5	0.000382
CURVE 2*	
310.9	0.000375
CURVE 3*	
310.9	0.0000447
CURVE 4*	
310.9	0.00450
CURVE 5*	
310.9	0.00320
CURVE 6*	
310.9	0.00215
CURVE 7*	
310.9	0.000537

*No graphical presentation

SPECIFICATION TABLE NO. 370 THERMAL CONDUCTIVITY OF PLUTON CLOTH

Curve No.	Ref. No.	Method Used	Year	Temp. Range, K	Reported Error, %	Name and Specimen Designation	Composition (weight percent), Specifications and Remarks
1	526	P	1962	455.8	± 7		Manufactured by Minn. Mining & Mfg. Co.; plain weave type with 26 yarns $in.^{-1}$ and 25 yarns $in.^{-1}$ filling; 8.0 oz yd^{-2}; 0.01304 in. ave thickness; 0.01056 in. thickness under load; measured under 16.2 psia pressure, 11.9 ppi tensile stress, and 3.9 psi compressive stress; thermal conductivity calculated from measured transient temperature changes.
2	526	P	1962	459.1	± 7		Similar to the above specimen except with 0.01069 in. thickness under load; measured under 16.5 psia pressure, 9.6 ppi tensile stress, and 3.2 psi compressive stress.
3	526	P	1962	456.3	± 7		Similar to the above specimen except with 0.01077 in. thickness under load; measured under 16.5 psia pressure, 8.4 ppi tensile stress, and 2.9 psi compressive stress.
4	526	P	1962	565.5	± 7		Similar to the above specimen except with 0.01056 in. thickness under load; measured under 16.9 psia pressure, 11.9 ppi tensile stress, and 3.9 psi compressive stress.
5	526	P	1962	563.6	± 7		Similar to the above specimen except with 0.01069 in. thickness under load; measured under 17.7 psia pressure, 9.6 ppi tensile stress, and 3.2 psi compressive stress.
6	526	P	1962	565.5	± 7		Similar to the above specimen except with 0.01077 in. thickness under load; measured under 17.3 psia pressure, 8.4 ppi tensile stress, and 2.9 psi compressive stress.
7	526	P	1962	674.1	± 7		Similar to the above specimen except with 0.01069 in. thickness under load; measured under 17.1 psia pressure, 9.6 ppi tensile stress, and 3.2 psi compressive stress.
8	526	P	1962	674.9	± 7		Similar to the above specimen except with 0.01077 in. thickness under load; measured under 16.4 psia pressure, 8.4 ppi tensile stress, and 2.9 psi compressive stress.
9	526	P	1962	682.6	± 7		Similar to the above specimen except with 0.01077 in. thickness under load; measured under 16.2 psia pressure, 8.3 ppi tensile stress, and 2.9 psi compressive stress.
10	526	P	1962	449.4	± 7		Similar to the above specimen except with 0.01069 in. thickness under load; measured in 1.5 mm Hg vacuum under 9.6 ppi tensile stress and 3.2 psi compressive stress.
11	526	P	1962	447.6	± 7		Similar to the above specimen except with 0.01077 in. thickness under load; measured in 2.0 mm Hg vacuum under 8.4 ppi tensile stress and 2.9 psi compressive stress.
12	526	P	1962	445.3	± 7		Similar to the above specimen except with 0.01077 in. thickness under load; measured in 1.0 mm Hg vacuum under 8.3 ppi tensile stress and 2.9 psi compressive stress.

SPECIFICATION TABLE NO. 370 (continued)

Curve No.	Ref. No.	Method Used	Year	Temp. Range, K	Reported Error, %	Name and Specimen Designation	Composition (weight percent), Specifications and Remarks
13	526	P	1962	551.8	± 7		Similar to the above specimen except with 0.01069 in. thickness under load; measured in 1.5 mm Hg vacuum under 9.6 ppi tensile stress and 3.2 psi compressive stress.
14	526	P	1962	551.2	± 7		Similar to the above specimen except with 0.01077 in. thickness under load; measured in 1.5 mm Hg vacuum under 8.4 ppi tensile stress and 2.9 psi compressive stress.
15	526	P	1962	550.9	± 7		Similar to the above specimen except with 0.01077 in. thickness under load; measured in 1.2 mm Hg vacuum under 8.3 ppi tensile stress and 2.9 psi compressive stress.
16	526	P	1962	663.3	± 7		Similar to the above specimen except with 0.01069 in. thickness under load; measured in 5.0 mm Hg vacuum under 9.6 ppi tensile stress and 3.2 psi compressive stress.
17	526	P	1962	665.1	± 7		Similar to the above specimen except with 0.01077 in. thickness under load; measured in 1.5 mm Hg vacuum under 8.4 ppi tensile stress and 2.9 psi compressive stress.
18	526	P	1962	666.7	± 7		Similar to the above specimen except with 0.01077 in. thickness under load; measured in 2.5 mm Hg vacuum under 8.3 ppi tensile stress and 2.9 psi compressive stress.

DATA TABLE NO. 370 THERMAL CONDUCTIVITY OF PLUTON CLOTH

[Temperature, T, K; Thermal Conductivity, k, Watt $cm^{-1}K^{-1}$]

T	k	T	k	T	k	T	k	T	k	T	k
CURVE 1*		CURVE 4*		CURVE 7*		CURVE 10*		CURVE 13*		CURVE 16*	
455.8	0.000450	565.5	0.000523	674.1	0.000666	449.4	0.00143	551.8	0.000291	663.3	0.000320
CURVE 2*		CURVE 5*		CURVE 8*		CURVE 11*		CURVE 14*		CURVE 17*	
459.1	0.000424	563.6	0.000507	674.9	0.000659	447.6	0.000208	551.2	0.000244	665.1	0.000346
CURVE 3*		CURVE 6*		CURVE 9*		CURVE 12*		CURVE 15*		CURVE 18*	
456.3	0.000424	565.5	0.000493	682.6	0.000651	445.3	0.000162	550.9	0.000260	666.7	0.000263

* No graphical presentation

SPECIFICATION TABLE NO. 371 THERMAL CONDUCTIVITY OF RENE 41 CLOTH

Curve No.	Ref. No.	Method Used	Year	Temp. Range, K	Reported Error, %	Name and Specimen Designation	Composition (weight percent), Specifications and Remarks
1	526	P	1962	445.9	±7		Plain type weave; avg thickness 0.00591 in.; manufactured by Avco Everett Res. Lab; 0.002779 in. thickness under load; measured under 17.5 psia pressure, 6.1 ppi tensile stress, and 15.6 psi compressive stress; thermal conductivity value calculated from measured transient temp changes.
2	526	P	1962	434.4	±7		Similar to the above specimen except with 5.3 ppi tensile stress.
3	526	P	1962	425.7	±7		Similar to the above specimen except with 5.7 ppi tensile stress.
4	526	P	1962	674.4	±7		Similar to the above specimen except with 18.0 psia pressure and 6.1 ppi tensile stress.
5	526	P	1962	607.3	±7		Similar to the above specimen except with 17.5 psia pressure and 5.7 ppi tensile stress.
6	526	P	1962	677.6	±7		Similar to the above specimen except with 0.00357 in. thickness under load; measured under 17.5 psia pressure, 6.1 ppi tensile stress, and 1.9 psi compressive stress.
7	526	P	1962	870.4	±7		Similar to the above specimen except with 0.002779 in. thickness under load; measured under 17.5 psia pressure, 6.1 ppi tensile stress, and 15.6 psi compressive stress.
8	526	P	1962	746.5	±7		Similar to the above specimen except with 17.6 psia pressure, 5.7 ppi tensile stress, and 15.6 psi compressive stress.
9	526	P	1962	891.5	±7		Similar to the above specimen except with 17.7 psia pressure, 6.2 ppi tensile stress, and 1.9 psi compressive stress.
10	526	P	1962	1065.5	±7		Similar to the above specimen except measured under 0.002779 in. thickness under load; 17.5 psia pressure, 5.0 ppi tensile stress, and 15.6 psi compressive stress.
11	526	P	1962	916.5	±7		Similar to the above specimen except with 18.3 psia pressure; 4.7 ppi tensile stress and 15.6 psi compressive stress.
12	526	P	1962	1042.1	±7		Similar to the above specimen except with 0.003570 in. thickness under load; measured under 17.7 psia pressure and 4.8 ppi tensile stress.
13	526	P	1962	459.8	±7		Similar to the above specimen except with 0.002779 in. thickness under load; measured in 18 mm Hg vacuum under 6.1 ppi tensile stress and 13.5 psi compressive stress.
14	526	P	1962	443.8	±7		Similar to the above specimen except with 11 mm Hg vacuum, 5.3 ppi tensile stress, and 13.5 psi compressive stress.
15	526	P	1962	433.3	±7		Similar to the above specimen except with 6 mm Hg vacuum, 5.7 ppi tensile stress, and 13.5 psi compressive stress.
16	526	P	1962	678.4	±7		Similar to the above specimen except with 24 mm Hg vacuum, 6.1 ppi tensile stress, and 13.5 psi compressive stress.
17	526	P	1962	607.2	±7		Similar to the above specimen except with 16 mm Hg vacuum, 5.7 ppi tensile stress, and 13.5 psi compressive stress.

SPECIFICATION TABLE NO. 371 (continued)

Curve No.	Ref. No.	Method Used	Year	Temp. Range, K	Reported Error, %	Name and Specimen Designation	Composition (weight percent), Specifications and Remarks
18	526	P	1962	679.1	±7		Similar to the above specimen except with 0.003524 in. thickness under load; measured in 17 mm Hg vacuum under 6.2 ppi tensile stress.
19	526	P	1962	892.6	±7		Similar to the above specimen except with 0.002779 in. thickness under load; measured in 24 mm Hg vacuum under 6.1 ppi tensile stress and 13.5 psi compressive stress.
20	526	P	1962	749.2	±7		Similar to the above specimen except with 20 mm Hg vacuum, 5.7 ppi tensile stress, and 13.5 psi compressive stress.
21	526	P	1962	868.2	±7		Similar to the above specimen except with 0.00352 in. thickness under load; measured in 8 mm Hg vacuum under 5.2 ppi tensile stress and 2.0 psi compressive stress.
22	526	P	1962	1070.4	±7		Similar to the above specimen except with 0.002779 in. thickness under load; measured in 67 mm Hg vacuum under 5.0 ppi tensile stress and 13.5 psi compressive stress.
23	526	P	1962	889.8	±7		Similar to the above specimen except with 27 mm Hg vacuum, 1.0 ppi tensile stress, and 13.5 psi compressive stress.
24	526	P	1962	432.9	±7		Plain type weave; CS-105 (Goodyear Acft. Co.) coating; 0.00877 in. avg thickness; manufactured by the Avco Everett Res. Lab; 0.004087 in. thickness under load; measured under 17.5 psia pressure, 5.3 ppi tensile stress, and 8.8 psi compressive stress.
25	526	P	1962	425.9	±7		Similar to the above specimen except with 17.4 psia pressure and 3.2 ppi tensile stress.
26	526	P	1962	427.1	±7		Similar to the above specimen except with 17.5 psia pressure and 4.0 ppi tensile stress.
27	526	P	1962	606.3	±7		Similar to the above specimen except with 17.5 psia pressure and 5.3 ppi tensile stress.
28	526	P	1962	594.3	±7		Similar to the above specimen except with 17.4 psia pressure and 3.2 ppi tensile stress.
29	526	P	1962	784.6	±7		Similar to the above specimen except with 18.4 psia pressure and 5.3 ppi tensile stress.
30	526	P	1962	784.5	±7		Similar to the above specimen except with 17.9 psia pressure and 3.2 ppi tensile stress.
31	526	P	1962	912.1	±7		Similar to the above specimen except with 18.4 psia pressure and 2.2 ppi tensile stress.
32	526	P	1962	916.5	±7		Similar to the above specimen except with 18.4 psia pressure and 2.6 ppi tensile stress.
33	526	P	1962	430.5	±7		Similar to the above specimen except with 3 mm Hg vacuum and 5.3 ppi tensile stress.
34	526	P	1962	435.4	±7		Similar to the above specimen except with 21 mm Hg vacuum and 3.2 ppi tensile stress.
35	526	P	1962	434.8	±7		Similar to the above specimen except with 9 mm Hg vacuum and 4.0 ppi tensile stress.
36	526	P	1962	601.5	±7		Similar to the above specimen except with 7 mm Hg vacuum and 5.3 ppi tensile stress.
37	526	P	1962	605.9	±7		Similar to the above specimen except with 30 mm Hg vacuum and 3.2 ppi tensile stress.
38	526	P	1962	785.2	±7		Similar to the above specimen except with 10 mm Hg vacuum and 4.6 ppi tensile stress.
39	526	P	1962	785.4	±7		Similar to the above specimen except with 40 mm Hg vacuum and 3.2 ppi tensile stress.
40	526	P	1962	904.8	±7		Similar to the above specimen except with 20 mm Hg vacuum and 2.2 ppi tensile stress.

DATA TABLE NO. 371 THERMAL CONDUCTIVITY OF RENE 41 CLOTH

[Temperature, T, K; Thermal Conductivity, k, Watt $cm^{-1} K^{-1}$]

Curve	T	k
CURVE 1*	445.9	0.000159
CURVE 2*	434.4	0.000173
CURVE 3*	425.7	0.000144
CURVE 4*	674.4	0.000292
CURVE 5*	607.3	0.000320
CURVE 6*	677.6	0.000339
CURVE 7*	870.4	0.000348
CURVE 8*	746.5	0.000502
CURVE 9*	891.5	0.000400
CURVE 10*	1065.5	0.000755
CURVE 11*	916.5	0.000999
CURVE 12*	1042.1	0.000317
CURVE 13*	459.8	0.000128
CURVE 14*	443.8	0.000267
CURVE 15*	433.3	0.000260
CURVE 16*	678.4	0.000239
CURVE 17*	607.2	0.000204
CURVE 18*	679.1	0.000216
CURVE 19*	892.6	0.000339
CURVE 20*	749.2	0.000379
CURVE 21*	868.2	0.000166
CURVE 22*	1070.4	0.000817
CURVE 23*	889.8	0.000194
CURVE 24*	432.9	0.000519
CURVE 25*	425.4	0.000978
CURVE 26*	427.1	0.000661
CURVE 27*	606.3	0.000789
CURVE 28*	594.3	0.000710
CURVE 29*	784.6	0.00101
CURVE 30*	784.5	0.000933
CURVE 31*	912.1	0.000824
CURVE 32*	916.5	0.000457
CURVE 33*	430.5	0.000161
CURVE 34*	435.4	0.000346
CURVE 35*	434.8	0.000247
CURVE 36*	601.5	0.000223
CURVE 37*	605.9	0.0000917
CURVE 38*	785.2	0.000254
CURVE 39*	785.4	0.000943
CURVE 40*	904.8	0.000431

*No graphical presentation

SPECIFICATION TABLE NO. 372 THERMAL CONDUCTIVITY OF SILK FABRIC

Curve No.	Ref. No.	Method Used	Year	Temp. Range, K	Reported Error, %	Name and Specimen Designation	Composition (weight percent), Specifications and Remarks
1	125	L	1923	313.2	1		Measured under a load of 21 psi.
2	125	L	1923	313.2	1		Second run of the above specimen.
3	125	L	1923	294.2	1		Waterproofed; 0.012 cm thick; measured under a load of 21 psi.
4	125	L	1923	294.2	1		Waterproofed; two layers; 0.025 cm thick; measured under a load of 21 psi.
5	125	L	1923	294.2	1		Waterproofed; six layers; 0.075 cm thick; measured under a load of 21 psi.
6	125	L	1923	294.2	1		Waterproofed; nine layers; 0.111 cm thick; measured under a load of 21 psi.

DATA TABLE NO. 372 THERMAL CONDUCTIVITY OF SILK FABRIC

[Temperature, T, K; Thermal Conductivity, k, Watt $cm^{-1}K^{-1}$]

T	k	T	k
CURVE 1*		CURVE 5*	
313.2	0.000418	294.2	0.00174
CURVE 2*		CURVE 6*	
313.2	0.000502	294.2	0.00184
CURVE 3*			
294.2	0.00142		
CURVE 4*			
294.2	0.00150		

* No graphical presentation

SPECIFICATION TABLE NO. 373 THERMAL CONDUCTIVITY OF ASBESTOS CEMENT BOARD (TRANSITE)

Curve No.	Ref. No.	Method Used	Year	Temp. Range, K	Reported Error, %	Name and Specimen Designation	Composition (weight percent), Specifications and Remarks
1	390	P	1961	339-589		Transite	Density 0.193-0.1918 g cm^{-3} between 339-589 K; thermal conductivity values calculated from thermal diffusivity, density, and specific heat of the specimen.

DATA TABLE NO. 373 THERMAL CONDUCTIVITY OF ASBESTOS CEMENT BOARD (TRANSITE)

[Temperature, T, K; Thermal Conductivity, k, Watt cm^{-1} K^{-1}]

T	k
CURVE 1*	
338.7	0.00770
366.5	0.00757
394.3	0.00749
422.1	0.00742
449.8	0.00739
477.6	0.00736
505.4	0.00736
533.2	0.00736
560.9	0.00733
588.7	0.00731

*No graphical presentation

SPECIFICATION TABLE NO. 374 THERMAL CONDUCTIVITY OF ASPHALT - GLASS WOOL PAD

Curve No.	Ref. No.	Method Used	Year	Temp. Range, K	Reported Error, %	Name and Specimen Designation	Composition (weight percent), Specifications and Remarks
1	109	L	1947	222-306			Thickness of specimen 1 in.; density 0.284 g cm^{-3}; as received.

DATA TABLE NO. 374 THERMAL CONDUCTIVITY OF ASPHALT - GLASS WOOL PAD

[Temperature, T, K; Thermal Conductivity, k, Watt $cm^{-1}K^{-1}$]

T	k
CURVE 1*	
222.0	0.000304
238.7	0.000326
255.4	0.000348
272.2	0.000369
288.9	0.000394
305.5	0.000414

* No graphical presentation

SPECIFICATION TABLE NO. 375 THERMAL CONDUCTIVITY OF CARDBOARD

Curve No.	Ref. No.	Method Used	Year	Temp. Range, K	Reported Error, %	Name and Specimen Designation	Composition (weight percent), Specifications and Remarks
1	125	L	1923	323.2	1		3.63 cm dia disk; measured under a load of 21 psi.
2	125	L	1923	323.2	1		Another run of the above specimen.
3	88	L	1926	273,293			Density 0.79 g cm^{-3}.

DATA TABLE NO. 375 THERMAL CONDUCTIVITY OF CARDBOARD

[Temperature, T, K; Thermal Conductivity, k, Watt $cm^{-1}K^{-1}$]

T	k
CURVE 1*	
323.2	0.00167
CURVE 2*	
323.2	0.00335
CURVE 3*	
293.2	0.00143
273.2	0.00138

* No graphical presentation

SPECIFICATION TABLE NO. 376 THERMAL CONDUCTIVITY OF CELLULOSE FIBERBOARD

Curve No.	Ref. No.	Method Used	Year	Temp. Range, K	Reported Error, %	Name and Specimen Designation	Composition (weight percent), Specifications and Remarks
1	109	L	1947	222-306			Lignin-impregnated fiberboard; 1.25 in. thick; density 1.37 g cm^{-3}; as received.

DATA TABLE NO. 376 THERMAL CONDUCTIVITY OF CELLULOSE FIBERBOARD

[Temperature, T, K; Thermal Conductivity, k, Watt cm^{-1} K^{-1}]

T	k
CURVE 1*	
222.0	0.00241
238.7	0.00247
255.4	0.00251
272.2	0.00255
288.9	0.00261
305.5	0.00265

*No graphical presentation

SPECIFICATION TABLE NO. 377 THERMAL CONDUCTIVITY OF CORNSTALK WALLBOARD

Curve No.	Ref. No.	Method Used	Year	Temp. Range, K	Reported Error, %	Name and Specimen Designation	Composition (weight percent), Specifications and Remarks
1	566	L	1929	322.8		No. 1	Made from cornstalks.
2	566	L	1929	322.8		No. 1	Second run of the above specimen.
3	566	L	1929	322.8		No. 2	Made from cornstalks.
4	566	L	1929	322.8		No. 2	Second run of the above specimen.
5	566	L	1929	322.8		No. 3	Made from cornstalks.
6	566	L	1929	322.8		No. 3	Second run of the above specimen.
7	566	L	1929	322.8		No. 4	Made from cornstalks.
8	566	L	1929	322.8		No. 4	Second run of the above specimen.
9	566	L	1929	322.8		No. 5	Made from cornstalks.
10	566	L	1929	322.8		No. 6	Made from cornstalks.

DATA TABLE NO. 377 THERMAL CONDUCTIVITY OF CORNSTALK WALLBOARD

[Temperature, T, K; Thermal Conductivity, k, Watt $cm^{-1}K^{-1}$]

T	k	T	k	T	k	T	k
CURVE 1*		CURVE 4*		CURVE 7*		CURVE 10*	
322.8	0.000593	322.8	0.000573	322.8	0.000544	322.8	0.000404
CURVE 2*		CURVE 5*		CURVE 8*			
322.8	0.000610	322.8	0.000513	322.8	0.000551		
CURVE 3*		CURVE 6*		CURVE 9*			
322.8	0.000554	322.8	0.000512	322.8	0.000624		

* No graphical presentation

SPECIFICATION TABLE NO. 378 THERMAL CONDUCTIVITY OF DIATOMITE AGGREGATE

Curve No.	Ref. No.	Method Used	Year	Temp. Range, K	Reported Error, %	Name and Specimen Designation	Composition (weight percent), Specifications and Remarks
1	186	L	1928	478-958	2.5	Sil-o-cel coarse grade; A	Granular specimen packed in a ring of 7.9 in. I.D. and 1.4 in. thick; oven dried thoroughly at 250 F; packing density 0.288 g cm^{-3}.
2	186	L	1928	467-958	2.5	Sil-o-cel coarse grade; B	Similar to the above specimen except packing density 0.352 g cm^{-3}.

DATA TABLE NO. 378 THERMAL CONDUCTIVITY OF DIATOMITE AGGREGATE

[Temperature, T, K; Thermal Conductivity, k, Watt $cm^{-1}K^{-1}$]

T	k
CURVE 1*	
477.6	0.000759
549.3	0.000838
597.6	0.000887
710.9	0.00101
772.1	0.00107
958.2	0.00127
CURVE 2*	
467.1	0.000837
549.8	0.000890
595.4	0.000935
718.7	0.00104
799.8	0.00107
957.6	0.00124

* No graphical presentation

SPECIFICATION TABLE NO. 379 THERMAL CONDUCTIVITY OF EXCELSIOR

Curve No.	Ref. No.	Method Used	Year	Temp. Range, K	Reported Error, %	Name and Specimen Designation	Composition (weight percent), Specifications and Remarks
1	109	L	1947	222-306			Thickness of specimen 1 in.; density 0.0295 g cm^{-3}; as received.
2	564	L	1944	297-325			Thickness of specimen 1 in.; density 0.0300 g cm^{-3}.

DATA TABLE NO. 379 THERMAL CONDUCTIVITY OF EXCELSIOR

[Temperature, T, K; Thermal Conductivity, k, Watt $cm^{-1} K^{-1}$]

T	k
CURVE 1*	
222.0	0.000307
238.7	0.000342
255.4	0.000376
272.2	0.000412
288.9	0.000451
305.5	0.000495
CURVE 2*	
297.1	0.000462
311	0.000498
324.8	0.000534

*No graphical presentation

SPECIFICATION TABLE NO. 380 THERMAL CONDUCTIVITY OF FIR PLYWOOD

Curve No.	Ref. No.	Method Used	Year	Temp. Range, K	Reported Error, %	Name and Specimen Designation	Composition (weight percent), Specifications and Remarks
1	109	L	1947	222-306			Thickness 0.75 in.; density 0.532 g cm^{-3}; as received.

DATA TABLE NO. 380 THERMAL CONDUCTIVITY OF FIR PLYWOOD

[Temperature, T, K; Thermal Conductivity, k, Watt $cm^{-1}K^{-1}$]

T	k
CURVE 1*	
222.0	0.00101
238.7	0.00104
255.4	0.00107
272.2	0.00111
288.9	0.00114
305.5	0.00117

* No graphical presentation

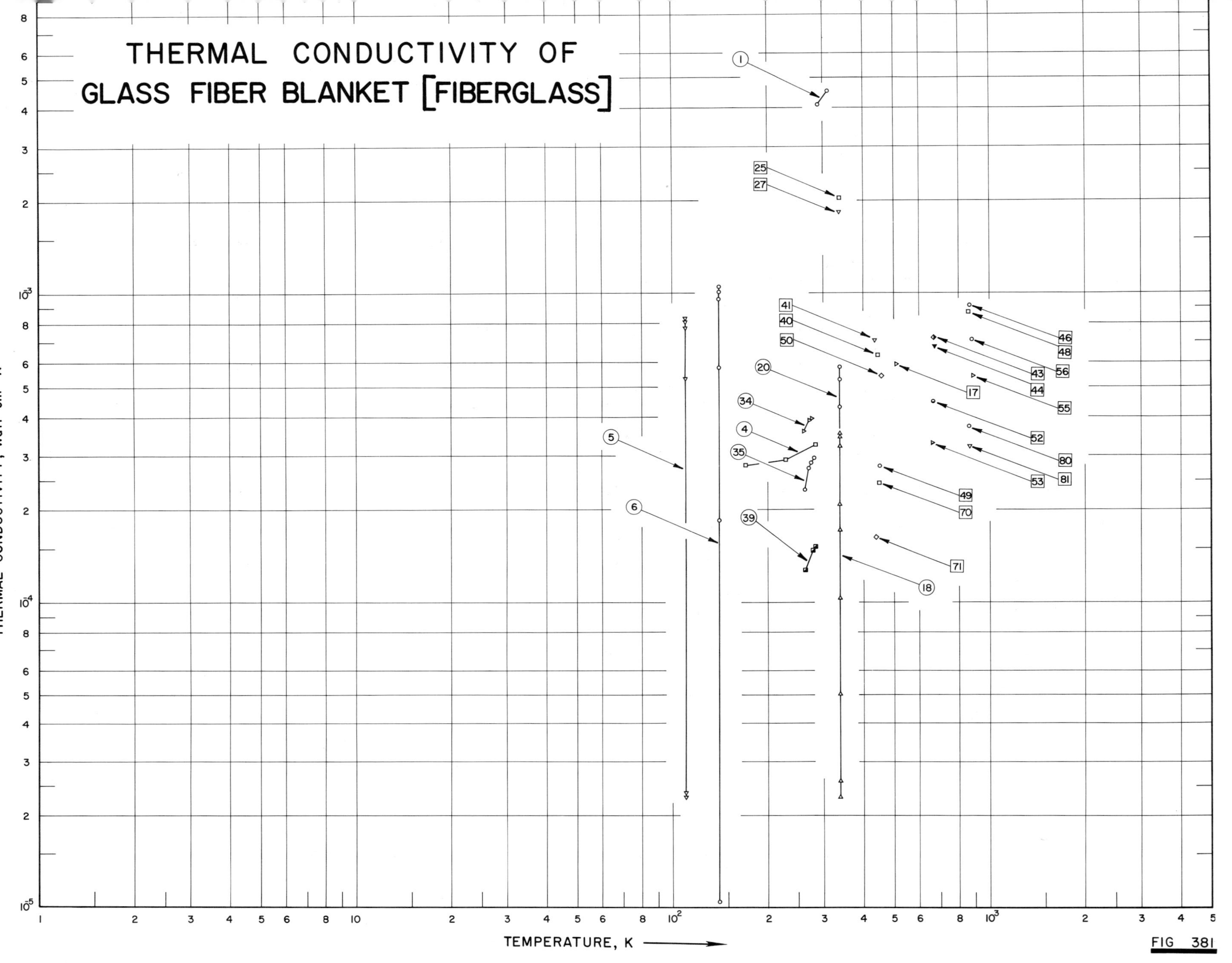
THERMAL CONDUCTIVITY OF
GLASS FIBER BLANKET [FIBERGLASS]
THERMAL CONDUCTIVITY, Watt cm⁻¹ K⁻¹
TEMPERATURE, K
FIG 381
1
25
27
41
40
50
20
34
4
35
5
6
39
46
48
56
43
44
17
55
52
80
81
53
49
70
71
18

SPECIFICATION TABLE NO. 381 THERMAL CONDUCTIVITY OF GLASS FIBER BLANKETS (FIBERGLASS)

[For Data Reported in Figure and Table No. 381]

Curve No.	Ref. No.	Method Used	Year	Temp. Range, K	Reported Error, %	Name and Specimen Designation	Composition (weight percent), Specifications and Remarks
1	559	L	1957	291, 312			Density 0.0096 g cm^{-3}.
2	559	L	1957	290, 311			Density 0.0681 g cm^{-3}.
3	559	L	1957	291, 311			Density 0.0157 g cm^{-3}.
4	567	L	1955	171-283			Fine glass fibre blanket - resin bonded; density 0.016 g cm^{-3}.
5	122	R	1953	110.0		Superfine PF	Blanket type bonded by an organic agent; about 0.5 in. thick; manufactured by Owens-Corning Fiberblas Corp; packing density 0.046 g cm^{-3}; filled with hydrogen; measured at various hydrogen pressures ranging from 0.0095 to 594 mm Hg.
6	122	R	1953	141.0		Superfine PF	Similar to the above specimen; filled with hydrogen; measured at various hydrogen pressures ranging from 0.0052 to 606 mm Hg.
7	122	R	1953	109.3		Superfine PF	Similar to the above specimen; filled with nitrogen; measured at various nitrogen pressures ranging from 0.0075 to 599 mm Hg.
8	122	R	1953	140.8		Superfine PF	Similar to the above specimen; filled with nitrogen; measured at various nitrogen pressures ranging from 0.0068 to 586 mm Hg.
9	122	R	1953	109.7		Superfine E	White wool type; supplied in bulk by Owen-Corning Fiberglas Corp; packing density 0.077 g cm^{-3}; filled with hydrogen; measured at various hydrogen pressures ranging from 0.0063 to to 602.5 mm Hg.
10	122	R	1953	140.8		Superfine E	Similar to the above specimen; filled with hydrogen; measured at various hydrogen pressures ranging from 0.0060 to 598.5 mm Hg.
11	122	R	1953	109.6		Superfine E	Similar to the above specimen; filled with nitrogen; measured at various nitrogen pressures ranging from 0.0080 to 599.0 mm Hg.
12	122	R	1953	140.6		Superfine E	Similar to the above specimen; filled with nitrogen; measured at various nitrogen pressures ranging from 0.0071 to 598.0 mm Hg.
13	122	R	1953	109.8		Superfine AA	Supplied by Owen-Corning Fiberglas Corp; fiber dia about 0.0005 in.; packing density 0.1629 g cm^{-3}; filled with hydrogen; measured at various hydrogen pressures ranging from 0.0064 to 593.0 mm Hg.
14	122	R	1953	141.3		Superfine AA	Similar to the above specimen; filled with hydrogen; measured at various hydrogen pressures ranging from 0.0054 to 600.0 mm Hg.
15	122	R	1953	109.2		Superfine AA	Similar to the above specimen; filled with nitrogen; measured at various nitrogen pressures ranging from 0.0058 to 788.0 mm Hg.
16	122	R	1953	140.5		Superfine AA	Similar to the above specimen; filled with nitrogen; measured at various nitrogen pressures ranging from 0.0052 to 577.0 mm Hg.
17	100	L	1957	508.7			Resin bonded; bulk density 0.0681 g cm^{-3}.

SPECIFICATION TABLE NO. 381 (continued)

Curve No.	Ref. No.	Method Used	Year	Temp. Range, K	Reported Error, %	Name and Specimen Designation	Composition (weight percent), Specifications and Remarks
18	568	L	1952	338.7		Insulation A3	Specimen thickness 1 in.; sp. gr. 2.01; avg fiber dia 2.58 μ; insulation density 0.0742 g cm^{-3}; vol fraction of fiber 0.0369; effective pore size 54.8 μ; filled with air; measured at various air pressures ranging from 0.00085 to 760 mm Hg.
19	568	L	1952	338.7		Insulation A2	Specimen thickness 1 in.; sp. gr. 2.01; avg fiber dia 2.58 μ; insulation density 0.024 g cm^{-3}; vol fraction of fiber 0.0120; effective pore size 169 μ; filled with air; measured at various air pressures ranging from 0.0003 to 760 mm Hg.
20	568	L	1952	338.7		Insulation A1	Specimen thickness 1 in.; sp. gr. 2.01; avg fiber dia 2.58 μ; insulation density 0.00375 g cm^{-3}; vol fraction of fiber 0.00436; effective pore size 465 μ; filled with air; measured at various air pressures ranging from 0.0013 to 760 mm Hg.
21	568	L	1952	338.7		Insulation A4	Specimen thickness 1 in.; sp. gr. 2.01; avg fiber dia 2.58 μ; insulation density 0.134 g cm^{-3}; vol fraction of fiber 0.00665; effective pore size 30.4 μ; filled with air; measured at various air pressures ranging from 0.00025 to 760 mm Hg.
22	568	L	1952	422.1		Insulation A4	Similar to the above specimen except measured with air at various pressures ranging from 0.0012 to 760 mm Hg.
23	568	L	1952	338.7		Insulation B	Thickness of specimen 1 in.; sp. gr. 2.5; avg fiber dia 1.51 μ; insulation density 0.119 g cm^{-3}; vol fraction of fiber 0.0474; effective pores size 25.0 μ; filled with air; measured at 760 mm Hg.
24	568	L	1952	422.1		Insulation B	Similar to the above specimen except measured at various air pressures ranging from 0.002 to 760 mm Hg.
25	568	L	1952	338.7		Insulation A1	Similar to the Insulation A1 except measured in helium at 760 mm Hg.
26	568	L	1952	338.7		Insulation A2	Similar to the Insulation A2 except measured in helium at 76 and 760 mm Hg.
27	568	L	1952	338.7		Insulation A3	Similar to the Insulation A3 except measured in helium at 760 mm Hg.
28	568	L	1952	338.7		Insulation A1	Similar to the Insulation A1 except measured in carbon dioxide at 760 mm Hg.
29	568	L	1952	338.7		Insulation A2	Similar to the Insulation A2 except measured in carbon dioxide at 76 and 760 mm Hg.
30	568	L	1952	338.7		Insulation A3	Similar to the Insulation A3 except measured in carbon dioxide at 760 mm Hg.
31	568	L	1952	338.7		Insulation A1	Similar to the Insulation A1 except measured in Freon-12 at 760 mm Hg.
32	568	L	1952	338.7		Insulation A2	Similar to the Insulation A2 except measured in Freon-12 at 76 and 760 mm Hg.
33	568	L	1952	338.7		Insulation A3	Similar to the Insulation A3 except measured in Freon-12 at 760 mm Hg.
34	569	L	1956	261-276		Pi 152	Thickness 5 cm; density 0.033 g cm^{-3}; fiber size about 12 μ; measured in air.
35	569	L	1956	263-280		Pi 152	The above specimen measured in CO_2.
36	569	L	1956	263-281		Pi 452	Thickness 5 cm; density 0.120 g cm^{-3}; fiber size about 12 μ; measured in air.
37	569	L	1956	261-286		Pi 352	Thickness 5 cm; density 0.088 g cm^{-3}; fiber size about 12 μ; measured in air.

SPECIFICATION TABLE NO. 381 (continued)

Curve No.	Ref. No.	Method Used	Year	Temp. Range, K	Reported Error, %	Name and Specimen Designation	Composition (weight percent), Specifications and Remarks
38	569	L	1956	263-286		Pi 352	The above specimen measured in CO_2.
39	569	L	1956	262-283		Pi 352	The above specimen measured in CF_2Cl_2.
40	526	P	1962	445.3	±7		Fiber glass cloth; 8 harness satin type, 57 ends in.$^{-1}$; 54 picks in.$^{-1}$; 8.9 oz yd^{-2}; avg thickness 0.01660 in.; 0.01270 in. thickness under load; measured under 17.5 psia pressure; 28.8 ppi tensile stress, and 9.5 psi compressive stress; thermal conductivity value calculated from measured transient temp changes.
41	526	P	1962	437.9	±7		Similar to the above specimen escept with 0.01341 in. thickness under load; measured under 16.9 psia pressure, 18.4 ppi tensile stress, and 5.9 psi compressive stress.
42	526	P	1962	463.8	±7		Similar to the above specimen except with 0.01321 in. thickness under load; measured under 17.2 psia pressure, 5.7 ppi tensile stress, and 6.8 psi compressive stress.
43	526	P	1962	667.4	±7		Similar to the above specimen except with 0.0134 in. thickness under load; measured under 17.6 psia pressure, 18.4 psi tensile stress, and 5.9 psi compressive stress.
44	526	P	1962	670.3	±7		Similar to the above specimen except with 0.0134 in. thickness under load; measured under 17.6 psia pressure, 15.8 psi tensile stress, and 5.9 psi compressive stress.
45	526	P	1962	671.6	±7		Similar to the above specimen except with 0.0132 in. thickness under load; measured under 17.7 psia pressure, 5.7 ppi tensile stress, and 6.8 psi compressive stress.
46	526	P	1962	868.3	±7		Similar to the above specimen except with 0.01270 in. thickness under load; measured under 18.0 psia pressure, 28.8 ppi tensile stress, and 9.5 psi compressive stress.
47	526	P	1962	872.2	±7		Similar to the above specimen except with 0.01341 in. thickness under load; measured under 17.3 psia pressure, 15.8 ppi tensile stress, and 5.9 psi compressive stress.
48	526	P	1962	864.7	±7		Similar to the above specimen except with 0.01321 in. thickness under load; measured under 17.3 psia pressure, 5.7 ppi tensile stress, and 6.8 psi compressive stress.
49	526	P	1962	454.8	±7		Similar to the above specimen except with 0.01270 in. thickness under load; measured in a vacuum of 7.0 mm Hg under 28.8 ppi tensile stress and 9.5 psi compressive stress.
50	526	P	1962	458.1	±7		Similar to the above specimen except with 0.01341 in. thickness under load; measured in a vacuum of 12.0 mm Hg under 18.4 ppi tensile stress and 5.9 psi compressive stress.
51	526	P	1962	453.1	±7		Similar to the above specimen except with 0.01202 in. thickness under load; measured in a vacuum of 3.0 mm Hg under 5.7 ppi tensile stress and 13.5 psi compressive stress.
52	526	P	1962	665.4	±7		Similar to the above specimen except with 0.01270 in. thickness under load; measured in a vacuum of 14.0 mm Hg under 28.8 ppi tensile stress and 9.5 psi compressive stress.

SPECIFICATION TABLE NO. 381 (continued)

Curve No.	Ref. No.	Method Used	Year	Temp. Range, K	Reported Error, %	Name and Specimen Designation	Composition (weight percent), Specifications and Remarks
53	526	P	1962	664.7	±7		Similar to the above specimen except with 0.01341 in. thickness under load; measured in a vacuum of 8.0 mm Hg under 15.8 ppi tensile stress and 5.9 psi compressive stress.
54	526	P	1962	664.8	±7		Similar to the above specimen except with 0.01202 in. thickness under load; measured in a vacuum of 7.0 mm Hg under 5.7 ppi tensile stress and 13.5 psi compressive stress.
55	526	P	1962	887.5	±7		Similar to the above specimen except with 0.01341 in. thickness under load; measured in a vacuum of 22.0 mm Hg under 18.4 ppi tensile stress and 5.9 psi compressive stress.
56	526	P	1962	884.4	±7		Similar to the above specimen except with 0.01341 in. thickness under load; measured in a vacuum of 17.0 mm Hg under 15.8 ppi tensile stress and 5.9 psi compressive stress.
57	526	P	1962	881.4	±7		Similar to the above specimen except with 0.01202 in. thickness under load; measured in a vacuum of 15.0 mm Hg under 5.7 ppi tensile stress and 13.5 psi compressive stress.
58	526	P	1962	443.1	±7		Aluminized fiberglass; plain weave; avg thickness 0.01021 in.; 0.00854 in. thickness under load; measured under 16.9 psia pressure, 26.0 ppi tensile stress, and 8.6 psi compressive stress.
59	526	P	1962	455.0	±7		Similar to the above specimen except with 0.00845 in. thickness under load; measured under 17.4 psia pressure, 28.5 ppi tensile stress, and 9.4 psi compressive stress.
60	526	P	1962	448.9	±7		Similar to the above specimen except with 0.00855 in. thickness under load; measured under 17.3 psia pressure, 26.0 ppi tensile stress, and 8.5 psi compressive stress.
61	526	P	1962	601.8	±7		Similar to the above specimen except with 0.00845 in. thickness under load; measured under 17.2 psia pressure, 28.8 ppi tensile stress, and 9.4 psi compressive stress.
62	526	P	1962	609.0	±7		Similar to the above specimen except with 0.00845 in. thickness under load; measured under 17.5 psia pressure, 28.5 ppi tensile stress, and 9.4 psi compressive stress.
63	526	P	1962	616.3	±7		Similar to the above specimen except with 0.00855 in. thickness under load; measured under 17.4 psia pressure, 26.0 ppi tensile stress, and 8.5 psi compressive stress.
64	526	P	1962	773.6	±7		Similar to the above specimen except with 0.00845 in. thickness under load; measured under 17.7 psia pressure, 28.8 ppi tensile stress, and 9.4 psi compressive stress.
65	526	P	1962	765.2	±7		Similar to the above specimen except with 0.00845 in. thickness under load; measured under 17.5 psia pressure, 28.5 ppi tensile stress, and 9.4 psi compressive stress.
66	526	P	1962	768.0	±7		Similar to the above specimen except with 0.00855 in. thickness under load; measured under 17.1 psia pressure, 26.0 ppi tensile stress, and 8.5 psi compressive stress.
67	526	P	1962	877.3	±7		Similar to the above specimen except with 0.00845 in. thickness under load; measured under 17.6 psia pressure, 28.8 ppi tensile stress, and 9.4 psi compressive stress.

SPECIFICATION TABLE NO. 381 (continued)

Curve No.	Ref. No.	Method Used	Year	Temp. Range, K	Reported Error, %	Name and Specimen Designation	Composition (weight percent), Specifications and Remarks
68	526	P	1962	876.7	±7		Similar to the above specimen except with 0.00845 in. thickness under load; measured under 17.3 psia pressure, 28.5 ppi tensile stress, and 9.4 psi compressive stress.
69	526	P	1962	875.1	±7		Similar to the above specimen except with 0.00855 in. thickness under load; measured under 17.5 psia pressure, 26.0 ppi tensile stress, and 8.5 psi compressive stress.
70	526	P	1962	448.3	±7		Similar to the above specimen except with 0.00844 in. thickness under load; measured in a vacuum of 6.0 mm Hg under 28.8 ppi tensile stress and 9.5 psi compressive stress.
71	526	P	1962	436.6	±7		Similar to the above specimen except with 0.00844 in. thickness under load; measured in a vacuum of 3.0 mm Hg under 28.5 ppi tensile stress and 9.5 psi compressive stress.
72	526	P	1962	415.4	±7		Similar to the above specimen except with 0.00845 in. thickness under load; measured in a vacuum of 2.7 mm Hg under 28.8 ppi tensile stress and 9.4 psi compressive stress.
73	526	P	1962	590.5	±7		Similar to the above specimen except with 0.00844 in. thickness under load; measured in a vacuum of 8.0 mm Hg under 28.8 ppi tensile stress and 9.5 psi compressive stress.
74	526	P	1962	560.9	±7		Similar to the above specimen except with 0.00844 in. thickness under load; measured in a vacuum of 5.0 mm Hg under 28.5 ppi tensile stress and 9.5 psi compressive stress.
75	526	P	1962	589.8	±7		Similar to the above specimen except with 0.00854 in. thickness under load; measured in a vacuum of 3.0 mm Hg under 26.0 ppi tensile stress and 8.6 psi compressive stress.
76	526	P	1962	760.7	±7		Similar to the above specimen except with 0.00844 in. thickness under load; measured in a vacuum of 15.0 mm Hg under 28.8 ppi tensile stress and 9.5 psi compressive stress.
77	526	P	1962	745.1	±7		Similar to the above specimen except with 0.00844 in. thickness under load; measured in a vacuum of 6.0 mm Hg under 28.5 ppi tensile stress and 9.5 psi compressive stress.
78	526	P	1962	755.2	±7		Similar to the above specimen except with 0.00854 in. thickness under laod; measured in a vacuum of 7.0 mm Hg under 26.0 ppi tensile stress and 8.6 psi compressive stress.
79	526	P	1962	878.7	±7		Similar to the above specimen except with 0.00844 in. thickness under load; measured in a vacuum of 15.0 mm Hg under 28.8 ppi tensile stress and 9.5 psi compressive stress.
80	526	P	1962	864.1	±7		Similar to the above specimen except with 0.00844 in. thickness under load; measured in a vacuum of 10.0 mm Hg under 28.5 ppi tensile stress and 9.5 psi compressive stress.

SPECIFICATION TABLE NO. 381 (continued)

Curve No.	Ref. No.	Method Used	Year	Temp. Range, K	Reported Error, %	Name and Specimen Designation	Composition (weight percent), Specifications and Remarks
81	526	P	1962	866.6	±7		Similar to the above specimen except with 0.00855 in. thickness under load; measured in a vacuum of 9.0 mm Hg under 26.0 ppi tensile stress and 8.6 psi compressive stress.
82	529	P	1956	290.4			Density 0.0641 g cm^{-3}; thermal conductivity value calculated from measured thermal diffusivity data.
83	529	P	1956	290.4			Similar to the above specimen.

DATA TABLE NO. 381 THERMAL CONDUCTIVITY OF GLASS FIBER BLANKET [FIBERGLASS]

[Temperature, T,K; Thermal Conductivity, k, Watt $cm^{-1}K^{-1}$]

CURVE 1

T	k
290.7	0.00407
311.7	0.00453

CURVE 2*

T	k
290.1	0.00412
310.8	0.00452

CURVE 3*

T	k
290.8	0.00402
311.2	0.00446

CURVE 4

T	k
170.7	0.000279
228.2	0.000291
283.2	0.000326

CURVE 5, T = 110K

p(mm)	k
0.0095	0.0000230
0.207	0.000237
1.16	0.000534
15.5	0.000774
52.5	0.000807
207	0.000823*
398	0.000828
594	0.000831*

CURVE 6, T = 141K

p(mm)	k
0.0052	0.0000105
0.128	0.000185
1.11	0.000579
.14.0	0.000955
52.5	0.00101
207.5	0.00104*
400	0.00105
606	0.00105*

CURVE 7*, T = 109.3K

p(mm)	k
0.0075	0.0000130
0.118	0.0000611
1.08	0.000111
13.5	0.000132
51.5	0.000134
200	0.000136
400	0.000140
599	0.000146

CURVE 8*, T = 140.8K

p(mm)	k
0.0068	0.0000130
0.118	0.0000649
0.968	0.000128
12.0	0.000162
50.0	0.000168
200.5	0.000171
395.0	0.000173
586.0	0.000177

CURVE 9*, T = 109.7K

p(mm)	k
0.0063	0.0000243
0.123	0.000232
0.886	0.000558
15.0	0.000792
51.5	0.000822
202.0	0.000846
367.0	0.000853
602.5	0.000858

CURVE 10*, T = 140.8K

p(mm)	k
0.0060	0.0000213
0.151	0.000272
1.17	0.000682
12.0	0.000986
52.5	0.00104
210.5	0.00107
402.0	0.00108
598.5	0.00109

CURVE 11*, T = 109.6K

p(mm)	k
0.0080	0.0000201
0.098	0.0000783
0.927	0.000133
13.0	0.000153
54.5	0.000156
205.0	0.000158
397.0	0.000161
599.0	0.0000171

CURVE 12*, T = 140.6K

p(mm)	k
0.0071	0.0000180
0.113	0.0000883
1.43	0.000163
11.5	0.000186
53.5	0.000191
195.5	0.000194
398.0	0.000196
598.0	0.000204

CURVE 13*, T = 109.8K

p(mm)	k
0.0064	0.0000113
0.123	0.000109
0.88	0.000345
17.0	0.000752
53.5	0.000807
201.5	0.000840
409.5	0.000851
593.0	0.000859

CURVE 14*, T = 141.3K

p(mm)	k
0.0054	0.0000130
0.089	0.0000891
1.06	0.000409
15.0	0.000904
56.0	0.00101
212.0	0.00105
399.5	0.00107
600.0	0.00108

CURVE 15*, T = 109.2K

p(mm)	k
0.0058	0.00000753
0.228	0.0000657
1.19	0.000110
13.0	0.000144
51.0	0.000151
220.5	0.000156
418.5	0.000157
602.5	0.000158
788.0	0.000159

CURVE 16*, T = 140.5K

p(mm)	k
0.0052	0.0000100
0.175	0.0000644
0.971	0.000119
12.0	0.000178
55.0	0.000189
220.5	0.000194
404.0	0.000196
577.0	0.000197

CURVE 17

T	k
508.7	0.000591

CURVE 18, T = 338.7K

p(mm)	k
0.00085	0.0000231
0.00175	0.0000231*
0.0065	0.0000231*
0.015	0.0000260
0.10	0.0000505
0.35	0.000104
1.0	0.000173
4.5	0.000209
17.5	0.000324
65	0.000346
200	0.000355
760	0.000356*

CURVE 19*, T = 338.7K

p(mm)	k
0.0003	0.0000793
0.0011	0.0000793
0.003	0.0000808
0.01	0.0000865
0.027	0.0000995
0.095	0.000144
0.2	0.000187
0.5	0.000252
1	0.000322
5	0.000398
20	0.000418
76	0.000423
200	0.000424
760	0.000425

CURVE 20, T = 338.7K

p(mm)	k
0.0013	0.000180*
0.003	0.000187*
0.01	0.000202*
0.027	0.000231*
0.1	0.000322*
0.35	0.000433
1	0.000526
3.5	0.000577
13	0.000581*
65	0.000583*
200	0.000584*
760	0.000584*

CURVE 21*, T = 338.7K

p(mm)	k
0.00025	0.0000101
760	0.000340

CURVE 22*, T = 422.1K

p(mm)	k
0.0012	0.0000346
760	0.000421

CURVE 23*, T = 338.7K

p(mm)	k
760	0.000320

CURVE 24*, T = 422.1K

p(mm)	k
0.002	0.00000577
1	0.0000707
10	0.000254
760	0.000392

CURVE 25, T = 338.7K

p(mm)	k
760	0.00204

CURVE 26*, T = 338.7K

p(mm)	k
76	0.00189
760	0.00193

CURVE 27, T = 338.7K

p(mm)	k
760	0.00182

CURVE 28*, T = 338.7K

p(mm)	k
760	0.000470

CURVE 29*, T = 338.7K

p(mm)	k
76	0.000326
760	0.000326

CURVE 30*, T = 338.7K

p(mm)	k
760	0.000255

*Not shown on plot

DATA TABLE NO. 381 (continued)

p(mm)	k
CURVE 31* T = 338.7K	
760	0.000376
CURVE 32* T = 338.7K	
76	0.000234
760	0.000234
CURVE 33* T = 338.7K	
760	0.000167

T	k
CURVE 34	
260.7	0.000360
270.2	0.000389
275.7	0.000395
CURVE 35	
263.2	0.000232
269.2	0.000273
274.7	0.000285
280.2	0.000296
CURVE 36*	
263.2	0.000291
270.7	0.000300
275.2	0.000307
281.2	0.000314
CURVE 37*	
260.7	0.000291
271.7	0.000325
277.2	0.000316
281.6	0.000322
286.2	0.000331

T	k
CURVE 38*	
263.2	0.000203
273.2	0.000215
278.7	0.000221
286.2	0.000227
CURVE 39	
262.2	0.000128
279.2	0.000149
283.2	0.000152
CURVE 40	
445.3	0.000633
CURVE 41	
437.9	0.000703
CURVE 42*	
463.8	0.000620
CURVE 43	
667.4	0.000718
CURVE 44	
670.3	0.000670
CURVE 45*	
671.6	0.000675
CURVE 46	
868.3	0.000907
CURVE 47*	
872.2	0.000891
CURVE 48	
864.7	0.000869

T	k
CURVE 49	
454.8	0.000277
CURVE 50	
458.1	0.000543
CURVE 51*	
453.1	0.000590
CURVE 52	
665.4	0.000448
CURVE 53	
664.7	0.000329
CURVE 54*	
664.8	0.000343
CURVE 55	
887.5	0.000543
CURVE 56	
884.4	0.000708
CURVE 57*	
881.4	0.000396
CURVE 58*	
443.1	0.000687
CURVE 59*	
455.0	0.000587
CURVE 60*	
448.9	0.000737

T	k
CURVE 61*	
601.8	0.000531
CURVE 62*	
609.0	0.000637
CURVE 63*	
616.3	0.000613
CURVE 64*	
773.6	0.000749
CURVE 65*	
765.2	0.000720
CURVE 66*	
768.0	0.000642
CURVE 67*	
877.3	0.000798
CURVE 68*	
876.7	0.000737
CURVE 69*	
875.1	0.000796
CURVE 70	
448.3	0.000244
CURVE 71	
436.6	0.000163
CURVE 72*	
415.4	0.000166

T	k
CURVE 73*	
590.5	0.000353
CURVE 74*	
560.9	0.000351
CURVE 75*	
589.8	0.000247
CURVE 76*	
760.7	0.000422
CURVE 77*	
745.1	0.000324
CURVE 78*	
755.2	0.000336
CURVE 79*	
878.7	0.000479
CURVE 80	
864.1	0.000372
CURVE 81	
866.6	0.000320
CURVE 82*	
290.4	0.000317
CURVE 83*	
290.4	0.000274

* Not shown on plot

SPECIFICATION TABLE NO. 382 THERMAL CONDUCTIVITY OF GLASS FIBER BOARD

Curve No.	Ref. No.	Method Used	Year	Temp. Range, K	Reported Error, %	Name and Specimen Designation	Composition (weight percent), Specifications and Remarks
1	128	L	1946	321.5			2.02 in. thick; density 0.176 g cm^{-3}.
2	128	L	1946	194, 247			Similar to above except 1.00 in. thick.
3	559	L	1957	289-316			Density 0.215 g cm^{-3}.

DATA TABLE NO. 382 THERMAL CONDUCTIVITY OF GLASS FIBER BOARD

[Temperature, T, K; Thermal Conductivity, k, Watt $cm^{-1} K^{-1}$]

T	k
CURVE 1*	
321.5	0.0000972
CURVE 2*	
193.7	0.0000579
246.5	0.0000734
CURVE 3*	
289.1	0.000499
289.7	0.000495
290.1	0.000494
311.2	0.000524
311.6	0.000524
311.7	0.000518
315.7	0.000533
315.7	0.000530

*No graphical presentation

SPECIFICATION TABLE NO. 383 THERMAL CONDUCTIVITY OF KOLDBOARD

Curve No.	Ref. No.	Method Used	Year	Temp. Range, K	Reported Error, %	Name and Specimen Designation	Composition (weight percent), Specifications and Remarks
1	189		1950	200-288			Supplied by Baldwin-Hill Co., density 0.256-0.272 g cm^{-3}; water absorption (24 hrs) 14.1 vol %.

DATA TABLE NO. 383 THERMAL CONDUCTIVITY OF KOLDBOARD

[Temperature, T, K; Thermal Conductivity, k, Watt $cm^{-1}K^{-1}$]

T	k
CURVE 1*	
200.0	0.000361
222.2	0.000389
239.2	0.000404
255.5	0.000433
272.1	0.000447
288.1	0.000476

* No graphical presentation

SPECIFICATION TABLE NO. 384 THERMAL CONDUCTIVITY OF MONOLITHIC WALL

Curve No.	Ref. No.	Method Used	Year	Temp. Range, K	Reported Error, %	Name and Specimen Designation	Composition (weight percent), Specifications and Remarks
1	553	L	1937	297.1		30a	39.3 sand, 60.7 crushed limestone concrete, plastic mix; 4.32 in. in thickness; density of concrete in wall 2.247 g cm^{-3}.
2	553	L	1937	297.1		31a	36.5 sand, 63.5 coarse gravel concrete, plastic mix; 4.168 in. in thickness; density of concrete in wall 2.295 g cm^{-3}.
3	553	L	1937	297.1		32a	36.5 sand, 63.5 coarse gravel concrete, dry tamp mix; 4.05 in. in thickness; density of concrete in wall 2.384 g cm^{-3}.
4	553	L	1937	297.1		33a	39. 8 fine cinders, 60. 2 coarse cinders, concrete, plastic mix; 3. 902 in. in thickness; density of concrete in wall 1.512 g cm^{-3}.
5	553	L	1937	297.1		34a	42.6 fine Haydite, 57.4 coarse Haydite plastic mix; 3.960 in. in thickness; density of concrete in wall 1.245 g cm^{-3}.
6	553	L	1937	297.1		35a	52.6 sand,47.4 coarse cinders, dry tamp mix; 3.958 in. in thickness; density of concrete in wall 1.901 g cm^{-3}.

DATA TABLE NO. 384 THERMAL CONDUCTIVITY OF MONOLITHIC WALL

[Temperature, T, K; Thermal Conductivity, k, Watt $cm^{-1}K^{-1}$]

T	k	T	k
CURVE 1*		CURVE 4*	
297.1	0.0175	297.1	0.00829
CURVE 2*		CURVE 5*	
297.1	0.0179	297.1	0.00538
CURVE 3*		CURVE 6*	
297.1	0.0189	297.1	0.0123

* No graphical presentation

SPECIFICATION TABLE NO. 385 THERMAL CONDUCTIVITY OF PAPER

Curve No.	Ref. No.	Method Used	Year	Temp. Range, K	Reported Error, %	Name and Specimen Designation	Composition (weight percent), Specifications and Remarks
1	59	L	1953	295-385			Oil impregnated paper; approx 0.05 in. thick; measured under a load of about 2 psi.
2	125	L	1923	313.2			Rice paper; thickness 0.003 cm; measured under a load of 21 psi.

DATA TABLE NO. 385 THERMAL CONDUCTIVITY OF PAPER

[Temperature, T, K; Thermal Conductivity, k, Watt $cm^{-1} K^{-1}$]

T	k
CURVE 1*	
294.7	0.00180
297.2	0.00180
310.2	0.00184
321.2	0.00184
335.2	0.00186
339.2	0.00180
353.2	0.00186
359.2	0.00182
360.2	0.00184
361.7	0.00186
365.2	0.00184
385.2	0.00180
CURVE 2*	
313.2	0.000460

*No graphical presentation

SPECIFICATION TABLE NO. 386 THERMAL CONDUCTIVITY OF SEA-WEED PRODUCT

Curve No.	Ref. No.	Method Used	Year	Temp. Range, K	Reported Error, %	Name and Specimen Designation	Composition (weight percent), Specifications and Remarks
1	561	R	1948	92-270			Powder form; contained in a hollow cylinder of 2 in. I. D. and 10 in. O. D.; bulk density 0. 130 g cm^{-3}.
2	561	R	1948	190			50. 0 block, 49. 5 powder; bulk density 0. 117 g cm^{-3}.

DATA TABLE NO. 386 THERMAL CONDUCTIVITY OF SEA-WEED PRODUCT

[Temperature, T, K; Thermal Conductivity, k, Watt $cm^{-1} K^{-1}$]

T	k
CURVE 1*	
92. 2	0. 000147
135. 5	0. 000197
162. 7	0. 000234
183. 2	0. 000258
200. 2	0. 000277
213. 2	0. 000293
223. 7	0. 000305
234. 2	0. 000315
243. 0	0. 000327
250. 2	0. 000337
256. 7	0. 000346
263. 7	0. 000350
269. 7	0. 000358
CURVE 2*	
190. 0	0. 000348

*No graphical presentation

SPECIFICATION TABLE NO. 387 THERMAL CONDUCTIVITY OF VEGETABLE FIBERBOARDS, MISCELLANEOUS

Curve No.	Ref. No.	Method Used	Year	Temp. Range, K	Reported Error, %	Name and Specimen Designation	Composition (weight percent), Specifications and Remarks
1	109	L	1947	222-306			0. 75 in. thick; density 0. 251 g cm^{-3}; moisture content 24. 1%.
2	109	L	1947	222-306			Similar to the above specimen but moisture content 3. 1%.
3	109	L	1947	222-306			Similar to the above specimen but zero moisture content.
4	558	L	1951	310. 9			0. 75 in. thick; density 0. 248 g cm^{-3} at 25 C; measured in air.
5	558	L	1951	310. 9			The above specimen measured in a vacuum of 0. 05 μ Hg.
6	563	L	1945	222-305			0. 75 in. thick; density 0. 248 g cm^{-3}.
7	564	L	1944	297-325			0. 75 in. thick; density 0. 251 g cm^{-3}.
8	560	L	1954	230-311			0. 75 in. thick; density 0. 248 g cm^{-3}; measured in air.
9	560	L	1954	230-311			The above specimen measured in CO_2.
10	560	L	1954	283, 311			The above specimen measured in F-12.
11	560	L	1954	230-311			The above specimen measured in vacuum.
12	128	L	1946	316. 5			1. 97 thick; density 0. 231 g cm^{-3}.
13	128	L	1946	194, 242			Similar to the above specimen except 1. 02 in. thick.

DATA TABLE NO. 387 THERMAL CONDUCTIVITY OF VEGETABLE FIBERBOARDS, MISCELLANEOUS

[Temperature, T, K; Thermal Conductivity, k, Watt $cm^{-1} K^{-1}$]

T	k
CURVE 1*	
222.0	0.000495
238.7	0.000516
255.4	0.000542
272.2	0.000565
288.9	0.000588
305.5	0.000611
CURVE 2*	
222.0	0.000424
238.7	0.000441
255.4	0.000460
272.2	0.000477
288.9	0.000495
305.5	0.000513
CURVE 3*	
222.0	0.000411
238.7	0.000430
255.4	0.000447
272.2	0.000466
288.9	0.000483
305.5	0.000502
CURVE 4*	
310.9	0.000509
CURVE 5*	
310.9	0.000108
CURVE 6*	
222.2	0.000411
239.2	0.000428
255.5	0.000447
272.1	0.000466
288.8	0.000483
305.3	0.000502
CURVE 7*	
297.1	0.000492
311.3	0.000509
324.8	0.000524
CURVE 8*	
310.9	0.00613
283.2	0.00578
255.4	0.00540
230.4	0.00507
CURVE 9*	
310.9	0.00483
283.2	0.00448
255.4	0.00410
230.4	0.00377
CURVE 10*	
310.9	0.00367
283.2	0.00339
CURVE 11*	
310.9	0.00126
283.2	0.00118
255.4	0.00107
230.4	0.00100
CURVE 12*	
316.5	0.000117
CURVE 13*	
194.3	0.0000855
242.1	0.0000979

*No graphical presentation

SPECIFICATION TABLE NO. 388 THERMAL CONDUCTIVITY OF WALLBOARD, MISCELLANEOUS

Curve No.	Ref. No.	Method Used	Year	Temp. Range, K	Reported Error, %	Name and Specimen Designation	Composition (weight percent), Specifications and Remarks
1	566	L	1929	322.8		No. 7	Commercial wallboard.
2	566	L	1929	322.8		No. 8	Commercial wallboard.
3	566	L	1929	322.8		No. 8	Second run of the above specimen.
4	566	L	1929	322.8		No. 9	Commercial wallboard.
5	566	L	1929	322.8		No. 9	Second run of the above specimen.

DATA TABLE NO. 388 THERMAL CONDUCTIVITY OF WALLBOARD, MISCELLANEOUS

[Temperature, T, K; Thermal Conductivity, k, Watt $cm^{-1}K^{-1}$]

T	k
CURVE 1*	
322.8	0.000640
CURVE 2*	
322.8	0.000581
CURVE 3*	
322.8	0.000594
CURVE 4*	
322.8	0.000633
CURVE 5*	
322.8	0.000587

* No graphical presentation

SPECIFICATION TABLE NO. 389 THERMAL CONDUCTIVITY OF WOOD PRODUCTS, MISCELLANEOUS

Curve No.	Ref. No.	Method Used	Year	Temp. Range, K	Reported Error, %	Name and Specimen Designation	Composition (weight percent), Specifications and Remarks
1	109	L	1947	222-306			Wood fiber blanket; 1 in. thick; density 0.056 g cm^{-3}; as received.
2	109	L	1947	222-306			Wood fiber mat; 1 in. thick; aerated; density 0.027 g cm^{-3}; as received.

DATA TABLE NO. 389 THERMAL CONDUCTIVITY OF WOOD PRODUCTS, MISCELLANEOUS

[Temperature, T, K; Thermal Conductivity, k, Watt $cm^{-1} K^{-1}$]

T	k
CURVE 1*	
222.0	0.000286
238.7	0.000304
255.4	0.000322
272.2	0.000340
288.9	0.000361
305.5	0.000384
CURVE 2*	
222.0	0.000270
238.7	0.000293
255.4	0.000316
272.2	0.000340
288.9	0.000363
305.5	0.000392

* No graphical presentation

SPECIFICATION TABLE NO. 390 THERMAL CONDUCTIVITY OF WOOL FELT

Curve No.	Ref. No.	Method Used	Year	Temp. Range, K	Reported Error, %	Name and Specimen Designation	Composition (weight percent), Specifications and Remarks
1	108	L	1920	313,343			9 in. dia x 0.96 in. thick; specific gravity 0.15.

DATA TABLE NO. 390 THERMAL CONDUCTIVITY OF WOOL FELT

[Temperature, T, K; Thermal Conductivity, k, Watt $cm^{-1}K^{-1}$]

T	k
CURVE 1*	
313.2	0.000623
343.2	0.000732

* No graphical presentation

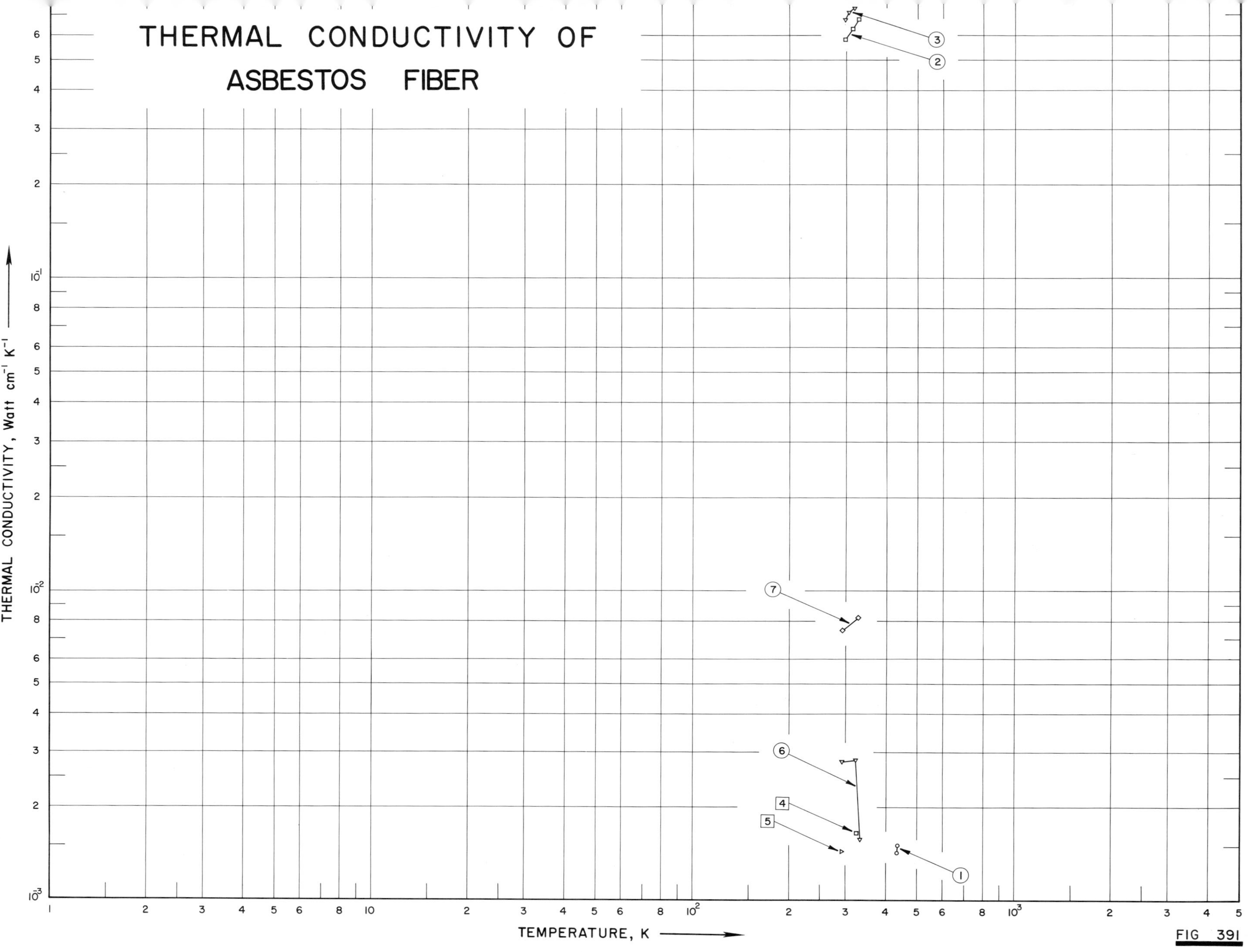

THERMAL CONDUCTIVITY OF
ASBESTOS FIBER
THERMAL CONDUCTIVITY, Watt cm⁻¹ K⁻¹
TEMPERATURE, K
1
2
3
4
5
6
7
FIG 391

SPECIFICATION TABLE NO. 391 THERMAL CONDUCTIVITY OF ASBESTOS FIBER

[For Data Reported in Figure and Table No. 391]

Curve No.	Ref. No.	Method Used	Year	Temp. Range, K	Reported Error, %	Name and Specimen Designation	Composition (weight percent), Specifications and Remarks
1	240	R	1919	435. 8, 436. 1			Fiber specimen packed in a hollow cylinder of size 12. 2 mm I. D. , 19 mm O. D. , and and 8 cm long.
2	570	L	1930	297-328		Asbestos fibre III	Average thickness 0. 0805 m; density 0. 151 g cm^{-3}.
3	570	L	1930	297-318		Asbestos fibre II	Average thickness 0. 0434 m; density 0. 202 g cm^{-3}.
4	108	L	1920	324. 2			Dia 9 in. , thickness 0. 344 in. cut from 1 in. sheet; sp. gr. 0. 894; transverse heat flow.
5	108	L	1920	293. 2			Dia 9 in. , thickness 0. 306 in. ; made from 0. 025 in. paper; sp. gr. 0. 98; transverse heat flow.
6	108	L	1920	293-333			Similar to the above specimen but with thickness 0. 356 in. and made from 0. 035 in. paper.
7	108	L	1920	293, 328			Dia 9 in. , thickness 0. 507 in. ; sp. gr. 1. 93; transverse heat flow.

DATA TABLE NO. 391 THERMAL CONDUCTIVITY OF ASBESTOS FIBER

[Temperature, T,K; Thermal Conductivity, k, Watt $cm^{-1}K^{-1}$]

T	k
CURVE 1	
435.8	0.00142
436.1	0.00151
CURVE 2	
297.2	0.581
314.7	0.628
328.2	0.674
CURVE 3	
296.7	0.674
306.7	0.709
318.2	0.732
CURVE 4	
324.2	0.00165
CURVE 5	
293.2	0.00144
CURVE 6	
293.2	0.00279
323.2	0.00282
333.2	0.00157
CURVE 7	
293.2	0.00745
328.2	0.00816

SPECIFICATION TABLE NO. 392 THERMAL CONDUCTIVITY OF MICANITE

Curve No.	Ref. No.	Method Used	Year	Temp. Range, K	Reported Error, %	Name and Specimen Designation	Composition (weight percent), Specifications and Remarks
1	125	L	1923	303.2	1	Commercial	Measured under a load of 21 psi.
2	125	L	1923	303.2	1	Commercial	Similar to above specimen.

DATA TABLE NO. 392 THERMAL CONDUCTIVITY OF MICANITE

[Temperature, T, K; Thermal Conductivity, k, Watt $cm^{-1}K^{-1}$]

T	k
CURVE 1*	
303.2	0.00209
CURVE 2*	
303.2	0.00418

* No graphical presentation

SPECIFICATION TABLE NO. 393 THERMAL CONDUCTIVITY OF MINERAL FIBER

Curve No.	Ref. No.	Method Used	Year	Temp. Range, K	Reported Error, %	Name and Specimen Designation	Composition (weight percent), Specifications and Remarks
1	564	L	1944	297-325		B	1 in. thick; packing density 0.0264 g cm^{-3}.
2	564	L	1944	297-325		A	1 in. thick; packing density 0.0561 g cm^{-3}.
3	564	L	1944	297-325		B	1 in. thick; packing density 0.0396 g cm^{-3}.
4	564	L	1944	297-325		B	1 in. thick; packing density 0.0529 g cm^{-3}.
5	564	L	1944	297-325		A	1 in. thick; packing density 0.0681 g cm^{-3}.
6	564	L	1944	297-325		A	1 in. thick; packing density 0.112 g cm^{-3}.

DATA TABLE NO. 393 THERMAL CONDUCTIVITY OF MINERAL FIBER

[Temperature, T, K; Thermal Conductivity, k, Watt $cm^{-1} K^{-1}$]

T	k	T	k
CURVE 1*		CURVE 4*	
297.1	0.000420	297.1	0.000361
311	0.000453	311	0.000389
324.8	0.000493	324.8	0.000418
CURVE 2*		CURVE 5*	
297.1	0.000398	297.1	0.000378
311	0.000428	310.9	0.000401
325.1	0.000463	324.8	0.000433
CURVE 3*		CURVE 6*	
297.0	0.000375	297.1	0.000362
311	0.000404	311	0.000388
324.8	0.000438	324.8	0.000412

*No graphical presentation

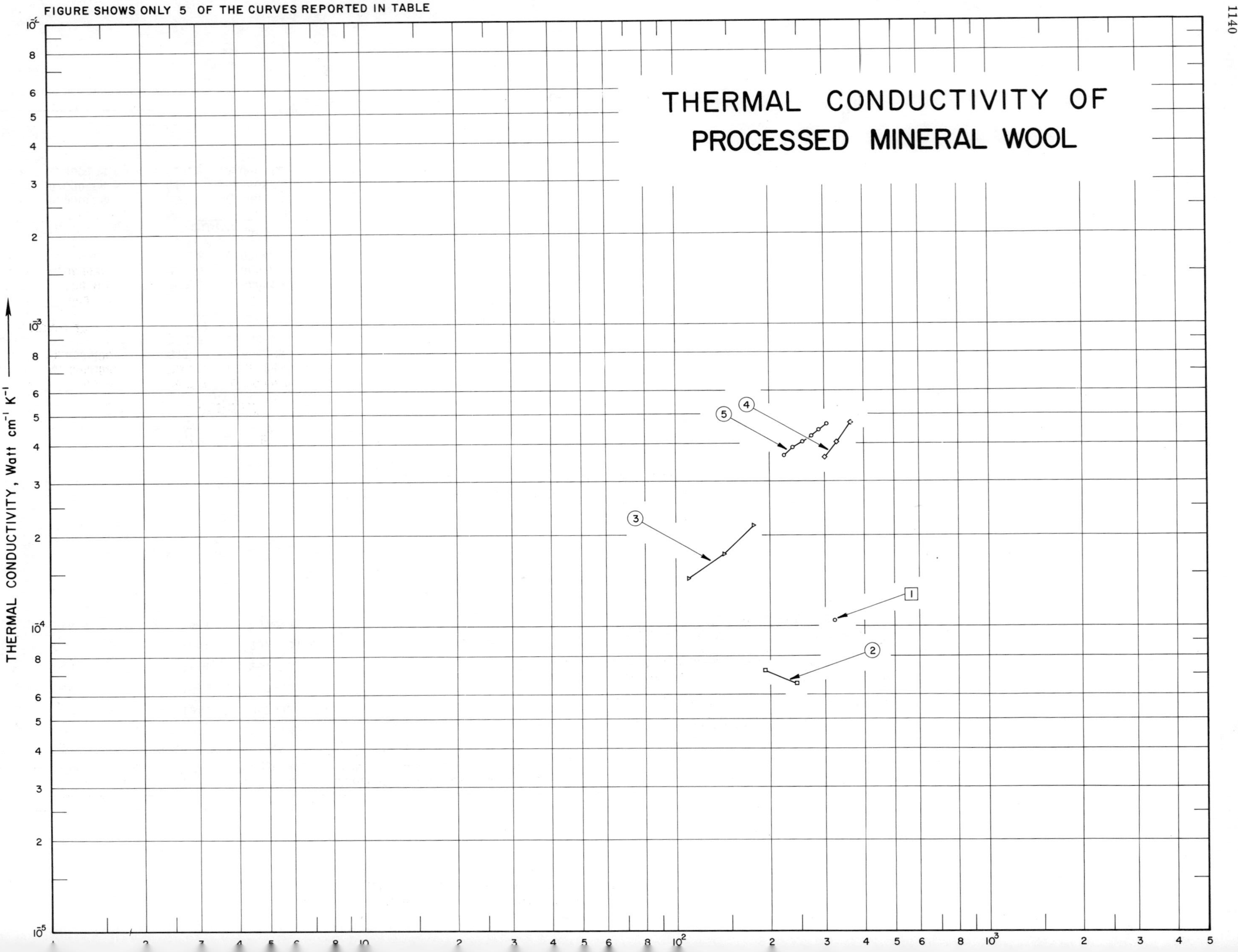
FIGURE SHOWS ONLY 5 OF THE CURVES REPORTED IN TABLE
THERMAL CONDUCTIVITY OF
PROCESSED MINERAL WOOL
THERMAL CONDUCTIVITY, Watt cm⁻¹ K⁻¹
1
2
3
4
5

SPECIFICATION TABLE NO. 394 THERMAL CONDUCTIVITY OF MINERAL WOOL, PROCESSED

[For Data Reported in Figure and Table No. 394]

Curve No.	Ref. No.	Method Used	Year	Temp. Range, K	Reported Error, %	Name and Specimen Designation	Composition (weight percent), Specifications and Remarks
1	128	L	1946	323.7		Mineral wool board	2.02 in. thick; density 0.229 g cm^{-3}.
2	128	L	1946	193, 244		Mineral wool board	Similar to the above specimen except 0.99 in. thick.
3	43	L	1954	110-177		Mineral wool felt	Density 0.064-0.128 g cm^{-3}.
4	43	L	1954	300-362		Mineral wool felt	Density 0.070-0.128 g cm^{-3}.
5	109	L	1947	222-306		Mineral wool board	1 in. thick; density 0.252 g cm^{-3}; as received.
6	189		1950	200-289		Mineral wool board	Density 0.232-0.264 g cm^{-3}; water absorption (24 hrs) 3.3 vol %.

DATA TABLE NO. 394 THERMAL CONDUCTIVITY OF PROCESSED MINERAL WOOL

[Temperature, T, K; Thermal Conductivity, k, Watt $cm^{-1} K^{-1}$]

T	k
CURVE 1	
323.7	0.000104
CURVE 2	
193.2	0.0000717
243.7	0.0000651
CURVE 3	
110	0.000144
143	0.000173
177	0.000216
CURVE 4	
300	0.000361
328	0.000404
362	0.000468
CURVE 5	
222.0	0.000366
238.7	0.000387
255.4	0.000405
272.2	0.000424
288.9	0.000444
305.5	0.000464
CURVE 6*	
200.0	0.000346
222.2	0.000361
239.2	0.000389
255.5	0.000404
272.1	0.000418
288.8	0.000447

*Not shown on plot

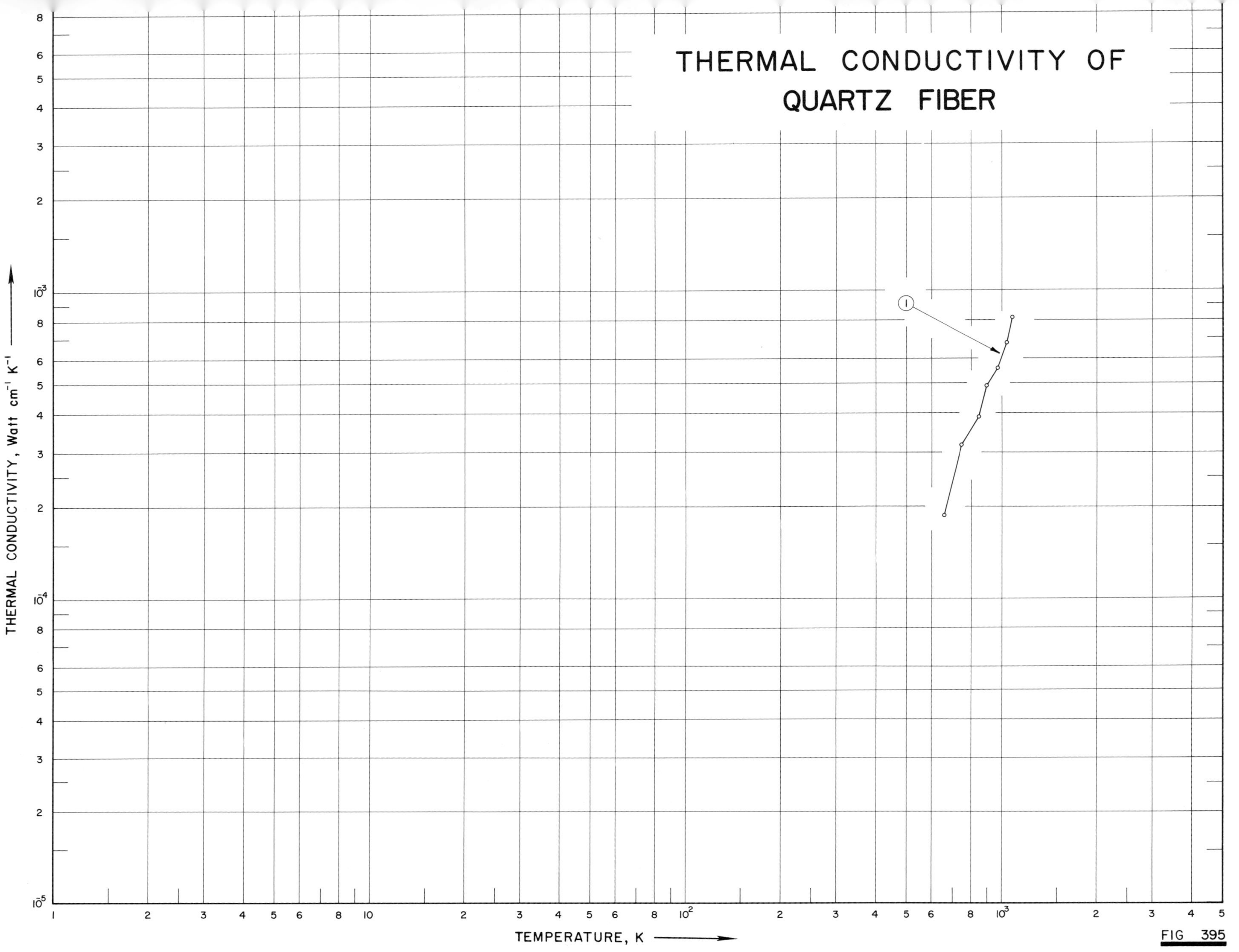
THERMAL CONDUCTIVITY OF
QUARTZ FIBER
1
THERMAL CONDUCTIVITY, Watt cm⁻¹ K⁻¹
TEMPERATURE, K
FIG 395

SPECIFICATION TABLE NO. 395 THERMAL CONDUCTIVITY OF QUARTZ FIBER

[For Data Reported in Figure and Table No. 395]

Curve No.	Ref. No.	Method Used	Year	Temp. Range, K	Reported Error, %	Name and Specimen Designation	Composition (weight percent), Specifications and Remarks
1	472	R	1963	661-1083		Dyna-quartz fiber	Specimen packed in a hollow cylinder of O. D. 2. 75 in. and I. D. 0. 75 in. ; density 0. 107 g cm^{-3}; supplied by Johns-Manville; measured in a vacuum of 5 x 10^{-6} to 1 x 10^{-4} mm Hg.

DATA TABLE NO. 395 THERMAL CONDUCTIVITY OF QUARTZ FIBERS

[Temperature, T, K; Thermal Conductivity, k, Watt cm^{-1} K^{-1}]

T	k
CURVE 1	
660.9	0.000187
747.1	0.000317
844.3	0.000389
894.3	0.000490
972.1	0.000555
1033.2	0.000678
1083.2	0.000815

SPECIFICATION TABLE NO. 396 THERMAL CONDUCTIVITY OF ROCK CORK

Curve No.	Ref. No.	Method Used	Year	Temp. Range, K	Reported Error, %	Name and Specimen Designation	Composition (weight percent), Specifications and Remarks
1	189		1950	200-288			Density range 0.22 ~ 0.26 g cm^{-3}; water absorption (24 hrs) 1.4 - 2.3 vol %.

DATA TABLE NO. 396 THERMAL CONDUCTIVITY OF ROCK CORK

[Temperature, T, K; Thermal Conductivity, k, Watt $cm^{-1}K^{-1}$]

T	k
CURVE 1*	
200.0	0.000346
222.2	0.000375
239.2	0.000389
255.5	0.000411
272.1	0.000433
288.1	0.000462

* No graphical presentation

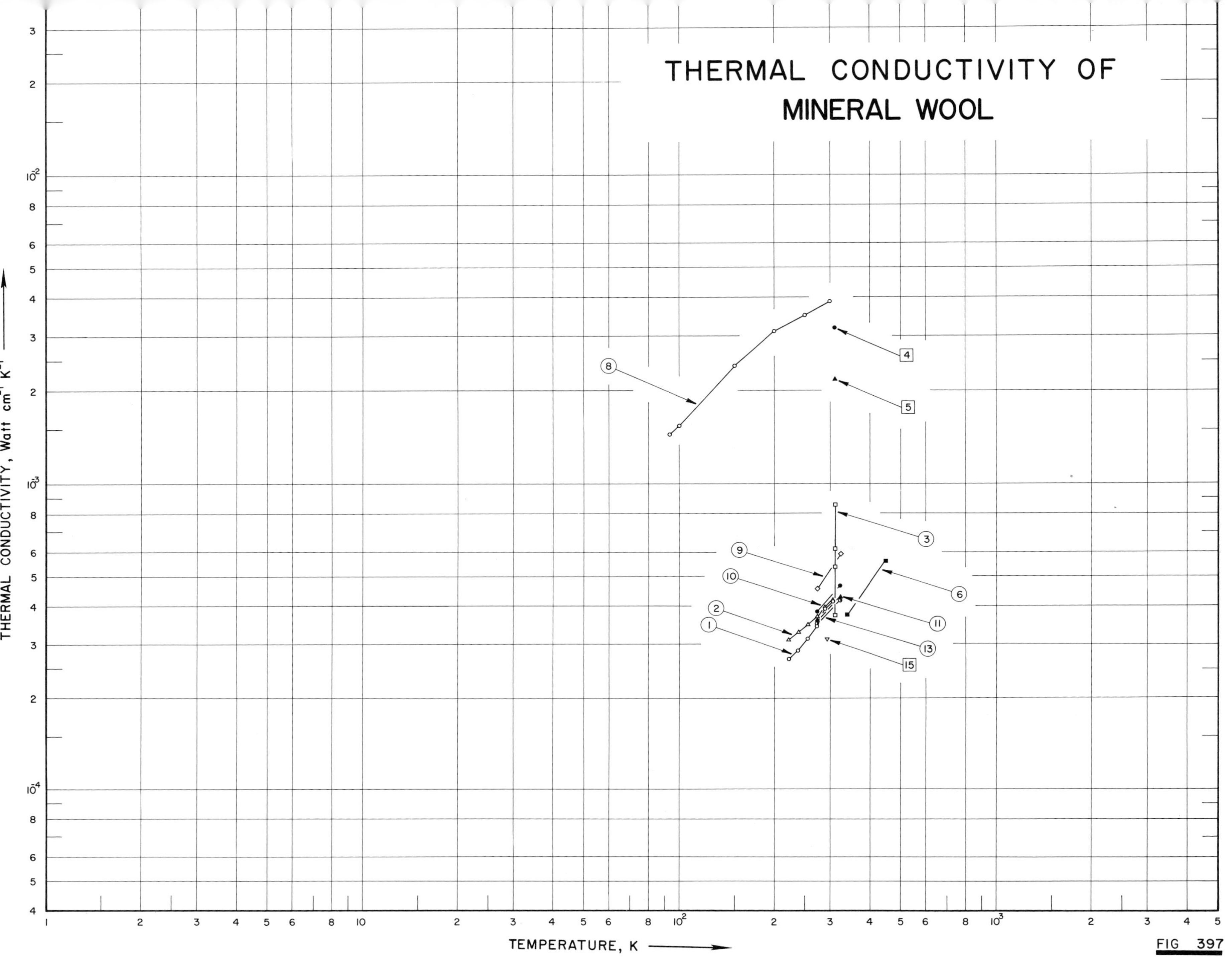
THERMAL CONDUCTIVITY OF
MINERAL WOOL
THERMAL CONDUCTIVITY, Watt cm⁻¹ K⁻¹
TEMPERATURE, K
1
2
3
4
5
6
8
9
10
11
13
15

FIG 397

SPECIFICATION TABLE NO. 397 THERMAL CONDUCTIVITY OF MINERAL WOOL

[For Data Reported in Figure and Table No. 397]

Curve No.	Ref. No.	Method Used	Year	Temp. Range, K	Reported Error, %	Name and Specimen Designation	Composition (weight percent), Specifications and Remarks
1	109	L	1947	222-306			1 in. thick; density 0.056 g cm^{-3}; as received.
2	109	L	1947	222-306			Nodulated; 1.5 in. thick; density 0.16 g cm^{-3}; as received.
3	558	L	1951	310.9			1 in. thick; density 0.123 g cm^{-3}; measured in air under various air pressures ranging from 0.02 μ Hg to 1 atm.
4	560	L	1954	310.9			1 in. thick; density 0.123 g cm^{-3}; measured in CO_2.
5	560	L	1954	310.9			The above specimen measured in F-12.
6	573	L	1954	339, 450			No details reported.
7	563	L	1945	222-305		Rock wool	Thickness of specimen 1 in.; density 0.056 g cm^{-3} at 25 C.
8	139	R	1950	93-300		Rock wool	Density 0.250 g cm^{-3}.
9	571		1950	273, 323		Rock wool	Density 0.030 g cm^{-3}.
10	571		1950	273, 323		Rock wool	Density 0.060 g cm^{-3}.
11	571		1950	273, 323		Rock wool	Density 0.090 g cm^{-3}.
12	571		1950	273, 323		Rock wool	Density 0.120 g cm^{-3}.
13	571		1950	273, 323		Rock wool	Density 0.145 g cm^{-3}.
14	571		1950	273, 323		Rock wool	Density 0.165 g cm^{-3}.
15	457	C	1954	293.2		Rock wool	Cork used as comparative material.

DATA TABLE NO. 397 THERMAL CONDUCTIVITY OF MINERAL WOOL

[Temperature, T, K; Thermal Conductivity, k, Watt $cm^{-1} K^{-1}$]

T	k
CURVE 1	
222.0	0.000268
238.7	0.000286
255.4	0.000312
272.2	0.000343
288.9	0.000391
305.5	0.000414
CURVE 2	
222.0	0.000307
238.7	0.000328
255.4	0.000349
272.2	0.000369
288.9	0.000392
305.5	0.000417
p(mm Hg)	k
CURVE 3 T = 310.9K	
atm.	0.000372
9.0	0.0000851
3.0	0.0000613
0.02	0.0000534
T	k
CURVE 4	
310.9	0.00318
CURVE 5	
310.9	0.00218
CURVE 6	
339.0	0.000375
450.2	0.000562

T	k
CURVE 7*	
222.2	0.000268
239.2	0.000286
255.5	0.000312
272.1	0.000343
288.8	0.000376
305.3	0.000414
CURVE 8	
93.2	0.0145
100	0.0155
150	0.0240
200	0.0310
250	0.0350
300	0.0390
CURVE 9	
273.2	0.000453
323.2	0.000593
CURVE 10	
273.2	0.000384
323.2	0.000465
CURVE 11	
273.2	0.000360
323.2	0.000430
CURVE 12*	
273.2	0.000360
323.2	0.000430
CURVE 13	
273.2	0.000349
323.2	0.000418

T	k
CURVE 14*	
273.2	0.000372
323.2	0.000430
CURVE 15	
293.2	0.00031

* Not shown on plot

SPECIFICATION TABLE NO. 398 THERMAL CONDUCTIVITY OF MYSTIC SLAG

Curve No.	Ref. No.	Method Used	Year	Temp. Range, K	Reported Error, %	Name and Specimen Designation	Composition (weight percent), Specifications and Remarks
1	545	P	1947	287.3		3D-1	Specimen consisted of basic residue from blast furnace located at Everett, Mass.; packed in a hollow cylinder of 5.36 in. dia and 10.68 in. long; material placed in container in five approximately equal layers and compacted; using increasing number of blows on each layer to obtain a uniform density through the specimen; specific gravity 2.45; density 1.382 g cm^{-3}; moisture content 9.1% of dry weight; thermal conductivity value calculated from measured transient temperature change.
2	545	P	1947	287.3		3D-2	Similar to the above specimen except density 1.736 g cm^{-3}; moisture content 33.5% of dry weight.
3	545	P	1947	287.3		3D-6	Similar to the above specimen except density 1.488 g cm^{-3}; moisture content 0.6% of dry weight.
4	545	P	1947	261.8		3D-3	Similar to the above specimen except density 1.471 g cm^{-3}; moisture content 5.5% of dry weight.
5	545	P	1947	261.8		3D-4	Similar to the above specimen except density 1.784 g cm^{-3}; moisture content 27.7% of of dry weight.
6	545	P	1947	261.8		3D-5	Similar to the above specimen except density 1.434 g cm^{-3}; moisture content 0.21% of dry weight.

DATA TABLE NO. 398 THERMAL CONDUCTIVITY OF MYSTIC SLAG

[Temperature, T, K; Thermal Conductivity, k, Watt $cm^{-1}K^{-1}$]

T	k	T	k	T	k
CURVE 1*		CURVE 3*		CURVE 5*	
287.3	0.00325	287.3	0.00253	261.8	0.0116
CURVE 2*		CURVE 4*		CURVE 6*	
287.3	0.00957	261.8	0.00424	261.8	0.00211

* No graphical presentation

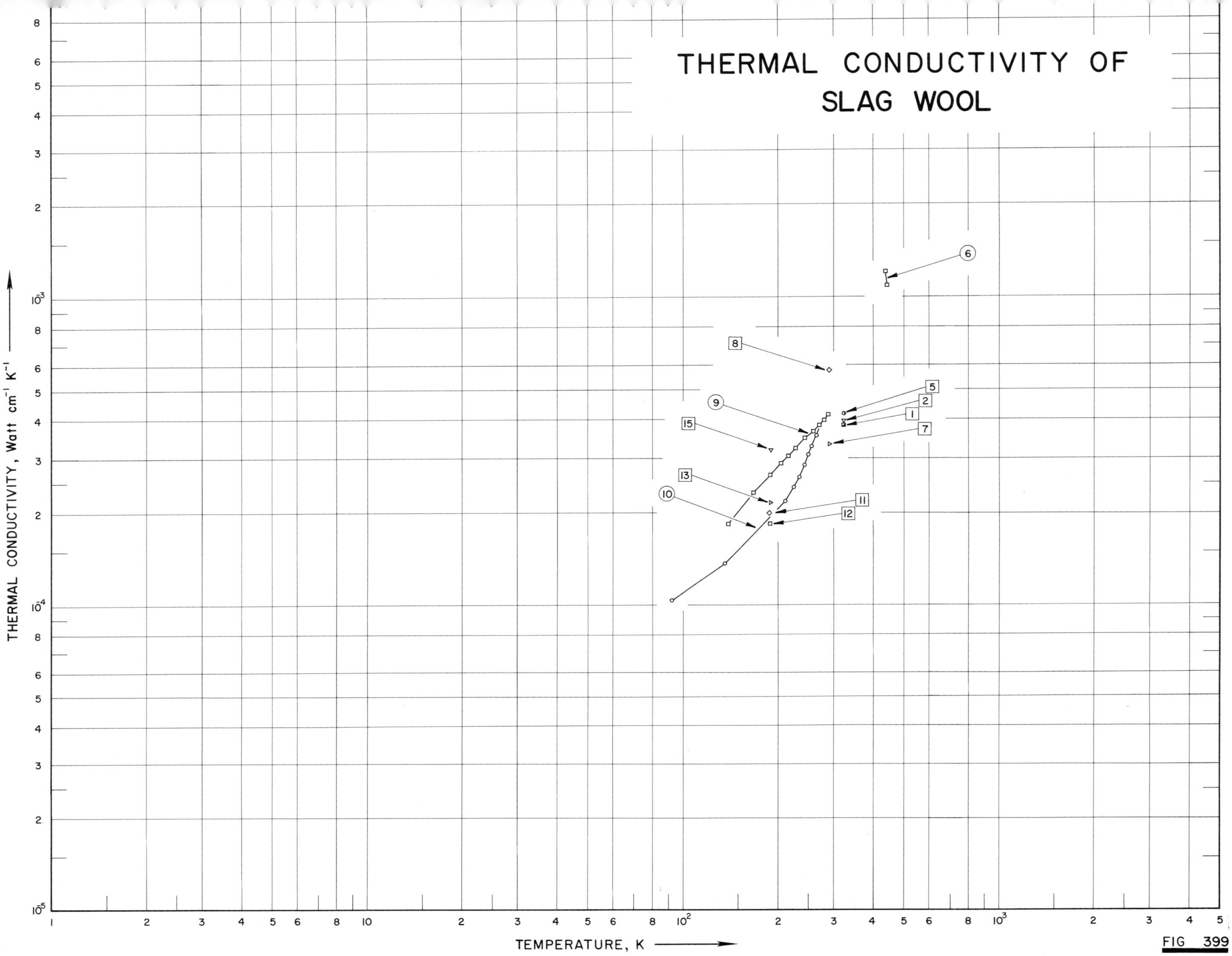

THERMAL CONDUCTIVITY OF SLAG WOOL

THERMAL CONDUCTIVITY, Watt cm^{-1} K^{-1}

TEMPERATURE, K

FIG 399

SPECIFICATION TABLE NO. 399 THERMAL CONDUCTIVITY OF SLAG WOOL

[For Data Reported in Figure and Table No. 399]

Curve No.	Ref. No.	Method Used	Year	Temp. Range, K	Reported Error, %	Name and Specimen Designation	Composition (weight percent), Specifications and Remarks
1	571		1950	323.2			Specimen consisted mainly of SiO_2, CaO and oxides of Al, Fe, and Mn; density 0.080 g cm^{-3}.
2	571		1950	323.2			Similar to the above specimen except density 0.100 g cm^{-3}.
3	571		1950	323.2			Similar to the above specimen except density 0.125 g cm^{-3}.
4	571		1950	323.2			Similar to the above specimen except density 0.150 g cm^{-3}.
5	571		1950	323.2			Similar to the above specimen except density 0.175 g cm^{-3}.
6	240	R	1919	438, 444			Specimen packed in a hollow cylinder of 12.2 mm I.D., 19 mm O.D., and 8 cm long.
7	529	P	1956	290.4			Density 0.128 g cm^{-3}; thermal conductivity value calculated from measured transient temp changes.
8	529	P	1956	290.4			Similar to the above specimen except density 0.320 g cm^{-3}.
9	561	R	1948	139-289		6A	Packed in a hollow cylinder of 2 in. I.D. and 10 in. O.D.; bulk density 0.197 g cm^{-3}.
10	561	R	1948	92-284		6B	Similar to the above specimen but bulk density 0.112 g cm^{-3}.
11	561	R	1948	187.3		6C	Similar to the above specimen but bulk density 0.147 g cm^{-3}.
12	561	R	1948	188.3		6D	Similar to the above specimen but bulk density 0.130 g cm^{-3}.
13	561	R	1948	189.3		6E	Similar to the above specimen but bulk density 0.095 g cm^{-3}.
14	561	R	1948	190.2		6F	Similar to the above specimen but bulk density 0.080 g cm^{-3}.
15	561	R	1948	190.2		6G	Similar to the above specimen but bulk density 0.059 g cm^{-3}.

DATA TABLE NO. 399 THERMAL CONDUCTIVITY OF SLAG WOOL

[Temperature, T, K; Thermal Conductivity, k, Watt $cm^{-1} K^{-1}$]

T	k
CURVE 1	
323.2	0.000384
CURVE 2	
323.2	0.000395
CURVE 3*	
323.2	0.000395
CURVE 4*	
323.2	0.000407
CURVE 5	
323.2	0.000418
CURVE 6	
438.1	0.00121
443.8	0.00109
CURVE 7	
290.4	0.000332
CURVE 8	
290.4	0.000577
CURVE 9	
139.2	0.000184
167.8	0.000233
189.2	0.000265
204.2	0.000289
216.0	0.000305
227.4	0.000322
243.7	0.000348
259.7	0.000365
270.7	0.000382
280.4	0.000397
288.7	0.000415

T	k
CURVE 10	
92.2	0.000104
136.5	0.000138
211.2	0.000218
224.5	0.000242
234.2	0.000261
243.2	0.000286
249.6	0.000307
256.2	0.000328
265.2	0.000355
274.2	0.000384*
279.7	0.000405*
284.2	0.000426*
CURVE 11	
187.3	0.000199
CURVE 12	
188.3	0.000185
CURVE 13	
189.3	0.000216
CURVE 14*	
190.2	0.000232
CURVE 15	
190.2	0.000317

* Not shown on plot

SPECIFICATION TABLE NO. 400 THERMAL CONDUCTIVITY OF BITUMEN

Curve No.	Ref. No.	Method Used	Year	Temp. Range, K	Reported Error, %	Name and Specimen Designation	Composition (weight percent), Specifications and Remarks
1	88	L	1926	292,356		I	Specimen thickness 5.5 mm; mean density 1.05 g cm^{-3}.
2	88	L	1926	294,349		II	Specimen thickness 5.0 mm; mean density 1.05 g cm^{-3}.

DATA TABLE NO. 400 THERMAL CONDUCTIVITY OF BITUMEN

[Temperature, T, K; Thermal Conductivity, k, Watt $cm^{-1}K^{-1}$]

T	k
CURVE 1*	
292.1	0.00166
355.9	0.00173
CURVE 2*	
294.1	0.00169
349.2	0.00171

* No graphical presentation

SPECIFICATION TABLE NO. 401 THERMAL CONDUCTIVITY OF BONE CHAR

Curve No.	Ref. No.	Method Used	Year	Temp. Range, K	Reported Error, %	Name and Specimen Designation	Composition (weight percent), Specifications and Remarks
1	549	L	1953	337.7			Bulk density 0.700 g cm^{-3} at 25 C; moisture content 2-4%.
2	549	L	1953	337.7			Similar to the above specimen except bulk density 0.791 g cm^{-3} at 25 C.
3	549	L	1953	337.7			Similar to the above specimen except bulk density 0.857 g cm^{-3} at 25 C.
4	549	L	1953	337.7			Similar to the above specimen except bulk density 0.823 g cm^{-3} at 25 C.

DATA TABLE NO. 401 THERMAL CONDUCTIVITY OF BONE CHAR

[Temperature, T, K; Thermal Conductivity, k, Watt $cm^{-1}K^{-1}$]

T	k
CURVE 1*	
337.7	0.00178
CURVE 2*	
337.7	0.00167
CURVE 3*	
337.7	0.00181
CURVE 4*	
337.7	0.00169

* No graphical presentation

SPECIFICATION TABLE NO. 402 THERMAL CONDUCTIVITY OF CHARCOAL

Curve No.	Ref. No.	Method Used	Year	Temp. Range, K	Reported Error, %	Name and Specimen Designation	Composition (weight percent), Specifications and Remarks
1	491	L	1958	293			Sutcliffe Speakman 208 C charcoal bed contained in a cylindrical annular space of 1.5 cm I.D., 3.9 cm O.D., and 5 cm long; filled with helium.
2	491	L	1958	293			Similar to the above specimen except filled with air.

DATA TABLE NO. 402 THERMAL CONDUCTIVITY OF CHARCOAL

[Temperature, T, K; Thermal Conductivity, k, Watt $cm^{-1}K^{-1}$]

T	k
CURVE 1*	
293	0.0036
CURVE 2*	
293	0.0020

* No graphical presentation

SPECIFICATION TABLE NO. 403 THERMAL CONDUCTIVITY OF COAL TAR FRACTIONS

Curve No.	Ref. No.	Method Used	Year	Temp. Range, K	Reported Error, %	Name and Specimen Designation	Composition (weight percent), Specifications and Remarks
1	572	F	1953	393-603	<1		4.52% free carbon and 1.6% humidity; from Kizelovski coal (USSR); sp weight (20 C) = 1.1673.
2	572	F	1953	453-613	<1		2.88% free carbon and 3.7% humidity; from Kuznetski coal (USSR); sp weight (20 C) = 1.1859.
3	572	F	1953	423-533	<1		9.95% free carbon and 3.9% humidity; from a mixture of Kuznetski and Karagandunski coals; sp weight (20 C) = 1.2195.

DATA TABLE NO. 403 THERMAL CONDUCTIVITY OF COAL TAR FRACTIONS

[Temperature, T, K; Thermal Conductivity, k, Watt $cm^{-1} K^{-1}$]

T	k	T	k
CURVE 1*		CURVE 3*	
393.2	0.00139	423.2	0.00130
468.2	0.00128	478.2	0.00119
488.2	0.00119	488.2	0.00115
528.2	0.00116	533.2	0.00105
553.2	0.00103		
598.2	0.00105		
603.2	0.00103		
CURVE 2*			
453.2	0.00129		
483.2	0.00123		
498.2	0.00114		
553.2	0.00102		
613.2	0.00096		

*No graphical presentation

REFERENCES TO DATA SOURCES

Ref. No.	TPRC No.	
1	16816	Norton, F. H., Fellows, D. M., Adams, M., and McQuarrie, M., USAEC Rept. NYO-594, 1-22, 1950.
2	16821	Norton, F. H., Kingery W. D., Fellows, D. M., Adams, M., McQuarrie, M. C., Coble, R. L., Francl, J., Vasilos, T., and Anderson, H. H., USAEC Rept. NYO-597, 1-15, 1951.
3	9404	Koenig, J. H., Rutgers Univ. N. J. Ceram. Research Sta. Progr. Rept. 2, 1-117, 1953. [AD 131 54]
4	16818	Norton, F. H., USAEC Rept. NYO-598, 1-10, 1951.
5	2441	Kingery, W. D., J. Am. Ceram. Soc., 37, 88-90, 1954.
6	35859	Neshpor, V. S., Ordanyan, S. S., Avgustinik, A. I., and Khusidman, M. B., NASA TTF-9350, 1-15, 1965.
7	16826	Norton, F. H., Fellows, D. M., Adams, M., McQuarrie, M., and Fullerton, C. P., USAEC NYOO-96, Progr. Rept. 2, 1-43, 1949.
8	10013	Sutton, W. H., J. Am. Ceram. Soc., 43 (2), 81-6, 1960.
9	7102	Koenig, J. H. (Director), Rutgers Univ., N. J. Ceram. Research Sta., Progr. Rept. 4, 1-35, 1953. [AD 293 35]
10	9316	Norton, F. H., Kingery, W. D., et al., USAEC Rept. NYO-602, 1-40, 1952.
11	7690	Powers, R. W., Schwartz, D., and Johnson, H. L., Cryogenic Lab., OSU, USAF TR264-5, 1-19, 1951.
12	7067	Koenig, J. H. (Director), Rutgers Univ. N. J. Ceram. Research Sta. Progr. Rept. 3, 1-25, 1953. [AD 198 33]
13	6200	Weeks, J. L. and Seifert, R. L., Rev. Sci, Instr., 24, 1054-7, 1953.
14	24366	Vasilos, T., Kingery, W. D., and Norton, F. H., USAEC Rept. NYO-3649, 1-33, 1953.
15	6364	Jamieson, C. P. and Mrozowski, S., Proc. Conf. on Carbon, Univ. Buffalo, 155-66, 1956.
16	9857	Worthing, A. G., Phys. Rev., 4, 535-43, 1914.
17	6739	Burr, A. C., Canad. J. Technol., 29, 451-7, 1951.
18	6225	Powell, R. W. and Schofield, F. H., Proc. Phys. Soc. (London), 51 (1), 153-72, 1939.
19	43	Brown, A. R. G., Watt, W., Powell, R. W., and Tye, R. P., Brit. J. Appl. Phys., 7, 73-6, 1956.
20	3556	Buerschaper, R. A., J. Appl. Phys., 15, 452-4, 1944.
21	9914	Kaye, G. W. C. and Higgins, W. F., Proc. Roy. Soc. (London), A122, 633-46, 1929.
22	6708	Eucken, A., Ann. Physik, 34 (2), 185-221, 1911.
23	9410	Koenig, J. H., Rutgers Univ. N. J. Ceram. Research Sta. Progr. Rept. 6, 1-98, 1952. [ATI 163 519]
24	8195	Ploetz, G. L., Muccigrosso, A. T., and Krystyniak, C. W., USAEC Rept. KAPL-1918, 1-24, 1958.
25	266	Berman, R., Simon, F. E., and Ziman, J. M., Proc. Roy. Soc. (London), A220, 171-83, 1953.
26	179	Ioffe, A. V. and Ioffe, A. F., Doklady Akad. Nauk USSR, 97 (5), 821-2, 1954.
27	7200	Berman, R., Simon, F. E., and Wilks, J., Nature, 168, 277-80, 1951.
28	1161	Berman, R., Foster, E. L., and Ziman, J. M., Proc. Roy. Soc. (London), A237, 344-54, 1956.
29	6944	Hedge, J. C. and Fieldhouse, I. B., AECU-3381, 1-10, 1956.
30	15942	Paprocki, S. J. (Editor), Keller, D. L., Hodge, E. S., Boyer, C. B., Fox, J. B., Kizer, D. E., and Porembka, S. W., USAEC Rept. BMI-1475, 1-92, 1960.
31	6478	de Vries, D. A. and Peck, A. J., Australian J. Phys., 11 (2), 255-71, 1958.
32	9289	de Haas, W. J. and Biermasz, Th., Physica, 4 (8), 752-6, 1937.
33	6565	Lucks, C. F., Matolich, J., and Van Velzor, J. A., USAF TR 6145, Pt, III, 1-71, 1954. [AD 954 06]
34	1287	Knapp, W. J., J. Am. Ceram. Soc., 26, 48-55, 1943.
35	107	Colosky, B. P., Am. Ceram. Soc. Bull., 31 (1), 465-6, 1952.
36	6544	NBS, NBS Rept. 4761, Progr. Rept. 1, 1-6, 1956. [AD 101 080]

Ref. No.	TPRC No.	
37	6970	Fieldhouse, I.B., Hedge, J.C., Lang, J.I., and Waterman, T.E., WADC TR 57-487, 1-88, 1958. [AD 150 954] [PB 131 718]
38	22625	Yoshida, I., J. Phys. Soc. Japan, 15 (12), 2211-9, 1960.
39	10947	Lucks, C.F., Deem, H.W., and Woods, W.D., Am. Ceram. Soc. Bull., 39 (6), 313-9, 1960.
40	16415	Ward, F.A., Phil. Mag., 48, 971-7, 1924.
41	8370	Johnston, H.L., Ohio State Univ. Research Foundation, USAF Proj. MX-588, Rept. 20, 1-10, 1948. [ATI 45 490]
42	7036	Koenig, J.H. (Director), Rutgers Univ. N.J. Ceram. Res. Sta. Progr. Rept. 1, 1-33, 1953. [AD 5 552]
43	6325	Verschoor, J.D., Refrig. Eng., 62 (9), 35-7 and 98, 1954.
44	16313	Barratt, T., Proc. Phys. Soc. (London), 27, 81-93, 1914.
45	10164	Norton, F.H., Kingery, W.D., et al., USAEC Rept. NYO-3646, 1-8, 1953. [AD 16 492]
46	163	Slack, G.A., Phys. Rev., 105, 832-42, 1957.
47	9972	Raeth, C.H., USAEC Rept. CP-2332, 1-25, 1944.
48	6502	Fieldhouse, I.B., Hedge, J.C., Lang, J.L., Takata, A.N., and Waterman, T.E., WADC TR 55-495, Pt. I, 20-48, 1956. [AD 110 404]
49	258	Tyler, W.W. and Wilson, A.C., Jr., Phys. Rev., 89 (4), 870-5, 1953.
50	820	Smith, A.W. and Rasor, N.S., USAEC Rept. NAA-SR-1590, 1-25, 1956.
51	734	Deegan, G.E., USAEC Rept. NAA-SR-1716, 1077, 1956.
52	166	Euler, J., Ann. Physik, 6, 18, 345-69, 1956.
53	7130	Rasor, N.S., USAEC Rept. NAA-SR-1061, 1-23, 1955. [AD 50 086]
54	190	Euler, J., Naturwissenschaften, 39, 568-9, 1952.
55	6560	Lucks, C.F. and Deem, H.W., WADC TR 55-496, 1-65, 1956. [AD 97 185]
56	8104	Fletcher, J.F. and Snyder, W.A., Am. Ceram. Soc. Bull., 36 (3), 101-4, 1957.
57	16820	Norton, F.H., Kingery, W.D., Fellows, D.M., Adams, M., McQuarrie, M.C., and Coble, R.L., USAEC Rept. NYO-596, 1-9, 1950.
58	6431	Doolittle, J.S., N. Carolina State College Quarterly Rept. 3, 1-4, 1953. [AD 122 141]
59	6210	Marshall, T.A., Brit. J. Appl. Phys., 4, 112-4, 1953.
60	23906	Tautz, H., Exp. Tech. der Physik, 8 (1), 34-7, 1960.
61	13785	Jaffee, R.I. and Maykuth, D.J., Aero-Space Eng., 19, 39-44, 1960.
62	23952	Pohl, R.O., Phys. Rev., 118 (6), 1499-508, 1960.
63	18030	Taylor, R.E., J. Am. Ceram. Soc., 44 (10), 525, 1961.
64	9880	Slack, G.A. and Newman, R., Phys. Rev. Letters, 1 (10), 359-60, 1958.
65	10682	Francis, R.K., Brown, R., McNamara, E.P., and Tinklepaugh, J.R., Wright Air Development Center, WADC TR 58-600, PB-151664, 1-69, 1958. [AD 205 549]
66	9287	Lees, C.H., Phil. Trans. Roy. Soc. (London), A204, 433-66, 1905.
67	1172	Ueberreiter, K. and Orthmann, H.J., Z. Naturforsch., A5, 101-8, 1950.
68	7104	Koenig, J.H. (Director), Rutgers Univ. N.J. Ceram. Res. Sta. Progr. Rept. 5, 1-24, 1954.
69	1411	Berman, R., Proc. Roy. Soc. (London), A208, 90-108, 1951.
70	1108	Berman, R., Foster, E.L., and Ziman, J.M., Proc. Roy. Soc. (London), A231, 130-44, 1955.
71	25435	McCarthy, K.A. and Ballard, S.S., J. Opt. Soc. Am., 41, 1062-3, 1951.
72	10101	Kingery, W.D. and Norton, F.H., USAEC Rept. NYO-6447, 1-14, 1955. [AD 55 595]
73	10192	Norton, F.H. and Kingery, W.D., USAEC Rept. NYO-6441, 1-7, 1954. [AD 27 184]
74	25427	Quirk, J. and Harman, C.G., J. Am. Ceram. Soc., 37, 24-6, 1954.
75	1639	Pengelly, A.E., Brit. J. Appl. Phys., 6, 18-20, 1955.
76	6991	McCreight, L.R., USAEC Rept. TID-10062, 1-19, 1952.
77	1425	Garrett, C.G.B., Phil. Mag., 41, 621-30, 1950.
78	26	Bijl, D., Physica, 14, 684-93, 1948.
79	5258	Griffiths, E., Powell, R.W., and Hickman, M.J., J. Inst. Fuel, 15, 107-20, 1942.
80	7180	Wilkes, G.B., J. Am. Ceram. Soc., 17, 173-7, 1934.

Ref. No.	TPRC No.	
81	20589	Buckner, O.S., J. Am. Ceram. Soc., 7, 19-28, 1924.
82	26012	Lowrance, D.T., Chance Vought Corp., CVC Rept. 2-53420/2R375, 1-95, 1962. [AD 273 802]
83	9247	Sheriff, I.I. and Bakr, M.Y., Silicates Ind., 21 (6-7), 287-9, 1956.
84	20757	Hartmann, M.L. and Westmont, O.B., J. Am. Ceram. Soc., 8, 92-95, 1925.
85	801	Kolechkova, A.F. and Goncharov, V.V., Ogneupory, 14, 445-53, 1949.
86	4492	Kolechkova, A.F. and Goncharov, V.V., Ogneupory, 13, 401-7, 1948.
87	110	Ruh, E., J. Am. Ceram. Soc., 37 (5), 224-9, 1954.
88	21109	Jakob, M., Z. tech. Physik, 7, 475-81, 1926.
89	10851	Norton, F.H. and Kingery, W.D., USAEC Rept. NYO-601, 1-52, 1952.
90	3610	Patton, T.C. and Norton, C.L., Jr., J. Am. Ceram. Soc., 26, 350-8, 1943.
91	617	Dudavskii, I.E., Zavodskaya Lab., 13, 710-5, 1947.
92	838	Plotnikov, L.A., Zavodskaya Lab., 16, 1136-9, 1950.
93	74	Wester, A. and Aggeryd, B., Das Papier, 11 (5-6), 98-104, 1957.
94	1070	Cammerer, J.S., Stahl und Eisen, 62, 503-10, 1942.
95	7182	Finck, J.L., J. Am. Ceram. Soc., 18, 6-12, 1935.
96	7181	Weinland, C.E., J. Am. Ceram. Soc., 17, 194-202, 1934.
97	8552	Shakhtin, D.M., UKAEA Rept. IGRL-T/C-67, 1-4, 1958.
98	3671	Norton, C.L., Jr., J. Am. Ceram. Soc., 25 (15), 451-9, 1942.
99	1967	Duplin, V.J., Jr., and Fitzsimmons, E.S., J. Am. Ceram. Soc., 35, 226-9, 1952.
100	10271	Bishui, B.M. and Prasad, J., Central Glass and Ceram. Research Inst. Bull. (India), 4, 144-9, 1957.
101	7185	Austin, J.B. and Pierce, R.H.H., Jr., J. Am. Ceram. Soc., 18, 48-54, 1935.
102	220	Suzuki, H., Kuwayama, N., and Yamauchi, T., Yogyo Kyokai Shi, 65, 206-11, 1957.
103	13367	Finck, J.L., J. Am. Ceram. Soc., 20, 378-82, 1937.
104	1991	Suzuki, H., Kuwayama, N., and Yamauchi, T., Yogyo Kyokai Shi, 64 (726), 161-6, 1956.
105	23960	Nuvoli, A., Ricerca Scientifica, 28 (10), 2095-101, 1958.
106	40	Noguchi, T. and Miyazaki, Y., Tanso, 5, 36-9, 1956.
107	21969	Hersey, M.D. and Butzler, E.W., J. Washington Acad. Sci., 14, 147-51, 1924.
108	22166	Taylor, T. S., Mech. Eng., 42, 8-10, 1920.
109	6359	Rowley, F.B., Jordan, R.C., and Lander, R.M., Refrigeration Engineering, 53, 35-9, 1947.
110	5402	Keesom, P.H., Physica, 11 (4), 339-42, 1945.
111	6567	Lucks, C.F., Bing, G.F., Matolich, J., Deem, H.W., and Thompson, H.B., USAF TR 6145, 1-39, 1952. [AD 95 239]
112	7695	Melonas, J.V., Covington, P.C., and Pears, C.D., Southern Res. Inst., WADC TR 58-129, 1-15, 1958. [AD 155 816]
113	206	Kunugi, M., Tanaka, S., Shimizu, S., and Kitamura, K., J. Ceram. Assoc. (Japan), 62 (703), 780-4, 1954.
114	10100	Kingery, W.D. and Norton, F.H., USAEC Rept. NYO-6446, 1-6, 1954. [AD 53 808]
115	56	Berman, R., Phys. Rev., 76, 315-6, 1949.
116	7477	Wilkinson, K.R. and Wilks, J., J. Sci. Instr., 26, 19-20, 1949.
117	1125	Krischer, O. and Esdorn, H., VDI Forschungsheft 450 Suppl. to Forsch. Gebiete Ingenieurw., B, (21), 28-39, 1955.
118	145	Geffcken, I.W., Glastechn. Ber., 25, 392-6, 1952.
119	7028	Eucken, A., Z. Tech. Physik, 6, 689-94, 1925.
120	768	Marathe, M.N. and Tendolkar, G.S., Trans. Indian Inst. Chem. Engrs., 6, 90-104, 1953.
121	21495	Carman, A.P. and Nelson, R.A., Univ. Illinois Exp. Sta. Bull. 122, 1-32, 1921.
122	7099	Powers, R.W., Johnston, H.L., Hansen, R.H., and Ziegler, J.B., Ohio State Univ. Cryogenic Lab. USAF TR-264-16, 1-29, 1953. [AD 27 569]
123	88	Ueberreiter, K. and Otto-Laupenmuhlen, E., Z. Naturforsch, II, 8A, 664-73, 1953.
124	17269	Hager, N.E., Jr., Rev. Sci. Instr., 31 (2), 177-85, 1960.

Ref. No.	TPRC No.	
125	7648	Griffiths, E. and Kaye, G.W.C., Proc. Roy. Soc. (London), A104, 71-98, 1923.
126	3265	Schallamach, A., Proc. Phys. Soc. (London), 53, 214-8, 1941.
127	20037	Clarke, L.N. and Kingston, R.S.T., Australian J. Appl. Sci., 1 (2), 172-87, 1950.
128	6321	Wilkes, G.B., Refrig. Eng., 52, 37-42, 1946.
129	21022	Hansen, C.A., Trans. Am. Electrochem. Soc., 16, 329-62, 1909.
130	2438	McQuarrie, M., J. Am. Ceram. Soc., 37 (2), 84-8, 1954.
131	2434	Francl, J. and Kingery, W.D., J. Am. Ceram. Soc., 37, 80-4, 1954.
132	7689	Fieldhouse, I.B., Hedge, J.C., and Lang, J.I., WADC TR 58-274, 1-79, 1958. [AD 206 892]
133	2429	Francl, J. and Kingery, W.D., J. Am. Ceram. Soc., 37, 99-107, 1954.
134	1115	Shul'man, A.R., Fedorov, V.N., and Shepsenvol, M.A., Zhur. tekh. Fiz., 22, 1271-80, 1952.
135	7095	Shakhtin, D.M. and Vishnevskii, I.I., Zavodskaya Lab., 23, 927-9, 1957.
136	10052	Kingery, W.D., J. Am. Ceram. Soc., 42 (12), 617-27, 1959.
137	6941	Soxman, E.J., Study of Heat Transfer of Ceramic Materials, Alfred Univ., ONR, I, 1-36, 1955.
138	1166	Ditmars, D.A. and Ginnings, D.C., J. Research NBS, 59 (2), 93-9, 1957.
139	76	Codegone, C., La Ricerca Sci., 20, 68-70, 1950.
140	22427	Kaye, G.W.C. and Higgins, W.F., Proc. Roy. Soc. (London), A113, 335-51, 1926.
141	1180	Scholes, W.A., J. Am. Ceram. Soc., 33 (4), 111-7, 1950.
142	9398	Berman, R., Foster, E.L., and Rosenberg, H.M., Brit. J. Appl. Phys., 6 (5), 181-2, 1955.
143	757	Kolechkova, A.F. and Goncharov, V.V., Ogneupory, 20, 39-44, 1955.
144	26008	Pears, C.D. (Project director), Southern Res. Inst. Tech. Documentary Rept. ASD-TDR-62-765, 20-402, 1963. [AD 298 061]
145	10110	Kingery, W.D. and Norton, F.H., USAEC Rept. NYO-6450, 1-16, 1955. [AD 73 173]
146	10045	Kingery, W.D. and Norton, F.H., USAEC Rept. NYO-6451, 1-16, 1955. [AD 80 699]
147	15618	Ratcliffe, E.H., Phys. and Chem. of Glasses, 1 (3), 103-4, 1960.
148	24174	Sujak, B., ACTA Physica Polonica, 20 (2), 167-74, 1961.
149	19423	Boettcher, A. and Schneider, G., Proc. of 2nd U.N. Conference on Peaceful Uses of Atomic Energy, 6, 561-3, 1958.
150	26079	Secrest, A.C., Jr., Foster, E.L., and Dickerson, R.F., USAEC Rept. BMI-1309, 1-16, 1959.
151	26078	Noe, A., Eldridge, E.A., and Deem, H.W., USAEC Rept. BMI-1377, 74-6, 1959.
152	25109	Bopp, C.D., Towns, R.L., and Sisman, O., USAEC Rept. ORNL-1945, 35-6, 1956.
153	10857	Norton, F.H., Kingery, W.D., et al., USAEC Rept. NYO-600, 1-20, 1951.
154	6249	Powell, R.W., Phil. Mag., 7, 27 (185), 677-86, 1939.
155	22782	Mannchen, W., Z. Metallk., 23, 193-6, 1931.
156	8330	Grard, C. and Villey, J., Compt. Rend., 185, 856-8, 1927.
157	124	de Nobel, J., Bull. Inst. Intern. Froid, Annexe 1956-2, 97-109, 1956.
158	824	Berman, R., Proc. Phys. Soc. (London), A65, 1029-40, 1952.
159	16782	Snyder, T.M. and Kamm, R.L., USAEC Rept. C-192, A-230, 1-56, 1955.
160	10690	Mrozowski, S., Andrew, J.F., Repetski, J., Strauss, H.E., and Wobschall, D.C., WADC TR-58-360, Pt. I, 1-59, 1958. [AD 206 664]
161	25359	Kasatochkin, V.I., Zamoluev, V.K., Kaverov, A.T., and Usenbaev, K., Proc. Acad. Sci. USSR Phys. Chem. Sect., 135 (1), 1009-12, 1960.
162	16829	Tyler, W.W. and Wilson, A.C., Jr., USAEC Rept. KAPL-789, 1-54, 1952.
163	10087	Rasor, N.S. and Smith, A.W., USAEC Rept. NAA-SR-862, 1-37, 1954. [AD 85 006]
164	204	Williams, W.S., Bull. Inst. Intern. Froid, Annexe 1956-2, 119-24, 1956.
165	29663	Ross, A.M., USAEC Rept. CRFD-817, 1-55, 1960.
166	9319	Eian, C.S. and Deissler, R.G., NACA RM E53G03, 1-18, 1953.
167	6968	Boegli, J.S. and Deissler, R.G., NACA RM E54L10, 1-20, 1955. [AD 56 099]
168	24212	Bethoux, O., Thomas, P., and Weil, L., Compt. Rend., 253 (19), 2043-5, 1961.
169	21298	Cohen, I., Lustman, B., and Eichenberg, J.D., J. Nuclear Materials, 3 (3), 331-53, 1961.

Ref. No.	TPRC No.	
170	10030	Norton, F.H. and Kingery, W.D., USAEC Rept. NYO-3644, 1-18, 1953. [AD 11 816]
171	10063	Scott, R., UKAEA Rept. AERE M/R 2526, 1-11, 1958. [AD 209 258 L]
172	14147	Howard, V.C. and Gulvin, T.F., UKAEA IG Rept. 51 (RD/C), 1-23, 1961.
173	27838	Daniel, J.L., Matolich, J., Jr., and Deem, H.W., USAEC Rept. HW-69945, 1-42, 1962.
174	14833	Reiswig, R.D., J. Am. Ceram. Soc., 44 (1), 48-9, 1961.
175	6288	Powell, R.W., Proc. Phys. Soc. (London), 49 (4), 419-25, 1937.
176	6978	Fieldhouse, I.B., Hedge, J.C., and Waterman, T.E., WADC TR 55-495, Pt. III, 1956. [AD 110 526]
177	6980	Rasor, N.S. and McClelland, J.D., WADC TR 56-400, I, 1-53, 1957. [AD 118 144]
178	10846	Rasor, N.S., North American Aviation, NAA-SR-43, 1-36, 1950.
179	20403	Icole, M., Ann de Chimie et de Physique, 25, 134-44, 1912.
180	20777	Mrozowski, S., Andrew, J.F., Juul, N., Strauss, H.E., and Wobschall, D.C., WADC TR 58-360, Pt. III, 1-54, 1961. [AD 258 531]
181	26075	National Carbon Co., USAEC ORO-240, 1, P1-2, 1959.
182	22833	Van Sant, J.H., USAEC Rept. GAMD-2089, 1-21, 1961.
183	25348	Gumenyuk, V.S. and Lebedev, V.V., Phys. Metals and Metallog, USSR, 11 (1), 30-5, 1961. [AD 262 622]
184	25074	Childers, H.M. and Cerceo, J.M., WADD TR 60-190, 1-66, 1961. [AD 272 691]
185	15792	Atomic International Div., North American Aviation, Inc., NAA-SR-5350, Progr. Repts., 1960.
186	20652	Hartmann, M.L., Westmont, O.B., and Weinland, C.E., Proc. ASTM, 28 (2), 820-47, 1928.
187	16179	Stephens, R.W.B., Phil. Mag., 14, 897-914, 1932.
188	27955	Crawford, J.H., Jr., and Cohen, A.F., USAEC Rept. ORNL-2614, 43-6, 1958. [TID-4500]
189	1097	Stone, J.F., Petroleum Engr., 22 (1), C11-C16, 1950.
190	1630	Hubscher, M., Silikattech, 5, 243-7, 1954.
191	777	Dauphinee, T.M., Ivey, D.G., and Smith, H.D., Can. J. Res., A28, 596-615, 1950.
192	1737	Hall, G.L. and Prettyman, I.B., India Rubber World, 113, 222-35, 1945.
193	9149	Eldred, V.W. and Curtis, G.C., Nature, 179, 910, 1957.
194	7681	Melonas, J.V., Covington, P.C., and Pears, C.D., Southern Res. Inst. WADC Tech. Rept. 58-179, 1-54, 1958. [AD 204 795]
195	8136	Wygant, J.F. and Crowley, M.S., J. Am. Ceram. Soc., 41 (5), 183-8, 1958.
196	8113	O'Brien, F.R. and Oglesby, S., Jr., WADC TR-54-306, Pt. 2, 16-22, 67-9, 1955. [AD 91 221]
197	1422	Freiling, J., Eckert, R.E., and Westwater, J.W., Ind. Eng. Chem., 44, 906-10, 1952.
198	6617	O'Brien, F.R. and Oglesby, S., Jr., Southern Res. Inst., WADC TR-54-306, Pt. 1, 1-133, 1955. [AD 81 159]
199	10146	O'Brien, F.R., Covington, P.C., Harlan, W.J., and Oglesby, S., Jr., Southern Res. Inst. Rept. 1864-523-IV, 1-40, 1954. [AD 65 211]
200	20020	Schoenborn, E.M., Armstrong, A.A., Jr., and Beatty, K.O., Jr., Bull. Am. Soc. Testing Mat'ls., No. 174, 54-9, 1951.
201	1629	Nomura, S., Rept. Inst. Sci. and Technol., Univ. Tokyo, 6, 117-9, 1952.
202	25307	Kodzhespirov, F.F., Soviet Phys.-Solid state, 3 (3), 567-70, 1961.
203	18522	Yoshida, I., Nomura, S., and Sawada, S., J. Phys. Soc. Japan, 13, 1550-1, 1958.
204	6945	Armour Research Foundation, ARF Project No. G-025, Final Rept. for Oct. 1956 to Sept. 1957, 19-28, 1957.
205	1632	Fahey, D.J., Dunlap, M.E., and Seidl, R.J., Southern Pulp Paper Manufacturer, 16 (9), 40-50, 1953.
206	22803	Seidl, R.J., Forest Products Lab. Rept. 1918, 1-27, 1956. [AD 110 961]
207	6520	Urey, H.B., Jr., Master thesis Oklahoma A & M College, 1-40, 1956. [AD 108 261]
208	399	Mark, M., Modern Plastics, 34 (9), 168, 247, 1957.
209	28922	Peterson, J.J., CVC Rept. 2-53420/2R374, 1-77, 1962. [AD 273 801]
210	1458	Whittemore, O.J., Jr. and Marshall, D.W., J. Am. Ceram. Soc., 35 (4), 85-9, 1952.
211	10481	Norton, F.H. and Kingery, W.D., USAEC Rept. NYO-6442, 1-8, 1954. [AD 40 872]
212	7005	Deem, H.W. and Lucks, C.F., BMI-713, 1-8, 1951.

Ref. No.	TPRC No.	
213	59	Tanaka, M., Tsubaki, T., Ueshima, T., and Kamiike, O., J. Ceram. Assoc. Japan, 64, 83-9, 1956.
214	15620	McCarthy, K.A. and Ballard, S.S., J. Appl. Phys., 31 (8), 1410-2, 1960.
215	16756	Berman, R., Foster, E.L., Schneidmesser, B., and Tirmizi, S.M.A., J. Appl. Phys., 31 (12), 2156-9, 1960.
216	16767	Salmon, D.F. and Bailey, J.F., USAEC Rept. ANP-71 (Del.) 1-15, 1951.
217	16036	Truesdale, R.S., Swica, J.J., and Tinklepaugh, J.R., College of Ceramics, State Univ. of N.Y. at Alfred Univ., Progr. Rept. 11, Nov. 1, 1959 to Apr. 30, 1960, 1-67, 1960. [AD 240 278]
218	16022	Taylor, R.E., USAEC Rept. NAA-SR-4905, 1-19, 1960.
219	26077	Schofield, H.Z., Duckworth, W.H., and Long, R.E., USAEC BMI-T 13, 1-24, 1949.
220	26076	Nelson, H.R., AEC Res. and Dev. Rept. BMI-HRN-Memo No. 10, 4-5, 1947.
221	12256	Birtwistle, S. and Shaw, R.A., UKAEA Industrial Group R &DB (S)-TN-2134, 1-15, 1959.
222	7013	Ratcliffe, E.H., Brit. J. Appl. Phys., 10, 22-5, 1959.
223	9861	Seemann, H.E., Phys. Rev., 31, 119-29, 1928.
224	9294	de Haas, W.J. and Biermasz, T., Physica, 5 (7), 619-24, 1938.
225	19203	de Haas, W.J. and Biermasz, T., Physica, 5 (4), 320-4, 1938.
226	25453	Berman, R., Klemens, P.G., Simon, F.E., and Fry, T.M., Nature (London), 166, 864-6, 1950.
227	8245	de Haas, W.J. and Biermasz, T., Comm. K. Onnes Lab. Univ. Leiden, Suppt. No. 82 to Nos. 241-252, 1-13, 1936.
228	15248	Cutler, M., Snodgrass, H.R., Cheney, G.T., Appel, J., Mallon, C.E., and Meyer, C.H., Jr., USAEC Rept. GA-1939, 1-99, 1961.
229	27954	Cohen, A.F., USAEC ORNL-2614, 39-42, 1958. [TID-4500]
230	784	Devyatkova, E.D. and Stilbans, L.S., Zhur. Tekh. Fiz., 22, 968-72, 1952.
231	11913	de Haas, W.J. and Biermasz, T., Physica, 5, 47-53, 1938.
232	10483	Norton, F.H. and Kingery, W.D., USAEC Rept. NYO-6445, 1-8, 1954. [AD 40 874]
233	16823	Norton, F.H. and Kingery, W.D., M.I.T., USAEC Rept. NYO-6449, 1-16, 1955.
234	16029	Armour Res. Foundation, APEX-516, 1-6, 1959.
235	26006	Beck, R.L., Trans. ASM, 55 (3), 556-64, 1962.
236	26080	Baskin, Y. and Handwerk, J.H., USAEC ANL 6529, 1-29, 1962.
237	14955	Wray, K.L. and Connolly, T.J., J. Appl. Phys., 30 (11), 1702-5, 1959.
238	20009	Green, A.T., Trans. Brit. Ceram. Soc., 21, 394-414, 1921.
239	9142	Deissler, R.G. and Eian, C.S., NACA RM E52 CO5, 1-44, 1952.
240	21929	Thomas, R., J. Soc. Chem. Ind., 38, 357T-360T, 1919.
241	24061	Yoshida, I. and Sawada, S., J. Phys. Soc. Japan, 15 (1), 199-200, 1960.
242	11743	Turnbull, A.G., Australian J. of Appl. Sci., 12 (3), 324-9, 1961.
243	26074	Neel, D.S., Pears, C.D., and Oglesby, S., Jr., WADD TR 60-924, Southern Res. Inst., 58-201, 1962. [AD 275 360]
244	26949	Grossman, L.N., Hoyt, E.W., Ingold, J.H., Kaznoff, A.I., and Sanderson, M.J., USAEC Rept. No. GEST-2009, 1-208, 1962. [AD 296 577]
245	23829	Taylor, R.E. and Nakata, M.M., WADD-TR-60-581, Pt. III, 1-21, 1962. [AD 285 236]
246	20113	Kamilov, I.K., Soviet Phys.-Solid State, 4 (9), 1693-7, 1963.
247	16913	Kanz, A., Mitt. Forsch. Inst., 2 (10), 223-34, 1932.
248	25063	Ewing, C.T., Spann, J.R., and Miller, R.R., NRL Memo Rept. 909, 1-10, 1959.
249	192	Mikashima, H. and Nakao, Y., Tetsu-To-Hagane, 39, 1229-33, 1953.
250	7140	Bidwell, C.C., Phys. Rev., 2, 10 (6), 756-66, 1917.
251	25961	Hedge, J.C., Kopec, J.W., Kostenko, C., and Lang, J.I., ASD-TDR-63-597, Wright-Patterson Air Force Base, Ohio, 1-128, 1963. [AD 424 375]
252	14416	Wilkes, G.B., J. Am. Ceram. Soc., 16 (3), 125-30, 1933.
253	15658	Flieger, H.W., Jr., and Ginnings, D.C., NBS Rept. 5642, 1-26, 1957.
254	790	Skanavi, G.I. and Kashtanova, A.M., Zhur. Tekh. Fiz., 25, 1883-92, 1955.
255	189	Neely, J.J., Teeter, C.E., Jr., and Trice, J.B., J. Am. Ceram. Soc., 33, 363-4, 1950.

Ref. No.	TPRC No.	
256	786	Chizhikov, D. M., Gulyanitskaya, Z. F., and Bogovarova, N. N., Izvest. Akad. Nauk SSSR, Otdel. Tekh. Nauk No. 6, 109-13, 1955.
257	22824	Claiborne, S. J., ORNL Central Files No. CF 53-1-233, 1-2, 1953.
258	26439	Taylor, R. E. and Nakata, M. M., Atomics International, 2nd ARPA Quarterly Progr. Rept., 1-16, 1963. [AD 297 874]
259	21939	Sturley, K. R., J. Soc. Chem. Ind., 51 (1), 271T-273T, 1932.
260	14083	Burton, J. T. A. and Ziebland, H., Ministry of Aviation (British) Explosives Res. and Dev. Establishment Rept. 17/R/60, 1-17, 1961. [AD 251 642]
261	1128	Gillam, D. G., Romben, L., Nissen, H. E., and Lamm, O., Acta Chemica Scand., 9, 641-56, 1955.
262	20657	Codegone, C., Ricerca Scientifica e Ricostruzione, 10, 701-6, 1939.
263	22822	Claiborne, S. J., ORNL CF-52-11-72, 1-2, 1952.
264	7531	Redmond, J. C. and Smith, E. N., Trans., AIME, 185, 987-93, 1949.
265	16037	Mrozowski, S., Andrew, J. F., Juul, N., Okada, J., Strauss, H. E., and Wobschall, D. C., WADC Tech. Rept. 58-360, II, 1960. [AD 236 663]
266	27009	Juul, N., Sato, S., and Strauss, H. E., Progr. Rept., State Univ. of N. Y. at Buffalo, No. 17, 1-11, 1963. [AD 405 540]
267	26684	Hartunian, R. A. and Varwig, R. L., Phys. Fluids, 5 (2), 169-74, 1962.
268	21703	Codegone, C., Ricerca Scientifica e Ricostruzione, 16 (3-4), 286-9, 1946.
269	16893	Flinta, J. E., USAEC Rept. TID-7546, (Book 2), 516-25, 1958.
270	29600	Taylor, R. E. and Nakata, M. M., AI-8058, 8-12, 1962. [AD 293 802]
271	16849	Johnson, J. R., Doney, L. M., Fulkerson, S. D., Taylor, A. J., Warde, J. M., and White, G. D., ORNL-2011, 7, 1956.
272	32169	Mischke, R. A. and Smith, J. M., Ind. Eng. Chem. Fundamentals, 1 (4), 288-92, 1962.
273	29585	Flynn, D. R., J. of Research of NBS, 67C (2), 129-37, 1963.
274	10083	Norton, F. H., Kingery, W. D., and others, MIT, USAEC Rept. NYO-3643, 1-124, 1953. [AD 13 940]
275	24384	Rudkin, R. L., ASD-TDR-62-24, II, 1-16, 1963. [AD 413 005]
276	24810	Booker, J., Paine, R. M., and Stonehouse, A. J., WADD TR 60-889, 1-133, 1961. [AD 265625]
277	26405	Taylor, R. E. and Nakata, M. M., Atomic Intern., 8494, 1-11, 1963. [AD 406 098]
278	30599	Burk, M., Mater. Res. Std., 3 (1), 25-8, 1963.
279	10022	Swann, R. T., NASA-TN-D-171, 1-24, 1959.
280	23153	Haskins, J. F., Jones, H., and Percy, J. L., MRG-323, 1-18, 1962. [AD 291 518]
281	29558	Minges, M. L. and Meiselman, J. M., Tech. Memo ASRCE TM 63-7, Wright-Patterson Air Force Base, Ohio, 1-13, 1963.
282	28067	Thielke, N. R., Penn. State Univ., Univ. Park, WADC TR 54-467, 1-88, 1955. [AD 88 128]
283	16094	Martin Co., Nuclear Div., Baltimore, USAEC Rept. MND-SR-1674, 1-72, 1959.
284	31786	Francis, R. K., McNamara, E. P., and Tinklepaugh, J. R., State Univ. of N. Y., College of Ceram. at Alfred Univ., Aeronautical Res. Lab., WADC, 1958. [AD 154 872]
285	25901	Makarounis, O. and Jenkins, R. J., USNRDL-TR-599, 1-16, 1962. [AD 295 887]
286	31788	Truesdale, R. S., Swica, J. J., and Tinklepaugh, J. R., State Univ. of N. Y., College of Ceram. at Alfred Univ., WADC TR 58-452, 1-36, 1958. [AD 207 079]
287	32520	Nishijima, T., Kawada, T., and Ishihata, A., J. Am. Ceram. Soc., 48 (1), 31-4, 1965.
288	33776	Gen. Elec. Co., Lamp Glass Dept., Circle No. 600, Lucalox, 1-11, 1963.
289	28655	Glower, D. D. and Wallace, D. C., Sandia Corp., USAEC Rept. TID-16906, SCDC-2886, 1-16, 1962.
290	20313	Burk, M., J. Am. Ceram. Soc., 46 (3), 150-1, 1963.
291	26233	Feith, A. D., Gen. Elec. Co., Adv. Tech. Service, USAEC Rept. GEMP-296, 1-25, 1964.
292	29231	Fitzsimmons, E. S., Gen. Elec. Co., Aircraft Nuclear Propulsion Dept., DC-61-6-4, 1-9, 1961.
293	209	Charvat, F. R. and Kingery, W. D., J. Am. Ceram. Soc., 40 (9), 306-15, 1957.
294	25618	Laubitz, M. J., Canadian, J. Phys., 41 (10), 1663-78, 1963.
295	18644	Lee, D. W. and Kingery, W. D., J. Am. Ceram. Soc., 43 (11), 594-607, 1960.
296	28870	Sibley, L. B., Allen, C. M., Zielenbach, W. J., Peterson, C. L., and Goldthwaite, W. H., WADC TR 58-299, 1-52, 1958. [AD 203 787]
297	103	Powell, R. W., Trans. Brit. Ceram. Soc., 53 (7), 389-97, 1954.

Ref. No.	TPRC No.	
298	24812	Fieldhouse, I. B. and Lang, J. I., WADD TR 60-904, 1-119, 1961. [AD 268 304]
299	8133	Karpinski, J. M., Hasselman, D. P. H., Tervo, R., and Fetterley, G. H., Am. Ceram. Soc. Bull., 37 (7), 329-33, 1958.
300	10615	Slack, G. A., Phys. Rev., 126 (2), 427-41, 1962.
301	32772	Flynn, D. R., NBS Rept. 7836, 1-22, 1963. [AD 407 802]
302	27922	Plummer, W. A., Campbell, D. E., and Comstock, A. A., J. Am. Ceram. Soc., 45 (7), 310-6, 1962.
303	20780	Birch, F. and Clark, H., Am. J. Sci., 238 (8), 529-58, 1940.
304	29152	Flynn, D. R. and Robinson, H. E., NBS Rept. No. 7135, 1-36, 1961. [AD 277 034]
305	24315	Flynn, D. R., NBS Rept. 7740, 1962. [AD 411 157]
306	26836	Chang, G. K. and Jones, R. E., Phys. Rev., 126 (6), 2055-8, 1962.
307	32611	Adams, M., J. Am. Ceram. Soc., 37 (2), 74-9, 1954.
308	9293	de Haas, W. J. and Biermasz, T., Physica, 2, 673-82, 1935.
309	26060	Devyatkova, E. D., Petrov, A. V., Smirnov, I. A., and Moizhes, B. Ya., Soviet Phys. -Solid State, 2 (4), 681-8, 1960.
310	35072	Thurber, W. R. and Mante, A. J. H., Phys. Rev., 139 (5A), 1655-65, 1965.
311	10809	Mason, C. R., Walton, J. D., Bowen, M. D., and Teague, W. T., Summary Rept. No. 3, Proj. No. A-212, Eng. Expt. Sta., Georgia Inst. Tech., Atlanta, 1-120, 1959. [AD 230 200]
312	10432	Ballard, S. S., McCarthy, K. A., and Davis, W. C., Rev. Sci. Instr., 21 (11), 905-7, 1950.
313	25602	Mason, C. R., Walton, J. D., Bowen, M. D., and Teague, W. T., Quarterly Rept. No. 14, Proj. No. A-212, Eng. Expt. Sta., Georgia Inst. Tech., Atlanta, 1-28, 1959. [AD 216 618]
314	10898	Charlesworth, D. H., AECL No. 828, 1-24, 1959. [AD 218 621]
315	12708	Nonken, G. C., Steel Processing, 34, 377-81, 1948.
316	18353	Norton, F. H. and Kingery, W. D., USAEC NYO-599, 1-31, 1951.
317	28929	JPL Research Summary No. 36-5, Pasadena, Calif., II, 12-5, 1960.
318	18625	McClelland, J. D. and Zehms, E. H., J. Am. Ceram. Soc., 43 (1), 54, 1960.
319	29130	General Electric Co., GEMP-16A, 1-33, 1962.
320	24817	General Electric Co., GEMP-9A, 1-71, 1962.
321	27117	Hartunian, R. A. and Varwig, R. L., USAF TDR-594 (1217-01) TN-2, 1-46, 1961. [AD 606 036]
322	17008	Cohen, A. F., USAEC ORNL-2413, 70-1, 1957.
323	22544	Gallo, C., Thermoelectricity Quarterly Rept. No. 4, 1-7, 1961. [AD 277 122]
324	28778	Tuchschmid, A., Ann. Phys. Chem., Beiblatter, 8, 490-2, 1884.
325	16421	Lees, C. H., Phil. Trans. Roy. Soc. (London), 183, 481-509, 1892.
326	21255	Kozak, M. I., Zhur. Tekh. Fiz., 22 (1), 73-6, 1952.
327	9254	Sinel'nikov, N. N. and Filipovich, V. N., Soviet Phys. Tech. Phys., 3, 193-6, 1958.
328	9238	Gafner, G., Brit. J. Appl. Phys., 8 (10), 393-7, 1957.
329	7082	Norton, F. H., Kingery, W. D., Loeb, A. L., Francl, J., Coble, R. L., and Vasilos, T., USAEC NYO-3647, 1-61, 1953. [AD 23 561]
330	7288	Powell, R. W., J. Sci. Instr., 34 (12), 485-92, 1957.
331	31783	Pears, C. D., Proc. 3rd Thermal Conductivity Conference, 453-79, 1963.
332	30255	Buessem, W. R. and Bush, E. A., J. Am. Ceram. Soc., 38 (1), 27-32, 1955.
333	35983	Sugawara, A., J. Appl. Phys., 36 (8), 2375-7, 1965.
334	28106	Juul, N., Sato, S., and Strauss, H. E., USAEC NP-12878, 1-11, 1963. [AD 416 276]
335	27395	Pike, J. N., Research Rept. No. C-16, Parma Research Lab., Union Carbide Corp., Parma, Ohio, 1-27, 1963. [AD 404 453]
336	38399	Powell, R. W., Tye, R. P., and Metcalf, S. C., 3rd Symposium on Thermophysical Properties, March 22-5, 1965, 289-95.
337	7663	Barratt, T., Proc. Phys. Soc. (London), 26, 347-71, 1914.
338	34428	National Carbon Co. (Carbon Products Div.), Union Carbide Corp., Industrial Graphite Engineering Handbook, published in a series of revised editions.
339	5	Zavaritskii, N. V. and Zeldovich, A. C., Zhur. Tekh. Fiz., 26, 2032-6, 1956.
340	29509	Hoch, M. and Vardi, J., Tech. Doc. Rept. No. ASD-TDR-62-608, Part I, Wright-Patterson Air Force Base, Ohio, 1-19, 1962.

Ref. No.	TPRC No.	
341	8196	Wagner, P., Driesner, A.R., and Kmetko, E.A., Second Intern. Conf. on the Peaceful Uses of Atomic Energy, Geneva, 7, 379-88, 1958.
342	16422	Crary, A.P., Physics, 4, 332-3, 1933.
343	10953	Union Carbide Corp. Thermoelectric Materials, Bi-Monthly Prog. Rept. No. 4, 1-33, 1959. [AD 231 650]
344	28899	Mrozowski, S., Andrew, J.F., Juul, N., Strauss, H.E., Tsuzuku, T., and Wobschall, D.C., WADC-TR-58-360, IV, 1-45, 1962.
345	15598	Bowman, J.C., Krumhansl, J.A., and Meers, J.T., Proc. 1957 Industrial Carbon and Graphite Conf. Soc. Chem. Ind. (London), 52-9, 1958.
346	10906	Durand, R.E. and Klein, D.J., U.S. At. Energy Comm. NAA-SR-1520, 1-33, 1956.
347	37379	Dull, R.B., WADD TR 61-72, Vol. XXVI, 1-448, 1964. [AD 602 607]
348	16590	Fieldhouse, I.B., Land, J.I., and Blau, H.H., Jr., WADC TR-59-744, 4, 1960.
349	33847	Holland, M.G., Klein, C.A., and Straub, W.D., J. Phys. Chem. Solids, 27(5), 903-6, 1966.
350	36698	Taylor, R., Brit. J. Appl. Phys., 16(4), 509-15, 1965.
351	36697	Mills, J.J., Morant, R.A., and Wright, D.A., Brit. J. Appl. Phys., 16(4), 479-85, 1965.
352	26843	Slack, G.A., Phys. Rev., 127(3), 694-701, 1962.
353	38041	Klein, C.A. and Holland, M.G., Phys. Rev., 136(2A), A575-90, 1964.
354	26574	Mason, I.B. and Knibbs, R.H., AERE-R 3973 (Gt. Brit.), 1-22, 1962.
355	36699	Goldsmid, H.J. and Lacklison, D.E., Brit. J. Appl. Phys., 16, 573-5, 1965.
356	17773	Snyder, T.M. and Kamm, R.L., USAEC Rept. No. C-96, 1, 1942.
357	21500	Thielke, N.R. (compiler), Thermoelectric Materials Bi-monthly Prog. Rept. No. 1 Nat. Carbon Co., 1-29, 1959.
358	113	Smith, A.W., Phys. Rev., 95(2), 1095-6, 1954.
359	16775	Garth, R.C. and Sailor, V.L., USAEC Rept. BNL-69, 1949.
360	33806	de Combarieu, A., Bull. Inst. Intern. Froid, Annexe 1965-2, 63-72, 1965.
361	22948	Pappis, J. and Blum, S.L., J. Am. Ceram. Soc., 44(12), 592-7, 1961.
362	36175	Hooker, C.N., Ubbelohde, A.R., and Young, D.A., Proc. Roy. Soc. (London), 284 (1396), 17-31, 1965.
363	28174	Wheeler, M.J., Brit. J. Appl. Phys., 16(3), 365-76, 1965.
364	2428	Kingery, W.D., Francl, J., Coble, R.L., and Vasilos, T., J. Am. Ceram. Soc., 2, 37(2), 107-10, 1954.
365	31160	Johnson, W. and Watt, W., Spec. Ceram., Proc. Symp. Brit. Ceram. Res. Assoc. 1962 (England), 237-59, 1963.
366	25034	Breckenridge, R.G. (compiler), Union Carbide Corp., 1-20, 1960. [AD 246 217]
367	22836	Mrozowski, S., Andrew, J.F., Juul, N., Sato, S., Strauss, H.E., and Tsuzuku, T., WADC TR 58-360, Part V, 1-68, 1963.
368	25973	Jain, S.C. and Krishnan, K.S., F.R.S., Proc. Roy. Soc. (London), Part II-IV, 225A, 1-32, 1954.
369	34346	Taylor, R., Phil, Mag., 8, 13(121), 157-66, 1966.
370	31576	Kraemer, H. and Schmeiser, K., Z. Physik. Chem., 35, 1-9, 1962.
371	35243	Slack, G.A., Phys. Rev., A139(2), 507-15, 1965.
372	8345	Donaldson, J.W., Foundary Trade J., 63, 141-4, 1940.
373	12215	Laubitz, M.J., Can. J. Phys., 37, 798-808, 1959.
374	35041	Mirkovich, V.V., J. Am. Ceram. Soc., 48(8), 387-91, 1965.
375	28172	Truesdale, R.S., M.S. Thesis, Alfed Univ., 1-91, 1960.
376	17449	McGill, R.C. and Smith, J.A.G., At. Energy Research Estab. (Gt. Brit.), R-3019, 1-2, 1959.
377	35248	Pryor, A.W., Tainsh, R.J., and White, G.K., J. Nuclear Materials, 14, 208-19, 1964.
378	27252	Ewing, C.T., Walker, B.E., Jr., Spann, J.R., Steinkuller, E.W., and Miller, R.R., J. Chem. Eng. Data, 7, 251-6, 1962.
379	25113	Slack, G.A., Proc. Intern. Conf. Semiconductor Phys., 630-3, 1961.
380	15610	Eucken, A. and Huhn, G., Z. Physik. Chem., 134, 193-219, 1928.
381	26750	Bosch, G., Philips Research Repts., 16, 455-61, 1961.

Ref. No.	TPRC No.	
382	38217	Slack, G.A., J. Appl. Phys., 35(12), 3460-6, 1964.
383	20929	Norton, F.H., J. Am. Ceram. Soc., 10, 30-52, 1927.
384	25061	Powell, R.W. and Tye, R.P., Special Ceramics, Proc. Symp. Brit. Ceram. Res. Assoc., 261-80, 1963.
385	26694	Cunningham, G.W., Kizer, D.E., and Paprocki, S.J., Plansee Proc. 4th Seminar, 483-509, 1962.
386	8486	Rauch, W.G., USAEC ANL-5268, 1-13, 1954.
387	27403	Eiermann, K. and Hellwege, K.H., J. Polymer Sci., 57, 97-104, 1962.
388	24016	Hattori, M., Bull. Univ. Osaka Prefecture, A9(1), 51-8, 1960.
389	6973	Covington, P.C. and Oglesby, S., Jr., WADC TR 57-10, 1-62, 1957. [AD 131 032]
390	16222	Smith, W.K., NOTS TP2624, 1-10, 1961. [AD 263 771]
391	33250	Bocquet, M. and Micaud, G., Bull. Inform. Sci. Tech. (Paris), 79, 83-96, 1964.
392	29272	Schweitzer, D. and Singer, R., BNL 696 (S-59), 56-8, 1962.
393	43747	Meyer, R.A. and Koyama, K., USAEC GA-4621, 1-12, 1963.
394	35533	Digesu, F.L. and Pears, C.D., AFML-TR-65-142, 1-279, 1965. [AD 471 337]
395	30973	Hooker, C.N., Ubbelohde, A.R., and Young, D.A., Proc. Roy. Soc., 276(1364), 83-95, 1963.
396	42488	Kaspar, J., SSD-TR-67-56, 1-43, 1967. [AD 813 821]
397	42736	Bortz, S.A. and Connors, C.L., IIT Res. Inst., 1-67, 1966. [AD 804 035]
398	44182	Kozlov, F.A. and Antonov, I.N., Soviet J. Atomic Energy, 19(4), 1333-4, 1965.
399	42392	Kozlov, F.A. and Antonov, I.N., Atomnaya Energiya, 19(4), 391-2, 1965.
400	34483	Godbee, H.W. and Ziegler, W.T., J. Appl. Phys., 37(1), 40-55, 1966.
401	43505	Kelly, P.N., Masters Thesis, Univ. of Idaho, 1-40, 1960.
402	21046	Kelly, P.N., USAEC IDO-14592, 1-25, 1962.
403	41346	Zadworny, F., Compt. Rend., B264(8), 569-72, 1967.
404	33179	Perelotov, I.I., TRC-Tras. 1068, 1-12, 1962.
405	33178	Perelotov, I.I., Teploenergetika, 7(2), 77-80, 1960.
406	38834	Brodie, D.E. and Mate, C.F., Can. J. Phys., 43(12), 2344-60, 1965.
407	30225	Snyder, N.H., Smoke, E.J., Wisely, H.R., and Ruh, E., 1-89, 1949. [AD 89 089]
408	28916	Powers, D.J., BLR61-15(M), 1-21, 1962. [AD 276 983]
409	33807	DeGōer, A.M., Bull. Inst. Intern. Froid, Annexe, 2, 73-82, 1965.
410	42732	Collins, C.G., USAEC GEMP-61, 157-169, 1966.
411	43749	Wood, W.D., Lucks, C.F., and Deem, H.W. (Dayton, R.W. and Tipton, C.R., editors), BMI-1448, 31-4, 1960.
412	31430	Emanuelson, R.C., Pratt and Whitney Aircraft, APR1048 (Suppl. 1), 1-8, 1963.
413	33540	Taylor, R.E. and Nakata, M.M., WADD-TR-60-581 (Pt. 4), 1-109, 1963. [AD 428 669] [AD 441 079]
414	32247	Taylor, R.E. and Morreale, J., J. Am. Ceram. Soc., 47(2), 69-73, 1964.
415	32912	Milnes, M.V., North American Aviation Inc., SDL408, 1-14, 1963.
416	33662	Bowen, M.D., M.S. Thesis, Georgia Inst. Technology, 1-32, 1959.
417	33440	Godfrey, T.G., Fulkerson, W., Kollie, T.G., Moore, J.P., and McElroy, D.L., USAEC ORNL-3556, 1-67, 1964.
418	33506	Amelinckx, S., Blank, H., DeConinck, R., Denayer, M., Devreese, J., Knaepen, F., Nagels, P., Penninckx, R., Strumane, R., and Van Lierde, W., EURAEC-524, 1-24, 1963.
419	31246	Lyons, M.F., Straley, R.L., Coplin, D.H., Weidenbaum, B., and Pashos, T.J., Trans. Am. Nucl. Soc., 6, 152, 1963.
420	37438	Daniel, R.C. and Cohen, I., USAEC WAPD-246, 1-151, 1964.
421	33408	Chernock, W.P., Burdg, C.E., Veil, E.I., and Zuromisky, G., USAEC CEND-204, 1-32, 1964.
422	33975	Bates, J.L., BNWL-150, 1.8-1.20, 1965.
423	28443	Daniel, J.L. and Bates, J.L., USAEC HW-76303, 2.19-.21, 1963.
424	31479	Deem, H.W. and Lucks, C.F., BMI 1324, 7-8, 1959.
425	31963	Deem, H.W. and Matolich, J., Jr., BMI-1644 (Del), J1-4, 1963.

Ref. No.	TPRC No.	
426	10850	Eichenberg, J.D., USAEC WAPD-200, 1-9, 1958.
427	35825	Feith, A.D., USAEC TID-21668 GE-TM63-9-5, 1-25, 1963.
428	41063	Christensen, J.A., USAEC HW-76302, 2.13-.15, 1963.
429	19982	Brenden, B.B. and Newkirk, H.W., USAEC HW-59574, 1-45, 1959.
430	39869	Swift, D.L., Intern. J. Heat Mass Transfer, 9(10), 1061-74, 1966.
431	40368	Lyons, M.F., Coplin, D.H., Hausner, H., Weidenbaum, B., and Pashos, T.J., USAEC GEAP-51000-1, 1-76, 1966.
432	29388	Godfrey, T.G. and McElroy, D.L. (Manly, W.D., Program Director), USAEC ORNL-3254, 264-6, 1962.
433	21545	Godfrey, T.G. and McElroy, D.L., (Manly, W.D., Program Director), USAEC ORNL-3210, 175-6, 1962.
434	29390	Godfrey, T.G. and McElroy, D.L., (Manly, W.D., Program Director), USAEC ORNL-3372, 294-6, 1962.
435	26868	Berg, K., Flinta, J.E., and Seltorp, L., Proc. U.N. 2nd Intern. Conf. Peaceful Uses Atomic Energy, 691-6, 1958.
436	23979	Bogaievski, M., Caillat, R., Delmas, R., Janvier, J.C., and Robertson, J.A.L., New Nucl. Mater. Including Non-Metal. Fuels, Proc. Conf., 307-22, 1963.
437	26424	Dean, R.A., USAEC CVNA-127, 1-34, 1962.
438	19992	Shapiro, H. and Powers, R.M., USAEC SCNC-294, 1-43, 1959.
439	25775	Bemden, V. et al., EURAEC-491, 1-42, 1962.
440	38108	Sievers, A.J. and Pohl, R.O., CFSTI NYO-2391-5, CONF-764-4, 1-12, 1964.
441	43420	Porneuf, A., USAEC CEND-153, 2, 131-4, 1962.
442	33090	Yoshizawa, Y., Sugawara, A., and Yamada, E., J. Appl. Phys., 35(4), 1354-5, 1964.
443	9285	Lees, C.H., Phil. Trans. Roy. Soc. (London), A191, 399-440, 1898.
444	36241	Mogilevskii, B.M. and Chudnovskii, A.F., Inzh.-Fiz. Zh., Akad. Nauk Belorus. SSR, 7(12), 23-31, 1964.
445	39345	Mogilevskii, B.M. and Chudnovskii, A.F., J. Engineering Physics, 7(12), 1-15, 1966.
446	35222	Green, S.E., Proc. Phys. Soc. (London), 44(243), Pt. 3, 295-313, 1932.
447	39890	Turnbull, A.G., Ph.D. Thesis, Imperial College of Science and Technology, (London), 1-146, 1959.
448	28731	Hecht, H., Ann. Physik, 14, 1008-30, 1904.
449	31897	Niven, C. and Geddes, A.E.M., Proc. Roy. Soc. (London), A87, 535-9, 1912.
450	36865	Pochettino, A. and Fulcheris, G., Atti. Accad. Sci. Torino, Pt. I Classe Sci. Fis. Mat. E. Nat., 58(14), 311-20, 1923.
451	21254	Ioffe, A.V. and Ioffe, A.F., Zh. Tekh. Fiz., 22(12), 2005-13, 1952.
452	26529	Devyatkova, E.D. and Smirnov, I.A., Fiz. Tverdogo Tela, 4(7), 1972-5, 1962.
453	26528	Devyatkova, E.D. and Smirnov, I.A., Soviet Phys.-Solid State, 4(7), 1445-6, 1963. [AD 270 841]
454	26819	Pohl, R.O., Phys. Rev. Letters, 8(12), 481-3, 1962.
455	43438	Peech, J.M., Bower, D.A., and Pohl, R.O., J. Appl. Phys., 38(5), 2166-71, 1967.
456	36978	Taylor, A., Albers, H.R., and Pohl, R.O., J. Appl. Phys., 36(7), 2270-8, 1965.
457	16072	Codegone, C., Ricerca Sci., 24(12), 2623-7, 1954.
458	22633	Tadokoro, Y., Science Repts. Tohoku Imp. Univ., 10, 339-410, 1921.
459	33125	Fritts, R.W., Ph.D. Thesis, Northwestern Univ., 1-68, 1950.
460	43681	Klein, P.H., J. Appl. Phys., 38(4), 1598-603, 1967.
461	43682	Klein, P.H. and Croft, W.J., J. Appl. Phys., 38(4), 1603-7, 1967.
462	31868	Dougill, G., Hodsman, H.J., and Cobb, J.W., J. Soc. Chem. Ind. (London), 34(9), 465-70, 1915.
463	26561	Pustovalov, V.V., Steklo i Keram., 18(12), 17-9, 1961.
464	27771	Pustovalov, V.V., Glass and Ceramics (Moscow), 18 (12), 618-20, 1961.
465	36785	Suemune, Y., J. Phys. Soc. (Japan), 20(1), 174-5, 1965.
466	17426	Douthett, D. and Friedberg, S.A., Phys. Rev., 121(6), 1662-7, 1961.
467	40937	Mante, A.J. H. and Volger, J., Phys. Letters, A24(3), 139-40, 1967.
468	31538	Snyder, N.H., Smole, E.J., Wisely, H.R., and Ruh, E., ATI 86345, 1-90, 1950.

Ref. No.	TPRC No.	
469	34386	Shanks, H.R. and Redin, R.D., J. Phys. Chem. Solids, 27(1), 75-8, 1966.
470	16093	Martin Co., USAEC MND-SR-1673, 1-85, 1959.
471	812	Weininger, J.L. and Schneider, W.G., Indust. Engng. Chem., 43(5), 1229-33, 1951.
472	25157	Wechsler, A.E. and Glaser, P.E., ASD-TDR-63-574, 1-171, 1963. [AD 420 193]
473	23316	Kitzes, A.S. and Hullings, W.Q., AECD 3625, 25-41, 1954.
474	22191	Tiller, W.A., Johansen, H.A., and Holstein, T.,1-97, 1958. [AD 215964] [PB 160 748-8]
475	8097	Griffiths, E. and Challoner, A.R., Trans. Brit. Ceram. Soc., 40, 40-53, 1941.
476	23408	Marchal, M. and Trouve, J., AEC-TR-5174, -131, 1962.
477	27176	Brown, D.J. and Stobo, J.J., Plansee Proceedings, 4th Seminar, 279-93, 1962.
478	28261	Howard, V.C., UKAEA IGR-TM/C-0164, 1-5, 1958.
479	28880	Taylor, R.E. and Nakata, M.M.,Atomics International, AI-7034, 1-14, 1962. [AD 270 841]
480	31184	Crane, J. and Gordon, E., USAEC UNC-5080, 1-67, 1964.
481	31996	Carniglia, S.C., NAA-SR-MEMO-9015 Conf-206-5, 1-41, 1963.
482	33116	Meerson, G.A., Kotel'nikov, R.B., and Bashlykov, S.N., Atomnaya Energiya (USSR), 9(5), 387-91, 1960.
483	30998	Meerson, G.A., Kotel'nikov, R.B., and Bashlykov, S.N., At. Energy (USSR), 9(5), 927-31, 1961.
484	27924	Taylor, R.E., J. Am. Ceram. Soc., 45(7), 353-4, 1962.
485	27379	Sedillo, L., Castonguay, T.T., and Donaldson, W.E., NAVWEPS Rept. 7918 (Pt. 2) NOTS TP2938, 1-28, 1963. [AD 407 515]
486	24980	Smallen, H., Bondesen, A.J., and Romaine, R.P., NOR-60-45, 1-25, 1961. [AD 270 421]
487	31889	Ratcliffe, E.H., Plastics (London), 22(55), 233, 1957.
488	18235	Howse, P.T., Jr., Pears, C.D., and Oglesby, S., Jr., WADD TR60-657, 1-137, 1961. [AD 260 065]
489	23836	Howse, P.T., Jr., and Pears, C.D., Modern Plastics, 39(1), 140-53, 246-8, 1961.
490	23490	Taylor, T.S., Elec. World, 76(24), 1159-62, 1920.
491	10860	Gluekauf, E. and Watts, R.E., UKAEA AERE C/M337, 1-3, 1958. [AD 158 977L]
492	16831	Shackleford, M.H., USAEC KAPL-M-MHS-23, 1-11, 1954.
493	6256	Fukuroi, T., Sci. Rept. Research Inst. Tohoku Univ., A1(2), 107-10, 1949.
494	14136	Francis, R.K. and Tinklepaugh, J.R., J. Am. Ceram. Soc., 43(11), 560-3, 1960.
495	36797	Engel, N.N. and McElroy, D.L., USAEC 2797 Conf-764-9, 1-11, 1964.
496	39765	Klein, M.V. and Caldwell, R.F., Rev. Sci. Instr., 37 (10), 1291-7, 1966.
497	24394	Lemanov, V.V. and Smirnov, I.A., Fiz. Tverd. Tela, 4(9), 2611-3, 1962.
498	28177	Lemanov, V.V. and Smirnov, I.A., Soviet Phys.-Solid State, 4(9), 1914-6, 1963.
499	35940	Moss, M., J. Appl. Phys., 36(10), 3308-19, 1965.
500	18374	Taylor, K.M., and Lenie, C., J. Electrochem. Soc., 107(4), 308-14, 1960.
501	20321	Long, G. and Foster, L.M., J. Am. Ceram. Soc., 42(2), 53-9, 1959.
502	28259	Keller, D.L., BMI-X-10027 EURAEC 10027, 1-10, 1963.
503	35040	Endebrock, R.W., Foster, E.L., and Keller, D., USAEC BMI-1690, 1-22, 1964.
504	30353	Kollie, T.G. and Moore, J.P., USAEC ORNL-3670, 142-3, 1964.
505	16624	Kurnick, S.W., Appel, J.C., and Cutler, M., GAMD-1529, 1-25, 1960. [AD 245 978]
506	39991	Golubkov, A.V., Devyatkova, E.D., Zhuze, V.P., Sergeeva, V.M., and Smirnov, I.A., Fiz. Tverd. Tela, 8(6), 1761-71, 1966.
507	39992	Golubkov, A.V., Devyotkova, E.D., Zhuze, V.P., Sergeeva, V.M., and Smirnov, I.A., Soviet Phys.-Solid State, 8(6), 1403-10, 1966.
508	27917	Niven, C.D., Can. J. Research, A18, 132-7, 1940.
509	35542	Rosser, W.A., Inami, S.H., and Wise, H., 1-16, 1964. [AD 614 081] [AD 464 840]
510	41597	Masumoto, K., Isomura, S., and Goto, W., J. Phys. Chem. Solids, 27(11/12), 1939-47, 1966.
511	33076	Cutler, M., Advanced Energy Conversion, 2, 29-43, 1962.
512	21241	McKee, D.J., USAEC DC-61-1-22, 1-8, 1961.
513	16138	Hamill, C.W., Waldrop, F.B., and Kite, H.T., USAEC Y-1366, 1-15, 1961.
514	33785	Suemune, Y., J. Phys. Soc. (Japan), 21(4), 802, 1966.

Ref. No.	TPRC No.	
515	41265	Suemune, Y., J. Phys. Soc. (Japan), 22(3), 735-43, 1967.
516	28919	Feigelson, R.S., ASTIA WAL-TR 853/1, 1-39, 1962. [AD 277 686]
517	26783	Luthi, B., Phys. Chem. Solids, 23, 35-8, 1962.
518	20510	Powers, R.M., Cavallaro, Y., and Mathern, J.P., USAEC SCNC-317, 1-128, 1960.
519	31964	Wright, T.R., Fackelmann, J.M., Kizer, D.E., and Keller, D.L., USAEC BMI-1644 (Del), U1-3, 1963.
520	32812	Pasquino, A.D. and Pilsworth, M.N., Jr., J. Polymer Sci., B2(3), 253-5, 1964.
521	577	Read, J.H. and Lloyd, D.M.G., Trans. Faraday Soc., 44, 721-9, 1948.
522	31136	Burdick, R.B. and Hoskyns, W.R., USAF ARL 63-170, 1-72, 1963. [AD 420 569]
523	33351	Arias, A., NASA TN-D-2464, 1-72, 1964.
524	23107	Ruh, R., J. Am. Ceram. Soc., 46(7), 301-7, 1963.
525	856	Fradkov, A.B., Doklady Akad. Nauk SSSR, 81, 549-51, 1951.
526	27214	Engholm, G., Lis, S.J., and Baschiere, R.J., ASD-TDR-62-810, 1-157, 1962. [AD 407 663]
527	29	Cămpan, T.I., Anghelache, D., and Belous, V., Bull. Inst. Politeh. Iasi., 2(6), 321-30, 1956.
528	91	Ueberreiter, K. and Otto-Laupenmühlen, E., Kolloid-Z., 133(1), 26-32, 1953.
529	6324	Mann, G. and Forsyth, F.G.E., Mod. Refrign., 59, 188-91, 1956.
530	24969	Myncke, H., Van Itterbeek, A., and deGreve, L., Bull. Inst. Intern. Froid, Annexe, 2, 241-9, 1956.
531	16507	Schultz, A.W. and Wong, A.K., ASTIA WAL TR 397/10, 1-22, 1958. [AD 154 351]
532	6680	Powell, R.L., Rogers, W.M., and Coffin, D.O., J. Res. Natl. Bur. Std., 59(5), 349-55, 1957.
533	29514	Chandler, H.H., USAF FTDM-2210, 1-3, 1962. [AD 285 365]
534	22621	Stops, D.W., J. Sci. Instr., 38(5), 221, 1961.
535	20374	Patten, G.A. and Skochdopole, R.E., Modern Plastics, 39(11), 149-52, 191, 1962.
536	20940	Wolkenstein, W.S., Technik, 8(9), 593-6, 1953.
537	16038	Hayes, R.A., Smith, F.M., Kidder, G.A., Henning, J.C., Rigby, J.D., and Hall, G.L., WADC TR 56-331 (Pt. 4), 1-157, 1960. [AD 240 212]
538	1102	Urbancová, L., Chem. Prumysl, 5, 338-40, 1955.
539	29491	Bernett, E.C., Wood, H.L., Jaffe, L.D., and Martens, H.E., NASA JPL-TR-32-368, 1-19, 1962.
540	20999	Breckenridge, R.G., 1-20, 1960. [AD 245 092]
541	29118	Pandorf, R.C., Chen, C.Y., and Daunt, J.G., 1-7, 1962. [AD 275 287]
542	42558	Rafalowicz, J., Acta Phys. Polon., 30(8), (Pt. 2), 205-22, 1966.
543	1913	Franck, E.U., Z. Elektrochem., 55(7), 636-43, 1951.
544	38106	Turnbull, A.G., Z. Physik. Chem., 42(3-4), 243-6, 1964.
545	25437	Shannon, W.L. and Wells, W.A., Proc. Am. Soc. Testing Mater., 47, 1044-55, 1947.
546	25145	Ingersoll, L.R. and Koepp, O.A., Phys. Rev., 24(Pt. 2), 92-3, 1924.
547	14860	Oliver, H., Trans. Ceram. Soc., 37, 49-59, 1938.
548	6990	Weeks, J.L. and Seifert, R.L., USAEC ANL-4938, 1-14, 1952. [AD 1929]
549	257	Barrett, E.P., Jonnard, A., and Missmer, J.H., Ind. Eng. Chem., 45(7), 1524-6, 1953.
550	23121	Hasselman, D.P.H. and Crandall, W.B., J. Am. Ceram. Soc., 46(9), 434-7, 1963.
551	20765	Clement, J.K. and Egy, W.L., Univ. Ill. Eng. Expt. Sta. Bull. No. 36, 1-31, 1909.
552	24268	Reinsdorf, S., Silikattech., 11(7), 312-8, 1960.
553	14744	Rowley, F.B. and Algren, A.B., Univ. Minn. Eng. Expt. Sta. Bull. No. 12, 1-134,1937.
554	9412	Powers, R.W. and Hansen, R.H. (Johnston, H.L., Editor), ATI 58926, (PR-24), 1-16, 1949.
555	9413	Powers, R.W. and Hansen, R.H. (Johnston, H.L., Editor), ATI 58926, (PR-25), 1-12, 1949.
556	268	Budnikov, P.P. and Al'perovich, I.A., J. Appl. Chem. (USSR), 25, 665-73, 1952.
557	785	Budnikov, P.P. and Al'perovich, I.A., Zhur. Priklad. Khim., 25, 582-91, 1952.
558	6357	Rowley, F.B., Jordan, R.C., Lund, C.E., and Lander, R.M., Heating, Piping and Air Conditioning, 23(12), 103-9, 1951.
559	6936	Zabawsky, Z., ASTM Spec. Tech. Publ. 217, 3-16, 1957.

Ref. No.	TPRC No.	
560	7020	Lander, R.M., Heating, Piping and Air Conditioning, 26 (12), 121-6, 1954.
561	19642	Chow, C.S., Proc. Phys. Soc. (London), 61, 206-16, 1948.
562	98	Lapshin, A., Myasnaya Ind. SSSR, 25 (2), 55-6, 1954.
563	6308	Rowley, F.B., Jordan, R.C. and Lander, R.M., Refrig. Eng., 50, 541-4, 1945.
564	7018	Lander, R.M., Univ. Minn. Inst. Technol. Eng. Exp. Sta. Tech. Paper 49, 1-34, 1944.
565	24258	Bettini, T.M., Ric. Sci., 20 (4), 464-6, 1950.
566	8326	Stiles, H., Chem. Met. Eng., 36, 625-6, 1929.
567	7333	Hickman, M.J. and Ratcliffe, E.H., 9th Intern. Congr. Refrign., Ag-S, 1955.
568	1939	Verschoor, J.D. and Greebler, P., Trans. Am. Soc. Mech. Engrs., 74, 961-8, 1952.
569	24968	Bartoli, R. and Laine, P., Bull. Inst. Intern. Froid, Annexe, 233-40, 1956.
570	21748	Codegone, C., Industria (Milan), 44 (15) 401-3, 1930.
571	1628	Braun, G., Bauwirtschaft, No. 47, 3-10, 1950.
572	850	Liplavk, I.L., Zhur. Priklad. Khim., 26, 178-84, 1953.
573	25445	Verschoor, J.D. and Wilber, A., Heating, Piping and Air Conditioning, 26 (7), 125-30, 1954.
574	29516	Chandler, H.H. and Hancock, F.E., USAF FTDM-2435, 1-9, 1962. [AD 285153]
575	193	Bock, H., Chem. Tech. (Berlin), 5 (7), 387-90, 1953.
576	31	Brown, W.G., Nature, 179, 1187, 1957.
577	13404	Powell, R.W. and Griffiths, E., Proc. Roy. Soc. (London), A163 (913), 189-98, 1937.
578	28116	Stephens, D.R., USAEC UCRL-7605, 1-19, 1963.
579	25431	Lewis, D.W., J. Am. Concrete Inst., 30 (5), 619-33, 1958.
580	787	Epshtein, A.S., Stroitel. Prom., 33 (9), 33-5, 1955.
581	8134	Hansen, W.C. and Livovich, A.F., Am. Ceram. Soc. Bull., 37 (7), 322-8, 1958.
582	104	Hansen, W.C. and Livovich, A.F., J. Am. Ceram. Soc., 36 (11), 356-62, 1953.
583	174	Gallaher, R.B. and Kitzes, A.S., USAEC ORNL 1414, 1-30, 1953.
584	33346	LaMarre, D.A. Simpson, G.R., and Thorburn, M.R. USAF, 1-41, 1962. [AD 437864]
585	456	Pochapsky, T.E., J. Chem. Phys., 21 (9), 1539-40, 1953.
586	16320	Nancarrow, H.A., Proc. Phys. Soc. (London), 45, 447-61, 1933.
587	25356	Zamoluev, V.K., Mukhanova, L.N., and Taits, E.M., Proc. Acad. Sci. USSR, Chemical Techn. Sect., 133 (5), 127-9, 1960.
588	29617	Dietz, J.L. and Hangen, W.J., ASTIA MT-M23, 1-38, 1956. [AD 289592]
589	9266	deVries, D.A. and Peck, A.J., Australian J. Phys. 11 (3), 409-23, 1958.
590	20264	Woodside, W. and Messmer, J.H., J. Appl. Phys., 32 (9), 1688-99, 1961.
591	20265	Woodside, W. and Messmer, J.H., J. Appl. Phys., 32 (9), 1699-706, 1961.
592	9416	Powers, R.W. and Hansen R.H., (Johnston, H.L., Editor), ATI 52496, 1-61, 1949.
593	6919	Woodside, W., Heating, Piping and Air Conditioning, 30 (9), 163-70, 1958.
594	1553	Howling, D.H. Mendoza, E., and Zimmerman, J.E., Proc. Roy. Soc. (London), A229, 86-109, 1955.
595	24809	Powers, D.J., USAF BLR61-20(M), 1-14, 1962. [AD 284355]
596	6940	Lucks, C.F., Thompson, H.B., Smith, A.R., Curry, F.P., Deem, H.W., and Bing, G.F., AFTR6145 Pt. 1, ATI 117715, 1-127, 1951.
597	32441	Ioffe, A.V., Fiz. Tverd. Tela, 5(11), 3336-8, 1963.
598	25014	Ioffe, A.V., Soviet Physics - Solid State, 5(11), 2446-7, 1964.

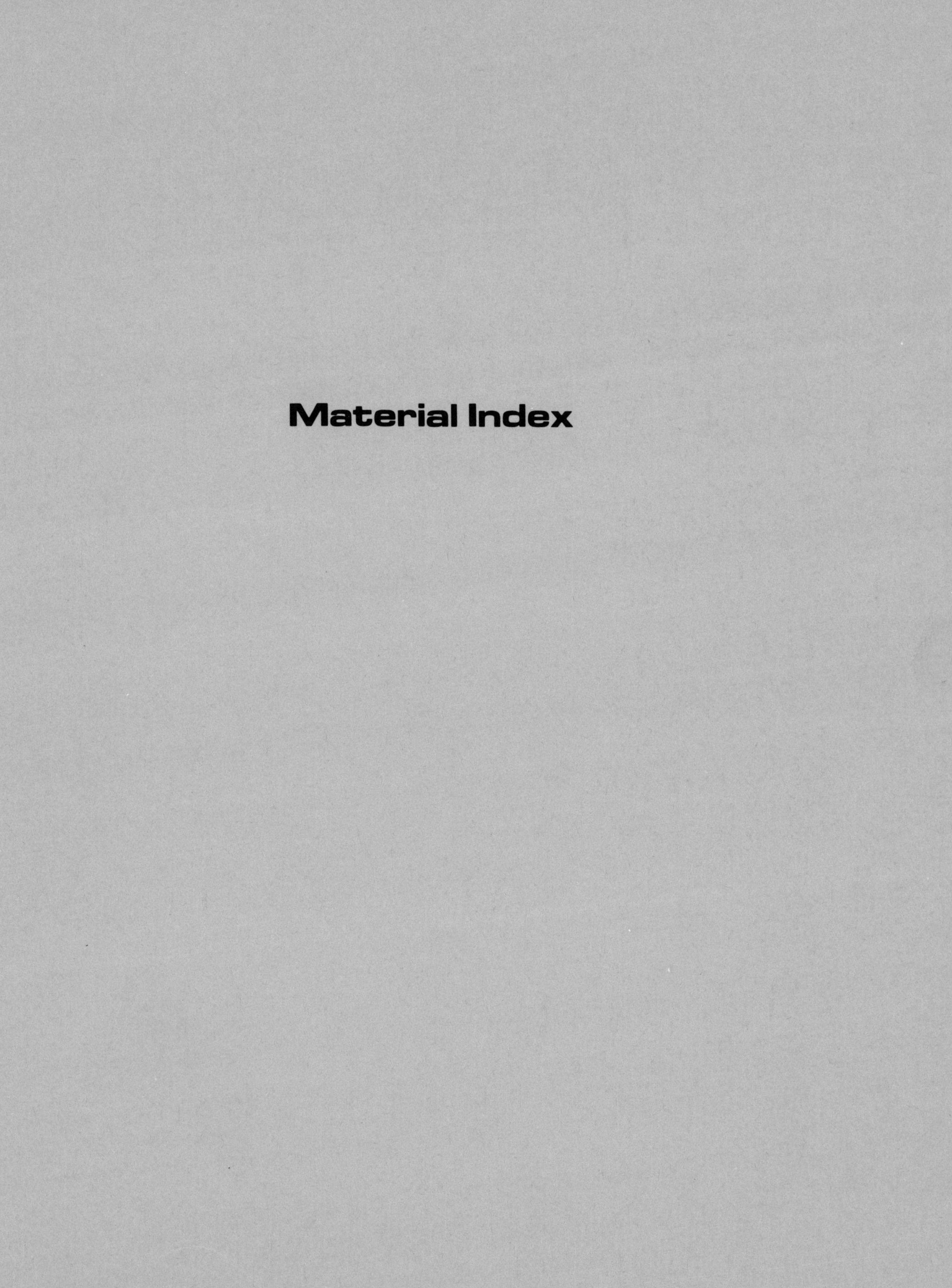

Material Index

MATERIAL INDEX TO THERMAL CONDUCTIVITY COMPANION VOLUMES 1, 2, AND 3